ISSUES IN ROCK MECHANICS

Proceedings
Twenty-Third Symposium on Rock Mechanics
The University of California
Berkeley, California
August 25-27, 1982

Edited by
Richard E. Goodman
University of California, Berkeley
Francois E. Heuze
Lawrence Livermore National Laboratory

Symposium Sponsored by
The University of California, Berkeley
and
US National Committee on Rock Mechanics

Published by the
Society of Mining Engineers
of the
American Institute of Mining, Metallurgical and Petroleum Engineers, Inc.
New York, New York•1982

Copyright © 1982 by the
American Institute of Mining, Metallurgical,
and Petroleum Engineers, Inc.

Printed in the United States of America
by Port City Press, Inc., Baltimore, MD

All rights reserved. This book, or parts thereof, may not be
reproduced in any form without permission of the publisher.

Library of Congress Catalog Card Number 82-71989
ISBN 0-89520-297-2

ISBN 0-89520-297-2

PREFACE

The publication of this volume by the Society of Mining Engineers revives a partnership between professional societies, universities, and government.

The Proceedings of the US Symposia record important developments and achievements in Rock Mechanics.

These developments, in 1982, embrace exploration of rock masses by geophysics, application of statistical methods, analysis and description of rock fracture, and the measurement of rock properties and attributes at all scales. Areas of application include not only excavation and drilling for construction, mining, and petroleum engineering, but also unprecedented endeavors for disposal of nuclear wastes, conversion and storage of energy in the ground, and in-situ mineral processing. Thus it is hardly surprising that the Symposia are attracting an increasing number of papers, and a growing participation from high technology organizations. Attendance by foreign visitors suggests the far-reaching importance of some of these topics. Indeed we are fortunate that these foreign guests enrich our meetings with their contributions.

The report on "Rock Mechanics Research Requirements" published in 1981 by the National Academy of Sciences has highlighted many unresolved questions in our field. The theme "Issues in Rock Mechanics" was selected to underscore our collective responsibility to address these questions. The Proceedings were printed before the meeting to promote thoughtful discussions at the Symposium. We hope that this conference will stimulate self-assessment and guide future activities.

Richard E. Goodman
Francois E. Heuze

Berkeley, California
May 17, 1982

Organizing Committee

Richard E. Goodman
Francois E. Heuze
Thomas W. Doe
Michael Hood
Michael S. King
John E. O'Rourke
Nanette Pike
Nicholas Sitar
Wilbur H. Somerton
Hans-Rudolf Wenk

Advisory Committee

Tor L. Brekke
Michael M. Carroll
Hugh C. Heard
Richard M. Nolting
Jahandar Noorishad
Walter A. Palen
P.N. Sundaram
Kent S. Udell
Lionel E. Weiss
Charles R. Wilson
Paul A. Witherspoon

Sponsors

The Organizing Committee for the 23rd US Symposium on Rock Mechanics gratefully acknowledges financial contributions from the following:

Dames & Moore
Washington, D.C.

Geotechnical Consultants, Inc.
San Francisco, California

Occidental Research Corporation
Irvine, California

Sohio Petroleum Company
San Francisco, California

Union Oil Company of California
Brea, California

Dowell Division, Dow Chemical USA
Tulsa, Oklahoma

Kaiser Engineers, Inc.
Oakland, California

MTS Systems Corporation
Minneapolis, Minnesota

Science Applications, Inc.
La Jolla, California

Terra Tek, Inc.
Salt Lake City, Utah

Utah International Inc.
San Francisco, California

Woodward-Clyde Consultants
San Francisco, California

In addition, the National Science Foundation sponsored two workshops during the Symposium: "Relationship Between Geophysics and Engineering Rock Mechanics", and "Deformation Mechanisms and Development of Textures in Rock".

US National Committee for Rock Mechanics

CHAIRMEN

Chairman
Kate H. Hadley
Exxon Company, USA
Harahan, Louisiana

Vice-Chairman
James H. Coulson
Tennessee Valley Authority
Knoxville, Tennessee

Immediate Past Chairman
Neville G. W. Cook
University of California
Berkeley, California

At-Large Members
Z.T. Bieniawski
Penn State University
University Park, Pennsylvania

John D. Bredehoeft
US Geological Survey
Menlo Park, California

Bruce R. Clark
Leighton & Associates
Irvine, California

Melvin Friedman
Texas A & M University
College Station, Texas

David T. Varnes
US Geological Survey
Denver, Colorado

Hilmar von Schonfeldt
Occidental Research Corp.
Irvine, California

SOCIETY REPRESENTATIVES

SOCIETY OF MINING ENGINEERS (SME)
Francis S. Kendorski
Engineers International
Downer's Grove, Illinois

TRANSPORTATION RESEARCH BOARD (TRB)
C. William Lovell
Purdue University
Lafayette, Indiana

GEOLOGICAL SOCIETY OF AMERICA (GSA)
George A. Kiersch
Cornell University
Ithaca, New York

AMERICAN SOCIETY OF CIVIL ENGINEERS (ASCE)
Fred H. Kulhawy
Cornell University
Ithaca, New York

SOCIETY OF PETROLEUM ENGINEERS (SPE)
Kenneth E. Gray
University of Texas at Austin
Austin, Texas

AMERICAN SOCIETY FOR TESTING & MATERIALS (ASTM)
Howard J. Pincus
Departments of Geological Sciences and
    Civil Engineering
University of Wisconsin, Milwaukee
Milwaukee, Wisconsin

AMERICAN GEOPHYSICAL UNION (AGU)
Earl R. Hoskins
Texas A & M University
College Station, Texas

AMERICAN SOCIETY OF MECHANICAL ENGINEERS (ASME)
Arfon H. Jones
Terra Tek, Inc.
Salt Lake City, Utah

US National Committee for Rock Mechanics Cont.

EX-OFFICIO MEMBERS

Thomas C. Atchison
University of Minnesota
Minneapolis, Minnesota

Herbert Friedman
National Research Council
Washington, D.C.

Robert M. White
National Research Council
Washington, D.C.

Don C. Banks
US Army Corps of Engineers
Vicksburg, Mississippi

ASSOCIATION OF ENGINEERING GEOLOGISTS (AEG)

Arthur B. Arnold
Bechtel, Inc.
San Francisco, California

SOCIETY OF EXPLORATION GEOPHYSICISTS (SEG)

Gordon P. Eaton
Texas A & M University
College Station, Texas

SECRETARIAT STAFF

EXECUTIVE SECRETARY
John E. Wagner
National Research Council
Washington, D.C.

ASSISTANT EXECUTIVE SECRETARY
Susan V. Heisler
National Research Council
Washington, D.C.

ADMINISTRATIVE ASSISTANT
Barbara S. Adams
National Research Council
Washington, D.C.

Published Proceedings of the Previous US Rock Mechanics Symposia

FIRST SYMPOSIUM ON ROCK MECHANICS, *Quarterly*, Colorado School of Mines, Vol. 51, No. 3, April 1956. (Out of print)

SECOND SYMPOSIUM ON ROCK MECHANICS, *Quarterly*, Colorado School of Mines, Vol. 52, No. 3, April 1957. (Out of print)

THIRD SYMPOSIUM ON ROCK MECHANICS, *Quarterly*, Colorado School of Mines, Vol. 54, No. 3, July 1959. (Out of print)

FOURTH SYMPOSIUM ON ROCK MECHANICS, *Bulletin 76*, Mineral Industries Experiment Station, The Pennsylvania State University, November 1961.

FIFTH SYMPOSIUM ON ROCK MECHANICS, University of Minnesota, May 1962. Published by Pergamon Press. (Out of print)

SIXTH SYMPOSIUM ON ROCK MECHANICS, University of Missouri, October 1964. Preprints only.

SEVENTH SYMPOSIUM ON ROCK MECHANICS, The Pennsylvania State University, June 1965. Most papers published in *Transactions,* Society Of Mining Engineers of AIME (1965-1966). (Out of print)

Failure of Breakage of Rock, EIGHTH SYMPOSIUM ON ROCK MECHANICS, University of Minnesota, September 1966. Published by The American Institute of Mining, Metallurgical, and Petroleum Engineers, Inc. (Out of print)

Status of Practical Rock Mechanics, NINTH SYMPOSIUM ON ROCK MECHANICS, Colorado School of Mines, April 1967. Published by the Society of Mining Engineers of AIME, 1972. (Out of print)

Basic and Applied Rock Mechanics, TENTH SYMPOSIUM ON ROCK MECHANICS, University of Texas at Austin, May 1968. Published by the Society of Mining Engineers of AIME, 1972. (Out of print)

Rock Mechanics — Theory and Practice, ELEVENTH SYMPOSIUM ON ROCK MECHANICS, University of California, June 1969. Published by the Society of Mining Engineers of AIME, 1970. (Out of print)

Dynamic Rock Mechanics, TWELFTH SYMPOSIUM ON ROCK MECHANICS, University of Missouri-Rolla, November 1970. Published by the American Society of Civil Engineers, 1972. (Out of print)

Stability of Rock Slopes, THIRTEENTH SYMPOSIUM ON ROCK MECHANICS, University of Illinois at Urbana, August-September 1971. Published by the American Society of Civil Engineers, 1972.

Published Proceedings Cont.

New Horizons in Rock Mechanics, FOURTEENTH SYMPOSIUM ON ROCK MECHANICS, The Pennsylvania State University, June 1972. Published by the American Society of Civil Engineers, 1973.

Applications of Rock Mechanics, FIFTEENTH SYMPOSIUM ON ROCK MECHANICS, South Dakota School of Mines and Technology, September 1973. Published by the American Society of Civil Engineers, 1975.

Design Methods in Rock Mechanics, SIXTEENTH SYMPOSIUM ON ROCK MECHANICS, University of Minnesota, September 1975. Published by the American Society of Civil Engineers, 1977.

Site-Characterization, SEVENTEENTH SYMPOSIUM ON ROCK MECHANICS University of Utah, August 1976. Preprint-proceedings published by the Utah Engineering Experiment Station, 1976. (Also, *Monograph on Rock Mechanics Applications in Mining,* Published by the Society of Mining Engineers of AIME, 1977.)

Energy Resources and Excavation Technology, EIGHTEENTH SYMPOSIUM ON ROCK MECHANICS, Colorado School of Mines, June 1977. Published by Johnson Publishing Company, 1977.

NINETEENTH SYMPOSIUM ON ROCK MECHANICS, Mackay School of Mines, University of Nevada, May 1978. Published by Conferences & Institutes, Extended Programs and Continuing Education, 1978.

TWENTIETH SYMPOSIUM ON ROCK MECHANICS, University of Texas at Austin, June 1979. Published by the Center for Earth Sciences and Engineering, University of Texas at Austin, 1979.

The State of the Art in Rock Mechanics, TWENTY-FIRST SYMPOSIUM ON ROCK MECHANICS, University of Missouri-Rolla, May 1980. Published by the University of Missouri-Rolla, 1980.

Rock Mechanics from Research to Applications, TWENTY-SECOND SYMPOSIUM ON ROCK MECHANICS, Massachusetts Institute of Technology, Cambridge, Massachusetts, July 1981. Published by the Massachusetts Institute of Technology, 1981.

NOTE: Those desiring to obtain out-of-print proceedings of earlier symposia are advised to refer to the Engineering Societies Library, 345 E. 47th St., New York, NY 10017, or University Microfilms International, 300 N. Zeeb Rd., Ann Arbor, MI 48106, regarding microfilm or photocopies of such publications.

TABLE OF CONTENTS

Preface • Richard E. Goodman, Francois E. Heuze................... iii
Organizing Committee iv
Advisory Committee iv
Sponsors... iv
US National Committee for Rock Mechanics v
Published Proceedings of the Previous Rock Mechanics Symposia vii

Chapter		
	Conference Keynote Address	1
1	Issues in Rock Mechanics: A Personal View.	
	B. Ladanyi..	3
	Exploration	15
2	Responsibilities, Opportunities and Challenges in Geophysical Exploration.	
	R.J. Lytle...	17
3	Standard Program for Site Selection Studies in Sweden for a High-Level Nuclear Waste Repository.	
	H.S. Carlsson..	23
4	Approaches to Evaluating the Permeability and Porosity of Fractured Rock Masses.	
	T.W. Doe, J.C.S. Long, H.K. Endo, C.R. Wilson..............	30
5	Acoustic Borehole Logging in a Granitic Rock Mass Subjected to Heating.	
	M.S. King, B.N.P. Paulsson.................................	39
6	Spatial Distribution of Permeability Around CSM/ONWI Room, Edgar Mine, Idaho Springs, Colorado.	
	P. Montazer, G. Chitombo, R. King, W. Ubbes................	47
7	Cross-Borehole Fracture Mapping Using Electromagnetic Geotomography.	
	A.L. Ramirez, F.J. Deadrick, R.J. Lytle....................	57
	Statistics	65
8	Playing the Odds in Rock Mechanics.	
	G.B. Baecher...	67
9	Permeability, Percolation and Statistical Crack Mechanics.	
	J.K. Dienes...	86

ISSUES IN ROCK MECHANICS

10	Effect of Correlation on Rock Slope Stability Analyses. E.F. Glynn, S. Ghosh.	95
11	A New Technique for Domain Delineation of Rock Mass Discontinuities. H.R. Hume, T.R. West, W.R. Judd.	104
12	A Rejection Criterion for Definition of Clusters in Orientation Data. M.A. Mahtab, T.M. Yegulalp.	116
13	Fourier Analysis for Estimating Probability of Sliding for the Plane Shear Failure Mode. Stanley M. Miller.	124
14	Statistical Characterization of Complex Crack and Petrographic Texture: Application to Predicting Bulk Physical Properties. N. Warren.	132

Stresses   141

15	Status of the Hydraulic Fracturing Method for In-Situ Stress Measurements. M.D. Zoback, B.C. Haimson.	143
16	The Influence of Rock Anisotropy on Stress Measurements by Overcoring Techniques. B. Amadei, R.E. Goodman.	157
17	Experience with the Solid Inclusion Stress Measurement Cell in Coal in Australia. R.L. Blackwood.	168
18	In-Situ Stresses in the Klerksdorp Gold Mining District, South Africa—A Correlation Between Geological Structure and Seismicity. N.C. Gay, P.J. Van der Heever.	176
19	Determination of the Instantaneous Shut-In Pressure from Hydraulic Fracturing Data and Its Reliability as a Measure of the Minimum Principal Stress. J.M. Gronseth.	183
20	Comparing Hydrofracturing Deep Measurements with Overcoring Near-Surface Tests in Three Quarries in Western Ohio. B.C. Haimson.	190
21	High Stress Occurrences in the Canadian Shield. G. Herget.	203
22	Behavior of a Rigid Inclusion Stressmeter in an Anisotropic Stress Field. G.W. Jaworski, B.C. Dorwart, W.F. White, W.R. Beloff.	211

CONTENTS

23 In-Situ Stress Determinations in Northeastern Ohio.
J.-C. Roegiers, J.D. McLennan, L.D. Schultz. 219
24 Stress Measurement of Rock Mass In-Situ and the Law of Stress Distribution in a Large Dam Site.
B. Shiwie, L. Guangyu. 230
25 Prediction of Hydraulic Fracture Azimuth from Anelastic Strain Recovery Measurements of Oriented Core.
L.W. Teufel. 238

Laboratory Tests 247

26 Mechanical, Thermal, and Fluid Transport Properties of Rock at Depth.
H.C. Heard. 249
27 The Effect of Water on the Mechanical Properties and Microstructures of Granitic Rocks at High Pressures and High Temperatures.
O. Alm. 261
28 Time-Dependent Volumetric Constitutive Relation for Fault Gouge and Clay at High Pressure.
C.-I. Chu, C.-Y. Wang. 270
29 Deformation Mechanisms in Granodiorite at Effective Pressures to 100 MPa and Temperatures to Partial Melting.
M. Friedman, J. Handin, S.J. Bauer. 279
30 The Effects of Fracture Type (Induced versus Natural) on the Stress-Fracture Closure-Fracture Permeability Relationships.
J.E. Gale. 290
31 A Work Hardening/Recovery Model of Transient Creep of Salt During Stress Loading and Unloading.
D.E. Munson, P.R. Dawson. 299
32 Steady-State Creep of Rock Salt in Geoengineering.
T.W. Pfeifle, P.E. Senseny. 307
33 An Experimental Investigation of the Combined Effects of Strain Rate and Moisture Content on Shale.
T.G. Richard, S.H. Advani. 315
34 Rock Mechanics Testing of Large Diameter Core at the Crandon Deposit.
R.G. Rowe. 324
35 The Influence of Stress Level on the Creep of Unfilled Rock Joints.
C.W. Schwartz, S. Kolluru. 333

36 Triaxial Creep of Oil Shale and Deformation of Pillars in the In Situ Retorting Environment.
K.P. Sinha, T.F. Borschel, J.F. Schatz, S. Demou............ 341
37 Behaviour of a Brittle Sandstone in Plane-Strain Loading Conditions.
V.G. Stavropoulou...................................... 351

Tectonophysics   359

38 Seismic Style in Relation to Heat Flow Along the San Andreas Fault System.
R.H. Sibson.. 361
39 Experimental Study of Cataclastic Deformation of a Quartzite.
J. Hadizadeh, E.H. Rutter............................... 372
40 Compression Experiments on Natural Magnetite Crystals at 200°C and 400°C at 400 MPa Confining Pressure.
C. Hennig-Michaeli, H. Siemes.......................... 380
41 Microstructures, Mineral Chemistry and Oxygen Isotopes of Two Adjacent Mylonite Zones: A Comparative Study.
T.E. LaTour, R. Kerrich................................. 389
42 Continuous Formation of Gouge and Breccia During Fault Displacement.
E.C. Robertson... 397
43 Workshop on Deformation Mechanisms and Texture Development in Rocks.
H.R. Wenk... 405

Fracture   421

44 An Examination of the Tensile Strength of Brittle Rock.
J.L. Ratigan.. 423
45 Experimental Investigation of Fragmentation Enhancement Due to Salvo Impact.
D.B. Barker, D.C. Holloway, J.F. Cardenas-Garcia.......... 441
46 Variables in Fracture Energy and Toughness Testing of Rock.
C.C. Barton.. 449
47 A Fracture Toughness Testing System for Prediction of Tunnel Boring Machine Performance.
A.R. Ingraffea, K.L. Gunsallus, J.F. Beech, P. Nelson........ 463
48 A Study of Fracture Toughness for an Anisotropic Shale.
V.H. Kenner, S.H. Advani, T.G. Richard.................. 471
49 Experimental Modeling of Micro-Crack Formation in Rocks.
T. Kobayashi... 480

CONTENTS

50 An Experimental Investigation of the Projectile Penetration into Soft, Porous Rock Under Dry and Liquid-Filled Conditions.
    A. Kumano, W. Goldsmith. 488
51 Statistics in Aid of Interpreting Fracture Data.
    E.Z. Lajtai. 496
52 Use of Laboratory-Derived Data to Predict Fracture and Permeability Enhancement in Explosive-Pulse Tailored Field Tests.
    S. McHugh, D. Keough. 504
53 A Simple R-Curve Approach to Fracture Toughness Testing of Rock Core Specimens.
    F. Ouchterlony. 515
54 Pre-Splitting and Stress Waves: A Dynamic Photoelastic Evaluation.
    K.R.Y. Simha, D.C. Holloway, W.L. Fourney. 523
55 The Mechanism, and Some Parameters Controlling, the Water Jet Cutting of Rock.
    D.A. Summers, S. McGroarty. 531
56 Sub-Critical Crack Growth in Stripa Granite: Direct Observations.
    G. Swan, O. Alm. 542
57 Step Cracks: Theory, Experiment, and Field Observation.
    R.K. Thorpe, M.E. Hanson, G.D. Anderson, R.J. Shaffer. 551
58 Correlation Between Fracture Roughness Characteristics and Fracture Mechanical and Fluid Flow Properties.
    Y.W. Tsang, P.A. Witherspoon. 560

Models   569

59 The Role and Credibility of Computational Methods in Engineering Rock Mechanics.
    B.H.G. Brady, C.M. St. John. 571
60 Finite Element Evaluations of Thermo-Elastic Consolidation.
    B.L. Aboustit, S.H. Advani, J.K. Lee, R.S. Sandhu. 587
61 Time and Temperature Dependent Stress and Displacement Fields in Salt Domes.
    H.W. Duddeck, H.-K. Nipp. 596
62 Implementation of Finite Element—Boundary Integral Linkage Algorithms for Rock Mechanics Applications.
    W.S. Dunbar. 604
63 On the Modeling of Nuclear Waste Disposal by Rock Melting.
    F.E. Heuze. 612

ISSUES IN ROCK MECHANICS

64   Nonlinear Thermo-Mechanical Behaviour and Stress Analysis in Rocks.
    K.Y. Lo, R.S.C. Wai, R.K. Rowe, L. Tham................... 620
65   A Hybrid Discrete Element-Boundary Element Method of Stress Analysis.
    L.J. Lorig, B.H.G. Brady...................................... 628
66   Numerical Simulation of Fracture.
    L.G. Margolin, T.F. Adams................................... 637
67   Numerical Simulation of Fluid Injection into Deformable Fractures.
    J. Noorishad, T.W. Doe....................................... 645
68   Verification of Finite Element Methods Used to Predict Creep Response of Leached Salt Caverns.
    D.S. Preece, C.M. Stone...................................... 655
69   Elastic Bending of Thick Rock Plates.
    J.G. Singh, P.C. Upadhyay.................................... 664
70   A Numerical Study of Excavation Support Loads in Jointed Rock Masses.
    M.D. Voegele, C. Fairhurst................................... 673
71   Influence of Creep Law Form on Predicted Deformations in Salt.
    R.A. Wagner, K.D. Mellegard, P.E. Senseny.................. 684
72   A Hybrid Quadratic Isoparametric Finite Element-Boundary Element Code for Underground Excavation Analysis.
    D. Yeung, B.H.G. Brady...................................... 692
73   Boundary Element Methods for Viscoelastic Media.
    W. Yongjia, S.L. Crouch...................................... 704
74   Compressibilities and Effective Stress Coefficients for Linear Elastic Porous Solids: Lower Bounds and Results for the Case of Randomly Distributed Spheroidal Pores.
    R.W. Zimmerman... 712

Field Tests   721

75   Questions in Experimental Rock Mechanics.
    N. Cook.. 723
76   Effects of Block Size on the Shear Behavior of Jointed Rock.
    N. Barton, S. Bandis... 739
77   The Corejacking Test: An Analysis of the Corejack Loading System.
    D.A. Blankenship, R.G. Stickney............................. 761
78   Evaluation of Opening and Hydraulic Conductivity of Rock Discontinuities.
    P.T. Cruz, E.F. Quadros, D. Correa F⁰, A. Marrrano.......... 769

CONTENTS

79 External Displacement Method for Determining the In-Situ Deformability of Rock Masses.
R.V. de la Cruz. 778
80 Spatial Distribution of Deformation Moduli Around the CSM/ONWI Room, Edgar Mine, Idaho Springs, Colorado.
A. Wadood M.A. El Rabaa, W.A. Hustrulid, W.F. Ubbes. 790
81 Measuring the Thermomechanical and Transport Properties of a Rockmass Using the Heated Block Test.
E. Hardin, N. Barton, M. Voegele, M. Board, R. Lingle, H. Pratt, W. Ubbes. 802
82 The Influence of Test Plate Flexibility on the Results of Cable Jacking Tests.
J.K. Jeyapalan, A.P.S. Selvadurai. 814
83 Tunnel Response in Modeled Jointed Rock.
H.E. Lindberg. 824
84 In Situ Measurements of Stress Change Induced by Thermal Load: A Case History in Granitic Rock.
R. Lingle, P.H. Nelson. 837
85 Geotechnical Monitoring of High-Level Nuclear Waste Repository Performance.
C.M. St. John, M.P. Hardy. 846
86 Calculated and Measured Drift Closure During the Spent Fuel Test in Climax Granite.
J.L. Yow, Jr., T.R. Butkovich. 855
87 Some Consideration of In-Situ Testing on Mechanical Properties of Rock Mass.
Z. Weishen, X. Dongjun. 864
88 Issues Related to Field Testing in Tuff.
R.M. Zimmerman. 872

Reinforcement 881

89 Calculation of Support for Hard, Jointed Rock Using the Keyblock Principle.
R.E. Goodman, G.-h. Shi, W. Boyle. 883
90 Case Studies of Rock Slope Reinforcement.
P.N. Calder. 899
91 Rock Support at Pine Flat, A Case History.
J. Cogan, P.M. Gomez. 912
92 Generalization of the Ground Reaction Curve Concept.
E. Detournay, C. Fairhurst. 924
93 Stabilization of Rock Excavations Using Rock Reinforcement.
T.A. Lang, J.A. Bischoff. 935

94 Rock Slope Reinforcement with Passive Anchors.
D.P. Moore, M.R. Lewis. 945
95 Revising Terzaghi's Tunnel Rock Load Coefficients.
D. Rose. 953
96 Rock Bolt Reinforcement System to Stabilise Shaft Intersections and Pit Bottom Roadways during Underground Reconstruction.
R. N. Singh, A.M. Heidarieh Zadeh. 961
97 Equipment, Automation, Rock Mechanics Principles and Safety Interfaces in the Control of Roof and Ribs of Mines.
J. J. Scott. 971
98 New Laboratory Instrumentation for the Evaluation of Rock Bolt Behavior.
E. Unal, H. R. Hardy, Jr., Z.T. Bieniawski. 985

Case Histories   **997**

99 Engineering Experience with Weak Rocks in Japan.
C. Tanimoto. 999
100 Some Applications of Rock Engineering to Geotechnical Practice.
I.S. Oweis, W.W. Lilly. 1015
101 Toppling Induced Movements in Large, Relatively Flat Rock Slopes.
A. Brown. 1035
102 Subsidence Monitoring—Case History.
P.J. Conroy. 1048
103 Rock Wedge Stability.
A.M. Crawford. 1057
104 Fireflood Microseismic Monitoring: Rock Mechanics Implications.
M.B. Dusseault, E. Nyland. 1065
105 Shear Stability of Mine Pillars in Dipping Seams.
W.G. Pariseau. 1077
106 Factors Governing the Stability of Rock Slopes in British Surface Coal Mines.
M.J. Scoble, W.J.P. Leigh. 1091
107 Preliminary Foundation Studies for Raising a Gravity-Arch Dam.
G.A. Scott, K.J. Dreher, C.C. Hennig. 1099
108 Interaction Effects Associated with Longwall Coal Mining.
A.N. Styler, R.K. Dunham. 1107
109 Complementary Influence Functions for Predicting Subsidence Caused by Mining.
H.J. Sutherland, D.E. Munson. 1115
110 Design and Analysis of a Circular Underground Powerhouse.
D. Zayakov, G. Yoshikado, P.R. Kneitz. 1122

Authors' Index. 1131

CONFERENCE KEYNOTE ADDRESS
Issues in Rock Mechanics

Branko Ladanyi
Ecole Polytechnique, Montreal, Canada

Chapter 1

ISSUES IN ROCK MECHANICS: A PERSONAL VIEW

by Branko Ladanyi

Professor of Civil Engineering
Ecole Polytechnique
Montreal, Canada

ABSTRACT

In reviewing the current issues in rock mechanics, the author puts an emphasis on problems related with the needs of extrapolation in scale and time, and of establishing proper conceptual models, especially in connection with the most recent large-scale and long-term projects such as underground disposal of nuclear waste. In such projects, where valid case histories will not be available for many years to come, the observational method can only be successful if combined with a proper framework approach, such as general rock mass classification systems and behavioral models.

INTRODUCTION

First of all, I wish to express my appreciation to the Organizing Committee for inviting me to give this Keynote Address to the 23rd U.S. Rock Mechanics Symposium. Certainly, I feel highly honored to be asked to address this distinguished audience.

Although my first feelings after receiving the invitation were ones of appreciation, these turned soon into the feelings of apprehension. I wondered what I could say to such an audience that would keep their attention for 45 minutes and would not be just a re-statement of the obvious.

After some reflection, I came to the conclusion that I had been invited, not because of my nearly twenty years of activity in rock mechanics, by which I do not differ from many of my colleagues, but more probably because of my trend towards geomechanical multi-disciplinarity. This trend has certainly slowed down my advances in any particular area, but has given me many times a great pleasure in finding how a study of any particular portion of geomechanics can help in understanding better the whole discipline.

So, I decided to give you some of my own impressions on the current issues in rock mechanics, with some emphasis on the multi-disciplinarity and the observational approach.

FIRST ISSUE: ROCK MASS BEHAVIOR

It may seem like an oversimplification, but if I were asked by a journalist what are the main issues in rock mechanics at present, I would be quick to answer that the main problem we have resides in the fact that we cannot measure directly certain fundamental properties of rock masses needed in design because of constraints imposed by the scale, the time and the money. To the next question, which would probably be: "So, what do you people do about it?", my answer would be something like: "You see, we cannot measure, but if we are patient enough to observe, develop conceptual models, back-calculate, check assumptions, make sensitivity analyses, establish frameworks and classification systems, and keep observing and improving them for a sufficiently long time, we hope that we shall eventually be able to estimate these properties of rock masses sufficiently closely for the design purposes".

As stated by Bieniawski (1979a) in his recent thorough review of several proposed rock mass classifications, the design of tunnels (and of most other structures in rock) uses three main approaches: analytical, observational, and empirical. The analytical approach is still little used in the present engineering practice, mainly because of the inability to obtain the necessary input data, as the ground conditions are rarely adequately explored. The observational approach, represents essentially a design-as-you-go philosophy, based on observations and monitoring. Although this approach is sound, it usually requires special contractual provisions not easily adaptable to the established contracting procedures. The empirical approach relates the experience encountered at previous projects to the conditions anticipated at a proposed site. If an empirical design is backed by a systematic approach to ground classification, it can effectively utilize the valuable practical experience gained at many projects. Rock mass classification can, if fulfilling certain conditions, effectively combine the findings from observation, experience, and engineering judgment for providing a quantitative assessment of rock mass conditions.

Most classification systems developed during the last four decades, i.e., since the Terzaghi's (1946) classification for tunnels, have been oriented towards certain special applications. While this was a logical and useful trend, it is nevertheless important to note that, at least some of them, have attempted to be less specific by furnishing a rating system that enables to estimate true rock mass properties based on large-scale observations of deformations and failure of rock masses. One such example is the RMR system (Bieniawski, 1975), recently enlarged to include the estimate of rock mass deformability (Bieniawski, 1979b) and shear strength (Hoek and Brown, 1980). Another example is the Q-system proposed by Barton et al. (1974).

This kind of a sound framework, although necessarily imperfect at the beginning, may become the basis of parameter evaluation for design, if it is continuously improved by adding new observational data.

In tunnelling, a hundred years of experience and many well-documented case histories have enabled a reasonable rock mass behavior prediction system to be established, with some general validity. In other cases, where valid case histories are still lacking, such classification systems may serve as basic frameworks for properly using the observational data when they become available.

SECOND ISSUE: EFFECT OF TIME

Another problem in rock mechanics one has to cope with more and more, is the difficulty in evaluating the effect of time on the behavior and safety of rock structures.

About ten years ago, within the Canadian Open Pit Slope Project 1972-77, we were asked to prove the validity of a rock joint shear strength concept (Ladanyi and Archambault, 1969) by back-analyzing a failed rock slope in an open pit mine. Not surprisingly, the back-analysis showed, that the slope which failed, would have been quite safe had the original joint system interlock been conserved. However, at the moment of slope failure, nearly all of the original interlock disappeared, and the overall shear resistance of the mass was close to the residual level. In such a case the prediction of the risk of failure is impossible without properly evaluating the probable effect of time on rock mass properties.

The effect of time on the stability of rock slopes was very clearly shown by Bjerrum and Jörstad (1963), who made a survey of 300 sites of rock falls and rock slides in Norway and concluded that a theoretical prediction of rock slides in rock slopes with sheeting joints is impossible, mainly because the failure is due to a progressive slope deterioration which may take centuries to develop. The failure of the slope may finally occur due to a minor effect such as a rise in the cleft-water pressure which "may be the last straw to break the camel's back".

While we use to think of the time effects in geomechanics in terms of "creep" or "consolidation", the fact is that the physical phenomena behind these overall manifestations may be quite different in various rock materials and environments. In blocky rock masses, in particular, both in slopes and behind tunnel linings, it is often observed that the rock mass tends to move continuously because it "deteriorates with time". This deterioration may be due to various causes, such as weathering, erosion, freeze and thaw, joint water pressure fluctuation, joint opening due to rebound, progressive failure of rock at stress concentration points, and others. Most of these phenomena and their combinations are difficult to predict, so that in many cases a continuous monitoring at the most characteristic points of a structure or a slope may be the only answer to a successful failure prediction (Miller and Hilts, 1969). Much too few such cases have been monitored up to now, and if so, the time of observation was generally too short for determining properly the time effect on the rock mass even for the design of ordinary rock structures, such as slopes, tunnels and dam foundations, not to mention the needs related to the design of underground repositories for the nuclear fuel waste, where the time scale is of the order of several thousands of years.

THIRD ISSUE: MODELLING

Behavioral Models

As it was well expressed by Gerrard (1977), the mathematical (or physical) modelling can provide the key to orderly and scientific advances in geomechanics when applied correctly as one component in an integrated cycle. The main features of such a cycle are the establishment of hypotheses of material behavior (constitutive relationships), the quantification of the various types of input data, the prediction and observation of full-scale performance of the engineering structures and feed-back to check the validity of the physical laws, input data and modelling process (Aitchison, 1973).

Models are by their very nature approximate representations of material response. Thus, the predictions can be expected to be also only approximate. As to the tolerable inaccuracy of prediction, the view of an engineer will usually differ from that of a scientist. As well stated by Gerrard (1977), the engineer's approach is to employ, within the time-scale of the project, the level of sophistication in modelling that will minimize the total cost of investigation, design, construction and maintenance, by reserving sophisticated predictive models only for the areas of the highest risk. On the other hand, the scientist is not constrained to a time-scale and seeks to understand all aspects of behavior to a highest possible level of accuracy of prediction.

Also according to Gerrard (1977), the development of "scientific" mathematical models for practical geomechanics problems should be based on the accumulation of data relating to the mechanical response of earth materials when subjected to the range of practical stress paths. It should include (a) the definition of input data in stochastic, rather than deterministic form, (b) the application of probabilistics to achieve the required degree of certainty in the prediction, and (c) the consideration of the geological processes relevant to the formation of the deposit and their correlation with observation of "inherent" fabric, prior stress history, current stress patterns and anisotropies in the mechanical properties.

Whoever has tried to describe and put into an understandable mathematical form the behavior of any natural earth material, has had to realize the immense complexity of the task, even for the simplest materials. Necessarily therefore, most models of material behavior cover only a very limited area of observed or observable conditions and give only a very narrow image of the material behavior.

A classical case of this problem was presented by Markus Reiner (1969) in his well-known address given at the 4th International Congress on Rheology. Although we are using the science of rheology in studying the behavior of both solids and liquids, everybody knows that the term "rheology" is based on the well-known Heraclitus' premise that "everything flows". But there are solids in rheology even if they may show relaxation and creep. According to Reiner the way out of this difficulty had been shown by the prophetess Deborah, even before Heraclitus. In her famous song after the victory over the Philistines, she sang "The mountains flowed before the Lord". Deborah knew two things. First that the mountains flow, as everything flows. But, secondly, that they flowed before the Lord, and not before man, for the simple reason that man in his short lifetime cannot see them flowing, while the time of observation of God is infinite. Markus Reiner then suggested that we define as a non-dimensional number the Deborah number:

$$D = \text{time of relaxation/time of observation}$$

The difference between solids and fluids is then defined by the magnitude of D: The greater the Deborah number, the more solid the material appears, the smaller the number, the more fluid it is.

In classical rock engineering, as long as we are concerned with the problems involving the human time scale, which we usually call "the service life", the rock appears as a solid, and a reasonable extrapolation of the observational data is possible. However, in some new types of problems, especially those involving underground storage of nuclear fuel waste, the time scale has been extended far beyond the length of a human life. The extrapolation in such a situation of any rock mechanics data obtained even for humanly long periods of time becomes uncertain and risky.

When the knowledge on a material behavior is only fragmentary, a framework approach may be of same help. As an example, some aspects of the behavior of a solid rock within a wide scope of pressures, temperatures and time scales, may be conveniently represented by a phase diagram which may cover either deformation or fracture phenomena. Such deformation and fracture maps have been recently established for many metals and non-metals including ceramics and ice (Figs. 1 and 2) (Stocker and Ashby, 1973; Ashby and Frost, 1975; Gandhi and Ashby, 1979).

Establishment of such maps for typical rock materials would be of a great help for extrapolating the test information far beyond the scope covered by the tests. Clearly, in case of rocks, one would like to add the third dimension to such maps: the hydrostatic pressure, which would transform them into deformation or fracture cubes.

Similar attempts towards establishing a framework for a material behavior have been made several times in the last thirty years. One of them is the well-known Cambridge model of the behavior of soils, (Roscoe et al, 1958) which has found some limited area of application also in rock mechanics. More than ten years ago, we have tried to use that approach to develop a unified failure concept for dense and porous rocks (Nguyen Don, 1972). Although in that work, we did not quite succeed in showing the applicability of the Cambridge model to porous rocks, we have found, nevertheless, a practical validity for dense rocks of another model of material behavior, i.e., that of the associated flow rule proposed by Drucker (Ladanyi and Nguyen Don, 1970). Although that proved to be a good method for predicting the dilatancy rate at failure in such rocks, this finding could not yet have been extended to the estimation of dilatancy rates in rock masses, due to the lack of valid observational proof (Brizzolari, 1981).

Conceptual models

When trying to solve a rock engineering problem, it is usually easier to handle problems not involving failure or fracture than those in which the mode of failure has to be guessed in advance.

In the pre-FEM period, photoelastic models were used extensively for studying stresses and strains in slopes and around underground openings in the elastic domain, while certain fracture problems were studied at a reduced scale on models made of some rock-like materials.

With the development of the FEM, the physical models have fallen into disfavor, at least in rock mechanics (this is less so in soil mechanics, where centrifuge model testing has found a wide application in recent years). However, while the FEM as a powerful problem solution tool can handle a wide variety of problems both in continua and in discontinua, it is clear that its results depend strongly on the quality of the input data and on the ability of the modeller to guess a correct mode of failure or succession of failure phenomena. Some typical cases difficult to solve by the FEM alone are the interaction between the foundations and the rock mass beyond the elastic domain, the behavior of rock mass around shallow tunnels, and the stability of rock slopes. In any of these cases, when failure is approached, a proper mode, based on observations should be guessed, and a proper limit analysis solution, simulating the observed failure mode, has to be found.

Classification of probable failure modes, or establishment of conceptual models, represents a useful basis for problem solving in that case. Some typical examples are the Terzaghi's (1946) classification of rock failure around shallow tunnels, the classification of slope failures by Heuze and Goodman (1971), and the selection of possible failure modes beneath foundations on rock, shown by Goodman (1980).

Risks in transfer of conceptual models

It is a great pleasure when one discovers, in a completely different field from one's own a ready-made solution of a problem one has been trying to solve for many years. Typical examples of such solution transfers can be found in the analogy among the diffusion phenomena in electricity, heat transfer and water seepage through porous media, which resulted in the development of the theory of consolidation in soil mechanics. One such lucky discovery happened to me about ten years ago, when I found that a solution for a needle viscosimeter, published by Nadai (1963), was directly applicable to the problem of designing a pile embedded in permafrost.

Another promising solution transfer which is presently being considered is in the area of borehole and shaft sealing in connection with the underground storage of nuclear waste. The fact is that a lot about this problem can be learned just by looking into the literature on the behavior of rock-socketed concrete piers, where a vast amount of information has been accumulated in the last ten years (South and Daemen, 1981).

However, sometimes, direct solution transfer may also be deceiving. As an example, in the 1960-ies some rock mechanics workers tried to apply the Prandtl's plasticity solution to solving punching problems in brittle rock. Although numerical results looked convincing, it is difficult to accept that a model based on shear failure in plastic domain would be applicable to a problem involving brittle crack propagation.

Another example of the risk in concept transfer is in the area of tunnel lining design. I think that by now everybody is familiar with the Rabcewicz (1963) concept of fracture sequence around a tunnel in rock. First, fracture is assumed to start at the midheight on both sides of the tunnel. This enables a vertical yield of the rock mass above the tunnel and the two lateral fracture zones to occur creating a large elliptic gravity flow zone above the tunnel (Fig. 3). This is essentially what we were taught as civil engineering students in Europe many years ago. We were told that this concept was based on the experience obtained in tunnelling through the Alps, but nobody mentioned to us that this failure sequence can only happen if the tunnel is driven through the ground where the vertical ground stress is much higher than the horizontal one. If opposite is the case, as in the Canadian Shield, the first fracture zones due to ground stress will develop not laterally, but above and below the tunnel (Lombardi, 1970) and the assumed sequence of events will never occur.

CONCLUSION

Although many problems in rock engineering have been successfully solved in the last twenty years, we are still uncertain about the main issue: How to design structures in large-scale rock masses and involving very long periods of time on the basis of limited experimental and observational data. The answer may come, although slowly, by a systematic use of the observational method, combined with the development of frameworks, classification systems, theories and conceptual models. However, for the latter, it may be useful to quote Manfred Eigen (1982): "A theory has only the alternative of being right or wrong. A model has a third possibility: It may be right, but irrelevant".

REFERENCES

Aitchison, G.D., 1973, "General Report, Session 4, Proc. 8th Int. Conf. Soil Mech. Found. Engrg, Moscow, URSS, Vol. 3, pp. 161-190.

Ashby, M.F. and Frost, H.J., 1975, "Deformation Mechanism Maps Applied to the Creep of Elements and Simple Inorganic Compounds", Frontiers in Materials Science, Murr, L.E. and Stein, C., eds, Marcel Dekker, New York, pp. 319-419.

Barton, N., Lien, R. and Lunde, J., 1974, "Engineering Classification of Rock Masses for the Design of Tunnel Support", Rock Mechanics, Vol. 6, pp. 183-236.

Bieniawski, Z.T., 1975, "Prediction of Rock Mass Behaviour by the Geomechanics Classification", Proc. 2nd Australia - New Zealand Conference on Geomechanics, Brisbane, pp. 36-41.

Bieniawski, Z.T., 1979a, "Tunnel Design by Rock Mass Classifications", Tech. Rep. GL-79-19 for U.S. Army Engineer Waterways Experiment Station, Vicksburg, Miss., NTIS No. AD/A-076540, 127 p.

Bieniawski, Z.T., 1979b, "The Geomechanics Classification in Rock Engineering Applications", Proc. 4th Congress of the Int. Soc. for Rock Mechanics, Montreux, Vol. 2, pp. 41-48.

Bjerrum, L. and Jörstad, F., 1963, Discussion of paper "An Approach to Rock Mechanics" by K.W. John, Proc. ASCE, Vol. 80, SM1, pp. 300-302.

Brizzolari, E., 1981, "Miniseismic Investigations in Tunnels: Methodology and Results", Geoexploration, Vol. 18, pp. 259-267.

Eigen, M., 1982, A Quote in SciQuest, Vol. 55, No. 3, March.

Gandhi, C. and Ashby, M.F., 1979, "Fracture-Mechanism Maps for Materials which Cleave: F.C.C. and H.C.P. Metals and Ceramics", *Acta Metallurgica*, Vol. 27, pp. 1505-1602.

Gerrard, C.M., 1977, "Background to Mathematical Modelling in Geomechanics: The Roles of Fabric and Stress History", *Finite Elements in Geomechanics*, Gudehus, G., ed., Wiley, New York, pp. 33-115.

Goodman, R.E., 1980, "*Introduction to Rock Mechanics*", Wiley, New York, 478 p.

Heuzé, F.E. and Goodman, R.E., 1971, "Three-Dimensional Approach for Design of Cuts in Jointed Rock", *Stability of Rock Slopes*, Proc. 13th Symp. on Rock Mechanics, Urbana, Ill., Cording, E.J., ed., ASCE, New York, pp. 397 - 441.

Hoek, E. and Brown, E.T., 1980, "Empirical Strength Criterion for Rock Masses", *Proc. ASCE*, Vol. 106, GT9, pp. 1013-1035.

Ladanyi, B. and Archambault, G., 1970, "Simulation of Shear Behavior of a Jointed Rock Mass", *Rock Mechanics - Theory and Practice*, (Proc. 11th U.S. Symp. on Rock Mech., Berkeley), W.H. Somerton, ed., AIME, New York, pp. 105-125.

Ladanyi, B. and Nguyen Don, 1970, "Study of Strains Associated with Brittle Failure", *Proc. 6th Can. Symp. on Rock Mech.*, Montreal, pp. 49-64.

Lombardi, G., 1970, "The Influence of Rock Characteristics on the Stability of Rock Cavities", *Tunnels and Tunnelling*, May, Vol. 2 pp. 104-109.

Miller, R.P., and Hilts, D.E., 1970, "Experimental Open-Pit Mine Slope Stability Study", *Rock Mechanics - Theory and Practice*, W.H. Somerton, ed., AIME, New York, pp. 147-167.

Nadai, A., 1963, "*Theory of Flow and Fracture of Solids*", Vol. 2, McGraw-Hill, New York.

Nguyen Don, 1972, "Un concept de rupture unifié pour les matériaux rocheux denses et poreux", *Ph.D. Thesis*, Ecole Polytechnique, Montréal, 296 p.

Rabcewicz, L.V., 1963, "Bemessung von Hohlraumbauten", *Rock Mechanics and Engineering Geology*, Vol. 1/3-4, Springer, Wien, pp. 223-243.

Reiner, M., 1969, "The Deborah Number", *Physics Today*, Vol. 17, p. 62.

Roscoe, K.H., Schofield, A.N., and Wroth, C.P. 1958, "On the Yielding of Soils", *Géotechnique*, Vol. 8, pp. 22-53.

South, D.L. and Daemen, J.J.K., 1981, "Laboratory Experiments of Borehole Plug Performance", Proc. 22nd U.S. Symp. on Rock Mechanics, M.I.T., Cambridge, Mass., pp. 110-114.

Stocker, R.L. and Ashby, M.F., 1973, "On the Rheology of the Upper Mantle", Reviews of Geophysics and Space Physics, Vol. 11, No. 2, pp. 391.

Terzaghi, K., 1946, "Rock Defects and Loads on Tunnel Supports", Rock Tunnelling with Steel Supports, Proctor and White, eds., The Commercial Shearing and Stamping Co., Youngston, Ohio, pp. 47-85.

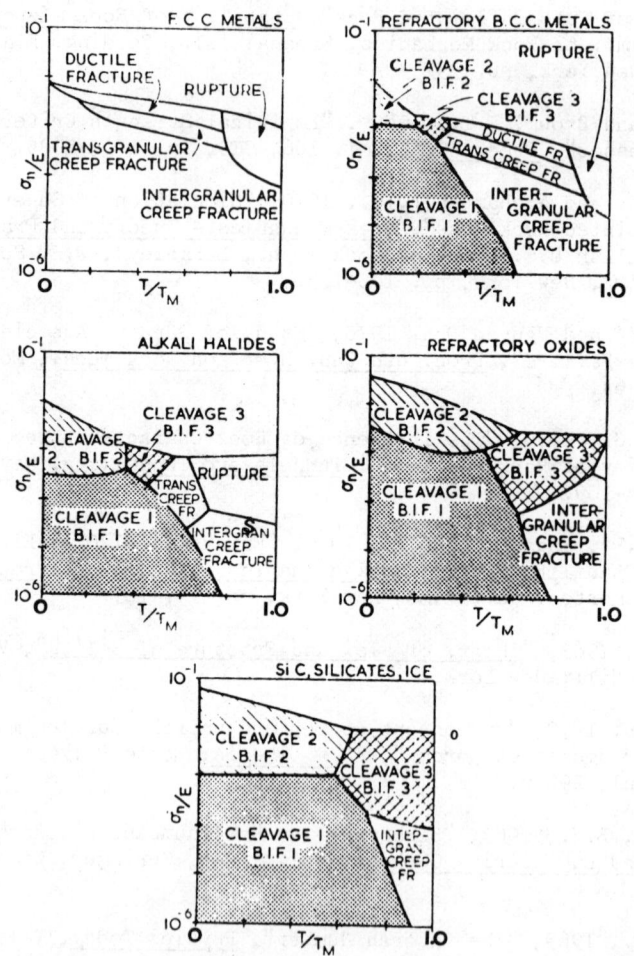

Fig. 1. Fracture mechanism maps for materials with different types of bonds (After Gandhi and Ashby, 1979).

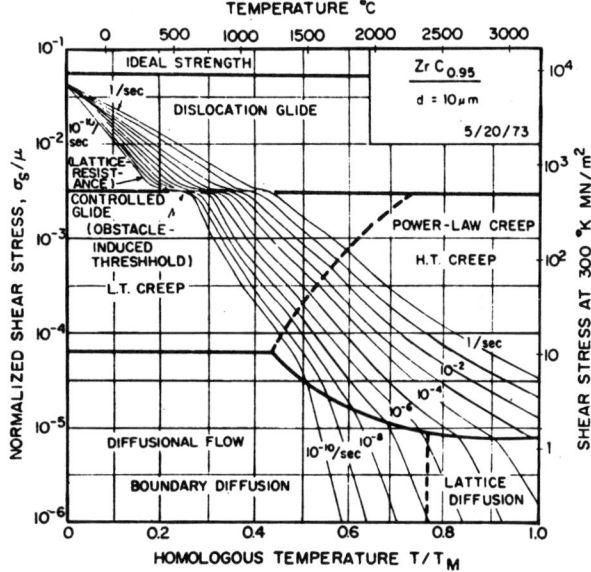

Fig. 2. Deformation mechanism map for ZrC (After Ashby and Frost, 1975).

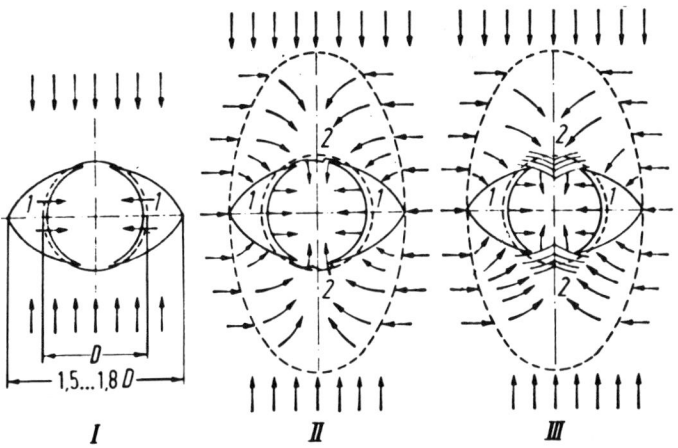

Fig. 3. Schematic representation of failure sequence around a tunnel (After Rabcewicz, 1963).

EXPLORATION

Chairman
Michael S. King
University of California
Berkeley, California

Co-Chairman
H. Reginald Hardy, Jr.
University of Pennsylvania
University Park, Pennsylvania

Keynote Speaker
Jeff R. Lytle
Lawrence Livermore National Laboratory
Livermore, California

require far greater precision and fineness of detail than is necessary for petroleum and mineral exploration. The petroleum and mineral explorationists are not working to develop techniques needed for engineering purposes. Hence, if significant advances are to be made in the state-of-the-art for construction site definition, engineering geophysicists need to respond to the responsibilities, opportunities and challenges before us. A subset of problems and possible future advances is given below.

Site Characterization Needs

Individuals well acquainted with the special needs of underground construction recently met and ranked which rock-mechanics problems related to underground construction need research. This effort was sponsored and documented[1] by the National Research Council. The participating individuals' attention was devoted to problems that had practical, day-to-day importance rather than to unanswered, fundamental research questions of uncertain impact. They specified that the research which would have the most important impact would include: 1) A summary and general description of the methodology used by practitioners; 2) an education of practitioners to raise the "average practice" closer to current "best practice;" 3) improved communication and sharing of experience; and, 4) documentation, quantification, and correlation of observables with geologic setting.

A second workshop independently organized by the National Science Foundation on site characterization and exploration also came to similar conclusions[2] as the above cited National Research Council study. This second group of workers emphasized: 1) The strong need to document case studies so that prior experience could be used to improve future performance; 2) the data base of in situ results needs to be reconciled with the data base of laboratory determined material properties. Both of these recognized needs create large bodies of data that can be handled effectively only by using computers. A call for an effective way of codifying this vast array of data was made. One possible way of solving part of this problem is given later.

In view of the above problems, certain of the responsibilities, opportunities and challenges facing the practitioners of geophysical diagnostics for discerning construction site conditions are addressed below.

Responsibilities

At certain times, crucial national programs may have the need for input from those with rock mechanics expertise. One such current program is the national program for geologic disposal of nuclear waste. This effort is overseen by the National Waste

Chapter 2

RESPONSIBILITIES OPPORTUNITIES AND CHALLENGES
IN GEOPHYSICAL EXPLORATION

R. J. Lytle

Lawrence Livermore National Laboratory
P. O. Box 5504, L-156
Livermore, California 94550

Abstract

   Geophysical exploration for engineering purposes is conducted to decrease the risk in encountering site uncertainties. Such studies are needed in construction of underground facilities. Current responsibilities, opportunities and challenges for those with geophysical expertise are defined. These include: replacing the squiggly line format, developing verification sites for method evaluations, applying knowledge engineering and assuming responsibility for crucial national problems involving rock mechanics expertise.

Introduction

   Designing, constructing and operating varied facilities (e.g., tunnels, geologic nuclear waste repositories, dams) requires a knowledge of the physical properties of the subsurface environment. Dependent upon the risk one is willing to accept, one's degree of confidence in his assessment of the actual site properties can assume a wide range. The lower the acceptable risk is, the higher the degree of confidence is desired. This paper addresses one means of attempting to lower the risk of unexpected site conditions--the use of geophysical methods for attaining knowledge of subsurface conditions.

   Geophysical exploration for engineering purposes is still in its relative infancy. Geophysical exploration for petroleum and minerals has seen extensive use. However, the scale and objectives of petroleum and mineral exploration and exploration for engineering purposes are greatly different. The construction site problem can

CHALLENGES IN GEOPHYSICAL EXPLORATION   19

Terminal Storage program and the Nuclear Regulatory Commission. Geotechnical instrumentation is a critical path item in the development of licensable waste repositories. An informal working group, consisting mainly of experimenters in nuclear waste isolation programs,[3] met in December 1981 to discuss experiences in using geotechnical instrumentation under conditions similar to those anticipated for repositories.

Quoting from the working group observations: "The consensus of the working group was that currently available instrumentation is inadequate for meeting the goals of the national program in Nuclear Waste Isolation. The extent of the inadequacy depends upon the objectives of the monitoring program. For instance, monitoring for model validation and verification requires precision and accuracy beyond that required in normal engineering applications of geotechnical instrumentation. Long term repository performance monitoring requires precision similar to that in other civil engineering programs, but requires life expectancy of instrumentation well beyond that of currently available instruments." The working group believes that development of geotechnical instrumentation has been neglected in the national programs. The lead time necessary to develop instrumentation, by and large, has not been recognized nor have the inadequacies of current instrumentation been fully appreciated. All of the programs reported a recurring common problem of having to use basically off-the-shelf equipment. While many objectives of the studies have been achieved using off-the-shelf equipment, there are many failures and difficulties in data analysis.

The working group believes that instrumentation development is a critical path item in the development of licensable waste repositories. While it may be possible to over-design a repository so that monitoring is minimized, nevertheless, they believe that long term monitoring, regardless of the over-design of the repository, will be required in the licensing process.

The working group recommends that federal agencies begin funding as soon as possible. There will be no time to field check the performance of newly developed instruments for long term monitoring other than during the execution of the at-depth test (ADT) program. It is critical, therefore, that these instruments be available at the beginning of the ADT so that their performance can be evaluated under field conditions in the likely repository medium before requiring long term monitoring in actual repositories. If we are to meet the requirements for instrumentation testing early in the ADT program (1983-1984), instrument development must begin immediately.

The working group cannot over-emphasize the need to begin this research program now. For the past three or four years, efforts have been made to modify off-the-shelf equipment with only minimal progress being made in developing the instruments needed. This three to four year hiatus has severely cut into the time needed for research. We feel the recommendations of this working group are legitimate, honest appraisals and estimates of work that needs to be done. We do not feel that monitoring all parameters of waste repositories for long periods of time is possible merely by using modified existing instrumentation."

Does the rock mechanics community at-large agree with this assessment of the state-of-the-art for instrumentation applied to geologic nuclear waste respositories? If it does, input from the society likely would have more impact that that of the working group alone. Action is needed now.

Opportunities

With the advent of new technology (e.g., fiber-optic sensors, improved computational and display capabilities), means of measuring, interpreting and displaying the site character are becoming feasible which were unthinkable a decade ago. The significant studies being made in computer technology are enormous, with even more rapid strides predicted. Within three years, for less than $5000, a computational capability will be available which this year costs $50,000. This capability will also be packaged in a smaller frame than required before. Thus, the field data processing of complicated character will be routine. What implication does this have? Consider the following.

Varied site characterization techniques exist and provide diagnostic information about the subsurface. However, a well-founded, well-accepted systematic procedure does not exist for deciding: 1) Which characterization techniques should be deployed for a particular site; 2) the sequential order of deployment of techniques; 3) when to terminate using site characterization techniques; and 4) how to integrate all available data to provide an effective diagnosis of the site character. Fortunately, prior work has been directed towards solving a similar problem.

I submit that suitable methods may exist for successfully implementing much of the desired codification and data handling via which one can make rational decisions. These goals may be achieved if the site characterization community shares it expertise and experimental data base.

Site characterization involves planning, executing, integrating and interpreting the results of an ordered sequence of observations and measurements pertinent to a geological site. Medical diagnosis involves planning, executing, integrating and interpreting the results of an ordered sequence of observations and mesurements pertinent to a human being. These different problems have important similarities. Both the site characterization and medical diagnosis procedures involve using data obtained for the particular site or patient. In addition, both procedures incorporate data learned from prior situations. This previously learned data consists of laboratory data, theoretical relationships, and case history data from prior sites or patients.

By adopting the knowledge engineering approach being successfully applied in medical diagnosis,[4] the following benefits may be accrued in diagnosis of a site.

1. There would be coordination among site characterizers of their case histories. This would permit the expansion of the codified data base, which would yield more accurate estimates of the relative likelihood of undesirable occurrences.

2. The present inadequate feedback loops for improvement of site characterization performance will be eliminated and a mechanism for providing continual updated improvements will be provided.

3. Communication pertinent to a particular site will become well organized, complete and available to all involved. This helps coordinate the efforts of each member of the site characterization team.

4. Use of this information can raise the expertise level of site characterizers in those areas where their expertise is not state-of-the art. This information thus serves as a very effective consultant.

5. To develop the data base and rationale for interpreting site data, the disparate groups that have knowledge of and interest in particular subsets of data are required to work together, to jointly define the suitable rationale, and to identify gaps in areas of needed knowledge. This communication between theorists, laboratory researchers, instrumentation engineers, data processors and site managers will define and update the current state-of-the-art and utility of particular measurements or observations.

These benefits are ones already identified by national experts in rock mechanics. Perhaps knowledge engineering may help us advance the state-of-the-art.

Challenges

National test sites where techniques can be tried and evaluated against varied targets of known character are needed. By developing such sites for varied conditions, the plethora of techniques can be evaluated under realistic conditions. By thus garnering such data, an unbiased set of observations can be made available as to the relative utility of various methods for detecting/defining various features of interest. By this means, organizations can more economically and realistically define how many and which techniques are needed to evaluate their site. In this way, the tradeoffs of risk versus the suite of techniques used can be better defined. Federal funding for establishing and maintaining such sites should be sought by the rock mechanics community. Similar sites run by hydrologists have had success.

One complaint regarding geophysicists' communication is their use of squiggly line data displays. This format seems to create communication barriers with non-geophysicists (e.g., contractors, geologists, civil engineers). It behooves us to develop alternative data display mechanisms that convey the essence of the squiggly line display. By suitably combining the experimental data, physical interaction mathematical description and data display mechanisms, it may be feasible to more effectively convey the meaning. Such improvements are needed now.

References

1. "Report for 1977 - U.S. National Committee for Rock Mechanics, A Summary of the Work Conducted during Calendar Year 1977," by the U.S. National Committee for Rock Mechanics and its Panels, published by the Assembly of Mathematical and Physical Sciences, National Research Council for the National Academy of Sciences, Washington, D.C., 1978.

2. C. H. Dowding, editor, "Site Characterization and Exploration," Proceedings of a Workshop sponsored by the National Science Foundation at Northwestern University, June 12-14, 1978, available from the American Society of Civil Engineers, 345 East 47th Street, New York, New York 10017.

3. D. G. Wilder, F. Rogue, W. R. Beloff, E. Binnall, E. C. Gregory, 1982, "Executive Committee Report Geotechnical Instrumentation Working Group Meeting," UCRL-87183, April 26, 1982.

4. E. H. Shortliffe, B. G. Buchanan, and E. A. Feigenbaum, "Knowledge Engineering for Medical Decision Making: A Review of Computer-Based Clinical Decision Aids," Proc. of the IEEE, September 1979, pp. 1207-1233.

Work performed under the auspices of the U.S. Department of Energy by Lawrence Livermore Laboratory under contract number W-7405-ENG-48.

Chapter 3

STANDARD PROGRAM FOR SITE SELECTION STUDIES IN SWEDEN
FOR A HIGH-LEVEL NUCLEAR WASTE REPOSITORY

compiled by Hans S Carlsson

Swedish Nuclear Fuel Supply Company/
Division Nuclear Fuel Safety
SKBF/KBS
Stockholm, Sweden

ABSTRACT

Like most industrial processes, nuclear power production creates wastes, which have to be handled and ultimately disposed of in a safe manner. Some of the wastes contain very long-lived radioactive substances and have to be isolated for very long periods of time before they have decayed to harmless levels.

Sweden, like many other countries with similar geology, intends to use crystalline rock for isolating the high-level nuclear waste from the biosphere.

The suitability of a rock mass for a repository needs to be investigated extensively. Apart from the political and practical considerations for a site selection, a detailed study has to be performed in terms of geology, hydrogeology, geophysics, geochemistry and rock mechanics.

The responsibility of the site selection studies in Sweden is entrusted to the Swedish Nuclear Fuel Supply Company, Division KBS (SKBF/KBS).

The paper gives a brief presentation of the "Standard Program" for the studies that will be carried out during 1980-1990. Modifications of the program are justified on the basis of gained experience or specific circumstances for each site. The program is compiled by the Geological Survey of Sweden (SGU) in cooperation with KBS.

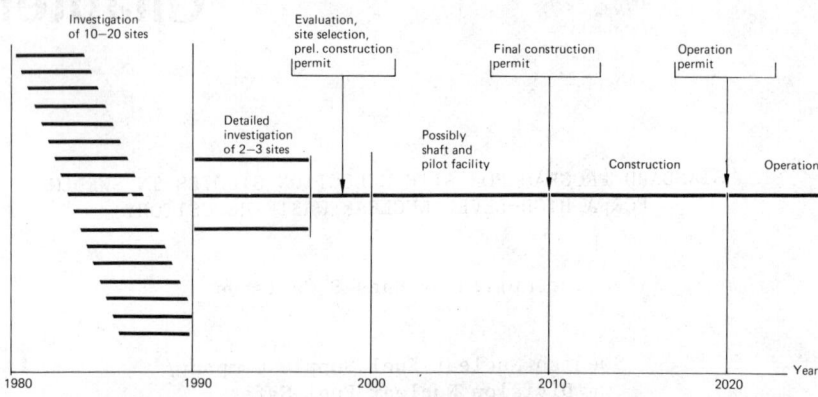

Fig. 1. General time schedule for a high-level nuclear waste repository in Sweden

INTRODUCTION

The current time-schedule for the selection of a site for a high-level nuclear waste repository in crystalline rock is given in the figure above. As can be seen from Figure 1, 10-20 sites will be investigated during 1980-1990 followed by a detailed study of 2-3 sites.

The overall investigations during this decade have been divided into four phases:

- selection of sites to be studied
- surface investigations
- borehole investigations
- final evaluation and modelling

Given below is a general description of the investigations within each phase.

It is important to note that the objective of these investigations is to find a site that fulfills the requirements of a safe disposal of high-level waste. That requirement is not the same as selecting the absolutely best site in Sweden.

PHASE 1 - SELECTION OF SITES TO BE STUDIED

The existing literature and maps in the fields of geology, hydrology and geophysics are used as a base for this study. The following main aspects are considered

SITE SELECTION STUDIES IN SWEDEN

- flat topography
- great distance between larger tectonic zones
- homogeneous structure of the rock
- low seismic activity
- low waterbearing capacity
- non-existence of valuable minerals and ores

Other factors influencing the selection of sites are landowners; availability; population; cultural and nature reservs etc.

A brief geological field study will follow if the response from the landowner is positive.

Diamond drilling of one vertical borehole with a diameter of 56 mm and to a depth of 1000 meters will follow the field study. The core is mapped and the hydraulic conductivity of 25-meter sections in the borehole is determined. Furthermore, a geophysical log of the hole is carried out. The following measurements are made:

- borehole deviation
- resistivity
- spontaneous potential
- natural gamma
- differential potential
- temperature, Eh and pH
- electrical conductivity of the borehole liquid

Three to five sites a year are expected to remain after completion of the above-mentioned studies. One or two sites out of these will be selected for further investigations each year.

PHASE 2 - SURFACE INVESTIGATIONS

This phase includes geological and geophysical mapping and the definition of a drilling program.

Geological Mapping

The geological surface investigations start with a regional mapping of an area bounded by topographic features. The size of the area is normally in the range of 100-500 km^2. Existing geological maps and a cooperation with other geologists familiar with the area in question form the base of this study.

The detailed geological mapping is concentrated to a local area of 4-6 km^2 within the regional area. Types of rocks, intrusions, tectonic features such as foliation, zones of weakness, fracture distribution, fracture fillings etc. are recorded.

Geophysical Surface Measurements

The geophysical surface measurements cover the entire local area (4-6 km^2) but profiles extending into the regional area are also performed. The data from the measurements should give indications of thickness of the overburden, shallow waterbearing discontinuities and mineralizations.

The following methods are used

- magnetic
- resistivity
- induced polarization
- electromagnetic measurements
- seismic measurements

The investigations mentioned above form the base for the definition of the drilling program. The location and the number of boreholes are dependent upon the distribution and the orientation of the existing fracture zones as interpreted from the surface investigations.

PHASE 3 - BOREHOLE INVESTIGATIONS

This phase includes percussion and diamond drilling, geological, geophysical and hydrological investigations in boreholes. The investigations will end up in a descriptive model of the area.

Drilling

Percussion drilling with a diameter of 115 mm and to a maximum depth of 200 meters is performed in order to define the strike and dip of the shallow fracture zones. The water flow in these zones is also recorded as well as the groundwater level.

The diamond drilled boreholes, with a diameter of 56 mm and a depth of about 600-800 meters are normally inclined and intentionally located to intersect the continuation of the previously observed fracture zones. The collar and dip of the holes are thus determined by the geological surface mapping and the shallow percussion drilling. The drilling water is filtered and a tracer is added.

The cores from the drilling are logged using standard logging procedures. Samples are taken for physical determinations and chemical analyses of the fissure fillings.

Cleaning of the Borehole

The drilling water used for cooling the diamond bit is normally taken from one of the percussion drilled holes. The contamination of the deep groundwater from the shallow drilling water is reduced by pumping nitrogen gas through a hose to the bottom of the diamond drilled hole. The nitrogen bubble will force the contaminated water towards the collar of the hole. Remaining drilling mud will also be forced out of the hole.

Geophysical Measurements

The geophysical measurements described in phase 1 are carried out in all diamond drilled holes. Resistivity and gamma logs are also performed in the percussion drilled holes. Crosshole techniques, currently under development in Sweden, are used when found appropriate and possible.

The following characteristics of the core are measured

- density, porosity, resistivity
- induced polarization, susceptibility
- remanent magnetism

Hydrogeological Investigations

The objective of the hydrogeological investigations is to determine the hydraulic conductivity, piezometric pressure and the porosity of the rock. These parameters are essential for the description of the groundwater flow in the rock mass.

The determination of the hydraulic conductivity is done by using transient injection tests under constant pressure. The entire holes are tested along 25-meter sections and thereafter in specially selected 5-meter sections. A conventional testing equipment and a new designed equipment are used. The old equipment uses steel rods whereas the new uses a multihose to keep the packers in place. The new equipment is highly automated and calculates the k-value immediately after the test by the help of a desk-computer at each site.

The piezometric pressure is determined both in conjunction with the determination of the k-values in each diamond drilled borehole as well as in separate tests in selected holes and sections.

Interference tests (pump tests) are performed in percussion drilled holes in order to get a better determination of the k-value in highly conductive zones.

Hydrochemical Investigations

The objective of these investigations is to document the chemical composition and age of the groundwater.

The sampling at each site is made in two diamond drilled boreholes each located in assumed groundwater inflow and outflow areas. Each borehole is sampled in four sections located at different depths along the hole and analyses with respect to the following constituents are performed:

- chemical-physical composition
- sulphide content
- tritium
- O^{18}/O^{16}
- C-14
- C^{13}/C^{12}
- helium
- radon-radium
- uranium-thorium
- strontium
- total organic content

In addition, pH, Eh, pS, pO_2, the electrical conductivity and the temperature, will be determined at the site and at each section of the holes.

PHASE 4 - FINAL EVALUATION AND MODELLING

The main objective of this last phase is to describe the groundwater flow through a tentative repository. Numerical models describing flow-times and nuclide release-curves to the biosphere through the investigated rock mass will be presented.

Descriptive Model

A descriptive model based on the previous investigations of the geological, hydrological, geophysical and geochemical conditions is presented. A special emphasize is made on the characterization of the hydraulic units in the rock. The descriptive model should end up in a definition of the location of the repository within the rock mass.

Numerical Models

A hydrothermal model, based on the descriptive model, will describe the groundwater flow in the investigated rock mass. The following parameters are used as input data

- groundwater level
- distribution and orientation of fracture zones
- porosity
- hydraulic conductivity for the homogeneous rock mass and for the fracture zones

Other input data, such as changes in viscosity of the fluid due to a change in temperature, will also be taken into account. Work on adding the changes in stresses and thus a change in aperture of existing fractures due to a change in temperature is in progress.

The nuclide migration model, using input data from the hydrothermal model, will take the fracture surface sorption into consideration as well as the diffusion through microfissures into the rock matrix. The calculations will thus end up in the final interpretation of the transportation of nuclides from the repository to the biosphere.

REFERENCES

Brotzen, O., "Site Investigations for a Nuclear Waste Repository in Crystalline Rock", Proceeding of OECD/NEA Workshop on Siting of Radioactive Waste Repositories in Geological Formations, Paris, France 19th-22nd May 1981.

Rasmuson, A., Narasimhan, T.N., Neretnieks, I., "Chemical Transport in a Fissured Rock: Validation of a Numerical Model". Accepted for publication in Water Resources Research.

Thoregren, U., "Standardprogram för Geologiska Områdesundersökningar avseende Förvaring av Högaktivt Avfall", Geological Survey of Sweden, Uppsala, Sweden, December 1981.

Thunvik, R., Braester, C., "Hydrothermal Conditions Around a Radioactive Waste Repository", SKBF/KBS Technical Report 80-19, Stockholm, Sweden, December 1980.

Chapter 4

APPROACHES TO EVALUATING THE PERMEABILITY AND POROSITY OF
FRACTURED ROCK MASSES

Thomas W. Doe, Jane C.S. Long, Howard K. Endo, and Charles R. Wilson

Earth Sciences Division
Lawrence Berkeley Laboratory
University of California
Berkeley, CA 94720

ABSTRACT

An approach to treating flow through fractured rocks is presented which involves (1) determining statistical distributions for fracture area, density, orientation, and aperture from field data, (2) computer generation of fracture systems from the statistical distributions, and (3) analysis of the permeability and porosity of the fracture systems. The fracture density and orientation data may come from core logging or mapping. The hydraulically appropriate apertures and areas of the fractures come from transient flow analyses of single fracture well tests. These tests may also yield information on fracture deformability. Analysis of flow through computer generated fracture systems shows that an equivalent porous medium permeability can be defined for some fracture systems if a large number of fractures is taken into account.

INTRODUCTION

The heterogeneity of fracures introduces a major element of uncertainty to rock mass exploration for hydrologic or mechanical purposes. This paper presents a stochastic approach to characterizing fluid flow in fractured rock masses. The basic philosophy is also applicable to mechanical characterization. The approach consists of (1) determining probability density functions for the parameters of the fracture system, (2) generating by computer fracture systems consistent with the field data, and (3) evaluating the permeability and porosity of the fracture systems. The fracture systems generated by the computer are based on the probability density functions of the geometric parameters. Spacing and length functions have negatively skewed distributions (lognormal or negative exponential) as shown by Priest and Hudson (1976) and Baecher and others (1977). Snow (1970) has suggested a lognormal form for aperture distributions. The computer generated fracture networks may contain fractures which are

PERMEABILITY, POROSITY OF FRACTURED ROCK

more permeable or more continuous than any of the fractures measured in the field. In this way the model can take into account large fractures which have been overlooked during the site exploration. In this study the model is used to determine representative values of hydrologic parameters; however, the model may also be used to analyze the effects of rare, very large fractures.

DETERMINING SIZE AND INTERCONNECTION OF FRACTURES FROM WELL TESTS

The aperture, extent, and interconnection of fractures are key factors in the transmission of groundwater through crystalline rocks. Unfortunately, the tools available for fracture characterization from boreholes are generally insensitive to these parameters. Visual or optical inspections of the core or the borehole walls and conventional geophysical logging techniques can indicate the location and orientation of fractures in a borehole, but they cannot measure the open area of the fractures or the hydraulic effective aperture. These two values can only be obtained through flow tests.

Over the last two years LBL has developed an approach to permeability testing which may be practical for gaining field information on the extent, interconnection, and possibly the deformability of fractures in situ. We consider single fractures to behave like individual confined aquifers with a transmissivity, T, related to the aperture by the cubic relationship of flow

$$T = \rho g \frac{e^3}{12\mu}$$

where ρ is fluid density, g is gravitational acceleration, μ is kinematic viscosity, and e is fracture aperture. Storativity, S, is related to the fracture's normal stiffness by

$$S = \rho g \left(\frac{1}{k_n} + e\beta \right)$$

where k_n is fracture normal stiffness and β is fluid compressibility. The transmissivity and storativity can be obtained from well tests. Of the available well techniques, we have concentrated on the constant pressure injection test for several reasons. It is the same test as the packer or Lugeon test used in civil engineering, except that steady flow is not assumed and careful attention is paid to the transient flow rate. Type curves used to analyze the data are similar to those used for pump tests; however, flow is the transient factor instead of pressure (Hantush, 1959). Constant pressure tests have a major advantage over transient pressure tests, such as the pump test or the pulse test, in that wellbore storage and the effects of deformability of the test equipment are virtually eliminated because the well and the test equipment are maintained at constant pressure. Another advantage of the constant head test is that it can be performed rapidly. Provided that low flow rate meters are available, measurements can be made within a few hours regardless of permeability, and the only reason for running longer term tests is to affect a larger volume of rock.

Mapping studies have generally shown that fractures are finite in extent. Figure 1 shows how finite fractures might intersect a well. The transient flow rate curve for finite fractures may have up to three major segments which are shown in Figure 1. The initial segment we call the _infinite_ portion because this segment follows the type curve for an infinite fracture. The curve match for this segment gives information on the transmissivity (or aperture) and the storativity (or stiffness) of the fracture. The second portion we call _depletion_ by analogy with the terminal phase of production from a well. During this phase of the test the pressure front from the injection has reached the boundary of the fracture and the flow is rapidly declining. The time at which the flow deviates from the infinite curve can be used to estimate the size of the fracture. The third stage of the test we call _leakage,_ which is a leveling out of the flow rate due to either leakage into other fractures, or, if the fracture is sufficiently extensive, to leakage into the intact rock. The three stages are not all necessarily present in a given well test. If the fracture is very large or if it intersects a highly conductive fracture near the well, the depletion portion may be absent as the flow goes directly from gradual transient decline to a steady state.

To measure low permeabilities using the constant head test, LBL has developed a field system for performing tests with high resolution of flow rate. The system (Figure 2) includes (1) turbine flowmeters for flow rates of 5.0 gpm down to .001 gpm, (2) positive displacement flow meters for flow of .001 gpm down to .00001 gpm and lower, (3) a downhole valve for shutting in the test zone and instantaneous starting of tests, and (4) a minicomputer system for immediate data analysis and graphic display.

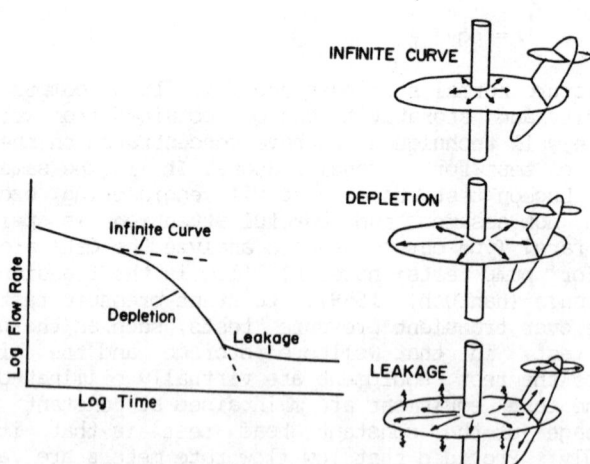

Figure 1.   Idealized flow rate versus time curve for constant pressure injection test into a single fracture.

PERMEABILITY, POROSITY OF FRACTURED ROCK

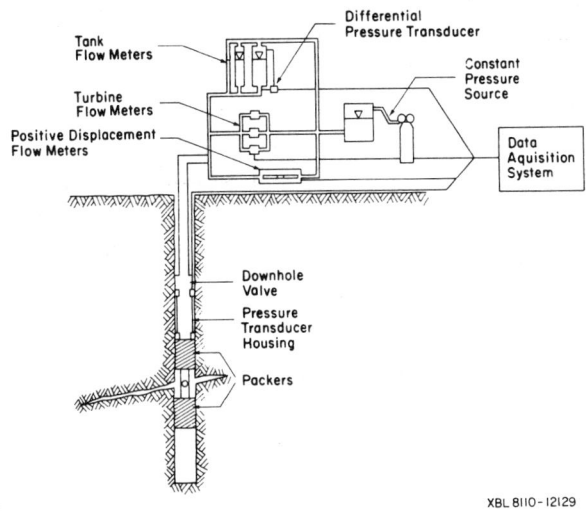

Figure 2. Schematic diagram of equipment for constant pressure well testing.

A series of twelve constant pressure tests was run in a 90 foot deep borehole in granite in the Sierra Nevada foothills in the spring of 1981. Three types of flow rate curves were found in the field tests. Figure 3A shows a curve for a finite fracture with leakage. Figure 3B possibly represents a fracture with a major connection to another fracture, as the depletion portion of the curve is absent. The third curve, Figure 3C, shows steady flow followed by transient decline. This curve does not fit any of the type curves developed. Calculated effective apertures ranged from .013 to .12 millimeters. Storativity values ranged from 4×10^{-8} to 8×10^{-6}, which corresponds to fracture normal stiffness values of 10^3 to 10^5 MPa/mm. The sizes of the fracture ranged from 0.4 in radius to 400 meters. Data such as these can be used in the numerical analysis described in the following section.

HYDROLOGIC ANALYSIS OF FRACTURE NETWORKS

A previous paper (Witherspoon, and others, 1981) discussed the theory of flow through fractured rock and homogeneous anisotropic porous media. That paper described how to determine when a fractured rock behaves as a continuum. A fractured rock behaves like an equivalent porous medium when: (1) there is an insignificant change in the value of the hydraulic conductivity, K, with the small addition or subtraction to the test volume, and (2) an equivalent conductivity tensor exists which predicts the correct flux when the direction of the constant gradient is changed. Fracture systems are generated by computer using the following procedure. For each fracture set, the centers of fractures were randomly selected in a two-dimensional

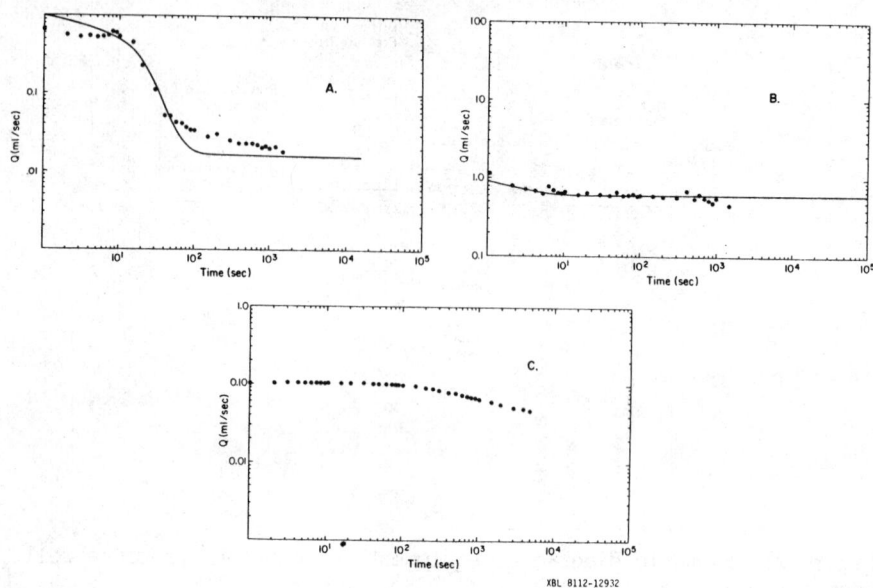

Figure 3. Flow rate curves from single fracture tests at Raymond, Calif.

region. The fracture passing through each center is assigned a length and an aperture selected from negatively skewed probability density functions (for example, negative exponential or lognormal). The orientation can also be varied randomly about the mean orientation of the fracture set. Once all fracture sets are generated, the networks are converted to line element meshes for numerical flow analysis. Square regions of the fracture mesh are selected for flow tests, which consist of imposing a hydraulic gradient across the square region. The square regions can be rotated to give conductivity in different directions. The conductivity is defined on the basis of the quantity of inflow to the square region. The results of the flow tests are then plotted as $1/\sqrt{K}$ versus direction in polar coordinates. If the conductivity plots are ellipses then the fractured rock is behaving on this scale as an approximate, homogeneous, anisotropic porous medium. The details of this method are discussed in Long, and others (1982). Some further applications of this method are described below.

In order to determine how the density of fractures affects the hydraulic behavior, three examples were analyzed in which the same fractures were squeezed into successively smaller areas. The directions of the principal permeabilities did not change as the fracture density was increased, however the overall conductivity ellipse became more regular with increased fracture density and larger numbers of fracture intersections. That is, the hydraulic behavior of the fracture systems becomes more like that of a homogeneous, anisotropic material as the fracture density increases.

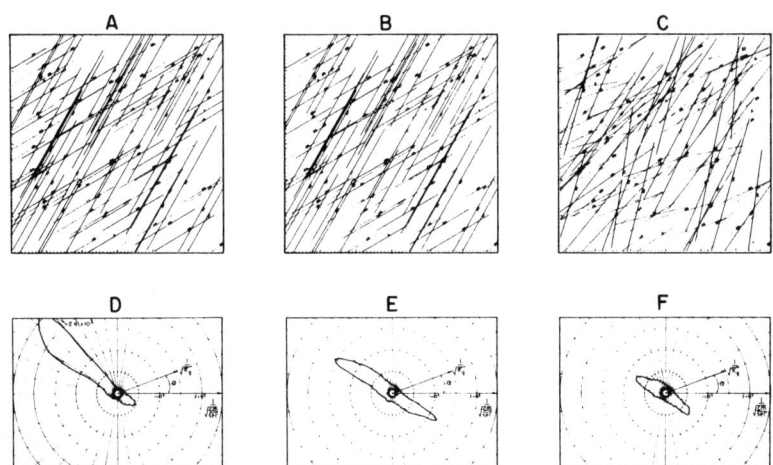

Figure 4. Effect of orientation or aperture distribution on conductivity ellipse.

The effect of distributing aperture or orientation is illustrated by comparing systems in which only aperture (Figure 4A) or only orientation (Figure 4C) had been distributed with systems in which all parameters are uniform (Figure 4B). The conductivity plot was less elliptical for the case in which only aperture was allowed to vary (Figure 4D). This is because not all the conductors are of equal strength. Varying the orientation of the fractures results in a more elliptical conductivity plot (Figure 4F). In this case, the number of fracture intersections increases because fractures of the same set are no longer parallel and can now intersect each other. The number of hydrualic connections is thus increased, and the conductivity plot becomes more symmetric and regular. Fracture systems with distributed orientations behave more like homogeneous porous media than do systems with uniform orientations. Fracture systems with distributed apertures behave less like homogeneous media than uniform aperture systems.

In order to illustrate the scale effect, a system of fractures was generated with two perpendicular sets of fractures, all with the same aperture and length. The orientation distribution for each set was the same. The resulting fracture system is shown in Figure 5A. Theoretically, two orthogonal sets with equal characteristics should have a roughly circular conductivity plot. Random variations from the circle can be due only to insufficient sample size. Figures 5 B to D show the conductivity ellipses for a flow region with steadily increasing size.

For Figure 5B, the results are erratic. Only a few fractures are included in the sample. Although in this figure most of the fractures transect the flow region, for certain values of the rotation angle, no fractures intersect the upstream side of the square region. Then the

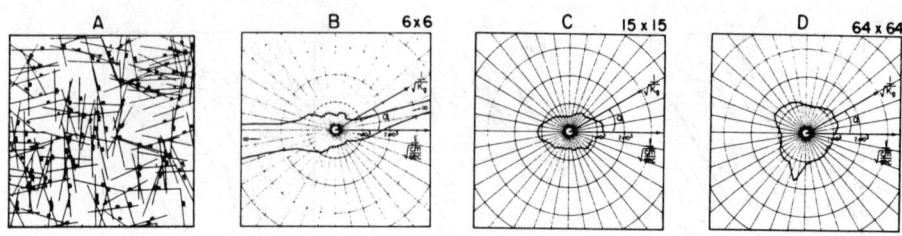

Figure 5. Scale effect study on an orthogonal fracture system.

conductivity is zero, and $1/\sqrt{K}$ is infinite. In Figure 5C, a 15x15 mesh, a fairly regular symmetric ellipse is obtained. For this flow region size, many of the fractures transect the entire flow region. However, the ellipse in Figure 5C is not circular as expected. Further increase in sample size causes oscillation in the form of the conductivity ellipse. As more fractures are gradually added to the sample, the effect of each fracture is to deform the ellipse in some way. However, the trend from B to D is toward the expected circular ellipse.

POROSITY IN FRACTURED MEDIA

Considering only the fracture void volume, three types of porosities can be defined: secondary porosity, rock effective porosity, and hydraulic effective porosity. The void volume per unit sample volume is known as the secondary porosity. The rock effective porosity considers only those pores which are interconnected, eliminating dead end and isolated void regions. The continuity and connectivity of the fracture network is associated with the rock effective porosity. The hydraulic effective porosity is the ratio of the specific discharge (q) to the average linear velocity (v) and governs the rate at which solute will migrate in the fracture system.

The secondary porosity and rock effective porosity can be determined by sequentially examining each fracture in the system. The hydraulic effective porosity requires the determination of the two flow properties v and q. The average linear velocity is the mean velocity of a fluid particle has traversing a linearized flow path through the medium (Figure 6) and is defined as $v=L/T$, where L is the sample length and T is the average travel time of the fluid particles.

In order to determine the average travel time, the movement of fluid particles in the network must be studied. The mechanical transport model devised for tracing fluid particle movement assumes that each fracture behaves hydraulically as a parallel plate with laminar flow prevailing. Instead of following individual fluid particles, packets of fluid particles contained in a streamtube are monitored. A streamtube traces the movement of a given quantity of

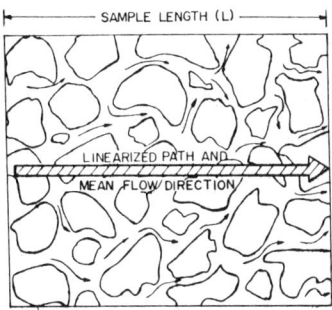 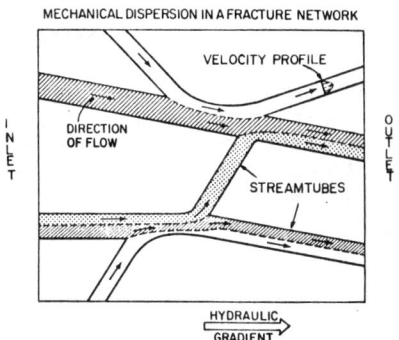

Figure 6. Left, fluid particle transport in a porous medium; right, mechanical transport in a fracture network.

flux through the system. The parabolic Poiseuille velocity profile is used to determine the width occupied by a streamtube in the fracture. Also by employing the fact that streamtubes cannot cross laminar flowlines, the flowpaths and travel times of the streamtubes can be determined for the fracture system (Figure 6) and the average linear velocity computed. The other flow property, specific discharge, is obtained by dividing the average total flux by the flow area. These two flow properties enable one to determine the hydraulic effective porosity (q/v).

In porous media, it is generally assumed that the three porosities are equal due to the complexity of pore structure and geometry. Clearly, each porosity relates to a unique rock property and will not necessarily be equivalent. Work has been initiated to examine the relationship between each type of porosity for stochastically generated fracture systems.

CONCLUSIONS

Constant head injection tests on single fractures can be used to obtain data for the aperture, area, and interconnection of fractures, provided a transient analysis of the data is used. Storativity values may be helpful in deducing the normal stiffnesses of fractures. The computer studies on synthetic fracture systems can use such data to examine the hydraulic behavior of fracture networks. The models can be used to identify critical fracture parameters which control equivalent porous medium permeability and porosity.

ACKNOWLEDGMENTS

This work was supported by the Assistant Secretary for Nuclear Energy, Office of Waste Isolation of the U.S. Department of Energy under contract number DE-AC03-76SF0098. Funding for this project was administered by the Office of Nuclear Waste Isolation at the Battelle Memorial Institute.

REFERENCES

Beacher, G., N.A. Lanney, and H. Einstein, 1977, "Statistical Descriptions of Rock Properties and Sampling,", Proceedings 18th U.S. Rock Mechanics Symposium, University of Colorado.

Hantush, M., 1959, "Nonsteady Flow to Flowing Wells in Leaky Aquifers", Journal of Geophysical Research, v. 64, no. 8.

Long, J.C.S., J. Remer, C. Wilson, and P. Witherspoon, 1982, "Porous Media Equivalents for Networks of Discontinuous Fractures," Water Resources Research, in press.

Priest, S.D., and J. Hudson, 1976, "Discontinuity Spacings in Rock," Int. Jour. Rock Mech. & Min. Sci., v. 13, p. 135-148.

Snow, D.T., 1970, "The Frequency and Apertures of Fractures in Rock," Int. Jour. Rock Mech. & Min Sci., v. 7, p. 23.

Witherspoon, P.A., Y. Tsang, J. Long, and J. Noorishad, 1981, "New Approaches to Problems of Fluid Flow in Fractured Rock Masses," Proceedings 22d U.S. Rock Mechanics Symposium, M.I.T.

Chapter 5

ACOUSTIC BOREHOLE LOGGING IN A GRANITIC
ROCK MASS SUBJECTED TO HEATING

M. S. King and B. N. P. Paulsson

Lawrence Berkeley Laboratory
University of California
Berkeley, California 94720

ABSTRACT

Four vertical boreholes in the vicinity of an electrical heater simulating a canister of nuclear waste in a granitic rock mass have been logged with an acoustic borehole sonde before and after thirteen months' heating. Excellent control of the structural geology and fractures present in the rock mass was provided by the considerable amount of oriented core recovered from the four boreholes used for acoustic logging and the large number of others used for a variety of rock mechanics instrumentation. Compressional and shear-wave velocities were measured as a function of stress on some twenty core samples, water-saturated and dry, which were recovered from the four acoustic logging boreholes. Further control was provided by a crosshole acoustic monitoring program conducted before, during and after heating between the same four boreholes, and by measurements of acoustic velocities in an isolated block of granite in the same heater drift, which was subjected to an increase in temperature and to changes in uniaxial stress. It is concluded from the acoustic logging tests, taken in conjunction with the acoustic cross-hole and heated block experiments, that acoustic techniques:
1. Provide a means for locating and delineating systems of weakly-bonded or open fractures within the rock mass;
2. Provide a sensitive means for monitoring a rock mass for changes in moisture content or structural damage to the rock fabric induced by thermal or mechanical loading;
3. Together with the rock bulk density yield the elastic constants of the rock mass.

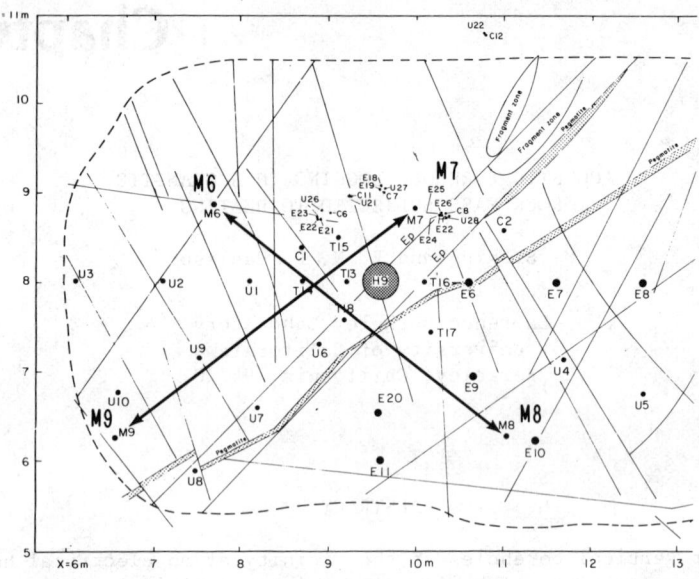

Figure 1. Sites of test boreholes in heater drift

INTRODUCTION

The propagation of elastic waves provides a convenient means for studying the elastic properties of rock. Since the presence of discontinuities such as fissures, fractures and joints can strongly affect the elastic properties of a rock mass, clearly the elastic wave velocities also will reflect their presence. The acoustic borehole log is employed to measure the elastic wave velocities over short distances of rock adjacent to the borehole; its response will therefore be governed by the elastic properties of the rock substance and by the presence of discontinuities intersecting the borehole over the interval measured.

In this paper are described the results obtained from acoustic borehole logging performed in four vertical, dry boreholes located in the vicinity of a vertical, cylindrical electrical heater embedded in a granitic rock mass. The boreholes, each 56 mm diameter and 10 m in depth, marked M6, M7, M8 and M9 in Figure 1, were logged before and after thirteen months heating of the rock mass. The electrical heater was 400 mm in diameter and 3 m long, with its centerpoint 4.2m below the drift floor. The study reported here was performed as part of a comprehensive rock mechanics and geophysics research program (Witherspoon et al, 1979) associated with large-scale heater tests conducted in an abandoned iron-ore mine in central Sweden.

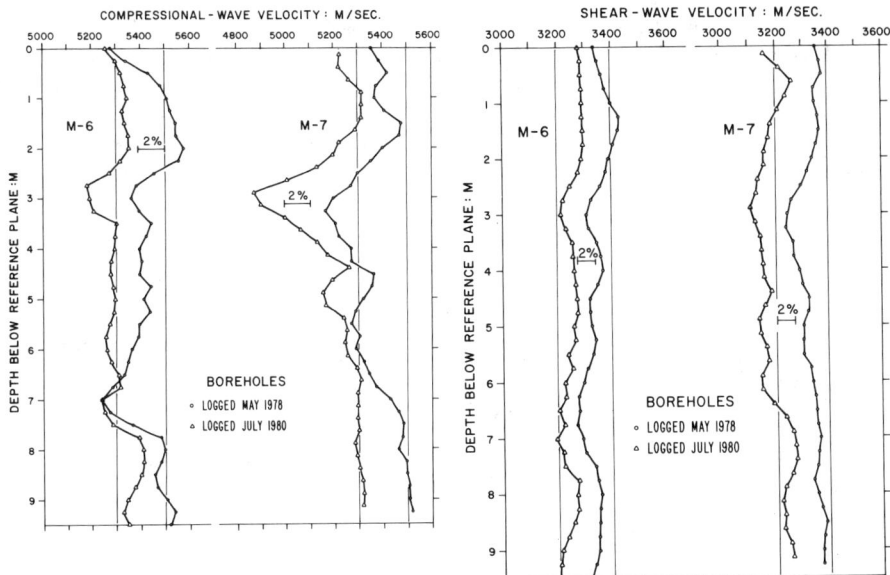

Figure 2. Compressional-wave velocities M6 and M7

Figure 3. Shear-wave velocities M6 and M7

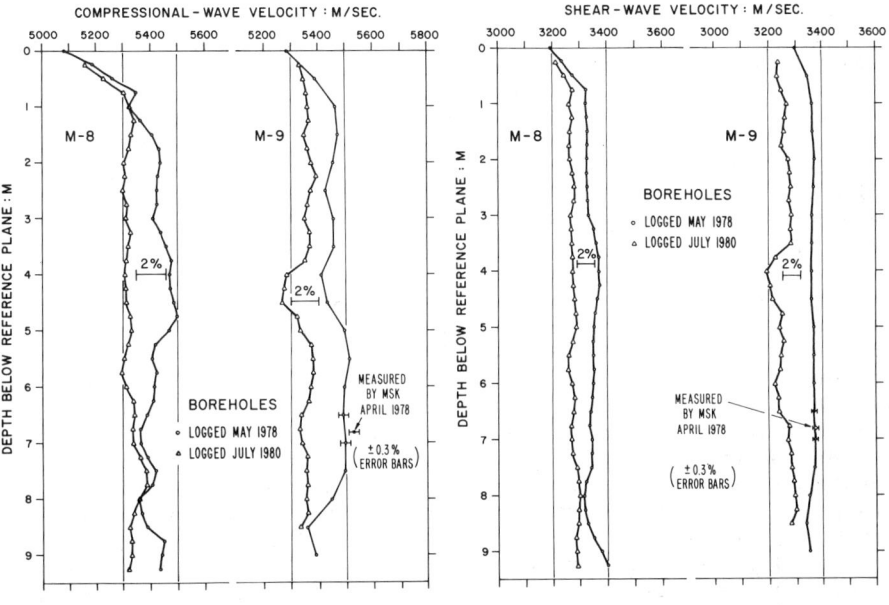

Figure 4. Compressional-wave velocities M8 and M9

Figure 5. Shear-wave velocities M8 and M9

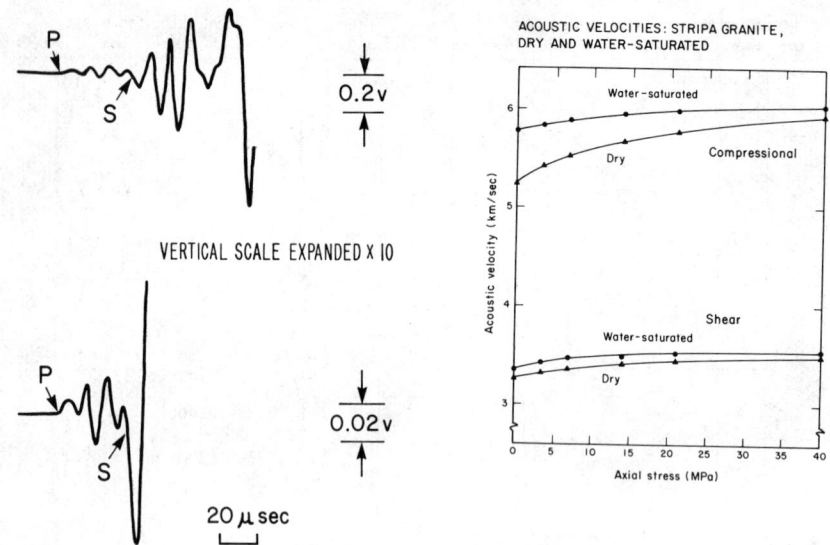

Figure 6. Typical oscilloscope trace of received signal in granite

Figure 7. Laboratory compressional and shear-wave velocities

The acoustic borehole logging system used in this study was originally developed (King et al, 1974) for assessing the quality of rock in dry boreholes adjacent to underground openings, and had been used for this purpose in a number of underground mines in Canada. The logging sonde consists of a pair of hydraulically-actuated transmitter and receiver shoes which are forced out against the borehole wall when measurements are to be made (King et al, 1974; 1978). An oscilloscope trace of a received signal typical of those obtained while logging dry boreholes in granite is shown in Figure 6. It will be observed that compressional and shear-wave arrivals are easily identified.

Excellent control of the structural geology and discontinuities present in the rock mass was provided by the considerable amount of oriented core recovered from the four boreholes used for acoustic logging and the large number of others used for a variety of rock mechanics instrumentation (Paulsson and Kurfurst, 1980). Compressional and shear-wave velocities were measured as a function of stress on some twenty intact core samples, water-saturated and dry, which were recovered from the four acoustic logging boreholes (Paulsson and King, 1980a). Further control was provided by a cross-hole acoustic monitoring program conducted before, during and after heating, between the same four boreholes (Paulsson and King 1980); and by measurements of acoustic velocities in an isolated block of granite located in the same heater drift (King and Paulsson, 1981), which was subjected to an increase in temperature and to changes in uniaxial stress.

ACOUSTIC BOREHOLE LOGGING

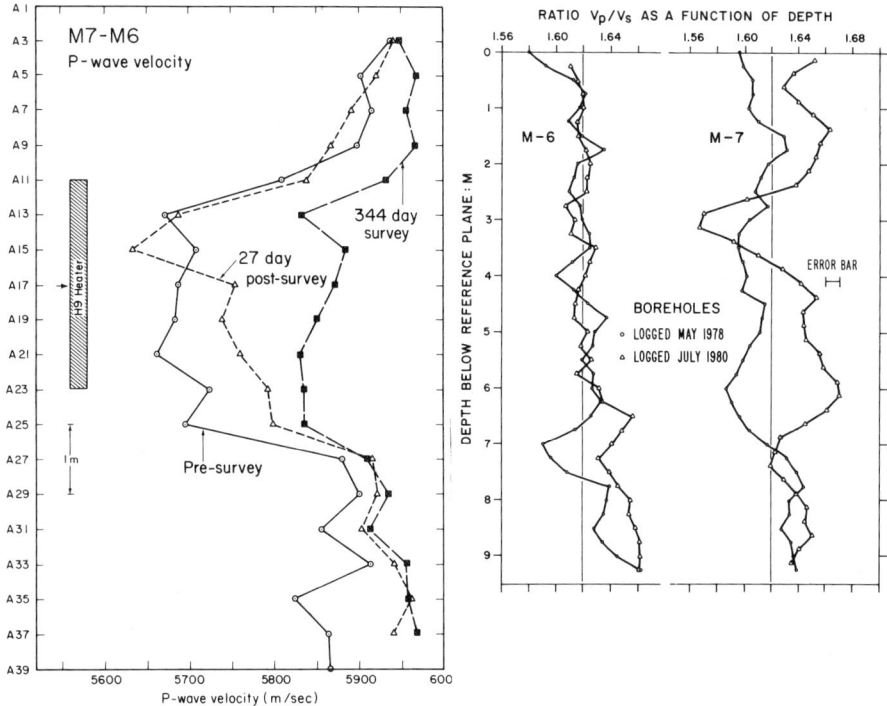

Figure 8. Cross-hole survey M6 and M7
(note the increase in velocity as rock mass was heated)

Figure 9. Ratio V_p/V_s for M6 and M7

DISCUSSION OF RESULTS

The acoustic velocity logs for the four boreholes, made 3 months before the heater was first turned on and 10 months after it was turned off, are shown in Figures 2, 3, 4 and 5. An oscilloscope trace of a received signal typical of those obtained in the granitic rock mass is shown in Figure 6. The sharp breaks associated with compressional and shear-wave arrivals enable calculations of compressional and shear-wave velocities to $\pm$ 0.3 percent. Compressional and shear-wave velocities measured on 21 intact core samples recovered from the heater drift as a function of axial stress, in their dry and water-saturated states, are shown in Figure 7 (Paulsson and King, 1980a).

Figures 2 to 5 indicate that there is an approximately 3-percent reduction in both the compressional and shear-wave velocities over the 26-month time interval between logging the boreholes. Water was found to seep into the boreholes to a varying degree throughout the test period. This water was collected at frequent intervals, and so conse-

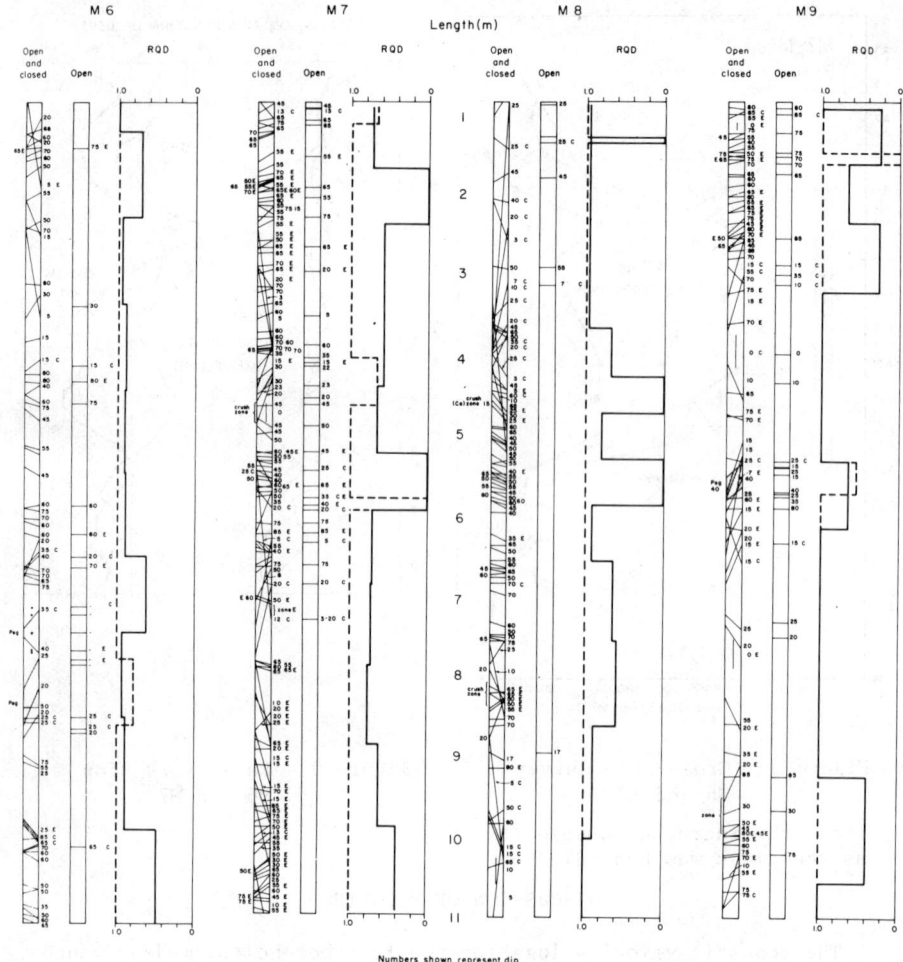

Figure 10. Fracture logs and RQD for test boreholes

quently the boreholes were logged in their dry state. A comparison of the maximum velocities recorded in Figures 2 to 5 in zones where there were few or no intersecting fractures (see Figure 10) and Figure 7 suggest that the rock immediately adjacent ot the boreholes was incompletely water-saturated during both periods they were logged. The decrease in velocity can probably be attributed partly to a general relaxation of the rock mass upon cooling and partly to a change in degree of water saturation in the rock adjacent to the boreholes. Similar behavior was noted in the cross-hole acoustic monitoring program (Paulsson and King, 1980).

The compressional wave velocities in boreholes M6 and M7 before heating both show lower velocities over the interval 2m to 8m below the drift floor. Through this zone passes a pegmatite dike and within it are found an abundance of weakly-bonded calcite-filled fractures, as shown in Figure 10. Results of the cross-hole acoustic surveys between M6 and M7, shown in Figure 8, indicate the same low-velocity behavior in this zone. The influence of calcite-filled fractures has been discussed by Paulsson and King (1980a).

Boreholes M6, M8 and M9, situated furthest from the heater, were at no point subjected to a temperature greater than 40°C during the 13-month heating period. Borehole M7, situated 0.9 m from the heater, was however subjected to a maximum temperature of approximately 90°C over a portion of its length opposite a point near the top of the heater. Over this particular length, a reduction of some 5 percent in compressional-wave velocity was observed upon conclusion of the heater tests, as shown in Figure 2. This behavior was consistent with that observed with the velocities measured within the isolated block of granite after it had been subjected to a temperature of approximately 100°C (King and Paulsson, 1981). The ratio of compressional to shear-wave velocities (V_p/V_s) over the lengths of boreholes M6 and M7 is shown in Figure 9. The reduction in the ratio V_p/V_s opposite the zone of maximum temperature is indicative of the increased microfracturing caused by the increase in temperature.

CONCLUSIONS

It is concluded from the acoustic borehole logging tests, taken in conjunction with the acoustic cross-hole and heated-block experiments, that acoustic techniques:
1. Provide a means for locating and delineating systems of weakly-bonded or open fractures within the rock mass;
2. Provide a sensitive means for monitoring a rock mass for changes in moisture contest or structural damage to the rock fabric induced by thermal or mechanical loading;
3. Together with the rock bulk density yield the dynamic elastic constants of the rock mass.

ACKNOWLEDGEMENTS

The research reported comprised part of an extensive rock mechanics and geophysics program to explore the possibilities of using a large crystalline rock mass as a geologic repository for nuclear waste, sponsored by the Swedish Nuclear Fuel Supply Company (SKBF) and the U.S. Department of Energy through the Office of Nuclear Waste Isolation under contract B511-0900-1. M.S.K. wishes to acknowledge support from the Geological Survey of Canada and the California Institute for Mining and Mineral Resources for partial support of the research. This work was also supported, in part, under U.S. Department of Energy contract DE-AC03-76SF00098.

REFERENCES

King, M.S., Pobran, V.S. and McConnell, B.V., 1974, "Acoustic borehole logging system", Proceedings 9th Canadian Rock Mechanics Symposium, Montreal, pp. 21-51.

King, M.S., Stauffer, M.R. and Pandit, B.I., 1978, "Quality of rock masses by acoustic borehole logging," Proceedings III International Congress Engineering Geology, Madrid, Section IV, Vol. 1, pp. 156-164.

King, M.S. and Paulsson, B.N.P., 1981, "Acoustic velocities in heated block of granite subjected to uniaxial stress", Geophysical Research Letters, Vol. 8, No. 7, pp. 699-702.

Paulsson, B.N.P. and King, M.S., 1980, "Between-hole acoustic surveying and monitoring of a granitic rock mass", International Journal Rock Mechanics Mining Sciences, Vol. 17, pp. 371-376.

Paulsson, B.N.P. and King, M.S., 1980a, "A cross-hole investigation of a rock mass subjected to heating", Proceedings Rockstore '80, Special Session on Nuclear Waste Disposal, Stockholm, Vol. 2, pp. 969-976.

Paulsson, B.N.P. and Kurfurst, P.J., 1980, "Characterization of the discontinuities in the Stripa granite - the full scale experiment drift", University of California, Lawrence Berkeley Laboratory, Report LBL-9063.

Witherspoon, P.A. et al, 1979, "Rock mass characterization for storage of nuclear waste in granite", Proceedings IV Congress International Society Rock Mechanics, Montreux, Vol. 2, pp. 711-718.

Chapter 6

SPATIAL DISTRIBUTION OF PERMEABILITY
AROUND CSM/ONWI ROOM
EDGAR MINE, IDAHO SPRINGS, COLORADO

by Parviz Montazer[1], Gideon Chitombo[2], Robert King[3], William Ubbes[4]

[1,2]Graduate Students, and [3]Professor, Department of Mining,
Colorado School of Mines, Golden, Colorado

[4]Program Manager, Office of Nuclear Waste Isolation (ONWI),
Battelle Memorial Laboratories, Columbus, Ohio

ABSTRACT

This paper describes the results of a detailed study that was undertaken to define the spatial distribution of permeability within a five meter thick envelope around the CSM/ONWI room. Detailed fracture orientations and distribution were determined through mapping of the walls of the opening, logging of oriented core, and visual examination of borehole walls with a borescope and a T.V. camera. Over 3,500 measurements were made from which the three dimensional distribution of fractures was determined. Two principally different testing equipment designs and various methods of air injection tests were used in all boreholes to identify and characterize conductive fractures. Through analizing the distribution of permeability and fracture index around the room a zone of blast damage of approximately .5 m thickness was delineated. The effect of stress modification seems to extend farther into the host rock to approximately 3.5 m depth.

INTRODUCTION

Igneous and metamorphic rocks are normally associated with geological discontinuities such as joints, faults and shear zones which are collectively termed as fractures. Their fluid transport properties are highly controlled by distribution, orientation, and characteristics of these features. When an underground opening is excavated in these rocks some of the characteristics of the fractures are modified in the vicinity of the opening due to blasting effects and stress modifications. Stress modification, depending on the orientation of the stress tensor with respect to the fracture orientation, may increase or decrease the permeability. Blasting can modify the permeability in the immediate vicinity of the opening by introduction of new fractures and opening of the pre-existing ones.

The purpose of this paper is to review some of the techniques used in characterizing permeability of the fractured metamorphic rocks around the CSM/ONWI room and to present the results with an attempt to relate the observed distribution of the permeability to the effects of the excavation.

The CSM/ONWI Room.

The experimental mine of the Colorado School of Mines (CSM) is located north-west of the town of Idaho Springs, about 40 km (25 miles) west of Golden, Colorado. Figure 1 is the plan view of the mine at the 2400 m (7880 ft.) level. The CSM/ONWI room, location of which is shown in this figure, was excavated during the summer of 1979 for installation of a thermomechanical test facility. Careful smooth

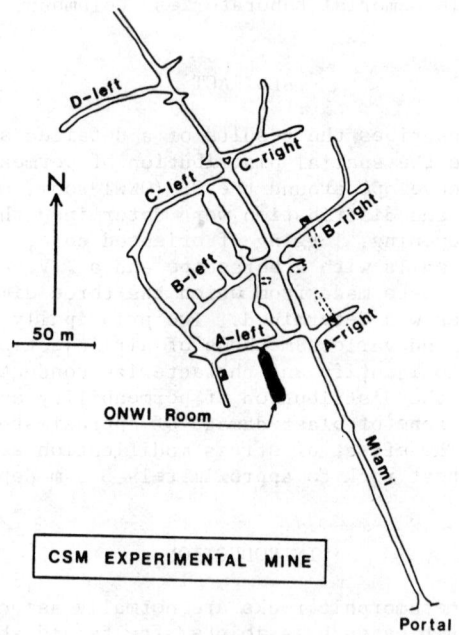

Figure 1. Site of Experimental Room. (Source of base map: Gary Van Huffel, CSM Senior Thesis, 1975).

wall blasting techniques were used to create the opening (Holmberg, 1981; Hustrulid, 1980). This room is approximately 100 m (300 ft.) below the ground surface and is 30 m (100 ft.) long, 3 m (10 ft.) high, and 5 m (15 ft.) wide (Figure 2). Forty-five NX boreholes were diamond drilled in and around the room shortly after its excavation. Forty two of these were drilled from inside the room in six

SPATIAL DISTRIBUTION AT EDGAR MINE, CO

radial sets. The seven holes of each set are arranged as shown in Figure 2. The west side of the room is paralleled by three longitudinal boreholes drilled horizontally from A-left that extend the room's entire length.

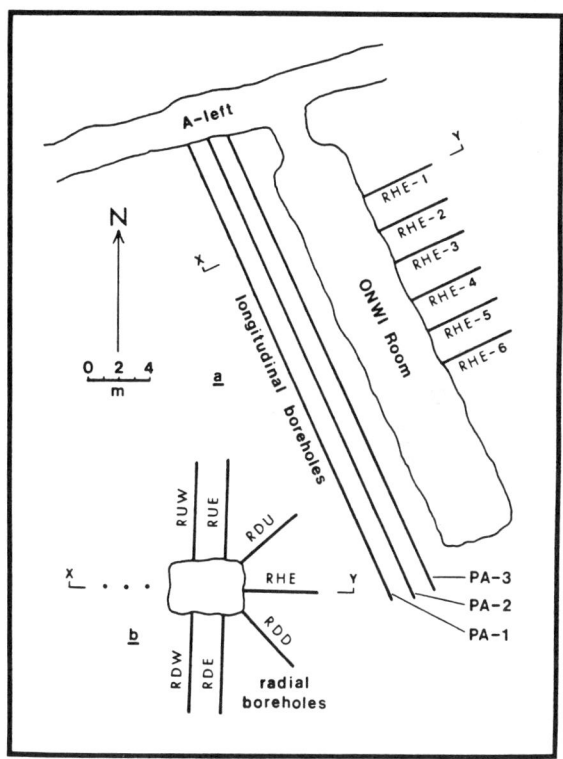

Figure 2. Plan view of the CSM/ONWI Room, with key to borehole designations.

Geology of the Room

Many small, tight folds associated with migmatization can be seen on the walls of the room. The rock around the room is strongly foliated with the foliation striking N70E and dipping 70NW. The main rock type is a medium to coarse and occasionally fine-grained migmatite biotite gneiss.

The main fracture in the room is a small shear zone of about .30 m width which crosses the room in the middle (Figure 3). This fracture has been intersected by all three longitudinal boreholes. Several other fractures cross the entire width of the room, the orientations of which are shown in Figure 3. Table 1 summarizes the orientations of the major fracture sets found by mapping of the walls

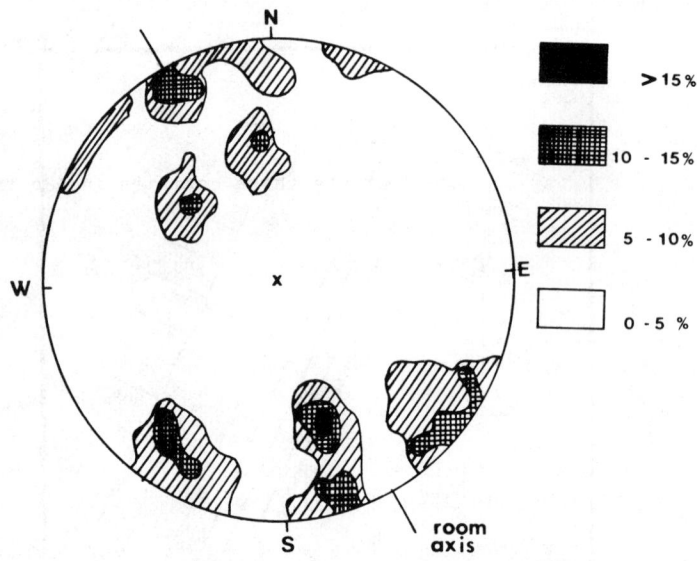

Figure 3. Lower hemisphere equal-area Schmidt net plot of fractures with continuity greater than 3 m.

of the room and borescope survey of the radial boreholes. Generally there are two major sets that are consistently seen around the room and within the boreholes. These are referred to as the foliation and diagonal fracture sets. Radial boreholes intersect only a few of the foliation fractures.

Table 1. Attitudes of the major fracture sets from wall mapping and borehole survey. (in decreasing order of significance).

	Strike/Dip			
	1	2	3	4
Wall mapping	N52W/90[a]	N75E/78NW[b]	N25E/68SE	N35E/90
Borescope survey	N50W/90[a]	N30E/55SE	N10E/78SE	N68E/70NW[b]

a) diagonal joints
b) foliation joints

SPATIAL DISTRIBUTION AT EDGAR MINE, CO

INJECTION TESTING

Steady-state air and water injection techniques in conjunction with pulse testing and pressure decay tests were employed to obtain data from which permeabilities were calculated for both the longitudinal and radial boreholes. The instrumentation and procedures used during testing of the longitudinal boreholes were more rigorous and complete than those used for the radial boreholes.

Instrumentation

Packer System. Figure 4 schematically shows equipment for testing of the boreholes. The main injection probe consists of four packers and a transducer housing spaced to create three cavities inside the borehole. Each cavity is connected to one pressure transducer through plastic or steel tubing. The central cavity can be pressurized with water or air; the pressure can be detected in all three chambers.

Water or air is injected into the central chamber through continuous tubing. The transducer for measuring pressure in the central cavity if housed within the cavity. Temperature of the injection fluid is measured by a thermocouple.

Each monitoring probe consists of two packers forming a single chamber. This chamber can be pressurized with water or air or it can serve as a pressure observation chamber.

Flowmetering System. Four devices were used for flow measurement and were placed in series to observe and compare the accuracy and applicability of each system. The flow tank system (Figure 4M1) relies on the differential pressure across the transducer caused by the head of the water in the tank to measure flow volume. The tracer line measures very small amounts of flow (Figure 4M2) by introducing a bubble or an electrolite into the flowline. The time required for the pulse to travel between the two electrodes can be converted to flow rate. Metering valves with a differential transducer (Figure 4M3) were used to measure the flow rates of both gas and water. Ratameters (Figure 4M4) give accurate flow data and are simple and easy to use. However, data collection is manual and cumbersome for long tests. Five ratameters were used to cover all possible ranges of flow rates (from 0.01 cc/min. to 25 lit./min of water and 1 cc/min. to 20 lit./min of air). The advantage of the metering valve and ratameters is that they can be used for air flow rate as well as for water flow rate measurements.

Data Acquisition System. The output signals from the transducers were input into a 16 channel programmable Kays Instrument Digistrip II data logger which was interfaced with a cassette tape recorder. Temperatures in the tanks, the injection zone, and the ratameters were measured using type T thermocouples. Conductivity meters, for the tracer line, were connected to a Schlumberger stripchart recorder.

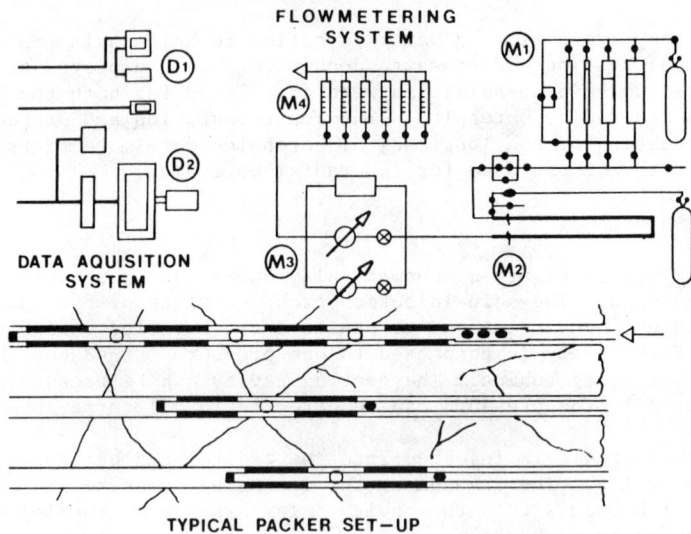

Figure 4. Schematic diagram of the instrumentation for injection testing of the longitudinal boreholes.

Systematic Injection Testing with Nitrogen

Longitudinal Boreholes. A relatively long injection zone (L=2.13 m or 7 ft.) was used to test the longitudinal boreholes. The long interval was chosen to reduce the number of tests. However, in order to precisely locate a fracture and to isolate fractures within a group, a special testing method was required. If the cavity length "L" (distance between packers) is longer than the apparent spacing of fractures, the permeability data from a single test may show the effects of several fractures, and their location within the tested interval cannot be determined precisely.

Consequently, the "sequentially overlapping interval" (SOI) method was used (Montazer and Hustrulid, 1981). In this method about 57% of every test interval was overlapped by subsequent interval. This allowed fracture location to within 14% and 25% of L in alternate test spacings. Therefore, testing effort was substantially reduced. In this case where L was 2.13 m and the packers were moved every 0.9 m, 30 tests (covering 28 m) provided information that would have required at least 90 tests employing side by side testing with an interval length of 0.3 m.

Radial Boreholes. The packers of a double packer monitoring assembly were shortened and the interval adjusted to .75 m (2.5 ft.) in order to obtain more complete coverage of the borehole. Boreholes were tested at intervals of .30 m (1.0 ft.) using the SOI method.

Cross-hole Testing

The main purpose of this set of tests was to determine the continuity and trends of conductivity variations along a few selected fractures. Because of the unsaturated nature of the more conductive fractures around the room, the trends delineated with single fluid could not be attributed to a specific cause. For this reason alternate injections of nitrogen and water were conducted to reduce the number of the factors causing the observed effects.

Fracture Index Method

A transient air injection method was employed to determine the fracture index (FI) for the radial boreholes. This method which was developed by Barrons (1978) is similar to pulse testing techniques, except that air is used instead of water and an external volume is provided to increase the resolution of the pressure decay in high permeability zones. The FI calculated from an approximate solution to the equation of flow of gas through porous media is inversely proportional to the permeability (of the porous medium). Therefore, test zones containing high conductivity fractures would show low FI and vice-versa.

RESULTS

For each interval, permeability versus inverse of pressure was plotted. This was done so that the comparison could be made between different fractures in adjacent holes on the basis of pressure-permeability trends rather than single values of permeabilities.

Single values of permeabilities extrapolated to infinite pressure were plotted along each borehole as shown in Figure . Correlation of this diagram with the fracture map revealed the conductive fractures which are traced on this map (note that strikes are distorted due to exaggeration in distance between boreholes in this diagram).

Three of the zones containing fractures show a strong diminishing trend in permeability towards the room (PA-3 is the borehole closest to the west wall of the room). These are zones containing fractures F6, the shear zone and F8. Figure 5 shows that the permeability of the interval containing the shear zone near PA-3 has reduced by more than one order of magnitude. F8 shows at least a one order of magnitude decline, and F6 shows a one order of magnitude decline. It can also be noted that the overall permeabilities along PA-3 are much smaller than along the other two boreholes.

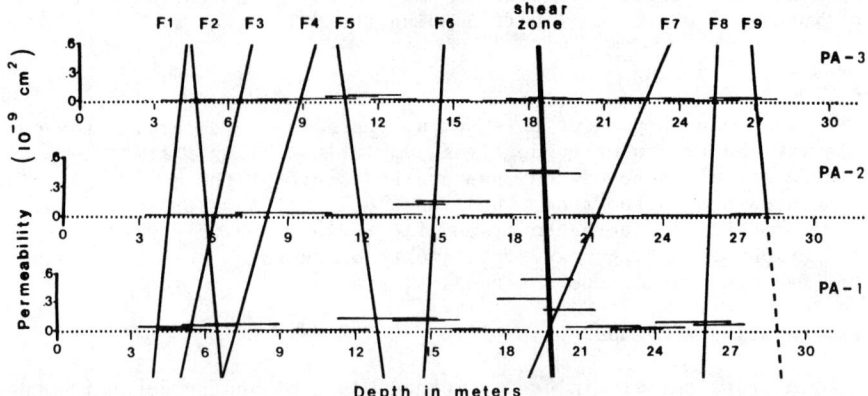

Figure 5. Infinite pressure permeabilities along the longitudinal boreholes. The boreholes are 1.2 m apart and PA-3 is 1.2 m away from the west wall of the room.

In radial boreholes for each testing interval only a single value of permeability was obtained. Although the rock matrix permeabilities observed during testing of the radial borehole are very low, the permeabilities of the fractured sections are several orders of magnitude larger than those calculated for fractures in the longitudinal boreholes.

The means of the logarithm of permeabilities and inverse of the fracture index for four groups of boreholes are plotted in Figure 6. From this figure it is evident that a zone of high permeability (or low FI) exist in all the boreholes within the first 0.5 meter. This is believed to be the damage envelop induced by blasting and stress concentration.

DISCUSSION

From data presented above and other evidence (Holmberg, 1981; Chitombo, et al. 1981) the blast damage zone seems to be limited to the first 0.5 m of the surface of the opening. The issue that is ambiguous is the reduction of the permeability of some of the vertical fractures near the opening. In particular the shear zone which shows at least one order of magnitude reduction in permeability in PA-3 (nearest the wall of the room). To seek the cause, a few possibilities are analyzed here. The reduction in permeability may be due to:

1. natural variations in fracture characteristics;
2. modification of the fracture traits due to high gas pressures during blasting (opening and closure of the fracture);

3. excessive shear displacement along the fractures near the opening, or
4. closure of the fracture due to increase in normal stress across it.

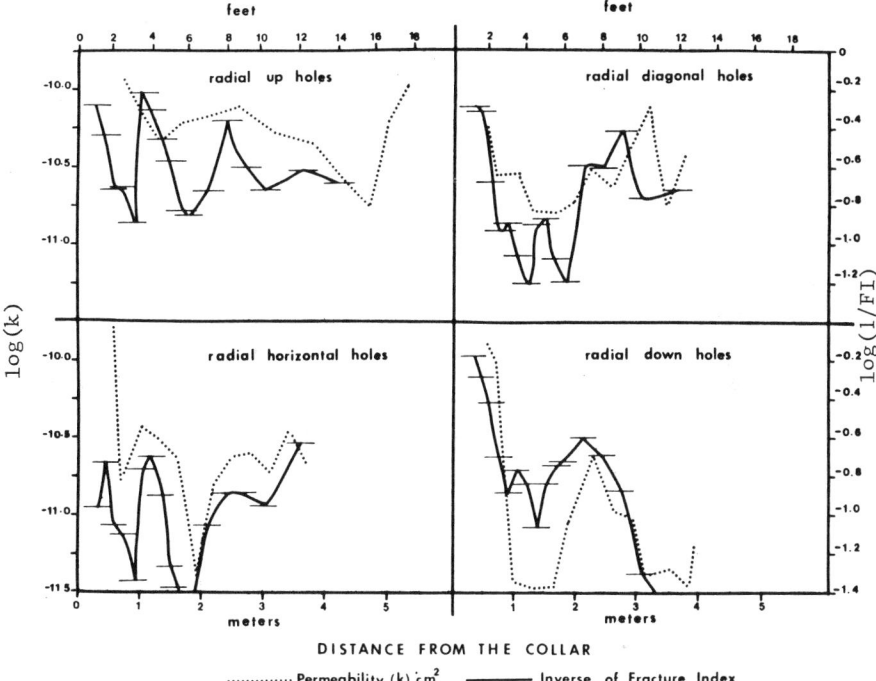

Figure 6. Distribution of the means of the logarithms of permeability and fracture index along four groups of boreholes.

Examination of the three boreholes with T.V. camera indicated that there are no significant changes in the geologic traits of the shear zone. The explosion gas effect does not explain the observed phenomena as there is no reason for it to act differently along the fractures with different orientation (those intersected by the radial boreholes). The last two possibilities are felt to be the most likely cause. They both could result in decrease in conductivity of the fractures that are perpendicular to the axis of the room and increase in conductivity of those oriented parallel to this axis. Preliminary analysis of the in situ stress measurements favors the 4th possibility.

CONCLUSIONS

Both methods of permeability and fracture index testing have been found to be accurate in locating conductive fractures. It is generally concluded that in the vicinity of the CSM/ONWI room significant alternation of the permeability due to the effects of the excavation are limited within a 1.5 m envelope around the opening. Furthermore, it is evident that the conductivity of the fractures, within this envelope is higher when they are oriented parallel to the axis of the room than when they are perpendicular to this axis.

NOTICE/DISCLAIMER

This report was prepared as an account of work sponsored by the United States Government. Neither the United States nor the Department of Energy, nor any of their employees, nor any of their contractors, subcontractors, or their employees, makes any warranty, express or implied, or assumes any legal liability or responsibility for the accuracy, completeness, or usefulness of any information, apparatus, product, or process disclosed, or represents that its use would not infringe privately-owned rights.

ACKNOWLEDGMENTS

The financial support for the work reported was through contract E512-04800 with the Project Management Division of the Battelle Memorial Institute, Columbus, Ohio.

REFERENCES

Barron, K., 1978, "An Air Injection Technique for Investigating the Integrity of Pillars and Ribs in Coal Mines", Int. Jour. Rock Mech. Sci. and Geomech. Abstr., vol. 15, pp. 69-76.

Chitombo, G. P., Hustrulid, W. A. and King, R., 1981, "Blast Damage Extent Around an Underground Opening", Battelle/ONWI subcontract no. E512-04800, Topical Report no. 140(6).

Holmberg, R., 1981, "Hard Rock Excavation at the CSM/ONWI Test Site Using Swedish Blast Design Techniques", Battelle/ONWI subcontract no. E512-04800, Topical Report no. 140(3), 140 p.

Hustrulid, W.A., 1981, "CSM/ONWI Hard Rock Test Facility at the CSM Experimental mine, Idaho Springs, Colorado: Program Summary", Battelle/ONWI subcontract no. E512-04800, Topical Report no. 140(1), 14 p.

Montazer, P. and Hustrulid, W.A., 1981, "An Investigation of Fracture Permeability Around an Underground Opening in Metamorphic Rocks," Battelle/ONWI subcontract no. E512-04800, Draft Topical Report no. 140(5), 200 p.

Chapter 7

CROSS-BOREHOLE FRACTURE MAPPING
USING ELECTROMAGNETIC GEOTOMOGRAPHY

A.L. Ramirez, F.J. Deadrick and R.J. Lytle

Lawrence Livermore National Laboratory
Livermore, California

ABSTRACT

This article describes the evaluation of a new geophysical technique used to map fractures between boreholes: electromagnetic geotomography used in conjunction with salt water tracers. An experiment has been performed in a granitic rock mass. Geotomographic images have been generated and compared with borehole geophysical data: neutron logs, acoustic velocity logs, caliper logs and acoustic televiewer records. Comparisons between the images and the geophysical logs indicate that clusters of fractures were detected but single fractures were not.

INTRODUCTION

The mapping of fractures within a rock mass is a three dimensional problem. It requires that fractures be detected and measured along exposed surfaces such as boreholes or tunnel walls. In many instances, this information can be successfully projected into the rock mass for short distances. However, as the projection distance increases, the uncertainty associated with the projections becomes larger and may become unacceptable.

There are borehole geophysical methods which offer a high level of resolution when used to detect fractures (e.g., acoustic televiewer). These methods, however, cannot penetrate deeply into the rock mass. Other methods (e.g., cross-borehole seismic) can probe deep into the rock and may provide evidence that fractures

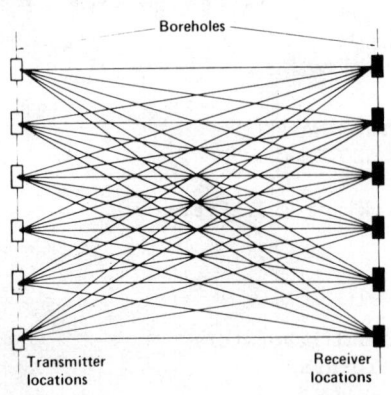

$$X_1 = {}_1D_{11}{}_1\alpha_{11} + {}_1D_{12}{}_1\alpha_{12} + {}_1D_{21}{}_1\alpha_{21} + {}_1D_{22}{}_1\alpha_{22}$$

$$X_2 = {}_2D_{11}{}_2\alpha_{11} + {}_2D_{12}{}_2\alpha_{12} + {}_2D_{21}{}_2\alpha_{21} + {}_2D_{22}{}_2\alpha_{22}$$

$$X_3 = {}_3D_{11}{}_3\alpha_{11} + {}_3D_{12}{}_3\alpha_{12} + {}_3D_{21}{}_3\alpha_{21} + {}_3D_{22}{}_3\alpha_{22}$$

Fig. 1. Method used to sample the rock mass. The antennas are moved along the boreholes to provide many different perspectives of the rock between boreholes.

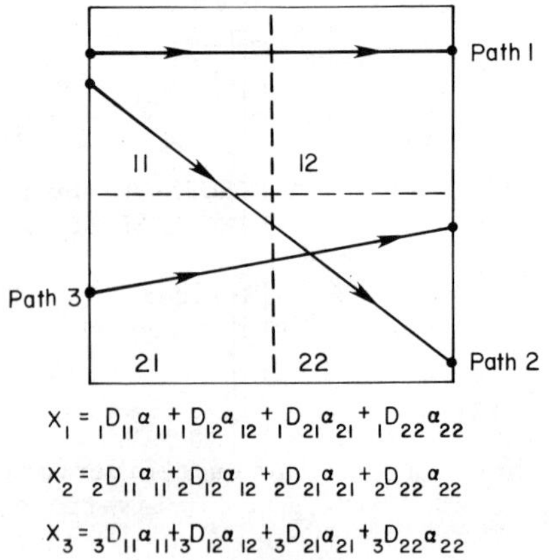

Fig. 2. Example of the method used to interpret the electromagnetic attenuation measurements.

exist but they lack the resolution necessary to map these fractures. Consequently, the Panel on Rock Mechanics Research Requirements (1981) has recommended that improvements be achieved in methods used to map fractures remotely (between boreholes).

GEOTOMOGRAPHY METHOD

Electromagnetic geotomography was combined with the use of salt water tracers to map fractures remotely. Geotomography was developed at LLNL and has been thoroughly described elsewhere (Lytle et al. 1978, Lytle et al., 1981). In this section we describe important characteristics of the geotomography method.

The geotomography method is a new way of collecting and interpreting geophysical data. The method is similar in concept to the data collection and interpretation procedures used in medical diagnostics, such as brain and body scans. Figure 1 shows how the geotomography method is used to collect subsurface geophysical data. By varying the depths of the transmitter and receiver in two boreholes, it is possible to detect regions having different properties using either electromagnetic waves or seismic waves. Multiple ray paths propagated along many different orientations are used to provide many different "views" of a region.

The geophysical data is interpreted using a modified version of the back projection technique described by Kuhl and Edwards

CROSS-BOREHOLE FRACTURE MAPPING

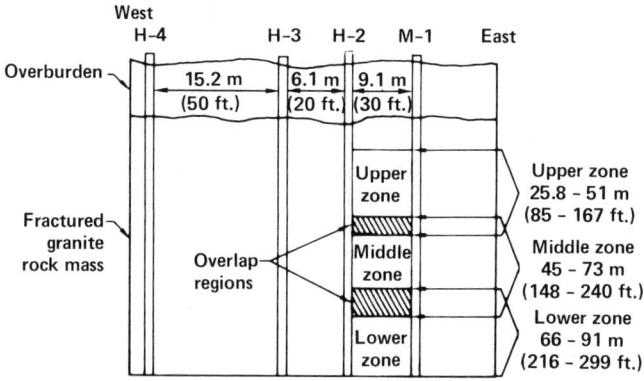

Fig. 3. Cross section of the experimental site showing the different rock mass regions probed between boreholes M-1 and H-2.

(1968). As a first step, the area between two boreholes is divided into a set of cells (i.e., smaller zones) by arbitrarily spaced vertical and horizontal lines. The objective is to infer the electromagnetic attenuation factor induced by the rock and the fluid-filled fractures within each of these. The attenuation factor is defined as α if the amplitude of a plane wave is reduced by the factor $\exp(-\alpha x)$ in traveling a distance of x meters (Sheriff, 1973). The resulting image will be a map showing the variations of the electromagnetic factor throughout all the cells. Fracture zones can be mapped on this image because they will appear as zones which create larger attenuation factors than those of intact rock. We use gray scale images to present the results of these calculations. In a gray scale image, a tone of gray is assigned to each cell. This tone represents the attenuation factor calculated for that cell.

Figure 2 shows a simplified example of the algorithm used to calculate the attenuation factor for each cell. For this example, the region between two boreholes was arbitrarily divided into four cells, and only three ray paths were used to sample the region. (For an actual geotomograph, many more cells and ray paths are used.) The attenuation X of rays, 1, 2, and 3 is expressed mathematically in Fig. 2: α_{11}, α_{12}, α_{21}, and α_{22} represent the attenuation factors characteristic of cells 11, 12, 21, and 22 respectively, and D_{11}, D_{12}, D_{21}, and D_{22} represent the distances which each ray path covered through these cells. Similar equations are constructed for ray paths 2 and 3 This set of linear equations is then solved iteratively for α_{11}, α_{12}, α_{21}, and α_{22}.

EXPERIMENT DESCRIPTION

An experiment has been performed at a site near the town of Oracle, Arizona, approximately 64 km north of the city of Tucson. The electromagnetic geotomography method was used in combination with salt water tracers which were forced into the granitic rock mass present at the site. The experimental site has been developed by the University of Arizona for the U.S. Nuclear Regulatory Commission. The granite rock mass contains an extensive network of fractures. These are quite numerous and exhibit large variability in orientation.

Figure 3 shows the four roughly coplanar boreholes available at the site during the experiment. Also shown are the various measurement zones for which geotomographic images have been constructed.

Measurements of the attenuation of electromagnetic waves were obtained in the field before and after the salt water was forced into the fracture network. We forced the salt water into the fractures by filling the full length of borehole M-1 with salt water while lowering the water level in borehole H-3 with a submersible pump. The salt water used had a conductivity ten times greater than that of the natural groundwater.

EXPERIMENTAL RESULTS

Geotomographs have been generated for all the measurement zones shown in figure 3. It has been assumed that boreholes M-1 and H-2 are vertical. Examples of these will be shown below; the remaining images are presented and analyzed by Ramirez et al. (1982).

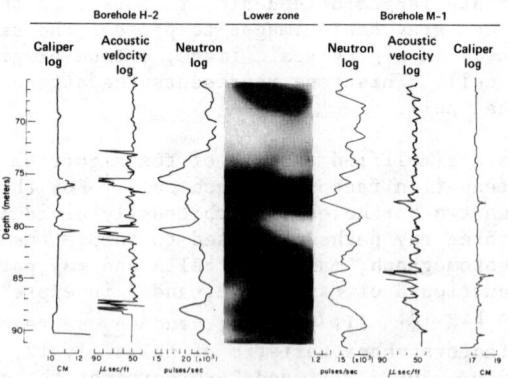

Fig. 4   Comparison of M-1 and H-2 geophysical logs with the lower measurement zone geotomograph. The data for this image was obtained after salt water was forced into the rock mass.

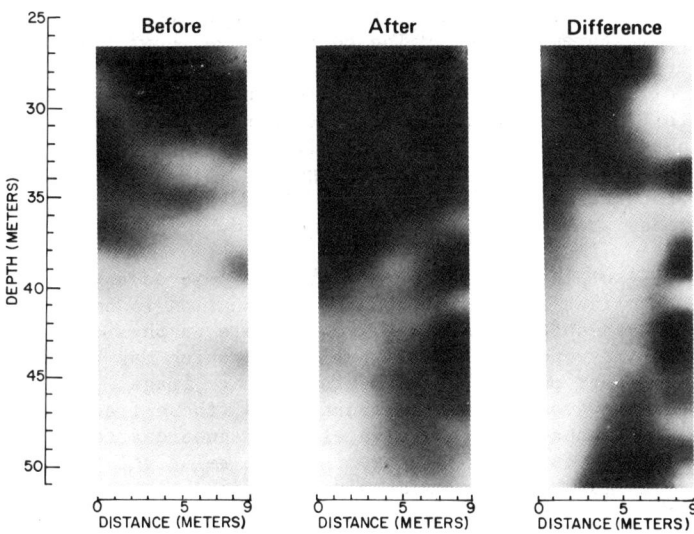

Fig. 5   Geotomographs of the upper measurement zone before and after salt water tracers were forced into the rock mass. A third geotomograph obtained by subtracting the "before" from the "after" is also shown.

Figure 4 shows a comparison of the lower measurement zone geotomograph (refer to figure 2) and geophysical logs of boreholes M-1 and H-2. The logs were recorded by W. Scott Keys of the U.S. Geological Survey. The right margin of the image coincides with borehole M-1 whereas the left margin coincides with borehole H-2. Zones of probable fracturing along the borehole wall are indicated by deflections in the logs.

Fractures should be indicated on the acoustic velocity and neutron logs by deflections to the left. On the caliper logs, most fractures should correspond with deflections to the right. The raw data for the geotomographic image was collected 2 days after salt water was introduced into borehole M-1.

This figure shows the most prominent correlations observed between the single borehole geophysical logs and the cross-borehole geotomographs. The darker colors in the image show the rock mass zones which caused the largest attenuation of the signal; hence, fractured zones should be represented by the darker colors. Good agreement is shown between logs and images from zones of inferred fracturing. The anomaly visible near the middle of the image at a depth of 76.5 m (251 ft) was tentatively identified by the U.S. Geological Survey as one of the most permeable and pervasive features identified on the geophysical logs (Keys, 1981).

Figure 5 shows a comparison of three geotomographs corresponding to the upper measurement zone. The center image shows the same region after salt water was forced into the rock. The image to the right is a difference geotomograph obtained by subtracting the "before" (left) image from the "after" (central) image. Note that the right hand image is substantially different from the other two. It shows the changes in electromagnetic attenuation factor created by the intrusion of salt water. The darker colors represent zones where maximum changes occurred.

Difference geotomographs appear to provide some advantages. These will highlight zones where signal attenuation is small but the change in attenuation is high. For example, a physically small permeable zone may induce small signal losses which may be difficult to detect on a standard geotomographic image. However, when a difference geotomograph is constructed, those fractures will be highlighted because the change in electromagnetic attenuation created by the intrusion of salt water can be relatively high. An example of this may be observed at a depth of 48.5 meters in figure 5.

FACTORS RELEVANT TO THE SUCCESSFUL APPLICATION OF THE METHOD

The electromagnetic geotomography method can be used to map the subsurface under certain conditions. Boreholes which penetrate and straddle the area of interest need to be available. They should be uncased or have casing made of non conductive casing materials, such as plastic. Otherwise the electromagnetic energy will travel up and down the casing only and will not penetrate the rock mass.

The minimum distance between the transmitter and receiver antennas should be much greater than $\lambda/2\pi$ (where λ is the wavelength of the waves in the rock mass). If this condition is met, the dominating electrical fields detected by the receiver antenna will attenuate linearly with distance and linear equations can be used in the reconstruction algorithms. Fortunately, the wave frequencies generally appropriate for probing have relatively small wavelengths, so the requirement is easily satisfied. For example, in Oracle $\lambda/2\pi$ was approximately 0.5m when 40MHz waves were used.

The probing geometry will control the resolution which can be achieved with geotomography. It will be defined by the locations of the transmitting and receiving antennas; in most applications, the total depths and alignments of the boreholes will determine the antennas' positions. Maximum resolution will be achieved when the area of interest can be totally encircled by the antennas. In most applications it will not be practical to do so but acceptable alternatives can be followed. At Oracle, for example, the antennas were moved along lines which extended far above and below the

CROSS-BOREHOLE FRACTURE MAPPING 63

center of the measurement zone. In this manner, the antennas effectively rotated roughly 300° around the center of the measurement area.

The geological features to be mapped using electromagnetic geotomography should have electrical properties different from those of the surrounding rock mass. Fractures filled with water offer good contrast relative to the surrounding rock when the rock itself contains little or no water within its pores and most of the water storage in the rock mass is along the fracture planes. These conditions are met by low porosity rocks, such as igneous rocks, which exist below the water table. Rocks such as sandstones or claystones, which store significant amounts of water in their pores, will be less desirable candidate media because the contrast between intact rock and fracture planes will be significantly smaller.

SUMMARY AND CONCLUSIONS

The potential of the electromagnetic geotomography method to map fractures remotely has been evaluated. A fractured granitic rock mass has been investigated and geotomographic images of the rock mass have been generated. These images were compared with borehole geophysical data (neutron logs, acoustic velocity logs, caliper logs, and acoustic televiewer records) and analyzed.

Comparisons between the geotomographic images and the borehole geophysical data suggest that geotomography has merit when used to map fractures in granite. In general, image anomalies coincide with geophysical log anomalies which can be indicative of fracturing along the borehole walls. Given the experimental conditions of the Oracle experiment, available data suggests that clusters of fractures (i.e., fracture zones) were detected but single fractures were not.

The results from Oracle indicate that salt water tracers create changes in electromagnetic attenuation which are useful to detect fractured zones filled with tracers. The contrast is artificially increased between fractures which accept the salt water and the intact rock. When salt water tracers are used and data is collected before and after tracers penetrate the fractures, a difference geotomographic image can be obtained. It appears to reveal smaller fractured zones, otherwise unrecognizable, which salt water has penetrated. A difference image will also tend to eliminate geological features which attenuate the electromagnetic signal but are not permeable fractures.

The conclusions presented herein should be considered tentative. They hinge upon the correlations of the geotomographic images and geophysical logs. The borehole geophysical methods only sample the rock mass along the borehole walls whereas the geotomography method samples the rock mass between the boreholes. These correlations are limited by the degree to which we can extrapolate the information provided by the geophysical logs into the rock mass between the boreholes. Exploratory coreholes will be drilled to verify the information provided by images.

ACKNOWLEDGMENTS

This work was funded by the U.S. Nuclear Regulatory Commission, Office of Nuclear Regulatory Research.

"Work performed under the auspices of the U.S. Department of Energy by the Lawrence Livermore National Laboratory under contract number W-7405-ENG-48."

REFERENCES

Keys, W. S., 1981, Private Communication, Technical Memorandum No. 69, U. S. Geological Survey, Denver, Colorado.

Kuhl, D. E., and Edwards, R. Q., 1968, "Reorganizing Data from Transverse Scans of the Brain Using Digital Processing, Radiology, Vol. 91, No. 5.

Lytle, R. J., Dines, L. A., Laine, E. F., and Lager, D. L., 1978, "Electromagnetic Cross-Borehole Survey of a Site Proposed for an Urban Transit Station", UCRL-52484, Lawrence Livermore Laboratory, Livermore, California.

Lytle, R. J., Lager, D. L., Laine, E. F., Salisbury, J. D., and Okada, J. T., 1981, "Fluid-FLow Monitoring Using Electromagnetic Probing", Geophysical Prospecting, Vol. 29, pp. 627-638.

Panel on Rock Mechanics Research Requirements, 1981, "Rock Mechanics Research Requirements for Resource Recovery, Construction, and Earthquake-Hazard Reduction," U. S. National Committee for Rock Mechanics, National Research Council, National Academy Press.

Ramirez, A. L., Deadrick, F. J., and Lytle, R. J., 1982, "Cross-Borehole Fracture Mapping Using Electromagnetic Geotomography", UCRL-53255 (in Preparation), Lawrence Livermore National Laboratory, Livermore, California.

Sheriff, R. E., 1973, The Encyclopedic Dictionary of Exploration Geophysicists, Society of Exploration Geophysicists, Tulsa.

STATISTICS

Chairman
Nicholas Sitar
University of California
Berkeley, California

Co-Chairman
Herbert H. Einstein
Massachusetts Institute of Technology
Cambridge, Massachusetts

Keynote Speaker
Gregory B. Baecher
Massachusetts Institute of Technology
Cambridge, Massachusetts

Chapter 8

PLAYING THE ODDS IN ROCK MECHANICS

by Gregory B. Baecher

Associate Professor of Civil Engineering
Massachusetts Institute of Technology

ABSTRACT

Rock engineering involves uncertainties which are large and difficult to quantify. The traditional design approach to these uncertainties has been conservatism, and has been satisfactory to the extent that failure rates are reasonably small. Yet, a price is paid for conservatism, and that price is not always small. Today, reliability and risk analysis are finding use in rock engineering. The introduction of such methods, usually under the name 'probabilistic design,' is controverisal. However, if the jingoism of both sides in this controversy is stripped away, it appears that new insights about uncertainty in rock engineering are emerging.

INTRODUCTION

The one sure-thing in rock engineering is uncertainty. Rock masses are poorly characterized, engineering mechanics provides imperfect models, and load conditions are inadequately specified. Yet, a frontal assault on this situation using probability theory and statistics has met opposition from the rock mechanics community. The strength of the opposition is surprising, particularly when compared with other technical disciplines. Even agronomists, members of a sister discipline, have based their experimental design on statistics since as early as the 1920's.

The shrillness of the opposition to formal methods of uncertainty analysis seems inexplicable against the background of technical issues alone. Yet, those urging the use of such methods have themselves been guilty of jingoism, and have done little to ease the acceptance of the new techniques. The present paper begins with a look at the various types of analyses that are now called 'probabilistic,' and then devotes one section each to three specific applications: Data analysis, reliability modeling, and risk assessment.

"PROBABILISTIC" TECHNIQUES

The set of techniques commonly called 'probabilistic' in fact comprises at least four different types, each of which uses methods of probability theory, statistics, or decision analysis. These might be described as, <u>data description</u> in which statistical measures are used to summarize a large number of data, <u>inference</u> in which statistical methodology is used to make estimates of engineering properties and conditions, <u>reliability analysis</u> in which probabilistic models are used to deduce the uncertainties associated with prodictions, and <u>risk analysis</u> in which uncertainties are combined with potential consequences to evaluate design alternatives. Data description is routine and thus ignored here. The remaining categories are each discussed in a separate section below.

Probability theory is an axiomized branch of mathematical logic, which is internally consistent and necessarily following from a limited set of axioms. Probability theory is used to deduce the uncertainties and relations among sets of variables, which follow necessarily from specified uncertainties and relations among other variables. Statistics on the other hand, is a set of techniques based on certain principles of inference, not on axioms. Statistical techniques are used to draw inferences about the real world from partial or limited observations, and these inferences do not follow necessarily. Therefore, the problems that probabilistic models and that statistical techniques are brought to bear against are different.

The term 'probability' is primative within the axioms of probability theory. This means that its properties are specified, but not its meaning. This has led to two schools of thought on the meaning of 'probability:' One holding 'probability' to be a frequency of occurrence within a long series of similar trials, the other holding it to be a degree of belief. The latter definintion is more common in geotechnical engineering, since many of the events and processes of interest are unique. They do not occur repetitively. In fact, one can view frequency and belief as two distinct concepts, each having its appropriate place, yet each obeying the laws of probability theory. This view is adopted here.

Because probability theory deals with deductive logic, there are certain uncertainties in geotechnical engineering that are not amenable to 'objective' probabilistic treatment. These have to do with the development of hypotheses to explain observations, and the assignment of initial (i.e., a priori) probabilities or credibilities to those hypotheses. These tasks are inductive and therefore intuitive.

A PROBABILITY 'PRIMER"

Consider an uncertain quantity X with realization x. The function $F(X)$, the cumulative density function (cdf) describes the probability that the realization of X is less than or equal to some value x,

PLAYING THE ODDS IN ROCK MECHANICS

$$F(x) = \Pr[X \leq x] \quad . \tag{1}$$

the derivative with respect to x is

$$f(X) = \frac{d}{dX} F(X) \quad , \tag{2}$$

the probaibltiy density function (pdf), and has the property that area under this function between two values x' and x" equals the probability of X lying in that interval. Clearly,

$$\int_{-\infty}^{+\infty} f(X) \, dX = 1.0 \quad . \tag{3}$$

The first moment of f(X) about the origin is said to be the expected value or mean of the distribution,

$$E[X] = \int_{-\infty}^{+\infty} X \, f(X) \, dX \quad , \tag{4}$$

and the second moment about the mean is said to be the variance,

$$V[X] = \int_{-\infty}^{+\infty} (X - E[X])^2 \, f(X) \, dX \quad . \tag{5}$$

The square root of V[X] is said to be the standard deviation, SD[X], and the ratio of the standard deviation to the mean is said to be the coefficient of variation, Cov[X].

As a first approximation, the first two moments (mean and variance) of a function of a random variable, Z=g(X), can be found by expanding f(X) in a Taylor series and truncating to two terms. This yields,

$$E[Z] \doteq g(E[X]) \quad , \tag{6}$$

$$V[Z] \doteq (\frac{dg}{dX})^2 V[X] \quad . \tag{7}$$

For vector $\underline{X}$, Eq. 6 remains unchanged, but Eq. 7 becomes

$$V[Z] \doteq \sum_{i=1}^{m} \sum_{j=1}^{m} \frac{\partial Z}{\partial X_i} \frac{\partial Z}{\partial X_j} C[X_i, X_j] \quad , \tag{8}$$

in which $C[X_i, X_j]$ is the covariance of X_i and X_j,

$$C[X_i, X_j] = E[(X_i - E[X_i])(X_j - E[X_j])] \quad . \tag{9}$$

For independent X_i, X_j, $i \neq j$, Eq. 8 reduces to the useful form,

$$V[Z] \doteq \sum_{i=1}^{m} (\frac{\partial Z}{\partial X_i})^2 V[X_i] \quad . \tag{10}$$

DATA ANALYSIS

Statistical methods are used to draw inferences about subsurface conditions from limited numbers of observations. As noted above, site characterization involves inductive as well as deductive uncertainties, and statistical methods of data analysis apply only to the latter. Geologists and even geotechnical engineers have long drawn inferences without the help of formal methods. The present interest in statistical methods springs from (1) a desire to derive the most information from increasingly expensive measurements, (2) the advent of data intensive in situ measuring devices (i.e, continuous profiling), and (3) to some extent, the introduction of physicists, chemists, and other professionals to traditionally geomechanical concerns (e.g., nuclear waste).

This section discusses selected problems in data analysis, starting with data scatter and errors of estimation. Brief attention is given to joint surveys, the most-worked-on statistical problem in rock mechanics. Then a few comments are made on exploration strategy.

Data Scatter

Fig. 1 shows field vane strength data from a "uniform" clay deposit measured in about 30 borings. Since data are scattered, the choice of a design profile and envelope is not obvious by inspection.

As a first approximation, the scatter may be divided into a spatial or 'real' part and a measurement error or 'noise' part:

$$\text{Data Scatter} = \begin{cases} \text{Spatial Variation} \\ + \\ \text{Measurement Noise} \end{cases} \qquad (11)$$

The spatial part is inherent to the soil or rock mass, and is therefore the variability the exploration program intends to characterize. The noise part is spurious and important only to the extent that it obscures the spatial variability.

The amount of measurement error in observed data scatter can be inferred in at least three ways, using replicate measurements, multiple profiling, or the spatial structure of the data scatter. The last is direct and inexpensive.

The spatial structure of data scatter about a trend is conveniently summarized by an autocovariance function (or equivalently, a variogram). Adopting the simple model

$$z(t) = x(t) + e(t) , \qquad (12)$$

in which $z(t)$ = the measurement at point t, $x(t)$ = the actual property, and $e(t)$ is measurement noise, the autocovarince function of $x(t)$ is

$$C_z[r] = E[(z(t)-\bar{z})(z(t+r)-\bar{z})] , \qquad (13)$$

in which $\bar{z}$ = the mean or trend of $z(t)$. Assuming stationarity, etc., $C_z[r]$ simply expresses the covariance of the observations as a function of their spatial separation. Typically, $C_z[r]$ is anisotropic, smaller in the vertical than horizontal direction.

Similar autocovariance functions can be defined for $x(t)$ and $e(t)$, and these are related to $C_z[r]$ by

$$C_z[r] = C_x[r] + C_e[r] \quad . \tag{14}$$

However, as $e(t)$ is presumably independent from one measurement to another, $C_e[r]$ must be a spike of height $V[e]$ at $r=0$, and zero elsewhere. Thus, $C_z[r]$ looks like Fig. 2, and the extrapolation of the observed $C_z[r]$ back to $r=0$ allows an estimate to be made of $V[e]$. For the data of Fig. 1, $V[e] \approx 40\%$ of the data scatter.

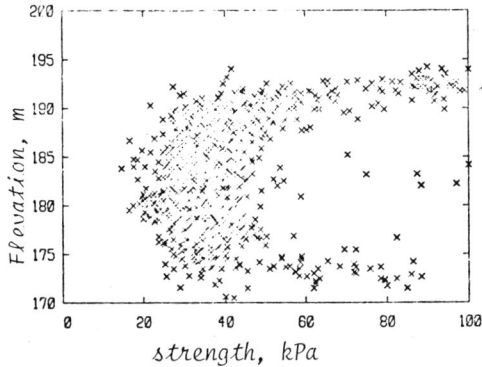

FIGURE 1 -- *Field vane strengths in 30 borings, Cov = 30%+*

In addition to data scatter, which is associated with variation about the mean trend, two systematic errors affect the estimation of the mean trend itself. First, the measurement procedure may introduce a systematic bias, B,

$$z(t) = B\, x(t) + e(t) \quad , \tag{15}$$

$$V[z(t)] = B^2\, V[x(t)] + V[e(t)] \quad . \tag{16}$$

Second, the total number of measurements is limited, thus statistical fluctuations introduce estimation errors. Combining, the total variance in an estimated profile becomes

$$V[\hat{x}(t)] = \begin{cases} \text{Variance about mean} \\ + \\ \text{Variance of mean} \end{cases} \tag{17}$$

in which "^" signifies an estimate. Mathematically,

$$V[\hat{x}(t)] = R\, V[x(t)] + V[B]\, \bar{z}(t)^2 + \bar{B}^2\, V[\bar{x}(t)] \quad , \tag{17}$$

in which R = a scale factor (below), $V[x(t)]$ is the spatial variance of the data estimated from Eqs. 14 and 16, $V[B]$ is the variance of the measurement bias, and $V[\bar{x}(t)]$ is the statistical estimation error on the mean trend. For n widely spaced measurements,

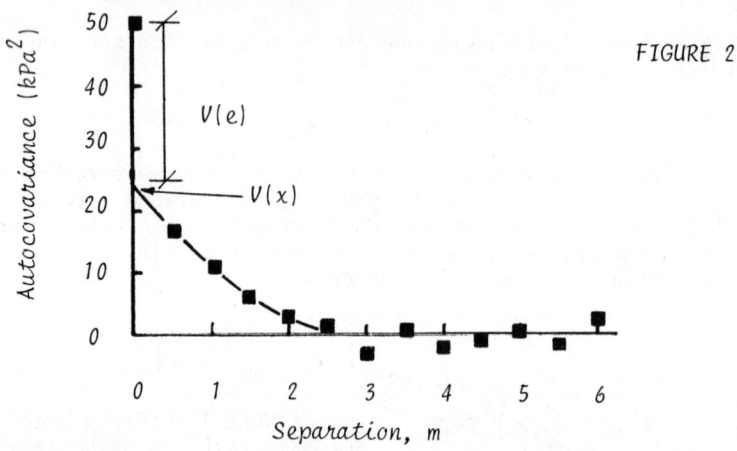

FIGURE 2

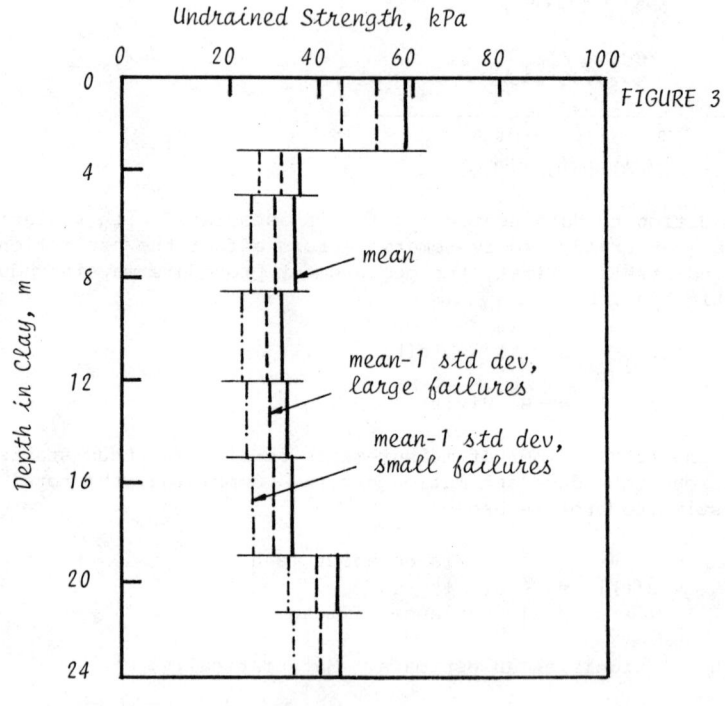

FIGURE 3

$$V[\bar{x}(t)] \pm \frac{V[x(t)] + V[e(t)]}{n} \quad . \quad (18)$$

For closely spaced measurements,

$$V[\bar{x}(t)] \pm \frac{(\underline{1}\,\underline{\underline{C}}_x\,\underline{1}^t) + V[e(t)]}{n} \quad (18)$$

in which $\underline{1} = \{1,\ldots,1\}_n$ and $\underline{\underline{C}}_x$ = the covariance matrix of $x(t)$, the ij^{th} element of which is the covariance of $x(t_i)$ and $x(t_j)$ taken from $C_x[r]$.

Based on Eq. 17, a mean profile and standard deviation envelopes are easily constructed. Note, however, that the standard deviation envelopes, which express the uncertainty in engineering parameters for analysis, must reflect mode of behavior and scale. This dependency is summarized in the factor R. For example, circular shear instability in a copper porphyry mine slope depends on total resistance over a surface of sliding. Thus, spatial variation in part averages out. Therefore, for very small instabilities $R \to 1.0$, but for very large ones $R \to 0$ (Fig. 3). Wedge instability, on the other hand, depends on the least favorable wedge. Therefore, for very small slopes (few wedges) $R \to 1.0$, but for large or long slopes $R \gg 1.0$. An analysis like that leading to Fig. 3, which discounts part of the data scatter and differentiates among uncertainties depending on modes of behavior and scale, can only be made statistically: Intuition is an inadequate substitute.

Joint Surveys

More work has been reported on the statistics of jointing and joint surveys than on perhaps all other rock engineering problems combined. Today, the character of spatial variations of joint populations is reasonabley well known. The statistical properties of joint survey designs are perhaps less well known, as are the relative merits of design alternatives. Work on joint surveys started early and has progressed far, probably because the frequency interpretation of joint populations is palatable to the deterministically minded.

A number of interesting problems remain in joint survey design. Primarily these involve the subtleties of geometric probability theory and the inherent biases of common survey designs. An important unanswered question remains how to design surveys for the collection of data pertinent to joint persistence or connectivity, important to strength and seepage. Questions also remain on the character of orientation data, which are poorly modelled analytically.

Exploration Strategy

One of the great promises of statistical methods is that they might be used to "optimize" exploration strategies, i.e., make exploration more efficient. They still hold this promise, but conceptual difficulties have hindered progress. None the less, this may yet become a fruitful area of work. Here, only two problems are considered, the simple problem of how to mix in situ tests, and the difficult problem of global strategies.

Simple Numbers and Mixes of Tests: The most useful results in coming years may derive from simple testing strategies of practical importance. The most obvious deals with numbers, types, and locations of in situ tests. While sophisticated techniques have appeared for sampling stochastic fields, practical tools are only now being developed.

Continuing the notation above, a test procedure may be characterized by its cost, bias, and noise: $\{C,B,e\}$. It seems reasonable to assume that cost and data quality are positively realted. More expensive tests should have both greater accuracy and precision than less expensive tests. Adopting Schmertmann's suggestion that cost might be functionally related to precision, (Engineering Foundation Conference, Sta Barbara, January 1982), assume that for test type i,

$$V[e_i] = K_e / C_i \tag{19}$$

in which K_e = constant. Then,

$$V[\bar{x}] \pm \frac{V[x] + K_e}{C_i n_i} \tag{20}$$

in which the number of tests of type i, n_i, is limited by the total cost of exploration C_T,

$$n_i = C_T / C_i . \tag{21}$$

Then, using only tests of type i, and with no loss of generality setting R=1,

$$V[\hat{x}] \pm V[x] + \frac{V[x] + K_e/C_i}{C_T/C_i} . \tag{22}$$

Thus, if Eq. 19 obtains, and if inaccuracy (i.e., bias) is ignored, then cheaper tests through larger numbers lead to smaller errors than do expensive tests (Fig. 4).

However, cheaper tests also contain more bias. If one assumes uncertainty in the bias error to be also related to cost, as

$$V[B_i] = K_b / C_i , \tag{23}$$

in which K_b = constant, then the relative merits of many cheap tests vs. few expensive tests is less clear. For limited programs, cheap tests may be better, but for extensive programs expensive tests may be better. However, a logical way of thinking about the question is available.

Obviously, a mix of high and low quality tests is normally used, but how should they be mixed? Consider a typical strategy: A few expensive test sections are built, and index tests are calibrated to their results. Then, a large number of index tests is used to characterize spatial variability. Presuming Eqs. 19 and 23 to obtain, how should the numbers of test sections, n_1, and index tests, n_2, be chosen?

If an index test is calibrated against a test section, then uncertainty in the calibration factor is controlled up to the bias of the more expensive tests. Thus, uncertainty in the bias reduces to

$$V[B_2] = V[B_1] / n_1 \quad . \tag{24}$$

Combining Eqs. 24, 22, and 17, and setting R=1.0,

$$V[\hat{x}] = V[x] + \frac{\overline{x}^2 \, K_b/C_1}{n_1} + \frac{(v[x] + K_e/C_2) \, \overline{B}^2}{n_2} \quad , \tag{25}$$

and

$$n_1/n_2 = (\overline{x}/\overline{B}) \sqrt{(K_b/C_1) / (V[x] + K_e/C_2)} \tag{26}$$

Here is a simple, easily understandable, and logical way to make exploration more efficient. A large number of related results are easily found.

Grand Schemes and "Global Optima:"
The complexities of site characterization make it unlikely that comrehensive optimizations will ever be possible. The ultimate stumbling block is that exploration data are used in accomplishing two tasks: Obtaining a geological concept of a site, and estimating engineering parameters. The first is inductive.

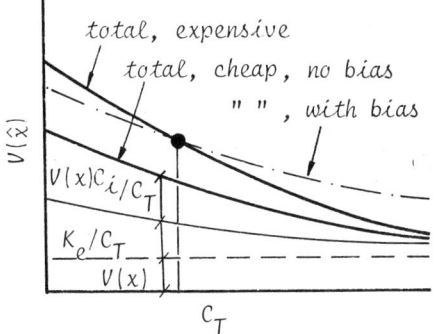

FIGURE 4

Most work reported in the literature on statistical exploration concentrates on engineering parameters. Little attention has been devoted to the parallel problems of mapping and detection of inhomogeneities. The neglect of mapping and search is not for lack of methodology. Techniques exist for both, and for search these are sophisticated. The increasing penetration of microcomputer systems in geotechnical engineering may increase the demand for such procedures.

Another topic to which surprising little attention has been paid is the use of subjective probability and decision analysis is exploration. Given the judgemental nature of rock engineering uncertainties, amounts of money at stake, and consequences of failures, decision analysis would be an obvious help. Increasing interest in risk analysis and trends toward risk-based regulatory criteria may provide motivation.

RELIABILITY MODELING

In current usage, reliability analysis refers to those systematic calculations through which uncertainties on parameters are propagated to uncertainties on predictions. The 'reliability' so analyzed is taken to be the degree to which an engineered facility can confidently be expected to perform as predicted. Thus, reliability must reflect many types of uncertainty, of both frequency and belief.

As used here, reliability analysis is deductive. A set of assumptions defining an engineering model is proposed, and uncertainties about the assumptions, parameter values, and boundary and initial conditions are propagated through the model to derive uncertainties on resulting predictions. The inductive parts of this analysis, specifically, defining a problem to analyze and building an engineering model, are ignored here.

Compounding Uncertainties

A very simple example of reliability analysis is the use of probabilistic relations in predicting rock mass deformability. The aggregate deformability of an idealized rock mass is a function of both (i) the discontinuities in the mass and their mechanicsl properties, and (ii) the intact material and its properties. Since the former often predominates, it alone will be considered.

For small stress increments the discontinuities will be assumed to display linear stress : deformation relationships both for shear and normal stresses. Thus, deformations normal to the discontinuities are approximated by

$$d_n = K_n \sigma_n \quad , \tag{27}$$

in which d_n = deformation, K_n = normal stiffness, and σ_n = normal stress. Analogously, shear deformations are approximated by

$$d_s = K_s \tau \quad , \tag{28}$$

in which τ = shear stress.

Assuming vertical deformation in the mass to be the simple sum of the vertical components of normal and shear deformations on all the discontinuities, and assuming the discontinuities not to interact, the uncertainty in vertical deformation of the mass resets on three things: The stiffnesses of each discontinuity, the increments of stress on each discontinuity, and the number and orientations of discontinuities. Were these known with certainty, the mass deformation could be precisely predicted.

The first reason these factors are not precisely known is that rock masses are variable. The geometry, physical properties, and spatial density of discontinuities varies from one location to another, and while in principle this variability could be mapped, in fact collecting

PLAYING THE ODDS IN ROCK MECHANICS

such detailed information in infeasible. Thus, the properties of discontinuities are described only up to frequency distributions, as discussed by many workers [2]. Spacings among discontinuities along sampling lines are known empirically to have a frequency function of the form $f(s) = \lambda \cdot \exp\{-\lambda s\}$, in which $f(s)$ = the frequency of spacings of size s, and $1/\lambda$ = the average spacing. Trace lengths appear to exhibit logNormal frequency functions. Limited study suggests that stiffnesses exhibit Normal frequency functions, and that K_n and K_s may exhibit correlations.

The combined effect of such spatial variation is that mass deformability also varies somewhat from one location to another. If the frequency functions of the factors are known--even if only as averages and standard deviations--then the corresponding frequency functions of mass deformability may be calculated through the chosen geomechanical model. Figure 5 shows the result, plotted against average RQD. This figure suggests that a large part of the scatter observed in rock mass deformation testing may be due to inherent variability within the mass. Knowing the fraction of data scattter due to spatial variation as opposed to measurement noise is important for design, for it strongly influences the precision of predictions and thus facility safety.

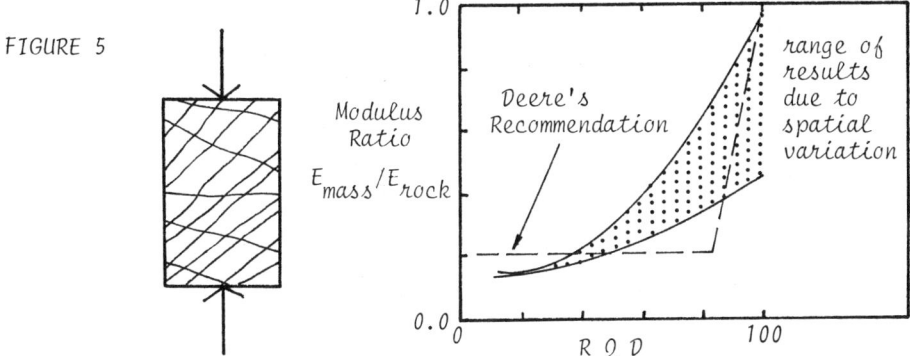

FIGURE 5

As discussed earlier, noise decreases the precision with which statistical estimates can be made, but is otherwise unimportant for predictions or designs. However, beyond data scatter, two other direct sources of uncertainty affect predictions. These are measurement bias and statistical estimation error, and both have been discussed. Each of these introduces uncertainty on predictions of the mean rock mass deformability, leading to a shipt of the entire prediction of Fig. 5, mean plus envelopes, either up or down. Such errors on mean predictions are more important than spatial variations when bulk response is at issue but less important when differential response is at issue. Errors on the mean produce systematic uncertainties, whereas spatial variations in many instances average out.

This simple example intends to illustrate that reliability theory methods provide a systematic framework for decomposing the significant contributions to uncertainty in engineering calculations. The word "calculations" is important, for reliability analyses seldom include all major uncertainties. Rather, they are used in parallel with the calculations normally made, to determine the implication of data scatter, limited numbers of tests and measurement and model bias. Analyses based only on 'best estimates,' even if conservatively chosen, do no allow the implications of differing sources of uncertainty, with differing influences on safety, to be logically and consistently evaluated.

Zen and the Selection of Factors-of-Safety

Factor of safety can be conveniently defined as the ratio of capacity to demand,

$$FS \equiv \frac{Capacity}{Demand} \quad . \quad (29)$$

Thus FS < 1 implies inadequate capacity and therefore 'failure.' This index is clear, concise, 'reasonable,' and wholly inadequate as a measure of safety. Despite its appearance, FS is only an ordinal measure of safety. Higher FS's imply more safety, but not how much more. For mechanically equivalent definitions of failure, FS can be changed arbitrarily by rearranging the terms in its equation, or by considering alternate load paths. Fortunately, reliability analysis provides an escape from this dilemma.

Slope stability calculations, in both soil and rock mechanics, have enjoyed by far the most attention from reliability analysts, and they provide a convenient example of factors of safety and alternative reliability indices. The simplest slope stability problem is block sliding on an individual fully persistent discontinuity (Fig. 6). Adopting Patton's approach, the resistance to sliding is contributed by mineral friction and by dilation of the discontinuity in over riding asperities. Denoting the mineral friction angle ϕ, and the maximum dilation angle i, two mechanically equivalent factors of safety might be defined as,

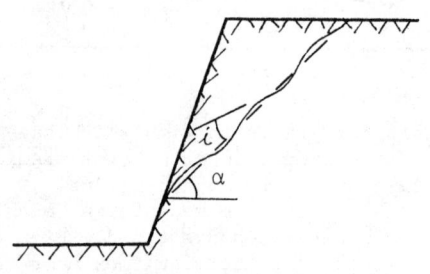

FIGURE 6

$$FS_I = \frac{\tan(\phi + i)}{\tan \alpha} \quad (30)$$

$$FS_{II} = \frac{\tan \phi}{\tan(\alpha - i)} \quad (31)$$

The first treats dilation as an increased capacity, the second as a decreased demand. The two are numerically equal for FS=1.0, but otherwise they may be quite different: For $\phi=30°$, $i=10°$, and $\alpha=20°$; FS_I=2.31 and FS_{II}=3.27. Similar non-invariance is easily found in more complex analyses. For example, total vs. effective stress analysis of undrained failures in shales may give different FS's even though the mechanical criteria of failure are the same.

PLAYING THE ODDS IN ROCK MECHANICS

An alternative to FS as a measure of safety is the first-order second moment reliability index

$$\beta_{FOSM} = \frac{E[FS] - 1}{SD[FS]} \tag{32}$$

in which E[FS] = the factor of safety based on mean capacity and mean demand (central FS). This measure includes information on uncertainties as well as best estimates, and is therefore more complete than FS alone. β_{FOSM} is based on second-moments of uncertain variables, that is, on means, variances, and convariances. This means that (i) no distributional assumptions are required, (ii) only rudimentary mathematics are needed, and (iii) computational demands are few. While an unique relation between β_{FOSM} and $Pr[FS<1]$ does not exist, for any arbitrary distributional form of FS, β_{FOSM} in monotonically related to $Pr[FS<1]$.

Unfortunately, β_{FOSM} also suffers non-invariance. Calculations using only second-moment information require approximations. Therefore changing the equation for FS, as for example from Eq. 30 to 31, may change β_{FOSM}, although the machanical model does not.

To overcome non-invariance, safety should be considered directly in parameter space. Fig. 7 shows the results of modified Bishop analysis for rotational sliding in a copper porphyry, plotted as a function of the Mohr-Coulomb parameters c, ϕ. The cross-hatched ellipse shows the best estimate $\hat{c}, \hat{\phi}$ for rock mass strength, and its one-standard-deviation envelope. Should the actual value of c, ϕ lie in the stable region of the plot, then within the limits of modified Bishop analysis the slope should be stable; and conversely, should the actual values lie to the left, the slope should be unstable. Therefore, an obvious measure of safety is "how far" the best estimate of $\hat{c}, \hat{\phi}$ lies away from the unstable boundary.

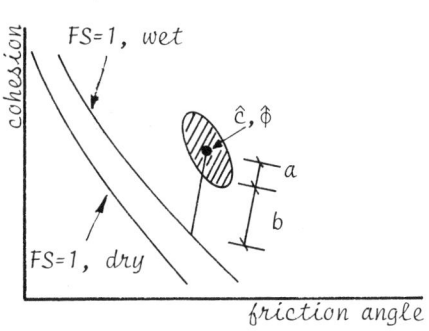

FIGURE 7

In analogy to β_{FOSM}, the distance from the best estimate $\hat{c}, \hat{\phi}$ to the unstable boundary should reflect uncertainty in c, ϕ, manifesting in the standard deviation envelope. A larger envelope about $\hat{c}, \hat{\phi}$ should imply lower safety. Because the standard deviation of c and ϕ need not be the same, and because the uncertainties may be correlated (i.e., the axes of the envelope may not be aligned with the c, ϕ axes), Hasover and Lind proposed measuring distance from $\hat{c}, \hat{\phi}$ to the unstable boundary in proportion to the standard deviaalong the direction of measurement, for example as the ratio $(b+a)/a$ in Fig. 7. They proposed using the shortest such distance as the reliability index now bearing their names, β_{HL}. For more than two uncertain parameters this minimum is easily found by linearly distorting the space to transform the standard deviation ellipsoid into a sphere, and then finding the minimum distance from $\hat{c}, \hat{\phi}$ to the transformed boundary.

Extremes and Averages

Spatial variability in a rock mass has an important consequence: The precision with which predictions can be made depends on the volume of rock mass mobilized. In averaging modes $0 \leq R \leq 1$, thus ignoring scale effects is conservative. This is not the case for failures that depend on extremes.

The best example of a failure due to extremes is wedge sliding in a slope. This is typically anlyzed using stereographic projection as in Fig. 8, where stable and unstable zones for sliding directions are shown. Also shown is the distribution of lines of intersections of pairs of joints. If for a particular pair of joints their line of intersection lies within an unstable region of the plot, then as a rough first approximation one often concludes that the wedge so defined is potentially unstable. The foregoing does not mean, however, that the distribution of lines of intersection can merely be integrated over the unstable region to obtain a probability of failure! The number so obtained, p, pertains to the probability that exactly two randomly selected joints in the rock mass would have a line of intersection, were they in proximity, that would define an unstable wedge. The probability of at least one such wedge failing becomes

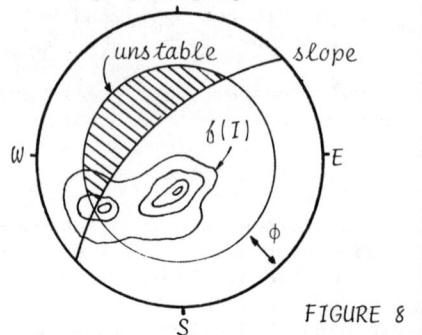

FIGURE 8

$$P_f = 1 - (1-p)^{C_n^2}, \qquad (33)$$

in which $n =$ number of joints, and $C_n^2 = n!/2!(n-2)!$. P_f rises rapidly with n; so rapidly, in fact, that were other factors (e.g., lack of persistence) not operative, few slopes in jointed masses would be stable.

"If We Do It Enough, Maybe We'll Get It Right"

Monte Carlo simulation has become popular, but is not without dangers. The principle danger is that it appears so easy, but may obfuscate important probabilistic relations among variables and assumptions. In practice, simulation frequently becomes little more than sensitivity analysis. Spatial variables are reduced to lumped parameters, and correlations are ignored for 'simplicity.' The potential for misuse of simulation is enormous, and one need know essentially nothing of probability theory to do so.

Dispite this warning, simulation does have a place. It may allow systematic risk analysis to be extrapolated from small analyses to large facilities. It allows complicated spatial interactions to be delt with. Fig. 9 shows results of simulation studies of joint persistence that could not be obtained analytically. However, herein lies a rule of thumb: If a problem can be solved analytically or through approximations, don't use simulation.

RISK ANALYSIS

Risk analysis attempts to combine data analysis and reliability models within an economic framework to reach decisions. Risk analysis intends to develop realistic probabilities, and therefore must incorporate subjective assessments. It also intends to balance direct investment against risk, and therefore must place values on consequences. A risk analysis is almost always project specific, and may be costly.

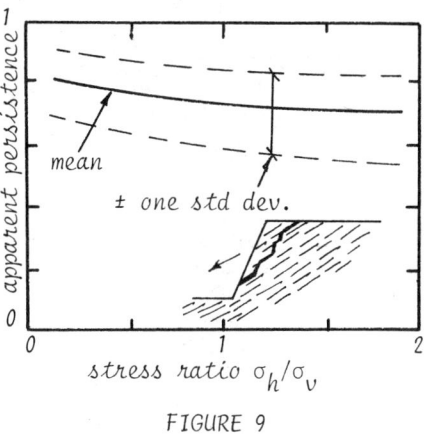

FIGURE 9

Tunnel Scheduling and Cost Estimation

Risk analyses have been performed on a number of projects, and each analysis proceeds in a slightly different way, tailored to the needs of the client, the uncertainties at issue, and the facility. Extensive event/fault tree analysis and simulations have been conducted for large chemical facilities, and on tailings storage and disposal plans. To the author's knowledge, modest scale risk analyses have been made for pit mine plans, on foundations for offshore structures, and on embankments for materials storage. However, the most extensive risk analyses of rock engineering in the public domain have probably been made on tunnel scheduling and construction.

The goal of risk analysis of tunnel operations is to obtain sound estimates of cost and advance rates--including uncertainty--and an evaluation of exploration data and construction procedure as they influence time and cost. The analysis procedes in three phases: (1) assessing tunnel geology and the uncertainty of engineering conditions, (2) assessing the performance of alternative construction techniques in various conditions, and (3) predicting effects of uncertainty on project cost and schedule. To date such analyses have been made on at least five projects.

In the first step, the tunnel is divided into units estimated to have similar but uncertain geological conditions, and each unit is divided into segments estimated to have similar but uncertain geotechnical conditions (fig. 10). For each segment, a set of geotechnical parameters is estimated, and uncertainties assigned to all assessments, using subjective probability techniques. These assessments are all made by the project geological and geotechnical staff, working with the risk analy-

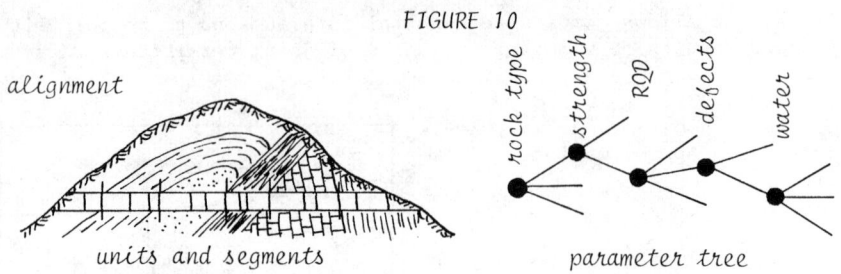

FIGURE 10

alignment

units and segments

parameter tree

sis team. For example, the parameters might include rock type, strength, jointing, defects, foliation, water inflows, etc; and each parameter might have several possible states and consequences for construction. The estimates are combined in a parameter tree, much like an even tree, which summarizes the state of information for one segment.

Exploration strategy is planned using statistical decision analysis. Options are identified, and evaluated by the changes they might lead to in the parameter trees. The 'worth' of an option is valued by its effect on expected construction cost, and the most cost effective options separated out from the rest. The prediction of potential changes in the parameter trees is made using an "exploration reliability matrix." This is a table summarizing the measurements one might make with a particular exploration tool, in a particular geotechnical condition. Bayes' Theorem is used to update probabilities in the parameter trees, using the exploration reliability matrix to provide likelihoods (i.e., conditional probabilities of the exploration results) and the original assessments as prior probabilities. Since all this information is numerical, a computer can be used to efficiently evaluate a large number of exploration options interactively.

Construction costs and schedules for alternative construction schemes are estimated by simulating a tunnel profile from the parameter trees, and using opitmal scheduling and control techniques to simulate tunnel construction. This two stage simulation is repeated many times to estimate advance rate and its uncertainty, and time streams of cost and its uncertainty. This may often be done on a microcomputer. An important result of these simulations is the guidance they provide for choosing among design-construction alternatives, and in selecting switching rules for adaptive tunneling methods.

"Regulatory Geology"[†]

To the surprise of many, geologist and engineer alike, regulatory decision in several sectors have become dependent on geological information. One result has been the finanical health of geotechnical consultants who provide seismic, surface faulting, geotechnical and other in-

---

[†] To the author's knowledge, this term is due to Richard Meehan, of Earth Science Associates, Palo Alto, California.

formation by which site safety might be demonstrated. Another, however, is a demand for risk-based safety studies, driven by a regulatory need to translate soft geological information into implications for public safety. This demand began first in nuclear siting, but spread within the energy sector (e.g., LNG facilities), and has even reached outside the "new technologies" (e.g., Auburn Dam). The prospect that risk analysis will be required to license hazardous waste sites, chemical and industrial facilities, and other projects appears likely.

Regulatory geology provides an interesting set of epistemological problems, in that geological data are sparce and qualitative, the events being predicted are rare, and the consequences of adverse facility performance may be serious. The General Electric Company's Test Reactor (GETR) at Vallecitos, California, presents a good example.

The GETR was the first licensed test reactor, and operated safely for 20 years until 1979. In 1979, at the time of GE's application for license renewal, trenching was conducted adjacent to the reactor at the foot of the Vallecitos Hills, and offsets were found in recent alluvium indicating the possible existance of surface faulting within meters of the reactor building. The NRC issued a "show cause" order and the plant was brought to a cold shut down.

Ignoring the subtleties of the GETR case, the issue of site safety resolved to a few simple things:

- What geological process caused the offsets (a fault, a surface slide, what?).
- Would it occur again under the reactor?

In essence, the "information base" upon which these questions would have to be answered consisted only of:

- The regional geology
- Observations of three offsets in recent sediments (Fig. 11)

Nevertheless, the decision to relicense GETR was important. Public safety was at issue, as was significant capital investment.

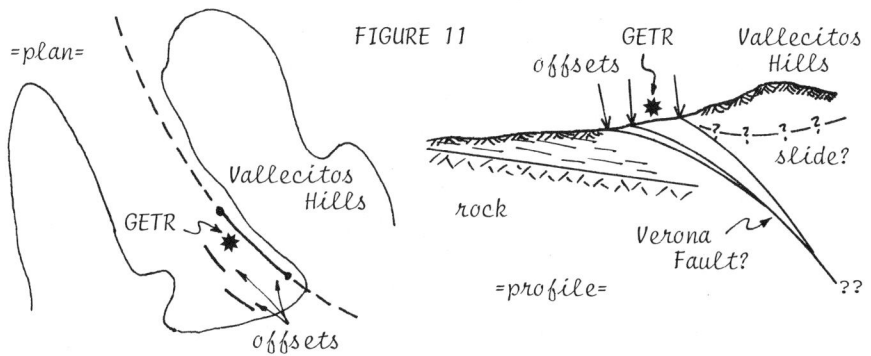

FIGURE 11

Risk analysis is being more frequently called upon to deal with problems such at the GETR, and to deal with them rationally. As a vehicle for explicitly organizing arguments--or for debunking spurious concerns --risk analysis plays a useful role. It allows a problem to be broken into parts, each to be analyzed in isolation, and then to be logically recombined to draw a conclusion. But risk analysis does not generate new information, nor is it value-free. As in any analysis, assumptions muct be made and boundaries set of the problem analyzed. Despite the extent of mathematics or geological reasoning, from a statistical point of view, if a small number of events, n, have occurred in T_o years, then the most that can be said about the future frequency of such events is summarized in Fig. 12. The data alone allow nothing more concrete.

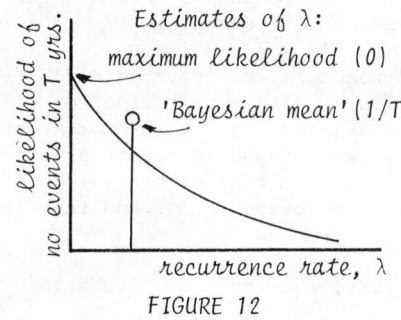
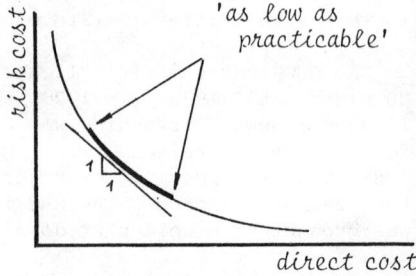

FIGURE 12

Acceptable Risk

FIGURE 13
(after Rowe, W., An Anatomy of Risk, John Wiley and Sons, 1977

Among the first questions asked by an owner or regulator when present with the results of a probabilistic risk analysis is, "What do I do with the numbers?" Traditional design has been based on factors of safety and few heuristics exist for deciding upon acceptable "probabilities of failure."

Risk is relative. An acceptable level of risk must usually be set in consideration of the marginal cost of mitigating it (Fig. 13). However, this is not always possible, and when it is not, the levels of ris now accepted in other civil facilities provide a useful reference. Fig. 14 shows approximate ranges of annual p_f and potential consequences for various facilities. Those ranges shown in full line are historical frequencies; those in broken line are a priori risk analyses. The data do not portray "acceptable" risks, only "accepted" ones.

CONCLUSIONS

Probabilistic modeling, statistical methods, and risk analysis have progressed far in recent years. For certain applications they now provide mature tools; for others they are in want of further development. The key to this work will be combining competent understanding of rock mechanics with equally competent understanding of inference and decision making theory. Demands for rational treatment of uncertainties is in-

FIGURE 14 -- "Accepted" risks for various civil facilities.

1. pit slopes
2. disease
3. merchant shipping
4. mobile drill rigs
5. fixed platforms
6. dams
7. dams (reactor safety st)
8. commercial aviation
9. landslide (specific site)
10. foundations and slopes
11. "CANVEY" oil facilities
12. "CANVEY" LNG facilities
13. 100 nuclear plants (RSS)

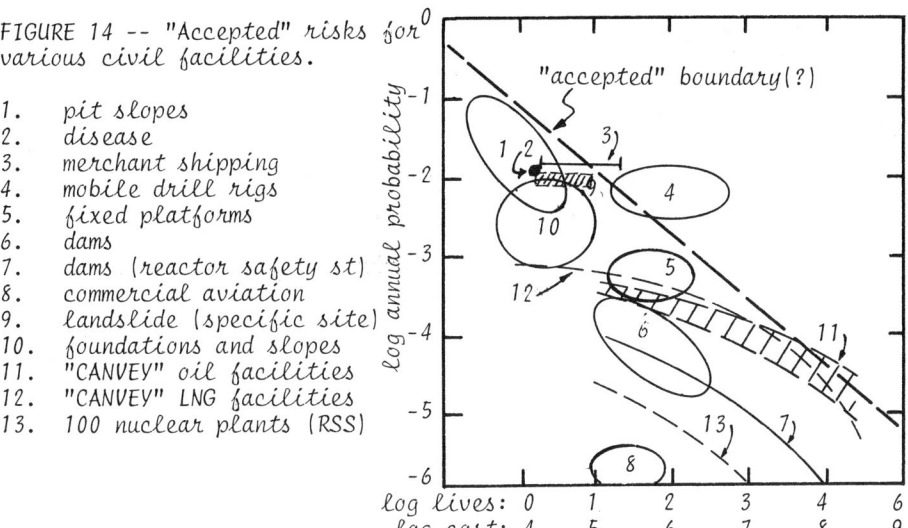

creasing, but it is incumbent upon those working on the new methodology to ensure that both geomechanical and probabilistic rigor are maintained.

REFERENCES

1. Einstein, H.H. and G.B. Baecher, "Probabilistic and statistical techniques in engineering geology," 30th Geomechanics Colloquy, Salzburg, 1981, 125pp.

2. Einstein, H.H., G.B. Baecher, D. Veneziano, and others, "Risk analysis of pit mine slopes," Report to the U.S. Bureau of Mines under Contract J 027 5015, 1979, 5 vols.

Chapter 9

PERMEABILITY, PERCOLATION AND STATISTICAL CRACK MECHANICS

John K. Dienes

Theoretical Division, Group T-3
University of California
Los Alamos National Laboratory
Los Alamos, New Mexico 87545

ABSTRACT

The permeability of sands and soils seems to be adequately described by Darcy's law, but the permeability of rocks is complicated by a number of factors related to the probability of crack intersections. During the last few years we have developed a statistical theory of fragmentation which has been successful in explaining the observed behavior of rocks. In this paper, a theory of permeability is developed which draws on the concepts employed in previous work. The permeability involves three factors, the average fluid flux per crack, the number of cracks per unit area, and the fraction of cracks that are not isolated. Probabilistic ideas are used in connection with the hydrodynamic theory of flow through a single crack to develop an integral expression for permeability. The result has the form of Darcy's law for anistropic media. Although the current work is motivated by the need to develop optimized oil-shale retorts, we believe that the theory can be applied to a variety of other problems.

INTRODUCTION

The permeability of rock plays an important role in many geophysical processes, in the recovery of minerals, and in the isolation of wastes. Still, it has received relatively little theoretical attention. Many features of transport in rocks have been discussed by Witherspoon et al. (1981). Some of the differences between transport in rocks and soils are noted here. In soils, the voids are generally all connected, whereas in rocks there may be many isolated cracks that do not contribute to the overall permeability. In soils, the voids tend to be compact in size, whereas in rocks they are often in the form of thin laminae, particularly following fragmentation processes. In soils, the void shapes do not change significantly under pressure, whereas in rocks it can significantly reduce crack openings, and since flow varies with the cube of thickness, the effect of closure can be very important. In addition, crack sets in rocks tend to have preferred orientations, whereas the porosity in soils is isotropic. Consequently, in rocks the permeability is a tensor, often with highly directional properties, whereas a scalar treatment is normally adequate for soils. Finally, we note that percolation theory (Broadbent and Hammersley, 1957) predicts a critical crack size below which rock is essentially impermeable because of the absence of connected paths for fluid convection. This critical property has no analogue in soil behavior.

In this paper, an expression is developed for the flux of fluid through a rock matrix which involves three factors; the average flux per crack, the number of crack traces per unit area, and the fraction of cracks connected in some continuous path. The first factor accounts for the flux through individual cracks, and is essentially Poiseuille's result adapted to flow in laminae. The second factor, the number of cracks per unit area, is found from the number of cracks per unit volume by an argument I presented at the 20th Symposium in this series. The third factor is investigated by percolation theory, which is extended to account for preferred cracking in the bedding planes herein. This paper presents a summary of results, rather than detailed derivations, which will be presented in related papers.

PERMEABILITY

Permeability of geological materials can result from convection between voids, as in sand or soil, from diffusion involving molecular processes, or from laminar transport through these cracks. In this paper, we consider only the last of these.

Consider an ensemble of thin laminae in each of which fluid flows down the pressure gradient following Poiseuille's law

$$Q_j = \theta(h^3 \ell/12\mu)(p_{,j} - n_k p_{,k} n_j) \tag{1}$$

where the Q_j are the components of flux, h is the maximum thickness of the crack, ℓ is its width, μ is the fluid viscosity, and the term in the right parentheses is the component of pressure gradient in the plane of the crack. The n_j are the components of the crack normal. Deviations of the crack shape from a uniform lamina are accounted for by θ, a parameter which is estimated to be about 1/2, but end effects and roughness may make it smaller. Now, let $\delta \tilde{N}_s^\alpha$ denote the number of crack intersections per unit area with the orientation and size of crack set α that intersect a plane, s. Let $\tilde{n}_k$ denote the components of the normal to s. Then the flux through s can be written

$$F_s = \sum_\alpha \delta\tilde{N}^\alpha_s Q^\alpha_k \tilde{n}_k = K_{jk} p_{,j} \tilde{n}_k/\mu \quad . \tag{2}$$

This defines the specific permeability tensor as

$$K_{jk} = \sum_\alpha B^\alpha \, \delta\tilde{N}^\alpha_s \, r^\alpha_{jk} \tag{3}$$

where

$$B^\alpha = \theta h^3_\alpha \ell_\alpha/12 \quad , \quad r^\alpha_{jk} = \delta_{jk} - n^\alpha_j n^\alpha_k \tag{4}$$

with δ_{jk} the Kronecker delta. The number of cracks per unit area is

$$\delta\tilde{N} = -\frac{\partial \tilde{N}}{\partial \ell} \delta\ell \, \delta\Omega \quad , \tag{5}$$

where $\delta\Omega$ is an element of solid angle and $\delta\ell$ is an element of trace length. It is convenient to consider the limit in which the crack sets are infinitely well resolved and the sum becomes an integral

$$K_{jk} = \int_\Omega d\Omega \int_0^\ell d\ell \left(-\frac{\partial \tilde{N}}{\partial \ell}\right) B(\ell,\Omega) \quad . \tag{6}$$

The integral on Ω is taken over the upper half of the unit sphere, and includes a solid angle of 2π sterradians. It can be shown by the method of Dienes (1979) that $\tilde{N}$ is related to the distribution of cracks per unit volume per 2π, $N_c(c,\Omega)$, by

$$\tilde{N} = -2 \int_{\ell/2}^\infty dc \sqrt{c^2 - \ell^2/4} \, N_c(c,\Omega) \sin\gamma_s \quad . \tag{7}$$

The quantity γ_s is the angle between s and crack set α, and can be found from

$$\cos\gamma = n_k \tilde{n}_k \quad , \tag{8}$$

Combining these results and writing the crack opening in terms of the aspect ratio

$$h = A(\Omega)\ell \quad , \tag{9}$$

we find

$$K_{jk} = \frac{64\theta}{45} \int_\Omega d\Omega \, A^3(\Omega) r_{ik} a_k G(\Omega) \tag{10}$$

The quantity $G(\Omega)$ is the fifth moment of crack size, c,

$$G(\Omega) = -\int_0^\infty dc \, \frac{\partial N(\Omega,c)}{\partial c} c^5 = N^o <c^5> . \tag{11}$$

where N^o denotes the number of cracks per unit volume that are not isolated. In the next sections, an expression for N based on a Liouville equation is derived, and subsequently a factor to account for crack isolation is developed using percolation theory. In finite difference calculations the integral in (10) would be done numerically. If the cracks are isotropically distributed, the integral in (10) can be done analytically with the result

$$K_{11} = K_{22} = K_{33} = \frac{8\pi^2}{15} \theta A^3 N_o <c^5> . \tag{12}$$

It is implicit in this result that all cracks have the same aspect ratio, A, a result that holds for the simplest stress fields, but not in general. If the distribution is taken to be the sum of isotropic and bedded parts, then G can be written

$$G(\Omega) = G_i + G_b \delta(\theta) \tag{13}$$

where δ is the Dirac delta function and θ is the angle of the crack normal with the z-axis. It follows that the permeability is the sum of two parts

$$K_{jk} = \frac{64\theta}{45} \left(\frac{3\pi^2}{8} A_i^3 G_i \delta_{jk} + A_b^3 G_b W_{jk} \right) \tag{14}$$

where $W_{11} = W_{22} = 1$, and otherwise $W_{jk} = 0$.

CRACK DISTRIBUTION

There is a substantial body of evidence that the distribution of crack sizes in competent rock and other materials is exponential. As cracks grow and intersect, this distribution will change. For simplicity, we assume that cracks are either active, that is capable of growth when unstable, or inactive, the result of having intersected a number, α, of other cracks such that growth is terminated. We take α to be 4, though this is not essential. When unstable, we assume that cracks grow at a rate, $\dot{c}$, which is assumed constant, and has a magnitude that is about a third the speed of sound. Again, this assumption is convenient, but not essential. It is shown by Dienes (1978) that crack distributions evolve in accord with a Liouville equation

$$\frac{\partial L}{\partial t} + \dot{c}\frac{\partial L}{\partial c} + \frac{\partial M}{\partial t} = 0 \tag{15}$$

where L and M denote the distributions of active and inactive cracks. For sparse distributions, we may write

$$\frac{\partial M}{\partial t} = kL \tag{16}$$

and the distributions can be determined analytically. We assume that crack sets with each orientation grow separately and that k is determined by the initial distribution. For a combination of isotropic and bedded cracks, an estimate for k is given by

$$k = (4\pi^2 \dot{c}/\alpha)\left(\pi^2 L_i^o \bar{c}_i^{-2}/2 + L_b^o \bar{c}_b^{-2} \sin\theta\right) \tag{17}$$

where L_i^o denotes the initial density of isotropic cracks per 2π and L_b^o denotes the initial density of bedded cracks. The distribution of active cracks, found by solving the Liouville equation (Dienes, 1978) assuming an initially exponential distribution is:

$$m = \frac{\partial M}{\partial c} = F(e^{-kc/\dot{c}} - e^{-c/\bar{c}}) \quad , \quad c \leqslant \dot{c}t \quad , \tag{18}$$

and

$$m = \frac{\partial M}{\partial c} = F(e^{\beta t} - 1)e^{-c/\bar{c}} \quad , \quad c \geqslant \dot{c}t \quad , \tag{19}$$

where

$$F = L_i^o k/\bar{c}\beta \quad , \quad \beta = \dot{c}/\bar{c} - k \quad . \tag{20}$$

The calculation of $G(\Omega)$, as given by (11) using this distribution function, is straightforward but lengthy. Here we note only its limiting value at late times,

$$G^\infty(\Omega) = 5! \, F(\dot{c}/k)^6 \quad . \tag{21}$$

If the crack distribution is represented as the sum of isotropic and bedded parts, then, as in (13), G^∞ will also be the sum of isotropic and bedded parts. If the bedded cracks dominate, as in oil shale, we then find from (17) that the average rate constant for formation of inactive cracks is

$$k_i = 8\pi \dot{c} L_b^o \bar{c}_b^2/\alpha \quad . \tag{22}$$

STATISTICAL CRACK MECHANICS

For bedded cracks, which cannot intersect each other because they are parallel, the rate of intersection is dominated by the isotropic cracks, so that

$$k_b = 2\pi^4 \dot{c} \, L_i \, \bar{c}_i^{-2}/\alpha \quad . \tag{23}$$

Then, for these two cases,

$$G_i^\infty = 5! \, L_i^0 (\alpha/8\pi L_b \, \bar{c}_b^{-2})^5 \quad , \quad G_b^\infty = 5! \, L_b^0 (\alpha/2\pi^4 L_i^0 \, \bar{c}_i^{-2})^5 \quad . \tag{24}$$

PERCOLATION THEORY

In their seminal paper on percolation theory, Broadbent and Hammersley (1957) refer to the wetting of stone as one of many interesting examples of critical phenomena, but they do not attempt a calculation of permeability. The fundamental idea is that if cracks, or any other objects, are sparsely connected, the probability of an infinite path connecting them is zero, but when the number of connections exceeds a critical value, infinite paths can occur, and the probability of a crack being isolated is no longer zero. This idea is quantified for isotropic cracks in the first part of this section, and extended to an anisotropic distribution in the second part.

Let q denote the probability that two cracks of an isotropic, homogeneous ensemble have not intersected. We require the probability, Q, that a path from one of the cracks (1) through a given adjacent crack (2) be finite, as illustrated in Fig. 1. The path from (1) may be finite either because it is not connected to (2), or, if it is connected to (2), because the $\sigma - 1$ paths from (2) are finite. Assuming $\sigma = 4$, this can be expressed algebraically as

$$Q = q + (1 - q)Q^3 \quad . \tag{25}$$

In the figure, and in the current analysis, it will be assumed that cracks can intersect up to four others, but then become inactive. The choice of four is thought to be physically plausible, and is mathematically convenient, but is not immutable.

Since Q is a probability, it must not exceed 1 or be less than 0. The two roots of (25) satisfying this restriction are

$$Q = 1, \quad \sqrt{1/(1-q) - 3/4} \, - 1/2 \quad . \tag{26}$$

If the probability of no intersection, q, exceeds 2/3, then Q exceeds 1, and the second root is no longer meaningful. In that case the branch $Q = 1$ is appropriate. At $q = 2/3$ the behavior is said to be at the critical point. The existence of critical points such as this

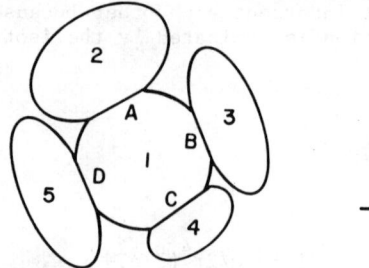 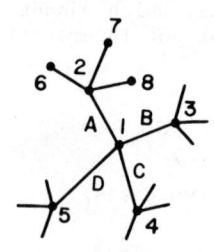

An inactive penny-shaped crack, 1, which has been arrested by intersection with four others, 2-5. The intersections are denoted by A, B, C, D.

A percolation-theory diagram representing the cracks and intersections of the figure on the left.

Fig. 1. Illustration of cracks, intersections, and their analysis by percolation theory.

is characteristic of percolation theory. In the calculation of permeability, the proportion, P, of cracks that are not isolated is required. We say that a crack is isolated when it is connected to one or zero others, and not isolated if it is connected to 2, 3 or 4 others. It is straightforward to show that

$$P = f(Q), \quad f(Q) = 1 - 4Q^3 + 3Q^4 \ . \qquad (27)$$

This relation is used in Fig. 2, in which the proportion of cracks that are not isolated is plotted against $p = 1 - q$, the probability of intersection.

We have undertaken to study the effects of anisotropy as an important part of our oil shale work, since it may be possible to enhance fragmentation by taking advantage of the weakness of its bedding planes. In pursuing this concept, we have also considered the generalization of percolation theory to the case where there are two families of cracks, one isotropic and one lying in the bedding planes. Here we will only quote the results of this analysis. It can be shown, by a generalization of the argument used for the isotropic distribution, that the proportion of isotropic cracks that are not isolated is given by $f_1 = f(R)$, where R is the solution to

$$(\nu_1 + \nu_2)R = \nu_1 \left(q_{11} + p_{11}R^3 \right) + \nu_2 p_{12} \left(q_{21} + p_{21}R^3 \right)^3 + \nu_2 q_{12} \ . \qquad (28)$$

The subscripts 1 and 2 refer to isotropic and bedded quantities, respectively. Thus, q_{ij} is the probability that a crack of type i does not intesect one of type j, and $p_{ij} = 1 - q_{ij}$. Similarly, Q_{ij} denotes the probability that a path starting from a crack of type i and

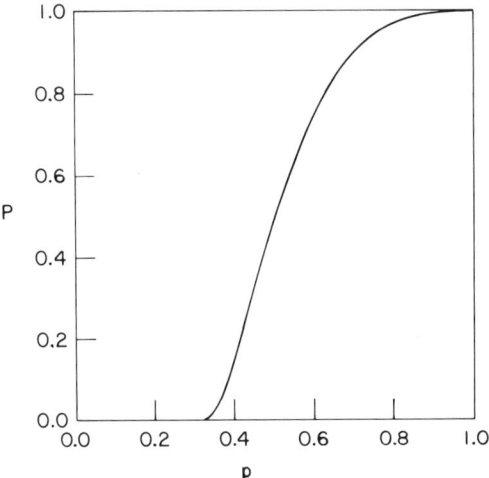

Fig. 2. The proportion, P, of cracks that are not isolated as a function of the probability, p, that two cracks have intersected.

intersecting a crack of type j is finite. The parameter ν_i denotes the frequency with which a line intersects cracks of type i. The proportion of bedded cracks that are not isolated, f_b, is given by $f(Q_{21})$ where f is given by (27) and

$$Q_{21} = q_{21} + p_{21}R^3 . \qquad (29)$$

In a complete numerical calculation, f_i multiplies G_i and f_b multiplies G_b in (13). There is a critical condition for anisotropic crack systems, just as for isotropic systems. The condition that (28) have a double root at R = 1 leads to the criterion

$$r + (\nu_2/\nu_1)s < 0 \qquad (30)$$

for all cracks to be isolated, where

$$r = 3p_{11} - 1 \quad , \quad s = 9p_{12}p_{21} - 1 . \qquad (31)$$

This condition is illustrated by the critical lines in Fig. 3. If the number of bedding cracks goes to zero, so that $\nu_2 = 0$, then the critical condition reduces to the condition obtained in the preceeding discussion of purely isotropic cracks.

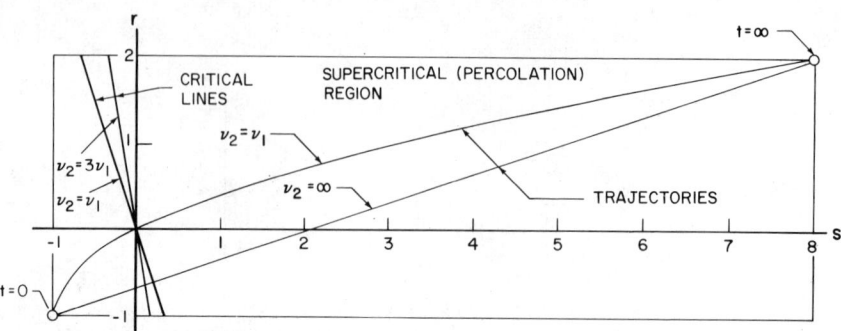

Fig. 3. Illustration of the critical condition for percolation. Percolation through a system of cracks is possible in the region to the right of the critical lines. Otherwise, the number of connected cracks is insufficient to allow for a continuous path. The trajectories show how the parameters r and s vary with duration of crack instability, t.

REFERENCES

Broadbent, S. R. and Hammersley, J. M., 1957, "Percolation Process I. Crystals and Mazes," Proc. Camb. Phil. Soc. 63, 629-641.

Dienes, J. K., 1978, "A Statistical Theory of Fragmentation," 19th U.S. Symposium on Rock Mechanics, Stateline, Nev.

Dienes, J. K., 1979, "On the Inference of Crack Statistics from Observations on an Outcropping," 20th U.S. Symposium on Rock Mechanics, Austin.

Dienes, J. K., 1982, "On the Stability of Shear Cracks and the Calculation of Compressive Strength," to appear in JGR.

Witherspoon, P. A., et al., 1981, "New Approaches to Problems of Fluid Flow in Fractured Rock Masses," 22nd U.S. Symposium on Rock Mechanics, M.I.T.

Chapter 10

EFFECT OF CORRELATION ON ROCK SLOPE STABILITY ANALYSES

by Edward F. Glynn and Satyajeet Ghosh

Assistant Professor, University of Pennsylvania
Geotechnical Engineer, Stone & Webster Engineering Corp.
Philadelphia, Pennsylvania

Engineer, United Engineers & Constructors, Inc.
Philadelphia, Pennsylvania

ABSTRACT

Many of the proposed techniques for calculating the reliability of rock slopes utilize Monte Carlo simulation. Monte Carlo approaches are attractive because they provide means of solving probabilistic problems with deterministic algorithms. One disadvantage shared by all simulation-based stability analyses is that they assume all the random variables are independently distributed. The paper focuses on the independence assumption and examines the effect of correlation on the probability of failure. In particular it considers the influence of possible (albeit hypothetical) correlation between joint attitudes (θ) and joint friction angles (ϕ). The study indicates that neglecting correlation can introduce significant errors into the reliability computations. With a modest positive correlation between θ and ϕ the error can lead to a 100% overestimate of the probability of failure.

INTRODUCTION

Reliability analyses provide an alternate approach to assessing the reliability of rock slopes. In reliability analyses the integrity of a slope is expressed as its probability of failure (P_f) rather than the traditional factor of safety (FS). One of the primary advantages of a probabilistic approach is that it explicitly considers the uncertainty associated with engineering parameters in that it characterizes parameters in terms of probability density functions rather than deterministic values.

At the present time there are a number of techniques available for examining the reliability of rock slopes. The techniques are remarkably varied with respect to sophistication and requisite computational effort. They range from graphical methods (McMahon, 1975) based on pole diagrams and steronets to stochastic models (Call and Nicholas, 1978; Veneziano, 1978) that treat jointing in rock masses as a analytically describable random process. The approaches which currently provide the most versatile means for estimating P_f rely on Monte Carlo simulation (Major et al., 1978; Glynn, 1978). In these approaches a deterministic algorithm that computes FS is mated with a routine that randomly selects sets of input parameters through Monte Carlo techniques. Each set of input parameters produces a single value of FS. P_f is the proportion of FS values that fall below unity.

Monte Carlo simulation is an attractive tool in reliability analyses because it can be applied to virtually any problem that can be modelled deterministically. As rock slope analyses become increasingly sophisticated with the addition of toppling and rotational failure modes the new analytical (albeit deterministic) models can be incorporated directly into Monte Carlo simulations. The weakness of Monte Carlo methods lies not so much in the FS algorithm, but rather in the selection of input parameters during simulation. Each random variable must have a rigorously defined probability density function (pdf). The paper discusses one of the problems encountered in selecting pdf's: are some (or perhaps many) of the random variables correlated and should their pdf's be defined as multivariate distributions?

THE CORRELATION PROBLEM

Consider the following scenario. Two random variables θ (joint attitude) and ϕ (joint friction angle) have a joint probability density function $f_{\theta\phi}(\theta,\phi)$. An individual interested in performing a reliability analysis would presumably be interested in identifying this function and could approximate it by fitting a curve over a statistically representative sample of "n" pair of (θ,ϕ) data points. These pairs of data points may represent the results of field and/or laboratory tests. The data (and their corresponding $f_{\theta\phi}(\theta,\phi)$) may reveal the existence of some correlation. For example, high friction angles may tend to be associated with joints that have steep dips. In any case, $f_{\theta\phi}(\theta,\phi)$ would be used (in conjunction with the pdf's of other random variables) to calculate P_f for the rock slope. The value of P_f would, of course, reflect the bivariate nature of $f_{\theta\phi}(\theta,\phi)$. Consider an alternate scenario. A second individual attempts to simplify the analysis by ignoring the fact that the data were measured in pairs. He determines $f_\theta(\theta)$ by merely examining the n values of θ and $f_\phi(\phi)$ by examining the n values of ϕ. In other words, he treats the two random variables as if they were unrelated or independent. Given this assumption of independence, the joint probability density function can be calculated as:

$$f_{\theta\phi}(\theta,\phi) = f_{\theta}(\theta) \cdot f_{\phi}(\phi). \tag{1}$$

This joint density function may be quite different from the one derived by examining pairs of data points despite the fact that both density functions are derived from the same data base. P_f's calculated from the two joint density functions may also be quite different.

Is there a significant difference between $(P_f)_j$, the value derived from the "correct" joint distribution, and $(P_f)_m$, the value derived from the two marginal distributions? Is one value consistently higher? Does the assumption of independence tend to make one over-estimate or under-estimate the reliability of the rock slope? These questions are important because the Monte Carlo models in current use actually compute $(P_f)_m$.

Monte Carlo models assume that all the random variables are independently distributed. The assumption eliminates the need for conditional distributions and reduces the volume of computations by a very considerable amount. At the present time it is difficult to produce hard evidence to suggest that correlation should be incorporated into the simulation. On the other hand, the burden of proof concerning independence rests on the individual who makes the assumption and, in many instances, the assumption cannot be justified a priori. There is not enough data available to develop conclusions, even tentative conclusions, about possible correlation. Correlation studies are somewhat beyond the state of the art in geologic data aquisition and reduction. Most of the current efforts in these areas are focused on establishing typical distributional forms i.e., marginal pdf's, for such variables as joint spacing, length and orientation. Einstein et al. (1979) summarized the work to date and noted how sampling biases can act as filters that obfuscate the underlying or true distributions. There is some information available on correlations among random variables. Dershowitz (1979) examined strike and dip data from numerous joint surveys and found that in many instances biaxially symmetric spherical distributions provided the best fit for pole diagrams. Marek and Savely (1978) and Gaziev and Rechitski (1979) suggest that joint friction angles and joint cohesion are correlated. Table 1 lists many of the parameters in rock slope analyses and categorizes the pairs with respect to correlation. The designations are largely speculative at the present time.

SENSITIVITY STUDY ON CORRELATION

The study presented in this paper examines one of the pairings in Table 1 - it considers the effect of correlation between θ and ϕ on P_f. The purpose of the study is to determine how closely correlated (as measured by ρ, the coefficient of correlation,) the two random variables must become before there is a significant difference between $(P_f)_j$ and $(P_f)_m$. Two cases were considered: a plane strain failure

	U	c	φ	l	p	s	λ	θ
Dip of Joint = θ	C	P	P	P	P	P	P	-
Strike of Joint = λ	P	P	P	P	P	P	-	
Joint Spacing = s	C	C	C	C	P	-		
Joint Persistence = p	C	C	P	C	-			
Joint Length = l	C	P	P	-				
Joint Friction Angle = φ	U	C	-					
Joint Cohesion = c	P	-						
Cleft Water Pressure = U	-							

C: Probably Correlated
P: Possibly Correlated
U: Probably Uncorrelated

Table 1 - Correlations Between Random Variables

with sliding on a single plane and a tetrahedral wedge failure with sliding on two planes. Both cases involved relatively simple conditions - continuous joints, no water pressures and no joint cohesion - in order to accenuate the influences of θ and φ.

Method of Analysis

The study assesses the influence of ρ by calculating both $(P_f)_j$ and $(P_f)_m$ for joint pdf's of θ and φ. Some "reasonable" joint distributions are assumed and $(P_f)_j$ and $(P_f)_m$ are calculated using conditional and marginal pdf's, respectively. In these simple slopes whose stability is controlled solely by θ and φ $(P_f)_j$ is the correct probability of failure. $(P_f)_m$ is the probability of failure that would be calculated by any technique that presumed θ and φ are independently distributed. The difference between $(P_f)_j$ and $(P_f)_m$ reflects the effect of correlation and represents the error incurred in a conventional Monte Carlo analysis. As discusses earlier, it is important to remember that both $(P_f)_j$ and $(P_f)_m$ are derived from the same underlying pdf i.e., data base. $(P_f)_j$ is merely a more refined interpretation of the data base.

Distributional Forms

Two types of joint distribution functions were used in the study: bivariate normal and ellipsoid.

The equation of a bivariate normal distribution centered at the origin is:

$$f_{XY}(x,y) = \frac{e^{-A}}{2\pi\sigma_x\sigma_y(1-\rho^2)} \quad (2)$$

where σ_x and σ_y are the standard deviations of X and Y and

$$A = -\frac{1}{2(1-\rho^2)}\left((x/\sigma_x)^2 - (2\pi xy)/(\sigma_x\sigma_y) + (y/\sigma_y)^2\right)$$

The bivariate normal distribution is unbounded with respect to X and Y. In the study this distribution models physical parameters which are bounded. Hence, the distribution has been truncated by slicing it with a plane parallel to the X-Y plane. The equation of the plane is $Z_0 = Z_{max}/C$ where Z_{max} is the maximum probability density which occurs at the origin and C is an arbitrary truncation coefficient. The trace of the distribution on the Z_0 plane is an ellipse. Once the distribution is truncated it is normalized to ensure that the volume under the curve is unity.

The ellipsoid distribution is an ellipsoid that is bisected by the X-Y plane. The equation of such a distribution centered at the origin is:

$$f_{XY}(x,y) = \frac{\left(1 - (x/a)^2 - (y/b)^2\right)^{1/2}}{2\pi ab/3} \qquad (3)$$

where a and b are the half-lengths of the major and minor axes

The ellipsoid distribution has a ρ equal to zero whenever the major and minor axes are coincident with the X and Y axes. However, ρ becomes non-zero if a does not equal b and if the major axis is rotated an angle β from the X axis. As indicated in Figure 1, ρ is a function of β and the ratio a/b.

Equations (2) and (3) represent joint pdf's of the random variables X and Y, both of which have mean values of 0.0. In the sensitivity study X corresponds to ϕ and Y corresponds to θ which have mean values of m_ϕ and m_θ, respectively. In general, both the bivariate normal distribution and the ellipsoid distribution will have five independent parameters: m_ϕ, m_θ, σ_ϕ, σ_θ and ρ in the former case; m_ϕ, m_θ, a, a/b, and β in the latter case. (The truncated bivariate normal distribution will also require C, the truncation coefficient.) As discussed above, the trace of a bivariate normal distribution on its truncation plane is an ellipse. The parameters of the ellipse i.e., a, a/b and β can be expressed in terms of σ_ϕ, σ_θ and ρ. Thus, the truncated bivatiate normal distribution can be uniquely identified by m_ϕ, m_θ a, a/b and β.

There is a similarity between truncated bivariate normal and ellipsoid distributions. Indeed, it is possible to find a member from each family that has the same basal ellipse i.e., the trace of the bivariate normal distribution on its trucation plane is identical to the trace of the ellipsoid distribution on the $f_{\phi\theta}(\phi,\theta) = 0$ plane. The two distributions will, of course, have different mathematical forms and, in general, will have different probability densities at any point on the ϕ-θ plane. On the other hand, both joint pdf's have elliptical traces (albeit different elliptical traces) on planes parallel to the

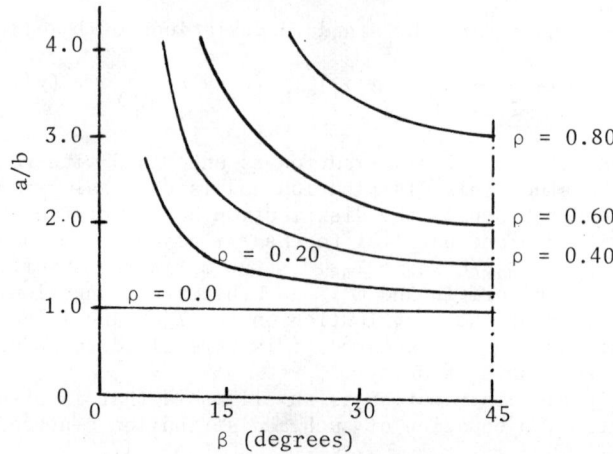

Figure 1 - Coefficient of Correlation: Ellipsoid Distribution

ϕ-θ plane. Furthermore, both joint pdf's have the same ρ. The ellipsoid distribution tends to be more disperse as the probability density decreases sharply at its boundaries.

RESULTS

Plane Strain Failure Mode

For a plane strain case wherein the strike of the discontinuity is parallel to the strike of the slope sliding occurs along a single plane. Under the simplest of conditions - continuous joints, no water pressures and no joint cohesion:

$$FS = \tan\phi/\tan\theta \qquad (4)$$

$(P_f)_j$ is that portion of $f_{\phi\theta}(\phi,\theta)$ in the region $\phi < \theta$.

Figures 2 and 3 show some typical results from the sensitivity study. Figure 2 compares $(P_f)_j$ and $(P_f)_m$ for bivariate normal

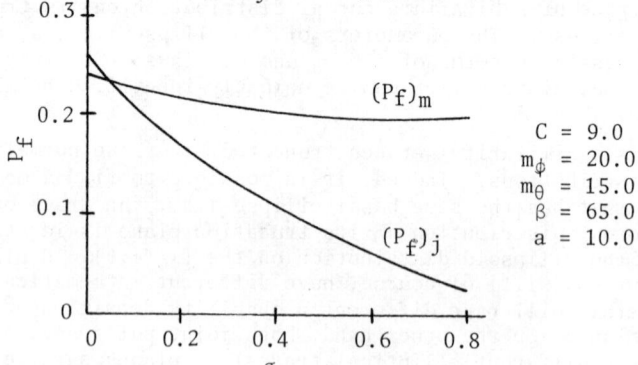

Figure 2 - $(P_f)_j$ vs. $(P_f)_m$: Bivariate Normal Distribution

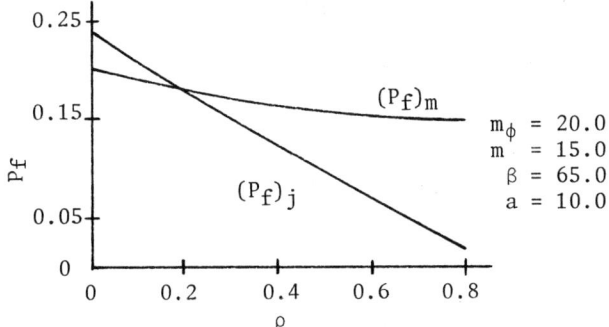

Figure 3 - $(P_f)_j$ vs. $(P_f)_m$: Ellipsoid Distribution

distributions with m_ϕ = 20.0, m_θ = 15.0, β = 65.0 and a = 10.0. Figure 3 is a similar plot for ellipsoid distributions. In both plots the independent variable is ρ. In other words, the a/b ratios of the basal ellipses were varied to achieve the requesite ρ.

Figures 2 and 3 clearly indicate that correlation can have a dramatic influence on the P_f calculations. $(P_f)_m$ is relatively unaffected by changes in ρ whereas $(P_f)_j$, the correct P_f, decreases with increasing ρ. In the bivariate normal case, $(P_f)_m$ overestimates P_f by 75% when ρ = 0.35.

Tetrahedral Wedge Failure Mode

The sensitivity study did investigate a limited number of cases that involved tetrahedral wedge failures in which sliding typically occurred on two planes. The analyses were based on the same simplifying assumptions - continuous joints, no water pressures and no joint cohesion - used in the plane strain cases. The study used Monte Carlo simulation to compare $(P_f)_j$ and $(P_f)_m$ when ϕ and θ (the dip of the joint) are jointly distributed. The data below summarizes the parameters in one of the cases.

Orientation of Slope: strike:N90W dip: 90S
Joint Plane 1 :   mean strike: N45W mean dip: 45SW
Joint Plane 2 :   mean strike: N45E mean dip: 45SE
Dispersion in Joint Poles   Fisher Distribution with a dispersion coefficient of 45.0
$f_{\phi\theta}(\phi,\theta)$:   Ellipsoid with m_ϕ = 45.0, m_θ = 45.0, β = 30.0, a/b = 1.33 and ρ = 0.26.

For this case, $(P_f)_j$ = 0.25 and $(P_f)_m$ = 0.30 .

CONCLUSIONS

Correlation between random variables can be an important consideration in reliability calculations for rock slopes. The correlations examined in the study are hypothetical; however, the results indicate that $(P_f)_m$, the P_f determined in Monte Carlo techniques, greatly overestimates $(P_f)_j$, the correct P_f, when there is even a mild positive correlation between ϕ and θ. The magnitude of the difference between $(P_f)_j$ and $(P_f)_m$ is only illustrative and is a function of the actual underlying joint pdf. Indeed, the difference would reverse if there were a negative correlation.

The study focussed on Monte Carlo analyses; however, the problem of identifying correlations is crucial to any reliability technique. This problem, along with the identification of typical pdf's, is perhaps the single most important step that remains before rock slope stability analyses can become a truly viable design tool.

ACKNOWLEDGEMENT

The authors would like to acknowledge the help of the National Science Foundation which funded this work under Grant ENG-7908096.

REFERENCES

Call, R.C. and Nicholas, D.E. (1978), "Prediction of Step Path Failure Geometry for Slope Stability Analysis," 19th U.S. Symposium on Rock Mechanics.

Dershowitz, W.S. (1979), "Probabilistic Model for Prediction of the Deformability of Jointed Rock Masses," S.M. Thesis, M.I.T., Cambridge, MA.

Einstein, H.H. et al. (1979), "Risk Analysis for Rock Slopes in Open Pit Mines," M.I.T. Research Report R80-22.

Gaziev, E.G. and Rechitski, V.I. (1979), "Method of Probability Analysis of Rock Slope Stability," 4th International Congress on Rock Mechanics, Vol. 1, pp 637-643.

Glynn, E.F. (1979), "A Probabilistic Approach to the Stability of Rock Slopes," Ph.D. Dissertation, M.I.T., Cambridge, MA.

Major, G., Ross-Brown, D. and Kim, H. (1978), "A General Probabilistic Analysis for Three-dimensional Wedge Failures," Supplement to 19th U.S. Symposium on Rock Mechanics pp 45-56.

Marek, J.M. and Savely, J.P. (1978), "Probabilistic Analysis of the Plane Shear Failure Mode," Supplement to 19th U.S. Symposium on Rock Mechanics, pp 40-44.

McMahon, B. (1975), "Probability of Failure and Expected Volume of Failure in High Rock Slopes," 2nd Australia-New Zealand Conference on Geomechanics, pp 308-313.

Veneziano, D. (1978), "Probabilistic Model of Joints in Rock," Unpublished manuscript, M.I.T., Cambridge, MA.

Chapter 11

A NEW TECHNIQUE FOR DOMAIN DELINEATION OF
ROCK MASS DISCONTINUITIES

Howard R. Hume, Terry R. West and William R. Judd

Graduate Student and Associate Professor
Department of Geosciences
Purdue University
West Lafayette, Indiana 47907

Professor
Geotechnical Engineering Department
Purdue University
West Lafayette, Indiana 47907

ABSTRACT

One of the outcomes of a sector slope design in a large open pit in the western United States has been the scrutiny of the structural domain concept. These domains essentially define the area of influence of particular structural characteristics. In the present instance these structural characteristics are joint discontinuities. One of the major drawbacks of defining a domain is that unless large quantities of discontinuity data are available this process of definition is largely qualitative in nature.

Viewed from a discontinuity analysis aspect the orientations of all the major joint sets at a given point are a concise descriptor of the structural geologic regime at that location. This is important to domain boundary definition, this in turn is extremely important for assigning which joint sets to use as input for slope stability analysis in a given discontinuity domain.

An attempt has been made to develop a quantitative method for defining these discontinuity domains based on limited amounts of orientation data thereby reducing the subjectivity normally involved in such operations. Subjectivity is not always a problem if large quantities of joint orientation data are available, since the mean and variance of joint clusters always can be compared and then grouped to define these domains. However, even this method suffers from the problem that only a single joint set orientation can be considered at a time during the grouping process. On the other hand a discontinuity domain requires that all joint sets at a point be considered simultaneously to truly define the structural discontinuity character of that point. This provides a far more complete

DOMAIN DELINEATION OF DISCONTINUITIES   105

description than would be the case if a single joint set per location were considered when formulating the domain boundaries.

With this in mind and realizing that economy of data collection is always desirable, joint set window mapping is the ideal vehicle for accurately describing a discontinuity domain. In joint set mapping the modal orientation of every joint set present at a given location is measured. This gives rise to one or more joint sets that define the structural joint discontinuity character for that location. Several joint set mapping locations obviously are needed to allow adequate resolution for defining the domain boundaries.

Since each joint set mapping location may contain a different number of joint sets, computational difficulties arise when trying to compare locations with each other.

Several comparison techniques were investigated but, some of them were too restrictive to make usable in practical situations. Pairwise comparison was achieved with relative ease on the computer and these results were used as a check of the validity of the methods which consequently were tested. Several joint set orientations, all measured in terms of dip angle and dip direction, and all from one fracture set mapping location, were represented in the form of a wave with a period of $360°$. The dip angle was treated as the dependent variable (y axis) and the dip direction as the independent variable (x axis). Modified forms of both cross correlation and trend surface analysis were attempted. In addition modified spline surface analysis was investigated in order to determine whether it could be applied usefully to describe and differentiate joint set mapping locations.

Once a measure of the similarity of these locations could be achieved, clustering of these locations into discontinuity domains is a simple step. There may be further uses for such a comparison method but as yet these have not been investigated.

INTRODUCTION

Structural domains define the zone of influence portraying particular structural characteristics. This paper discusses methods of quantifying joint discontinuity orientations but the method can be used for other applications. Joint discontinuity domains are of importance. Slope failure along these discontinuities is a possibility. In addition, some idea of the three-dimensional spatial distribution of the joints both areally and volumetrically is needed because dip and dip direction data collected at the surface may not be representative of the jointing regime at 100 feet or more into the slope. The method to be presented hopefully will fulfill these needs provided the data are collected on a slope face. Data collected at varying elevations on an inclined slope provide clues to the structural trends within the rock mass in a three-dimensional manner.

Delineation of these domains normally is difficult because:
1) they are a volumetric rather than an areal consideration, and
2) clear-cut boundaries do not exist but rather a gradual variation indicates a change has occurred.

In addition to accommodating their three-dimensional spatial character, all the associated joint sets at a given location should be considered collectively. To consider one joint set orientation and disregard the orientations of associated sets would provide an incomplete description of the joint orientations at a particular location.

Currently the methods for defining discontinuity domains involve amongst others:
1) purely visual assessment of boundaries,
2) Cusums analysis (Piteau and Russel, 1971), and
3) statistical analysis using the mean and variance of joint-set clusters plotted on an equal area net (CANMET, 1977).

The commonly used visual methods can provide reasonable estimates provided the total area of the rock mass under consideration is small and the data are not too voluminous or complex. The drawback is that visual assessments cannot treat adequately large quantities of information. They are not suitable in areas of limited exposure or masked characteristics because guesswork is required to fill the gaps where no information is available, i.e. these methods do not lend themselves to consistently accurate extrapolations.

The Cusums method is a much better procedure but requires an enormous amount of detailed orientation data collected along a scan line in a relatively limited area. Very good 3-dimensional coverage for the data collection is also necessary and this commonly is not possible with Cusums. The Cusums method can show the shift in orientation of a joint set with distance and effectively define the zone of influence of the set. It does however only consider one joint set at a time.

Statistical comparison of any appreciable quantity of clustered data can be done by interactive computer methods. The mean and variance of each joint-set cluster are used to determine the degree of similarity between joint sets. The statistical parameters are controlled by the designer doing the interactive search on the computer. However, this technique is also limited to just one joint set at a time and requires large volumes of orientation data. An added drawback is that it is more complicated than Cusums analysis.

The pseudowave method to be presented overcomes the aforementioned objections in that data collection can be made at point locations (joint-set locations).

DOMAIN DELINEATION OF DISCONTINUITIES 107

A small amount of information is sufficient to delineate discontinuity domains provided you select reasonably uniformly distributed locations separated by realistic distances. The reduction in data collection effort can be considerable. In addition the 3-dimensional spatial distribution is taken into account and the local character of the discontinuity domain is described uniquely because all joint sets at a given joint-set location are considered together. That is the contributions of the modal orientation values of each of the joint sets that appear at a given joint set location are considered collectively.

Carefully note that this method still is under development and extensive field verification cannot be provided at present. Additionally this technique may have further shape-related uses which as yet have not been identified.

INITIAL STUDIES

In evolving the pseudowave method, several other methods were evaluated. Initially a solution using a vector mean of the modal joint sets was considered but it was descarded because it proved to be a non-unique representation of a point location. Next a pairwise comparison with each of the other locations on a one-to-one basis proved satisfactory only as a check for further work. It suffered from the drawback that similarity was local because it was pair-specific and it could not be extrapolated to larger regions, on a more global* or regional basis i.e. the similarity measure it gives is true only for a given pair. It had the advantage however that it was adaptable to machine operations, thus it could handle data taken at a large number of point locations.

It was concluded finally that the most convenient and useful way of presenting dip and dip-direction data was in a waveform hereinafter referred to as a "pseudowave". If one considers that an equal area net has a maximum azimuth and zenith of $360°$ and $90°$ respectively, all one has to do is spread the net out in a linear fashion, analogous to rolling out a cylinder. The pseudowave therefore has a wavelength of $360°$ and a maximum possible amplitude of $90°$. An example is given in Figure 1.

This pseudowave is obviously only defined by a few points (the dip and dip direction values). It was decided to insert dummy values at $0°,0°$ and $0°,360°$ to give a closed shape. This would be common to all the pseudowaves and so its effect should be similar in each case.

---

* "Global" in certain parlance refers to a large area or region within an area of interest.

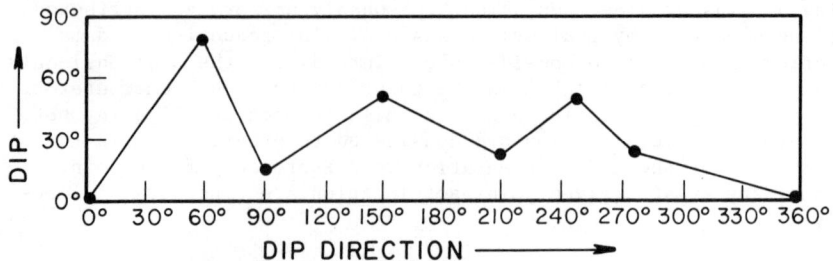

Figure 1. A pseudowave consisting of six joint orientation values (extra dummy values have not been inserted in this case).

A further modification in this regard was to insert extra dummy points at fixed equispaced distances along the pseudowave. Again, these would be common to all waves except where the dummy values conflicted with the actual data, then the data would take precedence. These dummy values take the form of a line of constant gradient and it is wise to verify that the gradient does not follow any general trend in the overall data. This easily can be checked by plotting all the joint-set orientation values from all the locations as one wave and then viewing the overall behavior. Again, since the dummy values were added to all the waves this is considered a reasonable compromise.

Once the pseudowave concept had been developed several analytical methods were attempted. Each fracture set location was represented by a pseudowave and it was compared with a standard reference wave by means of cross-correlation analysis. This method has been used effectively in geophysics for measuring the similarity between various waves. Success of this method was marginal in that it did not prove to be consistent. This was due largely to the fact that cross correlation uses the variance between corresponding (in the azimuth sense) pairs of points to obtain a measure of similarity. Because large numbers of dummy points would bias the waveform more than is justifiable, cross correlation is unable to provide a sufficiently sensitive measure of the slope change (shape change) of the wave at various points along the wave.

Two-dimensional trend surfaces were considered; however, exact fits of surfaces to the pseudowaves require high order polynomials. These polynomial surfaces have tremendous edge effects and rapid fluctuations between the data points and so they were discarded.

Finally, to overcome the disadvantages of high-order polynomial regression it was decided to fit low order spline functions to the data (both real and dummy) for each wave. New waveforms were produced using these linear piecewise polynomials; this involves joining each point to its neighboring point by a straight line (De Boor, 1979). Other possible spline functions were considered but the more sophisticated splines showed little advantage over the simple spline. Using this representation of the pseudowave, a shape description technique called Freeman chain-coding (Freeman, 1974) was adopted. Previously this method was used by electrical engineers involved in pattern recognition and artificial intelligence research. Recently, Bribresca and Guzman (1978) modified this method for new applications and the technique to be described was stimulated largely as a result of this work.

DESCRIPTION OF NEW METHOD

Stage 1

The pairwise comparison technique previously mentioned initially is used as a check on the behavior of the chain-coded data. This comparison basically gives an indication of the degree of similitude between any two data locations. As previously indicated this similarity is not global, i.e. the similarity measure it gives is true only for a given pair.

At Purdue University this phase of the work was achieved using a Fortran 77 program called FRACSRT. Essentially only the real data values are required and at this stage the pseudowave is not introduced. The dip and dip direction values for one fracture-set location are compared with those from another location. The number of matches between the two groups (allowing a certain user-specified tolerance for defining a "match") gives the similarity between the two groups. The number of matches is output in the form of a large symmetrical matrix that enables easy visual checking. This visual checking could be done by the computer but at present this has not been implemented.

Stage 2

This stage involves expressing the fracture set data as a pseudowave. This is the subjective part of the processing because every technique of expressing dip and dip direction data in the form of a continuous wave with wavelength and amplitude of 360° and 90° respectively has its drawbacks.

Resolution has its own problems. For example, optimizing the pseudowave so it is unique and representative of its fracture-set location requires insertion of dummy data. The effects of dummy data are minimized by introducing the same dummy data into all the fracture-set data except, of course, where this dummy information would conflict directly with the actual values. The low-order spline is then fitted by using straight lines to join all the points for a given joint set location.

The selection of the style and type of waveform is dependent on the character of the actual data and will vary depending on what is required. The generation of the dummy data (if required) is easily achieved on the computer.

Stage 3

Once the spline function has been fitted the pseudowave is partitioned by a grid. The grid fineness can be changed automatically by introducing this capability into the calculations.

Initially a coarse mesh is selected and the intersections of the wave with the vertical grid and horizontal grid lines are obtained. This is done using grid-intersection quantization (Freeman, 1974). An x, y coordinate system can be identified with a grid in such a way that every node can be described by coordinates (iD, jD) where i and j are integers and D represents the inter grid-line distance. For computer processing a suitable description of the wave can be achieved by sequentially giving the x and y coordinates of the nodes which lie closest to the curve. For practicality, D can be considered a scaling factor. There is more than one method of achieving this grid quantization but the method chosen is the grid-intersect quantization. This is applied by the following rules (Freeman, 1974):

whenever t and i are such that $x(t)-iD = 0$ then
j is such that $(j-\frac{1}{2})D < y(t) < (j+\frac{1}{2})D$

and whenever t and j are such that $y(t)-jD = 0$ then
i is such that $(i-\frac{1}{2})D < x(t) < (i+\frac{1}{2})D$

Referring to Figure 2, $x(t)-iD = 0$ and $y(t)-jD = 0$ represent the x and y values respectively where the wave crosses a grid line.

To obtain the chain code a set of reference directions is adopted. Usually a 4- or 8-direction system is adequate. Since the coordinates have to be in sequential x, y pairs, it is a relatively simple matter to assign these directional links in a clockwise manner provided one maintains the sequence. These links are all unit length with exception of the diagonal links that sometimes occur in the 8-direction system. These two-direction schemes together with an example of a chain-coded wave (8-direction scheme) are presented in Figure 3. The chain-code operation necessarily needs to be accomplished on a computer, so a Fortran program CHAIN was developed.

DOMAIN DELINEATION OF DISCONTINUITIES

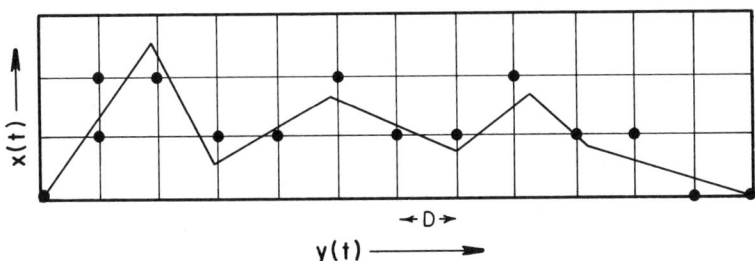

Figure 2. A grid superimposed on a pseudowave showing the position of the nodes once grid quantization has been achieved.

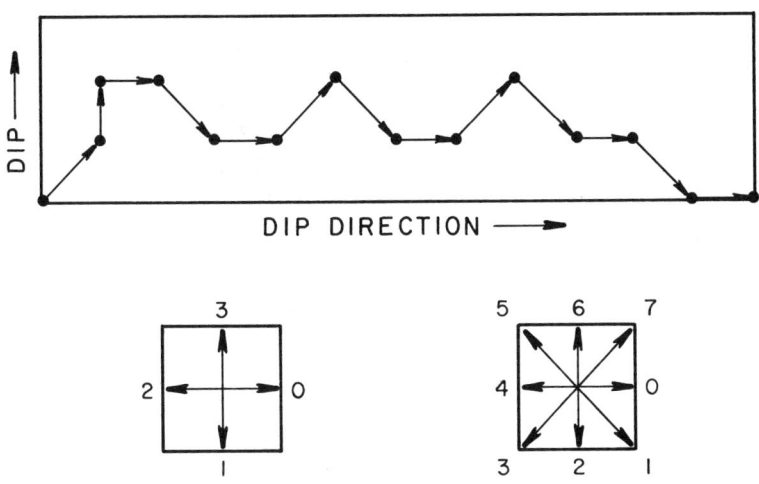

CHAIN CODE = 760107107 1010

Figure 3. This figure shows a pseudowave expressed in terms of directional links using an 8-direction system. A 4-direction and 8-direction schematic are also shown along with the 8-direction chain code of the pseudowave.

Stage 4

The pseudowave is now in the form of a chain code (which is really a sequence of unit-length directional links). Thus one really has only a unique shape descriptor. With minor changes this chain-code sequence could describe two waves that are moderately dissimilar in terms of the magnitude of the dip angles (yet these two waves might have fairly similar shapes). A technique that is shape-descriptive is not satisfactory on its own so some method of indicating the magnitudes of the shapes is necessary. To illustrate this, consider two circles with markedly different radii. Both circles have the same shape but their magnitudes are different so it would be incorrect to call them identical. Obviously the degree of similarity between any two waveforms is a function of both shape and magnitude.

The method adopted to handle this problem was to trace sequentially the unit length links and find where they were located in the vertical y direction. If each vertical segment of the pseudowave is treated as a column of cells the resulting histogram-like representation describes the wave just as the chain code did (Figure 4). The major difference however is that this cell code describes the amplitude and the shape of the wave in contrast to just providing information on the shape.

Once this is achieved the cell code of one wave can be compared column-to-column with another wave. This is continued until all the waves have been compared with one another. The column-by-column method is a direct comparison technique and provides a series of matches that have varying degrees of fit (also on a column-by-column basis). These varying degrees of fit for each column then have to be recombined to represent the overall degree of fit of the entire wave with respect to any reference wave. To achieve this, a form of distance coefficient was adopted as a similarity measure. The form of the distance coefficient is determined by trial and error. The cell coding and distance coefficient were obtained using another FORTRAN program called CHANSRT.

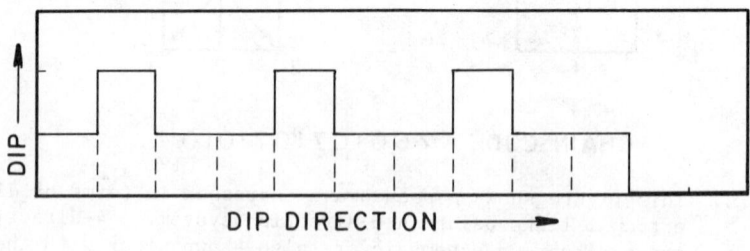

Figure 4. The cell coding obtained from the chain code shown in figure 3.

Stage 5

Once the distance coefficient for each fracture-set location has been calculated, the next step is to cluster these coefficients.

Since the similarities are now available on a global basis (i.e. they are now relative similarities) it is fairly easy to cluster these values. If the number of values is small, this can be done visually; for large quantities of values use a simple statistical clustering technique.

A more quantitative approach is to use a semivariogram which is the basis of the geostatistical method called Kriging (Blais and Carlier, 1968). Simply stated, the semivariogram is a plot of the semivariance of a variable (e.g. ½ x variance of assay values) with distance. When a factor no longer contributes to the semivariance function its zone of influence has been exceeded. This effectively defines a boundary for the zone of influence of that factor. The boundaries can be defined in any given geographical direction merely by calculating the requisite semivariogram in that direction and then observing the physical distance (indicated on the semivariogram as d) at which the semivariance function becomes constant (Figure 5).

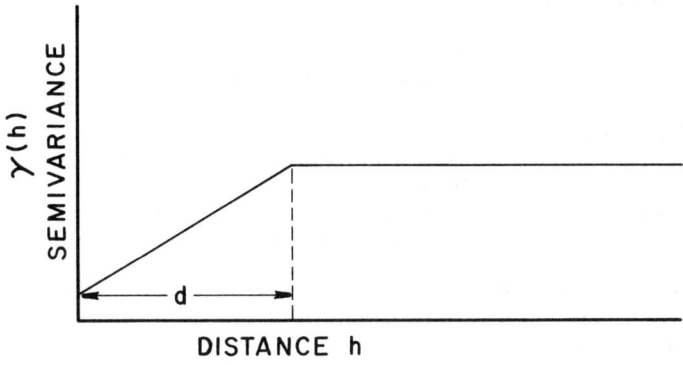

TRANSITIVE SEMIVARIOGRAM

Figure 5. An example of a semivariogram.

CONCLUSION

At the time of writing, this semivariogram concept has not been tested but it was to this end that the shape description method was developed. The semivariogram seems likely to lend itself well to defining the zone of influence of a geologic variable. In combination with a Kriging system, its capability can be extended even further. This includes an estimate of the error made by using Kriging to estimate the value of a variable at a given point.

It is not clear at present how successful the shape description method plus variogram will be with regard to domain delineation. It is certainly probable however that other geologic applications, in which shape description can be utilized, would benefit from such a method. Where shape alone is a deciding factor (magnitude not considered significant) the shape method can be used with very minor modifications. Programs CHAIN and CHANSRT have the ability to measure shape alone if necessary, this is why the 8-direction chain code has been retained. The 8-direction chain code gives a more faithful rendition of the shape than the 4-direction system. This extra contribution is not that significant however when using the cell coding scheme (i.e. when shape and magnitude are being considered simultaneously). While the results of this work are not conclusive at this time, they do present a direction for further thought and investigation.

ACKNOWLEDGEMENTS

The first author would like to express his thanks to his colleague Skip Watts for his encouragement and thoughtful comments during the development stages of this work. Very special mention and thanks are also due to Devan Citron without whose able assistance and programming it is doubtful whether this work would ever have been accomplished.

Financial support for this project was provided via the Indiana Mining and Mineral Resources Research Institute by U.S. Bureau of Mines grant number G1114019.

REFERENCES

Blais, R. and Carlier, A., 1968, "Application of Geostatistics in Ore Evaluation, Ore Reserve Estimation and Grade Control", in Special vol. No. 9, CIMM, Montreal, 48-61.

Bribiesca, E. and Guzman, A., 1978, "Shape Description and Shape Similarity Measurement for Two Dimensional Regions", Proc. 4th Int. Joint Conf. Pattern Recognition, Kyoto, Japan, 608-612.

CANMET, 1977, Pit Slope Manual, Supplement 2-2: Domain Analysis Programs, Canada Centre for Mineral and Energy Technology.

De Boor, C., 1978, A Practical Guide to Splines, Springer Press.

Freeman, H., 1974, "Computer Processing of Line Drawing Images", Computing Surveys vol. 6, No. 1, 58-97.

Piteau, D.R. and Russel, L., 1971, "Cumulative Sums Technique: A New Approach to Analyzing Joints in Rock", 13th U.S. Symposium on Rock Mechanics.

Chapter 12

A REJECTION CRITERION FOR DEFINITION OF CLUSTERS IN ORIENTATION DATA

by M. A. Mahtab and T. M. Yegulalp

Henry Krumb School of Mines, Columbia University, New York

ABSTRACT

This paper presents the development and application of an approach for clustering fracture orientation data. Data are projected on the surface of the unit upper hemisphere and the clustering approach is applied in three steps: (1) randomness test and identification of dense clusters, (2) estimation of cluster statistics, using Bingham distribution, and (3) rejection of extremal data based on Chauvenet's criterion.

INTRODUCTION

Geomechanical characterization of rock masses often requires analysis of data on orientations of natural fracures. On the scale of most engineering structures in rock, the fracture orientations may be clustered around one or more statistically preferred directions. A primary consideration is to identify these clusters such that their means and other parameters can be estimated for input to modeling and design.

An orientation measurement can be conveniently represented by a unit vector (defined by azimuth of dip and dip). It is, therefore, natural that cluster analysis techniques and probability distribution theory be applied to collections of data points on the unit hemisphere. The method and objectivity used in partitioning data affect the goodness-of-fit of a known spherical density function to the resulting clusters. The general practice (for example, Phillips, 1954) of defining clusters in orientation data by visual estimates of modes in the density contours plotted on equatorial projections is, clearly, too subjective. Computer-based methods presently used for partitioning orientation data (for instance, Shanley and Mahtab, 1976) are based on the assumption that the resulting partition must include all of the data points. These methods disregard the existence of errors (or 'noise') introduced by the measuring and sampling system and the assumed model for the geologic setting, thus creating problems of definition in clustering large sets of data and in fitting probability density functions to the resulting clusters.

What is needed is a rejection criterion that can reduce or eliminate the noise in the data while allowing an objective partitioning of the data. This paper presents the development of a rejection scheme, based on Chauvenet's criterion (see Gumbel, 1958), for objective partitioning of orientation data. The criterion is applied to the analysis of fracture orientation data from San Manuel mine in Arizona and Stripa mine in Sweden.

APPROACH FOR DEFINITION OF CLUSTERS

In what follows, we are concerned with partitioning a set of orientations A = [$(\theta_1, \phi_1), (\theta_2, \phi_2), \ldots, (\theta_n, \phi_n)$] into clusters. An observation, i, is defined by the azimuth (clockwise angle from North) of dip, $\theta_i (0 \leq \theta_i \leq 2\pi)$, and dip, $\phi_i (0 \leq \phi_i \leq \pi/2)$, of the normal to the fracture plane. We choose a left-handed, rectangular, Cartesian frame xyz such that x = (1, 0, 0) defines North, y = (0, 1, 0) defines East, and z = (0, 0, 1) defines the pole of the upper hemisphere of unit radius. The spherical coordinates (1, θ_i, ϕ_i) of observation i then define a unique point on the surface of the unit hemisphere.

The clustering algorithm proceeds in three steps: (1) randomness test and identification of dense points and "dense clusters," (2) definition of dense cluster statistics using Bingham's (1964) distribution, and (3) acceptance of (probable) unassigned points and rejection of extremal data.

Randomness Test

The randomness test used here is derived from Poisson distribution (see Molina, 1947). The test ensures that only significant concentrations (data modes) are selected as nuclei of clusters. We choose an angle $\alpha (0 \leq \alpha \leq \pi/2)$ defining a cone, with its apex at x=y=z=0, whose intersection with the surface of the unit hemisphere, $S(S=[(x, y, z) x^2 + y^2 + z^2 = 1, z \geq 0]$, outlines a fraction c of S(c=1-cos α). If n is the sample size (the number of points in A), then k is defined to be the smallest integer such that the expected number of fractional areas (of size c) containing k or more points is 1, that is,

$$\frac{1}{c}[1 - \sum_{j=0}^{k} (cn)^j e^{-cn}/j!] = 1 \tag{1}$$

All points in A that meet test (2) are called dense points (after Wishart, 1968). The dense points are now grouped to form dense clusters. The rule for determining whether two dense points, v and w say, belong to the same dense cluster is as follows: if there are dense points $x_0(=v)$, x_1, x_2, . . ., $x_m(=w)$, such that the angle between all consecutive pairs of points is less than or equal to α, that is, angle $(x_i, x_{i+1}) \leqslant \alpha$, $i = 1, 2, \ldots, m-1$, then v and w will belong to the same dense cluster. In this step all data points which were counted within α of dense points v and w are also included in the dense cluster containing v and w. The remainder of sample A, that is, the set of unassigned points, if any, is subjected to further tests for acceptance in the possible progressive growth of the already defined clusters, subject to the condition that no point may be removed from the existing clusters.

Cluster Statistics Using Bingham Distribution

Upto this point in our clustering algorithm, no 'a priori' judgment was made on the distribution that should describe the data. In order to construct a criterion for accepting or rejecting the unassigned points, we now seek to determine the statistics of the dense clusters using Bingham (1964) distribution on the sphere. Bingham distribution allows elliptical symmetry about the mean and has the advantage of fitting many sets of orientation data analyzed by the authors. Bingham distribution has the form

$$f(X,\mu) = \exp[\xi_1(\mu_1'x)^2 + \xi_2(\mu_2'x)^2 + \xi_3(\mu_3'x)^2]ds/4\pi_1F_1(1/2;3/2;Z) \quad (2)$$

where ξ_1, ξ_2, ξ_3 are dispersion parameters, and $_1F_1(1/2; 3/2; Z)$ is a hypergeometric function of matrix argument. Bingham imposes the constraint $\xi_3 = 0$ to render the maximum likelihood estimate of Z unique.

If the data are concentrated about a preferred orientation, both ξ_1 and ξ_2 will be negative. If $\xi_1 \neq \xi_2$, then $\xi_1 < \xi_2$ and the contours of the distribution will be elliptical in shape. In this example, the axis μ_1 would be parallel to the "minor axis" of the "ellipse," and the axis μ_2 would be parallel to the "major axis" of the "ellipse." The axis μ_3, that is, the best axis or the mean of the distribution, would be perpendicular to both μ_1 and μ_2. Estimation of parameters (ξ_i, μ_i, $i = 1, 2, 3$), discussed by Bingham (1964) and by Shanley and Mahtab (1976), is carried out in this step of the algorithm for each one of the dense clusters of sample A.

Criterion for Acceptance (or Rejection) of Unassigned Points

The third major step in the approach for clustering the points in sample A is to determine which of the unassigned points, if any, belong to which of the dense clusters. The acceptance test is carried out through an iterative process, with each iteration involving the following procedure:

Suppose j is a cluster containing n_j data points. We now compute a limiting angle $\bar{\phi}_j$. With reference to the mean μ_{3j} of cluster j such that the expected number of points beyond $\bar{\phi}_j$ is 2, that is,

CLUSTERS IN ORIENTATION DATA

$$n_j [1 - \int_0^{2\pi} \int_0^{\bar{\phi}_j} f_j(\theta, \phi) d\phi d\theta] = 2 \qquad (3)$$

where $f_j(\theta, \phi)$ is Bingham distribution with its parameters estimated by using data in cluster j.

Next we examine the closeness of an unassigned point $i(\theta_i, \phi_i)$ to each cluster. We choose the angle β_{ij} between i and μ_{3j} as the measure for closeness. The point i can be accepted in cluster j if $\beta_{ij} < \bar{\phi}_j$. If this condition is satisfied for one cluster only, i is assigned to that cluster. However, if $\beta_{ij} < \bar{\phi}_j$ is true for more than one cluster, the point i is assigned to the cluster j for which the probability P_{ij} is maximum; where

$$P_{ij} = 1 - \int_{\theta=0}^{2\pi} \int_{\phi=0}^{\beta_{ij}} f_j(\theta, \phi) d\phi d\theta \qquad (4)$$

At the end of each iteration, new values of cluster parameters, μ_{3j}, $\bar{\phi}_j$, are estimated using new n_j and Bingham distribution.

The iterative process continues until no additional points can be accepted into any of the existing clusters. The remaining points of A, if any, cannot be assigned and are rejected as extremal data.

APPLICATION

Application of the approach of definition of clusters is made to two sets of fracture orientation data: from San Manuel mine, Arizona (Shanley and Mahtab, 1976) and Stripa mine (Table 2), Sweden (Thorpe, 1979). [San Manuel data are not tabulated here, but are available from the authors or in the referenced material.] A summary of the results of analyses of these data is provided in Table 1.

TABLE 1. Summary of Results of Example Applications

	San Manuel		Stripa		
Total no. of data points	286		846		
Counting radius, α, deg.	15		10		
Poisson cutoff, k	16		21		
No. of iterations required	6		4		
No. of clusters defined	2		3		
No. of rejected points	104		0		
CLUSTER PARAMETERS					
Cluster number	1	2	1	2	3
No. of data points	80	102	231	328	287
Mean azimuth of dip, deg.	69	347	55	205	309
Mean dip, deg.	86	78	54	38	14
Theo. Chi-square at 95%	18.3	23.7	53.4	76.8	67.5
Actual Chi-square value	10.0	16.5	79.7	137.6	163.0

TABLE 2. Stripa Data: Azimuth of the Dip and Dip of Fractures

Polar equal-area projections of San Manuel data are presented as follows: Fig. 1A shows the entire (286) points with cluster boundaries and rejected points and Figures 1B and 1C show the data in clusters 1 and 2. The two clusters are significantly dense, as indicated by the point plots and the Chi-square values (Table 1). Therefore, a large number (104) of extremal points are rejected.

CLUSTERS IN ORIENTATION DATA 121

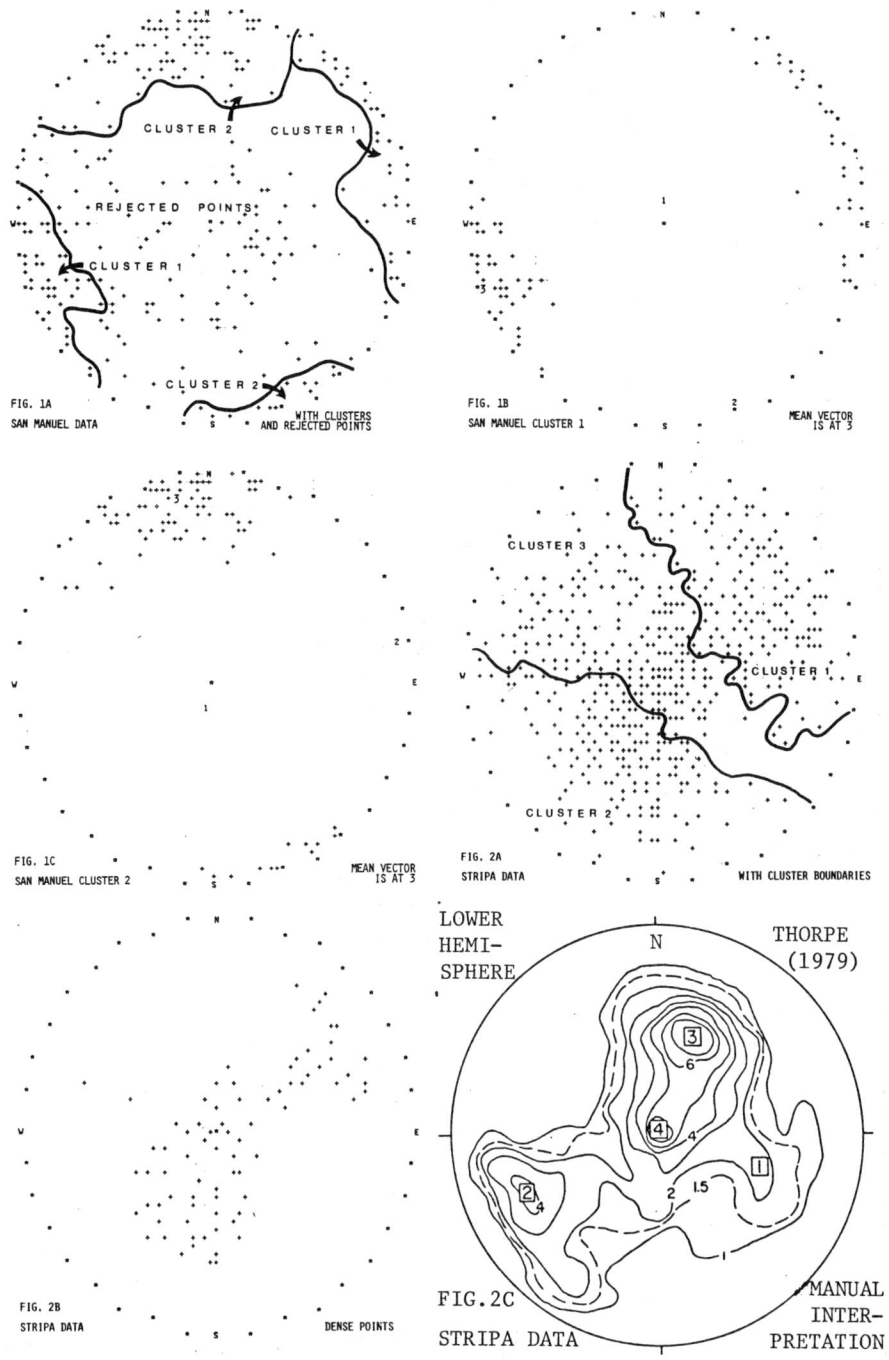

FIG. 1A SAN MANUEL DATA — WITH CLUSTERS AND REJECTED POINTS

FIG. 1B SAN MANUEL CLUSTER 1 — MEAN VECTOR IS AT 3

FIG. 1C SAN MANUEL CLUSTER 2 — MEAN VECTOR IS AT 3

FIG. 2A STRIPA DATA — WITH CLUSTER BOUNDARIES

FIG. 2B STRIPA DATA — DENSE POINTS

FIG. 2C STRIPA DATA — THORPE (1979) MANUAL INTERPRETATION

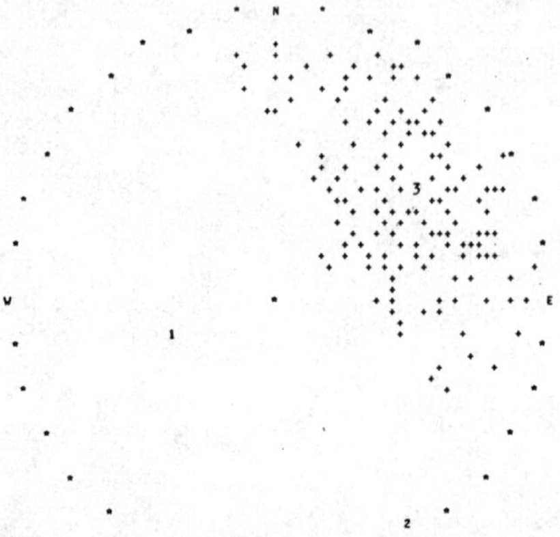

FIG. 3A
STRIPA CLUSTER 1   S   MEAN VECTOR IS AT 3

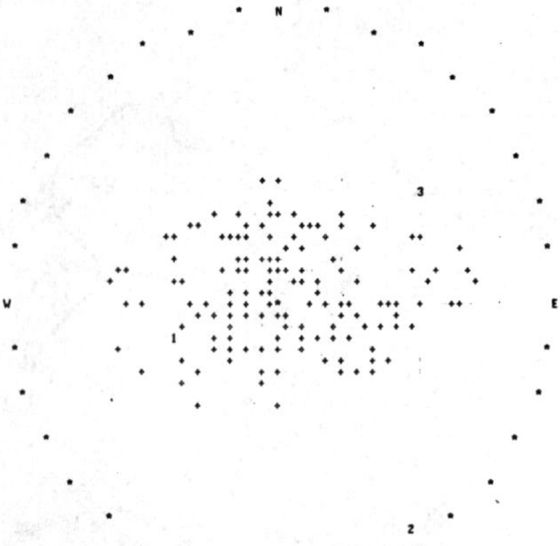

FIG. 3B
STRIPA CLUSTER 1   S   MEAN VECTOR IS AT THE POLE

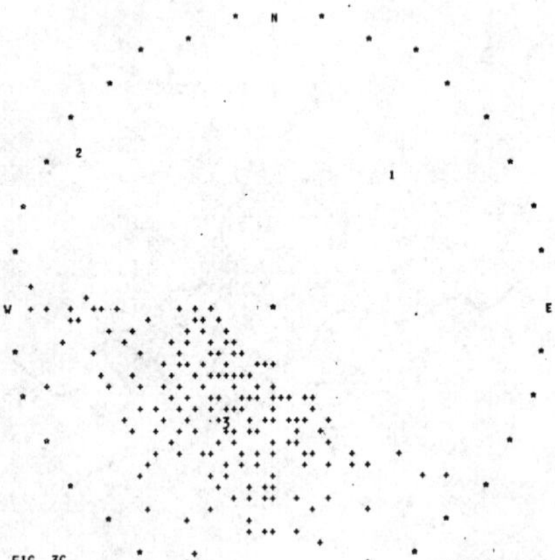

FIG. 3C
STRIPA CLUSTER 2   S   MEAN VECTOR IS AT 3

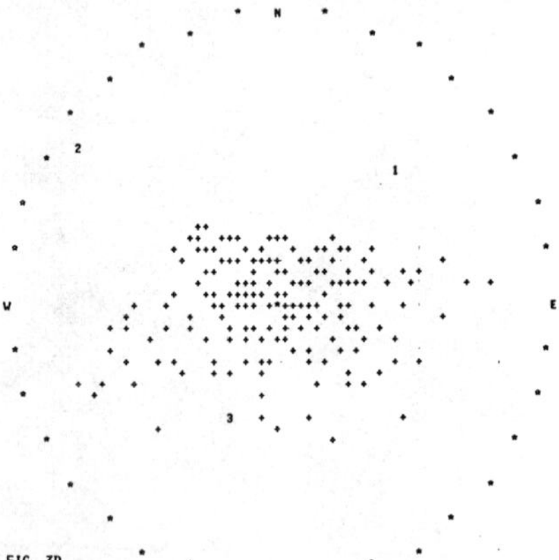

FIG. 3D
STRIPA CLUSTER 2   S   MEAN VECTOR IS AT THE POLE

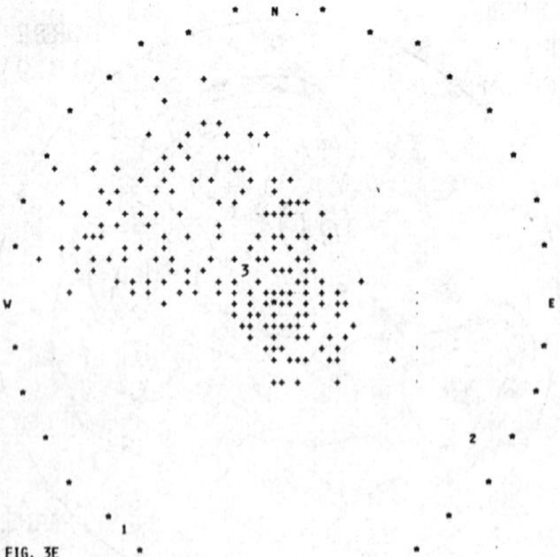

FIG. 3E
STRIPA CLUSTER 3   S   MEAN VECTOR IS AT 3

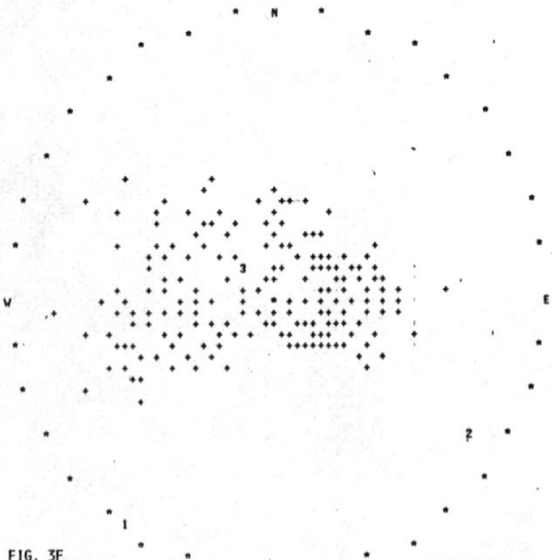

FIG. 3F
STRIPA CLUSTER 3   S   MEAN VECTOR IS AT THE POLE

Polar equal-area projections of Stripa data are presented as follows: Fig. 2A shows the entire (846) points with cluster boundaries, Fig. 2B shows the (80) dense points in the data, Fig. 2C is reproduced from Thorpe (1979) for comparison of his visual estimates (note projection of lower hemisphere) with our results, Figs. 3A to 3F present pairs of plots for the three clusters (for example, in Fig. 3A the pole of the upper hemisphere (0, 0, 1) is on the Z-axis of the Cartesian frame whereas in Fig. 3B the data have been rotated such that the pole is on μ_3 of Bingham distribution). No data point is rejected since the clusters are not significantly dense as seen from the plots and the Chi-square values (Table 1).

CONCLUSION

The approach presented here allows an objective selection of clusters and rejection of extremal points in orientation data. The selection of the counting radius, α, remains arbitrary and subject to judgment based on experience. It is suggested that the selection of α should be based on the precision of measurement, rationale for stratification or lumping of data sets, and the dispersion parameters, ξ_1, and ξ_2, of Bingham distribution for the clusters.

ACKNOWLEDGMENT

Partial support for the work reported here was provided by the U.S. Bureau of Mines. Stripa data were supplied by R. Thorpe of Lawrence Livermore Laboratory.

REFERENCES

Bingham, C., 1964, "Distribution on a Sphere and on the Projective Plane," Ph.D. Thesis, Yale University, 93pp.

Gumbel, E. J., 1958, Statistics of Extremes, Columbia University Press, New York, p. 83.

Molina, E. C., 1947, The Poisson Exponential Binomial Limit, D. Van Nostrand Co., New York, 92pp.

Phillips, F. C., 1954, The Use of Stereographic Projection in Structural Goelogy, Edward Arnold Ltd., London, 86pp.

Shanley, R. J. and Mahtab, M. A., 1976, "Delineation and Analysis of Clusters in Orientation Data," Mathematical Geology, vol. 8, no. 1, pp. 9-23.

Thorpe, R., 1979, "Characterization of Discontinuities in the Stripa Granite Time-Scale Heater Experiment," LBL-7083, July, University of California, Berkeley, 107pp.

Wishart, D., 1968, "Mode Analysis: A Generalization of Nearest Neighbour which Reduces Chaining Effects," Numerical Taxonomy, Academic Press, London and New York, pp. 282-309.

Chapter 13

FOURIER ANALYSIS FOR ESTIMATING PROBABILITY
OF SLIDING FOR THE PLANE SHEAR FAILURE MODE

Stanley M. Miller

Department of Civil Engineering
University of Wyoming, Laramie, Wyoming

ABSTRACT

The probabilistic nature of appropriate geologic variables can be included in analyzing the stability of potential slope failure modes. Random variables in the two-dimensional plane shear analysis of a specified structural discontinuity include the shear strength and waviness angle of the discontinuity and, in some cases, the rock mass density. Estimated probability density functions that describe these random variables are combined by Fourier analysis to produce an estimate of the safety factor probability density function, which appears to approximate a gamma distribution. The probability that sliding will occur along the specified structure is equal to the area under this density function where the safety factor is less than one.

INTRODUCTION

Slope failures in discontinuous rock masses are controlled by geologic structure, because displacements occur along zones or surfaces of weakness. Natural variabilities and measurement uncertainties associated with field mapping and laboratory testing prescribe that properties of geologic structure be treated in a statistical manner. They can be quantitatively described and then included in stability analyses to evaluate the risk of slope failures.

Plane shear failure is characterized by a potential failure mass capable of sliding along a semi-planar discontinuity that dips flatter than the slope angle. A recently developed probabilistic analysis of the plane shear failure mode is based on Monte Carlo techniques for predicting the probability of sliding (Marek and Savely, 1978). This probability is defined as the area under a simulated safety factor

FOURIER ANALYSIS FOR ESTIMATING SLIDING

distribution where the safety factor (a random variable) takes on values that are less than one. For a stability analysis that includes the waviness angle of the sliding surface (assumed to be exponentially distributed) and uses a power failure model for shear strength (assumed to be normally distributed), the resulting safety factor is approximately gamma distributed.

A Monte Carlo simulation provides only one possible realization of the true probability distribution of the safety factor. Also, a large number of sampling iterations (over 1000) is usually required to provide a reasonable estimate of the probability of sliding, making the associated computational time and costs objectionable and sometimes prohibitive. Therefore, fewer iterations are used, resulting in a poorer estimate of the probability of sliding (see Appendix).

Fourier analysis provides an alternative to Monte Carlo simulation in predicting the probability distribution of the safety factor. The sum of independent probability densities in space domain is analogous to the product of the Fourier transforms of the densities in frequency domain. An example of applying this principle to an engineering problem is presented by Borgman (1977). An efficient method for estimating the true probability density of the safety factor can be based on discrete Fourier procedures, which take advantage of the computational speed of the fast Fourier transform algorithm.

STABILITY ANALYSIS

Plane shear failure in fractured rock slopes is kinematically viable if the average dip of the sliding plane is less than that of the slope face and if the average strike is parallel or nearly parallel to that of the slope face. In addition, the assumption is made in a two-dimensional stability analysis that the potential sliding mass is laterally unconstrained ("side-release" assumption).

For the typical plane shear geometry shown in Figure 1, a two-dimensional stability analysis can be developed that includes the effect of the waviness of the sliding surface. Waviness is defined as the angle between the average dip and the flattest dip observed in a mappable geologic structure (Call, et.al., 1976).

The safety factor, s, is the ratio of resisting force to driving force. It can be algebraically expressed in the following form:

$$s = \frac{\tau L + W\cos(\alpha)\tan(r)}{W\sin(\alpha)} = \frac{2 L \sin(\beta)}{h^2 \sin(\beta-\alpha)} \left(\frac{\tau}{\gamma}\right) + \cot(\alpha)\tan(r) \qquad (1)$$

where: L = length of sliding surface,
h = height of failure mass,
W = weight of failure mass,

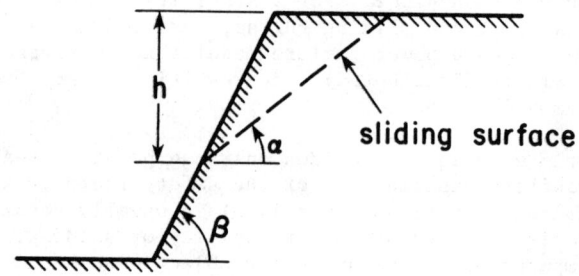

Figure 1. Typical Plane Shear Geometry

    β = average dip of slope face,
    α = average dip of sliding surface,
    γ = rock density,
    τ = shear strength (stress) of sliding surface,
    r = waviness of sliding surface.

Equation (1) is valid for dry conditions. Water in the potential failure mass causes pore pressures that reduce the effective normal stress acting on the sliding surface. This water pressure effect can be included in the stability analysis if warranted by field conditions.

For a specified slope geometry and sliding surface, the parameters L, h, β, and α are fixed. However, the shear strength and waviness are random variables whose probability distributions must be estimated by laboratory testing and structure mapping, respectively. The rock density can also be treated as a random variable (usually assumed to be normally distributed), although in most cases it has such a small coefficient of variation that it is assumed to be constant. For example, typical mean density values for crystalline rock range from 2.40 to 2.96 g/cm^3 (150 to 185 pcf), while typical standard deviations range from 0.032 to 0.192 g/cm^3 (2 to 12 pcf). These values result in coefficients of variation that are less than 0.08, indicating that rock density can typically be treated as a constant.

CHARACTERIZATION OF RANDOM VARIABLES

The statistical distribution of shear strength is a function of the applied normal stress. Laboratory direct shear tests of natural fracture surfaces or fault zone material provide plots of shear strength versus normal stress. After several specimens of the same rock type have been tested, a weighted least-squares regression procedure can be used to provide estimates of the mean and variance of the shear strength for the given rock type at specified values of normal stress (Miller, 1982). The general regression model is expressed as:

$$\tau = a\sigma^b + c \tag{2}$$

where: τ = predicted shear strength,
         σ = applied normal stress,
         a,b,c = parameters of regression model.

The distribution of τ can be considered a gamma probability density. This assumption allows τ to always be positive and also causes τ to be approximately normally distributed when its coefficient of variation is less than 0.10 (Andrew, 1981). Thus, the p.d.f. of shear strength at a given normal stress is given by:

$$f_T(\tau) = \begin{cases} \dfrac{e^{-\tau/\beta}\, \tau^{\alpha-1}}{\beta^\alpha\, \Gamma(\alpha)} & \text{, for } \tau > 0 \\ 0 & \text{, for } \tau \leq 0 \end{cases} \tag{3}$$

where: $\alpha = \mu_\tau^2 / \sigma_\tau^2$,

$\beta = \sigma_\tau^2 / \mu_\tau$,

$\Gamma(\alpha)$ = value of complete gamma function with argument α,
μ_τ = mean shear strength at the given normal stress,
σ_τ^2 = variance of shear strength at the given normal stress.

Numerous fracture mapping data have indicated that waviness tends to be exponentially distributed (Call, et.al., 1976). Therefore, the p.d.f. of waviness is given by:

$$f_R(r) = \begin{cases} \dfrac{e^{-r/\mu_r}}{\mu_r} & \text{, for } r \geq 0 \\ 0 & \text{, for } r < 0 \end{cases} \tag{4}$$

where: r = waviness of sliding surface,
        μ_r = mean waviness.

To estimate the safety factor p.d.f. by Fourier procedures, equation (1) is expressed in the following form:

$$S = AU + BV, \tag{5}$$

where A and B are constants, and U and V are random variables. For the general case in which the rock density is considered a random variable, A and B are defined as:

$$A = \frac{2 L \sin(\beta)}{h^2 \sin(\beta-\alpha)} \tag{6}$$

$$B = \cot(\alpha) \tag{7}$$

The new random variable U is the ratio of shear strength to rock density, and V is the tangent of the waviness angle. The normal stress must also be considered a variable, because it depends on the rock density according to:

$$\sigma = \frac{W\cos(\alpha)}{L} = D\gamma \tag{8}$$

where: $D = \dfrac{h \cos(\alpha) \sin(\beta-\alpha)}{2 \sin(\beta)}$.

By assuming that γ is normally distributed, then the normal stress σ is normally distributed with mean and variance given by:

$$\text{Mean}(\sigma) = D \cdot \text{Mean}(\gamma) \tag{9}$$

$$\text{Var}(\sigma) = D^2 \cdot \text{Var}(\gamma) \tag{10}$$

The cumulative distribution function of u is expressed as:

$$F_U(u) = P(U \leq u) = P\left(\frac{\tau D}{\sigma} \leq u\right) = P\left(\tau \leq \frac{u\sigma}{D}\right) \tag{11}$$

for $\sigma > 0$ and $D > 0$.

In integral form,

$$F_U(u) = \int_0^\infty \int_0^{ux/D} f_{\sigma,\tau}(x,y) \, dy \, dx$$

$$= \int_0^\infty f_\sigma(x) \left(\int_0^{ux/D} f_{\tau|\sigma=x}(y) \, dy \right) dx \tag{12}$$

The p.d.f. of U is determined by differentiating the cumulative distribution with respect to u:

FOURIER ANALYSIS FOR ESTIMATING SLIDING

$$f_U(u) = \int_0^\infty f_\sigma(x) \left[\frac{x}{D} f_\tau \Big|_{\sigma=x} \left(\frac{ux}{D}\right) \right] dx \tag{13}$$

The p.d.f. of r is determined by making a simple variable change. If v is set equal to tan(r), then r = arctan(v). The change in variable is made and the Jacobian included:

$$f_V(v) = f_R(r) = f_R(\arctan(v)) \left| \frac{\partial r}{\partial v} \right|$$

$$= f_r(\arctan(v)) \left[\frac{1}{1+v^2} \right]$$

$$f_V(v) = \begin{cases} \dfrac{e^{(-\arctan(v)/\mu_r)}}{\mu_r(1+v^2)} & \text{, for } v \geq 0 \\ \\ 0 & \text{, for } v < 0 \end{cases} \tag{14}$$

If the rock density is treated as a constant, then it is included in the constant A. The variable U has the same distribution as τ, and its gamma p.d.f. is given by equation (3). The p.d.f. of r is the same as that given for the general case, equation (14).

DISCRETE FOURIER ANALYSIS

The p.d.f. for the sum of two or more independent density functions can be expressed as the convolution of the functions with each other (Feller, 1966, p. 7). That is, the sum of independent densities in space domain is analogous to the product of their Fourier transforms (characteristic functions) in frequency space. Therefore, the safety factor, as expressed in equation (5), has a p.d.f. equal to the reverse Fourier transform of the product of the characteristic functions of AU and BU.

$$f_S(s) = F^- \left[F^+ \left(f_{AU}(Au) \right) \cdot F^+ \left(f_{BV}(Bv) \right) \right] \tag{15}$$

Procedures for estimating the probability of plane shear sliding by discrete Fourier analysis are summarized below. Use of the Fourier transform (FFT) algorithm provides a great deal of computational efficiency.

Step 1. Using the specified geometries of the rock slope and the potential plane shear failure mass, the constants A and B are calculated by equations (6) and (7). If the rock density is treated as a constant, then A must be divided by it.

Step 2. The probability densities of AU and BV are digitized at N increments of width Δx using equations (13) and (14), respectively. If the rock density is treated as a constant, then equation (3) is used instead of (13).

Step 3. These discretized probability densities are then transformed by FFT to produce two sets of complex Fourier coefficients, which represent discretized versions of the two characteristic functions. Because the same Δx increment was used to digitize both densities, the resulting discrete characteristic functions have the same Δf increments ($\Delta f = 1/N\Delta x$).

Step 4. At each frequency determined by Δf, the complex coefficients from the two transformed densities are multiplied together to produce a new set of complex coefficients, which represents the discrete characteristic function of the safety factor.

Step 5. This new set of complex Fourier coefficients is reverse transformed by FFT to produce a real-valued, discretized approximation of the safety factor p.d.f.

Step 6. The probability of sliding is determined by using numerical integration to calculate the area under the safety factor p.d.f. where the value of the safety factor is less than one.

For the typical case where rock density is treated as a constant, one would expect the safety factor to be approximately gamma distributed, especially when the mean waviness of the sliding surface is small (less than 10 degrees). For small values of the waviness r, the variable v (v=tan(r)) is approximately equal to r, making the distribution of BV approximately exponential (see equation (14)). Also, the shear strength τ is assumed to be gamma distributed. Thus, the safety factor is expected to be approximately gamma distributed. This result has been confirmed by Fourier analyses of plane shear stability for various slope geometries, sliding surfaces, and rock types.

CONCLUSIONS

The probability of plane shear sliding can be estimated with considerable computational efficiency by using Fourier analysis procedures based on the FFT algorithm. An estimate of the true p.d.f. of the safety factor is obtained, whereas Monte Carlo techniques provide only possible realizations of the safety factor distribution and are often prone to excessive amounts of computer time. Comparative studies have indicated that the Fourier method uses one-third to one-fifth the computer time required by the Monte Carlo method for typical plane shear geometries.

If the shear strength and waviness of the sliding surface are respectively assumed to be gamma and exponentially distributed, the resulting safety factor is approximately gamma distributed. It should be noted that application of the discrete Fourier procedure is not restricted to these particular types of statistical distributions of the input variables, but is valid for all mathematically defined probability distributions that adequately describe the random variables.

APPENDIX

The sampling variance of the probability of sliding in Monte Carlo simulation (Mihram, 1972, Chap. 4) is given by:

$$\text{Var}(P_S) = P_S(1 - P_S)/n \tag{A1}$$

where: P_S = probability of sliding,
n = number of Monte Carlo iterations.

Therefore, a simulation with 1000 iterations results in a standard deviation of P_S that is less than 0.016.

REFERENCES

Andrew, M., 1981. Comparisons of the Normal Distribution with Lognormal and Gamma Distributions by Simulation Examples. Unpubl. grad. seminar, Dept. of Stat., Univ. of Wyoming, Laramie, WY.

Borgman, L.E., 1977. Some New Techniques for Hurricane Risk Analysis. Paper presented at 9th Annual Offshore Technology Conf., Houston, TX, May 2-5.

Call, R.D., Savely, J.P., and Nicholas, D.E., 1976. Estimation of Joint Set Characteristics from Surface Mapping Data. Proc., 17th U.S. Symp. on Rock Mechanics, Salt Lake City, UT, p. 2B2.1-2B2.9.

Feller, W., 1966. An Introduction to Probability Theory and its Applications; vol. II. John Wiley & Sons, Inc., New York, 626 pp.

Marek, J.M. and Savely, J.P., 1978. Probabilistic Analysis of the Plane Shear Failure Mode. Paper presented at 19th U.S. Symp. on Rock Mechanics, Lake Tahoe, NV, May 1-3.

Mihram, G.A., 1972. Simulation: Statistical Foundations and Methodology. Academic Press, New York, 526 pp.

Miller, S.M., 1982. Statistical and Fourier Methods for Probabilistic Design of Rock Slopes. Ph.D. dissertation, Dept. of Geology and Geophysics, Univ. of Wyoming, Laramie, WY.

Chapter 14

STATISTICAL CHARACTERIZATION OF COMPLEX CRACK AND PETROGRAPHIC
TEXTURE: APPLICATION TO PREDICTING BULK PHYSICAL PROPERTIES

Nick Warren

Institute of Geophysics and Planetary Physics
University of California, Los Angeles, CA 90024

ABSTRACT

The problem considered is that of predicting rock physical properties from observations of the structure of the rock. The approach here departs from previous studies in that both the physical properties and structural features are taken as statistical data and the relations between them are cast as statistical functions. This approach I have termed a "quantitative petrostructure analysis". This paper deals with two aspects of making such a study; parameterizing patterns of topographic features of rock, and characterizing variables to allow correlation studies to be carried out. The principle of a quantitative petrostructure analysis is outlined and illustrated using maps from two samples of Conway granite.

INTRODUCTION

The Basic Problem

It has long been recognized that correlations exist between observable crack patterns, grain texture or fabric and bulk physical properties (examples of empirical and theoretical work on this include Brace, 1969; Tilmann and Bennett, 1973; Simmons, et al., 1974; O'Connell and Budiansky, 1974; and Warren, 1977). However, the problem of being able, in general and for over a large range of rock types, to quantitatively predict the pressure dependence of bulk physical properties from observation of rock structure has not been solved. This problem has two causes. The first is in the phenomenology of the problem, the second arises in the theory of the problem. Usual theoretical approaches are basically limited by the requirement of idealizing crack structure. The phenomenological approach is limited by the need to measure and quantify a very large set of possibly important petrographic variables.

Approach

The approach taken here is to treat this problem as a problem in statistics. Here then, three classes of variables are put on equal and quantitative footing. These are the bulk properties (e.g. moduli or resistivity), the compositional (scalar) variables which describe the rock makeup (e.g. mineral modes), and the topographic or geometrical variables (e.g. grain and crack geometries). This puts both the phenomenological and the theoretical approaches into the same frame and allows treatment of a large number of variables. In its broadest sense this approach is an application in pattern recognition and factor analysis (see for example Jareskog et al., 1976).

Feasibility

There are two tasks that must be accomplished to illustrate the feasibility of the approach. First is the demonstration that systematics exist in some class of bulk properties and that these occur at a sufficient level of resolution to infer that underlying statistical variables can be found. The second task is to invent a sufficiently general method to economically characterize and parameterize rock structure.

THE STUDY

A pilot study has been made using two samples of Conway granite (syenogranite) from Conway, New Hampshire. In previous work, a technique was developed for grouping the samples based on Moduli-pressure systematics (Warren and Tiernan, 1981). More recently, a technique for parameterizing and characterizing petrostructure has been developed (Warren, 1982). The study here illustrates this more recent development.

Moduli-Pressure Systematics

The two samples are part of a suite of ten samples from a borehole drilled by ERDA (now DOE) at an exploration Hot Dry Rock site (Conway, New Hampshire). Using measures of the dependence of ultrasonic velocity on pressure these ten samples were found to form three groups. The two samples used here (381.9 and 704.4) are from a single group, being well matched by their moduli-pressure dependence. Their similarities are demonstrated in Figures 1a and 1b. The figures show "crack spectra" inverted from the velocity data for these two samples. To illustrate the difference between these samples and a sample from another group, crack spectra for sample 448.1

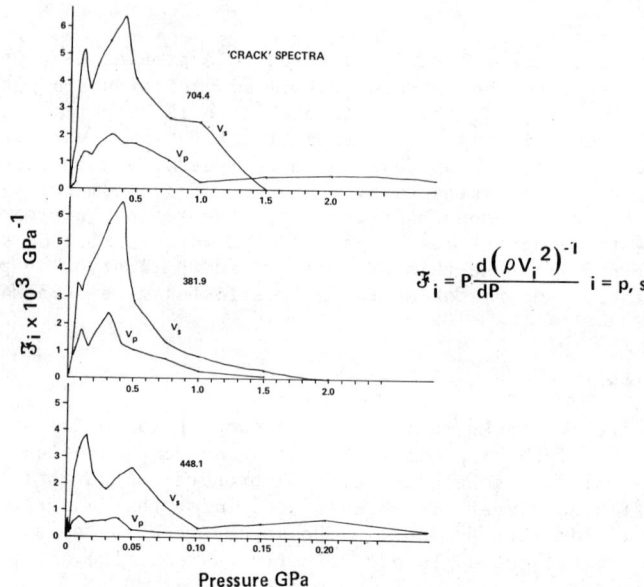

Figure 1. Spectra derived from velocity data for Conway samples.

are shown in Figure 1c. The crack spectra are obtained from the derivative of a spline curve fit to the data. In the inserted equation, ρ is density, P is pressure and V_i is velocity (i = p, and s denotes compressional and shear wave velocity respectively). Although they come from different depths (they are about 322 meters apart), both 381.9 and 704.4 are fine grained with similar textures. Both the samples and the groupings are described in detail in the referenced papers.

Petrostructure Analysis

Two microphotographs, one from each sample, were used in the second part of this study, the petrostructure analysis. By the term petrostructure is meant a descriptive space spanned by three axes: the mineralogical variables, the petrographic variables, and the microstructural variables. The variables fall broadly into two classes: 1) topographic and 2) compositional. The compositional variables (e.g. chemistry and mineralogy) are scalar fairly directly quantifiable. Topographic variables are <u>patterning</u> variables which must be quantified through mapping. They include grain orientation, mineral nearest-neighbor distances, crack patterns and crack heirarchies. These topographic variables are evaluated by a mapping

COMPLEX CRACK AND PETROGRAPHIC TEXTURE 135

technique which preserves the statistics of the geometrical relations in terms of a small fixed set of map parameters. Three parameters are used: 1) sets of coordinate points (pattern nodes), 2) listings of connected points and 3) flag values which specify the nature of connectivity between points, or identify or distinguish a group of points. The values of these three variables are specified through the process of digitizing a map or photo of the network.

In this pilot study, petrostructure was limited to "crack" structure (i.e. those visual features which showed crack-like visual contrast and morphology). The microphotographs were made at a magnification of 63X, mapped and digitized. The digitized maps are shown in Figure 2. A correlation between the two crack patterns was sought. The maps were selectively decomposed through a series of steps. Open networks were broken into length, orientation, spacing, and connectiveness distributions. Closed networks (e.g. crack boundaries that form grains) were additionally decomposed into area and shape distributions. From these basic distributions a large number of parameters were automatically evaluated using a computer program (SOCIOLOGY) which was developed as part of this analysis effort. The reader is referred to Warren (1982) for details.

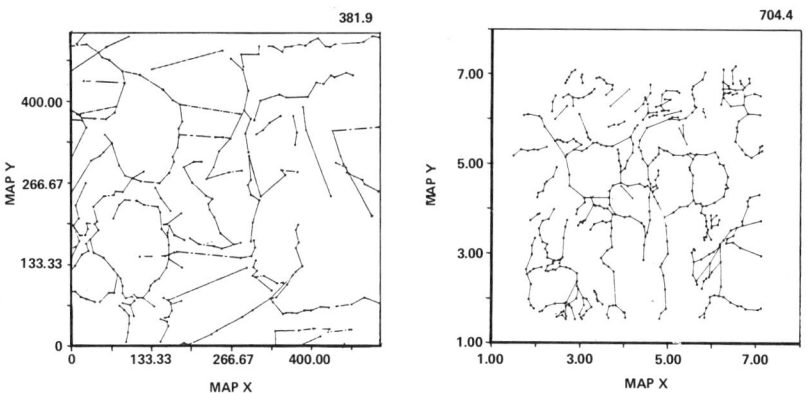

Figure 2. Computer generated maps.

Results

In this pilot study strong similarities in physical properties were shown to correlate with similarities in petrostructure. Exemplar results are shown in the figures. The first set of histograms (Figure 3) show a strong similarity between the two samples. Three

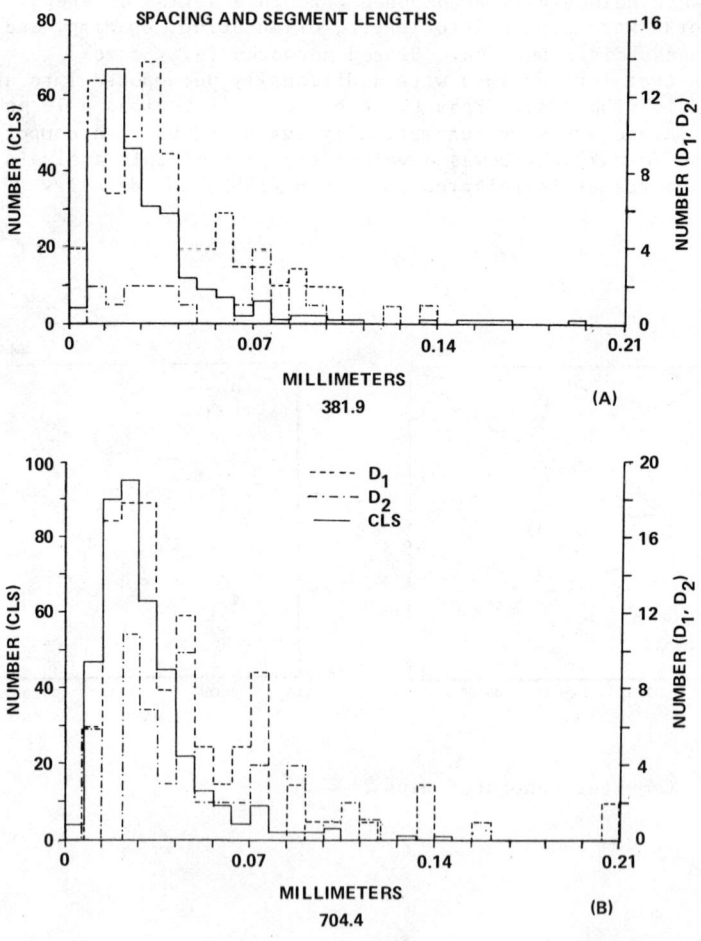

Figure 3. Histograms of crack feature. See text for discussion.

distributions are shown. These are the distributions of nearest-neighbor distances between crack tips (D_1), the equivalent distributions for crack branch points (D_2), and the directly digitized crack segment lengths (CSL). D_1 may be taken to be a measure of the distribution of a crack-stopping variable; namely either low stress regions or barriers. D_2 is a crack forking variable indicating the distribution of spacings at which cracks grow together or fork during growth. The CSL may be taken as a measure of the mean-free path of crack growth.

The results imply that for these samples the spacial properties of crack-stopping, kinking, and branching are equal within each rock and between each rock. The first part of this result would not be expected for a highly anisotropic or schistose rock. The second part, namely that the probability distributions controlling potential crack growth are equivalent in both rocks is in full agreement with the similarity predicted by the crack spectra obtained from the velocity data.

In Figure 4 a very striking agreement is found between these results and those of another study designed to determine actual crack length distributions in a granite (Hadley, 1976). Hadley used 400X SEM of Westerly granite to directly measure crack lengths and widths. Cracks were "any open flaw of aspect ratio less than 1 which did not change orientation by more than 20° over any significant portion of its length...". Her definition of crack length agrees well with the natural CSL determined for these Conway granite samples.

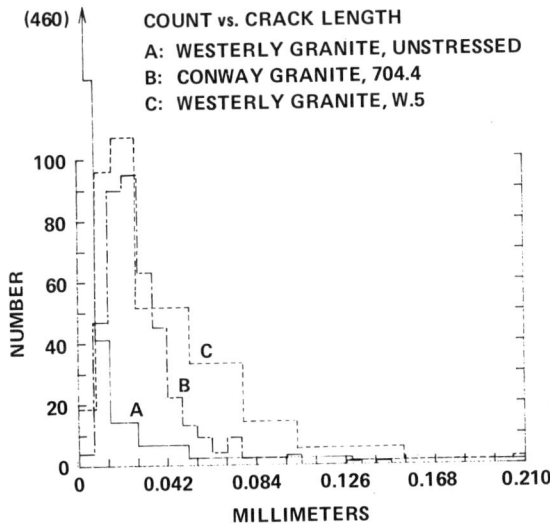

Figure 4. Superposition of crack length distributions.

A number of other variables distributions were also characterized in the pilot study. One was an alternative measure of crack length, namely the cumulated sum of CSL between crack tips and/or branch points. The distribution of this measure, as well as D_1, D_2, and CSL, are plotted along with the crack spectra amplitudes in Figure 5. This plot indicates the degree of correlation between the variables. The figure was simply generated by cross plotting the ordinate values of paired curves. If the curves are similar, then the diagram generated by cross plotting the y values should show a clustering along the 45° diagonal. For two samples with similar structures, but with a scale factor shift between them, the linear trend of the plot will fall off the 45° diagonal. In the case of little or no correlation, no linear trend will be seen. Figure 5 shows the basic similarity of the two samples is established over all the plotted variables.

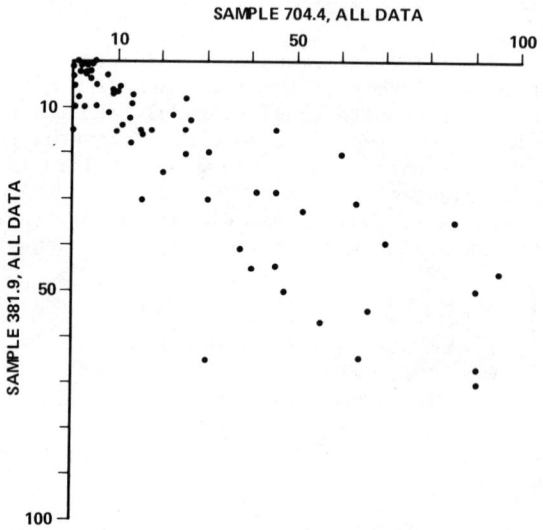

Figure 5. Correlation plot for four variables. Data are normalized to 100.

In a true factor analysis, studies of the data such as generated here would be searched for underlying and invisible factors which could serve to reduce the dimensionality of the data and maximize the predictive value of the data. Such techniques can be extremely valuable in rock geophysics.

COMPLEX CRACK AND PETROGRAPHIC TEXTURE

ACKNOWLEDGEMENTS

The Petrostructure Analysis Program has been developed at UCLA under DOE support (DE-AS03-76F00034) and in conjunction with the Hot Dry Rock Project at Los Alamos National Laboratory.

UCLA publication no. 2316.

REFERENCES

Brace, W.F., 1969, "Micromechanics in Rock Systems", Structure, Solid Mechanics and Engineering Design; The Proceedings of the Southhampton 1969 Civil Engineering Materials Conference, M. Te'eni, ed., Wiley-Interscience, pp. 187-204.

Hadley, K., 1976, "Comparison of Calculated and Observed Crack Densities and Seismic Velocities in Westerly Granite," Journal of Geophysical Research, Vol. 81, No. 20, July, pp. 3484-3493.

Jareskog, K.G., Klovan, J.E., and Reyment, R.A., 1976, Geological Factor Analysis (Methods in Geomathematics I), Elsevier Scientific Publishing Company, Amsterdam, pp. 174.

O'Connell, R.J. and Budiansky, B., 1974, "Seismic Velocities in Dry and Saturated Cracked Solids," Journal of Geophysical Research, Vol. 79, No. 35, December, pp. 5412-5426.

Simmons, G., Siegfried, R.W. III, and Feves, M., 1974, "Differential Strain Analysis: A New Method for Examining Cracks in Rocks," Journal of Geophysical Research, Vol. 79, No. 29, October, pp. 4383-4385.

Tilman, S.E. and Bennett, H.F., 1973, "A Sonic Method for Petrographic Analysis," Journal of Geophysical Research, Vol. 78, No. 35, pp. 8463-8469.

Warren, N., 1977, "Characterization of Modulus-Pressure Systematics of Rocks: Dependence on Microstructure," Geophysical Monograph 20, The Earth's Crust, American Geophysical Union, Washington, D.C., pp. 119-148.

Warren, N. and Tiernan, M., 1981, "Systematics of Crack Controlled Mechanical Properties for a Suite of Conway Granites from the White Mountains, New Hampshire," Tectonophysics, Vol. 73, pp. 295-322.

Warren, N., 1982, "Quantitative Petrostructure Analysis," Los Alamos Scientific Laboratory, New Mexico, in press.

STRESSES

Chairman
Thomas W. Doe
Lawrence Berkeley Laboratory
Berkeley, California

Co-Chairman
Bezalel C. Haimson
University of Wisconsin
Madison, Wisconsin

Keynote Speaker
Mark Zoback
US Geological Survey
Menlo Park, California

Chapter 15

STATUS OF THE HYDRAULIC FRACTURING METHOD
FOR IN-SITU STRESS MEASUREMENTS

by Mark D. Zoback and Bezalel C. Haimson

Office of Earthquake Studies
U.S. Geological Survey
Menlo Park, California

Department of Mineral Engineering
University of Wisconsin
Madison, Wisconsin

INTRODUCTION

A Workshop on Hydraulic Fracturing Stress Measurements was convened by the authors in December 1981 in Monterey, California under the auspices of the U.S. Geological Survey and the U.S. National Committee on Rock Mechanics. There are now over a dozen groups of investigators around the world who are actively using the hydraulic fracturing method for making in-situ stress measurements. Because of the extensive recent experience of these different groups, the Workshop enabled forty investigators from eight countries to comprehensively assess the status of the method. Based on the presentations and discussions at the Workshop, in this paper we briefly discuss the current status of the hydraulic fracturing stress measurement method concentrating on problem areas where further research needs to be done. A complete account of the presentations at the Workshop will be available through the Proceedings of the Workshop which will be published by the U.S. National Committee on Rock Mechanics in early 1983.

The most exciting aspect of the Workshop was the discovery by the various groups of investigators that others were having very similar experiences with the method. The consensus among the different groups of investigators at the Workshop was that they were recording very similar pressure-time data in the field, they were able to interpret them in similar ways, and they were achieving results that were quite encouraging because of several factors. First, the results at various locales were found to be internally quite consistent. This was reported by investigators from Germany (Rummel, Baumgartner, and Alheid, 1982), Japan (Tsukahara, 1982), Australia (Enever, 1982), and the U.S. (see, for example, Haimson, 1978, 1982a; Zoback et al. 1980). Second, the hydrofrac stress measurements generally agreed well with subsurface overcoring stress measurements wherever detailed comparisons could be made (Haimson, 1981, 1982; Li, et al. 1982; Doe, et al. 1982), and hydrofrac orientations seem to agree quite well with other stress measurement methods and stress field indicators (see Zoback and Zoback, 1980). Finally, the magnitudes of stresses determined by hydraulic fracturing seem to agree well with stress estimates based on the frictional strength of rock (Brace and Kohlstedt, 1980; Zoback and Hickman, 1982). Nevertheless, it was clear at the meeting that there are a number of areas where the method is more complex than is usually recognized, and other areas where further research is required to verify interpretation methods. There are discussed below.

DETERMINATION OF THE LEAST-PRINCIPAL STRESS

It has been well-demonstrated by laboratory and theoretical studies that hydraulic fractures propagate in a plane perpendicular to the least-principal stress (Hubbert and Willis, 1957; Haimson, 1970; Haimson and Avasthi, 1975; and others) and the magnitude of this stress can be determined simply from pressure in the fracture immediately after pumping has stopped (the shut-in pressure). Unless otherwise noted in this paper, we will usually refer to a vertical fracture propagating from a vertical well-bore and we assume that one principal stress is essentially vertical and results from the weight of the overlying rock. This should be true except at very shallow depth where topography or other effects (like excavations) alter the local stress field (see McGarr and Gay, 1978, and Zoback and Zoback, 1980). In this case the least-horizontal principal stress S_{hmin} is usually the most straightforward component of the stress field to measure. In this section, we investigate a number of factors that complicate the determination of S_{hmin}.

<u>Multiple Shut-In Pressures</u> - In cases where the overall least-principal stress is the vertical stress, S_v, a vertical fracture forms at the well bore at an azimuth perpendicular to S_{hmin} (as long as no sub-horizontal bedding planes or joints are present), and turns into a plane perpendicular to S_v as it propagates (see Haimson et al. 1976, Haimson, 1982a; Zoback et al. 1977; Roegiers and McClennan, 1982). As shown in Fig. 1, this results in multiple shut-in pressures which can be used to determine both S_{hmin} and S_v. In test #37, the fracture evidently initiated along a horizontal bedding plane. In test #39, the fracture was initiated vertically, but turned into a horizontal plane as it propagated. In general, this effect is quite common at shallow depths (≤ 300 m) and not particularly troublesome as long as one interprets the pressure-time history correctly.

<u>Decrease in Shut-In Pressure</u> - A distinctly different phenomena from multiple shut-in pressures that has been pointed out by several in-

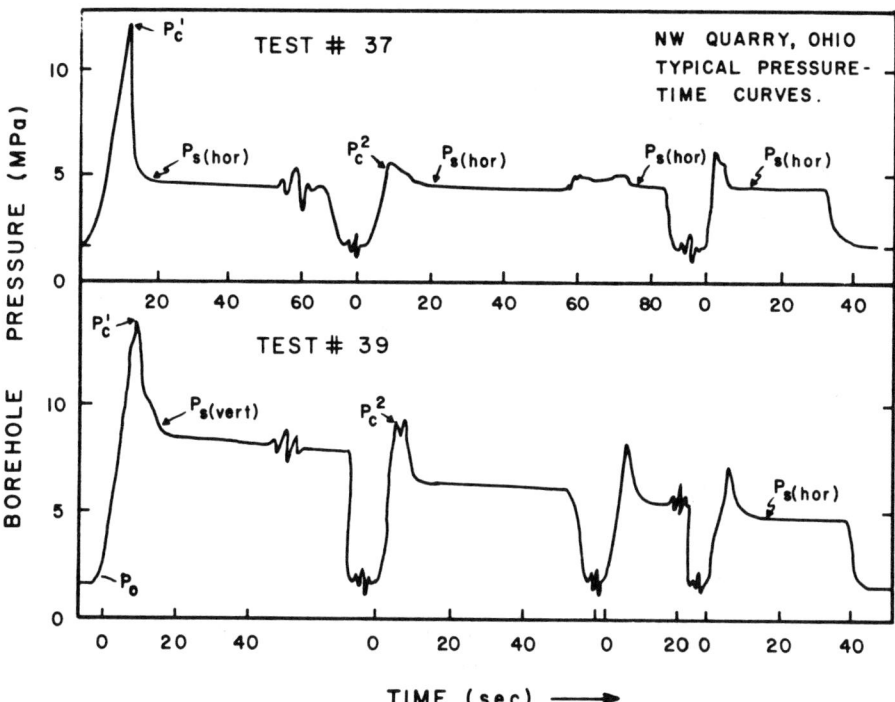

Figure 1 - Pressure-time records of two tests in shaly dolomite near Anna, Ohio. Test #37 represents initiation and extension of a horizontal bedding-plane fracture. In Test #39 a vertical fracture initiated at the well bore and turned into a horizontal plane. This interpretation was confirmed by an impression packer (Haimson, 1982b).

vestigators is a slight decrease in shut-in pressures as fractures are propagated (see for example Gronseth and Kry, 1982; Hickman and Zoback, 1982; Enever, 1982). An example of this behavior is shown in Figure 2 from a well drilled in granitic rock near Monticello, South Carolina. A noticeable decrease in shut-in pressure is often not observed (for example, the tests at 128 and 400 m). When it is observed, the shut-in pressure usually drops only a few bars unless the fracture is shut-in immediately after breakdown (as at 97 and 646 m), in which case the decrease is quite small after the second pressurization cycle. As discussed at length by Hickman and Zoback (1982), slight decreases in shut-in pressure as shown in Fig. 2 are apparently due to the decrease in importance of viscous effects in the fracture as it propagates. As concluded by all the investigators who have studied this effect, the shut-in pressure which should be used as a measure of S_{hmin} are the final values of the shut-in pressure. However, care must be taken not to mis-interpret hydraulic fractures which might intersect pre-existing fractures and joints, hydrofracs which break out around the packers, or vertical hydrofracs which turn into a horizontal plane thus exhibiting shut-in pressures related first to S_{hmin} and in the latter cycles to S_v (Figure 1).

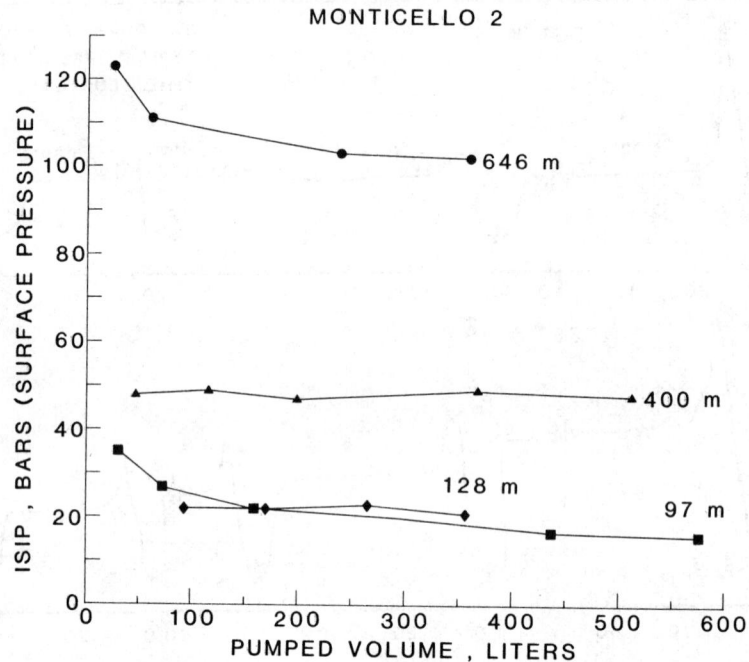

Figure 2 - Results from tests in a well in South Carolina which shows change in shut-in pressure as hydraulic fractures are propagated (Hickman and Zoback, 1982).

Gradual Pressure Changes Upon Shut-In - In some hydraulic fracturing tests the pressure change after the cessation of pumping is gradual and shut-in pressures are not distinct enough to be measured accurately. This phenomena is primarily due to leak-off from the straddled interval and fracture into the surrounding rock, leakage past the packers, or further fracture propagation after pumping stops. We are not, however, referring here to the marked viscous effects in massive hydraulic fractures propagated with gel and proppant (see Nolte, 1979 and Nolte and Smith, 1979) in which case fracture closure can follow shut-in by hours.

Various investigators have proposed analysis techniques to deal with indistinct shut-in pressures observed during stress measurement hydrofracs which are illustrated, in part, in Figure 3. Gronseth and Kry (1982) suggest using the inflection point in the pressure-time record (Fig. 3a), Doe et al. (1982) recommends plotting pressure as a function of log time (Fig. 3b), McLennan and Roegiers (1982) suggest plotting pressure as a function of log $\frac{t + \Delta t}{\Delta t}$ (Fig. 3c) where Δt is shut-in time and t is pumping time, Aamodt (1982) suggests using the Muskat method of plotting the log of pressure as a function of time and extrapolating back the linear curve to zero time (Fig. 3d), and

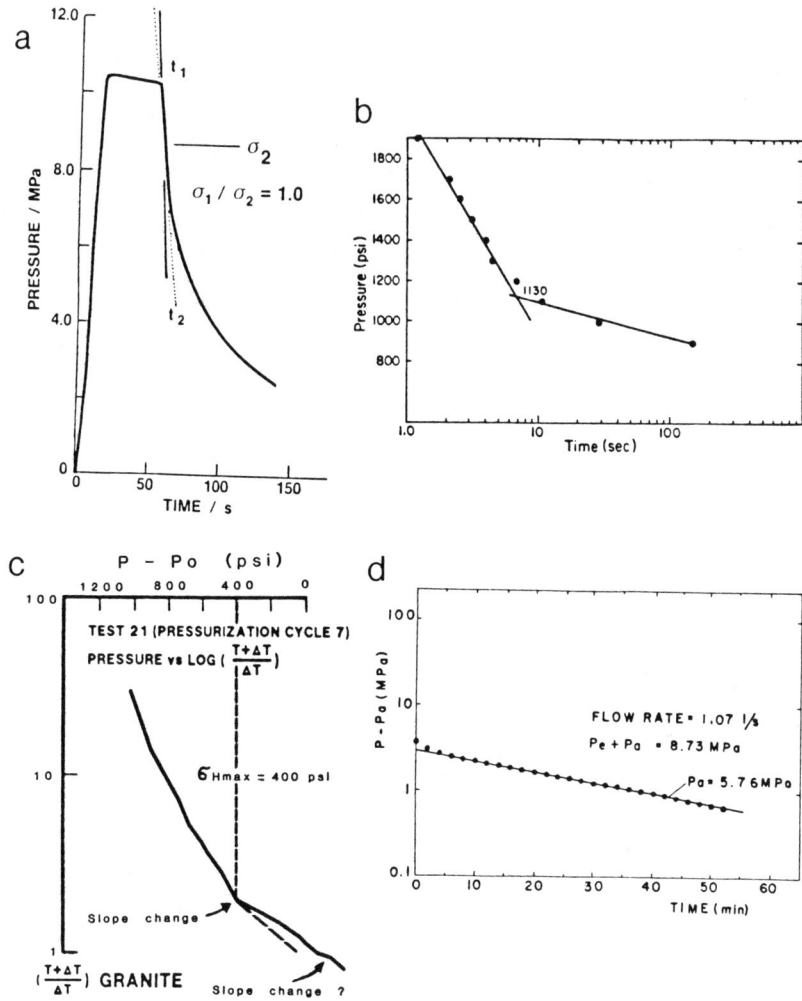

Figure 3 - Illustration of various methods for determining S_{hmin} when shut-in pressures are not distinct. a) inflection point method (Gronseth and Kry, 1982), b) pressure as a function of log time (Doe et al), c) pressure build-up method (McLennen and Roegiers, 1982), d) Muskat method (Aamodt and Kuriyagawa, 1982).

Haimson (1982c) has found that plotting log pressure as a function of log time gives the sharpest shut-ins.

These five methods seem to help identify the change in slope associated with fracture closure in at least the few tests examined by each investigator. However, as all the methods are based on different theoretical models (none of which have been rigorously developed or tested) considerable future work is needed in this area. Also, in cases in which this is especially troublesome it may be useful to consider measuring S_{hmin} by pumping at extremely slow flow rates with low viscosity fluid to determine the pressure necessary to hold the fracture open. In this case the viscous pressure drop in the fracture is so low that the pumping pressure is essentially equal to S_{hmin}.

DETERMINATION OF THE MAXIMUM HORIZONTAL PRINCIPAL STRESS

Since Hubbert and Willis (1957) derived the basic elastic solutions for hydraulic fracture initiation in a pressurized borehole in stressed rock, the basic applicability of this analysis and determination of the maximum horizontal principal stress (S_{hmax}) from the breakdown pressure and knowledge of the rocks' tensile strength has been discussed by Haimson and Fairhurst (1967) and others. In this section we discuss various issues concerning the determination of S_{hmax}.

Fracture Re-opening, or Secondary Breakdown, Pressures

The method described by Bredehoeft et al. (1976) of using the pressure necessary to re-open a hydraulic fracture (the fracture re-opening, or secondary breakdown, pressure) is being used widely. This is because either there is no core available to determine tensile strength (T), the appropriate T for the in-situ test is difficult to determine from laboratory tests (see below), or fracture initiation precedes breakdown by a considerable amount and the breakdown pressure is anomalously high (Zoback et al., 1976).

The basic assumption in using secondary breakdown pressures is that the fracture opens suddenly to accept fluid as the borehole is repressurized. The use of fracture re-opening pressures seems to yield very consistent results (see Zoback et al., 1980; Haimson and Rummel, 1982; Enever, 1982), but to determine the fracture re-opening pressure precisely, very careful test procedures are required, especially with respect to maintaining constant flow-rates (see Zoback et al., 1980). Hickman and Zoback (1982) discuss the method at length and show that for varying relative magnitudes of the horizontal principal stresses different forms of the pressure-time curve result (see Figs. 4, 5). Thus, pressure-time curves need not have "classic" shapes such as those shown in Fig. 1 or Fig. 5a in order to be a good hydrofrac measurement. The relative magnitudes of the horizontal principal stresses controls the shape of the pressure-time curve.

IN-SITU STRESS MEASUREMENTS 149

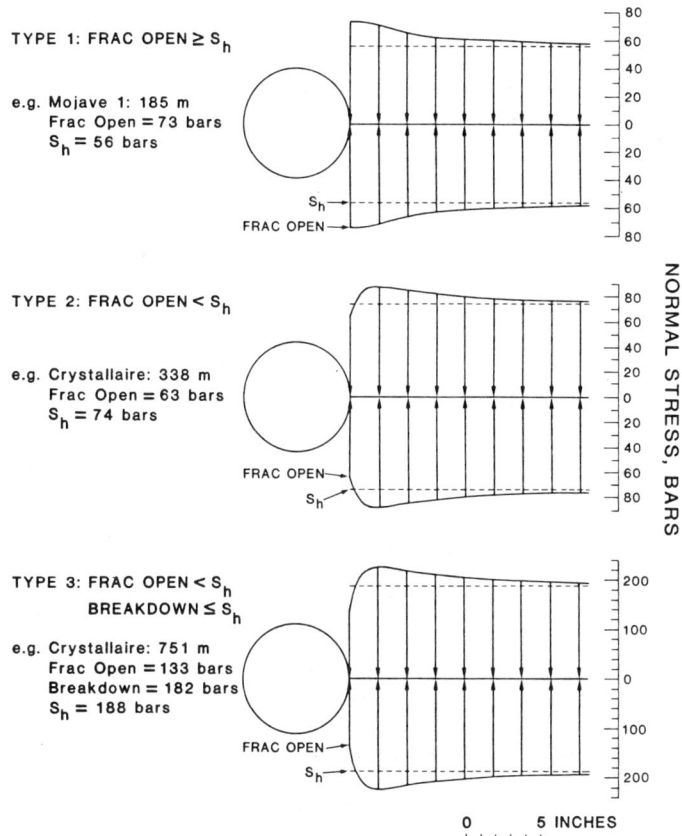

Figure 4 - Examples of the manner in which the relative magnitude of the horizontal principal stresses change the relative magnitude of the breakdown, fracture opening, and shut-in pressure (Hickman and Zoback, 1982). See Figure 5 for corresponding pressure-time histories.

In some cases secondary breakdown pressures cannot be used. For example, a case similar to the Type 2 (or 3) pressure-time curves shown in Figs. 4 and 5, is when the fracture re-opening pressure can be below the hydrostatic pressure. Also, in some tests constant flow-rate pumping may not be possible, or fluid infiltration may occur into a naturally propped fracture, or the packers may hold the fractures open (see below). All of these effects would basically negate the possible use of fracture re-opening pressures. A moderate decrease in the fracture reopening pressure is often seen that does not seem due to the processes mentioned above (Figure 6). Hickman and Zoback (1982) discuss this and conclude that the effect is probably due to fluid infiltration into the region surrounding the borehole (even in

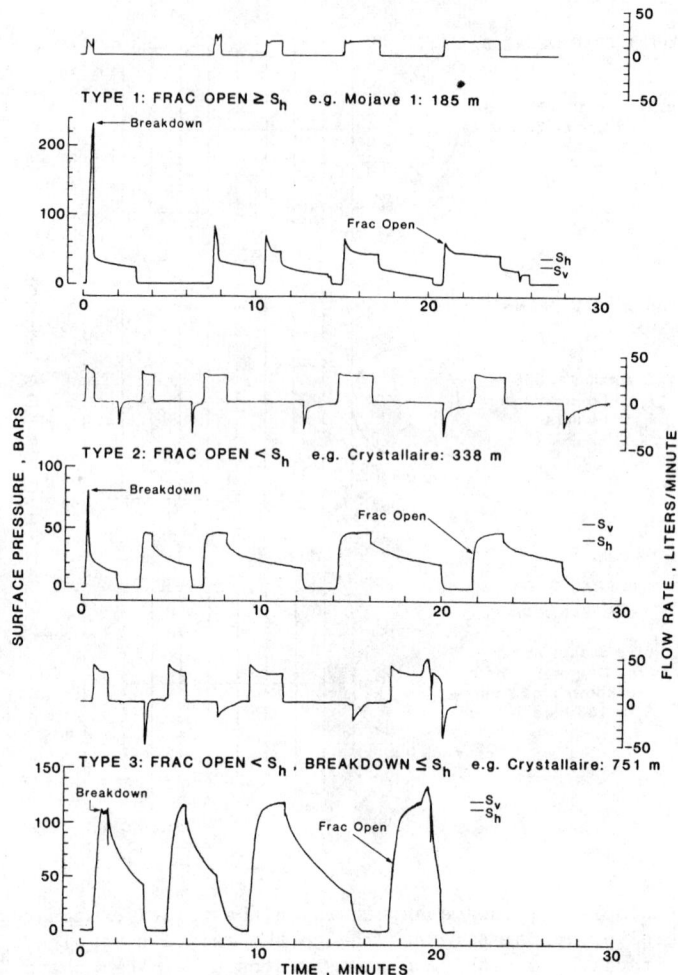

Figure 5 - Different types of pressure-time records which correspond to the relative stress magnitudes shown in Figure 4 (Hickman and Zoback, 1982).

relatively impermeable rock such as granite). They suggest that the appropriate fracture re-opening pressure to use for calculation of S_{hmax} is from early pressurization cycles before extensive fluid penetration occurs. In general, the use of secondary breakdown pressures for determination of S_{hmax} is an area which is extremely promising, seems to be yielding good results, but could benefit from more research.

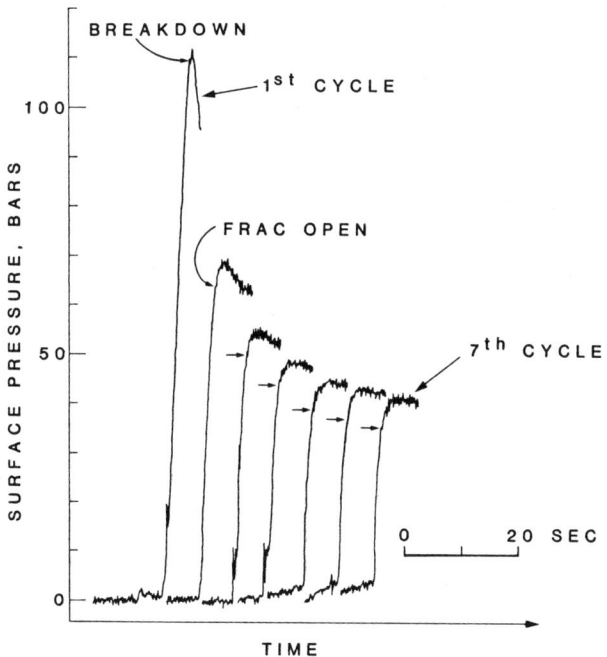

Figure 6 - Example of the decrease in fracture opening pressure (arrows) with repeated cycling (Hickman and Zoback, 1982).

The effect of fluid infiltration into the rock surrounding the test-hole is also discussed by Alexander (1982) with respect to the criterion for breakdown pressure in hydraulic fracturing. Haimson and Fairhurst (1967) extended the "no-infiltration" criterion (Hubbert and Willis, 1957), to include the effect of fluid flow from the pressurized test hole into the immediate rock. Extensive experimental testing has shown that in most cases this criterion predicts the magnitude of the pressure required to initiate (or reopen) hydraulic fractures remarkably well (Haimson and Fairhurst, 1970; Edl, 1973; Haimson and Rummel, 1982). However, based on these results, Haimson (1978) concluded that in the range of stresses of 0-50 MPa (most of field cases encountered to date) the "no-infiltration" criterion is a good indicator of the experimental values. The latter are typically off by no more than 20% from this criterion prediction. However, in order to use the "infiltration" criterion such stress-dependent parameters as Poisson's ratio, and jacketed and unjacketed compressibilities have to be determined first. Hence, the use of the much simpler "no-infiltration" criterion appears justified at this stage.

Tensile Strength - The primary breakdown pressure should be employed in addition to the secondary breakdown pressure or when the latter cannot be used provided a reliable estimate of in-situ tensile strength can be made. Laboratory studies often indicate values that are clearly higher than those in-situ (Haimson and Rummel, 1982);); and often the problem with tensile strengths can cause considerable uncertainty in the determination of S_{hmax}. However, substantial progress with this problem has been made by using a statistical fracture mechanics approach (Ratigan, 1982a) which accounts for such phenomena as the decrease in tensile strength with increase in sample size. The results of applying this method to tests conducted at Stripa Mine, Sweden (Ratigen, 1982) and Reydarfjordur, Iceland (Haimson and Rummel, 1982) suggest that T may be a more reliable parameter than usually assumed.

Inelastic and Anisotropic Materials - In cases in which the rock is considerably anisotropic or not strong enough to withstand the concentrated stresses around the borehole, there may be little chance of determining S_{hmax} or the direction of maximum horizontal compression through conventional means. However, in the case of anisotropy, Cornet (1982) recommends a new stress calculation technique that does not involve breakdown pressures. When the rock around the borehole fails under high stress concentrations, the azimuthal dependence of the stress concentration sometimes results in well-oriented 'breakouts' which can be used to determine the orientation of S_{hmax} (see Bell and Gough, 1979).

Packer Induced Stresses - The possibility that packer induced stresses effect the breakdown pressure has been suggested by several investigators (Roegiers, 1975; Warren, 1979). Problems would seem to be most severe when the breakdown pressure is substantially below the packer inflation pressure. This is borne out, in part, by experiments of Wilson (written comm., 1982) some of which are shown in Fig. 7 which shows the change in packer pressure as the pressure in the straddled region is changed. In these tests, it was shown that the straddle interval pressure is approximately equal to the packer pressure once the straddle interval pressure reaches the initial packer pressure (in this case 1000 psi). Note that this is true even when the ratio of hole diameter to uninflated packer diameter is as large as 1.6, which is about as large as in any field test. The inference one draws from these tests is that if the packer inflation pressure is less than or equal to the breakdown or fracture reopening pressure, it will have very little effect on the test. If, however, the breakdown or fracture reopening pressure is quite low, the packer inflation pressure could have an effect on the determination of S_{hmax}.

Stress Field Orientation - One of the most often asked questions about hydraulic fracturing is whether the orientation of the stress field at the borehole is indicative of the far-field stress or perhaps controlled by some other process (see Smith, 1979). Although it is very hard to test this problem directly, in every case studied by Zoback

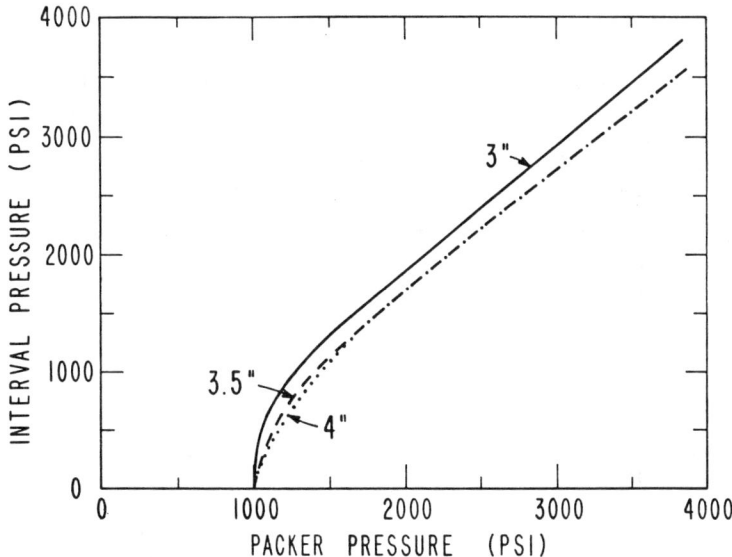

Figure 7 - Change in pressure in packers as pressure in straddled interval is raised for different hole diameters. Packer diameter is 2.36 inches (M. Wilson, written comm., 1982).

and Zoback (1980) and Haimson (1981, 1982b), excellent correlations were seen between hydraulic fracture orientations determined at the well bore and other stress-field measurements or indicators (see also Haimson, 1982b, for an updated stress map of the Upper Midwest showing the consistent stress directions in that large region of the continent.)

CONCLUSIONS

Further research is clearly needed in several important areas. For example, theoretical studies and laboratory (preferably "big-block") tests are needed to closely look at the fracture re-opening process as well as the fracture closing process during shut-in. Overall, the body of available hydrofrac data is still small. Many more tests are needed, especially where the results can be compared to overcoring data or other stress field indicators. Many investigators are attempting to develop more efficient field systems to make this possible. The consensus at the Workshop is that there is more reason than ever to believe that the hydraulic fracturing method is contributing substantially to our knowledge of the stress field in the earth's crust. With further refinement and improvement of field equipment and procedures, the method should gain even wider application and acceptance.

REFERENCES

Aamodt, L. and Kuriyagawa, M., 1982, Measurement of instantaneous shut-in pressure in crystalline rock, Proceedings of the Workshop on Hydraulic Fracturing Stress Measurements, U.S. Nat. Comm. on Rock Mech., Wash., D.C., in press.

Alexander, L.G., 1982, Note on effects of infiltration on the criteria for breakdown pressure in hydraulic fracturing stress measurements, Proceedings of the Workshop on Hydraulic Fracturing Stress Measurements, U.S. Nat. Comm. on Rock Mech., Wash., D.C., in press.

Bell, J.S., and Gough, D.I., 1979, Northeast-southwest compressive stress in Alberta: Evidence from oil wells, Earth Planet Sci. Lett., 45, 475-482.

Brace, W.F., and Kohlstedt, D.L., 1980, Limits on lithospheric stress imposed by laboratory experiments, Jour. Geophys. Res., 85, no. B11, 6248-6252.

Bredehoeft, J.D., Wolff, R.G., Keys, W.S., and Shuter, E., 1976, Hydraulic fracturing to determine the regional in-situ stress field Piceance Basin, Colorado, Geol. Soc. Amer. Bull., 87, 250-258.

Cornet, F.H., 1982, Analysis of injection tests for in-situ stress determination, Proceedings of the Workshop on Hydraulic Fracturing Stress Measurements, U.S. Nat. Comm. on Rock Mech., Wash., D.C., in press.

Doe, T. and others, 1982, Determination of the state of stress at the Stripa Mine, Sweden, Proceedings of the Workshop on Hydraulic Fracturing Stress Measurements, U.S. Nat. Comm. on Rock Mech., Wash., D.C., in press.

Edl, T.N., 1973, Mechanical instability of deep wells with particular reference to hydraulic fracturing, M.S. thesis, University of Wisconsin, Madison.

Enever, J.R., and Wooltorton, B.A., 1982, Experience with hydraulic fracturing as a means of estimating in-situ stress in Australian coal basin sediments, Proceedings of the Workshop on Hydraulic Fracturing Stress Measurements, U.S. Nat. Comm. on Rock Mech., Wash., D.C., in press.

Gronseth, J.M., and Kry, P.R., 1982, Instantaneous shut-in pressure and its relationship to the minimum in-situ stress, Proceedings of the Workshop on Hydraulic Fracturing Stress Measurements, in press.

Haimson, B.C., 1978, The hydrofracturing stress measuring method and recent field results, Inst. J. Rock Mech. Min. Sci. and Geomech., Abstr., vol. 15, 167-178.

Haimson, B.C., 1980, Near-surface and deep hydrofracturing stress measurements in the Waterloo quartzite, Inst. J. Rock Mech. Min. Sci. and Geomech., Abstr., vol. 17, 81-80.

Haimson, B.C., 1981, Confirmation of hydrofracturing results through comparisons with other stress measurements, Proc. of the 22nd U.S. Symp. on Rock Mechanics, 379-385.

Haimson, B.C., 1982a, Deep stress measurements in three Ohio quarries and their comparison to near-surface tests, Proc. of the 23rd U.S.

Symp. on Rock Mechanics, SME-AIME, this volume.

Haimson, B.C., 1982b, A comparative study of deep hydrofracturing and overcoring stress measurements at six locations with particular interest to the Nevada Test Site, Proceedings of the Workshop on Hydraulic Fracturing Stress Measurements, U.S. Nat. Comm. on Rock Mech., Wash., D.C., in press.

Haimson, B.C., 1982c, Stress measurements down to 1.5 km in the Precambrian granite of northern Illinois, J. Geophys. Res., in press.

Haimson, B.C., and Avasthi, J.M., 1975, Stress measurements in anisotropic rock by hydraulic fracturing, Proc. of the 15th Symp. on Rock Mech., Am. Soc. Civil Engrs., 135-156.

Haimson, B.C., Doe, T.W., Erbstoesser, S.R., and Fuh, G.F., 1975, Site characterization for tunnels housing energy storage magnets, Proc. of the 17th U.S. Symp. on Rock Mechanics, Salt Lake City, Utah, 4B4, 1-9.

Haimson, B.C., and Rummel, F., 1982, Hydrofracturing stress measurements in the IRDP drillhole at Reydarfjordur, Iceland, J. Geophys. Res., in press.

Haimson, B.C., and Fairhurst, C, 1967, Initiation and extension of hydraulic fractures in rock, Soc. Petr. Engrg. J., vol. 7, 310-318.

Haimson, B.C. and Fairhurst, C., 1970, In-situ stress determinations at great depth by means of hydraulic fracturing, Proc. 11th Symp. on Rock Mech., A.I.M.E., New York, 559-584.

Hickman, S., and Zoback, M.D., 1982, On the interpretation of hydraulic fracturing pressure-time data for in-situ stress determination, Proceedings of the Workshop on Hydraulic Fracturing Stress Measurements, U.S. Nat. Comm. on Rock Mech., Wash., D.C., in press.

Hubbert, M.K., and Willis, D.G., 1957, Mechanics of hydraulic fracturing, Trans. A.I.M.E., vo. 210, 153-162.

Li, F.-Q., and others, 1982, Experiments of in-situ stress measurements using the stress relieving and hydraulic fracturing techniques, Proceedings of the Workshop on Hydraulic Fracturing Stress Measurements, U.S. Nat. Comm. on Rock Mech., Wash., D.C., in press.

McGarr, A., and Gay, N.C., 1978, State of stress in the earth's crust, Ann. Rev. Earth Planet. Sci., vol. 6, 405-436.

McLennan, J.D., and Roegiers, J.-C., 1982, Do instantaneous shut-in pressures accurately represent the minimum principal stress, Proceedings of the Workshop on Hydraulic Fracturing Stress Measurements, U.S. Nat. Comm. on Rock Mech., Wash., D.C., in press.

Nolte, K.G., 1979, Determination of fracture parameters from fracturing pressure decline, Paper SPE 8341 presented at SPE-AIME 54th Annual Fall Technical Conference and Exhibition, Las Vegas, Nevada.

Nolte, K.G., and Smith, M.B., 1979, Interpretation of fracture pressures, paper SPE 82976 presented at SPE-AIME 54th Annual Fall Technical Conference and Exhibition, Las Vegas, Nevada.

Ratigan, J.L., 1982a, A statistical fracture mechanics determination of the apparent tensile strength in hydraulic fracture, Proceedings of the Workshop on Hydraulic Fracturing Stress Measure-

ments, U.S. Nat. Comm. on Rock Mech., Wash., D.C., in press.
Roegiers, J.-C., 1974, The development and evaluation of a field method for in-situ stress determination using hydraulic fracturing Ph.D. Thesis, University of Minnesota.
Roegiers, J.-C., and McLennan, J.D., 1982, Factors influencing the initial orientation of hydraulically induced fractures, Proceedings of the Workshop on Hydraulic Fracturing Stress Measurements, U.S. Nat. Comm. on Rock Mech., Wash., D.C., in press.
Rummel, F., 1982, Hydraulic fracturing stress measurements along the eastern boundary of the SW-German Block, Proceedings of the Workshop on Hydraulic Fracturing Stress Measurements, U.S. Nat. Comm. on Rock Mech., Wash., D.C., in press.
Smith, M.B., 1979, Effect of fracture azimuth on production with application to the Wattenberg gas field, Paper presented at 54th Annual Fall Technical Conference, Soc. of Petrol. Eng., Amer. Inst. of Mech. Eng., Las Vegas, Sept. 23-26.
Warren, W.E., 1979, Packer induced stresses during hydraulic well fracturing, Sandia Technical Report #79-1986.
Zoback, M.D., Rummel, F., Jung, R., and Raleigh, C.B., 1977, Laboratory hydraulic fracturing experiments in intact and pre-fractured rock, Int. J. Rock Mech. Min. Sci. & Geomech. Abstr., v. 14, 49-58.
Zoback, M.D., Tsukahara, H., and Hickman, S., 1980, Stress measurements at depth in the vicinity of the San Andreas fault: Implications for the magnitude of shear stress at depth, Journal of Geophysical Research, 85, no. B11, 6157-6173.
Zoback, M.D., and Hickman, S., 1982, In-situ study of the physical mechanisms controlling induced seismicity at Monticello Reservoir, South Carolina, Journal of Geophysical Research, in press.
Zoback, M.L., and Zoback, M.D., 1980, Faulting patterns in north-central Nevada and strength of the crust, J. Geophys. Res., 85, 275-284.

Chapter 16

THE INFLUENCE OF ROCK ANISOTROPY ON
STRESS MEASUREMENTS BY OVERCORING TECHNIQUES

By Bernard Amadei and Richard E. Goodman

Assistant Professor, University of Colorado
Boulder, Colorado

Professor of Geological Engineering, University of California
Berkeley, California

ABSTRACT

A medium is anisotropic if its properties vary with direction. This is the general characteristic of many rocks, for example, schists, slates, gneisses, phyllites and other metamorphic rocks. Bedded and regularly jointed rocks also display anisotropic behavior.

This paper is concerned with the influence of rock anisotropy on in-situ stress measurements. It is limited to stress measurements by overcoring techniques for which strains and displacements are recorded either on the walls of a pilot hole at the end of one or several boreholes or within instrumented solid or hollow inclusions perfectly bonded to the surface of the pilot hole. The rock is described as homogeneous, continuous, anisotropic and linearly elastic.

The following questions are answered with special emphasis on rocks that can be classed as transversely isotropic or orthotropic: the number of independent measurements obtainable in a single borehole; the number of boreholes required to determine the in-situ stress field; the influence of rock anisotropy on these numbers; the influence of the anisotropy type and the error involved by neglecting rock anisotropy.

INTRODUCTION

Stress measurements in rocks with oriented fabrics ought to take into account their anisotropic properties. This paper discusses the influence of anisotropy on interpretation of stress measurements by various overcoring techniques, in which strains or displacements are measured on the walls of a pilot hole drilled at the end of a borehole or within an instrumented inclusion bonded to the surface of the pilot hole. The discussion is based on a closed-form solution by Amadei (1982) that connects the six components of a general initial stress field with the six stresses, the six strains, and the three displacement components at any point along the walls of a circular hole or within a hollow isotropic inclusion perfectly bonded to the hole. The borehole is located within an infinite linearly elastic, anisotropic continuous and homogeneous medium. No restrictions are made on the type or orientation of the anisotropy or on the orientation of the applied stress field with respect to the hole.

These closed form solutions can be used to solve the inverse problem, that is to calculate 3D in-situ stress field components from measurements of strains and displacements associated with existing overcoring techniques in one or several boreholes. The only limitation is that low modulus, thin-walled hollow inclusions or very low modulus, solid inclusions must be used in order to neglect the finite character of the diameter of the overcored sample. The proposed theory is applicable to the following types of overcoring measurements in anisotropic media (Fig. 1):

(1) changes in diameter of a pilot hole, and changes in strain recorded on the walls of that hole;

(2) changes in length between two points located in different cross sections at the inner surface of a hollow inclusion or within a solid inclusion. The inclusion is perfectly bonded to the walls of a pilot hole. These changes in length reduce to changes in diameter when the two points are located within the same cross section;

(3) strain measurements using strain rosettes embedded at any point and in any direction within a hollow or solid inclusion perfectly bonded to the walls of a pilot hole; and

(4) combination of (2) and (3).

GENERAL RESULTS FOR OVERCORING IN ANISOTROPIC MEDIA

1. Introduction

In the closed form solution by Amadei, the strains and changes in length induced by overcoring in any borehole are linear functions of the six components of the global in-situ stress field. Therefore, determination of these components requires that we set up a system of six independent equations from the results of six measurements. Any additional measurement may improve the accuracy of this determination.

In practice, several questions often arise concerning the measurements of strains, and displacements in anisotropic rocks:

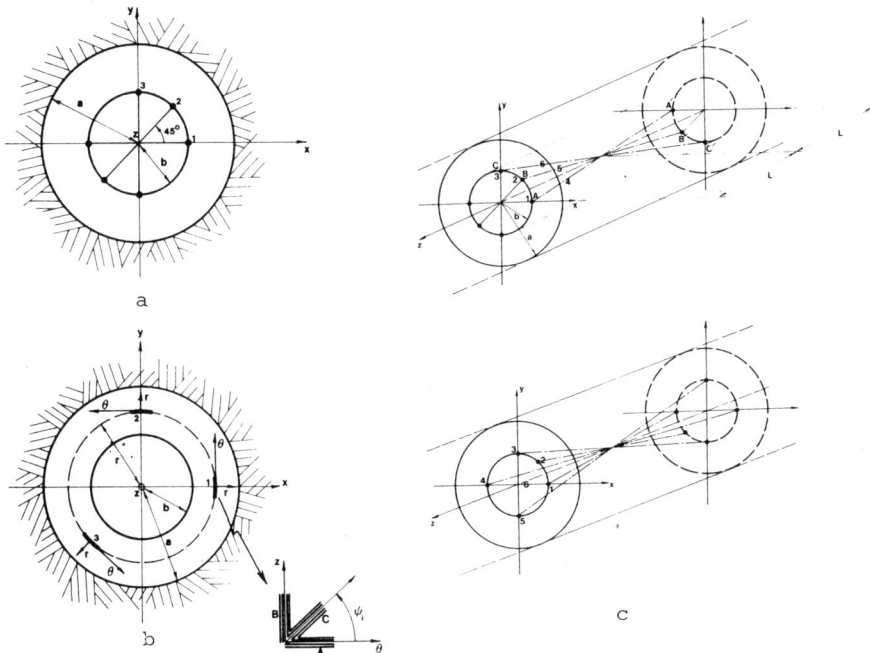

Fig. 1. (a) Diametral Measurements; (b) Strain Measurements; and (c) Combination of Measurements of Changes in Diameter and Changes in Length of Oblique Distances Within an Instrumented Hollow Inclusion

(i) How many independent measurements are in a single borehole and therefore how many boreholes do we need in order to determine the in-situ stress field?

(ii) Are the numbers influenced by the anisotropic character of the rock?

(iii) For a given set of measurements, how do anisotropy types and orientation influence the determination of the in-situ stress field? and,

(iv) How large an error is involved by neglecting rock anisotropy.

2. Anisotropic Solution

The influence of rock anisotropy on stress measurements has been addressed by several authors (Berry and Fairhurst, 1966; Berry, 1968; Becker and Hooker, 1967; Becker, 1968; Berry 1970; Niwa and

Hirashima, 1971; Hirashima and Koga, 1977). However, questions (i) to (iv) have never been answered completely for <u>anisotropic</u> media. The theory mentioned in the previous section is sufficient to address these questions.

Instead of focusing analysis on a specific method of overcoring, it will be assumed that strain and displacement measurements can always take place respectively within or at the inner surface of a <u>hollow inclusion</u> perfectly bonded to the surface of the pilot hole. This approach is very general since measurements on the rock surface or within a solid inclusion can then be considered as two limiting cases. The inclusion is also assumed to be soft in order to ignore the size of the overcoring diameter, and to reduce the tensile stresses at the rock/inclusion contact during overcoring.

In order to answer questions (i) to (iv), three computer programs were developed that make use of the closed form solutions mentioned in the Introduction. They calculate the least square estimates of the principal in-situ stress components and their orientation with respect to an arbitrary fixed global coordinate system XYZ from measurements of changes in strain within solid or hollow inclusions or changes in diameter and changes in length of oblique distances at the inner surface of hollow inclusions. These measurements take place in one or several boreholes inclined with respect to XYZ.

2.1 <u>Number of Independent Measurements in a Single Borehole</u>. Numerical examples using the computer programs mentioned previously have shown that if measurements take place on the inner surface of a hollow inclusion, there are at most <u>five</u> independent measurements of change in length of oblique distances such as AA' in Fig. 1c. This number is reduced to <u>three</u> if A and A' are within the same cross-section and changes in diameter are measured.

As far as measurements of change in strain are concerned, it is possible to obtain at least six independent measurements by orienting strain gages in different directions.

2.2 <u>Number of Boreholes</u>. The minimum number of boreholes required to calculate the complete state of stress depends primarily on the type of measurement used.

The complete state of stress can be determined within <u>a single borehole</u> by measuring changes in strain in six non-parallel strain gages or by combining measurements of change in length of oblique distances and diameters: for instance 5 oblique, 1 diametral; 4 oblique, 2 diametral; 3 oblique, 3 diametral. This is true regardless of the isotropic, anisotropic character of the rock and the orientation of the hole with respect to the planes and axes of elastic symmetry of the rock.

When changes in diameter are measured only, it would seem to be possible to use <u>two non-parallel boreholes</u> since there are at most

ROCK ANISOTROPY ON STRESS MEASUREMENTS

three independent measurements per hole as mentioned previously. Several numerical examples using the three computer programs have shown that the requirement for a third non-parallel borehole depends upon the following parameters:

- the angle between the two boreholes,
- the isotropic, anisotropic character of the rock,

and
- the orientation of both boreholes with respect to the planes and axes of elastic symmetry of the rock.

In these numerical examples, the rock was assumed to be either isotropic or transversely isotropic or orthotropic. The different cases are summarized in Table 1. It appears in this table that, in

TABLE 1. Number of Boreholes Required for the Determination of the Complete State of Stress from Diametral Measurements Only.

Angle between the two boreholes	Isotropic Anisotropic Character of the rock	Orientation of both boreholes with respect to the planes and axes of elastic symmetry of the rock	Requirement for a third borehole
any	Anisotropic	Each hole is not perpendicular to a plane of elastic symmetry	No
Non Perpendicular	Anisotropic	Only one hole is perpendicular to a plane of elastic symmetry	
any	isotropic	----	Yes
90°	Anisotropic	Only one hole is perpendicular to a plane of elastic symmetry	
90°	Anisotropic	Each hole is perpendicular to a plane of elastic symmetry	
any	Anisotropic	Both holes are within a plane of transverse isotropy	

comparison to the isotropic solution, the anisotropic character of the rock can reduce the number of boreholes to two. This applies when both boreholes are not perpendicular to any plane of elastic symmetry of the rock or when only one of the boreholes is perpendicular to a plane of elastic symmetry and is not perpendicular to the other.

When rock is isotropic, all planes and axes within the rock are now planes and axes of elastic symmetry and a third non parallel borehole is always required.

Amadei (1932) has shown that the possibility of reducing the number of boreholes from three in the isotropic solution to two in the anisotropic one is closely related to the anisotropic character of the rock. It was found that the use of two boreholes is limited to moderately or strongly anisotropic rocks only. For weakly anisotropic (almost isotropic) rocks, a third borehole is required. Practical applications and further discussion about the reduction of the number of boreholes from three to two in anisotropic rocks will be presented by the present authors in a future paper.

2.3 Influence of Rock Anisotropy on the Determination of the In-Situ Stress Field. Since it is difficult to draw general conclusions for all types of anisotropy, the present analysis is restricted to the transversely isotropic case only. One of several reasons for using this type of symmetry is that it can be visualized as a plane and can easily be represented geometrically.

In order to answer questions (iii) and (iv), let us consider the instrumented hollow inclusion shown in Fig. 2. The inclusion is assumed to have a ratio of outer radius to inner radius equal to 2, and a Young's modulus and a Poisson's ratio equal to 2.10^3 MPa and 0.3 respectively. Inclusion axes xyz are set parallel to XYZ of a global coordinate system. The inclusion is composed of three triple-gage strain rosettes embedded halfway within the inclusion. Each rosette consists of a gage A parallel to the tangential direction θ, a gage B parallel to the longitudinal direction z, and a gage C oriented at 225 degrees from A. Let us consider the following set of strain measurements in those gages.

Strain Rosettes	$10^6 \varepsilon_A$	$10^6 \varepsilon_B$	$10^6 \varepsilon_C$
1	-79	-86	44
2	-430	-85	-187
3	-52	-87	-210

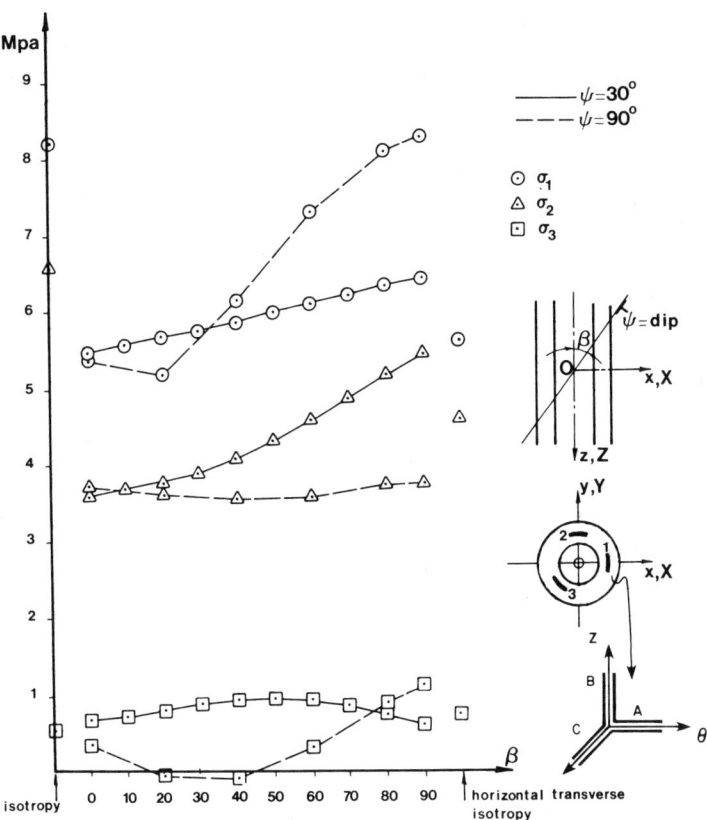

Fig. 2. Magnitude of the principal in-situ stress field components for different orientations of a plane of transverse isotropy.

The orientation of the anisotropy with respect to XYZ is defined by the strike β and the dip Ψ of the plane of transverse isotropy as shown in Fig. 2.

For the given set of strain measurements, the least square estimates of the principal components of the in-situ stress field $\sigma_1, \sigma_2, \sigma_3$ and their orientations with respect to X,Y,Z have been calculated for each of the following conditions

(1) The rock is isotropic with a Young's modulus and a Poisson's ratio equal to $4\ 10^4$ MPa and 0.25 respectively,

(2) The rock has a plane of transverse isotropy dipping at an angle Ψ of 30 or 90 degrees and striking at angles β varying from 0 to

90 degrees. Within the plane of transverse isotropy, the Young's modulus and Poisson's ratio of the rock are equal to $4\ 10^4$ MPa and 0.25 respectively. Perpendicular to this plane they are equal to $2\ 10^4$ MPa and 0.2 respectively with a shear modulus of $4\ 10^3$ MPa.*

(3) The plane of transverse isotropy is parallel to the plane XOZ. This case is termed "horizontal transverse isotropy" for sake of clarity.

The results are summarized in Fig. 2 as concerns the magnitude of $\sigma_1, \sigma_2, \sigma_3$ for Ψ equal 30 and 90 degrees, and in Fig. 3 for their orientation when Ψ is equal to 30 degrees only.

It appears through this numerical example, that the influence of the rock anisotropy on the determination of the in-situ stress field is non negligible especially for low values of the strike angle β. This influence is reduced as the plane of transverse isotropy turns and becomes perpendicular to the hole axis. When the plane of transverse isotropy is horizontal it also has a significant influence. For the present example, where the rock can be considered as strongly anisotropic, it appears that neglecting anisotropy by assuming that the rock is isotropic can create errors as large as 50 percent and up to 90 or 100 degrees for the magnitude and the orientation of the principal in-situ stress field, respectively. This could be largely reduced if the rock would be moderately or slightly anisotropic.

CONCLUSIONS

Several overcoring techniques have been proposed in the literature to measure in-situ stresses in rocks. For those techniques in which strains and displacements are recorded either on the walls of a pilot hole drilled at the end of one or several boreholes, or within instrumented hollow or solid inclusions perfectly bonded to the surface of the pilot hole, rock anisotropy can be included in the interpretation of measurements in-situ. Closed-form solutions proposed by Amadei (1982) allow one to do so. No restrictions are made on the type or the orientation of the anisotropy or on the orientation of the principal in-situ stress field with respect to the different boreholes where the measurements take place. These closed-form solutions are general enough to also apply to other types of overcoring measurements in anisotropic media. The only limitation is that the instrumented device within the pilot hole either does not interfere with the rock deformation or may be considered as a very low modulus solid inclusion or as a low modulus thin-walled hollow inclusion.

Although, the paper is concerned with the measurement of the absolute state of stress in-situ, the closed-form solutions previously referenced also apply for the determination of the components of any

---

* $\nu_{x'y'} = \nu_{x'z'} = 0.2$ where x'y'z' is a cartesian coordinate system attached to the anisotropy. The plane of transverse isotropy is perpendicular to the x' direction.

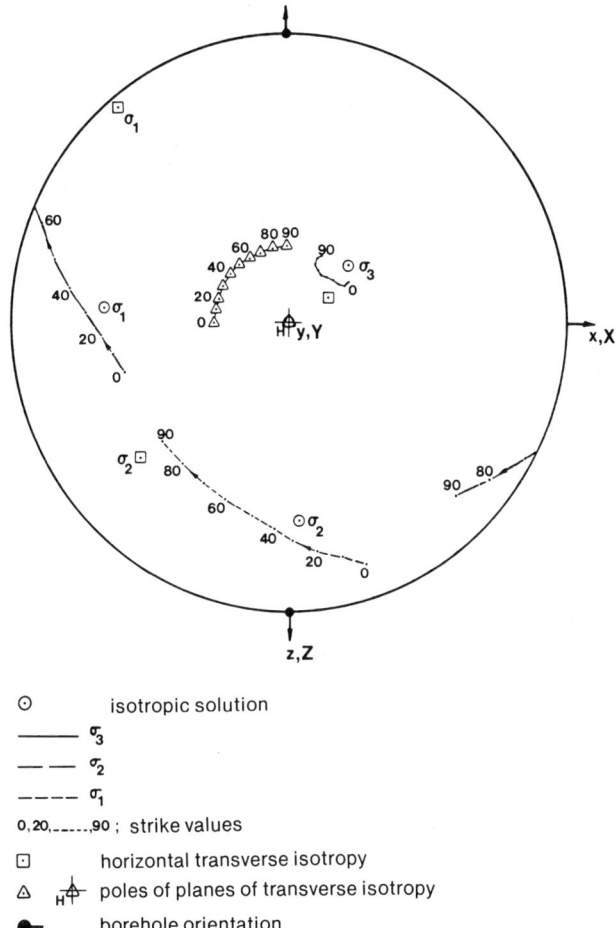

Fig. 3. Orientation of the principal in-situ stress field components for different strike angles β of a plane of transverse isotropy (Ψ = 30°) and for a horizontal transverse isotropy. Lower hemisphere stereographic projection.

change of the in-situ state of stress. In this case, the limitation previously mentioned is not required. In a single pilot hole, there are at most three independent measurements of diameter and five independent measurements of change in length of oblique distances recorded at the inner surface of a hollow inclusion or at the walls of the hole. It is possible to obtain six independent measurements of strain by orienting strain gages in different directions.

The minimum number of boreholes required to calculate the complete state of stress depends primarily on the type of measurement used. When changes in diameter are measured only, at least two non-parallel boreholes are needed. The requirement for a third non-parallel borehole depends upon the angle between the two boreholes, the isotropic anisotropic character of the rock and the orientation of both boreholes with respect to the planes and axes of elastic symmetry of the rock. Whenever those conditions are such that two boreholes can be used, the rock must be either moderately or strongly anisotropic.

Neglecting anisotropy by assuming that the rock is isotropic can create large errors when calculating the magnitude and the orientation of the in-situ stress field. An example dealing with a rock modelled as a strongly transversely isotropic medium has shown errors as large as 50 percent in the magnitude of the calculated principal stress field. Its orientation could be up to 100 degrees off from the isotropic solution.

A prerequisite for the applicability of the model used in this paper is that the directional character of the rock deformability must be known beforehand. If the latter is described using the constitutive relations of the theory of linear elasticity of anisotropic media, laboratory and in-situ testing is needed to determine the corresponding elastic constants.

REFERENCES

Amadei, B., 1982 "The Influence of Rock Anisotropy on Measurement of Stresses In-situ", Ph.D. Thesis, University of California, Berkeley, 472 pp.

Becker, R.M., 1968, "An Anisotropic Elastic Solution for Testing Stress Relief Cores", U.S. Bureau of Mines, RI. 7143.

Becker, R.M., and Hooker, V.E., 1967, "Some Anisotropic Considerations in Rock Stress Determination", U.S. Bureau of Mines, RI. 6965.

Berry, D.S., 1968, "The Theory of Stress Determination by Means of Stress Relief Techniques in Transversely Isotropic Medium", Tech. Rept. No. 5-68, Missouri River Division, Corps of Engineers, Omaha, Nebraska 68101, 36 pp.

Berry D.S., 1970, "The Theory of Determination of Stress Changes in a Transversely Isotropic Medium Using an Instrumented Cylindrical Inclusion", Tech. Rept. No. MRD-1-70, Missouri River Division, Corps of Engineers, Omaha, Nebraska 68101, 36 pp.

Berry, D.S. and Fairhurst, C., 1966, "Influence of Rock Anisotropy and Time Dependent Deformation on the Stress Relief and High Modulus Inclusion Techniques of In-Situ Stress Determination. Testing Techniques for Rock Mechanics, ASTM, STP. 402. Am. Soc. Testing Mats. pp. 190-206.

Hirashima, K. and Koga, A., 1977, "Determination of Stresses in Anisotropic Elastic Medium Unaffected by Boreholes from Measured Strains or Deformations", Proc. International Symposium on Field Measurements in Rock Mechanics, Zurich, Kovari K. (Editor), Vol. 1, pp. 173-182.

Niwa, Y. and Hirashima, K., 1971, "The Theory of the Determination of Stress in an Anisotropic Elastic Medium Using an Instrumented Cylindrical Inclusion", Memoirs of the Faculty of Eng., Kyoto Univ., Japan, Vol. 33, pp. 221-232.

Chapter 17

EXPERIENCE WITH THE SOLID INCLUSION STRESS

MEASUREMENT CELL IN COAL IN AUSTRALIA

by R. L. Blackwood

Senior Lecturer in Mining Engineering

The University of New South Wales

Sydney, New South Wales, Australia

ABSTRACT

The solid inclusion cell for absolute in situ stress measurement has been found to give reliable results in coal. The cell is described briefly, along with the methods of installing, overcoring and taking readings in underground coal mines. The validity of the cell performance is demonstrated by reference to the stress-relief curves obtained during overcoring and a three-dimensional B.I.E.M. analysis of the installation to test for stress anomalies deriving from the geometry of the overcoring groove-borehole-cell arrangement.

Case studies of two stress measurement projects using the cell are presented for coal mines in New South Wales, Australia.

INTRODUCTION

The Australian continent is subjected to an unusually large major horizontal stress trending approximately east-west. This is particularly important in the black coal districts of eastern Australia, where the horizontal stress causes severe strata control problems in many of the 93 underground mines throughout the states of New South Wales and Queensland. The horizontal stress also contributes to the occurrence of instantaneous outbursts of coal and gas in some of the mines. The problems are such that room and pillar mining is feasible only to a depth of 500 m, below which retreating longwalls are being increasingly used.

The magnitudes and directions of the regional stress field are perturbed locally by geological structure and so are not predictable at individual mine sites. Measurement of absolute stress has therefore been necessary, being carried out in a number of mines over the past 25 years using a range of instruments and methods. Most techniques require instrumentation to be installed in roof or floor rock.

The importance of measuring the stress in the coal rather than in the roof or floor rock is not generally appreciated. However it can readily be shown (Blackwood, 1976) that both the magnitude and direction of the normal stresses at the coal/rock interface are significantly altered when the principal virgin stresses are not orientated parallel and normal to the seam, for instance in dipping seams. It is thus desirable to measure the stress in the coal itself if it is the behaviour of the coal as a structural material that is to be studied. The only method which has been successfully used to measure three-dimensional stresses in coal is the solid inclusion cell, first described by Rocha and Silverio (1969) and subsequently developed for use in coal by the author.

CELL DESCRIPTION AND PERFORMANCE

Cell Description

The cell uses the stress-relief principle; in this case a solid epoxy cylinder is cemented into a borehole in coal and concentrically overcored by diamond drilling. The strain changes thus induced in the unstressed cylinder by the relaxing overcored annulus of coal are determined from a three-dimensional array of small biaxial strain gage rosettes (10 gages in all) encapsulated in the cylinder. The design details have been given in full elsewhere, along with the formulae for calculating the complete stress field in isotropic, elastic media from the observed strain changes (Blackwood, 1977). A diagrammatic view of the cell is shown in Fig. 1. All cells are routinely tested in a hydrostatic pressure cell before installation to check the integrity of electrical circuits and test the linearity of strain gage response.

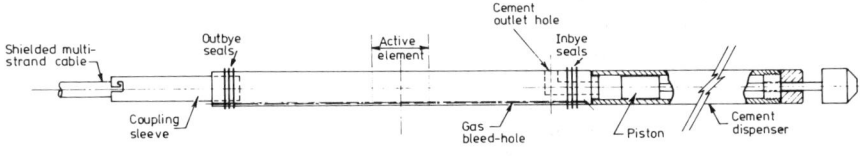

Fig. 1 General view of cell (not to scale).

Installation

The installation mechanism is generally similar to that adopted for other cemented inclusion devices. The cementing procedure bears further description here, however.

The cement used is a filled, two-component epoxy resin, Araldite

LC177N resin and HY956 hardener in proportions 10:1 by weight, which is pre-mixed immediately before installation. This takes about 16 hours to set hard. The cement is found to displace water, allowing a good bond to be achieved in very wet boreholes if necessary. The cement dispenser (Fig. 1) is filled and the piston placed on top of the cement so that the rod is fully extended. As the instrument is installed the leading edge eventually contacts the end of the borehole, depressing the piston and forcing cement through the outlet to fill the annulus between the seals at each end of the cell. The gas bleed-hole is included to allow any seam gas present to equilibrate to atmospheric pressure during the cement setting period.

The radial stress at the interface between the cell and the host coal (the "bond stress") may develop to magnitudes sufficient to cause tensile failure and loss of bonding unless the elastic modulus of the cell is much less than that of the coal (Duncan Fama, 1979). The modulus is therefore kept as low as possible with the result that bonding loss has not been found to be a problem in practice.

Overcoring

The technique developed for overcoring the cell in weak material such as coal involves the use of a stepped-crown diamond bit cutting a 105 mm diameter core, attached to a 165 mm diameter reamer and 1.52 m long thick-walled stationary inner tube core barrel. Although heavy and bulky it is necessary to use this type of barrel, since it is found that thin-walled trepanning bits and barrels used successfully in hard rock drilling cannot form a core in coal. Drilling speeds are kept as low as possible to avoid damaging the core.

Strain Measurement

It is important to monitor strain changes in the cell during overcoring to detect extraneous effects due to rock fracture, instrumentation failure and the like. The stress-relief curves obtained (graphs of strain change due only to stress-relief plotted against overcoring distance relative to each strain gage position) can be compared with theoretically derived curves for each strain gage orientation to test the validity of the stress-relief data before processing (Blackwood, 1978).

To do this, a shielded, multi-strand cable as shown in Fig. 1 carries 11 strain gage leads (10 active and one common) through the drill rods to be connected to the strain amplifier arrangement in half-bridge circuits with a common temperature compensation strain gage. In the two case histories described below, this arrangement consisted of a DC amplifier and switch box approved for operation in gassy coal mine atmospheres. The readings were taken on individual gages by the operator and recorded manually at 10 mm increments of overcoring distance, necessitating stopping the drill for about five minutes each time. The risk of jamming the core barrel in highly stressed coal was thus increased considerably. The strain amplifier used in present work gives instantaneous digital readout as each gage is selected, allowing strain changes to be monitored without stopping the drill. This improves the efficiency and reliability of the operation considerably.

Validation of Cell Performance

Laboratory Testing. A series of laboratory tests was carried out during the early stages of cell development (Blackwood, 1977). Uniaxial compressive testing of cells cemented into blocks of material ranging from steel to rock and rock-like mortars indicated that the cell performed as predicted.

Field Testing. In situ installations in coal gave indications of the reliability of operation of the cell, which was found to be excellent in view of the difficult material involved. No absolute stress values were available at that time so that determinations of precision were not possible. However, the stress-relief curves compared well with the theoretically predicted curves in virtually all cases, showing that the cell and the coal annulus were intact and the data valid.

B.I.E.M. Study. The plane strain analysis of the behaviour of the cell assumes that the cell is an infinitely long cylinder having an exactly continuous interface with the host material, which is an infinite elastic mass. Field stresses are assumed to be imposed at infinity.

A study was made of the three-dimensional geometry of the overcored cell, modelled on field experience, by the boundary integral equation method of stress analysis (B.I.E.M.) to determine the effects of the installation hole, the end effects of the finite cylinder, and the influence of the overcoring groove, on the plane strain assumptions (Blackwood, 1982).

It was found that the stress disturbance due to the installation hole end is negligible. The end effects of the cylindrical cell itself were found to extend to a distance of up to 5 radii in from the ends of the cell in certain stress fields. The strain gaged section at the centre of the cell is unaffected in the design adopted and it can be taken that a condition of plane strain applies there.

The modelled configuration showed the leading edge of the overcoring groove 2.5 radii beyond the inner end of the cell. In this position, the stress disturbance extended up to 6 radii into the cell, indicating that, in practice, overcoring should be extended a further distance before the cell is unaffected. Alternatively, the total length of the cell might be increased, keeping the strain gaged section at the central part of the cell length.

The effect of the outer end of the overcored annulus was found to have no effect on the stress field, with the practical implication that the instrument can be located close to the outer end of the installation borehole without affecting its performance. In addition, the pre-existing stress fields are removed completely in the annulus by the overcoring process, except for the small stresses retained by the reinforcing effect of the cell cemented in place.

CASE STUDY 1: APPIN COLLIERY, N.S.W.

Location

Appin Colliery is located some 60 km south-west of Sydney, the capital of New South Wales. Black, sub-bituminous coal of Permian age is mined from the Bulli seam, which averages 2.7 m thick in the test area. This is the uppermost of four seams worked in the region. The seam is relatively flat-lying and is overlain by about 485 m of Triassic sediments, mainly massive sandstones and shales.

Instrumentation

Two underground sites were located 540 m apart in a retreating longwall development road, Fig. 2. A total of 12 cells were installed in four horizontal boreholes, approximately at mid-seam height. (Two cells failed during overcoring, both due to poor installation technique. The failure rate has been reduced in subsequent work). The stress-relief curves were plotted for each strain gage in each cell as standard procedure. Results were only accepted on the basis of favourable interpretation of these curves.

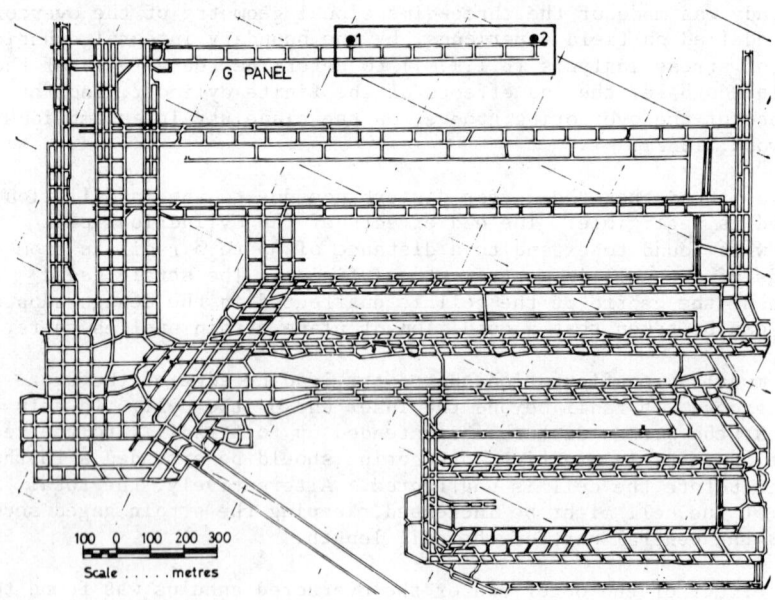

Fig. 2   Appin Colliery test site locations 1 and 2.

Measured Stresses

The stress profiles of all measured results are shown in Fig. 3. The major horizontal stress is 4.3 greater than the vertical stress, on average. This value has since been confirmed in stress measurements by others in an adjacent mine. The average directions of the horizontal stresses are shown as stress ellipses in Fig. 3d. The results from individual cells were reasonably consistent at each site and between sites.

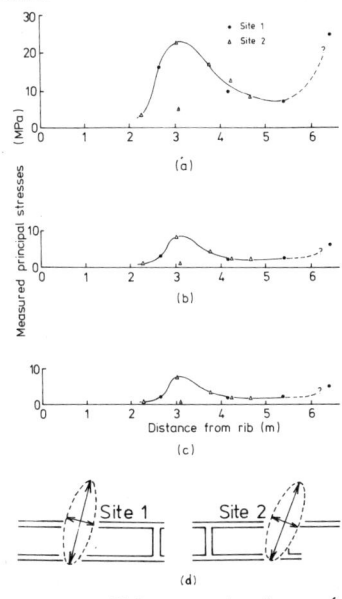

Fig. 3 Measured stress profiles at Appin: (a) Maximum horizontal (b) Vertical (c) Minimum horizontal (d) Horizontal stress ellipses (after Blackwood et al., 1976).

CASE STUDY 2: KANDOS No. 3 COLLIERY, N.S.W.

Location

Kandos No. 3 Colliery is a small mine located 162 km north-west of Sydney. It is included in the present discussion because of its unusual geological situation. As for Appin, the material is black, sub-bituminous, Permian coal, but of relatively inferior quality, mined for local specialized use as fuel in cement manufacture. The Lithgow seam, which is extensive throughout the region, at Kandos is perched above the surrounding country, occurring under a mesa-like formation such that it crops out around the periphery of the overlying mountain. As a result the seam is free of the lateral constraint that applies at most other Australian coal mine locations. The immediate depth of cover averages 160 m at the test sites, where the seam is 2.0 m in thickness.

Instrumentation

Two underground sites were located 970 m apart, where shown in Fig. 4, in a room and pillar mining layout which follows the cliff line of outcrop. Altogether nine cells were installed in a total of three horizontal boreholes at about mid-seam height. Stress-relief curves were plotted and checked against prototype curves for each strain gage, to be used as a selection/rejection criterion. One cell did not give consistent results on this basis and was disregarded accordingly.

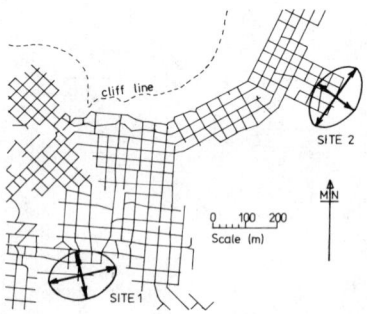

Fig. 4 Kandos Colliery test site locations and stress ellipses.

Measured Stresses

The stress profiles are shown in Fig. 5. The regionally dominant horizontal stress experienced elsewhere in the region is much reduced due to the stress relief at the outcrop such that the mean ratio of major horizontal stress to vertical stress is about 1.7. At a depth of 160 m the ratio would be expected to be much higher in fully constrained conditions.

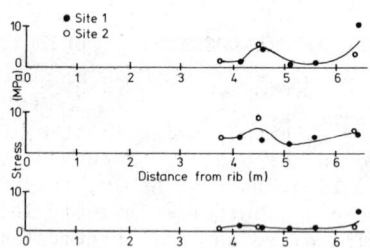

Fig. 5 Measured stress profiles at Kandos. (a),(b),(c) as for Fig. 3.

CONCLUSIONS

The limitations of the solid inclusion cell are appreciated, especially with regard to the application of an isotropic, elastic solution to rock mechanics problems in a clearly heterogeneous, possibly anisotropic and probably inelastic material such as coal.

Despite the shortcomings it has been demonstrated that the cell is useful as a practical device in routine coal mining geomechanical work. It is relatively quick and inexpensive to use, and displays few problems in the field.

The author is presently engaged in developments of the cell, in particular a program of controlled laboratory overcoring tests involving a range of commercially available instruments along with the solid inclusion cell to arrive at recommendations for the selection of suitable instrumentation in different rock conditions. Field site comparisons are proceeding with other instrument types. Trials of the application of the solid inclusion cell as a stress monitoring device is also in hand in a longwall coal block.

ACKNOWLEDGEMENTS

The author would like to thank the management of Kandos Colliery for permission to publish the results of work carried out as part of a research contract with The University of New South Wales.

REFERENCES

Blackwood, R.L., 1976. "The Development of a Method for Measuring In Situ Stress in Coal", Ph.D. thesis, Macquarie University, Sydney, 494 pp.

Blackwood, R.L., 1977, "An Instrument to Measure the Complete Stress Field in Soft Rock or Coal in a Single Operation", Proc. Int. Symp. on Field Measurements on Rock Mechanics, Vol. 1, Zurich, pp.137-150.

Blackwood, R.L., 1978, "Diagnostic Stress-Relief Curves in Stress Measurement by Overcoring", Int. J. Rock Mech. Min. Sci. & Geomech. Abstr., Vol. 15, pp. 205-209.

Blackwood, R.L., 1982, "A Three-Dimensional Study of an Overcored Solid Inclusion Rock Stress Instrument by the Boundary Integral Equation Method", Proc. Fourth Finite Element Conf. in Australia, Melbourne.

Blackwood, R.L., Hargraves, A.J. and McKay, J., 1976, "Absolute Stress Investigations in Coal at Appin Colliery, New South Wales". Proc. Int. Soc. Rock Mech. Symp. on Investigation of Stress in Rock - Advances in Stress Measurement, Sydney, pp. 17-22.

Duncan Fama, M.E., 1979, "Analysis of a Solid Inclusion In Situ Stress Measuring Device", Proc. Fourth Int. Soc. Rock Mech. Congr., Vol.2, Montreux, pp. 113-120.

Rocha, M. and Silverio, A., 1969, "A New Method for the Complete Determination of State of Stress in Rock Masses", Geotechnique, Vol. 19, pp. 116-132.

Chapter 18

IN SITU STRESSES IN THE KLERKSDORP GOLD MINING DISTRICT,
SOUTH AFRICA - A CORRELATION BETWEEN GEOLOGICAL STRUCTURE
AND SEISMICITY
by
N.C. GAY and P.J. VAN DER HEEVER

Assistant Director, Mining Operations Laboratory,
Chamber of Mines of South Africa,
Johannesburg, South Africa

Seismic Research Officer, Buffelsfontein G.M. Co. Ltd.,
Klerksdorp, South Africa

ABSTRACT

The Klerksdorp gold mining district, South Africa, experiences a relatively high level of seismicity with many seismic events of magnitude greater than 4. These large events appear to have foci close to fault planes, most of which displace the strata by several hundred metres over a wide area.

To help understand the mechanism of these large mining-associated seismic events, in situ stress measurements were made at 4 sites using "CSIR doorstopper" strain relief techniques. The sites were chosen close to, or some distance from fault planes and other structures. The results show that at all 4 sites a relatively large horizontal stress is oriented approximately northwest. The magnitude of this stress ranges from 0,78-1,0 of the vertical stress. A marked horizontal stress anisotropy, varying from 13-30 MPa is also present.

The dominant structural features are large, north-east striking normal faults which extend across the entire length of the goldfield. Displacements across these faults vary from a few centimetres to several hundred metres. Bedding plane faults across which large displacements may occur are also common as are north and north-west striking dykes. The majority of the faults strike at approximately 90° to the maximum horizontal stress and at all sites a large compressive stress plunges in the direction of dip of the strata and fault planes. Thus a significant shear stress may act across suitably oriented planes causing them to be in a state of potentially unstable equilibrium. This abnormal stress situation would exist along the entire length of the fault plane. As a result, disturbance due to mining activity at one point on the fault plane could result in the stored strain energy being released over a wide area precipitating a seismic event of significant magnitude.

INTRODUCTION

The Klerksdorp gold mining district (Figure 1) which comprises four large mines, covering a total area of 200 km^2 is situated approximately 160 km south-west of Johannesburg. The gold bearing horizons are confined to a fault bounded, north-east striking elliptical basin, situated on the northern limb of the main Witwatersrand syncline. Internally, the basin is severely disrupted by large normal faults, the majority of which strike parallel to the basin axis. Displacements across the faults range from a few centimetres to 1200 metres. Because of the predominance of normal faulting, graben and horst structures are very common, as are bedding plane faults.

Large scale mining activity began in the area in 1942 and for several years no untoward seismic activity was encountered. However as mining went deeper, increasing mine related seismicity occurred and by 1970, it was clear that the goldfield was experiencing an abnormal number of large events, compared to the other mining regions in South Africa.

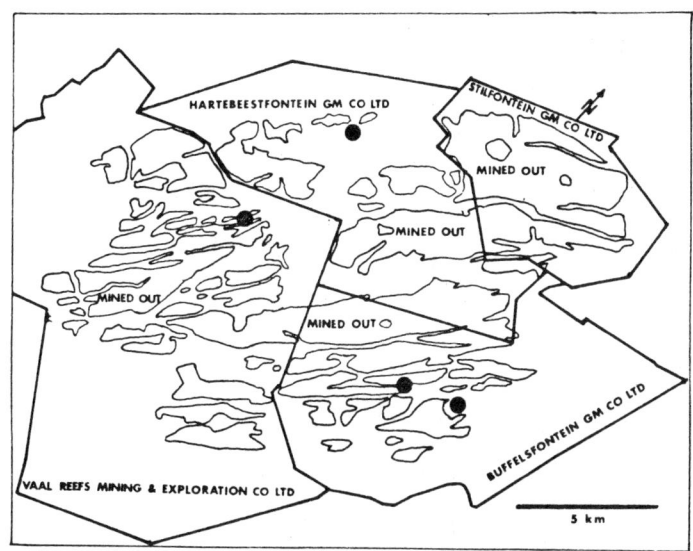

FIGURE 1 The Klerksdorp Mining District showing the four in situ stress measuring sites (black dots). The shallow site on Buffelsfontein is marked by the dot on the left of the mine area.

A seismic network, comprising 25 geophones, covers the district and is able to locate the foci of events to within ±50 metres. Thus, within this error, it is possible to correlate foci with seismically hazardous structures.

To investigate this correlation further, a programme of in situ stress measurements was undertaken. Four sites were chosen, two of which were close to (within 50 metres) large faults, two were close to dykes and one was in an area relatively far from any major structures. All sites were sufficiently far removed from areas of active mining to enable the virgin state of stress to be determined.

C.S.I.R. "doorstopper" strain cells were used to make strain relief measurements in three to five non-parallel boreholes at each site. Several measurements were made in each borehole at depths ranging from 6 to 15 metres. From these data, average strain reliefs were calculated for each borehole. In addition, representative samples were taken for the determination of the Young's Modulus (E) and Poisson's Ratio (ν). These elastic moduli were used to compute the stress concentration factors, following Coates and Yu's (1970) method, and, with the average strain reliefs, the components of the stress tensor.

RESULTS

The results of the measurement programme are summarized in Table 1 and plotted on lower hemisphere, equal area, stereographic projections together with relevant structural information in Figure 2. The orientations of the maximum and minimum principal horizontal stresses are also plotted.

TABLE 1 In situ stress data - Klerksdorp Mining District

Vaal Reefs   Depth 1868 m, overburden stress 52 MPa
                  Vertical and horizontal stresses:
                  σ_V = 53 MPa, σ_{H1} = 53 MPa, σ_{H2} = 39 MPa.

                  Principal stresses:
                  σ_1 = 56 MPa, plunges 44° towards 322°
                  σ_2 = 50 MPa, plunges 42° towards 140°
                  σ_3 = 39 MPa, plunges 1° towards 050°

Buffelsfontein   Depth 2166 m, overburden stress 61 MPa

                  σ_V = 67 MPa, σ_{H1} = 56 MPa, σ_{H2} = 26 MPa
                  σ_1 = 70 MPa, plunges 71° towards 148°
                  σ_2 = 54 MPa, plunges 19° towards 344°
                  σ_3 = 23 MPa, plunges 5° towards 252°

Hartebeestfontein Depth 2340 m, overburden stress = 64 MPa
σ_v = 66 MPa, σ_{H1} = 53 MPa, σ_{H2} = 40 MPa
σ_1 = 68 MPa, plunges 76° towards 144°
σ_2 = 52 MPa, plunges 11° towards 288°
σ_3 = 39 MPa, plunges 8° towards 020°

Buffelsfontein Depth 2560 m, overburden stress = 67 MPa
σ_v = 62 MPa, σ_{H1} = 48 MPa, σ_{H2} = 34 MPa
σ_1 = 67 MPa, plunges 67° towards 166°
σ_2 = 46 MPa, plunges 17° towards 300°
σ_3 = 31 MPa, plunges 16° towards 035°

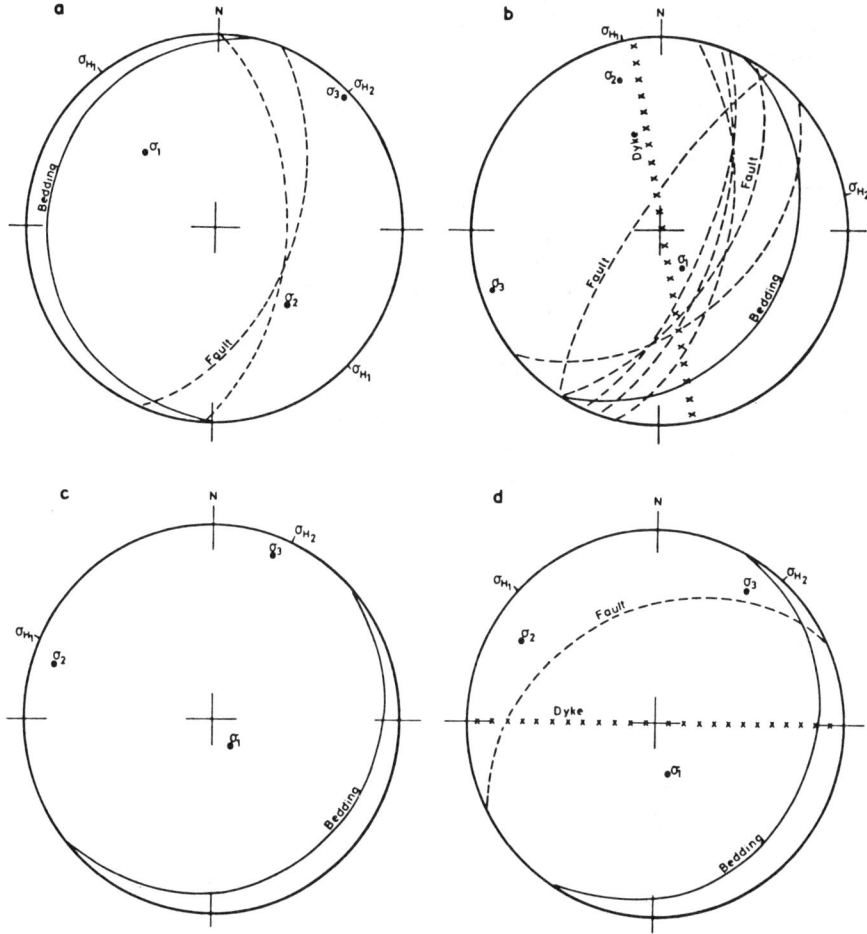

FIGURE 2 Equal area, lower hemisphere projections showing the orientations of the principal stresses and geological structures at each site: a) Vaal Reefs, b) Buffelsfontein (shallow), c) Hartebeestfontein, d) Buffelsfontein (deep)

DISCUSSION

The results at each site agree surprisingly well, particularly when one considers the different geological environments. Of the four sites, that on Hartebeestfontein (Figure 2c) has no significant structure near it. At Vaal Reefs (Figure 2a) measurements were made within 50 metres of two aseismic faults, one of which had a throw of 300 metres. The Buffelsfontein deep site (Figure 2d) was very close to a dyke which is displaced by a fault with a throw of 55 metres, neither structure being seismically hazardous. However, one of the faults close to the Buffelsfontein shallow site (Figure 2b) was known to be associated with seismic activity. Several other small faults and a dyke also occurred close to this site.

The most striking feature of the in situ stress data is the relative consistency in the orientations (c.f. Figure 3) with maximum principal stresses plunging steeply south south east, intermediate principal stresses plunging at low angles towards the north-west and minimum stresses being near horizontal and striking north-east south-west. This is not quite true since the Vaal Reefs data do not exactly conform, but there is very little difference between the magnitudes of the maximum and intermediate principal stresses. The maximum horizontal stress is oriented north-west (average strike 318°) and at all sites there are marked differences in horizontal stress magnitudes, the average value being 19 MPa with a maximum value of 30 MPa at the Buffelsfontein shallow site.

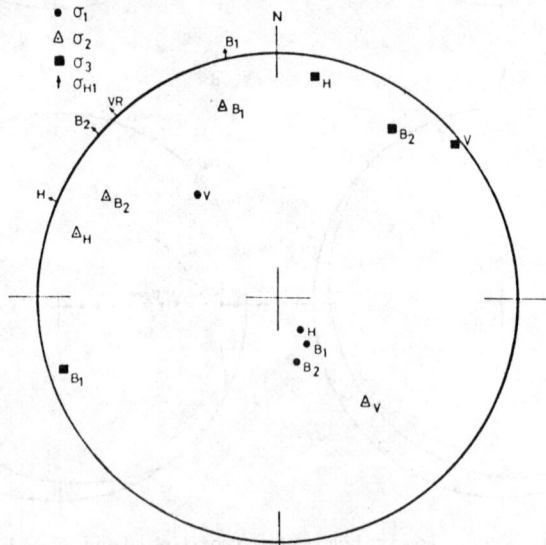

FIGURE 3   Equal area, lower hemisphere projection summarizing the in situ stress data for all four sites in the Klerksdorp area

Comparing Figures 2 and 3, it is seen that a large component of stress plunges in the direction of dip of both bedding and faults at all four sites. This implies that pre-mining shear and normal stresses may act on potential planes of weakness. The magnitudes of these stresses are given in Table 2, from which it appears that movement along either plane of weakness, is most likely to occur at the Buffelsfontein shallow site where the largest shear stress and smallest normal stress are found. This may explain the presence of a seismically active structure at this site but not at the other sites.

TABLE 2 Normal (σ_n) and shear (τ) stresses acting on bedding plane and normal fault surfaces, calculated using the Mohr construction

Site	Bedding plane σ_n	τ	Average fault plane σ_n	τ
Vaal Reefs	52 MPa	3 MPa	52 MPa	7 MPa
Buffelsfontein (shallow)	56 MPa	16 MPa	43 MPa	15 MPa
Hartebeestfontein	64 MPa	10 MPa	–	–
Buffelsfontein (deep)	60 MPa	12 MPa	66 MPa	7 MPa

However, the magnitudes of the measured stresses and, in particular, the stress differences and resultant shear stresses, are not sufficiently large to cause failure and need to be supplemented by mining induced stresses for this to occur. Even then, it is unlikely that this combined stress state would provide sufficient energy to the system to precipitate a large seismic event. In fact, Ryder et al. (1978) have shown theoretically, that slip along a fault plane located ahead of an advancing mining excavation, will only release sufficient energy to cause events up to magnitude 3.

How then do we envisage the generation of very large mining associated seismic events? Firstly, the large, regional horizontal stress acting at approximately right angles to the fault planes is capable of significantly enhancing the volume of strain energy stored adjacent to the fault plane over much of its strike length. This relatively highly stressed region is then disturbed by mining towards the fault plane at one or more localities along its strike, thus reducing the normal stress acting on the fault plane, to such an extent that the fault becomes unstable and sliding occurs at one or more isolated points. However, simultaneously, the strain energy stored within a much larger volume of rock is released precipitating a seismic event of significant magnitude.

The main factor in this scenario is the large, regional, northwest oriented horizontal stress, the existence of which may reflect the presence of residual and/or active tectonic stresses. Residual stresses associated with the main fault system, would

have a maximum stress near vertical and horizontal intermediate
and minimum stresses striking north-east and north-west
respectively, that is at right angles to the measured horizontal
stress components. However, the magnitudes of these horizontal
stresses are abnormally large compared to sites elsewhere in the
Witwatersrand goldfields and this may indicate the presence of
some residual stress due to faulting.

The other prominent structures in the region are dykes, of
which two main sets occur, striking approximately north and northwest respectively (Van der Heever, 1982). Gay (1979) has shown
that dykes may be very highly stressed bodies of rock in which
residual components of both tectonic and thermal palaeostresses
are present. In particular, the pre-mining state of stress within
the dyke would be lithostatic and that in the adjacent rock would
have a minimum principal stress oriented at right angles to the
plane of the dyke. Examination of Figure 2b, which shows the geological structure and orientations of the principal stresses at
Buffelsfontein shallow site, indicates that residual components of
stress associated with a dyke may well be reflected in the
measured state of stress.

However, the dominant factor in understanding the observed
state of stress in the Klerksdorp district is the existence over
much of Southern Africa of a system of horizontal stresses, with
the maximum principal stress oriented north-west (Gay, 1980).
These stresses presumably result from active forces generated
within the crust during plate tectonic movements and are represented by the north-west maximum horizontal stress reported here.

REFERENCES

Coates, D.F. and Yu, Y.S. 1970. "A note on the stress concentrations at the end of a cylindrical hole". *International Journal of Rock Mechanics and Mining Sciences*, Vol. 7, pp.585-588.

Gay, N.C. 1979. "The state of stress in a large dyke on E.R.P.M.,
Boksburg, South Africa". *International Journal of Rock
Mechanics and Mining Sciences*, Vol. 16, pp.179-185.

Gay, N.C. 1980. "The state of stress in the plates". *Dynamics of
Plate Interiors*. A.W. Bally et al. eds., pp.145-153.
American Geophysical Union, Washington D.C.

Ryder, J.A., Ling, T., and Wagner, H., 1978, Unpublished Report,
Chamber of Mines of South Africa, Johannesburg.

Van der Heever, P. 1982. M.Sc. thesis, Randse Afrikaans Universiteit, Johannesburg (in preparation).

Chapter 19

DETERMINATION OF THE INSTANTANEOUS SHUT IN PRESSURE
FROM HYDRAULIC FRACTURING DATA AND ITS RELIABILITY
AS A MEASURE OF THE MINIMUM PRINCIPAL STRESS.

by J. Mark Gronseth

Senior Research Engineer
Esso Resources Canada Limited
Calgary, Canada

ABSTRACT

The instantaneous shut in pressure often times is not a well defined feature of pressure-time records from in situ stress determinations by hydraulic fracturing. As the applications of in situ stress data become more sophisticated, the greater the need for an unambiguous method with which to identify the instantaneous shut in pressure becomes.

Controlled experiments suggest that the instantaneous shut in pressure should be equated with the pressure at the inflection point of the pressure-time record following shut in. A method for determining this value is presented. Often times the minimum instantaneous shut in pressure determined after several pressurization cycles is the best estimate of the stress acting perpendicular to the fracture plane.

INTRODUCTION

Over the years that hydraulic fracturing has been used as a means of in situ stress determination it has evolved from an experimental technique to an engineering tool. It is being used with increasing frequency for practical applications within the mining (Enever, 1982), heavy civil construction (Haimson, 1978), and petroleum industries (Kry and Gronseth, 1982), as well as for scientific investigations (Zoback et al., 1980), which are increasing our understanding of tectonic processes in the Earth's crust.

Regardless of the application, in situ stress determinations by hydraulic fracturing rely on the assumption that the instantaneous shut in pressure is equal in magnitude to the stress acting perpendicular to the plane of the induced fracture. While it can be shown experimentally that this assumption is justified, the instantaneous shut in pressure often times is not a well defined feature of pressure-time records from hydraulic fracturing tests. Identification of the instantaneous shut in pressure can be a highly subjective process. Several investigators interpreting the same pressure-time records can choose instantaneous shut in pressures differing by several MegaPascals.

It has been observed that instantaneous shut in pressures obtained from multiple pressurizations of a zone do not always remain constant from cycle to cycle. The instantaneous shut in pressure obtained from the first pressurization cycle of a zone can significantly overestimate the magnitude of the stress acting perpendicular to the fracture plane.

The purpose of this paper is to present the results of controlled laboratory and field tests which were performed to assess the reliability of the instantaneous shut in pressure as an estimate of the minimum stress acting perpendicular to the wellbore axis.

DETERMINATION OF THE INSTANTANEOUS SHUT IN PRESSURE

Experience reported in this paper suggests that for low flow rate hydraulic fracturing (<50 l/min), the instantaneous shut in pressure should be equated with the pressure at the inflection point in the pressure-time record following shut in.

A series of laboratory tests was performed to determine whether reliable estimates of the stress acting perpendicular to the plane of an induced fracture could be determined from hydraulic fracturing data. Details of the testing program are given in Gronseth and Detournay (1979).

Data was obtained by hydraulically fracturing a 38 cm x 38cm x 38 cm cube of Charcoal Gray Granite, loaded in biaxial compression. Specimen geometry is shown in Figure 1. Prior to testing, a fracture was propagated from the center hole into two fracture arrest holes located near the edges of the specimen. This was done so that multiple runs could be made on a single block. The internal stress distribution was modelled using the Displacement-Discontinuity method. It was found that the fracture arrest holes had no significant influence on the stress distribution at distances less than approximately 2.5 borehole diameters from center hole.

DETERMINATION OF SHUT IN PRESSURE

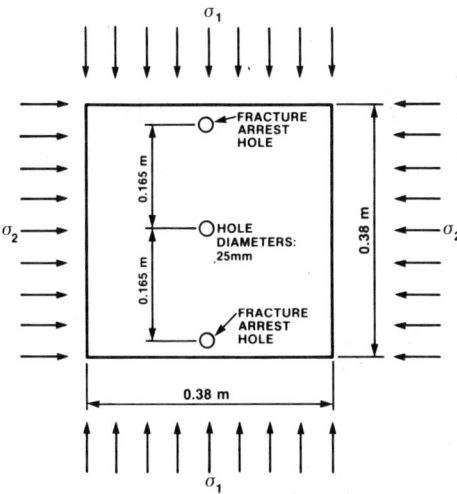

Fig. 1: Specimen Geometry

A simple graphical technique was used to determine the instantaneous shut in pressure. It consists of drawing one or two tangent lines to the pressure-time record immediately following shut in. The tangent lines allow the inflection point in the pressure-time record to be determined.

Two pressure-time records from these tests are shown in Figure 2. The tangent lines used to determine the instantaneous shut in pressure, along with the stresses which were applied perpendicularly to the fracture planes are shown in this figure. Line t_1 is tangent to the pressure-time record immediately following shut in. When the deviation of the pressure-time record from this tangent line is pronounced, the pressure at which the pressure-time record departs from the tangent line is the instantaneous shut in pressure. When this departure is not so well pronounced, a second line, t_2 drawn tangent to the pressure-time record at a slightly greater time can be used to determine the inflection point. The intersection of the two tangents gives the instantaneous shut in pressure.

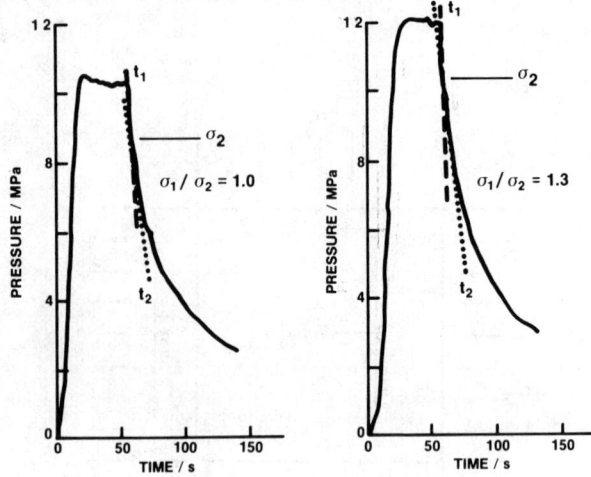

Fig. 2: Pressure-Time Records from Laboratory Tests Showing Determinations of the Instantaneous Shut In Pressure Using Tangent Lines.

Figure 3 is a plot of the minimum stress determined by the instantaneous shut in pressure vs. the applied minimum stress. Tests were performed with applied minimum stresses ranging from 1.7 MPa to 12.1 MPa. Ratios of the maximum to minimum applied stresses ranged from 1.0 to 6.0

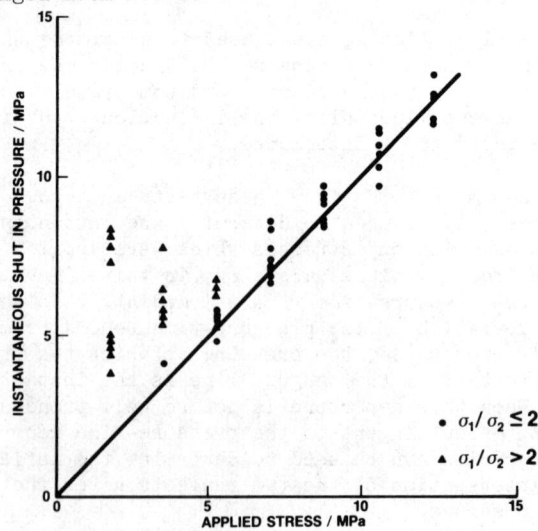

Fig. 3: Results of Laboratory Tests

Results shown in Figure 3 were obtained by re-opening a previously created fracture which was subjected to a variety of stress states. The fracture arrest holes were open to the atmosphere and very large pressure gradients existed along the fracture. These tests show that even under extreme conditions, the instantaneous shut in pressure, as defined by the inflection point in the pressure-time record following shut in, can give very reasonable estimates of the minimum stress.

ESTIMATION OF THE MINIMUM STRESS ACTING PERPENDICULAR TO THE WELLBORE AXIS.

Examination of many pressure-time records obtained from multi-cycle hydraulic fracturing shows that instantaneous shut in pressures do not always remain constant from cycle to cycle. In these cases measurable decreases in the instantaneous shut in pressure after each cycle is frequently observed. After a number of pressurization cycles the instantaneous shut in pressures usually approach a constant value.

Results obtained from tests performed at a location where stresses determined by hydraulic fracturing could be compared with stresses determined by other means, suggest that the minimum instantaneous shut in pressure provides the best estimate of the stress acting perpendicular to the fracture plane.

Tests were performed in a mine in Northeastern Minnesota at a depth of approximately 550 meters. The U.S. Bureau of Mines used the overcoring technique to determine the in situ stresses. These values, along with the calculated overburden stress, were used for comparison. Details of the testing program can be found in Gronseth and Detournay (1979).

TABLE 1

Comparison Between Minimum Stresses
Determined by Hydraulic Fracturing and Overcoring

HOLE	SECTION/m	PERCENT DIFFERENCE USING: First Shut in Press.	Min. Shut in Press.
10105	6.4	66.8	3.2
10105	7.6	38.6	7.7
10106	7.6	42.3	0
10106	11.6	50.5	19.1
10107	11.6	23.8	2.6

Table 1 lists the percent differences between the appropriate overcoring and hydraulic fracturing results. Results from the first pressurization cycles consistently overestimated the overcoring results, while those from the latter cycles were in much better agreement.

This behaviour is likely related to fracture size. It has been shown (Roegiers, 1975) that even in very low permeability rock masses rather significant volumes of fluid can leak off prior to fracture initiation as the tangential stresses acting around the wellbore become less compressive due to the applied pressure.

Not all of the fluid injected into the wellbore goes to the creation of new fracture surfaces. It takes a larger volume of fluid to create a fracture of sufficient size to be away from the influence of the wellbore than would be calculated assuming that the rockmass is impermeable.

The volume of fluid required to create a fracture large enough to allow reliable estimates of the stress acting perpendicular to the fracture plane to be made can rarely be predicted a priori. It is suggested that in situ stress determinations by hydraulic fracturing be performed at a constant flow rate of less than 50 l/min, using repeated low volume pressurization cycles. The zone should be repeatedly pressurized until changes in the instantaneous shut in pressure between cycles become small.

Instantaneous shut in pressures may not change between cycles, in which case the test can be terminated. Progressively propagating the fracture in this manner ensures that the created fracture will be large enough to be away from the influence of the wellbore. Judgement is required in interpreting the results since the observed pressure response could be associated with a bypassed packer, leaky check valve or tool joint, the fracture intersecting a pre-existing fracture or extending into a lower stress zone.

CONCLUSIONS

Controlled laboratory tests demonstrate that the instantaneous shut in pressure, when defined as the pressure corresponding to the inflection point in the pressure-time record following shut in, can be used to reliably estimate the minimum stress acting perpendicular to the wellbore axis.

Controlled field tests have shown that instantaneous shut in pressures are not necessarily reproduced by repeated pressurizations. The first instantaneous shut in pressure from a series of pressurization cycles can be significantly higher than those obtained from later cycles. Often times the minimum instantaneous shut in pressure from a number of pressurization cycles is the best estimate of the stress acting perpendicular to the fracture plane. Judgement is required in interpreting the results, since observed pressure responses could be associated with the fracture interacting with pre-existing fractures, lower stress zones or equipment malfunctions.

Low rate, low volume cyclic hydraulic fracturing is suggested as the appropriate method with which to perform in situ stress determinations by hydraulic fracturing.

ACKNOWLEDGEMENTS

The laboratory and mine experiments were performed by the author while in the Department of Civil and Mineral Engineering, Institute of Technology, University of Minnesota. This work was supported by the U.S. Department of the Interior, U.S. Geological Survey, under contract number 14-08-0001-16768. The support of the U.S.G.S. and the University of Minnesota, in particular Dr. W.D. Lacalanne and Dr. C. Fairhurst as well as the opportunity provided by Esso Resources Canada Limited to present this work is gratefully acknowledged.

REFERENCES

Enever, J.R. and Wooltorton, B.A., 1982, "Experience With Hydraulic Fracturing as a Means of Estimating In Situ Stress in Australian Coal Basin Sediments," U.S. National Committee For Rock Mechanics Special Report, In Press.

Gronseth, J.M. and Detournay, E., 1979, "Improved Stress Determination Procedures By Hydraulic Fracturing," Final Report, Contract No. 14-08-0001-16768, May, U.S. Geological Survey, Menlo Park, California.

Haimson, B.C., 1978, "The Hydrofracturing Stress Measuring Method and Recent Field Results," _International Journal of Rock Mechanics and Mining Sciences_, Vol. 15, No. 2, pp. 167-178.

Kry, P.R., and Gronseth, J.M., 1982, "In Situ Stresses and Hydraulic Fracturing in the Deep Basin," _Proceedings_, 33rd Annual Technical Meeting of the Petroleum Society of CIM, Calgary, Alberta, Canada.

Roegiers, J.C., 1975, "The Development and Evaluation of a Field Method for In Situ Stress Determination Using Hydraulic Fracturing," Technical Report MRO-1-75, March, Missouri River Division Corps of Engineers, Omaha, Nebraska.

Zoback, M.D., Tsukahara, H. and Hickman, S., 1980, "Stress Measurements at Depth in the Vicinity of the San Andreas Fault: Implications for the Magnitude of Shear Stress at Depth," _Journal of Geophysical Research_, Vol. 85, No. 811, Nov. pp. 6157-6173.

Chapter 20

DEEP STRESS MEASUREMENTS IN THREE OHIO QUARRIES
AND THEIR COMPARISON TO NEAR-SURFACE TESTS

Bezalel C. Haimson

University of Wisconsin
Madison, Wisconsin 53706

ABSTRACT

Anna, Ohio, at the junction of Cincinnati, Findlay and Kankakee arches, has been the site of repeated and sometimes damaging earthquakes. As part of a seismicity investigation near-surface (0.15-1 m depth) overcoring strain measurements were conducted in four limestone and dolomite quarries surrounding Anna, the results of which suggest a complicated regional stress pattern, with changes in the maximum horizontal stress direction from N45°E to N15°W. Since such abrupt stress fluctuations are rather atypical in the Midwest and since the reliability of near-surface tests in defining regional stress had not been established, we decided to remeasure the stresses at the same locations but at greater depths (0-200 m) using the hydrofracturing method. Owing to the weak horizontal bedding planes of the Silurian shaly limestones and shaly dolomites and the relatively shallow depths, many of the tests yielded horizontal fractures, providing information only on the vertical principal stress (~0.029 MPa/m depth). However, sufficient tests in each of three testholes resulted in vertical hydrofractures to enable the calculation of all three principal stresses. Within 50-170 m depth the horizontal stresses in all three quarries appear to belong to a uniform regional stress field defined by the following relations: $\sigma_{Hmin} = 5.1 + 0.014\ D$; $\sigma_{Hmax} = 10.1 + 0.014\ D$, where D is depth in meters. The direction of σ_{Hmax} within 50-170 m depth is uniform at N70°E ($\pm 15°$). The hydrofracturing results obtained at Anna, Ohio are in agreement with the prevailing stress regime in the midcontinent both with respect to magnitudes and directions. The results suggest a possible thrust mechanism for a proposed NW-SE fault along which most of the earthquake epicenters have been located, and a left lateral strike-slip at greater depths. The focal mechanism solution of a recent major earthquake near Sharpsburg, Kentucky, some 200 km south of Anna, also indicates strike-slip induced by a N60°E trending σ_{Hmax}. The discrepancies between the near-surface overcoring and the deeper hydrofracturing are attributed to a detachment or decoupling of the stress field at about 50 m depth which renders any shallow measurements unrepresentative of the regional state of stress.

INTRODUCTION

In a recent paper (Haimson, 1981) it was shown that in six case histories stress magnitudes and directions as determined by hydrofracturing agreed surprisingly well with those resulting from overcoring tests when both methods were employed at comparable depths. The study suggests that both methods measure the same quantity, namely the stress tensor, and that either can be used with some confidence provided the tests are correctly carried out. However, the conclusions of the paper or of any other study do not necessarily justify measuring the stress by either method at one depth in order to project its value to a different depth. For example, at Waterloo, Wisconsin, hydrofracturing tests in two adjacent drill holes in Precambrian quartzite revealed that the stresses in the upper 40 m were considerably different in both magnitudes and directions from the stress regime in the 50 m to 270 m depth (Haimson, 1980).

Similarly, at Darlington, Ontario, the state of stress in the Paleozoic shaly limestone (30 - 210 m) undergoes a shift both in direction and in magnitude as the rock changes to Precambrian granitic gneiss (220 - 300 m) (Haimson and Lee, 1980). Thus, measurements at 100 - 200 m depth cannot be used in this case to predict the state of stress at 250 m. Obviously, the decoupling that occurs here appears to be closely related to the change in rock type and age.

In the last several years, the increasing need to know the state of stress at depth over large areas has prompted several scientists to suggest that surface and near-surface overcoring measurements could be inexpensively conducted at a number of outcrop locations within a geologic unit and their results used to interpret regional trends of crustal stress. Eliminating the requirement for deep drill holes and their testing could save a substantial amount of money and time and makes near-surface testing very attractive. While the justification behind any extrapolation of near-surface results to greater depths has always been uncertain, work in shallow measurements has continued with various degrees of success (Engelder and Sbar, 1976; Froidevaux et al, 1980). One such near-surface program designed to determine the state of stress in the Anna, Ohio earthquake zone was recently undertaken as part of a seismicity investigation (Newman, 1977; Clark and Newman, 1977).

This paper presents their near-surface results at Anna and reports on our study of measuring deeper stresses at the same locations as the shallow tests. The deep-hole and near-surface tests are compared and conclusions are drawn regarding their relationship.

ANNA, OHIO EARTHQUAKE ZONE

Geologic Setting (after Newman, 1977)

The Anna seismic zone is located on the Indiana-Ohio platform near the junction of three mid-continent arches: Findlay Arch, Cincinnati Arch and Kankakee Arch (Figure 1). The anticlines separate three important Paleozoic basins: Michigan basin, Appalachian basin and Illinois basin. The area consists mainly of Silurian dolomites overlain by 5-15 m of Pleistocene glacial material. Bedrock is exposed in a few road cuts and in many dolomite and limestone quarries. A number of faults have been suggested for the area surrounding Anna, but none have been verified. In one of the many local quarries (Northwood Stone Co.) numerous small high angle reverse faults and monoclinal folds trending N27°W have been observed in the bedrock. In the quarries used for the stress measurement study two sets of joints are discernable. The predominant one strikes approximately N70°E to E-W; the other set trends N-S.

Seismicity

Anna, Ohio and its vicinity have been the site of repeated, often damaging, earthquakes. Unfortunately, instrumental recordings of these events have been rare until recently (Pollack et al, 1976). A compilation of the historic earthquakes was published by Bradley and Bennett (1965). At least two more earthquakes have been recorded since. The epicenters of the major ones are marked in Figure 5 (and 6). Some reached intensities of VIII on the modified Mercalli scale (Mauk et al, 1979).

The determination of the geologic structure related to the earthquakes around Anna, Ohio is difficult since surficial evidence is scarce. Some 'proposed' faults based mainly on borehole drilling and geophysical logging are shown in Figure 5 (and 6). The proposed fault most closely associated to the major earthquakes appears to be the one passing by Anna and trending northwest (Mauk et al, 1979).

NEAR-SURFACE OVERCORING MEASUREMENTS

Table 1 presents the results of the near-surface (0.15 - 1.0 m) doorstopper overcoring tests at four quarries near Anna, Ohio. The stress magnitude reasonableness is difficult to assess. However, the direction of σ_{Hmax} is definitely inconsistent between quarries although repeateable results were obtained within three of the four sites. The general conclusion was that either the stresses measured were decoupled from the regional stress field, or that a regional σ_{Hmax} direction did not exist (Newman, 1977).

Table 1. Calculated Near-Surface Stresses at Anna Ohio Based on Doorstopper Overcoring Measurements in Four Quarries (Newman, 1977).

Location	Test No.	σ_{Hmin} MPa	σ_{Hmax} MPa	σ_{Hmax} Direction
MRS	7	1.4	3.2	N35°E
MRS	9	2.1	2.2	N30°E
MRS	11	0.5	1.9	N59°E
MRS	12	0.1	5.0	N44°E
NLS	1	-2.9	0.1	N11°W
NLS	2	1.0	2.4	N20°W
NLS	3	0.9	4.6	N09°W
SD	2	0.2	1.6	N81°W
SD	3	0	3.0	N84°W
NW	2	Not calculated		N17°E
NW	4			N53°W

MRS-Miami River Stone Quarry SD-Standard Dolomite Quarry
NLS-National Lime and Stone Quarry NW-Northwood Stone Quarry

We became intrigued by the apparent complicated stress condition in the Anna area, since all of the previous deep measurements in the midcontinent implied a rather simple stress field with σ_{Hmax} oriented to the northeast (Sbar and Sykes, 1973; Haimson, 1977; see also Figure 8). Hence, we decided to conduct a series of downhole hydrofracturing stress measurements in the very same quarries in order to verify the persistence of the near-surface test results with depth, and answer the question of possible 'detachment' of the shallow stresses from the regional field.

HYDROFRACTURING STRESS MEASUREMENTS

Drilling

We selected three of the four quarries used for shallow stress tests and core-drilled in each a 200 m vertical NQ hole using our own Longyear 34 drillrig. The test holes were drilled in the proximity of the shallow overcoring sites so that the only variable in the intended comparison was depth. The extracted core was carefully treated so as to minimize mechanical breaking, was placed in boxes and logged for joints, bedding partings and lithology. The results were used in selecting unfractured intervals for testing. The core was also utilized for laboratory mechanical testing.

Testing

Testing procedure followed that described by Haimson (1974; 1978a). Hydraulic fracturing was conducted using two inflatable rubber packers straddled approximately 1 m. They were lowered to the pre-

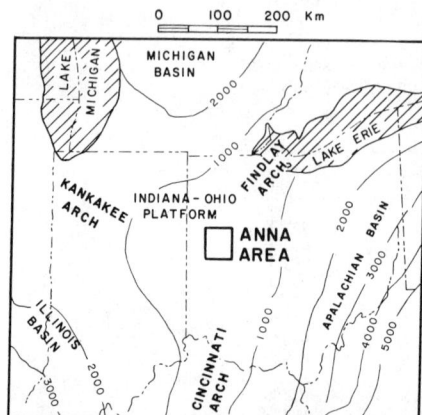

Figure 1. Regional map around Anna, Ohio showing major structural features and the Precambrian basement elevation contours.

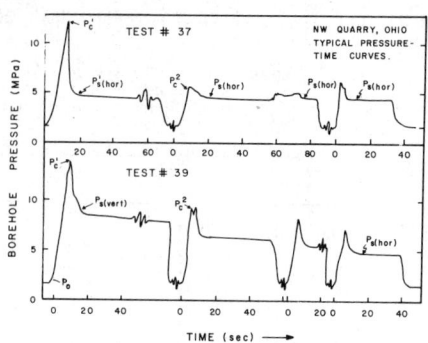

Figure 2. Typical pressure-time records. In test no. 37 repeatable constant shut-in pressure was indicative of horizontal hydrofracturing. In test no. 39 the initial vertical hydrofracturing shut-in pressure dropped from cycle to cycle until settling at a constant value corresponding to $P_{s(hor)}$. Both vertical and horizontal fracture impressions were obtained. Note: not all cycles shown.

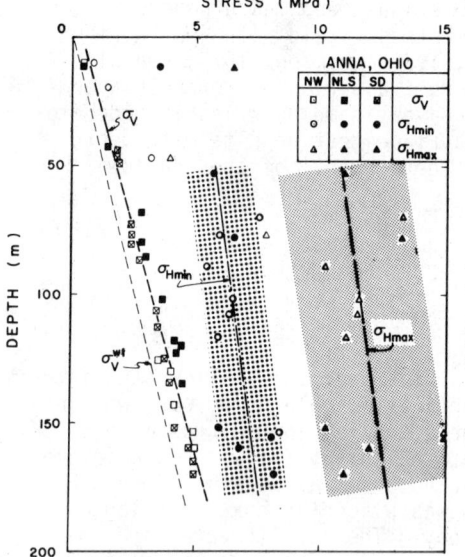

Figure 3. Variation of principal stresses with depth in three quarries near Anna, Ohio

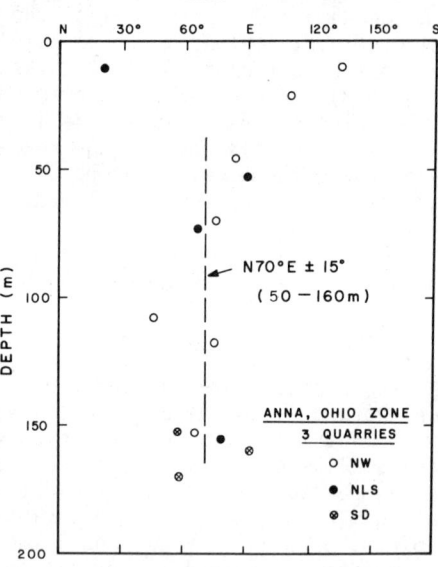

Figure 4. Variation of major horizontal principal stress direction with depth in three quarries near Anna, Ohio.

NW-Northwood Stone Quarry; NLS-National Lime and Stone Quarry; SD-Standard Dolomite Quarry

determined depth and were inflated to 7 MPa so that they adhered well to the borehole wall and sealed off the straddled interval. The latter was then pressurized, while maintaining higher packer pressure, until fracture occurred. The straddled interval pressure which was continuously recorded both at depth and on the surface using time-base recorders indicated hydrofracture initiation by a sudden drop in value (P_{c1} in Figure 2). The induced fracture was sometimes extended by continued pumping. Most often the pump was immediately shut off and the hydraulic line kept closed to allow the settling of the borehole pressure to an equilibrium level called the 'shut-in' value ($P_{s(hor)}$ or $P_{s(vert)}$ in Figure 2). The line was then bled off and several additional pressurization cycles conducted (Figure 2). These cycles provided values of the secondary breakdown pressure (P_{c2}) and additional values of P_s. When a test was completed the packers were deflated and moved to the next predetermined test depth.

Hydrofracture delineation was carried out using an impression packer, a 1m long inflatable packer wrapped around with a layer of very soft rubber. The impression packer was lowered to the depth of a hydrofractured interval and was inflated against the testhole wall and kept pressurized for about one hour. A down-hole magnetic survey-ing instrument was used to obtain the orientation of a mark on the packer with respect to north. When the packer was deflated and brought back to the surface, the 'impression' of the hydrofracture on the soft rubber was discerned and its attitude established.

Stress Calculations

Some 15 tests were conducted in each of the three testholes. Table 2 gives the average values of downhole hydrofracturing pressures P_{c1}, P_{c2} and P_s as obtained in the different pressur-ization cycles by both surface and down-hole transducers, the pore pressure (P_0) at each testing depth, and the vertical hydro-fracture directions in those tests were impression tests run. The pore pressure was calculated from the head of the water table in the testhole; the values of P_{c1}, P_{c2} and P_s were picked from the pressure-time curves as shown in Figure 2.

The vertical stress (σ_V) was obtained in two ways. Owing to the horizontally bedded nature of the rock and the fact that σ_V was generally the overall least principal stress horizontal hydrofractur-ing was obtained in most of the tests whether or not vertical frac-tures were also induced. In these tests:

$$\sigma_V = P_{s(hor)} \tag{1}$$

where $P_{s(hor)}$ is the shut-in pressure for horizontal hydrofractures. The other way of estimating the vertical stress was from the density of the rock (γ) and the depth D:

$$\sigma_V^{wt} = \gamma D \tag{2}$$

where σ_v^{wt} is the vertical stress resulting from the weight (wt) of the overlying rock.

The least horizontal compressive stress (σ_{Hmin}) was determined from

$$\sigma_{Hmin} = P_{s(vert)} \qquad (3)$$

where $P_{s(vert)}$ is the shut-in pressure for vertical hydrofractures.

The largest horizontal compressive stress (σ_{Hmax}) is usually obtained from the expression (Haimson, 1978a):

$$\sigma_{Hmax} = T + 3P_s - P_{c1} - P_o \qquad (4)$$

where T is the hydrofracturing tensile strength of the rock, or the resistance of the rock to hydrofracturing when the tangential stress at the test hole wall in the direction of σ_{Hmin} is zero. However, after hydrofracture initiation, the boundary conditions in subsequent pressurization cycles are basically the same as in the first cycle except that the rock surrounding the test interval has no tensile strength across the plane of the closed hydrofracture. Thus the secondary critical pressure P_{c2} can be substituted for $P_{c1} - T$ in equation (4):

$$\sigma_{Hmax} = 3 P_s - P_{c2} - P_o. \qquad (5)$$

We used equation (5) to calculate σ_{Hmax}.

Table 2 gives the calculated stress magnitudes and directions in the three quarries. The direction of σ_{Hmax} was taken as that of the vertical hydrofracture strike.

DISCUSSION OF RESULTS

Table 2 and Figures 3 and 4 give the individual principal stresses determined in each of the hydrofracturing tests. In two out of the three holes a majority of tests resulted in horizontal hydrofractures. The apparent reason for this result is the predominance of horizontal-bedding partings in the shale and dolomite of these holes. In some tests vertical fractures were accompanied by horizontal fractures. The latter case (such as test SD nos. 12, 28, 29, NW no. 39 and NLS no. 51) yields the most complete information, enabling the calculation of all three principal stresses from hydrofracturing data. Tests in which only horizontal hydrofractures are induced yield the least amount of information, namely the magnitude of σ_v. In each of the three holes there were sufficient vertical hydrofracturings resulting in consistent horizontal principal stresses. Thus, the state of stress could be established at each testhole. Figures 3 and 4 reveal, however, that the stress magnitudes and directions below a depth of approx-

Table 2 - Hydrofracturing Pressures and Resulting Stresses in 3 Quarries Near Anna, Ohio

Test No.	Depth m	P_o MPa	P_{c1} MPa	P_{c2} MPa	$P_{s(vert)}$ MPa	$P_{s(hor)}$ MPa	σ_V^{wt} MPa	σ_V MPa	σ_{Hmin} MPa	σ_{Hmax} MPa	σ_{Hmax} Direction
SD											
23	44	0.4	-	4.0		1.8	1.1	1.8			
21	47	0.5	21.5	8.8		1.8	1.2	1.8			
22	49	0.5	18.3	5.7		1.9	1.3	1.9			
15	73	0.7	14.2	8.1		2.4	1.9	2.4			
24	77	0.8	12.2	4.4		2.4	2.0	2.4			
9	81	0.8	18.7	4.9		2.5	2.1	2.4			
30	87	0.9	12.3	-		2.8	2.3	2.8			
25	107	1.0	15.1	5.1		3.4	2.8	3.4			
6	113	1.1	15.2	7.0		3.5	2.9	3.5			
11	125	1.2	9.1	5.0		3.8	3.3	3.8			
27	134	1.3	16.5	6.1		4.0	3.5	4.0			
28	152	1.5	9.1	6.3	6.0	4.4	4.0	4.4	6.0	10.2	N57°E
29	160	1.6	13.0	6.9	6.8	4.8	4.2	4.8	6.8	11.9	S88°E
7	165	1.6	17.1	6.8		5.0	4.3	5.0			
12	170	1.7	23.7	12.0	8.2	5.0	4.4	5.0	8.2	10.9	N58°E
NW											
44	10	0.1	7.3	2.9	0.8	0.4	0.3	0.4	0.8		S45°E
31	22	0.2	5.0	4.4	1.4		0.6		1.4		S70°E
43	47	0.4	9.4	5.2	3.2		1.2		3.2	3.9	N84°E
32	70	0.7	13.3	8.8	7.6		1.8		7.6	13.3	N75°E
42	77	0.8	14.2	9.4	6.0		2.0		6.0	7.8	
33	89	0.9	10.7	5.0	5.4		2.3		5.4	10.2	
34	102	1.0	12.4	6.9	6.5		2.7		6.5	11.6	
41	108	1.1	12.3	6.2	6.3		2.8		6.3	11.5	N45°E
35	117	1.1	12.4	5.6	5.9		3.0		5.9	11.1	N75°E
36	126	1.2	17.4	5.7		3.6	3.3	3.6			
40	131	1.3	14.1	5.1		4.1	3.3	4.1			
37	143	1.4	12.1	5.6		4.2	3.7	4.2			
38	150	1.5	12.3	5.6		4.6	3.9	4.6			
39	153	1.5	13.9	8.4	8.4	5.0	4.0	5.0	8.4	15.3	N65°E
45	160	1.6	11.1	8.2		5.1	4.2	5.1			
NLS											
51	11	0.1	13.5	4.2	3.6	0.4	0.3	0.4	3.6	6.5	N20°E
52	43	0.4	8.0	4.2		1.4	1.1	1.4			
65	53	0.5	15.0	5.7	5.7		1.4		5.7	10.9	N90°E
53	68	0.7	12.4	4.5		2.9	1.8	2.9			
64	73	0.7	10.7	5.9	6.6		1.9		6.6	13.2	N65°E
54	80	0.8	11.3	3.2		2.9	2.1	2.9			
63	86	0.9	12.6	3.3		3.1	2.3	3.1			
56	102	1.0	14.1	5.1		3.7	2.7	3.7			Horiz.
57	118	1.1	12.1	5.2		4.2	3.1	4.2			
62	120	1.2	11.3	4.3		4.5	3.2	4.5			
58	123	1.2	10.5	5.3		4.3	3.3	4.3			
61	135	1.3	10.3	5.8		4.6	3.6	4.6			
59	156	1.5	12.4	7.7	8.1		4.1		8.1	15.1	N78°E

P_o - pore-fluid pressure
P_{c1} - breakdown pressure
P_{c2} - secondary breakdown pressure
$P_{s(vert)}$ - shut-in pressure for vertical hydrofracture
$P_{s(hor)}$ - shut-in pressure for horizontal hydrofracture
σ_V - vertical stress based on weight of overlying rock
$\sigma_V, \sigma_{Hmin}, \sigma_{Hmax}$ - principal vertical and horizontal stresses as determined by hydrofracturing
SD - Standard Dolomite Quarry
NW - Northwood Stone Quarry
NLS - National Lime and Stone Quarry

imately 50 m are rather consistent from one hole to the next. It is quite obvious that all the testholes belong to a common stress field, for which the following stress-depth relationship was obtained using linear regression analysis:

$$\sigma_V^{wt} = 0.026\,D \qquad (6)$$

$$\sigma_V = 0.4 + 0.029\,D \qquad (7)$$

$$\sigma_{Hmin} = 5.1 + 0.014\,D \quad \text{at } N20°W \qquad (8)$$

$$\sigma_{Hmax} = 10.1 + 0.014\,D \quad \text{at } N70°E \qquad (9)$$

where stress is in MPa and D is depth in meters. The hydrofracturing vertical stress is slightly higher than the density-based σ_V^{wt} (by an average of 0.7 MPa) and confirms the reliability of estimating this magnitude from the weight of the overlying rock. As shown in Figure 3 below 50 m individual least horizontal stress results can deviate from the value predicted from equation (8) by ± 1.25 MPa; the largest horizontal stress results are more scattered and can deviate from equation (9) by ± 2.5 MPa. The direction of σ_{Hmax} in all three holes is particularly consistent at N70°E with a standard deviation of ± 15° (Figure 4).

The few results obtained between 10 m and 50 m depth are rather scattered but have one thing in common, namely the horizontal stresses appear to be of considerably smaller magnitude. This could be the result of some near-surface stress relaxation. A change is also evident in the direction of σ_{Hmax}. As noticed in Figure 4 in Northward Stone Quarry the direction undergoes a continuous rotation from S45°E at 10 m, to S70°E at 22 m, to N84°E at 47 m, to N75°E at 70 m. Below that it remains relatively constant are around N70°E. This kind of near-surface rotation and magnitude relaxation may have contributed to the discrepancy between the overcoring tests and the hydrofracturing results.

Unlike the near-surface tests (Figure 5), the hydrofracturing stress directions are very consistent between the different test quarries (Figure 6). With respect to magnitudes, the near-surface σ_{Hmin} and σ_{Hmax} values as extrapolated from equations (8) and (9) are approximately 5 MPa and 10 MPa respectively, considerably larger than the overcoring results (Table 1). The horizontal stress differential ($\sigma_{Hmax} - \sigma_{Hmin}$) is approximately constant within the hydrofracturing depth range, at 5 MPa; the average differential in the near-surface overcoring is much smaller (2 MPa). Overall, the comparison shows that near-surface tests in the Anna, Ohio quarries do not yield reliable information on the stress conditions at depth, owing to an apparent decoupling of stress fields occuring at about 50 m.

Hydrofracturing tests were conducted only to a depth of under 200 m. The question may be raised as to the reliability of these to

DEEP STRESS MEASUREMENTS IN OH   199

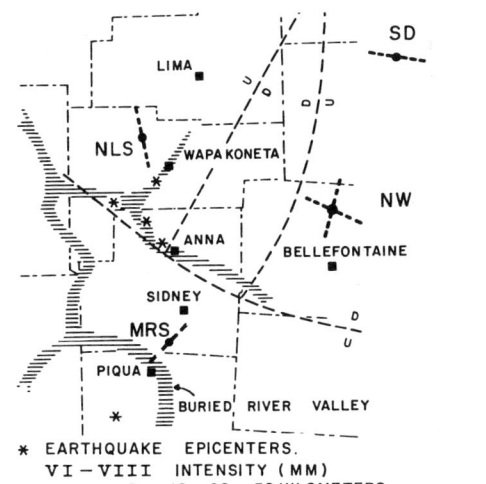

* EARTHQUAKE EPICENTERS.
VI – VIII INTENSITY (MM)
0 10 20 30 KILOMETERS

Figure 5. Map of Anna, Ohio area and the test sites (given by quarry initials) showing the direction of σ_{Hmax} resulting from near-surface overcoring tests (--•--) and proposed faults (– – –).

Figure 6. Map of Anna, Ohio area and the test sites (given by quarry initials) showing the direction of σ_{Hmax} resulting from down-hole hydrofracturing tests (—•—) and proposed faults (– – –).

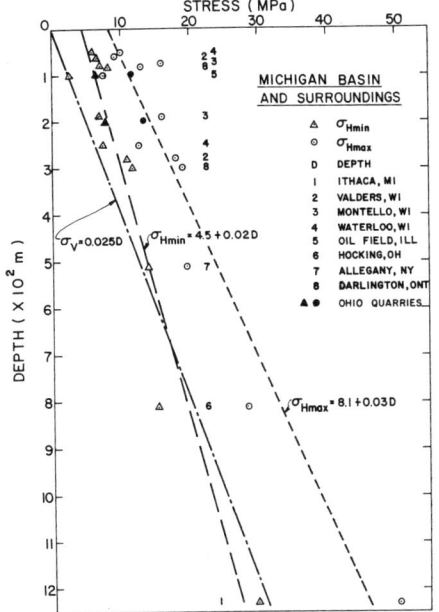

Figure 7. Variation of in situ stresses with depth in the Upper Midwest based on nine series of tests at as many locations. References given in Figure 8 (numbers in both figures refer to the same test sites).

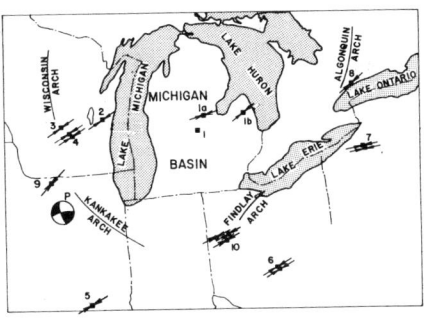

σ_{Hmax} DIRECTION IN AND AROUND MICHIGAN BASIN
FROM HYDROFRACTURING TESTS
● L.H. FOCAL MECH

Figure 8. Map of the Upper Midwest showing σ_{Hmax} direction at 10 test sites. References: tests no. 1, 3, 5, 6, 7. (Haimson, 1977); test no. 2 (Haimson, 1978b), test no. 4 (Haimson, 1980), test no. 8 (Haimson and Lee, 1980), test no. 10 (present paper), tests no. 1a, 1b and 9-unpublished data.

represent stresses at great depth. A direct answer can be provided only by measurement in deeper testholes. Indirect evidence, however, strongly supports the suggestion that around Anna the measurements in the top 200m are representative of the regional stress field. Figure 7 shows that stress magnitudes in the Anna, Ohio test holes blend well with those of many other tests in midcontinent. Figure 8 shows that σ_{Hmax} direction near Anna is also consistent with those in the midcontinent, which were obtained at a variety of depths reaching more than 1 km.

The hydrofracturing principal stress differentials at Anna were compared with shear stress levels predicted by McGarr (1980) from existing stress data. At 100m depth, for example, the maximum shear stress at Anna is $(\sigma_{Hmax} - \sigma_V)/2 = 4.5$ MPa. The same value based on McGarr's compilation for stresses in soft (sedimentary) rocks is 3.8 MPa + 0.004 MPa/m x 100m = 4.2 MPa, a rather remarkable coincidence.

The state of stress in the Anna testholes is indicative of thrust faulting ($\sigma_{Hmax} > \sigma_{Hmin} > \sigma_V$). The proposed northwest trending fault passing near Anna (Figure 6) could be a thrust fault based on its orientation with respect to the hydrofracturing principal stresses.

As seen in Figure 7 there is a possibility that at greater depths the vertical stress becomes intermediate in magnitude allowing for potential strike-slip faulting. The precise strike of the proposed fault near Anna has not been established but if it were, say, N70°W to N80°W a left lateral strike-slip mechanism resulting from the hydrofracturing stress field would be possible. Thus, the proposed fault along which most of the Anna earthquakes have been located and the hydrofracturing state of stress are compatible, with the constraint that either thrust or left lateral strike-slip faulting occurs.

It is interesting to note that on July 27, 1980 a major earthquake (magnitude 5.3) occurred at about 12-15 km below the town of Sharpsburg, Kentucky, some 200 km south of Anna (coordinates 38.17°N, 83.91°N). A focal mechanism solution of the event has indicated strike-slip faulting with a right-lateral slip along a fault plane striking N30°E (Herrmann et al., 1982). This solution suggests that the major horizontal principal stress acts at approximately 30° clockwise from the fault strike, i.e. at N60°E, and that σ_V is the intermediate stress value. Both suggestions are in close agreement with the Anna hydrofracturing measurements. The major difference between the Sharpsburg and Anna structures is that the fault in the former is a conjugate of the proposed fault in the latter. The general states of stress appear to be identical in principle.

ACKNOWLEDGEMENT

The hydrofracturing stress measurements in the Ohio quarries were supported by the National Science Foundation grant no. EAR78-15205. Thanks to Fred Mauk for providing information and reports related to seismicity studies at Anna, Ohio.

REFERENCES

Bradley, E.A. and Bennett, T.S., 1965, Earthquake history of Ohio, Bull. Seismol. Soc. Amer., Vol. 55, 745-752.

Clark, B.R. and Newman D.M. 1977, Modeling of non-tectonic factors in near-surface in situ stress measurements, In Proceedings 18th U.S. Symposium on Rock Mechanics (Edited by Fun-Den Wang & G.B. Clark), pp. 4C3 1-4C3 6. Colorado School of Mines Press, Golden, Colorado.

Engelder, J.T. and Sbar M.L. 1976, Evidence for uniform strain orientation in the Potsdam sandstone, Norther New York, from in situ measurements, J. Geophys. Res., Vol. 81, 3013-3017.

Froidevaux, C., Paquin C. and M. Souriau, 1980, Tectonic stresses in France: in situ measurements with a flat jack, J. Geophys. Res., Vol. 85-B, 6342-6346.

Haimson, B.C., 1974, A simple method for estimating in situ stresses at great depths. In Field Testing and Instrumentation of Rock, ASTM Spec. Tech. Publ. 554, pp. 156-182. American Society for Testing and Materials, Philadelphia.

Haimson, B.C., 1977, Crustal stress in the continental United States as derived from hydrofracturing tests. In The Earth's Crust, Geophysical Monograph 20 (Edited by J. C. Heacock), pp. 576-592. Am. Geophys. Union, Washington, D.C.

Haimson, B.C., 1978a, The hydrofracturing stress measuring method and recent field results. Int. J. Rock Mech. Min. Sci & Gromech. Abstr., Vol. 15, 167-178.

Haimson, B.C., 1978b, Additional stress measurements in the Michigan Basin. EOS Trans. (abstract), Am. Geophys. Union, Vol. 59, p. 1209.

Haimson, B.C., 1980, Near-surface and deep hydrofracturing stress measurements in the Waterloo quartzite, Int. J. Rock Mech. Min. Sci. & Geomech. Abstr., Vol. 17, 81-88.

Haimson, B.C., 1981, Confirmation of hydrofracturing results through comparisons with other stress measurements. Proceedings of the 23rd U.S. Rock Mechanics Symposium, p. 379-385, MIT press, Cambridge.

Haimson, B.C. and Lee C.F., 1980, Hydrofracturing stress determinations at Darlington, Ontario, in Underground Rock Engineering, 13th Canadian Rock Mechanics Symposium, pp. 42-50, the Canadian Institute of Mining and Metallurgy, Montreal.

Herrmann, R.B., Langston, C.A. and I.E. Zollweg, 1982, The Sharpsburg, Kentucky earthquake of July 27, 1980, Bull. Seismol. Soc. Amer. (in press).

Mauk, F.J., Coupland, M.D., Christewn, D., Kimball, J. and P. Ford, 1979 Geophysical investigations of the Anna, Ohio earthquake zone, Annual Progress Report NUREG/CB-1065 to the Nuclear Regulatory Commission, 182 p., the University of Michigan.

McGarr A., 1980, Some constraints on levels of shear stress in the crust from observations and theory, J. Geophys. Res., vol. 85-B, 6231-6238.

Newman, D.B., 1977, Near-surface in situ stress measurements in the Anna, Ohio earthquake zone, MS thesis, 74 p., the University of Michigan,

Pollack, H.N., et al., 1976, The Anna, Ohio earthquake zone seismic network, EOS Trans. (abstract), Am. Geophys. Union, vol. 57, p. 757.

Sbar, M.L. and Sykes L.R., 1973, Contemporary compressive stress and Seismicity in eastern North America. An example of intraplate tectonics, Geol. Soc. Amer. Bull., Vol. 84, 1861-1882.

Chapter 21

HIGH STRESS OCCURRENCES IN THE CANADIAN SHIELD

G. Herget

Associate Professor for Mining Engineering,
University of Kentucky, Lexington, Kentucky

SUMMARY

Three gradients have been identified in regard to the average horizontal ground stress increase with depth.

S_{Ha} (0-900 m) = 9.86 MPa + 0.0371 MPa/m
S_{Ha} (900-2200 m) = 33.41 MPa + 0.0111 MPa/m
S_{Hae} (extreme) = 12.36 MPa + 0.0586 MPa/m

The extreme stress values are not caused by mining geometry and are of limited extent only. In some cases geological disturbances like faults and rock type boundaries suggest an explanation but in other cases the unusually high stresses cannot be explained readily.

It is interesting to note that in the same locations the vertical stress components are very much higher than those derived from overburden weight and that the vertical stress gradient is similar to the gradient obtained from the extreme horizontal stress components.

INTRODUCTION

Ground stress determinations have been carried out in mines in Ontario and Manitoba, Canada, over more than a decade (Herget, 1980). Most of the sites are located in the Superior and Southern Tectonic Province of the Canadian Shield, which consist of Archaen and Proterozoic rocks comprising volcanics, metamorphosed sediments and granites. The youngest orogenic deformation occurred during the Grenville Orogeny (955 million years ago).

INSTRUMENTATION

Ground stress determinations in mines were carried out by overcoring methods, such as those using biaxial instruments, e.g., the USBM meter and CSIR doorstopper, and triaxial instruments, e.g., the triaxial strain cell developed by the CSIR (South Africa) and CSIRO (Australia).

Fortunately the rock material in the Canadian Shield is generally very strong. This is beneficial because the magnitude of ground stresses which can be determined depends on the range of the elastic behaviour of the rock under load. This has permitted the determination of ground stresses up to a magnitude of 130 MPa at a depth of 2100 m.

Discing of the drill core can make overcoring difficult at those stress magnitudes and the possibility exists that full strain recovery is not measured (Hast, 1979). In highly stressed ground it has been observed that the doorstopper method resulted in successful overcoring, whereas methods requiring the overcoring of an Ex annulus over 30-50 cm were unsuccessful.

Overcoring methods have been very successful when carried out in fine-grained, isotropic solid rock and within 10 to 20 m of excavation boundaries. For measurements in drill holes beyond this depth, overcoring procedures become time consuming and very costly.

In most of the locations where strain recovery was measured by overcoring, a redundancy of strain recovery data was available for each tensor determination, so that error determinations were possible with the method of least squares. Errors of $\pm$ 10-15% for the stress components are common. Details of overcoring methods, quality testing to obtain a high degree of reliability, and determination of physical parameters are described elsewhere, (Leeman, 1969; Gray and Barron, 1969; Herget, 1973).

INCREASE OF VERTICAL STRESS WITH DEPTH

Many investigators have observed that the vertical stress component (S_v) increases linearly with depth and that the increase is related to overburden weight. The density for quartz- and feldspar-rich rocks is about 2650 kg/m^3 and for basic and ultrabasic rocks 3300 kg/m^3. This results in a vertical stress gradient of 0.0260 to 0.0324 MPa/m respectively.

Figure 1 shows that many of the results obtained in the Canadian Shield are close to these gradients, but some results plot well outside this range.

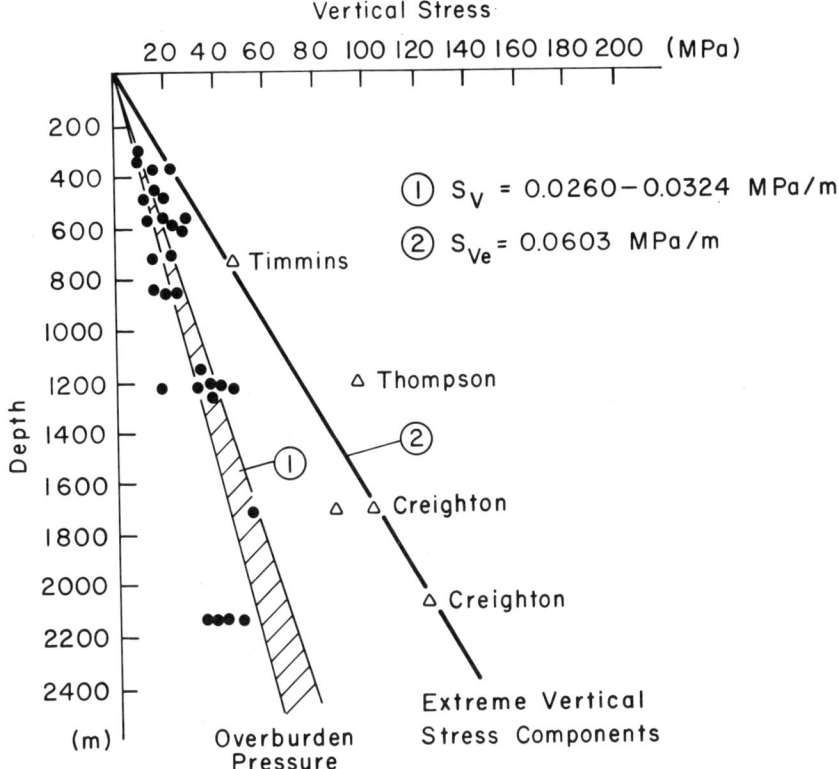

Fig.1: Increase of vertical stress with depth.

The high magnitude values were obtained at field sites where at the time of testing nothing extraordinary was observed in regard to the geological setting. With the benefit of hind sight one can mention in some cases a higher than usual frequency of quartz veins, or a shear zone being 50 m away, or the measurements having been made close to a completely healed contact between two metamorphic rock types with a change in elastic modulus of 20%. In other cases nothing offered an explanation for the higher than usual strain recovery.

The reported high stresses do not represent a measurement across a geological fracture as reported from the Sullivan Mine in British Columbia, where overcoring tests were carried out on both sides of a pronounced fracture and strain relief doubled close to the fracture surface (Royea, 1968). Discontinuities offer a straight forward explanation because the continuity of the rock material is interrupted and stress transfer occurs at local bridging points which therefore carry a high stress.

Generally, all stress determinations should be treated with caution where the calculated vertical stress component exceeds substantially the vertical stress component derived from overburden load. These higher than usual values are shown in Fig. 1 as triangles and they appear to follow a straight-line relationship going through the origin, with:

$$S_{Ve} = 0.0603 \text{ MPa/m} \qquad \text{Eq 1}$$

The regression yielded a correlation coefficient of 0.98.

From the observations of seismic activity around large hydro dams and the isostatic adjustments due to melting of ice sheets, it appears that the earth's crust is maintaining a delicate balance in a vertical direction and will adjust to changes in vertical loading of 0.5 to 1.0 MPa for large areas (Artyushkov, 1971). This identifies a probable outer limit for any extensive area exhibiting vertical stress components above overburden load. Therefore the high values in Fig. 1 are considered local stress disturbances.

INCREASE OF AVERAGE HORIZONTAL STRESS WITH DEPTH

It has been shown earlier that stresses can be found in the earth's crust where the average horizontal stress component is either equal to or larger or smaller than the overburden load (Herget, 1973). Around 1970, the majority of stress determinations showed the average horizontal stress component (S_{Ha}) to exceed the stress calculated from overburden weight, and a mean horizontal stress gradient of 0.0399 MPa/m provided a good fit to the data.

In the meantime, more information has become available and for measurements in the Canadian Shield a constant straight-line increase of the average horizontal stress with depth is not supported beyond a depth of about 900 m.

Figure 2 provides a plot of the ratio of average horizontal stress component to vertical stress component in relation to depth for Canadian data. This suggests that the ratio approaches unity with depth. On the basis of a reciprocal relationship, a regression was carried out which yielded:

$$S_{Ha}/S_V = 251.68/\text{depth (m)} + 1.14 \qquad \text{Eq 1}$$

with a correlation coefficient of 0.30. This regression line does not vary significantly whether outliers shown by triangles in Fig. 1 are included or excluded. Fig. 2 indicates clearly that the horizontal stress gradient changes with depth.

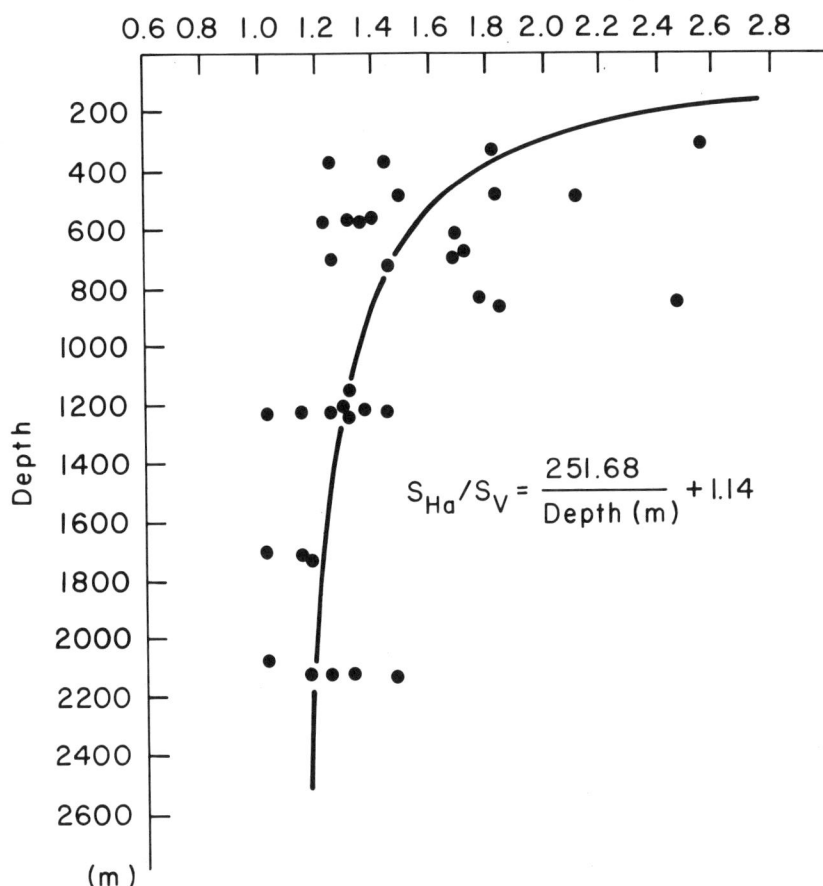

Fig.2: Change of ratio average horizontal stress/vertical stress with depth.

Fig. 3 shows a plot of the average horizontal stress component with depth and the values which are identified as extreme values in Fig. 1 also show up here as extreme values. This becomes really obvious if the first invariant of stress is calculated.

The average horizontal stress components (H_{ae}) identified as extreme are shown as triangles in Fig. 3 and follow a straight-line relationship:

$$S_{Hae} = 12.36 \text{ MPa} + 0.0586 \text{ MPa/m} \qquad \text{Eq 3}$$

with a correlation coefficient of 0.92. It is interesting to note that for the extreme stress tensors both the vertical gradient and the horizontal gradient are similar.

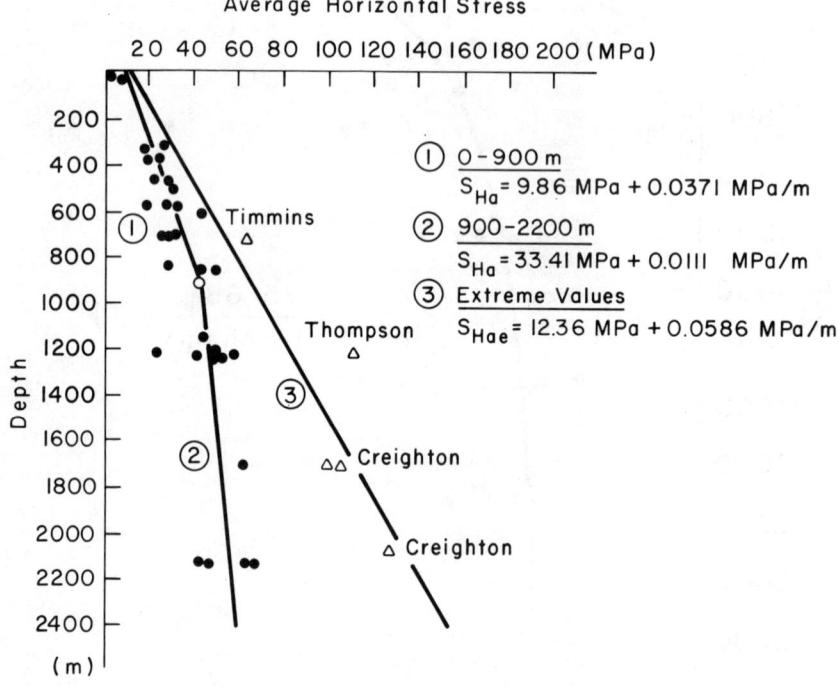

Fig.3: Increase of average horizontal stress with depth.

For the purpose of obtaining a common average horizontal stress gradient with depth, the extreme values were rejected and two straight lines were fitted to the remaining values, with a break at 900 m. This break in the gradient was found by adding and subtracting average horizontal stress components to the two populations until the best fit was obtained.

This provided the following:

1) S_{Ha} (0-900 m) = 9.86 MPa + 0.0371 MPa/m, r^2 = .59 Eq 4

2) S_{Ha} (900-2200 m) = 33.41 MPa + 0.0111 MPa/m, r^2 = 0.19 Eq 5

The change in gradient below 900 m is only supported by a few observations and the correlation coefficient (r^2) is low. Further observations are needed.

Changes in horizontal stress gradients with depth have been reported also from other locations (Brown and Hoek, 1978). Generally it can be said that predictions of ground stresses at depth cannot be made from near surface measurements.

CONCLUSION

It appears that the Canadian Shield possesses a relatively uniform stress field with occassional pockets of limited extent where ground stresses are above usual values.

ACKNOWLEDGEMENTS

The author is indebted to many mining companies who actively supported the difficult task of ground stress determinations at their properties. W. Zawadski, L. Tirrul, and J. Smith of the Elliot Lake Mining Research Laboratory, CANMET, EMR Canada assisted in obtaining the presented results.

REFERENCES

1. Artyushkov, E.V.,1971, "Rheological properties of the crust and upper mantel according to data on isostatic movements"; Geophysical Research, vol. 76, No. 5.

2. Brown, E.T. and Hoek, E., 1978, "Trends in relationships between measured in-situ stresses and depth." Int J Rock Mech Min Sci & Geom Abstr; vol. 15, pp.211-215.

3. Gray, W.M. and Barron, K.,1969, "Stress determinations from strain relief measurements on the ends of boreholes: planning, data evaluation and error assessment"; Proc Int Symp Determination of Stresses in Rock Masses,Lisbon, Portugal.

4. Hast, N. "Limit of stress measurements in the earth's crust"; Rock Mechanics, vol. 11, pp.143-150; 1979.

5. Herget, G.,1973, "Variations of rock stresses with depth at a Canadian iron mine"; Int J Rock Mech Min Sci; vol.10, pp.37-51.

6. Herget, G.,1973, "First experiences with the CSIR triaxial strain cell for stress determinations"; Int J Rock Mech Min Sci & Geom Abstr; vol. 10, pp.509-522.

7. Herget, G.,1980, "Regional stresses in the Canadian shield"; CIM Spec vol. 22, pp.9-16; 13th Canadian Rock Mechanics Symposium, Toronto.

8. Leeman, E.R.,1969, "The 'doorstopper' and triaxial rock stress measuring instruments developed by the CSIR"; J. South African Inst Min Met; vol. 69, pp.305-339.

9. Royea, M.J.,1968, "Rock stress measurement at the Sullivan Mine"; Proc 5th Can Rock Mech Symp; pp.59-74; Toronto.

Chapter 22

BEHAVIOR OF A RIGID INCLUSION STRESSMETER
IN AN ANISOTROPIC STRESS FIELD

by G. W. Jaworski[1], B. C. Dorwart[2], W. F. White[3], and W. R. Beloff[4]

[1]Assistant Professor, Department of Civil Engineering, University of New Hampshire, Durham, New Hampshire

[2]Geotechnical Engineer, Soil & Rock Instrumentation Division, Goldberg-Zoino & Associates, Inc., Newton Upper Falls, Massachusetts

[3]Manager, NSTF Test Engineering and Operation, Rockwell Hanford Operations, Richland, Washington

[4]Associate, Soil & Rock Instrumentation Division, Goldberg-Zoino & Associates, Inc., Newton Upper Falls, Massachusetts

INTRODUCTION

The rigid inclusion stressmeter is a relatively low-cost instrument developed by Hawkes and Bailey (1973) and was originally intended to monitor stress changes in mine pillars subjected to uniaxial stress changes. In this regard, this instrument has found considerable use by the mining industry. More recently, there have been attempts to use the stressmeter for determination of more general changes of the in situ rock stress. This paper reports on a study to characterize the performance of the rigid inclusion stressmeter for this application and presents results on the effect of anisotropic stress changes. These stress changes must be accounted for to properly analyze general changes of the in situ rock stress.

BACKGROUND INFORMATION

The rigid inclusion stressmeter is an instrument designed to measure variations of the in situ stress around a borehole. The stressmeter studied in this work, is manufactured by Irad Gage, Div. Creare Products, Inc. Lebanon, New Hampshire. It consists of a stiff hollow steel gage body that is preloaded across a borehole diameter. The preloading is activated by sliding a wedge between an upper platen and the gage body. Changes in the in situ stress result in slight deformations of the gage body. These deformations affect the stress in a highly tensioned wire that is in line with the rock contact surfaces. Stress changes in the wire are monitored with electronics that measure its natural frequency. The change in the wire stress is in turn related to the change in rock stress by calibration.

Hawkes and Bailey (1973) found that from changes in the wire stress, $\Delta\sigma_w$, relate to a corresponding rock stress change, $\Delta\sigma_r$, by a calibration constant, α, such that:

$$\Delta\sigma_r = \Delta\sigma_w/\alpha \qquad (EQ\ 1)$$

Using a relationship given by Hast (1958) for the radial deformation of a borehole, Hawkes and Bailey (1973) showed that the value of α given in EQ 1 would vary as a function of the magnitude and direction of the principal stress changes. It was further shown that for conditions where the stressmeter wire was parallel to the direction of a uniaxial stress application, the calibration constant would be three times as great and opposite in sign from that obtained when the stressmeter wire was normal to the direction of a uniaxial stress. An experimental study verified this result.

Babcock (1981) presented the results of a study that indicated the conditions under which stresses and strains normal to each other would carry over and affect measurements. These conditions were found to be dependent on the relative elastic properties of the rock mass and a rigid circular inclusion used for measurements.

At the Basalt Waste Isolation Project administered by Rockwell International, Rockwell Hanford Operations, clusters of three stressmeters are placed in numerous boreholes to monitor the in situ stress changes resulting from thermomechanical loading associated with nuclear waste isolation. To use the rigid inclusion stressmeter for this purpose, it was necessary to study the effects of rock-mass properties, anisotropic stress changes, and temperature on stressmeter behavior. This study consisted of an extensive laboratory calibration and characterization program coupled with an analytical study.

METHODOLOGY

<u>Analytical Study.</u> To study the behavior of a stressmeter in anisotropic stress fields, a number of linear elastic, finite element analyses were performed where the effects of anisotropic stress changes, elastic properties of the rock-mass, and preload stress settings could be investigated. These analyses assumed plane strain conditions in the plane normal to the axis of the borehole.

The contact area between the stressmeter and the borehole wall is believed to be a function of the initial stress setting of the stressmeter during installation. The contact area was varied by altering the connectivity of nodes between the borehole wall and the stressmeter assembly. No consideration was given to changes in the contact area with subsequent changes in rock stress; therefore, only trends resulting from these changes are discussed.

Calibration Program. The calibration program consisted of two parts. The first of these involved subjecting basalt and aluminum cylinders containing stressmeters to changes of stress equal in all directions and normal to the axis of cylinders. The second part of the calibration program involved setting a stressmeter in an aluminum slab that was subjected to a one dimensional stress change. By varying the orientation of the stressmeter wire with respect to the direction of the uniaxial stress change, various anisotropic stress states could be studied. Additionally, the thermal effects of the stressmeter behavior were also studied. Although the thermal effects are quite significant, the results of those studies are beyond the scope of this paper.

Samples of basalt were obtained from the Hanford Site. All cylindrical samples were 14.4 cm O.D. with a 4.8 cm I.D. The aluminum slab was 76.2 cm square and 10.2 cm thick with a 4.8 cm hole. Where stressmeters were set in the aluminum slab, the angle between the uniaxial loading and the stressmeter wire was incremented by 15° from 0° to 90°. The stress levels for all tests ranged from 0 MPa to 20.7 MPa in 2.1 MPa increments.

The aluminum was chosen to model the basalt because of similar elastic properties. It was assumed that by coupling the analytical study to the test results in aluminum, results obtained for aluminum could be extrapolated to basalt. The analytical study would define the extrapolation method involving the material properties.

RESULTS

Analytical Study. The result of previous work, the calibration program, and the finite element study indicated that changes in wire stress, $\Delta\sigma_w$, were significantly affected by the magnitude and direction of stress occurring in the rock mass. To account for more general changes in stress, various stress states were studied. It was found that wire stress varied linearly as a function of the stress normal and parallel to the stressmeter wire. These stresses can be expressed as functions of the major and minor principal stress changes, $\Delta\sigma_1$ and $\Delta\sigma_3$, respectively, and the angle, Θ between $\Delta\sigma_1$ and the orientation of the stressmeter wire. Thus, the change in wire stress, $\Delta\sigma_w$, is given by:

$$\Delta\sigma_w = K_1(\Delta\sigma_1 \sin^2\Theta + \Delta\sigma_3 \cos^2\Theta) + K_2(\Delta\sigma_1 \cos^2\Theta + \Delta\sigma_3 \sin^2\Theta) \quad \text{(EQ 2)}$$

where K_1 and K_2 are stressmeter calibration constants. It should be noted that the first part of EQ 2 refers to the stress normal to the stressmeter wire; whereas the second part refers to the stress parallel.

Given that $\Delta\sigma_w$ could be expressed as a function of $\Delta\sigma_1$, $\Delta\sigma_3$, and Θ as discussed above, the analytical study concentrated on the influence of material properties and contact area on calibration constants, K_1 and K_2. It was found that the effect of Poisson's ratio was small compared to that of modulus and contact area. The influence of modulus and contact area on calibration constants, K_1 and K_2, are shown in Figure 1. The larger values of K_2, along with its greater range, are indicative of the greater influence that the stress parallel to the stressmeter wire has on the change in the wire stress. However, if stresses normal to the stressmeter wire are not accounted for, considerable error could result when evaluating the change in the stress state, depending on the relative magnitude of the normal stress.

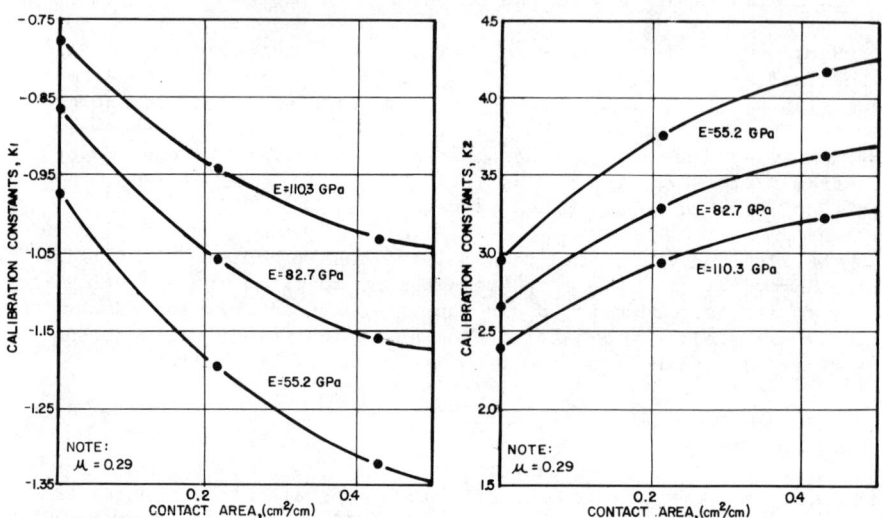

FIGURE 1 Calibration Constants, K_1 and K_2, as a function of rock mass modulus, E, and contact area

Calibration Program. The results of the calibration program are summarized in Table 1, where α values are listed for the tests performed for a wire stress during preload equal to 55.2 MPa. Typical results for tests performed using the aluminum slab are shown in Figure 2. Results obtained using basalt and aluminum cylinders are similar to those obtained for conditions where angles between the stressmeter wire and the stress application are less than or equal to 45°.

Table 1 illustrates a considerable amount of scatter in the

TABLE 1   SUMMARY OF RESULTS OF CALIBRATION PROGRAM

VWS SERIAL NO.	INITIAL WIRE STRESS (MPa)	TEST SAMPLE	ANGLE	VALUES		
				CYCLE 1	CYCLE 2	CYCLE 3
294	367	Basalt #5	C	1.26	1.50	1.50
		Basalt #3	Y	2.45	2.78	2.79
200	352	Basalt #6	L	1.42	1.67	1.69
		Basalt #4	I	1.76	1.96	1.97
258	303	Basalt #3	N	1.38	1.69	1.69
		Basalt #1	D	1.81	2.06	2.07
234	299	Basalt #4	E	0.97	1.18	1.18
		Basalt #2	R	2.31	2.51	2.52
107	274	Basalt #1		1.71	2.18	2.19
		Basalt #5	T	2.20	2.42	2.42
11	273	Basalt #2	E	2.19	2.47	2.48
		Basalt #6	S	1.84	2.04	2.05
268	254	Aluminum #1	T	3.17	3.39	3.39
			S	3.13	3.41	3.42
				3.20	3.48	3.49
268	254	Aluminum #2	0	5.38	5.77	5.79
			15°	4.77	5.20	5.28
			30°	3.49	3.66	3.66
			45°	1.36	1.46	1.45
			60°	-0.89	-0.90	-0.92
			75°	-1.60	-1.63	-1.63
			90°	-1.67	-1.65	-1.63

α values obtained with the basalt cylinders. This scatter may have been the result of one or a combination of the following

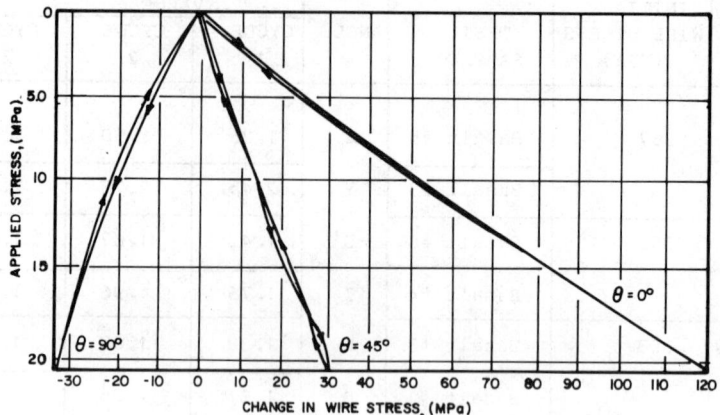

FIGURE 2   Test results with the stressmeter oriented at various angles, Θ, to an applied stress

causes: manufacturing tolerances, differences in the material properties among the test cylinders used, slight anisotropy with respect to the material properties of the test cylinders, inconsistent contact area between the stressmeter and the borehole wall for each test, and imperfect alignment of the stressmeter in the borehole.

The results of the aluminum cylinder tests show less scatter. In these tests, only one stressmeter and one test cylinder were used. The test results, using the aluminum slab subjected to a uniaxial stress increase, indicate that the values of α are significantly affected by the angle of stress application.

In all of the cylinder tests and the aluminum slab test at angles less than 60°, the α values calculated on the first cycle are always less than those calculated on the second and third cycles. Also, the α values obtained on the second and third cycles are generally in excellent agreement. This was attributed to a local failure of materials at contacts, resulting in a seating of the stressmeter in the borehole. When the uniaxial stress application was greater than 45°, there is little difference between the values of α obtained in the three cycles. Under these conditions, stresses at the contacts decrease with load application, indicating that a local failure of materials would not occur with cycling.

RIGID INCLUSION STRESSMETER BEHAVIOR

Figure 2 illustrates that when the angle between the stress application and the stressmeter wire is at 0° or 90°, the relationship is slightly curved. At 45°, however, the relationship is more linear. This is believed to result from a change in the contact area with stress application. The trends are consistent with the analytical study which indicate an increase in K_2 and a decrease in K_1 with increasing contact area. At 0°, stress application would cause an increase in contact area. Under these conditions, wire stress changes are a function of K_2 only. Hence, it should be expected that wire stress changes would increase with an increase in the stress level. Similarly, at 90°, the contact area decreases with stress application which increases the value of K_1. Under these conditions, wire stress changes are a function of K_1 only. However, at an angle of 45°, the contribution of K_1 and K_2 are equal and although the contact area would be expected to increase, the complimentary affects on K_1 and K_2 generally offset each other. However, the greater magnitude of K_2 compared to K_1, together with its greater range of values, would indicate that had the changes in contact area been greater, its influence would be noticed. In fact, in a separate study designed to investigate the influence of preload stress (which is directly related to contact area), it was found that α values generally increased with preload stress.

Discussion. The results of the analytical study were compared to those of the calibration program. It should be noted that for the cylinder tests, the value of α is equal to the sum of K_1 and K_2. In the tests on basalt, it was assumed that the contact area was equal to 0.426 cm^2/cm and that the modulus of elasticity was

TABLE 2 SUMMARY OF REGRESSION ANALYSIS OF

LABORATORY DATA FROM UNIAXIAL ALUMINUM SLAB TEST

ANGLE	AVERAGE MEASURED IN CYCLES 2 AND 3 (MPa)	PREDICTED BY REGRESSION OF MEASURED DATA (MPa)	ERROR (MPa)
0	119.0	116.0	3
15°	109.0	105.0	4
30°	75.7	75.7	0
45°	29.9	35.3	-5.4
60°	-18.8	-4.9	-13.9
75°	-33.7	-34.9	1.2
90°	-33.9	-45.7	11.8

81.4 GPa. Using these values, the value of α predicted by the analytical study is 2.4, which is within the range of the measured values. Assuming the same contact area, the predicted value of α for the aluminum cylinder test is 2.7, which is less than the average measured value of 3.4.

For comparison of results, a linear regression of the data was performed to obtain values for the constants, K_1 and K_2. These results, which are summarized in Table 2, indicate values of K_1 and K_2 to be -2.2 and 5.6 respectively, whereas the analytical study predicted values of -1.2 and 3.9. It should be noted that the values of K_1 and K_2 obtained from the aluminum slab test would indicate a value of α for the cylinder test of 3.4 which is in excellent agreement with that measured.

CONCLUSIONS

The most significant aspect of this paper is the influence of anisotropic stress changes on wire stress changes in a rigid inclusion stressmeter. This would indicate the need for instruments to be clustered in groups of three where general changes in stress state are of concern. Also, instruments must be calibrated for anisotropic stress changes to obtain calibration constants, K_1 and K_2. It was shown that this can be effectively accomplished by subjecting a cored slab containing a stressmeter to uniaxial stress changes and changing the orientation of the stressmeter with respect to the direction of the applied load.

The influence of installation conditions, manufacturing tolerances, and anisotropy with respect to material properties was inferred. However, the relative contribution could not be evaluated with the data available, though their combined influence can be significant. Therefore, additional study in this respect is now considered warranted.

REFERENCES

Hast, N., 1958, "The Measurement of Rock Pressure in Mines," Sveriges Geologiska Undersoknig, Arsbok Series C 3.1958.

Hawkes, I., and Bailey, F., 1973, "Design, Develop, Fabricate, Test, and Demonstrate Permissible Low Cost Stress Gages," Contract Report H0220050, U.S. Bureau of Mines.

Babcock, C.O., 1981, "Design Concepts for a Unidirectional Response Drill Hole Stress or Strain Measuring Gage," Proceedings of the 22nd U.S. Symposium on Rock Mechanics at M.I.T., ISRM.

Chapter 23

IN-SITU STRESS DETERMINATIONS IN NORTHEASTERN OHIO

by
Jean-Claude Roegiers, John D. McLennan and Lane D. Schultz

Senior Associate Scientist and Senior Research Engineer
Dowell Division of Dow Chemical U.S.A.
Tulsa, Oklahoma

Chief Geologist, Weston Geophysical, Westboro, Massachusetts

ABSTRACT

During construction of the intake tunnels for a nuclear power facility in Ohio, evidence of a potential geological discontinuity was discovered. Assuming that the discontinuity might still prevail underneath the site, it was decided to measure the stress profile across the discontinuity. This information would be used to assess future stability on the basis of predicted shear stresses acting along the fracture. Seven intervals were hydraulically fractured to determine the magnitude and orientation of the stress tensor, in order to infer the shear stresses acting along the discontinuity.

The formation consists of tightly interbedded shales. This, in combination with the shallow depths, would be expected to lead to horizontal hydraulic fractures -- a situation which generally provides limited information. Careful testing procedures, and meticulous interpretation, allowed the determination of the horizontal stress components as well as the vertical component, via the initiation and/or propagation of multiple fractures. The measured stresses fall within the range of stress magnitudes and orientation determined in other areas of the northeastern and north-central United States and in southern parts of Canada. In addition, a small stress gradient anomaly was recorded at depth.

On a different scale, the measured stress regime is also evaluated in light of the regional geological setting.

INTRODUCTION

A comprehensive investigative program evolved as a result of bedrock deformation exposed during the excavation phase of tunnel construction for the Perry Nuclear Power Plant bordering Lake Erie in Lake County, Ohio. Deterministic fault study objectives were realized as a consequence of interrelated geologic and geophysical research and engineering. Concurrent with and subsequent to tunnel excavations, the nature of the fault plane geometry, gouge and country rock mineralogy, as well as chemical constituency, was studied. After the necessary site specific data had been assembled, the localized anomalous deformation was interpreted in the context of its regional setting. The stress profile was measured across the extrapolated plane of deformation which also is a zone of lithologic variation controlled by sedimentation. The goal was to check for any anomalies that may exist in the neighborhood of a suspected fault.

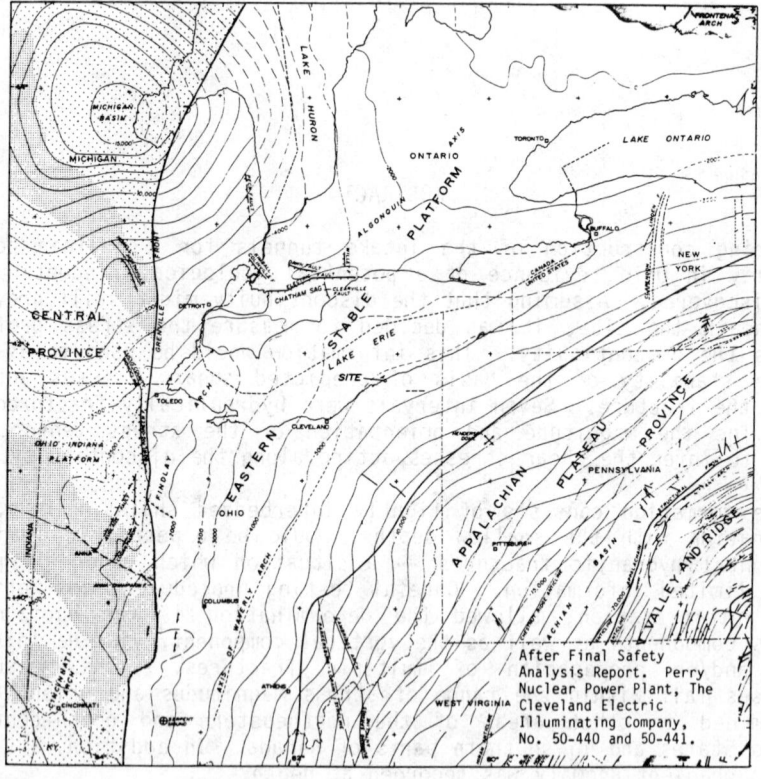

Figure 1. Site location.

GEOLOGIC SETTING

The Perry Nuclear Power Plant site is situated in the central part of the Eastern Stable Platform Tectonic Province, a wide region characterized by an Upper Precambrian crystalline basement complex, overlain unconformably by a sequence of Paleozoic sedimentary formations with little tectonic deformation (Fig. 1). Basement rocks in this province are comprised largely of high-grade, regionally metamorphosed schists, gneisses, marbles and calc-silicate granulites which were consolidated to a discrete crustal block during the Grenvillian Orogeny, 950 ± 150 million years ago.

Post-consolidation tectonic deformation in the site province is of minor extent. Activity was limited to the development of broad northeast-trending arches of epeirogenic origin along the western portion during Early to Middle Paleozoic time with a localized faulting activity on or near the arches in Middle to Late Paleozoic time. The only tectonic structure within the site province interpreted to be active is the Clarendon-Linden fault zone in western New York, about 160 miles northeast of the site.

The site province is bounded on the west by the Grenville Front; on the northeast (beyond the site region) by the Ottawa-Bonnechere graben structure; on the southeast by the moderately folded sedimentary rocks of the Appalachian Plateau; and on the south (beyond the site region) by the Kentucky River-Rome Trough fault system. Within this intraplate environment, there is no crustal manifestation of tectonic stress input. The Grenville Front is a profound tectonic boundary in the basement, separating the high-grade Grenvillian terrane of the site province to the east from essentially undeformed felsic intrusives, volcanic flows and sedimentary/pyroclastic rocks of the Keweenawan and Elosonian terranes to the west. Along much of the Grenville Front to the west and southwest of the site, the Precambrian basement rocks lie at depths less than 1.24 miles below ground surface.

Bedrock directly beneath the site is the Chagrin Shale member of the Ohio Shale formation (Upper Devonian). Regionally, these rocks dip gently to the southeast at a gradient of approximately 20 to 40 ft per mile. The Precambrian crystalline basement occurs at a depth slightly greater than 5,000 ft. To the south, the Devonian strata are overlain by successively younger Paleozoic sediments.

STRATIGRAPHY

A Devonian-Mississippian stratigraphic interval dominated by shale forms the bedrock surface of an 8- to 20-mile wide belt contiguous to Lake Erie westward from the Pennsylvania border to the vicinity of Sandusky, Ohio, and from there southward through central Ohio to the Ohio River. The subdivision of these Devonian-Mississippian shales in Ohio is based on their lithologic character according to Hoover (1960). The precise horizons separating the divisions are somewhat arbitrarily defined because of interfingering facies and their transitory nature vertically from one unit to the next. In the northeastern region, a complete columnar section through the shale sequence in stratigraphic order, oldest to youngest, includes the Plum Brook (logged in subsurface), Huron, Chagrin, Cleveland and Bedford shale members (Rector, 1950). Collectively, these members comprise approximately 1,500 ft of stratigraphic section of the site locale. The Huron, Chagrin and Cleveland shales together are also known as the Ohio Shale, representing most of the Devonian-Mississippian shale sequence in northeastern Ohio. The composite interval is underlain by Middle Devonian, predominantly nonargillaceous, carbonate rocks and capped by Berea Sandstone; the latter lies on a scoured Bedford Shale erosional surface.

Member subdivision of the Ohio Shale is accomplished mostly on the basis of color, primary structures and other physical criteria. The Huron Shale, stratigraphically averaging 410 ft throughout Ohio, is a black fissile shale containing conspicuous carbonate concretions. Its base is placed at the top of the highest gray shale (or limestone) bed of the underlying Plum Brook Shale. The top of the Huron is placed at the highest black shale where the gray, slightly arenaceous Chagrin begins. In some locales, the base of the Chagrin is conspicuous, beginning at the top of the uppermost layer of carbonate concretions (generally from one to six feet in diameter, but as large as 15 ft) or at the base of the lowermost Ohio Shale cone-in-cone structure.

The Chagrin Shale is essentially a noncarbonaceous, medium-gray, fissile, clay shale occupying an intermediary position between two highly carbonaceous fissile blue-black shales.

To the vertical extent of plant core drilling (730 ft), cores of the noncarbonaceous Chagrin Shale are identified as dark-gray to medium-gray silty or clayey shale with occasional light-gray sandy shale laminae. Huron Shale is black to dark brown, typically bituminous with lesser amounts of thinly bedded light-gray silty and sandy laminae but ubiquitous pyrite. A site stratigraphic column constructed from hydrofracture test boring TX-11 and Exploratory Well No. 202 logs, the latter located approximately five miles to the south, portrays the vertical transition from Chagrin to Huron lithology (Fig. 2). The Huron lithology is initially encountered at a depth of 463 ft and becomes more abundant with increasing depth.

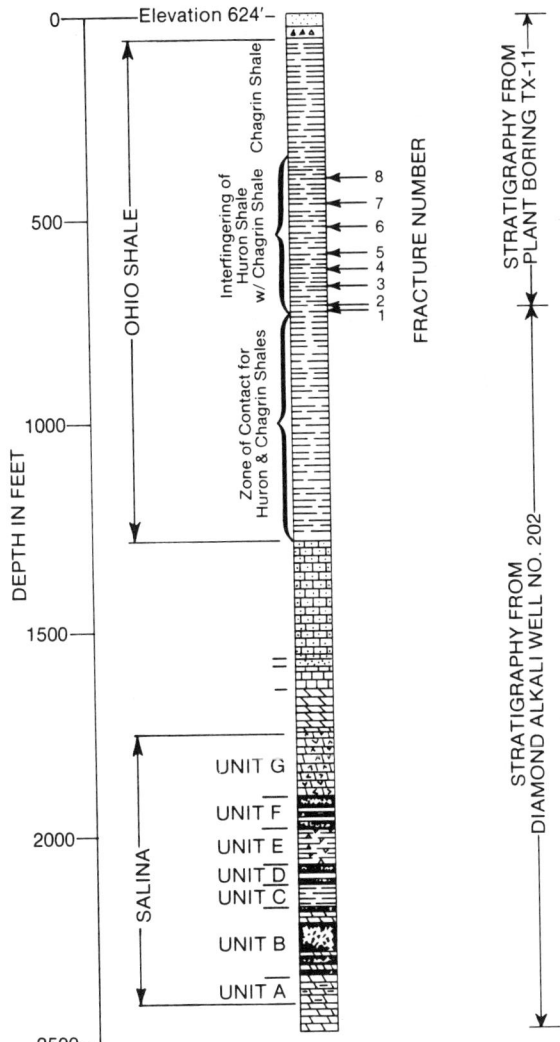

Figure 2. Stratigraphic column.

FIELD PROCEDURES

A total of seven horizons was successfully fractured between depths of 394 ft and 718 ft. The deepest horizon was fractured first in order to minimize alteration of the stress conditions at horizons to be fractured subsequently. Regular inflatable straddle packers were used to seal off the region of interest. The pressure recording was performed at the surface as well as downhole. Two different pumping systems were available to span a large range of flow rates. The fracture traces were recorded using impression packer elements and a single-shot survey instrument.

Due to the particular situation of shallow depth and a strongly bedded formation, usual test procedures were slightly modified in an attempt to obtain more usable information. First, the hydraulic fracture was initiated using a low pressurization rate. This often created a horizontal fracture from which the overburden load could be deduced. The complete stress sensor was determinable from low volume pumping at some of the horizons. After pressurization cycles, the well was shut-in for a relatively longer period to study the pressure decay curve. The system was then bled off before a second stage of repressurization using viscous fluids and high pumping rates. The consequence of this last procedure was the creation of an additional vertical fracture. The multiple fracturing system was then treated as has been described elsewhere (Roegiers and McLennan, 1981).

LABORATORY TESTING

Although the tensile strength was already determined from the field repressurization cycles, a number of burst tests were performed on available core (Bredehoeft et al., 1976; Zoback and Pollard, 1978). Further testing -- short rod tests (Barker, 1977), notched burst tests (Abou-Sayed, 1978) -- was also performed to estimate critical stress intensity factors which have been incorporated in several formulations for the determination of σ_{HMAX} (Abou-Sayed et al., 1977; Cleary, 1979). The following table gives the average values obtained for each depth range.

The general tendency is an increase in strength with depth. Surprisingly little anisotropy was detected despite the laminations.

DATA ANALYSIS

A summary of the field and laboratory in-situ stress measurement program results is shown in Table 2 and Fig. 3.

TABLE 1. Summary of Laboratory Data

Sample Depth (ft)	Tensile Strength (psi)	Critical Stress Intensity Factor (psi in.$^{1/2}$)	
		Burst Test	Short Rod Test
394	785	457 [V]	562 [V]
454	1,040	--	--
511	420	--	519 [H]
574	1,900	--	641 [V]
614	--	--	--
654	1,300	801 [V]	819 [H]
704	--	401 [V]	406 [H]
718	1,040	914 [V]	1,200 [V]; 877 [H]

V or H indicates vertical or horizontal fractures, respectively.

TABLE 2. Summary of Field Data Interpretation

Fracture Number	Depth (ft)	σ_{HMAX} (psi)	σ_{HMIN} (psi)	σ_V (psi)	Orientation of σ_{HMAX}
1	718	2,260	1,210	796	--
2*	704	--	--	--	--
3	654	1,460	1,020	730	80°
4	614	1,280	910	690	67°
5	574	810	810	630	100°
6	511	1,080	720	590	94°
7	454	850	660	580	37°
8	394	--	550	410	--

*Mechanical failure of surface iron precluded measurement at this depth.

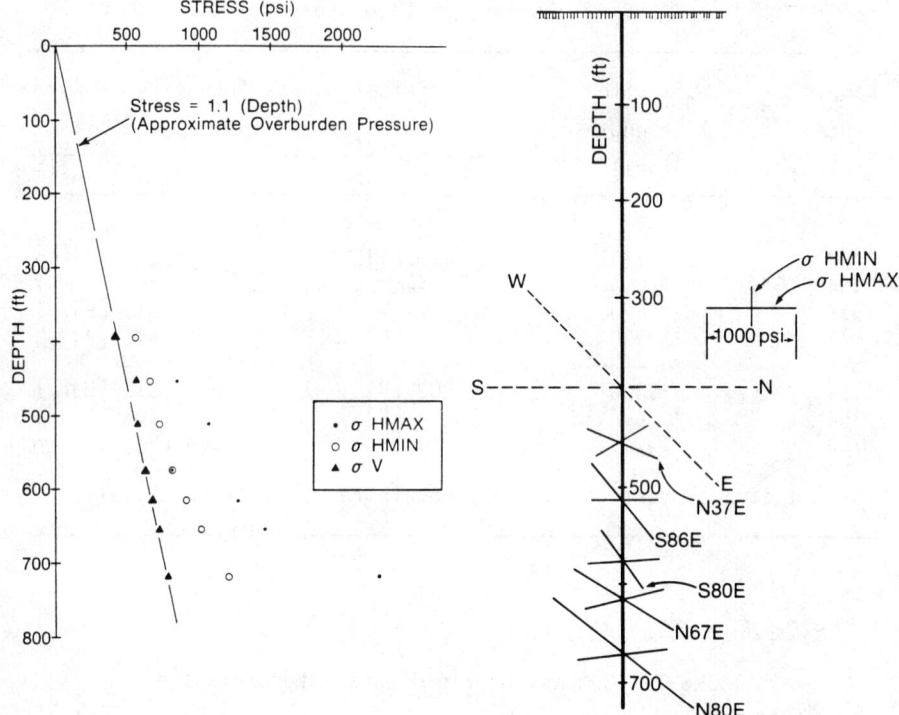

Figure 3A. Stress profile.

Figure 3B. Variation of stress and orientation with depth.

The general indications are as follows.

1. The orientation of σ_{HMAX} (σ_1) was measured to vary between N67°E and S80°E. This fits well with orientation of stress over a regional basis.

2. The stresses measured (the horizontal stresses are the maximum and intermediate principal stresses) fall within the limits of stresses measured in other parts of northeastern and north-central United States and southern Canada.

3. In all cases, except the uppermost interval, the complete stress tensor could be defined.

4. The vertical component, minimum principal stress gradient, corresponds closely to the anticipated overburden pressure.

5. The data interpretation, based on fracture mechanics considerations, can lead to a wide variation in the predicted values of

the maximum stress component (Fig. 4). Some of the reasons for this degree of uncertainty are that the following assumptions are usually made --

(a) symmetric bi-lobed fracture propagation;
(b) angle between σ_{HMAX} and fracture plane known;
(c) uniform pressure distribution in fracture (most formulas); and
(d) critical pre-existing fracture length.

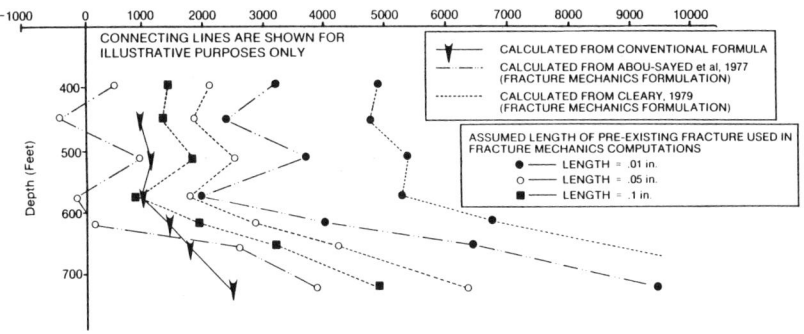

Figure 4. Fracture mechanics computations of σ_{HMAX}.

CONCLUSIONS

Hydraulic fracturing was conducted within borehole TX-11 in order to determine the magnitude and orientation of in-situ principal stresses for comparison with regional data.

At the shallower test depths, the tendency for $\sigma_1 \simeq \sigma_2 \simeq \sigma_3$ is well defined and gradient extrapolations of existing measurements to the surface are reasonable. High stress magnitudes were not experienced in either the tunnel or plant area excavations or concluded from measurements of extensometers installed in the bedrock walls of the emergency service water pumphouse. Conclusions regarding stresses from plant structure excavations are consistent with the gradient extrapolation of the deeper in-situ borehole measurements. Below a depth of approximately 600 ft, both σ_{HMAX} and σ_{HMIN} show an increase in gradient, with the gradient for σ_{HMAX} being larger. The gradient increase is attributed to changes in bedrock lithology rather than any structural discontinuities.

ACKNOWLEDGMENTS

The authors wish to express their gratitude to R. Wardrop of

Gilbert Associates, Inc., for his excellent field service, and to Cleveland Electric Illuminating Company for permission to publish this paper.

REFERENCES

Abou-Sayed, A. S., 1978, "An Experimental Technique for Measuring the Fracture Toughness of Rocks under Downhole Stress Conditions," VDI-berichte #313, pp. 819-824.

Abou-Sayed, A. S., Brechtel, C. E., and Clifton, R. J., 1977, "In-Situ Stress Determination by Hydrofracturing - A Fracture Mechanics Approach," Terra Tek Report TR-77-60.

Barker, L. M., 1977, "A Simplified Method for Measuring Plane Strain Fracture Toughness," Engineering Fracture Mechanics, Vol. 9, pp. 361-364.

Bayley, R. W., and Muehlberger, W. R. (compilers), 1968, Basement Rock Map of the United States, Exclusive of Alaska and Hawaii, U.S. Geological Survey, scale 1:2,500,000 (2 sheets).

Bredehoeft, J. D., Wolff, R. G., Keys, W. S. and Shuter, E., 1976, "Hydraulic Fracturing to Determine the Regional In-Situ Stress Field - Piceance Basin, Colorado," Geological Society of America Bulletin, Vol. 87, pp. 250-258.

Cleary, M. P., 1979, "Rate and Structure Sensitivity in Hydraulic Fracturing of Fluid-Saturated Porous Formations," 20th U.S. Rock Mechanics Symposium, Austin, Texas, pp. 127-142.

"Final Safety Analysis Report, Perry Nuclear Power Plant Units 1 and 2," Cleveland Electric Illuminating Company, Docket No. 40-440 and 50-441.

Heinze, W., Kellog, R. and O'Hara, N., 1975, "Geophysical Studies of Basement Geology of Southern Penninsula of Michigan," AAPG Bulletin, Vol. 59, pp. 1562-1584.

Hoover, K. V., 1960, "Devonian-Mississippian Shale Sequence in Ohio," Ohio Division of Geological Survey, Information Circular No. 27, 154 pp.

Owens, G. L., 1967. "The Precambrian Surface of Ohio," Ohio Division of Geological Survey, Report of Investigations No. 64, 30 p.

Rector, G. W., 1950, "Paleontology and Stratigraphy of a Well Core from Ashtabula, Ohio, M.Sc. Thesis, University of Michigan (unpub.) 37 pp.

Roegiers, J.-C., and McLennan, J. D., 1981, "Factors affecting the Initiation Orientation of Hydraulically Induced Fractures," Workshop on Hydraulic Fracturing Stress Measurements, Monterey, California.

Roegiers, J.-C., and McLennan, J. D., 1979, "Numerical Modelling of Pressurized Fractures," ISSN 0316-7968, Publication 79-02, University of Toronto, also EPB Open File Report 7208, Earth Physics Branch, Ottawa.

Chapter 24

STRESS MEASUREMENT OF ROCK MASS IN SITU AND
THE LAW OF STRESS DISTRIBUTION IN A LARGE DAM SITE

by Bai Shiwie and Li Guangyu

Institute of Rock and Soil Mechanics
Academia Sinica. Wuhan China

ABSTRACT

This paper presents the study in situ on stress field of rock mass in a large dam site. By means of two different measure instruments and the method of stress relief, a lot of measure in situ has been done, including stress measurement of three dimensions at ten points and two deep-hole stress measurements in the perpendicular boreholes which are 59.4m and 53m deep seperately, and the basic accordant results corresponding two methods have been got.

In the process of the deep-borehole stress measurement we got the stress value in critical state in which cylindrical core would become core-disking in the high stress area, which was first got in the world. The maximum principal stress was up to 650 kg/cm^2. The stress value obtained by test and the thickness of rock disks all basically accord to the analysis by mechanical mechanism of core-disking

The plane linear elastic finite element analysis has been completed. Results of inverse calculation all accord to the measurement results in four different positions. The whole outline of the stress field of dam site was delineated clearly.

LAW OF STRESS DISTRIBUTION

The investigating stress field of rock mass is one of crucial works to evaluate stability of rock engineering. The stress measurement of rock mass in situ and the finite element analysis is introduced in this paper.

METHODS FOR MEASURING ROCK STRESSES AND RESULTS

1. To clear up rock stresses around the dam, two different methods for measuring stresses in- situ were used: deep-hole stress-relief method with borehole deformeter as transducers, and stress-relief method with piezomagnetic stress units as transducers.

Borehole diameter deformation method uses type 36-2 borehole deformer as a transducer of whose sensitive elements are steel rings with good elasticity; and four strain gages, making up a double-bridge, are stuck on the ring. When deformation of borehole occurs, YJ-5 resistance strain instrument can record the message output from the double-bridge through rigid feelers; and then the change of diameter along feeler direction can be obtained (Fig 1.).

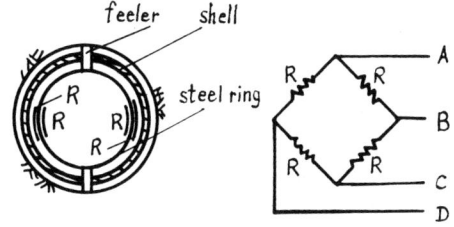

A. a sensitive element of deformers

B. measurement bridge

Fig. 1

The four steel rings fixed in a deformer can record at the same time deformations of a borehole along four orintations inclined at 45° each other. According to the deformations recorded, values of stresses and orientations in the plane normal to the axis of the hole would be calculated. Busing three intersected-boreholes method, values and orientations of cubic stresses can be obtained.

Undergoing a great number of tests for reliability and stability and engineering 36-2 borehole deformer has been proved stable and accurate enough to sense stress change up to 0.5 kg/cm , and has a relative error less than 10 % when measuring principal stress.

In piezomagnetic method, YJ-73 piezomagnetic stress gage is used as a transducer which is designed according to piezomagnetic effect.

When YJ-73 piezomagnetic stress instrument was put inside rock with E = 6×10^5 kg/cm² to measure stresses, it had such an accuracy that it could respond to a change of stresses of 0.2 kg/cm². Having been tested to inspect its reliability, YJ-73 has a relative error less than 10 % when measuring principal stress.

This paper would discuss the results measured from two methods in 1981-1982.

2. Measuring poits were mainly arranged inside rock mass where the axis of dam and underground building would be located. Cubic stress measurements with three intersected-boreholes method were performed in syenite in prospecting excavation 2 (pro. exc. 2), and in basalt in pro. exc.4. Two deep boreholes, 60m. in depth were arranged inside the dam base of the river bed (Fig. 2.).

Point 3, 4 and 5 were arranged inside syenite about 200m. in depth and 200-300m. horizontally away from the slope of the river bank There was overlay about 50m. above point 6 which was arranged into syenite. Point 7 and 8 did so inside basalt and point 9 did inside syenite. The results from point 10 and 11 are

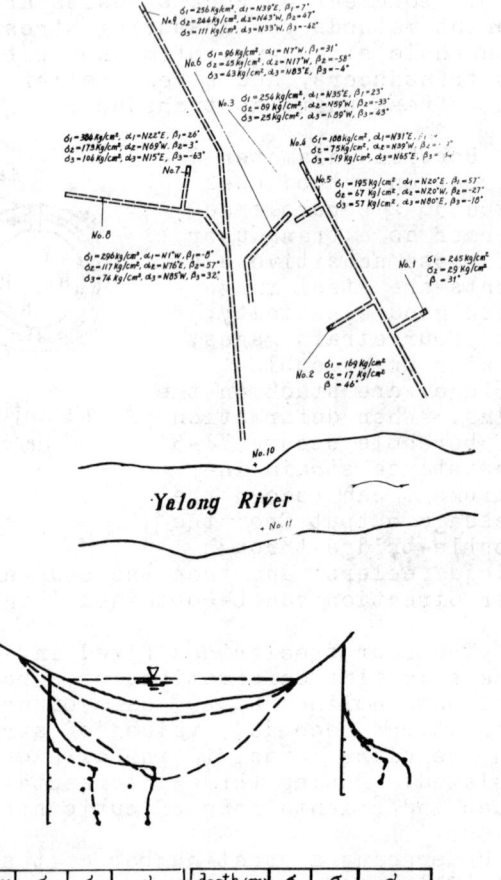

depth (m)	δ_1	δ_2	α	depth (m)	δ_1	δ_2	α
17.6	18	5	N78°W	21.9	15	-10	N6°E
24.5	50	12	N87°E	26.8	11	-20	N32°W
30.0	180	30	N80°W	31.4			
37.5*	650	291	N34°E	38.0	145	126	
40.5	659	259	N50°E	45.0	394	248	N12°E
45.0	494	202	N28°E	53.5	407	226	N32°E
55.3	600	315	N	* disking-core occured			
59.4	610	320	N 50°E	Fig.3			

shown in Fig. 3.

ANALYSIS OF RESULTS

1. Orientations of max. principal stress:

a) In general, stresses inside rock mass at both banks of incised river valley are intensely affected by valley slope topography. When there exists a larger horizontal stress , it is convenient for us to determine stresses in vertical borehole at the bottom of the bed. The orientation of axis of the borehole is basiclly consistent with that of min. stress; and orientation of max. stress obtained from this can basically represent real orientation of max. stress at the dam area. In analysing the results from point 10 and 11 (Fig. 3), it must be noted that the data obtained within less than 30m. can not sufficiently represent general characteristic orientation, because of some factors such as weathering, valley slope topography surface unloading. At the borehole more than 30m. in depth, all data show orientation of max. stress to be N-E, average N 40° E on the left bank, and N 22 E on the right. To determine the orientation of stress at the dam area does depend upon these accordant results.

b) At point 3, 4 and 5, the three units determined stresses respectively within less than 40m. at the same region by using two different measuring methods. The orientations of max. principal stresses measured range from N 20 E to N 35 E and accord basically to ones from deep borehole measurement.

Point 3, 4 and 5 were arranged into prospecting excavation 2, about 260-280m. far away from the slope of a bank. The results from these points show the orientations of initial stresses undisturbed.

Point 9 shows initial stress inside syenite deeply-coverd and its orientation to be N 39°E; point 7 shows the results inside basalt and the orientation of max. principal stress of N22 E accordant basically with above one.

c) The value of stress at point 6 inside prospecting excavation 24 is less than 100 kg/cm^2 and the orientation of max. stress is N 7°W. Elevation of excavation 24 is 100m. higher than that of excavation 2; therefore, at the point 6 above which thickness of overlay is only about 50m., thevalue of stress is a little lower. It is easier to understand. For the orientation of stress, it depends on local topography where the orientation of a mountain ridge is just N-S. Thus it can be seen the results at this point depend on particular topography.

point 8 was arranged inside basalt where there exists an alternation of basalt and syenite penetrating into the former; and initial stress field was disturbed; and later, adjustment of the stresses was affected by local stress cincentration due to the lithologic change. As a result, the orientation (N 1°W) of max. principal stress at this point is much more different from others, and it is impossible to infer general law of orientation from this datum.

d) In examining topography of the dam area, the geomorphic features clearly show that the region mainly made of syenite is just a V-shaped valley with strike S 60°E, near both banks of which contours are approximately parallel to the orientation of river bed; it is normal to the valley and the orientation along the ridge is just N 30°E which is accordant with that of max. principal stress.

It was seen in the engineering practic that most parts peeling off from the walls of adits took place in the branched excavations running at N 47 W, which proved the maximum principal stresses were from N-E orientation. From above, it is appropriate to regard the orientation of maximum principal stress as N 30°E.

2. The stress magnitude

a) Measure Points 3,4,5 were arranged in excavation 2, 260- 280m. deep covered with about 200m. where the rock mass was fresh and intact, without big tectonic rift, so the measured results of these 3 points are the important basis to determine initial stresses in undisturbed syenite. Points 3-5 showed maximum principal stress was about 200 kg/cm^2-250 kg/cm^2, and point 9 showed maximum principal stress was 256 kg/cm^2. Their magnitude orders are just the same.

b) Fig.3 shows deep hole stress measured results of Points 10 and 11. On the left bank when the hole depth was over 37m, the stress was over 600 kg/cm^2, and on the right bank when it was over 45m, the stress was over 400 kg/cm^2, which was the result of great stress concentration in the bottom of river valley.

The directions of stress were disorderly because of the effect of the bank slope and its value was smaller when the hole depth was less than 30m., which demonstrated that the rock mass near surface of bottom of river bed was faulted due to greater stress, the stress was released along the surface area and the stress concentration region moved into deeper portion.

c) During measuring at Point 10 with 37.5m in depth and there appeared 10 circle crackles which were distributed at 3cm intenals even in along 33cm core. The cracles did not completely open which represented the character of tention failure. Picture (Fig.4) showes the core obtained after the test.

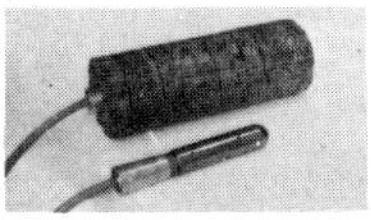

Cubic axi-symmetric finite element analysis demonstrated that under axi-symmetric lateral load the core bottom with central hole had tensile stress along symmetric axis during drill process. Using the mechanical parameter of rock in the dam site, Fig.5 showes the distrebution of tensile stress perpendicular to axial section. When relief deepth reaches to 3cm, the ratio of axial tensile stress to lateral compression stress was up to 0.163. In the dam site the tensile strength of syenite rock was 87 kg/cm², thus the lateral axi-symmetric compression stress 534 kg/cm² could lead to circle crack of core along the plane of maximum tensile stress, because the maximum value of stress concentration appeared on the outside surface of cylindrical core, crack began from the outside to form circle shape, and maximum interval of two crack was 3cm. When lateral stress got great enough, the core was cut completely. The more lateral stress increase, the thiner rock disk will become.

It was clear that over great horizontal stress would bring about axial tensile stress so as to form circle crack at the bottom of core though the horizontal stresses in the rock mass under the river bed were not axi- symmetrical.

d) Point 7,8 were in basalt and their other conditions were similar with Point 3,4,5,9, but it was apparent that the stress in basalt was greater than that in syenite (refer to Fig.2). The main reason bringing about the difference is due to the different rock mechanical properties of

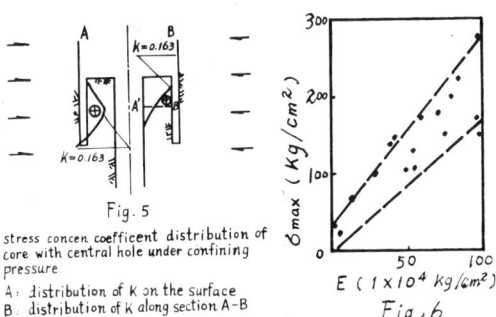

Fig. 5
stress concen coefficent distribution of core with central hole under confining pressure
A: distribution of k on the surface
B: distribution of k along section A-B

Fig.6

measure points. The compression strength and elastic module of basalt are higher than that of syenite and the rock mass with higher elastic module would store more elastic strain energy under the same strain condition. Fig.6 summed up the relation of a great many stresses obtained in in situ with rock elastic modulus. It is seen that stiff rock has better condition to store energy and its stress value would be higher than that of the rock with lower mechanical strength or elastic module.

3. Whole outline of stress state

On the basis of analysis of direction of principal stress, it was considered that the direction of maximum principal stress was perpendicular to the direction of valley N30°E. So the section along Excavation 4 and the calculation model involved 1000m in width was shown as fig.7 in order to form the whole outline of stress state. Simply, the calculation model of left bank was considered equilibrium at the central protion of valley , for the dam was located in nonsymmetrical V-shaped valley where, on the left, the rock properties were unitary and there were more aboundant data, but on the right the variation of the rock properties was not little and there was a little work on it. It is

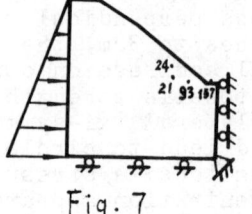

Fig. 7

well-known fact that horizontal stress increases linearly with the depth though there has been no mature theory about the distribution law of stress in superficial layer of crust. Therfore the finite element analysis considered effect of gravity and laterally applied triangular load increasing with depth.

The parameters used for calculation were $\gamma = 0.0027$ kg/cm^3, $E = 2.5 \times 10^5$ kg/cm^2, $\mu = 0.17$, let $\sigma_x = k\gamma H$, where k was coefficient of lateral pressure which was used to determine the grade of triangular load.

Table 1 showes the calculated value of stresses of some points compared with corresponding measured stresses.

Knot	σ_1	σ_2	α	Point	σ_1	σ_2	β
24	135		+33°	6	96		+31°
21	198		+25°	5	195		+57°
93	242	30	+25°	1	245	29	+31°
157	657		+7°		659		

Table 1

CONCLUSIONS

1. The direction of stress in rock mass near the bank slopes was controlled to a greater extent by topography. A great deal of work in situ showes the orientation of maximum stress is N30 E and perpendicular to the run of river, but the dip angle just directs the top mountain. This orientation has an important significance to determine axial orientation of underground building.

2. The dam site is a region with high stress. The maximum principal stress in undisturbed syenite is about 200 kg/cm^2, and that in basalt is about 300 kg/cm^2. The maximum principal stress intersects horizontal plane at small angle which shows that there is horizontal tectonic stress and its effect is much more than that of gravity.

Variation of stresses in different protions refer to the calculation result of the stress model.

3. The successful stress measurement in the core with cricle crack showes that crack begins from outside. The result from the finite element analysis accoeds with experamental evidence on thickness of cake-shaped core, critical stress, and crack mechanism.

4. Both results of calculation and measurement show serious stress concentration appeared in the bottom of river bed. The maximum principal stress reached 600 kg/cm^2. The estimation on the basis of calculation shows the stress concentration range is not less than 160m.

The stress concentration region of the bottom of river bed is not in surface of rock but after the depth of more than 37m. Stress concentration region was transfered into deeper portion.

5. The rock masses in different geological regions have different conditions to integrate stress because of different mechanical properties, so the rocks with higher elastic modulus would have higher stress.

The stress measuring work in the dam site region was together completed by Institute of Rock and Soil Mechanics Academia Sinica, Wuhan China, The Research Institute of Prospecting and Desinging Institute, Chengdu, and Earthquake Brigade of the State Bureau of Seismology, Chanhe, and support of borther organizations were important to carry out the study successfully. We hereby express our thanks.

REFERENCES

Li Guangyu, Bai Shiwei, 1979, "In-situ Study on Stress in Rock Mass", Mechanics of Rock and Siol, 1979, No.1.

Chapter 25

PREDICTION OF HYDRAULIC FRACTURE AZIMUTH FROM ANELASTIC STRAIN
RECOVERY MEASUREMENTS OF ORIENTED CORE*

by Lawrence W. Teufel

Sandia National Laboratories
Albuquerque, New Mexico

ABSTRACT

An anelastic strain recovery technique is presented as a method of determining the directions and ratio of the maximum and minimum horizontal in-situ stresses from oriented cores from deep wells. Predictions of hydraulic fracture azimuth inferred from principal stress directions determined from both elastic and time-dependent (anelastic) strain relaxation measurements of volcanic tuff are consistent with the observed orientations of hydraulic fractures produced in simulation and mineback experiments, underground at the Nevada Test Site. In addition, anelastic strain recovery measurements on oriented cores of shale and sandstone from well depths of about 1680 meters show consistent relaxation directions of the principal horizontal strains and have enabled a prediction of the principal horizontal stress directions and hydraulic fracture azimuth.

INTRODUCTION

As a result of an increased need for better recovery techniques to enhance gas production, massive hydraulic fracturing has been used in low permeability, gas-bearing sandstones of the western United States, where it is uneconomical to retrieve gas in the conventional manner. Massive hydraulic fractures are designed to extend as much as 1000 m radially from the well core and generally require up to 3×10^3 m^3 of fracture fluid. However, even with these large scale fracture-stimulation treatments much of the reservoir may be

---
* This work performed at Sandia National Laboratories supported by the U. S. Department of Energy under Contract # DE-AC04-76-DP00789.

left unrecovered because the drainage pattern around a fracture-stimulated well in tight reservoirs is elliptical rather than radial (Smith, 1979). This elliptical drainage pattern is enhanced as the fracture increases in length. Consequently, well location and spacing based on radial drainage may not be the optimum, economical gas recovery procedure in a particular field because of overlapping drainage at fracture tips and dead zones between fracture flanks. Accordingly, a reliable method of determining or predicting the fracture orientation is of great importance to the gas industry, because a knowledge of the fracture orientation would allow optimum spacing of wells in a developing field and thus could increase recoveries an additional 30% of the total reservoir (Smith, 1979).

Over the past several years much work has been done attempting to predict or measure in-situ fracture azimuth. Prediction efforts have been indirect and generally have related to either paleo-tectonic analysis (Smith, et al., 1978) or oriented core analysis to determine mechanical property anisotropies (Logan and Teufel, 1978). Direct measurements of in-situ hydraulic fracture propagation include tilt-meter surveys (Wood, et al., 1976), surface electric potential methods (Bartel, et al., 1976) and downhole passive seismic surveys (Schuster, 1978). The major problem with any of the fracture propagation surveys is depth. Resolution drops off rapidly as depth (and distance from sensors) increases. To date, the success of both indirect prediction efforts and direct measurements of fracture propagation have not always been satisfactory.

Another approach at predicting fracture orientation is to measure directly the in-situ stress field, since this is the dominant factor influencing fracture orientation (Hubbert and Willis, 1956; Warpinski, et al., 1980). Hydraulic fractures propagate perpendicular to the minimum principal compressive stress. Methods of determining the in-situ stress state in rock masses can be grouped into two general categories: hydraulic fracturing of open boreholes and stress relief methods.

Hydraulic fracturing may be a reliable method for directly determining the minimum principal compressive stress magnitude from the instantaneous shut-in pressure during the fracturing process, but a major shortcoming with this method is determining the stress directions from the fracture azimuth (Warpinski, et al., 1980). Fracture azimuth as detected at the borehole by fracture impression packers or borehole televiewers may be strongly influenced by borehole damage. Moreover, the reliability of these fracture detection methods is significantly reduced with increasing depth. In addition, there has been reluctance in industry to use this method in deep wells because a fracture is being produced in an uncased, open hole with attendant zone isolation and well control problems.

The stress relief method is generally accomplished by an over-coring process which allows the physical detachment of a body of rock from the rock mass, thus permitting the body to undergo differential relaxation in relief of in-situ stored strain energy (Obert and Duval, 1967). The recovered strain is measurable with strain gages and typically involves both an instantaneous, elastic component and a time-dependent (anelastic) component. Typically displacement or strain gages are installed at the bottom of a borehole prior to overcoring and stress relief, and the measured strain recovery can be related to the total stress components which acted prior to relief. A measure of the total strain relief (both elastic and anelastic) and the calculated total stress state is restricted to shallow depths. However, if gages are installed subsequent to stress relief, a partial component of anelastic strain recovery can still be determined. Voight (1967) first suggested that if the assumption is made that the partial anelastic recoverable strain is proportional to the total recoverable strain, then an approximate estimate of the in-situ stress state at depth can be made by instrumenting oriented drill cores upon removal from the borehole. Voight noted that there is empirical justification to consider that the recovered anelastic strain will be proportional to the total recoverable strain, and hence to the in-situ stress state in rheologically-isotropic rocks. Strain and stress ellipsoids calculated from anelastic strain recovery measurements will therefore be homothetic with ellipsoids which existed prior to relief. However, if nonlinear creep occurs there may be a rotation of the principal axes of strain relief with time. Then the correct orientations and ratios of the principal in-situ stresses may not be possible. An assessment of any nonlinear creep behavior can be made by monitoring the uniformity of the principal strain directions and ratios during anelastic recovery.

The purpose of this paper is to assess the feasibility of using partial anelastic strain recovery measurements on oriented cores from deep wells as a method of determining the directions and ratios of in-situ stresses and thus enable prediction of a hydraulic fracture azimuth. As part of Sandia National Laboratories stimulation and mineback experiment, an underground experiment in volcanic tuff at the Nevada Test Site, a direct comparison was made between observed hydraulic fracture azimuths and in-situ stress orientations determined from both elastic and anelastic strain recovery measurements. The principal directions of the elastic and anelastic strains were essentially coaxial, with the minimum strain (stress) perpendicular to the hydraulic fracture azimuth. In addition, anelastic strain recovery measurements on oriented core of Mesa Verde sandstone and shale from a deep well in the Piceance Creek Basin, Colorado were made and indicate that anelastic recovery occurs consistently over a sufficiently long time period to allow an approximate determination of the principal horizontal in-situ stress directions and ratio and therefore a prediction of the hydraulic fracture azimuth in deep wells.

PREDICTION OF HYDRAULIC FRACTURE AZIMUTH

PROCEDURE

In order to make an approximate determination of the in-situ principal stress ratios and directions at depth using anelastic strain recovery measurements, three strain measurements on three specified planes of an oriented core is required. However, if the stratigraphic beds are essentially horizontal and the assumption is made that either the maximum or intermediate principal stresses is vertical and equals the pressure due to the weight of the overburden, then only the directions of the principal horizontal strains are necessary to determine the azimuth of a vertical hydraulic fracture. To determine the directions and ratio of the principal horizontal strains only three measurements along known directions in that plane are required.

Anelastic strain recovery measurements have been made previously (Swolfs, 1975; Strickland, 1980) but the results have been inconsistent. The inconsistency of these measurements may be due, in part, to the procedure used in measuring strain relief. The previous studies have always used strain gages mounted on the surface of unsealed cores. Consequently, the effects of moisture evaporation and the problems of adequately bonding strain gages to the core surfaces which are saturated may have affected the strain relaxation measurements. Accordingly, in this study a set of clip-on disc gages (Schuler, 1979) are used rather than strain gage rosettes because they can be more easily placed on sealed core and within a considerably shorter time span. The disc-gages measure displacements across the entire diameter of the core. Displacements are converted to strain and for a 10.20 cm diameter core the sensitivity of the gages ranges from 2 to 8 microstrains. To determine the principal horizontal strains, three disc-gages are mounted at $45°$ to each other. The disc-gages are attached directly to the core surface by cutting slots in the polyurethane wrapping that seals the core and then resealing with epoxy. Data for each disc-gage is recorded manually every hour using a converted strain indicator box. The magnitude and orientation of the principal strains in the horizontal plane of the core are calculated following Frocht (p. 36, 1948). In addition, the measurements are made in a temperature controlled environment to avoid thermal effects.

RESULTS

Comparison of Elastic and Anelastic Strain Recovery

In order to compare the orientation and magnitude of the instantaneous elastic strain relief and the subsequent anelastic strain recovery of a core, a single horizon of volcanic tuff in G tunnel at the Nevada Test Site (depth 446 m) was selected for three overcoring strain relief measurements. Using a standard door-stopper technique overcoring measurements were made at the bottom of a 1.32 m hole, which was 22 cm in diameter. After overcoring, the elastic strain relief was measured and the cores (10.2 cm in diameter) were then extracted and sealed with polyurethane wrapping. Anelastic strain recovery measurements were made using disc-gages mounted to sealed cores following procedures previously described.

The instantaneous, elastic strain relaxation and the total time-dependent relaxation of three cores of volcanic tuff are given in Table 1. All of the strain relaxations were expansions. The elastic recovery constituted about 70 percent of the total relaxation. Shown in Figure 1 is a strain relief-time plot of the principal strains of the anelastic recovery for one core, determined from the strain relief measurements of three disc-gages over 4 hour time intervals. The recovery rate decreased rapidly with time and all of the relaxation took place within 30 hours. The anelastic recovery of the cores was fairly uniform with time as shown by the ratio of the principal strains. The ϵ_1/ϵ_2 ratio ranged from 1.3 to 1.8 during relaxation and the final strain ratio is within 10 percent of the elastic strain relief ratio (Table 1). For a given core the directions of the principal strain relief show limited variation with time, having a total variation of 28° or less during relaxation and a maximum difference of 13° between the principal directions of the mean anelastic strain relief and the elastic strain relief.

These strain relief measurements were made in conjunction with Sandia National Laboratories stimulation and mineback experiment. In this experiment dyed-filled hydraulic fractures were initiated in different horizons of volcanic tuff and then the vertical hydraulic fractures were exposed during subsequent mining to determine hydraulic fracture azimuth and geometry (Warpinski, et al., 1982). Twenty-two of these hydraulic fractures were near the overcoring site. The azimuth of these hydraulic fractures ranged from N34°E to N55°E. The direction of the maximum principal horizontal strain (stress) determined from both elastic and anelastic strain relief measurements ranged from 27°E to N61°E. Thus, there is excellent agreement between the stress direction determined from the strain relief measurements and the hydraulic fracture azimuth.

Table 1. Summary of Strain Relaxation Data from Volcanic Tuff

Core	Elastic Strain Relief				Anelastic Strain Relief					
	Final Principal Strain Magnitude		Ratio of ϵ_1/ϵ_2	Azimuth of ϵ_1^* (deg.)	Final Principal Strain Magnitude		Ratio of ϵ_1/ϵ_2		Azimuth of ϵ_1	
	ϵ_1 ($\mu\epsilon$)	ϵ_2 ($\mu\epsilon$)			ϵ_1 ($\mu\epsilon$)	ϵ_2 ($\mu\epsilon$)	Range	Mean	Range (deg.)	Mean (deg.)
1	984	641	1.54	50	358	238	1.35-1.62	1.49	45-61	55
2	1028	617	1.67	35	309	192	1.43-1.81	1.58	27-55	48
3	845	578	1.46	47	267	168	1.28-1.65	1.40	28-58	42
All Data†	952	612	1.56	44	311	199	1.28-1.81	1.49	27-61	48

*Azimuth is given as North being 0°, East being 90°, South being 180°, and West being 270°.
†Average ± one standard deviation, unless given as a range.

Anelastic Strain Recovery of Sandstone and Shale

Anelastic strain recovery measurements were made on cores from a well in the Piceance Creek Basin, Colorado, to determine if relaxation occurs over a sufficiently long time period to be applicable for estimating horizontal in-situ stress directions at depth. The cores were sandstone and shale from the Mesa Verde Formation and came from depths of 1682 to 1685 meters. The anelastic recovery of one of five instrumented cores is presented in a strain relief-time plot of the horizontal principal strains (Figure 2). Strain relief monitoring began 4 to 7 hours after the rock was cored. For a given core the principal strain directions did not vary over 26° during relaxation (Table 2). For all the cores the maximum principal strain direction, ϵ_1, had an average of N85W with a standard deviation of 8°.

SUMMARY

Anelastic strain recovery measurements have been made on oriented cores of volcanic tuff from an underground tunnel at the Nevada Test Site and oriented cores of Mesa Verde sandstone and shale from a deep well in Colorado. The horizontal anelastic recovery of the cores was fairly uniform with time, suggesting that the in-situ rock, in each case, may have been isotropic in the horizontal plane. Accordingly, the principal strains determined from any given time segment of these recovery curves are probably homothetic with the stress state prior to relief. Thus, measurements of a portion of the anelastic strain recovery of oriented cores may provide a good approximation of the principal in-situ stress directions at depth and enable a prediction of a hydraulic fracture azimuth in deep wells. However, it should be noted that stress orientations from partial anelastic

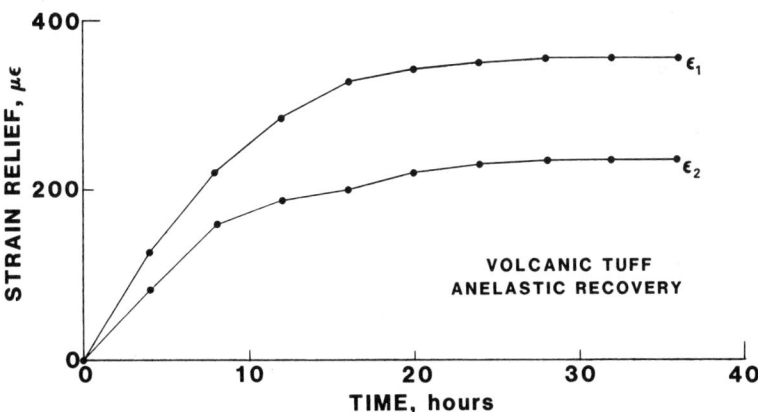

Fig. 1. Strain relief-time curves of the horizontal principal anelastic strains of a volcanic tuff core. Elastic strains (not shown) were 984 με and 641 με for ϵ_1 and ϵ_2.

strain recovery measurements assume that the in-situ rock is isotropic. A strong anisotropy in the rock prior to relief, such as one produced by a tectonic microcrack fabric, may bias the strain relaxation measurements, and thus could lead to an inaccurate determination of the in-situ stress directions. Detailed petrographic and directional petrophysical measurements on the relaxed oriented cores would provide evidence of anisotropy and if it exists at elevated confining pressure.

Table 2. Summary of Anelastic Strain Recovery Data from a Well in the Piceance Creek Basin, Colorado

Core Depth (m)	Rock Type	Final Principal Strain Magnitude		Ratio of ϵ_1/ϵ_2		Azimuth of ϵ_1*		Elapsed Time† (Hours)
		ϵ_1 ($\mu\epsilon$)	ϵ_2 ($\mu\epsilon$)	Range	Mean	Range (deg.)	Mean (deg.)	
1682.2	Shale	186	107	1.66-1.89	1.74	70-93	82	6.6
1682.5	Shale	171	94	1.72-1.95	1.82	79-103	91	6.1
1684.3	Sandstone	216	137	1.45-1.71	1.61	93-116	105	5.1
1684.6	Sandstone	235	148	1.39-1.75	1.57	86-111	100	4.7
1684.9	Sandstone	207	130	1.49-1.80	1.64	85-103	96	3.9
All Data**				1.39-1.95	1.67 ± .06	70-116	95 ± 8	

* Azimuth is given as North being 0°, East being 90°, South being 180°, and West, being 270°.
** Average ± one standard deviation, unless given as a range.
† Time interval between coring and onset of strain relief monitoring.

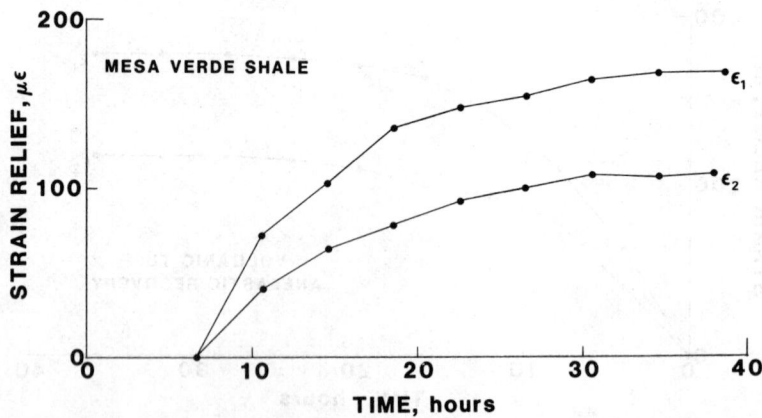

Fig. 2. Strain relief-time curves of the horizontal principal anelastic strains of a Mesa Verde shale core. Strain relief monitoring began 6.6 hours after coring.

REFERENCES

Bartel, L. C., McCann, R. P., and Keck, L. J., 1976, "Use of Potential Gradients in Massive Fracture Mapping and Characterization," SPE #6088.

Frocht, M. M., 1948, "Photoelasticity," John Wiley and Sons, New York, V. 1., 441 p.

Hubbert, M. K. and Willis, D. G., 1957, "Mechanics of Hydraulic Fracturing," Trans. A.I.M.E., 20, 153-168.

Logan, J. M. and Teufel, L. W., 1978, "The Prediction of Massive Hydraulic Fracturing from Analyses of Oriented Core," in 19th U.S. Rock Mechanics Symposium, Reno, Nevada, May 1-3, 340-344.

Obert, L. and Duvall, W. E., 1967, "Rock Mechanics and the Design of Structures in Rock," Wiley, New York, 558 p.

Schuster, C. L., 1978, "Seismic Activity Generated by Hydraulic Fracturing," presented at ASME Petroleum Division Conference, Houston, Texas.

Smith, M. B., 1979, "Effect of Fracture Azimuth on Production with Application to the Wattenberg Gas Field," SPE #8298.

Strickland, F. G. and Ren, N., 1980, "Predicting the In-Situ Stress for Deep Wells Using Differential Strain Curve Analysis," SPE #8954.

Swolfs, H. S., 1975, "Determination of In-Situ Stress Orientation in a Deep Gas Well by Strain Relief Techniques," unpublished Terra Tek Report TR 75-43.

Warpinski, N. R., Schmidt, R. A., and Northrop, D. A., 1980, "In-Situ Stresses: The Predominant Influence on Hydraulic Fracture Containment," SPE #8932.

Warpinski, N. R., Finely, S. J., Vollendorf, W. C., O'Brien, M., and Eshom, E., 1982, "The Interface Test Series: An In-Situ Study of Factors Affecting the Containment of Hydraulic Fractures," Sandia National Laboratories Report SAND81-2408.

Voight, B., 1968, "Determination of the Virgin State of Stress in the Vicinity of a Borehole from Measurements of a Partial Anelastic Strain Tensor in Drill Cores," Felsmechanik. V. Ingenieurgeol, 6, 201-215.

Wood, M. D., Pollard, D. D., and Raleigh, C. B., 1976, "Determination of In-Situ Geometry of Hydraulically Generated Fractures Using Tiltmeters, SPE #6091.

LABORATORY TESTS

Chairman
Wilbur H. Somerton
University of California
Berkeley, California

Co-Chairman
Aytekin Timur
Chevron Oil Field Research Co.
La Habra, California

Keynote Speaker
Hugh C. Heard
Lawrence Livermore National Laboratory
Livermore, California

Chapter 26

MECHANICAL, THERMAL, AND FLUID TRANSPORT
PROPERTIES OF ROCK AT DEPTH

H. C. Heard

Lawrence Livermore National Laboratory
University of California
P.O. Box 808
Livermore, California 94550

INTRODUCTION

As the world's population expands and nations struggle to better their relative position and standard of living, increased emphasis is being directed to the exploration and production of mineral resources, to the mitigation of natural hazards such as earthquakes and volcanic eruptions, and to geotechnical engineering projects such as high-level nuclear waste disposal, underground military facilities, and transportation tunnels. Several topics are common to all of these endeavors. One is the characterization of the rock mass in situ, including the definition of the initial environment (e.g., temperature, stress, etc.). A second is, depending upon the use of the underground or the desired result, the manner in which the boundary conditions or the environment are to be changed by man. Next, the physicochemical behavior of the rock mass must be measured, or inferred with uncertainty limits placed on this behavior, over the appropriate range of environmental conditions (e.g., temperature, chemistry, etc.) for the application proposed. And last, an appropriate, validated predictive model must be applied to estimate the rock mass response to the perturbation.

In the period preceding about 1955, only relatively simple testing of soils and rocks was commonplace, partly because the environment (depth) to which these data were applied was not so extreme as it is at present and partly because less precise predictions of material response were acceptable. Only limited and relatively simple test data seemed to be needed to guide the engineering. At the present time and certainly in the future, as we extend our interaction with the earth's crust to greater depths and impose more extreme perturbations on it, we need a larger body of data, determined under much more complex conditions, in order to make the required predictions. The present societal problem of high-level radioactive waste disposal is a case in point. We in the rock mechanics community are being asked to assure that heat-producing toxic waste will remain in a very localized region at depth for periods of 10^3 to 10^6 years. Before this assurance can be made with any degree of certainty, we must have test results (and validated models) in which the coupled time-dependent physical (mechanical)-thermal-chemical-hydrologic behavior of the waste-rock system is well-defined and understood. Earthquake prediction (time, place, and magnitude) is an equally complex problem.

Laboratory test data can be vital to understanding rock behavior under some set of mechanical-thermal-hydrologic-chemical conditions but in themselves cannot be directly substituted for field results. Many phenomena can be more easily or economically investigated in the laboratory because of the ability to control: (1) various boundary conditions; (2) pressure, temperature or pore pressure; (3) loading or thermal history; and (4) fluid flow rate or fluid chemistry. In addition, in situ environments simulated in the laboratory may be used to develop or calibrate field measurement techniques. Depending on the type of test or test conditions desired, the capability of common laboratory apparatus may accommodate samples ranging from 10^{-2} to 1 m--far smaller than is sometimes required to simulate some processes in nature: $\sim$ 1 m to in excess of 10^5 m. The purpose of this contribution is threefold: (1) to briefly identify those laboratory measurements which have been conducted in the past which are of use to predictive modeling of underground rock behavior; (2) to point out those types of laboratory tests which are desirable and are technologically feasible; and (3) to make recommendations for future research in coupled thermal, mechanical, and hydrologic laboratory testing--the theme of this session.

NATURAL IN SITU ENVIRONMENT

Among those environmental parameters identified as affecting rock behavior are: lithostatic pressure (P), temperature (T), pore fluid pressure (P_p), deviatoric stress field (σ), and chemical environment (X). Thermal, chemical, and hydrologic transport through the rock at depth also involve gradients in any of these with distance and variations with time. It is obvious that wide ranges in these parameters occur naturally in the crust. For example, local hot spring or volcanic activity indicates very much higher local temperatures than may exist elsewhere. Artesian springs or extreme fluid pressures in some oil fields, as compared to other regions, illustrate quite different fluid pressure gradients. The presence of naturally deformed rocks in mountain chains and nearby basins or local earthquake activity suggests high local stresses.

In order to compare the laboratory capability to simulate and control P, T, P_p, σ, and X which may be broadly "average" or "common" over large surface areas and depth in the crust, we must first estimate the natural environment. Lithostatic pressures in the crust may be calculated from density and the height of the overlying rock column. For granitic terrains, calculated pressures increase by about 26-MPa/km depth. Pressures from thick sedimentary basins can be expected to be slightly less and somewhat nonlinear with depth, especially near the surface (Dickinson, 1953). The in situ vertical stress determinations compiled by Hoek and Brown (1980) indicate large deviations from these average values in the shallow crust.

Crustal temperatures may be estimated from borehole temperatures and heat flow observations, coupled with thermal conductivity measurements and indirect evidence suggesting the chemical composition of the rocks at depth. In thick sedimentary basins in the continental crust such as the U.S. Gulf Coast, thermal gradients range from about 22°C km^{-1} for offshore Louisiana to about 36°C km^{-1} for onshore Texas (Nichols, 1947; Moses, 1961; Gretener, 1967). Linear extrapolation of these data to depths encountered in sedimentary basins would lead to widespread melting in shallow regions. No evidence exists for such melting and thus the gradient must decrease with depth. Lachenbruch (1968) has calculated temperature against depth for the Sierra Nevada batholith, a thickened region of the crust composed primarily of grandiorites. In addition, Herrin (1972) has calculated crustal temperatures for the Basin and Range provinces as well as the Canadian Shield with similar results. If we assume that these profiles are representative of the continental crust as a whole, then the temperatures that might be expected could extend from ambient values at the surface to 200-350°C at 10 km, 300-550°C at 20 km and perhaps 450-600°C at 30 km.

Pore fluid pressures at depth are strongly dependent upon the fluid density and viscosity, the rock permeability, the rate of deformation and compaction (or expansion) of the rock, as well as upon any dehydration or reaction of the mineral phases present. In many sedimentary basins, pore pressures at depths to several kilometers can be approximated by the pressure exerted at the base of a water column extending to the surface (Hubbert and Rubey, 1959). However, at greater depths, Hubbert and Rubey (1959), Rubey and Hubbert (1959), and Berry (1973) present data suggesting that pore fluid pressures in sediments often approach and occasionally are slightly greater than those exerted by the overburden. These locally high fluid pressures are presumably the result of lowered rock permeabilities coupled with relatively high compaction rates or of dehydration reactions of hydrous minerals during prograde metamorphism. Few deep borehole observations are available from crystalline igneous terranes; both normal and high values of pore pressure at depth have been reported in these rocks (Berry, 1973).

Estimates of maximum principal stress differences can be made from surface topography coupled with assumptions as to the crustal response. A calculation by Jeffreys (1970) showed that, for a 1-km surface load (density = 2.5 Mg m^{-3}) imposed on a low-density elastic crust floating on a denser substratum, stress differences can reach 300 MPa in the upper crust. If the crust is assumed to be elastic-plastic or if it possesses a discontinuity at the edge of the 1-km load, maximum values are nearer 200 MPa (Birch, 1964; Jeffreys, 1970). The very large gravity anomalies in Hawaii or in the Indonesian Arc suggest stress differences in the crust of about 200 MPa (Birch, 1964). The vertical and horizontal stresses compiled by Hoek and Brown (1980) from in situ measurements worldwide indicate deviations in the ratio of horizontal to vertical stress up to a factor of 6 in the uppermost 3 km.

The chemical environment of the pore fluid-rock system can vary widely. Mineral solution in pore fluids with subsequent redeposition, solid-solid or solid-melt phase reactions (see, for example, Griggs, 1940; Bridgman, 1945; Griggs and Handin, 1960), can enhance ductility and degrade strength in certain rocks. Trace amounts of H_2O affect the strength of silicate rocks (Griggs and Blacic, 1965; Griggs, 1967; Atkinson, 1979). Dissolved material in solution can enhance or degrade fluid flow and thermal transport through porous rocks.

When the initial environment is perturbed by a drillhole or creation of other free surface and the underground is then exploited, large changes may occur in any or all of the parameters P, T, P_p, σ, and X. For example, production of geothermal fluids, cryogenic gas storage, or injection of fluids for hydrocarbon reservoir stimulation alters local P, T, P_p, and σ. In situ combustion of coal or oil shale deposits will affect P, T, P_p, σ, and X. Mined storage cavities or deep mineral mining change local T, P_p, and σ. In estimating the most extreme environment likely to be encountered in rock after modification of the initial conditions in situ by man, we would somewhat subjectively set the range as: P = 0.1 to 1000 MPa, T = -160 to 1100°C, P_p = 0.1 to 1000 MPa, σ = 0 to the failure strength, while no limits may be placed on X.

LABORATORY MEASUREMENTS

Large amounts of physical (mechanical) data are available on most types of rocks tested over virtually the entire range of the above conditions. Most results have been accomplished in compression and on 10^{-2} m dimension samples. Less data are available in other stress/strain states: tension, extension, torsion, bending, one-dimensional strain loading, proportional strain loading, or hydrostatic compression. Limited results have been recently obtained under some of the above conditions for 0.1-m dimension samples (Trimmer et al., 1980). The entire range of elasticity, fracture, and flow has been covered for most rock types. Surface friction, internal friction, and plastic flow have been investigated. In addition, much rheological data are available, both as constant strain rate and creep tests. For detailed results, the reader is referred to Griggs and Handin (1960); Clark (1966); Heard et al. (1972); Vutukuri et al. (1974); Heard (1976); Lama and Vutukuri (1978a, b); Paterson (1978); Carter et al. (1981); Touloukian and Ho, (1981); and Heuze (1982).

Much more limited thermal properties data (thermal conductivity, thermal diffusivity, heat capacity, and thermal expansion) are available for rocks tested under controlled conditions of P, T, P_p, σ, or X. Most of these data have been determined on test samples within the range of dimensions noted above. Limited data have been reported on rocks under pressure: (Walsh and Decker, 1966; Lyubimova et al., 1979; Durham and Abey, 1981a,b). Although thermal expansion has been examined in more detail than other thermal properties, virtually no data have been measured at P, P_p, or σ in excess of 300 MPa. Several recent works include: Wong and Brace (1979); Van Der Molen (1981); and Page and Heard (1981). Most available data are summarized in Clark (1966), Touloukian and Ho (1981) and Heuze (1982).

Hydrologic test results as determined in the laboratory are even less common than those for thermal properties. Permeability and hydraulic conductivity observations have been made on fractured and intact igneous, sedimentary, and metamorphic rocks. Sample dimensions range from 10^{-2} to 1 m. Virtually all reported data have been measured at temperatures near 20°C except for the works of Summers et al. (1978) and Potter (1978). The effect of stress on permeability of granite has been reported by Zoback and Byerlee (1975). Recently, Brace (1977) has shown that electrical methods may be used to indirectly determine the hydrologic behavior of a variety of intact and jointed rocks. Available results have been reviewed and summarized in Martin (1979), Brace (1980), and Touloukian and Ho (1981).

Chemical transport, chemical (mineralogical) reactions or phase changes in the fluid-rock system at depth may also strongly influence the mechanical, thermal, and hydrologic behavior. Summers et al. (1978) and Potter (1978) have demonstrated that dissolved SiO_2, when transported by a H_2O-rich fluid phase in a P or T gradient, can either enhance or degrade permeability, depending on the crack geometry and the points of solution and deposition. Mineralogical reactions such as the hydration of brucite or the dehydration of gypsum or serpentine under stress alter the local temperature, pressure (and stress), and the amount of free water, thus perturbing the surrounding mechanical, thermal, and hydrological environment (Fyfe et al., 1958; Raleigh and Paterson, 1965; Heard and Rubey, 1966); Fyfe et al., 1978; and Chernosky, 1979. Van Der Molen (1981) has demonstrated that the volume changes preceding and associated with the $\alpha - \beta$ phase transition in quartz lead to local stress concentrations, cracking, and dilatancy in quartz-rich rocks. These effects in turn will influence the permeability and thermal conductivity (Walsh and Decker, 1966; Bauer and Handin, 1981; Durham and Abey, 1981a).

DIRECTIONS FOR FUTURE LABORATORY STUDIES

Although the availability of laboratory results under controlled conditions of P, P_p, T, σ, and X may vary considerably among mechanical, thermal, and hydrologic tests, the capability to accomplish these measurements presents no problems in technology, only in researcher interest and institutional or government support. Apparatus are currently in routine operation in many laboratories which can closely simulate the appropriate in situ environment for each of the broad areas noted above (e.g., Heard and Duba, 1978; Abey et al., 1982; Trimmer, 1982). However, the relatively small sample sizes, generally 10^{-2} to 10^{-1} m, can impose severe limitations on the extrapolation of these laboratory data to the desired in situ application. Where the laboratory test data have been reported on blocks as large as 1-2 m, the results have only a limited value to most applications because the test environment can simulate only very shallow conditions (e.g., Singh and Huck, 1973; Thorpe et al., 1980).

From the above brief discussions, many recommendations for future work should be obvious to all. To some, research directions may not be so clear or the suggested area to be emphasized by some may be perceived as too highly specialized, depending on the individual's philosophy of attack upon this type engineering/scientific problem. I believe that the main recommendations made previously concerning laboratory studies in support of a hard-rock, high-level nuclear waste repository are just as valid and useful today to the broader concept of man's structural use of the underground, to resource recovery or to the mitigation of natural hazards as when these recommendations were originally proposed (Heard, 1979; Brace, 1979; Handin, 1980; Handin and Heard, 1980). The same point holds equally well for similar recommendations made by the Panel on Research Requirements (USNCRM/NRC) for resource recovery, underground construction, and earthquake hazard reduction (US National Committee for Rock Mechanics, 1981). Several points have been selected from those recommendations for further emphasis here. No priority for implementation should be inferred from the ordering rank below.

1) More financial and personnel support is needed to perform the laboratory testing required. Laboratory testing provides the fastest and least expensive means for scoping physical-thermal-hydrologic-chemical properties and phenomenology. Perhaps more importantly, tests accomplished in the laboratory are the only means for acquiring an understanding of the fundamental physical/chemical processes which occur in situ. Moderate funding is needed to support or expand the current laboratory capabilities.

2) More laboratory data of all types are needed for rocks tested under conditions where all or most environmental conditions in situ are closely simulated, not just one or two. Data taken under shallow conditions cannot always be reliably extrapolated to the deep environment. In addition, more data

are needed in the sample scale range near 10^{-1} m.
3) Laboratory test data are needed where gradients in P, P_p, T, σ, and X are controlled and varied over the sample. One example would be investigation of fluid flow in a temperature and stress gradient. Another would be what is the time dependency of flow, fracture, and surficial friction when gradients in T, P_p, and σ are present. Another might address bulk compaction and dilatancy at moderate P_p, T when nontraditional boundary conditions are imposed (e.g., one-dimensional strain).
4) Additional effort must be directed towards the understanding of the physical/chemical processes involved--both in the laboratory and <u>in situ</u>. Without definition of the operative mechanisms, no truly predictive modeling is possible.
5) Some large scale ($\sim$ 1-m) laboratory testing appears to be feasible and worthwhile at this time. These experiments should be accomplished in close conjunction with field testing and should be done under the approropriate, carefully controlled <u>in situ</u> conditions. Since equilibration times (e.g., T, P_p) may be very long, these tests may tie up large apparatus for long periods, and thus the relatively few tests would be very expensive. Much careful planning would be required. It appears that the first priority for testing would be rocks with large-scale inhomogeneities (e.g., some discontinuities, porous zones or mineral segregations) or anisotropies (e.g., layering or regular jointing).
6) Increased emphasis must be directed towards scaling laboratory results on 10^{-2} to 1-m size test samples to the <u>in situ</u> case. Success in scaling here depends not only on points 1-5 (above) but also depends upon an understanding of the physical, thermal, and hydrologic properties of large scale, multiple discontinuities and layers.
7) Closer ties and better interfacing are needed between those researchers working in the laboratory, those developing mathematical models for coupled mechanical-thermal-hydrologic-chemical phenomena, and those working in the field with the natural system. Only when validated, coupled models are available can we ever have confidence in the prediction of rock behavior at depth at conditions impossible to measure because of cost, time, or scaling constraints.

ACKNOWLEDGEMENTS

The authors thank B. Bonner, A. Duba, W. Durham, and F. Heuze for their constructive comments.

This work was performed under the auspices of the U.S. Department of Energy by the Lawrence Livermore National Laboratory under contract number W-7405-ENG-48.

REFERENCES

Abey, A. E., Durham, W. B., Trimmer, D. T., and Dibley, L. L., 1982, "Apparatus for Determining the Thermal Properties of Large Geologic Samples at Pressure to 0.2 GPa and Temperatures to 750K," Review of Scientific Instruments, Vol. 53, (in press).

Atkinson, B. K., 1979, "Stress Corrosion and the Rate-Dependent Tensile Failure of a Fine-Grained Quartz Rock," Tectonophysics, Vol. 65, pp. 281-290.

Bauer, S. J. and Handin, J., 1981, "Thermal Expansion of Three Water-Saturated Igneous Rocks to 800°C at Effective Confining Pressures of 5 and 50 MPa," Transactions, American Geophysical Union, Vol. 62, p. 393.

Berry, F. A. F., 1973, "High Fluid Potentials in California Coast Ranges and Their Tectonic Significance," American Association of Petroleum Geologists Bulletin, Vol. 57, pp. 1219-1249.

Birch, F., 1964, State of Stress in the Earth's Crust, Judd, W. R., ed., American Elsevier, New York, N.Y.

Brace, W. F., 1977, "Permeability from Resistivity and Pore Shape," Journal of Geophysical Research, Vol. 82, pp. 3343-3349.

Brace, W. F., 1979, "Laboratory Measurements Workshop," pp. 19-22 in Proceedings of Workshop on Thermomechanical Modeling for a Hardrock Waste Repository, Ramspott, L. and Holzer, F., eds., Lawrence Livermore National Laboratory, UCAR-10043.

Brace, W. F., 1980, "Permeability of Crystalline and Argillaceous Rocks," International Journal of Rock Mechanics and Mining Science and Geomechanics Abstracts, Vol. 17, pp. 241-251.

Bridgman, P. W., 1945, "Polymorphic Transitions and Geologic Phenomena," American Journal of Science, Vol. 243a, pp. 90-97.

Carter, N. L., Friedman, M., Logan, J. M., and Stearns, D. W., 1981, Mechanical Behavior of Crustal Rocks, Geophys. Mono. 24, American Geophysical Union, Washington, D.C.

Chernosky, J. V., Jr., 1979, "Experimental Metamorphic Petrology," Reviews of Geophysics and Space Physics, Vol. 17, pp. 860-872.

Clark, S. P., 1966, Handbook of Physical Constants, Geol. Soc. America Mem. 97, Geological Society of America, New York, N.Y.

Dickinson, G., 1953, "Geological Aspects of Abnormal Reservoir Pressures in Gulf Coast Louisiana," American Association of Petroleum Geologists Bulletin, Vol. 37, pp. 410-432.

Durham, W. B. and Abey, A. E., 1981a, "The Effect of Pressure and Temperature on the Thermal Properties of a Salt and a Quartz Monzonite," pp. 79-84 in Proceedings 22nd US Symposium on Rock Mechanics, MIT, Cambridge, Mass.

Durham, W. B. and Abey, A. E., 1981b, "Thermal Conductivity and Diffusivity of Climax Stock Quartz Monzonite at High Pressure and Temperature," Proceedings of the 17th International Thermal Conductivity Conference, Gathersburg, Md., June 15-18 (in press).

Fyfe, W. S., Turner, F. J. and Verhoogen, J., 1958, Metamorphic Reactions and Metamorphic Facies, Geol. Soc. America Mem. 73, Geological Society of America, New York, N.Y.

Fyfe, W. S., Price, N. J., and Thompson, A. B., 1978, Fluids in the Earth's Crust, Elsevier, New York, N.Y.

Gretener, P. F., 1967, "On the Thermal Instability of Large Diameter Wells-An Observational Report," Geophysics, Vol. 4, pp. 727-738.

Griggs, D. T., 1940, "Experimental Flow of Rocks Under Conditions Favoring Recrystallization," Geological Society of America Bulletin, Vol. 51, pp. 1001-1022.

Griggs, D. T., 1967, "Hydrolytic Weakening of Quartz and Other Silicates," Geophysical Journal, Royal Astronomical Society, Vol. 14, pp. 19-31.

Griggs, D. T. and Handin, J., 1960, Rock Deformation, Geol. Soc. America Mem. 79, Geological Society of America, New York, N.Y.

Griggs, D. T. and Handin, J., 1960, "Observations on Fracture and a Hypothesis of Earthquakes," pp. 347-364 in Rock Deformation, Griggs, D. T., and Handin J., eds., Geol. Soc. America Mem. 79, Geological Society America, New York, N.Y.

Griggs, D. T., and Blacic, J. D., 1965, "Quartz: Anomalous Weakness of Single Crystals," Science, Vol. 147, pp. 292-295.

Handin, J., 1980, "Laboratory Investigations," pp. 10-15 in Proceedings of Workshop on Thermomechanical-Hydrochemical Modeling for a Hardrock Waste Repository, Lawrence Berkeley Laboratory Report 11204.

Handin, J. and Heard, H. C., 1980, "Laboratory Investigations," pp. 105-108 in Proceedings of Workshop on Thermomechanical-Hydrochemical Modeling for a Hardrock Waste Repository, Lawrence Berkeley Laboratory Report 11204.

Heard, H. C., 1976, "Comparison of the Flow Properties of Rocks at Crustal Conditions," Philosophical Transactions, Royal Society of London A, Vol. 283, pp. 173-186.

Heard, H. C., 1979, "Laboratory Measurements as Inputs to Modeling: Status and Needs," pp. 66-72 in Proceedings of Workshop on Thermomechanical Modeling for a Hardrock Waste Repository, Ramspott, L. and Holser, F., eds, Lawrence Livermore National Laboratory, UCAR-10043.

Heard, H. C. and Rubey, W. W., 1966, "Tectonic Implications of Gypsum Dehydration," Geological Society of America Bulletin, Vol. 77, pp. 741-760.

Heard, H. C., Borg, I. Y., Carter, N. L. and Raleigh, C. B., 1972, Flow and Fracture of Rocks, Geophys. Mono. 16, American Geophysical Union, Washington, D.C.

Heard, H. C. and Duba, A., 1978, "Capabilities for Measuring Physicochemical Properties at High Pressure," University of California, Lawrence Livermore National Laboratory Report UCRL-52420.

Herrin, E., 1972, "A Comparative Study of Upper Mantle Models: Canadian Shield and Basin and Range Provinces," pp. 216-231 in Nature of the Solid Earth, Robertson, E. C., ed., McGraw-Hill, New York, N.Y.

Heuze, F. E., 1982, "High Temperature Mechanical, Physical, and Thermal Properties of Granitic Rocks," International Journal of Rock Mechanics and Mining Science and Geomechanics Abstract (in press).

Hoek, E. and Brown, E. T., 1980, Underground Excavations in Rock, The Institution of Mining and Metallurgy, London.

Hubbert, M. K. and Rubey, W. W., 1959, "Role of Fluid Pressure in Mechanics of Overthrust Faulting: I. Mechanics of Fluid Filled Porous Solids and Its Application to Overthrust Faulting," Geological Society of America Bulletin, Vol. 70, pp. 115-166.

Jeffreys, H., 1970, The Earth, University Press, Cambridge England.

Lachenbruch, A. H., 1968, "Preliminary Geothermal Model of the Sierra Nevada," Journal of Geophysical Research, Vol. 73, pp. 6977-6989.

Lama, R. D. and Vutukuri, V. S., 1978a, Handbook on Mechanical Properties of Rocks, Vol. II, Trans. Tech. Pub., Clausthal, Germany.

Lama, R. D. and Vutukuri, V. S., 1978b, Handbook on Mechanical Properties of Rocks, Vol. III, Trans. Tech. Pub., Clausthal, Germany.

Lyubimova, A. I., Maslennikov, A. I., and Ganiev, Y.A., 1979, "Thermal Conductivity of Rocks at Elevated Temperatures and Pressures in the Water- and Oil-Saturated State," Physics of the Solid Earth, Vol. 15, pp. 358-361.

Martin, R. J., III, 1979, "Pore Pressure Effects in Crustal Processes," Reviews of Geophysics and Space Physics, Vol. 17, pp. 1132-1137.

Moses, P. L., 1961, "Geothermal Gradients Now Known in Greater Detail," World Oil, Vol. 152, pp. 79-82.

Nichols, E. A., 1947, "Geothermal Gradients in Mid-Continent and Gulf Coast Oil Fields," Petroleum Division, American Institute Mining, Metallurigical Petroleum Engineers, Vol. 170, pp. 44-50.

Page, L. and Heard, H. C., 1981, "Elastic Moduli, Thermal Expansion, and Inferred Permeability of Climax Quartz Monzonite and Sudbury Gabbro to 500°C and 55 MPa," pp. 97-104 in Proceedings 22nd US Symposium on Rock Mechanics, MIT, Cambridge, Mass.

Paterson, M. S., 1978, Experimental Rock Deformation-The Brittle Field, Springer-Verlag, New York, N.Y.

Potter, J. M., 1978, "Experimental Permeability Studies at Elevated Temperature and Pressure of Granitic Rocks," University of California, Los Alamos National Laboratory Report LA 7224 T.

Raleigh, C. B. and Paterson, M. S., 1965, "Experimental Deformation of Serpentinite and Its Tectonic Implications," Journal of Geophysical Research, Vol. 70, pp. 3965-3985.

Rubey, W. W. and Hubbert, M. K., 1959, "Role of Fluid Pressure in Mechanics of Overthrust Faulting: II. Overthrust Belt in Geosynclinal area of Western Wyoming in Light of Fluid Pressure Hypothesis," Geological Society of America Bulletin, Vol. 70, pp. 167-206.

Singh, M. M. and Huck, P. J., 1973, "Large Scale Triaxial Tests on Rock," pp. 35-60 in 14th Symp. on Rock Mechanics, Am. Soc. Civil Engineers, New York, N. Y.

Summers, R., Winkler, K., and Byerlee, J., 1978, "Permeability Changes During the Flow of Water Through Westerly Granite at Temperatures of 100-400°C," Journal of Geophysical Research, Vol. 83, pp. 339-344.

Thorpe, R., Watkins, D. J., Ralph, W. E., Hsu, R., and Flexser, S., 1980, "Strength and Permeability Tests on Ultra-Large Stripa Granite Core," Lawrence Berkeley Laboratory Report 11203.

Touloukian, Y. S. and Ho, C. Y., 1981, "Physical Properties of Rocks and Minerals," Vol. II-2, McGraw-Hill, New York, N.Y.

Trimmer, D. T., Bonner, B., Heard, H. C., and Duba, A., 1980, "Effect of Pressure and Stress on Water Transport in Intact and Fractured Gabbro and Granite," Journal of Geophysical Research Vol. 85, pp. 7059-7071.

Trimmer, D. T., 1982, "Laboratory Measurements of Ultra Low Permeability Geologic Materials," Review of Scientific Instruments (in press).

U.S. National Committee for Rock Mechanics, National Research Council, 1981, Rock Mechanics Research Requirements for Resource Recovery, Construction and Earthquake-Hazard Reduction, National Academy of Sciences, Washington, D.C.

Van Der Molen, I., 1981, "The Shift in the α-β Transition Temperature of Quartz Associated with the Thermal Expansion of Granite at High Pressure," Tectonophysics, Vol. 73, pp. 323-342.

Vutukuri, V. S., Lama, R. D., and Saluja, S. S., 1974, Handbook on Mechanical Properties of Rocks, Vol. I Trans. Tech. Pub., Clausthal, Germany.

Walsh, J. B. and Decker, E. R., 1966, "Effect of Pressure and Saturating Fluid on the Thermal Conductivity of Compact Rock," Journal of Geophysical Research, Vol. 71, pp. 3053-3061.

Wong, T. F. and Brace, W. F., 1979, "Thermal Expansion of Rocks: Some Measurements at High Pressure," Tectonophysics, Vol. 57, pp. 95-117.

Zoback, M. D. and Byerlee, J. D., 1975, "The Effect of Microcrack Dilatancy on the Permeability of Westerly Granite," Journal of Geophysical Research, Vol. 80, pp. 752-755.

Chapter 27

THE EFFECT OF WATER ON THE MECHANICAL PROPERTIES AND MICROSTRUCTURES
OF GRANITIC ROCKS AT HIGH PRESSURES AND HIGH TEMPERATURES

by Ove Alm

Research Scientist, Division of Rock Mechanics,
University of Luleå, Sweden

ABSTRACT

Wet and dry specimens of three rocks of approximately granitic composition were deformed at different experimental conditions. These experiments were carried out in order to study the extent to which water, either added or released from mineral reactions, may affect the mechanical properties and the microstructure of these rocks.

The experiments were conducted in a Griggs' hot creep apparatus at temperatures and pressures up to 900° C and 1000 MPa respectively; and most of them were performed in the constant strain rate mode ($1 \cdot 10^{-8} < \dot{\epsilon} < 1 \cdot 10^{-4}$ s^{-1}).

All three rocks were low porosity rocks, and the wet specimens never contained more than 0.9 wt % of added water. The experimental results are, however, quite unequivocal: They show that wet rocks are weaker than dry ones. A significant difference is even found for rocks tested in uniaxial compression at room temperature.

A few experiments carried out on specimens obtained from larger prestressed rock cores indicated that the central parts of these cores were more likely to contain more microcracks than the ends.

Optical microscope examinations revealed that extensive microcracking takes place even at the highest temperatures, where other mechanisms such as slip and dislocation creep also become important. Very little partial melting could be observed except for the specimens deformed at 900° C.

INTRODUCTION

It is believed that rocks of approximately granitic composition make

up a large part of the material in the Earth's crust, and therefore we
may assume that the mechanical properties of granitic rocks are on the
whole representative of those of the lithosphere, at least as regards
the upper part of it. We need to know more about these properties in
order to understand the mechanics of large crustal movements. The influence of water on these properties is of particular interest, since water
is by far the dominating pore fluid in the upper part of the crust.

Water is known to affect the mechanical properties of rocks in many
ways. A great number of reports (eg. Duba, Heard, and Santor, 1974)
argue that cracking and faulting are enhanced in the brittle regime,
provided that there is enough water present for the fluid pressure to
reduce the effective confining pressure. This rather vague and thus
unsatisfactory argument does not explain how the water actually affects
the rock. The results, however, support another important claim: That
the mechanical properties of rocks to a large extent depend on the
amount of pores and microcracks, and on the directional distribution
of these defects. A water pressure in the cracks and the pores will
oppose the external forces acting to close them, and may, at least in
places, prevent the cracks - and to a lesser extent even the pores -
from collapsing. The friction between the surfaces of the cracks will
thus be lost. In addition, water is also likely to affect the cohesion,
ie. the short range interaction, between the individual mineral grains.
Stress corrosion cracking suggested by Tullis and Yund (1980), and
reported by Atkinson and Meredith (1981), could just be some sort of
combination of the two effects mentioned above.

Broch (1974), invertigated the mechanical properties of a number of
rocks under both wet and dry conditions, and found that dry rocks were
stronger than water saturated rocks. His rocks were tested at room
temperature in uniaxial and triaxial compression, as well as in point
load testing; and all three methods gave consistant results regarding
the effect of water on the mechanical properties of rocks.

Minute quantities of water can also weaken the rock by hydrolysing
the strong covalent -Si-O-Si- bonds (eg. Griggs and Blacic, 1965, and
Tullis and Yund, 1980). Another interesting phenomenon can be observed
in multi-mineral rocks, for instance granite, where water seems to
lower the temperature for the onset of melting. Several recent reports
deal with such water-assisted partial melting in granites (eg. Van der
Molen and Paterson, 1979, and Paquet and François, 1980) or other
igneous rocks (Bauer et al., 1981). But a certain amount of caution
is advisable when we set out to apply the results of these reports to
geological models. The pressure, P, and temperature, T, conditions
assumed to exist in the lithosphere do not correspond to the ones
imposed by these authors. The experimental temperatures seem to be too
high in relation to the pressures used.

In a previous paper (Alm, 1979), the present author reported that
very little melting, next to nothing, could be detected in deformed
aplite at temperatures comparable to those used by Paquet and François
(1980), and Van der Molen and Paterson (1979). Thin sections of some

EFFECT OF WATER ON MECHANICAL PROPERTIES

of the wet aplite did, however, reveal a conspicuous band-like feature along a fairly large number of grain boundaries.

This paper presents some new data on the effect of water on mechanical properties and microstructures of granitic rocks at both high and low P-T conditions.

EXPERIMENT

Starting materials

Three rock types of approximately granitic composition were investigated; an aplite from Malmberget in Northern Sweden, a granite (the grey variety) from the Stripa test site for radioactive waste repository in Central Sweden, and an ultra-mylonite from the Tännäs Augen Gneiss Nappe in the Central Swedish Caledonides.

The three rock types differ both in grain size and mineral composition. The aplite is reasonably homogeneous and isotropic, even on a microscopic scale, with only minor variations in grain size and composition. The ultra-mylonite also displays a rather uniform grain size. The rock is banded and the grains are elongated with a preferred orientation subparallel to the foliation. By contrast, the granite exhibits a much more complicated microstructure (cf. Fig. 1A, 1B, and 1C). Fig. 1B indicates that we can expect fairly pronounced local variations in grain size in the granite specimens; and it can accordingly be difficult to achieve a homogeneous deformation in them. The granite was included in the investigation mainly because it contains more minerals that decompose at temperatures below the melting point (liquidus) of the rock and then release water which has been bound in the crystal structure.

Fig. 1.   Optical micrographs of undeformed aplite (A), granite (B), and mylonite (C). Scale bar is 1 mm in A, B and 0.1 mm in C.

Mineral and grain size data for the starting materials are summarized in Table 1.

TABLE 1. Mineral Composition (vol. %) and average grain size of investigated rocks.

Rock type / Mineral	Granite Stripa	Aplite Malmberget	Ultra-mylonite Tännäs
Quartz	35	35	24
Plagioclase	35	27	45
Microcline	24	33	25
Biotite			
Muscovite			
Chlorite	6	5	6
Epidote			
Accessories			
Average grain size (mm)	0,4-3	0,2	< 0,02

The extremely small grain size made it impossible to determine the mineral compositon of the ultra-mylonite by means of optical microscopy. The composition of this rock is thus calculated from chemical data according to the CIPW norm. - In the Stripa granite the average grain size varies a great deal from one place to another. A close look at Fig. 1B reveals that in our samples of this granite it is probably smaller than 0,7 mm.

Experimental details

The deformation experiments were carried out in a solid medium Griggs' hot creep apparatus operated in the constant strain-rate mode. The lay-out of the high-pressure cell used in the high-temperature experiments was identical to the one described by Alm et al. (1980). A thorough description of the experimental procedures is also given in that paper and in Alm (1979), and we shall here restrict ourselves to discussing certain details which are specific for these later experiments. One such detail is the type of pressure medium employed in the room temperature experiments. Here we found it pratical to use either teflon or lead. Deformation experiments on almost identical specimens gave approximately the same results for strength and fracture pattern for both these pressure media.

All experiments were performed at constant temperature, confining pressure, and average strain rate in the range 20-900°, 0.1-1000 MPa, and $1 \cdot 10^{-8} - 1 \cdot 10^4$ s^{-1} respectively. Furthermore, we found it preferable to use cylindrical specimens 8 mm in diameter and 10-11 mm in length in the high-pressure experiments, while larger ones, 15 mm in diameter and 30-35 mm length, were loaded to failure in conventional uniaxial compression tests. Only aplite specimens were tested in uniaxial compression.

In order to achieve a constant initial condition regarding the water content of the rocks we had all specimens oven dried for several hours at 60° C. The ones that were to remain dry were then stored in a desiccator, while those meant for wet tests were vacuum impregnated with deionized water several days before each experiment and then stored in a water bath until used. The porosity of all these rocks is quite small, and as a result very little water was absorbed by the specimens, in most cases less than 0,5 wt%. The amount of absorbed water was determined by weighing each specimen before and after water impregnation.

EFFECT OF WATER ON MECHANICAL PROPERTIES

Uniaxial compression tests were done on both dry and wet specimens. The wet runs were conducted with the specimen in a water bath during the whole run; but at high pressures it was necessary to encase the wet specimens in a copper jacket of 0,2 mm wall thickness. Because of the strength of this jacket we had to apply a correction to the stress-strain data obtained from the room temperature runs. The correction, -100 MPa, was determined by means of a separate experiment where a cylinder of lead was used instead of a rock specimen. The strength of the jacket decreased with increasing temperature, and no such correction was needed for the high-temperature runs.

In order to make comparisons of results from different runs worthwhile, we had to use the same procedure in all experiments when applying pressure and temperature. The procedures for wet runs were applied to all runs, ie. the pressure was increased before the temperature was raised. After reaching the final conditions we allowed 3-4 hours to elapse before we switched on the motor for the axial differential load.

The procedures discussed above are also used on a series of tests on prestressed specimens which we are now carrying out. These specimens were obtained from larger specimens loaded to about 50, 75 or 85 % of their compressive strength. The small specimens used in our investigation were cored from the central part of the big specimens or from their end parts. So far only aplite has been investigated in this way.

RESULTS

Mechanical results

Results in terms of fracture strength for the uniaxial compression tests are summarized in Table 2. Three runs where the specimens failed along distinct weakness planes are not included in the table.

TABLE 2. Results from uniaxial compression tests on dry (D) and wet (W) aplite specimens.

Spec. No.	Fracture Strength (MPa)	Strain Rate (s^{-1})	Density (kg/m^3)	Wet or Dry	Spec. No.	Fracture Strength (MPa)	Strain Rate (s^{-1})	Density (kg/m^3)	Wet or Dry
A06-D	206	$1,5 \cdot 10^{-6}$	2557	D	A05-D	182	$1,4 \cdot 10^{-8}$	2556	D
A07-D	200	$1,5 \cdot 10^{-6}$	2550	D	A09-D	162	$1,5 \cdot 10^{-8}$	2555	D
A08-D	215	$1,7 \cdot 10^{-5}$	2564	D	A10-D	173	$1,6 \cdot 10^{-8}$	2553	D
A32-D	218	$1,7 \cdot 10^{-5}$	2542	D	A20-W	150	$1,1 \cdot 10^{-8}$	2553	W
A19-W	175	$1,6 \cdot 10^{-5}$	2559	W	A24-W	159	$1,5 \cdot 10^{-8}$	2555	W
A21-W	199	$1,7 \cdot 10^{-5}$	2554	W	A29-W	156	$1,4 \cdot 10^{-8}$	2561	W
A22-W	196	$1,7 \cdot 10^{-5}$	2548	W	A30-W	196	$1,6 \cdot 10^{-8}$	2559	W
A31-W	175	$1,7 \cdot 10^{-5}$	2557	W	A34-W	169	$1,5 \cdot 10^{-8}$	2560	W
A33-W	182	$2,0 \cdot 10^{-5}$	2556	W					

In Table 2 we see that the wet aplite specimens were weaker than the the dry ones when tested in uniaxial compression. It also shows us that specimens deformed at the slowest strain rate were significantly weaker than those tested at a strain rate three orders of magnitude higher.

Fig. 2, 3, 4A and 4B display the stress-strain results from all our high pressure experiments. The strains are given as engineering strains, and a constant volume correction is applied to the stress data in order to correct for the gradual increase of the specimen diameter. The stresses in Fig. 2 are also corrected for the strength of the copper jacket. The curves show that there is a difference in the stress-strain results between the wet and the dry specimens for all three rock types. The wet ones display a more gradual increase of differential stress with increasing strain. The ultimate strength of the specimens was significantly reduced for the aplite (Fig. 2 and 3) and for the Stripa granite (Fig. 4A) because of the presence of water, but this effect was less pronounced for the ultra-mylonite (Fig. 4B), which is what could be expected, since the porosity in this last rock was so small that it hardly absorbed any water at all. In addition, we see that the strength of the rocks (aplite and granite) decreases with increasing temperature (Fig. 3 and 4A) and decreasing strain rate (Fig. 4A). Most of the "wet" curves exhibit a peak at about 10 % strain, although no throughgoing macroscopic fault can be seen. It seems as if the differential stress must exceed a certain threshold value before a new deformation mechanism becomes dominant. Also note in Fig. 2 that wet specimens from the central parts are weaker than those from the ends.

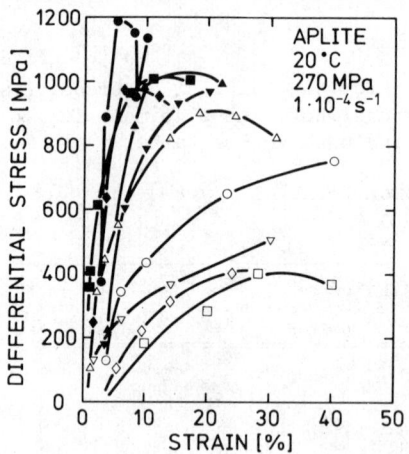

Fig. 2. Stress-strain data for wet (W) and dry (D) aplite specimens obtained from the central (M) or end (E) parts of larger prestressed cores. ○ W-M75; ● D-M75; △ W-E75; ▲ D-E75; □ W-M50; ■ D-M50; ▽ W-E50; ▼ D-E50; ◇ W-E85; ◆ D-E85.

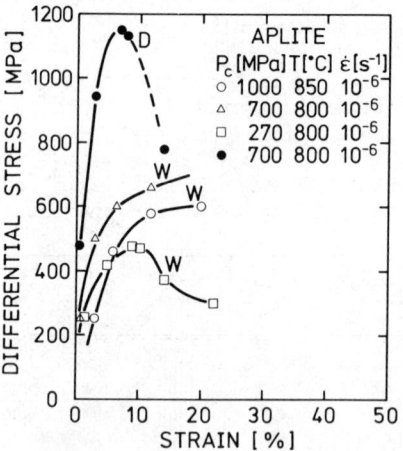

Fig. 3. Stress strain curves for wet (W) and dry (D) aplite specimens deformed at high temperatures and high pressures.

EFFECT OF WATER ON MECHANICAL PROPERTIES 267

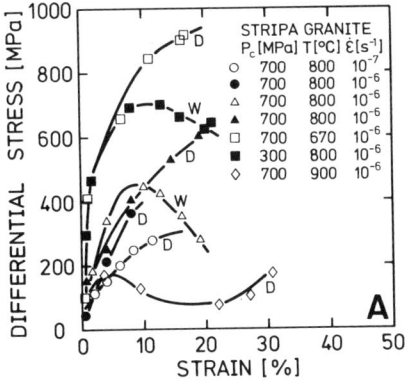
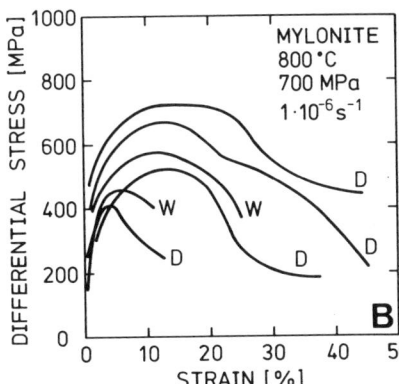

Fig. 4. Stress-strain curves for rocks deformed at high temperatures and pressures. A, Stripa granite; B, Tännäs ultra-mylonite deformed $\perp$ to foliation.

Microstructural Observations

No particular preference for any failure mode could be detected among the specimens tested in uniaxial compression. They looked much the same also in thin section, revealing more intense microfracturing close to macroscopic faults.

All prestressed aplite specimens deformed at room temperature failed by throughgoing, rather complex faults. On the microscopic scale we find that the wet specimens exhibit a more intense microfracturing than the dry ones. A typical example of this is shown in Fig. 5A-B. Direction of maximum applied stress is indicated by arrows in Fig. 5 and 6.

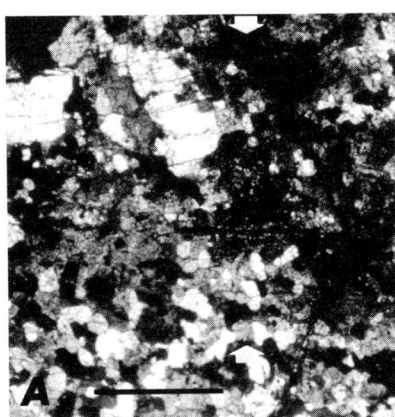

Fig. 5. Typical crack patterns in prestressed aplite specimens deformed at room temperature and 270 MPa confining pressure. A, wet conditions; B, dry conditions; Scale bar is 1 mm.

Brittle fracturing on the grain scale seems to play an important role even at the highest temperatures, though the ultra-mylonite may be an exception. Fig. 6A displays a portion of a thin section of a mylonite specimen deformed to about 45 % of its original length. By comparing Fig. 6A with 1C we can infer that the texture of this rock was not drastically altered by the superimposed experimental deformation. Furthermore, it should perhaps be pointed out that all our high-temperature experiments were done at temperatures high enough to promote partial melting in granitic rocks, if they contain water. Melt was indeed found in the 800° C wet granite specimens (Fig. 6C - Gl), and in the 900° C dry one. Trace amounts of melt could also be detected in wet aplite (Fig. 6B), whereas we can draw no definite conclusion for the mylonite due to insufficient resolution in the optical microscope. A strong undulatory extinction, recovery, and regions of intense deformation are other features observed in the minerals of the deformed rocks. Most of the mica was altered to a dark brown, almost opaque mineral (Fig. 6C - M).

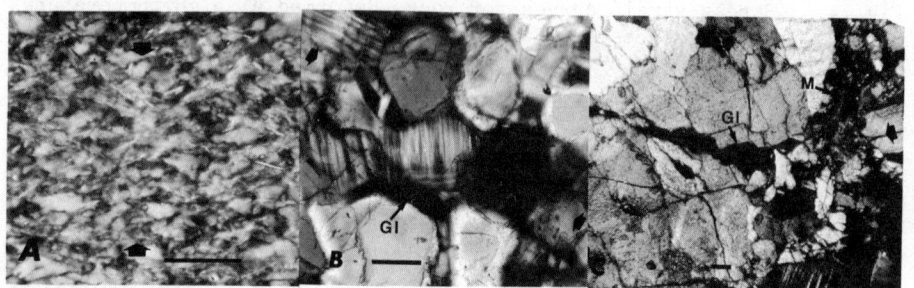

Fig. 6 Micrographs of wet mylonite (A), aplite (B), and granite (C) deformed at 800° C and 700 MPa. Region of altered mica is marked M and regions of glass - Gl; Scale bar is 0.1 mm.

CONCLUSIONS

Our experiments have shown that water may influence the deformation properties of rocks in several ways. A pure mechanical effect, where the closing of pores and microcracks is opposed by the pore-fluid pressure, was obviously active in the experiments on prestressed specimens. In addition, these results indicate a more intense microfracturing in the central part of a specimen loaded in uniaxial compression. It seems most likely that this "anomaly" is caused by the frictional forces acting at the ends of the specimens in such an experiment. Chemical effects causing stress corrosion can also play a significant role in these experiments, and in the uniaxial ones.

If the temperature is high enough, water will cause partial melting. Melt in the form of glass was also observed in the wet 800° C and the dry 900° C granite specimens - and in the wet 800° C aplite specimens, but here to a much lesser extent. A difference like this can result if the amount of initially absorbed water is insufficient to cause any appreciable amounts of melt, and if enough additional water is released in the granite from decomposition of hydrous minerals. This last argument seems to be supported by the results obtained from the 900° C granite specimen. This particular specimen contained a substantial amount of glass from partial melting although no water was added at all.

REFERENCES

Alm, O., 1979, "Influence of Water on the Strength and Deformation Properties of a Granitic Aplite at High Pressures and Temperatures," Bull. Minéral., Vol. 102, pp. 115-123.

Alm, O., Röshoff, K., and Stephansson, O., 1980, "Microstructures and Mechanical Characteristics of the Tännäs Augen Gneiss, Swedish Caledonides," Geologiska Föreningens i Stockholm Förhandlingar, Vol. 102, pp. 319-334.

Atkinson, B.K., and Meredith, P.G., 1981, "Stress Corrosion Cracking of Quartz: A Note on the Influence of Chemical Environment," Tectonophysics, Vol. 77, pp. T1-T11.

Bauer, S.J., Friedman, M., and Handin, J., 1981, "Effects of Water-Saturation on Strength and Ductility of Three Igneous Rocks at Effective Pressures to 50 MPa and Temperatures to Partial Melting," Proceedings, 22nd U.S. Symposium on Rock Mechanics, pp. 73-78.

Broch, E., 1974, "The Influence of Water on Some Rock Properties," Proceedings, 3rd Congress of the International Society for Rock Mechanics, Denver, Vol. 2, Part A, pp. 33-38.

Duba, A.G., Heard, H.C., and Santor, M.L., 1974, "Effect of Fluid Content on the Mechanical Properties of Westerly Granite," Lawrence Livermore Laboratory Report UCRL-51626, Livermore, California.

Griggs, D.T., and Blacic, J.D., 1965, "Quartz: Anomalous Weakness of Synthetic Crystals," Science, Vol. 147, pp. 292-295.

Paquet, J., and François, P., 1980, "Experimental Deformation of Partially Melted Granitic Rocks at 600 - 900° C and 250 MPa Confining Pressure," Tectonophysics, Vol. 68, pp. 131-146.

Tullis, J., and Yund, R.A., 1980, "Hydrolytic Weakening of Experimentally Deformed Westerly Granite and Hale Albite," J. Struct. Geol., Vol. 2, pp. 439-451.

Van der Molen, I., and Paterson, M.S., 1979, "Experimental Deformation of Partially-Melted Granite," Contr. Mineral. Petrol., Vol. 70, pp. 299-318.

Chapter 28

TIME-DEPENDENT VOLUMETRIC CONSTITUTIVE RELATION
FOR FAULT GOUGE AND CLAY AT HIGH PRESSURE

by Chaw-long Chu and Chi-Yuen Wang

Department of Geology and Geophysics
University of California, Berkeley, CA 94720

ABSTRACT

The time-dependent volumetric constitutive relation for a San Andreas fault gouge and a consolidated kaolinite are experimentally determined at confining pressures to 200 Mpa, under creep condition and at ultrasonic frequency. At any given pressure, the bulk modulus determined at 1 Mhz frequency is identical to the "elastic" modulus determined at a stepwise change of pressure. After the pressure change, there occur a time-dependent change of volume and thus the effective bulk modulus. The constitutive relations may be modeled by a standard linear viscoelastic medium with pressure-dependent moduli and viscosity. For the San Andreas fault gouge, viscosity increases from 2×10^{13} poise, at 17 Mpa, and 10^{14} poise, at 100 Mpa. For kaolinite sample, the corresponding values are 3×10^{12} poise and 10^{13} poise. These results imply that both stress and pore pressure in fault zones may show time-dependent variations in response to a change of state of stress, such as that occurring after an earthquake.

INTRODUCTION

Studies of the mechanical properties of fault gouge may be important for gaining an understanding of the mechanics of faulting. Factors which influence the fault gouge behavior include the confining pressure (e.g., Summers and Byerlee, 1977; Wang and Mao, 1979; Wang et al., 1980; Shimamoto and Logan, 1981; Morrow et al., 1982), strain rate (Logan, 1978; Solberg and Byerlee, 1981), hydration and dehydration (Logan et al., 1981), water saturation (Wang and Mao, 1979; Morrow et al., 1982), grain size and composition (Chu et al., 1981; Morrow et al., 1982). However, a general constitutive relation which can be used to adequately predict the time-dependent deformation of clayey materials under high pressure is still lacking. In the present study the time-dependent volumetric constitutive relations for a San Andreas fault gouge and an air-dired, consolidated kaolinite are

VOLUMETRIC CONSTITUTIVE RELATION

experimentally determined. The volumetric strain of fault gouge is important not only because it is a fundamental variable in the constitutive relation but also because it affects pore pressure in the media. The time-dependent volumetric constitutive relation of fault gouge, therefore, bears on the question of redistribution of pore pressure and effective stress in a fault zone after an earthquake and other types of change in the in situ state of stress.

SAMPLE PREPARATION AND EXPERIMENTAL PROCEDURE

Core samples of fault gouge were obtained through a USGS deep drilling program, from depths down to 400 m in the San Andreas fault zone in Dry Lake Valley near Hollister, California. The sample used for this study is #725 obtained from a depth of 725 feet. It was in the shape of a cylinder of 2.4 cm in diameter and about 4 cm long, with the ends of the cylinder cut into flat surfaces perpendicular to the axis of the cylinder, but otherwise left in the cored state with as little mechanical disturbance as possible. The sample was consolidated at a confining pressure of 200 Mpa for a period of 7 days, and was left in air for one and a half years before experimentation. Modal analysis of the sample using standard point counting technique showed that it contains 12.5% quartz, 5.1% feldspar, 6.4% calcite and 76% clay on a volume basis. X-ray analyses of the clay components in some gouge samples from the same location (Liechti and Zoback, 1979; Wu et al., 1979; Logan et al., 1981; Morrow et al., 1981) show that in general the clays consist largely of montmorillonite and kaolinite, with relatively small amounts of illite and chlorite. The sample was weighed and its dimension measured, yielding a bulk density of 2.25 g/cm^3. After completion of our experiment, the sample was heated in vacuum at a temperature of 105°C for thirty days. From the difference between the weights of the original and dried sample, we calculated a porosity of 0.218 and a degree of saturation of 0.89, assuming that the average density of the solid grains is 2.63 g/cm^3. The maximum volumetric strain of the fault gouge under pressure was 1.8% in our experiment. Thus, the sample was never saturated and the measured volumetric strain is due to the deformation of the solid framework of the fault gouge.

For the kaolinite, powdered sample (API standard) was seived to retain grains with sizes smaller that 124 micron, and was consolidated under confining pressure of 200 Mpa for 7 days, with a final bulk density of 2.22 g/cm^3. The consolidated sample was then machined into right cylinder of 6.3 cm in length and 2.54 cm in diameter. More detailed description of sample preparation was given in Wang et al. (1980).

For ultrasonic velocity measurements, piezoelectric transducers (lead zirconate ceramic) of either compression or shear mode vibration, with natural frequency of 1 Mhz, were mounted onto the ends of the sample. A standard pulse transient technique was used, with an accuracy of about 1% (Birch, 1960). For the measurement of sample

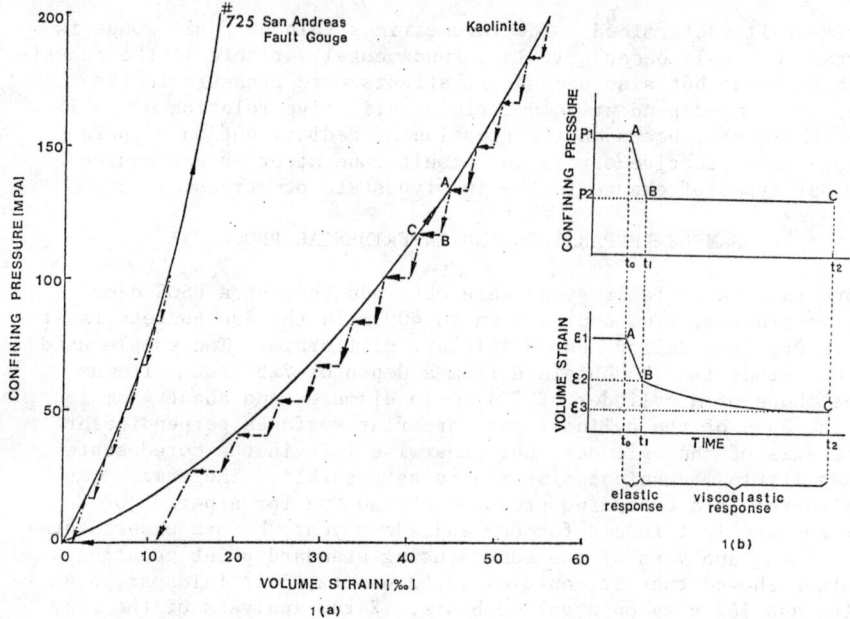

Fig. 1. (a) Confining pressure-volume strain relations of kaolinite and #725 fault gouge; solid curve represents quasi-static loading; dashed curve represents stepwise unloading.
(b) Illustration of stepwise unloading: as confining pressure decreased from P_1 at time t_o to P_2 at time t_1, the volume strain changed from ε_1 to ε_2; as pressure was kept at P_2 until time t_2, the sample was relaxed from ε_2 to ε_3.

deformation, two pairs of resistance strain gages (Micromeasurement EA-06-125TM-120) were epoxied onto the middle of the cylindrical wall of the sample; in each pair, one strain gage was along the axial direction of the sample, the other was perpendicular to this direction. The entire assemblage was then jacketed with a layer of liquid rubber to prevent the pressure medium from intruding into the sample during the experiment.

Standard technique for strain measurement was followed. However, the measurement procedure requires a brief explanation. Figure 1(a) shows the strain vs. pressure curves for our samples under various loading and unloading conditions. The solid curve represents the results of a quasi-static loading experiment; the dashed curve represents the results of stepwise unloading (from point A to B, say) followed by prolonged interval of time at constant pressure (from point B to C, say). Figure 1(b) gives a schematic illustration of the stepwise unloading process. At time t_o (point A), the confining pressure was suddenly released from P_1; it reached P_2 at time t_1 (point B),

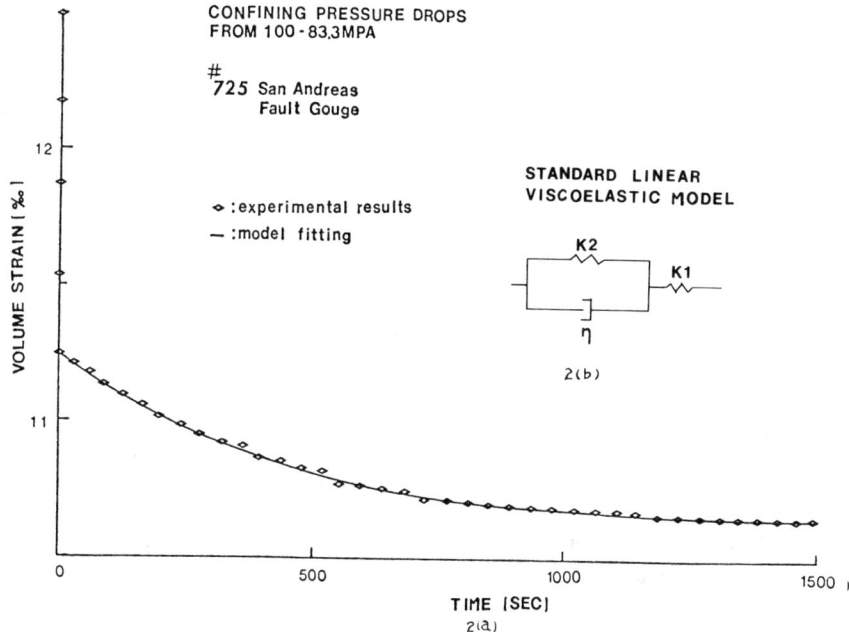

Fig. 2.(a) Typical experimental results of #725 fault gouge: volume strain was determined as a function of time after a stepwise decrease of confining pressures.
(b) A standard linear viscoelastic model used to fit experimental results.

and was kept constant for a prolonged interval of time until time t_2 (point C). The time interval between t_o and t_1 is less that 1 sec; the corresponding deformation between points A and B is labeled "elastic", during which the viscous component may be negligible because of the very short time interval. This point will be born out later by comparing the "elastic" moduli with the dynamic moduli determined by ultrasonic technique. For $t \geq t_1$, pressure was kept constant, deformation of the sample must be due to viscoelastic relaxation. The time interval between t_1 and t_2 was usually 1500 seconds or when no more volumetric change was detectable, whichever was longer. Data recording was carried out by using a minicomputer (PDP 11-10) via a 10-bit A/D converter (Wang et al., 1980). Sampling rate was 1 to 5 Hz during the first 20 seconds and 0.01 to 0.05 Hz afterwards.

RESULTS AND DISCUSSION

Figure 2(a) shows the results of a typical experiment in which the volumetric strain of the San Andreas fault gouge was determined as a function of time after a stepwise decrease of pressure. The first five points on the time axis are for the elastic response from which

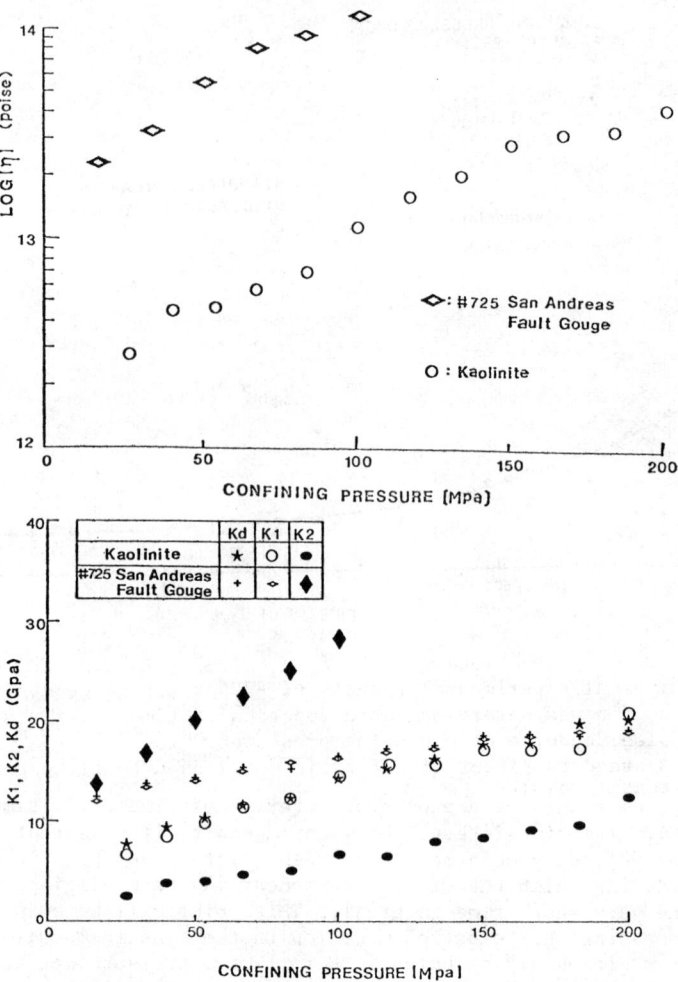

Fig. 3. Experimentally determined values of K_d, K_1, K_2 and η of kaolinite and #725 fault gouge as a function of confining pressure.

we calculated the "elastic" bulk moduli; those points after the first two were sampled at constant pressure and therefore represent changes due to relaxation. In Figure 3, the experimentally determined value of K_1 for both the San Andreas fault gouge and the consolidated kaolinite samples are plotted as a function of pressure up to 200 Mpa.

At ultrasonic frequencies, clayey fault gouges are effective in transmitting both compressional and shear waves (Wang et al., 1978). By measuring the velocities of these waves through the specimen we determined for the fault gouge the "dynamic" bulk moduli K_d at 1 Mhz as a function of pressure in Figure 3. It turns out that with the present experimental accuracy, these values are indistinguishable from the "elastic" moduli determined during the stepwise changes of pressure. The close resemblance of K_1 and K_d, together with the results from the following modeling study, strongly suggest that both K_1 and K_d may be considered as the unrelaxed moduli (i.e., moduli measured at "high-frequency") and there may be no major relaxation mechanism occurring between 1 and 10^6 Hz.

Rheological models are often proposed for mathematical description of the relations between stress, strain and time for soils. Although these models do not necessarily provide information on the physical mechanism for relaxation they are useful for visualization of the time-dependent deformation in terms of various arrangements of elastic and viscous components. For the data presented in Figure 2, the simplest model which provides an adequate description is a standard linear viscoelastic medium which consists of an elastic element and a Voigt element in series, as depicted in Figure 2(b). The volumetric strain (ε_v) of this model, in response to a stepwise change of pressure (δP), is given by

$$\varepsilon_v = \delta P \left(\frac{1}{K_1} + \frac{1}{K_2} (1 - e^{-K_2 t/\eta}) \right)$$

where K_1 is the unrelaxed moduli which may be identified with K_d; K_2 and η are the bulk moduli and viscosity in the Voigt element, which may be determined from the experimental data using a least square procedure. The solid curve in Figure 2(a) gives the best fitting model for the experimental data; it is evident that the chosen model closely describes the observation.

The rheological parameters are pressure-dependent; their variations with pressure are given in Figure 3. Viscosity of the San Andreas fault gouge, for example, increases from 2×10^{13} poise at 17 MPa to 1×10^{14} poise at 100 Mpa, at higher pressure, relaxation time for the fault gouge became excessively long that we did not make the necessary observation for the determination of the viscoelastic parameters.

For oscillatory phenomena, anelasticity of materials is commonly expressed in terms of the "quality factor", Q. Assuming the model discussed above for the volumetric deformation of the fault gouge, we may express the corresponding quality factor as

$$Q_K = Q_o (1 + \omega^2 \tau^2)/2\omega\tau$$

where ω is the angular frequency of the oscillation, τ is the relaxation time, Q_o is the minimum value of Q_K at $\omega\tau = 1$ and is related to the rheological parameters by

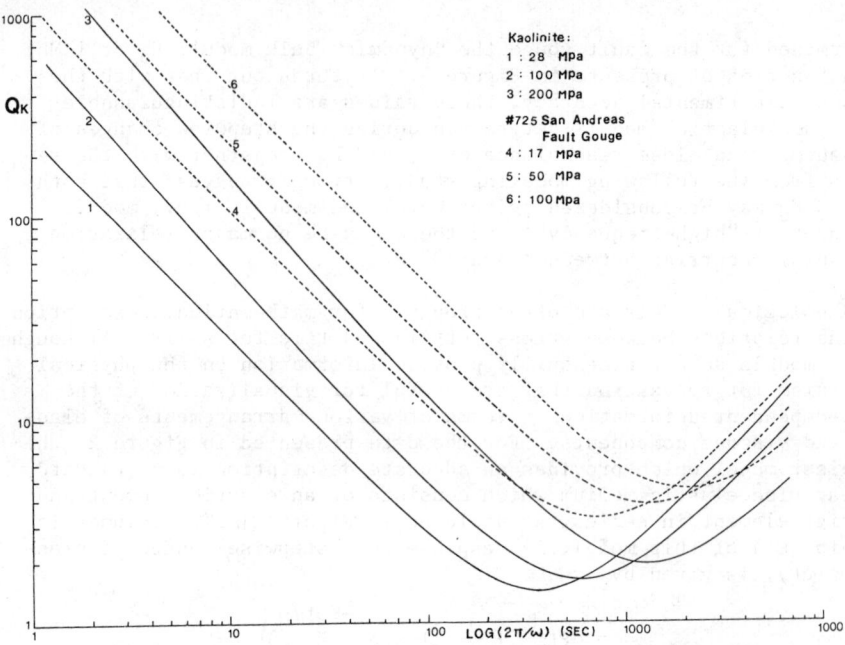

Fig. 4. "Quality factor" Q_K of kaolinite and #725 fault gouge as a function of time at various confining pressures.

$$Q_o = 2\left(\frac{K_2^2}{K_1^2} + \frac{K_2}{K_1}\right)^{1/2}$$

In Figure 4, we show the values of Q_K so determined. For the San Andreas fault gouge, the minimum value of Q_K (∼4) occurs at periods of about 1000 sec, and for consolidated kaolinite, $Q_K \cong 1.5$ occurs at periods of about 70 ∼ 100 sec. At periods away from the relaxation times, Q_K increases markedly both with decreasing and with increasing periods. For the San Andreas fault gouge, for example, Q_K reaches 10^2 to 10^3 at periods between 1 and 10 sec. The anelasticity of the fault gouge under shear deformation has not yet been determined. Thus, the results of the above study cannot be directly used for the interpretation of the attenuation of seismic waves in the San Andreas fault zone. If we assume that the quality factor for shear is of the same magnitude or greater as Q_K, as shown for partially saturated sandstones (e.g., Winkler and Nur, 1979), then our results imply that the San Andreas fault zone may be effective in transmitting seismic waves. Our data also suggest that in the San Andreas fault zone Q may exhibit strong frequency. Aki and Chouet (1975) showed that along the San Andreas fault in central California near (Stone Canyon), Q increases with frequency from 10^2 at 1 Hz to 10^3 at 20 Hz. Similar results were obtained in the San Andreas fault zone near Hollister in central California (McEvilly, 1982, pers. comm.). Although interpretations for this frequency-dependent Q are not unique (see Aki and Chouet, 1975), the interpretation that it may represent the material property of the fault zone is consistent with our experimental results.

ACKNOWLEDGEMENTS

We thank Mark Zoback for providing fault gouge samples from the San Andreas fault zone. Discussions with Yaolin Shi were helpful. This research was supported by a USGS Contract, USDI 14-08-0001-20530.

REFERENCES

Aki, K., and Chouet, B., Origin of coda waves: Source, Attenuation, and Scattering effects, J. Geophys. Res., 80, No. 23, 3322, 1975.

Birch, F., The velocity of compressional waves in rocks to 10 kilobars, part 1, J. Geophys. Res., 65, 1083, 1960.

Chu, C.L., Wang, C.Y., and Lin, W., Permeability and frictional properties of San Andreas fault gouges, Geophy. Res. Lett., 8, 565, 1981.

Liechti, R., and Zoback, M.D., Preliminary analysis of clay gouge from a well in the San Andreas fault zone in central California, Proc. Conf. VIII, Analysis of Actual Fault Zones in Bedrocks, USGS Open-file Report 79-1239, 269, 1979.

Logan, J.M., Creep stable sliding and premonitory slip, Pure Appl. Geophys., 116, 773, 1978.

Logan, J.M., Higgs, N.G., and Friedman, M., Laboratory studies on natural gouge from the USGS No. 1 Well in the San Andreas fault zone (to appear in AGU Monograph, the Handin Volume), 1981.

Morrow, C., Shi, L.Q., and Byerlee, J., Permeability and strength of San Andreas fault gouge under high pressure, Geophy. Res. Lett., 8, No. 4, 325, 1981.

Morrow, C., Shi, L.Q., and Byerlee, J., Strain hardening and strength of clay-rich fault gouges, Geophys. Res. Lett., (in press).

Shimamoto, T., and Logan, J.M., Effects of simulated fault gouge on the sliding behavior of Tennessee Sandstone: Non-clay gouges, J. Geophy. Res., 86, No. B4, 2902, 1981.

Solberg, P., and Byerlee, J., The effect of strain rate of frictional stress under moderate and elevated confining pressures, EOS Trans. AGU (Abst.), 62, No. 45, T3-1-B-6, 1981.

Summers, R., and Byerlee, J., A note on the stability of frictional sliding, Int. J. Rock Mech. Sci. and Geometh. (abs.), 14, 155, 1977.

Wang, C.Y., Lin, W., and Wu, F.T., Constitution of the San Andreas fault zone at depth, Geophy. Res. Lett., 5, 741, 1978.

Wang, C.Y., and Mao, N.H., Shearing in saturated clays in rock joints at high confining pressures, Geophys. Res. Lett., 6, 825, 1979.

Wang, C.Y., Mao, N., and Wu, F.T., Mechanical properties of clays at high pressure, J. Geophy. Res., 85, No. B3, 1462, 1980.

Winkler, K., and Nur, A., Pore fluids and seismic attenuation in rocks, Geophy. Res. Lett., 6, 1, 1979.

Wu, F.T., Roberson, H.E., Wang, C.Y., and Mao, N.H., Fault zones, gouge and mechanical properties of clays under high pressure, USGS Open-File Report 79-1239, 344, 1979.

Chapter 29

DEFORMATION MECHANISMS IN GRANODIORITE AT EFFECTIVE PRESSURES TO
100 MPA AND TEMPERATURES TO PARTIAL MELTING

by M. Friedman, J. Handin, and S. J. Bauer

Center for Tectonophysics
Texas A&M University
College Station, TX 77843

ABSTRACT

Deformation mechanisms in room-dry and water-saturated specimens of Charcoal Granodiorite, shortened at $10^{-4} s^{-1}$, at effective pressures (Pe) to 100 MPa and temperatures to partial melting ($\leq 1050°C$) are documented with a view toward providing criteria to recognize and characterize the deformation for geological and engineering applications. Above 800°C strength decreases dramatically at effective pressures ≥ 50 MPa and water-weakening reduces strength an additional 30 to $\overline{40}\%$ at Pe = 100 MPa. Strains at failure are only 0.1 to 2.2 percent with macroscopic ductility (within this range) increasing as the effective pressures are increased and in wet versus dry tests. Shattering (multiple faulting) gives way to faulting along a single zone to failure without macroscopic faulting as ductility increases. Microscopically, cataclasis (extension microfracturing and thermal cracking with rigid-body motions) predominates at all conditions. Dislocation gliding contributes little to the strain. Precursive extension microfractures coalesce to produce the throughgoing faults with gouge zones exhibiting possible Riedel shears. Incipient melting, particularly in wet tests, produces a distinctive texture along feldspar grain boundaries that suggests a grain-boundary-softening effect contributes to the weakening. In addition, it is demonstrated that the presence of water does not lead to more microfractures, but to a reduction in the stresses required to initiate and propagate them.

INTRODUCTION

Solutions of rock-mechanics problems associated with energy extraction from the geothermal regime above and in magma bodies and structural studies of exhumed paleomagma bodies require knowledge of the deformation mechanisms operative in rocks at temperatures to partial melting and at low to modest effective pressures. Recognition and documentation of these mechanisms in experimentally deformed rocks (1) serves as a basis for extrapolating laboratory data (strength, ductility, and flow laws) to the field (similar mechanisms must exist in both naturally and experimentally deformed materials for such extrapolations to be valid), and (2) provides the structural petrologist with fabric criteria potentially to recognize and interpret natural deformations. While extensive studies have been made of ductile mechanisms at high P and T (e.g., Carter, 1976; Tullis, 1979), comparatively little is known about mechanisms in the semi-brittle regime at temperatures to partial melting and at low to modest pressures and in wet as well as dry rocks (Carter and Kirby, 1978; Friedman et al., 1979; Van der Molen and Paterson, 1979; Bauer et al., 1981; and Bauer, in press).

A few studies have detailed the microstructures developed upon experimental deformation of igneous rocks at temperatures of partial melting (e.g., Van der Molen and Paterson, 1979; and Auer et al., 1981), but the minimum effective pressure was 300 MPa. In those tests axial (extension) microfractures and grain-boundary partings predominate, crystal-plastic behavior of the quartz and feldspar in slight, early melting of biotite and other hydrated minerals is evident, feldspar melts before quartz, and the melt occurs preferentially along the axial cracks. The latter was also observed at 1500 MPa by Ave Lallement and Carter (1970). The critical melt fraction for the transition from solid creep to viscous flow appears to range between 20 and 40% (Arzi, 1978; Van der Molen and Paterson, 1979; and Auer et al., 1981). Herein we confirm many of these earlier findings in room-dry and water-saturated 2 by 4-cm cylindrical specimens of Charcoal Granodiorite (St. Cloud Gray Granodiorite, see Kreck et al., 1974), shortened at $10^{-4}s^{-1}$, at effective confining pressures to 100 MPa, and at temperatures to partial melting ($\leq$ 1050°C). In addition, we emphasize the role of microfracturing and the very incipient stages of partial melting, both enhanced by the presence of water, as the mechanisms that account for the weakening of the rock with increasing temperatures, particularly between 800° and 1000°C.

EXPERIMENTAL RESULTS

Our specimen preparation, experimental methods, results for the granodiorite as well as for the Mount Hood Andesite and Cuerbio Basalt and the extrapolation of the data to an assessment of borehole stability are reported fully in Friedman et al. (1981), and in part in Friedman et al. (1979) and Bauer et al. (1981). Stress/strain curves for the granodiorite at temperatures to 400°C, wet or dry, typically show pronounced work-softening following a peak (ultimate) strength within the range of 0.1 to 2.2 percent shortening. The work softening (often 3 to 5 percent shortening where the test is terminated) reflects either movement along macroscopic faults or pervasive flow without macroscopic faulting. Within this narrow range of strains, ductility increases with increasing effective pressure, is greater in wet specimens, but is largely independent of temperature.

The ultimate strength of the rock decreases as temperature increases (Figure 1). That the strengths of the water-wet specimens are close to or less than those of dry counterparts at equal effective pressures suggests that the pore-fluid pressures are effective (corresponding pore pressures are pervasive), even at the strain rate of $10^{-4} s^{-1}$. Water-weakening (amounting to 30 to 40 percent) becomes apparent only at the highest effective pressure of 100 MPa and above 800°C.

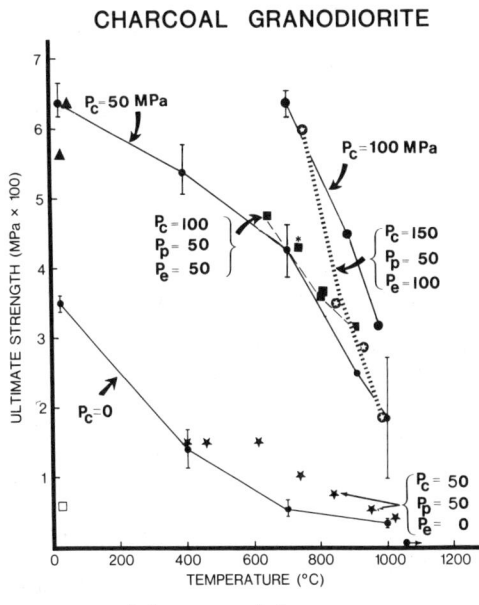

Figure 1. Strength versus temperature for room-dry and water-saturated granodiorite at effective pressures (Pe) to 100 MPa. Pc is confining pressure, and Pp is pore fluid pressure. Triangles represent dry specimens heated unconfined to 1000°C, then cooled and deformed at 25°C, Pe = 50 MPa. Open square represents three dry specimens heated unconfined to 900-1000°C, then cooled and deformed at 25°C.

We made no systematic investigation of melting temperatures per se. Twenty-five percent melt fraction of room-dry specimens is found at temperatures between 1050°C (melting point of copper jacket) and about 1250°C. Melting to this degree results in essentially total loss of strength so that a specimen shortens as much as 27 percent under the mere weight of a 2-kg steel spacer ($\Delta\sigma = 0.1$ MPa) with an equivalent viscosity in the order of 10^7 N·s/m². The incipient melting of feldspars discussed below occurs in water-saturated specimens at 900°C and in dry ones at $\geq$ 1000°C. While the incipient melting probably causes some strength reduction, strengths are still relatively high, e.g., 185 MPa at 1015°C and 100-MPa effective confining pressure (Figure 1).

PETROFABRIC RESULTS

Macroscopic

The macroscopic mode of deformation is determined prior to impregnation of the specimen with blue-staining epoxy for thin-section preparation and while it is still jacketed (Figure 2a). Relatively brittle behavior is characterized by the development of multiple faults of varying "strike" and with inclinations to the piston/cylinder axis (greatest principal compressive stress, σ_1) of 23° to 38° (Friedman et al., 1981, Table 1). This angle increases within this range as ductility increases. Multiple faulting tends to give way to faulting along a single zone as the effective confining pressure increases. Only one specimen failed by uniform flow without macroscopic faulting (925°C, 100-MPa, wet).

Microscopic

Microfractures are by far the most obvious deformation features in thin sections of all specimens. Included here are intragranular and transgranular microfractures, grain-boundary partings, and cleavages primarily in the feldspars, biotite, and hornblende. These features are made conspicuous and are separated from those possibly induced during thin-section preparation by impregnation under pressure of the blue-staining epoxy prior to removal of jackets. The microfractures result primarily from two stress states, one caused by the mechanical loading of the specimens during the triaxial-compression test, and the other due to the anisotropic thermal expansions induced by heating. The former are easily recognized by their preferred orientation. They are mainly extension "axial" fractures oriented, except for occasional jogs, subparallel ($\pm$ 10°) to cylinder axis, σ_1 (Figure 2b). Adjacent to shear zones they align as <u>en-echelon</u> axial cracks (Figure 2b) or as <u>en-echelon</u> ones (Figure 2c) that are either inclined shears or extension fractures, reflecting a locally inclined σ_1. Within shear zones (Figure 2d) these microfractures are either the initial precursive axial cracks or Riedel shears that reflect the shear deformation, probably the latter. The macroscopic faults appear to arise from the coalescence of microfractures.

The thermoelastic stresses account for some grain-boundary parting and for thermal cracks that are recognizable unambiguously only when they initiate at grain boundaries and taper inward (Friedman and Johnson, 1978). They add to the spacial and orientational scatter of the microfracture fabric, but do not significantly degrade triaxial-compressive strength (Figure 1).

A quantitative assessment of microfracturing, including all types except cleavage, is obtained by counting the number of cracks intersected along traverses across a thin section of each specimen. Between 1000 and 2200 microfractures are counted (at 250-power

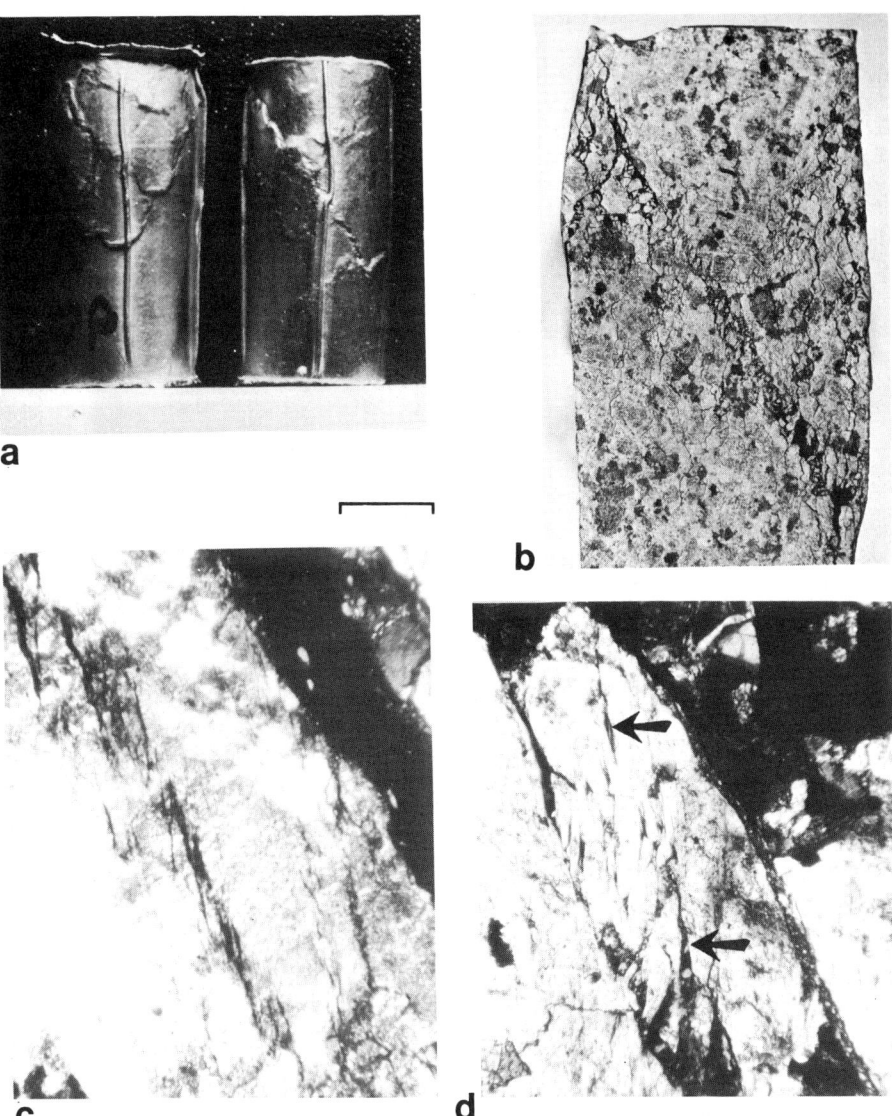

Figure 2. Brittle fracture in Charcoal Granodiorite. (a) Photograph of dry specimen, shortened at 920 and 1000°C at Pe = 50 MPa, shows imprint of faults on copper jackets. (b) Dry specimen, 1000°C and Pe=0, shows incipient fault is composed of coalesced extension fractures. (c) Photomicrograph shows an en echelon array of inclined microfractures in a zone parallel to the main fault, wet specimen, 750°C, Pe=50 MPa. (d) Detail of same fault zone shows closely spaced axial cracks or R_1-Riedel shears (arrow). In (a and b) specimen is 2 cm diameter; scale line is 0.15 mm for (c) and 0.3 mm for (d); plane polarized light.

magnification) along a total length of 8 to 10 cm. These lines are oriented perpendicular to σ_1. From these data the average number of microfractures intersected per mm is calculated and plotted against the temperature at which each specimen was deformed (Figure 3). This plot shows that (1) some cracks are due to heating only (at $P_c = 0$); (2) superposed on the thermal cracking, microfracturing induced by loading to failure is about the same regardless of temperature or effective confining pressure; and (3) the abundance of microfractures at a given temperature is the same in room-dry and water-saturated samples.

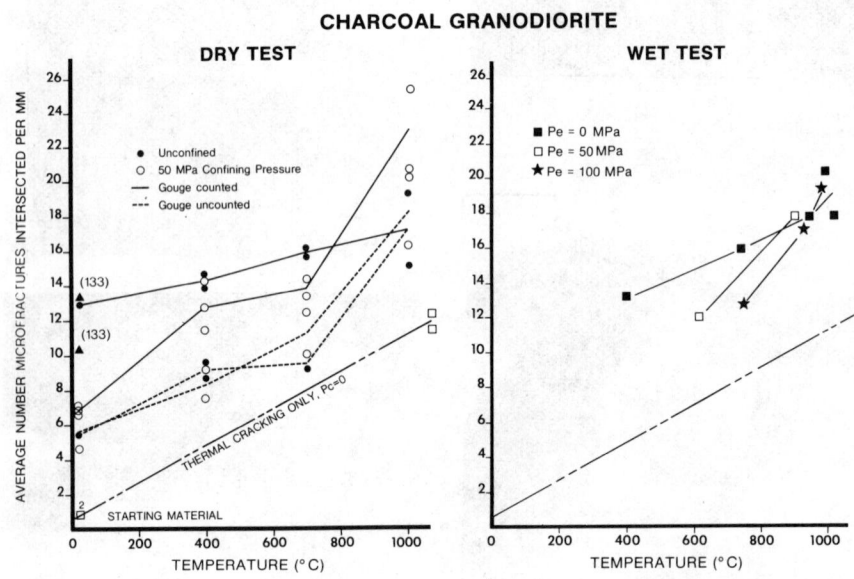

Figure 3. Microfracture abundance as functions of effective pressure and temperature in room-dry and water-saturated granodiorite. Specimen 133 (▲) was heated unconfined to 1000°C, then cooled, and deformed at 25°C, 50 MPa. Higher result include gouge, lower one does not.

Dislocation glide mainly is manifest as kink bands in favorably oriented biotite crystals at all temperatures. At $\geq$ 750°C quartz and feldspar exhibit more undulatory extinction than in the starting material or in specimens deformed at lower temperatures. At $\geq$ 900°C plagioclase crystals are bent (Figure 4a). While some of this bending is due to rotations across fractures (Tullis and Yund, 1977), some of the rotations are continuous and suggestive of bend gliding. In any event the contribution to total strain from crystal plasticity is minor in comparison to that from cataclasis.

DEFORMATION MECHANISMS IN GRANODIORITE 285

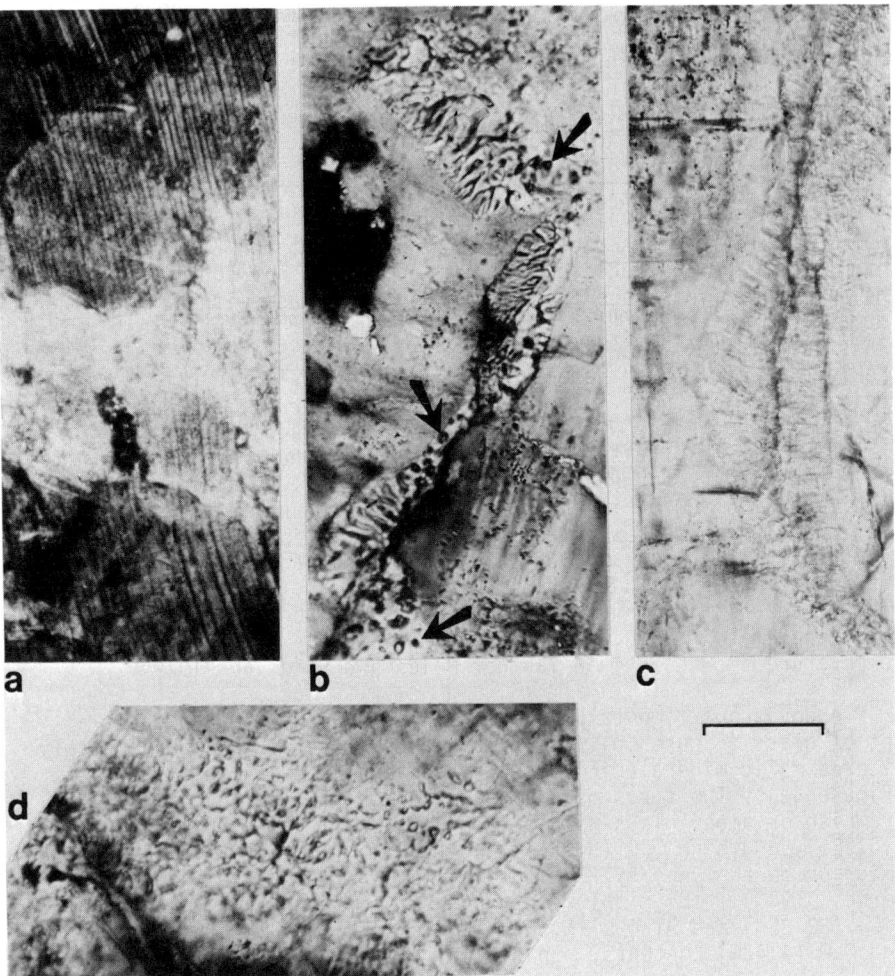

Figure 4. Photomicrographs of features reflecting dislocation glide (a) and possible grain-boundary softening (b-d). (a) Bent plagioclase crystal in wet specimen, 975°C, Pe=100 MPa. (b) Myrmakite-like texture along grain boundary between feldspar grains, wet specimen, 1015°C, Pe=0. Note high relief bubbles (straight arrows) and elongated bodies or tubes oriented nearly perpendicular to the boundary. (c) Similar textures along grain boundary between feldspars, wet specimen, 900°C, Pe=50 MPa. Note elongation is nearly normal to grain boundary. (d) Similar myrmakite-like texture within a feldspar grain exhibits radial configuration of worm-like bodies or tubes. Cross-polarized light for (a and b); plane-polarized light for (c and d); scale line is 0.3 mm for (a), 0.1 mm for (b), 0.05 mm for (c), and 0.02 mm for (d).

Incipient melting of the feldspars takes place at $\geq$ 900°C in wet specimens and at $\geq$ 1000°C in dry ones. In contrast to the obvious development of glass in the more advanced stages of melting described below, incipient melting is seen as a subtle myrmekite-like texture that initially forms along grain boundaries (Figures 4b, c) and later within grains (Figure 4d). Classical myrmekitic texture involves minute worm-like or finger-like bodies of quartz enclosed in sodic plagioclase (e.g., Williams et al., 1954, p. 20). The pseudo-myrmekitic texture developed here involves a complex of elongated (wormy) and nearly spherical bodies (bubbles) at the boundaries of both potash and plagioclase feldspars. It does not form in quartz. The high relief of these features suggests a phase with a much different index of refraction than that of the host grain, possibly glass or voids filled with volatiles. The fact that they lie within the host precludes the detection of their optical properties. At grain boundaries these features are oriented perpendicular to the boundary (Figure 4c); within grains they form radiating masses (Figure 4d). This texture does not appear in the starting material or in specimens deformed at $\leq$ 750°C. It is well developed in wet specimens at $\geq$ 900°C, in dry ones at $\geq$ 1000°C, and in the unmelted portions of specimens with about 25° melt fraction. Incipient melting of biotite occurs at 400°C and is manifest as small glassy bodies within and along the boundaries of the crystals (Figure 5a). The rims of the biotite (and hornblende) crystals become oxidized (blackened) at $\geq$ 400°C, and become progressively more altered as the temperature increases (see Van der Molen and Paterson, 1979).

Pseudoviscous flow takes place when partial melting produces a lot of glass (Figure 5b), glass-filled fractures (Figure 5c), highly altered mafic minerals (Figure 5d), and arborescent microlites (Figure 5e). Two types of glass are formed (Figure 5b). (1) A brownish or amber glass with relatively low SiO_2 (58.35%), Al_2O_3 (10.28%), and high FeO (9.52%), MgO (5.93%), and CaO (6.42%) contents (microprobe analyses by P. F. Hlava and E. J. Graber, Sandia Laboratories, 1979). (2) A colorless, highly vesicular glass with higher SiO_2 (62.38%) and Al_2O_3 (15.58%) and lower FeO (2.88%), MgO (1.87%), and CaO (4.42%) contents. Spacially, the former seems to originate from biotite and hornblende and to form first. The latter originates primarily from the feldspars as its high K_2O (7.05%) and Na_2O (4.79%) suggest. These melts appear to be immiscible, and show flow textures like banding (brownish glass) and elongated vesicles (colorless glass). In the early stages of melting the brownish glass tends preferentially to fill the axial extension fractures (Figure 5c) as Van der Molen and Paterson (1979) and Auer et al. (1981) also observed. At later stages most fractures and grain boundaries become filled with melt (glass). When melting occurs, the hornblendes and biotites become highly oxidized (blackened), and the hornblende alters to a curious texture (Figure 5d) that resembles spinifex (e.g., Arndt and Brooks, 1980).

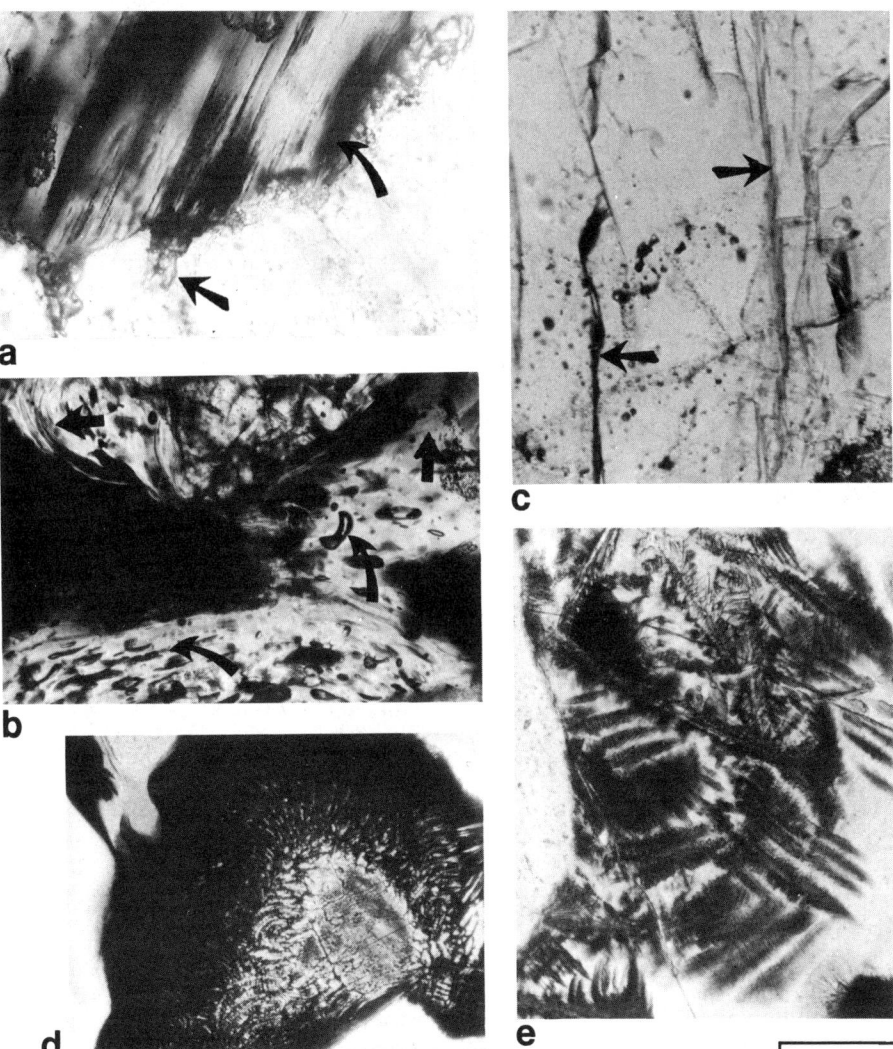

Figure 5. Photomicrographs of melt features in granodiorite. (a) Partially melted biotite crystal, dry specimen, 400°C and Pe=50 MPa shows blebs of glass (straight arrow) and oxidized zones (curved arrow). (b) Flow in colorless, vesicular glass (curved arrow) and in amber glass (straight arrow), dry specimen, shortened 27% at >1050°C and Pe=0. (b) Glass-filled axial cracks (arrow) in the incipient stages of melting, wet specimen, 1015°C and Pe=0. (d) Highly altered hornblende crystal with spinifex-like texture, dry specimen, 1000°C and Pe=50 MPa. (e) Arborescent microlites, dry specimen, 1000°C and Pe=50 MPa. Plane polarized light; scale line is 0.06 mm for (a and c), 0.14 mm for (b and e), and 0.38 mm for (d)

CONCLUSIONS

1. In triaxial compression at effective confining pressures to 100 MPa and all temperatures to partial melting, both dry and wet granodiorite deforms essentially by cataclasis (microfracturing) at least at the relatively high strain rate of $10^{-4} s^{-1}$. Extension microfractures coalesce to form macroscopic faults.

2. Strength reductions at $\geq 800°C$, including water-weakening apparent only at 100-MPa effective confining pressure (Figure 1), appear to result from microfracturing at lower differential stresses (not from an increase in fracture abundance, Figure 3), and from grain boundary softening where incipient melting of the feldspars occurs.

3. The myrmekite- and spinifex-like textures are symptomatic of partial melting and, therefore, of strength reduction, as are such previously recognized features as glass-filled fractures and oxidized mafic minerals.

REFERENCES

Arndt, N., and Brooks, C., 1980, "Komatiites," Geology, Vol. 8, pp. 155-156.

Arzi, A.A., 1978, "Critical Phenomena in the Rheology of Partially Melted Rocks," Tectonophysics, Vol. 44, pp. 173-184.

Auer, F., Berckhemer, H., and Oehlschlegel, G., 1981, "Steady State Creep of Fine Grain Granite at Partial Melting," Jour. of Geophysics, Vol. 49, pp. 89-92.

Avé Lallemant, H.G., and Carter, N.L., 1970, "Syntectonic Recrystallization of Olivine and Modes of Flow in the Upper Mantle," Geol. Soc. Amer. Bull., Vol. 81, pp. 2203-2220.

Bauer, S.J., Friedman, M., and Handin, J., 1981, "Effects of Water-Saturation on Strength and Ductility of Three Igneous Rocks at Effective Pressures to 50 MPa and Temperatures to Partial Melting," Proc. 22nd U.S. Symp. Rock Mech., Cambridge, MA, pp. 73-78.

Carter, N.L., 1976, "Steady-State Flow of Rocks," Revs. of Geophysics and Space Physics, Vol. 14, pp. 301-360.

Carter, N.L., and Kirby, S.H., 1978, "Transient Creep and Semibrittle Behavior of Crystalline Rocks," Pure and Applied Geophysics, Vol. 116, pp. 806-839.

Friedman, M., and Johnson, B., 1978, "Thermal cracks in unconfined Sioux Quartzite," Proc. 19th U.S. Symp. Rock Mechanics, Reno, pp. 423-430.

Friedman, M., Handin, J., Higgs, N.G., and Lantz, J.R., 1979, "Strength and Ductility of Four Dry Igneous ROcks at Low Pressures and Temperatures to Partial Melting," Proc. 20th U.S. Symp. on Rock Mech., Austin, Tx, pp. 35-50

Friedman, M., Handin, J., and Bauer, S. J., 1981, "Mechanical Properties of Rocks at High Temperatures and Pressures." Tech. Prog. Rept. No. 2, March - Nov., 1981, DOE/BES Contract: DE-AS05-79ER-10361, 67 pp.

Krech, W.W., Henderson, F.A., and Hjelmstad, K.E., 1974, "A Standard Rock Suite for Rapid Excavation Research." R.I. 7865, U. S. Bureau of Mines, 29 p.

Tullis, J.A., 1979, "High Temperature Deformation of Rocks and Minerals," Revs. Geophysics and Space Physics, Vol. 17, pp. 1137-1154.

Tullis, J.A., and Yund, R. A., 1977, "Experimental Deformation of Dry Westerly Granite." J. Geophy. Res., Vol. 82, pp. 5705-5718.

Williams, H., Turner, F.J., and Gilbert, C.M., 1954, "Petrography," W. H. Freeman, San Francisco, 406 p.

ACKNOWLEDGMENTS

We gratefully acknowledge support of this work by the U.S. DOE, Office of Basic Energy Sciences, Contract: DE-AS05-79-ER10361.

Chapter 30

THE EFFECTS OF FRACTURE TYPE (INDUCED VERSUS NATURAL) ON THE STRESS-FRACTURE CLOSURE-FRACTURE PERMEABILITY RELATIONSHIPS.

JOHN E. GALE

Department of Earth Sciences,
University of Waterloo,
Waterloo, Ontario, N2L 3G1.

ABSTRACT

Seven, 15 cm diameter, cores of gneissic granite, four containing induced fractures and three containing natural fractures all oriented normal to the core axis, were tested in an uniaxial compression mode over a range of 0 to 30 MPa. At given normal stress increments, over two or three loading and unloading cycles, the flowrates and changes in fracture aperture were measured. The induced fractures gave lower initial (at low stresses) and lower final flowrates (at maximum stress) than the natural fractures. Both types of fracture exhibited permanent fracture deformation between loading cycles as well as highly nonlinear, with distinct hystersis, loading and unloading flowrate-stress curves. A significant finding of this study is the breakdown of the cubic law for fracture flow in induced fractures subjected to normal stresses greater than 20 MPa, with the breakdown occuring at much lower stresses in natural fractures.

INTRODUCTION

With the growing need to isolate and contain toxic wastes, such as nuclear waste material, and the suggestion that this can be achieved by storing them in fractured crystalline or argillaceous rocks, it is essential that we understand how fluids and contaminants move, and the factors that control such movement, in fractured rocks.

Any attempt at understanding the factors that control the flow of fluids through fractured rocks must be based on a clear understanding of the structural nature of the rock mass. The main flow paths in fractured crystalline and argillaceous rocks are joints, fracture zones and shear zones. In this paper we are concerned with flow through the 'joint' member of the fracture family but the general term 'fracture' will be used throughout the rest of this paper. Gale (1982a) has discussed a number of factors that control flow through single and multiple fractures, such as stress, temperature, roughness, fracture geometry, etc, but as yet the relative importance of each factor has not been assessed. The discussion here will be restricted to the effects of changes in normal stress on fracture permeability as shown by the results of a laboratory study of induced and natural fractures in 15 cm diameter gneissic granite cores.

EFFECTS OF FRACTURE TYPE ON PERMEABILITY

EXPERIMENTAL PROCEDURES

Sample Collection and Preparation

Both induced and natural fractures were collected from surface exposures of granitic gneiss near Chalk River, Ontario, (Lumbers, 1974). Blocks of rock were obtained from outcrops by drilling overlapping boreholes. These blocks were cut to size, and a 1.5 cm diameter center hole was drilled along the length of the sample. A fracture was induced in each rock block using a modified form of the Brazilian test, with a rock bolt installed in the center hole to keep the fracture closed. The fracture was overcored, and cut to provide a 15 cm diameter by 30 cm long core. Natural fractures were collected by first installing a rock bolt in a 1.5 cm diameter hole drilled normal to the fracture plane and then overcoring with a 15 cm diameter thin walled core bit to provide a 30 cm long core with a natural fracture located about halfway along the core length and oriented perpendicular to the core axis. Final preparation of the cores followed that outlined by Gale and Raven (1982) and included either grinding the ends flat or adding sulphur caps to ensure end parallelism for testing purposes.

Sample Instrumentation and Testing Procedures.

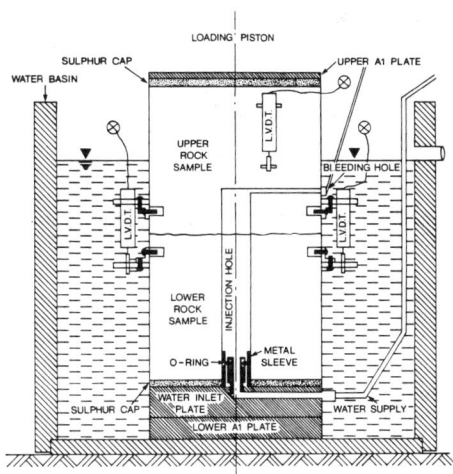

Figure 1. Schematic of rock sample ready for testing.

Figure 1 is a schematic of a fully instrumented sample placed within the loading frame and ready for testing. A metal sleeve was glued into the center hole at the base of the sample. A water inlet plate, with an O-ring tube attached, was fitted to the base of the sample. The O-ring tube fitted into the metal sleeve provided a water-tight connection for the water supply during injection testing. A .64 cm diameter hole was drilled from the outside of the sample to the top of the center hole. A swaglock fitting was glued into this hole which served to both bleed air pockets during the initial sample set-up as well as providing a means of measuring the fluid pressure in the center borehole during permeability tests.

The sample was then placed in a water reservoir in which the water level was maintained above the fracture plane. Three linear variable differential transducers, (LVDT's) were located across the fracture plane, spaced at 120 degrees, and one was located above the fracture plane. The three LVDT's mounted across the fracture plane measured the fracture and rock deformation and the LVDT mounted above the fracture plane measured the rock deformation. Hence with both sets of measurements the rock and fracture deformations can be separated.

Flowrates were determined by using compressed nitrogen as a driving force in positive displacement tanks, providing a constant fluid pressure for the injection tests. The loading frame was a 1.78 MN., MTS, servo controlled unit. The applied load was measured using a load cell that is built into the upper load platen. All of the samples were tested under load-control conditions.

After the sample had been placed in the MTS frame and instrumented the testing procedures, which are documented in Gale and Raven (1982) and Gale (1982b), followed those of Gale (1975) and Iwai (1976) and consisted of: 1) Increasing the load in increments up to 30 MPa and then decreasing the load until no load remained on the sample. 2) At each load increment four or five measurements were made of the flowrate, fluid pressures, applied load, temperatures and displacements. This represents one loading cycle. Each sample was subjected to two or three loading cycles.

EXPERIMENTAL RESULTS

Figures 2 and 3 show the rock and fracture deformations for each of the induced and natural fractures, respectively. The induced fractures (Figure 2) generally show greater permanent deformation between successive loading cycles than do the natural fractures (Figure 3). The considerable permanent deformation between the first and second loading cycle in three of the induced fractures and the lack of a similar deformation between the first and second loading cycles in Figure 3, the natural fractures, can be assumed to indicate that there was little disturbance of the natural fractures during the sampling process.

The flowrates, expressed as the flowrate normalized to a unit head, at different levels of normal stress are given in Figure 4 for the induced fractures and in Figure 5 for the natural fractures. These figures exhibit the typical nonlinear decrease in fracture flowrate with increasing normal stress. While there is a distinct hystersis in this flowrate-stress curve each of the induced fratures exhibits a similar pattern **even** with increasing number of loading cycles. However the cores containing natural fractures exhibit a more erratic behaviour. Sample 42 gave increasing values for the minimum flowrates at maximum stress with increasing number of loading cycles, possibly due to the opening of additional fractures in the sample. However samples 40 and 41 showed the more typical pattern of decreasing flowrates with increasing number of loading cycles.

Table 1 summarizes the magnitude of the fracture closure under maximum load conditions as well as the normalized flowrates at minimum and maximum normal stress levels for each loading and unloading cycle for which data was obtained. The decrease in total fracture closure with increase in number of loading cycles indicates the magnitude of the permenent fracture deformation produced by successive loading cycles, supporting the trend shown by Figures 2 and 3. Also the minimum and maximum flowrates given in Table 1 provide a ready comparison of both the induced and natural fractures. Table 1 clearly shows that, on average, the four induced fractures have lower initial and lower final flowrates than the natural fractures.

ANALYSIS AND DISCUSSION

In order to analyzed and compare the permeabilities of the two types of fractures, assumptions have to be made regarding the form of the flow law that is most applicable to flow in fractures. It has generally been assumed that the cubic law, derived from the Navier Stokes equation for laminar flow between two parallel plates which are not in contact, is applicable to flow in rough, deformable, fractures. This law assumes that the normalized

EFFECTS OF FRACTURE TYPE ON PERMEABILITY 293

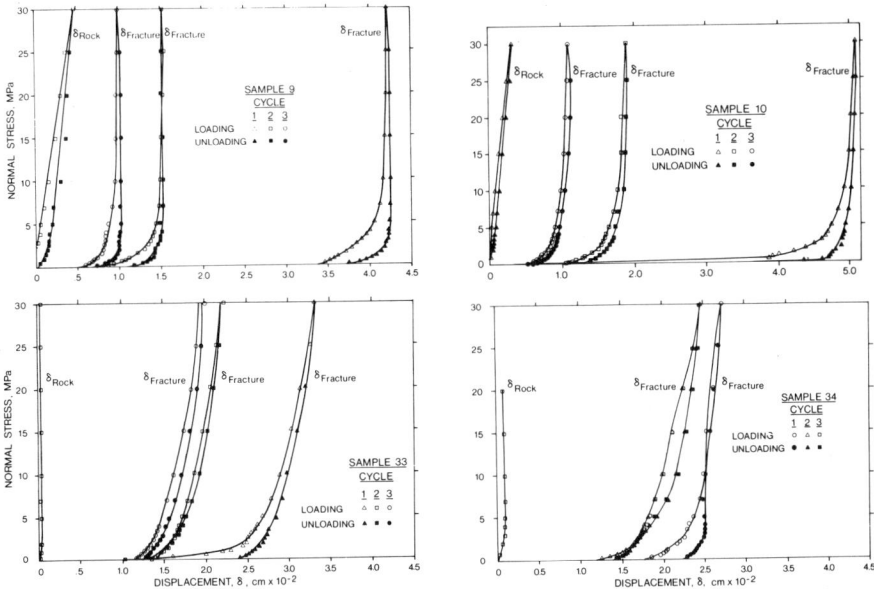

Figure 2. Rock and fracture deformation for induced fractures

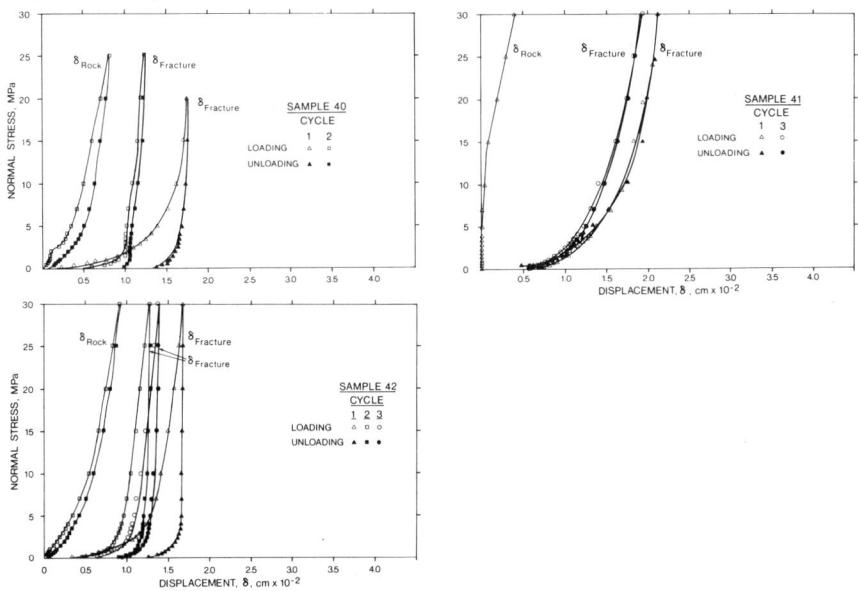

Figure 3. Rock and fracture deformation for natural fractures.

294 ISSUES IN ROCK MECHANICS

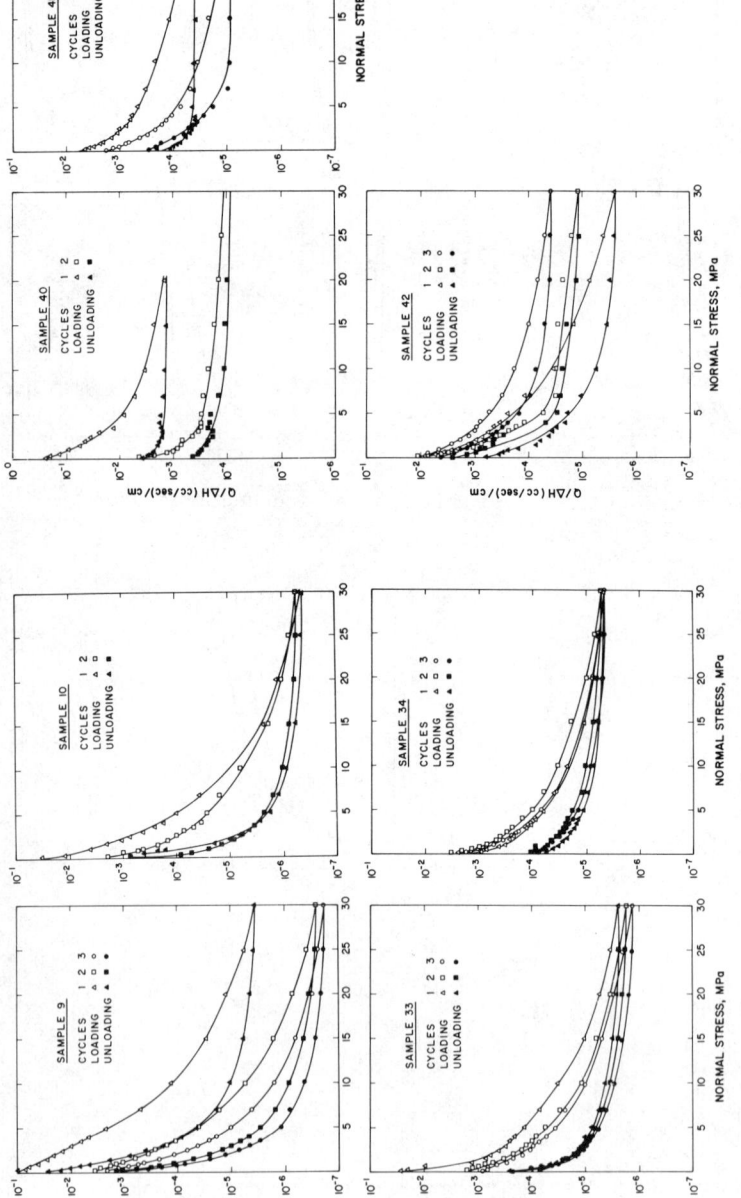

Figure 4. Normalized flowrates versus normal stress for induced fractures.

Figure 5. Normalized flowrates versus normal stress for natural fractures.

flowrate in a fracture is proportional to the third power of the apertures. In simplified form the cubic law can be written as;

$$Q/\Delta H = C(2b)^3 \qquad (1)$$

where $Q/\Delta H$ equals the flowrate per unit head through the fracture, $2b$ is the effective fracture aperture and C is a constant which in the case of radial flow is given by

$$C = \frac{2\pi}{\ln(r_e/r_w)} \frac{\rho g}{12\mu} \qquad (2)$$

Using this relationship we have calculated the equations relating fracture hydraulic conductivity (K_f) to stress in the form of

$$K_f = B\sigma^\alpha \qquad (3)$$

where β is the value of K_f at a normal stress (σ) of 1.0 MPa and α is the slope of the line. The log of K_f versus the log of normal stress (σ) is plotted in Figure 6. The values of β, α and the correlation cofficient for each curve are given in Table 2. As shown by the different slopes given in Table 2, and the actual curves in Figure 6, there is a distinct and very large difference in the stress permeability characteristics of induced versus natural fractures.

As mentioned above the K_f values were computed assuming that the cubic law applies to the fractures that were tested. However, since only fracture closure measurements were made, we do not have a direct measure of the effective fracture aperture. Thus to check the validity of the cubic law we recognize that the effective flow aperture (2b) consists of two components (Witherspoon et al, 1980): an unknown residual aperture which exists at the maximum applied normal stress ($2b_{res}$) and a measured value determined from L.V.D.T. closure measurements ($2b_m$). Thus,

$$2b = 2b_{res} + 2b_m. \qquad (4)$$

If the cubic law describes the relationship between fracture flowrate and aperture, then the measured flowrate and aperture data should plot as a straight line on a log-log plot with a slope equal to three. Thus to plot the experimental data one must assume that the cubic law is valid at maximum normal stress in order to calculate the residual fracture aperture ($2b_{res}$). In figures 7 and 8 the aperture versus flowrate data has been plotted for several of the induced and natural fractures. These plots clearly show that when we use the residual apertures ($2b_{res}$), computed at 30 MPa normal stress, the slope of the aperture versus flowrate curves for both the induced and natural fractures varies from one to greater than one in four. However as shown in Figure 8 choosing lower stress levels as the reference points in computing the residual apertures results in a curve that begins to approach a slope of one in three.

CONCLUSIONS

The presentation and brief discussion of these laboratory results suggests that natural fractures are much stiffer than induced fractures and that fracture closure at a given normal stress will be less for natural fractures. This concurs with the observation that flow through natural fractures at the higher stress levels is much larger than that through induced fractures. The experimentally determined aperture flowrate data show that the cubic law, or the approach to check the validity of the cubic law, is not applicable to rough, deformable, induced or natural fractures when the residual apertures are computed on the basis of flowrates measured at stress levels equal or greater than 30 MPa. However in most cases when lower normal stresses are used as the

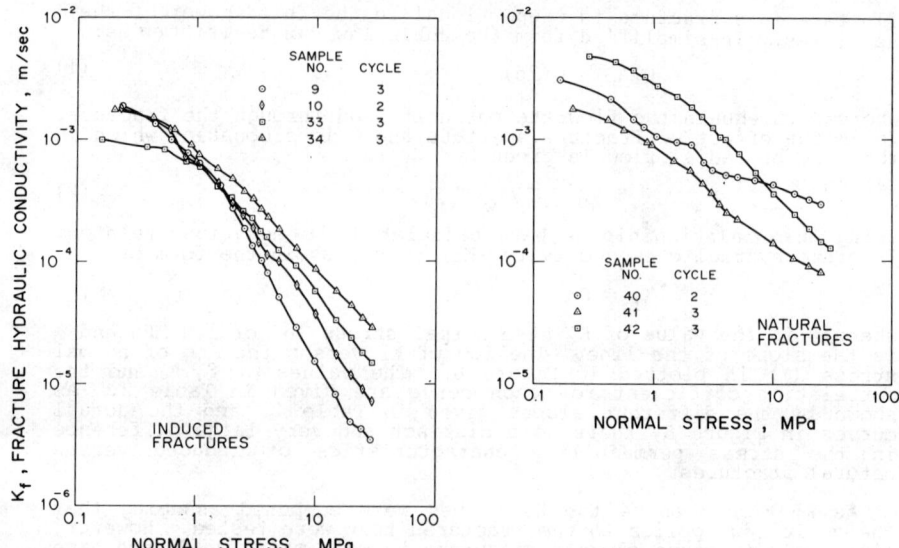

Figure 6. Log-log plots of fracture hydraulic conductivity versus normal stress for induced and natural fractures.

Table 1. Normalized flowrates at minimum and maximum stress conditions and maximum fracture closures for each loading cycle for both induced and natural fractures.

SAMPLE	INDUCED				NATURAL		
	9	10	33	34	40	41	42
CYCLE 1							
$Q/\Delta H$, max. (m^2/sec)	1.5×10^{-6}	4.7×10^{-6}	2.9×10^{-6}	4.0×10^{-7}	2.6×10^{-5}	5.2×10^{-7}	8.6×10^{-6}
$Q/\Delta H$, min. (m^2/sec)	3.0×10^{-11}	4.0×10^{-11}	2.5×10^{-10}	4.6×10^{-10}	1.7×10^{-7}	6.6×10^{-9}	3.9×10^{-9}
2b-cl (micrometers)	72	123	192	86	152	167	133
CYCLE 2							
$Q/\Delta H$, max. (m^2/sec)	3.6×10^{-7}	2.2×10^{-7}		3.2×10^{-7}	4.5×10^{-7}	5.9×10^{-7}	1.2×10^{-6}
$Q/\Delta H$, min. (m^2/sec)	2.6×10^{-11}	5.4×10^{-11}	missing data	5.2×10^{-10}	1.4×10^{-8}	1.5×10^{-8}	1.4×10^{-9}
2b-cl (micrometers)	63	90		122	43	133	79
CYCLE 3							
$Q/\Delta H$, max. (m^2/sec)	2.2×10^{-7}		8.1×10^{-8}	1.9×10^{-7}		2.4×10^{-7}	8.6×10^{-7}
$Q/\Delta H$, min. (m^2/sec)	1.8×10^{-11}	missing data	1.6×10^{-10}	4.2×10^{-10}	missing data	8.0×10^{-10}	4.0×10^{-9}
2b-cl (micrometers)	41		84	105		136	70

$Q/\Delta H$, max. equals the normalized flowrate at the minimum stress of approximately 0.25 MPa.
$Q/\Delta H$, min. equals the normalized flowrate at the maximum stress of approximately 30 MPa.
2b-cl maximum closure of fracture from minimum stress of $\simeq$ .25 MPa to the maximum stress of $\simeq$ 30 MPa.

EFFECTS OF FRACTURE TYPE ON PERMEABILITY

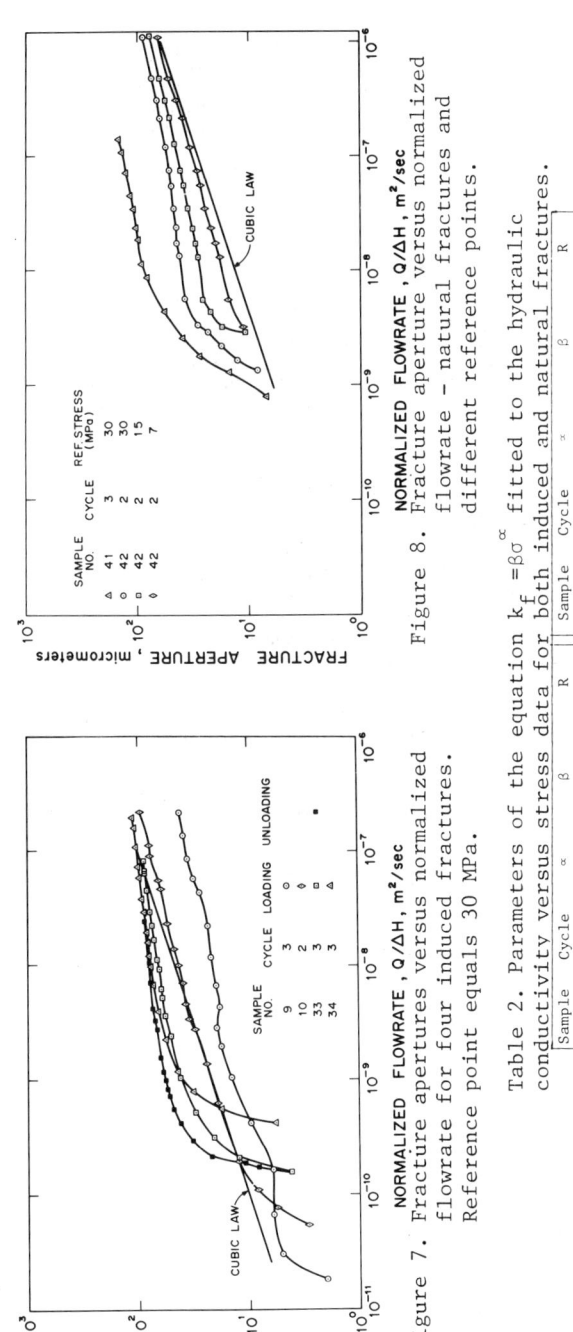

Figure 7. Fracture apertures versus normalized flowrate for four induced fractures. Reference point equals 30 MPa.

Figure 8. Fracture aperture versus normalized flowrate – natural fractures and different reference points.

Table 2. Parameters of the equation $k_f = \beta \sigma^\alpha$ fitted to the hydraulic conductivity versus stress data for both induced and natural fractures.

Sample No	Cycle No	α	β	R
INDUCED
9	1	-.276	2.81×10^{-3}	-.96
	2	-.215	9.25×10^{-4}	-.94
	3	-.203	4.28×10^{-4}	-.89
10	1			
	2	-.173	4.98×10^{-4}	-.91
33	1			
	2	-.139	6.72×10^{-4}	-.92
	3			
34	1			
	2	-.129	5.90×10^{-4}	-.90
	3			

Sample No	Cycle No	α	β	R
NATURAL
40	1			
	2	-.069	1.14×10^{-3}	-.77
	3			
41	1			
	2			
	3	-.104	7.38×10^{-4}	-.89
42	1	-.162	7.54×10^{-3}	-.95
	2	-.123	1.06×10^{-3}	-.80
	3	-.120	2.37×10^{-3}	-.92

α = Slope of K_f versus σ curve
β = K_f (m/sec) at a normal stress of 1 MPa
R = Correlation cofficient

reference point, for purposes of computing the residual aperture, the aperture-flowrate data approaches a cubic relationship. This observation is not in conflict with earlier conclusions by Witherspoon et al (1980) since their samples were not subjected to stresses greater than 15 to 20 MPa. However the implications of the results obtained from this study should be carefully considered when one is trying to predict groundwater velocities and groundwater residence times in fracture dominated flow systems.

ACKNOWLEDGEMENTS

Thanks are extended to B. Schmitke and R. Nadon for assistance with the laboratory measurements, to P. Fisher and N. Bahar for drafting and photographic work, and to S. Andrews for assistance with the data processing and manuscript preparation. The laboratory part of the work was funded by research funds provided by Environment Canada and a NSERC operating grant to the author.

REFERENCES

Gale, J.E., 1975, A numerical, field and laboratory study of flow in rocks with deformable fractures, Ph.D. Thesis, University of California, Berkeley, 255 pp.

Gale, J.E., 1982a, Assessing the permeability characteristics of fractured rock; in Recent Trends in Hydrology, Special Publication of the Geological Society of America, in press.

Gale, J.E., 1982b, A study of the effects of sample size on the stress permeability relationships for natural and induced fractures. (In press).

Gale J.E. and Raven, K., 1982, Effects of sample size on the stress permeability relationships of natural fractures, Law. Berk. Lab. technical report, (in press)

Iwai, K., 1976, Fundamental studies of fluid flow though a single fracture. Ph.D. Dissertation, Dept. of Civil Engineering, University of California, Berkeley, Ca.

Lumbers, S.B., 1974, Precambrian geology, Mattawa Deep River area, District of Nipissing and County of Renfrew, Ontario Dept. of Mines, Miscellaneous paper 59.

Witherspoon, P.A., Jang, J.S.Y., Iwai, K. and Gale, J.E., 1980, Validity of cubic law for fluid in a deformable rock fracture, Water Resources Research, Vol, 16, No. 6, pp. 1016-1024.

Chapter 31

A WORKHARDENING/RECOVERY MODEL OF TRANSIENT CREEP
OF SALT DURING STRESS LOADING AND UNLOADING*

By D. E. Munson and P. R. Dawson

Experimental Programs Division
Sandia National Laboratories, Albuquerque, NM 87185

Sibley School of Mechanical and Aerospace Engineering
Cornell University, Ithaca, NY 14853

ABSTRACT

An empirical model is developed that predicts accurately the transient response of salt creep to incremental and decremental changes in stress and temperature. Even though the model is empirical, it is derived from a firm theoretical framework for the micromechanical deformation processes of both steady-state and transient creep. In the model, the transient creep functions modify the steady-state creep behavior, which in turn is based upon creep mechanisms and the deformation-mechanism map. The model is applied successfully to experimental stress-drop results obtained for salt.

INTRODUCTION

Over the last few years an intensified national research and development program has been directed toward permanent disposal of radioactive waste. Prime candidates for geologic disposal are naturally occurring bodies of rock salt because, with time, the creep of salt encapsulates the waste. In order to predict the consequences of creep on encapsulation, as well as the short-term structural stability of the repository, a constitutive model is required for salt creep in the appropriate stress and temperature range. Because the assurance of public safety requires satisfactory results from numerical calculations of design and performance, there is considerable incentive to provide sophisticated models for these calculations. A primary obstacle in formulating a creep model is the difficulty describing creep

---

*This work supported by the U.S. Department of Energy.

during decrements in stress (unloading). In the repository such unloading occurs because of stress redistribution resulting from convergence of the storage rooms and cooling of the waste.

Creep of pure materials, such as salt, is a very complex process that has not yet fully yielded to theoretical analysis. Concerted efforts to understand the micromechanical aspects have provided a deformation mechanism framework for modeling certain parts of the creep process. In particular, models of micromechanical mechanisms, such as those proposed by Weertman (1968), are especially useful in understanding steady-state creep. Steady-state creep is an equilibrium condition between workhardening and recovery where the structure (internal dislocation and defect array) is constant. The individual mechanisms that control steady-state creep have been largely identified. These mechanisms and their temperature and stress regimes are commonly represented by deformation-mechanism maps (Munson, 1979).

Transient creep response for incremental and decremental stress changes and for primary creep is shown schematically in Figure 1. The theoretical analysis of transient creep is much less developed than for steady-state creep. Description of the transient response involves not only transient creep during the first stage of creep, but also transients that occur because of changes in applied stress and temperature. This complex response to changes in loading has proven difficult to model. Several empirical models have been advanced, with varying success, to simulate the transient response. Early empirical models that consisted of shifting segments, either in time or strain, of individual creep curves (at the new conditions of stress) to the time and strain condition of the initial creep curve (just before the stress change) are inadequate in decremental loading. More recently, the transient response has been treated according to first-order kinetics where the model causes a characteristic decay between the steady-state creep rate at the new stress and temperature and the creep rate at the time and strain of the stress change (Webster, et al., 1969 and Herrmann, et al., 1980). Unless special restrictions are enforced, this model can predict negative creep (strain recovery). A model given by Krieg (1980) that evokes workhardening and recovery is typical of another group of quite complicated models based on plasticity-creep concepts.

Perhaps the best current approach to the description of transient creep is based on micromechanical concepts. Dislocation mechanics is used to develop mechanism models which explain certain aspects of transient behavior. They have concentrated principally on transients observed during decremental stress changes and the concurrent changes in internal structure and backstress. Typical of these models are the anelastic backflow

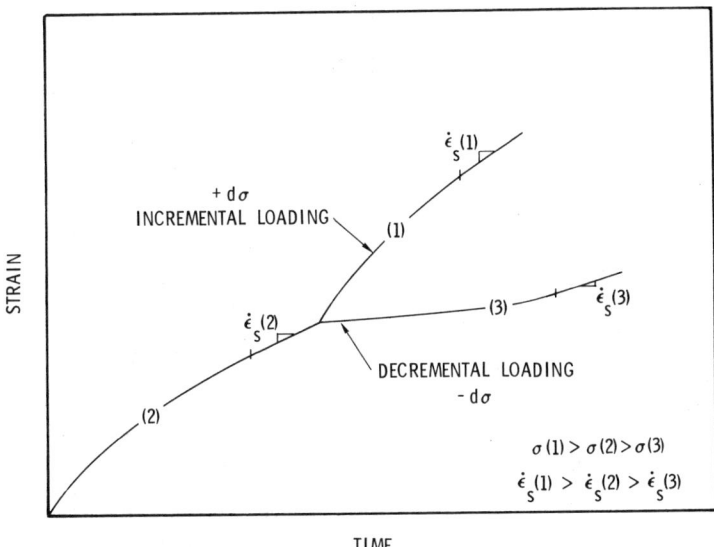

Fig. 1. Schematic of Stress Change Test.

models of Gibeling and Nix (1981). Such models are too detailed for our purposes; however, the experimental observations are instructive and form the foundation for the empirical model developed here.

CONSTITUTIVE MODEL DEVELOPMENT

In salt, internal substructures and clusters of dislocations (called <u>structure</u>) have been observed as a consequence of the creep process. These structures give rise to an internal backstress that opposes the applied stress. At any given applied stress, a structure develops that is in equilibrium. This equilibrium structure is observed experimentally as a steady-state creep rate. If we begin with a specimen that has less internal structure than the equilibrium structure, a potential exists that drives the structure toward equilibrium. By the same token, if we begin with a specimen that has more internal structure than the equilibrium structure, a potential exists that drives the structure toward equilibrium. Although the potential function is not known, schematically it must appear as in Figure 2.

The minimum in the potential function is the equilibrium condition. This state has some unique properties. First, the equilibrium or steady-state strain rate is a function of temperature and stress ($\sigma_s = f(T,\sigma)$) as demonstrated experimentally.

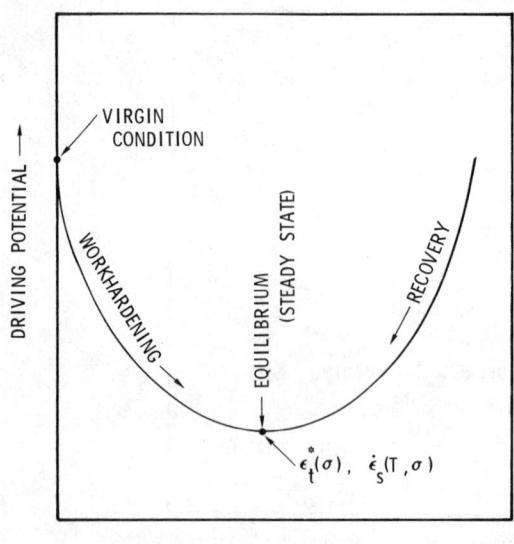

Fig. 2. Schematic Potential Function.

Second, at equilibrium, the total strain accumulates without additional transient strain. Third, as experimentally determined, the accumulated transient strain at equilibrium is a function only of the applied stress ($\epsilon_t^* = f(\sigma)$) (Robinson, et al., 1974).

Even though the potential function is unknown, a direct measure of this function is expressed through experimental strain rate vs time or strain rate vs strain relationships. In fact, the strain rate vs strain relationships for workhardening is empirically quite well known because the relationship is generated experimentally for primary creep in each standard creep test. This work-hardening branch of the strain rate vs strain relationship starting from the virgin condition for primary creep is indicated in Figure 3.

The strain rate vs strain relationship for the recovery branch is not well known because the relationship is not experimentally accessible using a standard creep test. This results from not being able to exceed the structure density at equilibrium in a standard creep test. Because of nonunique initial states for the recovery branch, a family of curves are obtained that start from the initial state and end at the equilibrium state. Two such curves are shown in Figure 3.

In the form shown in Figure 3, the recovery family of curves is difficult to treat. This is overcome by reframing the curves. These manipulations produce an <u>ab origine</u> (from the beginning) curve for transient creep. By introducing an internal state parameter, ζ, we can reduce the recovery family to a single curve that joins the workhardening curve at ϵ_t^*. The state parameter increases in workhardening, but decreases in recovery, always driving toward the equilibrium condition. <u>However, transient</u>

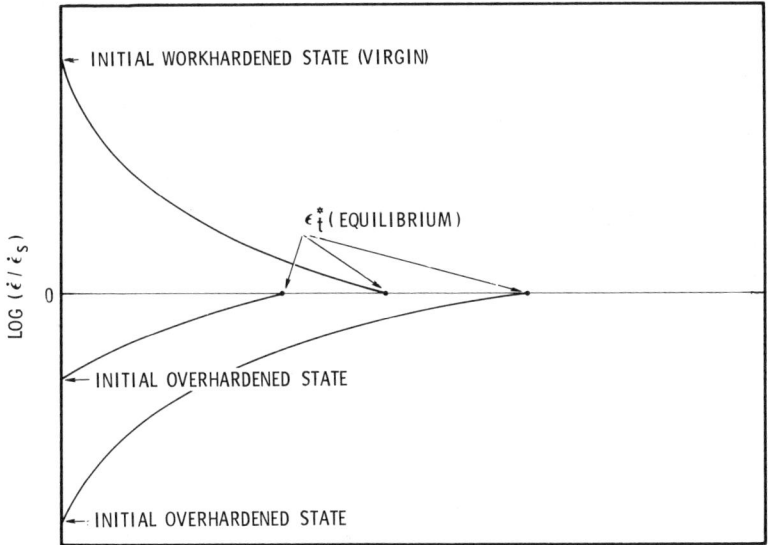

Fig. 3. Strain Rate-Strain Functions.

strain always increases in either workhardening or recovery. An ab origine construction of the transient creep curve in terms of the state parameter is shown in Figure 4.

CONSTITUTIVE MODEL

In developing the model, we follow and extend the earlier results of Munson and Dawson (1979). Steady-state creep of salt is the result of micromechanical mechanisms, as specified through the deformation-mechanism map, acting additively to produce the deformation (Munson, 1979). Thus, for n mechanisms, the steady-state rate is

$$\dot{\epsilon}_s = \sum_{i=1}^{n} \dot{\epsilon}_{s_i}(T, \sigma, S) \qquad (1)$$

where T is temperature, σ effective stress, and S a structure factor. The creep rate is simply related to the steady-state rate through

$$\dot{\epsilon} = F\dot{\epsilon}_s \qquad (2)$$

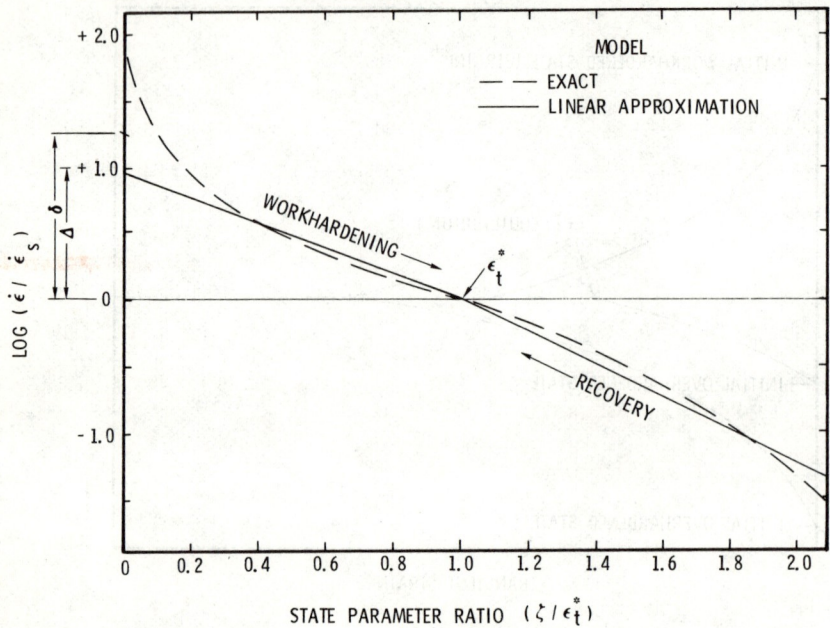

Fig. 4. Construction of the Transient Creep Curve.

where F is the normalized strain rate. The linear approximation, shown in Figure 4, to the exact function for F gives

$$F = \begin{cases} \exp\left[\Delta\left(1 - \dfrac{\zeta}{\epsilon_t^*}\right)\right], & \zeta < \epsilon_t^* \\ 1, & \zeta = \epsilon_t^* \\ \exp\left[\delta\left(1 - \dfrac{\zeta}{\epsilon_t^*}\right)\right], & \zeta > \epsilon_t^* \end{cases} \quad (3)$$

where Δ and δ are the workhardening and recovery parameters, respectively, and ϵ_t^* is the transient strain limit. The evolutionary equation for ζ is

$$\dot{\zeta} = \text{sign}(\epsilon_t^* - \zeta)(\dot{\epsilon}_t) \quad . \quad (4)$$

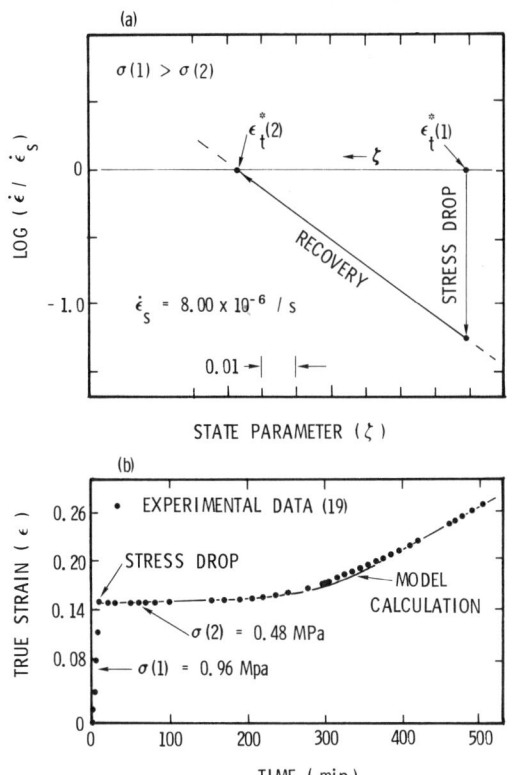

Fig. 5. Model Prediction and Experimental Stress-Drop Results.

Strategy for Stress-Change Analysis

The strategy for temperature and stress change is that a change in temperature or applied stress (either incremental or decremental) takes place at constant transient strain between the transient strain curves of Figure 4 representing the previous temperature and stress condition and the new condition. Any number of incremental or decremental changes in stress and temperature can be accommodated by progressing from one aboriginal curve to another. This strategy is also appropriate for continuous stress changes with time. The important features of this strategy are the use of transient strain accumulation, incorporation of both work-hardening and recovery processes, and the normalization of transient strain rates. While fundamentally correct, this strategy also is very appealing because it is quite simple.

Application to a Stress-Drop Experiment

Experimental data are comparatively rare for stress-drop or decremental stress-change tests that include complete recovery to a new steady-state rate. The experimental results reported by Robinson et al (1974) of one complete test of this type are shown in Figure 5b. Under the initial stress of 0.96 MPa, the test reached a steady-state creep rate of 2.8×10^{-4}/s. After the stress drops to 0.48 MPa, a new creep rate of 4.7×10^{-7}/s is observed. Upon recovery, this rate recovers a new steady-state rate of 8.0×10^{-6}/s.

Application of the appropriate branch of the model produces the predicted recovery transient and steady-state creep response shown in Figure 5. Comparison of the model calculation to the experimental data is quite accurate for this experiment.

CONCLUSIONS

Recent theoretical advances have begun to explain the underlying micromechanical processes that control transient creep response. These recent advances, when combined with earlier analyses of the micromechanical processes leading to steady-state creep, suggest a framework for the development of an empirical creep model for salt that treats workhardening, equilibrium, and recovery during creeping deformation in response to stress loadings and unloadings. Application of the model to a high-temperature decremental stress-change test that underwent complete recovery after the stress change showed very good agreement.

REFERENCES

Gibeling, J.C., and Nix, W.D., 1981, "Observations of an Elastic Backflow Following Stress Reductions During Creep of Pure Metals," Acta Met., $\underline{29}$, p. 1769.

Herrmann, W., Wawersik, W.R., and Lauson, H.S., 1980, "A Model for Transient Creep of Southeastern New Mexico Rock Salt," SAND80-2172, Sandia National Laboratories, Albuquerque, NM.

Krieg, R.D., 1980, "A Unified Creep-Plasticity Model for Halite," SAND80-1195, Sandia National Laboratories, Albuquerque, NM.

Munson, D.E., 1979, "Preliminary Deformation-Mechanism Map for Salt (With Application to WIPP)," SAND79-0076, Sandia National Laboratories, Albuquerque, NM.

Munson, D.E., and Dawson, P.R., 1979, "Constitutive Model for the Low-Temperature Creep of Salt (With Application to WIPP)," SAND79-1853, Sandia National Laboratories, Albuquerque, NM.

Robinson, S.L., Burke, P.M., and Sherby, O.D., 1974, "Activation Energy and Subgrain Size--Creep Rate Relations in Sodium Chloride," Phil. Mag., $\underline{29}$, p. 432.

Webster, G.A., Cox, A.P.D., and Dorn, J.E., 1969, "A Relationship Between Transient and Steady-State Creep at Elevated Temperatures," Metal Sci. J., $\underline{3}$, p. 221.

Weertman, J., 1968, "Dislocation Climb Theory of Steady-State Creep," ASM Tran. Quarterly, $\underline{61}$, p. 681.

Chapter 32

STEADY-STATE CREEP OF ROCK SALT IN GEOENGINEERING

by Tom W. Pfeifle and Paul E. Senseny

Staff Engineer, RE/SPEC Inc.
Rapid City, SD

Manager, Materials Laboratory, RE/SPEC Inc.
Rapid City, SD

INTRODUCTION

Engineered structures such as mines, shafts and tunnels, and storage caverns for hydrocarbons, chemicals and brine are being built in natural rock salt formations in increasing numbers. In addition, salt formations are being considered as potential hosts for mined, nuclear waste repositories. The popularity of these formations for engineering purposes stems from the many favorable commercial, physical, and thermal characteristics attributed to the material itself. Not only have the formations been stable and free of dissolution for hundreds of millions of years, but the salt is fairly easily mined and has very low permeability, low water content, and high thermal conductivity. Another characteristic of salt is its tendency to flow or creep when subjected to a shear stress. Although this behavior may be desirable in certain instances, it may be detrimental in other instances since the creep rate may be large enough to produce deformations that tend to reduce the volume of storage caverns or to restrict access to mining operations. In either case, the engineer is faced with the problem of accurately predicting the long-term structural deformations under various combinations of stress and temperature and is required to compare these deformations with long-term design criteria.

In the laboratory, deformation-versus-time curves obtained at constant stress and temperature show that creep of salt is similar to creep of many other crystalline solids. Figure 1 shows a typical deformation-versus-time curve for salt. When a shear stress is first applied, the rate of deformation is high, but decreases monotonically. This initial behavior is called transient or primary creep. At some time, the deformation rate no longer decreases but continues at a constant rate. This behavior is called steady-state or secondary creep. In some experiments, a third regime called tertiary creep follows the steady-state region. Tertiary creep usually occurs at low mean stress and low temperature and is characterized by accelerating creep rates and, ultimately, specimen failure (Wallner, 1981). The duration of transient creep is usually quite short, lasting only a month or two at most. The duration of steady-state creep can be very long, however, especially if low mean stresses and low temperatures are absent. Therefore, accurate prediction of long-term deformations requires that the steady-state creep rate be known precisely as a function of stress and temperature.

For cases when the load on the structure or temperature remain constant or change very slowly, the creep straining of salt can be approximated by a steady-state-only model. Elastic and transient creep strains associated with constant or slowly changing load and temperature are usually very small compared to strains that result from steady-state creep at the long times of interest for these structures. Constitutive models have been developed to relate the steady-state strain rate to the current stress and temperature. The values of the parameters in the models are determined by fitting the

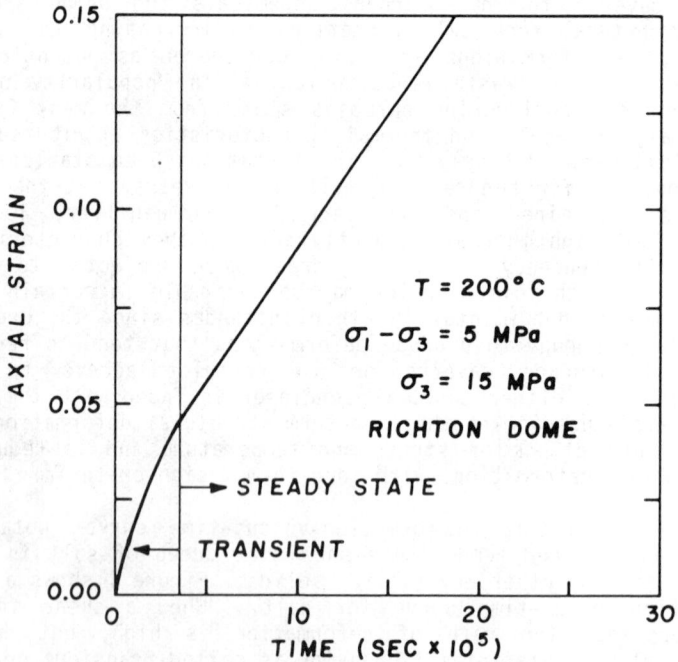

Fig. 1. Axial Strain as a Function of Time for Salt at Constant Stress and Temperature.

model to measured steady-state rates determined at different stresses and temperatures.

This paper compares two constitutive models that have been proposed for the steady-state creep of salt in the ranges of stress and temperature of interest to geoengineers. The models are fitted to laboratory data and their corresponding predictive capabilities are assessed. Although the more complex model, having more parameters, fits the data better than the simpler model, its predictive capability is less accurate.

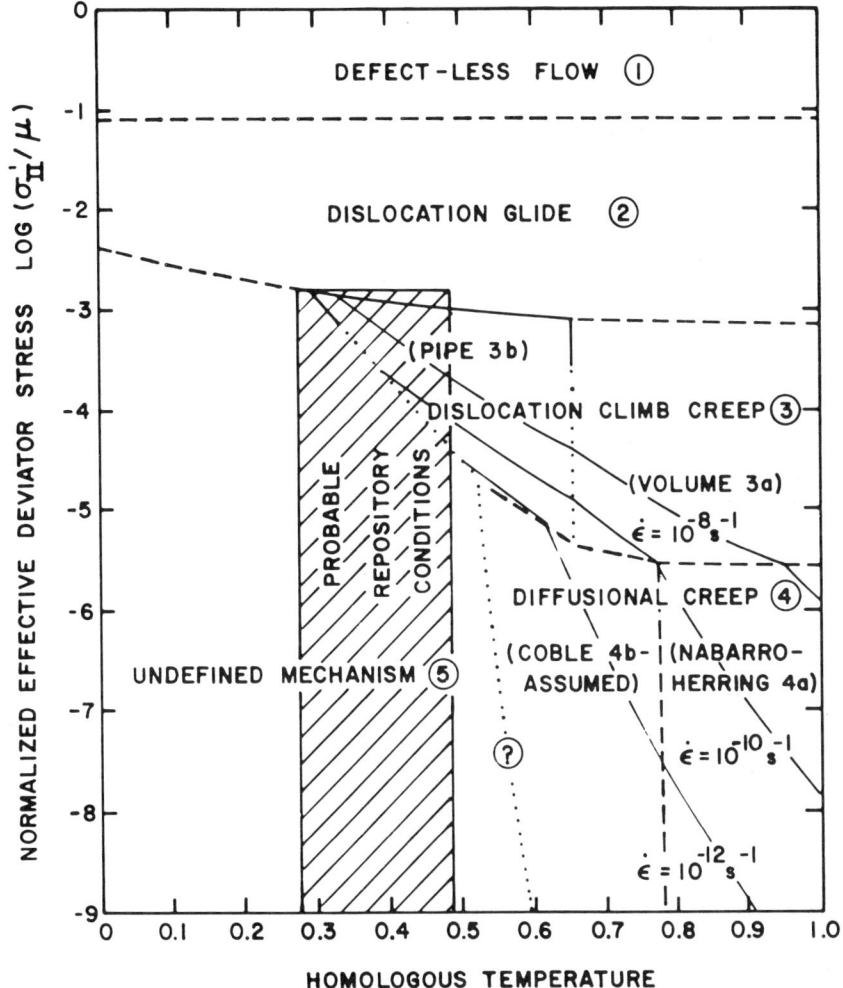

Fig. 2. Deformation-Mechanism Map for Salt (Munson, 1979).

CONSTITUTIVE MODELS

The constitutive model that is most often used to relate the steady-state creep rate to the stress and temperature is

$$\dot{\varepsilon}_{ss} = A\sigma^n \exp[-Q/RT] \qquad (1)$$

where $\dot{\varepsilon}_{ss}$ is the steady-state creep rate, σ is stress difference, T is absolute temperature, and A, n and Q/R are parameters whose values are determined by fitting the data. The functional form of Eq. 1 is obtained theoretically for deformation resulting from the action of a single micromechanism that involves thermally-activated motion of individual atoms. The parameter Q/R is then interpreted as the ratio between the activation energy for self-diffusion and the universal gas constant.

Typically, when Eq. 1 is fitted to data obtained in the ranges of stress and temperature encountered by geoengineers, the parameter values are not equal to those predicted theoretically. One possible explanation for this lack of agreement is that there are several deformation mechanisms that control creep in this stress and temperature regime.

Figure 2 shows a deformation map constructed for salt, which predicts that two mechanisms* operate in the region of interest (Munson, 1979). One deformation mechanism is expected to be dislocation climb-glide controlled by thermally-activated diffusion along dislocations. The second mechanism has not been identified. Physical arguments, based on deformation mechanisms, suggest that the one-term constitutive model given by Eq. 1 is not adequate to predict the stress and temperature dependence of the steady-state creep rate in this regime. Assuming that the second mechanism also involves thermally-activated motion of individual atoms and that it operates independently of the dislocation climb-glide mechanism, the constitutive model for steady-state creep is

$$\dot{\varepsilon}_{ss} = A_1 \sigma^{n_1} \exp\left[-\frac{Q_1}{RT}\right] + A_2 \sigma^{n_2} \exp\left[-\frac{Q_2}{RT}\right] \qquad (2)$$

where the subscripts 1 and 2 on the parmeters indicate the first and second mechanisms.

DATA BASE

To determine which constitutive model better represents the

---

*Dislocation glide may contribute at very high stress, but most of the structural geoengineering regime is outside the glide area.

steady-state creep of natural rock salt in the region of interest to geoengineers, the two models must be fitted to a specific data base. The data base used in this analysis was derived from eighty-eight triaxial creep experiments on natural rock salt from Avery Island, Louisiana (Hansen and Mellegard, 1980; Mellegard, Senseny and Hansen, 1981). The experiments were performed on two different sizes of right, circular cylinders to increase the size of the data base and to determine what influence, if any, specimen size has on the steady-state creep of rock salt. The smaller specimens were 50 mm diameter and 100 mm long, while the larger specimens were 100 mm in diameter and 200 mm long.

The creep strain-versus-time curve obtained from each test is similar to that shown in Figure 1. The transition between the transient and steady-state regimes is very smooth and consequently, difficult to locate. An algorithm has been developed that eliminates some of the subjectivity involved in determining where steady-state creep begins as well as the steady-state creep rate (Senseny, in Press). The procedure is to search the (log ε, log t) data for an interval in which a linear least squares fit gives (Δlog ε/Δlog t) = 1 $\pm$ η and a correlation coefficient $r^2 \geq r_0^2$. This interval is taken to be the steady-state regime. The steady-strain creep rate is determined from a linear least squares fit to the strain-versus-time data in the steady-state interval. For the present study, the values η = 0.01 and r_0^2 = 0.985 were used. This algorithm and acceptance criteria reduced the original data base to twenty-eight experiments. Table 1 gives the matrix of tests used and the steady-state creep rates measured at the corresponding stress and temperature.

STATISTICAL ANALYSES

Size Effects

The influence of specimen size was determined using two statistical procedures. The first procedure, called analysis of variance (Box, Hunter and Hunter, 1978), made use of a two-way classification where each of seven paired tests comprised a block. Specimen size is found to influence steady-state creep at only the 80 percent confidence level. A second procedure was to transform Eq. 1 into a linear form and to obtain parameter estimates and their corresponding 95 percent confidence intervals from separate fits to the smaller and larger specimen size data. The parameter estimates for one size are nearly identical to those for the other size and fall well within the respective confidence intervals. Therefore, specimen size does not have a significant influence on the steady-state creep of salt, at least as long as the ratio of specimen diameter to grain diameter is approximately ten, as recommended (ISRM, 1979). Consequently, constitutive models for the steady-state creep of salt that are based on laboratory data are adequate to predict field deformation. In the analyses that follow, data

TABLE 1. Steady-State Creep Rate* Data Base

STRESS DIFFERENCE (MPa)	TEMPERATURE (K)				
	298	343	373	423	473
2.5					2.71
3.5					14.8
5.0			1.19	12.3	42.6
5.5					52.6
6.0					98.1
6.9	0.250		3.86	38.6	155
8.6			8.24		
9.0			12.7	45.1	
10.3	1.29	6.15	16.3		
12.5			41.1		
12.8			36.7		
13.8		5.28	33.2		
15.0			62.7		
15.5			70.7		
17.2			133.		
17.5			147.		
18.1			154.		
20.7		43.6	283.		

*Creep rates given in table are x 10^{-9} s^{-1}

from tests on specimens of both sizes are pooled to provide the larger data base shown in Table 1.

Comparison of Models

The two models were fitted to the data using a nonlinear regression algorithm, ZXSSQ of the IMSL Library, a finite difference Levenburg-Marquardt routine (IMSL, 1980). Because of the orders-of-magnitude difference among the measured steady-state creep rates, the difference between the predicted and measured values is normalized by dividing by the measured value when computing the sum-of-squares error which is to be minimized. Table 2 gives the parameter values and sum-of-squares error (SSE) for the one-term and two-term model fits. The two-term model has a fifty percent lower sum-of-square error which indicates that the one-term model does not account for all the variation in the steady-state creep rate caused by variations in stress and temperature. Therefore, in addition to the physical evidence provided by the deformation map, this exercise provides statistical evidence that the one-term constitutive model does not completely describe the stress and temperature dependence of steady-state creep of salt for these stresses and temperatures.

As stated, the purpose of developing a constitutive model for steady-state creep is to predict the creep rate as a function of

TABLE 2. Data Fits For One- And Two-Term Constitutive Models

	A_1	n_1	Q_1/R	A_2	n_2	Q_2/R	SSE	PRESS
ONE-TERM	8.3×10^{-5}	3.57	6320	-	-	-	3.56	5.94
TWO-TERM	6.5×10^{-4}	4.60	8170	1.68×10^{-7}	1.87	2990	2.36	26.5

stress and temperature. The physical and statistical evidence presented for the inadequacy of the one-term model does not further imply that the two-term model is a better predictor than the one-term model. Although the two-term model fits the data better, it may not be able to predict it better. This stems from the distinction between two sources of error in a prediction: bias and variance.

Several statistical procedures are available to assess the predictive capability of mathematical models. The procedure used here is to calculate the PREdiction Sum of Squares, or PRESS statistic (Allen, 1971) for each model. To obtain PRESS for the one- and two-term models, each one of the twenty-eight observations is predicted in turn using the other twenty-seven. The normalized errors or residuals resulting from the successive predictions are squared and summed to form PRESS.

The values of the PRESS-statistic are given in the last column of Table 2. The PRESS value for the one-term model is only one fourth as large as that for the two-term model. Therefore, although both physical and statistical evidence indicate the need for a more sophisticated constitutive model than the one-term steady-state creep model given by Eq. 1, improvements in predictive capability will probably only be obtained by improved understanding of the nature of the unknown deformation mechanisms and its interaction with other deformation mechanisms.

CONCLUSION

Constitutive models for steady-state creep of natural rock salt at the stresses and temperatures encountered by geoengineers were evaluated using data obtained from triaxial compression creep experiments performed on Avery Island dome salt. Physical and statistical evidence indicates that a model based on one deformation mechanism involving thermally-activated diffusion of mass does not adequately represent the influence of stress and temperature on the steady-state creep of salt. However, a two-term model comprising two such mechanisms operating independently does not give as good of predictions of steady-state creep rate as does the one-term model.

Until improved understanding is obtained of the deformation mechanisms that operate in the stress and temperature regime commonly of interest, the one-term constitutive model should be used to predict the steady-state creep rates in the salt around a proposed structure.

ACKNOWLEDGEMENT

The work was funded through Contract E512-02300 with Battelle Memorial Institute, Project Management Division, under Contract EY-76-C-06-1830 with the Department of Energy. The contract was administered by the Office of Nuclear Waste Isolation.

REFERENCES

Allen, D. M., 1971, "The Prediction Sum of Squares as a Criterion for Selecting Predictor Variables", Technical Report No. 23, Dept. of Statistics, University of Kentucky.

Box, G. E. P., Hunter, W. G., and Hunter, J. S., 1978, Statistics for Experimenters, John Wiley and Sons, New York.

Hansen, F. D. and Mellegard, K. D., 1980, "Creep of 50-mm-Diameter Specimens of Dome Salt From Avery Island, Louisiana", RE/SPEC Inc., Rapid City, SD, Topical Report RSI-0118, ONWI-104, August, Prepared for Office of Nuclear Waste Isolation, Battelle Memorial Institute, Columbus, OH.

_____ : IMSL Library Reference Manual, Edition 8, IMSL, Houston.

ISRM, 1979, "Suggested Methods for Determining the Uniaxial Compressive Strength and Deformability of Rock Materials", Int. J. Rock Mech. Min. Sci. & Geomech. Abstr., Vol. 16, pp. 135-140.

Mellegard, K. D., Senseny, P. E., and Hansen, F. D., 1981, "Quasi-Static Strength and Creep Characteristics of 100-mm-Diameter Specimens of Salt From Avery Island, Louisiana," RE/SPEC Inc., Rapid City, SD, Topical Report RSI-0140, ONWI-250, March, Prepared for Office of Nuclear Waste Isolation, Battelle Memorial Institute, Columbus, OH.

Munson, D. E., 1979, "Preliminary Deformation Mechanism Map for Salt", SAND 79-0076, Sandia National Laboratories, Albuquerque, NM.

Senseny, P. E., in press, "Specimen Size and History Effects on Creep of Salt", Proceedings of the First Conference on the Mechanical Behavior of Salt, ed. H. R. Hardy and M. Langer, Trans Tech Publications, Clausthal-Zellerfeld.

Wallner, Manfred, 1981, Personal Communication.

Chapter 33

AN EXPERIMENTAL INVESTIGATION OF THE COMBINED EFFECTS
OF STRAIN RATE AND MOISTURE CONTENT ON SHALE

By Terry G. Richard and Sunder H. Advani

Associate Professor, Department of Engineering Mechanics
University of Wisconsin
Madison, Wisconsin

Professor & Chairman, Department of Engineering Mechanics
The Ohio State University
Columbus, Ohio

ABSTRACT

A sequence of 250 unconfined compression tests were completed on a common marine shale indicative of the overburden strata of east-central Ohio. The samples were of a light gray clay shale of Pennsylvanian age.

The shales considered are obvious layered materials with distinct bedding planes. For this reason there is sufficient evidence to consider this material as anisotropic in nature. With a properly chosen reference coordinate system this anisotropy can be reduced to a layered orthotropic characterization. This implies that the mechanical properties can be specified with regard to principal material directions.

With this nature in mind a systematic study of the combined influences of strain rate and moisture conditioning was initiated. Of particular interest is the variations in unconfined ultimate compressive strength with variations in the controlled parameters of strain rate, moisture conditioning level and direction of testing. Results of this particular phase of the program are presented.

INTRODUCTION

Mine roof collapse is a persistent problem frequently encountered in underground mining operations. Many environmental and mechanical factors contribute to this phenomena. Weakening of the overburden

due to inherent moisture content fluctuations has been noted
[Van Eeckhout, 1975, 1976]. The influence of rate of loading and
deformation has been addressed with regard to their influence on
various shale species [Chong, 1980; Lankford, 1976; Forrestal, 1978].
It was the intent of this study to present the combined parametric
influences of strain rate and moisture content on overburden shales
predominantly found in Ohio mining operations. It was also a primary
intent to facilitate ongoing analytical studies of mine-roof systems,
subsidence characterizations, hydraulic fracturing and overburden
crack penetration. To this end the mechanical properties of a
arenacious-clay shale are defined for a variety of strain rates and
moisture conditioning levels. This paper directs itself towards the
variations in ultimate unconfined compressive strength. Attempts are
made to illustrate the variation with and influence of the variable
parameters of the testing sequence.

MATERIAL DESCRIPTION

The material characterized is a Pennsylvanian age, clay-arenacious
shale which is dominant in the east and southeastern one-third of the
state of Ohio. This shale was collected from a cyclothemic unit
within the Conemaugh series, where the unit overlies the Middle
Kittanning (No. 6) coal. It is a marine shale medium gray in color,
massively bedded and appears somewhat slicken-sided in some areas.
The samples are composed of very fine grained material with a small
percentage of coarse grained silts. There is not a well defined
laminae in these shales but the bedding planes are detectable. X-ray
analysis indicates the primary components to be clays and quartz with
minor amounts of feldspar and possibly some galina. Primary clay
minerals appear to be chlorite, kaolinite and illite. It is evident
that some coarse grain quartz and silts add integrity and some
characterists of arenacious shales.

This clay shale has an average density of 2.54 gr/cc and moisture
content of 0.55% by weight in the as-received condition. By varying
the moisture conditioning environment the moisutre content of these
samples changed from 0.14% to 0.89% for specific test sequences.

SAMPLE PREPARATION

Shale is an obviously layered material with a somewhat distinct
orthotropic nature. This implies that the mechanical properties can
be specified with regard to principal material directions. Specimens
were cut from bulk shale smaples such that the longitudinal axis of
the specimen was normal to or in the plane of sedimentary bedding.
Prismatic specimens were fabricated measuring (19x19x38mm) and were
cut from bulk sample as soon as they were procured.

Of concern in this study was the combined influences of moisture
conditioning and rate of loading or deformation. To produce a

statistically significant result five specimens were tested for a particular combination of the three parameters under investigation - namely: moisture conditioning humidity level, strain rate and direction of applied load relative to the plane of bedding. This made it necessary to fabricate 250 coherent prismatic specimens.

Immediately after the specimens were cut, trimmed and ground they were either tested in the as-received condition or moisture conditioned at humidity levels of 33%, 47%, 63% or 81% R. H. in sealed chambers at 73°F. Those specimens which were conditioned were held within these chambers over saturated salt-solutions for a period of 4 to 5 months to insure moisture equilibrium within the specimen. Because of the extremely low diffusivity of these shales it was felt that periods of this length were necessary even though weight measurements of the specimens indicated stabilization within a period of half this length.

TEST PROCEDURE

Unconfined, monotonic compression tests were conducted at strain rates which varied from 1.33×10^{-4} s^{-1} to 1.33 s^{-1} in increments of one decade. Tests were conducted parallel and perpendicular to the plane of bedding on specimens in the as-received state and those conditioned at the humidity levels previously defined.

In the test sequence applied axial load, axial strain and transverse strain were measured continuously up to specimen failure. All tests were run in a strain controlled mode. From the data obtained it was possible to determine the unconfined compressive strength of the material, the axial and transverse strain at failure, Poisson's Ratio, the Secant Modulus at the Ultimate Stress and the strain energy density at failure. The variation of these mechanical properties with strain rate, direction of loading and moisture conditioning have been defined, however; at the higher strain rates only the axial parameters were monitored. For specimens where the axis of loading was in the plane of bedding the transverse expansion was alternately measured in orthogonal directions. This allows the description of Poisson's ratio parallel and perpendicular to the plane of bedding. Tables 1 and 2 define the results from the test sequence for variations in the direction of testing, strain rate and moisture conditioning level. In these tables only the influence of these parameters on the ultimate unconfined compressive strength is illustrated. The values of compressive stress represents the mean value of at least five specimen responses. The standard deviation (S.D.) is also given at the appropriate strain rate and moisture conditioning level.

TABLE 1
The Ultimate Compressive Strength of the Clay Shale Tested Parallel to the Plane of Bedding

Moisture Condition (% R.H.)	33	47	63	81	As Received
Strain Rate (s^{-1})	Ult.Stress (MPa)	Ult.Stress (MPa)	Ult.Stress (MPa)	Ult.Stress (MPa)	Ult.Stress (MPa)
1.33×10^{-4}	32.2 SD = (8.6)	27.6 SD = (11.0)	25.9 SD = (8.0)	22.1 SD = (4.4)	41.2 SD = (2.4)
1.33×10^{-3}	37.3 (10.5)	28.4 (8.4)	20.5 (5.1)	24.3 (2.2)	38.9 (7.3)
1.33×10^{-2}	26.8 (8.3)	25.8 (4.6)	30.1 (3.8)	24.1 (5.4)	48.4 (11.8)
1.33×10^{-1}	49.9 (11.4)	48.9 (12.8)	27.5 (11.7)	30.9 (4.9)	45.3 (6.9)
1.33	46.5 (10.7)	52.4 (19.8)	42.2 (6.4)	29.9 (3.0)	50.0 (11.2)

TABLE 2
The Ultimate Compressive Strength of the Clay Shale Tested Perpendicular to the Plane of Bedding

Moisture Condition (% R.H.)	33	47	63	81	As Received
Strain Rate (s^{-1})	Ult.Stress (MPa)	Ult.Stress (MPa)	Ult.Stress (MPa)	Ult.Stress (MPa)	Ult.Stress (MPa)
1.33×10^{-4}	50.8 SD = (8.3)	37.3 SD = (15.1)	27.4 SD = (4.3)	28.4 SD = (5.1)	57.0 SD = (11.3)
1.33×10^{-3}	49.6 (8.2)	36.2 (13.9)	24.4 (3.1)	30.7 (5.7)	52.8 (8.3)
1.33×10^{-2}	47.6 (3.9)	38.2 (14.3)	40.2 (4.3)	30.4 (3.3)	56.6 (12.1)
1.33×10^{-1}	57.2 (15.9)	56.6 (9.0)	42.8 (11.1)	35.8 (2.1)	56.6 (14.7)
1.33	62.1 (7.9)	57.5 (5.6)	47.8 (13.2)	41.3 (6.6)	68.7 (6.3)

TEST RESULTS AND CONCLUSIONS

The orthotropic nature of various shale species has previously been defined [Van Eeckhout, 1975]. The general results of this particular work and others of a similar nature have indicated a strengthening of this material with increases in strain rate and with a reduction of inherent moisture content [Chong, 1980]. This is not unlike the results of this testing program. Figure (1) illustrates the variation in unconfined ultimate compressive stress for this Pennsylvania age, clay shale with variations in the controlled parameters of moisture content and strain rate. This paraticular plot represents the results of 130 specimen tests where the axis of the applied load was in the plane or parallel to the plane of bedding. Each data point represents the mean value of at least five individual test results where the standard deviation ranged from 9% up to 38%. The curves drawn represent the fit of a least squares linear

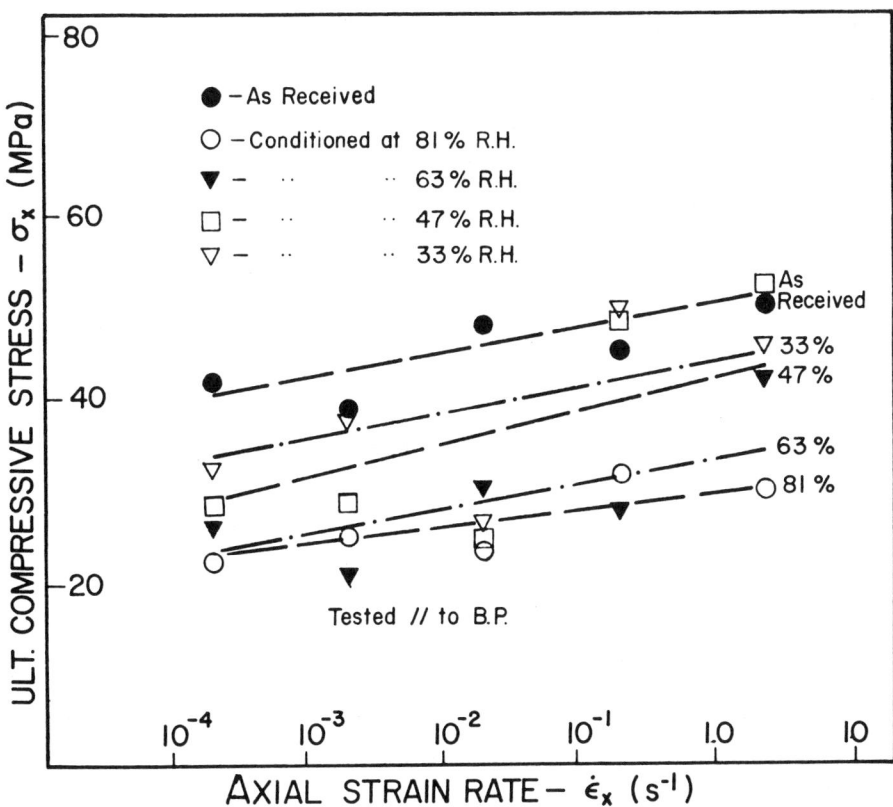

Fig. 1. Comparison of the Ultimate Compressive Strength Variation with Strain Rate and Conditioning Level

regression which typically had a correlation factor of 0.835. From this plot it is apparent that there is a persistent increase in the ultimate compressive stress of this material with increases in strain rate. This substantiates the general conclusion that shales of this nature are rate sensitivity, viscoelastic solids. The extend of this time dependent character and an accurate description of it was not within the scope of this program.

Also apparent from Figure (1) is that, in general, the higher the moisture level at which specimens were conditioned, the lower the resulting compressive strength. Specifically, those specimens conditioned at a relative humidity of 33% had a compressive strength 35% higher than those conditioned at 81% R. H. for the same period of time.

These trends of variation with strain rate and moisture content are observed when the axis of the compressive load is normal or perpendicular to the plane of bedding. Figure (2) illustrates this response. However, the magnitudes of the ultimate stress is typically 15% to 35% higher for shales tested perpendicular to the bedding plane compared to those tested parallel to the bedding plane.

Of particular concern in Figures (1) and (2) is the fact that the as-received material shows a significantly higher ultimate compressive strength than all conditioned samples regardless of the moisture level of conditioning. Figure (3) is presented to illustrate that specimens conditioned at a relative humidity level of 53% R.H. or above have a tendency for net moisutre absorption which increases with increasing R.H. At humidity levels below 53% R.H. the conditioning chambers act as desiccators producing a net reduction in specimen moisture content. This fact agrees very well with the

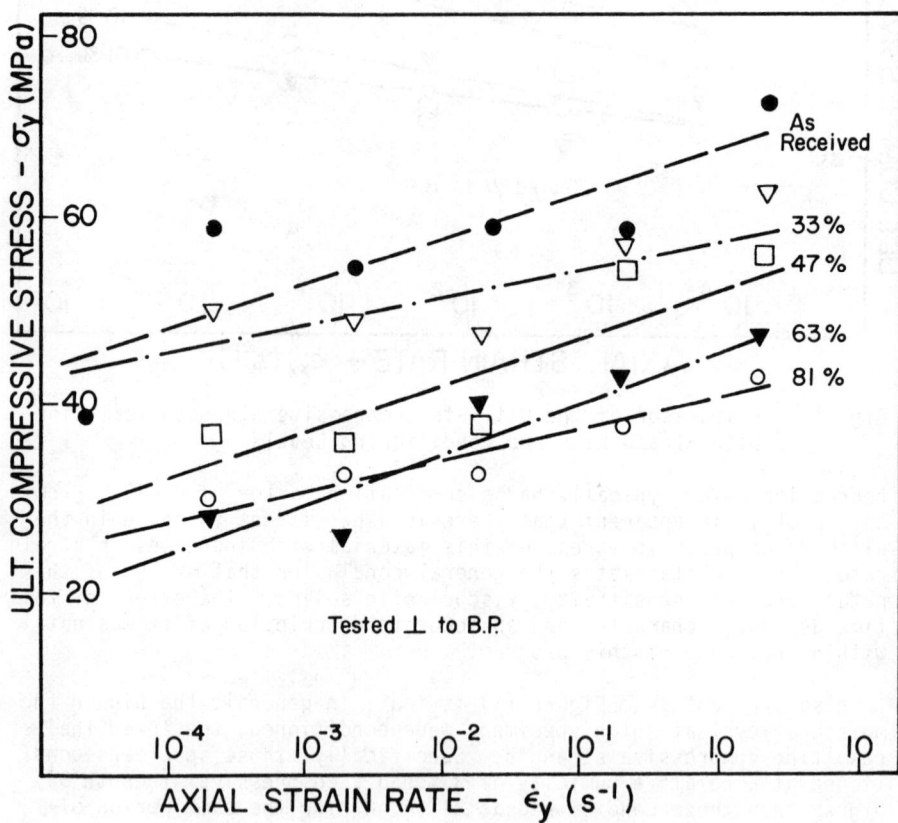

Fig. 2. Comparison of the Ultimate Compressive Strength Variation with Strain Rate and Conditioning Level

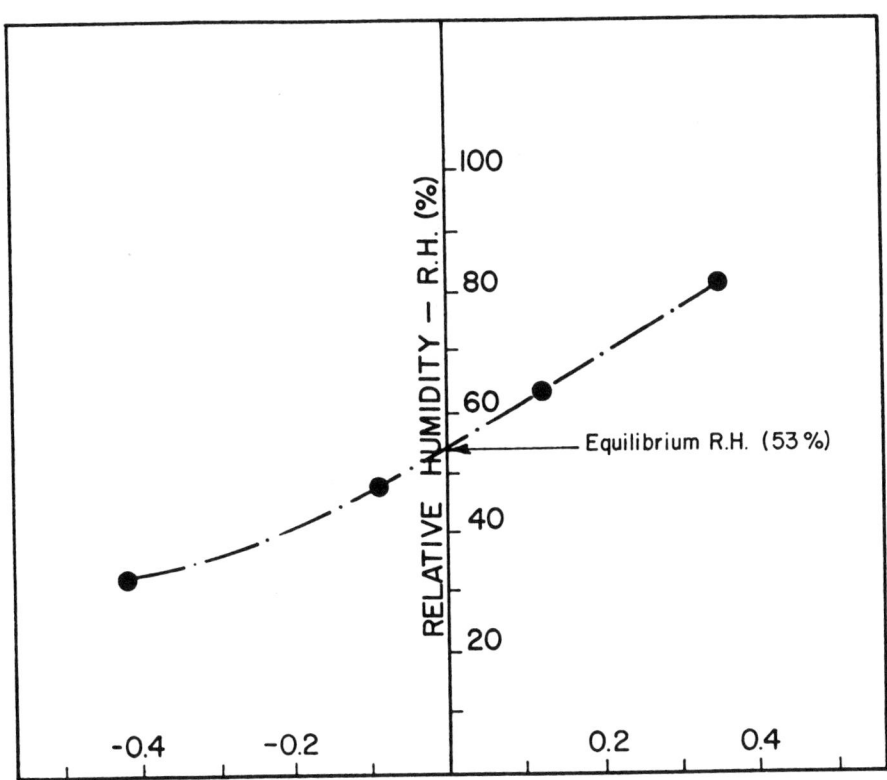

Fig. 3. Variation of Change in Moisture Content of Marine Shales Conditioned at Different Relative Humidities (76°F)

implied results of Van Eeckhout [Van Eeckhout, 1976]. It further states that the as-received specimens would sustain, in equilibrium, their inherent moisture if confined to an environment of 53% R.H. at 76°F.

However, the role of moisture content in shales and its influence on the mechanical properties of this material become unclear when recognizing that this material shows a generally lower strength when conditioned for a period of four months regardless of whether the material absorbs moisture or dehydrates. If increases in moisture content generally reduce the strength of highly layered and compacted shales then one would anticipate an increase in the strength of specimens conditioned at humidity levels below 53% R.H. relative to the as-received specimens. While specimens conditioned at humidity levels above this equilibrium level would generally be weaker in nature.

These test results strongly indicate that while the influence of moisture content is significant that there exists another strength reduction parameter which could be an aging phenomena stimulated upon release of in-situ stresses. This situation is not unlike the reversal of consolidation effects typically described for undisturbed clay strata [Vaid, 1977; Bjerrum, 1963].

In summary, the role of rate of deformation and moisture content on the mechanical response of shales is significant. These parameters control the variation of ultimate unconfined compressive stress in a somewhat well defined manner. When considering the mechanical properties such as stiffness and energy absorbing capacity the combined influence of strain rate and moisture content is extremely complex in nature. There appears to be a significant influence on the mechanical properties due to an aging phenomena which has yet been totally addressed.

ACKNOWLEDGMENTS

This research was supported by the U.S. Department of Interior, Bureau of Surface Mining and the Ohio Coal Research Laboratories Association. Their support is gratefully acknowledged and appreciated.

REFERENCES

Bjerrum, L. & Lo, K. Y., 1963, "Effect of Aging on the Shear Strength Properties of a Normally Consolidated Clay," <u>Geotechnique</u>, Vol. 13, No. 2, pp. 147-157.

Chong, K. P., Hoyt, P. M., Smith, J. W. & Paulsen, B. Y., 1980, "Effects of Strain Rate on Oil Shale Fracturing," <u>International Journal of Rock Mechanics, Mining Science and Geomechanics</u>, Vol. 17, pp. 35-43.

Forrestal, M. J., Grady, D. E. & Schuler, K. W., 1978, "An Experimental Method to Estimate the Dynamic Fracture Strength of Oil Shale in the 10^3 to 10^4 s^{-1} Strain Rate Regime," <u>International Journal of Rock Mechanics, Mining Science and Geomechanics</u>, Vol. 15, pp. 263-265.

Lankford Jr., J., 1976, "Dynamic Strength of Oil Shale," <u>Journal Society of Petroleum Engineers</u>, pp. 17-22.

Vaid, Y. P. & Campanella, R. G., 1977, "Time-Dependent Behavior of Undisturbed Clay," <u>Journal of Geotechnical Engineering - ASCE</u>, Vol. 103, pp. 693-709.

Van Eeckhout, E. M. & Ping, S. S., 1975, "The Effect of Humidity on the Compliance of Coal Mine Shales," International Journal of Rock Mechanics, Mining Science and Geomechanics, Vol. 12, pp. 335-340.

Van Eeckhout, E. M., 1976, "The Mechanisms of Strength Reduction due to Moisture in Coal Mine Shales," International Journal of Rock Mechanics, Mining Science and Geomechanics, Vol. 13, pp. 61-67.

Chapter 34

ROCK MECHANICS TESTING OF LARGE DIAMETER CORE
AT THE CRANDON DEPOSIT

Roger G. Rowe
Senior Minerals Geologist

Exxon Minerals Company
Rhinelander, Wisconsin

ABSTRACT

In late 1980, Exxon Minerals Company performed rock mechanics testing on oriented 150 mm drill core from their 75M tonne massive sulfide deposit near Crandon, Wisconsin. Instrumented samples from depths of 300 meters indicate premining horizontal stresses in a north-south direction (parallel to pillar faces) to average 12,300 kPa, with the perpendicular plane yielding an average stress of 14,630 kPa. Rock strength versus sample size was compared using uniaxial compressive strengths from 47 mm, 95 mm and 150 mm core. Massive sulfide averaged 153,300 kPa, 139,845 kPa and 75,330 kPa respectively, while the hanging wall and footwall averaged 92,000 kPa, 76,560 kPa and 57,960 kPa respectively. Point load compressive strengths were calculated using formula suggested by both Hassani et al., 1980 and Broch et al., 1972. Values calculated using the Hassani formula closely resembled preexisting point load and uniaxial compressive strengths, while values calculated using the Broch formula were 50 percent low.

INTRODUCTION

The Crandon massive sulfide deposit is located in northeastern Wisconsin, approximately 480 kilometers north of Madison, the state capital. Crandon, the nearest community, is located eight kilometers north of the proposed project site.

The orebody is a tabular deposit about 1,500 meters long, averages 38 meters wide, and dips approximately 80° to the north. The ore grade material penetrates to a depth of approximately 720 meters beneath the surface. Current published estimates of the probable tonnage and grade are 75 million metric tons that average 5.0 percent zinc, 1.1 percent copper, and 0.4 percent lead (Crandon Project team, 1980).

TESTING OF LARGE DIAMETER CORE AT CRANDON

It is proposed that the mining method will be sublevel blasthole open stoping utilizing mill tailings as the backfill materials. The stope block size, as dictated by rock mechanics studies, is currently estimated to be 45 meters wide measured along strike, and will extend from hanging wall to footwall with a vertical height of 120 meters.

ROCK MECHANICS PROGRAM

In late 1980, eight large diameter (150 mm) core holes were drilled to obtain a thirty ton bulk metallurgical sample. Rock mechanics testing was performed on rocks from three of those holes. Intermediate feasibility studies required more rock mechanics data and data of greater reliability than that which was available from earlier rock mechanics programs using NQ size (47 mm) core. The objectives of the recently completed large diameter rock mechanics program was to provide the quality of data from which could be better determined:

- Stope span calculations.
- Pillar strength dimensions and sequencing.
- Remnant pillar dimension analysis.
- Stope and pillar blasting procedures.
- Distribution of rock strengths.
- Stope and wedge stability analysis.

To accomplish these objectives, three types of rock mechanics testing were performed on the 150 mm core.

1) Inherent stress measurements.
2) Uniaxial compressive tests.
3) Point load tests.

Large Diameter Drilling and Logging

There have been very few large diameter (150 mm) core drilling programs in the base metals industry which involved both angle drilling and drilling to depths below 300 meters. The program was designed and expedited by Exxon drilling supervisors using contract personnel and equipment. The drilling contractor used a tophead drive Schramm T66B, which met operational criteria for size, speed and angle drilling capabilities. Drilling air was supplied by a rig-mounted compressor rated at .283 m^3/S @ 1.72 MPa (600 cfm @ 250 psi) and an auxiliary trailer-mounted compressor rated at .212 m^3/S @ 1.72 MPa (450 cfm @ 250 psi).

The 150 mm core was color-photographed in an undisturbed state for a permanent record of geologic structure and texture, and then immediately logged geologically, recording the following parameters:

- % recovery
- RQD
- bedding angle
- veining type and orientation
- fracture orientation
- fault zones and filling
- % base metal sulfides
- % pyrite
- weathering intensity (oxidation, leaching and clay development)
- % porosity
- rock type
- alteration

After logging, rock mechanics samples were identified and removed for processing.

Inherent Stress Testing

Electric strain gauge rosettes were installed on oriented hanging wall, ore, and footwall rocks in three drill holes. A total of 44 individual samples were instrumented. Two samples were instrumented from each sample location. The first sample was instrumented on a plane parallel to bedding. The second was prepared from the same rock specimen but was instrumented on a surface normal to bedding. These two surfaces are parallel to the proposed pillar faces.

Because the objective was to recover as much of the time-dependent strain energy as possible, strain gauges were installed on the core as quickly as possible. Several hours elapsed between the time the core was drilled and the time it arrived at the warehouse for processing. An additional time period of several hours was required to process and instrument the inherent stress sample.

After the instrumented samples were read for approximately one month, they were shipped to Queens University, Kingston, Ontario, where the geotechnical work was performed by John D. Smith Engineering Associates, Ltd. of Kingston, Ontario. There, the sample was prepared for overcoring of the strain gauge. A J. K. Smit skid-mounted hydraulic drill in the Mining Engineering Laboratory was used to overcore the gauge, using a 63 mm thinwall diamond impregnated bit. The gauge was read immediately prior to overcoring, and immediately after overcoring. Subsequent to overcoring, three readings were taken the first day, and once a day for several days thereafter, until the strain energy appeared to be fully released (Figure 1).

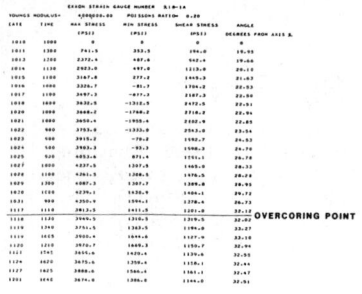

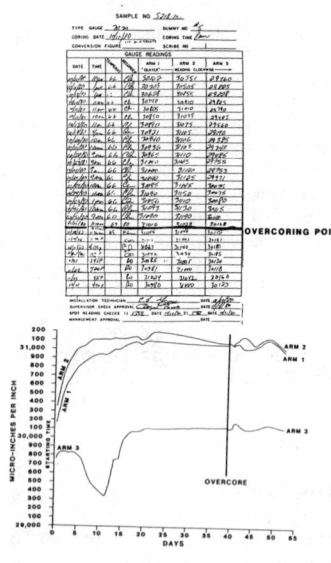

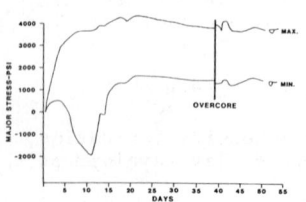

MAJOR AND MINOR PRINCIPAL STRESSES
(CALCULATED FROM SAMPLE 218-1a STRAINS)
FIGURE 1

STRAIN GAGE READINGS FOR SAMPLE S218-1a
FIGURE 2

TESTING OF LARGE DIAMETER CORE AT CRANDON

Strain readings were reduced via computer into principal stresses and direction (Figure 2). From this reduced data, the direction and magnitude of the major and minor principal stresses were determined for planes parallel and perpendicular to bedding. The theoretical method used to solve this problem is a modification of the Strain Relaxation Method, coupled with empirical data developed by Smith (1981).

Instances of tensile stresses are minor ($\sim$10%) and can be attributed to anomalous conditions. Some readings are out of the normal range of the average values, perhaps due to sample damage, and are likewise disregarded as anomalous and not used in calculation of average stress. For premining stresses parallel to bedding, the combined analysis listed in Table 1 yields a range of stresses (within one standard deviation) from 6,446 to 18,152 kPa (935 to 2,633 psi), with an average value of 12,300 kPa (1,784 psi) in the horizontal direction. For premining stresses perpendicular to bedding, the combined analysis listed in Table 2 yields a range of stresses (within one standard deviation) from 9,679 to 19,579 kPa (1,404 to 2,840 psi) with an average value of 14,629 kPa (2,122 psi).

MAJOR STRESS COMPONENT PARALLEL TO BEDDING (after Smith, 1981)

Gauge No.	Major Stress Direction	Stress Component // to Bedding in Horizontal Plane (psi)	Stress Component // to bedding on True Dip (psi)
209-1A	+28°W	1512	1522
-2A	+53°E	1281	1699
-3A	+12°W	1898	3079
-4A	+54°W	-1053	-1450
-5A	+11°E	3969	-1742
-6A	+25°W	-1416	-1743
-7A	+29°W	2956	2121
-8A	+21°E	1175	1467
-10A	+56°W	849	1259
218-1A	+35°E	3422	2396
-2A	+26°E	2082	1147
-3A	+17°E	-1582	1334
-4A	+60°W	1432	2480
-5A	+60°E	1326	2296
-6A	+17°E	-936	2494
-7A	+14°E	-2383	-1964
-8A	+43°W	1234	1151
-9A	+45°E	886	835
212-1A	+35°W	3060	2143
-2A	+13°W	1026	-1849
-3A	+44°W	2623	2532

Hole No.	Total No. of Gauges Used	Average Stress// to Bedding Horizontal (psi)	On True Dip (psi)
209	6	1612 ± 746	1858 ± 665
218	6	1730 ± 916	1767 ± 710
212	3	2236 ± 1071	2338 ± 275
COMBINED	15	1784 ± 849	1872 ± 650

TABLE 1

MAJOR STRESS COMPONENT PERPENDICULAR TO BEDDING (after Smith, 1981)

GAUGE NO.	(a) DIP OF BEDDING	(b) MAJOR STRESS DIRECTION	COMPONENT 1 TO BEDDING (SIN a-b) (psi)
209-1B	78°N	12°N	4278
-2B	78°N	-7°N	2430
-3B	68°N	28°N	1568
-4B	78°N	-75°N	1195
-5B	76°N	-60°N	2956
-6B	70°N	-14°N	2750
-7B	65°N	-75°N	2492
-8B	75°N	27°N	2221
-10B	85°N	-67°N	523
218-2B	84°N	-30°N	2567
-3B	84°N	-43°N	3832
-4B	84°N	-80°N	1417
-5B	84°N	-33°N	1901
-6B	84°N	-59°N	2377
-7B	88°N	-42°N	1940
-8B	88°N	-49°N	1239
-9B	88°N	18°N	-1264
212-1B	84°N	0°N	1591
-2B	84°N	-20°N	1486
-3B	84°N	-05°N	-1685

Hole No.	Total Number of Gauges Used	Average Stress 1 to Bedding (psi)
209	7	2230 ± 635
218	7	2181 ± 868
212	2	1539 ± 74
Combined	16	2122 ± 718

TABLE 2

Uniaxial Compressive Testing

Samples to be used for uniaxial compressive strength testing were moistened with water, wrapped in plastic to maintain a saturated environment, and packed in barrels for shipment to Queens University, Kingston, Ontario, where processing and testing would take place. Three sizes of core were tested, 150 mm, 95 mm, and 57 mm. The 95 mm and 57 mm diameter samples were drilled out of 150 mm samples. Drilling was performed using a J. K. Smit skid-mounted hydraulic drill located in the Mining Engineering Laboratory at Queens University. The following specifications were followed during final sample preparation and testing:

a) L:D nominally 2:1.
b) Ends of each sample cut parallel to each other and at right angles to the perpendicular axis.
c) Ends of the specimens lathed flat to .02 mm, and perpendicular to the axis within .001 radians.
d) Diameter and height of samples measured to + .001 in.
e) Loads applied to achieve failure from 5-15 minutes after initial loading, at constant stress rate.

After lathing, two PC-10-11 strain gauges were glued to each sample. Each gauge contained two Cu-Ni sensing elements oriented at right angles to each other. The gauges were placed midway along the length of each sample and diametrically opposite each other with one sensing element oriented to measure the longitudinal strain and the other to measure the transverse strain. The gauges were wired to a balancing bridge circuit so that recorded strain would actually be the average of the strains from each gauge. The objective of the uniaxial testing was to determine the uniaxial compressive strengths of 57 mm, 95 mm, and 150 mm diameter core, and values of Young's Modules and Poisson's Ratio for each sample.

Table 5 is a comparison between the compressive strengths of several hundred 47 mm pre-1980 core samples, and nineteen 150 mm samples generated in the current program. Compressive values are consistently lower in the current 150 mm program because larger core is inherently weaker due to discontinuities (Jaeger and Cook, 1976). Also, the core was taken at a shallower depth than earlier programs, and based upon geologic evaluation, has been weakened by supergene weathering.

An MTS SERVO-CONTROL unit capable of 2,068,000 kPa (300,000 pounds) ultimate load was used to perform the following tests:

a) deformation characteristics for all three sizes.
b) ultimate loads for the 57 mm and 95 mm sizes.

Because the ultimate loads on the 150 mm samples were generally in the 5,170,000 - 6,894,000 kPa (750,000 - 1,000,000 pound) range, they were tested on a 6,894,000 kPa (1,000,000 pound) press. Failure of some of the more competent samples was very violent, often being blown apart into small fragments. About 30% of the samples failed along joint planes. The remainder failed in compression. An estimate of the friction angle was made after failure.

The results of this testing were also used to determine the relationship between strength and size of sample. The Mining Research Center at Elliot Lake, Ontario, Canada has done considerable work on pillar design (Hedley, 1972). The Elliot Lake ore bearing strata is of similar age and has similar inherent stress properties as the Crandon deposit. The pillar strength equation produces a large reduction in compressive strength and is, therefore, relatively conservative. The Elliot Lake procedure was used during predevelopment for design purposes. The equation for pillar strength related to pillar dimensions is as follows:

$$Q_u = K \frac{W^a}{H^b}$$

where Qu = pillar strength
       K = strength of one foot cube
       W = pillar width (measured in strike direction)
       H = pillar height (H/W-F/W distance)
       a = 0.50   Average values quoted in the literature.
       b = 0.75   Determined empirically.

The current testing of strength versus size produces an average "K" value of 19,000 psi for 1 cubic foot of rock.

POINT LOAD TESTING

Point load testing has been used extensively on NQ (47 mm) cores from the Crandon deposit in the past four years for determining unconfined compressive strengths. Point load testing is an attractive alternative to conventional cylinder loading methods because of low cost, on site testing with portable equipment, and results that show less scatter than uniaxial compressive tests (Broch et al., 1971).

The purpose of including point load tests in this program was twofold:

1) To see how 150 mm point load data compares with point load data previously acquired from 47 mm core, and

2) To see the relationship between the 150 mm point load data generated during this program and the uniaxial compressive test data generated in this program.

Samples to be used for point load testing were moistened with water, wrapped in plastic to maintain a saturated environment, and shipped to Queens University, Kingston, Ontario. At Queens Laboratory, a 2,068,000 kPa (300,000 pound) MTS Servo-Control Unit (in manual control) with properly adapted platens was used to test the samples. The load was applied manually, at an approximate rate of 5000 lb/minute. At failure, the load P was read from a graph of load vs. displacement to an accuracy of ± 345 kPA (± 50 pounds).

Two types of tests were performed: diametral and axial. Since diametral tests were most consistent and more closely resemble uniaxial compressive test values, the following comparisons are between diametral point load tests and uniaxial compressive tests.

Table 3 lists the uniaxial compressive test values and diametral predictive compressive test values of 27 samples from the 150 mm program. The point load predictive compressive test values were calculated using both the method suggested by Broch (1972) and using the method suggested by Hassani (1980).

Table 4 lists the relationships between the point load predictive compressive strengths and the uniaxial compressive strengths. Correlation coefficients in the .4 range do not invalidate the point load values, they only indicate there is a poor relationship between the sample "pairs" selected for point load and uniaxial compressive testing. The low correlation coefficient appears to reflect our sample selection procedure in the laboratory, in which, due to heavy sample demand for other tests, individual samples for each sample "pair" were often taken up to 10 feet apart.

POINT LOAD PREDICTIVE COMPRESSIVE STRENGTHS AND UNIAXIAL COMPRESSIVE STRENGTHS (PSI)

Rock Type	Sample Number	Point Load Predictive Compressive Strengths ($S_{c//}$)		Uniaxial Compressive Strengths	
		After Hassani	After Broch	150 mm	95 mm
MASSIVE SULFIDE	218-1	23113	14790	17247	16182
	218-2	15979	9750	4555	25090
	218-2	23055	14760	4555	25090
	218-3	14703	8700	5749	6000
	209-5	5075	2888	14559	x
	209-6	19430	12258	1493	19090
	209-7	46255	31320	14298	18090
	209-8	31958	19836	18591	27545
	209-8	34510	22794	18591	27545
	209-9	28594	18096	16949	35727
ALL OTHERS (Hanging Wall, Footwall)	209-1	9671	5777	9744	7090
	209-2	4553	2610	2128	4027
	209-3	10005	5533	1045	x
	218-4	26477	18270	6477	6590
	218-5	17458	9744	9520	10818
	218-5	21025	12876	9520	10818
	218-6	19749	10440	5451	10273
	218-6	29290	18792	5451	10273
	218-7	12615	7308	5451	10273
	218-7	42050	26488	9445	11000
	218-7	33640	20880	9445	11000
	218-8	26100	16182	15903	9455
	209-10	9860	5377	11573	24182
	209-11	13659	7830	3995	15000
	218-9	26100	17574	20197	21182
	218-10	16820	9048	5413	11409
	218-10	19865	10512	5413	11409

TABLE 3

RELATIONSHIP BETWEEN POINT LOAD COMPRESSIVE STRENGTHS AND UNIAXIAL COMPRESSIVE STRENGTHS (PSI)

MASSIVE SULFIDE

Comparison	B	M	R	Relationship
150 mm - Broch	5951	.378	.44	$S_c = .378\ S_{c//} + 5951$
150 mm - Hassani	5361	.260	.45	$S_c = .260\ S_{c//} + 5361$
95 mm - Broch	15205	.405	.39	$S_c = .405\ S_{c//} + 15205$
95 mm - Hassani	14339	.295	.41	$S_c = .295\ S_{c//} + 14339$

ALL OTHERS (Hanging Wall and Footwall)

Comparison	B	M	R	Relationship
150 mm - Broch	4259	.311	.43	$S_c = .311\ S_{c//} + 4259$
150 mm - Hassani	4143	.194	.40	$S_c = .194\ S_{c//} + 4143$
95 mm - Broch	11506	.003	.005	$S_c = .003\ S_{c//} + 11506$
95 mm - Hassani	11464	.004	.01	$S_c = .004\ S_{c//} + 11464$

TABLE 4

In order to establish an accurate relationship between point load predictive compressive strengths and actual compressive strengths, regression analysis was performed on a suite of over 200 sample "pairs" of pre-1980 47 mm core, from which we had both point load tests and uniaxial compressive tests at the same sample location. The following relationship was developed:

Rock Type	Relationship
Massive Sulfide	$S_c = 1.14\ S_{c//} - 1482$ psi (R = .98)
All Others (Hanging Wall and Footwall)	$S_c = .452\ S_{c//} + 3663$ psi (R = .60)

Massive sulfide is a relatively uniform and homogeneous rock with a low anisotropic index. A high correlation coefficient can be expected in this type of rock. The hanging wall and footwall rocks, however, are strongly bedded, foliated and irregular, and commonly have a high anisotropic index. Because it is impossible to do a point load test and a uniaxial compressive test at precisely the same point, we believe a correlation coefficient of .60 in this type of heterogeneous rock is about the best that can be expected. Because the above relationships are considered good, they were used in converting 150 mm point load predictive compressive strengths ($S_{c//}$) to compressive strengths (S_c). Table 5 is a listing of both point load compressive strengths and uniaxial compressive strengths from pre-1980 (47 mm) test programs and our current 150 mm test program.

TABLE 5

POINT LOAD AND UNIAXIAL COMPARISONS (PSI)

POINT LOAD TESTING

Pre-1980 - 47 mm

Rock Type	No. Samples in Test	S_c
Massive Sulfide	6	17129**
All Others	136	12247

Current - 150 mm (After Hassani et al., 1980)

Rock Type	No. Samples in Test	S_c
Massive Sulfide	10	26182
All Others	17	12674

Current - 150 mm (After Broch et al., 1972)

Rock Type	No. Samples in Test	S_c
Massive Sulfide	10	16210
All Others	17	8858

UNIAXIAL COMPRESSIVE TESTING

Pre 1980 - 47 mm

Rock Type	No. Samples in Test	S_c
Massive Sulfide	20	22237
All Others	219	13345

Current - 95 mm

Rock Type	No. Samples in Test	S_c
Massive Sulfide	8	20285
All Others	12	11105

Current 150 mm

Rock Type	No. Samples in Test	S_c
Massive Sulfide	8	10927
All Others	12	8407

* All values after application in appropriate regression formula
Massive Sulfide $S_c = 1.14\ S_{c//} - 1482$ psi
All Others $S_c = .452\ S_{c//} + 3663$ psi

** Additional point load work on 47 mm core, since this study suggests this number to be considerably higher, perhaps double.

Point load testing of pre-1980 47 mm core indicates massive sulfide to have an average strength of 118,087 kPa (17,129 psi). (Work on 47 mm core since that time suggests this number to be about twice as high.) All other rock types have an average strength of 84,431 kPa (12,247 psi). Point load strengths from the 150 mm core calculated using the formula suggested by Hassani (1980) are much closer to these values than when calculated using the formula suggested by Broch (1972). The largest single factor which makes the Hassani approach more accurate is the chart for bringing the index value of 150 mm core back up to I_{s-50}. Hassani's curves are considerably steeper, which raises the index value considerably when compared to the Broch I_{s-50} curves. This will only have a significant effect when working with core much larger or smaller than 50 mm.

The comparisons of uniaxial compressive testing on Table 5 indicate the 150 mm core is weaker than 95 mm core, which is, in turn, weaker than 47 mm core. This corresponds to the strength vs. size concepts which were discussed in the uniaxial compressive test section of this paper.

While the correlation coefficient between 150 mm point load and 150 mm uniaxial compressive testing is poor due to poor sample pair selection, the individual values obtained by point load testing and uniaxial compressive testing are very good. Point load values, when calculated using the approach of Hassani, correspond well with previous data, and support the earlier numbers. Uniaxial compressive test data also compares well, when consideration is given for sample size.

CONCLUSIONS

The rock mechanics testing was a logistical and geotechnical success due to careful planning and quality control built into every aspect of the program. The data was applied to fulfill the engineering objectives previously outlined. Previous design efforts, based on pre-1980 rock mechanics efforts, have been verified and strengthened by the results of this program. All data generated to date will continue to be of use as mine design studies continue, or until the data is superseded by in situ measurements.

REFERENCES

Bieniawski, Z. T., Franklin, J. A., 1972, Suggested Methods for Determining the Uniaxial Compressive Strength of Rock Materials and the Point Load Strength Index: International Society for Rock Mechanics (ISRM), Committee on Laboratory Tests, Document No. 1.

Broch, E., and Franklin, J. A., 1972, The Point-Load Strength Test: Int. J. Rock Mech. Min. Sci., V. 9, p. 669-697.

Crandon Project Team, 1980, Preliminary Project Description - Crandon Project: Exxon Minerals Company public report prepared for Department of Natural Resources. Three Volumes.

Hedley, D. G. F., and Grant, F., 1972, Stope-and-Pillar Design for the Elliot Lake Uranium Mines: CIM, July, p. 37-44.

Jaeger, J. C., Cook, N. W., 1976, Fundamentals of Rock Mechanics: John Wiley and Sons, Inc., New York, N.Y.

Smith, John D., 1981, Rock Mechanics Testing and Engineering of Large Diameter Core: Exxon Minerals Company internal report.

Chapter 35

THE INFLUENCE OF STRESS LEVEL ON THE CREEP OF UNFILLED ROCK JOINTS

by Charles W. Schwartz and Subash Kolluru

Department of Civil Engineering
University of Maryland
College Park, Maryland

INTRODUCTION

Creep of rock in situ, like most rock mass behavior, will be largely governed by the behavior of the natural discontinuities -- bedding planes, faults, and joints, in particular. Several past studies have investigated various aspects of the time-dependent deformation of rock discontinuities; examples include Amadei and Curran (1980), Engelder and Scholz (1976), Johnson (1975), Kaiser and Morgenstern (1979), Solberg et al. (1978), and Wawersik (1974). These studies have provided much insight into the general phenomenon, but they have also produced some inconsistent conclusions and left many unanswered questions.

This paper will attempt to shed some light on one of these questions: What is the influence of stress level on the creep of unfilled rock joints? The discussion will follow a two-part approach to the problem: (1) laboratory creep experiments on small scale jointed specimens, (2) a simple theoretical mechanism to explain, at least in qualitative terms, some of the results from the experimental program.

EXPERIMENTAL PROGRAM

Uniaxial creep tests were conducted on intact and jointed samples of a specially formulated gypsum plaster "synthetic rock." Synthetic rock materials have several advantages over natural rock for long-term creep testing: (a) higher creep rates, (b) lower strength and therefore lower loading requirements, (c) simpler specimen preparation, and (d) more controlled and consistent properties. Gypsum based plasters have been used previously in studies of short-term strength and deformability of jointed samples (Patton, 1966; Einstein and Hirschfeld, 1973). Furthermore, Williams and Elizzi (1977) demonstrated that natural gypsum, at least, creeps in a manner similar to that of most other creep-susceptible natural rocks.

A gypsum plaster mix originally developed by Nelson (1968) for a study of short-term jointed rock behavior was used for all of the creep tests in our investigation. The mix consists of three ingredients: (a) water, (b) Hydrocal B-11 gypsum plaster (a product of the U.S. Gypsum Co.), and (c) Celite (a product of the Johns-Manville Co.), a fine-grained diatomaceous earth used as an admixture to prevent bleeding of the water during mixing and curing. By weight, the mix proportions are: water/Hydrocal = 0.45, water/Celite = 32. Based on extensive testing, Nelson (1968) determined that this mix produced a hardened plaster with appropriate low strength and brittle stress-strain properties which satisfied the standard dimensional similitude requirements for models. Short-term properties for this synthetic rock material, as determined by us from tests on 76 mm. (3 in.) by 152 mm. (6 in.) cylinders, are: unconfined compressive strength, σ_c = 26.5 MPa (3840 psi); Young's modulus, E = 8.1 GPa (1.17×10^6 psi); split-cylinder tensile strength, σ_t = 1.8 MPa (260 psi). The ratios of $\sigma_c/\sigma_t \cong 15$ and $E/\sigma_c \cong 300$ are typical of those for many natural rocks.

All creep tests were performed on 32 mm. (1.25 in.) by 32 mm. (1.25 in.) by 121 mm. (4.75 in.) tall rectangular prismatic specimens loaded in uniaxial compression. Specimen fabrication was carefully controlled to minimize variations in properties among samples. Jointed specimens were prepared by carefully sawing intact samples and then sanding the joint surfaces with 60 grit paper, producing a joint peak friction angle, ϕ_j, of 42°. The uniaxial constant-stress compressive loading was applied by dead weights using a specially modified soil consolidation frame. Axial deformations were measured by a conventional dial gauge readable to 0.0005 mm. (0.00002 in.). Thermal strains caused by ambient temperature changes were compensated using data from an unloaded dummy specimen. Test duration in most cases was approximately 50 hours.

TEST RESULTS

Intact Specimens

A series of sets of tests on intact specimens loaded to $0.2\sigma_c$, $0.4\sigma_c$, and $0.5\sigma_c$ stress levels was performed to provide information on the general nature of the creep behavior for the gypsum synthetic rock and its stress dependence. In order to minimize the influence of material variability between specimens in these and all other test series reported herein, three tests were run in each set and an average or "composite" power law creep function was fitted to the entire set of data using conventional least-squares regression analysis. The composite creep curves for the three stress levels are plotted in Figure 1. As is the case with natural rock (see, for example, Williams and Elizzi, 1977), the creep of the gypsum synthetic rock is significantly stress dependent, particularly at the higher stress levels.

Jointed Specimens

Joint Normal to Applied Stresses. By analogy with the effect of joints on short-term deformations, the presence of a single joint in a specimen would be expected to produce increased axial creep deformations due to, in the general case, the additional normal and shear deformations across the joint. The simplest case arises when the joint is perpendicular to the applied stress direction; then only normal joint movements occur. Figure 2 summarizes test data from two sets of tests at different stress levels for this simplest case. The joint creep strains, ε_{cj}, are a fictitious quantity obtained by subtracting the creep strains for an intact specimen from the total creep strains for the jointed sample at the same stress level; in other words, ε_{cj} represents the additional average creep strain in the sample due to the joint. The data in the figure clearly indicate that the joint does cause additional stress-dependent creep strains in the specimen.

Joint Inclined to Applied Stresses. Inclining the joint relative to the uniaxial stress direction will have two general effects: (a) both normal and shear deformations will develop across the joint, and (b) the relative contributions of the shear and normal joint deformations to the overall axial shortening of the specimen will vary with joint inclination. For a joint treated as an elastic seam with normal and shear stiffnesses K_n and K_s, the axial specimen displacement due to the joint, δ_j, for short-term conditions can be expressed as:

$$\delta_j = \frac{\sigma}{2K_s}[R_K(1+\cos 2\theta) \cos \theta + \sin 2\theta \sin \theta] \qquad (1)$$

in which σ is the applied uniaxial stress, θ is the angle between the joint pole and the uniaxial stress direction, and R_K is a joint stiffness factor equal to K_s/K_n. For low values of R_K, the axial displacement δ_j increases with increasing θ, reaching a peak value at $\theta \cong 45 - 50°$; after this point, δ_j decreases. The short-term displacement data from our tests follow the general trend of Equation (1) and suggest that $R_K \cong 0.05$ is appropriate for our joints.

Intuitively, it is reasonable to expect the axial creep strains, ε_{cj}, due to an inclined joint to follow the same trends as the short-term behavior given by Equation (1). Figure 3, which summarizes the results from two sets of creep tests on specimens loaded to a 0.4 σ_c stress level, suggests that this is in fact the case. Specimens with a joint inclined at $\theta = 15°$ produced ε_{cj} values up to 175% higher than those from specimens with joints perpendicular ($\theta=0°$) to the uniaxial stress direction.

Because rock creep is primarily a shear deformation phenomenon, the increase in axial displacements with increasing joint inclination might be expected to be more pronounced for the creep case than for the corresponding short-term case. Amadei (1979) hypothesized that the ratio τ_j/τ_p (τ_j = applied joint shear stress; τ_p = peak joint shear strength = $\sigma_n \tan \phi_j$) was the critical factor governing joint creep in shear; as τ_j/τ_p increases (i.e., as θ increases in our uniaxial tests), joint shear creep increases sharply. To verify this effect, we performed another set of creep tests in which jointed specimens were subjected to the same joint shear stress level as the $\theta = 15°$, $\sigma = 0.4\sigma_c$ tests described previously but loaded to a lower joint normal stress. This was accomplished by increasing θ to 30° and reducing σ to $0.231\sigma_c$; thus, while in both tests $\tau_j = 0.1\sigma_c$, τ_j/τ_p equalled 0.298 for the $\theta = 15°$ tests and 0.642 for the $\theta = 30°$ tests. By Amadei's hypothesis, the $\theta = 30°$ tests should exhibit much higher creep deformations both because τ_j/τ_p is higher and because the resulting joint shear creep is a larger component of the joint's contribution to the axial deformations. However, the data summarized in Figure 3 show just the opposite effect; at the larger value of θ, ε_{cj} is considerably reduced. Clearly the influence of joint inclination on specimen creep is more complex than initially expected. One possible explanation is that both the stress ratio τ_j/τ_p and the average absolute stress level exert a strong influence on joint creep.

To understand better the influence of stress level on joint creep, six sets of jointed specimen tests at the same uniaxial stress level but different inclinations (and therefore different τ_j/τ_p ratios) were conducted. These composite creep data for the six test sets ($\theta=0°$, 7.5°, 15°, 22.5°, 30°, 37.5°) are summarized in Figure 4, which plots ε_{cj} isochrones versus θ at various times. These test results are quite interesting and at first difficult to explain. At small joint inclinations (i.e., low τ_j/τ_p), increasing θ increased joint creep, as expected, reaching a peak at around θ of 15° to 20°. After this, increasing θ <u>decreased</u> joint creep. This again contradicts, at least in part, the hypothesis that joint shear creep increases monotonically with τ_j/τ_p; it also does not agree well with expectations based on short-term behavior, where the peak in δ_j occurs at much higher inclinations (Equation 1). However, the simple theoretical mechanism described in the next section may give some insight into this observed behavior.

ASPERITY CREEP MECHANISM

One possible explanation for the additional creep caused by rock joints is the accelerated creep of the intact rock caused by stress concentrations at joint asperities. A multistage simplification of the asperity geometry and the asperity creep behavior is depicted in Figure 5. The asperity is treated as a simple rectangular "pillar" spanning the rigid joint walls. Using the notation defined in Figure 5c, the average stresses in an individual asperity can be expressed as:

$$\sigma_a = 0.5\sigma A(1+\cos 2\theta) \tag{2}$$

$$\tau_a = 0.5\sigma A \sin 2\theta \tag{3}$$

Following the general formulation described by Bathe (1976) for creep under multiaxial stress conditions, the effects of σ_a and τ_a can be combined by considering s, the second invariant of the deviatoric stress tensor, producing:

$$s = \sigma A[1+0.5 \cos 2\theta - 0.5 \cos^2 2\theta]^{1/2} \tag{4}$$

assuming plane stress conditions for the asperity. The corresponding effective creep strain, e_c, is computed by substituting s into the general power law creep function with stress dependent coefficients:

$$e_c = a_0 s^{a_1} t^{a_2} s^{a_3} \tag{5}$$

in which a_0, a_1, a_2, and a_3 are material constants and t is time. The creep strain components in the normal and shear directions, ε_{cn} and ε_{cs}, can be expressed as:

$$\varepsilon_{cn} = 1.5\, e_c \left(\frac{s_n}{s}\right) = 0.5\, \sigma A (1+\cos 2\theta) a_0 s^{(a_1-1)} a_2 s^{a_3} t \tag{6}$$

$$\varepsilon_{cs} = 1.5\, e_c \left(\frac{s_s}{s}\right) = 0.75\, \sigma A (\sin 2\theta) a_0 s^{(a_1-1)} a_2 s^{a_3} t \tag{7}$$

in which s_n and s_s are the deviatoric stress components in the normal and shear directions. The axial creep displacement due to the joint, δ_{cj}, is then simply:

$$\delta_{cj} = \varepsilon_{cn} h \cos \theta + 2\varepsilon_{cs} h \sin \theta \tag{8a}$$

$$= 0.5\, \sigma A h a_0 s^{(a_1-1)} a_2 s^{a_3} t \,[(1+\cos 2\theta) \cos \theta + 3 \sin 2\theta \sin \theta] \tag{8b}$$

Equation (8b) can be used to estimate, at least qualitatively, the variation of axial creep displacement with joint inclination. Plots of normalized values for δ_{cj} calculated using reasonable values for the material parameters (based on our test data and physical reasoning) are given in Figure 6 for $\sigma = 0.4\sigma_c$ and t = 50 hours. Also shown on the figure are our experimental data and the theoretical plot of δ_j versus θ assuming elastic, time-independent joint behavior. It is clear from the figure that the asperity creep model is an improvement over the assumption that joint creep behavior mirrors the time-independent joint behavior; whereas the time-independent curve has a peak axial displacement at $\theta \simeq 50°$ (off the graph), the asperity creep model has a peak at $\theta \simeq 25 - 30°$, much closer to the observed value of $\theta \simeq 15 - 20°$.

Of course, the simple asperity creep model ignores many potentially important factors. One of these is the accelerated creep of the joint wall rock due to stress concentrations around the asperities. Finite element analyses are currently being conducted to study the significance of these effects.

CONCLUSIONS

Results from small scale laboratory creep tests of a "synthetic rock" gypsum material clearly show that clean, unfilled joints augment the axial creep displacements of rock specimens loaded in uniaxial compression. These increased creep deformations are due to both normal and shear creep displacements across the joint. Although joint creep, like intact rock creep, is strongly stress dependent, the nature of this stress dependence is more complex than has been assumed to this point. Joint creep appears to depend both upon the applied joint shear stress to shear strength ratio and upon the absolute stress level across the joint. A simple, theoretical asperity creep mechanism is proposed to explain, at least in part, the observed joint creep behavior.

ACKNOWLEDGMENTS

This material is based upon work supported by the National Science Foundation under Grant No. CME-8006685. Computer time for this project was supported in part by the Computer Science Center of the University of Maryland.

REFERENCES

Amadei, B. (1979), "Creep Behavior of Rock Joints," M.S. Thesis, University of Toronto, Toronto, Canada.

Amadei, B. and J.H. Curran (1980), "Creep Behavior of Rock Joints," Proceedings, 13th. Canadian Rock Mechanics Symposium, pp. 146-150.

Bathe, K.J. (1976), "Static and Dynamic Geometric and Material Nonlinear Analysis Using ADINA," Report 82448-2, Acoustics and Vibration Lab., Mechanical Engineering Dept., Massachusetts Institute of Technology, Cambridge, Mass.

Einstein, H.H. and Hirschfeld, R.C. (1973), "Model Studies on Mechanics of Jointed Rock," Journal of the Soil Mechanics and Foundations Division, ASCE, Vol. 99, No. SM3, March, pp. 229-248.

Engelder, J.T. and Scholz, C.H. (1976), "The Role of Asperity Indentation and Ploughing in Rock Friction-II," Int. J. Rock Mech. Min. Sci., Vol. 13, pp. 155-163.

Johnson, T.J. (1975), "A Comparison of Frictional Sliding on Granite and Dunite Surfaces," J. Geophys. Res., Vol. 80, No. 17, June, pp. 2600-2605.

Kaiser, P.K. and Morgenstern, N.R. (1979), "Time Dependent Deformation of Jointed Rock Near Failure," Proceedings, 4th International Congress on Rock Mechanics, ISRM, Montreux, Switzerland, Vol. I, pp. 195-202.

Nelson, R.A. (1968), "Modeling a Jointed Rock Mass," M.S. Thesis, Massachusetts Institute of Technology, Cambridge. Mass.

Patton, F.D. (1966), Multiple Modes of Shear Failure in Rock," Proceedings, 1st. Intl. Congress on Rock Mechanics, Lisbon, Vol. I, pp. 509-514.

Solberg, P.H., Lockner, D.A., Summers, R.S., Weeks, J.D., and Byerlee, J.D. (1978), "Experimental Fault Creep under Constant Differential Stress and High Confining Pressure," Proceedings, 19th U.S. Symposium on Rock Mechanics, Stateline, Nevada.

Wawersik, W.R. (1974), "Time-Dependent Behavior in Rock in Compression," Proc., 3rd Cong., ISRM, Denver, Vol. 2, Part A, pp. 357-363.

Williams, F.T., and Elizzi, M.A. (1977), "Creep Properties of Sherburn Gypsum under Triaxial Loading," Proceedings, Conference on Rock Engineering, Univ. of Newcastle Upon Tyne, England, pp. 71-83.

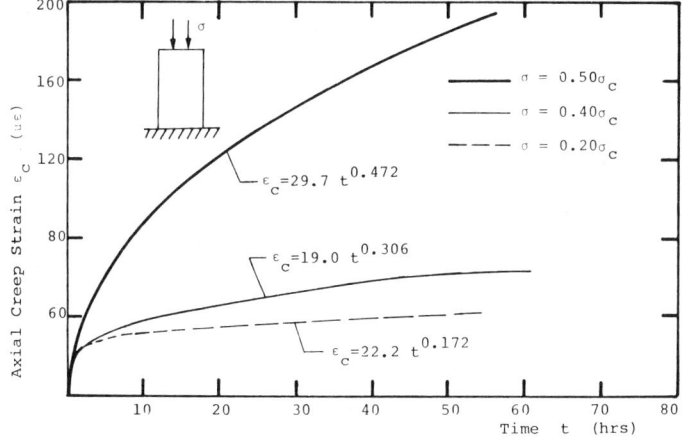

Figure 1. Intact Specimen Creep at Three Different Stress Levels

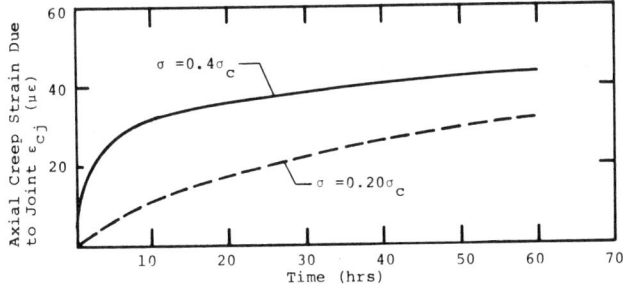

Figure 2. Effect of Single Joint Normal to Applied Stress

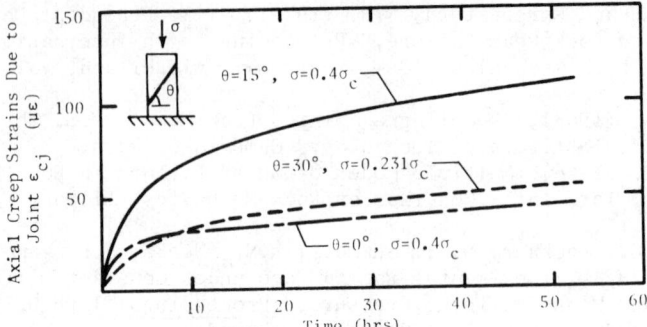

Figure 3. Effect of τ_j/τ_p on Axial Creep Strains

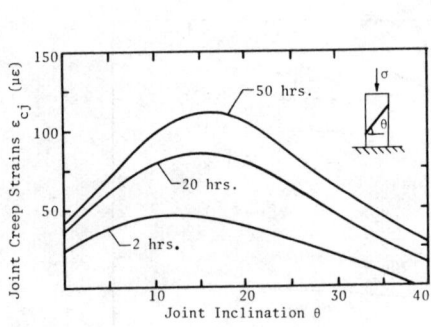

Figure 4. Effect of Joint Inclination on Axial Creep Strains

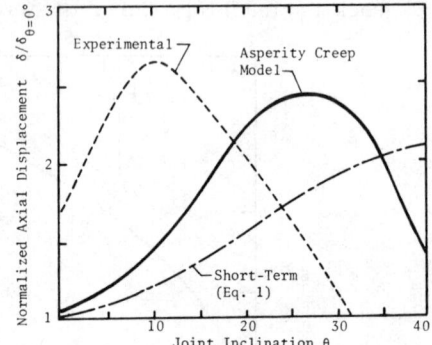

Figure 6. Results from Asperity Creep Model (A=2, a_0=6.9E-12, a_1=2, a_2=1.7E-07, a_3=2)

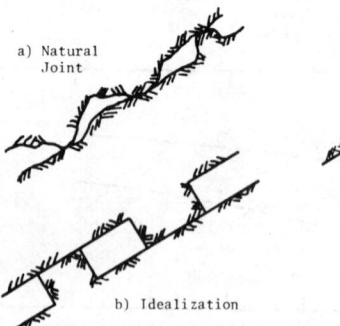
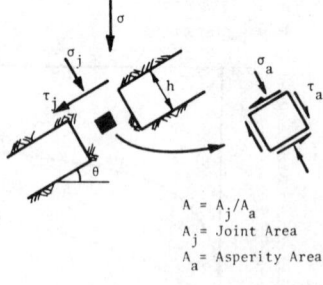

Figure 5. Idealization of Joint Asperities

Chapter 36

TRIAXIAL CREEP OF OIL SHALE AND DEFORMATION OF
PILLARS IN THE IN SITU RETORTING ENVIRONMENT

K.P. Sinha*, T.F. Borschel*, J.F. Schatz* and S. Demou†

*Terra Tek, Inc., Salt Lake City, Utah
†USBM, Minneapolis, Minnesota

ABSTRACT

Triaxial creep behavior of oil shale was investigated in the laboratory at simulated in-situ conditions. A range of temperature and stress conditions were chosen to represent those within the inter-chamber pillars of a modified in-situ retorting operation. Oil shale in grades up to 60 gallons per ton have been tested. Experimental data and a constitutive model have been presented here.

INTRODUCTION

Creep is a significant aspect of the mechanical behavior of oil shale. This has been confirmed by actual measurements on pillars in experimental underground oil shale mines (Agapito, 1972), and laboratory tests on oil shale samples (Miller, et al., 1979; Chu and Chang, 1980). It is, therefore, implied that time-dependent deformational behavior of oil shale will be a governing factor in all oil shale mine designs. In an in-situ, or modified in-situ retorting operation for recovering oil from oil shale, time dependent deformations are more pronounced due to elevated temperatures. In a modified in-situ retorting operation, part of the oil shale initially is conventionally mined and the retort chambers are prepared by rubblization of shale in place. Creep behavior of oil shale in this case may significantly affect the stability of the interchamber pillars and ground subsidence as well as the functioning of the retorts themselves due to resulting permeability loss. This paper describes a systematic investigation of the creep behavior of oil shale under temperature and stress conditions expected within an inter-chamber pillar. The investigation consists of performing a series of triaxial creep tests and constitutive modelling.

Interest in the mechanical properties and behavior of oil shale is quite recent. Sellars, et al. (1972) and Zambas, et al. (1972) seem to be the first to report a limited systematic investigation of time-dependent mechanical behavior in oil shale. They have given the results of long term uniaxial compression tests on oil shale samples from the Anvil Points, Colorado, site. Tests were conducted at a controlled temperature of 283K with samples held under constant uniaxial load until failure occurred or no deformation was measured during a period of five days. Maximum duration of tests was 1,000 hours. The main findings of these tests were that practically no creep occurred below a critical nominal stress of 13.79 MPa; very low initial creep strain rates occurred and rapidly reduced to zero (in only a few days) at stress levels between 13.79 MPa and 41.37 MPa (a logarithmic creep strain versus time behavior was observed during this period); creep rupture occurred in samples with oil content above 30 gallons per ton (GPT) at stress levels of 55.16 MPa (corresponding to approximately 80% of compressive strength).

Agapito and Page (1975) have described the long term vertical deformation behavior of actual oil shale pillars in the Colony Oil Shale Mine. They observed that the creep strain rate measured in a pillar was a function of the vertical pillar stress and proposed a linear relationship.

Chong, et al. (1978) performed creep and relaxation tests on oil shale samples from the Wyoming Green River formation subjected to uniaxial compression at room temperature. Samples with oil content from 10 GPT to 50 GPT were tested at stress levels of 25% to 75% of their ultimate compressive strength values and for times up to 80 hours. A standard linear visco-elastic model has had some success in describing the behavior.

Miller, et al. (1978) have indicated that at elevated temperature, creep exhibited by oil shale is very much augmented. Chu and Chang (1980) have described uniaxial compression creep tests on oil shale samples at temperatures ranging from 297K to 478K. They noted that creep behavior predicted on the basis of a generalized Kelvin-Voigt Model (Kelvin-Voigt unit in series with an elastic spring) with model parameters being functions of temperature, were in general agreement with experimental results. Olsson (1980) performed stress relaxation tests on oil shale samples of selected grades from the Anvil Points site and found that a linear visco-elastic constitutive law fitted his experimental results very well. Horino, et al. (1982) used a modified Burger's model in which the steady-state creep at room temperature was a non-linear function of stress. This paper describes a systematic investigation of the triaxial compression creep behavior of oil shale at elevated temperatures.

SAMPLE PREPARATION AND CHARACTERIZATION

Oil shale samples were collected from the Logan Wash Site, near Debeque, Colorado, on the property owned by the Occidental Oil Shale Company. Six inch diameter cores were drilled in directions both perpendicular and parallel to bedding planes. Three to four test specimens (5.08 cm diameter by 10.16 cm long) representing the same varve levels or the same sections were obtained from a single piece of 15.29 cm core. In addition, some test specimens were prepared from blocks of oil shale recovered from the muck pile in front of a blasted face. However, due to the blast damage, it was difficult to obtain, from the blocks, intact samples with no visible cracks. The oil content or the grade for each test specimen was estimated from a density versus grade correlation established on the basis of actual assay measurements in several samples. This correlation agrees very well (Figure 1) with a similar correlation established for oil shale at Anvil Points Site near Rifle, Colorado (Sellars, et al., 1972).

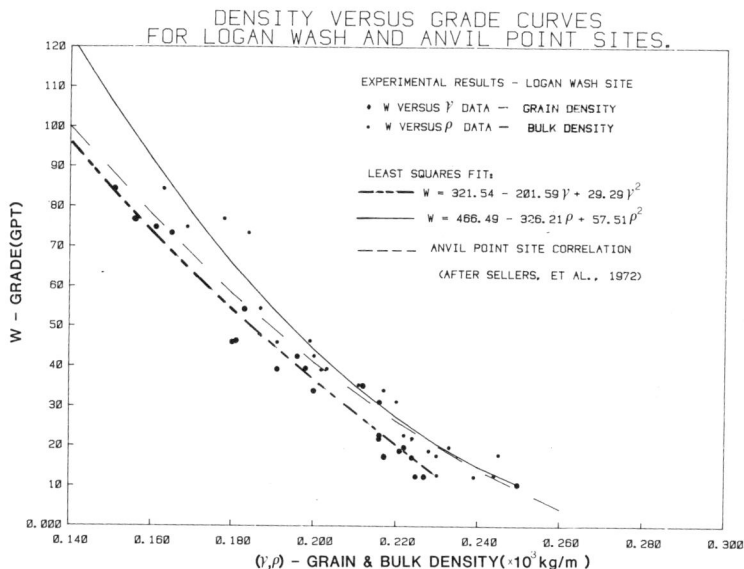

Figure 1. Grain and Bulk Density-Grade Relationships for Oil Shale at Logan Wash and Anvil Points Sites.

TEST EQUIPMENT AND EXPERIMENTAL TECHNIQUE

All the creep tests have been done using a system designed and built at Terra Tek for high-temperature creep measurements. The system uses gas-backed, thermally stabilized hydraulic accumulators for applying the axial and confining pressures. The gas-backed

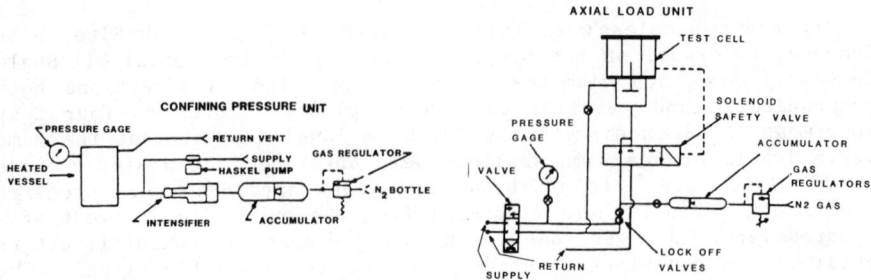

Figure 2. Schematic Diagram of the Axial Load and Confining Pressure Units in the Creep Test System.

accumulators are designed to maintain constant confining pressure and axial load over long periods. Figure 2 shows a schematic of the axial load and confining pressure units. The axial and transverse strain measurements are made using, respectively, linear variable differential transformers (LVDTs) and strain-gaged cantilever fixtures. The strain gages indicate some drift over prolonged measurement durations, especially at elevated temperature. The LVDTs, on the other hand, are extremely stable and have high enough sensitivity for creep measurements on most rocks. The heating of the speicmen is done internally within the cell by convection of the hot cell fluid within a ceramic shroud. Figure 3 shows a view of the entire testing system and the sample assembly.

The system is electronically linked with a PDP-11 computer system and all the pressure, temperature, stress and strain transducer outputs are acquired in real time.

The test procedure consists of the following steps. The prepared test specimen (5.08 cm diameter by 10.16 cm long) is accurately weighed and several measurements of its dimensions are taken to obtain the density and thereby estimate the grade. The specimen is jacketed with an elastomer material and then subjected to a nominal confining pressure in the test cell. The temperature inside the cell is then gradually raised until the thermocouples attached to the test specimen register the desired temperature. During the heating phase the specimen is allowed to expand freely and any excess pressure built in the cell is relieved. The specimen is allowed to equilibrate at the desired temperature for a brief period (1 to 3 hours) and the cell pressure and the axial load are adjusted to produce specified confining and deviatoric stresses. The creep phase begins at this stage and is continued for at least 24 hours under constant temperature and stress conditions. Table 1 shows the temperature and stress conditions under which the tests are being conducted. A maximum temperature of 523K was selected on the assumption that more than approximately 50% of an in-situ retort pillar will face a maxi-

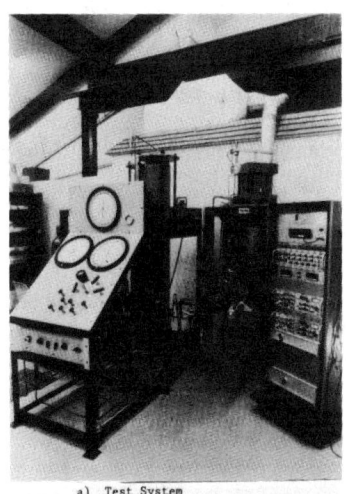 

a) Test System b) Specimen Assembly

Figure 3. Terra Tek High-Temperature Creep Testing System and a View of Test Specimen Assembly.

TABLE 1. Stress and Temperature Conditions During Creep Tests

Axial Stress - σ_1	6.89, 13.79, 20.68 MPa
Confining Pressure - σ_3	0, 1/3, 2/3 of σ_1
Temperature - T	323, 423, 523K

mum temperature less than 523K (Miller, et al., 1979). The test is completed by unloading and cooling in the reverse order.

Approximately 80 triaxial creep tests under different combinations of stress and temperature conditions implied by Table 1 have been conducted on oil shale specimens of grades up to 60 GPT.

RESULTS AND DISCUSSION

Elastic Properties

Elastic properties obtained from the loading portions of the stress-strain curves indicate some definite trends in spite of significant scatter in experimental data. Figure 4 shows the variation of Young's modulus with grade for temperatures of 323K and 523K. Trends suggest that:

o Young's modulus decreases with increasing temperature.

o An upper bound to Young's modulus values can be described by a relationship of the form,

$$E = A + \frac{B}{W}$$

in which E is Young's modulus, W is grade and A and B are constants depending upon temperature. Poisson's ratio, although generally increases with temperature, seems to have no definite correlation with grade.

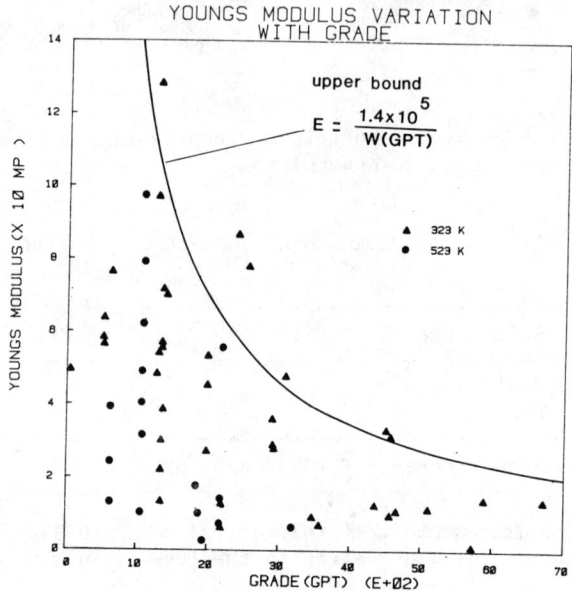

Figure 4. Variation of Young's Modulus With Grade.

Creep Characteristics

In general, classical creep behavior with primary and secondary creep phases is exhibited. In almost all of the tests, secondary creep phase started in less than 12 hours. Except for two of the tests in which the creep accelerated leading to failure in less than 6 hours, tertiary creep was not observed. This possibly was due to the short duration (24 to 48 hours) of the tests.

Figure 5 shows a typical set of creep curves for a high grade sample tested at low temperature in unconfined condition. In most cases the axial shortening was accompanied by radial extension resulting in volumetric dilatation. Some specimens exhibited creep compaction in which the sample volume (calculated on the basis of axial and radial strain measurements) decreased with time. Creep compaction generally is associated with high initial porosity and low deviatoric stress during creep.

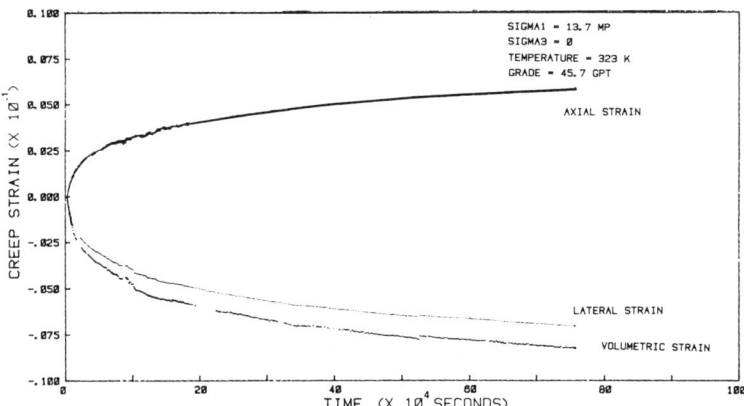

Figure 5. Creep Strain Curves for a High Grade Specimen Tested at Low Temperature in Unconfined Condition.

Figure 6 shows on a log-log plot selected points from the primary creep portion of creep curves for a few samples tested at 423K under different stress conditions. A linear regression to the points obtained for each individual test produces a power law relationship between the primary creep (ε_p) and time (t) given as

$$\varepsilon = at^b$$

The coefficients a and b along with the stress conditions for the tests represented in Figure 6 are shown in Table 2. Table 2 includes also the range of values for the coefficients a and b including the results for most of the tests conducted at all temperatures.

Figure 7 shows a composite log-log plot of secondary creep strain rate (ε_s) and deviatoric stress ($\sigma_1 - \sigma_3$) for all the tests. Most of the data points lie on this plot within a constant width band giving an upper and lower bound on steady-state creep strain rates. The resulting steady-state creep strain (ε_s) at time (t) therefore can be obtained by a relationship of the form

$$\varepsilon_s = C\,(\sigma_1 - \sigma_3)^D \cdot t$$

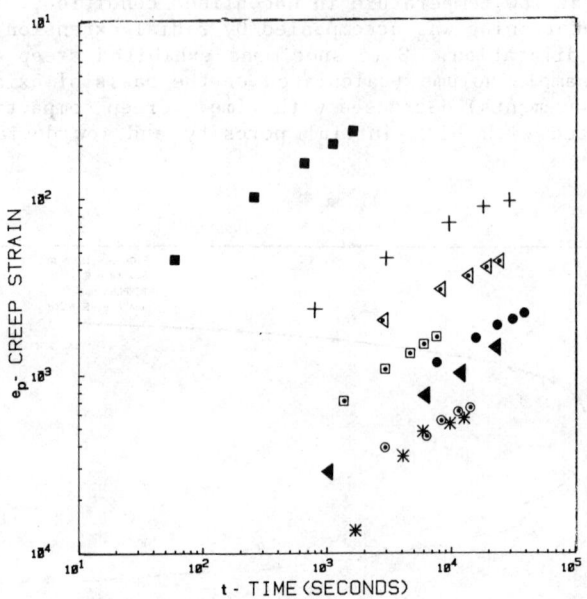

Figure 6. Primary Creep Correlation With Time for Specimens Tested at Temperature 323°K.

TABLE 2. Power Law Coefficients Describing Primary Creep for Tests Conducted at Temperature 323°K and Average Values for All the Tests

Tests at 323°K Represented in Figure 6	Grade (GPT)	$\sigma_1 - \sigma_3$ (MPa)	a	b
▫	21.8	13.79	0.486	2.10×10^{-5}
*	23.2	0.00	0.721	7.18×10^{-7}
□	34.7	4.48	0.499	5.99×10^{-4}
◁	45.7	6.89	0.519	7.90×10^{-6}
◀	45.7	13.79	0.352	1.19×10^{-4}
⊙	51.0	2.41	0.345	2.36×10^{-5}
●	57.3	9.31	0.387	3.65×10^{-5}
Range of values for most of the tests at all temperatures			0.329–0.519	1.19×10^{-4}–4.57×10^{-5}

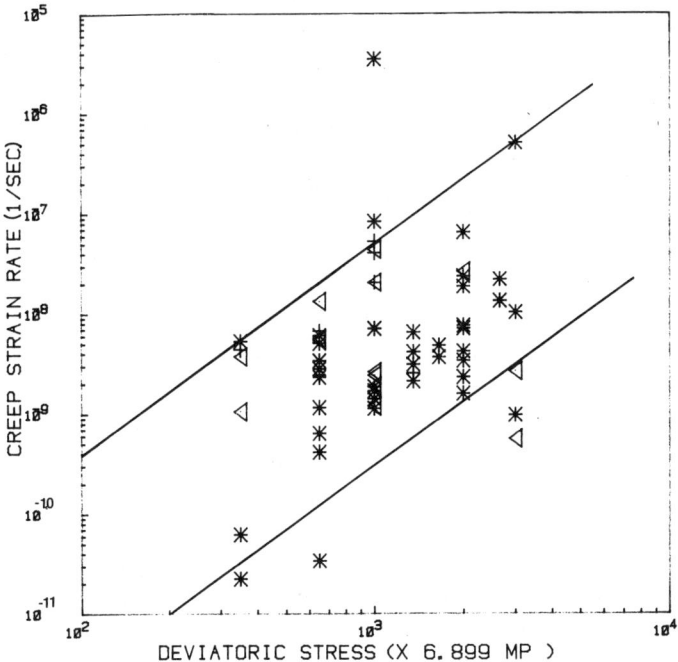

Figure 7. Steady-State Creep Correlation with Deviatoric Stress. $* = 323K$; $+ = 423K$; $\Delta = 523K$

With $(\sigma_1 - \sigma_3)$ measured in MPa and time in seconds, the constants C and D are given as

$D = 2.126$
$C = 3.798 \times 10^{-16}$ for upper bound
and 1.978×10^{-18} for lower bound

The total strain within the time frame of each test can therefore be described by a combination of thermal and rheological elements whose properties are derived from the coefficient of thermal expansion, Young's modulus and the creep coefficients. A general equation can be written as

$$\varepsilon = \alpha t + \frac{\sigma_o}{E} + \frac{\sigma_o}{\eta} t^b + \frac{\sigma_o D}{E_2} \cdot t$$

in which α is the thermal expansion coefficient, E is Young's modulus, σ_o is the stress difference and η and E_2 are derived from the creep coefficients a and C in the expression for primary and secondary creeps, respectively, described earlier.

The lateral strains have not been considered here for simplicity. Moreover, the lateral strain measurements, being localized, are more susceptible to variations due to inhomogeneous nature of oil shale. A suitable scaling of the data with resepct to stress ratio (σ_3/σ_1) in addition to the stress difference ($\sigma_3 - \sigma_1$) may provide a better model for the secondary creep strain rate. More complex and complete constitutive equations can be developed in terms of generalized stress and strain components and stress invariants. However, from the practical standpoint of pillar design, usefulness of any such model will be restricted.

REFERENCES

Agapito, J.F.T., 1972, "Pillar Design in Competent Bedded Formations," Ph.D. Thesis, Colorado School of Mines, Golden, CO, 195p.

Agapito, J.F.T. and Page, J.B., 1976, "A Case Study of Long-Term Stability in the Colony Oil Shale Mine, Piceance Creek Basin, Colorado," 17th U.S. Symposium Rock Mechanics, University of Utah, Salt Lake City, Utah, 138-143.

Chong, K.P., Smith, J.W. and Khaliki, B.A., 1978, "Creep and Relaxation of Oil Shale," Proc. 19th U.S. Rock Mechanics Symposium, University of Nevada, Reno, pp. 414-418.

Chu, M-S. and Chang, N-Y., 1980, "Uniaxial Creep of Oil Shale Under Elevated Temperatures," Presented at the 21st U.S. Rock Mechanics Symposium, University of Rolla, MO.

Horino, F.G., Dolinar, D.R. and Bickel, D.L., 1981, "Mechanical Properties, In-Situ Stress and Temperature Measurements - Logan Wash Oil Shale Site," Progress Report 10025 U.S.B.M. Denver Research Center.

Miller, R.J., Wung, F.D., Sladek, T. and Young, C., 1979, "The Effect of In-Situ Retorting on Oil Shale Pillars," Colorado School of Mines Contract No. H0262031 Interim Report, 194 p.

Olsson, W.A., 1980, "Stress-Relaxation in Oil Shale," Presented at the 21st U.S. Rock Mechanics Symposium, University of Rolla, MO.

Sellers, J.B., Haworth, G.R. and Zambas, P.G., 1972, "Rock Mechanics Research on Oil Shale Mining," Trans. of AIME, Soc. of Mng. Eng., 252, pp. 222-232.

Zambas, P.G., Howarth, G.R., Brackebush, F.W. and Sellars, J.B., 1972, "Large-Scale Experimentation in Oil Shale," Trans. Am. Inst. Min. Metall. and Pet. Eng., SME/AIME, 252, pp. 283-288.

Chapter 37

BEHAVIOUR OF A BRITTLE SANDSTONE
IN PLANE-STRAIN LOADING CONDITIONS

by Vassiliki G. Stavropoulou

Senior Research Officer, Chamber of Mines of South Africa
Research Organization, Johannesburg, R.S.A.

ABSTRACT

Although plane-strain is the preferred method of analysing many rock mecahnics problems, little is known about the behaviour of rock under this loading condition. A simple and inexpensive apparatus has been built for laboratory testing of rocks in plane-strain conditions. A thorough investigation of the specimen geometry and the friction end effects showed that the results can be considered as reliable.

The behaviour of a brittle sandstone in the simplest plane-strain condition ($\sigma_3 = 0$) is significantly different than that of the uni-axial compression ($\sigma_2 = \sigma_3 = 0$). The plane-strain values of the Young's Modulus and the Poisson's Ratio are decreased in comparison to the corresponding uniaxial values. Dilatancy is also reduced in plane-strain, whereas the pre-failure brittleness of the rock increases. The most dramatic effect of the plane-strain condition is that on the strength of the sandstone, which increases by 49 per cent in comparison to its uniaxial compression value.

INTRODUCTION

Plane-strain state in a rock mass is the state of strain in which all strain components normal to a certain plane are zero. This stage is often encountered in mining and civil engineering situations and in many more instances it can closely approximate the reality for design purposes. Its importance has been recognised by those involved in the theoretical analysis of engineering structures, with the result that almost every computer package incorporates a plane-strain analysis option. It is surprising that these theoretical developments have not been followed by any experimental work to assess the rock behaviour in plane-strain conditions. To the author's knowledge no plane-strain tests on solid rock have been reported yet.

This investigation attempts to assess the influence of the plane-strain loading condition on the linear elastic constants, on certain non-linear features, such as dilatancy, and finally on the strength of rock material. It is recognised that the plane-strain state is brought about by a suitable combination of the principal stresses, which generally are $\sigma_1 > \sigma_2 > \sigma_3$. If the rock behaviour were to be assessed in these general stress conditions, it would have involved "multiaxial" stress testing, which requires specially designed and expensive equipment with often questionable results because of ill-defined end conditions due to friction. Instead, it was decided to study at this stage only one simple plane-strain case, that of $\sigma_3 = 0$, and concentrate on the thorough investigation of the reliability of the test results.

A fine grain sandstone of the Karoo System has been selected as testing material. It is a fairly consistent rock with a uniaxial compressive strength 80,8 MPa.

THE PLANE-STRAIN TEST

Testing Procedure and Equipment

The plane-strain test described in this paper is essentially one of biaxial compression ($\sigma_3 = 0$) with the intermediate principal stress, σ_2, controlled in such a way as not to allow any deformation in its direction ($\varepsilon_2 = 0$). A configuration of this principle is shown in Figure 1. The implementation of this idea was achieved through a simple apparatus used in conjunction with the stiff compression machine of the Chamber of Mines Research Organisation. A perspective view of the testing set-up is shown in Figure 2. Its main design considerations were the application and checking of the plane-strain state, the elimination of all friction effects as far as possible and the accuracy in recording stresses and strains.

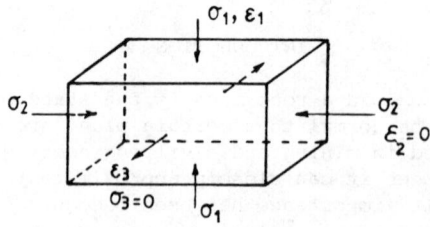

Fig. 1. Stress and strain configuration in the plane-strain test

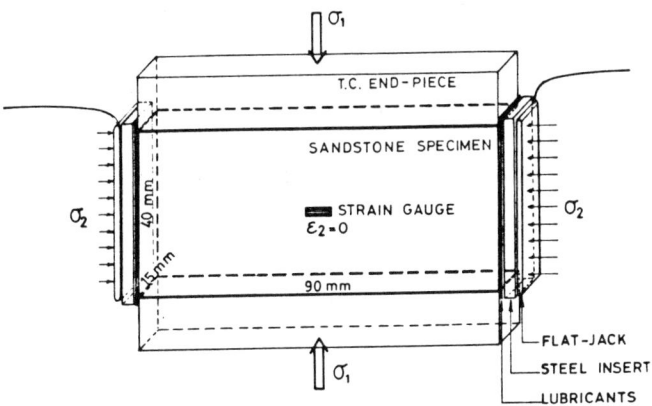

Fig. 2. The plane-strain testing set-up

The major principal stress, σ_1, is applied by the loading piston of the stiff testing machine. A pair of tungsten carbide end-pieces is used between the rock specimen and the platens of the machine in order to eliminate the friction effects as far as possible. The assembled set of the specimen end-pieces is placed in a solid rigid steel frame. The intermediate stress, σ_2, is applied through a pair of hydraulic flat jacks which are inserted between the specimen and the steel frame.

The testing procedure is very simple. While the specimen is loaded, the strain ε_2 is watched through a strain indicator. Simultaneously the stress σ_2 is increased at such a rate as to maintain $\varepsilon_2 = 0$ at all times. The σ_2 pressure control is presently manual, but servocontrol could easily be implemented with a few minor modifications of the equipment. The strain ε_2, which should be kept zero throughout the test, is checked by two strain gauges cemented on the specimen surface, as shown in Figure 2. The accuracy of the strain readings is 1×10^{-6} mm/mm. The lateral strain, ε_3, is measured by a caliper fitted with a set of strain gauges.

Specimen Size and Friction End Effects

The specimens are rectangular prisms 90 mm x 40 mm x 15 mm, as shown in Figure 2. These dimensions have finally been accepted after a series of studies involving various different combinations of dimensions. In order to minimize the friction effects of the end-pieces the breadth-to-height ratio was kept close to 3, as is the standard practice in the compression of cylindrical specimens.

A major problem of the plane-strain tests has been the development of friction between the flat jacks and the specimen side surfaces caused by the deformation in the direction of σ_1. This frictional force in fact generated a false increase of the failure stress.

Various lubricants have been used in order to eliminate these friction effects and the corresponding results of the peak strength are summarized in Table 1.

TABLE 1. Plane-strain Strength of Sandstone for Different Side-end Conditions

Mean Strength MPa	Standard Deviation %	Side Lubricant	Number of Specimens
118,1	3	Teflon	3
120,5	3	Teflon & Copper	9
128,5	5	No Lubricants	6

The use of lubricants appears to decrease considerably the side friction effects. When only teflon was used, it tended to extrude into the specimen and cause an erratic behaviour, although the ultimate strength was very similar to that resulting from the other lubricants. For this reason a combination of teflon and copper shims has been finally adopted.

Thorough investigation has shown a uniform distribution of stresses and a very satisfactory plane-strain state throughout the entire length of the specimen. A small discrepancy in the ε_2 strains has only been observed between the two opposite surfaces, due most probably to slight geometric irregularities. However, this never exceeded the magnitude of 100×10^{-6} mm/mm, which is virtually negligible. In order to eliminate this effect the specimens were machined to the exact dimensions with particular care being taken on the parallelity of the surfaces. The accepted tolerance limit was 0,02 mm between the two surfaces, on which σ_2 is applied.

EXPERIMENTAL RESULTS

The mean stress-strain curves of all the tests, which were free of side-end effects are presented in Figures 3 and 4.

The effects of the plane-strain loading are more lucidly assessed through a comparison with the uniaxial compression test results. The difference in the stress state of the two tests is in the intermediate principal stress, σ_2, which is zero in the uniaxial test, but non-zero in the plane-strain test. In both cases it is $\sigma_3 = 0$.

Linear Elastic Region

If the sandstone behaviour were assumed to be linear elastic, the following equations would be valid:

Uniaxial compression: $\sigma_1 = E \cdot \varepsilon_1$ (1)

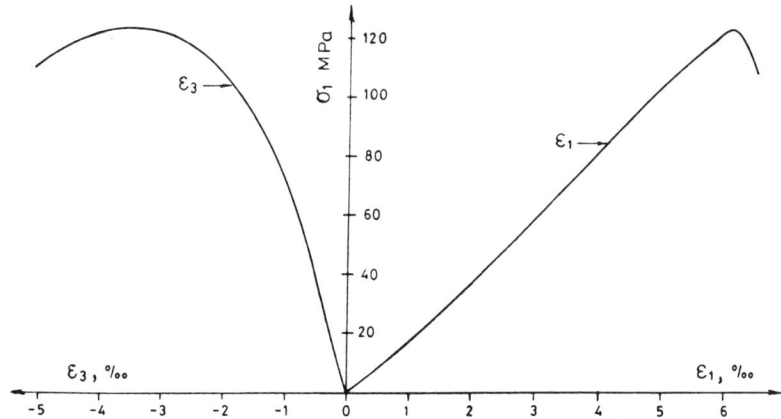

Fig. 3. Stress-strain curves of sandstone in plane-strain

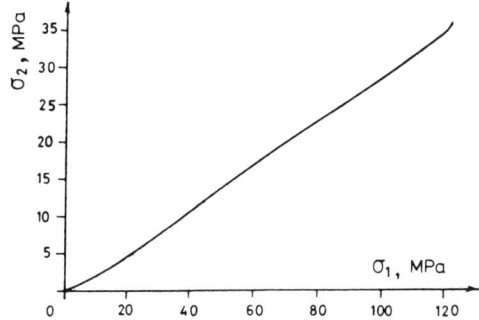

Fig. 4. Variation of σ_2 with σ_1 during plane-strain loading

Plane-strain:
$$\sigma_1 = \frac{1}{1-\nu^2} \cdot E \cdot \varepsilon_1 \quad (2)$$

and $\quad \nu = \dfrac{\sigma_2}{\sigma_1} \quad (3)$

According to equation (3) the slope of the $(\sigma_1 - \sigma_2)$ curve of Figure 4 gives the value of Poisson's Ratio. In this case it is measured $\nu = 0,30$, whereas the corresponding uniaxial value is $\nu = 0,34$, which means a 12 per cent reduction.

In order to assess the influence of the plane-strain condition on Young's Modulus, the $(\sigma_1 - \varepsilon_1)$ curve of Figure 3 was re-plotted, after substracting the Poisson's effect on σ_1, according to the equation (2), and it is shown in Figure 5 together with the uniaxial $\sigma_1 - \varepsilon_1$ curve. From the comparison of these two curves it is concluded that the Young's Modulus is reduced under plane-strain conditions. This observation seems to be even tenuously supported by Mogi's multiaxial stress-strain curves on Dunham dolomite (1979).

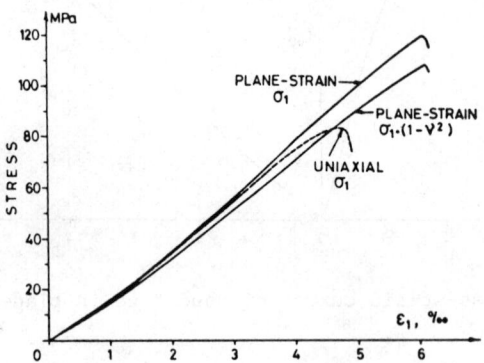

Fig. 5. Comparative stress-strain curves of sandstone in plane-strain and uniaxial compression

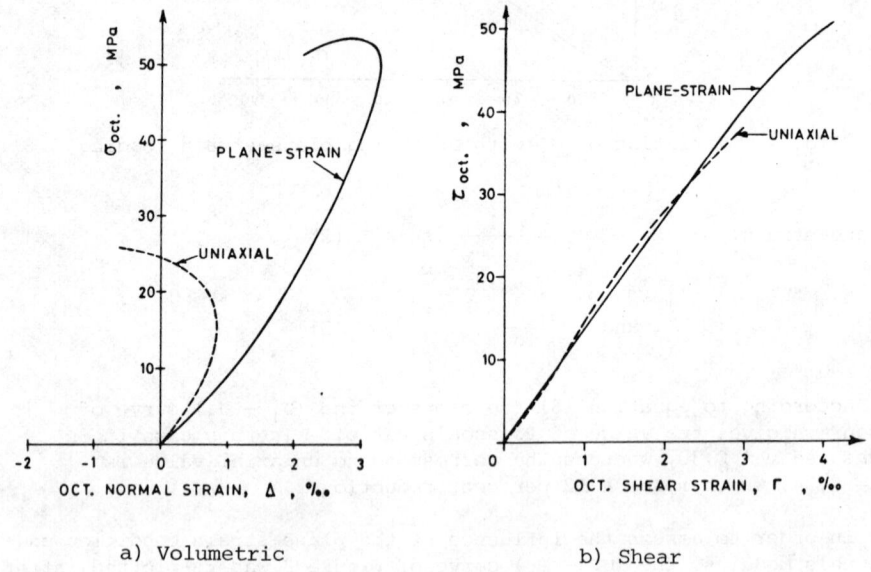

a) Volumetric   b) Shear

Fig. 6. Octahedral stress-strain curves of sandstone

A better correlation of the stress state with the bulk (volumetric) and the shear (deviatoric) rock behaviour is achieved when the normal and the shear components of the stress and the strain vectors on the octahedral plane are considered. Thus, from the original data of Figures 3 and 4, Figure 6 was derived. The slope of the linear region of these curves is directly proportional to the value of Bulk Modulus, K, and Shear Modulus, G. A comparison of the uniaxial and the plane-strain curves of Figures 6 a) and b) indicate a clear decrease of K in the plane-strain and a slight almost negligible decrease of G.

Non-Linear Region

The plane-strain state generally appears to suppress the non-linear behaviour of sandstone. As can be seen in Figures 3 and 5, the plane-strain ($\sigma_1 - \varepsilon_1$) curve deviates from linearity at failure by 3 per cent, whereas the uniaxial curve deviates by 7 per cent. This is in qualitative agreement with Mogi's results (1979) on the "ductility" of Dunham dolomite being decreased with increasing σ_2.

The ($\sigma_1 - \sigma_2$) curve of Figure 4 shows a constant slope almost until failure. This implies a constant value of Poisson's Ratio, whereas in the uniaxial compression the value of ν starts increasing at a stress level approximately 40 per cent of the failure stress. The explanation of this phenomenon lies with the mere presence of σ_2, which prevents the development of open cracks in its direction. The rock appears to behave as a solid material on the ($\sigma_1 - \sigma_2$) plane, allowing dilatancy only to take place in the unconfined direction.

The strong adverse effect, which the plane-strain state has on dilatancy, is clearly seen in Figure 6 a). Finn et al. (1967) studied theoretically and experimentally the behaviour of sand in plane-strain and triaxial conditions for the same σ_3 and arrived at the same conclusion.

Peak Strength

The plane-strain loading condition appears to cause a dramatic increase of the strength of sandstone by as much as 49 per cent in comparison to its uniaxial compression value, as can be seen in Figure 5. This effect could be attributed to the influence of the intermediate principal stress, σ_2, an observation already made by numerous researchers on results of multiaxial and biaxial tests. The biaxial strength of Dunham dolomite is, for example, 40 per cent higher than its corresponding uniaxial strength (Mogi). Masó and Lerau (1980) also reported a 38 per cent increase in the uniaxial strength of a sandstone when subjected to biaxial compression.

CONCLUSIONS

The behaviour of a brittle sandstone in plane-strain loading conditions with σ_3 being zero has been studied through a simple testing apparatus. A thorough investigation of the testing condition showed that the results are reliable.

The plane-strain state causes a 49 per cent increase of the failure strength in comparison to its corresponding uniaxial value and also a measurable decrease of the values of the elastic constants. A clear adverse effect of the plane-strain condition has been noticed on dilatancy and generally on the non-linear rock behaviour.

These observations have far-reaching effects in practical design applications. A linear elastic analysis can be carried out with a higher degree of confidence limit in a plane-strain domain than in a general stress state. A simultaneous readjustment of the elastic constants and the failure envelope (especially if it is of the Mohr-Coulamb type) derived from standard uniaxial and triaxial tests should also enhance the results of a linear elastic analysis.

ACKNOWLEDGEMENTS

The work reported in this paper is part of the research project CO1M10 of the Mining Operations Laboratory of the Chamber of Mines of South Africa Research Organisation. It also forms part of a Ph.D. thesis due to be submitted by the author to the University of the Witwatersrand, Johannesburg.

The assistance provided by the Engineering Branch in the manufacturing of the plane-strain apparatus is acknowledged with appreciation.

REFERENCES

Finn, W.D.L., Wade, N.H. and Lee, K.L., 1967, "Volume changes in triaxial and plane-strain tests". Proceedings, ASCE, Soil Mech. Div., SM6, pp. 297-308.

Masó, J.C. and Lerau, J., 1980, "Mechanical behaviour of Darney sandstone (Vosges, France) in biaxial compression". Int. J. Rock Mech. Min. Sci., Vol. 17, No. 2, pp. 109-115.

Mogi, K., 1979, "Flow and fracture of rocks under general triaxial compression". Proceedings, 20th U.S. Symposium on Rock Mechanics, University of Texas at Austin.

TECTONOPHYSICS

Chairman
Hans-Rudolf Wenk
University of California
Berkeley, California

Co-Chairman
Hugh C. Heard
Lawrence Livermore National Laboratory
Livermore, California

Keynote Speaker
Richard H. Sibson
University of California
Santa Barbara, California

Chapter 38

SEISMIC STYLE IN RELATION TO HEAT FLOW
ALONG THE SAN ANDREAS FAULT SYSTEM

Richard H. Sibson

Department of Geological Sciences
University of California
Santa Barbara
California 93106

ABSTRACT

Changes in seismic style along the San Andreas fault system appear to correlate with strike-parallel variations in regional heat flow. Large earthquake ruptures are associated with colder segments of the fault zone, while comparatively high levels of microearthquake activity and small to moderate shocks occur along the hotter segments, often in association with aseismic creep. Such behavior is consistent with geotherm and strain rate dependent profiles of shear resistance versus depth, constructed for quartz-bearing rocks. These profiles suggest that the transition from frictional to quasi-plastic behavior with depth should increase by a few kilometers passing from the hotter to the colder portions of the fault zone, giving rise to long-wavelength concentrations of distortional strain energy near the base of the seismogenic zone in the cold segments. Some support for this hypothesis comes from the observed deepening of microseismic activity passing northwest and southeast from the central Californian active region to the currently locked, ~400 km segments of the San Andreas fault that ruptured in 1906 and 1857.

INTRODUCTION

Observations gathered over the past 150 years or so suggest that alternating segments of the San Andreas fault system persistently exhibit different types of fault activity or seismic style (Fig. 1), and it seems probable that these different behavioral modes are long term characteristics of the fault system (Allen, 1968, 1981; Wallace, 1970; Scholz, 1977). The two ~400 km long segments that ruptured in the great (M~8) earthquakes of northern California (1906) and the "big bend" region (1857) now appear to be fully locked with rather low levels of microseismic activity and few moderate shocks. In contrast, segments in central and southern California are characterised by comparatively frequent small to moderate ruptures (3<M<7) and a high level of microseismic activity, sometimes accompanied by aseismic slip (fault creep). Aseismic slip rates reach 32 mm/yr along one 55 km section of the central California segment where geodetic studies indicate large scale rigid block motion without significant accumulation of elastic strain (Thatcher, 1979; Burford and Harsh, 1980). The complex activity northwest of the triple junction at Cape Mendocino results from transform faulting along the Mendocino Fracture Zone combined with the subduction and internal disruption of the Gorda plate (Smith and Knapp, 1980), and will not be discussed further.

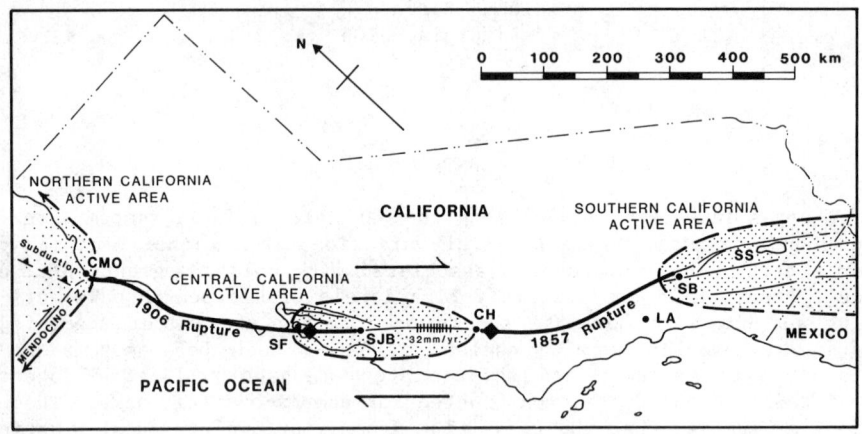

Fig. 1. Sketch map showing variations in seismic style along the San Andreas fault system, California (modified from Allen, 1968). Cross-barring indicates region of fast steady creep (~32 mm/yr); solid diamonds indicate possible epicenters for the 1857 and 1906 ruptures (CMO - Cape Mendocino, SF - San Francisco, SJB - San Juan Bautista, CH - Cholame, LA - Los Angeles, SB - San Bernardino, SS - Salton Sea).

Several explanations have been put forward to account for these variations in fault behavior. On geometrical grounds, Allen (1968) noted that both M8 ruptures were associated with "restraining" bends where the San Andreas fault trace lies oblique to the interplate slip vector (see also Scholz, 1977). Localisation of slip along the central San Andreas fault from San Juan Bautista to Cholame, and along the Hayward and Calaveras faults east of San Francisco Bay, has also been attributed to the intersection of these fault zones with serpentinites on the Coast Range thrust at the base of the Great Valley Sequence, a situation which may also induce high fluid pressures along the faults (Irwin and Barnes, 1975). This paper explores in some detail a further possibility, suggested by Sibson (1982), that the changes in seismic style may be related to variations in regional heat flow along strike.

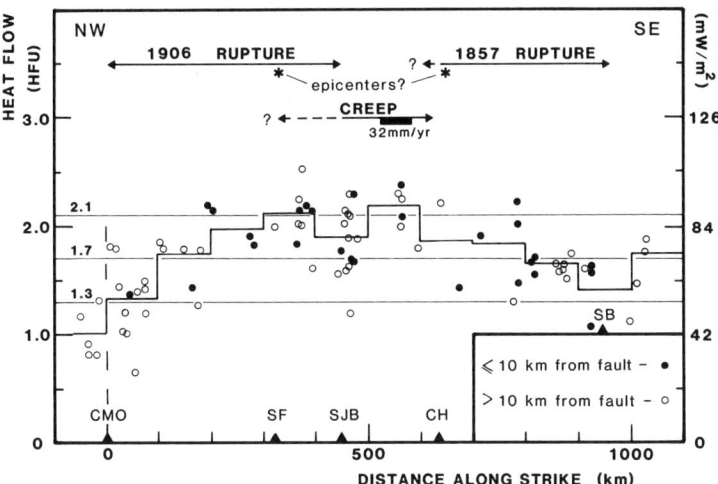

Fig. 2. Strike-parallel profile of heat flow measurements in relation to seismic style along the San Andreas fault (after Lachenbruch and Sass, 1980). Histogram plots the mean value of heat flow for each 100 km segment of the fault trace (see Fig. 1 for location key).

HEAT FLOW MEASUREMENTS

Measurements of heat flow in the dominantly strike-slip regime close to the San Andreas fault have been compiled by Lachenbruch and Sass (1980) into a strike-parallel profile which exhibits a broad maximum coincident with the central California segment of intense microearthquake activity and aseismic slip (Fig. 2). Comparatively few heat flow determinations are available for the southern active region southeast of the 1857 rupture, but aseismic slip and a high level of microseismic activity are also known in the region of high heat flow and geothermal activity southeast of the Salton Sea (Goulty et al. 1978). For the area covered in Fig. 2, the mean heat flow along the currently locked fault segments that ruptured in 1906 and 1857 is about 1.7 HFU (71 mW/m^2) while the average over the central California active region is about 2.0 HFU (84 mW/m^2), the mean value per 100 km segment peaking at over 2.1 HFU (88 mW/m^2) in the vicinity of the 55 km zone of most rapid, steady creep described by Burford and Harsh (1980). It is also interesting to note that the probable epicenters (rupture nucleation sites) for the 1906 and 1857 earthquakes (Boore, 1977; Sieh, 1978) lie near the ends of the high heat flow region associated with the creeping segment.

It has to be borne in mind that measurement and interpretation of conductive heat flow in tectonically active terranes, in particular estimation of geotherms, is fraught with difficulties (Oxburgh, 1980). However, while no simple correlation exists between measured heat flow values around the San Andreas fault and near-surface radioactive heat production, Lachenbruch and Sass (1980) note that the data are broadly comparable to those obtained from the Basin and Range province. On this basis, three geotherms estimated by Lachenbruch and Sass (1977) for the spread of Basin and Range heat flow values (BR1 - 2.1 HFU, BR2 - 1.7 HFU), and for the low heat flow characterising the stable cratonic crust of the eastern United States (SC - 1.3 HFU), are given in Fig. 3 and used to construct the shear resistance profiles below.

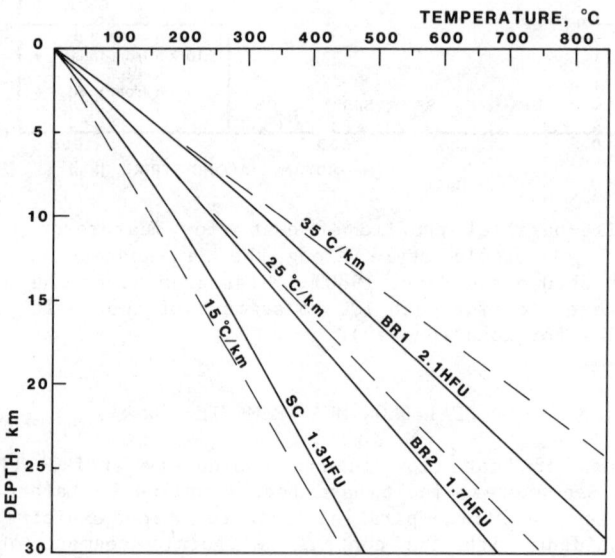

Fig. 3. Continental geotherms in the United States (after Lachenbruch and Sass, 1977).

FAULT RHEOLOGY AND SHEAR RESISTANCE PROFILES

Consideration of the varying rock deformation textures found in ancient continental fault zones, in relation to faulting style and metamorphic environment, has led to a general conceptual model for these structures (Sibson, 1977, 1982). A seismogenic frictional regime, dominated by discontinuous pressure-sensitive deformation involving cataclasis and frictional sliding, gives way with depth to a quasi-plastic regime where largely aseismic, continuous shearing is localised in mylonite belts developed under greenschist and greater grades of metamorphism. In quartzo-feldspathic crust, the changing response of quartz to deviatoric stress with increasing temperature appears to be the prime factor controlling the passage from frictional to quasi-plastic fault behavior, intracrystalline plasticity involving dislocation glide and climb becoming the dominant deformation mechanism at temperatures greater than 300 ± 50°C (Voll, 1976; Tullis and Yund, 1977). Deformation mechanisms involving diffusive mass transfer (in particular, pressure solution) are also believed to operate around this critical transition region over the temperature range 200-400°C (McClay, 1977), but their effect is difficult to assess.

This model has been quantified using the procedure employed by Sibson (1982) to provide a range of shear resistance profiles for the geotherms and strain rates appropriate to the San Andreas fault system (Fig. 4).

Frictional regime: The straight lines represent the minimum shear stresses needed to induce failure on existing thrust (T), strike-slip (S) and normal (N) faults through the frictional regime under hydrostatic fluid pressures, for a static frictional coefficient:

$$\mu = \tau/\sigma_n' = \tau/(\sigma_n - P) = 0.75 \qquad (1)$$

an adequate approximation to Byerlee's (1978) composite frictional criterion for rocks (τ and σ_n are respectively the shear and normal stresses acting on the fault, and P is the fluid pressure). The failure condition for strike-slip is plotted on the assumption that $\sigma_2 = (\sigma_1 + \sigma_3)/2$, where the three principal compressive stresses are $\sigma_1 > \sigma_2 > \sigma_3$. As $\sigma_2 \to \sigma_1$ or σ_3, the strike-slip criterion moves towards the failure lines for normal and thrust faulting respectively. Increasing fluid pressure above hydrostatic causes all failure lines to rotate towards the depth axis.

Quasi-plastic regime: Dislocation creep in quartz is assumed dominant during quasi-plastic mylonitization, so that the constitutive flow law is of the form:

$$\dot{\gamma} = A (\sigma_1 - \sigma_3)^n \exp(-Q/RT) \qquad (2)$$

where $\dot{\gamma}$ is strain rate, R is the gas constant, T is absolute temperature, and A, n and Q are material constants. Experimental deformation of nominally 'dry' Westerly Granite (average grain size

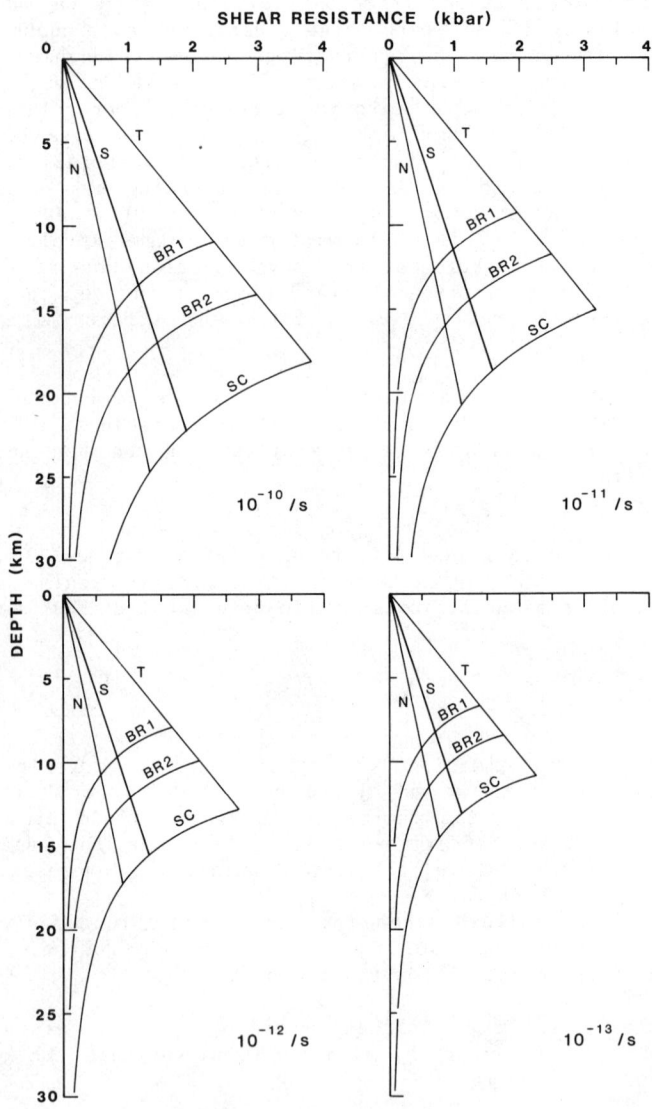

Fig. 4. Profiles of shear resistance versus depth for the three geotherms, at different shear strain rates in the quasi-plastic regime. Fluid pressure assumed hydrostatic in the frictional regime (0.36 x overburden pressure for an average crustal density of 2.8 g/cm^3).

~0.75 mm; composition about 30% each of quartz, oligoclase and microcline, plus 10% biotite and accessories) at high temperatures and pressures has yielded A = 8.83 x 10^{-4} $kbar^{-n}$ s^{-1}, n = 2.9, and Q = 25.3 kcal/mole for steady state creep, microstructural studies showing that quartz has accommodated most of the strain by crystal plastic flow combined with some microfracturing (Carter et al. 1981). Comparable flow laws have been obtained for 'wet' and 'dry' quartzite (Koch et al. 1980).

While caution must be exercised in extrapolating such flow laws to the natural environment (Tullis and Yund, 1980), the capacity for ready flow of quartz-bearing rocks under moderate to high grade metamorphic conditions at shear stresses below 1 kbar (as suggested by the plots in Fig. 4) is in reasonable accord both with field evidence for the onset of quartz plasticity at around 300°C, and with palaeostress estimates of 0.1 - 1 kbar obtained from quartz-bearing tectonites deformed in greenschist and amphibolite facies metamorphic environments (Kohlstedt and Weathers, 1980).

Flow shear resistance ($\tau = (\sigma_1-\sigma_3)/2$) calculated from this 'granitic' flow law is therefore believed appropriate to the mylonitization of quartz-bearing rocks and has been plotted for a range of strain rates, and for the three geotherms in Fig. 3, to provide the curved portions of the shear resistance profiles in Fig. 4.

Peak shear resistance and strain energy concentration: The rather sharp peak in shear resistance inferred for the transition from frictional to quasi-plastic behavior is unrealistic. The transition must in fact accommodate a mixture of brittle and ductile deformation, with pressure solution and allied deformation mechanisms acting to smooth out the peak. Nonetheless, the highest stress levels at failure are to be expected in this transition region. Since distortional strain energy density varies as the square of shear stress, a very pronounced concentration of strain energy should also develop around the frictional/quasi-plastic transition.

APPLICATION TO THE SAN ANDREAS FAULT SYSTEM

The need for a quartz-bearing middle and lower crust places lithological constraints on the applicability of these shear resistance profiles. However, Lin and Wang (1980) have interpreted the velocity structure along one side of the San Andreas fault in central and northern California in terms of a granitoid complex, with a significant quartz content, extending throughout the crust. Increasing temperature in the deep continental crust may also counteract diminishing quartz content by promoting plasticity of other important rock constituents such as feldspar.

Choice of strain rates: Shear strain rates, determined geodetically in the vicinity of the San Andreas fault and thought largely to represent the accumulation of elastic strain, typically lie between

10^{-13} and 10^{-14}/s (Prescott et al. 1981). However, rather higher localised shear strain rates may be inferred for quasi-plastic shear zones at depth. At the surface, the San Andreas fault zone is generally about 1 km wide (Allen, 1981), perhaps reflecting the width of a deep mylonite belt. An overall slip-rate of 30 mm/yr therefore yields a deep localised strain rate of 10^{-12}/s. However, seemingly continuous shear strain rates of around 10^{-11}/s have been obtained directly from surface measurements along the creeping segment (Burford and Harsh, 1980), so that a range from 10^{-11} to 10^{-12}/s seems the most likely.

Taking account of the heat flow data, the most appropriate shear resistance profiles for the San Andreas fault are therefore those for geotherms BR1 and BR2, at strain rates of 10^{-11} and 10^{-12}/s (Fig. 4). At constant strain rate, accepting the strike-slip frictional failure line as shown and neglecting the probable smoothing-out of the peak shear resistance, the depth of the frictional/quasi-plastic transition should increase by 2-3 km passing along strike from the hottest (2.1 HFU) to the colder (1.7 HFU) portions of the fault zone. With the same provisos, the peak values of shear resistance and distortional strain energy density would increase by factors of about 1.3 and 1.7 respectively. Note further that higher fluid pressures along the creeping segment should deepen the frictional/quasi-plastic transition according to our model.

MICROEARTHQUAKE DEPTH DISTRIBUTIONS

Shallow crustal earthquakes are thought to arise chiefly through frictional stick-slip (Brace and Byerlee, 1966), so that the depth of microseismic activity serves as a guide to the extent of the frictional regime. Sibson (1982) demonstrated that in continental terrane, the base of the seismogenic zone so defined could be modeled fairly satisfactorily as the frictional/quasi-plastic transition for quartz-bearing crust. Larger (M>5.5) shocks characteristically nucleate towards the base of the seismogenic zone, in the regions inferred to be capable of accumulating the highest strain energy concentrations.

Microearthquake activity along the hot creeping segment of the San Andreas fault is largely confined to the top 10 km of the crust (Wesson et al. 1973) in good accord with the modeled frictional/quasi-plastic transition for the BR1 geotherm at 10^{-12}/s. However, at both ends of the creeping segment in the vicinity of San Juan Bautista and Parkfield (~30 km northwest of Cholame), activity deepens fairly abruptly to around 15 km (Moths et al. 1981), somewhat more than predicted. Seismic activity is generally low along the locked 1906 rupture trace but M5 and M4.4 events have also been located at depths of 12-15 km near Corralitos and Pacifica in San Francisco Peninsula (McEvilly, 1966; Urhammer, 1981). In the Transverse Ranges adjacent to the 1857 rupture trace (Hanks and Brune, 1970), seismic activity locally extends to nearly 20 km depth, possibly because the geothermal gradient has been depressed by thrust stacking (Sibson, 1982). Observed depth distributions are therefore broadly consistent with the inference from the shear resistance profiles that the 1906 and 1857 ruptures are associated with deep cold sticking patches, capable of accumulating much higher strain energy concentrations before failure than the high heat flow region of central California.

DISCUSSION

Given the very considerable uncertainties involved in the estimation of geotherms and the construction of shear resistance profiles on the assumption of crustal homogeneity, the case for thermal control of faulting style along the San Andreas fault is far from proved. However, the depth distribution of earthquakes is in broad accord with the hypothesis. Thus while other factors (lithologic, geometric, etc.) may contribute to changes in seismic style, the probability of long-wavelength strain energy concentrations arising from variations in heat flow is hard to ignore.

REFERENCES

Allen, C.R., 1968, "The Tectonic Environments of Seismically Active and Inactive Areas along the San Andreas Fault System", Stanford Univ. Publ. Geol. Sci. 11, 70-82.

1981, "The Modern San Andreas Fault", in, The Geotectonic Development of California, W.G. Ernst, ed., Prentice Hall, N.J., 511-534.

Boore, D.M., 1977, "Strong Motion Recordings of the California Earthquake of April 18, 1906", Bull. Seism. Soc. Am. 67, 561-577.

Brace, W.F., and Byerlee, J.D., 1966, "Stick-slip as a Mechanism for Earthquakes", Science 153, 990-992.

Burford, R.O., and Harsh, P.W., 1980, "Slip on the San Andreas Fault in Central California from Alinement Array Surveys", Bull. Seism. Soc. Am. 70, 1233-1261.

Byerlee, J.D., 1978, "Friction of Rocks", Pure Appl. Geophys. 116, 615-626.

Carter, N.L., et al., 1981, "Creep and Creep Rupture of Granitic Rocks", Am. Geophys. Union Mon. 24, 61-82.

Goulty, N.R., et al., 1978, "Large Creep Events on the Imperial Fault, California", Bull. Seism. Soc. Am. 68, 517-521.

Hanks, T.C., and Brune, J.N., 1970, "Seismicity of the San Gorgonio Pass", (abstract), EOS, Trans. Am. Geophys. Union, 51, 352.

Irwin, W.P., and Barnes, I., 1975, "Effect of Geologic Structure and Metamorphic Fluids on Seismic Behavior of the San Andreas Fault System in Northern and Central California", Geology 3, 713-716.

Koch, P.S., Christie, J.M., and George, R.P., 1980, "Flow Law of Wet Quartzite in the α-Quartz Transition Field", (abstract), EOS, Trans. Am. Geophys. Union 61, 376.

Kohlstedt, D.L., and Weathers, M.S., 1980, "Deformation-Induced Microstructures, Palaeopiezometers and Differential Stresses in Deeply Eroded Fault Zones", J. Geophys. Res. 85, 6269-6285.

Lachenbruch, A.H., and Sass, J.H., 1977, "Heat Flow in the United States and the Thermal Regime of the Crust", Am. Geophys. Union Mon. 20, 626-675.

1980, "Heat Flow and Energetics of the San Andreas Fault Zone", J. Geophys. Res. 85, 6185-6222.

McClay, K., 1977, "Pressure Solution and Coble Creep in Rocks and Minerals: a Review", J. Geol. Soc. Lond. 134, 57-70.

McEvilly, T.V., 1966, "The Earthquake Sequence of November 1964 near Corralitos, California", Bull. Seism. Soc. Am. 56, 755-773.

Moths, B.L., et al., 1981, "Comparison between the Seismicity of the San Juan Bautista and Parkfield Regions, California", (abstract), EOS, Trans. Am. Geophys. Union, 62, 958.

Oxburgh, E.R., 1980, "Heat Flow and Magma Genesis", in, Physics of Magmatic Processes, R.B. Hargraves, ed., Princeton Univ. Press, Princeton, N.J., 161-199.

Prescott, W.H., Lisowski, M., and Savage, J.C., 1981, "Geodetic Measurement of Crustal Deformation on the San Andreas, Hayward, and Calaveras Faults near San Francisco, California", J. Geophys. Res. 86, 10853-10869.

Scholz, C.H., 1977, "Transform Fault Systems of California and New Zealand: Similarities in their Tectonic and Seismic Styles", J. Geol. Soc. Lond. 133, 215-230.

Sibson, R.H., 1977, "Fault Rocks and Fault Mechanisms", J. Geol. Soc. Lond. 133, 191-213.

1982, "Fault Zone Models, Heat Flow, and the Depth Distribution of Earthquakes in the Continental Crust of the United States", Bull. Seism. Soc. Am. 72, 151-163.

Sieh, K.A., 1978, "Central California Foreshocks of the Great 1857 Earthquake", Bull. Seism. Soc. Am. 68, 1731-1749.

Smith, S.W., and Knapp, J.S., 1980, "The Northern Termination of the San Andreas Fault", Cal. Div. Min. Geol. Spec. Rept. 140, 153-164.

Thatcher, W., 1979, "Systematic Inversion of Geodetic Data in Central California", J. Geophys. Res. 84, 2283-2295.

Tullis, J., and Yurd, R.A., 1977, "Experimental Deformation of Dry Westerly Granite", J. Geophys. Res. 82, 5705-5718.

1980, "Hydrolitic Weakening of Experimentally Deformed Westerly Granite and Hale Albite Rock", J. Struct. Geol. 2, 439-451.

Urhammer, R.A., 1981, "The Pacifica Earthquake of 28 April 1979", Bull. Seism. Soc. Am. 71, 1161-1172.

Voll, G., 1976, "Recrystallization of Quartz, Biotite and Feldspars from Erstfeld to the Levantina Nappe, Swiss Alps, and its Geological Significance", Schweiz. Miner. Petrogr. Mitt. 56, 641-647.

Wallace, R.E., 1970, "Earthquake Recurrence Intervals on the San Andreas Fault", Bull. Geol. Soc. Am. 81, 2875-2890.

Wesson, R.L., Burford, R.O., and Ellsworth, W.L., 1973, "Relationship between Seismicity, Fault Creep, and Crustal Loading along the Central San Andreas Fault", Stanford Univ. Publ. Geol. Sci. 13, 303-321.

Chapter 39

EXPERIMENTAL STUDY OF CATACLASTIC DEFORMATION OF A QUARTZITE.

by J.Hadizadeh* and E.H.Rutter

Geology Dept., Imperial College, London S.W.7., U.K.
(* Present address; Geology Dept., Garyounis University, Benghazi, Libya)

Abstract

The development of microcracking with progressive strain through failure in an orthoquartzite has been studied by means of optical and electron microscopy of specimens deformed experimentally at 200 MPa confining pressure, 20°C. Microcrack maps and statistics have been prepared.

The rock fails through the development of arrays of microcracks which form around grain boundaries and are initiated at notches on pore spaces. No special microstructural characteristic appears to be associated with the peak strength, and axially oriented cracks do not begin to develop until well past the strength peak. Axial microcracking is a precursor to shear fault localization.

Because most cracks form in highly shear stressed orientations, it is inferred that most of the work done in rupturing the rock is against the frictional forces on crack surfaces. An attempt is made to integrate the work done in this way, to compare it with the work done by the external forces.

Introduction

Despite a great deal of research over several decades, there is still no satisfactory treatment of the brittle failure of rocks, which adequately takes into account the microstructural changes which accompany failure. This is partly due to a lack of knowledge regarding the microstructural development through failure, and how it depends on the initial microstructural characteristics of the rock. This study was undertaken to determine the microstructural changes which occur in a pure orthoquartzite as it is loaded through strength failure at 200 MPa confining pressure, 20°C.

CATACLASTIC DEFORMATION OF QUARTZITE

Experimental

The rock used for these experiments was Oughtibridge Ganister, a mineralogically and texturally mature orthoquartzite (98% quartz) from the British Coal Measures. The grain size is 110 μm and the total porosity is 8%. Sub-spherical quartz grains are cemented by quartz overgrowths, which upon impingement enclose cuspate pores (fig.2.), within which the overgrowths are terminated by crystal faces. Triaxial experiments were performed on cylinders 6 mm diameter by 15 mm long of oven dried rock in a testing machine as described by Rutter (1972), at 20°C and at a strain rate of 10^{-4} sec^{-1}. Microcrack development was observed from samples loaded to various permanent strains both pre and post strength failure (fig.1.). Post failure, the load on this material falls in a stepwise fashion, so that it is often possible to offload the specimen manually from stable points in the post failure regime.

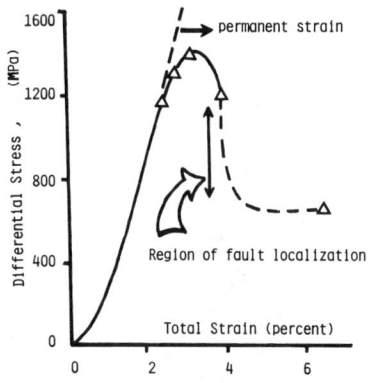

Fig. 1. Stress/strain curve for Oughtibridge Ganister, showing points at which samples were obtained for microstructural study.

Crack maps and microcrack statistics have been prepared from microscopic observations (figs. 2,3,4.), made using reflected light interference contrast microscopy, plus scanning electron microscopy of both polished and ion beam etched samples. Although ion etching enhanced crack visibility, it also confused them in the case of grain boundary microcracking, because the grain boundaries etched whether they were cracked or not. Compared with many rocks, microcracking of Oughtibridge Ganister was relatively well behaved, being uniformly distributed and of relatively low density, even after the strength peak. Through the use of the various microscopic techniques mentioned above, it is thought that most existing cracks are accounted for to within a factor of 2 or 3.

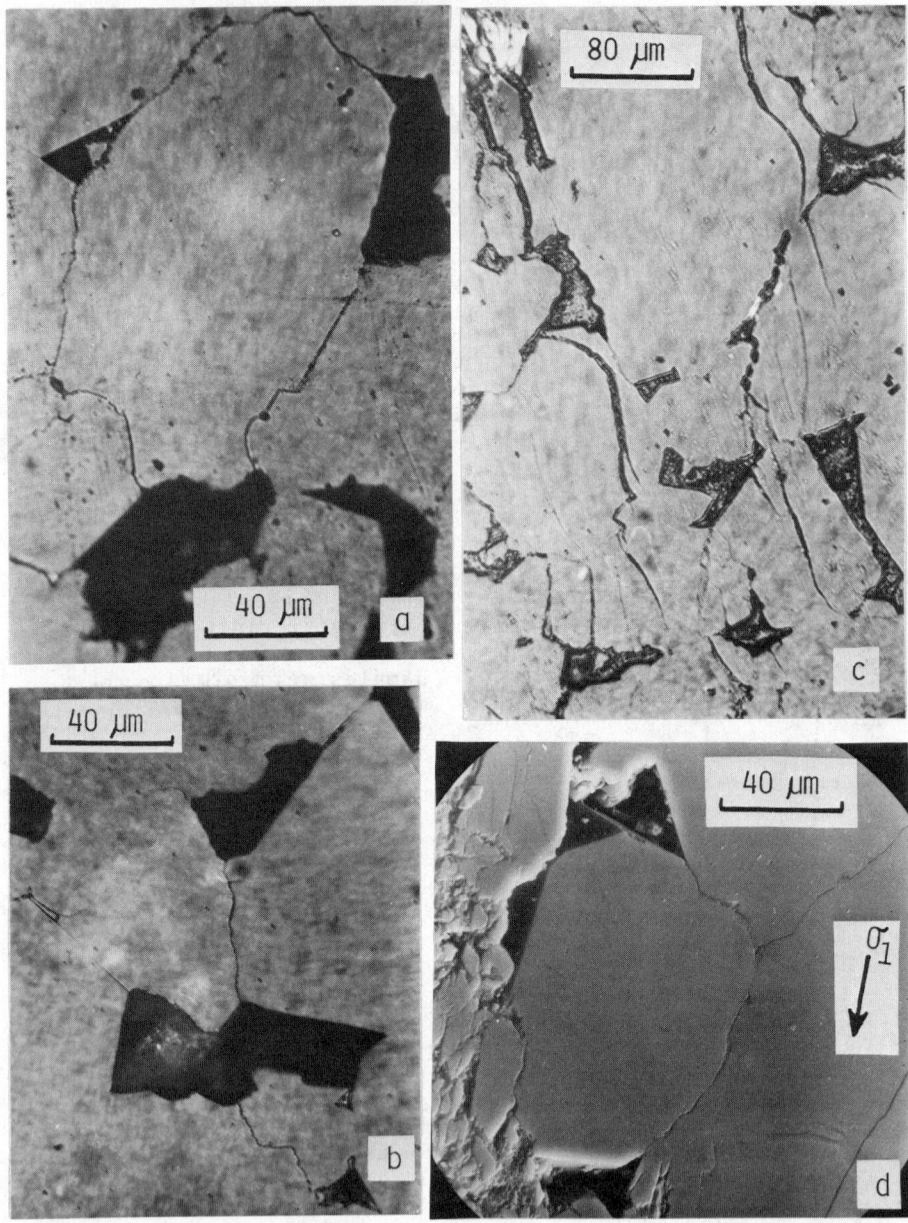

Fig.2. (a) and (b), Optical micrographs of polished surface of sample deformed to point prior to fault localization, showing grain boundary crack growth from pore notches. (c) Axial transgranular cracking in a faulted sample. (d) Grain boundary and trans-granular axial cracking adjacent to a fault zone (scanning electron micrograph).

CATACLASTIC DEFORMATION OF QUARTZITE 375

Microscopic Observations

From the crack maps it is apparent that this rock fails through the development of microcrack arrays, which propagate in shear mode, preferentially along the compromise grain boundaries formed by impingement of quartz cement overgrowths. For most of their lengths these cracks must grow with high shear tractions across them (fig.4). The cracks develop either from notches on pore space edges, or from pre-existing cracks already developed from pore space notches (fig.2). With increasing permanent strain grain boundary cracks begin to delineate individual grains and clusters of grains, as is apparent from fig.3. Only after almost all the grain boundaries are cracked, thereby loosening the structure until it is like a pile of interlocking glass beads, do axially oriented transgranular cracks propagate. The peak strength does not appear to be associated with any particular microstructural characteristic.

The development of axial cracks appears to lead to the localization of a fault zone and only occurs well after the strength peak. In contrast, other sandstones such as Tennessee sandstone, which has a mixed, clay plus quartz overgrowth cement, axial microcrack development is widespread prior to the strength peak. In the ganister, the development of grain boundary cracks is not expected to lead to the development of a marked induced elastic anisotropy. We consider that it is the development of such anisotropy by axial microcracking that leads to shear fault localization as an expression of internal instability in an anisotropic elastic medium (Cobbold, 1976; Rudnicki, 1977).

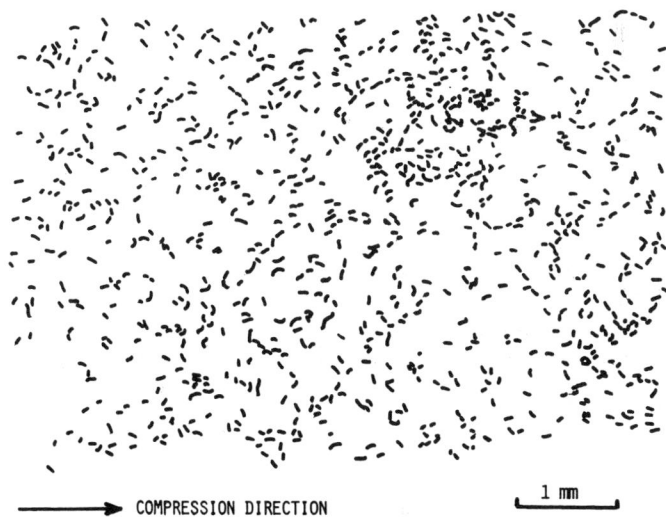

Fig.3. Crack map for a ganister specimen strained to the point of onset of axial transgranular cracking.

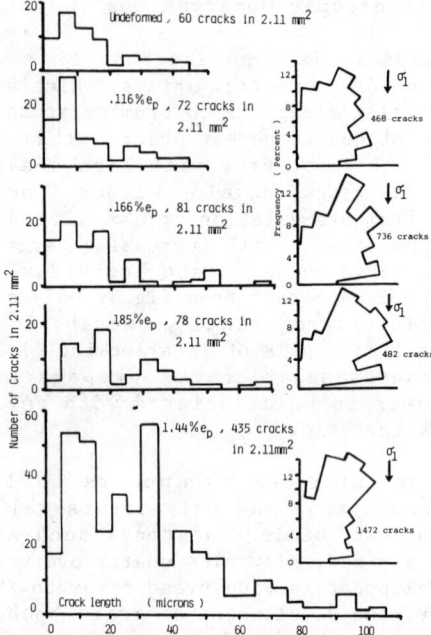

Fig.4. Histograms showing the development of crack length and density (left), and orientation distribution (right) with permanent strain.

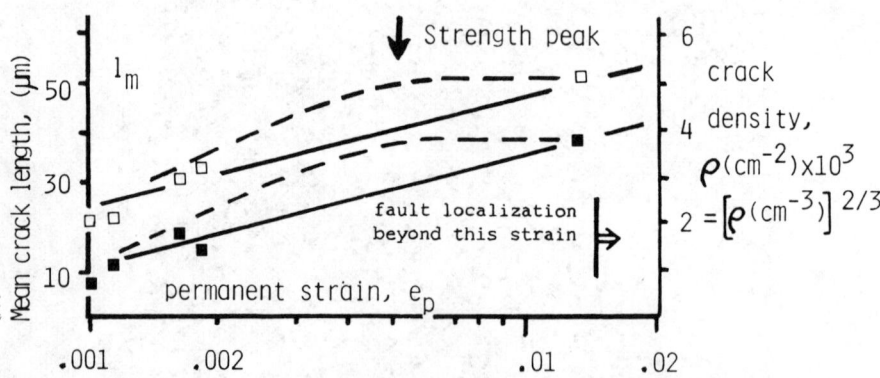

Fig.5. Graph showing the variation of mean crack length (open symbols) and crack density (closed symbols) with log permanent strain in the pre fault localization regime. The continuous and broken lines show respectively the functions describing mean crack length and density increase used to calculate the W_f curves shown in fig.6.

CATACLASTIC DEFORMATION OF QUARTZITE 377

Discussion

In fig.5 are shown plots of crack density, ρ, and mean crack length, l_m, against permanent strain, e_p, obtained from measurements on the two dimensional crack maps. Though these plots are not well constrained (there are large potential inaccuracies in the estimation of ρ, l_m and e_p) we can use them as a basis for rough calculations of energy dissipation through the growth of microcracks.

In the axisymmetric triaxial experiment the rate of work done, dW_e/de_p, by the external forces is

$$dW_e/de_p = \sigma' + p(de_v/de_p) \qquad (1)$$

where p is the confining pressure, e_v the permanent volume strain and σ' the differential stress. In this study volume changes could not be measured but based on behaviour observed in lower pressure experiments, permanent volume strains are expected to be roughly equal to $-e_p$. For the moment their contribution will be neglected, thus $dW_e/de_p = \sigma'$. For dW_e/de_p to have a maximum, d^2W_e/de_p^2 must be positive pre failure and negative post failure. Thus on a plot of log W_e vs. log e_p the slope at failure must be unity, greater than unity pre failure and less than unity post failure, in order that dW_e/de_p remains positive. These features are shown on a plot (fig.6) of integrated work done ($\int \sigma' de_p$) vs. log e_p obtained from stress/strain curves for the ganister. The effect of permanent dilatancy is to lower the level of this curve, but without changing its basic form.

The work done by the externally applied forces is dissipated through the formation of microcracks and frictional sliding along them. Making the simplifying assumptions that (a) all cracks are shear mode and are inclined at $45°$ to the loading direction and (b) all cracks are the same size at any instant, so that they are fully characterised by the crack density and the mean crack length, the contribution dW_c due to crack growth is

$$dW_c = \rho \; G \; \pi \; l_m \; dl_m \qquad (2)$$

where G is the crack extension force. The contribution dW_f due to friction is

$$dW_f = \tau_f \; \rho^{2/3} \; \pi \; \sqrt{2} \; l_m^2 \; de_p \; / \; 4 \qquad (3)$$

where τ_f is the resistance to sliding, and assuming that permanent strain is accumulated by equal shear displacements across all cracks. Displacement weakening in proportion to the amount of sliding on each crack can be incorporated by replacing

$$\tau_f = \tau_o \; (1 - \beta \; e_p \rho^{-1/3}) \qquad (4)$$

in (3), where β is chosen so that τ_f drops to the large displacement level of friction after 4% permanent strain. Based on fig.5 the simplest relations between ρ and l_m and e_p are

$$\rho = (a + b \log e_p) \quad ; \quad l_m = (c + d \log e_p) \tag{5}$$

Using these relations, (2) and (3) have been integrated. The contribution W_f is always greater than W_c, except for the very smallest cracks.

The integral of (3) together with (4) is plotted in fig.6. Without the displacement weakening effect, $\log W_f$ vs. $\log e_p$ approaches a slope of unity at high strains, and this remains true even taking into account the displacement weakening. The assumed exponential increase in e_p with l_m and ρ therefore overestimates the rate of increase of W_f.

On the other hand, if it is assumed that by about $e_p = 0.4$, crack density and length have stopped increasing, so that further strain depends on crack interactions and associated sliding, and so that the and l_m vs. $\log e_p$ curves become concave downwards as shown in fig.5, the forms of the $\log W_e$ vs. $\log e_p$ and $\log W_f$ vs. $\log e_p$ curves become similar. The curves can be made to correspond also in terms of energy magnitude by assuming that the crack density is underestimated by about x3 and l_m by about 40% (the two dimensional sections used to derive l_m would produce

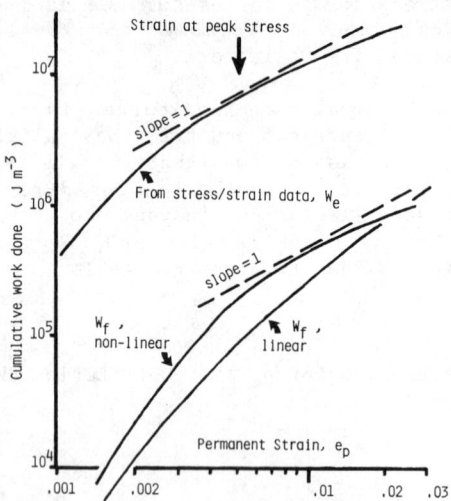

Fig.6. Plot of $\log W_f$ vs. $\log e_p$ obtained from stress/strain curves, neglecting the effect of permanent dilatancy. Calculated $\log W_f$ curves based on (a) linear increase in ρ and l_m with $\log e_p$ and (b) a non-linear rate, corresponding to the curves plotted in fig.4, are shown for comparison. The slope of the work curve equals unity at peak strength.

this effect), and taking into account the skewing of the crack length distributions. Crack growth always produces hardening through friction. To produce a stress drop at all it seems to be necessary to invoke some kind of reduction in friction stress as displacement per crack increases.

Conclusions

Most of the work done on this rock through loading through failure is dissipated through frictional sliding on microcracks. Pre and post peak-stress microcracking is concentrated around grain boundaries. The onset of transgranular axial cracking well past the stress peak leads to fault localization.

Crack length and number must increase rapidly with log e_p at first, then slow towards zero, and a crack displacement weakening process must be invoked in order to match the simple model described to the observed mechanical data for the pre fault-localization regime.

Though there have been attempts to describe rock failure in terms of a single microstructural process at peak strength, such as failure in flexure of column elements produced by axial microcracking (Peng and Johnson, 1972), it is clear that such an approach is not generally applicable, and that the failure characteristics of a particular rock type must depend strongly on its initial microstructural type.

Acknowledgements

The equipment used for this study was constructed through grant aid from the U.K. Natural Environment Research Council. P.R.Grant provided scanning electron microscopy facilities and R.F.Holloway assisted with rock deformation equipment. This work was carried out during the tenure of a research studentship (J.Hadizadeh) at Imperial College supported by the Government of Iran.

References

Cobbold, P.R., Mechanical effects of anisotropy during large finite deformations. Bull. Geol. Soc. France, 7(6), 1497-1510, 1976.

Peng, S.S. and **Johnson, A.M.**, Crack growth and faulting in cylindrical samples of Chelmsford Granite. Int. J. Rock Mech. Min. Sci., 9, 37-86, 1972.

Rudnicki, J.W., The effect of stress induced anisotropy on a model of brittle rock failure as a localization of deformation. In: Energy Resources and Excavation Technology, Proc. U.S. Symp. Rock Mech., 18, Ed. F.Wang and G.Clark., pp.3B4-1 to 3B4-8, 1977.

Rutter, E.H., Influence of interstitial water on the rheological behaviour of calcite rocks. Tectonophys. 14, 13-33, 1972.

Chapter 40

COMPRESSION EXPERIMENTS ON NATURAL MAGNETITE CRYSTALS AT 200 °C AND 400 °C AT 400 MPa CONFINING PRESSURE

by Christa Hennig-Michaeli and Heinrich Siemes

Institut für Mineralogie und Lagerstättenlehre der RWTH,
Aachen, West Germany

ABSTRACT

The effect of temperature and orientation on the plastic behaviour of an impure titaniferous magnetite from Palabora (South Africa) has been investigated. Prismatic specimens were axially compressed along [001], [110] and [111]. The strain rate was about $3 \cdot 10^{-6}$ s^{-1}.

Trace analyses of glide lines revealed a plastic anisotropy due to glide on several non-equivalent crystallographic planes. At 200 °C $\{111\}<11\bar{2}>$ twinning, $\{111\}<1\bar{1}0>$ slip and $\{100\}<011>$ slip have been operative. At 400 °C $\{111\}<1\bar{1}0>$ slip and $\{100\}<011>$ slip are primary glide modes. $\{110\}<1\bar{1}0>$ slip seems to be a secondary glide mode in the [001] specimens at 400 °C.

$\{111\}<11\bar{2}>$ twin glide generates isolated sharp lamellae. $\{111\}<1\bar{1}0>$ slip produces smooth surface striae. $\{100\}<011>$ slip is characterized by coarse and partially wavy slip bands. $\{110\}<1\bar{1}0>$ slip displays faint remarkably equidistant glide lines.

The critical resolved shear stresses (CRSS) decrease with rise of temperature. The CRSS ratio of slip on $\{100\}$ to slip on $\{111\}$ increases from 1.1 at 200 °C to 1.4 at 400 °C giving rise to a distinct flow anisotropy of the magnetite at 400 °C.

INTRODUCTION

Little is at present known about the plastic behaviour of magnetite especially about the intracrystalline glide mechanisms. Deformation twins in magnetite with $K_1 = (111)$ and $K_2 = (11\bar{1})$ have been observed in magnetite plates compressed at 25 °C (Grühn 1918), in octahedral crystals compressed along [001] at 25 °C (Hennig-Michaeli and Siemes 1973), in naturally deformed magnetites (Ramdohr 1969, p. 898) and in polycrystalline hematite magnetite ores compressed at 25 °C and 300 °C, whereas in specimens which had been deformed at 450 °C twin lamellae were lacking (Hennig-Michaeli and Siemes 1973). $\{111\}<1\bar{1}0>$ slip has been determined to be the glide mode in magnetite crystals compressed along [111] at room temperature (Müller and Siemes 1972).

Charpentier et al. tested the plasticity of synthetic magnetite crystals between 25 °C and 1000 °C by microhardness studies. At low temperatures glide lines in the vicinity of the indents were consistent with $\{111\}$ and $\{100\}$ glide planes. At higher temperatures the glide lines adopted a wavy nature.

Müller and Siemes (1972) showed that the strength of axially compressed polycrystalline magnetite decreases from 25 °C to 300 °C whereas the ductility increases. The degree of preferred orientation of the strain induced [110] fibre texture was essentially the same at all temperatures. First deformation maps for magnetite omitting the dislocation glide regime have been presented by Atkinson (1977). From a review of the data on the deformation behaviour of spinel structure crystals and the hardness data in question Veyssières et al. (1978) inferred that the transition from dislocation glide to dislocation climb presumably takes place between 0.3 T_m and 0.4 T_m (T_m = melting temperature).

Most experiments on $MgO \cdot (Al_2O_3)_n$ spinels have been performed at temperatures greater than 0.6 T_m. In stoichiometric spinels (n = 1) $\{111\}<1\bar{1}0>$ slip is the dominating glide mode (Radford and Newey 1967, Hwang, Heuer and Mitchell 1974). In spinels with n = 1.1 $\{111\}<1\bar{1}0>$

Table 1: Maximum Schmid factors (S) of potential glide modes in magnetite in relation to the compression directions [001], [110] and [111]. N_p = number of glide planes, N_s = number of glide systems.

	[001]			[110]			[111]		
	N_p	N_s	S	N_p	N_s	S	N_p	N_s	S
$\{111\}<1\bar{1}0>$ slip	4	8	.41	2	4	.41	3	6	.27
$\{110\}<1\bar{1}0>$ slip	4	4	.5	4	4	.25	0	0	0
$\{111\}<11\bar{2}>$ twinning	4	4	.47	2	4	.25	3	6	.15
$\{100\}<011>$ slip	0	0	0	2	4	.35	3	3	.47

slip and $\{110\}<1\bar{1}0>$ slip have similar CRSS (Duclos, Doukhan and Escaig 1979), whereas in n = 2 spinels $\{110\}<1\bar{1}0>$ slip exhibits considerably lower CRSS than $\{111\}<1\bar{1}0>$ slip (Radford and Newey 1967, Lewis 1968).

The present experiments on a rather impure material have been conducted to obtain first CRSS data of glide modes in magnetite. They have been performed at 200 °C (= .26 T_m) and at 400 °C (= .37 T_m). The Schmid factors of the potential glide modes in relation to the chosen directions of compression [001], [110] and [100] are listed in tab. 1.

EXPERIMENTS

Starting Material. The magnetite crystal from Palabora (South Africa) is a rounded specimen with diameters up to 80 mm. It contains numerous tabular ilmenites parallel to $\{111\}$ with diameters up to 3 mm and thicknesses up to 0.1 mm. Minute ulvite exsolution droplets are included in the magnetite except in areas close to the ilmenites. In these areas the mean magnetite composition is $(Fe_{7.1}Mg_{0.9})(Fe_{7.6}Ti_{0.3}Al_{0.1})_2O_3$ as determined by microprobe analysis assuming stoichiometry. In areas containing ulvite the magnesium content increases with increasing titanium content. Polycrystalline magnetite ilmenite bands up to 0.5 mm thickness pass through the crystal. In the magnetite as well as in bent ilmenites subgrains can be observed. X-ray reflexion pole figures exhibit broad reflexes indicating lattice distortions. The Vickers indentation hardness of microscopically homogeneous magnetite as well as that of ulvite carrying areas is 593 ± 10.

The magnetite crystal was sawn into plates parallel to $(1\bar{1}0)$ from which test specimens of size 7x7x14 mm³ were cut with compression axes along [001], [110] and [111]. The surfaces of the prismatic specimens were polished by hand and the orientations were checked by X-ray reflexion pole figures. Misorientations up to 10° occurred.

Experimental details. The compression experiments were conducted in a three axial high pressure apparatus with silicon oil as pressure medium. The specimens were sealed against the oil by aluminum jackets. In order to prevent reactions with the magnetite surfaces at 400 °C gold foils were inserted between the specimens and the aluminum jackets. During the 400 °C runs the magnetite surfaces had locally been oxidized. The high confining pressure of 400 MPa should prevent brittle failure. Nevertheless the experiments had to be stopped after small deformation since rupture was indicated. The strain rate was approximately kept constant at $3 \cdot 10^{-6}$ s^{-1}.

Determination of glide systems. The surface structures of the deformed specimens were studied microscopically by means of interference contrast objectives. Tiny glide lines are recognizable only on excellently polished surfaces which have not been obtained in all cases.

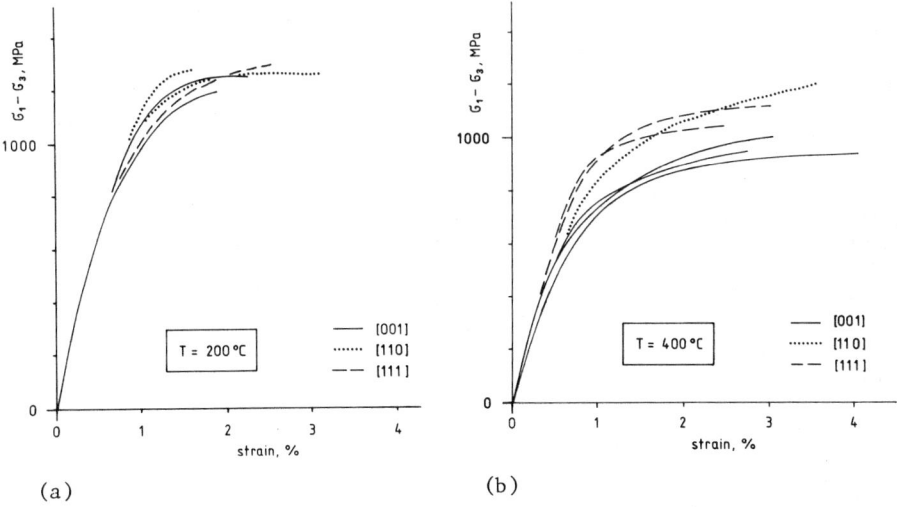

Fig. 1. Stress strain curves of Palabora magnetite crystals deformed at 200 °C (a) and 400 °C (b) at 400 MPa. Strain rate about $3 \cdot 10^{-6}$ s^{-1}.

Table 2: Compressing tests on Palabora magnetite crystals. 400 MPa confining pressure, strain rate about $3 \cdot 10^{-6}$ s^{-1}.

experiment	orientation	maximum strength MPa	shortening %	yield stress MPa	observed glide modes	CRSS MPa
200 °C tests						
MPA6	[001]	1210	1.0	1110	{111} twinning {111}<1$\bar{1}$0> slip	520
MPA7	[001]	1250	1.3	1180	{111} twinning {111}<1$\bar{1}$0> slip	550
MPB2	[110]	1280	0.8	1240	{111}<1$\bar{1}$0> slip {100}<011> slip	510
MPB4	[110]	1266	2.3	1160	{111}<1$\bar{1}$0> slip {100}<011> slip	480
MPC1	[111]	1300	1.4	1150	{100}<011> slip	540
400 °C tests						
MPA4	[001]	902	2.8	750	{111}<1$\bar{1}$0> slip {110}<1$\bar{1}$0> slip	310
MPA5	[001]	940	1.8	780	{111}<1$\bar{1}$0> slip {110}<1$\bar{1}$0> slip	320
MPA8	[001]	1000	2.2	780	{111}<1$\bar{1}$0> slip {110}<1$\bar{1}$0> slip	320
MPB8	[110]	1220	2.6	910	{111}<1$\bar{1}$0> slip	370
MPC2	[111]	1120	2.2	1000	{100}<011> slip	470
MPC4	[111]	1050	1.4	940	{100}<011> slip	440

Skratches might obscure the glide line engravings. Direction analyses of glide lines on the surfaces were used to determine the glide planes. The assumption was made that the glide planes are low index planes and that the glide directions are low index lattice directions with a small Burgers vector. The "observed" glide systems, therefore are tentative interpretations of the strain induced surface lines.

Measurements of the CRSS. The determination of the CRSS is subject to some uncertainties as in all experiments several glide systems have contributed to the strain. The deflection of the stress strain curves (SSC) from the elastic part into the plastic branch takes place gradually (fig. 1) extending over stress differences up to 200 MPa. The yield stresses were fixed somewhat arbitraryly at the midmost point of the deflection and the CRSS of the principal glide systems were calculated from that stress value.

RESULTS

Stress strain curves, strength. The results of the deformation experiments are summarized in tab. 2. In the 200 °C experiments the SSC are running closely together (fig. 1 a). A significant anisotropy of flow behaviour cannot be recognized. At 400 °C the [001] specimens are distinctly weaker than the [111] and [110] specimens respectively (fig. 1 b).

The crystals are only weakly deformed and initiating fracture is observed. The permanent strain is small, 0.8 % to 1.4 % at 200 °C and at most 2.8 % at 400 °C. An undefinable amount of deformation is due to displacements along fractures. Furthermore the strain is affected by the plastic deformation of the ilmenites which presumably starts in the "elastic" part of the SSC.

The strengths of the crystals are unexpectedly high with more than 1200 MPa at 200 °C and more than 900 MPa at 400 °C. They are higher than those of a coarse grained polycrystalline magnetite ore tested under the same conditions.

The Observed Glide Mechanisms in Magnetite. All specimens are deformed inhomogeneously. On the 200 °C specimens numerous glide lines are well exposed but only more indistinct linear structures may be observed on the undulating surfaces of the 400 °C samples.

COMPRESSION ALONG [001]: Prism surfaces ($1\bar{1}0$), (110). On the 200 °C samples two kinds of glide lines are developed. Sharp isolated lamellae along traces of {111} are regarded as {111} deformation twins (fig. 2 a). They can be closely arranged and twisted. Attempts to identify the lamellae as twins by X-ray goniometry provided only ambiguous reflexes of twin orientations since the twins occupy only a minute fraction of the crystal volume.

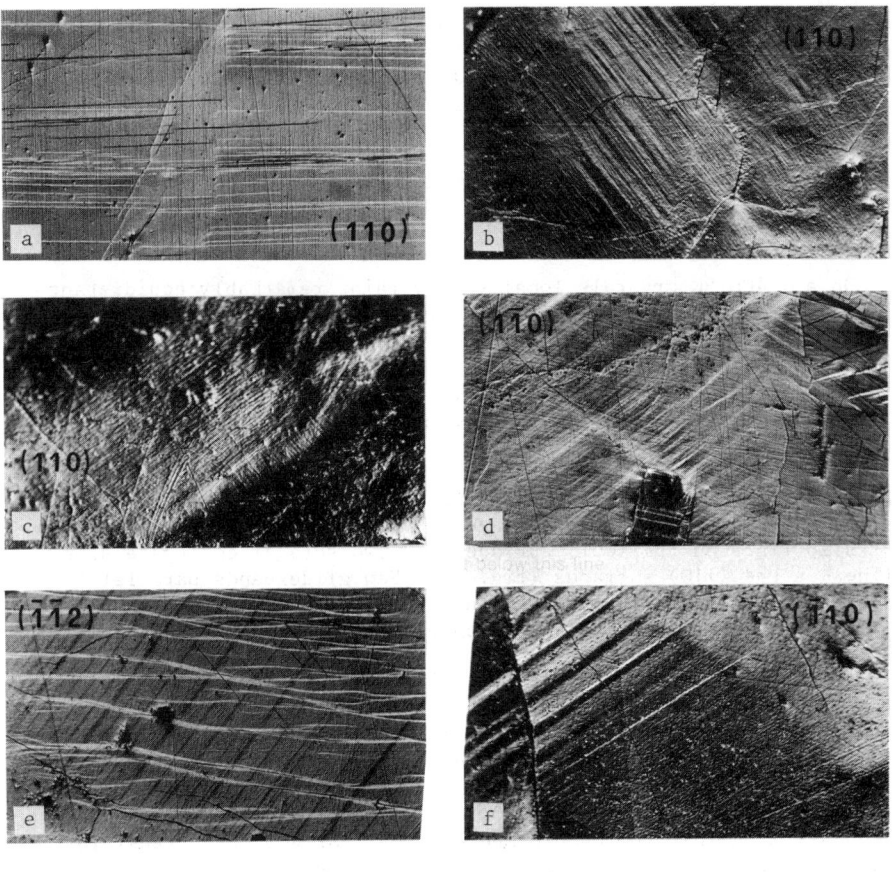

Fig. 2. Surface structures of the deformed Palabora magnetite crystals. Compression directions (CD) N-S. Interference contrast.

(a) {111} twin lamellae (E-W) along traces of (111) and (11$\bar{1}$) are stopped at a subgrain boundary. MPA7, 200 °C, CD [001].
(b) {111}<1$\bar{1}$0> slip lines (N 36° W) are traces of ($\bar{1}$11). MPA8, 400 °C, CD [001].
(c) {110}<1$\bar{1}$0> slip lines (N 54° E) are traces of (101) and {111}<1$\bar{1}$0> slip lines (N 36° W) are traces of ($\bar{1}$11). MPA5, 400 °C, CD [001].
(d) {111}<1$\bar{1}$0> slip lines are traces of (111) (N 54° E) and (11$\bar{1}$) (N 54° W), ilmenites with deformation twins. MPB4, 200 °C, CD [110].
(e) {100}<011> slip bands are wavy along traces of (001) (E-W) and straight along traces of (010) (N 38° E). MPC1, 200 °C, CD [117].
(f) {100}<011> slip bands along traces of (100) and (010) (N 54° E) dissolve into narrowly spaced slip lines. MPC2, 400 °C, CD [111].

Accumulations of tender lines and smooth surface striae like strokes of brushes with weak relief parallel to {111} traces have been referred to {111}<1$\bar{1}$0> slip. 1/2 <110> has been stated to be the slip vector in all slip modes observed in spinels (tab. 1).

On the undulating surfaces of the 400 °C specimens {111}<1$\bar{1}$0> slip lines of even weaker relief than at 200 °C are common (fig. 2b, c). The ripple of the surfaces may be generated by multiple slip in small crystal volumes.

On all 400 °C crystals locally very faint remarkably equidistant straight lines are noticeable on surface areas without {111} lines (fig. 2c). These lineations are consistent with {110} traces. Probably {110}<1$\bar{1}$0> slip has been operative in addition to {111}<1$\bar{1}$0> slip. At one place isolated thin lamellae like twins along two {111} traces have been observed in a 400 °C specimen.

COMPRESSION ALONG [110]: Prism surfaces (1$\bar{1}$0), (001). The plastic deformation predominantly results from {111}<1$\bar{1}$0> slip. (111) and (11$\bar{1}$) have been operative and have produced horizontal accumulations of slip lines on the (001) surfaces and two sets of conjugate slip lines on the (1$\bar{1}$0) surfaces (fig. 2d). Few glide bands parallel to {100} traces occur. The glide along {100} is characterized by coarse glide bands with a relief far stronger than that of the {111}<1$\bar{1}$0> slip lines.

COMPRESSION ALONG [111]: Prism surfaces (1$\bar{1}$0), (11$\bar{2}$). On all samples a lot of coarse glide bands parallel to {100} traces are visible. The horizontal (001) glide bands on the (11$\bar{2}$) surface are remarkably wavy and their ends are splitted (fig. 2e). The waviness is about the same at both temperatures. At 200 °C the {100} glide bands show sharp boundaries against the undeformed material. At 400 °C a transition from coarse glide bands into fine narrowly spaced glide lines can be observed (fig. 2f). It is suspected that the glide bands have been produced by {100}<011> slip.

At 400 °C in addition to {100}<011> slip locally in the vicinity of deformed ilmenites {111}<1$\bar{1}$0> slip lines occur.

The observed glide modes in ilmenite. The ilmenites on the whole are stronger strained than the magnetite and squeezed out of the magnetite surface. Sometimes weakly deformed ilmenites may be observed in strongly deformed magnetite. At both temperatures glide lines and glide bands parallel or subparallel to the base and deformation twins on a rhombohedron are developed. At 400 °C the (0001) slip lines are very fine and closely arranged and several ilmenite plates are wrinkled by kinkbands.

CRSS of the observed glide systems in the Palabora magnetite. The CRSS of the Palabora magnetite are high (tab. 2). In camparison to those values a pure magnetite from Kennedy Ranges (Australia) tested by compression along [111] at 200 °C displayed a CRSS of 270 MPa for {100}<011> slip. The structure of the Palabora magnetite (subgrains, bent ilmenites, broad X-ray reflexes) indicates that the crystal has been deformed before and thus probably been strengthened by dislocation pile-ups. The results show that the CRSS of the slip modes in the Palabora magnetite decrease with increase of temperature. The CRSS ratio of slip on {100} to slip on {111} increases from ~1.1 at 200 °C to ~1.4 at 400 °C giving rise to the distinct anisotropy of the flow behaviour at 400 °C.

DISCUSSIONS

The results of the deformation experiments on the Palabora magnetite are surprising inasmuch as glide lines existed for all glide systems so far observed in crystals of spinel structure. The glide modes which have been derived from the traces seem to be reasonable assumptions as they all together have high Schmid factors in relation to the respective compression directions. That applies especially to {100}<011> slip which has been operative during compression along [111] (S = 0.47) and to {110}<1$\bar{1}$0> slip which presumably has become active during compression along [001] (S = 0.5) at 400 °C. The occurrence of {100}<011> slip confirms the observations of Charpentier, Rabbe and Manenc (1967) who detected {100} glide lines next to indentations on synthetic magnetite crystals and of Veyssières et al. (1978) who observed them next to indents on nickel ferrite. Obviously this glide mode has considerable influence on the mechanical behaviour of magnetite in the low temperature deformation regime.

The results have to be regarded as preliminary as long as the glide modes have not been verified by TEM investigations.

Acknowledgements. This work was supported by the Minister of Science and Research (Minister für Wissenschaft und Forschung des Landes Nordrhein-Westfalen). Thanks are due to the staff of the Mineralogical Institute, Aachen.

REFERENCES

Atkinson, B.K., 1977, "The Kinetics of Ore Deformation: Its Illustration and Analysis by Means of Deformation-Mechanism Maps", Geologiska Föreningens Förhandlingar, Vol. 99, pp. 186-197.

Charpentier, Ph., Rabbe, P. and Manenc, J., 1968, "Mise en Évidence de la Plasticité de la Magnétite Mésure de la Dureté en Location de la Temperature", Materials Research Bulletin, Vol. 3, pp. 69-78.

Duclos, R., Doukhan, N. and Escaig, B., 1978, "High Temperature Creep Behaviour of Nearly Stoichiometric Alumina Spinel", Journal of Materials Science, Vol. 13, pp. 1740-1748.

Grühn, A., 1918, "Künstliche Zwillingsbildung des Magnetits", Neues Jahrbuch für Mineralogie, Geologie und Paläontologie, pp. 99-112.

Hennig-Michaeli, Ch. and Siemes, H., 1975, "Zwillingsgleitung beim Magnetit", Neues Jahrbuch für Mineralogie, Abhandlungen, Vol. 123, pp. 330-334.

Hwang, L., Heuer, A.H. and Mitchell, T.E., 1974, "Slip Systems in Stoichiometric $MgAl_2O_4$ Spinel", Deformation of Ceramic Materials, R.C. Bradt and R.E. Tressler, eds., Plenum Press, pp. 257-270.

Lewis, M.H., 1968, "The Defect Structure and Mechanical Properties of Spinel Single Crystals", Philosophical Magazine, Vol. 17, pp. 481-498.

Müller, P. and Siemes H., 1972, "Zur Festigkeit und Gefügeregelung von experimentell verformten Magnetiterzen", Neues Jahrbuch für Mineralogie, Abhandlungen, Vol. 117, pp. 39-60.

Radford, K.C. and Newey, C.W.A., 1967, "Plastic Deformation in Magnesium Aluminate Spinel", Proceedings of the British Ceramic Society, Vol. 9, pp. 131-145.

Ramdohr, P., 1969, "The Ore Minerals and Their Intergrowths", Pergamon Press.

Veyssière, P., Rabier, J., Garem, H. and Grilhé, J., 1978, "Influence of Temperature on Dissociation of Dislocations and Plastic Deformation in Spinel Oxides", Philosophical Magazine, Vol. 38, pp. 61-79.

Chapter 41

MICROSTRUCTURES, MINERAL CHEMISTRY AND OXYGEN ISOTOPES
OF TWO ADJACENT MYLONITE ZONES: A COMPARATIVE STUDY

by Timothy E. LaTour and Robert Kerrich

Geology Department, Georgia State University
Atlanta, Georgia, U.S.A. 30303

Geology Department, University of Western Ontario
London, Ontario, Canada N6A 5B7

ABSTRACT

Microstructures and mineral chemistry of amphiboles and plagioclases from two spatially related mylonite zones at the Grenville front are significantly different, suggesting different physical conditions (including temperature) for the formation of the two zones. Oxygen isotopes indicate multiple episodes of fluid-rock interaction, consistent with previously determined mylonitization temperatures and postulated tectonic evolution of the area.

INTRODUCTION

Two discrete mylonite zones have been distinguished at the Grenville front near Coniston, Ontario, Canada (Fig. 1). These zones have been designated MZ I and MZ II by La Tour (1981a) and have been differentiated on the basis of several criteria: First, the mesoscopic folds of MZ II are convolute and "swirled" in style, whereas those in MZ I are generally similar in style. Second, microstructures of rocks in MZ II exhibit evidence of having undergone much recrystallization and neocrystallization during deformation, whereas microstructures characteristic of MZ I are indicative of dynamic recovery dominant over recrystallization, plus brittle microfracturing. Third, metamorphic reactions associated with mylonitic deformation in MZ II are prograde in nature, whereas those in MZ I are exclusively retrograde. Fourth, garnet-biotite and garnet-amphibole geothermometry indicates that mylonitic deformation in MZ II occurred at approximately $610°C$, whereas mylonitic deformation in MZ I occurred at temperatures below $540°C$ (La Tour, 1981b). Fifth, oxygen isotope analysis indicates extensive low-temperature interaction of meteoric water with rocks of MZ I, but not with those of MZ II.

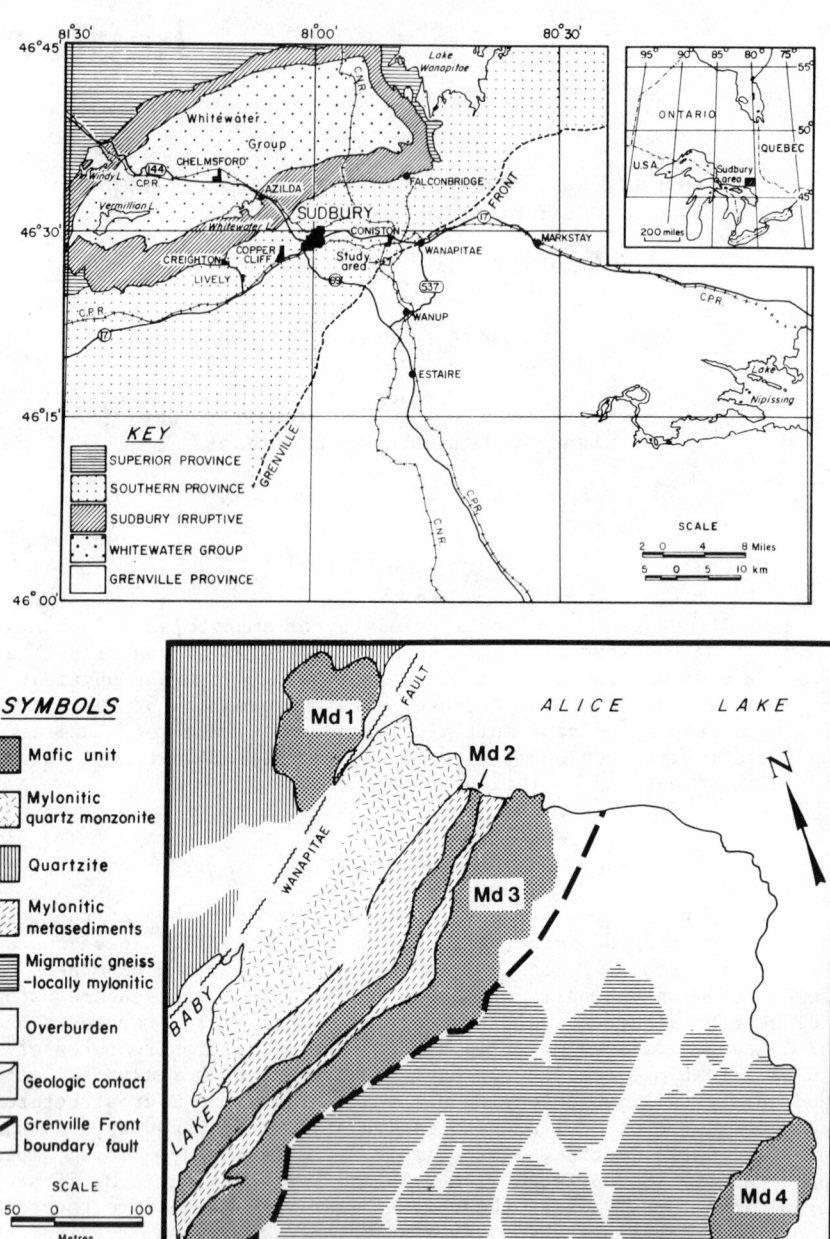

Fig. 1. Location map and lithologies at the Grenville front near Coniston. Metadiabases prefixed "Md."

LITHOLOGIES

Lithologies affected by mylonitization in MZ II include quartzofeldspathic layered gneisses (with or without amphibole), rare pelites, metadiabase unit Md-4, and amphibolites interlayered with the migmatitic gneiss (Fig. 1). Those mylonitically deformed in MZ I include gneisses and pelites, metadiabase units Md-2 and Md-3, amphibolites within the gneiss, metasediments and quartzites, and a narrow body of quartz monzonite (Fig. 1). Due to local variations in types and abundances of minerals present in the gneisses, only the amphibolites and metadiabases were appropriate for comparative microstructural and geochemical study. Specifically, amphibole and plagioclase microstructures and mineral chemistry have been the most instructive.

MICROSTRUCTURES AND MINERAL CHEMISTRY

Mylonite zone (MZ) II

All amphiboles from MZ II were recrystallized during mylonitic deformation, some thoroughly but others only along their margins. In the metadiabases, the high whole-rock Mg/Fe ratio of the rocks (thought to be originally pyroxene cumulates) enabled recrystallization to medium-grained magnesio-hornblende and actinolitic hornblende (Table 1), plus oligoclase and clinozoisite. The relatively high Al contents of such Mg-rich amphiboles, along with the Ca content of plagioclase (sodic andesine), attests to relatively high metamorphic grade during deformation. Amphibole prisms are moderately preferentially aligned, and plagioclase is equant and twinned.

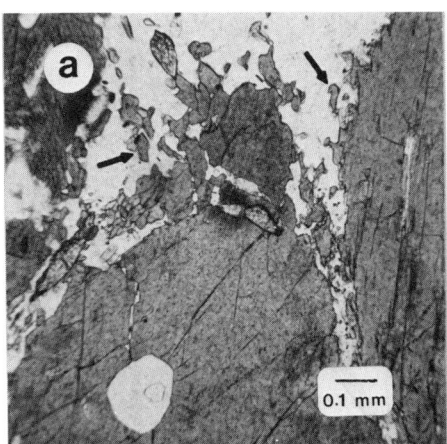

Fig. 2. a. "Satellite" grains of amphibole (arrows). b. Recrystallized plagioclase (p) between amphibole porphyroclasts.

TABLE 1. Representative amphibole analyses from MZ II

	metadiabase	amphibolite	
Sample no.	S-122	S-148-2 core	S-148-2 rim
No. analyses	6	7	1
SiO_2	50.84 (0.83)*	39.70 (0.63)	39.04
TiO_2	0.38 (0.03)	1.34 (0.10)	0.96
Al_2O_3	6.89 (0.33)	13.34 (0.20)	13.35
Fe_2O_3	0.539	1.547	4.753
FeO	8.535	23.368	22.863
MnO	0.15 (0.04)	0.29 (0.07)	0.26
MgO	16.14 (0.21)	4.06 (0.24)	3.33
CaO	12.55 (0.23)	11.28 (0.11)	10.88
Na_2O	0.77 (0.12)	1.48 (0.07)	1.25
K_2O	0.30 (0.01)	1.46 (0.08)	1.49
total	97.094	97.865	98.176
	number of ions based on 23 oxygens		
Si	7.2742	6.2097	6.1241
Al	.7258	1.7903	1.8759
Al	.4364	.6696	.5929
Fe 2+	1.0214	3.0569	2.9995
Fe 3+	.0580	.1821	.5611
Mg	3.4416	.9464	.7785
Mn	.0182	.0384	.0345
Ti	.0409	.1576	.1133
Ca	1.9241	1.8905	1.8287
Na-M4	.0594	.0584	.0915
Na-A	.1542	.3904	.2887
K	.0548	.2913	.2982

*Numbers in parentheses indicate one standard deviation.

Amphibolites that have much lower whole-rock Mg/Fe than the metadiabases, possess hornblende and andesine, with or without biotite or garnet. Deformation of the amphibole during mylonitization occurred primarily along pre-existing hornblende boundaries, although some hornblende was totally recrystallized. Deformation gave rise to highly embayed boundaries of hornblende porphyroclasts, surrounded by fine granoblastic to idioblastic "satellite" grains of hornblende, ilmenite and rutile (Fig. 2a), floating in fine, recrystallized andesine (Fig. 2b). Epidote is very rare. Optically distinct rims on the hornblende porphyroclasts are chemically indistinguishable from their tiny, floating "satellite" grains, but different from their cores. In some cases the rims and "satellite" grains have equal or higher Al contents than the cores (Table 1). This, coupled with (i) the almost complete absence of chemical breakdown of deformed garnet in the same rocks, and (ii) recrystallization of

TABLE 2. Representative amphibole analyses from MZ I

	metadiabase		amphibolite			
Sample no.	S-59		S-113A core		S-113A rim	
No. analyses	7		3		2	
SiO_2	54.13	(0.43)*	41.69	(0.21)	49.79	(0.94)
TiO_2	0.20	(0.05)	1.08	(0.14)	0.13	(0.01)
Al_2O_3	2.33	(0.39)	10.85	(0.21)	4.96	(0.47)
Fe_2O_3	0.215		1.860		1.688	
FeO	7.986		18.716		17.022	
MnO	0.22	(0.02)	0.38	(0.01)	0.39	(0.01)
MgO	18.14	(0.35)	8.02	(0.36)	11.23	(0.65)
CaO	13.04	(0.54)	11.46	(0.15)	12.11	(0.06)
Na_2O	0.30	(0.07)	1.75	(0.18)	0.80	(0.04)
K_2O	0.06	(0.01)	1.49	(0.05)	0.41	(0.06)
total	96.622		97.296		98.529	

number of ions based on 23 oxygens

Si	7.7268	6.4291	7.3518
Al	.2732	1.5709	.6482
Al	.1189	.4017	.2152
Fe 2+	.9534	2.4139	2.1020
Fe 3+	.0231	.2158	.1875
Mg	3.8591	1.8432	2.4712
Mn	.0266	.0496	.0488
Ti	.0215	.1253	.0144
Ca	1.9945	1.8936	1.9160
Na-M4	.0029	.0568	.0449
Na-A	.0801	.4665	.1842
K	.0109	.2931	.0772

*Numbers in parentheses indicate one standard deviation.

andesine to new andesine in these rocks, implies that metamorphic conditions during mylonitization were near those recorded by the amphibole cores.

Undulatory extinction of amphiboles and plagioclase is rare, and subgrain development or other evidence of dynamic recovery has not been observed. The mylonitically deformed amphibolites and metadiabases of MZ II are therefore blastomylonites, having formed at elevated temperatures (and pressures?) during high strain.

Mylonite zone (MZ) I

Microstructures of amphiboles in MZ I are more varied than those of MZ II, ranging from features indicative of brittle fracture, to subgrain development, to recrystallization. Recrystallization of the

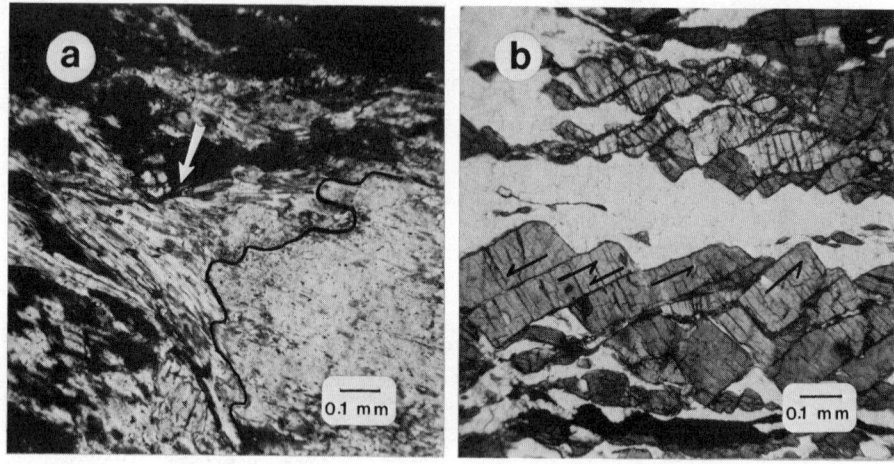

Fig. 3. a. Amphibole beard (arrow) in the pressure shadow of an amphibole porphyroclast (outlined in ink for clarity).
b. En echelon fragments of hornblende showing slip on a fracture subnormal to the c axis.

type seen in MZ II was not observed in MZ I, but rather occurs in two other forms: (i) amphibole beards in the pressure shadows of amphibole porphyroclasts (Fig. 3a), observed exclusively in the high Mg/Fe metadiabases near the Grenville front boundary fault (Fig. 1). Its origin is thought to be analogous to mica and chlorite beards formed during pressure solution. (ii) complete recrystallization to fine-grained epidote-amphibolite from the pre-existing coarse-grained low-Mg/Fe amphibolites. Conversion of andesine to sodic oligoclase (or albite), and profuse growth of epidote (clinozoisite) have accompanied this recrystallization. Recrystallized amphiboles have much lower Al contents than the presumed host grains, as do rims on larger porphyroclasts (Table 2). These new mineral assemblages are indicative of retrograde metamorphism, and considerable additional hydration of the rocks.

Subgrain development in hornblendes has been seen in only two samples, and appears similar to that described elsewhere. Brittle fracture is common, but probably accounts for little of the total reduction in grain size of most samples. In one sample, brittle fracture subnormal to the c axis is spectacularly developed, and en echelon fragment displacement appears to have been accomodated by ductile deformation of the enclosing quartz (Fig. 3b) (Allison and La Tour, 1977). Stress corrosion cracking may have also contributed to amphibole deformation in rocks of MZ I (Kerrich et al., 1981), and it is clearly related to retrograde metamorphic reactions.

MICROSTRUCTURES OF TWO MYLONITE ZONES

OXYGEN ISOTOPES AND FLUID REGIMES

Mylonite zone (MZ) II

Estimates of the ambient temperature of deformation in MZ II from quartz-muscovite fractionations (2.6-2.8 ^{o}/oo) are 590-620°C, in close accord with temperatures deduced from garnet-biotite geothermometry. The $\delta^{18}O$ of quartz in schists is relatively uniform at 11.3-11.8 ^{o}/oo; quartz veins in these schists have isotope compositions within the same range as the host rock, signifying growth under rock-dominated conditions. Fluids present during deformation and veining have a $\delta^{18}O$ of about +9 ^{o}/oo, based on the $\delta^{18}O$ of quartz in schists and veins, plus the quartz-water fractionation equation (Clayton et al., 1971).

A second population of quartz veins is present in MZ II, characterized by $\delta^{18}O$ quartz of 7.6-8.0 ^{o}/oo. These later veins are interpreted to have grown at about 500°C from fluids of +4.0 to +4.5 ^{o}/oo. Veining was unrelated to the ambient conditions or pore solutions present during evolution of MZ I or MZ II, and may represent a small late pulse of fluids inducing hydrofracturing, and derived from a deeper, cooler source external to the immediate rock units, as hot rocks moved along MZ II over cooler units at depth.

Mylonite zone (MZ) I

In deformed granites of MZ I, quartz-albite fractionations of 1.8-2.0 ^{o}/oo correspond to isotopic temperatures of 440-500°C, broadly commensurate with estimates from garnet-biotite geothermometry. The quartz in deformed granites has a $\delta^{18}O$ of 10.9-11.7 ^{o}/oo. As for the case of MZ II, the $\delta^{18}O$ of vein quartz closely corresponds to that of quartz in the host rock, signifying rock-dominated conditions, with a $\delta^{18}O$ fluid estimated at +8 to +9 ^{o}/oo.

The granite of MZ I also has a second population of veins related to late fractures, with $\delta^{18}O$ quartz of -1 ^{o}/oo. Preliminary evidence from fluid inclusions yields filling temperatures in the range of 200-250°C. The depth at which fracturing took place is unknown, but pressure corrections to temperatures may be about +20 to +50°C. Based on these results, fluids implicated in the late veins would have an isotopic composition of -9 ^{o}/oo (300°C) to -14 ^{o}/oo (200°C). Continental meteoric water is the only terrestrial fluid reservoir with a negative $\delta^{18}O$ (cf. Taylor, 1974); hence, this episode of veining is interpreted to represent penetration of isotopically light meteoric water deep into MZ I, perhaps derived from high altitudes in the Grenville mountains. This is consistent with post-orogenic formation of MZ I, related to isostatic rebound of the Grenville crust (La Tour, 1980).

CONCLUSIONS

Comparison of microstructures and mineral chemistry of amphiboles and plagioclases in both MZ I and MZ II indicates different dominant deformation mechanisms and temperatures of deformation for the two zones. Oxygen isotope analysis of minerals in mylonitic rocks of both zones has yielded temperatures consistent with those previously determined (La Tour, 1981b), and has revealed a complex history of rock-fluid interaction during various phases of deformation. The compatibility of diverse types of data, that is, microstructures, mineral chemistry and oxygen isotopes, strengthens previously suggested differences in origin for the two zones: MZ II is thought to have formed during crustal thickening (orogeny) and MZ I during crustal thinning, probably during isostatic rebound (La Tour, 1981b).

REFERENCES

Allison, I. and La Tour, T. E., 1977, "Brittle Deformation of Hornblende in a Mylonite: a Direct Geometrical Analogue of Ductile Deformation by Translation Gliding," Canadian Journal of Earth Science, Vol. 14, pp. 1953-1958.

Clayton, R. N., O'Neil, J. and Mayeda, T. K., 1972, "Oxygen Isotope Exchange Between Quartz and Water," Journal of Geophysical Research, Vol. 77, pp. 3057-3067.

Kerrich, R., La Tour, T. E. and Barnett, R. L., 1981, "Mineral Reactions Participating in Intragranular Fracture Propagation: Implications for Stress Corrosion Cracking," Journal of Structural Geology, Vol. 3, pp. 77-87.

La Tour, T. E., 1980, "Metamorphic Evidence for Discrete Mylonite Zones at the Grenville Front," (abstract), Geological Society of America Abstracts With Programs, Vol. 12, No. 7, p. 68.

La Tour, T. E., 1981a, "Significance of Folds and Mylonites at the Grenville Front in Ontario," Geological Society of America Bulletin, Vol. 92, Part II, pp. 997-1038.

La Tour, T. E., 1981b, "Metamorphism and Geothermometry Near Coniston, Ontario: a Clue to the Tectonic Evolution of the Grenville Front," Canadian Journal of Earth Science, Vol. 18, pp. 884-898.

Taylor, H. P., 1974, "The Application of Oxygen and Hydrogen Isotope Studies to Problems of Hydrothermal Alteration and Ore Deposition," Economic Geology, Vol. 69, pp. 843-883.

CONTINUOUS FORMATION OF GOUGE AND BRECCIA

DURING FAULT DISPLACEMENT

Chapter 42

by Eugene C. Robertson

Geophysicist
U. S. Geological Survey
Reston, Va. 22092

INTRODUCTION

A direct proportionality between the observed displacement of a fault and its thickness of breccia and gouge has been proposed recently (Robertson, in press). To validate this finding, an explanation of the mechanics of continuous breccia formation is needed and is offered in this paper. A summary of evidence for the finding will show why an explanation is needed.

The correlation between fault displacement d and accompanying breccia-plus-gouge thickness t seems to be linear and reasonably well established over 6 orders of magnitude, with a range in d from 1 mm to 1 km, on a trend of $d/t = 100$, shown in Figure 1. Most of the points are for faults observed underground in mines. The $d - t$ relation is limited to normal and reverse faults, that have a flat to steep dip and a displacement less than 5 km. It is not applied to large strike-slip and low-angle overthrust faults, whose displacements and breccia thicknesses are widely variable and uncertain. Undisplaced jointed rock is not included in the thickness measurement. A complete description of sources and analysis of the data is given in Robertson (in press).

One would intuitively expect that continued movement on a fault would grind up the previously formed breccia and gouge and not produce any more, but the evidence (Fig. 1) indicates otherwise. A physical model to explain continuous formation of gouge and breccia from the wall rocks can be obtained from laboratory experiments on friction.

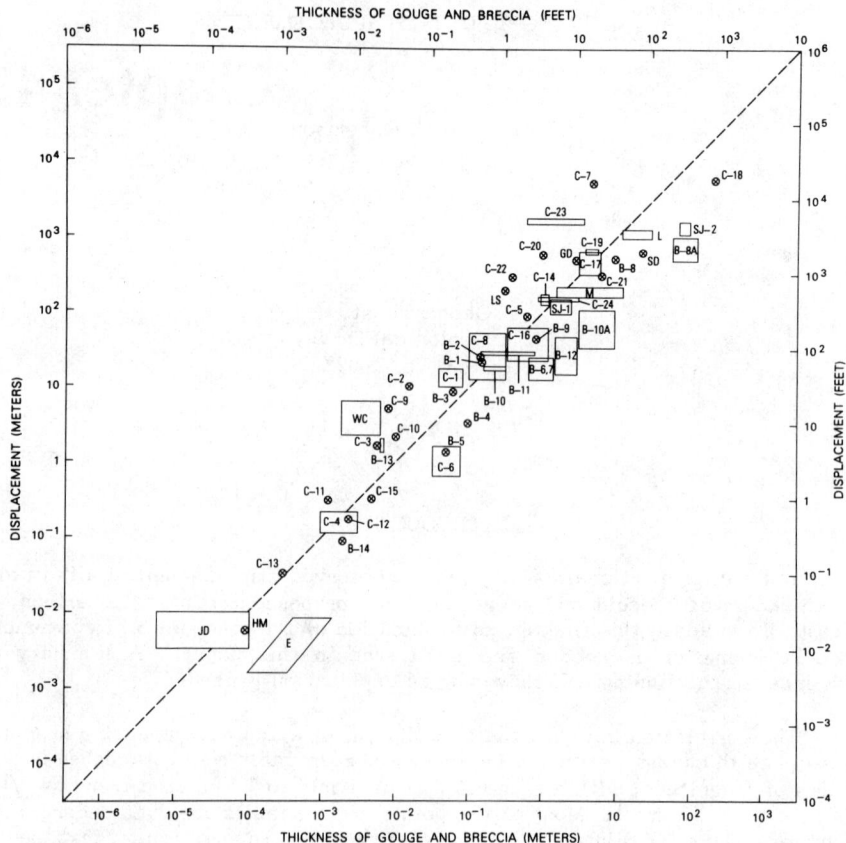

Figure 1. Fault displacement $\underline{d}$ plotted against thickness $\underline{t}$ of breccia and gouge of faults at the following localities: points "B", in mines at Butte, Montana; points "C", in mines at Coeur d'Alene, Idaho; points "M", "L", and "LS", in mines in Arizona, Colorado, and Ontario; points "JD", "HM", and "E", from laboratory data; remaining points, from surface observations in Montana and New Mexico (Robertson, in press, Fig. 1). A trend line (dashed) at $\underline{d} / \underline{t} = 100$ is shown.

BRECCIATION MODEL

Engelder (1978) reviewed the sliding and fracturing mechanisms in friction, citing the important work of Brace, Byerlee, Friedman, Logan, Bombolakis, Scholz, and Teufel. He concluded that "there is a common process accompanying frictional sliding on both fresh rock and on gouge." Sliding mechanisms cited are shearing through interlocked asperities, asperity indentation and ploughing, plastic flow by melting of asperity tips or by crystal gliding, and sliding within gouge or between intact rock and gouge. By these mechanisms breccia and gouge would be formed but not necessarily in increasing amounts as fault movement increases.

Continuous production of gouge and breccia as a fault moves can be explained by a physical model of Hundley-Goff and Moody (1980), who used a transmission electron microscope to study quartzite samples tested in frictional sliding by Hayes (1975). They found an "increased number of intragranular cracks with increased axial displacement"; the cracks formed mostly at $90°$ to the sliding surfaces and showed no shear displacement (Hundley-Goff and Moody, 1980, Figs. 1,3,5). They found that the cracks cut across dislocations and terminated in regions free of dislocations, so they concluded that dislocation mobility is not the controlling process. Their stages of surface destruction are summarized here in Figure 2; they ascribe destruction of the rock surfaces to intragranular cracking, asperity penetration and ploughing, and plucking of grains and fragments, a process which is independent of sliding direction.

The formation of intragranular cracks by asperity indentation and ploughing was described by Engelder and Scholz (1976); they found that both soft and hard minerals produced cracks when drawn over a quartz surface. Cracks are also produced in grains by stress concentrated at point contacts with other grains in a rock (Kranz, 1979, Fig. 5; Tapponier and Brace, 1976, Fig. 11). Of course, displaced breccia grains and grain fragments, entrained in the fault zone, as well as asperities will indent grains in the walls and will be dragged over the wall surfaces to form microcracks and to plough and comminute other grains (like the comminution process in the Hardinge grinding mill.)

BRECCIATION IN FAULT SAMPLES

Production of breccia and gouge during fault movement takes place ultimately by the same process of indentation, cracking, plucking, abrasion, and comminution of grains in rocks in the earth as it does in laboratory samples of rocks. Plucking of a grain from the slickensided surface of the quartz monzonite wall of a northwest fault in a Butte mine is shown in Figure 3; microcracks in the adjacent rock are also apparent in the electron photomicrograph. Figure 4 shows the head of a ploughed groove in the sandstone wall of a small overthrust fault. Whether an asperity or a grain fragment created the groove is not clear. A series of ploughed grooves and plucked grain cavities in the quartzite of a fault in the Bunker Hill Mine are shown in Figure 5. Figure 6 shows some plucked grains, gouge, and the

400 ISSUES IN ROCK MECHANICS

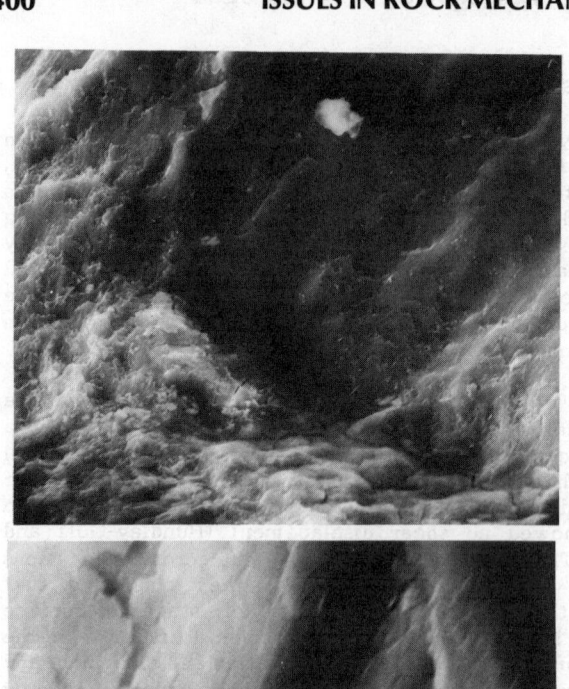

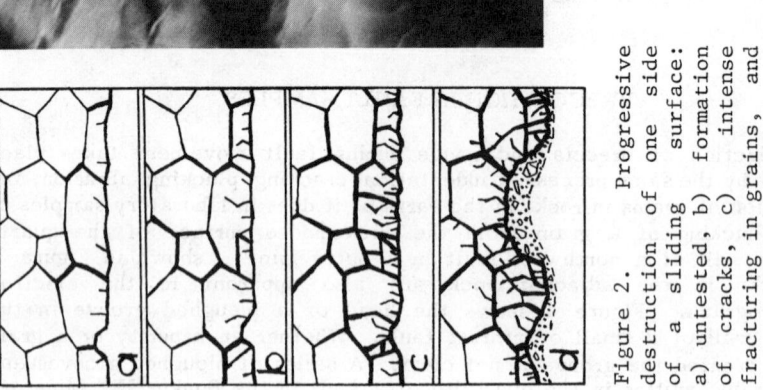

Figure 2. Progressive destruction of one side of a sliding surface: a) untested, b) formation of cracks, c) intense fracturing in grains, and d) pluckout and formation of gouge (Hundley-Goff and Moody 1980, Fig. 7).

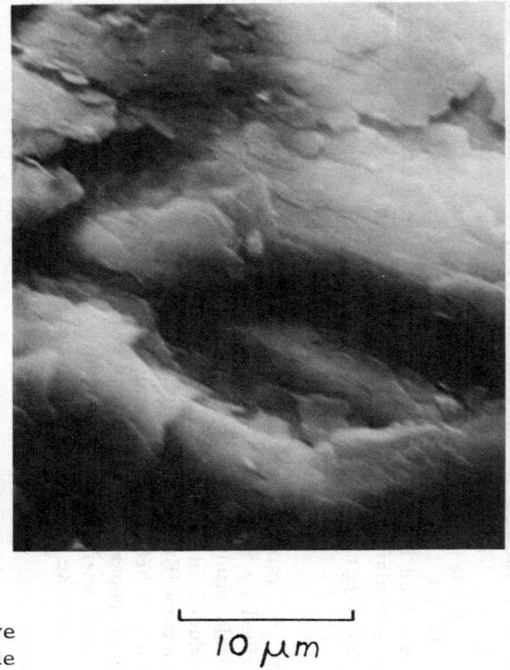

Figure 3. Plucked grain cavity in quartz monzonite, northwest fault on the 3400 level, of the Steward Mine,

Figure 4. Ploughed groove in sandstone wall of a fault in a roadcut, Wolf Creek, Montana.

FORMATION OF GOUGE AND BRECCIA 401

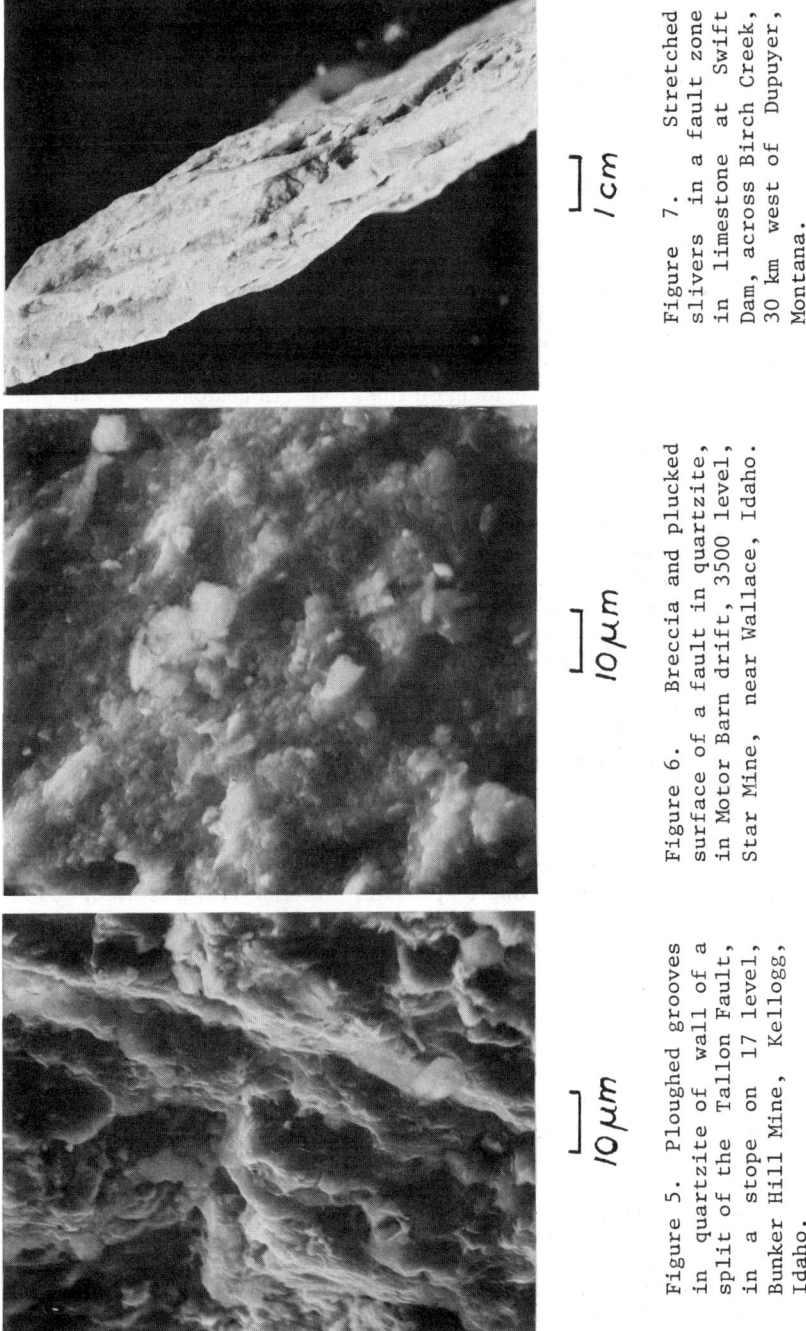

Figure 5. Ploughed grooves in quartzite of wall of a split of the Tallon Fault, in a stope on 17 level, Bunker Hill Mine, Kellogg, Idaho.

Figure 6. Breccia and plucked surface of a fault in quartzite, in Motor Barn drift, 3500 level, Star Mine, near Wallace, Idaho.

Figure 7. Stretched slivers in a fault zone in limestone at Swift Dam, across Birch Creek, 30 km west of Dupuyer, Montana.

slickensided surface of a fault in quartzite in the Star Mine. These features are like those described by Hundley-Goff and Moody (1980), and on the basis of all the microscopic evidence, I conclude that these authors' brecciation model is applicable to faults in the earth.

EFFECTS OF ROCK TYPE AND AMBIENT CONDITIONS

Rocks involved in the faults shown in Figure 1 include quartz monzonite, quartzite, shale, dacite, diabase, dolomite, graywacke, sandstone, and shale, all of which produce about the same proportion of gouge and breccia to fault movement. The independence of rock type of the brecciation process is corroborated by Byerlee's (1978) important finding that friction is also independent of rock type in that all common rocks have the same coefficient of friction at confining pressures of more than 1 kb. Limestone is an exception as it does not produce breccia because it is ductile and not brittle, as shown by the deformed slivers in Figure 7.

In their reviews, Stesky (1978) Engelder (1978), Rutter and Mainprice (1978), and Moody and Hudley-Goff (1980) described the effects of various ambient conditions on faulting. Water in a fault lowers friction strength, causes greater rounding and size reduction of rock fragments, and promotes stable sliding over stick slip. Increase in strength of rocks with increasing confining pressure is closely related to increase in friction strength as normal pressure increases (Byerlee, 1978). However, at low confining pressures, less than 1 kb, compressive and friction strength are lower, and the amount of microcracking (Moody and Hundley-Goff, 1980) and the amount of brecciation (Robertson, in press) are reduced. The history of faulting can be important in breccia formation; reactivation of a fault with a change from high to low confining pressure (or vice versa) would reduce (or increase) friction strength and breccia formation.

The effects of strain rate and of temperature on brecciation can be disregarded (Rutter and Mainprice, 1978). Stesky (1978) stated, "Friction strength of rocks is a rather remarkable property since, for a wide variety of rock types, it is largely independent of mineralogy, temperature to about $400^{\circ}C$, and loading or sliding rate. The dominant variable seems to be confining pressure or normal stress."

CONCLUSIONS

The production of breccia and gouge from the wall rocks in a fault will take place in the following succession as the walls move: (1) formation of microcracks at the boundaries and within grains by asperity and grain-fragment indentation and ploughing; (2) extension of cracks; (3) intense fracturing of grains and slight shearing; (4) plucking of grains and grain fragments to form breccia; and (5) shearing and comminution of fragments to form gouge. In the stick-slip mode of faulting, steps (1), (2), and (3) would take place during stick, accompanied by small displacement; steps (4) and (5) would take place during slip, when most of the displacement would occur. In a stable sliding mode, all steps would take place somewhere on the fault surface. The important conclusion is that these processes will lead to continuous production of gouge and breccia on a fault as displacement continues.

Ambient conditions can be expected to be similar among most faults in the earth at 1 - 10 km depth, and thus production of breccia and gouge would be alike. In general, brecciation would not be affected significantly by temperature, strain rate, and rock type, but water, confining pressure, and history of deformation could affect brecciation and account for the variation in observed thicknesses.

ACKNOWLEDGEMENTS

Electron micrographs, Figures 3, 4, 5, and 6, were made by Edward J. Dwornik, and a direct photo, Figure 7, was prepared by Deborah Dwornik, and I am very grateful to them both. Permission to use Figure 2 by Judith B. Moody and Emily M. Hundley-Goff is also gratefully acknowledged.

REFERENCES CITED

Byerlee, J.D., 1978, "Friction of Rock," in Rock Friction and Earthquake Prediction, J.D. Byerlee and Max Wyss, editors: Birkhauser, Basel, pp. 583-991.

Engelder, J.T., 1978, "Aspects of Asperity-Surface Interaction and Surface Damage of Rocks during Experimental Frictional Sliding," in Rock Friction and Earthquake Prediction, J.D. Byerlee and Max Wyss, editors, Birkhauser, Basel, pp. 705-716.

Engelder, J.T. and Scholz, C.H., 1976, "The Role of Asperity Indentation and Ploughing in Rock Friction-II," International Journal of Rock Mechanics and Mining Science, Vol. 13, pp. 155-163.

Hayes, M.J., 1975, "Effects of Surface Active Fluids on Sliding Behavior of Crab Orchard Sandstone," M. Sc. Thesis, University of North Carolina, Chapel Hill, NC, 58 p.

Hundley-Goff, E.M., and Moody, J.B., 1980, Microroscopic Characteristics of Ortho-quartzite from Sliding Experiments, I. Sliding surface," Tectonophysics, Vol. 62, pp. 279-299.

Kranz, R.L., 1979, "Crack Growth and Development during Creep of Barre Granite," International Journal of Rock Mechanics and Mining Science, Vol. 16, pp. 23-35.

Moody, J.B., and Hundley-Goff, E.M., 1980, "Microscopic Characteristics of Orthoquartzite from Sliding Friction Experiments, II. Gouge," Tectonophysics, Vol. 62, pp. 301-319.

Robertson, E.C., in press, "Relationship of Fault Displacement to Gouge and Breccia Thickness," Society of Mining Engeering, Trans. AIME.

Rutter, E.H., and Mainprice, D.H., 1978, "The Effect of Water on Stress Relaxation of Faulted and Unfaulted Sandstone," in Rock Friction and Earthquake Prediction," J.D. Byerlee and Max Wyss, editors, Birkhauser, Basel, pp. 634-654.

Stesky, R.M., 1978, "Rock Friction—Effect of Confining Pressure, Temperature, and Pore Pressure," in Rock Friction and Earthquake Prediction, J.D. Byerlee and Max Wyss, editors, Birkhauser, Basel, pp. 690-704.

Tapponier, P. and Brace, W.F., 1976, "Development of Stress-induced Microcracks in Westerly Granite," International Journal of Rock Mechanics and Mining Science, Vol. 13, pp. 103-112.

Chapter 43

WORKSHOP ON DEFORMATION MECHANISMS AND TEXTURE DEVELOPMENT IN ROCKS

By H. R. Wenk

Department of Geology and Geophysics

University of California at Berkeley

Berkeley, California 94720

ABSTRACT

This paper describes a workshop on deformation mechanisms and texture development in rocks. Empahsis is on relating macroscopic mechanical properties with microscopic structures and the methods of their quantitative analysis.

INTRODUCTION

In conjunction with the 23rd U.S. Symposium of Rock Mechanics at Berkeley a workshop addresses issues of the elementary structures which underly the macroscopic mechanical behavior and methods of analysis, emphasizing the study of textures. It was possible to attract an interdisciplinary group of scientists, specializing in material science, metallurgy, physics and geology to present the state of the art and also instruct in practical procedures of data collection and analysis.

The program includes the following contributions:

A. REVIEWS AND APPLICATIONS

D.J. Barber. Physics Department. University of Essex, Colchester (England): "The Role of Crystal Defects in the Development of Microstructures in Rocks."

T.G. Langdon. Department of Materials Sciences. University of Southern California, Los Angeles (Calif.): "Steady-State Creep in Rocks."

G. Oertel. Department of Earth and Planetary Sciences. University of California, Los Angeles (Calif.): "Preferred Orientation of Phyllosilicate Grains in Deformed Rocks."

P. Van Houtte. Department of Materials Science. Katholieke Universiteit, Leuven (Belgium): "Texture Interpretation using the Taylor Theory."

H. Mecking. Bereich Werkstoff physik. Technische Universitat Hamburg-Harburg, Hamburg (West Germany): "Texture Development in Metals - an Analog for Rocks."

L.E. Weiss. Department of Geology and Geophysics, University of California, Berkeley (Calif.): "Symmetry of Textures."

H.J. Bunge. Institut f. Metallkunde und Metallphysik. Technische Universitat Clausthal - Zellerfeld (West Germany): "The Harmonic Method of Texture Analysis Applied to Deformed Mineral Aggregates"

C. Esling. Laboratoire de Metallurgie Structurale. Université de Metz, (France): "The Statistical Significance of the Orientation Distribution Function."

H. Schaeben, H. Siemes. Institute Für Mineralogie und Lagerstättenkunde. Technische Universitat Aschen (West Germany): Comparison of the Harmonic and Vector Method of Quantitative Texture Analysis."

A. Vadon and R. Baro. Laboratoire de Metallurgie Structurale. Université de Metz (France): "Application of the Vector Method to Experimentally Deformed Calcite Limestone."

F. Wagner. Laboratoire de Metallurgie Structurale. Université de Metz (France): "Some Calcite Deformation Textures: Comparison of Textures Predicted by the Taylor Theory with Experimental Results."

H. Kern, H.R. Wenk, Institut für Mineralogie und Petrographie, Universitat Kiel (West Germany) and Department of Geology and Geophysics, University of California, Berkeley (Calif.): "Calcite Texture Development in Experimentally Induced Ductile Shear Zones."

G. Price. CSIRO, Mount Waverly (Australia): "Recent Progress in the Analysis and Interpretation of Quartzite Textures."

B. INTRODUCTION TO PRACICAL DATA ANALYSIS

- Interpretation of electron micrographs. D.J. Barber, University of Essex, England.
- Texture determination. Measurement and data reduction. H.R. Wenk, University of California at Berkeley.
- Texture representation. Harmonic analysis. E. Bechler - Ferry University of Metz, France.
- Texture representation. Vector method. A. Vadon. University of Metz, France.

- Texture simulation. F. Wagner. University of Metz, France.
P. Van Houtte, University of Leuven, Belgium.

In the following pages some of these methods and ideas are summarized, relying largely on work of the contributors to the workshop.

Metallurgists have long recognized the close relationship between macroscopic mechanical properties, particularly strength and failure, upon microscopic structures and crystallite orientation. In fact, a large section of metallurgy is concerned with establishing quantitative relationships between microstructure and properties. But while every metallurgist knows that a sheet of metal becomes more brittle by repeated bending because of dislocation strain hardening, rock mechanics has been mainly concerned with circumventing or stabilizing weak links and measuring properties rather than investigating their elementary causes.

This workshop addresses some of these basic principles and may add to a better understanding of mechanical properties of rocks. In contrast to metals rocks already exist and we can generally not modify them by an appropriate production process. Furthermore they are far more complex in structure and composition than metals but many basic principles are nevertheless very similar.

Physical properties are determined by the mineralogical composition, crystallite orientation and shape and by existing cavities. If a rock is subjected to stress, it fails in directions largely controlled by its physical properties which are often extremely unisotropic (e.g. Kern & Fakhimi, 1975, Johnson & Wenk, 1974). Joints for example are generally normal to the lineation and the foliation plane which is determined in a schist by the orientation of mica flakes. Most rocks cleave preferentially parallel to the foliation plane. Like in metals the strength of rocks is a function of texture and direction dependent. Any predictions of these properties have to rely on a quantitative texture analysis.

Mining engineers are mainly concerned with brittle processes, such as failure and fracturing but plastic deformation of rocks has considerable importance for other engineering applications. Plastic deformation in mudstones may result in landslides or in slow creep. Also, in composite rocks macroscopic fractures often develop from microcracks and plastic yield of weak constituent minerals. But plastic deformation also occurs in crystalline rocks. Marble slabs in Greek temples show significant bending as a result of intracrystalline glide processes while they were exposed to their own weight for two thousand years.

Such slow plastic deformation at strain rates of 10^{-10} to 10^{-15} sec.$^{-1}$ governs in most geological environments, both in the crust and in the mantle. Stresses are usually low (estimated at less than a kilobar e.g. Brace and Kohlstedt, 1980) but over the long geologic times sufficient to produce large strains.

SHEETSILICATES AND SCHISTOSITY

Most important for mechanical properties and the anisotropy of polymineralic rocks are phyllosilicates. Mechanisms which lead to the establishment of preferred orientation have received a lot of attention and are also fairly well-understood largely because of the systematic studies of G. Oertel (e.g. 1970, 1980, and particularly in his 1982 review, see also e.g. Bell, 1981, Means & Paterson, 1966, Weber, 1981, Williams, 1977). The calculation of strain from the preferred orientation of platy phyllosilicate grains relies on the simple but powerful model of March (1932) which assumes that rigid platelets reorient themselves in a plastic matrix when strained homogeneously. The starting distribution is usually assumed to be random which implies that the strain accumulated in a pelitic sediment is measured from the moment of first deposition and includes processes of sediment compaction, generally neglected by structural geologists (Oertel and Curtis, 1972).

Originally random distribution of clay flake orientations has been observed in recently deposited sediments in many instances. Mudrocks are almost always compacted by the weight of later deposits. The strain so induced is, as a rule, axially symmetric about the vertical and thus about the pole of the bedding planes. Tectonically caused deformation may start early, before the rock is fully compacted, or it may begin much later, and the mechanisms of deformation increase in complexity with the increasing temperature and hydrostatic stress of deeper burial. Inherited phyllosilicate grains seem to be very resistent if not as mineral species then at least by preservation of the essential structure of their constituent sheets of silica tetrahedra. Given this permanence of phyllosilicate grains, the mechanism of rock deformation, whether intergranular slip, cataclasis, or dissolution and reprecipitation and various intragranular processes in the non-phyllosilicate matrix, hardly matters as far as the recording of strain is concerned. In slates and other deformed mudrocks the average preferred orientations of mica or chlorite grains has been successfully used to estimate strain, and results are similar to those obtained from independent strain markers in the same rocks such as deformed fossils and iron reduction bodies. Also the anisotropy of texture in such rocks corresponds very closely to the anisotropy of physical properties (e.g. Wood et al., 1976). Oertel (1974) has demonstrated that similar principles apply to some metamorphic rocks where he was able to determine strain from texture and by reversing strain to unfold an antiform.

DEFORMATION MECHANISMS IN ROCKS 409

INTRACRYSTALLINE GLIDE

Other rocks particularly monomineralic rocks such as quartzites, marbles, dunites accomodate strain and develop preferred orientation by mechanisms which are similiar to those of metals and a comparison of the two materials is appropriate. While diffusion rates are $10^6 - 10^8$ times slower in rocks than in metals, times available in geologic deformations are not days as in metallurgic processes, but millions of years.

Texture development in metals is predominantly controlled by the forming and annealing operations. Therefore the understanding of the atomistic mechanisms underlying recrystallization and plastic flow is essential for the assessment and control of textures. At present the mechanisms of deformation-texture development are best understood. If deformation is carried out solely by simple dislocation movement, textures can be predicted in great detail on the basis of the continuum theories of polycrystal deformation (e.g. Mecking, 1981, Sevillano et al., 1980).

In the case of recrystallization the controlling mechanisms are less clear, due to the complexity of the physical processes involved in recrystallization. Here two simultaneously active mechanisms of quite different physical nature have to be considered, namely, nucleation and growth of recrystallizing grains (e.g. Gottstein, 1978). By varying the conditions of processing of a metal a great variety of textures can be achieved. In a reverse sense, it ought to be possible to trace the history of formation of a rock from its texture.

The concept of dislocations (line defects) in crystalline materials, first introduced to explain the observed low strengths of simple solids (in comparison with theoretical bond-breaking models) has been proven and supplemented by a wealth of research in the last thirty years. The present advanced state of materials science and many modern developments in metallurgy owe much to dislocation theory and to techniques such as transmission electron microscopy (TEM), by means of which dislocations and other crystal defects can be seen.

It is now clear that dislocation mechanisms cannot be ignored if proper understanding of rock mechanics is to be achieved. Dislocation and point defect mechanisms are central to the flow process of both crustal and mantle rocks, and concepts arising out of dislocation theory have relevance to fracture processes. Interaction between dislocations, and between dislocations and point defects control processes of static and dynamic recovery, creep, recrystallization. At low temperature dislocation glide, including cross slip dominate while at higher temperature diffusive processes become important boundaries and surfaces can act as sources and sinks for point defects and dislocations. The general shape changes in the grains of polycrystalline materials, which are necessary for avoidance of macroscopic fracture, can be achieved both if sufficient deformation systems can be activated, and with the assistance of diffusion mechanisms. Since slip leads to crystal re-orientation without change of crystal structure, both mechanisms can lead to the development of preferred orientations. Cataclasis will occur, even at elevated temperatures, in the absence of adequate slip mechanisms or mechanisms for relieving the build-up of local stress concentrations of the order of the fracture stress. At low temperatures slip will be accompanied by fracture, combined with grain sliding and rolling (Borrodaile, 1981) while increasing temperature will involve firstly a great contribution from grain boundary sliding (Langdon, 1970, 1981a) and secondly, increased diffusional flow. At low stresses and high temperatures, steady state creep can occur via both dislocation climb and diffusional flow mechanisms. The various processes lead to quite different and characteristic microstructures in metals and their counterparts are being increasingly recognized in minerals. They are indicated by particular dependences of strain rate on stress and temperature, as discussed by Ashby and Brown (1981) or Langdon (1981). There is a close parallel between the roles of deviatoric stress in influencing the diffusion paths of point defects and dissolution/deposition sites in models of pressure solution (Stocker & Ashby, 1973; Rutter, 1976). Creep is particularly important in tectonic deformations both in the crust and in the mantle and has been studied in some depth (e.g. Weertman 1970, Carter 1976, Heard 1976, Kirby 1977).

Since only limited information is available on the transient or primary stages of flow in high temperature deformation, most models deal with steady-state conditions.

At temperatures at or above $0.5\ T_m$, where T_m is the absolute melting point, the steady-state creep rate, $\dot{\varepsilon}$, is generally given by

$$\dot{\varepsilon} = \frac{AGb}{kT} \left(\frac{b}{d}\right)^p \left(\frac{\sigma}{G}\right)^n D_o \exp(-Q/RT)$$

where G is the shear modulus, b is the Burgers vector of the dislocations, k is the Boltzmann's constant, T is the absolute temperature, d is the grain size, σ is the applied stress, D_o is the frequency factor, Q is the activation energy, R is the gas constant, A is a dimensionless constant, and p and n are constants.

Steady-state creep behavior falls into three distinct regimes which depend on the value of the stress exponent, n. At high stresses, there is an exponential dependence on stress; at intermediate stresses, there is a power-law behavior and the value of n is typically within the range of $\sim$ 3-7; at low stresses, there is Newtonian viscous behavior and n = 1. The different regimes correlate

DEFORMATION MECHANISMS IN ROCKS

with distinct and clearly defined substructures. For example, subgrain formation occurs in the presence of dislocation climb, and it is now firmly established that, under steady-state conditions, there is an inverse relationship between the average subgrain size and the applied stress (e.g. Kohlstedt & Weathers 1980, Twiss 1977). The transmission electron microcsope is used to characterize these microstructures which can then be interpreted and give information about deformation conditions. This is often done by comparing microstructures in naturally deformed rocks with those obtained in experimental deformations under well-defined conditions (e.g. dolomite: Barber et al., 1981).

TEXTURE DEVELOPMENT

We have looked so far at microscopic processes within crystals and along grain boundaries. The expression of dislocation movements in a polycrystal is generally the formation of a texture. It relies on rotations of individual cyrstallites but - as a whole - can be treated as a continuum. If strain is accomodated by slip and mechanical twinning - rather than due to reorientation of rigid anisotropic platelets as discussed above - models proposed by Sachs (1928) and Taylor (1938) are applicable to predict the texture given the slip systems and the strainpath (see e.g. review by Sevillano et al. 1980). The Taylor theory particularly has become a very useful tool to predict textures in metals (e.g. Chin et al. 1969, Kocks 1970, Van Houtte 1978) and applies also to minerals (quartz: Lister et al. 1978, 1979; calcite: Lister 1978, Wagner et al. 1982). An excellent test for the Taylor theory are limestones which permit a quantitative comparison of theoretical simulatio and laboratory experiment. Deformation mechanisms in calcite are well known (e.g. Turner et al. 1954) and in the triaxial apparatus of Kern (1976) textures of low symmetry (plane strain with both pure and simple shear) can be produced which are less ambiguous than those attained in conventional uniaxial compression tests (e.g. Wenk et al. 1972). So far agreement of simulated and experimentally produced textures is very satisfactory both at low temperature where mechanical twinning dominates and at high temperature where slip is more important (Wagner et al. 1982). Low symmetry deformation is present in most geological environments, particularly plane strain in folding and on thrust planes. The large majority of experiments have been conducted in a geometry with fixed strain axes but rotational components such as in pure shear are most significant and contain more possibilities of interpreting the strain path. Friedman and Higgs (1981) and Kern and Wenk (1982) have obtained shear textures in carbonate rocks. In limestone very strong monoclinic textures were produced which can be derived from simple shear geometry (Kern and Wenk, 1982).

The close relationship between the microscopic elementary scale and the macroscopic properties of deformed rocks was pointed out. It is directly analogous to the concept of crystallographers who discovered a long time ago that one has to study the atomic structure in order to understand the external crystal morphology (e.g. genotype-phenotype concept of P. Niggli, 1924).

EXPERIMENTAL APPROACH

Different methods are used to establish this relationship in polycrystals than in single crystals. Electron microscopy plays an important role (e.g. Christie and Ardell, 1976). A complex methodology had been developed to measure, represent and interpret preferred orientation. The next paragraphs will describe some of the techniques and problems. At a practical session of the workshop Evelyne Bechler, Francis Wagner, Albert Vadon (Metz, France) will introduce the procedures which rely largely on sophisticated computer programs.

Preferred orientation of crystallites in a rock can be measured grain by grain on the universal stage microscope (e.g. Turner and Weiss, 1963) and fabric diagrams are constructed. More recently pole-figures have been measured by X-ray and neutron diffraction and by photometric optical methods (Price, 1973). Diffraction methods are most widely used in metallurgy and are becoming increasingly popular in geology. But there are problems with pole-figures. The correlation of individual crystal directions in various pole-figure is lost (compare Wenk et al 1981, Wagner et al. 1981). In X-ray diffraction, pole-figures are determined on slabs in reflection or in thin sections in transmission. Both geometries require rather involved intensity corrections (e.g. Baker, et al. 1969; Oertel, 1970). This is greatly facilitated with neutron diffraction where large spherical specimens can be used and absorption is minimal (Morris, 1978). In addition application of 2θ-sensitive continuous detectors allows to measure many pole-figures simultaneously which is particularly convenient for minerals with complex spectra of many peaks and frequent overlaps (Bunge et al. 1982).

TEXTURE REPRESENTATION

Traditionally textures of deformed rocks have been described with fabric diagrams depicting the orientation distribution of a single crystallographic direction. This is not very complete and therefore Roe (1965) and Bunge (1965) have suggested a three-dimensional representation of Euler angles ϕ_1, ϕ, ϕ_2, (ODF) to relate three-dimensional coordinates with the three-dimentional crystal coordinates with the three-dimensional specimen (or strain) coordinates. A problem with Euler angles arises because by definition they relate right handed coordinate systems while most crystals and specimen geometries are centrosymmetric (e.g. Mathies, 1980, 1981, Bunge et al. 1980, Weiss and Wenk, 1982). But the ODF has nevertheless many successful applications. It serves as the weight function when mean values of anisotropic properties of polycrystalline samples are to be calculated from the corresponding properties of their constituent crystals (e.g.

Young's modulus, sound velocity, yield stress, thermal expansion and so on).

There are two main methods to determine the ODF from measured pole-figures. The first and widely used method introduced by Bunge (1965, 1969) and Roe (1965) is the harmonic series expansion method. The experimentally determined pole-figures as well as the unknown ODF are developed into series of spherical harmonics or generalized harmonics respectively. The integral relation between both functions leads to a linear relation between their respective coefficients which can be solved for C_l^{mn} coefficients of the ODF if the F_l^n (hkl) coefficients of the pole-figure are known. Once the coefficients C_l^{mn} are known, the ODF can be calculated for any desired orientation ϕ_1, Φ, ϕ_2. Computer programs performing the calculations are available.

The second method of reproducing the ODF is a generalization of biaxial method (Williams 1968) and was recently developed by Ruer (1976) and Vadon (1981). It is known as the vector method (not to be mistaken for the vector method in crystallography). Both pole-figure and texture function are discretized into vectors of high dimensions, which are then coupled by a corresponding matrix. Establishing this matrix the geometrical interdependence of the direct pole-figure, the inverse pole-figure, and orientation space as well as the interdependence of the intensities at distinct points in one pole-figure are taken into account. The general pole-figure is no longer treated as a two dimensional projection of the ODF. As the vector method thus makes use of all available geometrical information and the mathematically required positiveness of the ODF it is capable to considerably diminish the number of required pole-figures and even to process incomplete pole figures (Ruer, Baro 1977; Vadon 1981). Programs for application of the vector method to minerals have been developed but they are not yet quite ready for routine use.

At present it is rather difficult to find summaries which give an introduction to topics of this workshop. Older textbooks such as those of Sander (1950), Turner and Weiss (1963) are no longer up-to-date and excellent compilations such as that of Nicolas and Poirier (1976) have a broader goal and only touch on some aspects. Hopefully an expanded version of reviews presented at the workshop and complemented by additional contributions will soon give a comprehensive introduction to texture analysis of geological materials.

Acknowledgments: The workshop became possible through enthusiastic collaboration of an interdisciplinary group of scientists. Since results will mainly help in interpreting the mechanical history of rocks, geologists are most obliged to materials scientists, physicists and mathematicians for lending their expertise to this endeavor. Acknowledged is the support of travel by various institutions which enabled many to attend the Berkeley conference and a grant from the National Science Foundations.

REFERENCES

Ashby, M.F., and Brown, A.M, 1981, "Flow in Polycrystals and the Scaling of Mechanical Properties," <u>Deformation of Polycrystals, Mechanisms and Microstructures</u>, Hansen et al. eds., Riso International Symposium, Roskilde, Denmark, pp. 1-13.

Baker, D.W., Wenk, H.R., & Christie, J.M., 1969, "X-ray Analysis of Preferred Orientation in Fine-Grained Quartz Aggregates," <u>Journal of Geology</u>, Vol 77, pp. 143-172

Barber, D.J., Heard, H., & Wenk, H.R., 1981, "Deformation of Dolomite Single Crystals from 20-800°C," <u>Physics and Chemistry of Minerals</u>, Vol. 7, pp. 271-286.

Brace, W.F., & Kohlstedt, D.L., 1980, "Limits on Lithospheric Stress Imposed by Laboratory Experiments," <u>Journal of Geophysical Research</u>, Vol. 85, pp. 6248-6252.

Bell, T.H., 1981, "Foliation Development - The Contribution, Geometry and Significance of Progressive, Bulk, Unhomogeneous Shortening," <u>Tectonophysics</u>, Vol.75, pp. 273-296.

Borradaile, G.H., 1981, "Particulate Flow of Rock and the Formation of Cleavage," <u>Tectonophysics</u>, Vol. 72, pp. 305-321.

Bunge, H.J., 1965, "Zur Darstellung Allgemeiner Texturen," <u>Zeitschr. Metallkunde</u>, Vol. 56, pp. 872-874.

Bunge, H.J., 1969, <u>Mathematische Methoden der Texturanalyse</u>, Akademie Verlag: Berlin. pp 330.

Bunge, H.J., Esling, C. & Müller, L., 1980, "The Role of the Inversion Centre in Texture Analysis," <u>Journal of Applied Crystallography</u>, Vol. 13, pp. 544-554.

Bunge, H.J., Wenk, H.R., & Pannetier, J., 1982, "Neutron Diffraction Texture Analysis using a 2Θ Position Sensitive Detector," <u>Texture and Microstructures</u> (submitted).

Carter, N.L., 1976, "Steady State Flow of Rocks," Review of <u>Geophysics and Space Physics</u>, Vol. 14, pp. 301-360.

Chin, G.Y., Hosford, W.F., and Mendorf, D.R., 1969, "Accomodation of Constrained Deformation in F.C.C. Metals by Slip and Twinning," <u>Royal Society Proceedings, A</u>, Vol. 309, pp. 433-456.

Christie, J.M., and Ardell, 1976, "Deformation Structures in Minerals," <u>Electron Microscopy in Mineralogy</u>, Wenk, H.R., et al., eds. Springer Verlag, Heidelberg, pp. 374-403.

Friedman, M. and Higgs, N.G., 1981, "Calcite Fabrics in Experimental Shear zones," Mechanical Behavior of Crustal Rocks, The Handin Volume, Carter, et al., eds., American Geophysical Union Monograph, Vol. 24, pp. 11-27.

Gottstein, G., 1978, "Recent Aspects in the Understanding of Recrystallization Texture Development," Textures of Materials, Gottstein, G., and Lücke, K., eds., Springer Verlag, Heidelberg, pp. 93-110.

Heard, H.C., 1976, "Comparison of the Flow Properties of Rocks and Crustal Conditions," Philosophical Transactions of the Royal Society of London, A, Vol. 283, pp. 173-186.

Johnson, L., and Wenk, H.R., 1974, Anistropy of Physical Constants in Metamorphic Rocks," Tectonophysics, Vol. 23, pp. 79-98.

Kern, H., and Fakhimi, 1975, "Effect of Fabric Anisotropy on Compressional-Wave Propagation in Various Metamorphic Rocks for the Range 20 - 700°C at 2 kbars," Tectonophysics, Vol. 28, 227-244.

Kern, H., 1976, "Preferred Orientation of Experimentally Deformed Limestone, Marble, Quartzite and Rock Salt at Different Temperatures and States of Stress," Tectonophysics, Vol. 39, pp. 103-120.

Kern, H., and Wenk, H.R., 1982, "Experimental Development of Preferred Orientation in a Ductile Shear Zone," Tectonophysics (submitted).

Kirby, S.H., 1977, "States of Stress in the Lithosphere: Inferences from the Flow Laws of Olivine," Pure and Applied Geophysics, Vol. 115, pp. 245-258.

Kocks, U.F., 1970, "The Relation Between Polycrystal Deformation and Single Crystal Deformation," Metallurgical Transactions, Vol. 1, pp. 1121-1143.

Kohlstedt, D.L., Weathers, M., 1980, "Deformation - Induced Microstructures, Paleopiezometers, and Differential Stresses in Deeply Eroded Fault Zones," Journal of Geophysical Research, Vol. 85, pp. 6269-6285.

Langdon, T.G., 1970, "Grain Boundary Sliding as a Deformation Mechanism During Creep," Philosophical Magazine, Vol. 22, pp. 689-700.

Langdon, T.G., 1981a, "Current Problems in Superplasticity," Creep and Fracture of Engineering Materials and Structures, Wilshire, B, and Owen, O.R.J., eds., Pinridge Press, Swansea, pp. 141-156.

Langdon, T.G., 1981b, "Deformation of Polycrystalline Materials at High Temperature," Deformation of Polycrystals.Mechanisms and Microstructures," Hansen, N., et al., eds. Riso International Symposium, Roskilde, Denmark, pp. 45-54.

Lister, G.S., 1978, "Texture Transition in Plastically Deformed Calcite Rocks," Proceedings of the Fifth International Conference on Texture of Materials, Aachen, Gottstein, G., and Lücke, K., eds., Springer Verlag, Vol 2, pp. 199-210.

Lister, G.S., Paterson, M.S. Paterson, and Hobbs, B.E., 1978,, "The Simulation of Fabric Development in Plastic Deformation and its Application to Quartzite: The Model," Tectonophysics, Vol. 45, pp. 107-158.

Lister, G.S., and Paterson, M.S., 1979, "The Similation of Fabric Development during Plastic Deformation and its Application to Quartzite: Fabric Transitions," Journal of Structural Geology, Vol. 1, pp. 99-115.

March, A., 1932, "Mathematische Theorie der Regelung nach der Korngestalt bei Affiner Deformation," Zeitschrift fuer Kristallographie, Vol. 81, pp. 285-297.

Mathies, S., 1980, 1981, "On the Reproducibility of the Orientation Distribution Function of Texture Samples from Pole Figures, I-VII Crystal Research and Technology, Vol. 15, pp. 431-444; pp. 601-614; pp. 823-835; pp. 1189-1195; pp. 1323-1328. Vol 16, pp. 513-520; pp. 1061-1071.

Means, W.D., and Paterson, M.S., 1966, "Experiments on Preferred Orientation of Platy Minerals," Contributions to Mineralogy and Petrology, Vol 13, pp. 108-133.

Mecking, H., 1981, "Low Temperature Deformation of Polycrystals. Deformation of Polycrystals.Mechanisms and Microstructures," Hansen, N., et al., eds., Riso International Symposium, Roskilde, Denmark, pp. 73-86.

Morris, P.R., 1978, "Recent Developments in Experimental Methods of Texture Measurement," Textures of Materials, Gottstein, G., and Lücke, K., eds., Springer Verlag, Heidelberg, pp. 17-24.

Nicolas, A., and Poirier, J.P., 1976, Crystalline Plasticity and Solid State Flow in Metamorphic Rocks, Wiley: New York, PP. 444.

Niggli, P., 1924, Lehrbuch der Mineralogie I. Allgemeine Mineralogie, Borntraeger, Berlin.

Oertel, G., 1970, "Deformation of a Slaty Lapillar Tuff in the Lake District, England," Geological Society of America Bulletin, Vol. 81, pp. 1173-1188.

Oertel, G., and Curtis, C.D., 1972, "Clay-Ironstone Concretion Preserving Fabrics due to Progressive Compaction," Geological Society of America Bulletin, Vol. 83, pp. 2597-2606.

Oertel, G., 1980, "Strain in Ductile Rocks on the Convex Side of a Folded Competent Bed," Tectonophysics, Vol. 66, pp. 15-34.

Oertel, G., 1982, "Preferred Orientation of Phyllosilicate Grains in Deformed Rocks," Bulletin of The Geological Society of America, (submitted).

Price, G.P., 1973, "The Photometric Method in Microstructural Analysis," American Journal of Science, Vol. 273, pp. 523-537.

Roe, R.J., 1965, "Description of Crystallite Orientation in Polycrystalline Materials. III. General Solution of Pole Figure Inversion," Journal of Applied Physics, Vol. 36, pp. 2024-2031.

Ruer, D. 1976, "Méthode Vectorielle d'Analyse de la Texture," Thèse, Université de Metz, pp. 163.

Ruer, D., and Baro, R., 1977, "A New Method for the Determination of the Texture of Material of Cubic Structure from Incomplete Reflection Pole Figures," Advanced X-ray Analysis, Vol. 20, pp. 187-200.

Rutter, E.H., 1976, "The Kinetics of Rock Deformation by Pressure Solution," Philosophical Transactions of the Royal Society of London, A, Vol. 283, p. 203-219.

Sachs, G., 1928, "Zur Ableitung einer Fliessbedingung," Zeitschrift Verein Deutscher Ingenieure, Vol. 72, pp. 734-736.

Sander, B., 1950, Einführung in die Gefügekunde der Geologischen Körper, II., Springer: Wien.

Sevillano, J., Van Houtte, P., and Aernoudt, E., 1980, "Large Strain Work Hardening and Textures," Progress in Materials Science, Vol. 25, pp. 69-412.

Stocker, R.L., and Ashby, M.F., 1973, "On the Rheology of Upper Mantle," Reviews of Geophysics and Space Physics, Vol. 11, pp. 391-426.

Taylor, G.I., 1938, "Plastic Strain in Metals," Journal for the Institute of Metals, Vol. 62, pp. 307-324.

Twiss, F.J. 1977, "Theory and Applicability of Recrystallized Grain Size Paleopiezometer," Pure and Applied Geophysics, Vol. 11, pp 227-244.

Turner, F.J., and Weiss, L.E., 1963, "Structural Analysis of Metamorphic Tectonites," McGraw Hill, New York, pp. 545.

Turner, F.J, Griggs, D.T., and Heard, H.C., 1954, "Experimental Deformation of Calcite Crystals," Geological Society of America Bulletin, Vol. 15, pp. 883-933.

Vadon, A., 1981, "Généralisation et Optimisation de la Méthode Vectorielle d'Analyse de la Texture," Thèse, Universite de Metz, pp. 298.

Van Houtte, P., 1978, "Simulation of the Rolling and Shear Texture of Brass by the Taylor Theory Adapted for Mechanical Twinning," Acta Metallurgica, Vol. 26, pp. 591-604.

Wagner, F., Wenk, H.R., Esling, C., and Bunge, H.J., 1981, "Importance of Odd Coefficients in Texture Calculations for Trigonal-Triclinic Symmetries," Physica Status Solidi,(a), Vol. 67, pp. 269-285.

Wagner, F., Wenk, H.R., Kern, H, Van Houtte, P., Esling, C., 1982, "Development of Preferred Orientation in Plane Strain Deformed Limestone. Experiment and Theory," Contributions to Mineralogy and Petrology, (submitted).

Weber, K., 1981, "Kinematic and Metamorphic Aspects of Cleavage Formation in Very Low-Grade Metamorphic Slates," Tectonophysics, Vol. 78, pp. 291-306.

Weertman, J., 1970, "The Creep Strength of the Earth's Mantle," Review of Geophysics and Space Physics," Vol. 8, pp. 145-168.

Weiss, L.E., and Wenk, H.R., 1982, "Symmetry Considerations in Texture Analysis, (in preparation).

Wenk, H.R., Venkitasubramanyan, C.S., Baker, D.W., Turner, F.J., 1972, Preferred Orientation in Experimentally Deformed Limestone," Contributions to Mineralogy and Petrology, Vol. 38, pp. 81-114.

Wenk, H.R., Wagner, F., Esling, C., Bunge, H.J., 1981, "Texture Representation of Deformed Dolomite Rocks, Tectonophysics, Vol. 78, pp. 119-138.

Wenk, H.R, and Wilde, W.R., 1972, "Orientation Distribution Diagrams for Three Yule Marble Fabrics," Flow and Fracture of Rocks, Heard et al., eds., Geophysical Monograph, American Geophysical Union, Wash., Vol. 16, pp. 83-94.

Williams, P.F., 1977,"Foliation: A Review and Discussion," Tectonophysics, Vol. 39, pp. 305-328.

Williams, R.O., 1968, "Analytical Methods for Representing Complex Textures by Biaxial Pole Figures," Journal of Applied Physics, Vol. 39, pp. 4329-4335.

Wood, D.S., Oertel, G, Singh, J., Bennett, H.F., 1976, Strain and Anisotropy in Rocks, Philosophical Transactions of the Royal Society of London, A, Vol. 283, pp. 27-42.

FRACTURE

Chairman
Michael Hood
University of California
Berkeley, California

Co-Chairman
Anthony R. Ingraffea
Cornell University
Ithaca, New York

Keynote Speaker
Joe L. Ratigan
RE/SPEC Inc.
Rapid City, South Dakota

Chapter 44

AN EXAMINATION OF THE TENSILE STRENGTH OF BRITTLE ROCK

by Joe L. Ratigan

Staff Scientist, RE/SPEC Inc.
Rapid City, SD

INTRODUCTION

Rock mechanics engineers are seldom concerned with obtaining the tensile or fracture strength of brittle rock at low mean stresses. The reason for this is two fold. Firstly, the behavior of many excavations in brittle rock is controlled by the inherent discontinuities in the rock mass. Because of the jointing and faulting, tensile strength is usually taken to be negligible. The second reason that tensile strength is seldom obtained in the laboratory is that the technology of utilizing this strength measure in design is undeveloped. Despite the fact that tensile strength of brittle rock is often ignored, there are a number of applications in rock mechanics wherein the knowledge of tensile strength is of fundamental importance. The apparent tensile strength must be known in a hydraulic fracture experiment if the state of in situ stress is to be determined from the initiation of the hydraulically induced fracture. In certain underground situations, the apparent tensile strength of intact rock beams defined by jointing or bedding planes is important in determining required rock bolting. Numerous other situations require a knowledge of apparent tensile strength including such high technology uses of underground space as geothermal energy extraction and LPG storage.

Three observations are invariably made when intact rock samples are taken into the laboratory and tested to determine tensile strength.

(1) The apparent tensile strength depends on the sample size (the larger the specimen, the lower the tensile strength)

(2) The apparent tensile strength depends upon the type of test being performed.

(3) With any given test and specimen size, a scatter (usually skewed) about the mean is obtained.

The first dilemma (commonly referred to as the <u>size effect</u>) is also observed with respect to strength at higher (<u>more compressive</u>) mean stresses, although to a lesser extent than with tensile strength. The observation of size effect has prompted many investigators in rock mechanics to recognize that the tensile strength measured at the usual laboratory scale is not a material property. The second observation noted above is often disregarded for lack of an explanation. Rather than explaining the differences in strength measured in different types of tests, we readily give adjectives from the test type to label the strength measure. For example, we refer to the Brazilian tensile strength, the direct tensile strength, the Modulus of Rupture (a strength measured in bending), etc. Although the rock mechanics community has not completely come to the realization that the same phenomena occurs in fracture toughness testing, it is becoming popular to refer to the fracture toughness of rock with an adjective taken from the test type (Ouchterlony (1980)). The third observation noted above is often totally neglected in the reporting of test results. Scatter about the mean is often attributed to testing methodology and/or sample-to-sample inhomogeneity. Statistical moments can be useful in understanding the strength characterization of brittle rock and should be reported.

The purpose of this paper is to discuss various theories of tensile or fracture strength of brittle rock and to examine critically how these theories address the laboratory observations noted above. A brief review of the foundations of each of the strength characterizations is presented. This review is followed by laboratory results for two granites and two marbles.

Portions of this paper (particularly that which relates to statistical fracture mechanics) have appeared in more extensive detail in other works of the author (Ratigan (1981), (1982)).

STRENGTH CHARACTERIZATIONS

In the discussion of the various theories of strength of brittle rock we shall refer to a global and a local failure criterion. The global criterion relates to the structural collapse or loss of load carrying ability of the entire structure being considered. The local failure criterion relates to the failure of a material region.

Continuum Strength

Much of the technology that we use in rock mechanics is <u>borrowed</u> from the <u>metals field</u>. Strength characterizations and testing are no exception. In a direct tension test of a metal specimen, the cross section that finally fractures will have on the order of one thousand grains across a diameter. However, in rock mechanics, we only suggest that more than ten grains be across any diameter in the same test.

The most commonly encountered method of determining or estimating failure of a structure in brittle rock is that in which the strength is considered to be a pointwise property of the rock. This continuum strength approach is commonly taught in the academic community in strength of materials courses and is the usual background of most graduate engineers. In the continuum strength approach, the global and local criteria are taken to be identical. Once a material region has satisfied some local criterion or has failed, the structure is considered to have lost its operation function. This method of limit design is often used, for example, in evaluating the potential for failure of a rock beam in a roof of an underground opening or for determining when a fracture might propagate in a hydraulic fracture experiment.

A great deal of effort has been expended in the rock mechanics community for determining the local failure criteria for brittle rock as it is used in the continuum strength approach. Perhaps the first recognizable effort in developing a local criterion for determining failure in tension in brittle materials was due to Griffith (1921). Numerous modifications of Griffith's criterion have appeared in the literature. For example, McClintock and Walsh (1962) considered friction on the hypothesized crack of a Griffith criterion to arrive at what is called the modified Griffith criterion. The research effort expended on the Griffith criterion took place despite the fact that the criterion cannot explain any of the three observations noted in the introduction to this paper. In fact, the continuum strength approach cannot explain any of the three observations, regardless of the local criterion.

Contemporary local criteria used in the continuum strength approach for general states of stress (compressive or tensile) often involve a pressure dependent shear criteria with a tension cutoff at some finite magnitude of tensile normal stress. The purpose of the tension cutoff is to remove all but the shear failures from a shear failure criterion. However, it is common to observe the results from a Brazilian test incorrectly plotted in a Mohr diagram with unconfined and triaxial test results. Further, it is not clear that an unconfined compression tests should be plotted on this figure if the unconfined test resulted in axial splitting of the specimen.

Deterministic Fracture Mechanics

Fracture mechanics has only been an entity in the engineering profession for about sixty years (following the introduction by Griffith in 1921). The field has however, only been actively researched in the last thirty years. This research has been mainly concerned with the fracture behavior of metals and ceramics. Research into the fracture mechanics of rock has only occurred over the last ten to fifteen years. In recent years, fracture toughness testing in rock has become quite popular despite the fact that this strength measure is not commonly used in any facet of rock mechanics design.

The fracture mechanics approach to the strength of brittle rock requires the delineation of a material defect or crack. In this regard, delineation includes the specification of the size and location of the crack within the structure in question. This is not always a severe restriction in the laboratory. However, the requirement of crack delineation almost always negates the use of deterministic fracture mechanics in field situations. Once the crack has been defined, the stress intensity at the crack tip can be stated from a knowledge of the boundary conditions and geometry on the structure. The stress intensity factor formulas which are available in many texts and references in the literature typically assume a one dimensional defect. That is to say, the crack is sufficiently flat so as to preclude any increase in stress intensity by any in-plane stresses. Some igneous rocks possess defects satisfying this requirement; however, few sedimentary rocks satisfy the restriction.

If the stress intensity factor at the crack tip becomes equal to the fracture toughness of the material, the crack will instantaneously propagate. Thus, the local criterion for failure in the deterministic fracture mechanics approach is that the stress intensity factor cannot exceed the fracture toughness of the material. After the stress intensity factor has reached the fracture toughness, the crack may propagate for a finite distance and stop or it may propagate until the structure in which the crack resides looses its load carrying ability. Whether the crack propagates or not depends on the change in the strain energy release rate with respect to an increase in crack length (Griffith (1921)). Thus, this is the global criterion for failure in the deterministic fracture mechanics approach to the strength of brittle rock.

Two of the observations which we note in the laboratory with respect to strength at low mean stress can be explained with deterministic fracture mechanics. The apparent tensile strength for a material with fixed defect size will depend on the specimen geometry. The apparent tensile strength will naturally depend on the type of test being performed since, again, the stress intensity factor depends on the geometry, boundary conditions and loading of the specimen. The deterministic fracture mechanics approach cannot be used to explain the third observation from the laboratory testing at low mean stress. Since the method is deterministic, only a singular value of strength can be predicted from a given test with a specific rock type.

Statistical Fracture Mechanics

The foundation of statistical fracture mechanics rests in the supposition that all materials inherently contain defects that eventually lead to structural collapse under increasing load. It is obvious that all geological materials inherently contain mechanical defects. These flaws or defects need not be physical voids, but may be soft minerals in contact with a significantly stiffer matrix material (Brace (1964)). Statistical fracture mechanics differs from a Griffith type criterion in that the material defect or crack need not be uniform in size or in strength. Statistical fracture mechanics also recognizes that more than one defect can exist in a material, whereas the Griffith criterion only requires a single crack. These are the three primary factors which differentiate statistical fracture mechanics from deterministic fracture mechanics.

The global failure criterion for a statistical fracture mechanics relates to the number of defects or flaws which must fail for structural collapse or loss of load carrying ability. The number of possibilities is infinite. However, a single defect criterion is usually selected. Single defect criteria are termed <u>Weakest Link Models</u> (Weibull (1939a)), for the collapse of the specimen or structure depends only on the failure of a single link in a series structure of links. The cumulative failure probability, G, for a weakest link model can be stated as

$$G = 1 - \exp \{- \int_R n(\sigma)dR\} \qquad (1)$$

where:

    R = the geometric domain of the structure where defects reside

    $n(\sigma)$ = material function (number of flaws per unit region with strength $< \sigma$)

    σ = stress (tension positive)

The geometric domain, R, may comprise the volume of the specimen or structure as well as the free surface. Thus, failure may be caused by surface flaws or defects or the failure may be caused by volumetrically distributed flaws depending upon the type of loading. If a material obeys the weakest link theory, Equation (1) is an exact characterization of the material strength. However, empirism will be introduced in the specification of the material function $n(\sigma)$, the local failure criterion.

Weibull defined a term B, which he referred to as the risk of rupture, viz:

$$B = \int_R n(\sigma)\, dR \tag{2}$$

The function $n(\sigma)$ may be determined in the laboratory for a specific test (Evans and Jones (1978)). This methodology will not be adopted, however, because of the lack of generality. Weibull (1939b) proposed a functional form for uniaxial tension;

$$n(\sigma) = \begin{cases} \left[\dfrac{x(\sigma) - x_u}{x_0}\right]^m & x(\sigma) > x_u \\ 0 & x(\sigma) \leq x_u \end{cases} \tag{3}$$

where:

x = some suitable function of stress

x_u = the value of x below which rupture does not occur

x_0 = scaling constant

m = Weibull modulus.

Weibull stated that the region R, could well be both a volumetric region in addition to a free surface region, so that in the general case, the risk of rupture can be stated as;

$$B = \int_S n_s(\sigma)\, dS + \int_V n_v(\sigma)\, dV \tag{4}$$

where:

$$n_s(\sigma) = \begin{cases} \left[\dfrac{x_s(\sigma) - x_{u_s}}{x_{0_s}}\right]^{m_s} & x_s(\sigma) > x_{u_s} \\ 0 & x_s(\sigma) \leq x_{u_s} \end{cases}$$

$$n_v(\sigma) = \begin{cases} \left[\dfrac{x_v(\sigma) - x_{u_v}}{x_{0_v}}\right]^{m_s} & x_v(\sigma) > x_{u_v} \\ 0 & x_v(\sigma) \leq x_{u_v} \end{cases}$$

Weibull's theory is often erroneously stated in the literature as;

TENSILE STRENGTH OF BRITTLE ROCK

$$T_a V^{1/m} = \text{constant} \tag{5}$$

where:

T_a = apparent tensile strength.

Equation (5) is a specific form of Weibull's theory that only arises following the assumptions that (1) $x_u = 0$ and (2) $n_s(\sigma) = 0$. Both of these assumptions must be verified in the laboratory before being discarded or adopted.

Weibull selected the function $x(\sigma)$ to be the tensile stress normal to a material defect or flaw. However, others have suggested that a more appropriate selection for the function $x(\sigma)$ may be the strain energy release rate associated with Mode I fracture. If we assume that a material contains flat, non-interacting cracks, the critical strain energy release rate associated with Mode I fracture can be shown to be;

$$\mathcal{G} = k \sigma^2 = x(\sigma) \tag{6}$$

where:

k = a proportionality constant

σ = the tensile stress normal to the crack.

With this selection for $x(\sigma)$, we may also make the assignments;

$$\begin{aligned} x_0 &= \mathcal{G}_0 \\ x_u &= \mathcal{G}_u \end{aligned} \tag{7}$$

For any tension test, uniaxial or multiaxial, the stress state in the specimen can be stated as;

$$\sigma(x_i) = T_a f(x_i) \quad i = 1,2,3 \tag{8}$$

where:

x_i = spatial coordinate

$f(x_i)$ = function whose value is ≤ 1 in the domain where $\sigma(x_i) > 0$.

Equation (8) permits the risk of rupture with $\mathcal{G}_u = 0$[†] to be determined as;

$$B = T_a^m \left[C_s + C_v \right] \tag{9}$$

[†]When $\mathcal{G}_u = 0$, the model is termed a two parameter model; otherwise it is called a three parameter model.

where:

C_s = a constant arising from integration over the surface

C_v = a constant arising from integration over the volume.

Thus, the cumulative probability of failure, $G(T_a)$ can be stated as;

$$G(T_a) = 1 - \exp\{-T_a C^m\} \qquad (10)$$

where:

C = a constant.

Rearranging and taking logarithms twice, Equation (10) becomes:

$$\ln \ln \left[\frac{1}{1 - G(T_a)}\right] = m \ln \left[T_a C^{\frac{1}{m}}\right]$$
$$= m \ln T_a + \ln C \qquad (11)$$

A multitude of methods exists for determining the parameters for a two parameter model whether the variant is stress or strain energy (e.g., Robinson (1970), Robinson and Finnie (1969)). However, we may categorize all of the methods into two general classes, viz;

Class I: Methods relying on the frequency histogram of measured strengths or statistical moments of the histogram for a given test series.

Class II: Methods relying on the relationship between statistical moments (usually, the mean) between two or more test series.

Despite the extreme difficulty in determining the histogram accurately, methods in Class I above appear to have received the greatest application in the literature. The most popular method in Class I is to employ the formulation of Equation (11). If a plot of experimental data deviates from linearity, the material is not well represented with Equation (10). If the curvature is concave downward, $\mathcal{G}_u \neq 0$ is suggested. Other potential material features can be identified in this functional space; for example, anisotropy.

A very significant feature of a two parameter model is that the coefficient of variation (standard deviation divided by the mean) depends only upon the parameter m. Thus, for any tensile test of any sample size, the coefficient of variation is constant if the material is well represented by a two parameter model.

TENSILE STRENGTH OF BRITTLE ROCK

STRENGTH MEASUREMENTS

In this section of the paper we shall examine several series of test data with respect to each of the tensile or fracture strength theories previously discussed with the objective of evaluating the aptness of each theory as applied to various rock types.

Uniaxial Tension

The uniaxial tension tests which will be discussed in this paper include the direct tension test and the four point bend test. The rock mechanics profession is well aware that direct tension tests are difficult to perform on rock. It is time consuming to machine the specimens to a "dogbone" shape, and thus, this is seldom done. Using a right circular cylinder and following the ASTM procedure (ASTM (1974)) however, it is impossible to perform a direct tension test on a sedimentary rock with a low modulus. For this reason, laboratory tests for direct tension usually deviate from the ASTM specifications.

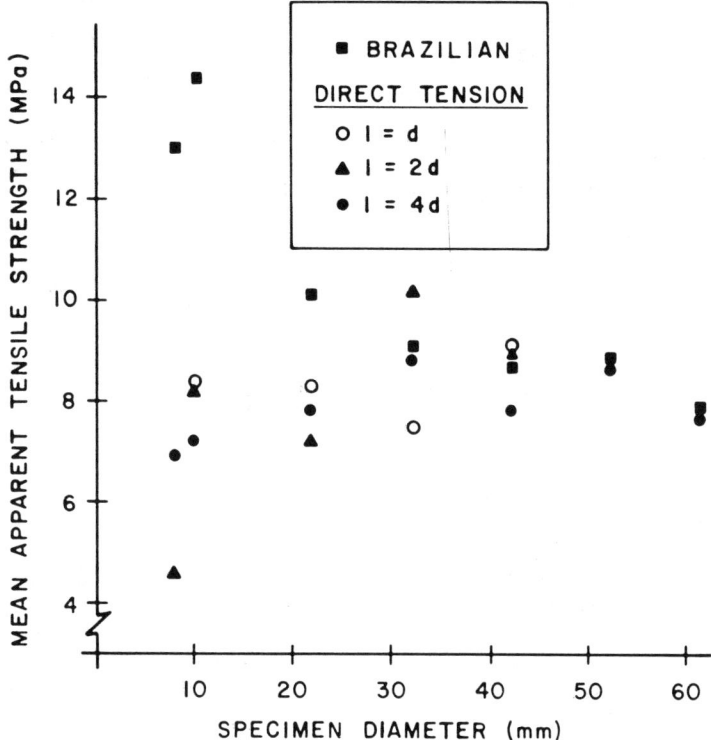

Fig. 1. Apparent Tensile Strength for Brazilian and Direct Tension Testing of Bohus Granite (Data From Wijk et al (1978)).

The continuum concept of strength dictates that the direct tension strength must remain constant regardless of the size of the test specimen. The deterministic fracture mechanics approach indicates that there will be a size effect which will depend on the relative dimensions of the defect with respect to the dimensions of the specimen. A two parameter statistical fracture mechanics model requires that there always will be a decrease in the apparent tensile strength with increasing sample size. Finally, a three parameter statistical fracture mechanics model indicates that there will be an asymptotically decreasing strength with increasing sample size.

The most significant study of direct tension on a single rock type known to the Author is due to Wijk et al (1978). Wijk et al performed a great number of direct tension tests on specimens of Bohus granite over a wide range of specimen sizes and found no size effect. Wijk et al (1978) also performed Brazilian tests on Bohus granite with various sized specimens and found a size effect (see Figure 1). The continuum strength theory is incapable of explaining these results as is the two parameter statistical fracture mechanics model. The deterministic fracture mechanics method can be used to explain the results for certain magnitudes of initial crack lengths, but cannot be used to explain the variation in results at a given sample size. The three parameter statistical fracture mechanics model can account for the results. The value of one of the three parameters of this model can be inferred from the results in Figure 1. Specifically, $\mathcal{G}_u$ can be estimated to be the square of the direct tension strength or the asymptote of the Brazilian tensile strength (approximately 64 MPa^2).

A four point bending test is schematically illustrated in Figure 2. Also shown in this figure are three specimen cross sections which will be discussed. Assuming linear elastic response, the stress state at rupture is;

$$\sigma = \begin{cases} \dfrac{2T_a y}{a} & 0 \leq x \leq \dfrac{L}{6} \\[2ex] \dfrac{3T_a y}{a}\left[1 - \dfrac{2x}{L}\right] & \dfrac{L}{6} \leq x \leq \dfrac{L}{2} \end{cases}$$

The continuum strength concept dictates that the strength measure T_a must be the same for each of the cross sections. The deterministic fracture mechanics theory would indicate that the strengths may be different for each of the cross sections depending on the location and size of the defect. Both the two and three parameter statistical fracture mechanics models would indicate that the strengths in each of the cross sections would be different.

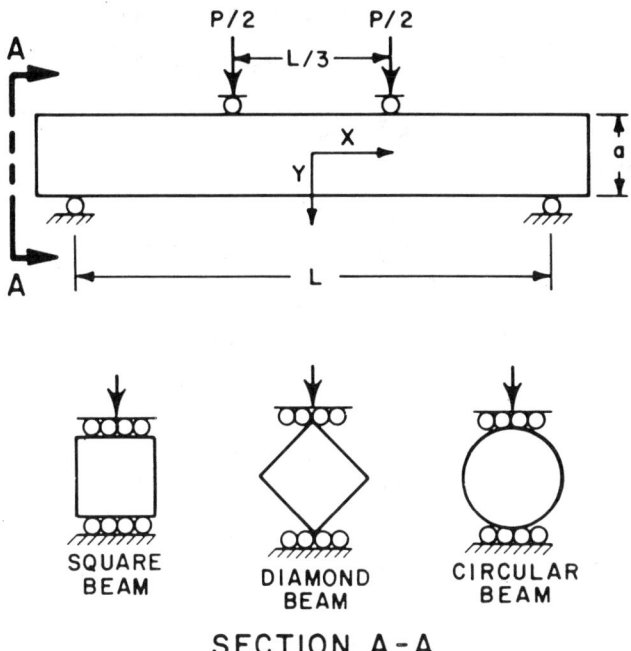

Fig. 2. Four Point Bending Test.

Results obtained by Ratigan (1981) are shown in Table 1 for Cararra marble, Stripa granite and Sierra White granite. Both the Sierra White granite and the Carrara marble display a distinct anisotropy with respect to the apparent tensile strength in this test.

If we assume that a weakest link model is appropriate for a rock in four point bending, we may use the space of Equation (11) to evaluate the parameter m (a qualitative measure of size effect) in the local failure criterion. The apparent tensile strength results for the diamond shaped beams of Carrara marble are shown in Figure 3. This figure illustrates a phenomenon which is perhaps more significant than the apparent strength anisotropy. Specifically, the size effect (as qualitatively indicated by the slope of the data) is different in each orthogonal direction. Thus, not only is the Carrara marble anisotropic, but the degree of anisotropy is dependent upon the specimen size. Conceivably, there exists sample sizes with which anisotropy would not be detectable.

TABLE 1. Apparent Tensile Strength in Four-Point Bending
(Data From Ratigan (1981))

Rock/ Orientation	Beam Cross Section		
	Square	Circular	Diamond
Carrara Marble			
X	17.7 (28)	17.6 (15)	20.7 (6)
Y	11.1 (3)	13.5 (17)	15.9 (6)
Z	-	18.6 (27)	25.3 (19)
Sierra White Granite			
X	-	13.7 (15)	-
Y	-	15.0 (16)	-
Z	19.7 (15)	19.7 (16)	23.2 (15)
Stripa Granite	21.1 (17)	25.3 (18)	28.0 (17)

Note: (1) All strengths in MPa
(2) Number of tests in parentheses
(3) Nominal beam size = 2 cm x 2 cm x 8 cm

Volkov (1962) states that Bartenev and Bovkunenko (1956) were the first to report anisotropy of size effect in glass filaments. The Author is unaware of any publication wherein the anisotropy of size effect in rock is discussed. If we for the moment assume that failure inducing defects are either mineral or crystal interfaces (or differential mineral locations) it becomes physically obvious that there should be an anisotropy of size effect since all rocks contain elongated minerals of some sort.

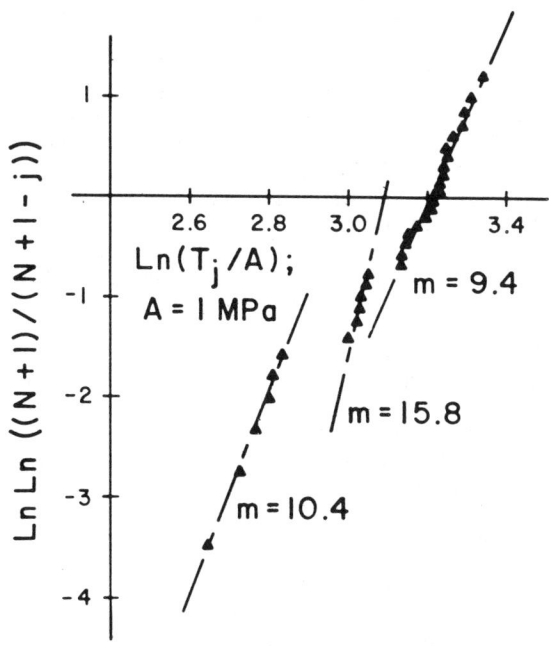

Fig. 3. Apparent Tensile Strength for Diamond Beams of Carrara Marble in the Space of Equation (11)(Data From Ratigan (1981)).

Multiaxial tests

A multiaxial tension test will be defined as a tension test in which non-zero stress components exist in directions which are orthogonal to the fracture surface. These tests are often referred to as indirect tension tests. The most common test of this type is the Brazilian test (Carniero and Barcellos (1953)) in which a thin disk (see Wijk (1978) for a definition of thin for this test) is diametrically loaded until rupture occurs on a plane coincident with the line of loading. Numerous other multiaxial tension tests are used in the rock mechanics community. Thin disks can be loaded in bending (Swan (1980)). Small hydraulic fracturing tests have become fairly common in the laboratory (e.g., Haimson (1968)). Ratigan (1981) has used a test similar to the laboratory hydraulic fracture test using a synthetic rubber for the pressurizing medium rather than water. This test enables one to use a significantly smaller rock specimen than in the hydraulic fracture test and also requires considerably less equipment. Another multiaxial tension test which does not receive a great deal of attention is the "pinching-off" test (Bridgeman (1912), Jaeger and Cook (1963), Sato and Naoyuki (1975)) in which a pressure is applied to the exterior of a rock core with or without any load along the core axis. This test is seldom used because of the fact that it is not understood. Matsuo (1979) has presented an analysis of this test with a two parameter statistical fracture mechanics model and has explained some laboratory observations reasonably well.

Size effects in brittle rock are much more apparent in multiaxial tension tests than in uniaxial tension tests. Statistical fracture mechanics indicates this is because less of a material region is at a tensile stress near the breaking stress in addition to the fact that fewer material defects are located in orientations which are acted upon by tensile stresses. The size effects observed in multiaxial tension tests cannot be explained through the use of the continuum concept of strength and therefore, the method will not be discussed in this section of the paper. Both the deterministic and the statistical fracture mechanics models are capable of reproducing or explaining many results observed in multiaxial tension tests. However, as was stated earlier, the deterministic fracture mechanics concept is incapable of reproducing any variance about the mean.

The two and three parameter statistical fracture mechanics concepts are less similar than they might initially appear. If both models are "fit" to a set of experimental data and these models are used to "predict" the apparent tensile strength in a different type of test, the two predictions can be significantly different. The fundamental difference between the two models is of course the magnitude of the parameter $\mathcal{G}_u$. Two physical interpretations are available for this parameter. Firstly, this is the strength measure below which there is zero probability of failure. Secondly, as specimen size is increased, the strength will asymptotically approach a value proportional to $\mathcal{G}_u^{1/2}$ for any test.

Using data from the literature, Ratigan (1981) has evaluated the parameter m (assuming a two parameter model) using both Class I and Class II methods (see section on statistical fracture mechanics). The m values determined from the Class I method are consistently greater in magnitude than the values of m determined from the Class II method. If one accepts the general form of the statistical fracture mechanics model presented in this paper, the results indicate that (1) rock is best characterized with a non-zero value of $\mathcal{G}_u$ and/or (2) volumetrically distributed flaws may not be the sole source of failure inducing defects. If a material is best characterized with a three parameter model, experimental data plotted in the space of Equation (11) will appear concave down (Weibull (1939b)). Results obtained by Ratigan (1981) for rubber fracturing tests on Stripa granite are shown in the space of Equation (11) in Figure 4.

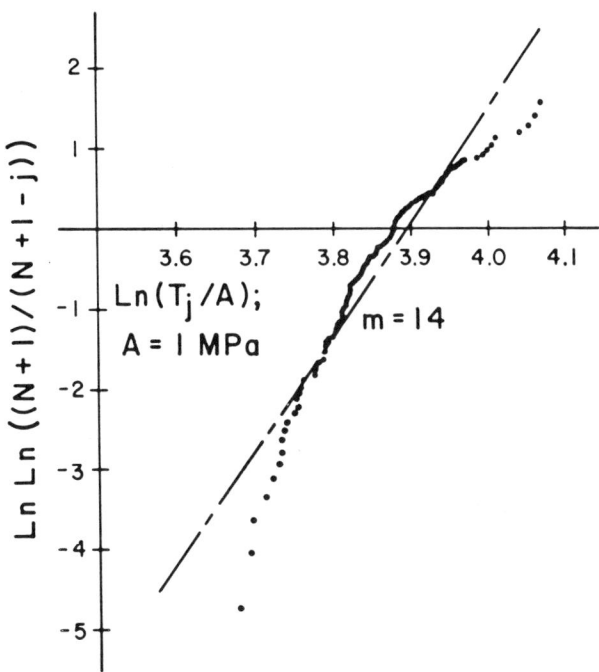

Fig. 4. Apparent Tensile Strength for Rubber Fracturing of Stripa Granite in the Space of Equation (11) (Data From Ratigan (1981)).

The behavior of the apparent tensile strength of brittle rock may be evaluated in a multiaxial, nonhomogeneous stress state in the laboratory using the hydraulic fracture test (Haimson (1968)). In this test, the two principal stresses in the plane (horizontal) orthogonal to the axis of the borehole can be varied independently. Haimson (1968) performed laboratory hydraulic fracture tests on several rock types. The results for Tennesse marble are shown in Figure 5. When the two horizontal stress components are equal, the apparent tensile strength decreases with an increase in the maximum horizontal stress. When the two horizontal stress components are equal, the apparent tensile strength decreases with an increase in

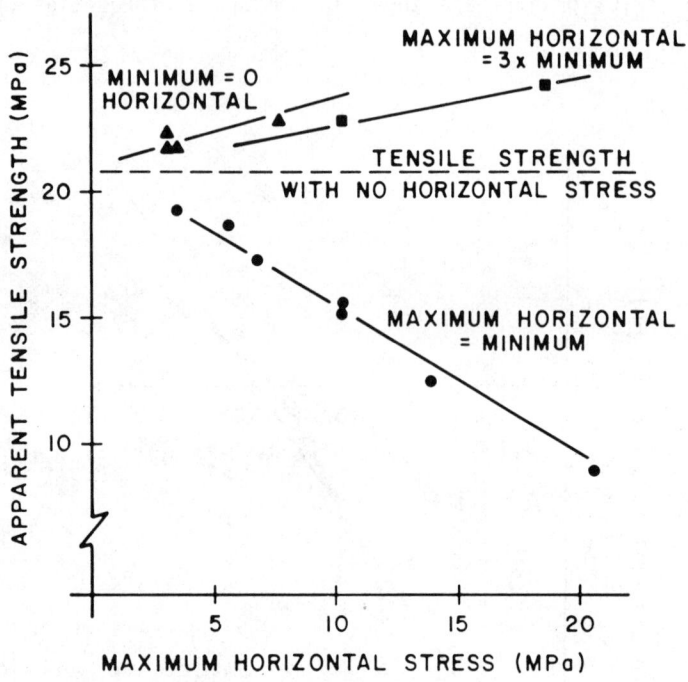

Fig. 5. Apparent Tensile Strength in Laboratory Hydrofrac for Tennessee Marble (Data From Haimson (1968)).

the maximum horizontal stress. When the maximum horizontal stress is three times the minimum (or if the minimum equals zero) the apparent tensile strength increases with an increase in the maximum horizontal stress. Haimson's results on other rock types also display this behavior. Deterministic fracture mechanics cannot be used to explain these results. Ratigan (1982) has shown that a three parameter statistical fracture mechanics model can reproduce the behavior. Ratigan (1982) has also shown that the apparent tensile strength will always decrease with an increase in the maximum horizontal stress if the minimum is greater than one half the maximum. Otherwise, the apparent tensile strength will increase. When the minimum is equal to one half the maximum, the apparent tensile strength does not change with an increase in horizontal stress.

CONCLUSIONS

Theories of tensile strength have been reviewed and critically examined with respect to various laboratory results. The continuum strength approach is found to be incapable of reproducing any of the observed testing results. Deterministic and statistical fracture mechanics can be used to explain much of the behavior noted in tensile testing. However, the statistical fracture mechanics approach appears capable of reproducing some test results that deterministic fracture mechanics is incapable of reproducing.

ACKNOWLEDGEMENTS

The author gratefully acknowledges the careful preparation of the manuscript by Ms. Judy Hey and the excellent drafting by Mr. Danny Nelson. Funds provided by RE/SPEC Inc. for the preparation of this paper are also acknowledged with appreciation.

REFERENCES

American Society for Testing and Materials (ASTM), 1974, "Standard Method of Test for Direct Tensile Strength of Rock Core Specimens", Test D2936-71.

Bartenev, G. M. and Bovkunenko, A. N., 1956, "Strength of Glass Fibers and the Effect of Various Factors Upon Their Strength", ZhTF, No. 11, 2058.

Brace, W. A., 1964, "Brittle Fracture of Rocks", In State of Stress in the Earth's Crust, Judd, W. R., ed., pp. 111-180.

Carniero, F. L. L. B. and Barcellos, A., 1953, Union of Testing and Research Laboratories for Materials and Structures, No. 13.

Evans, A. G. and Jones, R. L., 1978, "Evaluation of a Fundamental Approach for the Statistical Analysis of Fracture", Journal of the American Ceramic Society, 61, No. 3-4, pp. 156-160.

Finnie, I., 1977, "Waloddi Weibull - 90 Years Young on June 18, 1977", Journal of Engineering Materials and Technology, pp. 193.

Griffith, A. A., 1921, "The Phenomena of Rupture and Flow in Solids", Philosophical Transactions of the Royal Society, 221A, pp. 163-198.

Haimson, B. C., 1968, "Hydraulic Fracturing in Porous and Nonporous Rock and Its Potential for Determining In-Situ Stresses at Great Depth", Missouri River Division Corps of Engineers, Technical Report No. 4-68.

McClintock, F. A. and Walsh, J. B., 1962, "Friction on Griffith Cracks Under Pressure", Fourth U. S. National Congress of Applied Mechanics, pp. 1015-1021.

Matsuo, Y., 1979, "A Probabilistic Treatise on a Lateral Fluid Pressure Test of Brittle Materials", Bulletin of the Japanese Society of Mechanical Engineers, Vol. 22, No. 170.

Ouchterlony, F., 1980, "Review of Fracture Toughness Testing of Rock", Swedish Detonic Research Foundation, Report DS 1980:15, Stockholm.

Ratigan, J. L., 1981, "A Statistical Fracture Mechanics Approach to the Strength of Brittle Rock", Ph.D. Thesis, University of California, Berkeley.

Ratigan, J. L., 1982, "A Statistical Fracture Mechanics Determination of the Apparent Tensile Strength in Hydraulic Fracture", Workshop on Hydraulic Fracturing and Stress Measurements, United States Geological Survey, Monterey, CA, (To be published as a USGS Open File Report).

Robinson, E. Y., 1970, "The Statistical Nature of Fracture", UCRL-50622, Lawrence Livermore Laboratory.

Robinson, E. Y. and Finnie, I., 1969, "On the Statistical Interpretation of Laboratory Tests on Rock", Proceedings of Colloque de Geotechnique, Toulouse, France, pp 1.87-1.118.

Sato, Y. and Naoyuki, S., 1975, International Journal of Mechanical Science, Vol. 17, p. 705.

Swan, G., 1980, "Some Observations Concerning the Strength-Size Dependency of Rocks", Research Report TULEA 1980:01, Lulea, Sweden.

Volkov, S. D, 1962, Statistical Strength Theory (Translated from the Russian Edition), Gordon and Breach.

Weibull, W., 1939a, "A Statistical Theory of the Strength of Materials", Ingeniorsvetenskap Akademiens Handlingar, Nr. 151.

Weibull, W., 1939b, "The Phenomena of Rupture in Solids", Ingeniorsvetenskap Akademiens Handlingar, Nr. 153.

Wijk, G., Rehbinder, G. and Logdstrom, G., 1978, "The Relation Between the Uniaxial Tensile Strength and the Sample Size for Bohus Granite", Rock Mechanics, 10, pp. 201-219.

Wijk, G., 1978, "Some New Theoretical Aspects of Indirect Measurements of the Tensile Strength of Rocks", International Journal of Rock Mechanics and Mining Science, Vol. 15, pp. 149-160.

Chapter 45

EXPERIMENTAL INVESTIGATION OF FRAGMENTATION
ENHANCEMENT DUE TO SALVO IMPACT

D. B. Barker, D. C. Holloway and J. F. Cardenas-Garcia

Department of Mechanical Engineering
University of Maryland
College Park, MD 20742

ABSTRACT

Tround International Inc. is currently developing a hard rock drill that uses a salvo of three frangible projectiles to prefracture the rock ahead of a conventional tri-cone bit. This paper reports on a series of experiments conducted for Tround to study the fragmentation enhancement possible through the use of an optimally timed and spaced salvo of projectiles. In this study the complex three-dimensional problem was simplified to a two dimensional salvo impact problem in a transparent brittle plate of glass. The use of high speed photography and dynamic photoelasticity was then used to study the development of the impact damage zone and the various interactions of the stress waves.

INTRODUCTION

One of the most promising new techniques for rock excavation that has come out of some 30 or more novel hard rock excavation techniques studies under the old NSF/RANN program (Crouch, 1973 and Wang, 1975) is projectile impact. Projectile impact is a practical way to get large amounts of energy efficiently to the rock face. Two research groups, Sandia Laboratories (Newsom, et. al, 1976) and Physics International (Watson and Godfry, 1972) went beyond the normal laboratory confines and conducted field experiments in the early 70's demonstrating the feasibility and advantages of using projectile impact as a hard rock fragmentation or comminution technique.

All previous research, with the exception of the Sandia-Tround Tera-drill program, has been concerned with the fragmentation due to the impact of a single projectile, see for example Vanzant, 1973, Bauer and Calder, 1969, and Kabo, et. al., 1977. It is an accepted and proven fact that in hard rock blasting improved fragmentation and rock removal occurs with proper choice of delay between adjacent charges (Winzer, et. al., 1979). Simultaneous and very long delay times do not yield as good a fragmentation. Besides timing, charge spacing also affects fragmentation. With large amounts of energy being delivered to a rock face by projectile impact, it is expected that similar improvements in fragmentation as seen in blasting could occur by firing a salvo of projectiles with proper timing and spacing.

This paper reports on a series of experiments conducted for Tround International Inc. to study the fragmentation enhancement possible through the use of an optimally timed and spaced salvo of projectiles. Tround is using a salvo of three frangible projectiles to prefracture hard rock ahead of a conventional tri-cone bit for deep drilling application (Dardick, 1977). Sandia Labs found that 100% drilling rate increases could be expected with such a technique even with a non-optimized salvo (Newson, et. al., 1976). This paper discusses the fragmentation enhancement possible by optimizing the time delay and spacing between two projectiles in a salvo.

Past studies of projectile impact in rock have been primarily conducted on a post-mortem basis. In order to study the dynamic interaction between adjacent projectiles, it was decided in this study to simplify the complex three-dimensional problem and study edge impact on a transparent two-dimensional plate. By using high speed photography and the technique of dynamic photoelasticity, it was possible to follow the development of the damage zone and individual cracks as well as observing the stress waves within the material. Admittedly, the two-dimensional modeling of the actual problem magnifies some of the fragmentation damage as compared to the three-dimensional situation, but a good qualitative understanding of the problem can be obtained.

EXPERIMENTAL PROCEDURE

Twelve millimeter (1/2 inch) thick plate glass was used to simulate a hard brittle material such as rock. Admittedly rock is neither as brittle nor as homogeneous as plate glass, but a qualitative understanding of the phenomenology of the problem can be obtained by using a brittle transparent material such as glass. The plates of glass were impacted on one edge by two frangible ceramic projectiles 6mm (1/4 inch) in diameter and 6mm (1/4 inch long. The projectiles were fired into the edge of the plate at a velocity of about 500 m/s (1640 ft/s) by two independently fired explosive cannons. Two different projectile spacings of 76mm (3 inches) and 140mm (5-1/2 inches) and inter-projectile impact delays varying from almost simultaneous impact to more than 100 μs between impacts were studied. The glass plates were as large as practically possible, 915 mm (36 inches) square, to negate as much as possible the influences of reflected stress waves. All tests were photographed during the impact event with a multiple spark gap camera equipped to function as a dynamic light field polariscope. The camera has the capability to record 16 independently controlled frames of dynamic information with an exposure duration of less than 1 µs. The use of the Cranz-Schardin multiple spark gap camera to record stress waves and crack propagation has been previously described in detail, see for example Dally and Riley, 1967, or Dally, 1971. Even though glass is only a very weakly birefringent material, the technique of dynamic photoelasticity made the various stress waves created in the impact event visible. It was then possible to watch the fragmentation pattern develop and determine the influence of the various stress waves.

DISCUSSION OF RESULTS

A sequence of three out of the sixteen dynamic photographs for a typical test with a projectile spacing of 76mm (3 inches) is shown in Figure 1. The first photo in the figure shows the two small ceramic projectiles just before impacting the plate glass and the last photo shown was taken about 50 µs after impact. The glass has been marked with 25.4mm (1 inch) square grid lines. From the complete photographic record it was determined that the first projectile to impact was traveling at 501 m/s (1643 ft/s). The second projectile impacted 10.5 µs after the first and it was traveling at 483 m/s (1584 ft/s).

An enlargement of the middle photo in Figure 1 is shown in Figure 2. This particular photo was taken 27 µs after the first impact and 16.5 µs after the second impact. The outgoing dilitational or P-wave front from each projectile impact is indicated on the photo. The impact crater damage zone appears as a dark shadow. Surface cracks can also be seen traveling into the glass from the top free surface completing the appearance of a shallow conical damage area. These surface cracks, which actually form a large portion of the crater, are initiated by what appears to be the Rayleigh wave.

The test shown in Figures 1 and 2 show a 76mm (3 inch) spaced salvo impact where both projectiles impacted before any stress wave from the first impact reached the second impact site. The interprojectile impact delay was 10.5 µs and as close to simultaneous as experimentally obtained. For the test with a 140mm (5-1/2 inch) projectile spacing, a 8.5 µs interprojectile impact delay was as close to simultaneous as obtained. Various interprojectile delays were then tried for the two different salvo spacings. The longest delays tried included an infinite delay where the second projectile impacted after all dynamic influences of the first impact had disappeared, (experimentally this was on the order of minutes).

To evaluate the amount of fragmentation in the various tests, it was necessary to measure damage before the reflected stress waves from model boundaries arrived back in the impact zone. Careful study of the photographs of the dynamic event and comparison with post-mortem model photographs showed that the reflected stress waves re-initiated various cracks in the impact region. These re-initiated cracks caused additional damage that would not be present with a salvo impact on a true infinite half plane. Thus the salvo impact fragmentation damage was measured by analyzing a single photo from each test a constant time after the second impact, but before the arrival of the reflected wave from the first impact could re-initiate any cracks in the impact zone. The fragmented area or region that was densely cracked was carefully measured with a planimeter off the single selected photo for each test. This measured fragmented area between the two impact sites served as our quantitative measure of damage.

Figure 3 shows the measured fragmented area between the two impact sites as a function of impact delay. The relative damage as plotted on the absicissa is the measured salvo fragmentation of damage area normalized by the measured damage area from a single projectile. From the figure it is obvious that a salvo of two projectiles, no matter what inter-projectile impact delay, caused minimally three times the damage of a single projectile. An impact delay on the order of 60 to 70 µs caused over four times the damage of a single projectile.

The apparent data scatter in Figure 3 can be smoothed by taking into consideration the variations in projectile velocities from test to test. Average projectile velocities were 500 m/s (1640 ft/s), but maximum variations did occur up to plus or minus ten percent. Assuming that the fragmentation damage is directly proportional to total kinetic energy in the salvo, Figure 3 is replotted as Figure 4 where the damage area for the salvo is divided by first the kinetic energy of the salvo and then normalized by the damage area due to a single projectile divided by the kinetic energy of that single projectile.

Care must be used in interpreting data from Figure 4 since all the data has been normalized with respect to total kinetic energy in either the salvo or single projectile. Experimentally the salvo contained almost twice the kinetic energy of a single projectile. Figure 4 shows that a salvo minimally produces one and one-half the damage of a single projectile on a per kinetic energy basis. For the 76mm (3 inch) spaced salvo, delays on the order of 40 to 60 µs produced over two times the damage of a single projectile on a per kinetic energy basis, but in reality produced over four times the actual damage.

For both the 76mm (3 inch) and the 140mm (5-1/2 inch) spaced salvos, the nearly simultaneous impact delay produced the poorest damage. From a careful study of the dynamic photographs from each test and Figures 3 and 4, it became obvious that maximum salvo fragmentation enhancement occurred only when the second projectile

INVESTIGATION OF FRAGMENTATION 445

in the salvo was sufficiently delayed. Maximum damage consistantly occurred only when the second projectile impacted after the Rayleigh surface wave from the first projectile passed the site of the second impact. For 76mm (3 inch) spacing this corresponds to minimally about 31 µs and 57 µs for the 140mm (5-1/2 inch) spacing. Figure 4 dramatically shows this increase in fragmentation for the 76mm (3 inch) spacing. For the wider 140mm (5-1/2 inch) spacing the increase is not as dramatic.

The dynamic photographs showed that all the cracks in the impact zone were initiated by the time the Rayleigh wave passed. For maximum fragmentation the second projectile then impacts. The outgoing stress wave from this impact not only initiates its own new fractures but also reinitiates the already existing fractures from the first impact. These reinitiated fractures are driven greater distances. This stress wave crack interaction is the key to salvo fragmentation enhancement.

The maximum salvo enhancement for the 140mm (5-1/2 inch) spaced projectiles was not as great as the 76mm (3 inch) spaced projectiles. It is felt that the smaller maximum enhancement is due to the greater spacing and geometric attenuation of the stress waves. With the greater spacing and distance the stress waves have to travel, they do not have as much strength to reinitiate and drive previously initiated cracks in the region of the first impact.

Minimum projectile delays were relatively easy to find for maximum salvo enhancement, but maximum delays were more elusive. As previously mentioned stress wave reflections from model boundaries reinitiated cracks and created more damage than would be present in a true infinite half plane. Various attempts with wave traps were tried but were not completely successful. Data points are entered in both Figures 3 and 4 for a very long delay between projectiles. The measured damage is slightly greater than it would be for the true infinite half plane. Assuming for the moment that this additional damage is negligible, note that a very long delay is much better than a simultaneous delay, but probably not as good as an optimum delay.

CONCLUSIONS

An experimental investigation of fragmentation enhancement due to salvo impact has been conducted with a dual projectile salvo impacting the edge of a large plate of glass. Through the use of high speed photography and dynamic photoelasticity it was possible to study the development of the impact damage zones and stress waves for various projectile impact delays. It was found that a dual projectile salvo produces more than three times the damage of a single projectile. With an optimum inter-projectile impact delay the salvo enhancement is more than four times the damage of a single projectile.

The interactions between the stress waves from the second impact and the cracks created by the first impact were the reasons for salvo fragmentation enhancement. Maximum enhancement was found to occur when the second projectile impacts after the Rayleigh wave from the first projectile has passed the second impact site. With such a delay the cracks in the damage zone of the first projectile have had a chance to initiate and develop before the second projectile impacts. The stress waves from the second projectile then reinitiates and drives these fractures to greater lengths than if there were only a single projectile or if the projectiles in the salvo had impacted close to simultaneous.

REFERENCES

Bauer, A. and Calder, P.N., 1969 "Projectile Penetration in Rock", Proceedings of 5th Canadian Rock Mechanics Symposium, pp. 157-170.

Crouch, S.L., 1973, Conference on Research in Tunneling and Excavation Technology, NSF/RANN

Dally, J.W. and Riley, S.F., 1967, "Stress Wave Propagation in a Half Plane Due to a Transient Point Load", Developments in Theoretical and Applied Mechanics, Vol. 3, Pergamon Press, New York.

Dally, J.W., 1971, "Applications of Photoelasticity to Elastodynamics", Proceedings of Symposium on Dynamic Response of Solids and Structures, Stanford University.

Dardick, D., 1977, "Tround Terra Drill Process and Apparatus", U.S. Patent 4,004,642, Jan. 25

Kabo, M., Goldsmith, W. and Sackman, J.L., 1977, "Impact and Comminution Processes in Soft and Hard Rock", Rock Mechanics, Vol. 9, pp. 213-243.

Newsom, M.M., Alvis, R.L. and Dardick, D., 1976, "The Tera-Drill Program, A Progress Report and Program Plan", Report No. SANDIC-0228, June, Sandia Laboratory.

Vanzant, B.W., 1962, "Dynamic Rock Penetration Tests at Atmospheric Pressure", Proceedings 5th Symposium on Rock Mechanics, pp. 61-91.

Wang, F.D., 1975, Conference on Research on Excavation Technology, NSF/RANN.

Watson, J.D. and Godfrey, C.S., 1972, "REAM - A New Concept for Hard Rock Excavation", Proceedings of the 8th Canadian Rock Mechanics Symposium, pp. 141-158.

Winzer, S.R., Furth, W. and Ritter, A., 1979, "Initiator Firing Times and their Relationship to Blasting Performance", Proceedings of 20th U.S. Symposium on Rock Mechanics, pp. 461-470.

INVESTIGATION OF FRAGMENTATION 447

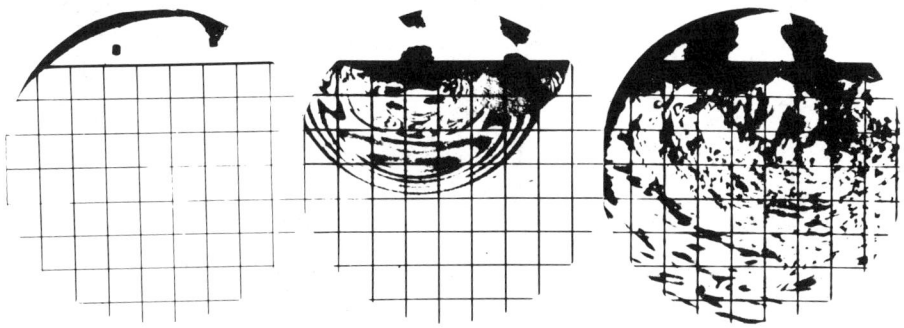

Figure 1. A sequence of three photos showing the salvo of two projectiles impacting the plate of glass.

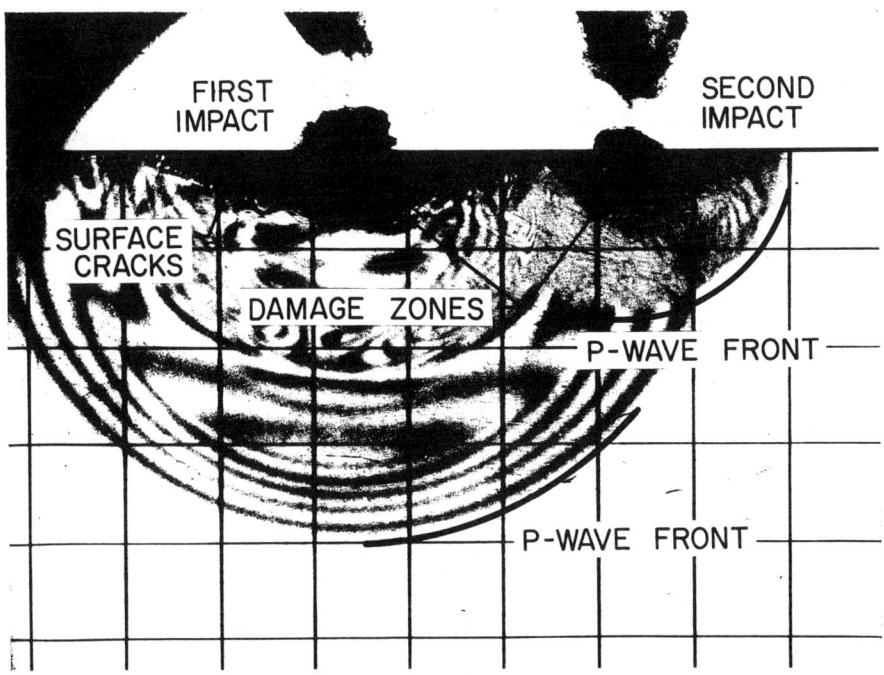

Figure 2. An enlarged photo of the impact region taken 27 μs after the first impact.

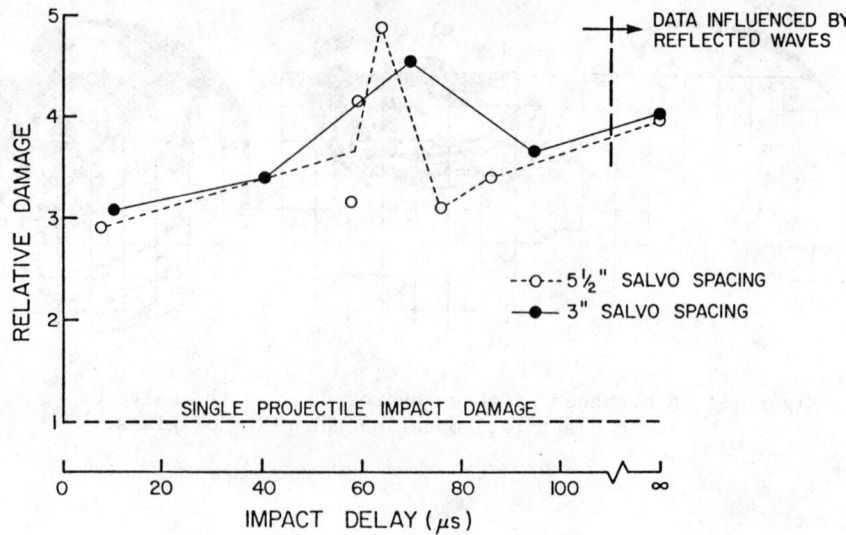

Figure 3. A plot showing salvo fragmentation damage normalized by single projectile damage as a function of inter-projectile impact delay.

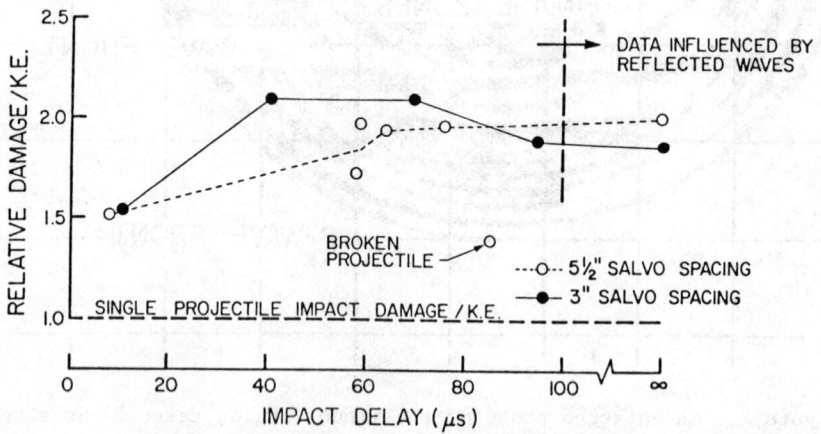

Figure 4. A plot showing the influence of inter-projectile delay on the salvo fragmentation damage normalized by both the single projectile damage and the total available kinetic energy in the impact.

Chapter 46

VARIABLES IN FRACTURE ENERGY AND TOUGHNESS TESTING OF ROCK

by CHRISTOPHER C. BARTON

Department of Materials Science and Mineral Engineering

University of California
Berkeley, California

ABSTRACT

Each variable known to affect laboratory measurement of fracture-energy and fracture-toughness is reviewed. Specific examples are cited where each of the variables have been isolated.

ACKNOWLEDGEMENT

I thank Lindamae Peck and Robert B. Gordon for helpful discussion. The research for this paper was done while the author was in the Department of Geology and Geophysics at Yale University. The research was supported by the Division of Engineering, Mathematical and Geoscience Office of Basic Energy Science, Department of Energy, under Contract No. GDE AC-02-79ER10 445.

INTRODUCTION

Fracture-energy (G) and fracture-toughness (K) are inter-related parameters that are a measure of a material's resistance to crack growth where:

$$G = K^2(1-m^2)/E$$

where m = Poisson's ratio

E = Young's modulus

Fracture-energy is the energy consumed in producing a unit of crack surface. Fracture-toughness is the value of the stress-intensity factor (also referred to as K) at which crack growth commences.

Compilations by Ouchterlony (1980) and Barton (1982) reveal considerable variation in measured values of fracture-energy and toughness, even for the same rock. For example, the fracture-energy of Indiana limestone measured in eight independent studies under STP conditions ranges from 16 to 230 J/m^2 (Barton, 1982). Yet within each one of the eight studies the variation is less than 20%.

Some variation is expected, for rock is a heterogeneous aggregate material, but the variation between studies is more likely due to the diversity of test conditions. The few studies where each of the variables has been isolated are cited below.

Effect of Specimen Configuration and Loading Geometry

Specimen configurations and loading geometries that have been used to measure fracture-energy or toughness of rock are shown in figure 1. For a linear-elastic, homogeneous, isotropic material, specimen configuration and loading geometry should not affect the measured fracture-energy and toughness. For rock, which is not such a material, fracture-energy and toughness may not be independent of specimen and loading geometry.

For example, in theory the angle of the wedge used to drive the crack in the double-cantilever beam specimen (see figure 1), should not influence these properties. However, the fracture energy of Sioux quartzite is found to depend on the angle of the wedge used to split a double cantilever beam test specimen (Peck & Gordon, 1982). The fracture energy increases 20 percent with decrease in wedge angle from 100° to 10° for cracks propagated at the same velocity. They suggest that the greater component of compressive load accompanying a high wedge angle may open microcracks and thereby reduce the energy needed to propagate the primary crack.

In contrast, the fracture toughness of Indiana limestone is unaffected (in burst tests) by axial stress up to 7 MPa (Abou-Sayed, 1978).

Effect of Crack Length and Specimen Size

The fracture-energy and fracture-toughness of rock increase by as much as 50% during the initial increments of crack growth and then become independent of crack length, see Hoagland et al. (1973a & b), Schmidt (1976), Schmidt and Lutz (1979), Ingraffea and Schmidt (1978), Barton (1981), and Peck (1982).

The increasing energy probably is related to the formation of the process zone of microcracks observed alongside of, and ahead of, the advancing crack, whether the crack grows from a notch or from a preexisting fatigue crack. Examples of the length of crack propagation beyond which fracture-energy or toughness is independent of length are cited in Table I. The term "apparent"

TABLE I

Length of crack propagation beyond which fracture energy or toughness is independent of length.

Rock Type	Crack Length (cm)	Reference
Salem (Indiana) Limestone	3-5	Schmidt (1976)
Indiana Limestone	3-5	Ingraffea and Schmidt (1978)
Portland Sandstone	3	Barton (1981)
Barre Granite	0.9	Hoagland et al. (1973b)
Westerly Granite	5-10	Schmidt and Lutz (1979)
Sioux Quartzite	1-3	Peck (1982)
Sioux Quartzite	1-3	Hoagland et al. (1973b)

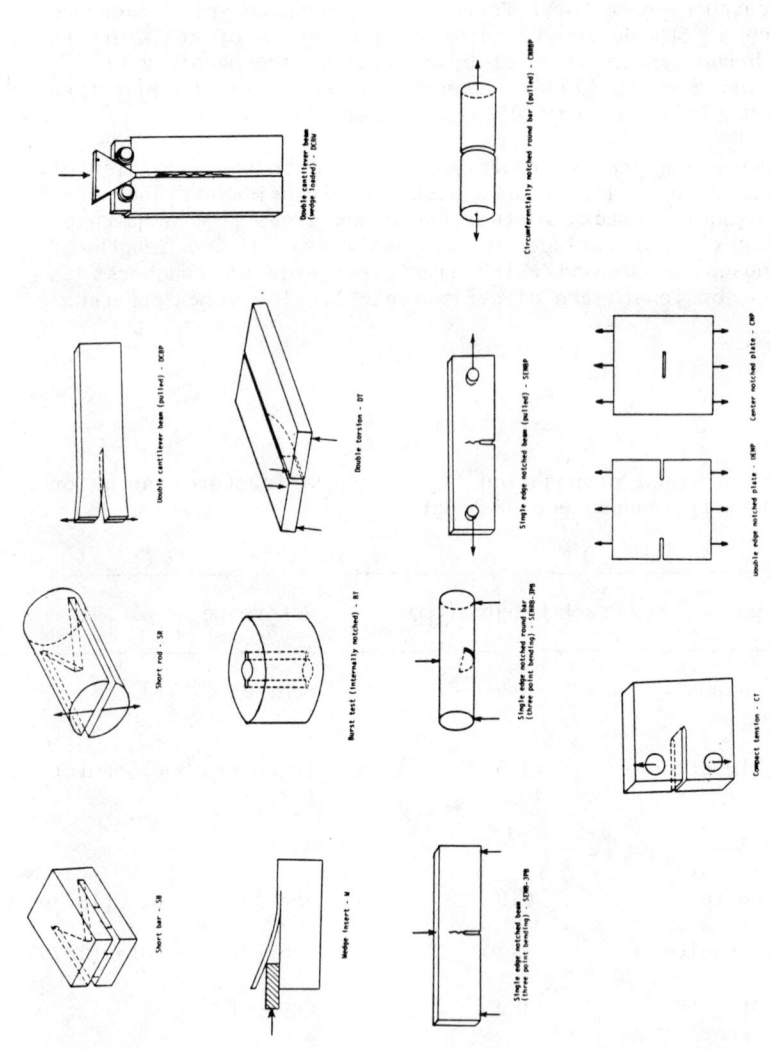

Fig. 1. Specimen geometries used to measure fracture-energy and toughness of rock

and the subscript Q are used by Schmidt (1976) and Schmidt and Lutz (1979) where the fracture-energy or toughness is not independent of crack length. The ASTM Standard E399 criterion for a valid test in metals accurately defines the crack length beyond which fracture-toughness measurements for Westerly granite are independent of crack length (Schmidt and Lutz, 1979). By the criterion, a test is valid when crack length is greater than or equal to $2.5(K_Q/\sigma_{ys})^2$, where K_Q is the fracture-toughness measured and σ_{ys} is the uniaxial tensile strength of the rock.

The same ASTM standard applies to the minimum specimen dimension (usually thickness). The fracture-toughness of Westerly granite is independent of specimen thickness over a range of 18 to 100 mm (see Schmidt and Lutz, 1979). For reproducible and potentially useful measurement of fracture-energy or toughness, specimen thickness should be greater than the width of the process zone of microcracking. Process zones 5 to 10 grain diameters wide are reported by Friedman et al. (1972), Hoagland et al. (1973a and b), Peck (1982), and Barton (1982).

Effect of Planar Anisotropy

Most rocks contain planar anisotropies, which may affect fracture-energy or toughness measurements; the most obvious are:

(1) primary bedding and lamination
(2) metamorphic banding and foliation due to compositional differences and the flattening of grains and alignment of platy minerals
(3) microjoint sets and planar microcrack sets.

The fracture-energy or toughness is lowest, in the following rocks, for cracks propagated in the bedding plane and in two mutually perpendicular planes orthogonal to the bedding plane is 30 and 43% higher than in the bedding plane in Anvil Points oil shale (Schmidt, 1977), 17 and 65% higher in Berea sandstone and in Salem limestone 27 and 86% higher (Hoagland et al. 1973a) and 20 and 27% higher (Schmidt, 1976). The lowest value for fracture-toughness in St. Pons marble is for cracks propagated in the metamorphic foliation plane. Values for this parameter in two mutually perpendicular planes orthogonal to this foliation plane are 23 and 37% higher, (Henry et al., 1977).

The fracture toughness of Berkeley granite is lowest for cracks propagated in the rift plane (the easiest splitting direction as determined by the experience of quarrymen), and is 40 and 81% higher in the grain and hardway planes (Halleck and Kumnick, 1980). For the Chelmsford granite the fracture-toughness in the rift and hardway planes is 4% higher than in the plane of the grain (Peng and Johnson, 1972). For Barre granite the fracture energy in the rift and grain is 4 and 32% higher than in the hardway plane (Hoagland et al., 1973b).

Effect of Environment - Water and Stress Corrosion Cracking

Water may affect crack propagation in rock chemically, by reacting with material at the crack tip and/or mechanically by reducing friction in the process zone. Fracture-energy or toughness in the presence of water has been reported lower than that measured in air by 10% in the Anvils Point oil shale (Schmidt, 1977), 33% in Berea sandstone, 34% in Salem limstone (Hoagland et al., 1973a), 15% in Sioux quartzite (Peck, 1982), and 60% in Portland sandstone (Barton, 1981).

The fracture-energy of Dresser basalt and Sioux quartzite was not reduced in the presence of water (Hoagland et al., 1973b), but this may be due to the experimental method. In these tests the specimens were soaked for ten hours but the experiment was run in air. Perhaps the low permeability of basalt and quartzite prevented water absorption and so despite soaking, the interiors of the specimens remained dry. Wilkening (1978) reports that water has no effect on the fracture-energy of Barre granite, but does not describe his technique for introducing water. Pre-soaking and cracking the specimen while submerged assures free access of water to the crack tip region during crack propagation (e.g. Schmidt, 1977; Peck, 1982; Barton, 1981, 1982).

The time to failure (static fatigue) for natural quartz loaded in uniaxial compression decreases with increasing partial pressure of water (Martin, 1972). Crack growth rate in natural quartz loaded in uniaxial compression increases with the partial pressure of water (Martin and Durham, 1975). Other studies, based on fracture-mechanics, employ double-torsion specimens to determine the relation between crack velocity, V, and stress-intensity factor.

Curve A in figure 2 shows the form of K-V curves for glass in a vapor-water environment (Wiederhorn, 1967; Waza et al., 1980) and for micrite and marble in a liquid-water environment (Henry et al., 1977). Curve B shows the form of the K-V curve for micrite in a vapor-water environment 75% humidity (Henry et al., 1977).

There are three distinct regions of curve A. Region I is thought to be controlled by the kinetics of the chemical reaction or dissolution at the crack tip, and Region II by the rate of transport of water from the surrounding environment to the crack tip or by the rate at which the dissolved species diffuses away

FRACTURE ENERGY, TOUGHNESS TESTING OF ROCK 455

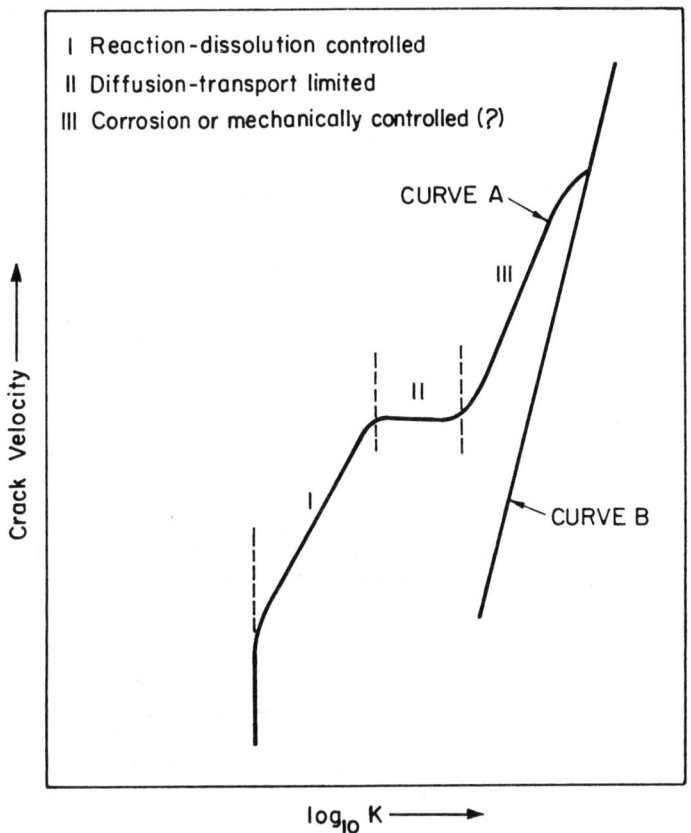

Fig. 2. The form of K-V curves in rock

from the saturated crack tip region (Anderson and Grew, 1977). For glass regions I and II depend on partial pressure (relative humidity) of water vapor while region III is apparently independent of environment (Wiederhorn, 1967). Alternatively in micrite and marble region III may represent a corrosion process (Henry et al., 1977).

The theoretical approaches suggested to explain regions I and II in terms of stress corrosion cracking are summarized by Atkinson (1979b) who follows Anderson and Grew (1977) in pointing out that the theoretical aspects are still in an active stage of development and who follows Evans (1972), and Wiederhorn (1974) and uses the power law relation developed by Charles (1958) as reformulated by Wiederhorn (1974) in terms of the stress-intensity factor. In this formulation the crack velocity is given by:

$$v = v_o \exp(-\Delta H/RT) K_I^n$$

ΔH = activation enthalpy
R = gas constant
T = absolute temperature
K_I = stress-intensity factor
v_o and n = experimentally determined constants
n is sometimes known as the stress corrosion index.

For silicates in a water environment stress corrosion is thought to be due to hydrolysis of the strong siloxane bonds to the weaker hydrogen bonded Si-OH-HO-Si group at the crack tip. No corrosion reaction for carbonates has been proposed.

Best fit curves to experimental data are shown in figure 3. Type A curves have been generated for micrite and marble and region II portions of type A curves for synthetic quartz, Arkansas novaculite, and esite and basalt. Type B curves have been generated for micrite in vapor-water (humidity 75%) and for andesite and basalt at humidities of 50% and 30% respectively. We may note that for rock, type A curves are generated in liquid-water environments and type B curves in vapor-water environments. For andesite and basalt the region I portion of type A curves is translated to higher values of K in vapor-water environments with little change in slope. Apparently B-curves for these rocks are dependent on environment. The crack velocity corresponding to the point of intersection of type A and B curves may be the velocity at which the crack outruns environmental effects. Alternatively, the B-curve may not become vertical (and thus environment independent) until some higher crack velocity and stress-intensity factor. The present data base does not permit resolution between the two possibilities and therefore the stress-intensity factor (i.e., the critical fracture-toughness of the rock). The critical fracture-toughness (K_{Ic}) reported by Henry et al. (1977) for micrite is higher than the stress-intensity factor at the A curve-B curve intersection. This may be because either the environment independent stress-intensity factor is greater than the stress-intensity factor corresponding to the intersection or because they used double-torsion specimens to generate the K-V curve and three point bend specimens to measure K_{Ic}.

Atkinson (1978b) infers a lower stress corrosion limit from the lower portion of A curves for synthetic quartz cracked in the a-plane normal to the z-plane at crack velocities of 10^{-9} m/sec. The lower limit at 25°C in liquid-water environment corresponds to a stress intensity factor that is approximately 0.52 K_{Ic} and in vapor-water environment, 0.64 K_{Ic}. A stress corrosion limit was not observed for synthetic quartz cracked in a liquid environment

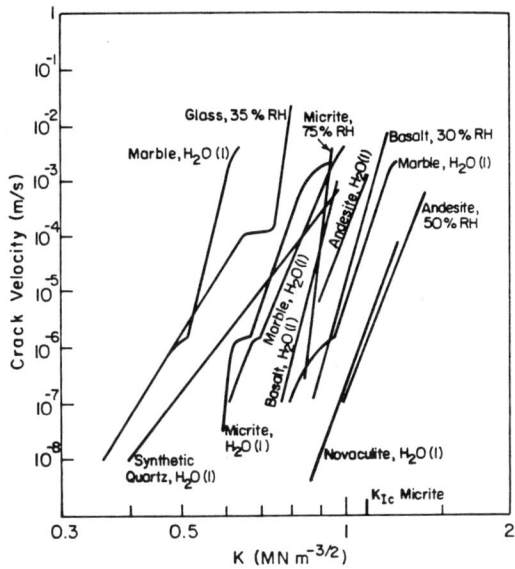

Fig. 3. K-V curves. Marble and micrite from Henry et al. (1977); glass, basalt and andesite from Waza et al. (1980); synthetic quartz from Atkinson (1979b); novaculite from Atkinson (1980).

in the a-plane normal to r at crack velocities down to 10^{-8} m/sec. He cites the stress corrosion limit in the soda-lime glass-water system at 0.25 K_{Ic} and that for porcelain-water system at 0.54 K_{Ic} at 25°C.

Effect of Temperature

Most fracture-energy and toughness measurements have been made at 20-25°C. The fracture-toughness of synthetic quartz cracked in a liquid-water environment is 20% lower at 80°C than at 20°C while that of novaculite is 16% lower at 80°C than at 20°C, (Figure 4). The fracture-energies of Berea sandstone, Salem (Indiana) limestone, and Barre granite are reported to be approximately 35%, 8%, and 21% lower, respectively, when measured at -196°C in liquid nitrogen than at 22-25°C in air. Dresser basalt and Sioux quartz-ite values fall within the lower end of the range of values at -196°C measured at 22-25°C in air (Hoagland et al. 1973a and b).

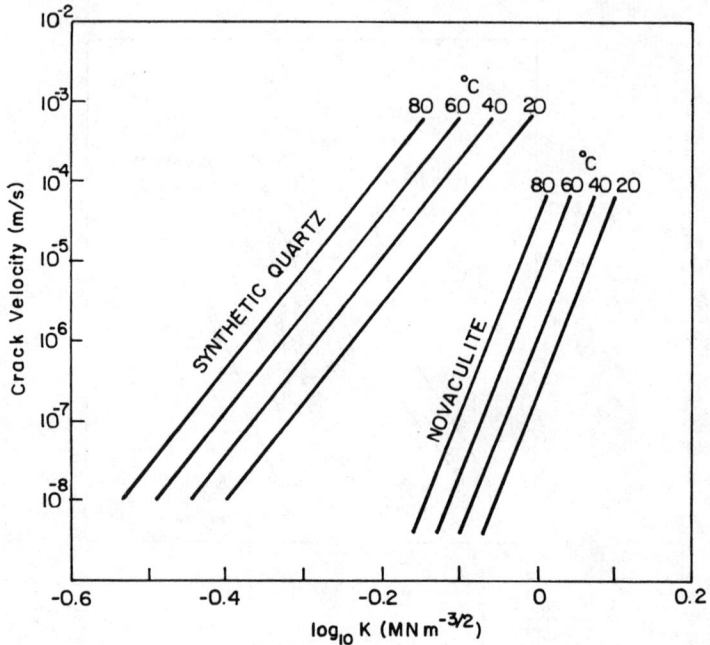

Fig. 4. K-V curves for synthetic quartz cracked on the a-plane in a direction normal to r and for Arkansas novaculite, in water at temperatures from 20°C to 80°C (Atkinson, 1979b, 1980).

Effect of Confining Pressure

The fracture-energy and toughness of rock increases with hydrostatic confining pressure (see figure 5). The Tennessee sandstone and Carthage and Lueders limestones double cantilever beam specimens were jacketed with a neoprene rubber paint, but once the crack began to propagate, its surface and tip were exposed to the confining fluids. The surface energies were not affected when the confining fluid was changed from an oil-base fluid containing a "leak-off control agent" and a low fluid-loss water base drilling mud.

The Indiana limestone single edge notched beam (pulled)specimens were jacketed with a flexible urethane coating which prevented the confining fluid from penetrating the crack surfaces. The fracture-energy of Indiana limestone measured in burst tests increases with confining pressure but is independent of pore-pressure [up to 7 MPa] (Abou-Sayed, 1978).

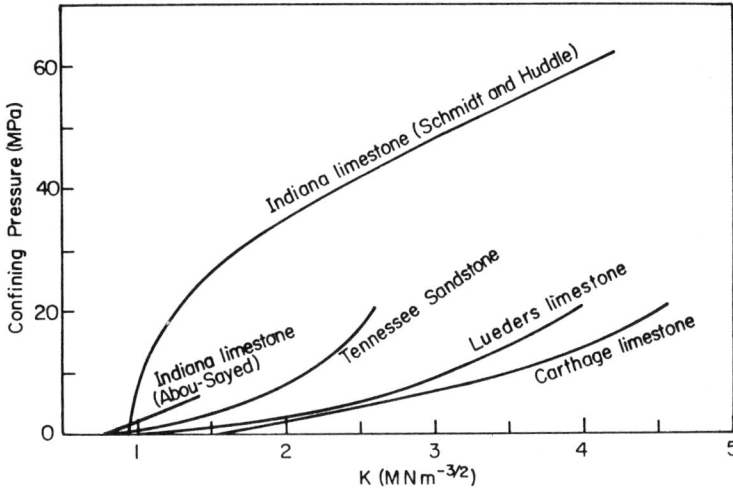

Fig. 5. Increase in fracture-toughness with hydrostatic confining pressure. Indiana limestone, Schmidt and Huddle (1976) and Abou-Sayed (1978); Tennessee sandstone and Carthage and Lueders limestone, Perkins and Kretch (1966).

CONCLUSION

Much of the variation in fracture-energy and fracture toughness between studies is due to the diversity of testing procedures. Because of incomplete records of experimental conditions most published measurements generally cannot be compared even in relative terms. Until standardized test methods are adopted complete details of the test conditions, and procedures, should be reported.

REFERENCES

Abou-Sayed, A.S., 1978. An experimental technique for measuring the fracture toughness of rocks under downhole stress conditions, VDI-Berichte, Nr.313, pp.819-824.

ASTM, 1981. "Standard test method for plane-strain fracture toughness of metallic materials, E399 -81", Annual Book of ASTM Standards, Part 10, published by the American Institute for Testing and Materials, Philadelphia, Pa., pp. 588-618.

Atkinson, B.K., 1979a. Fracture toughness of Tennessee sandstone and Carrara marble using the double torsion testing method, Int. J. Rock Mech. Min. Sci. and Geomech. Abstr., Vol. 16, pp.49-53.

Atkinson, B.K., 1979b. A fracture mechanics study of subcritical tensile cracking of quartz in wet environments, Pure and Applied Geophysics, Vol. 117, pp. 1011-1024.

Atkinson, B.K., 1980. Stress corrosion and the rate dependent tensile failure of fine grained quartz rock, Tectonophysics, Vol. 65, pp.281-290.

Barton, C.C., 1981. Fracture toughness and stress corrosion cracking of Portland sandstone, (abstract) EOS, Vol. 62, No. 17, p. 396.

Barton, C.C., 1982. Ph.D. dissertation, Yale Univ., Unpublished.

Charles, R.J., 1958. Dynamic fatigue in glass, J. Applied Phys., Vol. 29, pp. 1657-1662.

Evans, A.G., 1972. A method for evaluating the time-dependent failure characteristics of brittle materials - and its application to polycrystalline alumina, J. Mat. Sci., Vol. 7, pp.1137-1146.

Friedman, M., Handin, J., and Alani G.,1972. Fracture-surface energy of rocks, Int. J. Rock Mech. Min. Sci., Vol. 9, pp. 757-766.

Halleck, P.M. and Kumnick, A.J., 1980. "The influence of orientation on fracture toughness and tensile moduli in Berkeley granite," in Proc. 21st U.S. Symposium on Rock Mechanics, pp. 235-242.

Henry, J.P., Paquet, J., and Tancrez, J.P., 1977. Experimental study of crack propagation in calcite rocks, Int. J. Rock Mech. Min. Sci. and Geomech. Abstr., Vol. 14, pp. 85-91.

Hoagland, R.G., Hahn, R.G., and Rosenfield A.R., 1973a. Influence of microstructure on fracture propagation in rock, Rock Mechanics Vol. 5, pp. 77-106.

Hoagland, R.G., Hahn, G.T., and Rosenfield, A.R., 1973b. Influence of microstructure on fracture progagation in rock, Report to U.S. Bureau of Mines Twin Cities Research Center, 40 p.

Ingraffea, A.R. and Schmidt, R.A., 1978. "Experimental verification of a fracture mechanics model for tensile strength prediction of Indiana limestone," Proceedings, 19th U.S. Symposium on Rock Mechanics, pp. 247-253.

Martin, R.J., 1972. Time-dependent crack growth in quartz and its application to the creep of rocks, Jour. Geophys. Res., Vol. 77, No. 8, pp. 1406-1419.

Martin, R.J., III and Durham, W.B., 1975. Mechanisms of crack growth in quartz, J. Geophys. Res., Vol. 80, pp. 4837-4844. Deformed sedimentary rocks, Geol. Soc. Am. Bull., Vol. 53, pp. 381-408.

Ouchterlong, F., 1980. Review of fracture toughness testing of rock, Swedish Detonic Research Foundation Report DS 1980:15.

Peck, L. and Gordon, R.B., 1982. The effect of compressive stress on the fracture energy of Sioux Quartzite, Geophysical Research Letters, Volume 9, No. 3, pp. 186-189.

Peck, L., 1982. Stress Corrosion and Crack Propagation in Sioux Quartzite, Ph.D. dissertation, Yale Univ., unpublished.

Peng, S. and Johnson, A.M., 1972. Crack growth and faulting in cylindrical specimens of Chelmsford granite. Int. J. Rock Mech. Min. Sci., Vol. 9, pp. 37-86.

Perkins, T.K. and Bartlett, L.E., 1963. Surface energies of rocks measured during cleavage. J. Soc. Petroleum Engineers, Amer. Inst., Mining and Mech. engineers, Vol. 3, No. 4, pp. 307-313.

Perkins, T.K. and W.W. Kretch, 1966. Effects of cleavage rate and stress level on apparent surface energies of rocks. J. Soc. Petroleum Engineers, Amer. Inst. Mining and Mechanical Engineers, Vol. 6, No. 4, pp. 308-314.

Schmidt, R.A. 1976. Fracture toughness testing of limestone, Exp.

Mechanics, Vol. 16, No. 5, 161-167.

Schmidt, R.A. 1977. "Fracture mechanics of oil shale - unconfined fracture toughness, stress corrosion cracking, and tension test results," in Proc. 18th U.S. Symposium on Rock Mech., pp. 2A2-1 2A2-6.

Schmidt, R.A. and Huddle,   1976. "Effect of confining pressure on fracture toughness of Indiana limestone," in 17th U.S. Symposium on Rock Mechanics, compiled by W.S. Brown, S.J. Green, and W.A. Hustrulid, Utah Engineering Experiment Station, Univ. of Utah, Salt Lake City, Utah, pp. 503-1 to 503-6.

Schmidt, R.A. and Lutz, T.J., 1979. "K_{Ic} and J_{Ic} of Westerly granite - effects of thickness and in plane dimensions," in Fracture Mechanics Applied to Brittle Materials, Am. Soc. for Testing and Materials Publ. STP 678, Philadelphia, Pa., pp. 166-182.

Waza, T., Kurita, K., and Mizutani, H., 1980. The effect of water on the subcritical crack growth in silicate rocks, Tectonophysics, Vol. 67, pp. 25-34.

Wiederhorn, S.M., 1967. Influence of water vapor on crack propagation in soda-lime glass, J. Am. Ceramic Soc., Vol. 50, pp. 407-414.

Wiederhorn, S.M., 1974. "Subcritical crack growth in ceramics," in Fracture Mechanics in Ceramics, Vol. 2, Eds. R.C. Bradt, D.P.H. Hasselman, and F.F. Lange, Plenum, New York, pp. 613-646.

Wilkening, W.W., 1978. "J-integral measurement in geological materials," in Proc. 19th U.S. Symposium on Rock Mechanics, Vol. 1, pp. 254-258.

Chapter 47

A FRACTURE TOUGHNESS TESTING SYSTEM FOR PREDICTION
OF TUNNEL BORING MACHINE PERFORMANCE

by A.R. Ingraffea, K.L. Gunsallus, J.F. Beech, and P. Nelson

Assistant Professor and Manager of Experimental Research

Graduate Students
School of Civil and Environmental Engineering
Cornell University, Ithaca, New York

INTRODUCTION

Fracture toughness, K_{Ic}, is an intrinsic material property and is a measure of the energy required to create new surface area in a material. Fracture toughness measurements can be made for a wide range of rock types (Ouchterlony, 1980). As a measure of energy of comminution, fracture toughness might be used for more sensitive predictions of Tunnel Boring Machine (TBM) performance than are possible with other index measures in current use.

This paper describes part of an ongoing, comprehensive, analytical/experimental program aimed at evaluation of K_{Ic} in such a role. The program includes efforts to:

1. Provide a firm analytical basis for a test specimen geometry: the core-derived short-rod.
2. Develop simple, inexpensive testing equipment and techniques.
3. Evaluate the accuracy of the above by testing rock and other materials and comparing the measured fracture toughness with those obtained by other methods.
4. Measure fracture toughness of a substantial number of different rocks from three ongoing TBM projects.
5. Compare fracture toughness results with other strength measures.
6. Compare all test results with observed TBM performance in the three projects.

Presented here are summaries of all but the last of these efforts. Comparative evaluation of K_{Ic} as a more senstive rock property for TBM performance prediction will be the subject of the oral presentation and a subsequent paper.

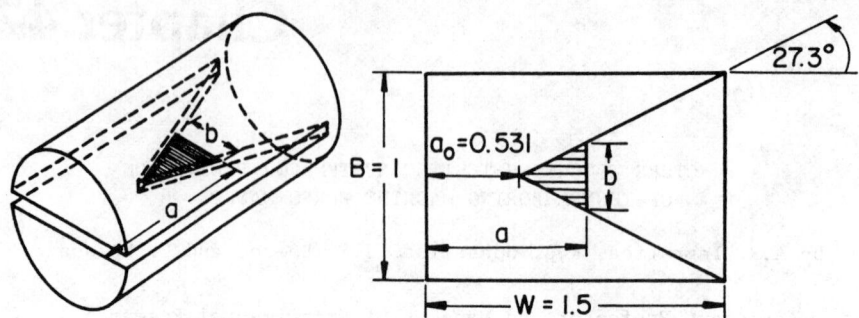

Fig. 1. Short rod configuration and relative dimensions

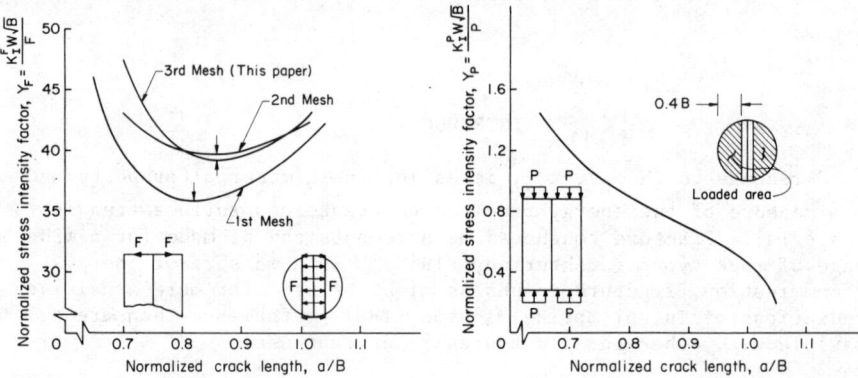

Fig. 2. Normalized stress intensity factor for wedge loading. Arrow denotes Y_{min}.

Fig. 3. Normalized stress intensity factor for axial loading.

SHORT-ROD DESCRIPTION AND ANALYSIS

The short-rod specimen is easily prepared from core. The geometry used in testing and analysis (Fig. 1) is the same as that studied by Beech and Ingraffea (1982). The purpose of the three-dimensional finite element analysis is to provide a rigorous calibration of the short-rod specimen. Barker (1977) showed that the critical stress intensity factor (SIF) could be determined from short-rod tests using the following equation:

$$K_{Ic} = AF_c / [B^{3/2}(1 - \nu^2)^{1/2}] \qquad (1)$$

where B is the specimen diameter, ν is Poisson's ratio, F_c is the load at crack instability and A is a calibration constant which is a function only of geometry. Note that, as indicated in equation (1), short-rod testing does not require a determination of crack length.

Beech and Ingraffea (1982) used the compliance calibration technique to determine the average SIF along the crack front, as a function of crack length. Their results are shown in Figure 2. The average SIF is computed using

$$K_I = \frac{F}{W\sqrt{B}} Y_F = \left[\frac{F^2 E'}{2b} \frac{dC}{da}\right]^{1/2} \qquad (2)$$

where $E' = E/(1 - \nu^2)$, for plane strain, E is Young's modulus, Y_F is the normalized average stress-intensity factor, b is the crack front length (see Fig. 1), and dC/da is the rate of change of compliance with crack length. The minimum value of this curve corresponds to $a = a_c$, $F_c = F_{max}$, and $K_I = K_{Ic}$. The constant A was determined by evaluating the right side of equation 2 at $a = a_c$ and setting it equal to equation 1. However, it was shown that a further mesh refinement would provide a more accurate value of A. Such an analysis was performed in the present investigation and a value of 25.4 was obtained for A. Updated values of Y_F are plotted in Figure 2. The shape of the crack front and the critical crack length predicted by the finite element analysis have been confirmed in tests on transparent polystyrene specimens.

As reported below, premature failure often developed in assocation with specimen discontinuities. To inhibit this mode of failure, an axial load was applied, as shown in Figure 3. An axial load, distributed as shown, results in an additional stress intensity along the crack front. This means a smaller value of F_c is required to reach instability and Equation 1 is no longer valid. The effect of axial loading was analyzed using the same compliance calibration technique mentioned above. The results of this study are shown in Figure 3.

The resultant critical SIF is obtained from superposition by,

$$K_{Ic} = \left[(F_c Y_F + P Y_p)/W\sqrt{B}\right]_{minimum}$$

TESTING PROCEDURE

As the analysis discussion suggests, short-rod testing procedure is straightforward. Aluminum end plates are epoxied to the top of the specimen to act as loading lines for the splitting force, F, and to distribute the axial load, P. The splitting force is applied with a simple testing apparatus designed for that purpose. Figure 4, center, shows the loading device in position on a specimen; in the lower left, a properly failed limestone specimen is seen.

Preliminary testing, however, showed that the application of the splitting force sometimes caused horizontal shearing failure in specimens with weak bedding (Fig. 4, lower right). Two methods were used in attempting to eliminate this phenomenon. The first one involved prenotching the specimen with a wire saw. The expected result of prenotching was to minimize the possibility of an excessively high pop-in or crack-initiating load. The second method, mentioned above, consisted in the application of axial load prior to testing. This force serves to give the bedding greater shearing resistance as well as to help restrict out of plane crack deviation. The calibrated clamps used to apply the axial force are shown attached to a specimen in Figure 4.

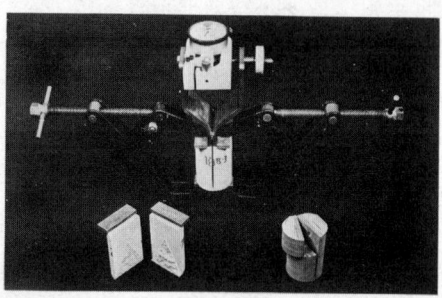

Fig. 4. Testing device, axial load clamps, and tested specimens.

VERIFICATION

Although the accuracy of the short-rod analyses has been independently confirmed (Bubsey et al, 1982), additional verification tests on polystyrene and Indiana limestone specimens were conducted during the present study. Five tests were performed on polystyrene giving a K_{Ic} of 2790 ± 140 MNm$^{-3/2}$. This value agrees well with the value of 2690 ± 50 MNm$^{-3/2}$ reported by Krenz et al (1976) for polystyrene with the same molecular weight and made by the same manufacturer.

A total of 31 tests were performed on specimens of Indiana limestone cored from beams previously used in three point bend (3PB) determination of K_{Ic} (Ingraffea, 1982). These tests comprised 9 with no modifications, 13 with prenotching, and 9 with axial load without prenotching. The values of K_{Ic} for these test series were 1050 ± 100, 1030 ± 90, 975 ± 60 MNm$^{-3/2}$, respectively. Comparison of the first two averages shows that the effect of prenotching was negligible, as expected. The K_{Ic} determined from the axially loaded specimens differed from the others by only 7%. In these tests, the SIF contributed by the axial load, approximately 6.7kN, is about 30% of the total and cannot be ignored. Figure 5 shows the comparison between 3PB and short-rod results. The single value of K_{Ic} from each 3PB test is compared to the average and range of 4-6 short-rod tests on specimens taken from the same beam. Short-rod values tend to be higher than those from 3PB tests. However, short-rod results fall well within the range of all previously reported toughness results for Indiana limestone (Ouchterlony, 1980).

TESTING RESULTS

Fracture toughness determination was made on sedimentary rocks from the Buffalo Light Rail Rapid Transit project, Culver-Goodman tunnel in Rochester, and the Chicago Tunnel and Reservoir Plan project. Because of space limitations, only the results from the Buffalo project will be presented here. A high incidence of horizontal shear failure with the Buffalo rocks led to the development of the prenotching and axial loading techniques discussed previously. It was found that prenotching did not solve the shear failure problem. However, use of the axial loading technique resulted in a test failure rate of less than 5%. To be consistent, all of the Buffalo rocks were both prenotched and axially loaded. As prenotching was found to be unnecessary and did not affect results, the remainder of the rocks were tested with axial load only. All results reported here were obtained with specimens prepared from vertical NX cores and, unless otherwise noted, with random fracture plane orientation.

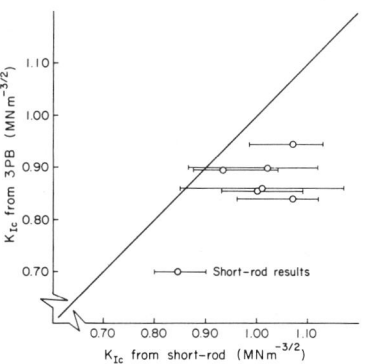

Fig. 5. Comparison of 3PB (Ingraffea 1981) and short-rod results for Indiana limestone.

The Buffalo Tunnel alignment is confined to the lower two members of the Silurian Bertie Formation, the Falkirk, and the Oatka. The Falkirk Member is a medium-bedded to massive, hard, grayish-tan, medium to coarse grained dolostone with undulating shale partings. The Oatka Member is a medium to thinly bedded, moderately hard, bluish-gray, shaley dolostone and dolomitic shale.

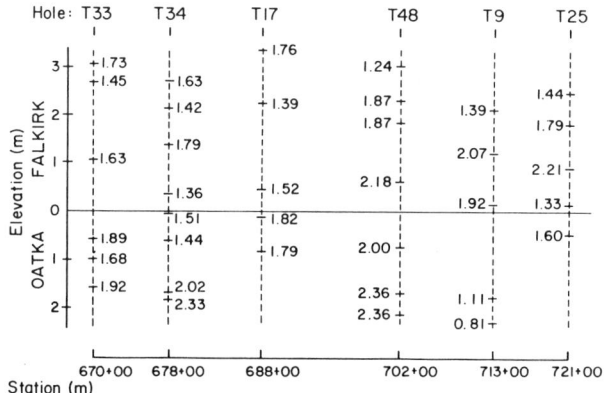

Fig. 6. Fracture toughness results ($MNm^{-3/2}$)

Figure 6 shows short-rod K_{Ic} results for specimens from 6 bore holes spaced along the tunnel. The vertical location of each specimen is referenced to the Falkirk/Oatka contact. Specimens were taken from the horizon of the 6 meter diameter tunnel whose springline was approximately coincident with the contact.

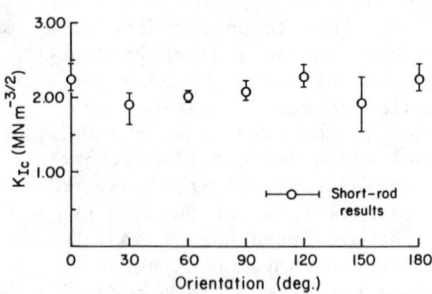

Fig. 7. Variation of fracture toughness in Falkirk member with relative fracture plane azimuth.

To determine further the spatial variability of K_{Ic} in a rock unit, an auxiliary test series was performed. To isolate longitudinal and vertical variability from the effect of fracture plane orientation, specimens were prepared from a block of Falkirk rock measuring approximately 300 mm on each side. Three specimens at each of six crack plane azimuths were tested. Results are shown in Figure 7.

COMPARISON OF K_{Ic} WITH OTHER STRENGTH MEASURES

In addition to the short rod tests, other currently popular measures of strength were evaluated. For each bore hole, point load, uniaxial compressive, and Brazil tests were performed on specimens also taken from the tunnel horizon. Variability of each of these measures along the tunnel axis as a percentage of its overall average value is compared to that of K_{Ic} in Figure 8.

DISCUSSION

The short-rod test is quick and inexpensive. Specimen preparation requires no machining or grinding operations. No complex instrumentation is needed to perform the test The test has, however, been shown to be as accurate as standard, more complex K_{Ic} measurement methods currently in standard use.

However, Figure 6 shows a wide variation in results for the Buffalo rock units tested. This variation can be attributed to two causes: rock property variation with location and fracture orientation, and precision of the testing system. The results of the orientation study (Fig. 7) show that a variation of 10% around the mean could be caused by random selection of fracture plane azimuth. Also, the maximum difference from the mean of any single datum in the orientation study is only 25%. Therefore, the large variation exhibited by the data in Figure 6, sometimes over short elevation changes, is most likely real.

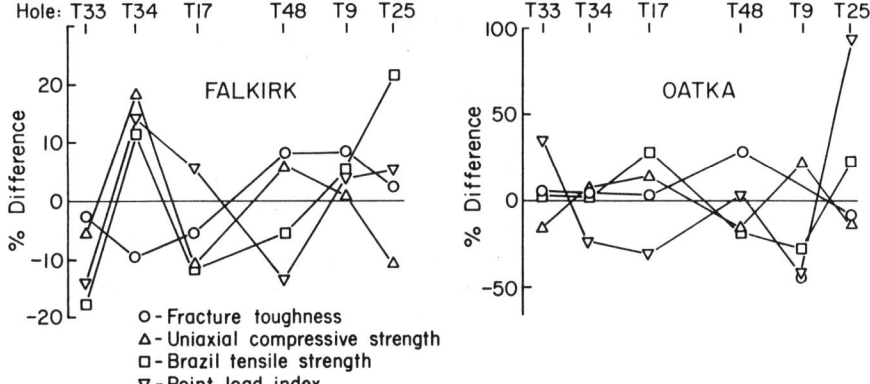

Fig. 8. Comparison of variability of strength measures

Moreover, Figure 8 indicates that, in general, the variation in K_{Ic} is less than that of the other measures. Since K_{Ic} is an intrinsic, rather than index, property, and since its measurement is direct and simple, without such complexities as multiple failure mechanisms and end effects, this should be expected. Figure 8 also shows that while some general trends exist between fracture toughness and the other measures, these trends are not consistent along the tunnel or, surprisingly, between units. Since these measures often vary out of phase, the question arises as to which property might best predict variations in TBM penetration rates. A future paper will address this question.

SUMMARY

A program to develop a more reliable TBM performance prediction system is in progress. Ideally, to relate TBM performance to lithology, the material property which controls a mechanism of rock chip formation under a roller cutter should be used. Fracture toughness appears to be an attractive candidate for this role. From the overall study, this paper reports on:

1. The development of a fracture toughness testing technique. This includes accurate K_I-calibration, loading device fabrication, verification studies, and modifications for testing weakly bedded sedimentary rocks.

2. The application of this technique to rocks from current TBM projects. Variability in K_{Ic} has been shown to be of the same order as that of other strength measures.

A data store of fracture toughness and TBM performance values from a range of projects is being accumulated. Subsequent papers will examine analytically the effect of varying fracture toughness on chip formation and statistically evaluate toughness as a predictor of penetration rate.

ACKNOWLEDGMENTS

This work is being partially supported by the Department of Transportation contract No. DTRS 57-80-C-00107. Our appreciation is extended to Goldberg-Zoino Associates of New York, P.C., the Niagara Frontier Transportation Authority and Richard Flanagan for their assistance in obtaining the rock cores. The authors also express special thanks to Professors Thomas O'Rourke and Fred Kulhawy for their continued enthusiastic support of this effort.

REFERENCES

Barker, L.M., 1977, "A Simplied Method for Measuring Plane Strain Fracture Toughness," Engineering Fracture Mechanics, Vol. 9, No. 2, pp. 361-369.

Beech, J.F. and Ingraffea, A.R., 1982, "Three-Dimensional Finite Element Calibration of the Short-Rod Specimen," International Journal of Fracture, Vol. 18, No. 3, Mar., pp. 217-229.

Bubsey, R.T., Munz, D., Pierce, W.S., and Shannon, J.L., Jr., 1982, "Compliance Calibration of the Short Rod Chevron-Notch for Fracture Toughness Testing of Brittle Materials," International Journal of Fracture, Vol. 18, No. 2, Feb., pp. 125-133.

Ingraffea, A.R., 1981, "Mixed-Mode Fracture Initiation in Indiana Limestone and Westerly Granite," Proceedings, 22nd U.S. Symposium on Rock Mechanics, Cambridge, MA, pp. 186-191.

Krenz, H.G., Ast, D.G., and Kramer, E.J., 1976, "Micro-Mechanics of Solvent Crazes in Polystyrene," Journal of Materials Science, Vol. 11, No. 12, Dec., pp. 2198-2210.

Ouchterlony, F., 1980, "Review of Fracture Toughness Testing of Rock," DS1980:15, Swedish Detonic Research Foundation, Stockholm.

Chapter 48

A STUDY OF FRACTURE TOUGHNESS FOR AN ANISOTROPIC SHALE

by Vernal H. Kenner and Sunder H. Advani
Department of Engineering Mechanics
The Ohio State University
Columbus, Ohio

and

Terry G. Richard
Department of Engineering Mechanics
University of Wisconsin
Madison, Wisconsin

ABSTRACT

This paper presents and discusses the results of thirty-one fracture toughness tests on Pennsylvanian Age Clay Shale. Two different orientations for this anisotropic material were studied and two series of tests were conducted, namely, 1. a test series to determine the effect of moisture content on fracture toughness, and 2. a test series to study mixed mode fracture. Increased absorbed moisture is found to reduce fracture toughness. Several analyses of mixed mode failure are found to give varying non-conservative predictions of failure stress intensity level.

INTRODUCTION

In a variety of energy resource recovery problems it is vital that the fracture characteristics of the involved rocks be known. In a combined experimental and theoretical effort to study several such problems associated with the utilization of coal, we have examined the fracture toughness of Pennsylvanian Age Clay Shales which form the overburdens for the Middle Kittaning Coal Seam in Ohio. Some data have been collected which pertain to 1. the effects of moisture content on fracture toughness, 2. the consequences of the anisotropy of this material for fracture toughness, and 3. the character of mixed-mode fracture in this material.

In this paper we present these results. The data on moisture dependence are examined in the light of studies of the dependence of

compressive strength on this variable which have been conducted for
the same shale (Advani and Richard, 1981). The examination of the
mixed-mode fracture results is made with an interest to comparing
results reported recently for similar tests involving much more
nearly isotropic rocks (Ingraffea, 1981).

EXPERIMENT

Rectangular shale samples nominally 152mm x 38mm x 19mm were cut
using a rock saw to produce fracture toughness specimens as described
in ASTM E399 (Anon, 1978) and depicted in Fig. 1. Specimens were cut
so that the bedding plane was oriented both parallel to and perpendicular to the plane of the loading (Fig. 1). Owing to the fragility
of the specimens, precracking was not attempted; the "crack" thus
consisted of a saw cut which had a nominal thickness of approximately
0.25mm. For four mode I specimens tested early in the program, however, the saw cut was 1.12mm thick. Specimens were conditioned
prior to testing by storage at ambient temperature (22-24°C) over
saturated salt solutions yielding relative humidity (RH) levels of
33%, 47%, 63% and 81%. Testing occurred well beyond the time when
the equilibrium specimen weight had been established. The "equilibrium" state of the unconditioned specimens was found to correspond
to a storage RH level of 53% (Advani and Richard, 1981).

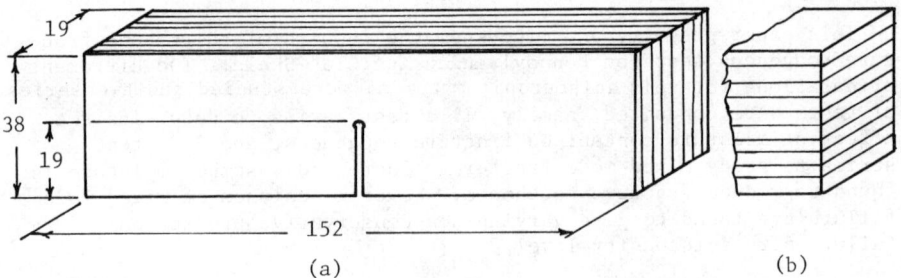

Fig. 1. Specimen dimensions (mm) and orientation relative to the
bedding planes. a. Parallel load test specimen.
b. Perpendicular load test specimen.

Specimens were loaded in a servohydraulic testing machine (MTS)
at a constant stroke rate of 0.05mm/minute. The four-point loading
configuration shown in Fig. 2 permitted either pure bending, pure
shear or combinations of shear and bending to be applied to the
specimen ligament; this produced mode I, mode II and mixed-mode
loading, respectively. Moisture dependence of the fracture data
was only determined for mode I loading. Load and load point displacement were always measured and, in the mode I tests, crack
opening displacements were also detected through the use of a
mimiature LVDT. For tests involving shear on the ligament the
relative displacement of specimen halves at failure was such that

the LVDT would have suffered damage and thus it was not employed in these tests.

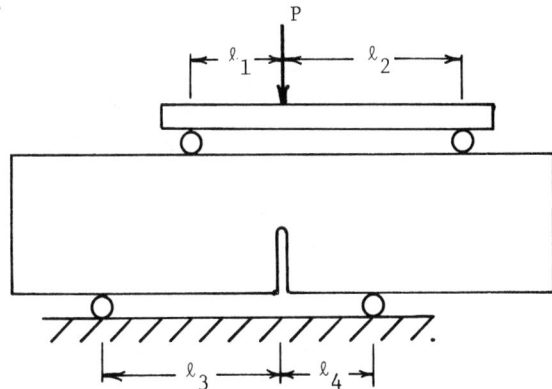

Fig. 2. Specimen loading arrangement. For mode I tests, $\ell_1 = \ell_2 = 38.1$mm; for mixed-mode tests, $\ell_1 = 16.9$mm, $\ell_2 = 59.3$mm, $\ell_3 = 50.8$mm and $\ell_4 = 25.4$mm; for mode II tests, $\ell_1 = \ell_4 = 25.4$mm and $\ell_2 = \ell_3 = 50.8$mm.

RESULTS

Fig. 3 presents typical load vs. load point displacement (P-δ) and load vs. crack opening displacement (P-COD) records from mode I tests. The relatively more pronounced nonlinearity of the P-COD trace for loading perpendicular as compared to loading parallel to the bedding plane is noted. For the parallel loading case, data reduction as specified by the ASTM standard (Anon. 1978) would produce what is there referred to as a valid K_{Ic}. There the appropriate critical load, P_Q, would be that associated with a secant stiffness equal to 95% of the initial tangent stiffness (Fig. 3). Furthermore, P_Q would satisfy a condition, imposed by the standard, that the maximum load not exceed $1.1P_Q$. For the case of perpendicular loading, Fig. 3 indicates that this condition would not be met. Observing that the ASTM data reduction described above is intended to eliminate test results (from metals) in which excessive plastic deformation occurs and that no such deformation is present in our results, here we (arbitrarily) calculate the stress intensity at break, K_{break} utilizing the maximum load encountered by the specimen.

Two-dimensional crack displacement finite element computations were conducted in order to calculate K_I and K_{II} for the several experimental loading conditions. The nominal specimen geometry (Fig. 1) was modeled by the mesh depicted in Fig. 4. Triangular isoparametric elements were used with one side of the original quadrilateral collapsed and the mid-side node placed at the quarter

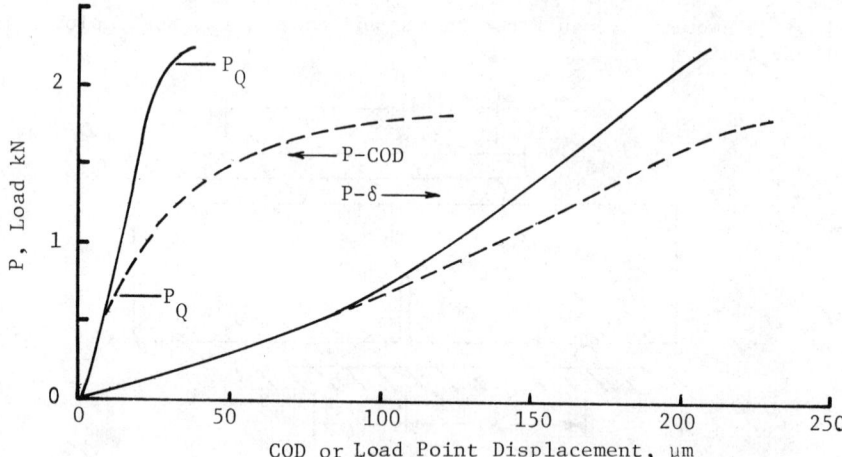

Fig. 3. P-δ and P-COD records. Solid curves represent results for parallel loading and dashed curves represent results for perpendicular loading. P_Q represents load level associated with 95% of the initial slope (see text).

points to produce the $\sqrt{r}$ singularity at the crack tip (Advani, 1980). Minor variations of actual specimen geometry from the nominal dimensions were accounted for by scaling consistent with results for edge cracked infinite beams subjected to shear or bending which are given by Tada et al. (1973). While always amounting to less than 2% for K_{II}, the correction for K_I ranged to as much as 22% of the nominal value.

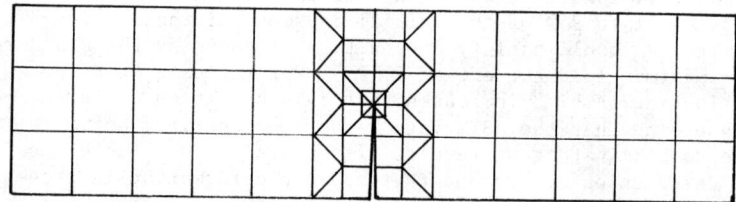

Fig. 4. Finite element mesh.

Fig. 5 presents the results of twenty-two mode I tests in the form of $K_{I\ break}$ vs. conditioning RH level. Fig. 6 gives the results of five mode II and four mixed-mode tests; in this figure normalization of break stress intensity values has been based on a mean value of $K_{I\ break}$ as described in the sequel.

FRACTURE TOUGHNESS FOR ANISOTROPIC SHALE

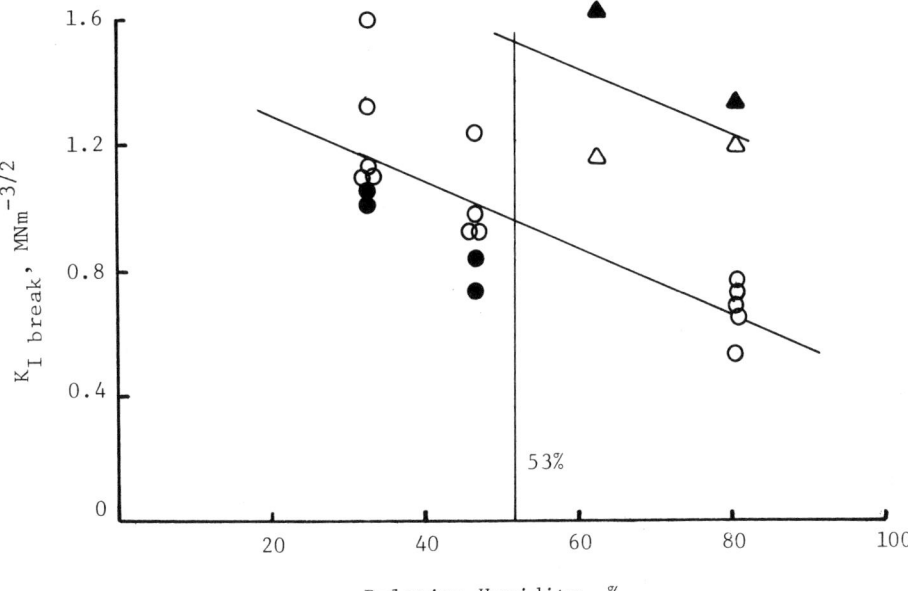

Fig. 5. Mode I fracture toughness as a function of conditioning relative humidity level. Triangles denote results from specimens for which saw-cut thickness was 1.12mm and circles represent specimens for which saw-cut thickness was 0.5mm. Open symbols represent parallel loading and closed symbols represent perpendicular loading.

DISCUSSION

As noted above, several specimens were fabricated with the final crack-simulating slot being approximately four times greater in thickness than that usually employed. These specimens, represented by triangular data points in Fig. 5 exhibited significantly increased $K_{I\,break}$ values. When these data points are excepted from Fig. 5, a clear decrease in $K_{I\,break}$ with increasing storage relative humidity (i.e., increasing moisture content) is observed. The straight line fitted through these data points (all of the circles) represents a least squares fit which has a correlation coefficient of 0.76; the probability that this value would be exceeded for unrelated variables is less than 0.1% (Young, 1962). The present experiments show an average lowering of $K_{I\,break}$ by approximately 42% for specimens conditioned at 81% as compared to those stored at 33% RH. This finding appears to be consistent with a decrease in ultimate compressive strengths reported to be typically 30% to 50% for tests of similarly conditioned samples of the same shale (Advani and Richard, 1981).

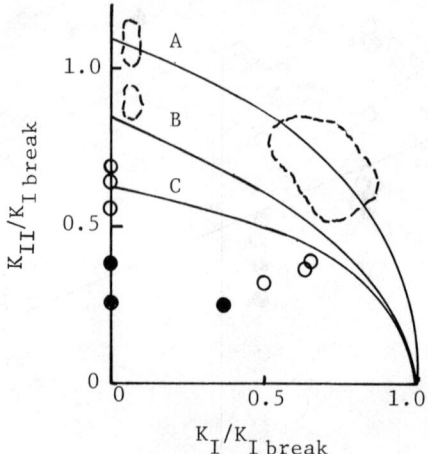

Fig. 6. Mixed-mode fracture results. Open and closed circles correspond to loading parallel or perpendicular to the bedding plane, respectively. The solid curves are theoretical predictions. Curve A is due to Sih (1974), curve B is due to Erdogan and Sih (1963) and curve C is due to Hussain et al. (1974). The dashed curves approximately enclose the domain of experimental points due to Ingraffea (1981).

The present tests did not indicate a clear difference in $K_{I\,break}$ values for loading perpendicular as opposed to loading parallel to the bedding planes. Although the fine-cut specimens (circles) produced somewhat lower $K_{I\,break}$ values at 33% RH and 47% RH there was no substantial difference at 81% RH. Furthermore, the few results for the wider cut are in the opposite direction. This ambiguity, reinforced by the observation that in neither loading case can a mode I crack propagate in or parallel to a bedding plane indicate that, at most, the difference is small enough to be obscured by statistical variations. Although likely more isotropic than the present shale, the limestone tested by Schmidt (1976) exhibited no obvious differences for mode I fracture toughness for the two configurations we have tested. The majority of mixed-mode or mode II tests were conducted on "as received" material. When this was not the case, the mode II data presented in Fig. 6 were corrected to "as received" conditions by adjusting $K_{I\,break}$ and $K_{II\,break}$ values to the 53% RH level according to the slope of the least squares fit of Fig. 5. All data presented in this plot are normalized by dividing by $949 \times 10^3 \mathrm{Nm}^{-3/2}$, which is the $K_{I\,break}$ value indicated by the (lower) fitted straight line at 53% RH.

FRACTURE TOUGHNESS FOR ANISOTROPIC SHALE

For loading parallel to the bedding plane the anisotropy of the shale should not be a factor insofar as the problem may be considered plane. Thus, for this case (open circles only) the experimental points may reasonably be compared to analytical predictions for mixed mode fracture, several of which are shown in Fig. 6 by the solid lines. Also presented here is the approximate domain of experimental results for Indiana limestone and Westerly granite due to Ingraffea (1981).

The present results are seen to fall below those previously reported and closest to (slightly below) the energy release rate theory of mixed-mode fracture proposed by Hussain et al. (1974). Since our results represent propagation from a very thin saw cut instead of an actual crack, the possibility of friction of mechanical interlocking at the crack surface to inhibit mode II motion is not present. The absence of such effects would seem to mitigate toward lower measured K_{II} values, consistent with the present observations. It is noted that limited mode II data due to Ingraffea (1981) for notches appear to be somewhat low compared to otherwise comparable data for mixed-mode propagation from the tips of natural cracks.

For mixed-mode and mode II loading the anisotropy of the present shale appears to be important. The data points for perpindicular loading are always below those for parallel loading. This is consistent with theoretical results (Palaniswamy and Knauss, 1978; Swedlow, 1976) which indicate that, for pure mode II loading, a crack will propagate in a direction perpendicular to its initial plane, i.e., in (or at least parallel to) the bedding plane for our perpendicular loading case. Indeed, in the one perpendicularly loaded specimen which exhibited highly pronounced bedding planes, (and, also, failed at the lowest load level) the crack was found to have propagated in a bedding plane for approximately 12mm before then jumping to the adjacent bedding plane.

In concluding discussion of the experimental results presented above, we consider the effect of using saw-cuts in lieu of actual cracks. By using the line fitted to experimental points for 1.12mm wide saw-cuts (Fig. 5), a normalized $K_{I\ break}$ value may be determined for 53% RH. Consistent with the notion of stress at a blunt-crack tip being inversely proportional to $\sqrt{\rho}$, where ρ is the radius of curvature (Tada et al., 1973), fit values of $K_{I\ break}$ have been plotted against the square root of saw-cut thickness, $\sqrt{t}$, in Fig. 7. Extrapolation to zero thickness indicates that the $K_{I\ break}$ value from a sharp crack would be approximately half that observed for the 0.25mm-wide crack, or approximately 425 $kNm^{-3/2}$. By way of comparison, fracture toughness values for Indiana Limestone have been reported in the range of 700 to 1000 $kNm^{-3/2}$ (Schmidt, 1976; Ingraffea, 1981).

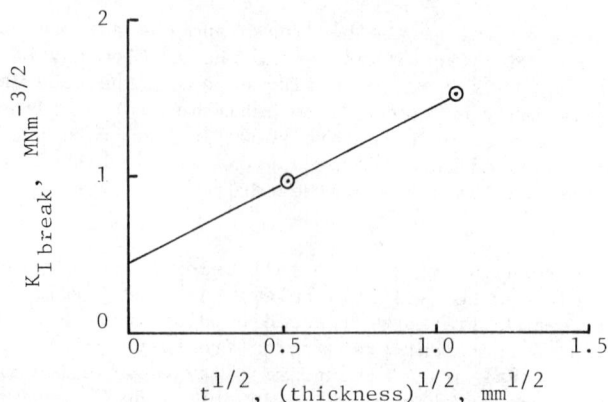

Fig. 7. Mode I fracture toughness vs. square root of crack thickness (extrapolated to zero thickness).

ACKNOWLEDGEMENT

This work was supported by U.S. Bureau of Mines Grant G1115393 and the Ohio Coal Research Laboratories Association Contract No. OCRLA-9. The authors also wish to acknowledge the efforts of Messrs. O. Gurdogan and I. Szauter, who assisted with the finite element calculations and the experiments, respectively.

REFERENCES

Advani, S.H., 1980, "Hydraulic Fracturing and Associated Stress Modeling for the Eastern Gas Shales Project," DOE/METC/10514-28 Final Report, U.S. Department of Energy.

Advani, S.H. and Richard, T.G., 1981, "Fracture Mechanics and Structural Response Investigations Associated With Energy Resource Recovery," R.F. Project 762367/713219 Final Report, The Ohio State University Research Foundation.

Anon., 1978, "Standard Test Method for Plane-Strain Fracture Toughness of Metallic Materials," ASTM E 399-78a, Part 10, Annual, Book of ASTM Standards, Americal Society for Testing and Materials Materials, Philadelphia, pp 540-561.

Erdogan, F. and Sih, G.C., 1963, "On the Crack Extension in Plates Under In-Plane Loading and Transverse Shear," Journal of Basic Engineering, Vol. 85, pp. 519-527

Hussain, M.A., Pu, S.L. and Underwood, J.H., 1974, "Strain Energy Release Rate for a Crack Under Combined Mode I and Mode II," Fracture Analysis, ASTM STP 560, Americal Society For Testing Materials, pp 2-28.

Ingraffea, A.R., 1981, "Mixed-mode Fracture Initiation in Indiana Limestone and Westerly Granite," Proceedings of the 22nd U.S. Symposium on Rock Mechanics, pp 186-191.

Palaniswamy, K. and Knauss, W.G., 1978, "On the Problem of Crack Extension in Brittle Solids Under General Loading," Mechanics Today, Vol. 4, pp 87-148.

Schmidt, R.A., 1976, "Fracture-toughness Testing of Limestone," Experimental Mechanics, Vol. 16, No. 5, pp 161-168.

Sih, G.C., 1974, "Strain-energy-density Factor Applied to Mixed-mode Crack Problems," International Journal of Fracture Mechanics, Vol. 10, pp 305-321.

Swedlow, J.L., 1976, "Criteria for Growth of the Angled Crack," Cracks and Fracture, ASTM STP 601, American Society for Testing and Materials, pp 506-521.

Tada, H., Paris, P.C. and Irwin, G., 1973, The Stress Analysis of Cracks Handbook, Del Research Corporation, Hellertown, PA.

Young, H.D., 1962, Statistical Treatment of Experimental Data, McGraw Hill.

Chapter 49

EXPERIMENTAL MODELING OF MICROCRACK FORMATION
IN ROCKS

by Takao Kobayashi

Department of Mechanical Engineering
University of Maryland
College Park, MD 20742

ABSTRACT

Micro-crack formation zones developed near a crack tip in rocks and ceramics play a significant role in fracture behavior of these materials. In order to study micro-crack formation near a crack tip and its effects on crack behavior, a mica-epoxy composite material was developed.

Multi-layer mica flakes were embedded in an epoxy matrix. Under high stress in the vicinity of a crack tip the mica delaminated, simulating the micro-cracking along grain boundaries or crystal planes in rocks and ceramics.

A compact tension specimen was fabricated from mica-epoxy composite material. During the test the load and crack opening displacement were measured following the ASTM E561 (R-curve determination) procedure. Through the compliance calibration method the effective crack length and the K_R were computed. The experiment clearly demonstrates the formation of micro-cracks and their effects on the overall behavior of the specimen. The paper discusses the significance of micro-crack formation on fracture toughness characterizations, and the advantages of the use of mica-epoxy composite for detailed studies of interaction between the micro-crack formation zone and the specimen geometry and size.

MODELING OF MICRO-CRACK FORMATION 481

INTRODUCTION

One of the major differences in fracture behavior of rocks and ceramics from that of metals is the formation of micro-cracks around a crack tip (1-5). This phenomenon poses an interesting problem in the application of linear elastic fracture mechanics to these materials. Linear elastic fracture mechanics assumes that a zone near a crack tip exists in which stresses and strains are characterized by a single parameter: the stress intensity factor, K. This zone is called a singularity dominated zone, and its size may depend upon such parameters as specimen dimensions, crack size and loading. Another parameter which has to be considered is the size of a fracture process zone such as micro-crack formation zone in rocks and ceramics. When the singularity dominated zone size is sufficiently large compared with the fracture process zone, a stress intensity factor characterization is quite appropriate; however, when the size of the fracture process zone approaches or exceeds that of the singularity dominated zone, a linear elastic fracture mechanics characterization may no longer be valid. It is, thus, imperative to understand the role of micro-crack formation in fracture mechancis characterization of rocks and ceramics.

In order to study micro-crack formation and its effect on crack behavior and to study the effects of specimen geometry, specimen size, and loading systems, a mica-epoxy composite specimen was developed. Multi-layered natural mica flakes were embedded in an epoxy matrix. Under high stress near a crack tip, delamination in mica flakes took place, simulating micro-crack formation in rocks. An advantage of this system is that the zone in which delamination of mica flakes occurred is clearly visible. Furthermore, the size and distribution density of mica flakes are controlling parameters in simulating specific materials. This paper describes the mica-epoxy composite fabrication technique and preliminary test results.

DESCRIPTION OF MICA-EPOXY COMPOSITE MATERIAL

The mica flakes employed in this work were purchased from Mica Products Co. of Balboa Island, California, as natural mica-sheet of 0.2 mm to 1.5 mm in thickness and approximately 75mm X 150mm in size. This mica-sheet was thinned down to 0.1 mm by delaminating the layers. The sheet of 0.1 mm thickness still consisted of three to

four layers. After delaminating the sheet, it was cut
into approximately 3mm X 3mm squares by a paper-cutter.

A matrix epoxy consisted of three components: an epoxy
resin (Epon 828[*]), a curing agent (Jeffamine D-400[**]),
and an accelerator (398[**]). The weight ratio of each
component is 100 pph of the epoxy resin, 35 pph of the
curing agent, and 5 pph of the accelerator. The
mechanical properties and photoelastic studies of the
fracture behavior of this epoxy have been reported
elsewhere (7). The mica-epoxy composite fabrication
procedure is: 1) throughly mix measured quantities of
epoxy resin, curing agent, and mica flakes; then de-gas
the mixture in a vacuum chamber, 2) mix the accelerater
and de-gas briefly, 3) pour the mixture into an open mold
and distribute mica flakes as uniformly as possible with
a stick, 4) place mold cover with care to avoid trapping
air in the mold, and 5) place the assembled mold in the
frame to rotate it so that the mica flakes are suspended
and uniformly distributed in the epoxy matrix.

During stages 1 through 4, it was observed that mica
flakes settled at the bottom of the container or the
mold; however, the successful suspension and uniform
distribution of mica flakes were achieved by slowly
rotating the mold. The mold rotating device and the
assembled mold in the device are shown in Fig.1.

The mold consisted of two 300mm X 300mm X 12.5mm glass
plates coated with polyvinyl alcohol as mold release
agent. Four 12.7mm square aluminum bars were used as a
spacer frame. A seal was provided by a 20mm diameter
vinyl tubing placed inside of the aluminum frame.

The specimen in the mold was kept rotating for 24hrs
for initial cure. Then it was removed from the mold and
placed on a flat plate at 35 °C for two weeks for final
cure. Room temperature cure was done to prevent an
introduction of residual stresses due to differences in
thermal expansion coefficents of mica and epoxy resin.

SPECIMEN GEOMETRY AND TEST PROCEDURE

A compact specimen specified by the ASTM E561-80
(R-curve determination) was utilized for this test with a
modification for crack opening displacement measurement.
Crack opening displacement measurement was made with a
crip-gage mounted on knife edges located at 0.1567W
(W=101.6 mm) away from the load-line. The specimen
dimensions are shown in Fig. 2.

The specimen was loaded step-wise in a loading machine under a displacement control mode. The load and crack opening displacement, $2v_1$, were recorded on an X-Y recorder. The specimen compliance was calculated from the slope of the load versus $2v_1$ curve. An effective crack length and resistance K_R were calculated from the result of the specimen compliance. The development of a zone in which mica delamination was observed was photographed in the circular polariscope.

EXPERIMENTAL RESULTS AND DISCUSSION

Fig. 3 shows the X-Y recorder trace of the load as a function of crack-opening-displacement $2v_1$. The resistance curve is shown in Fig. 4. Photographs of delaminated mica-chips around a crack tip at different stages are shown in Fig. 5.

Examination of load versus crack opening displacement record indicates that after development of delamination of mica chips the crack did not close completely upon unloading. This phenomenon was also observed during testing of rocks (7).

The resistance curve shown in Fig. 4 exhibits several large kinks; however, the sizes of kinks are determined by the size of mica flakes embedded in the epoxy resin. If smaller mica flakes are used, it is believed that a smoother resistance curve can be obtained.

Comparison of crack extension shown in Fig. 4 with the photographs of mica delamination zone shown in Fig. 5 is of particular interest. Fig. 4 shows that an effective crack length extension before instability was about 5 mm; however, the mica delamination zone observed was about 18-30 mm in size. It is interesting to note that the delamination zone appeared to stretch more in the direction perpendicular to the crack plane. The shape is similar to that of Schmidt's process zone model (2).

Finally, formation of delaminated mica flakes around a crack tip appeared to increase a resistance to crack propagation significantly. The K_{IC} value of the epoxy resin itself is of the order of 0.9 MPa$\sqrt{m}$. Furthermore, crack arrest was observed after 35 mm crack extension in mica-epoxy composite even though the crack instability took place at such a high value of K_R as 2.3 MPa$\sqrt{m}$.

CONCLUSION

The preminary results of crack behavior in mica-epoxy

composite indicate that with this material effects of micro-crack formation around a crack tip can be studied very effectively. The formation of micro-cracking around a crack tip increases crack initiation toughness and resistance to crack propagation significantly. Futhermore, the micro-cracking zone size observed was much larger than the effective crack size increase computed by the compliance method.

REFERENCES

1. Hoagland, R.G., Hahn, G.T., Rosenfield, A.R, 1973, "Influence of Microstructure on Fracture Propagation in Rock," Rock Mechanics, Vol. 5, pp77-106.

2. Schmidt, R.D., 1980, "A Microcrack Model and Its Significance to Hydraulic Fracturing and Fracture Toughness Testing," Proceedings of the 21st U.S. Symposium on Rock Mechanics.

3. Wu, C. Cm., Freidman, S.W., Rice, R.W., Mecholsky, J.J., 1978, "Microstructural Aspects of Crack Propagation in Ceramics," Journal of Material Science, 13, pp2659-2670.

4. Kobayashi, T, Fourney, W.L., 1978, "Experimental Characterization of the Development of Micro-Crack Process Zone at a Crack Tip in Rock under Load," Proceedings of the 19th U.S. Symposium on Rock Mechanics, pp243-246.

5. Kobayashi, T.,Fourney, W.L., Holloway, D.C., 1979, "Further Examination of Microprocess Zone in Pink Westerly Granite," presented at the 1979 Symposium of the American Society of Ceramics.

6. Kobayashi,T., Dally, J.W., 1977, "A System of Modified Epoxies for Dynamic Photoelastic Studies of Fracture," Experimental Mechanics, 17, No. 10, pp367-374.

7. Schmidt, R.A., Lutz, T.J., 1979, "K_{IC} and J_{IC} of Westerly Granite--Effects of Thickness and In-plane Dimensions," Fracture Mechanics Applied to Brittle Materials, ASTM STP 678, pp166-182.

MODELING OF MICRO-CRACK FORMATION

Fig. 1. Mica-Epoxy Composite Fablication Mold and Mold Rotating Device

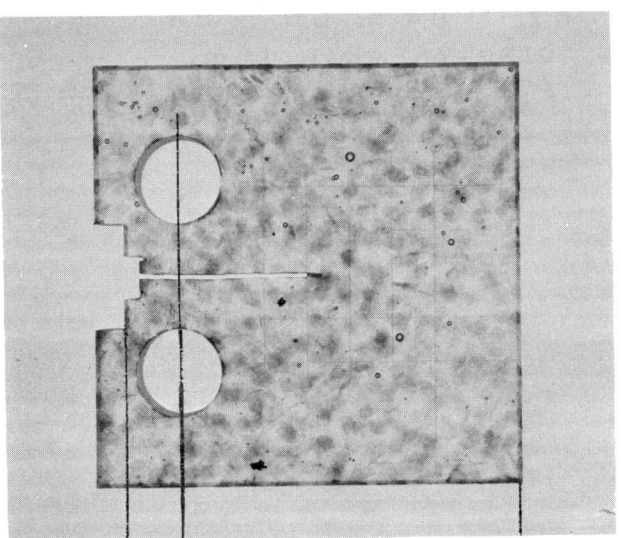

B=12.7mm

Fig. 2. Mica-Epoxy Composite Compact Tension Specimen and Its Dimensions

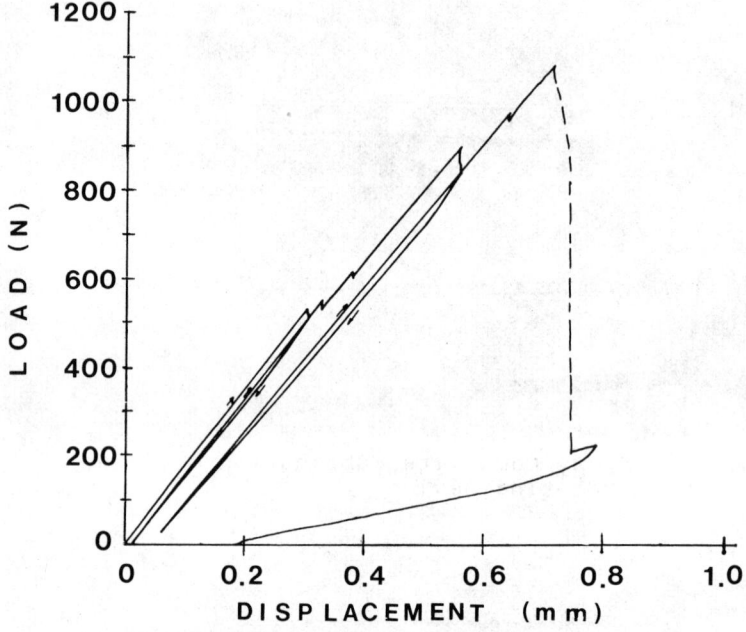

Fig. 3. X-Y Plot of Load versus Crack Opening Displacement

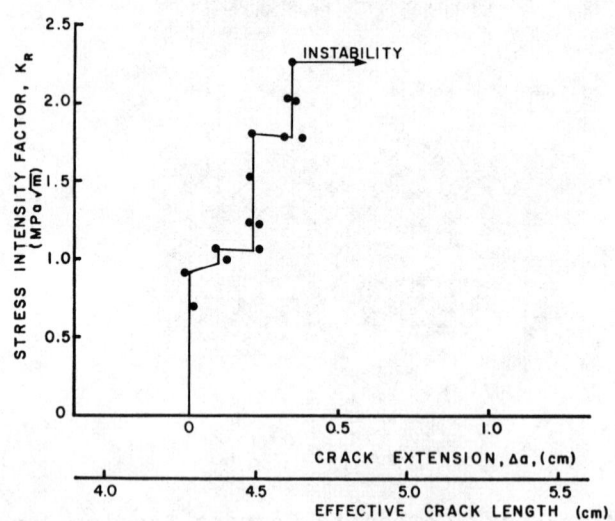

Fig. 4. Fracture Resistance Curve for Mica-Epoxy Composite Material

MODELING OF MICRO-CRACK FORMATION

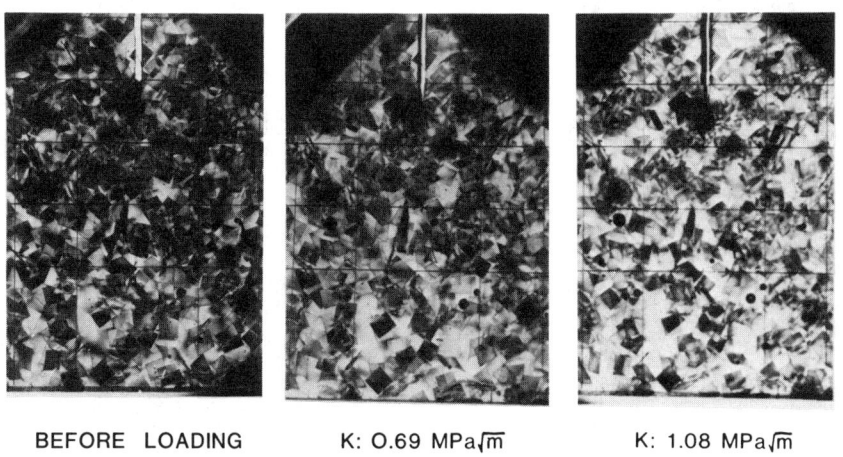

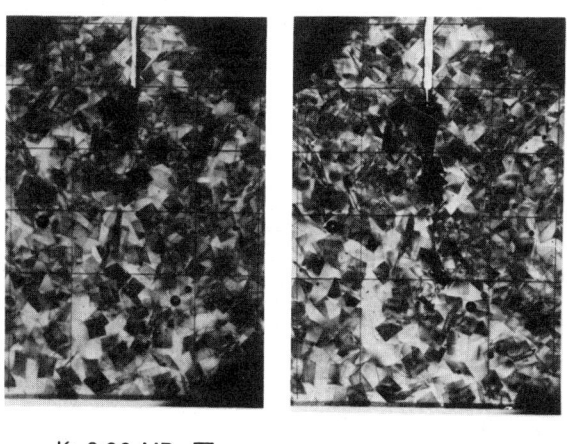

Fig. 5. Photographs of Development of Zone of Delaminated Mica Flakes around a Crack Tip

Chapter 50

AN EXPERIMENTAL INVESTIGATION OF THE PROJECTILE PENETRATION INTO SOFT, POROUS ROCK UNDER DRY AND LIQUID-FILLED CONDITIONS

by Akihiko Kumano and Werner Goldsmith

Consultant, Nutech Engineers
San Jose, California

Professor, Department of Mechanical Engineering
University of California, Berkeley, California

ABSTRACT

An experimental investigation was conducted to study the response of soft, porous rock, green shale, under dry and liquid-filled conditions to normal impact of hemispherically-tipped cylindrical steel projectiles with diameter and mass of 6.35 mm and 31 g, respectively. Disk-shaped green shale specimens with nominal thickness and diameter of 32 mm and 127 mm, respectively, were tested in the initial projectile velocity range of 18 to 42 m/s. For the liquid-filled tests, the green shale specimens were saturated with kerosene. A total of eight impact tests were conducted. Information on the extent and pattern of the damage in the rock and velocity histories of the projectiles were collected. It was found that while the penetration depth was quadratically related to the initial projectile kinetic energy for the dry case, the penetration depth for the liquid-filled case was directly proportional to the initial projectile kinetic energy. The results of the impact tests under dry and liquid-filled conditions conformed well to a linearly viscous model and a Coulomb frictional model, respectively. The major deformation processes in the dry specimens were found to be compaction and equivolumnial distortion of the material. In the liquid-filled case, however, formation of cracks and equivolumnial distortion with less material compaction were observed. Furthermore, craters were formed as a result of the crack formation in the liquid-filled specimens.

INTRODUCTION

Numerous studies have been conducted in an attempt to understand the behavior of soft, porous materials. The response of porous materials to load has been studied extensively, primarily from analytical considerations of continuum mechanics (Bhatt et al., 1975). In addition, the compressibility of porous rocks under quasi-static loading has been studied experiemntally (Stephens et al., 1970). Wave propagation problems in a porous material has also been investigated theoretically (Carroll et al., 1973). A series of studies on the penetration of projectiles into soft materials, in which large deformations are generally experienced, resulted in the development of finite-difference or finite-element procedures for either viscoelastic or elastic-plastic targets subjected to large deformations (Johnson, 1976).

In the present investigation, projectile impact on soft, porous green shale under dry and liquid-filled conditions was studied experimentally. This is a substantial extension of a previous effort in which the impact of 6.35 mm diameter spherical projectiles on dry green shale was studied (Kabo et al., 1977). The current effort aims, in part, to extend the available data base by measurement of the velocity histories and characterization of the damaged region in the green shale under both dry and liquid-filled conditions. The major objective here, however, is to study the effect of the presence of liquid in the rock on the response of green shale to impact. The purpose of the present study also includes the construction of simple models describing the relation between the final penetration depth and the initial projectile kinetic energy. Comparison of the test data obtained under dry condition to the predictions of a more comprehensive theoretical model was also included in the present investigation.

EXPERIMENTAL PROCEDURE

Sample Preparation

<u>Dry Samples</u> The specimens consisted of a green shale disk with a diameter of 140 mm and a thickness of 35 mm, generated by means of a standard rock core drill. This shale, obtained from the site of an open pit mine at Boron, California, contains feldspar, quartz, mica, biotite, clays, and other rock fragments. The measured mechanical properties of this material under dry condition are summarized in Table 1.

The disk-shaped shale specimens were initially sawed diametrically into two halves, and the cut surfaces were then finished with fine sandpaper. Grid lines were drawn on these surfaces with soft lead pencils and the two halves were reassembled as a disk upon insertion

Table 1 Mechanical Properties of the Green Shale

Density, kg/m^3	Young's Modulus, GPa	Shear Modulus, GPa	Compressive Strength, MPa	Tensile Strength, MPa	Porosity, % by Volume
1670	3.77	1.47	7.15	0.37	28

in a specially-fabricated metal shell. Two BLH FAE-12-12-S6L strain gages were mounted on the shell wall to monitor the tightness of the shell so that both gages exhibited the same strain of 0.0003 for all tests. This corresponds to an internal compressive stress of 2 MPa, based on thin-walled vessel analysis, which is approximately 5.4 times the tensile strength of green shale. For the tests of the dry condition, the assembled specimen was clamped directly to the end face of an aluminum cylinder, 152 mm in diameter and 203 mm long.

Liquid-Filled Samples As before, the disk-shaped shale specimens were first sectioned into two halves in the diametrical direction and the same surface treatment given. In the liquid-filled case, the grid lines were drawn using white paint as this color yielded the best contrast with the dark liquid-filled shale in photographs.

In the present study, kerosene was selected as the liquid for saturating the shale samples since (1) green shale is mechanically and chemically stable in kerosene and (2) green shale can be saturated with kerosene more readily than with a more dense liquid such as glycerin. The shale samples were saturated with kerosene by immersion in a pool of kerosene under a vacuum condition for at least two hours. The liquid-saturated halves were assembled in a manner identical to the dry case using the same shell. The assembled sample was secured in an aluminum casing. The front face was covered by a plastic plate with a slot through which the projectile impacted the specimen. The aluminum casing was filled with kerosene which completely immersed the assembled sample during the tests. The aluminum casing was again placed on one end of the same aluminum back stop cylinder previously mentioned.

Impact Tests

Table 2 summarizes the conditions for the eight impact tests conducted. Mild steel 6.35 mm diameter projectiles with a hemispherical tip were used. These projectiles were threaded on one end and were screwed onto a 12.7 mm diameter aluminum cylinder with a length of 57 mm whose surface was cut by thirty 0.79 mm wide grooves with a depth of 0.79 mm. The nominal total mass of the projectile was 31 g. These assembled projectiles were fired by a pneumatic gun with a 12.7 mm diameter bore in the velocity range from 18 to 42 m/s. In this velocity range, the projectiles suffered no observable damage.

The velocity histories of the projectile during the penetration was measured by a device at the gun muzzle consisting of a pin photo-diode and an optical fiber light source. When the beam is interrupted by the previously described grooves in the aluminum cylinder, a signal is generated which is monitored on oscilloscopes. The velocity history during penetration can be readily obtained from the system geometry and the oscilloscope sweep speed. The overall error for the velocity-penetration history measurement including the data reduction was estimated to be less than 8 percent.

Table 2 Summary of Impact Tests on Green Shale

Test Number	Sample Condition	Initial Projectile Velocity, m/s	Initial Projectile Kinetic Energy, J	Final Penetration, mm	Constants c kg/s and f N in Eqs. 3 and 4
1	dry	18.3	5.3	3.2	c = 177
2	dry	30.7	14.8	5.8	c = 164
3	dry	38.0	22.7	7.2	c = 164
4	liquid-filled	21.3	7.2	3.1	f = 2269
5	liquid-filled	26.8	11.3	4.7	f = 2369
6	liquid-filled	32.4	16.5	9.2	f = 1769
7	liquid-filled	37.1	21.7	9.4	f = 2270
8	liquid-filled	42.0	27.8	11.8	f = 2317

The grid lines of each specimen were photographed before and after the impact. This permitted the examination of the deformation pattern and damage degree in the specimen due to the projectile impact. Fig. 1 depicts typical final grid line patterns due to impact of hemispherically-tipped projectiles under dry and liquid-filled conditions.

RESULTS AND DISCUSSIONS

Table 2 contains the results of the eight impact tests conducted. The penetration depth is plotted logarithmically against the bullet initial kinetic energy data for the present tests in Fig. 2 together with the information obtained in Goldsmith, 1976 and 1979. The data points for each contact geometry and specimen condition cluster in the vicinity of a straight line. The impact of hemispherically-tipped and spherical projectiles on dry specimen resulted in a straight line with a slope of approximately one-half. Similar results were obtained by the impact of conically-tipped projectiles on dry specimens. However, the depth achieved by the conically-tipped strikers for a given energy is approximately 50 percent greater than the hemispherically-tipped and spherical projectiles due to a lower target resistance for this more acute penetrator. Impact of hemispherically-tipped projectiles

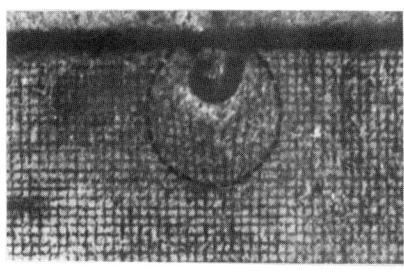
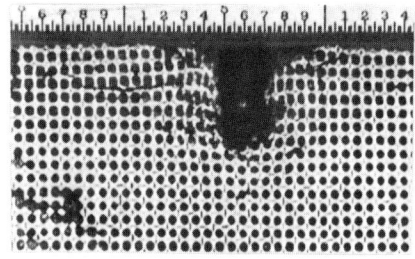

(a) Test No. 2, Dry   (b) Test No. 6, Liquid-Filled

Fig. 1 Damage Patterns in Green Shale Specimens

on liquid-filled specimens again resulted in a straight line in the range tested, but with a slope of approximately unity. From the plot, the relationships between final penetration p and the initial striker kinetic energy E may be approximated as:

$$p \propto E^{\frac{1}{2}} \quad \text{for the dry case and} \tag{1}$$

$$p \propto E \quad \text{for the liquid-filled case.} \tag{2}$$

Although the present observations are limited to the velocity range considered here, the difference in the p-E relation strongly suggests that additional mechanisms are involved when liquid is added to the shale.

In view of the relations described by Eqs. 1 and 2, the data for the two cases may be fitted to simple models involving either a linearly viscous resistive force $c\dot{x}$ or a Coulomb frictional resistive force f with equations of motion given respectively by

$$\ddot{x} + (c/m)\dot{x} = 0 \quad \text{and} \tag{3}$$

$$\ddot{x} + (f/m) = 0 \tag{4}$$

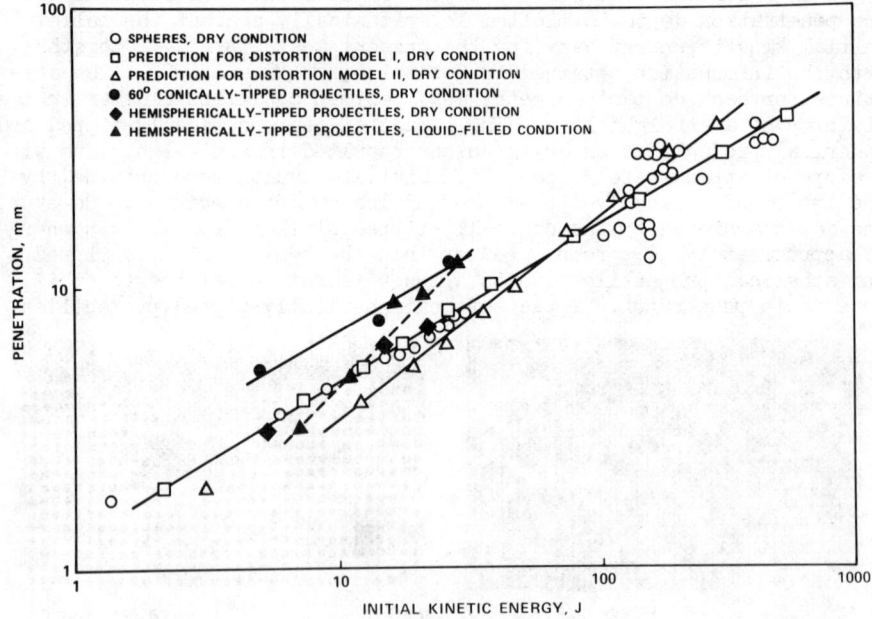

Fig. 2 Final Penetration Depth and Initial Kinetic Energy

Here, m is the mass, c the viscous constant, x the coordinate in the direction of motion, and a dot denotes differentiation with respect to time t. The solutions of Eqs. 3 and 4, subjected to initial and terminal conditions involving striking speed V and final position p, are given by

$$x = (V \cdot m/c) \cdot (1 - e^{-ct/m}) \quad \text{and} \tag{5}$$

$$x = V \cdot \{t - f \cdot t^2/(2m)\} \tag{6}$$

respectively. Constants c and f are given by

$$c = m \cdot V/p \tag{7}$$

$$f = (1/2) \cdot m \cdot V^2/p \tag{8}$$

The viscous model may be used to describe the data for the dry case and the Coulomb frictional model for the liquid-filled case. The values of c and f were calculated from the test data and are listed in Table 2. Consistent values of c and f were obtained. The average values of c and f were found to be 168 kg/s and 2200 N, respectively.

The photographs of Fig. 1 for Test No. 2 and 6 depict the damaged region of the impact areas. In Fig. 1 (a), the dry case, the region of permanent deformation is appreciable, particularly near the contact surfaces, bounded by an elliptical contour. In the dry cases the material near the impact area has been compacted without cracking or surface dishing. Distal deformed regions have been subjected to only small amounts of deformation, with a much smaller degree of compaction. In Fig. 1 (b), the liquid-filled case, the formation of cracks is observed. This feature was observed in all liquid-filled specimens and the general crack pattern was the same for all cases. Another different feature observed in the liquid-filled case is the deformation pattern of the grid lines. In the region closer to the free surface of the specimen and the projectile boundary, the horizontal grid lines are observed to have deformed in the surface direction. In the dry case, all the horizontal grid line deformation was toward the deeper area of the specimen. It appears that less compaction is involved in the liquid-filled case. In the photograph of liquid-filled specimen, the formation of crater around the impact point is observed. This is another feature that was not observed in the dry case. Large fragments of shale were found after impact which suggests that the crater was formed by chipping. The crater formation was observed in all liquid-filled samples except for Test No. 4.

Typical results of the velocity history measurements of the striker during penetration are shown in Fig. 3. In the dry case the velocity data points fit well with the best-fit curve. On the other hand, noticeable deviation of the velocity data points from the best-fit curve is observed in the liquid-filled case. This may be due to the formation of cracks and craters which will result in a sudden change of the boundary condition or sudden energy release.

A more comprehensive model to predict the velocity change as a function of penetration depth and the final depth for a given geometry and initial kinetic energy of the projectile for the dry specimens was constructed. The model prescribes a flow or deformation pattern of the rock based on the observed grid line distortion. And it is assumed that all of the kinetic energy of the projectile is consumed in (1) compaction and (2) equivolumnial distortion of the material. The energy associated with compaction was computed using an empirical relation between hydrostatic pressure and volumetric strain. For the computation of energy consumed in equivolumnial distortion, two representations for the effective stress-strain relation of perfectly-plastic comportment of the material were used. The first model, Model I, is represented by a perfectly-plastic material with the flow stress independent of volumetric strain. The second version, Model II, is similar but with the flow stress dependent on the confining pressure.

The final penetration was computed using the model for an initial projectile kinetic energy range of 5 to 500 J for the hemispherically-tipped projectile impact. The results are shown in Fig. 2. The computed projectile velocity history for Test No. 3 is shown in Fig. 3 for Models I and II. For Model I, both the final penetration and the velocity history are in exellent accord, indicating that a representation consisting of compaction and an effective stress with a constant flow stress hypothesis adequately describes the process. The agreement for Model II is not nearly good as in the previous case. Furthermore, a significant difference in the predicted and the observed final penetration depth exixts. It appeas that a stress-strain relation, as portrayed by Model I adequately delineates the process and that the assumption of constant flow stress seems to suffice in describing the energy associated with equivolumnial distortion.

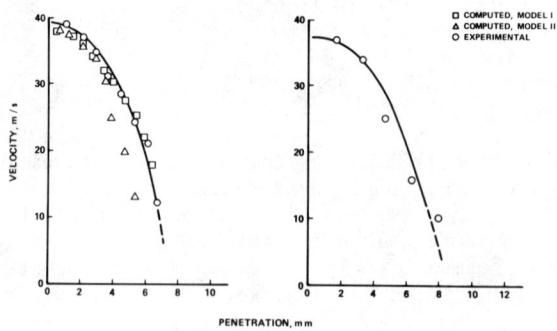

(a) Test No. 3, Dry   (b) Test No. 7, Liquid-Filled

Fig. 3 Velocity-Penetration Data

CONCLUSIONS

The following conclusions are reached for the range of test conditions considered in the present investigation:

(1) Under the dry condition the final penetration depth scaled well quadratically with the initial kinetic energy. The data conformed well to a linearly viscous model.
(2) Under the liquid-filled condition the final penetration depth scaled well linearly with the initial kinetic energy. The data conformed well to a Coulomb frictional model.
(3) The two major processes involved in the permanent deformation in dry shale are (a) compaction and (b) equivolumnial distortion.
(4) The observed damage in the liquid-filled specimen consisted of crack and crater formations in addition to those observed in the dry specimen.
(5) According to the observed grid line distortion, less compaction is involved in the liquid-filled case compared to the dry case.
(6) A computational model, which accounts for compaction and equivolumnial distortion, was constructed for the dry case. Good agreement was obtained between the observed and predicted final penetration depth and velocity history when the flow stress was considered to be independent of volumetric strain.

BIBLIOGRAPHY

Bhatt, J.J. et al., 1975, "A Spherical Model Calculation for Volumetric Response of Porous Rocks," J. of Applied Mechanics 42, pp. 363.

Carroll, M.M. and Holt, A.C., 1973, "Steady Waves in Ductile Porous Solids," J. of Applied Physics 44, 10, pp. 4388.

Goldsmith, W., 1976, "Fragmentation Processes in Soft Rock," DNA-001-73-C-0226, January, Defense Nuclear Agency.

Goldsmith, W., 1979, "Mechanics of Rock-Tool Interaction under Impact Loading," ENG-76-17989, August, National Science Foundation.

Johnson, G.R., 1976, "Analysis of Elastic Plastic Impact Involving Severe Distortions," J. of Applied Mechanics 98, pp. 439.

Kabo, J.M. et al., 1977, "Impact and Comminution Process in Soft and Hard Rock," Rock Mechanics 9, pp. 213.

Stephens, D.R. et al., 1970, "Pressure-Volume Equation of State of Consolidated and Fractured Rcok to 40 kb," Int. J. of Rock Mechanics Min. Sci. 7, pp. 257.

Chapter 51

STATISTICS IN AID OF INTERPRETING FRACTURE DATA

by E.Z. Lajtai
Departments of Civil and Geological Engineering
University of Manitoba
Winnipeg, Canada. R3T 2N2

ABSTRACT

The probability distribution of fracture data from a rock mechanics test contains information about the effectiveness of the testing procedure and/or about the nature of the operating failure mechanism. The tensile strength of Lac du Bonnet granite forms a bimodal probability distribution when determined through line loading as in the Brazilian test. The same data obtained from three point bends tests are unimodal. The bimodality can in part be removed by distributing the line load over a finite area.

Moisture has the anticipated effect in lowering both the average compressive and the average tensile strength of Lac du Bonnet granite. The reduction in strength however is not uniform within the probability distribution; the change is the largest at low and the smallest at high strength. Conversely, long-term loading is more damaging to the high strength members of the distribution.

The large scatter of failure times in static fatigue experiments can in part be removed or even utilized when the probability distribution of instantaneous strength is taken into account.

INTRODUCTION

Rock mechanics professionals are all aware of the probabilistic nature of geotechnical measurements and events. They routinely make use of statistical techniques in reducing data and often make statistical inferences about the means based on the assumption of a normal or "student t" distribution. In the process of analyzing fracture data relating to the Lac du Bonnet granite, the comparison of the distributions of fracture data has been yielding more useful information than statistical tests of population means.

THE BRAZILIAN TEST

The measurement of tensile strength by diametral compression of rock core slices has always been somewhat contraversial. The reason for this is that fracture may not start at the centre, i.e. under the influence of the tensile stress concentration, but at the platten-rock contact where high compressive stress concentrations may develop. In the testing of Lac du Bonnet granite, the use of a flat, ungrooved platten (concentrated line loading) results in a probability distribution that is clearly bimodal (Figure 1); the Weibull two-parameter cumulative distribution function (Weibull, 1951) gives a relatively poor fit at the low-strength tail. The source of the bimodal distribution is not the rock itself, because an alternate test, the three point bend test, gives a unimodal form (Figure 1). The use of the

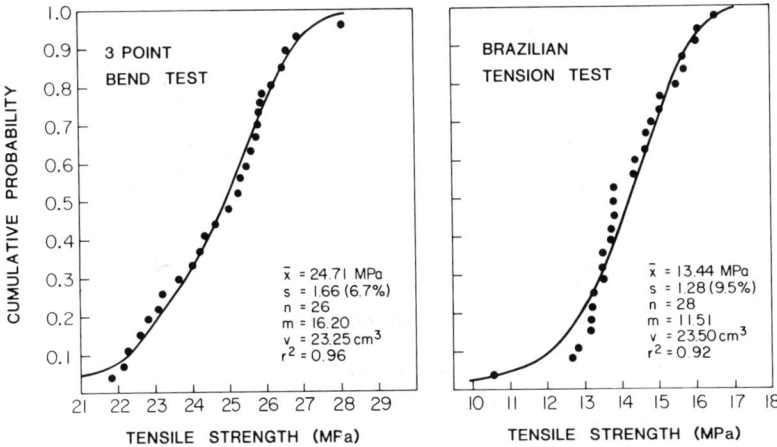

Fig. 1. The cumulative probability distribution of test results from the three point bend and Brazilian tension tests. The fitted curves are Weibull probability functions.

recommended finite area loading (Mellor and Hawks, 1971) seems to produce the desired result (Figure 2). The biomodality has not completely disappeared, but the unimodal Weibull fit is much better as quantified by the higher value of the coefficient of codetermination (r^2).

CRACK GROWTH AND FAILURE IN COMPRESSION

The growth of microcracks in a uniaxial compression test is usually demonstrated by tracking the course of the volumetric strain. Alternatively, the intensity of cracking at a certain value of compression may be characterized by the tensile strength of the test specimen with tension applied across diametral planes as in the Brazilian test. The resulting strength may then be compared with the tensile strength of a control specimen that has not been subjected to compressive loading. Although a large number of specimens are required for meaningful results, there are two advantages. First of all, the method complements the static fatigue experiments. What else can one do with a specimen

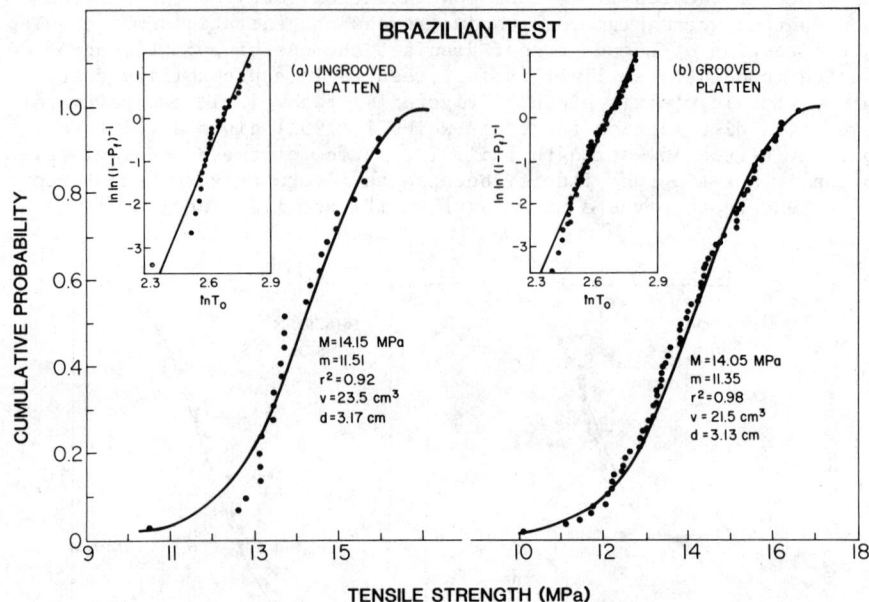

Fig. 2. Brazilian test data for (a) ungrooved and (b) grooved platten. The inserts are the linearized Weibull plots for the same data.

that refuses to fail in a static fatigue experiment? Last but not least, the characterization of crack growth in compression by tensile strength data opens the door to using the Weibull theory and other methods from the wide repertory of engineering fracture mechanics.

The Stress Effect

In the uniaxial compression of short cylinders of Lac du Bonnet granite (length to diameter ratio = 1, diameter = 31 mm), dilatancy starts in the 70 to 100 MPa stress range, (Lajtai, 1982 - in press). The tensile strength however is unaffected to about 200 MPa (Figure 3). Specimens compressed about 240 MPa show the effect of compression by a strong reduction in strength at the low-strength tail of the distribution. The nature of the transition is illustrated well by the test series representing compression at 216 MPa; more than one half of the samples were apparently unaffected.

Time and Environmental Effects

The effect of moisture (vacuum evacuated specimen tested submerged in water) on strength is shown in Figure 4 by corresponding distributions relating to both the tensile and the compressive strength of

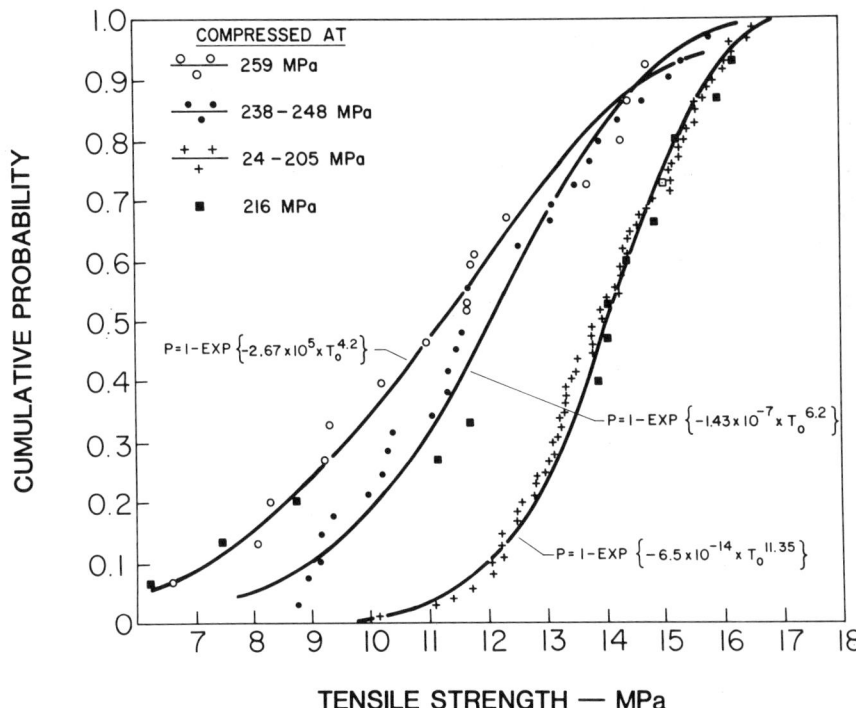

Fig. 3. Tensile strength of previously compressed granite. Note the convergence of the distributions at high strength.

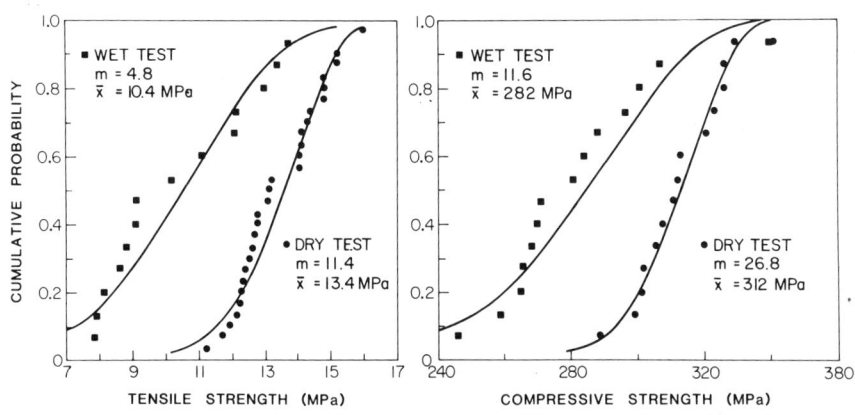

Fig. 4. The influence of moisture on tensile (left) and compressive (right) strength.

short cylinders of Lac du Bonnet granite. As expected there is a decrease in median strength, but the reduction in strength within the distribution is strongly skewed. In both sets, the Weibull cumulative distributions converge toward high strength, suggesting that some specimens were insensitive to the presence of moisture.

The only thing common to both the crack growth and the failure strength experiments is that loading was applied "instantaneously" (1 MPa and 0.1 MPa per second in compression and tension respectively). Indeed, the convergence at high strength can be eliminated by introducing the time factor through long-duration loading. Figure 5 illustrates the point. Three sets of specimen (14 in each set) were compressed at 140 MPa for one week but under different environmental conditions: dry, room temperature (22°C) and humidity ($\approx 40\%$); wet,

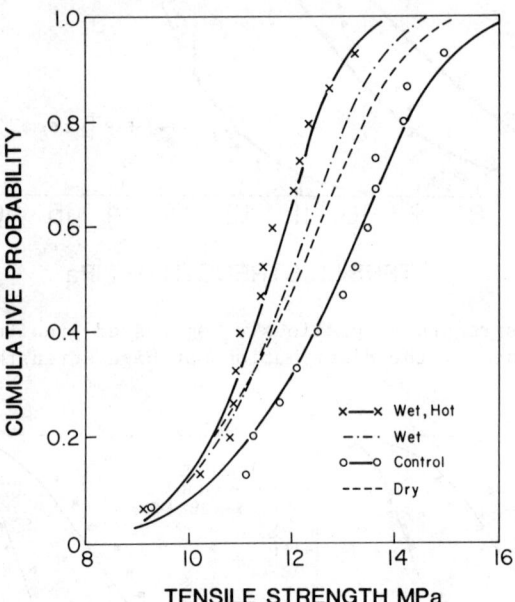

Fig. 5. The change in the tensile strength of granite that was subjected to 140 MPa compression for one week and to three different environments.

room temperature but submerged in water; hot-wet, submerged in water at 80-90°C. All specimens survived the one week compression following which they were tested in Brazilian tension. The resulting distributions are compared with the fourth set, the control group, whose members were not subject to compression. For the sake of clarity, the individual data points are shown only for the two extreme sets.

INTERPRETING FRACTURE DATA

Without help from independent investigations, the significance of changing data distribution patterns cannot be understood with any degree of certainty. Microscopic observations of crack patterns in Lac du Bonnet granite tested in double torsion (Svab and Lajtai, 1981 - in press) and compression (Lajtai, 1981 - in press) suggest that crack growth is controlled by the mineral constituents quartz and biotite. Quartz is known to be sensitive to time dependent stress corrosion cracking (Martin, 1972). One may speculate therefore that the high strength tail of the strength distribution is controlled through the stress and time dependence of fracture in quartz.

ELIMINATING KNOWN CAUSE OF SCATTER

Results from static fatigue experiments in which the load is kept constant at a fraction of strength and the "time to fracture" is measured are subject to wide scatter. For example, short cylinders of Lac du Bonnet granite compressed in water at 225 MPa fail in a range from a fraction of a second to an hour, presenting the interpreter with a scatter across four orders of magnitude (Figure 6). Part of the scatter can be removed through a method used in the design of ceramic components (e.g. Burke et al, 1971 or Wilkins, 1971). The method depends on the assumption that the distribution of instantaneous

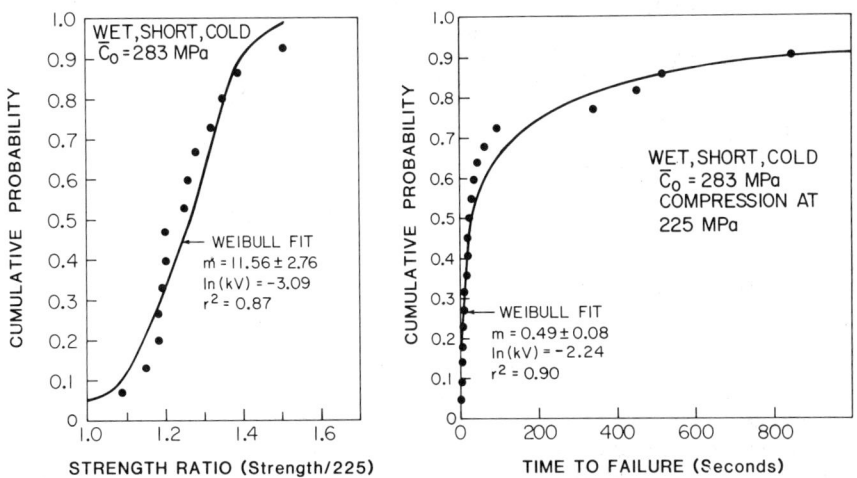

Fig. 6. Comparison of a normalized instantaneous strength with failure times at 225 MPa constant compression.

strength is related to the order of failure times measured under sustained loading at constant stress; the weakest specimen fails first and the strongest last. Pairing is simple when the number of tests in both the instantaneous strength and failure time distributions are the same; in this particular case interpolation of times for the 14 strength tests was necessary to obtain the 14 pairs. The static fatigue curve in which the logarithm of stress level (reciprocal of the strength ratio) is plotted against the logarithm of corresponding failure time, was constructed in Figure 7. The solid line is the linear least square approximation based on the 14 pairs. A more elegant way to proceed is by matching the fitted Weibull cumulative distribution curves for the constants shown in Figure 6.

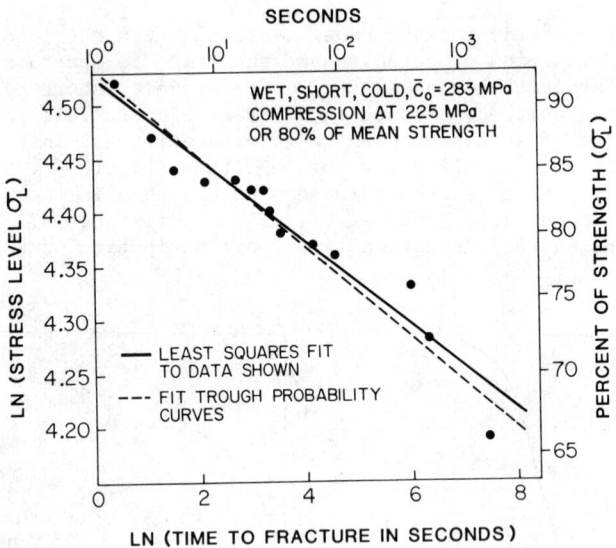

Fig. 7. The relationship between stress level and failure time.

ACKNOWLEDGEMENTS

The test data for the paper have been produced during the course of an experimental program in support of the Canadian nuclear fuel waste management program administered by the Whiteshell Nuclear Research Establishment. The project officer is Gary R. Simmons. Many of the ideas and procedures described here grew out of discussions with scientists at WNRE in particular with Gordon Bird, Hans Tammemagi and Brian Wilkins. Financial support comes from the Atomic Energy of Canada Limited and the National Science and Engineering Council of Canada through operating grant A8076.

REFERENCES

Burke, J.H., Doremus, R.H., Hillig, W.B., and Turkalo, A.M., 1971, "Static Fatigue in Glasses and Alumina", In: Ceramics in Severe Environments, editor W.W. Kriegel and H. Palmour, Plenum Press, New York, 435-444.

Jaeger, J.C., and Cook, N.G.W., 1979, "Fundamentals of Rock Mechanics", Third Edition, Chapman and Hall, London.

Lajtai, E.Z., (1982 - in press), "On the Mechanical Response of Granite to Conditions Associated with the Disposal of Nuclear Fuel Waste", International Society of Rock Mechanics Symposium Proceedings, Aachen, 1982.

Lajtai, E.Z., (1981 - in press), "Creep and Crack Growth in Lac du Bonnet Granite Due to Compressive Strength", Proceedings of the 5th Canadian Fracture Mechanics Conference, Winnipeg.

Martin, R.J., 1972, "Time-dependent Crack Growth in Quartz and its Application to the Creep of Rocks", J. Geophys. Res., 77, 1406-1419.

Mellor, M., and Hawkes, I., 1971, "Measurement of Tensile Strength by Diametral Compression of Discs and Annuli", Eng. Geol., 5, 173-255.

Svab, M. and Lajtai, E.Z., (1981 - in press), "Microstructural Control of Crack Growth in Lac du Bonnet Granite", Proceedings of the 5th Canadian Fracture Mechanics Conference, Winnipeg.

Weibull, W., 1951, "A Statistical Distribution Function of Wide Applicability", Appl. Mech. Rev., 5, 449-451.

Wilkins, B.J.S., 1971, "Static Fatigue of Graphite", J. Amer. Ceram. Soc. 54, 593-595.

Chapter 52

USE OF LABORATORY-DERIVED DATA TO PREDICT
FRACTURE AND PERMEABILITY ENHANCEMENT
IN EXPLOSIVE-PULSE TAILORED FIELD TESTS

by Stuart McHugh[*] and Douglas Keough

SRI International
333 Ravenswood Avenue
Menlo Park, CA 94025

INTRODUCTION AND BACKGROUND

Because of the potential for large reserves from increased production in the eastern Devonian shales, the Department of Energy (DOE), Morgantown Energy Technology Center (METC) selected these materials for the application and evaluation of existing and novel well stimulation techniques. One technique selected for further study was pulse tailoring, or tailored pulse loading, i.e., shaping the input borehole pressure history to produce a specific fracture pattern and to increase communication with the reservoir.

The complexity and expense of in-situ mineback experiments prohibits direct evaluation of pulse tailoring in various geologic formations. An alternative approach used by SRI International consisted of evaluating fracture enhancement by computational simulations with supporting laboratory experiments and using data from field tests by other agencies to validate the computational model (McHugh and Keough, 1980). Laboratory experiments were conducted on core samples to determine fracture phenomenology and parameters and to verify material properties. The fracture parameters were obtained from posttest measurements of fracture distributions and correlation with stresses calculated by a wave propagation code. Model development was guided by these experiments and similar experiments in various materials performed by other agencies. The validity of the modeling, parameters, and scaling was evaluated by comparing fracture patterns from field tests with calculational simulations.

---

[*]Now at Lockheed Missiles and Space Co., Bldg. 204, 0/52-35, 3251 Hanover Street, Palo Alto, CA 94304.

PREDICTING FRACTURE AND PERMEABILITY

Although it was originally proposed that field experiments be performed in Devonian shale, no sites were readily available in which it was possible to mineback around the explosive cavity and map the fracture pattern. Therefore METC decided to use Sandia's facilities at the Nevada Test Site (NTS) and perform the experiments in ashfall tuff. Because the program objective was initially to evaluate the pulse tailoring technique rather than to apply it to a specific geologic formation, the program continued with this material substitution.

In this paper we summarize the NAG-FRAG computational model, discuss the laboratory experiments and measurements used in the derivation of the fracture and material properties, and compare the computational results to the observations from the field tests.

NAG-FRAG COMPUTATIONAL MODEL

SRI developed a combined experimental and computational procedure (referred to as NAG-FRAG) based on observed stress-wave-induced activation of inherent flaws that form cracks in rocks and other materials. Crack growth caused by the internal gas pressure within cracks is also treated, and Figure 1 describes this process for a pulse applied to a borehole. The model (and computational subroutine, also called NAG-FRAG, based on the model) and fracture parameters (a subset of material properties and input to NAG-FRAG subroutines) treat populations of cracks and hence describe average crack behavior, rather than treat details of individual crack growth and coalescence. The procedure and the computational model (used as a subroutine within a stress wave propagation code) have shown predictive capabilities for a variety of nongeologic and geologic materials (Seaman, Curran, and Shockey, 1976). Details of the model are presented in reports by McHugh and Keough (1980) and Seaman (1980). Here only the main features are summarized.

The experimentally determined algorithms used in NAG-FRAG to characterize tensile fracturing in borehole experiments are listed below:

Fracture Distribution: $N_g = N_o \exp(-R/R_1)$ ... (1)

where N_g is the cumulative crack density, R is the crack radius, N_o is the crack density (the total number of cracks per unit volume), and R_1 gives the shape of the distribution. The inherent flaw distribution is specified in NAG-FRAG by the parameters T_3 (the value of R_1 at nucleation) and T_7 (the maximum flaw size at nucleation). These and other fracture parameters discussed later are listed in Table 1 for ashfall tuff. In computational simulations, relation (1) specifies the distribution within an individual fractured computational cell. The fracture distributions from each computational cell are weighted by cell volume and crack length and combined to represent the total or "global" calculated distribution for comparison with experiments.

$$\text{Crack Nucleation:} \quad \overset{\circ}{N} = \overset{\circ}{N}_o \exp[(\sigma - \sigma_{no})/\sigma_1] \quad \ldots \quad (2)$$

where $\overset{\circ}{N}$ is the nucleation rate, $\overset{\circ}{N}_o$ ($=T_4$, Table 1) is the threshold nucleation rate, σ is the tensile stress normal to the plane of the crack, σ_{no} ($=T_5$) is the threshold nucleation stress, and σ_1 ($=T_6$) governs the sensitivity of nucleation rate to stress level.

$$\text{Crack Growth:} \quad \frac{dR}{dt} = T_1(\sigma + P_c - \sigma_{go}) R \quad \ldots \quad (3)$$

where R is the crack radius, dR/dt is the growth velocity, P_c is the fluid pressure (if any) in the crack, and T_1 is the growth coefficient. The growth threshold ($\sigma_{go} = T_2$) may depend on fracture toughness and crack size (McHugh and Keough, 1980; Seaman, 1980). Use of a constant value for σ_{go} mainly skews the calculated crack size distribution toward smaller cracks, but the general results in either case are similar. NAG-FRAG contains a gas penetration model to determine when the gas penetrates the cracks and the resulting magnitude of P_c. (T_{14} is a "switch" in the model that determines whether the gas penetration model is used.) The limiting growth velocity (dR/dt) is the Rayleigh velocity.

$$\text{Stress Relaxation due to Crack Interaction:} \quad \frac{t_n}{T_9} = \frac{N_o^{-1/3}}{c_\ell} \quad (4)$$

where c_ℓ is the longitudinal sound speed and T_9 is a dimensionless time constant assumed to be unity in this study. For the rock fracture problem treated here, there were so few cracks at such large distances that the crack opening did not immediately relax the stresses. Therefore the time constant in relation (4) was used to delay the crack opening, and the material softening caused by damage was delayed until the material had been unloaded by stress-relief waves originating at the fracture.

$$\text{Spallation:} \quad t_p = \beta \gamma^3 \tau_z T_f \quad (5)$$

where the criterion for spall is $t_p > T_g$ (t_p is the fragmented fraction, $\beta(=T_{10})$ is the ratio of number of fragments to number of cracks (set to 0.25), $\gamma(=T_{11})$ is the ratio of fragment size to crack size (set to 1.00), $\tau_z = \Sigma NR^3$ (where the summation Σ is over the total crack density, N, and crack size, R), and $T_f(=T_{13})$ is a proportionality constant such that the value of a fragment is $T_f \gamma^3 R^3$. Relation (5) for spallation is used to statistically describe fragmentation and separation in a computational cell. (NAG-FRAG also provides for consolidation following spallation.)

Permeability Index:
$$k^* = N_o \{2R_1 \exp(\overline{\ell}/2R_1) [R_1 + \overline{\ell}/2] - \exp(R_{max}/R_1) [(R_{max} + R_1)^2 - (R_1^2 - (\overline{\ell}/2)^2)]\} \quad (6)$$

where $R_{max} (> \overline{\ell}/2)$ is the maximum crack size (k^* is set to zero if $R_{max} < \overline{\ell}/2$), and $\overline{\ell}$ is the dimension of the region through which significant fluid flow occurs ($\overline{\ell}$ must be specified before a permeability measurement or calculation can be made). The permeability index is also linked to the gas penetration model and P_c [relation (3)]. For example, if $k^* = 0$, there is no gas flow (i.e., $P_c = 0$). In the derivation of k^*, a cylindrical annulus of length $\overline{\ell}$ is chosen, and the average crack area exposed at the boundary of the annulus is computed. This approach assumes that only cracks larger than $\overline{\ell}$ can coalesce to allow fluid flow from one annulus to another. The permeability index (k^*) is used as a figure of merit to assess the ease of fluid flow through regions of interest, and to show the expected permeability enhancement due to fracturing; k^* also allows fracture patterns to be assessed for specific applications. [Details of the permeability model are given by McHugh and Keough (1980)].

In summary, the fracture and permeability enhancement caused by a specific borehole pulse shape can be calculated once the appropriate material and fracture properties have been determined.

LABORATORY EXPERIMENTS

The bulk material properties (Table 1) were determined from tests on samples of ashfall tuff provided to SRI by Sandia Laboratories. Fracture parameters T_3 and T_7 were determined from an analysis of the grain size distribution, and T_6 was determined from the fracture toughness. Fracture parameters T_8 through T_{13} were fixed from an analysis of the relationship of fragment geometry to crack size, and T_{14} was set to allow gas penetration. Thus parameters T_3 and T_6 through T_{14} were fixed for all the calculational simulations of fracturing and are not considered further.

However, fracture parameters T_1, T_2, T_4, and T_5 had to be determined from laboratory springing experiments on 0.3-m-diam (12-in.) by 0.3-m-long (12-in.) cores of ashfall tuff (Figure 2). A

1.3×10^{-2} m-diam (0.5-in.) borehole was drilled along the core axis, gages, and an explosive charge of pentaerythritol tetranitrate (PETN) was emplaced. The borehole was stemmed to prevent venting of explosive gases, the cores were put into a pressure vessel, which was subjected to a hydrostatic confining pressure [6.9×10^6 Pa (1000 psi), to simulate the overburden at the depth of the field tests], then the explosive was detonated. After the tests, the cores were removed from the pressure vessel and sectioned to examine the cracking.

Two types of experiments were conducted. In the first set of experiments, borehole pressure gages and in-material stress and particle velocity gages were emplaced at various locations in the cores [Figure 2(b)]. Then computational predictions of stress and particle velocity were compared with these gage measurements to ensure that the one-dimensional stress wave propagation code PUFF (McHugh and Keough, 1980) used in these simulations was predicting the correct stress distribution (Figure 3). Because the finite detonation velocity of the PETN was not simulated in the one-dimensional calculations, the calculated and observed arrival times of the shock wave were not expected to agree, and only the peak values and durations are compared in Figure 3. The calculated peak stress and particle velocity are within 10% of the measurements. The single peak in radial stress at 1.5 μs (rather than the calculated double peak) occurs because the finite detonation velocity of the PETN, and resultant gage motion, was not simulated. The baseline for the stress measurement was arbitrarily chosen so that the stress minima at 8 to 9 μs are probably in better agreement than shown. The agreement between calculated and measured histories is satisfactory considering the approximations used in the calculations and the difficulty of the measurements. Since the calculated fracture distributions depend on the stress histories, this agreement also indicates that the conclusions from the calculational simulations of fracturing are applicable to these experiments in tuff.

In the second set of experiments, only borehole pressure gages were used. We performed sheathed borehole tests in which a thin-walled, 5×10^{-4}-m (0.02-in.) steel liner was used to prevent the explosive gases from entering the fractures and unsheathed borehole tests in which the explosive gases were free to enter the cracks. After each test, the cores were sectioned and the cracks were counted and converted to a fracture distribution (a cumulative number of cracks per unit volume versus crack length) for comparison with calculations. Calculational simulations of fracturing in these laboratory tests were performed with different values of T_1, T_2, T_4, and T_5 until predicted and observed fracture distributions were in agreement. [The results from this procedure on the laboratory experiments are presented in McHugh and Keough (1980)].

At this point, all the fracture parameters for tuff, T_1 through T_{14} (Table 1), were fixed from the laboratory measurements and

PREDICTING FRACTURE AND PERMEABILITY 509

tests on the cores. Their values were not changed during any subsequent calculational simulations of fracturing in the field tests.

FIELD TESTS

Data from two sets of field experiments (performed by Sandia Laboratories at NTS approximately 1400 feet underground) were used for comparison with calculational simulations. The first set was from the 1978 Gas Frac series with three tests: GF1, GF2, and GF3. The second set was from the 1980 MultiFrac series with five tests: MF1 through MF5. In each test, a cylindrical propellant and/or explosive charge was used for the pressure source, and borehole pressure and radial and tangential stresses and accelerations were recorded during the tests. Preshot and postshot flow measurements provided information about permeability enhancement in the Multi-Frac series. (No flow measurements were made in the Gas Frac series.) The fracture patterns were mapped during mineback after the tests. Details of all the field tests are given in Warpinski et al. (1978) and Schmidt et al. (1981).

The borehole pressure histories and associated fracture patterns for Gas Frac and MultiFrac are shown in Figures 4 and 5, respectively. These borehole pressure histories were used as a pressure boundary condition in PUFF, and the material properties and fracture paramters from Table 1 were used as input to NAG-FRAG. The numbers and lengths of fractures in each test [Figures 4(b) and 5(b)] were tabulated and converted to fracture distributions for comparison with the calculated simulations. Here only the fracture results from GF2, MF4, and MF5 and the permeability results are considered [See McHugh and Keough (1981) for complete results]. Because of difficulties in measuring the stresses and mapping the fractures in the ashfall tuff field tests, the following results indicate the ability of the NAG-FRAG technique to rank fracture and permeability enhancement rather than fully reproduce all details of the distributions.

RESULTS

The calculated fracture and permeability results are compared with the observations in Figures 6 through 9. As stated before, the material and fracture properties (Table 1) used in the calculational simulations were derived from laboratory data only, and no attempt was made to adjust these properties to obtain better agreement between the calculational results and observations from the field tests in these figures.

GF2 (Figure 6)

The calculations underestimated the number of small cracks (i.e.,

<25 cm) by a factor of two to five; however, the calculated crack densities for crack lengths of 25 to 250 cm are identical to those observed. The calculated maximum crack lengths are about 25% longer than those observed.

MF4 (Figure 7)

As shown in Figure 5(b), only part of the fractures were excavated and mapped. Consequently, only a lower bound on crack length and an upper bound on crack density could be determined from the observations (McHugh and Keough, 1981). The calculated density of small cracks is only slightly less than that observed, and the remainder of the distribution is essentially correct.

MF5 (Figure 8)

In MF5 extensive fracturing occurred around the borehole, and counting all the cracks was difficult. Therefore, the observed crack densities for crack lengths between 20 and 120 cm are probably a minimum bound on the actual crack densities.

The calculated crack densities for these crack lengths are two to three times greater than the minimum bound. The two largest cracks at 550 and 670 cm [Figure 5(b)] were not calculationally reproduced because of a restriction on calculated maximum crack size, which can probably be removed in subsequent simulations. The calculated and observed total crack densities for small crack lengths are identical.

In summary, the calculational fracture simulations of GF2, MF4, and MF5 reproduced the main features of the fracture observations although some differences in details need to be resolved.

The calculated and observed permeability for MF1, MF4, and MF5 are shown in Figure 9. (There were no permeability results in the Gas Frac tests.) From Figure 5(b) and Figure 9, we see that when the crack length increases by about a factor of two, the permeability also increases by two to three orders of magnitude. MF1 and MF4 have comparable calculated and observed permeability enhancements. However, the extent of the calculated and the measured permeability enhancements in MF5 is much larger than those in MF1 and MF4. Thus there is reasonable agreement between the ranking of permeability enhancement in both calculations and observations.

Although not discussed in this paper, as part of this same DOE/METC program, SRI also examined the effect of input pulse shape (i.e., borehole pressure history), the effect of gas flow into the fractures, and the effect of material compaction around the borehole on the resultant fracture pattern. These pulse tailoring results are described in detail in McHugh and Keough (1981) and summarized in McHugh and Keough (1982). In addition, SRI calculationally examined the effect of the material properties (e.g., density, moduli) on fracture and the significance of shear versus tensile fracturing. Application of these pulse tailoring results from ashfall tuff to an actual Devonian shale reservoir is also discussed in McHugh and Keough (1981).

CONCLUSION

In general, there is reasonable agreement between calculated and observed fracture and permeability enhancement. Therefore we conclude that using laboratory-derived parameters, calculational simulations of field tests in an isotropic medium can reproduce the fracture and permeability enhancement sufficiently to permit a computational assessment of the effect of various pulse shapes.

REFERENCES

McHugh, S. L., and Keough, D., 1980, "Small-Scale Experiments with an Analysis to Evaluate the Effect of Tailored Pulse Loading on Fracture and Permeability," Final Reports for Phase I (June 1980) and (June 1981) on SRI International Project PYU 8621, to the Department of Energy, Morgantown Energy Technology Center, Morgantown, WV.

McHugh, S. L., and Keough, D., 1982, "Fracture and Permeability Enhancement with Pulse Tailoring," Paper 10844 to be presented at the SPE/DOE Unconventional Gas Recovery Symposium, May, Pittsburgh, PA.

Schmidt, R. A., et al., 1981, "MultiFrac Test Series Final Report," Sandia Report SAND81-1239, Sandia National Laboratories, Albuquerque, NM.

Seaman, L., Curran, D. R., and Shockey, D. A., 1976, "Computational Models for Ductile and Brittle Fracture," Journal of Applied Physics, Vol. 47, No. 11, pp. 4814-4826.

Seaman, S. L., 1980, "A Computational Model for Dynamic Tensile Microfracture and Damage in One Plane," SRI Poulter Laboratory Technical Report 001-80, February, Menlo Park, CA.

Warpinski, N. R., et al., 1978, "High Energy Gas Frac," Sandia Laboratories Report SAND78-2342, December, Sandia Laboratories, Albuquerque, NM.

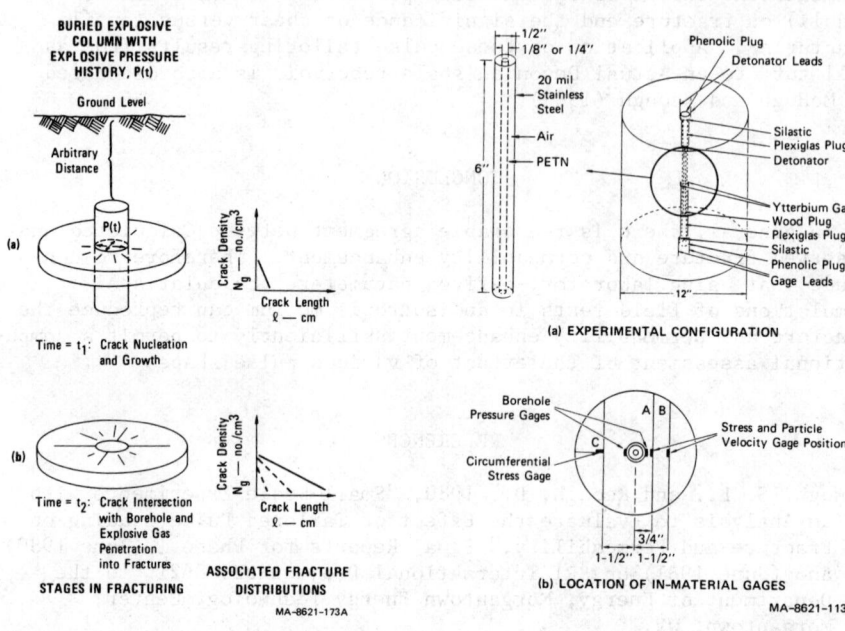

FIGURE 1 CONCEPTUAL BASIS FOR NAG-FRAG MODEL

FIGURE 2 LABORATORY EXPERIMENT

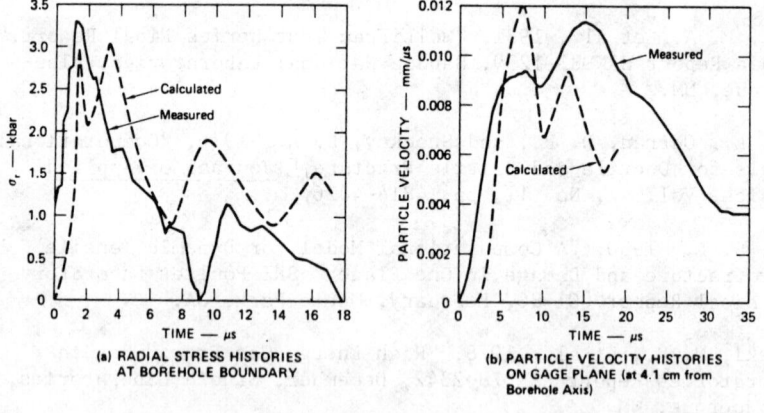

FIGURE 3 MEASURED AND CALCULATED STRESS AND PARTICLE VELOCITY HISTORIES FROM A LABORATORY EXPERIMENT

PREDICTING FRACTURE AND PERMEABILITY

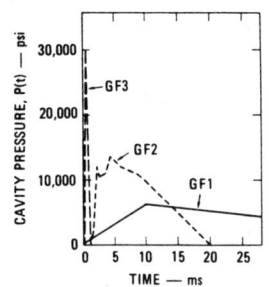

(a) PRESSURE HISTORIES
(position of zero arbitrary)

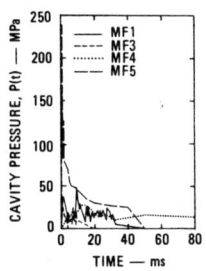

(a) PRESSURE HISTORIES
(position of zero arbitrary)

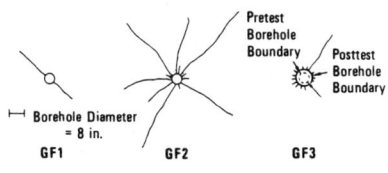

(b) FRACTURE PATTERNS

MA-8621-172B

FIGURE 4 GAS FRAC TESTS

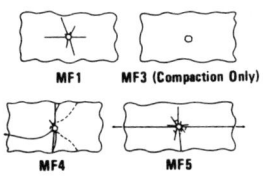

(b) FRACTURE PATTERNS

MA-8621-143B

FIGURE 5 MULTIFRAC TESTS

FIGURE 6 GF2

FIGURE 7 MF4

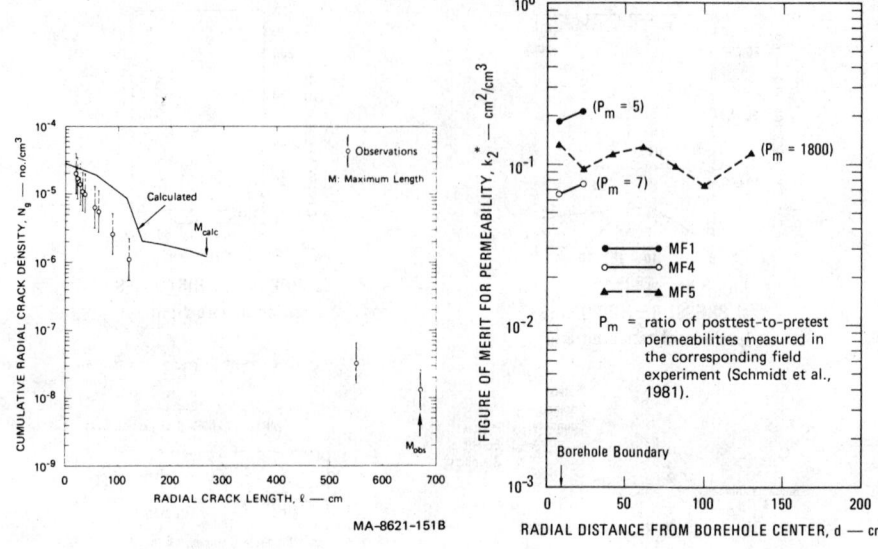

FIGURE 8 MF5

FIGURE 9 PERMEABILITY RESULTS

TABLE 1

MATERIAL AND FRACTURE PROPERTIES OF TUFF USED
IN NAG-FRAG CALCULATIONAL SIMULATIONS
OF LABORATORY AND FIELD EXPERIMENTS*

	Description	Value Used in Laboratory Springing and MultiFrac Field Experiments
ρ_o	Initial density (g/cm^3)	1.96
C	Bulk modulus (dyn/cm^2)	9.15 × 10^{10}
G	Shear modulus (dyn/cm^2)	2.37 × 10^{10}
Y	Yield strength (dyn/cm^2)	1.00 × 10^9
T_1	Growth coefficient (cm^2/dyn-s)	− 3.00 × 10^{-3}
T_2	Growth threshold (dyn/cm^2)	− 5.00 × 10^6
T_3	Nucleation size parameter (cm)	1.20 × 10^{-1}
T_4	Nucleation rate coefficient (no./cm^3-s)	3.00 × 10^{-1}
T_5	Nucleation threshold (dyn/cm^2)	− 3.50 × 10^7
T_6	Stress sensitivity factor (dyn/cm^2)	− 2.00 × 10^8
T_7	Upper size nucleation cutoff (cm)	1.00
T_8	Spall criterion	1.00
T_9	Coefficient in stress relaxation time constant	1.00
T_{10}	Ratio of number of fragments to number of cracks	0.25
T_{11}	Ratio of fragment radius to radius crack	1.00
T_{12}	ΣNR^3 at coalescence	0.20
T_{13}	Coefficient for calculating fragment volume	0.13
T_{14}	Gas penetration switch	1.00

*For more details on how the values were obtained, see McHugh and Keough (1980).

Chapter 53

A SIMPLE R-CURVE APPROACH TO FRACTURE TOUGHNESS
TESTING OF ROCK CORE SPECIMENS

by Finn Ouchterlony

Swedish Detonic Research Foundation
S-12611 Stockholm, Sweden

ABSTRACT

A simple bilinear R-curve description of crack extension resistance is applied to the testing of sub-size rock core specimens. The R-curve consists of a linear sub-critical part and a flat post-critical part. Formulas for the apparent energy rate resistance quantities G_m, J_m, and $\bar{R}$ are developed. Experiments on notched cores of Ekeberg marble, Bohus granite, and Stripa granite are compared with these predictions. The R-curve model gives reasonable results for G_m-data but fails essentially to describe the J_m and $\bar{R}$-data. Another model with a sharply increasing linear post-critical part could possibly account for the $\bar{R}$-data, but direct R-curve measurements seem a better approach.

INTRODUCTION

Proper K_{Ic} testing of rock requires certain minimum specimen dimensions (Ouchterlony,1982). Yet rock testing is preferably made with core specimens which nearly always will be subsize in comparison and yield erroneously low toughness results if subcritical cracking is disregarded.

It is thus important to investigate whether K_{Ic}-values can be deduced from sub-size specimens. J-integral methods offer one relatively complicated possibility (Schmidt and Lutz,1979; Costin,1981). Specimens that permit a sufficient amount of crack growth from a notch before the evaluation is made may offer simpler one. Examples are the Short Rod (Barker,1977) and the Chevron Edge Notch Round Bar in Bending (Ouchterlony,1980).

The Short Rod toughness value is based on an assumedly flat R-curve of the test material. Since non-linear correction is needed to make it compatible with K_{Ic} (Costin,1981) this assumption would seem insufficient. This paper provides added evidence to this effect by comparing theoretical results with measurements on core bend specimens with straight edge notches, the SECRBB specimen by Ouchterlony (1981). It is based on the report by Ouchterlony (1981a).

SPECIMEN GEOMETRY AND EVALUATION PARAMETERS

The SECRBB specimen is shown in figure 1 together with notation. Formulas for isotropic linear elasticity are given by Ouchterlony (1981). The stress intensity factor is given as

$$K_I = 0.25(S/D) \cdot Y' \cdot F/D^{1.5}. \tag{1}$$

Setting $\alpha = a/D$, the dimensionless stress intensity factor follows as

$$Y' = 12.7527\alpha^{0.5}(1+19.646\alpha^{4.5})^{0.5}/(1-\alpha)^{0.25}, \tag{2}$$

valid for $S/D = 3.33$ when $0 < \alpha < 0.6$. Y' is derived from the dimensionless compliance $\bar{g}$, which is given by

$$g = 15.6719[1+0.1372(1+\nu)+11.5073(1-\nu^2)\alpha^{2.5}(1+7.0165\alpha^{4.5})], \tag{3}$$

valid in the same interval. Its dimensional form is $\lambda = g/ED$, where λ is the inverted slope in a force-displacement diagram. E is the Young's modulus of the material and ν its Poisson's ratio.

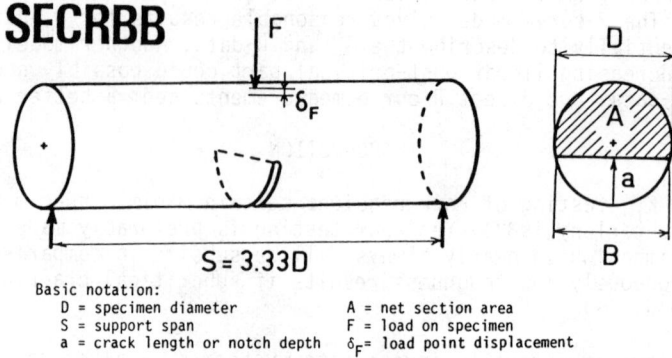

Basic notation:
D = specimen diameter
S = support span
a = crack length or notch depth
A = net section area
F = load on specimen
δ_F = load point displacement

Fig.1: Single edge crack round bar in bending, or SECRBB specimen.

The relation between J-integral and work done W is

$$J_I = \eta_{eq} \cdot W/A_{eq} = \gamma \cdot W/D^2. \tag{4}$$

Here $A_{eq} = B \cdot (D-a)$ is an equivalent uncracked area and

$$\eta_{eq} = \begin{cases} (1-\alpha) \cdot d\ell n g/d\alpha & \text{when } 0 < \alpha \leq 0.5 \text{ and} \\ 2.5 & \text{when } 0.5 < \alpha < 1. \end{cases} \tag{5}$$

FRACTURE TOUGHNESS TESTING OF ROCK

See Ouchterlony (1981) equations 24 to 26 and figure 11.

The data was obtained from complete failure curves $F(\delta_F)$, see figure 2. Testing details are given by Ouchterlony (1981). One crack resistance parameter obtained was the specific work of fracture

$$\bar{R} = W_f/A = \frac{1}{A} \int_0^\infty F \cdot d\delta_F. \tag{6}$$

Further an approximate fracture toughness K_m and an approximate J-integral resistance J_m were evaluated. If the material behavior is sufficiently ideal elastic-brittle then one expects that $\bar{R} \approx G_m = (1-\nu^2)K_m^2/E \approx J_m$. Usually this isn't found in practise. Both K_m and J_m may give erroneously low values (Ouchterlony, 1982).

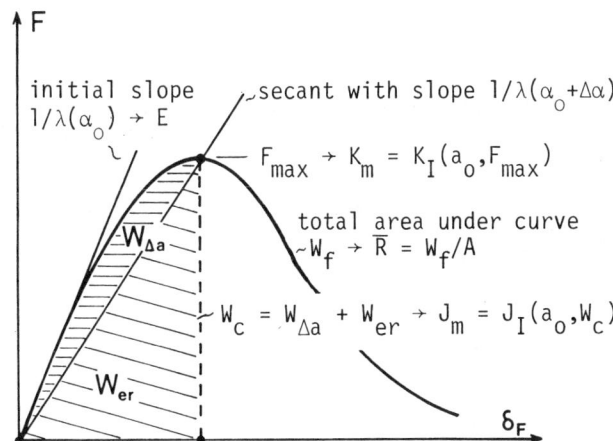

Fig.2: Schematic of complete failure curve for SECRBB specimen of elastic-brittle material with sub-critical crack growth.

K_m is obtained from the failure load and the initial notch length in the appropriate K_I-formula,

$$K_m = K_I(a_0, F_{max}). \tag{7}$$

It should not be confused with K_Q or K_{IC} as per the ASTM standard E399. In this paper K_{IC} will however mean an ideal one parameter characterization of a material's resistance to crack growth which presumably can be measured on sufficiently large specimens.

J_m in turn is obtained from the work done up to F_{max} and a_0 as

$$J_m = J_I(a_0, W_c), \tag{8}$$

with W_c defined in figure 2. A reliable modulus value, required to compare J_m with K_m, may be determined from the compliance $\lambda_0 = \lambda(a_0)$

of the notched specimen. The ν-value may probably be estimated with sufficient accuracy.

The proposition is that R-curves are regarded as material properties, independent of starting crack length and the specimen in which they are developed. Thus the crack-extension resistance in stress intensity terms may be written $K_R(a-a_0)$. To predict crack instability this R-curve is compared with the crack-extension force or $K_I(a,F)$-curves of the specimen. The unique K_I-curve that develops tangency with the R-curve defines the critical load.

Here two linear elements are joined to a simple R-curve with a well defined knee and a flat post-critical part,

$$K_R = \begin{cases} (a-a_0)/\Delta a \cdot K_{Ic} & \text{when } a_0 \leq a \leq a_0+\Delta a \\ K_{Ic} & \text{when } a > a_0+\Delta a. \end{cases} \quad (9)$$

Here K_{Ic} and Δa are material properties. Swan (1980) supports the linear sub-critical part and the knee but not necessarily the flat part which is a prerequisite for the Short Rod evaluation (Ouchterlony,1982).

This curve and the critical crack-extension force curve of the SECRBB specimen are plotted in figure 3. The toughness K_m and its energy rate version follow as

$$\frac{G_m}{G_{Ic}} = \left[\frac{K_m}{K_{Ic}}\right]^2 = \left[\frac{Y'(\alpha_0)}{Y'(\alpha_0+\Delta\alpha)}\right]^2 = \frac{g(\alpha_0)}{g(\alpha_0+\Delta\alpha)} \cdot \frac{\gamma(\alpha_0)}{\gamma(\alpha_0+\Delta\alpha)} \quad (10)$$

Here $\alpha_0 = a_0/D$, $\Delta\alpha = \Delta a/D$, and $G_{Ic} = (1-\nu^2)K_{Ic}^2/E$ have been used.

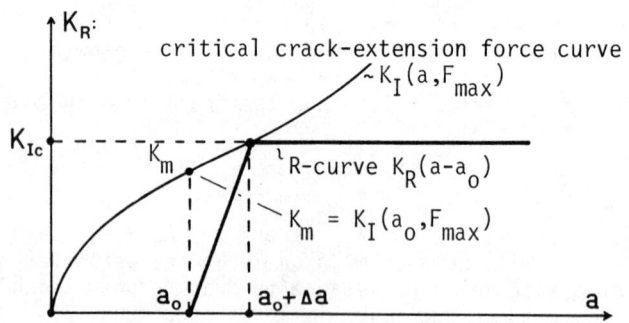

Fig.3: Simple bilinear R-curve and critical crack-extension force curve illustrate evaluation of approximate toughness K_m.

The approximate J-integral J_m requires more analysis, see figure 2. W_c consists of two parts, the recoverable energy W_{er} and the consumed fracture energy $W_{\Delta a}$. There follows that

FRACTURE TOUGHNESS TESTING OF ROCK

$$W_{er} = \frac{1}{2}\lambda(\alpha_0+\Delta\alpha)F_{max}^2 = G_{Ic} \cdot D^2/\gamma(\alpha_0+\Delta\alpha)$$

$$W_{\Delta a} = \frac{1-\nu^2}{E} \int_{a_0}^{a_0+\Delta a} K_R^2(a-a_0) \cdot B da = G_{Ic} \cdot 2D^2 \cdot \int_{\alpha_0}^{\alpha_0+\Delta\alpha}(\alpha-\alpha_0)^2\sqrt{\alpha-\alpha^2}d\alpha/\Delta\alpha^2.$$
(11)

Let $I_2(\alpha_0,\Delta\alpha)$ denote the integral factor in the last member, index 2 refering the exponents of $(\alpha-\alpha_0)$ and $\Delta\alpha$. Then

$$J_m/G_{Ic} = \gamma(\alpha_0)/\gamma(\alpha_0+\Delta\alpha)+2I_2 \cdot \gamma(\alpha_0).$$
(12)

Being a semi-integrated measure it should give a better prediction of K_{Ic} than K_m does.

The specific work of fracture $\overline{R}$, finally, is given by

$$A \cdot \overline{R} = \frac{1-\nu^2}{E}\int_{a_0}^{D} K_R^2(a-a_0) \cdot B da.$$
(13)

With the subcritically swept section area being $2D^2 \cdot I_0(\alpha_0,\Delta\alpha)$

$$\overline{R}/G_{Ic} = 1-2(I_0-I_2)D^2/A.$$
(14)

It doesn't depend on the elastic properties of the specimen.

A comparison of G_m, J_m, and $\overline{R}$ is given in figure 4. Both $\overline{R}$ and J_m are superior to G_m as G_{Ic} predictors. G_m may be as much as 50% lower than G_{Ic}. Since a well established crack is desirable, J_m is the best one. For $0.3 \leq \alpha_0 \leq 0.6$ it falls within 10% of G_{Ic}.

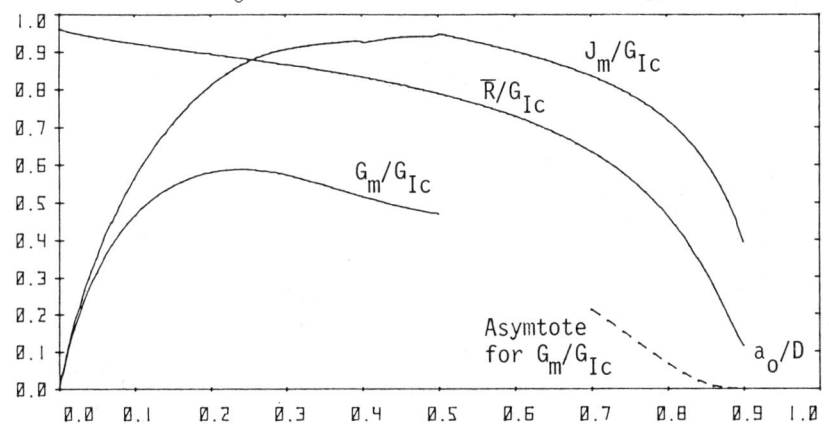

Fig.4: Comparison of energy rate crack resistance measures G_m, J_m, and $\overline{R}$ in non-dimensional form for $\Delta a/D = 0.10$.

ENERGY RATE CRACK RESISTANCE DATA

Cores with $D \approx 41$ mm and 1.6 mm notch width were used. 55 tests on Ekeberg marble and 22 on Bohus granite were made, each with its own modulus from λ_0. These moduli are independent of notch depth and agree closely with strain-gage values from uniaxial tests (Ouchterlony 1981; 1981a). Swan's (1980) data for Stripa granite was also reinterpreted. The elastic properties, E ± standard deviation, are
  $E = 85.7 \pm 3.3$ GPa and $\nu \approx 0.22$ for Ekeberg marble,
  $E = 41.4 \pm 2.7$ GPa and $\nu \approx 0.10$ for Bohus granite, and
  $E = 57 \pm 4$ GPa and $\nu \approx 0.12$ for Stripa granite.

The energy rate crack resistance data is plotted in figures 5 to 7. All results are similar. $\bar{R}$ is always the largest and on average $\bar{R}:J_m: G_m$ is roughly 4:2:1. $\bar{R}$ depends considerably on a_0/D, G_m depends on a_0/D, and so does J_m for Bohus granite. This dependence and their numerical incompatibility make all three quantities unfit as single parameter material properties. Use of the present two parameter R-curve model to interpret the data instead yields that:

The best least squares fit G_m-curves are obtained when
  $G_{Ic} = 39.2$ J/m^2 and $\Delta a/D = 0.10$ for Ekeberg marble,
  $G_{Ic} = 66.8$ J/m^2 and $\Delta a/D = 0.11$ for Bohus granite, and
  $G_{Ic} = 93.2$ J/m^2 and $\Delta a/D = 0.10$ for Stripa granite.

The $\Delta a/D$-values are practically equal. On one hand $\Delta a/D$ constant and material independent points to a geometrical explanation. On the other the time independent R-curve approach requires Δa constant and material dependent. There is also time dependent sub-critical crack growth to contend with but no definite conclusion can be drawn from the present material.

The G_{Ic}-results are clearly reasonable, if somewhat low compared to the J_{Ic} levels. To achieve a good fit for the G_m-curve is the least demanding test of the simple R-curve model since only the sharp knee is involved. An assessment of the minimum required crack length yields ≈ 130 mm for all three rocks which emphasizes the need to predict K_{Ic} from sub-size specimens.

A J_m-curve depends both on the sub-critical part of the R-curve and on the knee. The curve fit is actually worst around $\Delta a/D \approx 0.10$. The predicted G_{Ic}-values are not bad but the simple R-curve model essentially fails to account for the J_m-data (Ouchterlony,1981a).

The trend of the $\bar{R}$-data with a_0/D is correctly reproduced by the $\bar{R}$-curve in figure 4 whose level is roughly 50% too low. The $\bar{R}$-curve integrates the whole R-curve. A bilinear model with sharply increasing post-critical part would raise the predicted $\bar{R}$-curve, possibly to the level of the data. Such an R-curve could maybe also explain why uncorrected Short Rod values are too low (Ouchterlony, 1981a).

FRACTURE TOUGHNESS TESTING OF ROCK

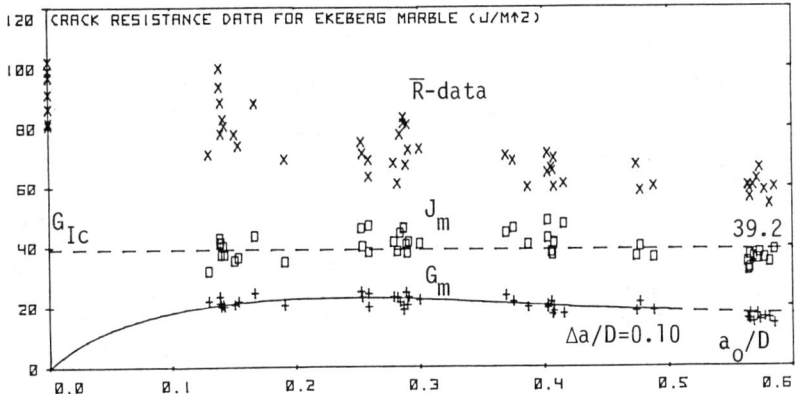

Fig.5: Energy rate crack resistance data for Ekeberg marble.

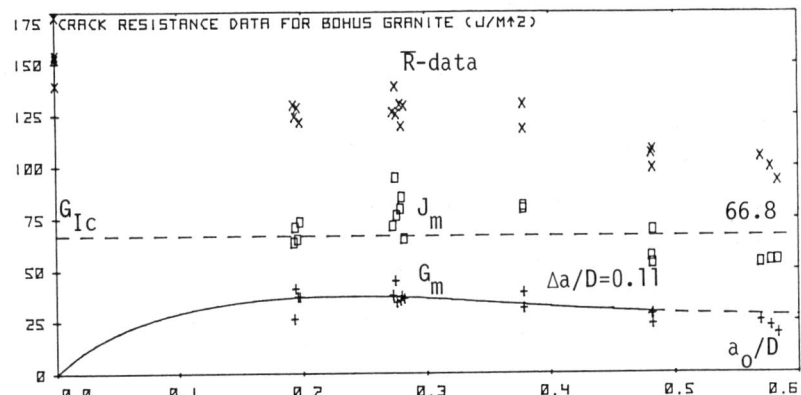

Fig.6: Energy rate crack resistance data for Bohus granite.

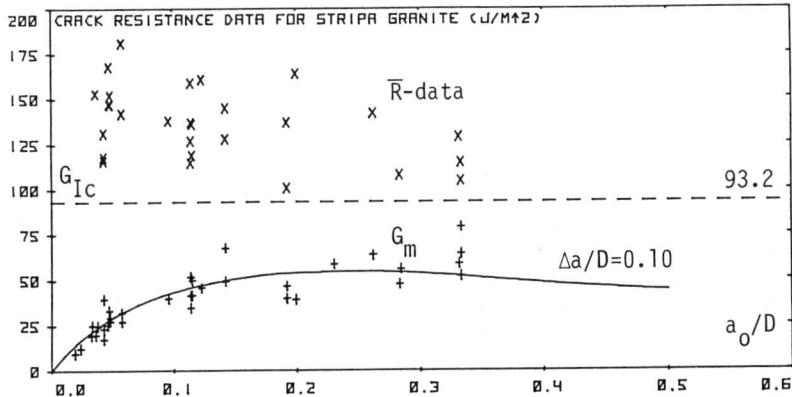

Fig.7: Energy rate crack resistance data for Stripa granite.

CONCLUSIONS

The simple R-curve model with a flat post-critical part is in general inconsistent with the experimental energy rate resistance data. The knee of the R-curve is supported by the G_m-data, which is less demanding than the J_m and $\bar{R}$-data. The simple R-curve model fails with the latter. One with a sharply increasing linear post-critical part can possibly explain the $\bar{R}$-data.

Since the factual R-curves of rock aren't so clear-cut, a direct measurement of them seems better than indirect conclusions based on their assumed form. Unfortunately this points away from simple K_{Ic}-evaluations with sub-size SECRBB and Short Rod core specimens.

REFERENCES

Barker, L.M., 1977, "A Simplified Method for Measuring Plane Strain Fracture Toughness," *Engineering Fracture Mechanics*, Vol. 9, pp. 361-369.

Costin, L.S., 1981, "Static and Dynamic Fracture Behavior of Oil Shale," *Fracture Mechanics for Ceramics, Rocks, and Concrete*, Freiman, S.W. and Fuller Jr., E.R., eds., STP 745, ASTM, Philadelphia, pp. 169-184.

Ouchterlony, F., 1980, "A New Core Specimen for the Fracture Toughness Testing of Rock," DS 1980:17, Swedish Detonic Research Foundation (SveDeFo), Stockholm, Sweden, 18 pp.

Ouchterlony, F., 1981, "Extension of the Compliance and Stress Intensity Formulas for the Single Edge Crack Round Bar in Bending," *Fracture Mechanics for Ceramics, Rocks, and Concrete*, Freiman, S.W. and Fuller Jr., E.R., eds., STP 745, ASTM, Philadelphia, pp. 237-256.

Ouchterlony, F., 1981a, "A Simple R-Curve Approach to Fracture Toughness Testing of Rock with Sub-Size SECRBB Specimens," DS 1981:18, SveDeFo, Stockholm, Sweden, 41 pp.

Ouchterlony, F., 1982, "Review of Fracture Toughness Testing of Rock," *Solid Mechanics Archive*, in press, 80 pp.

Schmidt, R.A. and Lutz, T.J., 1979, "K_{Ic} and J_{Ic} of Westerly Granite - Effects of Thickness and In-Plane Dimensions," *Fracture Mechanics Applied to Brittle Materials*, Freiman, S.W., ed., STP 678, ASTM, Philadelphia, pp. 166-182.

Swan, G., 1980, "Fracture Stress Scale Effects for Rocks in Bending," *International Journal of Rock Mechanics and Mining Sciences*, Vol. 17, pp. 317-324.

Chapter 54

PRE-SPLITTING AND STRESS WAVES: A DYNAMIC
PHOTOELASTIC EVALUATION

K. R. Y. Simha, D. C. Holloway and W. L. Fourney

Department of Mechanical Engineering
University of Maryland
College Park, MD 20742

ABSTRACT

 An experimental investigation was performed to evaluate the role of stress waves in the pre-splitting operation. 3D birefringent Plexiglas models and the dynamic photoelastic technique were used to visualize the stress waves generated by the detonation. Simultaneous as well as sequential detonation of the explosive charges were investigated. The development of fractures and stress waves were recorded using a high speed multiple spark gap camera of the Cranz-Shardin type. The experimental data was analyzed and displayed on a Lagrangian diagram to present an overall picture of the entire dynamic event. It was repeatedly observed in the tests that the time required for the completion of the pre-splitting operation as practiced in the field is of the same order as that for the transit time for the stress waves between the blastholes thereby revealing the highly elastodynamic nature of the event.

INTRODUCTION

 Pre-splitting denotes the preliminary operation of fixing the blasting limits. Pre-splitting is achieved by employing mild, decoupled charges with the sole intention of connecting the blastholes with minimal damage and vibration. Usually the blastholes in pre-splitting are fired simultaneously while some times millisecond delays are employed between successive blastholes. Pre-split blastholes are typically 5-10 cms in diameter and are spaced about 10-12 diameters from each other. Typically, the decoupling ratio; i.e., the ratio of blasthole diameter to charge diameter, varies from 2-3 (Dupont Blasters' Handbook, 1977).
 Existing literature on pre-splitting (Konya, 1980, and Kihlstrom, 1970) and related mining operations accomplished by the detonation of

explosives (Porter, et. al., 1970, and Ito, et. al., 1970) indicate a general lack of agreement on the basic mechanisms of dynamic fracture. Consequently, several experimental observations have been often misinterpreted and misunderstood. In particular, the influence of stress waves on the dynamic event of pre-splitting has not been investigated in detail. It is erroneously believed that the gas pressure is entirely responsible for the dynamic event. Although it is a fact that the gas pressure generates the stress wave systems, the unstable propagation of fractures that ensues upon the detonation is completely decided by the instantaneous dynamic state of stress that develops in the wake of the propagating stress waves. A complete description of the pre-splitting operation must therefore rely on elastodynamic consideration rather than the conventionally adopted elastostatic approaches.

Several investigators have addressed the problem of dynamic fracture due to multiple explosions using small scale models of plastics, limestone, etc. In 1973, Dally, et. al., examined some aspects of stress wave effects on explosively induced fractures. In these experiments a brittle photoelastic polyester called Homalite 100 was employed as the model material. The model thicknesses were 6.35mm (1/4") and 4.76mm (3/16"). In 1974 Pederson, et. al., conducted a photoelastic evaluation of the conventional pre-splitting procedure using once again 6.35mm (1/4") Homalite models. In 1978 Swift, et. al., employed 3D samples of polystyrene to study the effects of delay on fracturing. However in this investigation the results were analyzed on a post mortem basis to quantify fracture. Although post mortem analyses can reveal important features, the results are at best qualitative due to the influence of stress wave reflections. In 1980, Konya performed an experimental investigation of the pre-splitting operation using Plexiglas and Berea sandstone models. Even though the dynamic events were observed using high speed cameras, no attempt was made to visualize the interaction of stress waves with fractures. The purpose of the present paper is to delineate the role played by the stress waves in the pre-splitting operation and to emphasize the elasto-dynamic nature of the event.

EXPERIMENTAL PROCEDURE

The dynamic event of pre-splitting was simulated in the laboratory by using 3D Plexiglas models of 100.8mm (4") thickness. The visualization of stress waves was achieved by exploiting the birefringent property of Plexiglas and the techniques of dynamic photoelasticity. The fringe patterns were recorded by a high speed multiple spark gap camera of the Cranz-Schardin type. Three different model configurations were tested as shown in Figure 1 but only the results of configurations α and β are reported in this paper. In configuration α, two holes of 5.55mm (7/32") diameter and 5014mm (2") apart were drilled in the middle of a 100x200x300mm (4"x8"x12") Plexiglas block. In configuration β three holes were drilled to simulate the pre-splitting process better. The blasthole diameter and spacing were obtained by scaling down the values used in the field. In configuration γ the blasthole diameter was increased to 11.9mm (15/32") and grooved at the sides to investigate the benefits of notched boreholes. In all the cases PETN was used as the explosive. The explosive charge was contained in a 3.17mm (1/8") shrink tube. The detonation was accomplished by discharging a 10μf capacitor from a potential of 2KV. A small amount of lead azide was placed on the end of a No. 28 miniature coaxial cable which in turn was placed in the PETN. For the purpose of data analysis the stress wave velocities in Plexiglas were established separately and these values are: Dilatational or (Primary) P wave velocity = 2540 m/sec (100,000 in/sec); Distortional or (Secondary) S wave velocity = 4150 m/sec (45,300 in/sec).

EXPERIMENTAL ANALYSIS

In all the tests, sixteen pictures of the dynamic event were taken at predetermined time intervals from the time of explosion to the time of completion of the pre-split. Attention was also given to study the influence of stress wave reflections from the model boundaries. Other work conducted in the laboratory on 97.54x97.5cm (3ft x 3ft) glass plates have shown that reflected stress waves can have a significant effect on the final fracture pattern, (1982, Barker, et. al.,). Figure 2 shows some typical dynamic isochromatic fringe patterns at two distinct times for three different tests. In the following, a typical test involving the simultaneous detonation of two charges is analyzed in detail.

a. Simultaneous Pre-split Blasting: Figures 3, 4 and 5 show the procedure for analyzing the experimental data of a test pertaining to configuration α. Two 200mg PETN charges explaced in 5.55mm (7/32') diameter holes 50.4mm (2") apart were detonated simultaneously. Figure 4 illustrates the development of the fractures from the two blastholes A and B. In this figure there are two distinct phases labeled, I and II. Phase I denotes the time from the beginning of the dynamic event until the time the radial fractures from one of the blastholes begins interacting with the stress wave systems arriving from the other blasthole. Seven to eight radial fractures appear randomly but the one connecting the blastholes will eventually dominate the event to complete the pre-splitting operation. Phase II constitutes the period of fracture growth due to the combined action of the stress waves originating from both the blastholes. Crack propagation during this phase depends on the instantaneous stresses that develop due to the dynamic superposition of stresses transients. It is also during this phase that a complex interplay between the moving cracks and stress waves ensues giving rise to several interesting phenomena like crack branching, crack arrest and crack curvature. It can be seen from Figure 4 that the cracks AB and BA between the blastholes are about four times faster than the cracks AO and BO propagating away from each other. A similar trend was also observed by Konya, 1980.

In order to gain a better understanding of the interaction of stress waves and fractures, the temporal development of fractures and stress waves was displayed on a Lagrangian basis as shown in Fig. 5. This diagram presents an overall picture of the entire dynamic event and serves as a viable platform for speculating on the complex interactions of stress waves and fractures with regard to their effect on the pre-splitting operation. In this diagram P_A^+ and P_B^+ are the primary wave front positions in the positive direction, AB, the line of join of blastholes A and B while S_A^+ and S_B^+ are the corresponding secondary waves. P_A^-, P_B^-, S_A^- and S_B^- represent the propagation in the negative direction BA. This distinction serves a purpose in studying pre-split blasting with delayed charges. Also in Fig. 5 it is possible to identify several distinct events marked A through I that lead to the completion of the pre-splitting operation. A and B represent the detonation; C (10µs) represents the interaction of P waves on the line joining the blastholes between A and B; D (14µs) and E (16µs) represent the beginning of the interaction of fracture propagating from one of the blastholes with the P waves generated from the other blasthole. F (22µs) is the intersection of S waves. G (25.5µs) and H (27µs) mark the beginning of the interaction between the S waves and fractures. Finally I (32.5µs) marks the completion of the dynamic event. It can be noted that the time required for the completion of the pre-splitting event is comparable with the transit time of the stress waves between the blastholes. It can also be noted that the fracture growth between the blastholes is not uniform due to the complex interaction of stress waves and fractures. Post mortem examination of the model indicated an intensely strained fracture had connected the blastholes. Further radial cracks were noted at both the blastholes with a few of them extending to 75mm (3") due to stress wave reflections from the boundaries. The model was cleaved in two by the combined action of stress waves extending the pre-splitting crack and the eventual gas pressurization of the cracks. Other work has shown that there can be a very long delay before the cracks are pressurized (Fourney, et. al., 1981). Fig. 6 shows the final pre-split model.

  b. Delayed Pre-split Blasting: Tests were conducted to investigate the effect of sequential detonation of pre-splitting. Tests were conducted for configuration A as well as B and the data reduction was done on the same basis as described earlier. Fig. 7 shows the results of a typical delayed pre-splitting test conducted with a delay of 10µs between three successive detonations of 125mg of PETN charges each.

CONCLUSIONS

An experimental investigation has been conducted to evaluate the role of stress waves on the pre-splitting operation. On the basis of tests conducted on scaled down 3D Plexiglas models, it can be said that a complete description of the pre-splitting operation must rely on elastodynamic considerations wherein the effect of stress waves becomes dominating. It was repeatedly observed that the time for the completion of the pre-splitting operation was of the same order as that of the transit time for the stress waves between the blastholes. The dynamic event of pre-splitting furnished an excellent example of a mining operation in which the stress waves play a significant role.

ACKNOWLEDGMENTS

The work reported herein in a part of the Ph.D. requirement of K.R.Y. Simha and was sponsored by the University of Maryland Minta Martin Fund.

REFERENCES

1. Blasters' Handbook, 1977, Dupont Co., pp. 374-378.
2. Konya, J.J., 1980, "Pre-split Blasting: Theory and Practice" Proc. AIME Annual Meeting.
3. Kihlstrom, B., 1970, "The Technique of Smooth Blasting and Pre-splitting with Reference to Completed Projects" Nitro Nobel Report No. B672.
4. Porter D.D. and Fairhurst, C., 1970, "A Study of Crack Propagation Produced by the Sustained Borehole Pressure in Blasting", Proc. 12th U.S. Symposium on Rock Mechanics.
5. Ito, I., Sassa, K., Tanimoto and Katsuyamu, K., 1970, "Rock Breakage by Smooth Blasting", Proc. Second Congress of the Intl. Soc. for Rock Mechanics.
6. Dally, J.W., Fourney, W.L. and Holloway, D.C., 1973, "A Dynamic Photoelastic Investigation of Explosively Induced Fracture", Formal Report to U.S. Bureau of Mines Contract No. 40220010.
7. Pederson, A.L., Fourney, W.L. and Dally, J.W., 1974, "Investigation of Pre-splitting and Smooth Blasting Techniques in Construction Blasting", Report to the National Science Foundation.
8. Swift, R.D., Schatz, J.F., Dusham, W.B., Hearst, J.R. and Kusbov, A., 1973, "Effect of Simultaneous and Sequential Detonation on Explosively Induced Fracture", Proc. 19th U.S. Symp. on Rock Mechanics.
9. Barker, D.B., Holloway, D.C. and Cardenas-Garcia, J.F., 1982, "Experimental Investigation of Fragmentation Enhancement Due to Salvo Impact", Proc. 23rd U.S. Symposium on Rock Mechanics.
10. Fourney, W.L., Holloway, D.C. and Barker, D.B., 1980, "Pressure Decay in Propagating Cracks" DOE Report Contract DE-AP21-79MC42577.

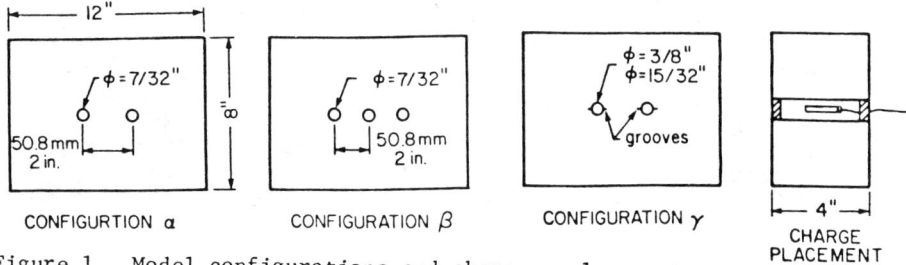

Figure 1. Model configurations and charge emplacement

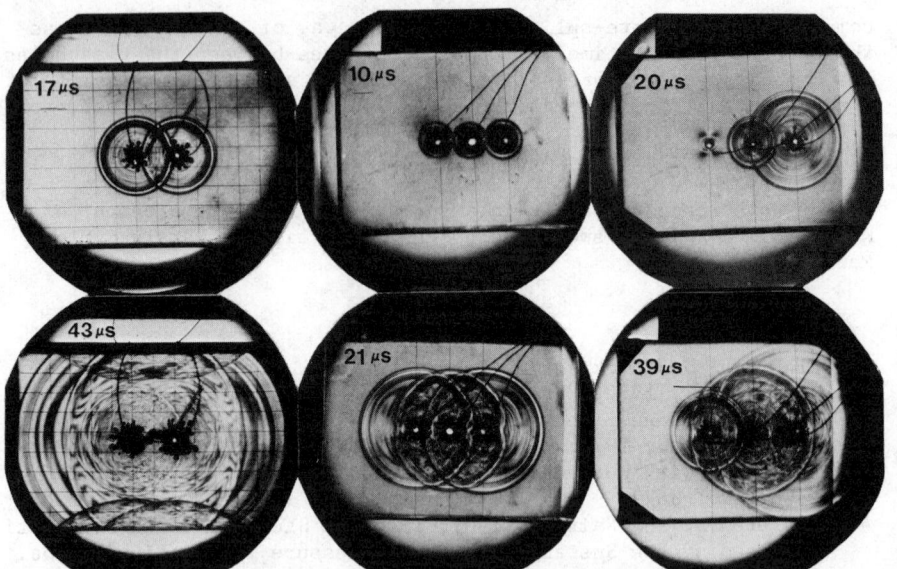

Figure 2. Pre-splitting process: a. Simultaneous detonation of two charges; b. Simultaneous detonation of three charges; c. Delayed detonation of three charges

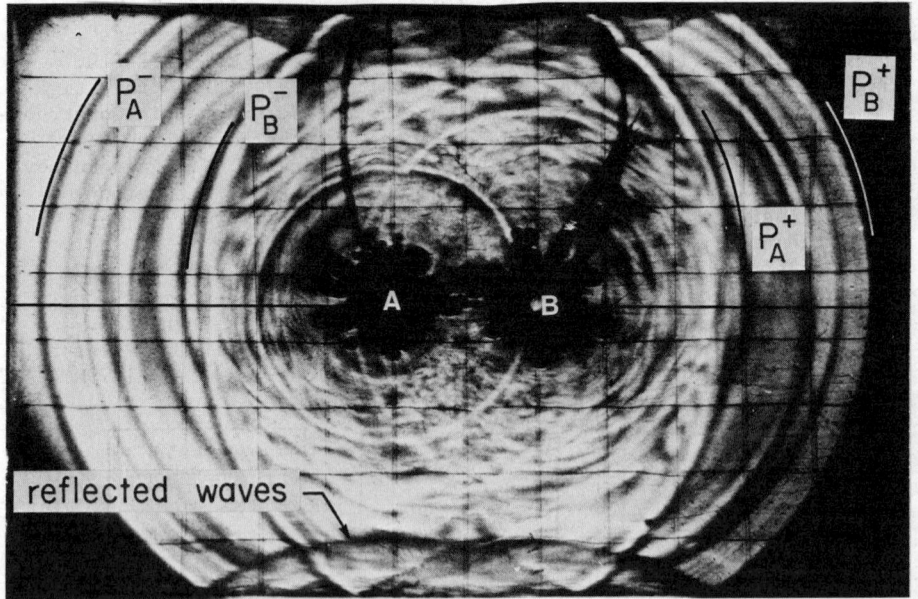

Figure 3. Fractures and stress wave systems 43μs after detonation

PRE-SPLITTING AND STRESS WAVES 529

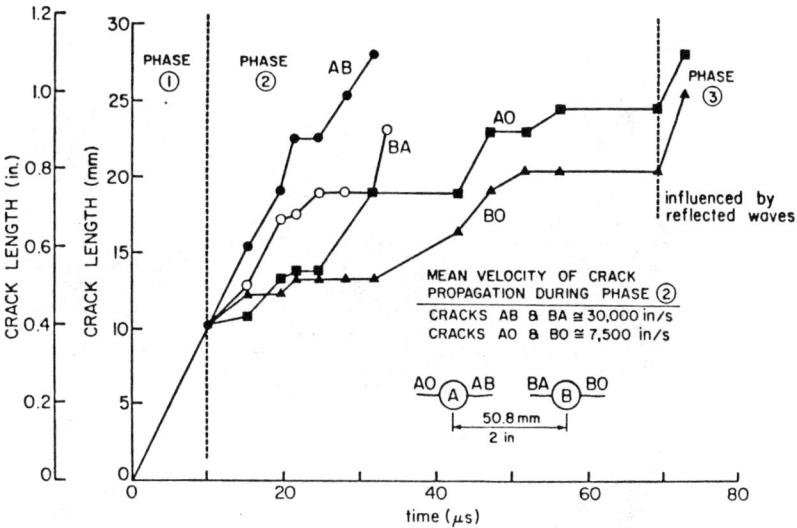

Figure 4. Crack growth along the line connecting blastholes (simultaneous detonation of two charges)

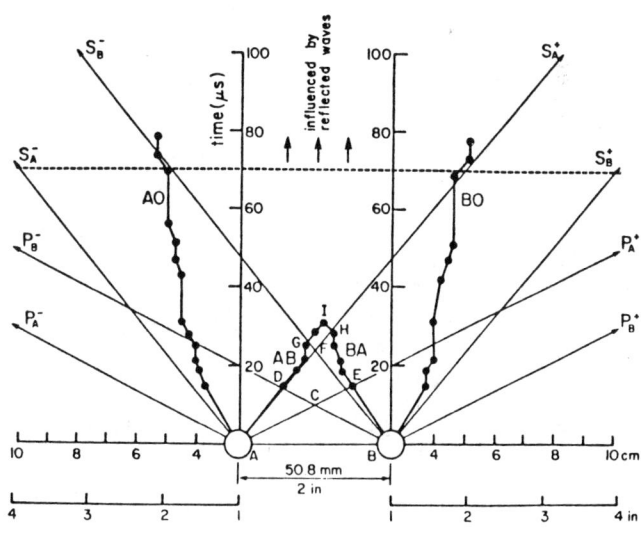

Figure 5. Lagrangian diagram of crack growth and stress waves (simultaneous detonation of two charges)

530　　　　　　　　　ISSUES IN ROCK MECHANICS

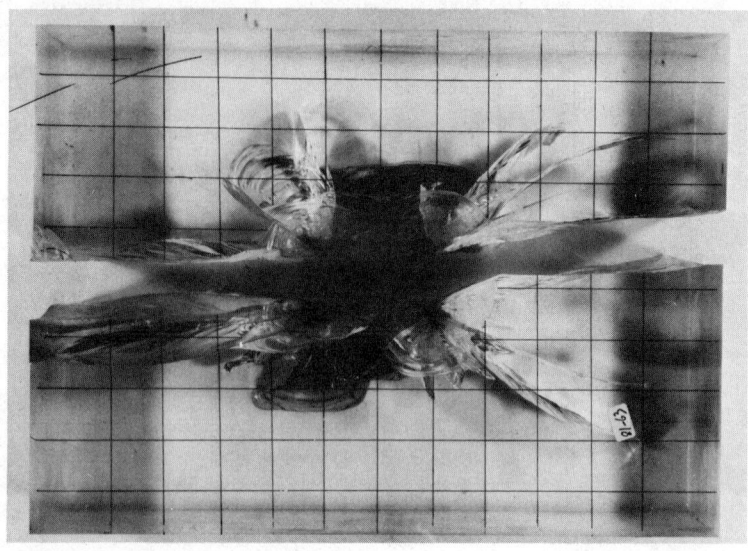

Figure 6. Pre-split model (simultaneous detonation of two char

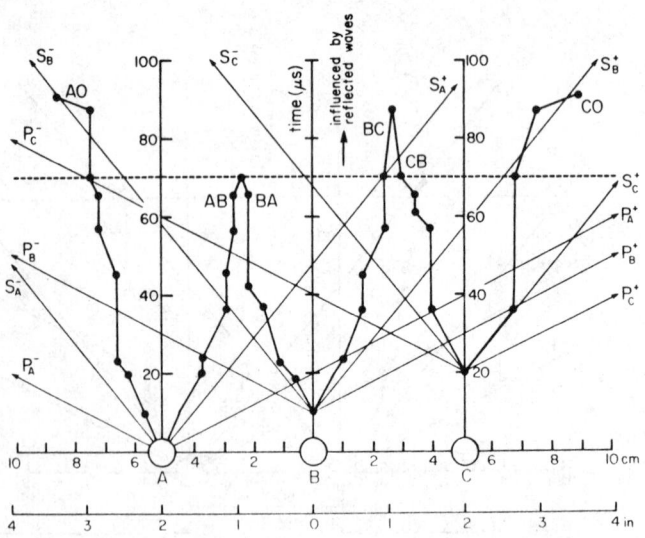

Figure 7. Lagrangian diagram (delayed detonation of three charges with 10μs delay)

Chapter 55

THE MECHANISM, AND SOME PARAMETERS CONTROLLING,
THE WATER JET CUTTING OF ROCK

by David A. Summers and Steven McGroarty

Curators' Distinguished Professor of Mining Engineering;
Director, Rock Mechanics & Explosives Research
University of Missouri-Rolla
Rolla, Missouri

Graduate Research Assistant
Rock Mechanics & Explosives Research Center
University of Missouri-Rolla
Rolla, Missouri

ABSTRACT

The last ten years has witnessed the change in water jet cutting studies, on rock, from laboratory testing to prototype use of field equipment.

Unfortunately, while equipment and technological developments have occurred, the understanding of the exact mechanisms of water jet erosion have been less well developed. Commonly, theoretical evaluations have used the uniaxial compressive strength of the target rock as the major parameter of rock erosion resistance.

It is considered that the structural properties of the rock, and most particularly, the flaw density and heterogeneity of the rock are much more important parameters. This is demonstrated and discussed.

Locating the jet and target under water and pressurizing the containing vessel makes it is possible to induce cavitation in and around the water jet and create a much slower erosion process.

INTRODUCTION

The use of high pressure water jets as a tool for cutting geotechnical material has become only seriously accepted within the last ten years. Part of this acceptance has been brought about by a change in the understanding of the way in which a water jet cuts material and the conditions therefore necessary for such cutting to be effective. In order for ultimate optimization of an application to be ensured, however, an adequate theoretical modeling of the process is required. A pre-condition for such a model is that an appreciation of the events which occur during the process, and the parameters which control them, exists.

It is not the purpose of this paper to describe a model, the technology as yet has had insufficient study that all the events which go on in the cutting zone can be adequately identified for such a step. It is, however, possible to identify certain signposts which have become evident and to make certain suggestions, based on observation, which might hasten such a model development.

THE ERRONEOUS USE OF COMPRESSIVE STRENGTH

A typical approach to a difficult problem is to simplify and idealize initial conditions in a first evaluation of the problem, and then to gradually add the more complex conditions of reality. This approach has been tried by several investigators, as a means of tackling the water jet problem.

A paper written by Powell and Simpson (Powell and Simpson, 1969) illustrates this procedure. The stress distribution was first calculated for a homogeneous, linearly elastic half-space impacted by a pressurized water jet, whose normal pressure profile on the surface had been found to approximate the equation

$$F(r) = \frac{1}{2} \rho u^2 [1 - 3(\frac{r}{b})^2 + 2(\frac{r}{b})^3]$$

in the region, radius b, over which the jet loads the surface, and where the jet has radius a, velocity v, density , and b = a 20/3 and where r is the radial distance from the center of the inpact zone. It was presumed that the jet imposed no shear stress on the surface.

Once this stress distribution was calculated, then the half space was presumed filled with randomly oriented and distributed incipient cracks, which otherwise had no effect on the stress distribution. Based on a prior calculation of the maximum principal stress (σ_1) and minimum principal stress (σ_3), then the criterion for fracture growth was that

$$-\frac{(\sigma_1 - \sigma_3)^2}{8(\sigma_1 + \sigma_3)} = \sigma_t$$

where σ_t is the tensile strength of the material.

Based on this evaluation, Powell and Simpson indicated that no fracture should occur in the material until the incident jet pressure exceeded 20 times the rock tensile strength, and plotted a projected relationship between likely depth of cut and multiples of this tensile strength (k) (Fig. 1.)

It is unfortunate that this initial theoretical review was followed by other research which seemed to indicate that high jet pressures were required for effective cutting. Cooley, for example, found, using a single pulse water cannon, that specific energy of excavation reduced as jet pressure was increased from 1.43 to 3.95 times the rock compressive strength, with suggested optimal values at above 10 times the rock compressive strength (Cooley and Brockett, 1972). As a result there began a school of thought in writing theoretical predictive equations for water jet action that a meaningful dimensionless parameter could be achieved by dividing jet pressure by rock compressive strength (Cooley, 1974; Singh, Finlayson, and Huck, 1972). From this it has become common to consider that the uniaxial compressive strength is a meaningful parameter in assessing cuttability of a rock by water jet. Such is an erroneous assumption, as can be shown by drilling through two rocks of approximately equivalent compressive strength (Fig. 2).

In the initial impact of a water jet on a smooth surface, erosion will occur, close to the nozzle, around the perimeter of the impact zone (Fig. 3). This is because of possibly three separate factors. Investigators at Cambridge, have shown that round-ended water jets will create a stress, as it impacts a surface (which can exceed 2.5 times water hammer) (Rockester and Brunton, 1979). It is considered that this is generated because until this ring zone is reached the surface of the water drop is collapsing on the surface faster than the water already on the surface can escape. In consequence no shear is measured within the zone, and only relatively low values outside it (ibid.).

This phenomenon of high pressure impact stress is extremely transient and requires a controlled shape to the jet leading edge. Attempts to develop rock cutting devices based on this have so far failed because of the difficulty in controlling the latter (Ripkin and Wetzel, 1972).

As impact pressure is increased, however, circumferential cracks are generated around the impact zone, (Field, et al, 1979). Field relates these to the Hertzian cracks which occur when a sphere impacts on a flat surface. Because of the stepped nature of the surface at these cracks the subsequent flow of fluid over the surface will initiate erosion at this location.

The third possible factor controlling this erosion is that,

close to the nozzle the central core of the jet is still solid and moving at relatively constant velocity (Summers and Zakin, 1975) it is only on the edges of the jet that there is the rapid decay in pressure discussed by Powell and Simpson (Ref. 1). This pressure decay will reduce the loading on the surface such that there is a differential stress across the particles concurrent with the curvature of the surface from the unloaded area into the zone of impact. This possibility is illustrated by the fact that as the target plate is moved further from the jet, the central core disappears in the jet structure and on the target surface (Fig. 4).

THE USE OF TENSILE STRENGTH

Failure on a smooth, homogeneous, surface is thus initiated as surface fractures. However the presence of a thin surface layer of soft material will inhibit the creation of the high water hammer level pressures on the surface (Matthewson, 1979). This could, in part, explain why in the experiments carried out by Foreman and Secor (Foreman and Secor, 1973) no damage was seen on the surface if a thin copper plate was placed between the impacting jet and the underlying rock surface.

The major point of that paper was, however, that cutting was enhanced by water pressurization of the pores in the rock, and that this created failure under Griffith crack growth at a water pressure of approximately 3 times the rock tensile strength. The validity of this correlation is again challenged in so far as it is easier to cut granite with a water jet than it is to cut marble. Concurrently, it is suggested that the concept of internal failure within the rock mass by growth of "concealed" Griffith cracks is also not valid. During the first authors early research, a block of granite was placed under a water jet operating at 70 MPa. No discernable cutting of the rock ensued. When, however, the rock was rotated around an axis eccentric to that of the jet a hole was drilled through the 20 cm thick sample (Summers, 1968).

Water jet cutting of rock is therefore considered to occur by a process of crack growth initiating from surface cracks which are impacted by the water jet. This viewpoint is enhanced by Fields' result that eching the surface of a glass plate prior to impact, which removes surface flaws, dramatically reduces impact damage (Field, 1966). This, however, is not, in itself, sufficient to explain some of the data described above - particularly the marble:granite cutting situation. It is, however possible to accept that this explains the increase in difficulty found in cutting finer grained materials and the restriction of the cutting path to a line directly under the jet in granular materials. This is in contrast to spallation around the traverse line which can occur in crystalline materials. It is considered that this difference is due to the relative crack lengths, established by

WATER JET CUTTING OF ROCK 535

Brace (Brace, 1964) as being equivalent to grain size, and the greater density of crack arrest locations in the granular material. These, in turn, will control the rate and extent of crack growth, according to Bieniawski (Bieniawski, 1967) and thus the cutting rate of the rock.

The controlling physical parameters of the rock therefore are likely to be flaw density (inversely correlated to grain size and the surface energy of the rock. It is however incorrect to use an average value as being pertinent.

CRACK GROWTH CONSIDERATIONS

In seeking to resolve the problem as to why the water jet cut granite more easily than marble, the hypothesis was propounded that the jet was differentially compressing two adjacent mineral components of the rock, that this in turn caused a weakening of the bond at the interface between the crystals and this could be exploited by the water jet.

Given however that the water jet operates at a pressure of 100 MPa to cut granite, and that the mineral modulus is on the order of 64 GPa, then the relative displacement of a crystal 2 mm long would be on the order of 0.003 mm at maximum. This did not seem an adequate initial amount to justify hypothesis.

It was then considered that the multi-phase materials exist as a combination of soft and harder minerals and that, were the softer materials preferentially cut out before the harder ones, then it would not be necessary for the jet to cut the hard components. Such a process was developed, demonstrated (Summers and Raether, 1982) and is now commercially available for the cutting of concrete (Hi-Tech, Inc.). In this use of water jets, the pressure used (80 MPa) is sufficient to cut the cement paste, but does not cut the aggregate which is however removed by a lack of surrounding support (Fig. 5).

An examination of the section through a water jet cut in granite does not however show the same phenomenon occuring in that rock (Fig. 6). Rather the hole shows that there is jet penetration and material removal occuring along the crystal boundaries. The clear evidence is that the water jets do work by preferentially attacking existing crystal boundaries, first. It might be, however,, be that the lower surface energy of the mica components (0.0132 cm kg/cm^2) relative to theat of the quartz and feldspar components (0.0306 cm kg/cm^2) leads to greater crack growth in those regions.

Two stages, however, can be hypothesized for this during jet attack. In the first instance the water jets will preferentially attack the crack areas, causing crack growth to the point that in the second stage the crack is long enough to be grown to material failure under the main jet impact pressure. Evidence for this is suggested by a study of cavitation damage of a dolomite sample.

CAVITATION EROSION

A test chamber has been built and previously described (Summers and Mazurkiewicz, 1981) wherein a 5 cm x 2.5 cm x 2.5 cm sample of rock is slowly traversed under a water jet at pressure. The test is carried out under water in the chamber which is pressurized, by valving the drain pipe. By adjustment of the cell pressure, relative to the jet pressure (a ratio which simply approximates to the cavitation number) cavitation can be induced around the flowing jet and caused to collapse on the rock surface. By control of the feed rate of the rock, the residence time of the rock under the jet can be changed. As the rock speed is slowed from 5 cm to 1 cm/min a change in the cut surface can be discerned (Fig. 7).

Initially, the entire surface is pitted by the collapse of cavitation bubbles. As the residence time increases, however, this even attack does not. Rather, the cavitation attack starts to concentrate at weakness points and from these larger cracks develop. The point suddenly is reached where these cracks reach a critical length and there is a substantial change in the depth of cut, as the main jet pressure can now exploit the cracks and remove larger fragments of material.

It is conjectured that this demonstrated activity is similar to that which occurs, but at a much more rapid rate, with conventional water jets at pressure. If such is the case, then there is indeed no correlation possible between conventionally measured rock properties and jet cuttability and some new method of assessing viable rock properties is required.

As a concluding remark, it might be mentioned that this affinity for water jets to attack existing grain boundaries has proven effective in breaking down sedimentary lead ores into their component minerals at the grain size level. This has given the opportunity for a simple mechanical separation or, in the case of one ore where the galena and host rock are in different size ranges, a simple mechanical sieving to separate most of the ore.

ACKNOWLEDGEMENTS

This work was funded under the Mining and Minerals Research and Resources Institute Program of the U.S. Bureau of Mines. We are pleased to acknowledge this support and that of the students, Mr. R. Brandom and Mr. T. Fort III, and staff Mr. B. Hale and Mr. J. Tyler, who made this report possible.

REFERENCES

Bieniawski, Z.T., "Mechanism of Brittle Fracture of Rock," Int. J. Rock Mech. Min. Sci., Vol. 4, Oct. 1967, p 395-430.

Brace, W.F., "Brittle Fracture of Rocks," State of Stress in the Earth's Crust, Elseview, New York, 1964, p 111-180.

Cooley, W.C. and Brockett, P.E., "Rock Disintegration by Pulsed Liquid Jets," Annual Report on ARPA Order 1579, Bureau of Mines Contract H0210012, Terraspace, Inc., Maryland, 1972.

Cooley, W.C., "Correlation of Data on Jet Cutting by Water Jets Using Dimensionless Parameters," Paper H4, Proc. 2nd Int. Symp. Jet Cutting Tech., Coventry, UK, 1972.

Field, J.E., "Stress Waves, Deformation and Fracture Caused by Liquid Impact, Phil. Trans. Roy Soc. (London) 260A, 1966, p 86-93.

Field, J.E., et al., "Liquid Jet Impact and Damage Assessment for Brittle Solids," Paper 13, Proc. 5th Int. Conf. on Erosion by Solid and Liquid Impact, Cambridge, UK, 1979.

Foreman, S.E. and Secor, G.A., "The Mechanics of Rock Failure Due to Water Jet Impact," SPE 4247, Proc. 6th Conf. on Drilling, Rock Mech., Austin, Texas, 1973, p 163-174.

Hi-Tech, Inc., Milbank, South Dakota.

Matthewson, M.J., "Theoretical Aspects of Thin Protective Coatings," Paper 73, Proc. 5th Int. Conf. on Erosion by Solid and Liquid Impact, Cambridge, UK, 1979.

Powell, J.H. and Simpson, S.P., "Theoretical Study of the Mechanical Effects of Water Jets Impinging on a Semi-Infinite Elastic Solid, Int. J. Rock. Mech. Min. Sci., Vol. 6, p 353-364, 1969.

Ripkin, J.F. and Wetzel, J.M., "A Study of the Fragmentation of Rock by Impingement with Water and Solid Impactors," Project Report on ARPA, Order 159, Bureau of Mines Contract H0210021, 1972.

Rockester, M.C. and Brunton, J.H., "Pressure Distribution During Drop Impact," Paper 6, Proc. 5th Int. Conf. on Erosion by Solid and Liquid Impact, Cambridge, UK, 1979.

Singh, M.M., Finlayson, L.A., and Huch, P.J., "Rock Cutting by High Pressure Water Jets," Paper B8, Proc. 1st Int. Symp. Jet Cutting Tech., Coventry, UK, 1972.

Summers, D.A., "Disintegration of Rock by High Pressure Jets," Ph.D. Thesis, Univ. of Leeds, UK, 1968.

Summers, D.A. and Zakin, J.L., "The Structure of High Speed Fluid Jets and Their Use in Cutting Various Soil and Material Types," Final Report on U.S. Army Contract Daak01-74-C-0006, 1975.

Summers, D.A. and Mazurkiewicz, M., "The Further Development of a Cavitation Test Cell," Proc. 1st U.S. Water Jet Symp., Golden, Co., 1981.

Summers, D.A. and Raether, R.J., "Comparative Use of Intermediate Pressure Jets for Slotting and Removing Concrete," Proc. 6th Int. Symp. Jet Cutting Tech., Guildford, UK, BHRA, 1982, p 387-396.

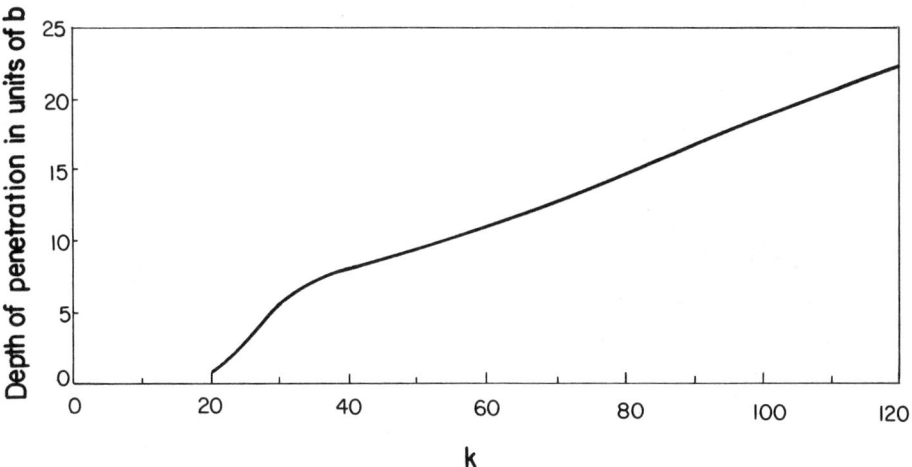

Fig. 1. Proposed relationship between hole depth and tensile strength (after Powell and Simpson).

Fig. 2. Variation in hole diameter as a 70 MPa water jet drill passes from Berea sandstone to Indiana limestone to Berea sandstone.

Fig. 3. Damage created by a 70 MPa jet on an aluminum plate held 2.5 cm from the nozzle.

Fig. 4. Damage created by a 70 MPa jet on an aluminum plate held 20 cm from the nozzle.

Fig. 5. Slot cut through a 20 cm concrete wall showing that the cement paste is removed and the aggregate is not damaged.

Fig. 6. Microscope section of a cut made in granite showing the preferential attack on crystal boundaries.

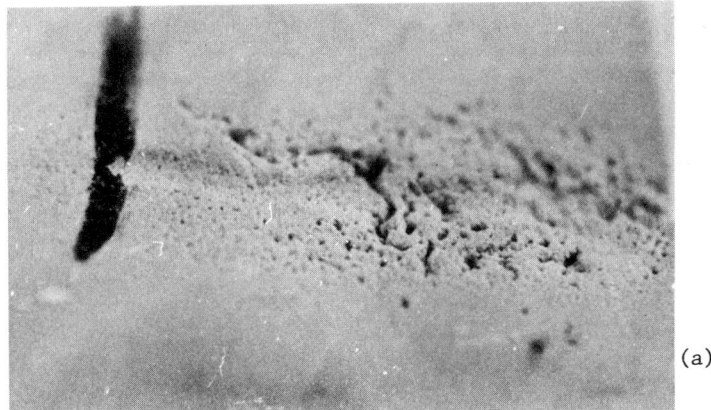

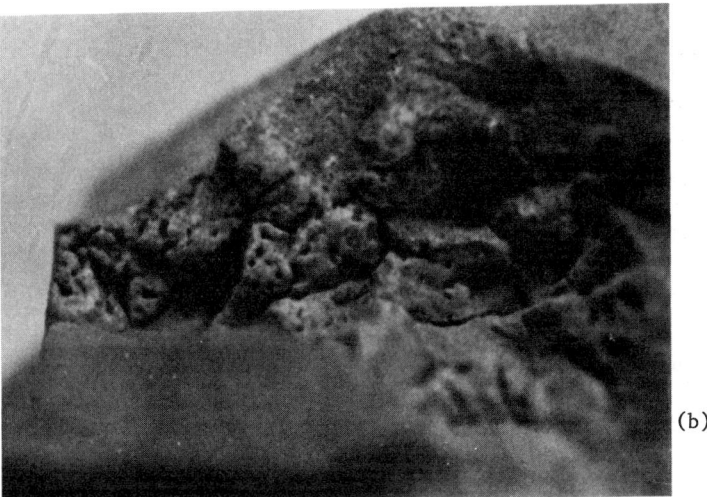

Fig. 7. Samples of dolomite cut by a cavitating water jet (a) at a traverse velocity of 1.25 cm/min; (b) at a traverse velocity of 1 cm/min. The change in erosion pattern can be seen as the jet acts on (a) and the increase in cutting length is shown in (b).

Chapter 56

SUB-CRITICAL CRACK GROWTH IN STRIPA GRANITE: DIRECT OBSERVATIONS

by Graham Swan and Ove Alm

Research Scientists, Division of Rock Mechanics,
University of Luleå, Sweden

ABSTRACT

In-situ fracture toughness experiments have been performed with a scanning electron-microscope. The rock, Stripa granite, was loaded both in tension and three-point bending using the same size notched plates. Observations, which included crack extension measurements as a function of load, are presented in the form of R-curves. The data does not support the proposition that R-curves can be regarded as independent of specimen configuration for this granite. Typically crack extension was seen to begin at loads of 70-80 % failure load. Tensile specimens showed considerably greater sub-critical crack growth than the bend specimens. Evaluation of the approximate fracture toughness K_m for both specimen types produced as expected erroneously low estimates of K_{Ic}, being worst in the case of the tensile specimen.

Some crack extension data for Stripa granite using the indirect methods of compliance calibration, fluorescent dye penetrant and acoustic emission are compared. In a material such as Stripa granite it is considered that no single method is without its limitations.

INTRODUCTION

From both practical and economic considerations, the advantages gained by the use of standard rock core specimens in fracture toughness measurements for geomechanical use are considerable. To-date it is well-understood that valid K_{Ic} determinations only exist for a limited number of rock types (nominally fine-grained) and geometries (typically the SENB standard according to ASTM E-399). While the problems of using sub-size specimens in <u>un-notched</u> testing have been effectively ignored in the standard procedures by the expedient of subsequently evoking some size effect model, this has not been the approach adopted by fracture mechanics workers. The question has thereby directly arisen among those attempting to standardize rock fracture toughness tests, if and how valid K_{Ic} values can be deduced from sub-size specimens? By sub-size is here meant that notch depth a_o must be at least equal to a_{min}, where

$$a_{min} = 2.5 \, (K_{Ic}/\sigma_t)^2 \tag{1}$$

σ_t being the uniaxial tensile strength.

Possible answers to this question will not be discussed explicitly in this paper. Rather, the intention is to contribute generally to present understanding with regard to what is variously called sub-critical crack growth, environment-enhanced crack growth, static fatigue, stress corrosion cracking and pre-failure micro-cracking. The significance of what will be here termed sub-critical crack growth to the rock fracture process is considered fundamental. The upper scale at which it ceases to be so is most likely to be dictated by the appearance of large structural weaknesses such as intact joints, fissures, etc. Certainly it is true to state that whether the problem of using sub-size specimens is approached using J-integral methods (eg. Schmidt and Lutz, 1979), specially notched specimens (eg. the short Rod of Barker, 1977) or direct R-curve measurements (Ouchterlony, 1981), a description of the associated micro-crack geometries is implicit.

In this paper some direct observations using a scanning electron-microscope (SEM) are reported and compared with results using conventional indirect techniques. The SEM observations were taken from in-situ bending and direct tension experiments. The rock chosen was Stripa granite, being one for which a great variety of physical/mechanical data already exists in the open literature. On the basis of this data a realistic value for a_{min} using (1) is 100 mm. Such a value only serves to emphasise the practical desirability of using sub-size specimens if fracture toughness testing is ever to be accepted within the mining rock mechanics community.

IN-SITU SEM EXPERIMENTS

Since it was recognised by various workers in the seventies that complications arise when attempting to measure crack extension lengths in rock, surface observations have become less popular. It is now understood that on the scale of grain size, cracks coalesce in a more or less discontinuous fashion with increasing load. This presents a conceptual problem of identifying a well-defined crack tip from which to measure length. Conventional grid techniques (Swan, 1975) being indirect, suffer by not being able to provide enough discriminating information. The use of in-situ SEM techniques, while limited to relatively slow crack growth rates, offer improved prospects for making measurements. This was recently demonstrated by the work of Batzle et al (1980).

Experimental Procedures

Ten single-edge-notched plates where prepared with the nominal dimensions given in Fig. 1. After carefully cutting to oversize, these dimensions were arrived at by grinding and polishing down to 1 μm. Finally the surface to be observed was coated under vacuum with carbon. This coating was of sufficient thickness to prevent charging during the experiment while permitting the visibility of grain boundaries at accelerating voltages between 20-30 kV. Furthermore, perforation of this layer due to crack extension was made immediately visible through the contrast obtained as a result of charging.

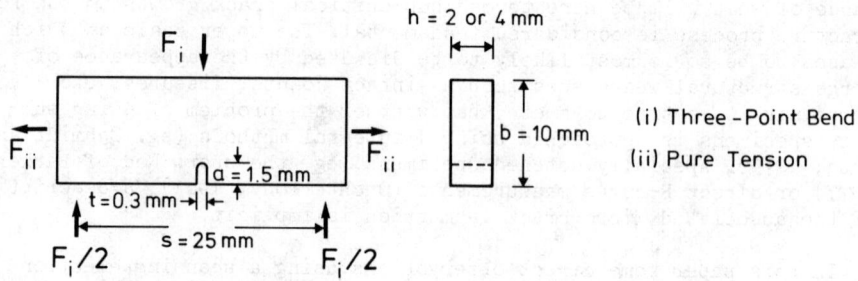

Fig. 1. Dimensions of SEN specimens used in SEM experiments.

The specimens were placed into the 2 kN loading stage of a JEOL 50A SEM. In the case of tensile testing, specimens were glued in-situ so making for a stiff loading configuration. The progress of the subcritical crack growth was monitored as a function of applied load and in some cases crack opening displacement. Both direct photography and TV video viewing was used for recording purposes.

The data reduction to R-curves followed the procedure outlined in ASTM E561-78T, the crack growth measured directly from photographic plates. For the bend tests the non-standard dimensions $(S/b = 2.5)$ required that an estimate of the dimensionless stress intensity factor $Y'(a/b)$ be made. The result

$$Y'(a/b) = 1.90 - 3.34\ (a/b) + 15.30\ (a/b)^2 - 26.25\ (a/b)^3 + 26.40\ (a/b)^4 \qquad (2)$$

was used in

$$K_I = 3FS\ \sqrt{a}/2hb^2 \qquad (3)$$

The crack opening displacement measurements, which were made optically on tensile specimens, were used via a standard compliance calibration to infer crack growth.

Results

Of the ten specimens prepared the test results from only five were deemed valid. These are summarised in Table 1 and Fig. 2. The approximate fracture toughness K_m is simply obtained by inserting the failure load F_{max} and the initial notch depth a_o in the appropriate formula $K_I(a, F)$.

TABLE 1. Summarised data from SEM experiments.

Specimen, Type	a_o (mm)	F_{max} (N)	K_m (MN/m$^{3/2}$)	$\Delta a/b$
5, SENB	1.49	215.6	1.122	0.218
6, SENB	1.49	229.7	1.393	0.153
16, SENB	1.44	98.5	1.066	0.182
21, SENT	1.28	317.5	0.623	0.331
22, SENB	1.64	169.8	1.101	0.194

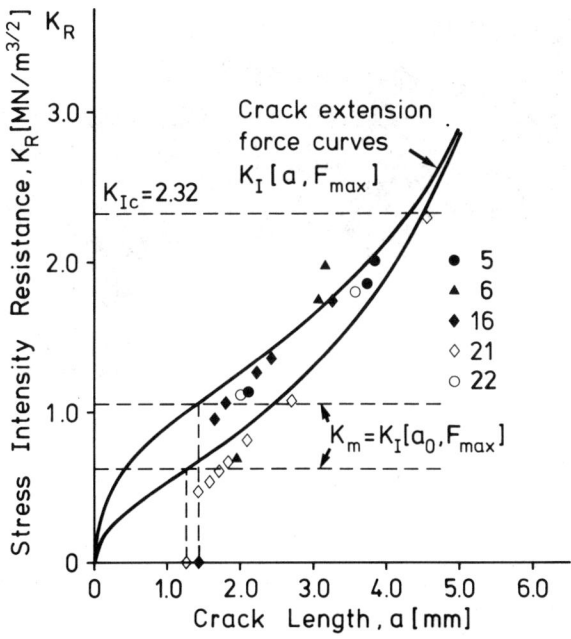

Fig. 2. R-curve data from SEM in-situ experiments on SEN specimens of Stripa granite. The K_{Ic} value is an estimate due to Ouchterlony (1981).

An advanced state of crack extension development for specimens 5 and 21 as observed in the SEM is shown in Fig. 3a and 3b respectively. A frequent difficulty with this superficial method of extension monitoring is illustrated in Fig. 4. The sequence shows a regular notch extension which is suddenly interupted by the appearance of an adjacent growth. Such instantaneous development implies the existance of subsurface cracking which goes unobserved, suspected by Kobayashi and Fourney (1978).

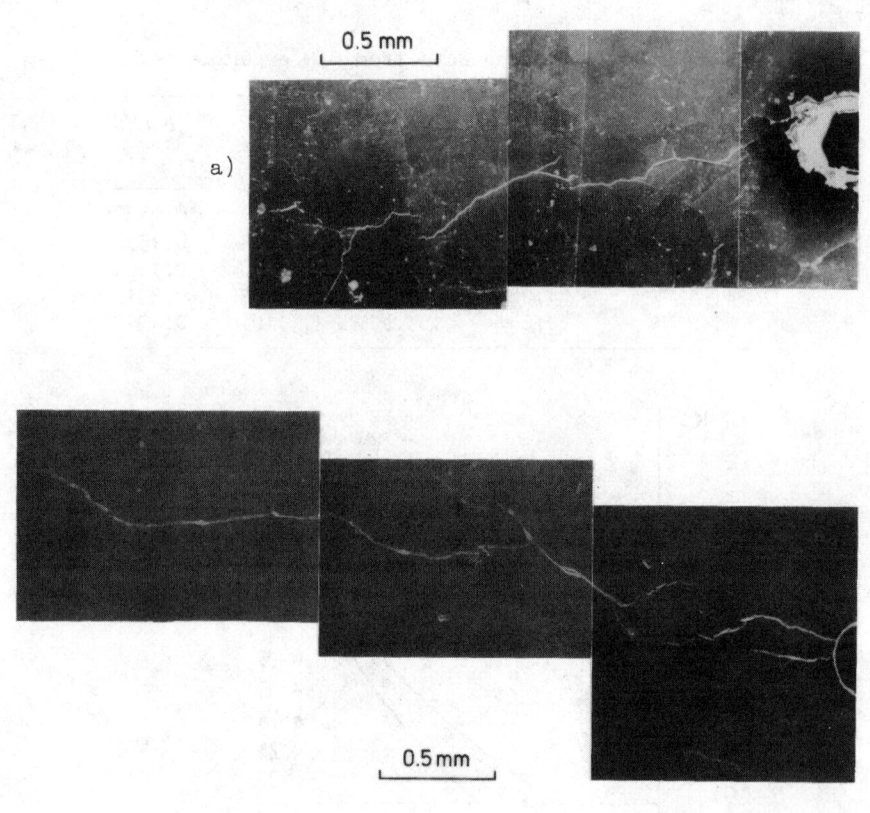

Fig. 3. Superficial crack patterns immediately prior to specimen failure in the SEM.
a) SENB, specimen 5. b) SENT, specimen 21.

Some difficulties were also experienced with crack mouth displacement measurements. Results from tensile specimen 21, Table 2, show a discontinuous behaviour. Crack lengths as calculated from dimensionless compliance CEh are not comparable to those from direct SEM observations. The reason for this is probably due to contributions from occasional local grain motions on the pure bulk opening.

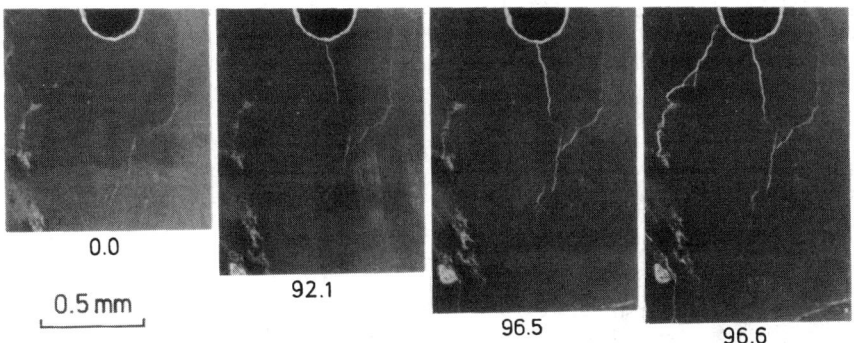

Fig. 4. Crack growth sequence from a SENB specimen for which subsurface cracking is evident. The figures refer to instantaneous load values in Newtons.

TABLE 2. Summarised data from CMD measurements for specimen 21.

Load, F (N)	CMD (μm)	CEh	Crack Length, $(\Delta a + a_o)/b$ Compliance Calibration	Direct Method
239.4	0.272	0.260	0.227	0.159
256.4	0.426	0.380	0.264	0.169
273.3	0.734	0.615	0.314	0.179
298.8	1.093	0.837	0.349	0.209
300.5	0.580	0.442	0.279	0.273
307.3	0.631	0.470	0.285	0.460

INDIRECT CRACK GROWTH STUDIES

A number of recently published papers have given rise to a conflicting understanding with regard to rock strength size effects in granitic rocks. While there are those who generally advocate the suitability of Weibull type models (eg. Ratigan, 1981), experiments both with bending and pure tension loads have not provided unequivocal evidence supporting this (compare Wijk et al, 1978 and Swan, 1980). Certainly the observation of sub-critical crack growth as reported above is expected to restrict the usefulness of statistical fracture models to Stripa granite. It may also be used to account for certain irregularities in sub-size fracture toughness determinations using SENRBB specimens. The results from some indirect studies it is hoped will add credence to these remarks.

Acoustic Emission Methods

The evidence that sub-critical crack growth is accompanied by acoustic emission in granitic rocks is considerable. In fracture toughness testing the chevron notch geometry has been deliberately used to produce prescribed amounts of precracking. For the Short Rod specimen this is evidenced with Stripa granite by records typically of the form shown, Fig. 5. Interestingly enough, while the specimens were rated as subsize, the approximate toughness K_m value of 2.18 MN/m$^{3/2}$ is an improvement upon SENRBB data (see Swan, 1980). As to the study itself no attempt was made to correlate actual extension with emissivity.

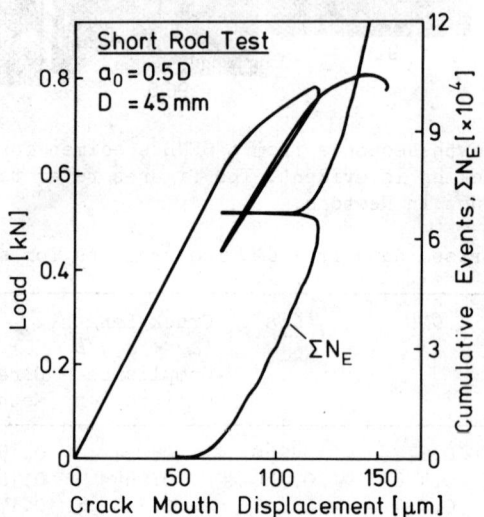

Fig. 5. Acoustic emission record for Stripa granite, Short Rod specimen.

Dye Penetrant Methods

Inasmuch as dye penetrant techniques do not give crack extension data simultaneous with the specimen in a loaded state they may be regarded as indirect. While there exists disagreement over which penetrant technique might give best results for granitic rocks (Schmidt and Lutz, 1979), the present authors favour a sectioning technique. This involves unloading the specimen after some growth has occurred, cutting sections through the notch front, Fig. 6a, grinding the section surface, ultrasonic cleaning for 30 secs and finally dye immersion for 30 mins. The resulting specimen (shown dotted in Fig. 6a) is then observed on surface xx and yy before completing the notch half-section's fracture surface and observing the depth of penetration. A result from this procedure is shown in Fig. 6. It suggests that the extended crack front is not straight as required for a valid toughness determination. This aparently answers a recent suspicion of Ouchterlony (1981). Quantitatively, the dye estimates mean length agrees quite well with compliance calibration data (Swan, 1980). In general the edge penetration gave more decisive results than the depth penetration.

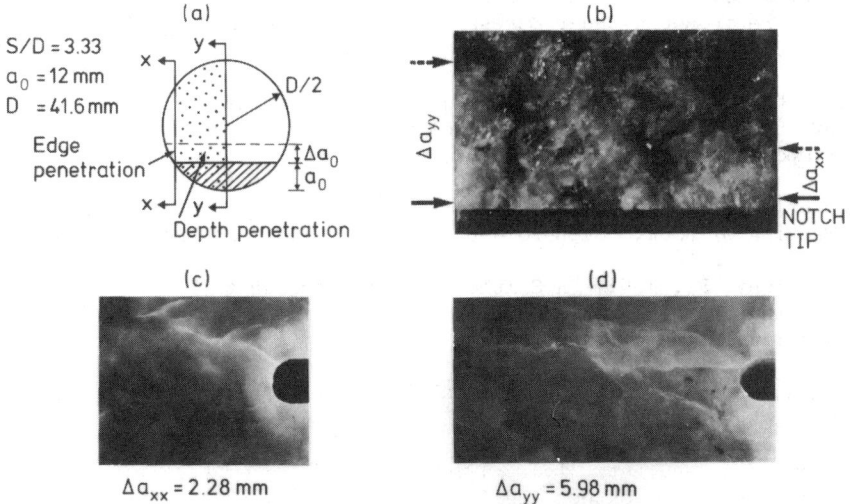

Fig. 6. Dye penetrant assisted crack length measurements from a sectioned SENRBB specimen, Stripa granite.
a) SENRBB notch section geometry. b) Notch plane section.
c) Section xx. d) Section yy.

CONCLUSIONS

The direct method of observing sub-critical crack growth using a SEM can provide useful data in connection with the application of sub-size specimens for fracture toughness estimates. The method itself requires considerable specimen preparation and care in execution as well as data evaluation. It has however been possible to obtain convincing R-curve data prior to the event of failure for a number of SEN Stripa granite specimens. While evidence for subsurface activity has been observed, the method apparently is sufficiently sensitive as to limit this effect. Optical CMD measurements in the SEM would seem to require some averaging technique.

The cheaper and less sophisticated indirect techniques of acoustic emission and dye penetrant both confirm the existence of sub-critical crack growth in Stripa granite. For a material as complex as Stripa granite it is considered that no single crack extension monitoring technique is without its limitations.

REFERENCES

Barker, L.M., 1977, "A Simplified Method for Measuring Plane Strain Fracture Toughness," Engng. Fracture Mechanics, 9, pp. 361-369.

Batzle, M.L., Simmons, G. and Siegfried, R.W., 1980, "Microcrack Closure in Rocks Under Stress: Direct Observation," J. Geop. Res., 85, B12, pp. 7072-7090.

Kobayashi, T. and Fourney, W.L., 1978, "Experimental Characterization of the Development of the Micro-Crack Process Zone," 19th U.S. Rock Symposium, Nevada, pp. 243-250.

Ouchterlony, F., 1980, "A New Core Specimen for the Fracture Toughness Testing of Rock," SveDeFo Report, DS 1980:17, Stockholm.

Ouchterlony, F., 1981, "A Simple R-Curve Approach to Fracture Toughness Testing of Rock with Sub-Size SECRBB Specimens," SveDeFo Report DS 1981:18, Stockholm.

Ratigan, J.L., 1981, "A Statistical Fracture Mechanics Approach to the Strength of Brittle Rock," Ph D Thesis, Univ. California, Berkeley.

Schmidt, R.A. and Lutz, T.J., 1979, "K_{Ic} and J_{Ic} of Westerly Granite - Effects of Thickness and In-Plane Dimensions," Fracture Mechanics Applied to Brittle Materials, A.S.T.M. - S.T.P. 678, S.W. Freiman Ed., pp. 166-182.

Swan, G., 1975, "The Observation of Cracks Propagating in Rock Plates," Int. J. Rock Mech. Min. Sci. & Geomech. Abstr., 12, pp. 329-334.

Swan, G., 1980, "Fracture Stress Scale Effects for Rocks in Bending," Int. J. Rock Mech. Min. Sci. & Geomech. Abstr., 17, pp. 317-324.

Wijk, G., et al, 1978, "The Relation between the Uniaxial Tensile Strength and Sample Size for Bohus Granite," Rock Mech., 10, pp. 210-219.

Chapter 57

STEP CRACKS: THEORY, EXPERIMENT, AND FIELD OBSERVATION*

Richard K. Thorpe, Merle E. Hanson, Gordon D. Anderson, and Ronald J. Shaffer

Earth Sciences Department
Lawrence Livermore National Laboratory
University of California
Livermore, CA 94550

ABSTRACT

The propagation of pressurized fractures across a frictional interface is discussed, with emphasis on the case where an offset, or step, in the crack is produced. Theoretically, the steps can occur at regions of reduced shear strength along the interface. As a fracture is propagated toward a weak zone, extensional strain is concentrated at the edges of the zone where higher shear stresses can be sustained. Thus, when the fracture intersects the weak zone, it can be reinitiated at some small distance away on the opposite wall of the interface. The phenomenon has been studied through numerical modeling and laboratory experiments of the hydraulic fracturing process. Field observation of natural step cracks indicates that the mechanism is applicable to the genesis of jointing patterns in rock. Such features should not be confused, therefore, with steps caused by shear displacement of an interface.

INTRODUCTION

Discontinuities in fractures propagated across an interface, or "step cracks", can often be observed in naturally or artificially fractured rock, as shown schematically in Figure 1. Such naturally occuring features are of interest because they may be indicative of shear displacement along the interface, which would be important in terms of rock mass deformability and fluid flow behavior. Similarly, step cracks in artificially fractured rock masses would influence

---

* Work performed under the auspices of the U.S. Department of Energy by the Lawrence Livermore National Laboratory under contract number W-7405-ENG-48.

the effectiveness of permeability enhancement measures, such as massive hydraulic fracturing, by increasing the tortuosity of the flow path through which proppant material must move. As part of the Department of Energy's gas stimulation program, recent work at Lawrence Livermore National Laboratory has focused on the propagation of a hydraulic fracture across an interface, and the mechanics of step crack formation are now well understood both theoretically and experimentally (Hanson, et al. 1981, and Hanson et al. 1980).

Results of this work can be incorporated into the process of interpreting natural joint patterns, with important consequences in describing the genesis and present kinematic and fluid flow aspects of fracture systems. This is especially relevant to nuclear waste disposal, for which considerable effort has been spent on characterizing fracture systems in crystalline rocks (for example, Thorpe, 1979, Olkiewicz, et al. 1979, and Wilder and Yow, 1981). Misinterpretation of natural step cracks as evidence of shearing along an intersected interface could lead to overestimation of its continuity and hydraulic conductivity.

THEORETICAL MODELING

Studies on the hydraulic fracturing process using two-dimensional numerical models have been reported by Shaffer, et al. (1980) and Hanson et al. (1978). Results show that as a pressurized crack approaches a perpendicular interface with uniform friction, it is drawn toward the interface if slippage occurs there. That is, the tendency for further propagation, as indicated by the Mode I stress intensity factor, increases as the crack tip is advanced toward the interface. The geometry of the crack changes, and the stress intensity increases. Obviously, with lower shear strength along the interface, the tendency to propagate the crack toward the interface will be greater. Once the crack intersects it, however, the chance of penetrating the opposite side is reduced when the interface is unbonded.

The effect of varying the frictional coefficient along an interface has more recently been analyzed (Hanson, et al. 1981). In this case the pressurized fracture is propagated into an interface where the frictional coefficient is smaller on the fracture path than a short distance away (Fig. 2). One result noted from the analysis was that as the fracture intersected the interface, motion was induced on that side of the interface as shown in Fig. 2. The resultant of this motion was an increased strain parallel to the interface, ε_y, near where the friction changes and on the side opposite to the fracture (right side of interface on Fig. 1). This parallel strain opposite the pressurized fracture is shown in Fig. 3 as a function of distance along the interface (δ) in the y direction. The range of δ from 11 to 17 on Fig. 3 is the region of reduced friction shown on Fig. 2. The fracture intersects the interface at $\delta = 14$. The coefficient of friction is shown by the dashed lines and has a value of 0.1 where the fracture intersects the interface, and jumps to 0.7 approximately 1/2 crack length from the intersection point. The strain increases significantly in the region where the frictional coefficient increases. Hence one would expect the fracture path to have an abrupt stop at the interface and reinitiate at one of the points where the parallel strain is highest. Figure 4 displays the strain ε_y when a fracture intersects the region of decreased friction off the symmetry axis. In this case, the increased friction region nearest the pressurized fracture axis displays the highest strain. The fracture thus would reinitiate in this region, producing a step pattern.

EXPERIMENTAL RESULTS

Laboratory experiments have been performed to qualitatively demonstrate the phenomena described above (Hanson, et al. 1980 and Anderson, 1979). First, the frictional properties for several surface finishes on Indiana limestone and Nugget sandstone were studied (Anderson, 1979). These data were then used to estimate the shear strength of an interface across which a hydraulic fracture was directed. The apparatus for the hydraulic fracturing experiments is shown in Fig. 5, where several 5x10x10 cm blocks of the test rock are stacked between the platens of a vertical press. The sides of the blocks were unconfined so that only a uniform normal stress was applied to the interfaces. Usually the stack consisted of three blocks, with the fracturing fluid injected into the middle of the center block through a tube cemented into it. With load applied to the interfaces, the fluid was pressurized until breakdown and crack initiation occurred.

Results of these experiments indicate that for the Indiana limestone and Nugget sandstone there exist critical threshold shear stresses across the interface below which the crack will not propagate into the adjacent block. When the coefficient of friction was lowered uniformly over the interfaces in another set of tests, the normal stress had to be increased substantially to provide the necessary shear resistance to allow the crack to cross the interface.

A change in the frictional properties of an interface such as one region having a lower coefficient of friction than an adjacent region can also influence crack propagation across the interface, and possibly cause a step crack to form. Fig. 6 shows the result of an experiment in which a 3/4-inch strip of lubricant was coated on an otherwise smooth, dry limestone interface parallel to the fluid injection tube. The opposite interface contained no lubricant whatever. The three-block stack was placed in a press and a load in excess of the threshold for cracks to cross the interface was applied. As can be seen in the figure, at the interface with the lubricated strip the crack reached the interface within the strip and then continued into the adjacent block laterally displaced to the edge of the lubricated zone. At the interface containing no lubricant the crack continued directly across the interface. These results are consistent with the theoretical modeling discussed above.

FIELD OBSERVATIONS

As mentioned previously, step cracks have been observed in the field as a result of hydraulic fracturing. Examples of the phenomenon are provided by Warpinski, et al., 1982 from their mine-back experiments in tuff at the Nevada Test Site. Similar step cracks have been produced by hydrofracing in coal seams to drain methane (Hanson, et al. 1981).

Although natural fractures or joints are not necessarily produced by internal pressurization, the presence of naturally stepped or truncated cracks suggests that similar principles govern the propagation of any extensional fracture across a frictional interface. The interface could be another extensional feature or a fault, and a step could occur then at any region of reduced friction along the interface. Of course, the geometry of the crack would also be affected by other factors such as nearby discontinuities or inhomogeneities in the rock and local variations in stress orientation and magnitude. The important issue is that an apparent offset of a natural joint should not be interpreted solely as a result of shearing along the interface.

To illustrate this point further, some examples can be given. As part of Lawrence Berkeley Laboratory's nuclear waste storage research project at Stripa, Sweden, field mapping of fractures in granite yielded a number of cases of natural step cracks (Thorpe, 1979, and Olkiewicz, et al. 1979). Most of these formed where high-angle, NW-SE striking joints crossed subhorizontal fractures, and inference could be made that the subhorizontal set offsets and thus postdates the other (Thorpe, 1979). This conjecture would be weakened, however, by absence of clear signs of shearing on the subhorizontal set. Recognizing the step crack phenomenon alleviates the confusion.

A detailed illustration of these features is given by Fig. 7, which is a circumferential diagram of natural fractures on the surface of a large 1.0x1.6 m cylindrical sample of Stripa granite. The specimen was used for uniaxial strength and fluid flow studies which have been reported by Thorpe, et al. (1980). Its axis was oriented nearly horizontally in the field such that the fractures labeled A, B, and C in the figure were from the high-angle set, while fractures D, E, and F represent the subhorizontal set. Offsets of up to 1 or 2 cm can be seen in fracture B of the former set where it intersects fractures D, E, and F. During its preparation the sample was parted along fracture B, and the step pattern was clearly exposed, as shown by Fig. 8. Because of the rough or irregular surfaces and lack of continuity of fractures D, E, and F through the specimen, it is unlikely that the steps are the result of shearing. Instead, it appears that fracture B was propagated in step-wise fashion across the three other joints, and thus postdates them. This chronology is consistent with observation that the high-angle joints at Stripa (fractures A, B, and C) are more persistent, or longer, than other discontinuities, excluding true shear fractures (Thorpe, 1979). They are also the most pervasive and closely spaced set (Olkiewicz et al. 1979), so correctly interpreting them as extensional, rather than shear fractures, is important for rock mechanics and hydrologic considerations.

CONCLUSIONS

Step cracks can be the result of fractures propagating across an interface at areas of reduced frictional resistance. While the offset could also be caused by shearing along the interface, the presence of a step crack is not necessarily indicative of shear displacement. The proper interpretation of step cracks is thus important for understanding the genesis and chronology of jointing in a rock mass.

From a more practical standpoint, step cracks can reduce the effectiveness of hydraulic fracturing measures to improve permeability of a formation. The steps can act as obstructions in the flow path, restricting movement of both proppants and fluids. It is, of course, virtually impossible to predict or eliminate step cracks; however, their likelihood of formation might be diminished if measures can be taken to increase the frictional resistance of existing interfaces, such as by reducing pore pressure in the medium.

REFERENCES

Anderson, G. D., 1979, "The Effects of Mechanical and Frictional Rock Properties on Hydraulic Fracture Growth Near Unbonded Interfaces," SPE Paper 8347 presented at 54th Annual AIME Society of Petroleum Engineers Meeting, Las Vegas, NV, September.

Hanson, M.E., G. D. Anderson, and R. J. Shaffer, 1978, "Theoretical and Experimental Research on Hydraulic Fracturing," paper 78-PET-49, presented at ASME Energy and Technology Conference, Houston, Texas, November. Also Lawrence Livermore National Laboratory Report UCRL-80558.

Hanson, M. E., G. D. Anderson, R. J. Shaffer, and L. D. Thorson, 1981, "Some Effects of Stress, Friction, and Fluid Flow on Hydraulic Fracturing," SPE/DOE Paper 9831 presented at Symposium on Unconventional Gas Recovery, Denver, CO., March. Also Lawrence Livermore National Laboratory Report UCRL-85003.

Hanson, M.E., G. D. Anderson, and R. J. Shaffer, 1980, "Effects of Various Parameters on Hydraulic Fracture Geometry," SPE Paper 8942 presented at Symposium on Unconventional Gas Recovery, Pittsburgh, PA, May.

Olkiewicz, A., J. E. Gale, R. K. Thorpe, and B. Paulsson, 1979, "Geology and Fracture System at Stripa," Lawrence Livermore National Laboratory Report LBL-8907, SAC-21, University of California, Berkeley, CA.

Shaffer, R. J., M. E. Hanson, and G. D. Anderson, 1980, "Hydraulic Fracturing Near Interfaces," presented at ASME Energy and Technology Conference, New Orleans, February 24, 1980. Also Lawrence Livermore National Laboratory Report UCRL-83419.

Thorpe, R.K., 1979, "Characterization of Discontinuities in the Stripa Granite - Time Scale Experiment," Lawrence Berkeley Laboratory Report LBL-7083, SAC-20, University of California, Berkeley, CA.

Thorpe, R.K., D. J. Watkins, W. E. Ralph, R. Hsu, and S. Flexser, 1980, "Strength and Permeability Tests on Ultra-Large Stripa Granite Core," Lawrence Berkeley Laboratory Report LBL-11203, SAC-31, University of California, Berkeley, CA, September.

Warpinski, N.R., S. J. Finley, W. C. Vollendorf, M. O'Brien, and E. Eshom, 1982, "The Interface Test Services: An In-situ Study of Factors Affecting the Containment of Hydraulic Fractures," Sandia National Laboratories Report SAND81-2408, February.

Wilder, D.G., an J. L. Yow, Jr., 1981, "Fracture Mapping at the Spent Fuel Test-Climax," Lawrence Livermore National Laboratory Report UCRL-53201, May.

STEP CRACKS

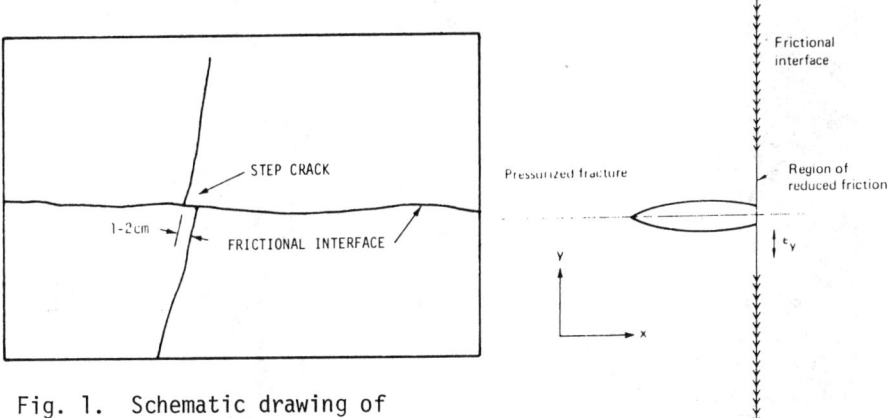

Fig. 1. Schematic drawing of step crack.

Fig. 2. Pressurized fracture intersects frictional interface.

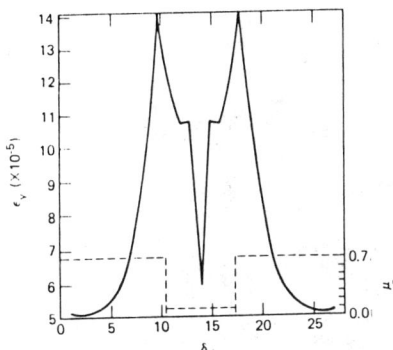

Fig. 3. Parallel strain for symmetric case.

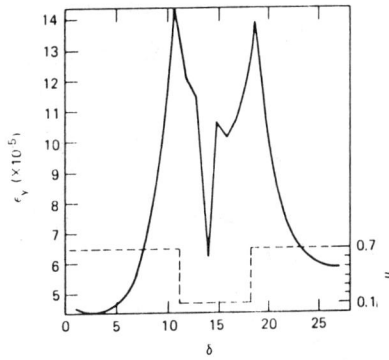

Fig. 4. Parallel strain for asymmetric case.

558 ISSUES IN ROCK MECHANICS

Fig. 6. Effect of a low friction region on crack growth across an unbonded, loaded interface.

Fig. 5. Hydraulic fracture - interface experimental setup.

STEP CRACKS   559

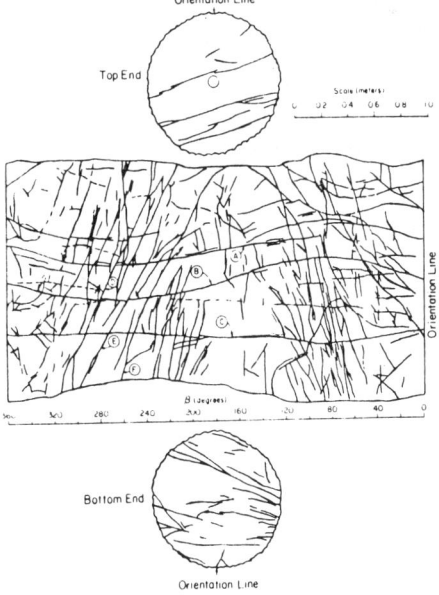

Fig. 7. Layout map of natural fractures on surface of 1 x 1.6 m granite cylinder.

Fig. 8. Surface of fracture B showing steps at intersection with fractures D, E, and F.

Chapter 58

CORRELATIONS BETWEEN FRACTURE ROUGHNESS CHARACTERISTICS AND
FRACTURE MECHANICAL AND FLUID FLOW PROPERTIES

Y. W. Tsang and P. A. Witherspoon

Earth Sciences Division
Lawrence Berkeley Laboratory
University of California
Berkeley, California   94720

ABSTRACT

Normal stress-fracture closure variation and stress-dependent fluid flow rates were calculated for rock fractures of different aperture distributions. This study shows that both the mechanical and hydraulic properties of the fracture are controlled by the large-scale roughness of the rock fracture. Hence the typical large-scale undulation wavelength of the rock joint dictates the suitable sample size to be used in normal stress and fluid-flow measurements, if results due purely to sample size are to be avoided.

INTRODUCTION

We present a study in which the fracture closure and fluid flow properties of the fracture when subjected to normal stress were correlated to the roughness characteristics of the fracture walls. The calculations for this investigation were based on an earlier model of ours (Tsang and Witherspoon, 1981) in which a mathematical expression was derived to relate the nonlinear effective Young's modulus of a rough-walled fracture to the fracture-roughness profile. In that earlier work we deduced from normal stress-displacement data of fractures, the fracture-roughness profile and in turn the fluid flow through such a fracture as a function of normal stress. The agreement obtained between the predicted flow and the measured flow lends validity to the theory. In the present study, we started with different fractures of known roughness profile and calculated their stress-displacement and stress-flow properties. The aim is to find correlations between the geometrical characteristics of the fracture roughness and the fracture mechanical and fluid-flow properties. Our investigation led to rather interesting conclusions regarding size effects.

FRACTURE ROUGHNESS CHARACTERISTICS

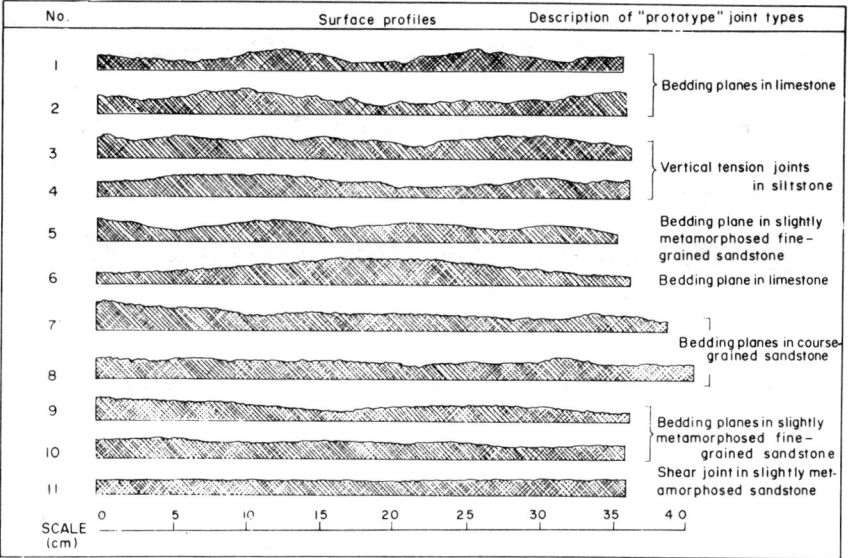

Figure 1. Selection of typical joint surface profiles (after Bandis, Lumsden, and Barton, 1981).

METHOD

Figure 1 is reproduced from Figure 9 of Bandis, Lumsden, and Barton (1981). It shows a selection of joint surface profiles from natural exposures of sandstone, siltstone, and limestone. Note that the profile of each joint surface is characterized by a large-scale undulation on which is superposed a small-scale roughness, whose average amplitude and wavelength are both much smaller than that of the large-scale undulation. These profiles range from rough undulation to almost smooth and planar. If each joint surface profile were to represent both the top and bottom halves of a fracture, then different fracture apertures may be simulated from each profile when the upper and lower joints are mismatched in varying degrees.

Figure 2 shows four different variations that were generated from the same joint surface profile (1 in Fig. 1). The fractures in Figure 2a,b,c,d were constructed by a horizontal displacement to the right of the top joint with respect to the lower joint by fractions varying from 0.013, 0.03, 0.056, 0.17, respectively, of the entire profile length; plus enough upward displacement of the top joint to ensure that there was no overlap between the upper and lower profiles.

In the earlier work (Tsang and Witherspoon, 1981), we used a void description of the fracture, that is, each fracture such as shown in Figure 2 can be considered as a collection of elongated voids, each

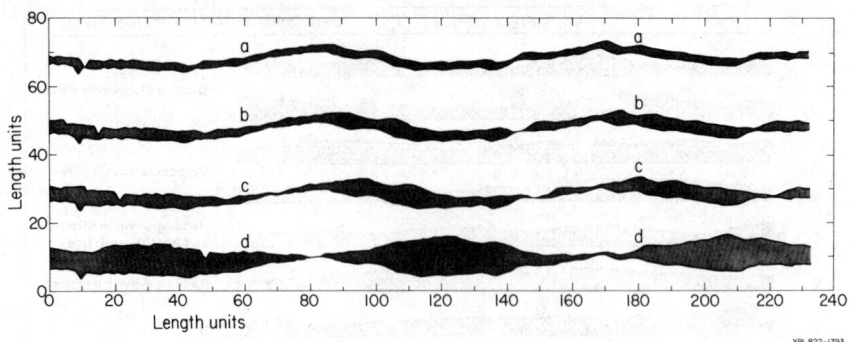

Figure 2. Four fractures generated from the surface profile number 1 of Figure 1. [XBL 822-1793]

with crack length 2d. The fracture closes under increasing normal stress, causing the crack lengths of the voids to shorten. The effective Young's modulus of the fractured jointed rock, E_{eff}, on the other hand, increases with increased normal stress on the fracture to approach the value of the constant Young's modulus, E, of the intact rock. A relationship between the average crack length over all the voids and the effective Young's modulus at each stage of applied stress was proposed as:

$$1 - \frac{E_{eff}}{E} \propto \langle d \rangle. \tag{1}$$

When measurements are made on intact rock over a thickness ℓ, then by definition,

$$E = \ell \frac{d\sigma}{d(\Delta V_r)} \tag{2}$$

and for a rock with a single fracture, one has

$$E_{eff} = \ell \frac{d\sigma}{d(\Delta V + \Delta V_r)} \tag{3}$$

where ΔV_r is the deformation of the intact rock and ΔV, the closure of the fracture. Now, if the modulus E of the intact rock is a constant, equations (2) and (3) reduce to

$$\int d\sigma = \frac{E}{\ell} \int \frac{E_{eff}/E \; d(\Delta V)}{(1 - E_{eff}/E)}. \tag{4}$$

FRACTURE ROUGHNESS CHARACTERISTICS

The integrand on the right-hand side of (4) can be computed from the geometry of the fracture profile. First, the fracture aperture and the profile length are both discretized into the same length units. Then the aperture density n(b), the number of discretized units with the aperture b, is computed. Initially, when fracture closure ΔV is zero, the upper and lower joints of the fracture have only one point of contact as shown in Figure 2. As ΔV is allowed to increase, more areas of the upper and lower joints come into contact, giving rise to shorter crack lengths since crack length is defined by the distance between two consecutive points of contact. At each increment of closure, all the length units of the fracture with apertures equal to or less than ΔV will have come into contact; consequently, for each ΔV the resultant shortened crack lengths and their average can be computed from n(b). The proportionality sign in (1) implies that a reference crack length at zero stress is needed. Since the entire length of profile 1 shown in Figure 1 contains approximately three large-scale undulation wavelengths, it is reasonable to choose for the reference the average crack length when three length units of the upper and lower joints are in contact. With this reference chosen, E_{eff}/E is calculated and the integration in (4) may be carried out. The presence of the arbitrary proportionality constant in (1) and E/ℓ outside the integrand in (4), implies that the above computation procedures can yield the dependence of normal stress σ versus fracture closure but not the absolute value of the stress level.

APERTURE DENSITY AND STRESS-DISPLACEMENT CURVES

The aperture density n(b) versus aperture b, and the normal stress σ versus fracture closure ΔV are plotted in Figures 3-6, corresponding respectively to the fractures shown in Figures 2a through 2d.

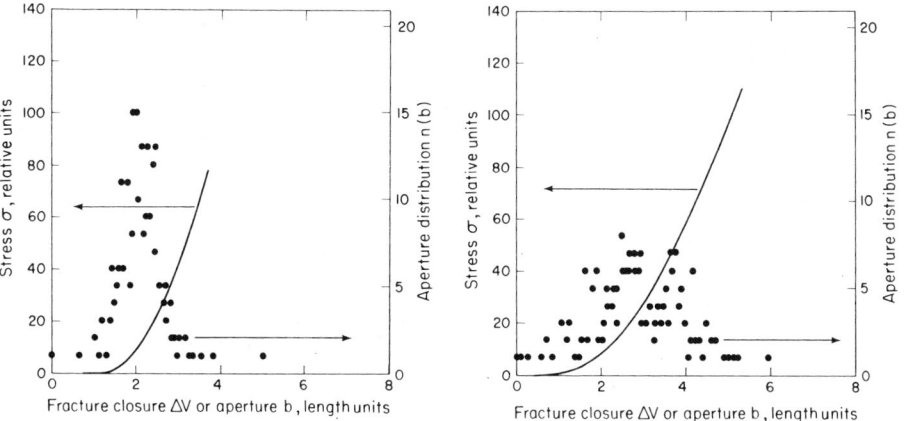

Figures 3 and 4. Fracture aperture distribution and stress-fracture closure variation for fracture 2a and fracture 2b.

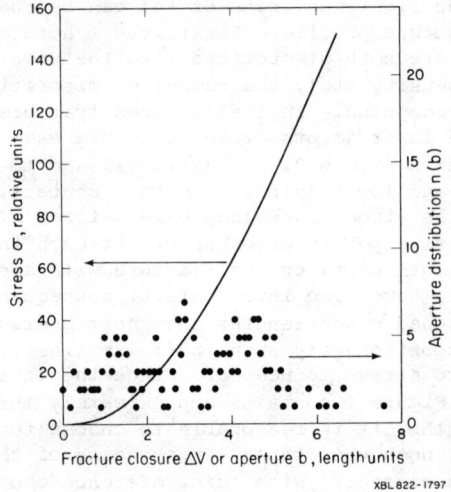

Figure 5. Fracture aperture distribution and stress-fracture closure variation for fracture 2c.

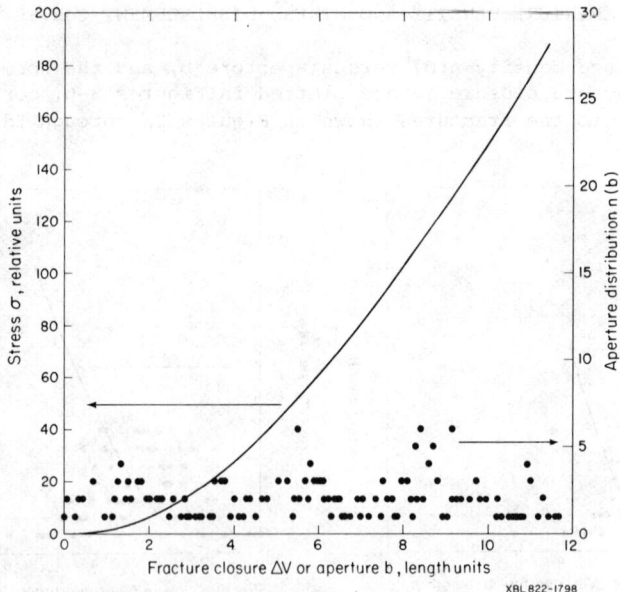

Figure 6. Fracture aperture distribution and stress-fracture closure variation for fracture 2d.

The same horizontal and vertical scales are used for the four figures so that a meaningful comparison may be made among them. Each aperture distribution shows a noisy background plus a discernable overall envelope. The noise originates from the small-scale roughness of the joint surface, but the overall shape of the envelope is controlled by the large-scale undulation of the joint. The envelopes range from a narrow peaked distribution on Figure 3 for the most well-matched fracture shown in Figure 2a to a flat, almost uniform distribution on Figure 6 for the most mismatched fracture shown in Figure 2d. The characteristics of the stress versus fracture closure curves may be correlated to the shape of the aperture distribution. The flat tail of the σ versus ΔV curve at small ΔV (Fig. 3) corresponds to low aperture density $n(b)$ at small b. Therefore, when the aperture density is appreciable at small b (Fig. 5), the flat tail of the σ versus ΔV curve, so prominent in Figure 3 has all but disappeared. As $n(b)$ peaks, with increasing ΔV, the fracture contact area increases, causing the average crack length to decrease and resulting in the steady increase of the slope $d\sigma/d\Delta V$. In the progression from Figure 3 through Figure 6, as the mismatch between the top and bottom joints increases, the aperture distribution broadens and the variation of σ with ΔV becomes less steep; or, in other words, softer mechanical property is predicted for the more mismatched fractures.

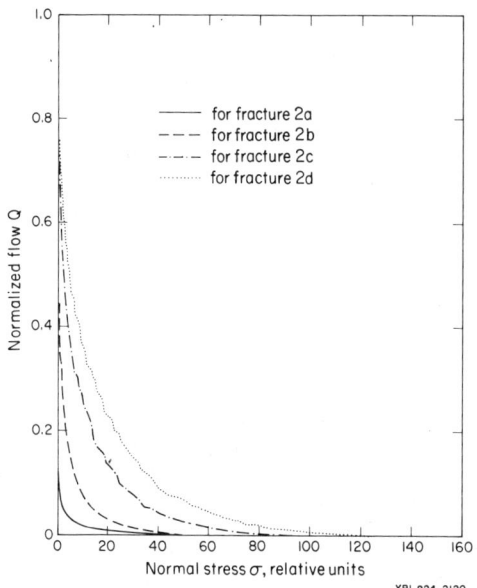

Figure 7. Flow-stress variation for the fractures 2a, 2b, 2c, and 2d.

FLOW VERSUS NORMAL STRESS

We plot in Figure 7 the fluid flow Q as a function of stress for the four fractures of Figure 2. The flow is normalized to 1 at zero stress and varies as (Tsang and Witherspoon, 1981)

$$\langle b^3(\Delta V)\rangle = \int_{\Delta V}^{b_o} n(b)(b - \Delta V)^3 db \qquad (5)$$

at each stress level corresponding to fracture closure ΔV; b_o denotes the maximum aperture of the fracture at zero stress. The sharp drop of the flow at small stresses corresponds to both the flat tail and the sharp rise in the aperture distribution $n(b)$. Therefore the flow drop with stress becomes progressively slower from case (a) to (d) as the rise in the aperture distribution becomes less steep from Figure 3 to Figure 6. The wiggles in the curves arise from the small-scale roughness of the fracture walls. The flow approaches zero asymtotically as the fracture closure becomes complete. Such asymtotic behavior had been observed (e.g., Iwai, 1976; Kranz et al., 1979)

CONCLUSION

This study leads to an understanding of the relationship between fracture roughness and the fracture mechanical and hydraulic properties when the fracture is subjected to normal stress. The large-scale roughness of a well-matched fracture is characterized mathematically by a narrow and peaked aperture distribution whereas that of an ill-matched fracture is characterized by an aperture distribution that is broad and flat. The small-scale roughness of the fracture wall contributes to the background noise of the overall shape of the aperture distribution. Since the features in σ versus ΔV and Q versus σ curves have been shown to correlate to the shape of the aperture distribution, it is the large-scale roughness of the fracture walls that controls the mechanical and hydraulic behavior of the fracture. Therefore, rock samples smaller than a typical large-scale undulation wavelength do not represent the fracture roughness properly, then spurious results due purely to 'size' can occur when such small samples are used in stress and flow measurements. Figure 1 shows that the undulation wavelengths range from about 12 cm for profiles 1, 5 and 10 to perhaps 70 cm in profile 6. Most other profiles have wavelengths on the order of 30 cm. If Figure 1 is a good sampling of typical rock profiles, then our study seems to suggest that in order to get a reliable database from which laboratory experimental results may be extrapolated to field situations, rock samples much larger than those normally employed in laboratory stress-strain and stress-flow measurements are called for.

ACKNOWLEDGMENTS

This work was supported by the Director, Office of Energy Research, Office of Basic Energy Sciences, Division of Engineering, Mathematics, and Geosciences of the U. S. Department of Energy under Contract DE-AC03-76SF00098.

REFERENCES

Bandis, S., Lumsden, A. C., Barton, N. R., 1981, "Experimental Studies of Scale Effects on the Shear Behavior of Rock Joints," *International Journal of Rock Mechanics and Mining Sciences*, Vol. 18, pp. 1-21.

Iwai, K., 1976, "Fundamental Studies of the Fluid Flow Through a Single Fracture," Ph.D. Thesis, University of California, Berkeley, 208 pp.

Kranz, R. L., Frankel, A. D., Engelder, T., and Scholz, C. H., 1979, "The Permeability of Whole and Jointed Barre Granite," *International Journal of Rock Mechanics and Mining Sciences*, Vol. 16, pp. 225-234.

Tsang, Y. W. and Witherspoon, P. A., 1981, "Hydromechanical Behavior of a Deformable Rock Fracture Subject to Normal Stress," *Journal of Geophysical Research*, Vol. 86, No. B10, Oct., pp. 9287-9298.

NUMERICAL MODELS

Chairman
Walter A. Palen
Sohio Petroleum Co.
San Francisco, California

Co-Chairman
Paul F. Gnirk
RE/SPEC Inc.
Rapid City, South Dakota

Keynote Speakers
Barry Brady
University of Minnesota
Minneapolis, Minnesota

Christopher St. John
Applied Mechanics Inc.
Grand Junction, Colorado

Chapter 59

THE ROLE AND CREDIBILITY OF COMPUTATIONAL METHODS IN
ENGINEERING ROCK MECHANICS

by B.H.G. Brady and C.M. St. John

Associate Professor, Dept. of Civil and Mineral Engineering
University of Minnesota, Minneapolis, MN 55455

Consultant, Los Angeles, California

ABSTRACT

Computational schemes for analysis of rock mass response to excavation, loading and other imposed changes, are employed pervasively in rock mechanics practice. Applications range in complexity from determination of stress and displacement distributions around openings, to prediction of the thermo-hydromechanical behaviour of a saturated fissured mass. While this may be taken as apparent sophistication in rock mechanics design activity, the indiscriminate use of computer methods may sometimes conceal inferior and inadequate engineering procedures. These inadequacies may not always be recognized, due to the general uncertainties which are inherent in rock mechanics investigations. This paper seeks to identify some common deficiencies in rock mechanics applications of computational schemes, and to propose ways of eliminating them. It also indicates some specific developments and applications of computational procedures, which are required to address persistent design problems in geomechanics practice.

INTRODUCTION

Rock mechanics has evolved rapidly as an engineering discipline in the last twenty-five years. To some extent, this has been a response to the needs posed by the increasing physical scale and complexity of surface and underground rock structures, and the financial risks associated with these engineering ventures. Notable examples of such motivating ventures are large gravity and arch dam projects, deep level mining, hazardous waste isolation and comprehensive hydroelectric and stream diversion projects. In these and other cases, improved engineering practice has been based on sound scientific investigation and greater understanding of the fundamental processes controlling the load-deformation behaviour of rock masses.

The emergence of rock mechanics as a coherent body of engineering science has been virtually contemporaneous with the widespread application of computers in scientific investigation and routine engineering activity. Thus, the period of increasing industrial demand for engineering rock mechanics skills has coincided with the dramatic expansion of analytical capacity associated with developments in computational mechanics. It is not surprising, therefore, that the techniques of numerical analysis are firmly embedded in design aspects of engineering rock mechanics. Certainly, the stage has been reached where a rock mechanics engineer, intimately involved in excavation design practice, can play a very limited role if not computationally literate. This implies, as a minimum, an understanding of the engineering principles exploited in various computational schemes, the valid range of conditions under which a particular scheme might be applied, and possible numerical difficulties that might be associated with particular algorithms, whether related to general numerical instability or machine dependence of the source code.

Even though the engineer must understand the principles and limitations of the computational tools applied, this need not be to a level expected of a specialist in computational mechanics. It must, however, be sufficient to provide an informed and rational basis for defining computational activities required to support any particular investigation, and for interpretation of the meaning and significance of any analytical results. Definition of appropriate computational activities requires assessments of the engineering significance of the work to be done and the quality of the engineering data available to support any analytical investigations. These assessments can only be made by a rock mechanics engineer since a specialist in computational mechanics will seldom possess sufficient understanding of the problem or physical insight to make the necessary judgements.

Finally, it may be observed that application of computational techniques poses certain practical problems in the result-orientated arena of engineering practice. Results can be obtained with relative ease, and at moderate cost, using various computational tools. These results are often of an unspecified quality, since limitations in the modelling activity or the engineering data on which it is based, are seldom well defined, or even understood. Indeed, it would be most unusual to find a set of performance objectives defined, with evidence provided that these objectives have been met. Lack of endeavor in these areas inevitably leads to loss of credibility for any numerical modelling effort, even where the tools of computational mechanics are applied correctly and appropriately.

COMPUTATIONAL METHODS IN ROCK MECHANICS 573

ROCK MECHANICS DESIGN METHODOLOGIES

Observational Method

Logical methodologies for excavation design in rock have developed in response to recognition of the basic problems which arise in design of structures in geologic media. These problems include: scale effects--the dependence of rock mass mechanical response to engineering activity on the geometric scale of applied loading; the difficulty of obtaining and testing representative, undisturbed specimens of the elements of a rock mass; and definition of ambient conditions, including the initial state of stress. Sampling of intact rock material is now reasonably routine, but sampling and testing a major structural feature is not. Similarly, the ambient state of stress in a rock mass is not as readily determined as a superficial evaluation might suggest. The general consequence of these and other limitations is that practical rock mechanics problems are frequently rather ill defined. The quality of the data base and limitations of analytical tools available for design calculations have been implicitly recognized in the traditional design methodology used in rock mechanics. Such a methodology is listed schematically in Fig. 1. This scheme identifies the various assumptions and simplifications which are necessary to render a design problem tractable. The logical steps include the development, from the site exploration data, of a geotechnical model of the problem domain. In this process, particular mechanical properties are assigned to lithologic units and rock structural features, geometric properties are ascribed to the structural features, and the in-situ state of stress is approximated by a local volume average of the stress tensor.

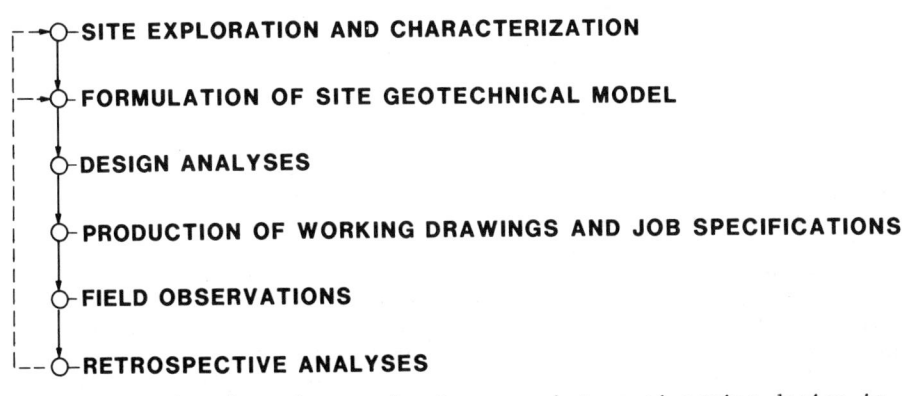

Fig. 1. Logic of an observational approach to engineering design in rock.

In design analyses using the established site model, the analytical schemes were, almost without exception, based on assumptions of simple constitutive behaviour, and simple excavation geometry. Thus, it was convenient to consider isotropic or transversely isotropic elastic behaviour of the rock mass. Plane strain solutions for excavations with circular, elliptical, or approximately square cross-sections, or which could be represented as parallel-sided slits, were then used to estimate stress and displacement distributions in the medium. These solutions were obtained from sources such as Sneddon (1946), Muskhelishvili (1953), Savin (1963), and Lekhnitskii (1963). More recently, these standard solutions were augmented by elementary computational techniques that allowed more complex geometrical configurations and somewhat more realistic constitutive descriptions to be incorporated within the model.

The mode of application of classical solutions and simple numerical schemes was mainly in parameter studies. These were used for determining the most advantageous shape, orientation and location of an opening, and to examine issues such as the effect of excavation sequence on the probable performance of an excavation or set of excavations. The latter topic is a particular concern in the design of a mining layout. Since the various types of solutions exploited continuous behaviour of the rock medium, implicit account could not be taken of the presence of discontinuities. However, it was possible to check to determine if the elastic stress distribution could be sustained throughout the problem domain. If the analysis suggested the slip could occur on a plane of weakness, qualitative but significant inferences could be drawn about the consequences of the inelastic response. For example, for the circular excavation shown in Fig. 2, extensive slip on the flat-lying feature reasonably could be expected to result in cracking in the crown of the opening.

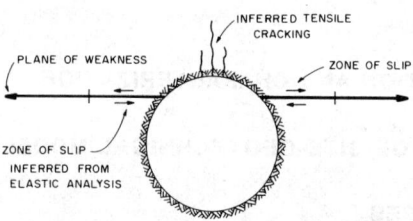

Fig. 2. Application of a simple analytical scheme for qualitative determination of effects of rock-mass non-linearities.

The essential component of the observational methodology was field observation of the practical performance of a rock structure, followed by retrospective analysis of the measured, in-situ rock mass response. These activities provided information for reformulating the geotechnical model of the problem domain, and a basis for unification of prototype performance with the analytical predictions generated from the geotechnical model. In particular, the key features and parameters controlling the deformation of the host rock medium induced by excavation could be identified, and their role incorporated in subsequent design analyses.

It is recognized that the logical scheme discussed above is little more than an implementation of the observational method suggested for soil mechanics practice. It is particularly appropriate in mining rock mechanics practice, where excavation activity may involve the repetitive development of structural units, such as pillars, stope spans and stope crowns, as mining proceeds through an orebody. Application of the method in non-mining rock mechanics is illustrated by Starfield and McClain (1973), in preliminary studies of nuclear waste isolation in salt. In this case, deficiencies in site characterisation and the analytical scheme for excavation design were mitigated by execution of pilot scale in-situ tests. These were used to generate salt mass creep parameters by retrospective analysis, which might have been confirmed subsequently by monitoring the performance of a full-scale facility at the site of the in-situ tests.

Appropriate Methodology for Advanced Computational Methods

The observational method generally employs deliberately simple analytical schemes, and a design engineer is therefore forced to maintain an awareness of fundamental limitations in the results of design analyses. Within the last five years, computer codes have become widely available which incorporate more general constitutive laws for a rock mass. These include various modes of non-linear response, several types of time-dependent behaviour, various levels of coupling of mechanical, thermal and hydraulic response, and non-dilatant and dilatant slip on discontinuities (National Academy of Sciences, 1981). The inference is that the level of complexity that can be introduced to represent the deformational response of a loaded rock mass, can provide a capacity, in principle at least, to undertake more significant design analyses. This assertion is not demonstrably true. In fact, it is likely that advanced computational techniques have had, to date, relatively little impact on the quality of the data generated in any design exercise. The reasons for this assertion are as follows.

A logical methodology which would justify the design application of an advanced computational scheme, and properly exploit its capacity, is illustrated in Fig. 3. It is seen that it is necessary to retain the six key steps of the observational method in this new strategy. This is so since the introduction of a more sophisticated analytical system does nothing to eliminate the need to develop a simplified geotechnical description of the actual site. In other words, the rock mechanics engineer retains responsibility for defining the problem in a tractable form. The rationale advanced in Fig. 3 also indicates that computer codes, which can incorporate a wide range of constitutive relationships, are developed without any consideration of specific site conditions, rock mass properties, and the like. Indeed, most computer codes should be regarded as general purpose tools that may be used as a basis for development of geotechnical models of particular sites or systems.

The process of code development follows a well established routine. It involves, firstly, construction of an algorithm relating problem

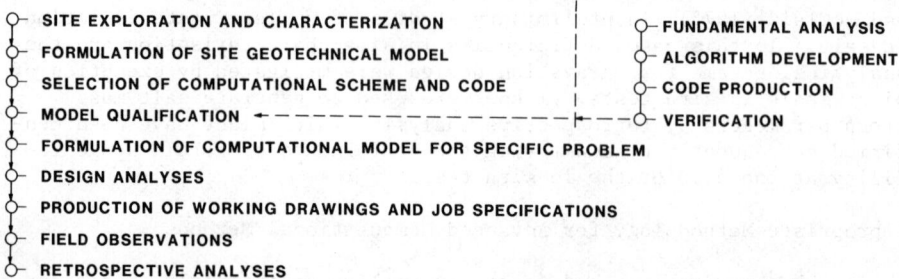

Fig. 3. A logical methodology for semi-quantitative application of advanced computational schemes in rock mechanics.

geometry, initial conditions, boundary conditions, constitutive behaviour, and stress and displacements fields in the medium. This is followed by implementation of the algorithm as a sequence of computer instructions, and elimination of coding errors and errors in logic from the solution procedure. Production of user information, consisting of code listing, input data specification and specimen problems, is often considered to constitute completion of code development. In such a state, a computer code has no value in rock mechanics practice. It should not be used at all until every effort has been made to ensure that the code is computationally correct for all circumstances under which it will be used. This process, which includes comparison with analytical solutions and results obtained using other codes, preferably based on different computational principles, is termed code verification.

Once the code has been verified, it must be qualified for application to particular classes of problems. The process of qualification involves comparison with observed behaviour of real geologic materials and typical geologic systems. Interestingly, it is not only the code which needs to be qualified, since the code cannot be used in isolation from the methodology that leads from site investigation through development of the model incorporating a physical description of the system and a quantitative constitutive description. Accordingly, the term model, rather than code, qualification will be used in the following discussion of activities constituting verification and qualification.

CODE VERIFICATION

The objective in code verification is to demonstrate, to a suitable level of tolerance, correspondence between the code's solution to a particular problem and an independent solution to the same problem. The level of effort required for confirmation of satisfactory code performance is clearly related to the complexity of the constitutive behaviour incorporated in the computational scheme. For example, if a scheme purports to represent material creep, the only credible verification of code performance is the analysis of a range of problems, subject to various types of loading, for which the solutions have been obtained by some completely unrelated procedure. The aim is to ensure that the verification tests result, in the computer analysis, in execution of all segments of the code, and evaluation of all components of the logic. The definition of suitable test problems for code verification must therefore be based on detailed understanding of the code structure. It is unfortunate that it is frequently only the writer of a code who is sufficiently familiar with the subtleties of a code's construction, to design or undertake a comprehensive program of code verification tests. There is also seldom any recognition by the user community of the positive, indispensable contribution such work makes to the functional validity of subsequent code applications.

The simplest codes to verify are those based on linear constitutive behaviour, such as linear elasticity. Uniqueness of solution (independence of solution path) and the existence of an ample number of classical solutions for simple problem geometries, present the possibility of simple verification tests. Even in this case, problems arise. One problem is associated with the implied necessity to demonstrate close correspondence between the computational and independent solutions to a problem. For example, suppose a verification study for a Boundary Element code indicated that 100 boundary elements disposed around a circular hole produced satisfactory agreement with the Kirsch solution. A reasonable interpretation of this result is that the code does not perform satisfactorily, if it requires such a large number of elements to handle such a simple problem geometry. Demonstration of tolerable correspondence with an analytical solution, with a sensible number of boundary elements, constitutes in this case a more valid confirmation of satisfactory code performance.

A second problem in verifying linear codes is that the geometries of problems for which closed form analytical solutions exist are symmetric. When these are used to assess the performance of a code, correspondence between the numerical and independent solutions is a necessary but not sufficient condition for verification. The risk is that symmetric problems can be processed in a code with self-cancelling, internal errors. Undetected errors would then appear when the code was applied to unsymmetric problems. There appears to be no simple solution to this problem, other than to employ another code capable of solving the same problem, but based on an entirely different analytical procedure.

Verification of codes incorporating non-linear constitutive behaviour presents acute difficulties. Although classical analytical solu-

solutions are available for rigid-perfectly plastic media, these materials may bear little resemblance to the complex hardening/softening media encountered in rock mechanics. Studies of these simple problems are therefore merely interim steps in a process that is required to be much more comprehensive. In the absence of classical analytical solutions for the more complex types of media, the only logical approach is to conduct controlled physical tests. The test specimen should obviously be constructed from a material for which a generalised form of the constitutive behaviour has been incorporated in the subject code. The execution of these tests would involve the application of problem initial, boundary, and loading conditions which could be duplicated precisely in the computational analysis. Verification of the satisfactory performance of a code would then require correspondence, to an agreed tolerance, between the macroscopic deformation behaviour of the test specimen and that determined from the computational scheme. Careful attention would need to be given to application of an exhaustive stress path, including components such as cycles of loading and unloading. Tests would also need to be devised to determine the significance of computation path on the solution to particular types of problems which involve, in the computational scheme, imposition of an arbitrary route to the final state.

It is reasonable at this point to question whether field tests can constitute a basis for code verification. It is our position that field experiments can make virtually no contribution in code verification because they represent an extra level of complexity, compared with laboratory experiments, due to the poor definition of experimental parameters. Also, apart from basic questions of control of operating variables, there are physical limitations on reproducibility. Any rationally designed experimental program generates information on the reproducibility of the product data, and its relationship to the fundamental properties describing system behaviour. The virtual uniqueness of any field experimental site would preclude the replicative experiments which are needed to assure the intrinsic and absolute quality of the product data set.

The preceding discussion indicates that the role and value of physical experiments for code verification is not generally understood in the rock mechanics community. For the codes which model non-linear behaviour, where the need for verification experiments is most acute, there has been very little work which recognizes the extent of the problem. Some useful examples which contradict this assertion are given by Stewart (1981) and Hart (1981).

MODEL QUALIFICATION

The issue of model qualification is concerned with establishment of the acceptability of a particular computational scheme and modelling methodology for a definite design exercise. Since the scheme is to be applied to a site-specific problem, the obvious first requirement is that the features to be incorporated in the site model be represented in properly verified segments of the candidate code. The decisive requirement is that, for a given set of properly determined site parameters, the model can predict the response of the rock medium to some controlled perturbations, to some prescribed tolerance. As an example, in the preliminary phases of the design of a hydroelectric machine hall, measurements and estimates would be made of the rock mass properties and the in-situ state of stress. The qualification exercise would involve computational and experimental determinations of stresses and displacements around a trial excavation in the rock mass. Correspondence between the observations and predictions, within the tolerance produced by input data uncertainty, would constitute qualification of the modelling approach, for the subsequent, principal design activity.

In the execution of a model qualification experiment, there can be no question of using the qualification experimental data to generate site data for use in subsequent computational analysis--in fact the opposite is true. All site data, from site characterisation and from qualification experiment response, must be rigorously exclusive. Levels of confidence should be ascribed to each of the rock mass parameters determining its response to the imposed, experimental perturbation, and bounds of possible response estimated using the model. If these bounds bracket the measured response, as shown in Fig. 4, the modelling methodology and computational scheme can be considered to be qualified for application in circumstances where important system characteristics, such as medium type and degree and nature of fracturing, are similar to those of the qualification experiment.

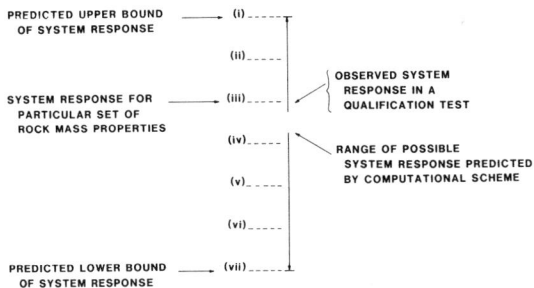

Fig. 4. Criterion for qualification of a computational methodology for a specific site, by predictions bounding the observed response of a test site.

Clearly, there are many circumstances in which the nature of the problem under investigation precludes comprehensive experimental qualification. For example, the time scale of concern when considering geologic isolation of hazardous wastes may preclude addressing some aspects of system performance within any realizable experimental program. In such circumstances, qualification should include demonstration that the results of modelling investigation are consistent with historical experience in similar or analogous situations.

SOME PERSISTENT ROCK MECHANICS PROBLEMS RELATED TO COMPUTATIONAL MECHANICS

The preceeding discussion was concerned with general issues concerning the application philosophy, formulation and credible validation of computational schemes for design in rock. There are several other computational mechanics issues which currently remain either unresolved in principle or unconsidered in practice, which deserve immediate attention.

Problem Geometry and Principal Stress Orientations

Two-dimensional problem geometry is a useful working hypothesis in many design exercises, due to the efficiency and economy of the associated analytical work. Most codes for the analysis of stress and displacement distributions around excavations which can be represented by two-dimensional geometry assume that the long axis of the excavation is parallel to a field principal stress direction. Such an assumption is neither justified nor necessary. For the general problem illustrated in Fig. 5, the long axis of the excavation is non-coincident with a field principal stress direction, and any point in the medium is subject to excavation-induced displacements u_x, u_y, u_z. The essential notion in the assumption of two-dimensional geometry is that conditions are identical in all planes perpendicular to the long axis of the excavation; i.e., for the chosen reference axes,

$$u_x = u_x(x, z), \quad u_y = u_y(x, z), \quad u_z = u_z(x, z) .$$

It follows that all six induced stress components are, in general, non-zero.

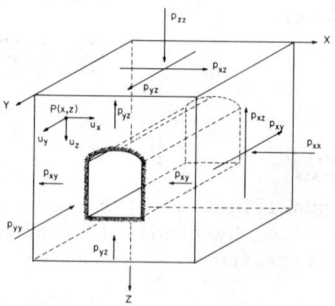

Fig. 5. For a long opening with axis non-parallel to a field principal stress, displacements at any point P consist of plane components u_x, u_z, and antiplane component u_y.

The problem usually solved in typical two-dimensional stress analysis is the plane problem, involving displacements u_x, u_z in the plane perpendicular to the excavation axis. The other component of the problem, called the antiplane problem by Filon (1937), involves the displacement u_y parallel to the excavation axis. The combination of the plane and antiplane problems was called the Complete Plane Strain by Bray (1976) (see Brady and Bray, 1978). For isotropic elasticity, and transversely isotropic elasticity where the excavation lies in, or is perpendicular to, the plane of isotropy, the plane and antiplane problems are decoupled, and can be analysed separately. For all other elastic problems (Goodman and Amadei, 1981), and for all problems involving non-linear constitutive behaviour, the plane and antiplane problems are coupled. In these cases, grossly misleading results are to be expected if the antiplane component of a problem is arbitrarily ignored. However, the antiplane component can be incorporated by simple modification of conventional two-dimensional codes (St. John, 1977) or by suppression of appropriate degrees of freedom in three-dimensional representations (Wittke, 1977).

To illustrate the significance of the antiplane problem, it is useful to consider the relative effects of ignoring and including the antiplane component in a two-dimensional analysis, for a simple excavation geometry. Fig. 6(a) shows a circular excavation with its axis inclined at various angles to a field principal stress axis. Fig. 6(b) indicates the magnitude of sidewall stress, and crown stress, for a particular field principal stress ratio. When the excavation axis is within about 10° of the principal stress axis, boundary stresses obtained from plane strain and complete strain analyses are virtually identical. Thereafter, the simple plane strain analysis greatly underestimates the boundary stresses. A reasonable conclusion is that it is rarely valid to ignore the antiplane problem in any two-dimensional analysis.

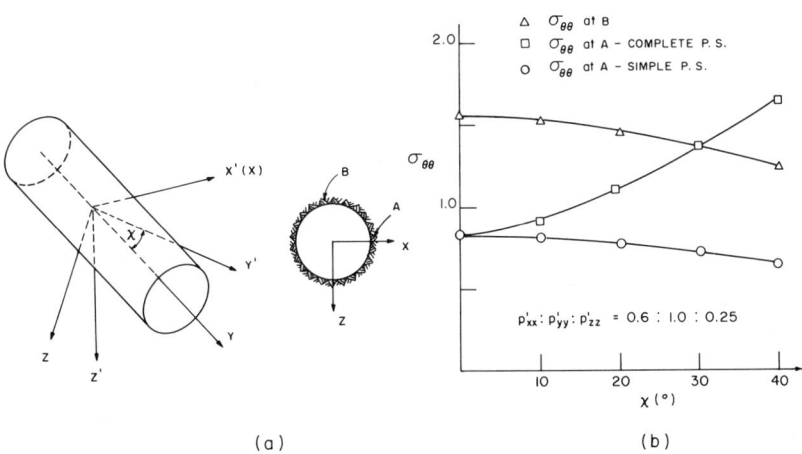

Fig. 6. Problem geometry and maximum and minimum boundary stresses around a circular hole in a triaxial stress field, for complete plane strain and simple plane strain analyses.

Hybrid Computational Schemes

It has been observed (NAS, 1981) that Boundary Element methods of stress analysis offer computational efficiency, but are effectively limited to linear material behaviour. Differential methods, on the other hand, model complex constitutive behaviour, but the physical size of a problem domain that can be accommodated is necessarily restricted. Hybrid computational schemes, involving a domain modelled by some differential method embedded in a Boundary Element domain, eliminate the individual disadvantages of the separate methods, and offer positive advantages. These include, firstly, elimination of uncertainties about the location of, and applied conditions on, the outer boundary of the differential domain. Secondly, far-field behaviour can be represented appropriately and efficiently as a linear continuum. Finally, zones of complex constitutive behaviour in a rock structure are usually small and localized, so that these zones only require the versatility conferred by a differential formulation. This implies that computational efficiency is promoted directly by limiting the extent of the differential domain. The overall product of hybrid codes is more realistic computational representation of a site model, and greater credibility of the results of predictive analyses of the performance of the prototype.

An example of a problem for which a hybrid scheme is most appropriate is the design of a subsurface nuclear waste repository in an evaporite medium. The scale of the analytical problem in this case is such that even simple design exercises using a Finite Element code can lead to development of a completely absurd computational model of the prototype. Some F.E. analyses of this problem conducted to date have required that the pillars between repository galleries be each represented by a single finite element. This is equivalent to imposing linear variation of stress in a domain which is to be subject to the concentrated thermomechanical loading associated with waste canister emplacement. Such a crude representation cannot be expected to yield any but the coarsest predictive data on the response of the medium to repository construction and operation. Indeed, development of such a model might legitimately be considered as an inappropriate application for a finite element code.

The solution to the large-scale computational problem presented by repository design in salt would appear to be the development of a hybrid Finite Element-Boundary Element scheme. The far-field rock would be represented by a Boundary Element formulation for a linear, thermo-viscoelastic half space. Selected zones of the repository horizon and its immediate environs, constituting the repository near-field, would be represented by a Finite Element or Finite Difference formulation, incorporating large-strain, geometrically non-linear creep of the salt. Embedment of the differential domain in the subsurface of the half space, as shown in Fig. 7, would be based on conditions assuring continuity of stress, displacement, temperature and flux across the interface between the two domains.

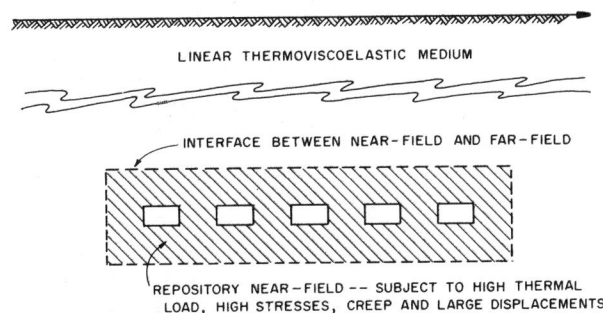

Fig. 7. Conceptual scheme for effective computational treatment of problems posed by design of a nuclear waste repository in salt.

In spite of the engineering value, algorithmic simplicity and increased public credibility to be derived from hybrid computational schemes, there does not appear to be a very determined effort to realize these advantages. However, there are exceptions (Kelly, et al. 1979; Brady and Wassyng, 1981). Some work is known to be in progress on hybrid schemes for a boundary element half space, linear problems and with inclusions modelling complex constitutive behaviour of various types.

Mine Structural Stability

A mine structure rarely resembles the orderly layout of rooms and pillars usually depicted for mining stratiform orebodies. The design of these complex mining layouts demonstrates a clear need for the hybrid codes described previously. An additional need in mine design is to assure the stability of the rock structure generated by mining activity. The specific problem in mining, not encountered in other aspects of rock mechanics practice, is that realization of the maximum economic potential of an orebody inevitably leads to extensive failure in the host rock mass. The problem of instability arises when a small perturbation in the equilibrium condition in a structure can provoke a sudden and large change in its geometry.

Mine instability is not a problem uniquely related to deep mining, as is apparently assumed. It occurs naturally at any depth when high extraction ratios result in stresses that exceed the rock mass strength. Some examples are given by Tincelin and Sinou (1960), Bryan, et al. (1964), and Blake (1972). Many other examples are known. Currently, the principles of a computational approach to determination of mine stability are well established, but there is little motivation for its implementation. The particular requirement is to formulate a scheme for handling the bifurcation/localization mode of rock mass rupture, such as that described by Rudnicki and Rice (1975).

Slip on Planes of Weakness

Slip on a structural feature, such as a fault, is a natural mode of response which must be incorporated in codes modelling rock mass response to some perturbation. Most existing Boundary Element and Finite Element Difference codes can model slip of only elastic orders of magnitude. This presents no conceptual difficulty, since, in massive rock with a limited number of transgressive features, slip magnitudes of elastic orders of displacements are sufficient to eliminate excess shear stress. A study by Holman (1982) suggests that more basic problems exist. Most Boundary Element formulations for slip do not satisfy the boundary conditions around an excavation intersected by a fault, and underestimate the extent of slip.

Current Finite Element methods of modelling slip do not suffer from the basic deficiencies of Boundary Element methods. Their main disadvantage seems to be the concentration of elements required to achieve sufficient mechanical freedom near a plane of weakness to accept a high local displacement gradient.

Site Characterisation

Determination of the in-situ mechanical properties of a rock mass usually proceeds by measuring the response of the medium to some local, applied load, and applying an analytical solution derived from a simple model of rock mass behaviour. This usually results in unduly restrictive assumptions concerning the properties of the rock mass, such as homogeneity, anisotropy and constitutive behaviour. Frequently, isotropic elastic behaviour is assumed in the retrospective analysis of site characterisation tests.

The application of computational methods to characterisation problems is well established in structural mechanics (Kavanagh, 1973; Maier, et al., 1977). In rock mechanics, Cividini, et al. (1981) demonstrated the application of a Finite Element code, coupled with a Simplex algorithm, to the analysis of a large scale surface loading experiment. The obvious value of this type of approach is that a more realistic site model can be formulated. It also means that some level of confidence can be attached to the calculated values of the site parameters, if the number of measurements in the characterisation test is sufficient to produce an overdetermined system.

CONCLUSIONS

The evolution of rock mechanics as an engineering discipline has been accompanied by significant improvement in the capacity to analyse rock response to imposed loading. This capacity is mainly due to the development of powerful computer methods. Exploitation of the maximum value of these enhanced computational schemes requires a sound appreciation of conditions for their valid application in engineering practice. A logical philosophy needs to be established for precise verification of the performance of codes, particularly those which incorporate non-linear constitutive behaviour or coupled modes of response. Criteria must also be defined which qualify a particular modelling procedure for application to a specific design problem. Such criteria must be related to the degree of certainty with which the rock mass conditions at the site of interest can be determined by site characterisation procedures. The development of such a logical philosophy must surely be of paramount importance if numerical models are to play a credible role in rock mechanics practice.

It has been suggested that there are a number of areas in which further development and application of computational schemes are justified immediately. In none of the cases discussed does there appear currently to be sufficient motivation and support to undertake the required work to achieve the related improvements in rock mechanics design practice. The list of issues for further attention is obviously not definitive, and there are undoubtedly others that need to be resolved to meet the current engineering needs for computational schemes.

REFERENCES

Blake, W., 1972, "Rock-burst Mechanics," Quarterly of Colorado School of Mines, Vol. 67, No. 1, 64 pp.

Brady, B.H.G., and Bray, J.W., 1978, "The Boundary Element Method for Determination of Stresses and Displacements around Long Openings in a Triaxial Stress Field," Int. J. Rock Mech. Min. Sci. & Geomech. Abstr., Vol. 15, pp. 21-28.

Brady, B.H.G., and Wassyng, A., 1981, "A Coupled Finite Element-Boundary Element Method of Stress Analysis," Int. J. Rock Mech. Min. Sci. & Geomech. Abstr., Vol. 18, pp. 475-485.

Bryan, A., Bryan, J.G., and Fouche, J., 1964, "Some Problems of Strata Control and Support in Pillar Workings," Mining. Eng., Vol. 123, pp. 238-254.

Cividini, A., Jurina, L., and Gioda, G., 1981, "Some Aspects of 'Characterization' Problems in Geomechanics,'" Int. J. Rock Mech. Min. Sci. & Geomech. Abstr., Vol. 18, pp. 487-503.

Goodman, R.E., and Amadei, B., 1981, "Formulation of Complete Plane Strain Problems for Regularly Jointed Rocks," Proc. 22nd U.S. Rock Mechanics Symposium, M.I.T., pp. 245-251.

Hart, R.D., 1981, "A Fully-Coupled Thermal-Mechanical-Fluid Flow Model for Non-Linear Geologic Systems," Ph.D. Thesis, Univ. of Minn.

Holman, W.E., 1982, M.Sc. Thesis, Univ. of Minn. (in preparation).

Kavanagh, K.T., 1973, "Experiment Versus Analysis: Computational Techniques for the Description of Static Material Response," Int. J. Numer. Meth. Eng., Vol. 5, pp. 505-515.

Kelly, D.W., Mustoe, G.G.W., and Zienkiewicz, O.C., 1979, "Coupling Boundary Elements with Other Numerical Methods," Chapter 10 of Recent Developments in Boundary Element Methods I, Applied Science Publishers, Essex, U.K.

Leknitskii, S.G., 1963, Theory of Elasticity of an Anisotropic Elastic Body, Holden-Day, San Francisco.

Maier, G., Grierson, D.E., and Bast, M.J., 1977, "Mathematical Programming Methods for Deformation Analysis at Plastic Collapse," Comput. Struct., Vol. 7, pp. 599-612.

Muskhelishvili, N.I., 1953, Some Basic Problems in the Mathematical Theory of Elasticity, Nordhoff, Holland.

National Academy of Sciences, 1981, "Numerical Modeling," in Rock Mechanics Research Requirements for Resource Recovery, Construction, and Earthquake Hazard Reduction, National Academy Press, Washington, D.C.

Rudnicki, J.W., and Rice, J.R., 1975, "Conditions for the Localization of Deformation in Pressure-Sensitive Dilatant Materials," J. Mech. Phys. Solids, Vol. 23, pp. 371-394.

Sneddon, I.N., 1946, "The Distribution of Stress in the Neighbourhood of a Crack in an Elastic Solid," Proc. Roy. Soc. A, Vol. 187, pp. 229-260.

Starfield, A.M., and McClain, W.C., 1973, "A Case Study in Rock Mechanics--Laboratory and Large-Scale In-Situ Tests and a Theoretical Model for the Feasibility of Radioactive Waste Disposal," Int. J. Rock Mech. Min. Sci. & Geomech. Abstr., Vol. 10, pp. 641-656.

Stewart, I.J., 1981, "Numerical and Physical Modelling of Underground Excavations in Discontinuous Rock," Ph.D. Thesis, Univ. of London.

St. John, C.M., 1977, "An Analysis of Deep Boreholes in Hot, Dry, Crystalline Rock," Proc. 18th U.S. Symposium on Rock Mechanics.

Tincellin, E., and Sinou, P., 1960, "Collapse of Areas Worked by the Small Pillar Method. Practical Conclusions and an Attempt to Formulate Laws for the Phenomena Observed," Proc. 3rd Int. Conf. Strata Control, pp. 571-588, Paris.

Wittke, W., 1977, "Static Analysis for Underground Openings in Jointed Rock," in Numerical Methods in Geotechnical Engineering, Ed. C.S. Desai, J.T. Christian, McGraw-Hill.

Chapter 60

FINITE ELEMENT EVALUATIONS OF
THERMO-ELASTIC CONSOLIDATION

by B.L. Aboustit, S.H. Advani, J.K. Lee and R.S. Sandhu

College of Engineering
The Ohio State University
Columbus, Ohio 43210

ABSTRACT

A comprehensive finite element formulation for thermo-elastic consolidation based on a general variational principle is presented. Sample one dimensional problems associated with isothermal consolidation, heat conduction, and thermo-elastic consolidation are presented. In addition, evaluations of thermo-elastic consolidation for an underground coal conversion field site are conducted.

INTRODUCTION

The finite element method has been extensively used in solving problems of isothermal consolidation (Sandhu and Wilson, 1969; Hwang et al, 1971). For various energy extraction applications such as underground coal conversion and oil shale retortion, however, the formation is subjected to both in situ stresses and thermal gradients. The associated thermo-elastic consolidation modeling with coupled pore pressure and temperature effects is still in its infancy. Schiffman(1971) used the dynamical theory of interacting continua developed by Green and Nagdhi(1965) and consolidation theory by Biot(1955) to present a simplified theory of thermo-elastic consolidation. Weres(1979) considered the concentration of the solute in lieu of the pore pressure in his finite element solution of the thermo-mechanical mixture problem. Witherspoon et al(1981) employed the variational principle for the poroelastic component of the hydrothermo-elastic phenomenon in which the temperature effects were introduced physically, and the Galerkin formulation was employed to solve the energy equation.

In this paper, a finite element model framework for thermo-elastic consolidation response simulation is presented. The de-

developed variational principle represents a generalization of the results presented by Sandhu(1976) for isothermal consolidation and formulations on thermo-elasticity developed by Nickell and Sackman (1968). Numerical results for idealized one dimensional models and a field problem are presented.

GOVERNING FIELD EQUATIONS AND VARIATIONAL PRINCIPLE

The governing field equations are characterized by

(i) the equilibrium equations for the solid-fluid mixture

$$(\sigma_{ij} + \Pi\delta_{ij} - \beta T\delta_{ij})_{,j} + \rho f_i = 0 \qquad (1)$$

(ii) the stress-strain and strain-displacement relations for the linear, homogeneous elastic solid matrix

$$\sigma_{ij} = E_{ijkl}\,\varepsilon_{kl} \qquad (2)$$

$$\varepsilon_{ij} = 1/2(u_{i,j} + u_{j,i}) \qquad (3)$$

(iii) Darcy's law for irrotational fluid flow

$$q_i = K_{ij}(\Pi_{,j} + \rho_2 f_j) \qquad (4)$$

(iv) the continuity equation for a non-chemically reacting continuum with the solid skeleton fully saturated by an incompressible fluid

$$q_{i,i} = -\dot{\varepsilon}_{ii} \qquad (5)$$

(v) Fourier's law of heat conduction

$$h_i = -k_{ij}\,T_{,j} \qquad (6)$$

(vi) an energy equation for the solid-fluid mixture with convection ignored

$$h_{i,i} + \rho C_v \dot{T} + \beta \tau_o \dot{\varepsilon}_{ii} = 0 \qquad (7)$$

A variational principle of the form, obtained by Gurtin(1964), equivalent to the preceding equations with appropriate boundary and initial conditions, is

$$J[u,\Pi,T] = \int_R [-2u_i*\rho f_i + \varepsilon_{ij}*\sigma_{ij} - K_{ij}(\Pi_{,j}+\rho_2 f_j)*g'*(\Pi_{,i}+\rho_2 f_i)$$

$$-\frac{k_{ij}}{\tau_o} T_{,j}*g'*T_{,i} + 2\Pi*(u_{i,i} - u_{i,i}(0)) - 2\beta T*(u_{i,i}-u_{i,i}(0)) \qquad (8)$$

$$- T*\frac{\rho c_v}{\tau_o}(T-T_o)]dR - 2\int_{S_2} u_i*\hat{t}_i dS + 2\int_{S_4} \Pi*g'*\hat{Q}dS - 2\int_{S_6} T*g'*\frac{\hat{H}}{\tau_o} ds$$

with the forced boundary conditions $u_i(x,t) = \hat{u}_i(x,t)$ on S_1, $\Pi(x,t) = \hat{\Pi}(x,t)$ on S_3 and $T(x,t) = \hat{T}(x,t)$ on S_5. Vanishing of the first variation of Eq. 8 implies the satisfaction of the field equations (1),(5),(7) along with the natural conditions $\hat{t}_i = (\sigma_{ij} + \Pi\delta_{ij} - \beta T\delta_{ij})n_j$ on S_2, $\hat{Q} = q_i n_i$ on S_4, and $\hat{H} = h_i n_i$ on S_6.

FINITE ELEMENT FORMULATIONS

The element displacements, pore pressure, and temperature are defined in terms of a set of generalized coordinates. Vanishing of the spatial variation results in a set of first order linear differential equations in time and further discretization in the time domain yields a set of linear algebraic equations. The spatial discretization is expressed in the form

$$u_i^m = [N_u^m(x)]\{u(t)\}, \quad \Pi^m = \{N_\Pi^m(x)\}\{\Pi(t)\}, \quad T^m = \{N_T^m(x)\}\{T(t)\} \quad (9)$$

The corresponding strains, volumetric strain, pressure and temperature gradients are

$$\varepsilon^m(x,t) = [N_\varepsilon^m(x)]\{u(t)\}, \quad \Delta^m = \{N_\Delta^m(x)\}\{u(t)\} \quad (10a)$$

$$\nabla\Pi^m(x,t) = [N_q^m(x)]\{\Pi(t)\}, \quad \nabla T^m = [N_h^m(x)]\{T(t)\} \quad (10b)$$

Substituting Eqs. 9 and 10 in Eq. 8 and applying the extremization principle with respect to u, Π, T, we obtain the discretized form of the equilibrium, continuity and energy equations respectively

$$[K_{uu}]\{u\} + [K_{pp}]\{\Pi\} - [K_{Tu}]\{T\} = \{M_1\} + \{M_3\} \quad (11)$$

$$[K_{pu}]\{u\} - g'*[K_{pp}]\{\Pi\} = g'*\{M_2\} - g'*\{M_4\} \quad (12)$$

$$-[K_{Tu}]\{u\} - [C_{TT} + g'*K_{TT}]\{T\} = g'*\{M_5\} \quad (13)$$

where

$$[K_{uu}] = \sum^M \int [N_e^m][E^m][N_e^m]^T dR_m, \quad [K_{pu}] = \sum^M \int \{N_\Delta^m\}\{N_\Pi^m\}^T dR_m$$

$$[K_{Tu}] = \sum^M \int \beta \{N_\Delta^m\}\{N_T^m\}^T dR_m, \quad [K_{pp}] = \sum^M \int \{N_q^m\}[K^m]\{N_q^m\}^T dR_m$$

$$[K_{TT}] = \sum^M \int \{N_h^m\}[k^m/\tau_0]\{N_h^m\}^T dR_m, \quad [C_{TT}] = \sum^M \int (\rho C_v/\tau_0)\{N_T^m\}\{N_T^m\}^T dR_m$$

$$\{M_1\} = \sum^M \int [N_u^m]\{\rho f^m\}^T dR_m, \quad \{M_2\} = \sum^M \int [N_q^m][K^m]\{\rho_2 f^m\} dR_m$$

$$\{M_3\} = \sum^M \int [N_u^m]\{\hat{t}_m\} dS_2^m, \quad \{M_4\} = \sum^M \int \{N_\Pi^m\} \hat{Q}_m dS_4^m$$

and
$$\{M_5\} = \sum^M \int \{N_T^m\}(\hat{H}_m/\tau_o)dS_6^m$$

For the temporal discretization, a forward time integration scheme is adopted with temporal functions. Eqs. (11),(12), and (13) can be written as

$$\begin{bmatrix} K_{uu} & K_{pu} & -K_{Tu} \\ K_{pu}^T & -\alpha\Delta t K_{pp} & 0 \\ -K_{Tu}^T & 0 & -(C_{TT}+\alpha\Delta t K_{TT}) \end{bmatrix} \begin{Bmatrix} u(t_n) \\ \Pi(t_n) \\ T(t_n) \end{Bmatrix} = \begin{Bmatrix} R_u(t_n) \\ R_\Pi(t_n) \\ R_T(t_n) \end{Bmatrix} \quad (14)$$

where $\Delta t = t_n - t_{n-1}$, $\{R_u(t_n)\} = \{M_1(t_n)\} + \{M_3(t_n)\}$

$$\{R_\Pi(t_n)\} = [K_{pu}]^T\{u(t_{n-1})\} + (1-\alpha)\Delta t[K_{pp}]\{\Pi(t_{n-1})\}$$
$$+ \alpha\Delta t(\{M_2(t_n)\} - \{M_4(t_n)\}) + (1-\alpha)\Delta t(\{M_2(t_{n-1})\}-\{M_4(t_{n-1})\})$$

and
$$\{R_T(t_n)\} = -[K_{Tu}]^T\{u(t_{n-1})\} + [-C_{TT} + (1-\alpha)\Delta t K_{TT}]\{T(t_{n-1})\}$$
$$+ \alpha\Delta t\{M_5(t_n)\} + (1-\alpha)\Delta t\{M_5(t_{n-1})\}$$

FINITE ELEMENT SIMULATIONS AND RESULTS

Based on the preceding formulations, one dimensional problems associated with thermoelastic consolidation are initially simulated for response calibration and code validation. A field problem, applicable to in situ coal gasification, is subsequently investigated.

The one dimensional simulations entail numerical evaluation of the isothermal poro-elastic, heat conduction and thermo-elastic problems (Fig. 1a). The (8-4-4) isoparametric composite element is used in the analysis (Fig. 1b). A comparison of the exact and finite element solutions for the pressure and termperature profiles at different time steps is shown in Fig. 2. The corresponding surface settlement histories for the cases of isothermal and thermoelastic consolidation are illustrated in Fig. 3. To the authors' best knowledge, an exact solution for the latter case is unavailable.

A field simulation of the plane strain thermal consolidation characteristics of the Hanna II, phase 2 in situ gasification experiments with the determined cavity shape (Youngberg and Sinks 1981) and specified boundary conditions is illustrated in Fig. 4. The selected material properties are listed in Table I.

THERMO-ELASTIC CONSOLIDATION

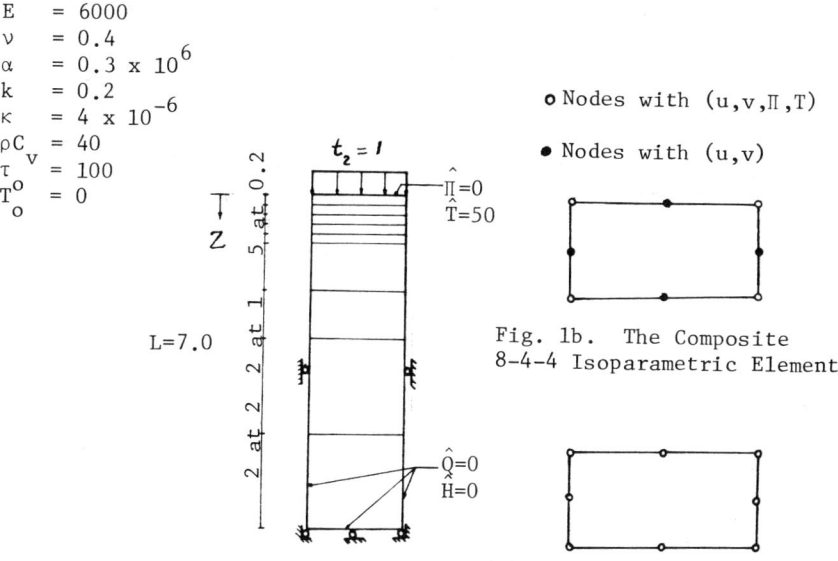

$E = 6000$
$\nu = 0.4$
$\alpha = 0.3 \times 10^6$
$k = 0.2$
$\kappa = 4 \times 10^{-6}$
$\rho C_v = 40$
$\tau_v = 100$
$T^o_o = 0$

○ Nodes with (u, v, Π, T)
● Nodes with (u, v)

Fig. 1a. 1-D Problem

Fig. 1b. The Composite 8-4-4 Isoparametric Element

Fig. 1c. The Composite 8-8-8 Isoparametric Element

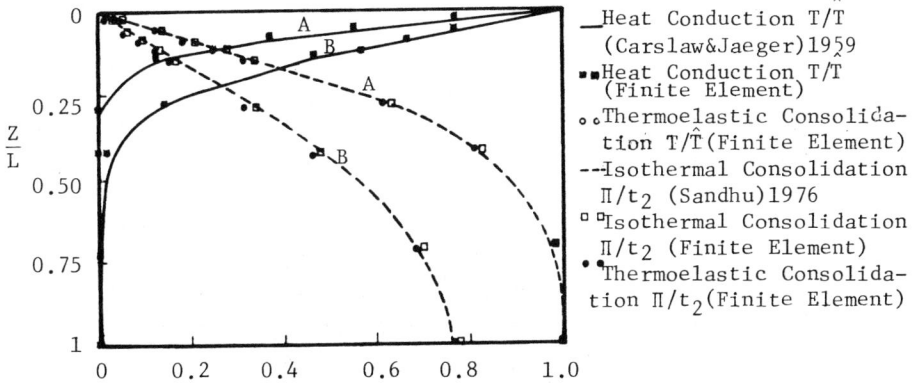

— Heat Conduction $T/\hat{T}$ (Carslaw & Jaeger) 1959
■■ Heat Conduction $T/\hat{T}$ (Finite Element)
○○ Thermoelastic Consolidation $T/\hat{T}$ (Finite Element)
--- Isothermal Consolidation Π/t_2 (Sandhu) 1976
□□ Isothermal Consolidation Π/t_2 (Finite Element)
●● Thermoelastic Consolidation Π/t_2 (Finite Element)

Fig. 2. Pore Pressure and Temperature Distribution:
$A(\frac{kt}{\rho_C{_v}L^2} = 0.005)$; $B(\frac{kt}{\rho_C{_v}L^2} = 0.02)$

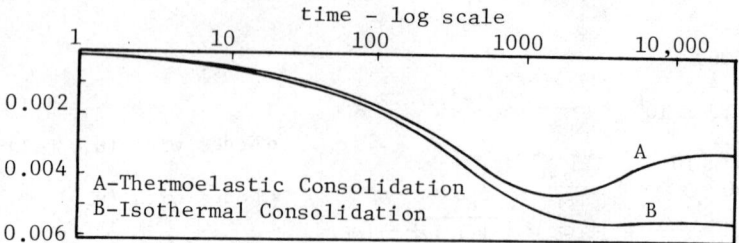

Fig. 3. Surface Settlement History for the 1-D Problem Using Finite Element Method

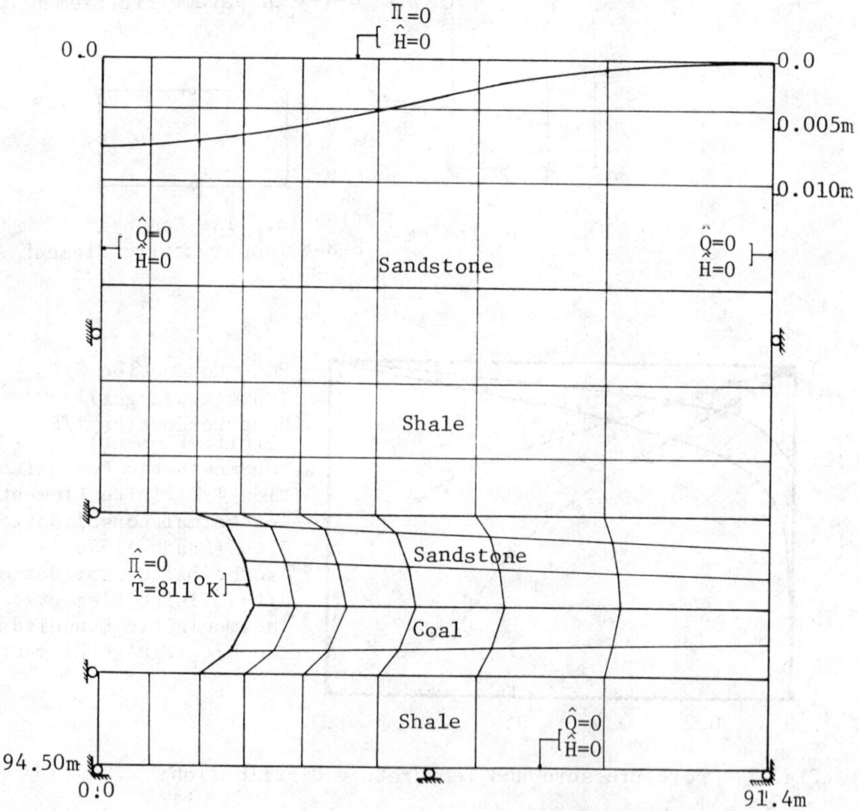

Fig. 4. Finite Element Discretization and Surface Settlement After 6 Months for Field Simulation

Table I. Material Properties for Hanna Experiment Simulation

Material Property	Units	Shale	Coal	Sandstone	Water
E	Pa	10.7×10^9	12.4×10^8	13.8×10^9	-
ν		0.33	0.33	0.33	-
α	/°K	8.1×10^{-6}	5.0×10^{-6}	8.1×10^{-6}	-
k	W/m°K	0.99	0.21	0.99	-
C_v	J/Kg°K	670	724	921	-
K	m/skgm^{-3}	9.9×10^{-11}	12.1×10^{-11}	12.1×10^{-11}	-
ρ	Kg/m^3	2003	2003	2292	-
ρ_2	Kg/m^3	-	-	-	1000

The (8-8-8) composite element illustrated in Fig. 1c is employed in the spatial discretization. The predicted surface settlement after 6 months is shown in Fig. 4. If a clay overburden is used in lieu of shale, the corresponding maximum surface settlement is about 0.5 m. In addition to the consolidation evaluations, temperature, pore pressure, effective stress, flow intensity, and thermal flux magnitudes have been computed at each time step.

CONCLUSIONS AND RECOMMENDATIONS

The presented methodology provides a comprehensive framework for the simulation of thermo-poro-elastic responses with potential applications in underground coal conversion, oil shale retortion, geothermal energy recovery, and nuclear waste disposal. Incorporation of temperature-moisture dependent properties and inelastic material behavior is recommended for model refinement.

ACKNOWLEDGEMENT

This research was supported by the U.S. Department of Energy under Contract No. DE-AS20-80LC10335.

NOMENCLATURE

σ_{ij} = stress tensor
Π = pore pressure
T = temperature change from reference state τ_o
α = coefficient of linear thermal expansion of the solid
β = thermoelastic coefficient
E_{ijkl} = elastic property tensor

ν = Poisson's ratio
u_i = displacement component
ε_{ij} = strain tensor
q_i = fluid velocity component
K_{ij} = permeability tensor
h_i = heat flux component
k_{ij} = heat conductivity tensor
C_v = heat capacity coefficient
ρ, ρ_2 = mass density of mixture and fluid, respectively
Δ = dilatation
f_i = body force component
$(\ )*(\ )$ = convolution
g' = unity $\forall\ t \in [0, \infty)$

REFERENCES

Biot, M.A., 1955, "Theory of Elasticity and Consolidation for a Porous Anisotropic Solid," J. Appl. Phys., vol. 26, 182-185.

Carslaw, H.S. and Jaeger, J.C., 1959, <u>Conduction of Heat in Solids</u>, D.R. Hillman and Sons Ltd., Great Britain.

Green, A.E. and Nagdhi, P.M., 1965, "A Dynamical Theory of Interacting Continua," Int. J. of Eng. Sci., vol. 3, 231-241.

Gurtin, M.E., 1964, "Variational Principles for Linear Initial Value Problems," Quart. J. Appl. Math., vol. 22, 252-256.

Hwang, C.T. et al., 1971, "On Solution of Plane Strain Consolidation Problems by Finite Element Methods," Can.Geo.J., vol. 8, 109-118.

Nickell, R.E. and Sackman, J.L., 1968, "Variational Principles for Linear Coupled Thermoelasticity," Quart. J. Appl. Math., vol. 26, 11-26.

Sandhu, R.S. and Wilson, E.L., 1969, "Finite Element Analysis of Seepage in Elastic Media," ASCE, J. Eng. Mech. Div., 95, 641-652.

Sandhu, R.S., 1976, "Finite Element Analysis of Soil Consolidation," NSF Grant 72-04110-A01 Report No. 6, Department of Civil Engineering, The Ohio State University, Columbus, Ohio.

Schiffman, R.L., 1971, "A Thermoelastic Theory of Consolidation," Environmental and Geophysical Heat Transfer, ASME, Heat Transfer Div., vol. 4, 78-84.

Weres, J., 1979, "Coupled Mechanical, Thermal, and Diffusive Fields in Solids - A Finite Element Approach," <u>Numerical Methods in Thermal Problems</u>, Ed. Lewis, R. and Morgan, K., Pine Ridge Press.

Witherspoon, P.A., et al., 1981, "New Approach to Problems of Fluid Flow in Fractured Rock Masses," Proc. 22nd U.S. Symposium on Rock Mechanics, 1-20.

Youngberg, A.D. and Sinks, D.J., 1981, "Postburn Study Results for Hanna II, Phases 2 and 3 Underground Coal Gasification Experiment," Proc. 7th Annual UCC Symposium, CONF810923.

Chapter 61

TIME AND TEMPERATURE DEPENDENT STRESS AND DISPLACEMENT FIELDS IN SALT DOMES[+]

by Heinz W. Duddeck and Hans-Konstantin Nipp

Professor Dr.-Ing., Institut für Statik, Technische
  Universität Braunschweig, Federal Republic of Germany
Dr.-Ing., Research associate at the same institute

ABSTRACT

The design of storage space or deep mining in rock salt should take into account not only time dependency of the rock salt behavior but also temperature effects. For deep mining of potassium salt the natural increase of temperature with depth may no longer be negligible. For the storage of cryogenic liquid gas in salt caverns more knowledge is needed about the propagation of deep temperatures and their effects on the stress-strain-creep behavior of the surrounding salt. High temperature problems arise for nuclear waste storages in caverns or drill holes emitting radiation. A numerical model is proposed for the analysis of displacements and stresses including their time and temperature induced change. For the material behavior of salt constitutive equations are chosen composed of a combination of Hooke-Kelvin-Bingham-Newton-model based on experimental results (excluding at present deep cooling), Langer 1980. The parameters involved are depending on the stress level and the temperature. The finite-element-method is applied to rotationally symmetrical three-dimensional problems. The non-linear calculation method is based on an implicit method with respect to time. A modified Newton-Raphson-method and the initial strain approach are employed. Along with some experiences on suitable numerical procedures, examples are shown for the characteristical effects of time-temperature dependency. Having achieved convergency and reliability a method is now available for practical design and parametric sensibility studies.

[+] The paper is a report on a more comprehensive investigation (Nipp, 1982), sponsored by the Volkswagen-Stiftung.

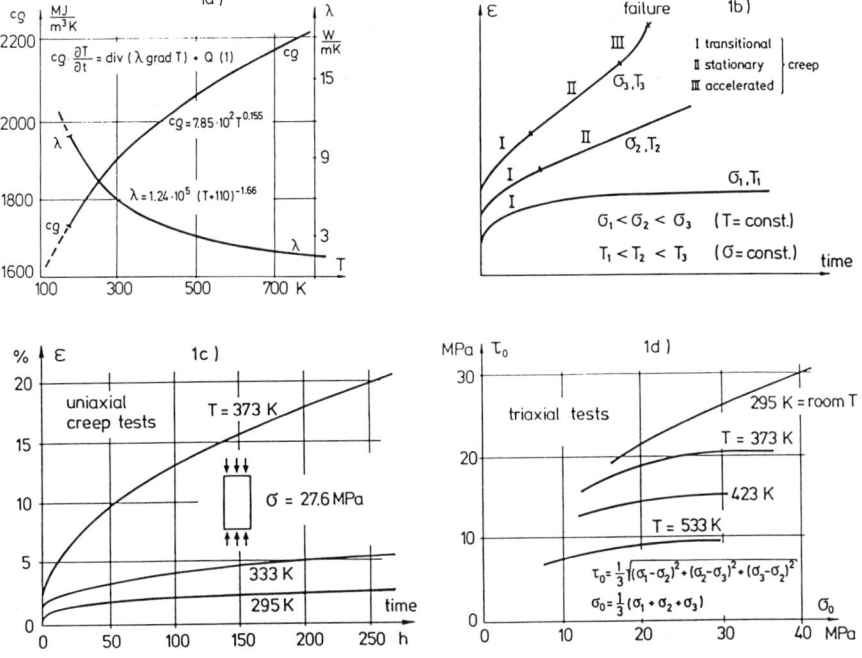

Fig. 1.   Effects of temperature T on the properties of rock salt.
  1a)   Thermal conductivity λ and specific heat capacity cϱ.
  1b)   Generalized creep phases.
  1c)   Example for results of uniaxial creep tests (Langer 1980).
  1d)   Example for max τ_o declining with rising temperature T (Langer 1980).

THERMAL AND TIME BEHAVIOR OF ROCK SALT

Fig. 1a is the result of the evaluation of experimental results given by different authors for specific temperature regions. A potential function is chosen for the thermal conductivity λ, covering the entire temperature range. In the same way a functional for the heat capacity cϱ is derived. Thus, the temperature propagation can be calculated from the given specific heat energy Q applying the equation in Fig. 1a. For the modulus of elasticity E(T) and the heat expansion coefficient linearized functions with regard to temperature T or even constant values may be chosen.

It is well known that creep is accelerated by increased temperature (Fig. 1b and 1c), whereas failure expressed e. g. by the 2nd invariant of the stress deviator (Fig. 1d) is decreasing with higher temperature. For more discussions on time-temperature-behavior of salt the reader is referred e.g. to Langer 1979 and Nipp 1982.

After careful consideration of the different alternatives for describing the stress-strain-time-temperature-behavior of an average rock salt, preference was given to the rheological model of Fig. 2.

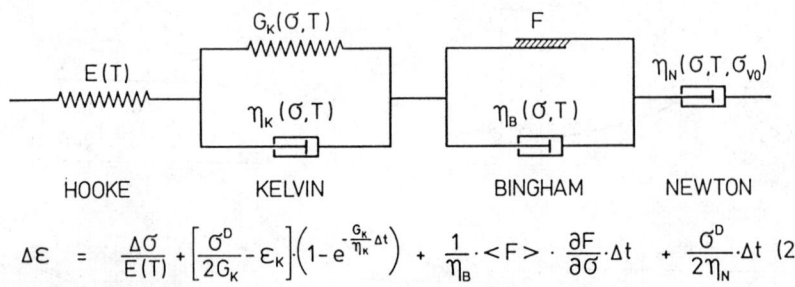

$$\Delta\varepsilon = \frac{\Delta\sigma}{E(T)} + \left[\frac{\sigma^D}{2G_K} - \varepsilon_K\right]\left(1 - e^{-\frac{G_K}{\eta_K}\Delta t}\right) + \frac{1}{\eta_B} \cdot <F> \cdot \frac{\partial F}{\partial \sigma} \cdot \Delta t + \frac{\sigma^D}{2\eta_N} \cdot \Delta t \quad (2)$$

Fig. 2. Symbolic rheological model and incremental material law, assumed here for rock salt. σ^D = deviator of the stress tensor.

The non-linear model covers:
- temperature dependent elasticity E(T) (Hooke),
- transitional creep, shear modulus G_K and viscosity η_K,
- stationary creep beyond a creep limit F corresponding to dislocation climb (Bingham), for F a limit in analogy e.g. to the condition of Drucker/Prager or that of von Mises may be chosen,
- stationary creep corresponding to dislocation glide which is the dominant creep source at a higher stress-level starting with σ_{vo} (Newton-model).

The model does not include the phase III in Fig. 1b of accelerated creep and creep failure. Research and laboratory tests on phase III are at present still in a preliminary state and have not yet resulted in a consistent theory which verifies hypotheses. For the three-dimensional analysis equ. (2) has, of course, to be written in generalized stress-strain-matrices. Poissons's ratio is assumed to be independent of temperature.

The coefficients for the Kelvin- model follow from:

$$\dot{\varepsilon} = 1.5 a_1 e^{-Q/RT} \cdot \sigma_v^{b_1 - 1} \cdot m e^{-mt} \cdot \sigma^D; \quad m = \frac{G_K}{\eta_K} = \text{const.} \quad (3)$$

yielding: $G_K(\sigma,T) = \left[3 a_1 e^{-Q/RT} \cdot \sigma_v^{b_1 - 1}\right]^{-1}.$ \quad (4)

σ_v is the one dimensional equivalent of the three-dimensional stress level. For the Newton model strain due to dislocation glide will start only if σ_{vo} is surpassed. The viscosity coefficients for the Bingham part η_B and the Newton model η_N are derived by comparison with the material equations given by Langer et. al. (1980): (5)

$$\frac{1}{\eta_B}=3a_2 e^{-Q/RT}\sigma_v^{b_2-1} \; ; \; \frac{1}{\eta_N}=3a_3 e^{-Q/RT}\sinh\left[b_3 <\sigma_v-\sigma_{vo}>\right].$$

a_i and b_i are parameters of constant value free for the adjustment to experimental results.

ANALYSIS AND NUMERICAL PROCEDURE

It is evidently allowed to assume only a single-sided coupling of the thermodynamical behavior of rock salt. The propagation of the temperature field is obviously independent of the stress-strain field, whereas the opposite is not true, Fig. 3. Hence, the temperature field may be evaluated by the equation given in Fig. 1a independently in the same steps of time as required for the incremental procedure of the mechanical analysis. Applying the finite-element-theory the temperature field is calculated for the nodal points. An example for temperature propagation is shown in Fig. 4. The implicit time-step numerical solution of equ. (6) follows Zienkiewicz (1977). Numerical stability and accuracy are provided by choosing $\xi = 2/3$ (Galerkin).

$$[\xi W_L + W_S/\Delta t_{j+1}]\Delta T_{j+1} = (1-\xi)\cdot H_j + \xi H_{j+1} - W_L T_j . \quad (6)$$

W_L = conductivity matrix, W_S = heat capacity matrix, $H(t)$ = temperature load.

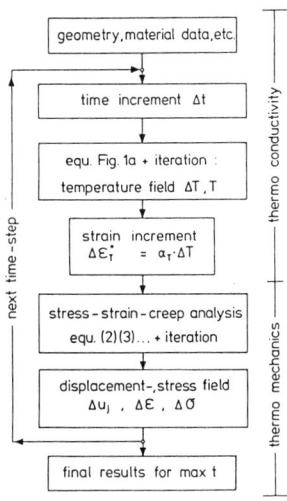

Fig. 3 . Procedure of analysis

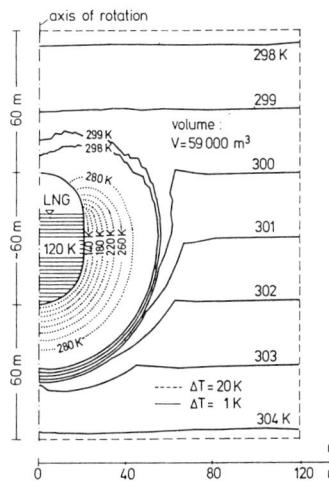

Fig. 4. Calculated distribution of temperature around a cavern partially filled with cryogenic liquid gas LNG, 365 days after applying 120 Kelvin to the wall.

The implicit method is also applied for the thermo-mechanical part of Fig. 3, see equ. (7),(8) and also Hughes and Taylor (1978).

FEM displacement method: $K^i_{j+\xi} \cdot \Delta u^i_{j+1} = \Delta P^i_{j+\xi}$, (7)

element matrix: $k^i_{j+\xi} = 2\pi r \int B^T \cdot D^i_{j+\xi} B \cdot dA$, (8)

with $D^i_{j+\xi} = \left[E^{-1}_{j+1} + \Delta t_{j+1} \cdot \left(\dfrac{\partial \dot{\varepsilon}}{\partial \sigma} \right)^i_{j+\xi} \cdot \xi \right]^{-1}$.

EXAMPLE 1: CAVERN FOR NATURAL GAS.

Here only few results of those given in (Nipp 1982) are presented. The displacement fields are not shown. Neglecting for simplicity reason the excavation phases of the cavern in Fig. 5, constant gas pressure of 15 MPa is

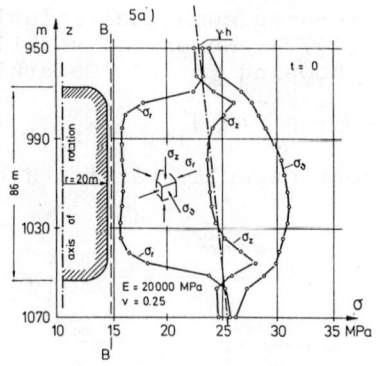
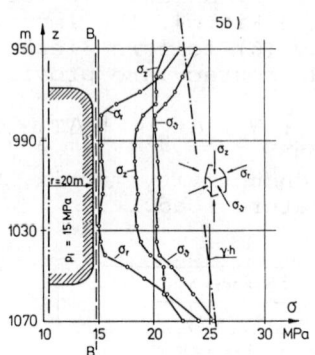
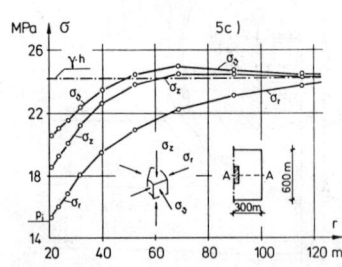
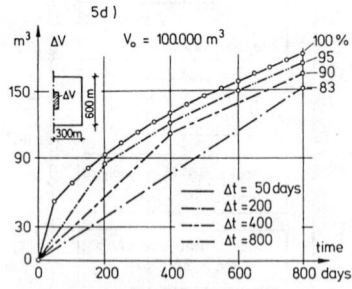

Fig. 5. Cavern of rotational symmetry in salt dome filled with gas.
5a) Calculated stress field along vertical section B-B (close to wall) without creep (t = 0).
5b) Stress field along section B-B after t = 10 years.
5c) Stress field along the horizontal plane A-A after t = 10 years.
Stress level before the excavation: $\sigma_r = \sigma_z = \sigma_\vartheta = \gamma \cdot h$ = dead weight of cover.
5d) Loss of open volume ΔV for the gas filled cavern within 800 days, showing the dependency of the results on the value of the increment Δt chosen for the iteration procedure.

applied and assumed that creep may start at this moment. On the basis of experimental results (Langer 1980 and Wallner 1981) the following functions are derived for the terms in equ. (2) to (5), $E(T)$ = const as in Fig. 5a (dimensions: MPa, days, Kelvin):

$$1/G_K = 0.63\ e^{-5390/T} \cdot \sigma_v^{4.0} \qquad F = \sqrt{J_2^D} - 0.577 = 0,$$

$$1/\eta_K = 0.22\ e^{-5390/T} \cdot \sigma_v^{4.0} \tag{9}$$

$$1/\eta_B = 0.54\ e^{-6495/T} \cdot \sigma_v^{4.0}.$$

The stress field at time $t = 0$ (Fig. 5a) is everywhere smaller than the stress level σ_{vo} in Fig. 2. Therefore, the Newton part is without any significance. Triangular finite ring-elements are chosen for the rotationally symmetric salt dome section of 600 m hight and 600 m diameter. The effect of creep on the stress field is quite pronounced, see Fig. 5a and 5b. The stress peaks for σ_z are vanishing with time. The stresses around the cavern are reduced by 16 % for σ_z to 35 % for σ_ϑ. The convergence of the open cavern, Fig. 5d shows that the deformations are growing and the stress field changing beyond the analyzed 800 days as long as the stress deviator is not zero.

EXAMPLE 2: INSTATIONARY HEAT SOURCE

In order to prove the capacity of the method a more complex example is analyzed simulating an arbitrarily chosen heat source, Fig. 6. Because of the non-availability of reliable experimental investigations the input data and the results of this example should not be taken as relevant to a real nuclear waste deposal. Assuming rotational symmetry with regard to the vertical axis and for simplicity sake with regard to the A-A plane, a ground segment as in Fig. 6a is analyzed. Each of the ground layers are supposed to be homogenous and isotropic. The specific heat emission of the central core may follow the time depending function, see Fig. 1a, also for λ and $c\varrho$:

$$Q = 0.81 \cdot e^{-6,61\ \cdot\ 10^{-5}t}. \quad [W/m^3] \tag{10}$$

The temperature calculated increases to its maximum at 54 years, Fig. 6b, and decreases thereafter due to heat transfer to the surface and the larger surrounding. By applying constant values for λ and $c\varrho$ instead of the functions given in Fig. 1a proved that the temperature rose 15 % less.

The temperature induced stresses at the time of max T (54 years) are shown in Fig. 6c to 6f. For the different

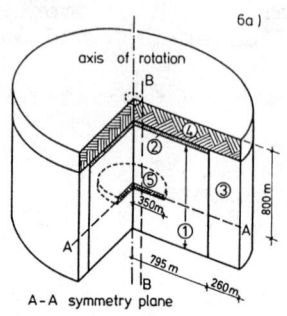
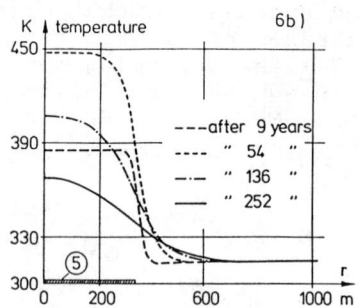
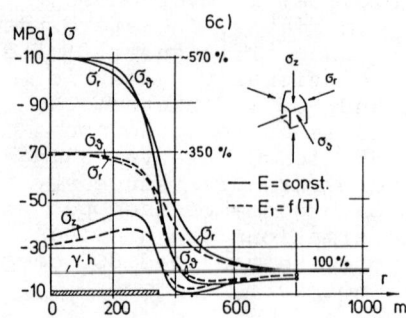
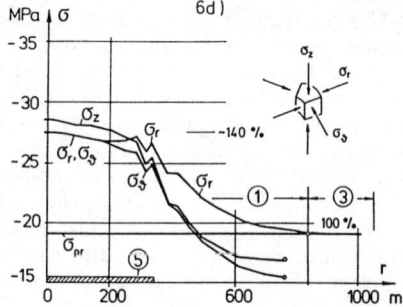
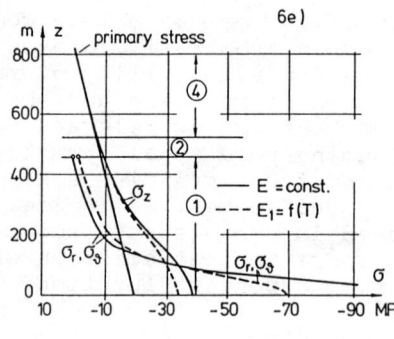
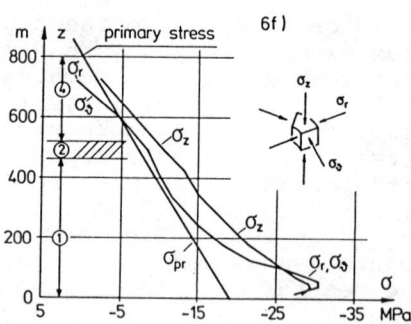

Fig. 6. Effects of a heat source on a salt dome and the adjacent ground.
6a) Analyzed ground segment consisting of (1) rock salt, (2) gypsum head (60 m deep), (3) adjacent rock, (4) soil cover (280 m thick) and (5) heat source (50 m deep).
6b) Rise and fall of the temperature along the horizontal plane A-A (see 6a) caused by time dependent heat emission of the source (5).
6c) Stress fields in plane A-A after 54 years, showing the considerable relaxation if a modulus of elasticity is applied which is temperature dependent $E_1 = f(T)$. Results without applying creep analysis.
6d) Stress field as in Fig. 6c) with additional relaxation due to applying the rheological model Fig. 2.
6e) Deviation from the dead weight primary stress along the vertical section B-B after t = 54 years without applying creep analysis, but showing the effect of a temperature dependent modulus of elasticity $E_1 = f(T)$.
6f) Stress field as in Fig. 6e) due to relaxation applying the rheological model Fig. 2.

types of ground in Fig. 6a the following moduli of elasticity are chosen: E_1 = 26.000 - 40 T; E_2 = 500; E_3 = 5.000; E_4 = 250 MPa. Rheological behavior is assumed for the rock salt only. For long time investigations neglecting the primary creep (Kelvin part in Fig. 2) is allowed. Hence, the following material coefficients are chosen, η_B and F as in equ. (9), $G_K = \infty$, $\eta_K = \infty$: (11)

$$\frac{1}{\eta_N} = (7.8 \cdot 10^{12} \cdot e^{-\frac{13590}{T}} + 1.9 \cdot 10^4 \cdot e^{-\frac{6495}{T}}) \cdot \frac{1}{\sigma_v} \cdot \sinh[0.37 < \sigma_v - 20 >]$$

Here it is assumed that only the deviatoric part of the stress tensor is causing creep, not the hydrostatical part of it. The values given in percentages of the primary stress (without a temperature rise) demonstrate the strong effects of the temperature dependent E_1 and the creep behavior. As expected, this is much more pronounced for the case of directly induced deformations than for the case of acting loads (see also Fig. 5). The analysis, implicit with regard of time, allowed for rather large time steps, nevertheless providing stability and sufficient accuracy. The numerical procedure exhausted the capacity of the central computer of the University. For further progress, finding realistic and appropriate material laws for rock salt behavior is more urgent than refinement of the analysis.

REFERENCES

Hughes, T.R.J., and Taylor, R.L., 1978, "Unconditionally stable algorithms for quasi-static elasto/visco-plastic finite element analysis", Computers & Structures, p.169.

Langer, M. 1979, "Rheological Behaviour of rock masses", General report, Proc. 4. Inst. Congress Rock Mechanics, Vol. 3, Montreux.

Langer, M., et. al., 1980, "Das Verformungs- und Bruchverhalten von Steinsalz, "Bundesanstalt für Geowissenschaften und Rohstoffe, Hannover.

Nipp, H.-K. 1982, "Temperatureinflüsse auf rheologische Spannungszustände im Salzgebirge," Bericht Nr. 82-36, Institut für Statik, Technische Universität Braunschwg.

Wallner, M., 1981, "Berechnung thermomechanischer Vorgänge bei der Endlagerung hochradioaktiver Abfälle im Salzgestein unter Verwendung eines optimierten Finite Element Programms," Bericht zum BMFT-Forschungsvorhaben KWA 2070-8, BGR, Hannover.

Zienkiewicz, O.C., 1977, "The finite element method", 3rd ed. McGraw-Hill, London.

Chapter 62

IMPLEMENTATION OF FINITE ELEMENT--
BOUNDARY INTEGRAL LINKAGE ALGORITHMS
FOR ROCK MECHANICS APPLICATIONS

by W. Scott Dunbar

Senior Staff Engineer
Woodward-Clyde Consultants
San Francisco, California

INTRODUCTION

Methods of linking boundary integral (BI) solutions with finite element (FE) solutions have been well described theoretically in other publications (e.g., Zienkiewicz, et al., 1977). The purpose of this paper is to point out some practical aspects of implementing these linkage algorithms for rock mechanics applications and to describe an alternative linkage algorithm.

A not so obvious motivation for using FE-BI linkages is illustrated in Figure 1. The cost of modeling the excavation of the crusher station by the BI method can equal or exceed the cost of an equivalent FE model. This is due to the large surface area which must be modeled by the BI method. In such cases, a more economical approach might be to model the region immediately adjacent to the crusher station by finite elements and link the solution at the periphery of the FE model with a solution for the exterior region. Further economies can be realized if some flexibility in the type of exterior solution is possible. For example, it would be desirable to use a coarse discretization of the FE-BI interface for a BI solution, which would be less demanding of computer storage and time, or to use an analytical solution for the exterior region. The alternative linkage algorithm to be discussed allows such flexibility.

BOUNDARY INTEGRAL METHODS

There are basically two formulations of BI methods: the direct and the indirect. FE-BI linkage algorithms depend on which formulation is used. The purpose of this section is to briefly describe each method.

BOUNDARY INTEGRAL LINKAGE ALGORITHMS

Given displacements, u, and/or tractions, t, at N nodal points on a surface S, the direct method solves for the unknown displacements and/or tractions by means of the following relationship

$$Au = Bt \qquad (1)$$

where, in m dimensions, the matrices A and B are of order mN. (See Rizzo, 1967; Cruse, 1969, 1974; Lachat and Watson, 1976.)

Indirect BI methods are formulated in terms of fictitious quantities defined on S. In the traction discontinuity method (TDM), displacements and tractions are given by

$$u = Ba \qquad t = Ca \qquad (2)$$

where a is a vector of dimension mN composed of traction discontinuities on S. B is the same matrix as in Equation (1) and C is composed of combinations of derivatives of the elements of B. (See Banerjee and Butterfield, 1977).

Another indirect method is the displacement discontinuity method (DDM) where the fictitious quantities are displacement discontinuities on S. The formulation and solution procedures of the DDM are similar to those of the TDM. (See Crouch, 1976a, b; Dunbar and Anderson, 1981).

VARIATIONAL PRINCIPLE

Zienkiewicz et al (1977) showed that the potential energy of the regions labelled H in Figure 2 is given by

$$P = \frac{1}{2}\int_S u't\,dS - \int_{S_u} t'u^*\,dS - \int_{S_t} u't^*\,dS \qquad (3)$$

where u^* and t^* are, respectively, the displacements and tractions given on S_u and S_t. The prime denotes a transpose. Equation (3) may be used together with matrix equations of the direct and indirect methods to derive FE stiffness matrices for the region H.

FE-BI LINKAGE

In the direct BI method, the tractions at N nodal points on the surface S are given by Equation (1)

$$t = B^{-1}Au = Zu.$$

Substitution of this expression into Equation (3) and minimizing with respect to u results in the following stiffness relation for u at the N nodal points on S:

$$Ku - f = 0$$

$$K = \frac{1}{2}\int_S (Z' + Z)\,dS \qquad f = \int_{S_u} Z'u^*\,dS + \int_{S_t} t^*\,dS.$$

The units of K are FORCE/LENGTH so that K may be directly added to the stiffness matrix of the FE region. However, to obtain K it is necessary to explicitly invert B which is unsymmetric and of the same order as K. Generally, one would like to avoid this as B can become large, especially in three-dimensional problems involving large surface areas.

A slightly different approach is required for indirect BI methods. The displacements u* are given in terms of the FE nodal displacements w on S.

$$u^* = Pw$$

where P is the FE shape function. Substituting this and Equation (2) into Equation (3) and minimizing with respect to a results in

$$Fa + Gw = h \qquad (4)$$

Minimizing with respect to w gives

$$G'a = 0 \qquad (5)$$

where

$$F = \frac{1}{2}\int_S (A'D + D'A)\,dS \qquad G = \int_S D'P\,dS$$

$$h = \int_S A't^*\,dS$$

Eliminating a between Equations (4) and (5) results in the following stiffness relation for w

$$G'F^{-1}Gw - G'F^{-1}h = 0. \qquad (6)$$

The units of the matrix $G'F^{-1}G$ are FORCE/LENGTH.

The matrix F is symmetric and may be decomposed into the Cholesky product

$$F = R'R$$

where R is an upper triangular matrix. Letting $X = R^{-1}G$ and $y = R^{-1}h$, Equation (6) may be written as

$$X'Xw - X'y = 0.$$

BOUNDARY INTEGRAL LINKAGE ALGORITHMS

The matrix X may be computed column by column by repeated back substitution

$$RX_j = G_j$$

where X_j and G_j are the jth columns of X and G respectively. The solution X_j may overwrite the jth column of G. The vector y can be computed in a similar manner, and overwrite the vector h. All of the above can be implemented by simple, compact and readily available computer codes.

The stiffness matrix $G'F^{-1}G$ is symmetric but fully populated. Consequently, it will increase the bandwidth of the global FE stiffness matrix. The associated increase in computer storage requirements may be at least partially avoided by the use of "skyline" solution algorithms (Bathe and Wilson, 1976; Zienkiewicz, 1977). The effect on solution time of several large "skyscrapers" in the column skyline of the global stiffness matrix has apparently not been investigated.

Another aspect of the above FE-BI linkage algorithm is that the discretization of the interface S is the same for both the FE and BI solutions. In the interests of economy and practicality, it may be desirable to use a coarser discretization for the BI solution or, even better, an analytical solution, if possible. The alternating algorithm described in the next section allows such flexibility in the solution for the region H of Figure 2.

ALTERNATING ALGORITHM

In the alternating algorithm, the FE and BI solutions are treated separately. Displacements and tractions produced by the two solutions at the interface S between the FE region and the region H are matched in an iterative manner.

Suppose the FE and BI problems are posed as:

A: FE solution, $t = t^*$ on S

B: BI or analytical solution, $u = u^*$ on S.

The alternating algorithm may then be described as follows:

(1) Assume a value of t on S, say t^*,

(2) Given t^*, compute desired solution and $u = u^*$ on S from A,

(3) Given u^*, compute $t = t^*$ on S from B,

(4) Go to 2.

The above procedure is continued until the difference between two successive desired solutions reaches an acceptable limit.

A test example of the application of this algorithm to a two-dimensional problem in plane strain is shown in Figure 3. The FE model is intended to approximate the problem of a hole of radius a(=1) in an infinite medium with a compressive stress at infinity and zero stress at r = a. The solution for the region exterior to the FE region is the classical Lame solution of a hole of radius b(=3) in an infinite medium with a displacement boundary condition at r = b and a stress boundary condition at infinity.

For the first iteration the traction on S was assumed to be the applied compressive stress at infinity. The FE solution for the radial stress corresponding to this assumption is shown in Figure 4. The average of the radial displacements computed on S from the FE solution was used as the boundary condition for the analytical solution for the exterior region. Ten iterations of the algorithm were required to reduce the difference between successive solutions to 10^{-3}. The results of the first and second iterations are shown in Figure 4. Successive solutions tended to oscillate about the final solution. (The final solution is greater than the exact solution because a low order element was used and the stiffness of the mesh is over estimated. The error is approximately 5 percent.)

It appears that the alternating algorithm is potentially a simple and effective method of linking FE and BI solutions. It remains to investigate methods of accelerating convergence and to test the method on geometrically unsymmetric problems.

ACKNOWLEDGEMENTS

I would like to thank Dr. Nick Sitar and Dr. Richard Goodman of the Department of Geotechnical Engineering at Berkeley for their help with this paper. Dr. Gerald Mavko of the U. S. Geological Survey at Menlo Park provided some insight to the nature of the alternating algorithm.

REFERENCES

Banerjee, P. K. and Butterfield, R., 1977. Finite Elements in Geomechanics, Gudehus, G., ed., pp. 529-570, Wiley.

Bathe, K. J. and Wilson, E. L., 1976. Numerical Methods in Finite Element Analysis, Prentice-Hall.

Crouch, S. L., 1976a. Solution of Plane Elasticity Problems by the Displacement Discontiniuty Method. Int. J. Num. Meth. Engng., Vol. 10, pp. 301-343.

Crouch, S. L., 1976b. Analysis of Stresses and Displacements Around Underground Excavations: an Application of the Displacement Discontinuity Method. V. of Minn. Geomech. Rep't., U. of Minn.

Cruse, T. A., 1969. Numerical Solutions in Three-Dimensional Elastostatics. Int. J. Solids Structures, Vol. 5, pp. 1259-1274.

Cruse, T. A., 1974. An Improved Boundary Integral Method for Three-Dimensional Elastic Stress Analysis. Comput. Struct., Vol. 4, pp. 741-754.

Dunbar, W. S. and Anderson, D. L., 1981. The Displacement Discontinuity Method in Three-Dimensions. <u>Boundary Element Methods</u>, Brebbia, C. A., ed. pp. 153-173, Springer-Verlag.

Lachat, J. C. and Watson, J. O., 1976. Effective Numerical Treatment of Boundary Integral Equations: A Formulation for Three-Dimensional Elastostatics. Int. J. Num. Meth. Engng, Vol. 10, pp. 991-1005.

Rizzo, F. J., 1967. An Integral Equation Approach to Boundary Value Problems of Classical Elastostatics. Q. Appl. Math, Vol. 25, pp. 83-95.

Zienkiewicz, O. C., 1977. <u>The Finite Element Method</u>. McGraw-Hill.

Zienkiewicz, O. C., Kelly, D. W., and Bettess, P., 1977. The Coupling of the Finite Element Method and Boundary Solution Procedures. Int. J. Num. Meth. Engng., Vol. 11, pp. 355-375.

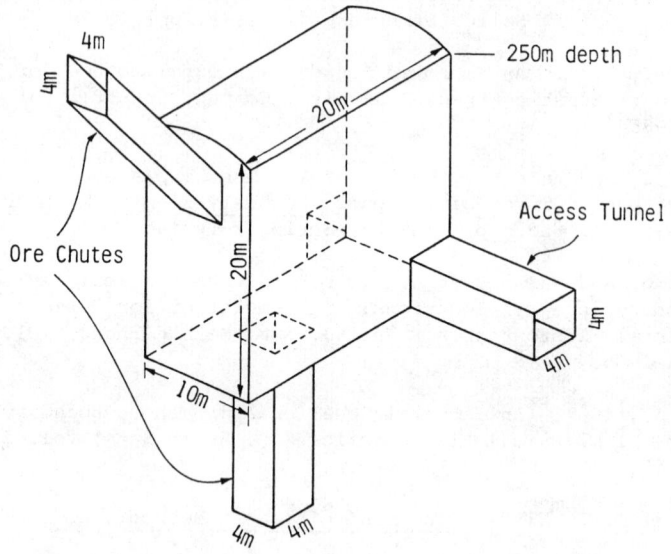

Figure 1: CRUSHER STATION GEOMETRY

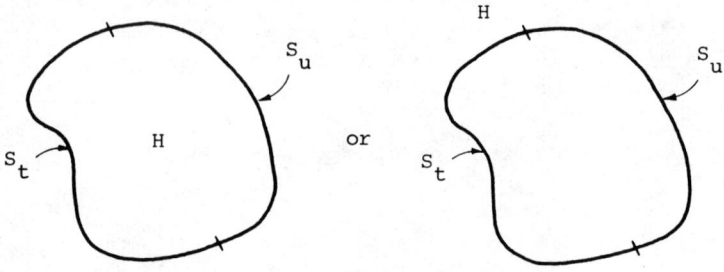

Displacements prescribed on S_u

Tractions prescribed on S_t

$$S = S_u + S_t$$

Figure 2

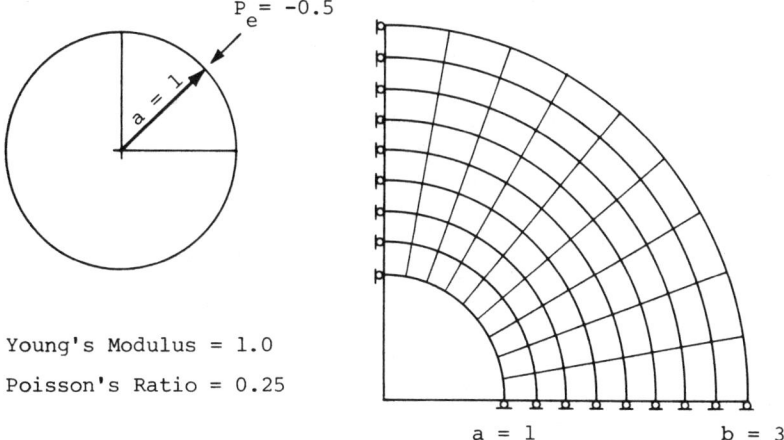

Figure 3: TEST PROBLEM

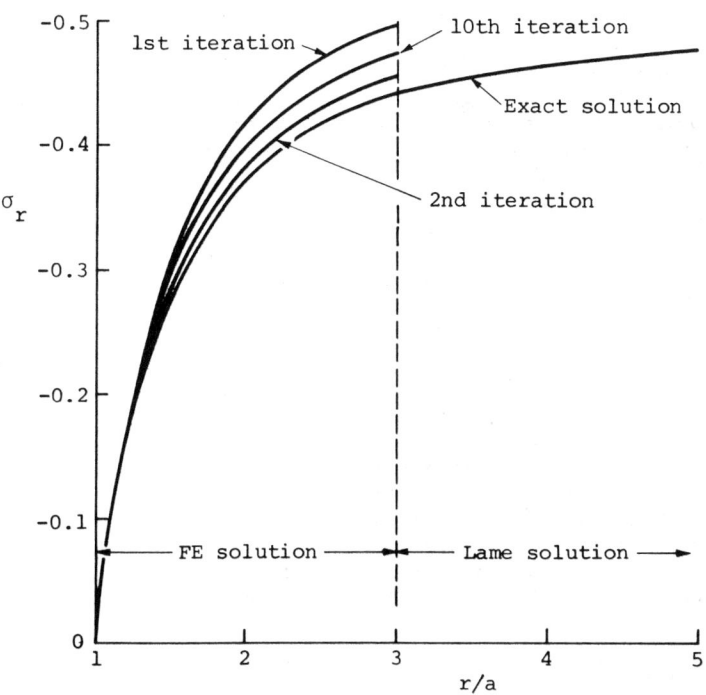

Figure 4: RESULTS OF TEST PROBLEM

Chapter 63

ON THE MODELING OF NUCLEAR WASTE DISPOSAL BY ROCK MELTING

Francois E. Heuze
Lawrence Livermore National Laboratory
Livermore, Ca.   94550

ABSTRACT

   Today, the favored option for disposal of high-level nuclear wastes is their burial in mined caverns. As an alternative, the concept of deep disposal by rock melting (DRM) also has received some attention. DRM entails the injection of waste, in a cavity or borehole, 2 to 3 kilometers down in the earth crust. Granitic rocks are the prime candidate medium. The high thermal loading initially will melt the rock surrounding the waste. Following resolidification, a rock/waste matrix is formed, which should provide isolation for many years. The complex thermal, mechanical, and hydraulic aspects of DRM can be studied best by means of numerical models. The models must accommodate the coupling of the physical processes involved, and the temperature dependency of the granite properties, some of which are subject to abrupt discontinuities, during α-β phase transition and melting. This paper outlines a strategy for such complex modeling.

THE DRM CONCEPT*

   The basic idea behind DRM is to deposit the nuclear waste into the rock mass with such a power output that the rock initially will melt. Whether the waste is encapsulated or not, it will eventually be held in a matrix formed upon resolidification of the rock. Such a matrix would have a high resistance to water leaching. In addition, the high temperatures created in DRM would tend to keep the water away for many years, allowing a strong decay of the radioactive elements before water access. This mitigates the fact that waste retrieval after emplacement would be very difficult, if not impossible.

---

*Because of space limitation, many references are not included in this paper. They can be found in Heuze, 1981 (Ref. 1) which contains 121 references. The report is available from the author or from the Lawrence Livermore National Laboratory.

Four variations of DRM have been proposed so far:[1]

- The Deep Underground Melt concept (DUMP) was first introduced by Lawrence Livermore National Laboratory[2] (Fig. 1) in 1971, and was later refined. In the latest version (1978) the waste is injected in a liquid form for one year, and cooling water is provided. When the water supply is shut-off, after a year, the water is allowed to boil off and the cavity is sealed. Then, the melting phase starts. Purely conductive thermal calculations were performed in the mid 70's, assuming a maximum power output of 23 MW at 1 year, decreasing to 2 MW at 200 years. The melt radius grows to 80 m at 60 years, and slowly decays thereafter. At 200 years, the temperature is 100° at 200 m from the source.
- Angelo, at the University of Arizona (1976) proposed the Solidified Waste In-Situ Melting concept in which waste canisters are dispersed in rock rubble inside a shallower cavity. The lower thermal loading limits the extent of melting, and the rubble voids can accommodate a large thermal expansion (Fig. 2).
- Deep Self Burial and Deep Rock Disposal were two concepts proposed by Sandia Laboratories. They involve disposal in deep boreholes, rather than cavities, with or without encapsulation. In the Deep Self Burial case, the canisters are self-burying, by melting their way down from the bottom of the hole.(Figures 3 and 4).

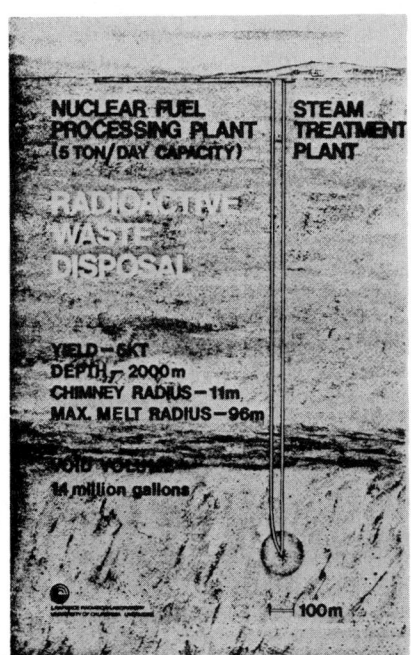

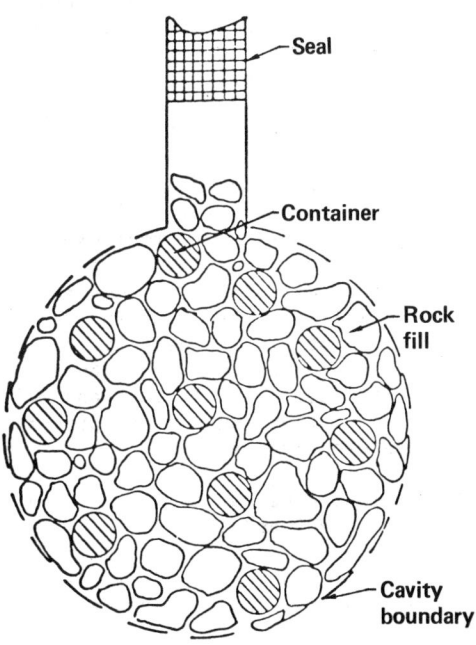

Figure 2: Solidified Waste In-Situ Melting Concept.

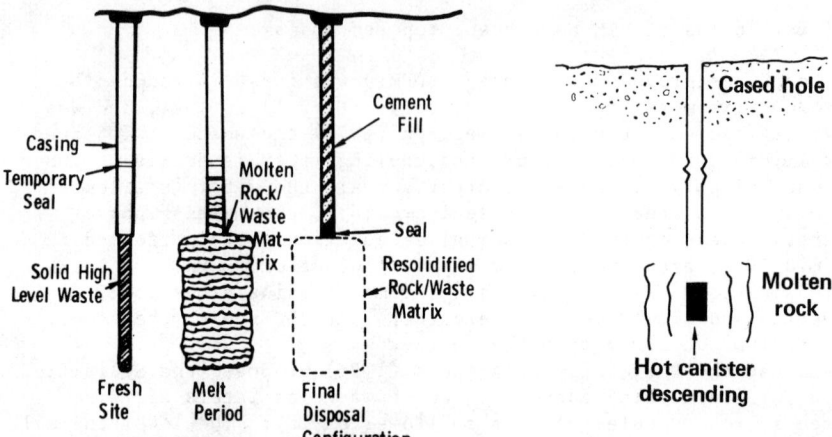

Figure 3: Deep Rock Disposal (Klett, 1974).

Figure 4: Deep Self Burial (Logan, 1973)

We focus upon granite as a candidate host rock and evaluate the state of knowledge of relevant granite properties over a wide range of temperature and pressure. Other candidate rocks which also have been mentioned include basalt, salt, and shale. Carbonate rocks have been rejected because of their capacity to generate large quantities of gas when melting. Such gases would be undesirable because they would provide an additional vehicle by which radioactivity can escape into the biosphere.

TEMPERATURE DEPENDENT PROPERTIES OF GRANITE

The following mechanical and transport properties are required to model DRM: modulus, Poisson's ratio, tensile strength, compressive strength, viscosity, thermal expansion, density, permeability, melting temperature, heat of fusion, specific heat, thermal conductivity and thermal diffusivity. Temperatures in the crust, around a DRM cavity, will vary from ambient, at a large distance, to beyond the melting point, in the near field. Over the range of 40° to 1200°C, granite undergoes two temperature phase changes: the α-β transition of quartz (573°C, at 1 atm), and melting (1050°C, dry). Because of space limitation, only those properties which vary drastically with temperature are discussed:

- Modulus: goes to zero at melting,[3] with a fairly regular decay between 20° and 1050°C

- Poisson's ratio goes from less than 0.25 to a nearly incompressible value at melting (0.49999 in the calculations)

- tensile,[3] and compressive[4,5] strengths go to zero at melting, with an accelerated decay after 400°C. Decay of cohesion and friction normalized to the values at 20°C is shown in Fig. 5.

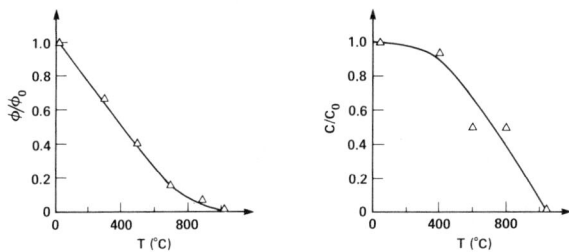

Figure 5: Relative Decrease of Friction (ϕ) and Cohesion (c), with Temperature, for dry Westerly Granite.[4,5]

- thermal expansion undergoes two drastic discontinuities,[6] as shown in Fig. 6.

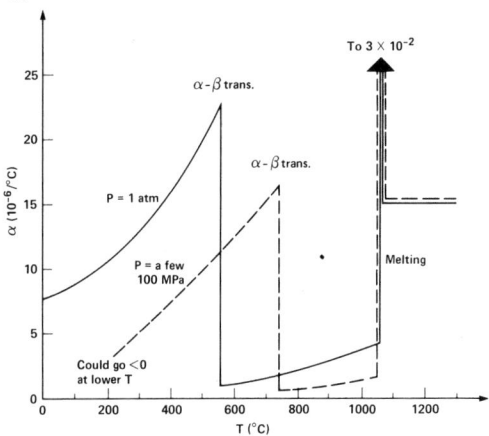

Figure 6: Schematic Variation of Granite Linear Thermal Expansion, Based on Results for Delegate Aplite.[6]

Thermal expansion is of particular interest because the large expansion of the host rock upon melting may create new fractures in the rock mass. Such fractures can then provide a path for rock melt and gas migration.

- viscosity: the few published results[7,8] indicate a dry granite viscosity of 10^{11} g.s/cm^2 at 1000°C, decreasing to 5.10^5 at 1600°C. For granite with 1% water by weight, the viscosity decreases from 2.10^{10} at 700°C to 10^6 at 1200°C.

- creep law: because of the durations and temperatures involved in DRM, the steady state creep component will overshadow the transient creep effects. The accepted secondary creep law for rocks is Weertman's law.[9]

$$\dot{\varepsilon} = A \cdot \sigma^N \cdot e^{-Q/RT}$$

where A and N are constants, Q is the activation energy, R the universal gas constant, and T the absolute temperature. The only granite for which parameters were found in this research is Westerly granite,[10] for which: $A = 4.10^{-27}$ (Pa^{-N}/s); $N = 2.9$; $R = 8.31$ j/mole. deg; $Q = 1.058 * 10^5$ j/mole and σ is in Pascals.

- specific heat: specific heat varies only between about 800 and 1200 j/kg. deg from 20°C to melting; but, in the models, one accounts for the latent heat of fusion of $3.344 * 10^5$ j/kg by raising sharply the specific heat in a small temperature interval about the melting point.[11]

- thermal conductivity and thermal diffusivity decay by about 40% between 20°C and the α-β transition, and increase thereafter at approximately the same rate.

MODELING DRM

Desirable Features of Models

The analysis of any DRM scenario will require that complex coupled phenomena be simulated. It seems that only numerical models can perform such calculations. The most desirable and necessary features of models are:

- finite element or finite difference models should be used. Boundary methods are not sophisticated enough at this time.

- elastic response, creep and fractures should be accounted for.

- free convection will take place; the thermo-mechanical solutions should be strongly coupled.

- slow viscous motion over long times indicates the need for implicit schemes and capability for large deformations.

- code input should accommodate the sometimes unusual variation of material properties with temperature.

- for analysis of flow of gases and free surface rock melt propagation in fractures, Eulerian models seem preferable.

- pre-processing, rezoning, and post-processing capabilities would be most useful.

Candidate Codes

An extensive survey of modeling in a wide range of geological, hydraulic, and mechanical process indicated that 3 codes are prime candidates for use in DRM modeling: SANGRE,[12] COUPLEFLO,[13] and KRAK.[14] Another three codes, NACHOS, TACO/NIKE-2D, and ADINA/ADINAT would require significant enhancement to be used effectively. The relevant characteristics of the 6 codes are summarized in Table 1.

MODELING OF NUCLEAR WASTE DISPOSAL

Table 1. Summary of capabilities for selected codes to model DRM thermo-mechanics and thermo-hydraulics.

	SANGRE	COUPLEFLO	KRAK	NACHOS	TACO/ NIKE-2D	ADINA/ ADINAT
Finite element	x	x		x	x	x
Finite difference			x			
Plane	x	x		x	x	x
Axisymmetric	x	x	x	x	x	x
Lagrange	x	x^a	x		x	x
Euler		x	x	x		
Transient	x	x	x	x	x	x
Implicit	x	x	x^b	x	x	x
Coupled	x				x	x
Elastic solids	x				x	x
Fracture mech.			x			
Weertman creep	x	x				
Melting^c	x	x			x	x
Incompressible	x	x		x		x
Convection	x	x		x		
Newtonian	x	x	x	x		
Viscous dissipation	x	x				
Free surface	x	x				
Time-dep. heat source		x	x	x		x
Multi-material	x	x	x		x	x
Anisotropy	x	x			x	x
Temp.-dependent properties	x	x	x	x	x	x
Pre-post-processor	x	x		x	x	x
Origin	LANL	SANDIA	LANL	LANL	LLNL	MIT
	Codes usable without major modifications.			Codes need significant enhancement.		

^{a} Pseudo-Lagrangian: new system geometry advanced through Euler integration of the velocity field.
^{b} In KRAK, the crack flow and crack propagation are solved implicitly, whereas the porous medium flow is solved explicitly.
^{c} Programs will be said to handle melting if they can accommodate the related variations in thermal and mechanical properties as well as handle latent heat.

<u>Modeling Strategy</u>

Because no single code can model all the geotechnical aspects of DRM, the problem is decomposed so that the analysis can be performed in parts, using the best features of the three prime codes. Thus, SANGRE, COUPLEFLO and KRAK would be applied to four separate analyses as shown in Table 2.

Table 2: Relationships between rock melt scenario, types of analysis, and candidate codes.

Types of analysis and code used	Rock melting scenario			
	DUMP	Deep Self-Burial	Deep Rock Disposal	In-Situ Melting
┌ Cavity melt/SANGRE	x		x	x
│ Melt migration in └▶ cracks/KRAK	x		x	x
┊ Gas migration in └▶ cracks/KRAK	x		x	x
Self burial/COUPLEFLO		x		

⟶ Strong coupling: SANGRE output defines KRAK input for pressure driving the melt.

--▶ Partial coupling: SANGRE output provides temperatures in the solid medium but not vapor pressures for KRAK input.

As of early 1982, modeling of DUMP has started, with SANGRE.

SUMMARY

A strategy was developed to model the thermal, mechanical, and hydraulic aspects of nuclear waste disposal by rock melting (DRM). This was based on an examination of current knowledge of high temperature properties of granite, and on an evaluation of state-of-the-art numerical programs. This research also may benefit studies of other complex geological processes such as folding, plate tectonics, diapirism and magma migration, as well as studies of DRM in rocks other than granite.

REFERENCES

1. Heuze, F.E., 1981, "On the Geotechnical Modeling of High-Level Nuclear Waste Disposal by Rock Melting", Lawrence Livermore National Laboratory, UCRL-53183. December.

2. Cohen, J.J., Lewis, A.E., and Braun, R.L., 1971, "In-Situ Incorporation of Nuclear Waste in Deep Molten Silicate Rocks", Lawrence Livermore National Laboratory, UCRL-73320, July.

3. Bauer, S.J., and Johnson, B., 1979, "Effect of Slow Uniform Heating on the Physical Properties of the Westerly and Charcoal Granites", Proc. 20th Symp. on Rock Mechanids, Austin, Texas (Am. Soc. Civ. Eng., New York), pp 7-18.

4. Tullis, J. and Yund, R.A., 1977, "Experimental Deformation of Dry Westerly Granite", J. Geophys. Res., v. 82, pp 5705-5718.

MODELING OF NUCLEAR WASTE DISPOSAL

5. Friedman, M., Handin, J., Higgs, N.G. and Lantz, J.R., 1979, "Strength and Ductility of Four Dry Igneous Rocks at Low Pressures, and Temperatures to Partial Melting", Proc. 20th Symp. on Rock Mechanics, Austin, Texas (Am. Soc. Civ. Eng., New York), pp 35-50.

6. Van der Molen, I., 1981, "The Shift of the α-β Transition Temperature of Quartz Associated with the Thermal Expansion of Granite at High Pressures", Tectonophysics, v. 73, pp 323-342.

7. Rowley, J.C., 1974, "Rock Melting Applied to Excavation and Tunneling", Proc. 3rd Congress Int. Soc. Rock Mechanics, Denver, CO (Nat. Acad. Sci., Washington, D.C.), v. II-B, pp 1447-1453.

8. Shaw, H.R., 1965, "Comments on Viscosity, Crystal Settling, and Convection in Granitic Magmas", Am. J. Sci., v. 263, pp 121-152.

9. Weertman, J. and Weertman, J.R., 1975, "High-Temperature Creep of Rock, and Mantle Viscosity", Annual Rev. of Earth and Planetary Sci., v. 3, pp 293-315.

10. Carter, N.L., Anderson, D.A., Hansen, F.D., and Kranz, R.L., 1981, "Creep and Creep Rupture of Granitic Rocks", Geophysical Monograph 24 (American Geophysical Union, Washington, D.C.), pp 61-82.

11. Comini, G., Del Guidice, S., Lewis, R.W., and Zienkiewicz, O.C., 1974, "Finite Element Solution of Non-Linear Heat Conduction Problems, with Special Reference to Phase Change", Int. J. Num. Meth. in Eng., v. 8, pp 613-624.

12. Anderson, C.A., and Bridwell, R.J., 1980, "A Finite Element Method for Studying the Transient Non-Linear Thermal Creep of Geological Structures", Int. J. Num. Analyt. Meth. in Geomech., v. 4, pp 255-276.

13. Dawson, P.R. and Chavez, P.F., 1978 "COUPLEFLO: A Computer Program for Coupled Creeping Viscous Flow, and Conductive-Convective Heat Transfer", Sandia National Laboratories, Albuquerque, Part I: Theoretical Background (SAND-78-1406), and Part II: User's Manual (SAND-78-1407).

14. Travis, B.J., and Davis, A.H., 1980, "Calculation of Gas-Driven Fracture Propagation in Rocks", Proc. 21st Symp. on Rock Mech., Rolla, Miss. (U. of Missouri, Rolla), pp 356-361.

ACKNOWLEDGMENTS

This work was supported by the U.S. Department of Energy under contract W-7405-ENG-48 with the Lawrence Livermore National Laboratory. The author expresses his appreciation to H. C. Heard, and L. L. Schwartz for many useful discussions, and to L. D. Burrow for her fine typing of the manuscript.

Chapter 64

NONLINEAR THERMO-MECHANICAL BEHAVIOUR AND
STRESS ANALYSIS IN ROCKS

K.Y. Lo[1], R.S.C. Wai[2], R.K. Rowe[3] and L. Tham[4]

[1,3,4] Professor, Associate Professor and Postdoctoral Fellow
Faculty of Engineering Science
The University of Western Ontario
London, Canada N6A 5B9

[2] Geotechnical Engineer, Ontario Hydro
700 University Avenue
Toronto, Ontario, Canada

ABSTRACT

Laboratory tests for the determination of thermo-mechanical properties of three rock types were performed at temperatures up to $400°C$. Results showed that for the medium and coarse Granitic Gneisses, the coefficient of thermal expansion and Young's modulus are strongly temperature and temperature history dependent. A threshold temperature exists for each rock type above which material nonlinearity becomes important, and is associated with the progress of thermal cracking. Finite element computations were performed to illustrate the influence of material nonlinearity and boundary constraint on thermal stresses. Finally, a case history of thermal spalling was analyzed and it is shown that the results of analysis are consistent with field observations.

INTRODUCTION

Design of engineering facilities for energy and resource development such as underground nuclear power plants, high level radioactive waste repositories, hydrocarbon storage and in situ extraction of bitumen from oilsands and shales involve consideration of thermal loading. The operating temperatures for these facilities may vary from a low of $50°C$ for energy storage to $250°C$ for extraction of bitumen. The thermo-mechanical properties of the host rock are therefore of direct interest to the engineer, and a method of analysis for thermal stresses incorporating the thermal behaviour of the rock will be useful in the design and performance evaluation of these projects.

Laboratory tests for the determination of thermo-mechanical properties were performed on rock cores of a medium grained, a coarse

THERMO—MECHANICAL BEHAVIOR IN ROCKS 621

grained Granitic Gneiss, and a Limestone from a 303 m deep borehole. The measurements include (a) linear coefficient of thermal expansion, α, (b) diffusivity, κ, (c) uniaxial compressive strength, σ_o, (d) Young's modulus E and (e) Poisson's ratio ν in the temperature range of 20°C to 400°C. Cycles of first heating, cooling and reheating were performed so that the effects of temperature history on each parameter may be studied. The change in the crack porosity was also assessed by wave velocity measurements, and the role of thermal cracking in influencing the rock behaviour explored.

THERMO-MECHANICAL PROPERTIES

The coefficients of linear thermal expansion, α, were determined to an accuracy of approximately 0.5×10^{-6} per °C by means of fused quartz dilatometers using (i) a water bath and (ii) an electric furnace to maximum temperature of up to 400°C. The specimen size was generally 14 mm diameter and 75 mm long. The rate of heating employed was 2°C per minute so as to avoid premature thermal cracking (Richter and Simmons, 1974).

The results of three tests on Granitic Gneiss B at depth of 262 m are shown in Fig. 1. It may be observed that
(a) The coefficient of thermal expansion generally increases with temperature.
(b) Below a temperature of 110°C, there is no significant difference in α between first heating, cooling and reheating.
(c) Above 110°C, results of first heating lie above those from cooling and reheating.

The difference in behaviour between first heating and subsequent cooling/reheating may be attributed to thermal cracking which occurs after a threshold temperature is exceeded. This phenomenon will be further explored. It is significant to note, however, that α is a nonlinear material property. The rate of increase for α for reheating and cooling is about 4×10^{-6} per 100°C. The results of the coarse grained Granitic Gneiss are similar, the threshold temperature is about 130°C and increase in α in cooling and reheating is also approximately 4×10^{-6} per 100°C.

The results of two tests on the Gull River Limestone indicated a similar general behaviour to that of the Gneiss with a higher threshold temperature of about 210°C and a rate of $\Delta\alpha/100°C$ of 2×10^{-6}.

The test arrangement for thermal diffusivity, κ, consists of embedding a rock specimen 47.6 mm diameter and 15 mm thick in a cubical asbestos block of 180 mm. Thermocouples are installed at the top and bottom of the specimen. The entire assembly is heated up in a furnace and the value of κ is computed from the temperature histories recorded by the thermocouples using one dimensional heat conduction theory. The accuracy of the results is estimated to be about 0.1×10^{-6} m^2/s.

A study of the test results on medium grained Granitic Gneiss shows that (a) κ decreases with temperature increase. The rate of change, however, decreases with temperature. (b) The temperature history has a discernible effect on κ at lower temperatures, although the magnitude of this effect is probably within the natural variation of κ in a rock formation. Results of tests on the coarse grained Granitic Gneiss and Limestone showed similar behaviour.

The strength and deformation properties of the rocks were investigated by performing uniaxial compression tests on cylindrical specimens of 32.5 mm diameter and 65 mm in height. A MTS servo controlled testing machine (220 kN capacity) and a stiff testing frame were employed. The rate of load application was 4 kN/min. The deformations were measured by high temperature strain gauges, adhesives and fibreglass insulated leadwires. The arrangement was found to perform satisfactorily to a temperature of 370°C. A total of 37 tests were performed.

The influence of temperature history on the deformation modulus for medium-grained Granitic Gneiss is shown in Fig. 2 in which the ratio of modulus at any temperature to that at room temperature is plotted against temperature. It is obvious that at a given temperature, the value of the modulus depends on the temperature history to which the sample has been subjected. It is also significant to note that decreases in modulus occur at the same temperature independent of temperature history and path. This threshold temperature agrees well with that depicted in the measurements of coefficient of thermal expansion.

The Poisson's ratio for Granitic Gneiss also decreases with temperature increase. For Limestone, the effect is less marked; E remains essentially constant (<10%). The decrease in ν amounts to less than 20% for temperature increase up to 350°C.

An examination of the test results for the Granitic Gneisses shows that above about 120°C, there is a definite trend for the strength of the materials to decrease with temperature at the rate of approximately 30 MPa per 100 deg C. Below 120°C, the results are more erratic. The scatter in the data may be due to sample variations in terms of both material and moisture conditions. Although the specimens were vacuum dried, various amounts of moisture might still remain or be reabsorbed from the atmosphere during the setting up of the experiments. The weakening effect of moisture on the strength of rocks is well documented (Colback and Wild, 1965; Ballivy et al., 1976). There is evidence that upon heating, the strength of the rocks increases slightly at first (probably due to drying) before starting to decrease.

From a limited number of tests performed (8 tests), it seems that the uniaxial compression strength of the Limestone is insensitive to heating (at least up to about 300°C). This observation is consistent with previous results that E for Limestone is also essentially constant with temperature.

It has been shown above that a threshold temperature exists above

THERMO—MECHANICAL BEHAVIOR IN ROCKS

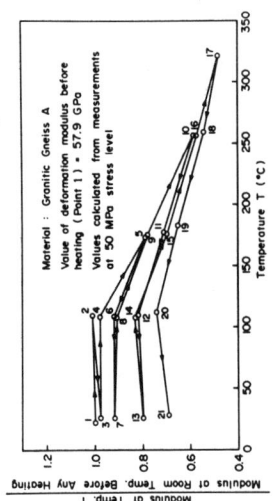

Fig. 2 Variation of Deformation Modulus of Specimen DAR/UCF/226.68 Under Cyclic Heating

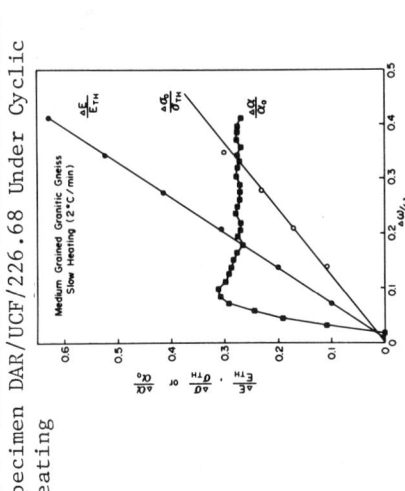

Fig. 4 Relationship Between Change in Modulus, Strength, Expansion Coefficient and Wave Velocity

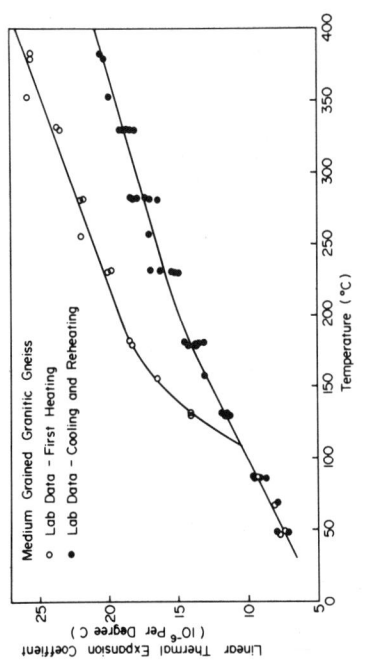

Fig. 1 Variation of Thermal Expansion Coefficient With Temperature

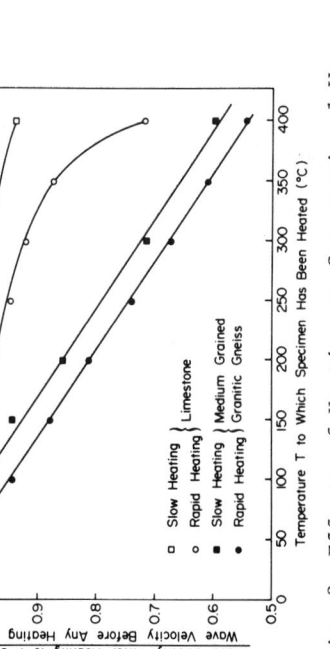

Fig. 3 Effects of Heating on Compressional Wave Velocity

which thermal cracking occurs and the process is an important factor in controlling the material behaviour of the rocks at elevated temperatures. To explore further this process, measurements of compressional wave velocity (ω), which was shown by Nur and Simmons (1970) to be indicative of crack porosity, were carried out.

Pairs of "identical" specimens (47.6 mm diameter, 50 mm thick) were vacuum dried. One specimen was subjected to "slow" heating at a rate of $2°C$ per minute to a specified temperature. The other specimen was subjected to "instantaneous" heating by placing the specimen in a preheated furnace. Measurements of ω were then carried out. The results for Limestone and medium grain Gneiss are shown in Fig. 3. Data for coarse grain Gneiss are very similar to those for medium grain Gneiss and are omitted for clarity. From Fig. 3, the following observations may be made: (a) The velocity with which compressional waves travel through a rock is reduced after heating, indicating an increase in crack porosity. (b) For a given rate of heating, the crack porosity increases with the magnitude of the temperature. (c) For a given temperature, the crack porosity increases with the rate of heating. This may be attributed to the thermal stress generated by higher temperature gradient. (d) The threshold temperatures for slow heating for rocks are consistent with those values deduced from results of α and E measurements at the same rate of heating. (e) The Limestone is much less susceptible to thermal cracking, as also indicated by results of tests for α and E.

From the results of these tests, an interpretation of the thermal behaviour of the rocks and the interrelationship of the thermal parameters at elevated temperatures may be made. Thermal stresses arise from two physical causes; thermal gradient and thermal "mismatch" of the constituent minerals. For the slow rate of heating employed in these experiments, it may be shown, from the classical theory of thermal elasticity that the tensile stress generated by thermal gradient is far below the tensile strength of the rock specimens, and thus may be neglected for the purpose of discussion.

"Thermal mismatch" results from different adjacent minerals having different coefficients of thermal expansion. With temperature increase, the grains tend to undergo different amounts of expansion. Compatibility of strains along the grain boundaries will restrict expansion of those grains having larger expansion coefficients while grains with lower expansion coefficients will be forced to elongate by an amount in excess of its own thermal expansion. This process depends on the magnitude of temperature increase, but not on the rate and will continue with temperature increase until the grain boundaries separate under internal tensile stress resulting in thermal cracking. This explains why cracking occurs even for slow rate of heating and a threshold temperature exists for each rock type. The same argument holds for homogeneous materials composed of anisotropic minerals, unless all grains are perfectly aligned in one direction. The presence of existing cracks (Griffith cracks) may enhance the process of thermal cracking but is not the cause by itself.

The influence of increase in crack porosity on E and α is shown in Fig. 4 for the medium grained Granitic Gneiss for results of tests above the threshold temperature. In Fig. 4, ΔE is the change in modulus at a higher temperature from E_{TH} at threshold temperature (110°C), $\Delta\sigma$ is the change in uniaxial strength from that at threshold temperature and $\Delta\alpha$ is the increase in α in excess of that value (α_o) due to intrinsic nonlinearity as determined from cooling and reheating, and $\Delta\omega$ is the change of compressional wave velocity from the initial wave velocity (ω_o). It may be seen that as $\Delta\omega/\omega_o$ increases, $\Delta E/E_{TH}$ and $\Delta\sigma_o/\sigma_{TH}$ increase linearly due to the weakening effect of thermal cracks. However, $\Delta\alpha/\alpha_o$ increases rapidly but tends to a constant value as $\Delta\omega/\omega_o$ increases.

THERMAL STRESS ANALYSIS

Using a finite element solution incorporating the nonlinear thermal-mechanical properties (Wai et al., 1981), computations were performed to study the effects of (a) material nonlinearity and (b) boundary constraints on thermal stresses for a 30 m diameter tunnel excavated deep into Granitic Gneiss and subjected to a sudden temperature rise of 180°C. Fig. 5 illustrates that, while each of the time-dependent parameters exerts significant influence on the computed thermal stresses, the effect of α_ψ dominates. The maximum stress at the tunnel surface may differ by a factor of 2 if constant parameters are employed. It appears, therefore, that for field cases where the temperature rise exceeds the threshold value, material nonlinearity should be incorporated into thermal stress analysis.

For the case of free thermal expansion in the z-direction (axial) due to presence of compressible sub-vertical joints, results of analysis show that there is a general reduction of stresses in the long term. Under short term conditions, however, high compressive stresses occur within a thin region at the surface followed by a zone of high tensile stresses. If cooling takes place instead of heating, the sign of stresses will be reversed and the surface of the rock cavern experiences very high tensile stresses and may lead to thermal spalling. Careful assessment must therefore be given to the design provisions such as stress relief slots and cooling a previously heated cavern after "accident" condition in underground nuclear plant design.

A CASE HISTORY OF THERMAL SPALLING

Thermal spalling has been observed in several field cases. Gray (1965) reported mass spalling of a 750 mm square, 3 m long test passage when subjected to a temperature rise of 61°C within a duration of 8.5 hours. The passage was excavated by drill and blast in a competent Granitic Paragneiss. No rock mechanics data were reported for the rock. However, it may be of some interest to examine whether spalling may be predicted in this case by assuming the data of Granitic Gneiss reported in this paper as applicable. To study the effect of geometry, two circular passages with diameters of 750 mm and 1000 mm inscribing and circumscribing the experimental passage were also analyzed.

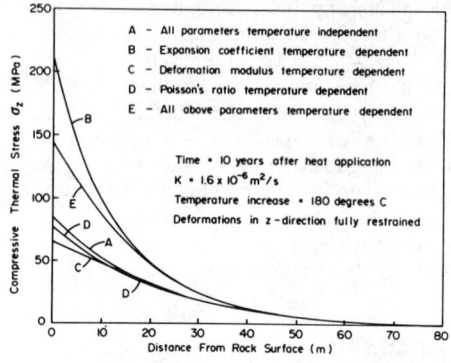
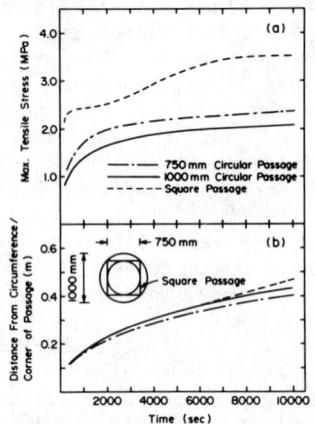

Fig. 5 Effects of Temperature Dependent Parameters on Thermal Stresses

Fig. 6 Variation in (a) Maximum Tensile Stress; and (b) Location of Maximum Tensile Stress With Time

The variation in the maximum tensile stress, and the position of this stress, with time is shown in Fig. 6 for all three passages. The maximum tensile stress approaches a peak value at approximately 2.8 hrs after the start of heating. The magnitude of this peak value is highly dependent on the passage geometry, having values of 2.1 and 2.4 for the 1000 mm and 750 mm circular passages and 3.5 MPa for the square passage. This suggests that the likelihood of spalling will depend upon the geometry of the passage or cavern as well as the temperature change and rock properties. The location of the point of maximum tensile stress advances with the temperature front and the peak tensile stress occurs at between 400-460 mm from the face of the passage.

The results of the thermal stress analysis show that the transient thermal tensile stresses computed would be of the order of the mass tensile strength expected, suggesting thermal fracturing would occur. The time span of occurrence and the smooth surface that developed also appear to be consistent with the results of the analysis.

For lack of experimental data, it is not possible to draw any firm quantitative conclusion regarding the observed spalling and the results of the theoretical analysis. It is suggested however that the theoretical analysis would provide a designer with a means of assessing the effects of geometry changes, uncertainty regarding the rock parameters and the magnitude of any temperature change on the likelihood of spalling.

CONCLUSIONS

(a) Results of tests show that, both for the medium and coarse grain

Granitic Gneiss, the thermal expansion coefficient increases, the diffusivity decreases, the Young's modulus decreases, the uniaxial compression decreases and the Poisson's ratio decreases as the test temperature is raised. For Limestone, α increases, κ decreases but E, σ_o and ν remain relatively unchanged.

(b) In general, the thermo-mechanical parameters are dependent both on temperature and temperature history.

(c) For a given rate of heating, a threshold temperature exists beyond which thermal cracking occurs. The temperature-dependent behaviour is, to a large extent, due to the progress of thermal cracking.

(d) Results of computation by a finite element analysis show that both the material nonlinearity and discontinuities in the rock mass exert significant influence on the thermal stresses and these factors should be incorporated in mathematical modelling.

(e) Significant transient tensile stress may develop during heating (or cooling) and due consideration should be paid in the design of underground cavities in rock to account for this effect.

ACKNOWLEDGEMENT

The research performed is supported by the Natural Sciences and Engineering Research Council under Grant No. A7745.

REFERENCES

Ballivy, G., Ladanyi, B. and Gill, D.E., 1976, Effect of Water Saturation History on the Strength of Low-Porosity Rocks. ASTM STP 599, American Society for Testing and Materials, pp. 4-20.

Colback, P.S.B. and Wild, B.L., 1965, The Influence of Moisture Content on the Compressive Strength of Rocks. Proc. Rock Mech. Symp., University of Toronto, pp. 65-83.

Gray, W.M., 1965, Surface Spalling by Thermal Stresses in Rocks. Proc. Rock Mech. Symp., Toronto, pp. 85-106.

Nur, A. and Simmons, G., 1970, The Origin of Small Cracks in Igneous Rocks. Int. J. Rock Mech. Min. Sci., Vol. 7, pp. 307-314.

Richter, D. and Simmons, G., 1974, Thermal Expansion Behaviour of Igneous Rocks. Int. J. Rock Mech. Min. Sci. & Geomech. Abstr., Vol. 11, pp. 403-411.

Wai, R.S.C., 1981, Rock Behaviour Under Elevated Temperatures, Ph.D. Thesis, Faculty of Engineering Science, University of Western Ontario.

Wai, R.S.C., Lo, K.Y. and Rowe, R.K., 1981, Thermal Stress in Rocks With Nonlinear Properties. Research Report, Faculty of Engineering Science, The University of Western Ontario, GEOT-8-81.

Chapter 65

A HYBRID DISCRETE ELEMENT-BOUNDARY ELEMENT METHOD OF
STRESS ANALYSIS

by L.J. Lorig and B.H.G. Brady

Graduate Student, Dept. of Civil and Mineral Engineering
University of Minnesota, Minneapolis, MN 55455

Associate Professor, Dept. of Civil and Mineral Engineering
University of Minnesota, Minneapolis, MN 55455

ABSTRACT

The Discrete Element Method is a numerical technique suitable for use in modeling the discontinuum behavior of jointed rock. The disadvantage of this method, in its application to analysis of underground excavations, is the necessity to discretize the complete problem domain. Since the far-field rock responds to the excavation as an elastic continuum, it may be appropriately represented by the Boundary Element Method. This exploits the inherent advantage of the Boundary Element Method in representing a quasi-infinite domain in terms of its internal surface geometry and associated boundary conditions. The procedures employed in developing a first-generation coupled Discrete Element-Boundary Element algorithm are described and the solution to a simple problem verifying the performance of the coupled code is presented. A procedure for numerical treatment of installed support systems is also presented.

INTRODUCTION

The design of underground excavations requires the capacity to determine support requirements and to assess the adequacy and timeliness of various types of installed support and reinforcement. The interactive nature of support mechanics is often explained in terms of a reaction curve for the rock medium, and a force-displacement curve for the support system. Thus, the mechanical behavior of both the local rock mass and the support is used to determine the point at which load applied to the peripheral rock by the support system re-establishes equilibrium in the rock mass. For the case of a circular excavation made in a homogeneous isotropic mass in a hydrostatic pre-mining stress field, the load-deformation curve for the rock mass can be

DISCRETE ELEMENT—BOUNDARY ELEMENT

determined and used in calculating the support required to achieve control of boundary displacements.

Two implicit assumptions in the application of the ground reaction curve--radial symmetry and rock mass continuum behavior--limit the ability of the model to analyze many real situations. In particular, the assumption of radial symmetry cannot take into account the flexural properties of the support. The assumption of rock mass continuum behavior does not permit consideration of cases where slip or separation occurs on discontinuity surfaces. The hybrid computational schemes described here are being developed to eliminate these difficulties. They will provide a numerical model of rock-support structure interaction suitable for use in designing support for arbitrarily shaped excavations in a jointed rock mass.

A HYBRID DISCRETE ELEMENT-BOUNDARY ELEMENT METHOD

Discontinuities in the rock mass frequently exercise a significant role in the local response of the rock to excavation development. For hard rock, or low field stresses, the main modes of local rock mass response involve separation and slip on discontinuity surfaces. Numerical models of such media must account for displacements which are orders of magnitude greater than block elastic deformations. The computational scheme must also model the progressive loosening behavior of the jointed rock mass. The only procedure which fulfills these requirements is the Discrete Element Method (also known as the Rigid Block or Distinct Element Method). Use of this method in the analysis of underground excavations requires the introduction of arbitrary external boundaries and boundary conditions. Another limitation of the Discrete Element Method is that the size of the numerical problem is related to the number of elements in the problem domain, and therefore to its volume.

Field measurements of displacements indicate that the far-field rock, which may be distant only a few equivalent radii from the excavation boundary, responds to the excavation as an elastic continuum. The Boundary Element Method has been frequently used to analyze problems in the design of rock excavations where there is no tendency for slip or separation. In such cases, the rock mass is modeled as an infinite or semi-infinite elastic body.

A hybrid Discrete Element-Boundary Element Method eliminates the individual disadvantages of the two methods and offers the following advantages: (1) elimination of arbitrary location and description of the conditions of the outer boundary of the blocky (Discrete Element) domain; (2) valid representation of elastic far-field behavior with the Boundary Element formulation; (3) only the excavation near-field region requires modeling with discrete elements, thereby reducing the size of the domain with complex constitutive behavior, and ensuring computational efficiency.

The nature of the coupled Discrete Element-Boundary Element system is shown in Fig. 1, with a set of discrete elements embedded in a boundary element domain. The medium is subject to known in-situ initial stresses prior to excavation, and excavation eliminates tractions on the surfaces defining the opening boundary. The condition to be satisfied on the interface between the Boundary Element domain and the Discrete Element domain is continuity of excavation induced displacements and equilibrium of the final stress state. It is assumed that it is sufficient if these conditions are satisfied at nodes defining the interface between the two domains.

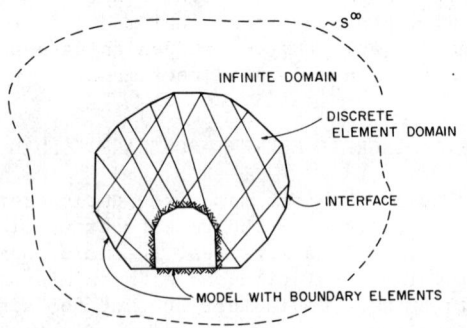

Fig. 1. Schematic representation of coupled Discrete Element-Boundary Element domains and resolution into component problems.

DISCRETE ELEMENT METHOD

This method of analysis, when applied to a discontinuous rock mass, represents the rock medium as a set of discrete blocks, each block capable of connecting mathematically with adjacent blocks. The elasticity of the blocks, the deformability of discontinuities and the frictional properties are represented by spring-slider systems with prescribed force-displacement relationships. In the numerical model, spring-slider systems are located at contact points between a block corner and the adjacent block edge. Blocks are rigid in the sense that the amount of penetration or overlap between adjacent blocks can be determined directly from block geometry and block centroidal translation and rotation. Normal forces which develop at block contacts are determined from the amount of notional penetration. Similarly, shear forces are determined by the amount of relative shear displacement with plastic shear failure developing according to the Mohr-Coulomb criterion. Elastic conditions may be re-established if the failure law is not satisfied at a subsequent stage. The Discrete Element Method sets shear and normal forces to zero if a tension condition exists. Again, shear and normal forces may redevelop with subsequent compressive interaction. Most discrete element formulations use elastic-perfectly plastic shear load deformation behavior as well as constant normal stiffness. Due to the explicit nature of the Discrete Element Method formulation, more complex constitutive behavior is

DISCRETE ELEMENT—BOUNDARY ELEMENT

possible with little increase in computational effort.

Both static and dynamic relaxation methods have been used in Discrete Element formulations. The terms "static" and "dynamic" refer to the motion law used in the solution procedure. Dynamic relaxation as applied to a jointed rock mass was introduced by Cundall (1971). The method uses the unbalanced forces and moments to compute block accelerations according to Newton's second law of motion. Accelerations are numerically integrated twice to compute displacements. Damping factors are required to solve static equilibrium problems. Stewart (1981) introduced a static relaxation routine developed from conventional relaxation principles similar to those presented by Hardy Cross (1932) for determination of moment distribution in continuous beams.

Performance of Discrete Element Routine

Since the elasticity of discrete elements is represented completely in terms of nodal springs, the Discrete Element Method may be used to model an elastic continuum if slip and separation are not allowed and if nodal spring stiffnesses are selected in accordance with elastic theory. For equal size rectangular elements with Poisson's ratio equal to zero, these constants are given by

$$K_{nx} = \frac{Ebc}{2a}, \quad K_{ny} = \frac{Eac}{2b}, \quad K_{sx} = \frac{Eac}{2b}, \quad K_{sy} = \frac{Ebc}{2a} \tag{1}$$

where E is Young's Modulus and a, b and c are the element half length, half height and thickness, respectively.

The elastic continuum formulation was used to determine stresses and displacements around a hole in a finite elastic plate subject to hydrostatic compression. Restricting the discrete elements to rectangular shapes meant that the hole only crudely represented a circle. In the numerical model, one quadrant was discretized into 108 square elements. Shear springs were eliminated along symmetry lines and normal springs along symmetry lines were set artificially high. Fig. 2 compares the analytical (Kirsch) solution for displacements and stresses around a circular opening in an infinite elastic plate with results obtained using the Discrete Element Method. Since the discrete elements form an irregular hole boundary, the radius shown in Fig. 2 was selected for use in the Kirsch solution. Differences between analytical and numerical results probably arise from approximating an infinite elastic plate by a finite plate, and from representing a circular hole by the shape lacking the smooth boundary curvature of the prototype.

BOUNDARY ELEMENT FORMULATION

Production of a stiffness matrix to define the elastic force-displacement behavior of the interior of the Boundary Element domain

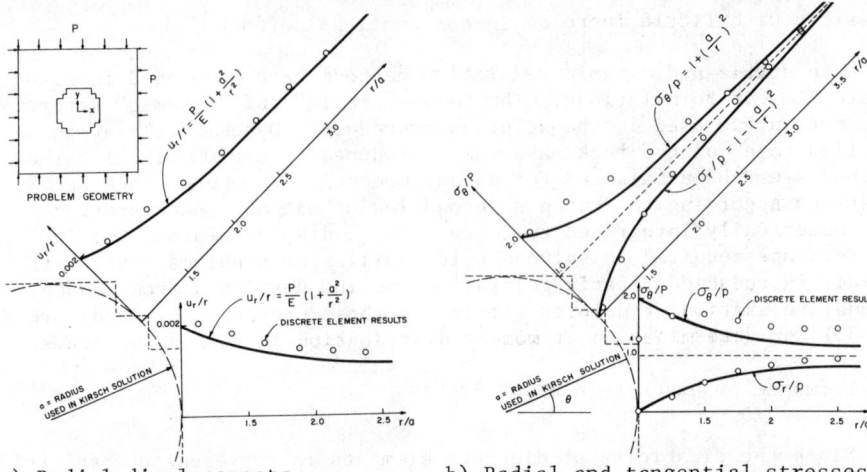

a) Radial displacements.   b) Radial and tangential stresses.

FIG. 2. Comparison of analytical solution and discrete element results for opening in an elastic plate.

is a primary requirement in coupling discrete elements and boundary elements. Brady and Wassyng (1981) demonstrated how such a stiffness matrix is established and used in coupling finite elements and boundary elements. In the current procedure, a linear isoparametric formulation of problem geometry and functional variation is used to construct a boundary constraint equation for the interface between domains. Development of the equation is initiated by applying Betti's Reciprocal Theorem to the two load cases in Fig. 3, yielding

$$\int_S (T_x^{xi} u_x + T_y^{xi} u_y) dS = \int_S (t_x U_x^{xi} + t_y U_y^{xi}) dS \tag{2}$$

The kernel functions T_x^{xi}, T_y^{xi}, U_x^{xi}, and U_y^{xi} are given by the Kelvin solution. Following division of surface S into n elements and integration over all elements, equation (2) can be recast in terms of nodal values of traction and displacement, yielding

$$\sum_{j=1}^{n} (T_{xj}^{xi} u_{xj} + T_{yj}^{xi} u_{yj}) = \sum_{j=1}^{n} (t_{xj} U_{xj}^{xi} + t_{yj} U_{yj}^{xi}) \tag{3}$$

where T_{xj}^{xi}, etc., represent the result of integration of the kernel-shape function products over the range of each element partitioned into components associated with each node. An identical procedure is followed for a Y-directed unit line load applied at the particular boundary node, to establish an expression similar to equation (3). Repetition of these procedures for loads applied at all nodes i of the surface S results in a set of 2n simultaneous equations, written in the form

DISCRETE ELEMENT—BOUNDARY ELEMENT

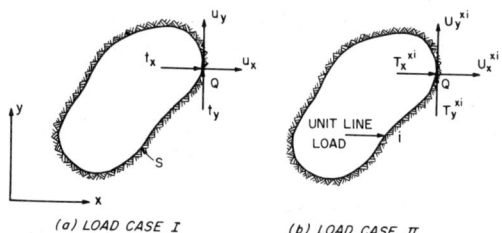

Fig. 3. Load cases used in boundary element formulation.

$$[T][u] = [U][t] \qquad (4)$$

Equation (4) can be recast in terms of nodal forces, q , by multiplying columns of [u] by appropriate factors obtained from the geometry of adjacent elements, yielding

$$[T][u] = [U'][q] \qquad . \qquad (5)$$

Rearrangement of equation (5) yields

$$[q] = [U']^{-1}[T][u] \qquad (6)$$

or $\qquad [q] = [K_b][u] \qquad (7)$

where $[K_b]$ represents the required stiffness matrix. The matrix inversion required by equation (6) can be achieved efficiently using the Transpose Elimination technique (Wassyng, 1982).

PRINCIPLES OF INTERFACING BOUNDARY ELEMENTS AND DISCRETE ELEMENTS

The interface between the two domains contains points common to both regions. Determination of induced displacements at each boundary element node permits direct determination of induced nodal forces [q], from equation (7). These are related to the total nodal forces $[q^t]$ by the expression

$$[q^t] = [q^o] + [q] \qquad (8)$$

where $[q^o]$ represents the initial nodal forces. Thus, equation (7) may be rewritten

$$[q^t] = [K_b][u] + [q^o] \qquad . \qquad (9)$$

The total nodal forces determined in this way can then be applied to appropriate block corners located on the periphery of the Discrete domain in the subsequent relaxation step of the iteration. Thus, the relaxation process continues, with nodal displacements and forces being updated and used in the subsequent relaxation cycle.

PERFORMANCE OF FIRST GENERATION HYBRID DISCRETE ELEMENT-BOUNDARY ELEMENT CODE

A coupled Discrete Element-Boundary Element computer code has been written using a static relaxation version of the discrete element formulation and the direct boundary element formulation described previously. Performance of the code has been assessed by investigating the problem of a square hole excavated in an infinite elastic plate subjected to hydrostatic stress. Fig. 4a shows that finer discretization gives a fair estimate of the boundary stresses, when one notes that the discrete element formulation imposes locally constant stress distribution in a block. Thus, the high stress gradients near the hole corner are reduced by an averaging process in the computational treatment. Fig. 4b illustrates the stress distribution in the Boundary Element domain, showing the hybrid code results are in accord with the independent elastic solution for the problem.

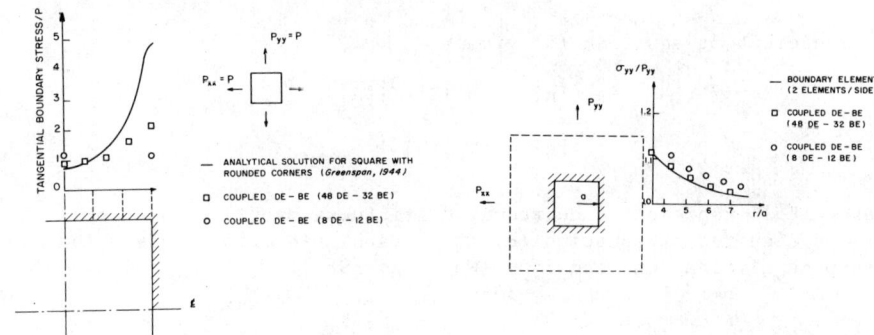

a) Tangential boundary stresses (Boundary Element domain).

b) Stresses in boundary element domain.

Fig. 4. Stress distributions around a square hole in infinite elastic medium.

PRINCIPLES OF INTERFACING DISCRETE ELEMENTS AND BEAM ELEMENTS

Beam elements, a member of the family of finite elements, have been used in the analysis (Brierley, 1975) and design (Monsees, 1977) of liner type support systems for underground excavations. In these cases, the rock mass was modeled by a series of springs oriented radially and tangentially relative to the support system. Key requirements in developing a hybrid Discrete Element-Beam Element code for modeling the interaction of near-field rock and the support system

DISCRETE ELEMENT—BOUNDARY ELEMENT

are: (1) determination of a stiffness matrix to define the force-displacement behavior of the support system, and (2) satisfaction of continuity requirements for force and displacement at the interface between the discrete elements and beam elements.

In analyzing the interaction problem, the size of the stiffness matrix for the support system can be reduced by deleting free nodes, i.e., those nodes which are not located at the rock-support interface. This is the case since these nodes are subjected to neither directly imposed external loads nor displacements by the surrounding medium. Suitable matrix reduction methods are described in such texts on structural analysis as Martin (1966). Displacements of interface nodes are obtained directly from the discrete element analysis and are used to calculate nodal forces mobilized in the support system. The calculated nodal forces are applied to the discrete elements during the subsequent iteration cycle.

CONCLUSIONS

The behavior of jointed rock around the periphery of an underground excavation is controlled by its interaction with the exterior quasi-elastic rock mass, and the interior, installed support system. It has been shown that stiffness matrices can be developed, from a standard boundary element formulation and a structural element representation of support components, for each of the elastic bodies which constrain the jointed assembly. When the jointed rock in the excavation near-field is modeled with discrete elements, it is a straightforward matter to calculate the load-deformation behavior of the complete system through a relaxation routine. This exploits simple force-displacement laws for joint surfaces, and takes account of the forces mobilized in the support system and in the surrounding rock medium by displacements of rock blocks.

A logical extension of the computational scheme described here would incorporate three-dimensional problem geometry. This would allow more realistic analysis of such problems as deformation behavior near a tunnel development end, and a tunnel intersection with a larger excavation.

ACKNOWLEDGEMENTS

The authors are grateful to Dr. A. Wassyng for helpful discussions related to this work and for providing the efficient equation solvers used in the codes. This work was partially supported by the University of Minnesota Computer Center.

REFERENCES

Brady, B.H.G. and Wassyng, A., 1981, "A Coupled Finite Element-Boundary Element Method of Stress Analysis," *International Journal of Rock Mechanics, Mining Sciences and Geomechanics Abstracts*, Vol. 18, pp. 475-485.

Brierley, G., 1975, "The Performance during Construction of the Liner of a Large, Shallow Underground Opening in the Rock," Ph.D. Thesis, University of Illinois at Urbana-Champaign.

Cross, Hardy, 1932, "Analysis of Continuous Frames by Distributing Fixed-End Moments," Paper No. 1793, *Transactions, American Society of Civil Engineers*, Vol. 96, pp. 1-10.

Cundall, P.A., 1971, "A Computer Model for Simulating Progressive Large Scale Movements in Blocky Rock Systems," *Proceedings, International Symposium on Rock Fracture*, Nancy (ISRM), Paper II-8.

Greenspan, M., 1944, "Effect of Small Hole on the Stresses in a Uniformly Loaded Plate," *Quarterly of Applied Mathematics*, Vol. 11, April, pp. 60-71.

Martin, H.C., 1966, *Introduction to Matrix Methods of Structural Analsis*, McGraw-Hill Book Co., pp. 173-175.

Monsees, J.E., 1977, "Station Design for the Washington Metro System," *Proceedings of the Engineering Foundation Conference--Shotcrete for Ground Support*, ACI Publication SP-54.

Stewart, I.J., 1981, "Numerical and Physical Modelling of Underground Excavations in Discontinuous Rock," Ph.D. Thesis, Imperial College, University of London.

Wassyng, A., 1982, "Solving $Ax = b$: A Method with Reduced Storage Requirements," *Society for Industrial and Applied Mathematics Journal of Numerical Analysis*, Vol. 19, No. 1, Feb., pp. 197-204.

Chapter 66

NUMERICAL SIMULATION OF FRACTURE

by L. G. Margolin and T. F. Adams

Earth and Space Science Division
Los Alamos National Laboratory
Los Alamos, New Mexico

ABSTRACT

The Bedded Crack Model (BCM) is a constitutive model for brittle materials. It is based on effective modulus theory and makes use of a generalized Griffith criterion for crack growth. It is used in a solid dynamic computer code to simulate stress wave propagation and fracture in rock. A general description of the model is given and then the theoretical basis for it is presented. Some effects of finite cell size in numerical simulations are discussed. The use of the BCM is illustrated in simulations of explosive fracture of oil shale. There is generally good agreement between the calculations and data from field experiments.

INTRODUCTION

Numerical simulation of stress wave propagation and fracture in rock is a topic of great current interest and importance. Applications of numerical programs range from *in situ* techniques for recovery of energy and mineral resources to nuclear weapons testing, and even to the study of earthquakes.

Numerical simulation requires a solid dynamic computer code to simulate stress wave propagation and a constitutive model to represent the material response, including fracture. The Bedded Crack Model (BCM) is a constitutive model that has been developed for brittle materials. It is based on a microphysical picture in which the evolution of a statistical distribution of penny shaped cracks is calculated.

The BCM addresses two questions. For a material containing penny shaped cracks:

1) how does the stress field affect the cracks - this is, when can cracks grow?

2) how do the cracks affect the material properties - that is, what are the effective elastic moduli of a cracked material?

Intrinsic to the model is the statistical framework used to describe the distribution of cracks as a function of size.

In the next section, we describe the theoretical basis of the model. We then investigate some effects of finite cell size in numerical simulation of stress wave propagation. Finally, we illustrate the use of the BCM in a stress wave code to simulate blasting in oil shale. In general, the calculations agree with data from field experiments.

THEORETICAL BASIS

Griffith Criterion

The question of when a crack can grow is answered by a generalized Griffith criterion. The criterion says that a crack will grow whenever that growth reduces the potential energy of the crack and the material that contains it. Griffith (1920) applied this criterion to the case of a two-dimensional crack (slit) in normal tension. We have generalized Griffith's analysis to three-dimensional (penny shaped) cracks in the x-y plane. A crack of radius c in normal tension will grow if

$$\sigma_{zz}^2 + \left(\frac{2}{2-\nu}\right)\left(\sigma_{xz}^2 + \sigma_{yz}^2\right) \geq \frac{4\pi TE}{c} . \tag{1}$$

Here ν is Poisson's ratio, E is Young's modulus and T is the surface tension. Equation 1 shows that in a given stress field, there is a critical crack size. Cracks bigger than the critical size grow, while smaller cracks are stable. The equation also shows that the effect of shear stress is to reduce the critical crack size.

The criterion can also be applied to closed cracks (normal compression). In this case, friction between the crack faces becomes important (McClintock and Walsh, 1962). The criterion for closed cracks has the form

$$\left(\frac{2}{2-\nu}\right)\left[\sigma_{xz}^2 + \sigma_{yz}^2 - (\tau_o - \mu\sigma_{zz})^2\right] \geq \frac{4\pi TE}{c} . \tag{2}$$

Here μ is the dynamic coefficient of friction and τ_o represents a cohesion.

Effective Elastic Moduli

The effective moduli are found from static solutions for strain as a function of applied stress in a randomly cracked body.

The basic assumption is that the total strain can be written as the sum of the strain in the material of the body plus the additional strain due to opening and sliding of the cracks. The additional strain due to the cracks is linear in the applied stress if crack interactions are ignored (Hoenig, 1979). Thus we can write

$$\varepsilon_{ij} = (M_{ijkl} + \tilde{M}_{ijkl}) \sigma_{kl} . \tag{3}$$

Here M_{ijkl} is the modulus of the uncracked material. $\tilde{M}_{ijkl}$ is proportional to the third moment of the crack distribution.

The crack size distribution changes with time as the cracks grow. The constitutive law then is

$$\frac{d\sigma_{kl}}{dt} = C_{ijkl} \left(\frac{d\varepsilon_{ij}}{dt} - \frac{d\tilde{M}_{ijmn}}{dt} \right) \sigma_{mn} , \tag{4}$$

where $C = (M + \tilde{M})^{-1}$. Thus, the constitutive relation has the form of a Maxwell solid with a variable relaxation time.

Crack Statistics

The initial distribution of cracks is assumed to be exponential: the number of cracks with radius greater than c is $N_0 \exp(-c/\bar{c})$. The constant $\bar{c}$ is a characteristic length scale of the initial distribution and N_0 is the total number of cracks per unit volume. The exponential dependence is not crucial to the model, but is convenient and is consistent with data for many rocks (Shockey et al., 1974).

At the beginning of each computational cycle, we use the Griffith criterion (equation 1 or 2) to determine which cracks, if any, may grow. In each cell, based on the stress, there is a critical size. In principle then, all cracks larger than the critical size grow at the asymptotic crack velocity for the duration of the cycle. Even for relatively simple stress histories, the critical crack size would have to be saved for each cell for each cycle. In the BCM, this problem is overcome by using a two parameter fit to represent the distribution function. Crack growth is not allowed until the critical size (c_{min}) reaches its minimum value. At this point, all cracks larger than this value of c_{min} are assumed to grow. The unstable cracks continue to grow until the smallest active cracks no longer satisfy the Griffith criterion. The time interval of active growth is one parameter and c_{min} is the second.

MESH EFFECTS

The finite size of a computational cell leads to two problems, both associated with numerical diffusion. First, there is the diffusion of fracture ahead of the wave. As a steep wave propagates through a cell, the cracks behind the peak cause a degradation of the effective elastic moduli of the material. Since there is only one set of moduli for the cell, the wave begins to propagate through partially fractured material, leading to too much attenuation. The problem can be characterized in terms of two time scales. One scale is physical, representing the time for the material to suffer significant reduction of the effective moduli. The second scale is numerical, the transit time of the wave through the cell, i.e., the cell size divided by a sound speed. When the numerical time scale becomes comparable to the physical time scale, numerical diffusion of the effects of fracture becomes significant.

The second problem is associated with the use of artificial viscosity to represent shock waves in the mesh (Wilkins, 1980). The artificial viscosity smears the numerical precursor to the shock over three or four computational cells. Because all cracks grow with the same asymptotic speed (Dunlaney and Brace, 1960), the shape of the precursor plays an important role in determining the amount of fracture ahead of the wave peak. The real rise time of the pulse is probably much less than is simulated with aritifical viscosity.

In the BCM, we prevent fracture in the precursor by not allowing crack growth until the wave peak is detected. This corresponds to assuming a very sharp wave front, so it underestimates slightly the attenuation. This treatment reduces the effect of the first problem, the diffusion of fracture. When the cells are large the attenuation may be greatly underestimated.

The effect of fracture in one-dimensional wave propagation is to cause an exponential attenuation (Piau, 1979). However, the attenuation coefficient is sensitive to mesh spacing if the cells are too large. This is demonstrated in Fig. 1, where we plot the calculated attenuation coefficient, μ, against cell size for a one-dimensional problem. For small cells, the attenuation asymptotically approaches a constant value. The figure shows that a critical point occurs when $\mu \Delta x \approx 1$. This is consistent with the criterion for numerical diffusion since $(\mu c_s)^{-1}$ is a time for fracture and $(\Delta x/c_s)$ is the transit time of a cell. Here c_s is a sound speed. If it is not possible to use sufficiently small cells, a shock fitting technique that allows subgrid resolution of the wave must be employed.

APPLICATION TO OIL SHALE

The BCM has been used with the two-dimensional stress wave code, YAQUI, to simulate fracture of oil shale by high explosives.

YAQUI is a finite difference "ALE" (Arbitrary Lagrangian-Eulerian) code (Amsden, Ruppel, and Hirt 1980). The ALE formulation allows the oil shale to be followed in Lagrangian coordinates, preserving material interfaces, while the high explosive gases are followed in nearly Eulerian coordinates as they rush up the borehole past the oil shale.

The version of the BCM used for the calculations assumes that the penny shaped cracks all lie in planes parallel to the bedding planes. This is a reasonable first approximation, since the bedding planes in oil shale are planes of weakness (Youash, 1969). Reasonable values are assumed for the initial crack density and mean crack size. These quantities are measurable, in principle, in the laboratory, although studies at SRI (Murri et al., 1977) have shown that it is difficult to identify the pre-existing cracks and flaws in direct microscopic observations in oil shale. The elastic constants for the rock matrix in which the cracks are embedded are taken from published fits to laboratory data (Johnson, 1979).

The YAQUI code with the BCM constitutive model has been used to simulate a series of oil shale blasting experiments that was conducted in the Colony Mine near Parachute, CO. These experiments were done in cooperation with the Colony Development Corporation and Atlantic Richfield. Experiment 79.10 involved 24.7 kg of a commercial ammonium nitrate/fuel oil explosive (ANFO) emplaced in a 0.15 m-diameter borehole drilled vertically into the mine floor. The charge was 1.7 m in length and the depth to the bottom of the charge was 3.3 m. The charge was detonated from the bottom.

Experiment 79.10 produced a crater filled with loose rubble. The crater was subsequently excavated and profiles of the crater were measured for comparison with the calculations. The predicted fracture distribution 3.0 m-sec after the firing of the detonator is shown in Fig. 2. A typical measured cross section for the 79.10 crater is also shown in this figure. The predicted extent of fracture at the free surface agrees well with the width of the crater in the field. The code also predicts a large amount of fracture beneath the observed crater. However, the crater is not just the region where cracks have grown, but where complete fragmentation and tumbling of the rubble has occurred as well. Thus, we expect the observed crater to be shallower, since the broken rock at depth is locked in place and does not acquire upward momentum.

It is instructive to compare the predicted fracture for experiment 79.10 with the observed crater profile for experiment 79.12. That experiment consisted of four charges, each of which was approximately the same in size and depth of burial as the single charge in experiment 79.10. The four charges were arranged in a 3.2 m square pattern. A typical profile of the experiment 79.12 crater (along one side of the pattern) is shown in Fig. 3, along with the predicted fracture pattern, centered on one of the charges.

There is a good agreement between the predicted fracture and the observed crater from the surface down to about the level of the bottom of the charge. The effect of having four charges, spaced as in experiment 79.12, therefore appears not to be an increase in the extent of breakage, but rather in the total amount of loose and tumbled rubble. This probably has to do with the occurrence of multiple shocks and the way the high pressure explosive product gases penetrate into the shock-induced fracture network. The net effect is to cause the rubble-filled crater to more closely match the extent of the fractured rock. The calculation shows predicted fracture below the explosive borehole. This fractured rock will not be tumbled, even in a multiple borehole experiment, because of its location.

Code calculations can also be compared with field data from acceleration, velocity, and stress gauges. Measurements of peak vertical velocity at several locations on the free surface were made in a recent field experiment. That experiment was similar to experiment 79.10, except that low-density TNT replaced ANFO as the explosive. The predicted and observed peak velocities are plotted against range from ground zero in Fig. 4. This figure shows good agreement between the calculations and the observations.

ACKNOWLEDGEMENTS

This work was supported by the U. S. Department of Energy, Office of the Assistant Secretary for Fossil Energy.

REFERENCES

Amsden, A. A., Ruppel, H. M., and Hirt, C. W., 1980, "SALE: A Simplified ALE Computer Program for Fluid Flow at All Speeds," LA-8095, June, Los Alamos, NM, available from National Technical Information Service, Springfield, VA.

Dunlaney, E. N., and Brace, W. F., 1960, "Velocity Behavior of a Growing Crack," Journal of Applied Physics, Vol. 31, No. 12, December, pp. 2233-2236.

Griffith, A. A., 1920, "The Phenomena of Rupture and Flow in Solids," Philosophical Transactions of the Royal Society A, Vol. 221, No. 6, October, pp. 163-198.

Hoenig, A., 1979, "Elastic Moduli of a Non-Randomly Cracked Body," International Journal of Solids and Structures, Vol. 15, No. 2, February, pp. 137-154.

Johnson, J. N., 1979, "Calculation of Explosive Rock Breakage: Oil Shale," Proceedings of the 20th U. S. Symposium on Rock Mechanics, June, U. S. National Committee for Rock Mechanics, pp. 109-118.

McClintock, F. A., and Walsh, J. B. 1962, "Friction on Griffith Cracks in Rocks Under Pressure," Proceedings of the 4th U. S. National Congress of Applied Mechanics, Berkeley, CA, pp. 1015-1021.

Murri, W. J., et al., 1977, "Determination of Dynamic Fracture Parameters for Oil Shale, Final Report Covering the Period 1 June 1976 to 30 September 1976, Contract 03-4487 Under AT(29-1)-789," Stanford Research Institute, Menlo Park, CA, 42 pp.

Piau, M., 1979, "Attenuation of a Plane Compressional Wave by a Random Distribution of Thin Circular Cracks," International Journal of Engineering Science, Vol. 17, No. 2, February, pp. 151-167.

Shockey, et al., 1974, "Fragmentation of Rock Under Dynamic Loads," International Journal of Rock Mechanics and Mining Science and Geomechanics Abstracts, Vol. 11, No. 8, August, pp. 303-317.

Wilkins, M. L., 1980, "Use of Artificial Viscosity in Multidimensional Fluid Dynamic Calculations," Journal of Computational Physics, Vol. 36, No. 2, July, pp. 281-303.

Youash, Y., 1969, "Tension Tests on Layered Rocks," Geological Society of America Bulletin, Vol. 80, No. 2, February, pp. 303-306.

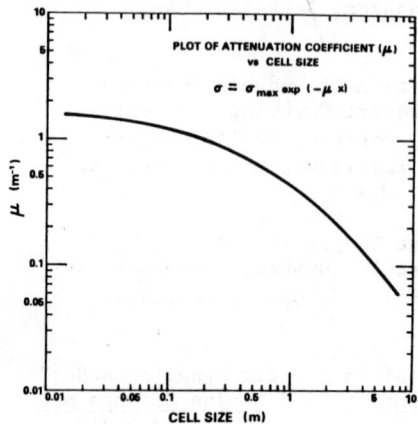

Fig. 1. Attenuation coefficient versus cell size.

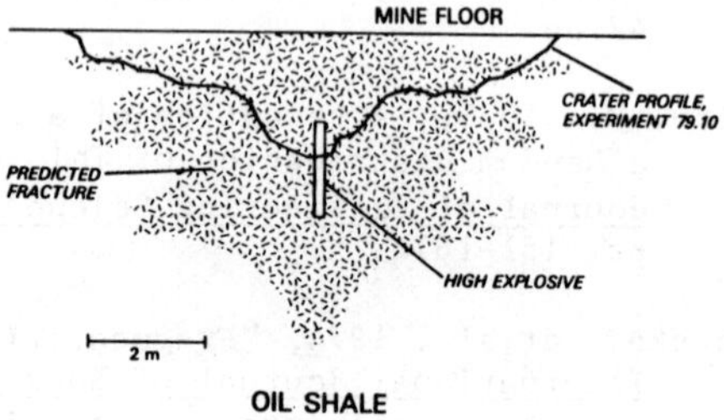

Fig. 2. Predicted fracture and observed crater for experiment 79.10.

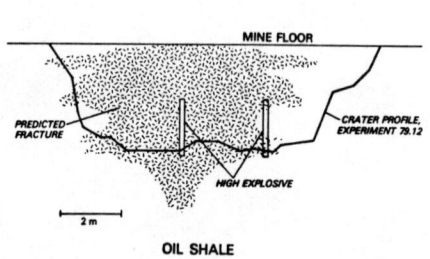

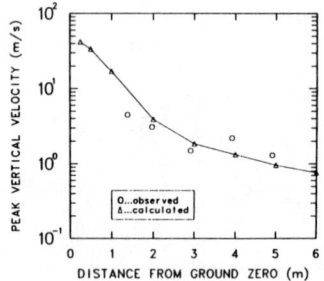

Fig. 3. Predicted fracture for experiment 79.10 and observed crater for experiment 79.12.

Fig. 4. Peak vertical velocity versus distance from ground zero.

Chapter 67

NUMERICAL SIMULATION OF FLUID INJECTION INTO DEFORMABLE FRACTURES

J. Noorishad and T. W. Doe
Earth Sciences Division
Lawrence Berkeley Laboratory
University of California
Berkeley, California   94720

ABSTRACT

The problem of fluid flow from a well into a horizontal fracture has been studied using a nonlinear coupled hydromechanical finite element model. Such an approach is required due to the strong nonlinearity and anisotropy of deformation and fluid flow moduli introduced by the presence of the fracture. The resulting nonuniform total stress field invalidates the application of the conventional fluid flow techniques to the problem. Stable convergent solution from the above analysis are verified by a coupled steady state algorithm.

INTRODUCTION

Injection of fluids into fractured rocks is a common practical concern in many phases of civil, mining, and petroleum engineering. The traditional means of predicting injection flow rate or well pressure behavior is by use of solutions of the general transient equations of fluid flow in porous or equivalent porous media (Dewiest, 1968; Snow 1968). Availability of numerical techniques and computers has extended this analysis capability and made the simulation of fluid flow in fractured and/or porous media a possible task. In these treatments the deformability of the medium is thought to be represented through the concept of specific storage. This assumption, inherently requires the existence of a uniform total stress field unaffected by temporal and spacial variations of the fluid pressure field (Terzaghi, 1925). Such conditions generally, prevail for regional fluid flow problems in saturated porous continua within acceptable range of approximation. However, the requirements cannot be met where rapid variations of pressure field are taking place or when the media possess strong nonlinearity and anisotropy with regard to different deformation moduli and fluid flow properties (Snow, 1968; Noorishad, 1971). Therefore, the traditional fluid flow treatments may no longer be applicable in such situations. Development of the general theory of consolidation by Biot (1941) has provided the basis on which more realistic attempts of predicting fluid flow behavior of deformable media have been made (e.g. Ghaboussi and Wilson, 1971). In this study a new coupled hydromechanical finite element technique, developed by Ayatollahi et al. (1982) and generalized by Noorishad et al. (1982), is used to simulate a more realistic behavior of a deep lying fracture subject to fluid injection under constant head and constant flow conditions. In verifying these results the fluid flow equation of the coupled phenomena is reformulated in an approximate manner and a finite element algorithm for the coupled steady state phenomena is worked out.

FIELD EQUATIONS OF COUPLED HYDROMECHANICAL PHENOMENA

Field equations originally formulated by Biot (1941) in his general theory of consolidation are the basis for the hydromechanical analysis of the fractured porous media (Ayatollahi et al. 1982). These relationships for the isotropic continuum portions of the medium include the stress-strain equation, the fluid flow law, and the law of static equilibrium as follows:

$$\tau_{ij} = 2\mu e_{ij} + \lambda \delta_{ij} \delta_{k\ell} e_{k\ell} + \alpha \delta_{ij} P \tag{1}$$

$$\zeta = -\alpha \delta_{ij} e_{ij} + \frac{1}{M} P = -\alpha e + \frac{1}{M} P$$

$$\frac{\partial \zeta}{\partial t} = \nabla \cdot \frac{k}{\eta_\ell} (\nabla P + \rho_\ell g \nabla z) \tag{2}$$

$$\frac{\partial \tau_{ij}}{\partial x_j} + \rho_s f_i = 0 \tag{3}$$

where τ_{ij} = solid stress tensor (tension positive), e_{ij} = solid strain tensor, e = bulk dilatation equal to $e_{11} + e_{22} + e_{33}$, P = fluid pressure, δ_{ij} = Kronecker delta function, α = Biot's coupling coefficient, M = Biot's storativity coefficient, ζ = fluid volume strain, $\underline{k}$ = intrinsic permeability tensor of porous parts, η_ℓ = liquid dynamic viscosity, ρ_ℓ = fluid mass density, ρ_s = average porous solid density, f_i = body force components, and μ, λ = Lame's elasticity constants. Similar equations for the fracture portion of the medium have been developed which are not given here for the sake of brevity. The reader is referred to Noorishad et al. (1982). The field equations for the porous solid and the fracture along with initial and boundary conditions completely define the mixed initial boundary value problem of fluid flow in a deformable fractured porous rocks.

SIMULATION OF FLUID INJECTION 647

SOLUTION APPROACH

Complexity of the phenomena of flow of fluids in deformable fractured rocks practically inhibits analytical solution attempts. Only simple problems involving linear isotropic porous media have been solved. Extension of the existing variational finite-element method for linear elastic porous media (Ghaboussi and Wilson, 1971) provides the numerical formulation of the problem as follows:

$$\underline{K}\,\underline{U} + \underline{C}\,\underline{P} = \underline{F}$$

$$\underline{C}^T\underline{U} - \underline{E}\,\underline{P} - 1*\underline{H}\,\underline{P} = 1*\underline{Q}$$

(4)

where $\underline{K}$ = structural stiffness matrix of the fractured media, $\underline{C}$ = coupling matrix, $\underline{E}$ = storativity matrix, $\underline{H}$ = fluid conductivity matrix, $\underline{F}$ = nodal force vector, and $\underline{Q}$ = nodal flow vector. In equation (4), $\underline{U}$ and $\underline{P}$ represent the nodal displacement and pressure vectors and 1* stands for time integration.

Transient Solution

A predictor-corrector scheme is used to obtain the time marching solution of equation (4). The development of the above equations and the details of the solution are explained in Noorishad et al. (1982).

Steady State Solution

Simple modification of equation (4) offers a way of directly determining steady-state solutions for coupled hydromechanical problems. Considering the basic assumptions regarding the nature of the dependent variables (Finlayson, 1972), the term $1*\underline{H}\,\underline{P}$ in equation (4) could be replaced by $t_{st}\underline{H}\,\underline{P}_{st}$ for linear problems. In the latter expression t_{st} equals a very large time value and $\underline{P}_{st}$ represents the steady-state pressure to be found. Implementing this operation in equation (4) results in the following equation

$$\underline{K}\,\underline{U}_{st} + \underline{C}\,\underline{P}_{st} = \underline{F}$$

$$\underline{C}^T\underline{U}_{st} - (\underline{E} + t_{st}\underline{H})\underline{P}_{st} = t_{st}\underline{Q}$$

(5)

A single direct solution of equation (5) provides the steady state results for coupled hydromechanical problems in porous elastic solids.

For nonlinear materials, such as fractured rocks, equation (4) is nonlinear on the account of the dependency of $\underline{K}$ and $\underline{H}$ matrices on the dependent variables. Therefore, the term $1*\underline{H}\,\underline{P}$ can only be replaced by $t_{st}\underline{H}_{st}\underline{P}_{st}$ which appears to make the solution complicated. However, the same formulation as in equation (5) in which $\underline{K}$ and $\underline{H}$, representing the first approximations of final values of $\underline{K}_{st}$ and $\underline{H}_{st}$ can be used to start an iteration loop in which all nonlinear quantities are updated until convergence to true value of P_{st} and U_{st} is achieved. At this point, the nonlinear structural stiffness matrix $\underline{K}$ and the system conductivity matrix $\underline{H}$ will have taken their final steady state values.

This steady state modification, implemented in the computer code, provides a useful option for prompt assessment of the late time results of the transient problems such as the ones considered below.

APPLICATION TO FLUID INJECTION PROBLEMS

The problem considered here, as sketched in Figure 1 is that of a horizontal fracture in a low permeability, rigid rock (e.g. granite) located at a depth of 100 meters. A 0.05 m radius well intersects the fracture at its center. It is assumed that the fracture encounters a region of high permeability capable of maintaining a constant head at a radial distance of 150 m from the well. The rock has a density of 2600 kg/m^3. The hydrologic system is at ground level hydrostatic equilibrium before fluid injection takes place. Other system properties are shown in Table 1. To simulate this problem by a two-dimensional model certain assumptions have to be made. The overburden is replaced by a 10 m slab of rock with a load on its upper surface equal to the remaining 90 m of overburden. Due to the very small deformations involved not much accuracy is lost by ignoring the tangential tractions which might be mobilized at the top of the modeled slab. To insure static equilibrium, corresponding in-situ stress and initial pressures are input in the model. The problem is assumed to have axial symmetry, and its outer boundary is restricted to vertical movement only. The fracture is modeled by a number of joint elements (Goodman et al., 1968) with initial stiffnesses of K_s = 0.5 GPa/m, K_n = 1.6 GPa/m and is assumed to have a linear behavior in the load-deformation range under consideration. To represent the rock rigidity, the Young's modulus is assumed to have a value of 70 GPa. The rock is of sufficiently low permeability (10^{-20} m^2), such that under the hydraulic conditions and the time periods considered the flow into the rock itself would be confined to the 10 m slab of the model.

Constant Head Injection

Steady state analysis. The system is pressurized by injection under 50 m of constant differential head. Note that the injection head is less than one quarter of the overburden stress on the fracture. The steady-state pressure profile in the fracture is plotted in Figure 2. The resulting steady-state flow rate is 0.381 m^3/sec.

In a second run the calculated flow rate above is used to simulate a constant flow rate injection problem. The results of this run plots precisely on the pressure profile curve of Figure 2 as is expected, thus adding confidence to the soundness of the steady-state algorithm.

Transient analysis. To simulate the transient behavior of the system in the above problem, the coupled finite-element technique

SIMULATION OF FLUID INJECTION

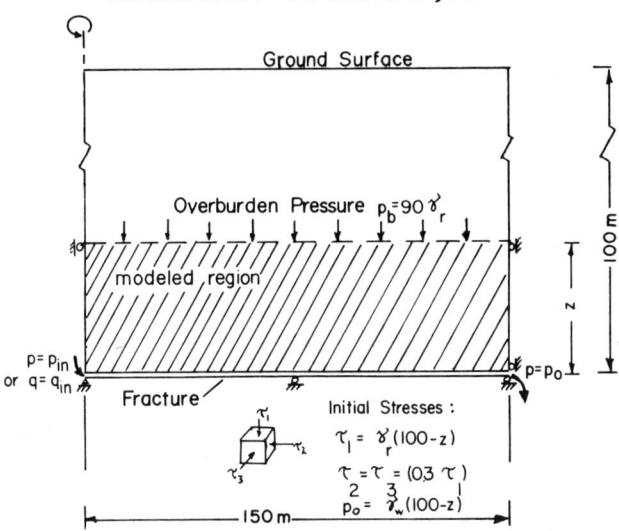

Figure 1. Schematic geometry of the model used in injection problem.

Table 1. Data Used for Various Analysis of Fracture Injection Problem.

Material	Property	Value
Fluid	Mass density, ρ_ℓ Compressibility, β_p Dynamic viscosity, η_ℓ	9.80×10^2 kg/m^3 5.13×10^{-1} GPa^{-1} 2.80×10^{-4} N sec/m^2
Rock	Young's modulus, E_s Poisson's ratio, ν_s Mass density, ρ_s Porosity, θ Intrinsic Permeability, k Biot's constant, M Biot's coupling constant, α	70.0 GPa, 0.7 GPA 0.25 2.5×10^3 kg/m^3 0.015 10^{-20} m^2 1.47 GPa, 14.0 GPa* 1.0, 0.0*
Fractures	Initial normal stiffness, K_n Initial tangential stiffness, K_s Cohesion, C_o Friction angle, δ Initial aperture, b Porosity, θ Biot's constant, M Biot's constant, α	1.60 GPa/m 0.50 GPa/m 0.0 30° 10^{-4} m 0.50 1.47 GPa, 14.0 GPa* 1.0, 0.0*

* Used in the uncoupled case.

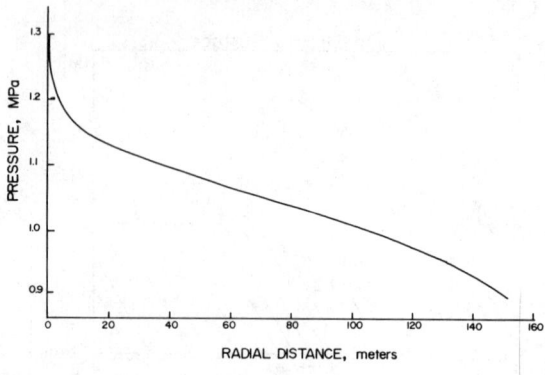

Figure 2. Steady-state pressure distribution along the fracture.

(Noorishad et al. 1982) is used. The resulting values of the well flow rates are plotted in Figure 3. Curve B exhibits an initial exponential decline lasting about 10 seconds followed by a lengthy slow rise continuing for about a 1000 seconds until a steady state trend of much longer duration takes over. The final steady state flow rate obtained here closely approximates the results of the one-step steady state analysis performed earlier. The early time behavior follows the familiar pattern of the traditional constant injection head fluid flow problems shown in curve C. To delineate this behavior, comparison needs to be made between fluid flow equations of the coupled (hydromechanical) problem and that of the uncoupled or equivalent nondeformable problem. In the coupled theory, using the assumption of incompressible solid grains, the flow equation as expressed in equation (2) is of the following form:

$$\nabla \cdot \underline{k} \nabla (P + \rho_\ell gz) = \frac{\partial}{\partial t} (-\alpha e + \theta \beta_p P) \qquad (6)$$

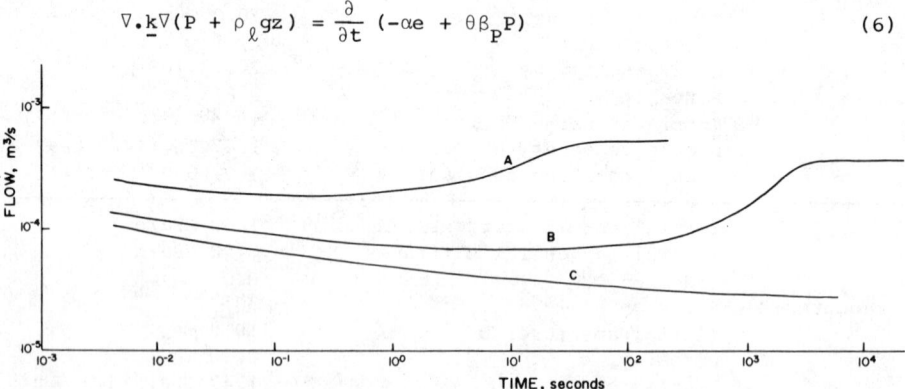

Figure 3. Transient well flow rate versus time for (a) nondeformable fracture C, (b) deformable fracture overlain by rigid rock (E = 7.0 GPa) B, and (c) deformable fracture overlain by soft rock (E = 0.7 GPa) A.

SIMULATION OF FLUID INJECTION

Under the condition of uniform total stresses in the fluid flow region, such as those assumed in one-dimensional theory of consolidation, it is possible to write $\tau' = -P$, where τ' is the effective normal stress in the fracture. Using the definition of compressibility one easily finds $\beta_r = -e/P$. Notice that τ', P and e are incremental in nature and represent deviations from initial state. Therefore,

$$\nabla \cdot \underline{k} \nabla (P + \rho_\ell gz) = \frac{\partial}{\partial t} (\beta_r + \theta \beta_p) P \qquad (7)$$

which is the same as the familiar fluid flow equation with $(\beta_r + \theta \beta_p)$ representing the constant specific storage parameters S_s. Thus, under <u>specific conditions</u>, the true fluid flow behavior of a deformable porous elastic continua may comparably be obtained by a fluid flow analysis alone using the conventional fluid flow equation. The traditional approach has also been employed for analysis of fluid flow behavior of nonlinear material, such as fractured rocks, by using an equivalent specific storage for fractures defined as $S_s^f = 1/(2bK_n)$ (e.g. Snow 1968). Such extensions of the concept of specific storage may hold valid for extensively fractured rocks with frictionless fractures under special circumstances. However, the combination of anisotropy and nonlinearity due to the presence of fractures and the consequent pressure-coupled total stress field, invalidates application to fluid flow problems in most fractured rocks. To demonstrate this point, the problem under consideration is solved in uncoupled manner (i.e., fluid flow analysis alone) using an equivalent storage value of $S_s^f = 1/2bK_n$ equal to 0.06 MPa^{-1} (neglecting the much smaller fluid compressibility contribution). The results, checked also by Jacob-Lohman (1951) analytic solution, plot much higher than the coupled analysis curve indicating need for a much smaller specific storage value in the analysis. With some trial and error, using a specific storage of 2.0 GPa^{-1}, an uncoupled solution, curve c, is obtained. These solutions and comparison with the coupled analysis results point to the fact that although the fracture tends to behave in regard to fluid flow <u>in the very early period</u>, like an equivalent porous material, its storage is not represented by the $1/2bK_n$, but rather by the deformation behavior of the rock-fracture system. This porous-media-like behavior is rapidly violated in later time by the effects of deformation on the intrinsic fracture permeability, necessitating a coupled stress-flow analysis of such problems.

The above coupled transient analysis has been repeated for a much softer rock having a Young's modulus of 0.7 GPa. The same general behavior, though much less striking as shown in Figure 3 curve A, is observed.

Constant flow rate injection

A transient wellbore pressure analysis for a constant flow rate injection problem has been made using the same system geometries as described above for the constant-head case. The analyses are made using a flow rate of 0.381×10^{-3} m^3/sec, and both deformable and nondeformable conditions are applied for the fracture. In the latter case, a storage coefficient of 2.0 GPa^{-1} (as in the earlier problem) is used. The transient pressure curves are shown in Figure 4. As with the transient flow rate case, the results of the conventional fluid flow and the coupled stress and fluid flow analysis follow the same pattern for the very early behavior but deviate as pressure front advances radially in the fracture.

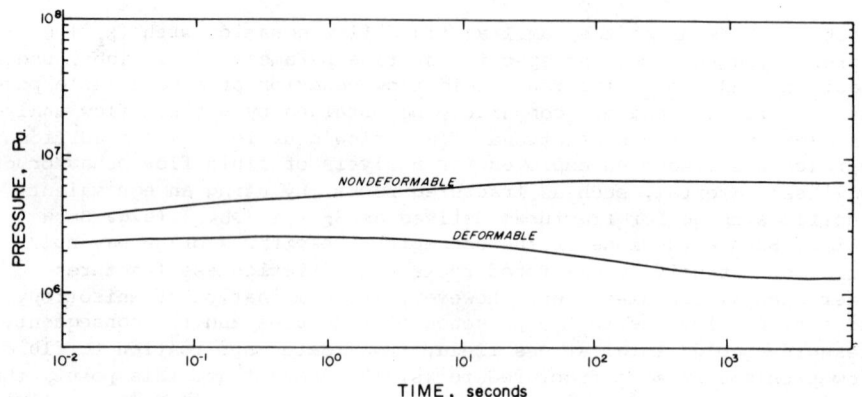

Figure 4. Transient well pressure versus time for (a) nondeformable fracture and (b) deformable fracture overlain by rigid rock (E = 70. Pa).

CONCLUSION

The primary conclusion of this work is that pressure-dependence of permeability through fracture deformation may have a major effect on the behavior of fluid-injection phenomena. In well testing simulations, the inclusion of deformation results in drastic changes in the form of transient flow rate or pressure curves that may invalidate use of conventional well-test type curves. The deformation of the fracture depends on the state of the effective stress tensor in the rock rather than the pressure alone.

Unfortunately, there is not any solution technique available for verifying the coupled stress-flow model. However, using an approximate reformulation of the flow equation in the coupled method, an explanation for early time behavior of the flow rate, in the constant head injection problem, is presented. An important by-product of this attempt is that fluid flow analysis of fractured rocks can not generally be performed by employing the conventional methods which use the concept of specific storage. Definition of fracture specific storage, based on its deformation moduli (Snow, 1963), may not be realistic even under favorable conditions.

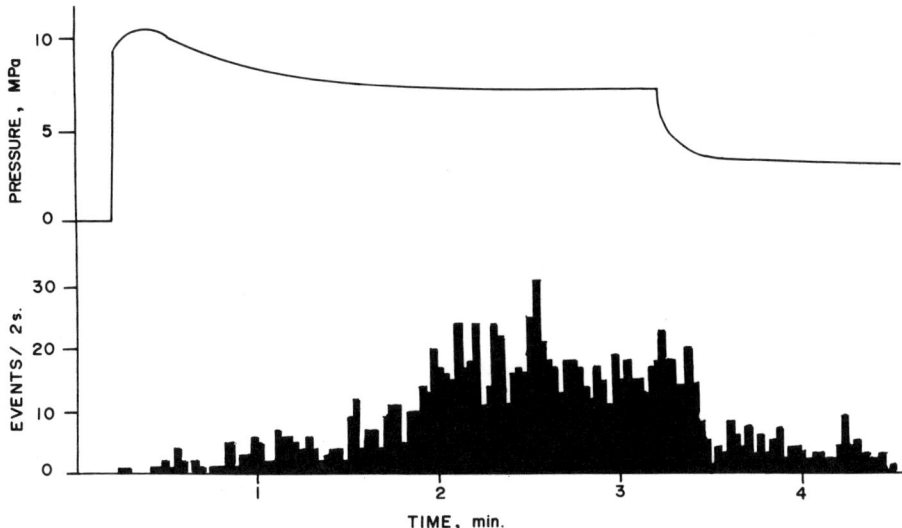

Figure 5. Pressure-time record for reinjection of a hydraulic fracture shown with acoustic emission record (Majer and McEvilly, 1982).

Employing an algorithm developed here, the steady-state coupled hydromechanical problem is solved directly and the same steady flow rate and pressure profile as those resulting from the transient analyses are obtained.

The general form of the transient pressure curves predicted by the coupled model for the constant wellbore flow case have been obtained from hydraulic fracturing experiments at the Stripa mine in Sweden (Majer and McEvilly, 1982). A reinjection of a previously generated fracture at constant flow (Figure 5) showed the initial increase in pressure followed by a gradual pressure decline. Such behavior has in the past been interpreted as fracture extension, however, acoustic emission data obtained during the pumping showed that breakdown of the rock near the fracture tip did not occur until well into the test.

Existence of mixed and geometrically complex boundary conditions and variable rock properties will naturally alter the behavior of pressure and flow rate from those described in the simple models presented in this paper. Nonetheless, we are confident that coupled stress-flow models can provide valuable insight into the phenomena accompanying fluid injection.

ACKNOWLEDGEMENTS

This work was supported by the Assistant Secretary for Nuclear Energy, Office of Waste Isolation of the U. S. Department of Energy under Contract DE-AC03-76SF00098. Funding for this project is administered by the Office of Nuclear Waste Isolation at Battelle Memorial Institute.

REFERENCES

Ayatollahi, M. S., Noorishad, J., and Witherspoon, P. A., 1982, "Stress-Fluid Flow Analysis in Fractured Rock Masses," Journal of Engineering Mechanics Div., ASCE, in press.

Biot, M. A., 1941, "General Theory of Three-dimensional Consolidation," Journal Applied Physics, Vol. 12, pp. 155-164.

Dewiest, R., 1966, "On the Storage Coefficient and the Equation of Groundwater Flow," Journal of Geophysical Research, Vol. 71, pp. 1117-1122.

Jacob, C. E. and Lohman, S., 1952, "Nonsteady Flow to a Well of Constant Drawdown in an Extensive Aquifer," Transactions, Am. Geophys. Union, Vol. 33, No. 4.

Ghaboussi, J. and Wilson, E. L., 1971, "Flow of Compressible Fluids in Porous Media," SESM Report No. 72-12, University of California, Berkeley.

Goodman, R. E., Taylor, R. L., and Brekke, T., 1968, "A Model for the Mechanics of Jointed Rock," Journal of Soil Mechanics and Foundation Division, ASCE, Vol. 94, No. SM3.

Finlayson, B. A., 1972, The Method of Weighted Residual and Variational Principles, Academic Press, New York.

Majer, E. and McEvilly, T., 1982, "Stripa Acoustic Emission Experiment," Proceedings, Workshop on Hydraulic Fracturing Stress Measurements, December, 1981, USGS Open File Report, in press.

Narasimhan, T. N., and Witherspoon, P. A., 1977, "Numerical Model for Saturated-Unsaturated Flow in Deformable Porous Media, 1. Theory," Water Resources Research, Vol. 13, No. 3, pp. 657-664.

Noorishad, J., Ayatollahi, M. S., and Witherspoon, P. A., 1982, "Coupled stress and Fluid Flow Analysis of Fractured Rocks," International Journal of Rock Mechanics and Mining Sciences, in press.

Noorishad, J., Witherspoon, P. A., and Brekke, T. L., 1971, "A Method for Coupled Stress and Flow Analysis of Fractured Rock Masses," Geotechnical Engineering Publication No. 71-6, University of California, Berkeley.

Snow, D. T., "Fracture Deformation and Changes of Permeability and Storage upon Changes of Fluid Pressure," Quarterly Colorado School of Mines, Vol. 63, No. 1, p. 201.

Terzaghi, K., 1925, Erdbaumechanik auf Bodenphsikalischer Grundlage, Leipzig, F. Deuticke.

Chapter 68

Verification of Finite Element Methods Used to Predict
Creep Response of Leached Salt Caverns*

Dale S. Preece and Charles M. Stone

Sandia National Laboratories
Albuquerque, New Mexico 87185

Introduction

The Strategic Petroleum Reserves (SPR) is a national program dedicated to storage of large quantities of crude oil in leached salt caverns in the Texas-Louisiana gulf coast area. The program has required storage in existing caverns at each site and includes plans to leach new ones. The structural stability of each cavern depends on the behavior of the surrounding salt. In the past cavern stability has been predicted based on experience with other caverns in the vicinity. With the development of good non-linear finite element structural computer programs, it is possible to predict stability before the cavern is created.

An important step in verifying the applicability of these predictive methods is comparison of analytical and field data. The analytical data come from finite element analyses of a specific cavern where the laboratory determined material properties and an approximation to the cavern geometry are used as input. These comparisons have been made for caverns in the Bayou Choctaw Dome (Louisiana) and Eminence Dome (Mississippi) with reasonably good correlation being obtained between measured and predicted volumetric response of the caverns.

This paper discusses 1) the acquisition of material properties, 2) the finite element program SANCHO, and 3) the comparison of analytical and field data for each of the caverns.

---
* This work was performed at Sandia National Laboratories supported by the U. S. Department of Energy under Contract Number DE-AC0476-DP00789.

Laboratory Triaxial Creep Testing

The creep response of rock salt is usually somewhat site specific even though similarities have been shown to exist among test samples from many locations in the United States (Herrmann and Lauson, 1981). Because material differences are possible salt samples must be obtained from each site and tested for elastic properties and creep characteristics. The elastic properties are obtained using standard triaxial test procedures and the creep characteristics are obtained using two specially constructed triaxial test machines which were designed for long term (3 months to 1 year) creep testing of cylindrical rock samples.

The data obtained from each test are recorded on magnetic tape for later computer processing. This processing involves fitting the data from many tests to a number of different creep models. The creep model used in this work is one which relates effective secondary creep strain rate to effective stress (Herrmann and Lauson, 1981):

$$\dot{e}_s = A\bar{\sigma}^n$$

where

$\dot{e}_s$ = effective secondary creep strain rate

A = constant for a given temperature

$\bar{\sigma}$ = effective stress

n = stress exponent

Finite Element Analysis

SANCHO is a developmental structural finite element code which was derived from HONDO II (Key, Beisinger and Krieg, 1978). SANCHO has most of the important features associated with HONDO II such as large strain, large deformation, and a large selection of constitutive models. Dynamic relaxation is used to find static solutions at user specified time steps. A dynamic relaxation solution involves adding an acceleration term to the equilibrium equation, which converts a static problem into a dynamic one involving a pseudo-time measure. An internally computed "optimum" damping value is used to follow the "transient" response out in pseudo-time until a converged static solution is obtained. Convergence of this iterative procedure is based on the satisfaction of global equilibrium at a given load step. The magnitude of the residual force vector is compared to the magnitude of the applied load vector to determine when global equilibrium has been reached.

PREDICTING CREEEP RESPONSE   657

The creeping material model is currently restricted to secondary creep expressed in a power law form. The integration of the model is done "semi-analytically" and has been shown in tests to be quite accurate. There are no stability or time step restrictions as usually associated with classical Euler integration involved with this method. The only consideration is that the strain rate is constant within the time step so that accuracy of the solution is the dominant concern (Key, Stone and Krieg, 1980).

SANCHO was a participant in the recent WIPP (Waste Isolation Pilot Plant) Benchmark II exercise where a generic waste isolation drift in bedded salt was analyzed (Morgan, Krieg, and Matalucci, 1981). SANCHO results compared very well with results from the eight other structural codes which were exercised in the benchmark study.

Volumetric Measurements and Computations

In a geologic media the properties may vary from point to point. Because of this, it was decided that correlation between field data and finite element analyses would be more meaningful on a volumetric basis since the integrated nature of the volumetric response (decrease in cavern volume due to creep closure) tends to reflect an average of material properties. Volumetric response was also more easily obtained in the field because flow rate and pressure could be measured at the wellhead whereas creep displacement measurements would have to be made underground.

Bayou Choctaw Cavern Number 2

Bayou Choctaw Number 2 was drilled to a depth of 562 meters in 1934. Brine production eventually produced the 9.02 million barrel cavern shown in Figure 1. The United States Department of Energy (DOE) purchased the cavern in 1976 along with others in the dome but this cavern was judged unsuitable for oil storage because of the thin roof salt and has been used recently as a test cavern (Hogan, 1980).

The wellhead of this cavern is typically sealed for several months at a time with the brine pressure gradually building due to creep closure of the cavern. When the pressure reaches approximately 689 kPa, the wellhead is opened and brine is allowed to flow until the pressure has been significiantly reduced. It is possible to convert pressure changes at the sealed wellhead to volume changes due to creep. The relationship can be obtained by observing the pressure change when a measured amount of fluid is pumped into or out of the cavern.

658 ISSUES IN ROCK MECHANICS

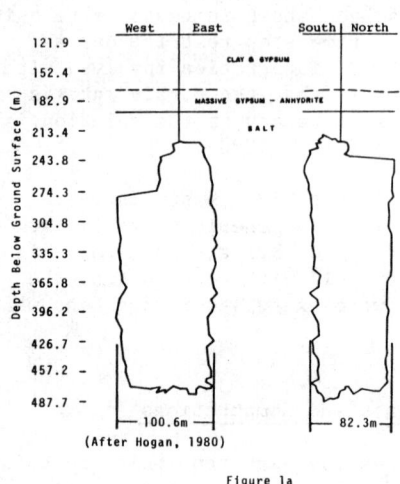

Figure 1a

Geometry of Bayou Choctaw Cavern Number 2

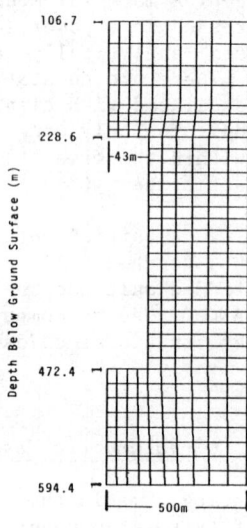

Figure 1b

Finite Element Model of Bayou Choctaw Cavern Number 2

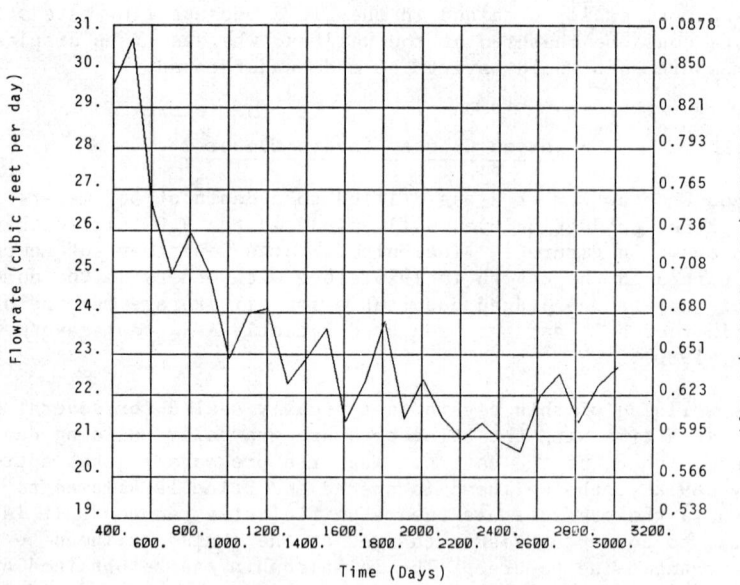

Figure 2

Flowrate Versus Time

PREDICTING CREEEP RESPONSE 659

The axisymmetric finite element approximation of this cavern is shown in Figure 1. The cavern model has a volume of 9.025 million barrels, a height of 244 meters and a radius of 43 meters (chosen to give the measured sonar volume). The boundary conditions include geostatic pressure across the top of the mesh and pressure inside the cavern corresponding to brine head from the ground surface. The boundary conditions on the right and left sides of the mesh allow vertical but no horizontal displacement. On the bottom of the mesh the horizontal displacements are free and the vertical are fixed. All of the caverns surrounding cavern 2 are at approximately the same depth except cavern 15, whose top is well below the bottom of cavern 2. Since there are no other caverns in a northeasterly direction from cavern 2 the pillar distance required to simulate an infinite boundary was included in the average of pillar distances used to obtained the width of the mesh. The pillar distance required to simulate an infinite boundary was determined to be eight times the cavern radius by analyzing a generic cavern with successively wider meshes until a minimal change in stress was observed on the right boundary at 30 years.

The salt properties used were from the West Hackberry Dome since those data were available at the time of the analysis (Wawersik, Hannum and Lauson, 1980). These properties were compared with elastic coefficients measured at Bayou Choctaw and found to be in close agreement. The creep properties were determined using a temperature at the mid-height of the cavern of 41°C. This was estimated from borehole temperature logs. The elastic properties of the shale in the caprock were taken from previous analyses involving shale layers (Benzley, 1980). The material properties chosen are given in Table 1.

Table 1

Material Properties

Shale

Youngs Modulus = 1.48×10^8 psf
Poissons Ratio = 0.29

Salt

Youngs Modulus = 4.61×10^8 psf
Poissons Ratio = 0.26

Coefficients for Secondary Creep Equation

Stress Exponent n = 4.9
A (22°C) = 2.54×10^{-31} (psf and days)
A (41°C) = 9.02×10^{-31} (psf and days)
A (47°C) = 1.10×10^{-30} (psf and days)
A (60°C) = 1.769×10^{-30} (psf and days)

The volumetric response of the cavern obtained from the finite element model is plotted as a function of time in Figure 2. There is some transient behavior evident on the plot since the rate of change of the total volume is gradually decreasing. The flow rate computed at 3000 days where the tranisent behavior has diminished is about 0.62 cubic meters per day. Comparing this with the measured flow rate of 0.85 cubic meters per day indicates that reasonably good correlation (for field events of this size and uncertainty) between field and calculated data has been obtained.

Eminence Cavern Number 1

In 1970 Transcontinental Gas Pipe Line Corporation (TRANSCO) completed cavern number 1 as the first of four natural gas storage caverns (Allen, 1972). The geometrical description of this cavern is shown in Figure 3. This is an interestng cavern to test the predictive methods on because: 1) it is very deep in comparison with Bayou Choctaw Number 2 providing an indication of the range over which calculations are valid, 2) it has a relatively simple geometry making creation of a finite element mesh easy, 3) good data were available from TRANSCO on the volumetric response of the cavern with time, and 4) the large difference between internal pressure and geostatic stress caused rapid closure of the cavern.

The initial volume of the cavern was determined using the volume of fresh leaching water and the salinity of the returning brine. Sonar and metering of input and output volumes of brine or fresh water were used to make subsequent measurements of cavern volume. The natural gas storage pressure in the cavern varied between a maximum of 26.61 MPa and a minimum of 8.96 MPa. A time weighted average of internal cavern pressure taken from TRANSCO data (Fenix and Scisson, 1975) was 20.68 MPa. This value was used in the analysis.

The axisymmetric finite element model of this cavern is shown in Figure 3. The finite element model has a cavern volume of 1.0788 million barrels compared to a volume measured from leach and sonar data to be approximately 1.1008 million barrels. The mesh is made up of four node quadrilateral elements with geostatic pressure across the top and a constant pressure of 20.68 MPa inside the cavern. Boundary conditions similar to the Bayou Choctaw 2 model were placed on the right and left sides and along the bottom of the mesh.

Since a temperature log of Eminence was not available the temperature of the salt around the cavern was taken from a Bayou Choctaw temperature log (Hogan, 1980) to be 60°C. This temperature was used in the selection of the creep properties given in Table 1. The width of the pillar between caverns was also not available from TRANSCO. However, judging from the fact that there are only four caverns total in Eminence dome and the dome width is

PREDICTING CREEEP RESPONSE 661

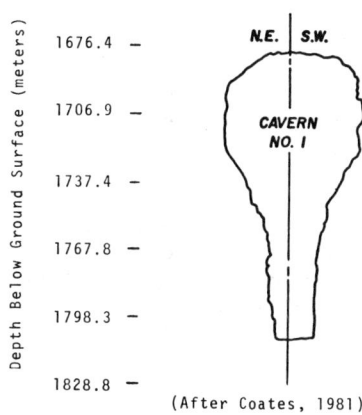

(After Coates, 1981)

Figure 3a

Geometry of Eminence Cavern Number 1

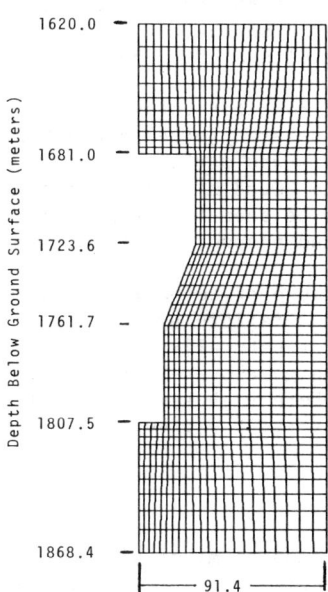

Figure 3b

Finite Element Model of Eminence Cavern Number 1

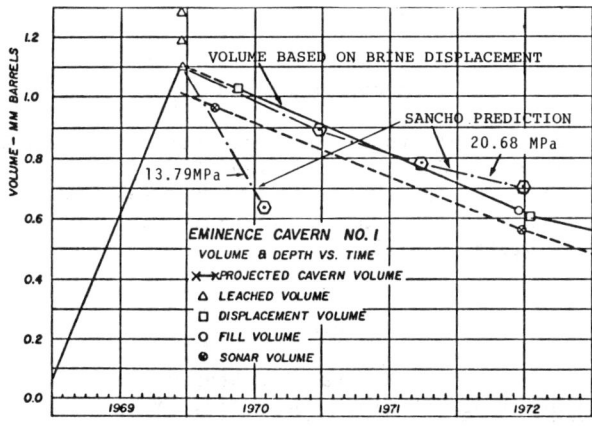

Figure 4
Total Volume Versus Time
(After Fenix & Scisson, 1975)

approximately one mile in diameter (Allen, 1972), a spacing of 183 meters was chosen with a pillar of 128 meters. Previous calculations had shown that widening the mesh any further did not change the results significantly.

The computed volumetric response of the cavern and its measured response are plotted against time in Figure 4 (Coates, 1981). The analysis was performed at two different internal pressures, 13.79 MPa and 20.68 MPa, to give an indication of pressure sensitivity. It can be seen from Figure 4 that the computed response tracks the measured response quite closely.

Because of severe creep closure the caverns at Eminence were termed a failure by some (Coates, 1981) though the caverns were never structurally unstable and did not collapse. It appears that this large closure rate was the result of the caverns being too deep where high temperature and large stress differences accelerated the creep closure of the caverns.

Summary

The use of material properties from triaxial creep tests in conjunction with finite element analyses provides a reasonably reliable method for predicting the volumetric response of salt caverns due to creep. The material testing procedures and the analytical approach used here have been shown to be valid by comparison of analytical results with flow rate and total volume data obtained in the field. These methods are being used to predict the response of new and existing caverns which will be employed in the SPR program.

Acknowledgements

The rock salt material properties were obtained from triaxial experiments performed under the direction of Wolfgang Wawersik, Sandia National Labortories, Geomechanics Division - 5532. All field data were obtained by members of Sandia National Laboratories, SPR Geotechnical Division 4773.

References

Allen, K., "Eminence Dome - Natural-Gas Storage in Salt Comes of Age," Journal of Petroleum Technology, November 1972.

Benzley, S. E., "Structural Analysis of the West Hackberry #6 SPR Storage Cavern," Sandia National Laboratories, SAND80-1904, August 1980.

Coates, G. K.; Lee, C. A.; McClain, W. C.; Senseny, P. E., "Failure of Man-Made Cavities in Salt and Surface Subsidence Due to Sulphur Mining," RE/SPEC Inc., Rapid City, S. D., January 1981.

Fenix and Scisson, Inc., "Analysis of Data from Eminence Dome Storage Caverns," unpublished report prepared for Transcontinental Gas Pipe Line Corporation, 1975.

Herrmann, W. and Lauson, H. S., "Analysis of Creep Data for Various Natural Rock Salts," Sandia National Laboratories, SAND81-2567.

Herrmann, W.; and Lauson, H. S., "Review and Comparison of Transient Creep Laws Used for Natural Rock Salt," Sandia National Laboratories, SAND81-0738, April 1981.

Hogan, R. G., "Geological Site Characterization Report, Bayou Choctaw Salt Dome," Sandia National Labortories, SAND80-7140, December 1980.

Key, S. W.; Beisinger, Z. E.; and Krieg, R. D., "HONDO II - A Finite Element Program for the Large Deformation Response of Axisymmetric Solids," Sandia National Laboratories, SAND78-0422, October 1978.

Key, S. W.; Stone, C. M.; and Krieg, R. D., "A Solution Strategy for the Quasi-Static Large Deformation, Inelastic Response of Axisymmetric Solids," presented at U. S. European Workshop Nonlinear Finite Element Analysis in Structural Mechanics, Ruhr-Universitat; Bochum, W. Germany, July 1980.

Morgan, H. S.; Krieg, R. D.; and Matalucci, R. V., "Comparative Analysis of Nine Structural Codes Used in the Second WIPP Benchmark Problem," Sandia National Laboratories, SAND81-1389, December 1981.

Wawersik, W. R.; Hannum, D. W.; and Lauson, H. S., "Compression and Extension Data for Dome Salt from West Hackberry, Louisiana," Sandia National Laboratories, SAND79-0669, September 1980.

Chapter 69

ELASTIC BENDING OF THICK ROCK PLATES

by J. G. Singh and P. C. Upadhyay

Department of Mining Engineering

Department of Mechanical Engineering
Institute of Technology, B.H.U.
Varanasi 221005, India

ABSTRACT

Reissner's thick plate bending theory has been recast for rocks which exhibit the property of double elasticity: different Young's modulus and Poisson's ratio in compression and tension. The governing differential equation and other relevent expressions have been presented in a form compatible to the applied mechanics results so that they can readily be used. Effect of the ratio of Young's modulii in compression and tension, (E_c/E_t), on the maximum plate deflection and the maximum bending moments has been evaluated over a wide range. It has been found that the inclusion of bi-modulus property through a parameter ($\beta = E_c/E_t$), changes the sheer deformation of the plate to such an extent that the bi-modular effect of rock cannot be overlooked in the thick plate analysis, especially when β is expected to be greater than 2.

INTRODUCTION

When the resultant stress field around an opening in a stratified ground (e.g., coal measure rocks) overcomes the cohesion between the rock layers in the neighborhood, the phenomenon of bed-separation takes place overhead. Structurally these unstuck rock layers are "plates," whose boundaries are defined by the lateral extent of the bed-separation. Since in many mining situations the lateral extent of bed-separation, and hence the span of the plate is limited, the thickness to span ratio of such plates is relatively higher than the permissible range in the thin plate theory. Therefore, it is always desirable to use thick plate formulations while analyzing the stability and deflection of such rock plates; a thin plate treatement is likely to introduce serious errors.

For the elastic bending of rock beams, it has been shown (Jaeger and Cook, 1971) that the inclusion of the double elastic property (unequal Young's modulus in tension and compression) of rocks affects the bending results significantly. The same conclusion has been arrived at while investigating into the effects of double elastic property of rocks in the bending of thin rock plates (Singh, Upadhyay and Saluja, 1980). It has been shown that, as the difference between the Young's modulii in tension and compression increases, the maximum plate deflection changes signficantly in comparison to the one obtained on the basis of equal values of the two modulii. In fact, in the above reference it has been pointed out that many of the field observations which could not be explained properly within the framework of the elasticity theory, can very well be explained if the bi-modulus property of rock is duly incorporated in the analysis.

In light of these results, therefore, it was felt to be of interest to re-cast the applied mechanics theory of thick plates, incorporating the bi-modulus property of rocks. The motivation for undertaking this work is twofold: Firstly, to get an estimate of the quantitative difference in the results (maximum deflection and bending moments, etc.); and, secondly, to provide alternate expressions compatible to the applied mechanics results which could readily be used in the analysis of thick rock plate bending problems, like the one pointed out in the opening paragraph. It has been found that the shear deflection, which plays a significant role in case of thick plates, is so grossly changed with the inclusion of the double elastic property that a direct adoption of the applied mechanics results, valid only for steel like mono-elastic materials, should be considered prohibitive.

RECASTING OF THICK PLATE EQUATIONS

Referring to Fig. 1, which shows the plate geometry, let the neutral surface (NS) divide the plate height h into ht and hc, corresponding, respectively, to the thicknesses of tensile and compressive zones on the two sides of the neutral surface. The cartesian coordinate system OXYZ, then, has its axes (OX,OY) on the undeflected NS, and OZ directed downwards.

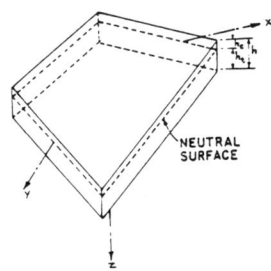 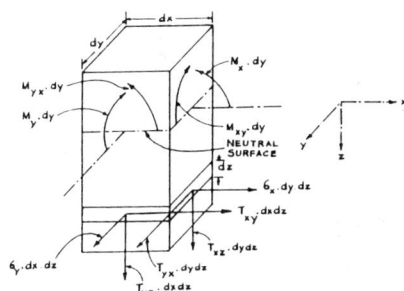

Fig. 1. Plate geometry   Fig. 2. Forces on a Plate Element

A normal traction q acts on the upper face of the plate; plate is devoid of any body force. Rock is assumed to be homogeneous, isotropic and linearly elastic.

In the development of thick plate formulation by Reissner's variational approach (Reissner, 1945, 1947), stresses are needed to evalulate the energy functional. These are borrowed from the thin plate theory, to start with, and are expressed as functions of force and moment intensities (per unit length along X and Y axes). For rocks, following the relevent (incorporating bi-modus property) thin plate treatement (Singh, Upadhyay and Saluja, 1980), the required stress expressions may be written as

$$\sigma_x = (A,A') M_x \cdot z,$$
$$\sigma_y = (A,A') M_y \cdot z,$$
$$\tau_{xy} = (A,A') M_{xy} \cdot z,$$
$$\tau_{xz} = 1/2 (A,A') Q_x [(h_c^2, h_t^2) - z^2],$$
$$\tau_{yz} = 1/2 (A,A') Q_y [(h_c^2, h_t^2) - z^2], \text{ and}$$
$$\sigma_z = -1/6 (A, A') q [z^3 - 3 (h_c^2, h_t^2)z + 2(-h_c^3, h_t^3)] - (1,0) q;$$

$$(z<0, z>0). \qquad (1)$$

where,

$$(A,A') = \frac{3(1+\beta^{\frac{1}{2}})^2}{(1,\beta)h^3}, \text{ and} \qquad (2)$$

(Q_x, Q_y), (M_x, M_y) and (M_{xy}, M_{yx}) are, respectively, the transverse shear force, bending moment and the twisting moment, per unit length along a line parallel to (OY, OX); and are defined as

$$(Q_x, Q_y) = \int_{-h_c}^{h_t} (\tau_{zx}, \tau_{zy}) \, dz,$$

$$(M_x, M_y, M_{xy}) = \int_{-h_c}^{h_t} (\sigma_x, \sigma_y, \tau_{xy}) z \, dz. \qquad (3)$$

Figure 2 shows the various forces acting on a rectangular plate element.

The thicknesses h_c and h_t are obtained from the relation (Singh, Upadhyay and Saluja, 1980),

ELASTIC BENDING OF THICK ROCK PLATES

$$(h_c, h_t) = \frac{(1, \beta^{\frac{1}{2}})h}{(1 + \beta^{\frac{1}{2}})} \tag{4}$$

in which, $\beta = (E_c/E_t)$ represents the ratio of the Young's modulii of rock in compression and tension.

Equations (1-4) are all based on the plane stress assumption in thin plates, wherein all the strains and the normal stress on the planes parallel to the NS are taken to be zero, i.e., $\varepsilon_z = \varepsilon_{zy} = \varepsilon_{zx} = 0$ and $\sigma_z = 0$. But in the thick plates, where the shear strains cannot be ignored, the conditions are $\varepsilon_z = 0$, $\varepsilon_{zy} \neq 0$, $\varepsilon_{zx} \neq 0$, and $\sigma_z \neq 0$. Hence, the strain displacement relations are given as

$$(\varepsilon_x, \varepsilon_y, \varepsilon_{xy}) = z(k_x, k_y, k_{xy}), \quad \varepsilon_z = 0 \quad \text{and}$$

$$(\varepsilon_{zy}, \varepsilon_{zx}) = (\dot{\varepsilon}_{zy}, \dot{\varepsilon}_{zx}) \tag{5}$$

where,

$$(k_x, k_y, k_{xy}) = -[\frac{\partial \phi_x}{\partial x}, \frac{\partial \phi_y}{\partial y}, \frac{1}{2}(\frac{\partial \phi_y}{\partial x} + \frac{\partial \phi_x}{\partial y})], \quad \text{and}$$

$$(\dot{\varepsilon}_{zy}, \dot{\varepsilon}_{zx}) = \frac{1}{2}[(\frac{\partial w}{\partial y}, \frac{\partial w}{\partial x}) - (\phi_y, \phi_x)]. \tag{6}$$

In eq. (6), (ϕ_x, ϕ_y) represent the bending slopes in the (OX,OY) direction and w is the normal deflection in the Z direction, of a point on the NS. Thus, eq. (5) gives the strains at a point of distance z from NS, in terms of the quantities defined at the corresponding point on the NS.

Now, to derive the governing equation for the deflection w, that includes the contribution of the shear strains ε_{zy} and ε_{zx}, Reissner's variational equation

$$\delta[\iiint (\sigma_x \cdot \varepsilon_x + \sigma_y \cdot \varepsilon_y + 2\tau_{xy} \cdot \varepsilon_{xy} + 2\tau_{yz} \cdot \varepsilon_{yz}$$

$$+ 2\tau_{zx} \cdot \varepsilon_{zx} - \overline{W}) \, dxdydz - \iint q \cdot w \cdot dxdy] = 0 \tag{7}$$

where,

$$\overline{W} = \frac{1}{2E}[\sigma_x^2 + \sigma_y^2 + \sigma_z^2 - 2\nu(\sigma_x \cdot \sigma_y + \sigma_y \cdot \sigma_z + \sigma_z \cdot \sigma_x)] \tag{8}$$

is used; which on substitution of the stressses from eq. (1) and the strains from eqs. (5,6), and on integration, gives

$$\int(-M_y \cdot \delta\phi_y - M_{xy} \cdot \delta\phi_x + Q_y \cdot \delta w)dx + \int(-M_x \cdot \delta\phi_x - M_{xy} \cdot \delta\phi_y + Q_x \cdot \delta w)dy$$

$$+ \iint(\frac{\partial M_x}{\partial x} \cdot \delta\phi_x + \frac{\partial M_y}{\partial y} \cdot \delta\phi_y + \frac{\partial M_{xy}}{\partial x} \cdot \delta\phi_y + \frac{\partial M_{xy}}{\partial y} \cdot \delta\phi_x - Q_x \cdot \delta\phi_x$$

$$- Q_y \cdot \delta\phi_y - \frac{\partial Q_x}{\partial x} \delta w - \frac{\partial Q_y}{\partial y} \delta w - q \cdot \delta w) \, dxdy + \iint [k_x \cdot \delta M_x + k_y \cdot \delta M_y$$

$$+ 2k_{xy} \cdot \delta M_{xy} + 2\dot{\varepsilon}_{zx} \cdot \delta Q_x - 2K_1(M_x \cdot \delta M_x + M_y \cdot \delta M_y) + 2\nu k_1 (M_x \cdot \delta M_y$$

$$+ M_y \cdot \delta M_x) + K_2 \, q \cdot (\delta M_x + \delta M_y) - 4(1+\nu) \, K_1 \, M_{xy} \cdot \delta M_{xy} + 2\dot{\varepsilon}_{yz} \cdot \delta Q_y$$

$$-2K_3 (Q_x \cdot \delta Q_x + Q_y \cdot \delta Q_y)] \, dxdy = 0 \qquad (9)$$

The constants K_1, K_2 and K_3, in eq. (9), are defined as

$$K_1 = \frac{3(1+\beta^{\frac{1}{2}})^2}{2 \, E \cdot h^3},$$

$$K_2 = \frac{3\nu(1+\beta^{3/2})}{10 \, E \cdot h \, (1+\beta^{\frac{1}{2}})} + \frac{3\nu}{2 \, E \cdot h}, \text{ and}$$

$$K_3 = \frac{6(1+\nu)(1+\beta^{3/2})}{5 \, E \cdot h \, (1+\beta^{\frac{1}{2}})} \qquad (10)$$

The first two integrals in eq. (9) furnish the boundary conditions

$$(Q_y \cdot w \, , \, M_{xy} \cdot \phi_x \, , \, M_y \cdot \phi_y) = 0,$$

and

$$(Q_x \cdot w \, , \, M_x \cdot \phi_x \, , \, M_{xy} \cdot \phi_y) = 0, \qquad (11)$$

whereas, the third integral gives the equilibrium equations

$$\frac{\partial}{\partial x}(Q_x, M_x, M_{xy}) + \frac{\partial}{\partial y}(Q_y, M_{yx}, M_y) - (-q, Q_x, Q_y) = 0. \qquad (12)$$

From the fourth integral, the relations obtained are

$$k_x - 2 \, K_1 \cdot M_x + 2\nu \, K_1 \cdot M_y + K_2 \cdot q = 0 \quad ,$$

$$k_y - 2K_1 \cdot M_y + 2\nu K_1 \cdot M_x + K_2 \cdot q = 0,$$

$$2k_{xy} - 4(1+\nu) K_1 \cdot M_{xy} = 0$$

$$2\dot{\varepsilon}_{yx} - 2 \cdot K_3 \cdot Q_y = 0, \text{ and}$$

$$2\dot{\varepsilon}_{zx} - 2 \cdot K_3 \cdot Q_x = 0 \tag{13}$$

Now, the solution of the first two equations for M_x and M_y, and the third equation for M_{xy}, in eq. (13), gives

$$(M_x, M_y, M_{xy}) = D[(k_x, k_y, k_{xy}) + (k_y, k_x, -k_{xy})] + (1,1,0) \; S \cdot q \cdot \tag{14}$$

where,

$$D = \frac{E \cdot h^3}{3(1-\nu^2)(1+\beta^{\frac{1}{2}})^2}, \text{ and } S = \frac{\nu[5(1+\beta^{\frac{1}{2}}) - (1+\beta^{\frac{3}{2}})]}{10(1-\nu) \; (1+\beta^{\frac{1}{2}})^3} \cdot h^2 \tag{15}$$

The fourth and fifth equations in (13) yield

$$(Q_y, Q_x) = \frac{1}{K_3} (\dot{\varepsilon}_{yz}, \dot{\varepsilon}_{zx}) \tag{16}$$

which, on combining with eqs. (6), gives

$$(\phi_y, \phi_x) = - B \cdot (Q_y, Q_x) + (\frac{\partial w}{\partial y}, \frac{\partial w}{\partial x})$$

and $\quad (k_x, k_y, k_{xy}) = B [\frac{\partial Q_x}{\partial x}, \frac{\partial Q_y}{\partial y}, \frac{1}{2}(\frac{\partial Q_y}{\partial x} + \frac{\partial Q_x}{\partial y})]$

$$- (\frac{\partial^2 w}{\partial x^2}, \frac{\partial^2 w}{\partial y^2}, \frac{\partial^2 w}{\partial x \partial y}) \tag{17}$$

where

$$B = \frac{12(1+\nu) \; (1+\beta^{3/2})}{5 \; E \cdot h \cdot (1+\beta^{\frac{1}{2}})} \tag{18}$$

At this stage, following the standard treatment of thick plate analysis (Reissner, 1945) using eqs. (10-18), the final form of the governing differential equation for the deflection can easily be obtained as

$$D \nabla^4 w = q - K \nabla^2 q, \tag{19}$$

where D is the same as in eqn. (15), and

$$K = \frac{(8+\nu)(1+\beta^{3/2}) - 5\nu(1+\beta^{1/2})}{10(1-\nu)(1+\beta^{1/2})^3} \cdot h^2 \qquad (20)$$

In eq. (19) $\nabla^4 \equiv \nabla^2\nabla^2$ while ∇^2 is the two dimensional (x and y) Laplacian operator.

It may be observed that the final form of the governing differential equation (eq. 19), for the deflection, is the same as for the mono-elastic materials like steels, etc., excepting that the flexural rigidity D and the constant K are different. In fact, in $\beta = 1$ we get the same values of D and K as reported for the mono-elastic materials,

$$D = \frac{Eh^3}{12(1-\nu^2)}, \text{ and } K = \frac{(2-\nu)}{10(1-\nu)} \cdot h^2 \qquad (21)$$

If the values of D and K, as obtained for rocks (eqs. (15,20)) are expressed in a form compatible to eqs. (21), it appears as

$$D = \frac{E(\lambda_1 h)^3}{12(1-\nu^2)} \text{ and } K = \frac{(2-\nu)}{10(1-\nu)} (\lambda_2 h)^2 \qquad (22)$$

where

$$\lambda_1 = \left[\frac{4}{(1+\beta^{1/2})^2} \right]^{1/3}, \text{ and } \lambda_2 = \left[\frac{(8+\nu)(1+\beta^{3/2}) - 5\nu(1+\beta^{1/2})}{(2-\nu)(1+\beta^{1/2})^3} \right]^{1/2} \qquad (23)$$

Thus, equations (22) and (23) enable us to use all the existing thick plate results of mono-elastic materials just by multiplying h by factors λ_1 and λ_2 in the expressions of D and K, respectively.

For mono-elastic materials. the maximum deflection and the maximum bending moments of a thick rectangular plate, loaded by a constant normal traction "q" can always be expressed as (Salerno and Goldberg, 1960),

$$(w)\text{max.} = \left[\alpha + \alpha' \left(\frac{h}{a}\right)^2 \right] \frac{qa^4}{Eh^3}, \text{ and}$$

$$(M_x, M_y)\text{max.} = \left[(\phi,\psi) + (\phi',\psi') \left(\frac{h}{a}\right)^2 \right] \cdot q a^2 \qquad (24)$$

where, "h" and "a" are, respectively, the thickness and the smaller span of the plate. The values of α, α', ϕ, ϕ', ψ and ψ' depend on the plate boundary conditions. The first term on the right-hand side of eqs. (24) represent the bending contribution, and the second term is due to the shear deformation of the plate which vanishes as h/a tends to zero for a thin plate.

For rocks, equations analogous to (24) may be written as

$$(w)_{max.} = [(\xi_1 \cdot \alpha) + (\xi_2 \cdot \alpha')(\frac{h}{a})^2] \frac{qa^4}{Eh^3} \qquad (25)$$

and

$$(M_x, M_y)_{max.} = [(\phi,\psi) + \eta \cdot (\phi',\psi')(\frac{h}{a})^2] qa^2 \qquad (26)$$

where, $\xi_1 = \dfrac{(1+\beta^{\frac{1}{2}})^2}{4}$, $\xi_2 = \dfrac{(8+\nu)(1+\beta^{3/2}) - 5\nu(1+\beta^{\frac{1}{2}})}{4(2-\nu)(1+\beta^{\frac{1}{2}})}$

$$\eta = \frac{(8+\nu)(1+\beta^{3/2}) - 5\nu(1+\beta^{\frac{1}{2}})}{(2-\nu)(1+\beta^{\frac{1}{2}})^3} = \xi_2/\xi_1$$

RESULTS AND DISCUSSION

Since the effect of factor ξ_1, which accounts for the changes in the pure bending part of the deflection only, is the same as discussed in the previous work (Singh, Upadhyay and Saluja, 1980) of the authors, while discussing the percentage change the flexual rigidity due to β, it is not being shown here. Plots of ξ_2 and η, as function of β, are shown in Figs. 3 and 4, respectively.

The main conbribution of this paper lies in bringing out the effect of β on the shear deflections, through ξ_2. It may be seen from Fig. 3 that for the values of β in the range of 1.5 to 4.0 (for sand stones) the value of ξ_2 varies from 1.35 to 3.25, for $\nu = 0.2$. This shows that the shear part of the deformation can be enhanced by as much as 30 to 325 percent, depending on the value of β for the particular rock. Therefore, it is realized that the shear contribution to the deflection of the plate is much more significant in case of rocks than for materials with $\beta = 1$. As the role of shear becomes still more important in the dynamic performance of the structures, it is felt that where the rock structures are likely to be subjected to the dynamic loading conditions due to blast loading, etc., consideration of the shear effect cannot be overlooked.

Figure 4 shows the influence of β on the maximum bending moments, (M_x, M_y) max., through the η vs. β plots for different ν. It may be seen that, though the effect of β is not so pronounced on η, as it is on ξ_1 and ξ_2, it does change the values of the maximum bending moments quite appreciably. This effect is worth accounting for, especially, when the criterion of the performance of a rock plate is based on the development of maximum bending moments inside the plate (fracture initiation, etc.).

Thus, it may finally be concluded that the rock parameter β influences the maximum plate deflection to such an extent that without incorporating the property of double elastic constants of rocks, any theoretical assessment of the stability of rock plate structures undergoing bending is likely to be highly erroneous, especially when

β is expected to be greater than 2.

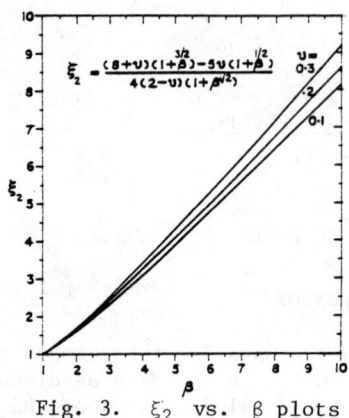

Fig. 3. ξ_2 vs. β plots

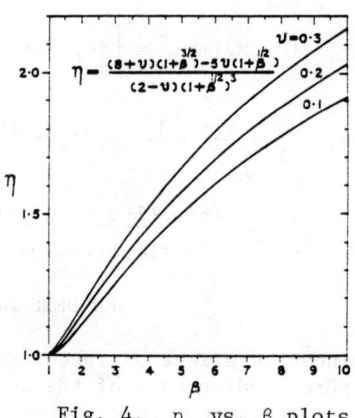

Fig. 4. η vs. β plots

REFERENCES

Jaeger, J.C., and Cook, N.G.W., 1971, <u>Fundamentals of Rock Mechanics</u>, Chapman and Hall Ltd., London.

Reissner, E., 1945, "The Effect of Transverse Shear Deformation on the Bending of Elastic Plates," Journal of Applied Mechanics, Vol. 12, pp. 69-77.

Reissner, E., 1947, "On Bending of Elastic Plates," Quarterly of Applied Mathematics, Vol. 5, pp. 55-68.

Salerno, V.L., and Goldberg, M.A., 1960, "Effect of Shear Deformation of the Bending of Rectangular Plates," Journal of Applied Mechanics, Vol. 27, pp. 54-58.

Singh, J.G., Upadhyay, P.C., and Saluja, S.S., 1980, "The Bending of Rock Plates," International Journal of Rock Mechanics, Mining Sciences & Geomechanic Abstracts, Vol. 17, pp. 377-381.

Chapter 70

A NUMERICAL STUDY OF EXCAVATION SUPPORT LOADS IN JOINTED ROCK MASSES

M. D. Voegele* and C. Fairhurst**

*Science Applications, Inc.
Salt Lake City, Utah

**University of Minnesota
Minneapolis, Minnesota

INTRODUCTION

A computer study of excavation support loads in tunnels and other excavations in discontinuous rock masses was undertaken with the distinct element method introduced by Cundall (1971). In the distinct element method the rock mass is modeled as a composition of individual blocks of rock defined by the joint planes. Normal and shear deformation laws are ascribed to the joint surfaces; the blocks of rock themselves are considered to be rigid. The particular version of the program utilized in this study ran a mini computer and used interactive techniques for the input of geometrical data such as joint orientation and spacing and subsequent display of rock mass deformation and other data.

Two major aspects of the behavior of excavations in jointed rock masses are examined in this paper: i) the behavior of unsupported excavations in rock masses with varying joint configuration, joint properties and in situ stress fields; and ii) the modification of the behavior of similar excavations when support is provided. In the latter case the excavation behavior was examined through the use of ground reaction curves.

THE DISTINCT ELEMENT METHOD

The Distinct Element method introduced by Cundall (1971) is a computer based analysis that simulates the behavior of a system of discrete, semi-rigid rock blocks. Block interactions are governed by realistic friction and stiffness laws; each block may undergo unlimited displacement and rotation while progressive failure is modeled.

The program described herein was run in an interactive mode on a dedicated mini-computer coupled to a cathode ray tube (CRT) graphic output device. The CRT was used both for the input of geometric and material information as well as for the output data which consisted of drawing the movements of the blocks as a function of time.

The program calculation cycle comprises force-displacement relations for the block contacts and laws of motion for the block centroids. Realistic relationships are used to relate normal force to normal displacement and shear force to shear displacement. The normal force-displacement relationship owes its simplicity to the assumption that the normal stiffness of a joint plays a small role in the failure process of the rock mass and that shear force does not affect normal force. Thus normal force is assumed proportional to the overlap between two blocks. Diagramatically,

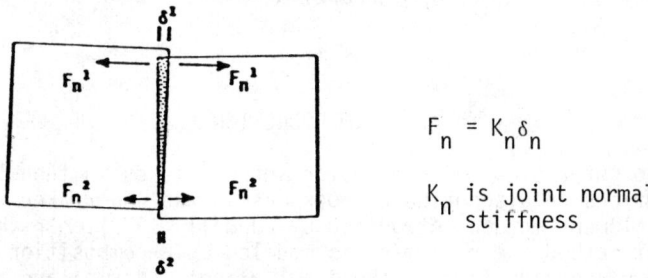

$$F_n = K_n \delta_n$$

K_n is joint normal stiffness

An argument for the validity of representing a joint by two point contacts is that owing to irregularities present on a real joint, contact will occur only at discrete points, quite possibly only two.

The shear force-displacement relationship cannot be described by such a simple formulation because the shear force depends upon the past history of movement of the blocks as well as the amount of normal force. To account for this, the shear force is calculated incrementally with the incremental amount of shearing force assumed proportional to the relative movement of a block corner along another block face. The incremental shear force is then added, noting the sense of movement, to the shear force already existing between the two blocks. Diagramatically:

EXCAVATION SUPPORT LOADS

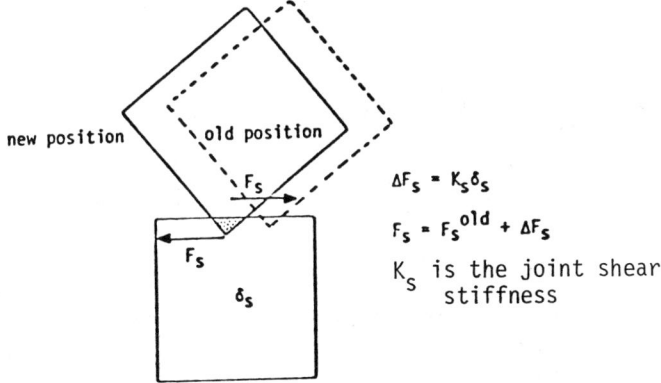

$\Delta F_s = K_s \delta_s$

$F_s = F_s^{old} + \Delta F_s$

K_s is the joint shear stiffness

Although not strictly necessary from a physical standpoint, the normal force is also calculated incrementally in the program so that all forces are derived from incremental displacements. This formulation does, however, simplify the task of incorporating nonlinear phenomena, such as dilatation, associated with the normal stress.

Two failure laws are incorporated in the program. Since it is probably unrealistic to have tensional resistance across a joint, a "no tension" criterion is adopted at each time step, by simply setting normal forces that become negative to zero. The criterion governing shear failure is the Mohr-Coulomb-Navier law. At every time step, the shear force at each contact point is tested and limited to a maximum force, which is dependent upon the normal force.

The force-displacement relations are thus used to calculate the set of forces acting on each block solely due to the geometric position of each block relative to its neighbors. The forces acting on each block may be resolved into an equivalent force vector and a moment acting on the block centroid. If a law of motion is now implemented (in this case Newton's second law) the linear acceleration vector can be calculated as the quotient of the resultant force and the mass of the block. Similarly, the rotational acceleration is the quotient of the resultant moment and the rotational moment of inertia of the block. By choosing a suitable time step, these accelerations may be numerically integrated twice to give the displacement of the block.

The complete calculation cycle can be summarized as:

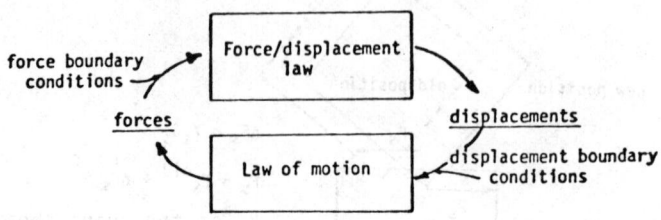

In addition to the main calculation cycle, routines are needed to keep track of the coordinates of contacts; the use of arbitrarily large displacements and the attendant large number of possible contact points requires the implementation of a dynamic memory allocation scheme.

The Distinct Element formulation is oriented toward the behavior of each block as an individual mass. The kinematic behavior of each block is independently calculated using Newton's law of motion; each block senses the blocks surrounding it only as boundary conditions. If the movement of a block leads to penetration or relative movement along the surface of another block then the normal and shear stiffness will lead to interblock contact forces by a simple application of Hooke's law with an upper limit to the forces set by the Mohr-Coulomb relation. These forces are simply treated as boundary conditions for the first block. When a contact is broken by a relative displacement between the two blocks involved, there is no longer a need to consider the effect that these blocks have upon each other.

ROOF FAILURE IN UNDERGROUND ROOMS

The conditions preceeding failure in multilayer roofs were seen to be characterized by two common features. First, loss of force transmittal across the lower contact of the midspan joint was not indicative of failure. Frequently, significant horizontal force reduction after the joint opened was required before failure occurred. The second general behavior pattern that was recognized concerns the distribution of contact forces in the immediate roof. Figure 1 presents a typical multilayer model and a section of its contact force distribution. The blocks are in equilibrium but a reduction in the horizontal thrust of approximately 10% would lead to failure; this is a typical force distribution of a multilayer model at stress conditions slightly greater than those at which failure occurs. Three characteristics of the force distribution in multilayer models were noted in all models tested and are indicated in Figure 1 by the letters A, B, and C. The characteristics are:

    A) absence of force transmittal across the lower contact of the mid span joint

B) minimal vertical force transmittal within the suspended zone, especially to the lower row of blocks

C) the development of an additional contact force where the blocks adjacent to the abutment rotate into the next upward level of blocks

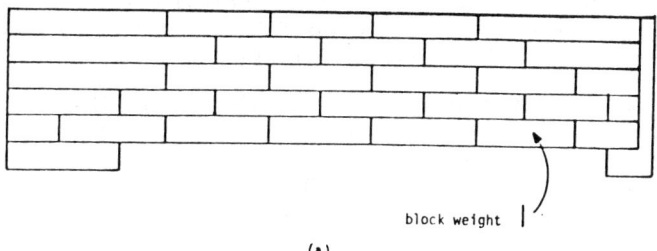

(a)

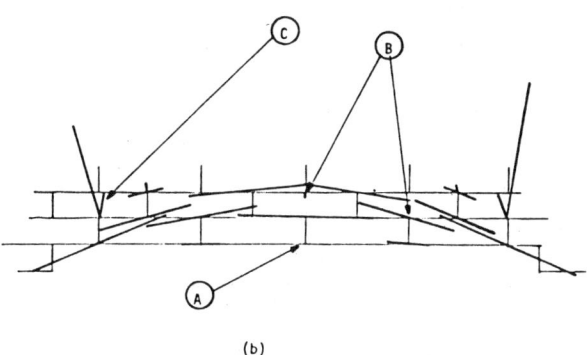

(b)

Figure 1. Contact force distribution in lower rows of multilayer model.

The second characteristic is to be expected in light of the model; the corbelling effect of the blocks outside of the suspended zone acts to lessen the span over which the next row of blocks must be supported. The other two observations, A and C, are closely related and provide a reasonable explanation as to why the behavior of the multilayer models depart from the linear arch model. The linear arch model is based upon a simple contact force distribution; the model used for the development of the linear arch equation is not valid for the multilayer cases. As the lower row of the multilayer model deflects some rotation of the blocks occurs and leads to the development of a shearing resistance along the top of the block. The presence of an additional shearing resistance also explains how stable conditions can be maintained even though the lower contact of the mid span joint is broken.

The presence of the additional force acting on the block tends to maintain equilibrium in a manner not accounted for by the linear arch model.

The results from Distinct Element runs can be expressed in a way that may be useful for design purposes. The example presented below utilizes the data to illustrate an empirical relationship between parameters. These relationships are characterized by errors in the order of 4% rather than the 40% error experienced when using linear arch theory to predict the horizontal thrust.

The example illustrates a relationship between the horizontal force required for stability, the number of blocks in the bottom row, (a factor which is analogous to joint spacing) and the friction angle of the joints, in models similar to those shown in Figure 2. The excavation width and the block thickness are constant in this analysis.

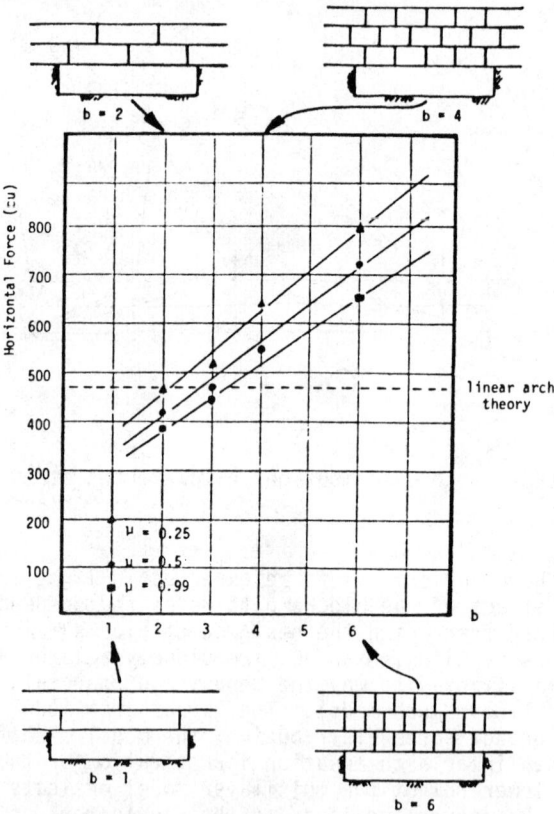

Figure 2. Linear relationship between horizontal force, number of blocks in the lower row and joint friction angle (constant span and block thickness).

EXCAVATION SUPPORT LOADS

The data points, which represent the failure conditions for 11 test models, and the associated linear trends are plotted in Figure 2. Also included in the figure is a horizontal dashed line which represents the value of horizontal force necessary to maintain roof stability as calculated by linear arch theory. The data points corresponding to a monolithic lower roof (b=1) are included on the plot and are seen to deviate from the trend; a frictional resistance relationship predicts these values correctly. For a constant span and block thickness, linear arch theory predicts that the value of horizontal thrust should be a constant and does not consider the effect of friction. The actual data indicated that a linear relationship exists between horizontal thrust, joint spacing in the roof and friction angle of the joints. The data values indicated that the side force required for stability increases both as the joint spacing decreases and as the friction coefficient of the joints decreases.

SUPPORT REQUIREMENTS OF UNDERGROUND EXCAVATIONS

The Distinct Element method was also used to study the support requirements of numerous excavation roofs which posses the joint pattern characteristics of the basic model with regular, continuous jointing in the horizontal direction and regular, discontinuous jointing in the vertical direction. This is a plane strain model and the aspect ratio of the blocks for a given problem is constant. The results of this investigation are presented in this section through ground reaction curves which are representative of the observed responses.

The results presented previously indicated that the stability of the roof of an excavation in jointed rock was most sensitive to the magnitude of the horizontal stress. It follows logically, therefore, that an investigation of the support requirements of excavations in jointed media should be concerned with the effect of horizontal stress on the ground behavior as expressed by a ground reaction curve relating the total load acting on the support to the vertical deflection of the support. The joints are modeled as planar and do not posses cohesion. The tendency of construction procedures such as blasting is to destroy the cohesion of the joint surfaces near the excavation. This, coupled with the fact that the models portray the behavior of failing masses lead to the conclusion that the analyses are valid in terms of the cohesive strength of the joints. The fact that the joints are considered to be planar, however, does detract somewhat from the validity of the analyses. Real joints are non-planar; perfectly mating rough surfaces can only be forced to slide relative to one another if they are free to move apart. This dilatancy leads to increased mass strength for if the the joint separates two confined blocks, the only way relative movement can occur is if shearing of the rock mass takes place.

Figure 3 presents two ground reaction curves for the six meter wide excavation illustrated in the figure. Part (a) of the figure

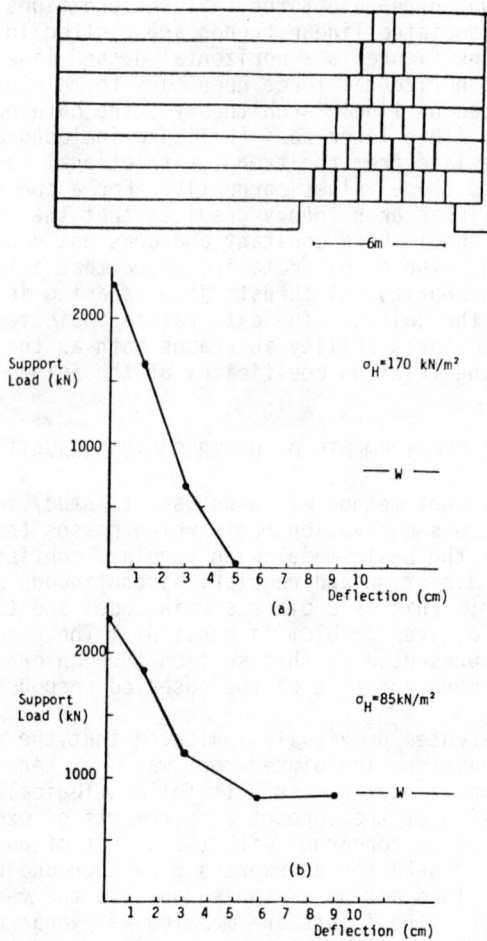

Figure 3. Ground Reaction Curves for 6m Wide Excavation:
(a) High Horizontal Stress; (b) Low Stress.

illustrates the ground reaction curve for a case where sufficient horizontal stress exists to stabilize the mass in the absence of externally applied support. The ground reaction curve reflects this fact indicating that at a value of the roof deflection of approximately five centimeters, the load acting on the support is zero. The second ground reaction curve illustrated in the figure represents a situation where the magnitude of the horizontal stress field is insufficient to stabilize the mass without the introduction of external support. The parameter W, indicated on the ground reaction curve, is the total weight of the material within the zone of moving material. The form of the ground reaction curve suggests that as

EXCAVATION SUPPORT LOADS

deflection of the roof continues the required support force approaches a constant value, and that this value is given by the roof load W. The tendency for the ground reaction to indicate a constant value of the required support force was observed in the majority of the cases examined.

The modeling of jointed excavation roofs led to the conclusion that the ultimate load to be resisted by the support system could be predicted, in the majority of cases, by the roof load W. A relationship between the ultimate support load and the span of the excavation can be seen. This relationship was found to be a function of the aspect ratio of the blocks, but relatively insensitive to the friction coefficient of the joints. The relationship between the support load required and span is given approximately by:

$$L = n B^2$$

where

$n = 2 + 5A$; and A is the block aspect ratio. (a)

THE USE OF THE DISTINCT ELEMENT METHOD IN THE DESIGN OF SUPPORT SYSTEMS FOR EXCAVATION IN JOINTED MASSES

In response to the idealized assumptions of joint behavior utilized in the analyses, the support force required for stability was seen typically to be a function of the geometric properties of the excavation. In particular, the ultimate resisting force was found to be given approximately by the roof load, which could be calculated. This data was also utilized in a comparison with several of the empirical schemes to see if a correlation existed. The primary purpose of this investigation was to see if the Distinct Element calculated response of an excavation in jointed rock, taking account of mass/support interaction, could be correlated to "dead weight" load schemes such as those proposed by Terzaghi (1946), Cording, et al. (1971), Stini (1950), and Barton, et al. (1974). Figure 4 presents a summary of the required support force as a function of span for those masses investigated by the Distinct Element method; also included in the figure is actual design data summarized. The trend of the data have, in this instance, been calculated using equation a. The presented curves fit the data as well as those suggested by Cording, et al. (1971); however, in this case the curves are a function of the aspect ratio of the blocks formed by the jointing. It is not immediately clear that there should be a correlation between RQD and aspect ratio of the blocks. It certainly would be feasible to estimate the block aspect ratio if directionally biased RQD data were available, but RQD data is not typically recorded in this manner. The properties of the basic model chosen for investigation indicated that a reasonable estimate of the upper limit to the amount of load to be resisted by the support system could be calculated in terms of the geometric parameters of the rock mass and excavation. The eventual results indicate that this upper limit, the roof load, was actually the value for which the support should be designed.

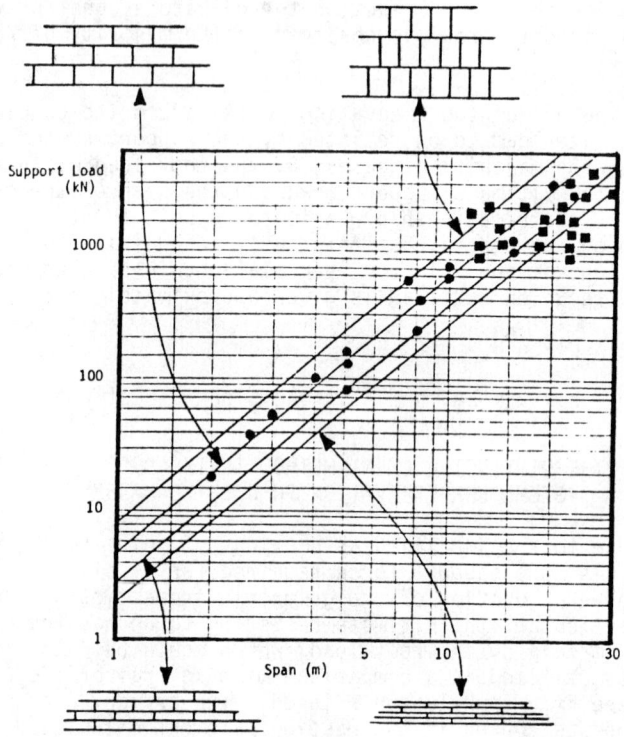

Figure 4. Summary of Distinct Element calculated required support loads and design data presented by Cording et al. (1971); also illustrated are the various aspect ratios.

The implications of the results presented in this section can be interpreted in one of two ways. By neglecting dilatancy a correlation was found between the required support force and the roof load. This support force was also found to correlate fairly well with empirical methods, particularly those of Stini (1950) and Cording, et al. (1971). If it can be inferred that the failure to incorporate the dilatancy properties of real joints in the analysis leads to a value of the mass strength that is too low, then it can be concluded that the roof load and thus the empirical methods represent a conservative value of design load. The second interpretation also follows from the properties of the joints. It is reasonable to expect that the dilatancy properties of joints would play a minor role in situations of relatively low stress. It can thus be concluded that dimensioning the supports to resist the roof load, or using one of the empirical schemes should give the best results in problems involving low stresses.

ACKNOWLEDGEMENTS

The research described herein was a portion of the senior author's doctoral dissertation. Dr. Peter Cundall made numerous significant suggestions throughout the program as well as providing the basic computer model. Support for the research was provided by the U.S. Army Corps of Engineers program of Rational Design of Tunnel Supports.

REFERENCES

Barton, N.R., R. Lien and J. Lunde, 1974, "Engineering Classification of Rock Masses for the Design of Tunnel Support," Rock Mechanics V6 N4, December.

Cording, E.J., A.J. Hendron and D.V. Deere, 1971, "Rock Engineering for Underground Caverns," Proc. Symp. Underground Rock Chambers ASCE, Phoenix, pp. 567-600.

Cundall, P.A., 1971, "A Computer Model for Simulating Progressive Large Scale Movements in Blocky Rock Systems," Proc. Symp. Rock Fracture (ISRM) Nancy, paper II-8.

Stini, J., 1950, Tunnelbaugeologie Wien, Springer-Verlay, pp. 215-221.

Terzaghi, K., 1946, "Introduction to Tunnel Geology," in Rock Tunneling with Steel Supports, by Proctor and White, Commercial Shearing and Stamping, Youngstown, Ohio.

Chapter 71

INFLUENCE OF CREEP LAW FORM ON PREDICTED DEFORMATIONS IN SALT

by Ralph A. Wagner, Kirby D. Mellegard and Paul E. Senseny

RE/SPEC Inc.
P. O. Box 725
Rapid City, SD 57709

ABSTRACT

Six creep laws for salt, each fitted to the same laboratory data base, are used in a numerical model of the deformation of an opening in salt to determine the influence of creep law form on the predicted closures. Results show that predicted closures can be very sensitive to the functional form of the creep law chosen to model the laboratory data.

INTRODUCTION

This study assesses the influence of creep law form on the prediction of time-dependent deformation. This is accomplished by considering six commonly used creep laws in the numerical analyses of a nuclear waste disposal room in salt. Each creep law is fitted to a common set of laboratory data. Therefore, any variations in the predicted deformations are a function of the creep law form and not variation in the laboratory data. Results show that prediction of long-term deformation of structures can vary substantially, depending on which creep law form is used to represent the laboratory-measured time-dependent deformation in the numerical analysis.

BACKGROUND ON THE CREEP LAWS

Creep laws are one kind of the many constitutive laws that model the rate-dependent deformation of materials. In rock mechanics studies, rate-dependent laws for salt are applied in the design of underground storage structures, radioactive waste repositories, and salt mines where the combination of stress, temperature, and time give rise to significant time-dependent deformations.

The creep law forms discussed below are fitted to laboratory tests that involve constant stress and temperature. Temporal changes in stress and temperature are accommodated by using either time-hardening or strain-hardening creep accumulation strategies (Finnie, 1960) for transient strains and assuming steady-state strain rate is independent of prior stress and temperature. Neither creep accumulation strategy for the transient strains has a physical basis, but both are used extensively. Four creep laws considered in this study involve a time-hardening strategy, whereas a strain-hardening strategy is used in the other two. Table 1 lists these six creep laws in a tensorial form that gives the creep strain rate as a function of stress, temperature and time or accumulated creep strain. The table also provides the parameter values determined for these laws by fitting them to a common data base.

Exponential-Time Creep Law

An exponential-time creep law was selected as the baseline creep law (Senseny, 1981) for use in numerical analyses of potential nuclear waste repositories in salt. This form was selected because it fits the data well, incorporates steady-state deformation, and can account for complex thermomechanical history. To account for temporal changes in stress and temperature, a strain-hardening strategy was adopted. Development of the exponential-time law for salt is based on first-order kinetics and parallels development of a similar law for high-temperature creep of steel (Webster et al, 1969).

Power Creep Law

The power law is an empirical formula that is commonly used to relate transient creep strain to stress, temperature, and time. Stress and temperature histories are incorporated by using a time-hardening creep formulation in this paper, although a strain-hardening approach is often used.

Exponential-Temperature Creep Law

The exponential-temperature law is another empirical formula that relates transient creep strain to stress, temperature, and time. The law is very similar to the power law except that the strain varies exponentially with temperature rather than as a power of temperature (Table 1). This law is preferred by some analysts because the temperature dependence is similar to that obtained for thermally-activated rate processes. The exponential temperature law is usually fitted to creep data obtained at high temperatures (greater than half the melting temperature), but it is also used to describe the tran-

TABLE 1

DESCRIPTION OF SIX CONSTITUTIVE RELATIONS FOR SALT CREEP

TYPE	FORM	PARAMETER VALUES
POWER (Time-Hardening)	$\dot{\varepsilon}_{ij} = A m t^{m-1} (3J_2)^{\frac{n-1}{2}} (3/2\, S_{ij}) T^p$	$A = 6.00$ E-6 (MPa^{-n}hr^{-m}) $m = 0.47$ $n = 2.21$ $p = 6.00$
MODIFIED POWER (Time-Hardening)	$\dot{\varepsilon}_{ij} = A m t^{m-1} (3J_2)^{\frac{n-1}{2}} (3/2\, S_{ij}) T^p\ +$ $B(3J_2)^{\frac{q-1}{2}} (3/2\, S_{ij}) \exp(-\lambda/T)$	$A = 4.44$ E-8 (MPa^{-n}hr^{-m}) $m = 0.36$ $n = 3.90$ $p = 11.5$ $B = 3.85$ E-6 (MPa^{-q}hr^{-1}) $q = 6.20$ $\lambda = 5104$ (K)
MODIFIED POWER (Strain-Hardening)	$\dot{\varepsilon}_{ij} = m \left(\dfrac{A(3J_2)^{\frac{n}{2}} T^p}{\varepsilon_e^{1-m}} \right)^{\frac{1}{m}} \dfrac{3/2\, S_{ij}}{(3J_2)^{\frac{1}{2}}} +$ $B(3J_2)^{\frac{q-1}{2}} (3/2\, S_{ij}) \exp(-\lambda/T)$	$A = 4.44$ E-8 (MPa^{-n}hr^{-m}) $m = 0.36$ $n = 3.90$ $p = 11.5$ $B = 3.85$ E-6 (MPa^{-q}hr^{-1}) $q = 6.20$ $\lambda = 5104$ (K)
EXPONENTIAL TEMPERATURE (Time-Hardening)	$\dot{\varepsilon}_{ij} = A m t^{m-1} (3J_2)^{\frac{n-1}{2}} (3/2\, S_{ij}) \exp(-\lambda/T)$	$A = 9.62$ E-3 (MPa^{-n}hr^{-m}) $m = 0.47$ $n = 2.17$ $\lambda = 2200$ (K)
MODIFIED EXPONENTIAL TEMPERATURE (Time-Hardening)	$\dot{\varepsilon}_{ij} = A m t^{m-1} (3J_2)^{\frac{n-1}{2}} (3/2\, S_{ij}) \exp(-\lambda/T)\ +$ $B(3J_2)^{\frac{q-1}{2}} (3/2\, S_{ij}) \exp(-\beta/T)$	$A = 3.87$ E-1 (MPa^{-n}hr^{-m}) $m = 0.36$ $n = 3.60$ $\lambda = 4723$ (K) $B = 3.85$ E-6 (MPa^{-q}hr^{-1}) $q = 6.20$ $\beta = 5104$ (K)
EXPONENTIAL TIME (Strain-Hardening)	$\dot{\varepsilon}_{ij} = A(3J_2)^{\frac{n+1}{2}} \exp(-\lambda/T) \left\{ 1 + B\varepsilon_a - \dfrac{B\varepsilon_a\, \xi}{(J_2)^{n/2} \exp(-\lambda/T)} \right.$ $\left. \cdot \int_0^t (J_2)^{n/2} \exp(-\lambda/T) \exp\left[-\int_{t''}^{t} \xi dt'\right] dt'' \right\} \dfrac{S_{ij}}{2J_2}$ where: $\xi = \begin{cases} BA(3J_2)^{n/2} \exp(-\lambda/T) & \dot{\varepsilon}_{ss} > \dot{\varepsilon}_{ss}^* \\ B\dot{\varepsilon}_{ss}^* & \dot{\varepsilon}_{ss} < \dot{\varepsilon}_{ss}^* \end{cases}$	$A = 1.42$ E+1 (MPa^{-n}hr^{-1}) $n = 4.04$ $\lambda = 8103$ (K) $B = 200$ $\varepsilon_a = 1.05$ E-1 $\dot{\varepsilon}_{ss}^* = 1.44$ E-4 (hr^{-1})

t, t', t'' = Time (hr)
T = Temperature (K)
$\dot{\varepsilon}$ = Creep Strain Rate (hr^{-1})
ε_e = Total Effective Transient Creep Strain

J_2 = Second Invariant of Stress Deviator (MPa^2)
S_{ij} = Deviatoric Stress Tensor (MPa)
$\dot{\varepsilon}_{ss} = A(3J_2)^{\frac{n-1}{2}} (3/2\, S_{ij}) \exp(-\lambda/T)$

sient creep of salt in the range of temperatures expected in a nuclear waste repository. A time-hardening formulation is used in this paper to accommodate temporal changes in temperature and stress.

Modified Power and Modified Exponential-Temperature Creep Laws

The power and exponential-temperature laws model only transient creep. These transient laws can be modified by combining a steady-state expression as shown in Table 1. The data are fitted better by the modified power and modified exponential-temperature laws than the two transient laws because these modified laws have additional parameters. However, negative values for the steady-state strain rate are often obtained when these modified laws are fitted to laboratory data that do not contain a substantial amount of steady-state deformation. Both strain-hardening and time-hardening formulations for transient creep are used for the modified power law; whereas, only a time-hardening formulation is used for the modified exponential-temperature law.

DATA BASE

The six constitutive forms for salt creep considered in this study are fitted to a data base comprising laboratory creep tests performed in triaxial compression. The creep tests used 100-mm-diameter cylindrical specimens of dome salt from Avery Island, Louisiana (Mellegard et al, (1981).

A total of 23 experiments were included in the matrix of creep tests. The matrix covers a temperature range of 24°C to 200°C, confining pressures ($\sigma_2 = \sigma_3$) of 0.7 MPa to 20.7 MPa, and axial stress differences ($\sigma_1 - \sigma_2$) of 5.4 MPa to 20.7 MPa. The length of time for these experiments averaged 28 days with variations between 2 and 78 days. The laws were fitted to the data by first fitting each test to a constant stress and temperature form of the law. The stress and temperature dependence of the parameters obtained from these fits was then determined by a subsequent least-squares regression.

All laws fit the data base equally well (comparable Sum of Squares Error), although the relative amount of creep strain may vary considerably at different temperatures. To illustrate these variations, creep strain was calculated as a function of time at constant stress and temperature for two laws that produce extreme deformations in the subsequent example problem. Creep strain calculated from the exponential-temperature and the exponential-time creep laws are shown in Figure 1 for times up to 25 years at a stress of 9 MPa and temperatures of 25°C, 50°C, 75°C and 100°C. This plot shows that for temperatures below 50°C, the exponential-temperature law predicts larger creep strains. At temperatures above 75°C, however, the exponential-time law predicts larger creep strains for most of the time considered.

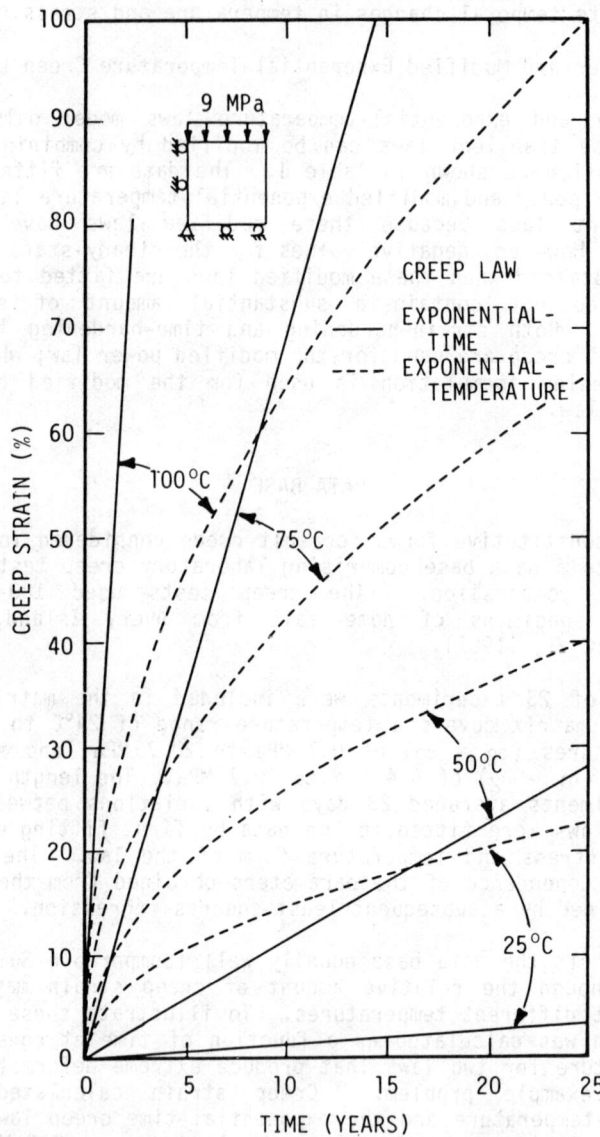

Figure 1. Comparison of two creep laws used to evaluate a problem with constant loading conditions.

NUMERICAL ANALYSES

The six constitutive forms for salt creep were compared by using them in a finite element analyses of the time-dependent deformation of a representative nuclear waste disposal room in salt. An areal thermal loading of 25 W/m^2 is assumed to be generated from 10 year old commercial high-level waste. The depth of the disposal room (610 m), initial temperature (30°C), and material properties of salt are characteristic of a repository situation adopted in a recent study (Wagner et al, 1982). The exponential-time creep law form was used exclusively in that study. An initial lithostatic state of stress is assumed before the instantaneous excavation of the disposal room. The initial stress state is proportional to the repository depth and overburden density. Temperature and stress at any spatial location was assumed to be constant for each time step of the creep analyses.

Separate numerical analyses of the thermal and thermomechanical responses are performed because these two responses are not fully coupled. The thermal analysis provided identical temperature distributions which were used in the thermomechanical analysis of each creep law form.

RESULTS

Centerline roof-to-floor closure from the numerical analyses using each of the six creep laws are shown in Figure 2. The closure-versus-time curves vary substantially, showing that the creep law form is very important when predicting long-term deformation of structures in salt. In this particular example problem, the power law predicts the most room closure while the exponential-time law predicts the least.

The two transient-only laws, the power law and the exponential-temperature law, predict that long-term closure occurs at a monotonically decreasing rate. These two closure curves are nearly parallel because both have the same time dependence (like t^m) and the change in stress and temperature is small in later time. The very large closure predicted using these laws results from the poor fits to the data at the low temperatures (~50°C) experienced in this problem.

The modified power and modified exponential-temperature laws predict that the long-term closure-versus-time curves are nearly linear because the steady state terms dominate in later time. These three curves are nearly parallel after about two years when the stress and temperature change slowly because they also have identical time dependence ($At^m + \dot{\varepsilon}_{ss}t$). The strain-hardening formulation of the modified power law predicts greater closure than does the time-hardening formulation. This is because the actual rate of hardening is less with the strain-hardening strategy for the nearly constant stress and increasing temperature histories encountered in our analyses of a nuclear waste disposal room.

The exponential-time creep law predicts smaller deformations than the modified laws because the transient strain rates predicted by this law become vanishingly small after a few weeks, whereas in the three

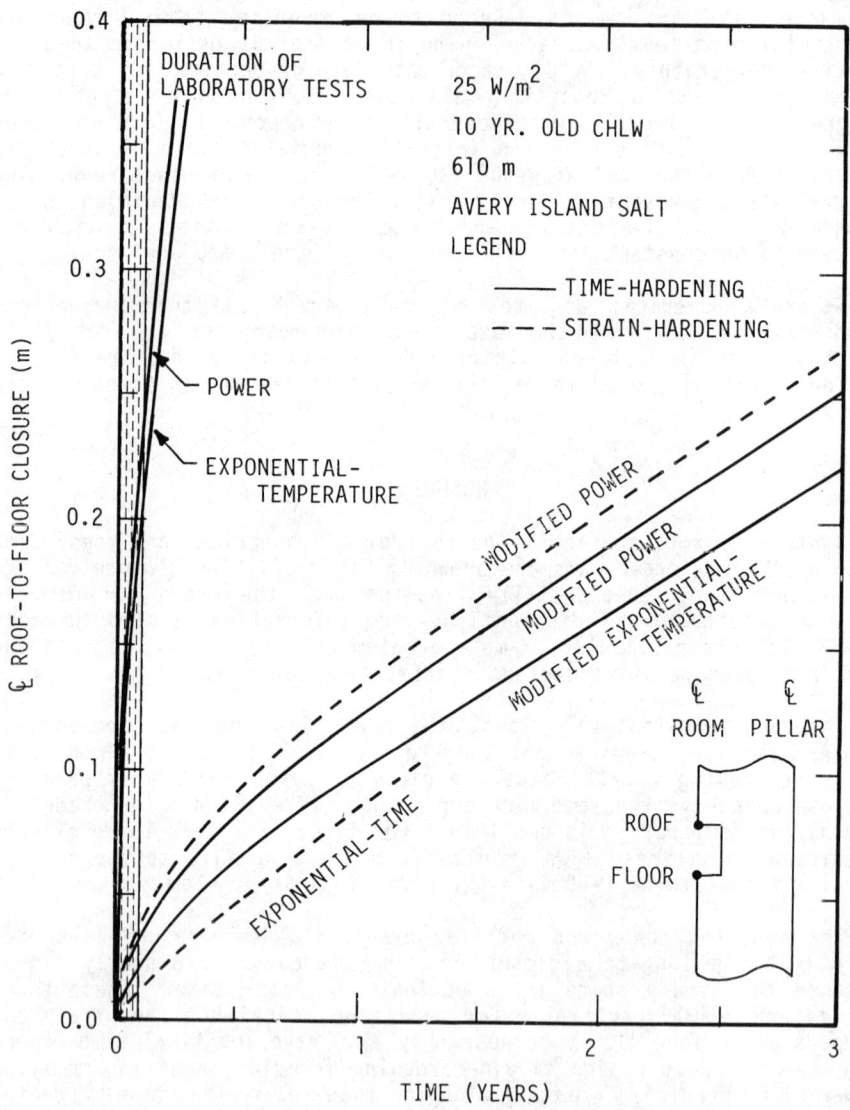

Figure 2. Centerline roof-to-floor closure predicted using six different creep laws.

modified laws the transient strain rates decrease very slowly and contribute significantly to the deformation for a long time.

CONCLUSION

This study shows the importance of carefully selecting a creep law form for salt when predicting structural deformations at times that lie outside the data base. Although not thoroughly investigated in this study, creep deformation is also found to be sensitive to loading conditions, length of time considered, and the laboratory test matrix. Six commonly used creep laws were fitted to a common data base of laboratory triaxial compression creep tests. These laws were then used to predict deformation of a nuclear waste disposal room. Substantially different deformation-versus-time curves were obtained. Although it is beyond the scope of this paper to evaluate these creep laws, it was stated that the exponential-time law has been selected by the National Waste Terminal Storage program for modeling of nuclear waste repositories in salt.

ACKNOWLEDGEMENTS

This work was performed under a subcontract with Battelle Memorial Institute, a DOE contractor. The subcontract was administered by the Office of Nuclear Waste Isolation (ONWI) and is part of the National Waste Terminal Storage (NWTS) Program. The authors acknowledge the technical contributions of Dr. Joe L. Ratigan and Dr. Gary D. Callahan. We also thank Ms. Jean M. Wilson for typing this manuscript.

REFERENCES

Finnie, I., 1960, "Stress Analysis in the Presence of Creep", Applied Mechanics Reviews, Vol. 13, No. 10, October, pp. 705-712.

Mellegard, K. D., Senseny, P. E., and Hansen, F. D., 1981, "Quasi-Static Strength and Creep Characteristics of 100-mm-diameter Specimens of Salt from Avery Island, Louisiana", ONWI-250, March, Office of Nuclear Waste Isolation, Battelle Memorial Institute, Columbus, OH.

Senseny, P. E., 1981, "Review of Constitutive Laws Used to Describe the Creep of Salt", ONWI-295, August, Office of Nuclear Waste Isolation, Battelle Memorial Institute, Columbus, OH.

Wagner, R. A., Loken, M. C., and Tammemagi, H. Y., In Preparation, "Preliminary Thermomechanical Analyses of a Potential Nuclear Waste Repository at Four Salt Sites", ONWI-   , Office of Nuclear Waste Isolation, Battelle Memorial Institute, Columbus, OH.

Webster, G. A., Cox, A. P. D., and Dorn, J. E., 1969, "A Relationship Between Transient and Steady-State Creep at Elevated Temperatures", Metal Sci. J., Vol. 3, pp. 221-225.

Chapter 72

A HYBRID QUADRATIC ISOPARAMETRIC FINITE ELEMENT-BOUNDARY ELEMENT
CODE FOR UNDERGROUND EXCAVATION ANALYSIS

by Denis Yeung and B.H.G. Brady

Graduate Student, Dept. of Civil and Mineral Engineering
University of Minnesota, Minneapolis, MN 55455

Associate Professor, Dept. of Civil and Mineral Engineering
University of Minnesota, Minneapolis, MN 55455

ABSTRACT

A hybrid finite element-boundary element code with eight noded isoparametric finite elements and three noded boundary elements is described for plane strain and linear isotropic elasticity. In the direct formulation of the Boundary Element Method, quadratic variation of geometry and displacements, and linear variation of tractions with respect to the element intrinsic coordinates are used. The boundary element region is treated as a "super-element" in the finite element sense.

INTRODUCTION

Numerous problems in the design of underground excavations for mining and civil engineering purposes require consideration of the displacements induced and total stresses resulting from the excavation process in an initially stressed medium. The determination of these quantities by complete discretization of the solution domain as required by the Finite Element Method is usually not economically feasible. In practice, an infinite or semi-infinite domain is approximated by a finite one, whose dimensions are a matter of subjective judgement derived by trial-and-error procedures.

A coupling of the Finite Element and Boundary Element Methods of stress analysis eliminates this problem. The Finite Element Method can be used to model the near-field around the excavation, where complex constitutive behavior may be assumed to exist due to the higher stress concentrations. The Boundary Element Method can be used to model the far-field remote from the excavation surface, where linear elastic behavior may be assumed.

An example of a hybrid code using triangular finite elements and two noded boundary elements was recently published by Brady and Wassyng (1981). For the practical solution of more realistic problems, the use of more sophisticated finite and boundary elements would result in an increase in accuracy and a decrease in the amount of data preparation. This paper describes a hybrid code utilizing eight noded isoparametric quadratic finite elements and three noded boundary elements, and may be considered an extension of the work of Brady and Wassyng (1981).

In this paper, summation is implied on repeated subscripts and superscripts. The range of the repeated indices will be from 1 to 2 unless explicitly stated.

BOUNDARY CONDITIONS AND F.E. FORMULATION

In underground excavation analysis, the induced displacements and stresses are obtained from the solution of the exterior problem (zero initial stress and zero body forces) with boundary conditions given by

$$\underline{t}(\underline{x}) = - \underline{P}(\underline{x}) \, \hat{n}(\underline{x}) + \underline{f}(\underline{x}) \qquad \underline{x} \in S \qquad (1)$$

and $\qquad \underline{u}(\underline{x}) = \underline{t}(\underline{x}) = 0 \qquad \underline{x} \in S_\infty \qquad (2)$

where $\underline{t}$ is the traction vector, $\underline{P}$ the pre-excavation stress tensor, $\hat{n}$ the unit outer normal, $\underline{f}$ the applied force vector and $\underline{u}$ is the displacement vector. The total stresses are obtained from the algebraic sum of the induced and pre-excavation stresses. This implies that the solution algorithm for underground excavation analysis may concentrate on the "induced" problem.

The displacement formulation of the Finite Element Method is well documented (Zienkiewicz, 1977). The F.E. Method consists of deriving for each F.E. the relationship

$$\underline{K}^e \, \underline{u}^e = \underline{f}^e \qquad (3)$$

where $\underline{K}^e$ is the stiffness matrix, $\underline{u}^e$ the nodal displacement vector and $\underline{f}^e$ is the nodal point force vector, from the known geometrical and material properties of the element. A coupling of the F.E. and B.E. Methods of stress analysis therefore requires the determination of a similar relationship for the Boundary Element region.

BOUNDARY ELEMENT FORMULATION

The direct formulation of the Boundary Element Method is based on Maxwell-Betti's reciprocal theorem and any fundamental solution for a point force in an infinite, linear elastic, homogeneous and isotropic medium. For convenience, we have used Kelvin's solution. This gives Somigliana's identity

$$u_k(\xi) = \int_S \{U_{ik}(x,\xi) t_i(x) - T_{ik}(x,\xi) u_i(x)\} dS(x) \qquad x \in S, \; \xi \in V \qquad (4)$$

where u_k is the displacement, t_i the traction, S the near-field surface of the B.E. region, V the solution domain of the B.E. region, and U_{ik} and T_{ik} are obtained from Kelvin's solution.

By taking ξ to a point $x_o \in S$ and taking account of the $(1/r)$ singularity in T_{ik}, we obtain the boundary integral equation

$$(\delta_{ik} + c_{ik}) u_i(x_o) = \int_S U_{ik}(x,x_o) t_i(x) \, dS(x)$$

$$- \int_S T_{ik}(x,x_o) u_i(x) \, dS(x) \qquad x, \; x_o \in S \qquad (5)$$

relating the tractions and displacements on S. The second integral in equation (5) is interpreted as a Cauchy Principal Value integral, and $c_{ik} = -\tfrac{1}{2} \delta_{ik}$ if x_o is on a smooth boundary. Explicit derivations of equations (4) and (5) are given by Brebbia (1978), and Banerjee and Butterfield (1981).

In discretized form, equation (5) becomes

$$(\delta_{ik} + c_{ik}) u_i(x_o) = \sum_{j=1}^{n} \int_{S_j} U_{ik}(x,x_o) t_i(x) \, dS_j(x)$$

$$- \sum_{j=1}^{n} \int_{S_j} T_{ik}(x,x_o) u_i(x) \, dS_j(x) \qquad (6)$$

where n is the number of boundary elements.

We are now concerned with a description of (a) the geometry of the boundary element surface, S_j, (b) the surface variation of the tractions, t_i and (c) the surface variation of the displacements, u_i. These approximations are now described in turn.

ISOPARAMETRIC FINITE ELEMENT

Quadratic Description of Problem Geometry

The geometry of a boundary element, S_j, is approximated as follows:

$$x(\xi) = x^1 N^1(\xi) + x^2 N^2(\xi) + x^3 N^3(\xi) = x^i N^i(\xi) \quad i = 1,2,3, \quad -1 \leq \xi \leq 1$$
$$y(\xi) = y^1 N^1(\xi) + y^2 N^2(\xi) + y^3 N^3(\xi) = y^i N^i(\xi) \tag{7}$$

where x^i, y^i are the nodal coordinates, $x(\xi)$, $y(\xi)$ the coordinates of a point within S_j and $N^i(\xi)$ are the Lagrangian interpolation functions (Fig. 1).

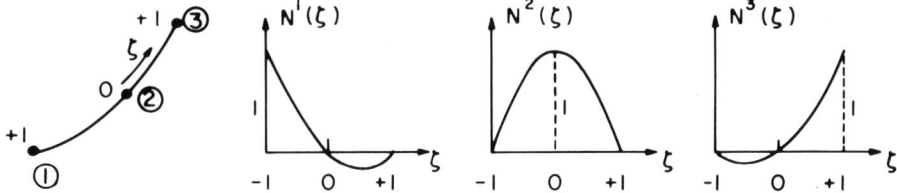

Fig. 1. A quadratic boundary element and the quadratic Lagrangian shape functions.

For numerical integration by Gaussian quadrature, i.e.,

$$\int_{-1}^{1} f(x) \, dx \doteq w_i \, f(x_i) \quad i = 1,2,\cdots$$

we need an expression for the Jacobian, J_j, of the transformation

$$dS_j(\underset{\sim}{x}) = \left\{ \left(\frac{dx}{d\xi}\right)^2 + \left(\frac{dy}{d\xi}\right)^2 \right\}^{\frac{1}{2}} d\xi \equiv J_j(\xi) \, d\xi \ . \tag{8}$$

Equation (6) becomes

$$(\delta_{ik} + c_{ik}) u_i(\underset{\sim}{x}_o) = \sum_{j=1}^{n} \int_{-1}^{1} U_{ik}(\xi,\underset{\sim}{x}_o) \, t_i(\xi) \, J_j \, d\xi$$
$$- \sum_{j=1}^{n} \int_{-1}^{1} T_{ik}(\xi,\underset{\sim}{x}_o) \, u_i(\xi) \, J_j \, d\xi \ . \tag{9}$$

Linear Variation of Tractions

We may rewrite the first integral of equation (9) as

$$\int_{-1}^{1} U_{ik}(\xi,\underset{\sim}{x}_o) \; t_i(\xi) \; J_j \; d\xi = \int_{-1}^{0} U_{ik}(\xi,\underset{\sim}{x}_o) \; t_i(\xi) \; J_j(\xi) \; d\xi$$

$$+ \int_{0}^{1} U_{ik}(\xi,\underset{\sim}{x}_o) \; t_i(\xi) \; J_j(\xi) \; d\xi = \int_{-1}^{1} U_{ik}(\eta,\underset{\sim}{x}_o) \; t_i(\eta) \; J_j(\eta) \; \frac{d\xi}{d\eta} \; d\eta$$

$$+ \int_{-1}^{1} U_{ik}(\beta,\underset{\sim}{x}_o) \; t_i(\beta) \; J_j(\beta) \; \frac{d\xi}{d\beta} \; d\beta \tag{10}$$

where η and β are suitable changes of variable of integration. We now approximate the tractions along S_j as follows

$$\left.\begin{array}{l} t_x(\eta) = t_x^1 \; \overline{N}^1(\eta) + t_x^2 \; \overline{N}^2(\eta) \\ t_y(\eta) = t_y^1 \; \overline{N}^1(\eta) + t_y^2 \; \overline{N}^2(\eta) \end{array}\right\} \; t_i(\eta) = t_i^p \; \overline{N}^p(\eta)$$

$$-1 \leq \eta, \beta \leq +1$$

$$\left.\begin{array}{l} t_x(\beta) = t_x^2 \; \overline{N}^1(\beta) + t_x^3 \; \overline{N}^2(\beta) \\ t_y(\beta) = t_y^2 \; \overline{N}^1(\beta) + t_y^3 \; \overline{N}^2(\beta) \end{array}\right\} \; t_i(\beta) = t_i^{p+1} \; \overline{N}^p(\beta) \tag{11}$$

where t_i^p are the nodal tractions and $\overline{N}^i$ are the linear Lagrangian shape functions $(\overline{N}^1(\alpha) = \tfrac{1}{2}(1 - \alpha), \; \overline{N}^2(\alpha) = \tfrac{1}{2}(1 + \alpha), \; -1 \leq \alpha \leq 1)$.

Equation (10) may thus be written as

$$\int_{-1}^{1} U_{ik}(\xi,\underset{\sim}{x}_o) \; t_i(\xi) \; J_j(\xi) \; d\xi = \int_{-1}^{1} U_{ik}(\eta,\underset{\sim}{x}_o) \; t_i^p \; \overline{N}^p(\eta) \; J_j(\eta) \; \frac{d\xi}{d\eta} \; d\eta$$

$$+ \int_{-1}^{1} U_{ik}(\beta,\underset{\sim}{x}_o) \; t_i^{p+1} \; \overline{N}^p(\beta) \; J_j(\beta) \; \frac{d\xi}{d\beta} \; d\beta$$

$$j = 1, n \quad -1 \leq \eta, \beta \leq 1 \tag{12}$$

These integrals may be evaluated by Gaussian quadrature, except for three possible cases when the term U_{ik} has a logarithmic singularity. In these cases, it is necessary to perform a change of variable of integration and utilize the special numerical integration formula,

$$\int_{0}^{1} \ln(x) \; f(x) \; dx \doteq w_i \; f(x_i) \qquad i = 1, 2, \cdots$$

Quadratic Variation of Displacements

We may rewrite the second integral of equation (9) as

$$\int_{-1}^{1} T_{ik}(\xi,\underset{\sim}{x}_o) \, u_i(\xi) \, J_j(\xi) \, d\xi = \int_{-1}^{1} T_{ik}(\xi,\underset{\sim}{x}_o) \, u_i^p \, N^p(\xi) \, J_j(\xi) \, d\xi \qquad (13)$$

where we have assumed that

$$u_i(\xi) = u_i^p \, N^p(\xi) \qquad i = 1,2 \qquad p = 1,2,3 \qquad -1 \le \xi \le 1 \quad . \qquad (14)$$

That is, u_i^p are the nodal displacements, $u_i(\xi)$ are the displacements of a point within S_j, and $N^p(\xi)$ are the quadratic shape functions of Fig. 1. These integrals may be evaluated by Gaussian quadrature except for three possible cases when the integrand has a singularity of (1/r). The Cauchy Principal value of these integrals and the "free term" c_{ik} of equation (9) can be determined by rigid body considerations (Cruse, 1974 and Watson, 1979).

By evaluating equations (12) and (13), and taking $\underset{\sim}{x}_o$ successively to each boundary element node, we have

$$\underset{\sim}{T}\underset{\sim}{u} = \underset{\sim}{U}\underset{\sim}{t} \qquad (15)$$

where $\underset{\sim}{t}$ is the vector of nodal tractions, $\underset{\sim}{u}$ the vector of nodal displacements, and $\underset{\sim}{T}, \underset{\sim}{U}$ are coefficient matrices, whose terms are obtained as integrals of kernel-shape function products over the range of an element. Equation (15) may be written as

$$\underset{\sim}{U}^{-1}\underset{\sim}{T}\underset{\sim}{u} = \underset{\sim}{t} \qquad (16)$$

or

$$\left\{\underset{\sim}{C} \, \underset{\sim}{U}^{-1} \, \underset{\sim}{T}\right\}\underset{\sim}{u} \equiv \underset{\sim}{K}^b \, \underset{\sim}{u} = \underset{\sim}{C} \, \underset{\sim}{t} \equiv \underset{\sim}{f}^b \qquad (17)$$

where $\underset{\sim}{C}$ is the conversion matrix of distributed tractions to statically equivalent nodal point forces, $\underset{\sim}{K}^b$ the "super-element" stiffness matrix for the B.E. region and $\underset{\sim}{f}^b$ the "super-element" nodal point force vector.

Conversion of Distributed Tractions to Nodal Point Forces

We now describe the determination of the $\underset{\sim}{C}$ matrix of equation (17), expressing the relationship between the distributed tractions to the statically equivalent nodal point forces. Considering a boundary element S_k (Fig. 1) defined by nodes i(= 1,2,3), along which the tractions, t_j (j = 1,2) are governed by equation (11) and the displacements by equation (14), we may apply a unit virtual displacement in the jth coordinate direction at node i only. Then, by the principle of virtual work, we have

$$f^i_j = \int_{S_k} t_j \, N^i(\xi) \, dS_k$$

where f^i_j is the equivalent nodal point force in the jth coordinate direction at node i, and N^i is the quadratic shape function of Fig. 1. By identical reasoning as in equations (10) and (12), we have

$$f^i_j = \int_{-1}^{1} \left\{ t^p_j \overline{N}^p(\eta) \right\} N^i(\xi)\big|_\eta \, J_k(\xi)\big|_\eta \, \frac{d\xi}{d\eta} \, d\eta + \int_{-1}^{1} \left\{ t^{p+1}_j \overline{N}^p \right\} N^i(\xi)\big|_\beta \, J_k(\xi)\big|_\beta \, \frac{d\xi}{d\beta} \, d\beta$$

where $\overline{N}^p$ (p = 1,2) are the linear shape functions. By evaluating these integrals over all boundary elements (k = 1, n), we obtain $\underset{\sim}{C}$ which is sparsely populated and whose coefficients are integrals of the products of the linear and quadratic shape functions.

COUPLING PROCEDURES AND STRESS DETERMINATION

Equations (3) and (17) may now be assembled to form the global system of equations, thereby satisfying continuity of nodal displacements and equilibrium of nodal forces (Hinton and Owen, 1979). The B.E. nodal tractions may then be determined from equation (16) and the induced stresses and displacements within the B.E. region determined by differentiation and discretization of equation (4). The stresses at points along the boundary elements can be determined by using a local coordinate system as described by Brady (1979), but with the tangential strain given by

$$\varepsilon_{\ell\ell} = \frac{du/d\xi}{dS/d\xi} \frac{dx/d\xi}{dS/d\xi} + \frac{dv/d\xi}{dS/d\xi} \frac{dy/d\xi}{dS/d\xi}$$

CODE VERIFICATION

The preceding principles were implemented in a code for underground excavation analysis. Results are now presented to verify the performance of the code and to demonstrate its practical potential. Fig. 2 shows a circular hole excavated in a homogeneous uniaxial compressive stress field. The excavation surface was approximated by 16 nodes (i.e., 8 boundary elements), and the solution domain treated as a single "super-element." Table 1 compares the computed results with Kirsch's analytical solution.

Fig. 3 shows a circular hole also excavated in a homogeneous, uniaxial compressive stress field. The solution domain is approximated by two rings of finite elements around the excavation surface, and a "super-element." Table 2 compares the computed results with Kirsch's analytical solution.

Fig. 4 shows two circular holes excavated in a homogeneous, uniaxial compressive stress field. Table 3 compares the computed results for simultaneous and sequential excavation of the two holes. Each hole was approximated by two boundary elements (i.e., 4 B.E. nodes). In these exercises, we have used 3 point numerical integration, and tension is taken as positive.

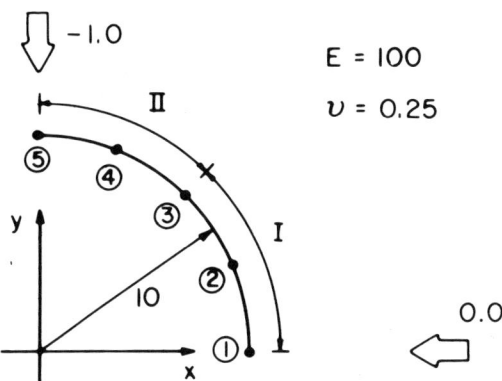

Fig. 2. Circular hole in an infinite, homogeneous, isotropic elastic medium.

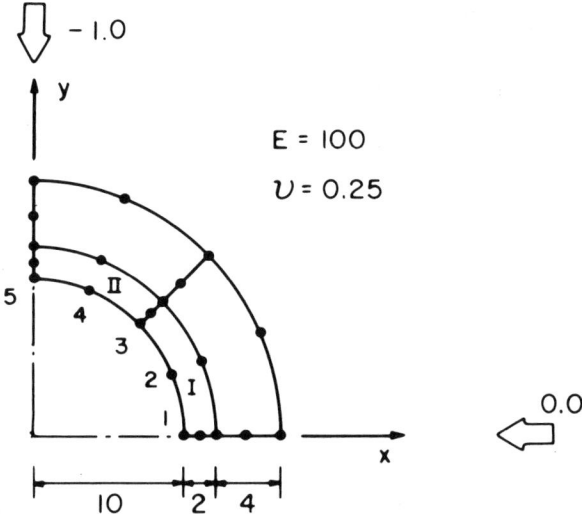

Fig. 3. Approximation of solution domain by finite and boundary elements.

TABLE 1. Comparison of "Super-element" Results with Kirsch's Solution

Induced Nodal Displacements				
Node	Computed		Analytical	
No.	u_x	u_y	u_x	u_y
1	0.06246	0	0.06250	0
2	0.05770	-0.07173	0.05773	-0.07175
3	0.04416	-0.13254	0.04421	-0.13262
4	0.02385	-0.17320	0.02389	-0.17318
5	0	-0.18742	0	-0.18750

Boundary Element Stresses						
B.E.	Computed			Analytical		
No.	σ_{xx}	σ_{yy}	σ_{xy}	σ_{xx}	σ_{yy}	σ_{xy}
I	-0.1078	-2.7385	0.5434	-0.1125	-2.7290	0.5542
	-0.3557	-2.0545	0.8583	-0.3537	-2.0592	0.8526
	-0.5444	-1.2222	0.8128	-0.5452	-1.2354	0.8221
II	-0.1631	-0.0787	0.1092	-0.1549	-0.0645	0.1002
	0.3549	0.0768	-0.1471	0.3527	0.0601	-0.1455
	0.8123	0.0252	-0.1612	0.8083	0.0333	-0.1641

TABLE 2. Comparison of Coupled F.E.-B.E. Results with Kirsch's Solution

Induced Nodal Displacements			
Node	Computed		Analytical	
No.	u_x	u_y	u_x	u_y
1	0.06095	0	0.06250	0
2	0.05672	-0.07092	0.05773	-0.07175
3	0.04430	-0.13268	0.04421	-0.13262
4	0.02305	-0.17219	0.02390	-0.17318
5	0	-0.18594	0	-0.18750

Finite Element Stresses					
F.E.	Computed			Analytical		
No.	σ_{xx}	σ_{yy}	σ_{xy}	σ_{xx}	σ_{yy}	σ_{xy}
I	-0.02645	-2.7606	0.21634	-0.0818	-2.7994	0.2375
	-0.42350	-2.1194	0.69549	-0.3383	-2.0115	0.7519
	-0.50078	-0.9332	0.61732	-0.4752	-0.8725	0.6189
	-0.17767	-2.3708	0.13105	-0.2210	-2.4051	0.1472
	-0.30554	-1.8834	0.52186	-0.2918	-1.8760	0.4882
	-0.21998	-0.9928	0.48466	-0.2619	-1.0392	0.4784
	-0.37215	-2.0911	0.07631	-0.2991	-2.1204	0.0874
	-0.24949	-1.7420	0.38732	-0.2552	-1.7661	0.3140
	-0.01173	-1.1194	0.38798	-0.1210	-1.1414	0.3759
II	-0.24437	-0.3291	0.32115	-0.3013	-0.3509	0.3216
	0.43731	0.0875	-0.02239	0.3366	0.0131	-0.0756
	0.81799	-0.0390	-0.04187	0.8588	0.0224	-0.0623
	-0.15178	-0.6221	0.33095	-0.1113	-0.5875	0.3354
	0.30437	-0.1002	0.06482	0.2921	-0.1243	0.0951
	0.56100	0	0.02037	0.5921	0.0339	0.0041
	-0.02030	-0.8639	0.31669	0.0102	-0.7478	0.3335
	0.23184	-0.2425	0.12188	0.2554	-0.2342	0.1956
	0.37803	0.0698	0.05547	0.4107	0.0088	0.0444

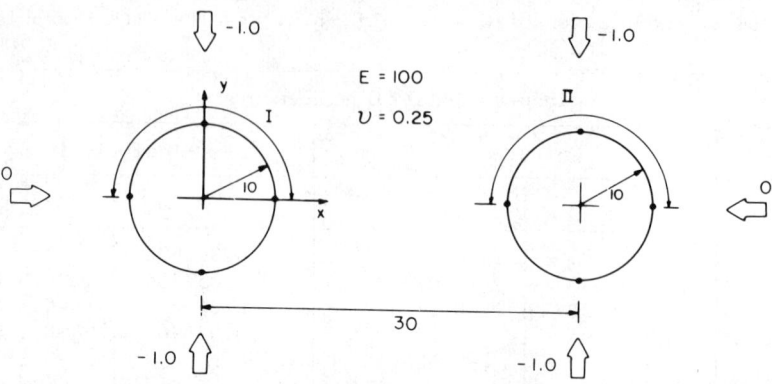

Fig. 4. Two circular holes in an infinite, homogeneous, isotropic elastic medium.

TABLE 3. Comparison of Results of Simultaneous and Sequential Excavation Sequences

B.E. No.	Boundary Element Stresses					
	Simultaneous Excavation			Sequential Excavation		
	σ_{xx}	σ_{yy}	σ_{xy}	σ_{xx}	σ_{yy}	σ_{xy}
I	-0.5975 0.9017 -0.5072	-0.6522 0.3368 -0.5620	0.5975 0 -0.5072	-0.6015 0.8898 -0.5060	-0.6562 0.3368 -0.5607	0.6015 0 -0.5060
II	-0.5072 0.9017 -0.5975	-0.5620 0.3368 -0.6522	0.5072 0 -0.5975	-0.5064 1.0250 -0.5554	-0.5690 0.3533 -0.6423	0.5021 0.0095 -0.6001

CONCLUSIONS

In this paper, we have described a procedure for determination of induced displacements and total stresses resulting from the excavation process in an infinite and initially stressed medium. We have treated the far-field as a boundary element region and derived a "super-element" in the finite element sense. An alternative procedure is to treat the finite element region as an equivalent boundary element domain (Brebbia and Georgiou, 1979). However, this has limitations for modeling complex constitutive behavior around the excavation.

We have made no attempt to ensure the required symmetry of the "super-element" stiffness matrix. The results of our debugging exercises did not justify any such modification.

We have used a lower (i.e., linear) order variation of tractions along the element in contrast to Beer and Meek (1981), who used a quadratic variation of tractions and obtained appreciable inaccuracy in their results. A lower order variation of tractions is necessary, since the stresses and tractions are related to the first derivative of the displacements.

This code may be modified to incorporate complex constitutive behavior in the finite element domain.

REFERENCES

Banerjee, P.K., and Butterfield, R., 1981, *Boundary Element Methods in Engineering Science*, McGraw-Hill, U.K.

Beer, G., and Meek, J.L., 1981, "The Coupling of Boundary and Finite Element Methods for Infinite Domain Problems in Elasto-Plasticity," *Boundary Element Methods*, Brebbia, C.A., ed., Springer-Verlag, New York, pp. 575-592.

Brady, B.H.G., 1979, "A Direct Formulation of the Boundary Element Method of Stress Analysis for Complete Plane Strain," *International Journal of Rock Mechanics and Mining Sciences*, Vol. 16, pp. 235-244.

Brady, B.H.G., and Wassyng, A., 1981, "A Coupled Finite Element-Boundary Element Method of Stress Analysis," *International Journal of Rock Mechanics and Mining Sciences*, Vol. 18, pp. 475-485.

Brebbia, C.A., 1978, *The Boundary Element Method for Engineers*, John Wiley, New York.

Brebbia, C.A., and Georgiou, P., 1979, "Combination of Boundary and Finite Elements in Elastostatics," *Applied Mathematical Modelling*, Vol. 3, pp. 212-220.

Cruse, T.A., 1974, "An Improved Boundary-Integral Equation Method for Three Dimensional Elastic Stress Analysis," *Computers and Structures*, Vol. 4, pp. 741-754.

Hinton, E., and Owen, D.R.J., 1979, *Finite Element Programming*, Academic Press, New York.

Watson, J.O., 1979, "Advanced Implementation of the Boundary Element Method for Two and Three Dimensional Elastostatics," *Developments in Boundary Element Methods*, Banerjee, P.K., and Butterfield, R., eds., Applied Science Publishers, London.

Zienkiewicz, O.C., 1977, *The Finite Element Method*, McGraw-Hill, New York.

Chapter 73

BOUNDARY ELEMENT METHODS FOR VISCOELASTIC MEDIA

by Wang Yongjia[1] and Steven L. Crouch[2]

[1] Graduate Student, Department of Civil and Mineral Engineering
University of Minnesota, and Associate Professor
Department of Mining Engineering, Northeast Institute of Technology
People's Republic of China

[2] Professor, Department of Civil and Mineral Engineering
University of Minnesota

ABSTRACT

This paper describes a numerical method for computing time-dependent displacements and stresses in linear viscoelastic media. For such materials, the correspondence principle can be used to obtain the time-dependent solution directly from the solution to an associated elastic problem in Laplace transform space. The elastic solution, in turn, can easily be obtained by boundary element methods. The work presented in this paper is based on the displacement discontinuity method, adapted for the case in which the region of interest consists of two half-planes bonded together. A practical example of the method is given for a problem involving a panel of rooms and pillars in a thick layer of salt that is overlain by limestone. Viscoelastic behavior of the salt causes time-dependent displacements and stresses in both rock types.

INTRODUCTION

Determination of time-dependent displacements and stresses is a subject of great practical interest in mining engineering. In-situ measurements of closure of underground openings and contact stresses in tunnel linings indicate that many rocks exhibit creep behavior. In salt and potash mines, time-dependent displacements and stresses must be considered in practically all mine planning decisions.

The theory of linear viscoelasticty is the simplest continuum theory for modeling time-dependent effects (Fung, 1965). The term viscoelasticity is used to describe rheological models built of dashpots (which represent viscosity) and springs (which represent elasticity). The theory of linear viscoelasticity has been used to obtain solutions to a few special problems involving underground excavations (see, for example, Berry, 1977). However, analytic solutions can only be obtained for a limited class of problems, and for most situations one has to use numerical methods.

The numerical method used in this paper is comparable to the work of Rizzo and Shippy (1971) and is based on the correspondence principle of linear viscoelasticity (Lee, 1955). The correspondence principle allows one to construct the solution to a viscoelastic problem from the solution to an associated elastic problem. The elastic problem can easily be solved by boundary element methods. Rizzo and Shippy (1971) used the direct boundary integral method for this purpose. The work presented in this paper uses the displacement discontinuity method, adapted for the case in which the region of interest consists of two half-planes bonded together (Crouch and Starfield, 1982). Either, or both, of the half-planes can be viscoelastic.

CONSTITUTIVE EQUATIONS FOR LINEAR VISCOELASTIC MEDIA

The governing equations for a problem in quasi-static viscoelasticity are the static equilibrium equations and the kinematic (strain-displacement) relations, together with a set of constitutive equations. The constitutive equations for a linear viscoelastic material can be written as (Alfrey, 1944; Tsien, 1950)

$$P_s(D) \, s_{ij}(\underline{x},t) = Q_s(D) \, e_{ij}(\underline{x},t)$$

$$P_v(D) \, \sigma_{kk}(\underline{x},t) = Q_v(D) \, \varepsilon_{ij}(\underline{x},t) \quad (1)$$

where $\underline{x}$ is the position vector, t is time, $D = \frac{\partial}{\partial t}$ and $P_s(D)$, $Q_s(D)$, $P_v(D)$ and $Q_v(D)$ are the differential operations defined by

$$P_s(D) = \sum_{k=0}^{n_1} a_k D^k \; ; \; Q_s(D) = \sum_{k=0}^{m_1} b_k D^k$$

$$P_v(D) = \sum_{k=0}^{n_2} c_k D^k \; ; \; Q_v(D) = \sum_{k=0}^{m_2} d_k D^k \quad (2)$$

In (1) s_{ij} and e_{ij} are the deviatoric components of the stress and strain tensors σ_{ij} and ε_{ij}:

$$s_{ij} = \sigma_{ij} - \frac{1}{3} \delta_{ij} \sigma_{kk}$$

$$e_{ij} = \varepsilon_{ij} - \frac{1}{3} \delta_{ij} \varepsilon_{kk} \quad (3)$$

If we consider the stress-strain relationship of a network of springs and dashpots such as the Burgers model (Reiner, 1969) in Fig. 1, it can be shown that stress σ (or s) and strain ε (or e) are related by an equation of the form of (1). For this reason, a viscoelastic body is often represented by a mechanical model.

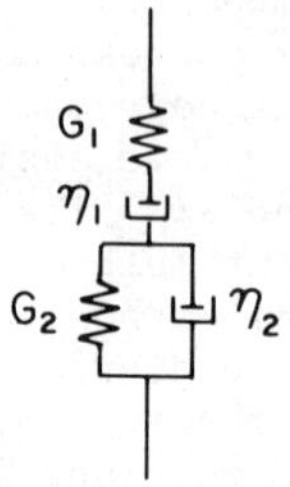

$$D^2\sigma + \left(\frac{G_1}{\eta_1} + \frac{G_1}{\eta_2} + \frac{G_2}{\eta_2}\right)D\sigma + \frac{G_1 G_2}{\eta_1 \eta_2}\sigma$$

$$= G_1 D^2 \epsilon + \frac{G_1 G_2}{\eta_2} D\epsilon$$

Fig. 1. Burgers model and its response relation

CORRESPONDENCE PRINCIPLE

The Laplace transform of a function $f(t)$ is defined as

$$\bar{f}(s) = \int_0^\infty f(t) e^{-st} dt \qquad (4)$$

where s is called the transform parameter. According to Lee (1955) the Laplace transform can be used to remove the time variable from a problem in quasi-static linear viscoelasticity. The resulting problem, defined in Laplace transform space, is formally equivalent to a problem in linear elasticity. The 'elastic constants', however, are functions of the transform parameter s, as are the transformed boundary conditions for the problem. This procedure can only be employed if (i) the material is initially "dead", i.e the stress is applied suddenly at time t=0 when the system is unstrained, and (ii) the body shape does not change during loading. If these conditions are met, then the solution to a viscoelastic problem can be found directly from the solution to the 'corresponding' elastic problem. The elastic solution (in Laplace transform space) can be accurately obtained using boundary element methods, as discussed below. The time-dependent solution itself is obtained by finding the inverse Laplace transform.

The s-varying elastic contents are found from the constitutive equations (1). The Laplace transforms of these equations are

$$P_s(s) \bar{s}_{ij}(\underline{x},s) = Q_s(s) \bar{e}_{ij}(\underline{x},s)$$

$$P_v(s) \bar{\sigma}_{kk}(\underline{x},s) = Q_v(s) \bar{\epsilon}_{kk}(\underline{x},s) \qquad (5)$$

The constitutive equations for the elastic case are

$$s_{ij} = 2G e_{ij} \quad ; \quad \sigma_{kk} = 3K \epsilon_{kk} \qquad (6)$$

and comparison of (5) and (6) yields

$$\bar{G}(s) = \frac{Q_s(s)}{2P_s(s)} \quad ; \quad \bar{K}(s) = \frac{Q_v(s)}{3P_v(s)} \tag{7}$$

The s-varying constants $\bar{E}(s)$ and $\bar{\nu}(s)$ can thus be expressed as

$$\bar{E}(s) = \frac{9\bar{K}(s)\,\bar{G}(s)}{3\bar{K}(s) + \bar{G}(s)} = \frac{3Q_v(s)\,Q_s(s)}{Q_s(s)\,P_v(s) + 2P_s(s)\,Q_v(s)}$$

$$\bar{\nu}(s) = \frac{1}{2} \cdot \frac{3\bar{K}(s) - 2\bar{G}(s)}{3\bar{K}(s) + 2\bar{G}(s)} = \frac{P_s(s)\,Q_v(s) - Q_s(s)\,P_v(s)}{Q_s(s)\,P_v(s) + 2P_s(s)\,Q_v(s)} \tag{8}$$

Suppose now that induced tractions $(t_x)_o$ and $(t_y)_o$ are imposed on the boundary at time t=0 and remain constant for t > 0. Then, according to the correspondence principle, we can transfer the problem to the s-domain merely by letting

$$(t_x)_o \rightarrow (t_x)_o/s \quad ; \quad (t_y)_o \rightarrow (t_y)_o/s$$

$$E \rightarrow \bar{E}(s) \quad ; \quad \nu \rightarrow \bar{\nu}(s) \tag{9}$$

where s is the transform parameter and P_v, Q_v, P_s, Q_s are polynomials in s as defined by (2).

It is known that the stresses are independent of the elastic constants for a two-dimensional stress boundary value problem in homogeneous, isotropic linear elasticity (Timoshenko and Goodier, 1970). According to the correspondence principle, therefore, the stresses are independent of time for a comparable problem in homogeneous, isotropic linear viscoelasticity. If the material is anisotropic or inhomogeneous, however, the stresses for the elastic problem will depend on the elastic constants and hence the stresses will vary with time in the viscoelastic solution. The displacements, of course, will always be functions of time in a viscoelastic material.

INVERSE LAPLACE TRANSFORM TECHNIQUE

Schapery (1962) developed a collocation method of numerical Laplace transform inversion. In this method it is assumed that, at any point, each component of stress and displacement can be represented by the series

$$f(t) = A + Bt + \sum_{j=1}^{m} a_j\, e^{-b_j t} \tag{10}$$

where A, B, a_j and b_j are constants. Taking the Laplace transform of equation (10) and multiplying by the transform parameter s gives

$$s\bar{f}(s) = A + \frac{B}{s} + \sum_{j=1}^{m} \frac{a_j}{1 + b_j/s} \tag{11}$$

In order to find the constants in this equation a value of m and a sequence of values of s must be selected, i.e.

$$s = s_n, \quad n = 1, 2, \ldots, M \tag{11}$$

where $M = m + 2$. Shapery suggested the relationship between s and t as $s = 0.5/t$, so a range of s from 0.001 to 100 is equivalent to a range of t from 500 to 0.005. The m values of b are taken to be the first m values of s. Equation (11) then can be written

$$s_n \bar{f}(s_n) = A + \frac{B}{s_n} + \sum_{j=1}^{m} \frac{a_j}{1 + b_j/s_n}, \quad n = 1, 2 \ldots, M \tag{12}$$

which is a set of M linear algebraic equations in the M unknowns A, B and a_j, and can be solved using standard procedures. Guidelines for selecting the discrete values of s are given by Shapery (1962), and also by Rizzo and Shippy (1971).

AN APPLICATION FOR A BOUNDARY ELEMENT METHOD

A boundary element method is a numerical technique for solving a problem in terms of boundary data (e.g. displacements and tractions) only. Several different kinds of boundary element methods have been developed for linear elasticity (see, for example, Crouch and Starfield, 1982), and any of them can be used to solve problems in linear viscoelasticity. The solution procedure can be summarized as follows:

(1) choose an appropriate rheological model, for example the Burgers model of Fig. 1;
(2) according to the required time range, choose a number (M) of different values of s;
(3) compute the s-varying elastic moduli and transform the boundary conditions to the s-domain for each value of s_j (j=1 to M), and then solve the M modified boundary value problems using a boundary element method;
(4) find the inverse Laplace transform of the M solutions in the s domain to obtain the time-dependent solution.

As an illustration of this approach, consider the problem shown in Fig. 2. This problem represents a panel of rooms and pillars in a thick salt layer overlain by limestone. The far-field stresses are hydrostatic and equal to -10MPa (compression). The limestone is considered elastic with constants $E^* = 5 \times 10^4$MPa and $\nu^* = 0.2$. The rock salt is incompressible, i.e. $\nu = 0.5$, but exhibits viscoelastic behavior in shear, which is represented using the Burgers model of Fig. 1. The parameters in the model are chosen as $G_1 = 4 \times 10^3$MPa, $\eta_1 = 8 \times 10^4$MPa-day, $G_2 = 2 \times 10^4$MPa, $\eta_2 = 10^6$MPa-day. In addition, seven values of s (10^{-4}, 10^{-3}, 10^{-2}, 10^{-1}, 1, 10 and 10^{-5}) are selected for the numerical inversion technique.

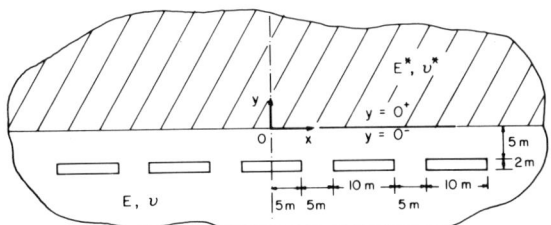

Fig. 2. Five uniformly spaced excavations.

Crouch and Starfield (1982) have given the analytic solution for a displacement discontinuity in one of two bonded isotropic elastic half-planes. This solution was used to solve the viscoelastic problem of Fig. 2 by the displacement discontinuity method, according to the procedure outlined above. Details concerning the solution of the associated elastic problem by the displacement discontinuity method are given by Crouch and Starfield (1982).

The time-dependent closure D_y at the mid-span of the central room is shown in Fig. 3. The instantaneous elastic deformation and the transient and steady-state creep of the salt are evident. The amount of deformation and the rate of creep, of course, can be altered by choosing different parameters in the Burgers model. In practice, one would want to choose these parameters to match measured room closure rates.

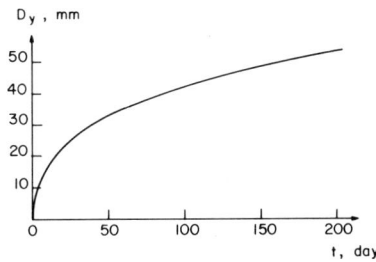

Fig. 3. Time-dependent closure at the mid-span of the central room.

Representative results for the time-dependent stresses along the interface between the salt and limestone are given in Fig. 4. Figures

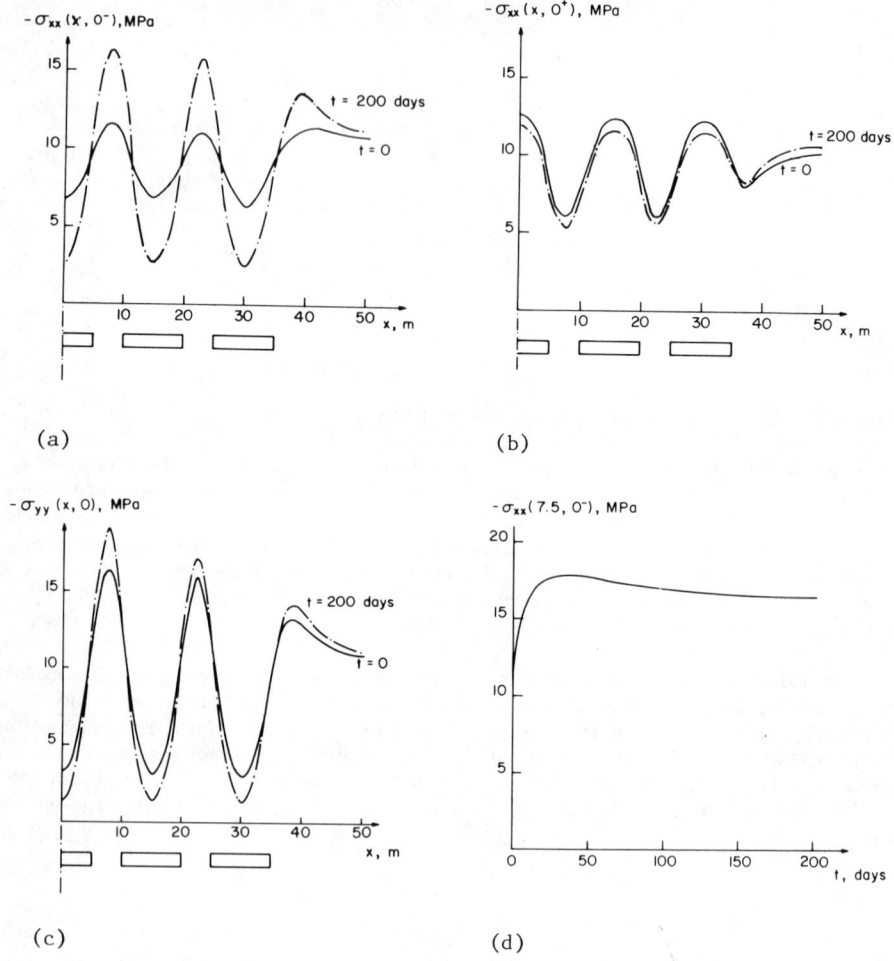

Fig. 4. Time-dependent stresses along the interface between the salt and limestone.

4a and 4b show the horizontal compressive stress $-\sigma_{xx}$ in the salt just below the interface (Fig. 4a) and in the limestone just above it (Fig. 4b). The horizontal stress is clearly discontinuous along the interface, and the amount of the discontinuity appears to increase with time. The maximum horizontal compressive stress in the salt occurs at x = 7.5m. The horizontal stress in the limestone is a minimum at this same location (Fig. 4b).

The vertical compressive stress $-\sigma_{yy}$ along the interface is shown in Fig. 4c. This stress component is continuous across the interface. It can be seen that the stress above the pillars increases with time. Above the rooms, however, it decreases.

Figure 4d shows that the time-dependent stress $-\sigma_{xx}$ at $(7.5,0^-)$ has an extreme value somewhere between the initial and final times. Additional analysis indicates that other stress components exhibit a similar time-dependent response. Thus, it appears that an optimization of time-dependent stress concentrations can be achieved for inhomogeneous viscoelastic rocks. Further efforts in this direction are currently being made to include moving boundaries and sequential excavation, where time-dependent effects occur even more prominently.

The authors realize that there is a lack of experimental data for viscoelastic rocks. Qualitative analysis by computer modeling, however, allows one to obtain a better understanding of the physics of complex problems. The sophistication of the modeling can, if necessary, be improved when reliable field data are available. Indeed, the computer model itself can be used to plan the instrumentation program, thus reducing the risk of making meaningless measurements.

REFERENCES

Alfrey, T., 1944, "Non-homogeneous Stresses in Viscoelastic Media," Quart. Appl. Math., Vol. 2, pp.113-119.

Berry, D.S., 1977, "Progress in the Analysis of Ground Movements due to Mining," in: J.D. Geddes (ed.), Large Ground Movements and Structures, Wiley, New York, pp. 187-208.

Crouch, S.L., and Starfield, A.M., 1982, Boundary Element Methods in Solid Mechanics, George Allen & Unwin, London.

Fung, Y.C., 1965, Foundations of Solid Mechanics, Prentice-Hall, Englewood Cliffs, New Jersey.

Lee, E.H., 1955, "Stress Analysis in Viscoelastic Bodies," Quart. Appl. Math., Vol. 13, pp. 183-190.

Reiner, M., 1969, Deformation, Strain and Flow, 3rd ed., Lewis, London.

Rizzo, F.J., and Shippy, D.J., 1971, "An Application of the Correspondence Principle of Linear Viscoelasticity Theory," SIAM J. Appl. Math., Vol. 21, No. 2, pp. 321-330.

Shapery, R.A., 1962, "Approximate Methods of Transform Inversion for Viscoelastic Stress Analysis," Proceedings, 4th U.S. National Congress of Applied Mechanics, pp. 1075-1085.

Timoshenko, S.P., and Goodier, J.N., 1970, Theory of Elasticity, 3rd ed., McGraw-Hill, New York.

Tsien, H.S., 1950, "A Generalization of Alfrey's Theorem for Viscoelastic Media, Quart. Appl. Math., Vol. 8, pp. 104-106.

Chapter 74

COMPRESSIBILITIES AND EFFECTIVE STRESS COEFFICIENTS FOR
LINEAR ELASTIC POROUS SOLIDS: LOWER BOUNDS AND RESULTS
FOR THE CASE OF RANDOMLY DISTRIBUTED SPHEROIDAL PORES

by Robert W. Zimmerman

Research Assistant
Department of Mechanical Engineering
University of California
Berkeley, California

POROUS SOLID COMPRESSIBILITIES

There are four different compressibilities associated with porous solids, each relating the fractional change in either bulk volume or pore volume with the change in either pore pressure or confining pressure. These will be referred to as C_{bc}, C_{bp}, C_{pc}, and C_{pp}, with the first subscript denoting the relevant volume, and the second subscript denoting the relevant pressure. They are defined as:

$$C_{bc} = \frac{-1}{V_b}\left(\frac{dV_b}{dP_c}\right)_{P_p} \qquad C_{bp} = \frac{1}{V_b}\left(\frac{dV_b}{dP_p}\right)_{P_c} \qquad (1)$$
$$C_{pc} = \frac{-1}{V_p}\left(\frac{dV_p}{dP_c}\right)_{P_p} \qquad C_{pp} = \frac{1}{V_p}\left(\frac{dV_p}{dP_p}\right)_{P_c}$$

Note that minus signs are included in two of the definitions so that each of the compressibilities will always have a positive value. In general, each compressibility depends on the elastic moduli of the matrix material, the porosity, and the exact geometry of the pores. Taking the latter factor into account calls for separate analysis for each different pore shape, which of course can only be attempted for a few relatively simple cases. But lower bounds for each of the four compressibilities can be found, independant of pore structure, in terms of only the porosity ($\emptyset$) and the elastic moduli of the matrix material (any pair chosen from K, E, G and ν).

The first step is to find relationships between the four compressibilities, starting with the following observation: If a porous solid made of a linear elastic, isotropic and homogeneous matrix material is subjected to uniform hydrostatic pressure of magnitude P on its external and all of its internal (pore) surfaces, the state of stress in the solid will be exactly the same as that which would occur if the pores were filled up with the matrix material, keeping the same external tractions (Nur and Byerlee, 1971). The change in bulk volume in this latter case is $-(P/K)V_b$, where henceforth all elastic moduli without subscripts will refer to the matrix material. But by superposition, this change in bulk volume must be the sum of the changes due to confining pressure P alone ($-PC_{bc}V_b$) and due to pore pressure P alone ($PC_{bp}V_b$). Equating these and cancelling out the common term PV_b leads to the relation $C_{bp} = C_{bc} - (1/K)$. A similar argument, concentrating on the pore volume, leads to the relation $C_{pp} = C_{pc} - (1/K)$. Finally, application of the Betti reciprocal theorem reveals that $V_p C_{pc} = V_b C_{bp}$, hence $C_{bp} = \emptyset C_{pc}$. If the reciprocal of C_{bc} is denoted by K', the above relations can be used successively to express all four compressibilities in terms of K, K', and $\emptyset$:

$$C_{bc} = \frac{1}{K'} \qquad C_{pc} = (\frac{1}{K'} - \frac{1}{K})\frac{1}{\emptyset}$$

$$C_{bp} = \frac{1}{K'} - \frac{1}{K} \qquad C_{pp} = (\frac{1}{K'} - \frac{1+\emptyset}{K})\frac{1}{\emptyset} \qquad (2)$$

It must be born in mind that K' will depend on K, ν, $\emptyset$ and the pore geometry.

Variational principles have been used (Hashin and Shtrikman, 1961) to find an upper bound on the ratio (K'/K) which involves only the Poisson ratio of the matrix material and the porosity. Examination of eqs. (2) shows that this bound can be used to provide lower bounds on all four compressibilities. The results, when expressed in terms of E, ν, and $\emptyset$ are:

$$C_{bc} = \frac{6(1-2\nu) + 3(1+\nu)\emptyset}{2E(1-\emptyset)} \qquad C_{bp} = \frac{9(1-\nu)\emptyset}{2E(1-\emptyset)}$$

$$C_{pp} = \frac{3(1+\nu) + 6(1-2\nu)\emptyset}{2E(1-\emptyset)} \qquad C_{pc} = \frac{9(1-\nu)}{2E(1-\emptyset)} \qquad (3)$$

These lower bounds are plotted in Figs. 1-4 for various values of ν, and for porosity values of up to .4 (although the bounds are in fact valid for all values of $\emptyset$ from zero to one). For fixed E, ν, and $\emptyset$, these are lower bounds over all possible pore geometries. It is known that the bound on C_{bc} is the sharpest possible, since it is in fact obtained for the case of a hollow spherical shell. It follows from eqs. (2) that all four of these bounds are the sharpest possible. While it is perhaps obvious that a spherical shell will exhibit the minimum possible bulk compressibility for given E, ν, and $\emptyset$, it does not seem to be geometrically transparent that the spherical shell will also have the minimum possible value of C_{bp}, for instance. Note that all of the above relations required no assumptions as to the sizes or shapes of the pores, nor any assumptions concerning macroscopic homogeneity.

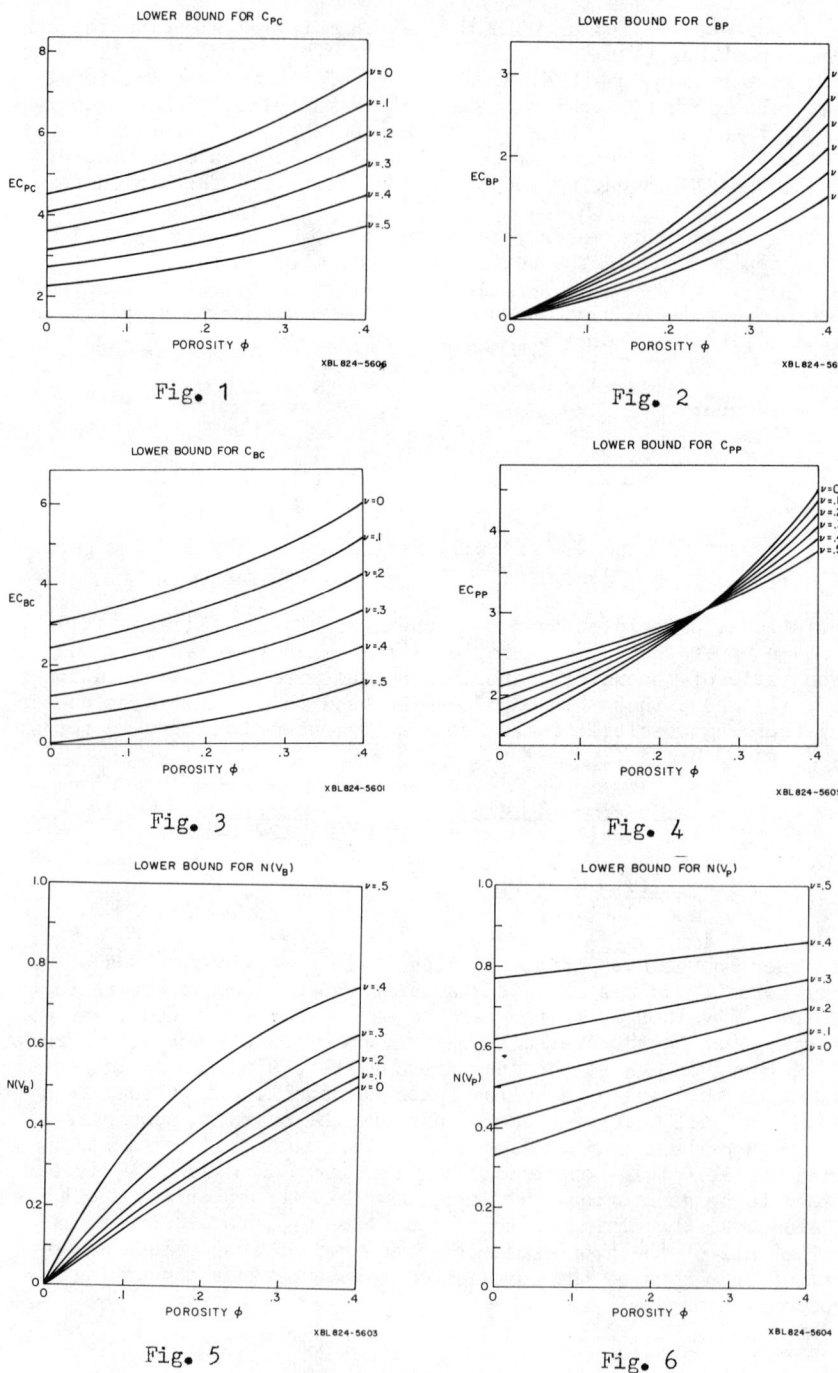

Fig. 1

Fig. 2

Fig. 3

Fig. 4

Fig. 5

Fig. 6

LINEAR ELASTIC POROUS SOLIDS

EFFECTIVE STRESS COEFFICIENTS

Changes in quantities such as V_p, V_b or $\emptyset$ will always turn out to be expressed as linear combinations of P_p and P_c. For example, for some quantity F, ΔF will equal $C_1 P_c + C_2 P_p$, where C_1 and C_2 are constants. This can be written as $\Delta F = C_1(P_c + (C_2/C_1)P_p) = C_1(P_c - N(F)P_p)$, where $N(F) = -(C_2/C_1)$ is the effective stress coefficient for F. Comparison of the partial derivatives of F with respect to P_c and P_p leads to the following useful interpretation of $N(F)$:

$$N(F) = -\left(\frac{dF}{dP_p}\right)_{P_c} \bigg/ \left(\frac{dF}{dP_c}\right)_{P_p} \tag{4}$$

Nur and Byerlee used this definition to find that $N(V_b) = 1-(K'/K)$, which immediately shows that this coefficient will always lie between zero and one. In fact, the upper bound on (K'/K) leads to a lower bound on $N(V_b)$.

The effective stress coefficient for pore volume changes can similarly be found to be equal to:

$$N(V_p) = \frac{1-(K'/K)(1+\emptyset)}{1-(K'/K)} \tag{5}$$

This is most conveniently studied by first expanding it in a power series in terms of (K'/K):

$$N(V_p) = 1 - \emptyset(x + x^2 + x^3 + \ldots) \qquad x = (K'/K) \tag{6}$$

In this form it is clear that $N(V_p)$ is also never greater than one. Dividing the two effective stress coefficients and invoking the bound on (K'/K) leads to the conclusion that $N(V_p)$ will in general be greater than $N(V_b)$.

Now $N(V_p)$ is equal to (C_{pp}/C_{pc}), but it is not immediately obvious that the ratio of the two lower bounds for the compressibilities will lead to a lower bound for $N(V_p)$ itself, since this is not algebraically true in general. But if the ratio of the two lower bounds is subtracted from an arbitrary value of this ratio, and then put over a common denominator, it is seen that all terms cancel out except one which is inherently non-negative. Hence this ratio does give a lower bound for $N(V_p)$. The lower bounds on the two effective stress coefficients can be expressed in terms of ν and $\emptyset$ as follows:

$$N(V_b) = \frac{3(1-\nu)\emptyset}{2(1-2\nu)+(1+\nu)\emptyset} \qquad N(V_p) = \frac{(1+\nu)+2(1-2\nu)\emptyset}{3(1-\nu)} \tag{7}$$

These bounds are plotted in Figs. 5 and 6, for various values of ν, and for $\emptyset$ ranging from 0 to .4 (although they are valid for all $\emptyset$). Note that in the limit of an incompressible matrix material, that is as ν approaches .5, both coefficients will approach 1, regardless of the porosity.

Since $\emptyset = (V_p/V_b)$, eq. (4) can be used in conjunction with the rules for differentiation of a quotient, and eqs. (2), to yield an expression for $N(\emptyset)$. The final result is that $N(\emptyset)$ will be equal to one, for all values of ν, $\emptyset$, and regardless of the pore structure. In other words, the porosity changes will simply depend on the difference between the confining pressure and the pore pressure.

APPLICATION TO THE CASE OF SPHEROIDAL PORES

For a solid having pores of a specified shape, it is in principle possible to derive expressions for the compressibilities which should be very accurate at low porosities, and at the same time provide lower bounds which are specific to pore shape. This is done by first calculating C_{pp} for an isolated pore in an infinite medium, and then using eqs. (2) to find the other three compressibilities. Since the stresses around a pressurized cavity generally decrease quite rapidly as the distance from the cavity increases, such a calculation should involve only slight errors if the porosity is low and the pores are well-separated. Yet since this method assumes that the pore must expand into a non-porous surrounding material which is stiffer than the actual porous material, it is also clear that this method will underestimate C_{pp}. If eqs. (2) are manipulated to express all of the compressibilities in terms of C_{pp}, it is seen that the above method leads to lower bounds for each. It can also be shown that the resulting values for $N(V_b)$ and $N(V_p)$ will be lower bounds. For fixed E, ν, and $\emptyset$, these bounds are over all possible spatial distributions of pores of the given shape.

One fairly general shape of pore is the spheroid, an ellipsoid with two of its axes having equal length, which in limiting cases can represent a sphere, cylinder, or penny-shaped crack. The Boussinesq stress function approach (Sadowsky and Sternberg, 1947; Edwards, 1951) can be used to solve the problem of an isolated spheroidal pore under hydrostatic pressure in an infinite medium. The pore volume change is found by integrating the normal displacement over the surface of the cavity, which leads directly to C_{pp}.

Prolate Spheroids

A prolate spheroid is cigar-shaped, with the two equal axes (length $\bar{q}$) being shorter than the third axis (length q, with $q^2 - \bar{q}^2 = 1$). The aspect ratio r is defined to be the length of the major axis divided by the length of the minor axis, hence $r = (q/\bar{q})$. After solving for the displacements and integrating over the pore surface as described above, the resulting expression for C_{pp} is:

$$C_{pp} = \frac{(1+3R)[1-2(1-2\nu)R - 3q^2] - 2(1-2\nu)(1+2R)}{4G[(1+3R)q^2 - (1+R)(\nu + \nu R + R)]}$$

(8)

where $R = \bar{q}^2 Q$, $Q = 1 + (q/2)\text{Ln}[(q-1)/(q+1)]$

This equation has been evaluated for the case $\nu = .25$ over the entire range of r, with the results shown in Table 1. Spherical pores are represented by the limiting case of $r = 1$, while the other limit of $r = \infty$ represents needle-shaped pores. It seems that the result for needle-shaped pores can be used whenever the aspect ratio is greater than about 10, with an error of less than 3%, while the result for spherical pores can be used for all aspect ratios up to about 1.5, with an error of less than 4%. This latter observation is important for cases, such as certain glasses or ceramics, for which one would expect the pores to deviate only slightly from true sphericity. The relative insensitivity of C_{pp} to aspect ratio should be noted, with only a 33% increase over the entire range of r.

r	1	1.1	1.5	2	5	10	50	100
KC_{pp}	1.250	1.254	1.299	1.368	1.561	1.628	1.664	1.667

Table 1: Pore volume compressibility for a solid with a low concentration of prolate spheroidal pores.

Oblate Spheroids

The results for prolate spheroids can, by a slight change of variables, be rendered applicable to oblate spheroids, where the two equal axes are longer than the third. This entails replacing q with $i\bar{q}$, and $\bar{q}$ with iq, which results in $Q = 1 - \bar{q} \operatorname{arc cot} \bar{q}$. The ensuing expression for C_{pp} is so formally similar to eq. (8) that it need not be written out explicitly. The results are plotted in Fig. 7 on a logarithmic scale, for $\nu = .25$, and a range of aspect ratios from 1 to 100. The limiting value of $r = 1$ again reproduces the spherical pore solution, while the limit of $r = \infty$ corresponds to what is referred to as a penny-shaped crack. There is no limiting value for C_{pp} in this case, but the leading terms of the asymptotic expansion can be found by starting with the first two terms of the Taylor series for R, for small values of $\bar{q}$, which is $R \doteq (\bar{q}/2) - 1$. Insertion of this into eq. (8), with the above-mentioned modifications, results in:

$$KC_{pp} = \frac{4(1-\nu^2)r}{3\pi(1-2\nu)} - 1 + \text{terms of order } (1/r) \qquad (9)$$

The first two terms of this expansion coincide with the result which can be inferred from the work of Sack (Sack, 1946), who passed to the limit of an infinitely thin crack before solving the equations. It turns out that the Sack approximation will be within less than 1% of the exact value, for aspect ratios which are greater than 25. In fact the error is only 15% for a value of r as low as 5.

Manipulation of eqs. (2) and (5) leads to $N(V_p) = KC_{pp}/(1+KC_{pp})$, which for the present case is shown on a semilogarithmic plot in Fig. 7. It is clear that a solid with cracks of aspect ratio on the order of 100 or more will have an effective stress coefficient for pore volume which is not sensibly less than 1. The bulk volume behaviour is described by $KC_{bc} = 1+(1+KC_{pp})\phi$, and $N(V_b) = 1-(1/KC_{bc})$. These are plotted for oblate spheroidal pores in Fig. 8, for $\nu=.25$, and ϕ from 0 to .3. In contrast with prolate spheroidal pores, in this case the aspect ratio has a marked effect on bulk volume behaviour.

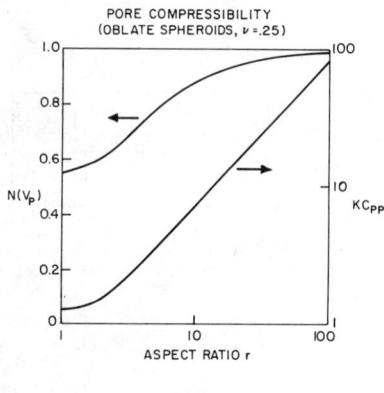
Fig. 7

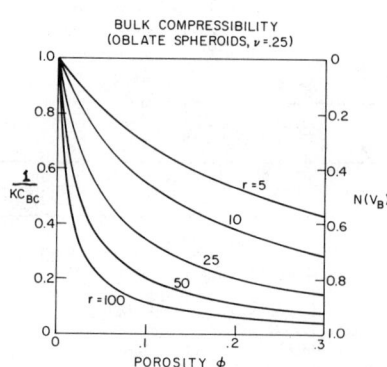
Fig. 8

It must be remembered that these results for spheroidal pores do not take into account so-called pore-pore interactions. But after C_{pp} was found for an isolated pore, eqs. (2), which are exact, were used to determine the other three compressibilities. This leads to an overestimation of $(1/KC_{bc}) = (K'/K)$, but gives the exact first-order in ϕ perturbation. Note that other methods, such as wave-scattering, (Kuster and Toksoz, 1974) have so far had the disadvantage of predicting negative values of K' for all ϕ greater than some critical value, which in the case of oblate spheroids turns out to be extremely low. Such results are useless past this critical value, while the results derived above at least serve as a non-trivial upper bound on (K'/K) for all ϕ.

ACKNOWLEDGMENTS

This work has been supported in part by United States Department of Energy Contract W-7405-ENG-48, administered by the Lawrence Berkeley Laboratories, and in part by a United States Department of Energy Domestic Mining and Mineral Fuel Conservation Fellowship.

REFERENCES

Edwards, R.H., 1951, "Stress Concentrations Around Spheroidal Inclusions and Cavities," Journal of Applied Mechanics, Vol. 18, No. 1, pp. 19-30.

Hashin, Z. and Shtrikman, S., 1961, "Note on a Variational Approach to the Theory of Composite Elastic Materials," Journal of the Franklin Institute, No. 271, pp. 336-341.

Kuster, G.T. and Toksoz, M.N., 1974, "Velocity and Attenuation of Seismic Waves in Two-Phase Media: Part 1- Theoretical Formulations," Geophysics, Vol. 39, No. 5, pp. 587-606.

Nur, A. and Byerlee, J.D., "An Exact Effective Stress Law for Elastic Deformation of Rock with Fluids," Journal of Geophysical Research, Vol. 76, No. 26, pp. 6414-6419.

Sack, R.A., 1946, "Extension of Griffith's Theory of Rupture to Three Dimensions," Proceedings of the Physical Society of London, Vol. 58, pp. 729-736.

Sadowsky, M.A. and Sternberg, E., 1947, "Stress Concentrations Around an Ellipsoidal Cavity in an Infinite Body...," Journal of Applied Mechanics, Vol. 14, No. 3, pp. A191-A201.

FIELD TESTS

Chairman
Francois E. Heuze
Lawrence Livermore National Laboratory
Livermore, California

Co-Chairman
Z.T. Bieniawski
Pennsylvania State University
University Park, Pennsylvania

Keynote Speakers
Neville G.W. Cook
University of California
Berkeley, California

Nick R. Barton
Terra Tek Inc.
Salt Lake City, Utah

Chapter 75

QUESTIONS IN EXPERIMENTAL ROCK MECHANICS

Neville G. W. Cook

Department of Materials Science and Mineral Engineering
University of California
Berkeley, California 94720

ABSTRACT

The ultimate goal in rock mechanics is to make quantitative predictions of the response of rock to changes. Measured and calculated values of changes in displacements or temperatures in the rock for six different field experiments are compared using linear regression. When values of the mean compressive stresses are large compared with the values of the deviatorial stresses, linear elasticity, using values for Young's modulus from laboratory tests, provides a good predictive model. However, if the values of the mean compressive stresses are small compared with those of the deviatorial stresses the behavior and response of the rock, though systematic, is not linear. Linear heat conduction provides good predictions of changes in rock temperatures but thermal displacements are not predicted well by linear thermoelasticity.

INTRODUCTION

The ultimate goal in rock mechanics is to develop an understanding of the behavior and properties of rock sufficient to enable quantitative predictions to be made concerning the response of rock masses to changes wrought by man or nature.

The practical experiences and observations of civil and mining engineers and of geologists and geophysicists are of invaluable help in interpreting field observations and laboratory test results but are themselves insufficiently complete to arrive at a quantitative predictive model. Laboratory tests on relatively small samples of rock, made over wide ranges of stresses, temperatures and pore fluid pressures, have provided a wealth of qualitative information and quantitative data about the behavior and properties of rocks (Griggs and Handin, 1960; and Carter et al, 1981). In principle, the use of this information and these data in models based on sound principles should provide the predictive capabilities about rock masses that are required. Field experiments, in which some of the responses of rock masses to known changes are measured, provide data against which the validity of predictions using models can be checked.

Unfortunately, it is seldom, if ever, practicable to make sufficient measurements in the field to define completely the response of a rock mass. Nevertheless, if the behavior and properties of rock are understood sufficiently well, measurements that are made of the response of rock masses to changes should correlate well with predictions based on a sound model. If the correlation between measurement and prediction is poor: the model is inapplicable or incomplete; the behavior and properties of the rock masses are not understood sufficiently well; the changes have not been defined properly, or the measurements are not good. The disparities between measurement and prediction seldom provide sufficient information to resolve these uncertainties, because of the incomplete nature of field measurements.

Field experiments are difficult, expensive and time consuming. The number of good field experiments is, therefore, small. Two quantities that can be measured with confidence in the field are changes in relative displacement between points and changes in temperature at specific points in a rock mass

In this paper, measured and calculated changes in relative displacements and temperatures are compared, using linear regression, for six different field experiments done over the past two decades. These particular six experiments have been selected because their results are well documented and because the author is familiar with each of them.

Three of these field experiments were done in different gold mines of the Witwatersrand System, namely, E.R.P.M., Harmony and W.D.L. The fourth was done at a coal mine in South Africa. One of the remaining two was done at Stripa, Sweden and the other is still underway at Climax, Nevada Test Site.

ANALYSIS OF DATA

The analyses consist of performing a linear regression between calculated, x, and measured, y, values of the displacements or temperatures. The values of b and m in the equation $y = mx + b$ and the value of the coefficient of correlation, r, provide different measures of the applicability of the model used to make the calculations. In general, a value of r near unity suggests that the model is applicable, but a value of m different from unity suggests that the values of the coefficients used to describe the properties of the rock mass, such as Young's modulus or the coefficient of thermal expansion, are inappropriate. For an applicable model, the value of b should be zero if the measurements also are good. All the models used in this paper are linear and most of the measurements are relatively good. The value of b relative to the actual values of the measured changes, b', is, therefore, indicative of non linear behavior and response of the rock.

E. R. P. M. Displacement

At E. R. P. M. 23 benchmarks were established along 600m of 58 haulage situated between 100m and 200m above the mined out areas and the intervening unmined, gold-bearing reef between two longwall faces being mined toward one another (Ortlepp and Cook, 1964). This haulage, at 2600m below surface traverses strata of the Witwatersrand System that extend to surface save for a conformable diabase sill 60m thick about 300m above the reef. These benchmarks comprised vertical rockbolts 2.4m in length anchored only at their upper ends into the roof of the haulage. Precise levelling of the benchmarks was done initially in February, 1962 and, subsequently, in February, 1963 and in December, 1963 (Cook et al., 1966). Vertical displacements of these benchmarks occurred as the geometry of the longwall stopes was changed by mining. Using an electrical resistance analog to model the complex stope geometry (Salamon et al., 1964) and a value for Young's modulus of about 76GPa, derived from laboratory tests on small specimens of rock, the changes in the elastic displacements of 10 benchmarks as a result of the changes in stope geometries at these dates were calculated.

Calculated and measured values of the displacements together with a linear, least squares regression fit to them are as is illustrated in Figure 1. From this Figure and the values for the regression line in Table 1, it appears that the value of Young's modulus is slightly low, $m = 0.86$, and that the linear, elastic model is applicable, $r = 0.97$ and $b' = 0.05$.

Harmony Displacements

A vertical, wire extensometer extending from 7m in the footwall to 50m in the hanging wall was installed 20m ahead of an advancing stope face at a depth of 1677m below surface. The strata include a shale layer overlying the reef about 1m into the hanging, and 460m of lava 600m into the hanging, followed by sandstones and shales of the Karoo System. The dip of the stope was only 5° and its geometry was simple, so that the displacements of the anchor points of the extensometer, as a result of the advance of the stope face toward it, could be predicted using a solution in the theory of elasticity for a long, open horizontal slit and a value for Young's modulus derived from laboratoary tests on small specimens of rock (Ortlepp and Cook, 1964).

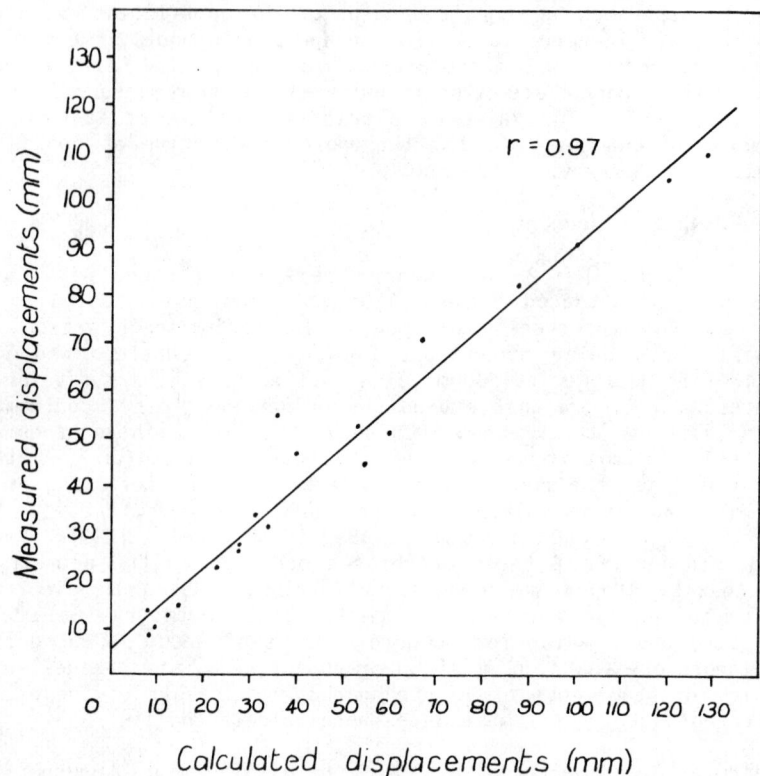

Figure 1. The relationship between calculated and measured changes in the elevation of a hanging wall haulage at E. R. P. M. resulting from mining.

Calculated and measured values of displacements, together with a linear, least squares regression fit to them, are as is illustrated in Figure 2. Values for the coefficients b', m and r are summarized in Table 1. From Figure 2 and Table 1 it appears that the rock is linear, b' = 0.05, and the predictive model is good, r = 0.93, but that the laboratory value for Young's modulus is significantly too low, m = 0.72.

W. D. L. Displacements

Benchmarks, similar to those used at E.R.P.M., were installed in a horizontal haulage over a length of about 750m at a depth below surface 1590m, from about zero to about 300m above the stope. The hanging consists of about 1000m of lava succeeded by

Table I
Linear regression coefficients

	b	m	r
Figure 1, E. R. P. M.	0.05	0.86	0.97
Figure 2, Harmony	0.05	0.72	0.93
Figure 3, W. D. L.	-0.04	1.12	0.97
Figure 4, Coal Mine	0.03	0.98	0.99
Figure 5, Stripa Temperatures	-0.02	0.93	0.998
Figure 6, Stripa Displacements	-0.15	0.39	0.93
Figure 7, Climax All Displacements	0.28	-0.25	0.13
Negative displacements only	0.10	0.79	0.43
Figure 8, Climax Measured displacements only	0.01	0.55	0.82
Figure 9, Climax Temperatures	0.08	0.85	0.98

strata of the Transvaal System including dolomite and Karst formations near surface. The benchmarks were first levelled in May, 1962 and subsequently in January, 1964 (Cook et al., 1966). Changes in elevation at these benchmarks as a result of changes in stope geometry produced by mining, were calculated using an electrical resistance analog, the theory of elasticity and a value for Young's modulus derived from measurements on small, laboratory specimens of rock.

Calculated and measured changes in elevation of these benchmarks together with a linear, least squares regression fit to them are as is illustrated in Figure 3. Values for the coefficients b', m and r are summarized in Table 1. From Figure 3 and Table 1 it appears that the rock is linear, b' = -0.04, and that the predictive model is good, r = 0.97, but that the laboratory value for Young's modulus is slightly too high, m = 1.12.

Coal Mine Displacements

Surface subsidence and vertical displacements in the strata above and below a coal seam were measured using precise surface leveling and multiple point, vertical wire extensometers, respectively (Salamon and Oravecz, 1970). The coal seam, at a depth of about 63m below surface was mined in a room and pillar layout. The electrical resistance analog (Cook et al., 1966) and numerical integration were used to calculate the subsidence and vertical displacements

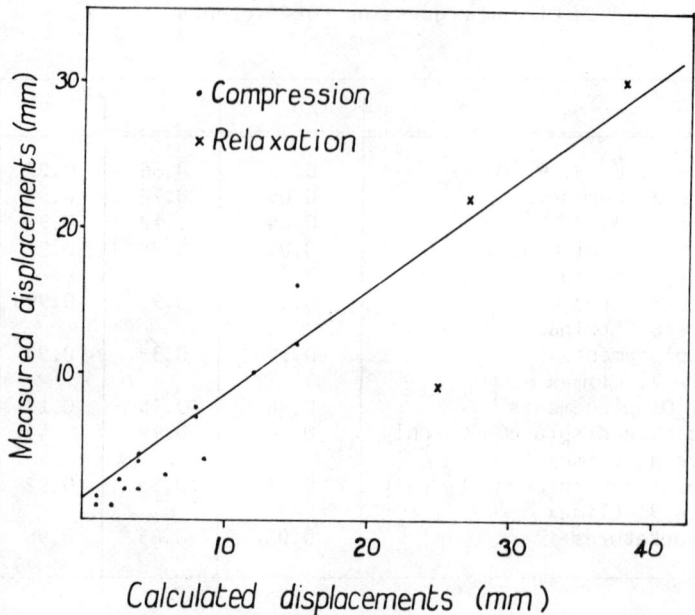

Figure 2. The relationship between calculated and measured changes in the anchor positions of a vertical extensometer as a result of stoping toward and through the extensometer at Harmony (r = 0.93).

resulting from mining. However, the values for the Young's moduli of the coal and the overlying strata were chosen so as to achieve a least squares fit between calculated and measured displacements. The resulting value for Young's modulus of the coal was 3.09GPa, that is, similar to values obtained from laboratory tests on coal specimens, and that for the overburden was 4.92GPa. This latter value is about a fifth of the value for Young's modulus measured in laboratory tests on small specimens of the overburden rock. Presumably, the reduction in the value of Young's modulus arises from the transversely isotropic and stratified nature of the overburden (Salamon, 1964). Calculated and measured displacements are as illustrated in Figure 4 and values for the regression line are summarized in Table 1. The rock is linear, b' = 0.03 and the model based on the theory of elasticity and fitted values of Young's moduli is good, r = 0.99 and, as would be expected, m = 0.98.

Stripa Measurements

Measurements of changes in displacements and temperatures around electrical heaters simulating the heat output of canisters of radioactive waste were made in granite adjacent to a defunct iron

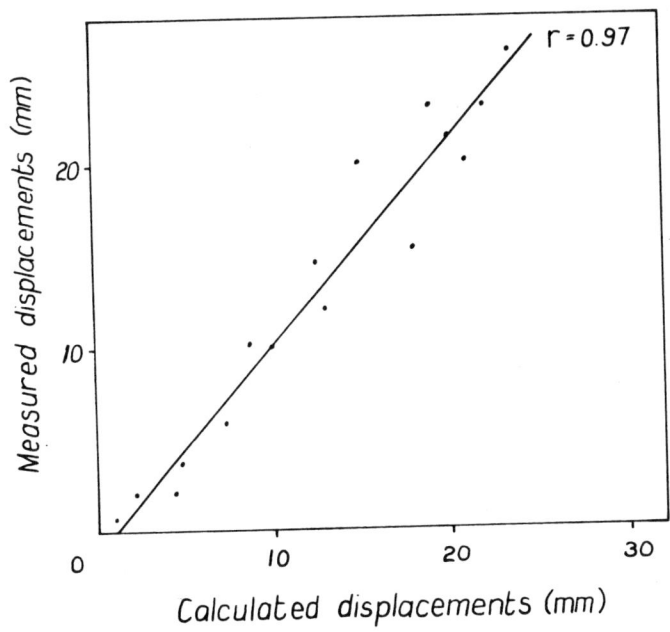

Figure 3. The relationship between calculated and measured changes in the elevation of a hangingwall haulage at W. D. L. resulting from stoping.

ore mine at a depth of 340 m below surface near Stripa, Sweden (Witherspoon et al., 1981). The granite at Stripa is intersected by four sets of pervasive joints; the joint spacing in each set is predominantly less than 1 m (Thorpe, 1981).

Changes in rock temperature were measured at several elevations in vertical boreholes at different radial distances and azimuths around the vertical hole containing the heater. Predictions of changes in temperature as a function of position and time were made using the linear theory of heat conduction, with values for the thermal conductivity and diffusivity determined from laboratory measurements (Chan et al., 1978). Temperatures measured in the mid-plane of the 5kW heater 190 days after heating commenced and predictions of these values, together with a linear, least square regression fit to them are as is illustrated in Figure 5. From this Figure and the values in Table 1 it can be seen that the rock is linear, b' = -0.02, the predictive model is excellent, r = 0.998, and that the values for the thermal conductivity and diffusivity of the granite are marginally too high, m = 0.93.

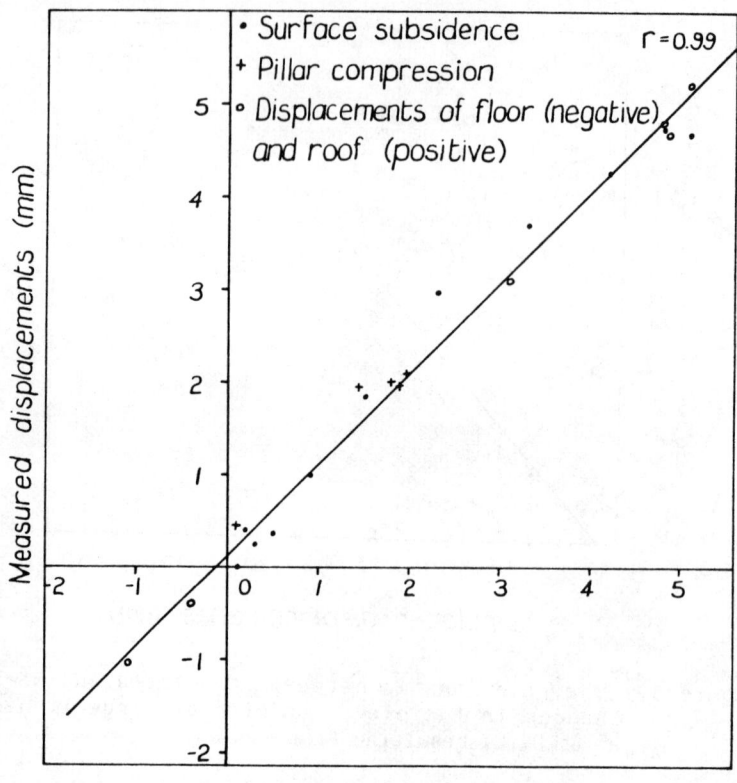

Figure 4. The relationship between calculated and measured surface subsidence and vertical strata movement resulting from excavation of a room and pillar coal mine.

Thermally induced displacements were measured at 5 axial positions along vertical and horizontal boreholes at various distances and azimuths around the heaters (Witherspoon et al., 1981). Changes in displacement of the anchor positions as a result of heating the rock were predicted using the theory of linear thermoelasticity and typical values for Young's modulus and the coefficient of linear thermal expansion. Calculated values of relative displacement were found to be consistently much less than those measured. Calculated and measured values of displacements between anchors above and below the midplane of the heater and across a diameter, together with a linear, least squares regression fit to them, are as is illustrated in Figure 6. From this figure and the values shown in Table 1 it appears that, although r = 0.93, the rock is non linear, b' = 0.15,

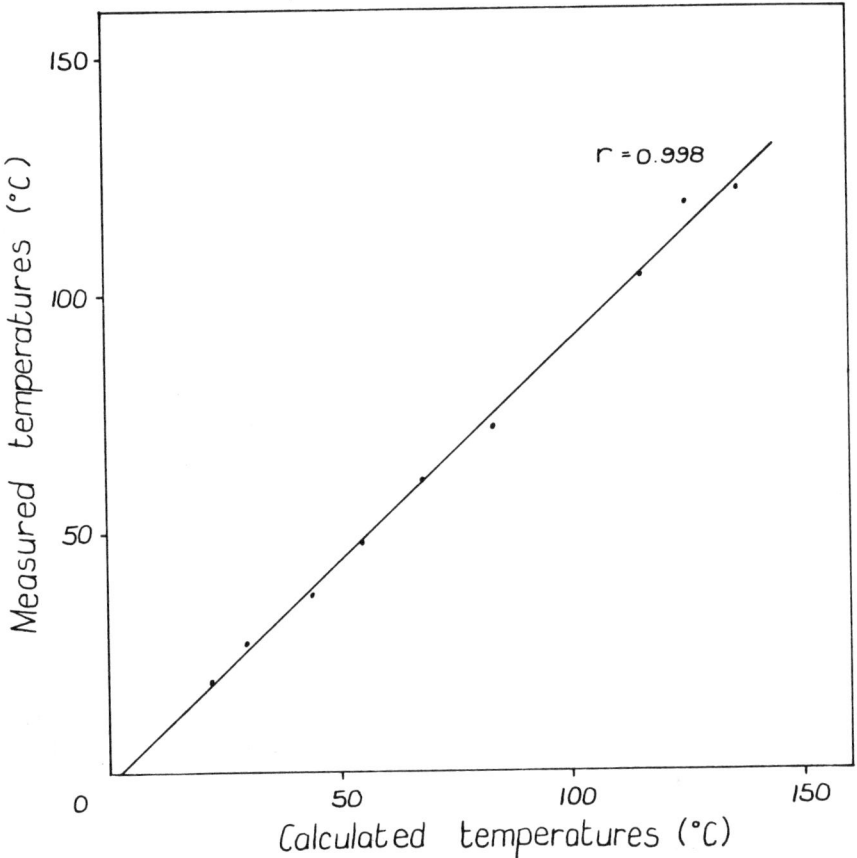

Figure 5. The relationship between calculated and measured thermal displacements in the rock adjacent to a 5kW heater at Stripa.

and that the typical typical value for the coefficient of thermal expansion is wrong, m = 0.39.

Climax Measurements

The Spent Fuel Test at Climax is located at a depth of 420m below surface in the Climax granite of the Nevada Test Site (Ramspott et al., 1982). The excavations comprise three parallel drifts over 64m in length on centers of 9.7m. The central drift has the largest cross section, about 6.1m high by 4.6m wide. Vertical holes in its floor contain 11 spent fuel assemblies and six electrical heaters. This central drift was excavated after the two outside drifts had been developed. Before excavating the central drift, multiple point extensometers were installed from the outside drifts

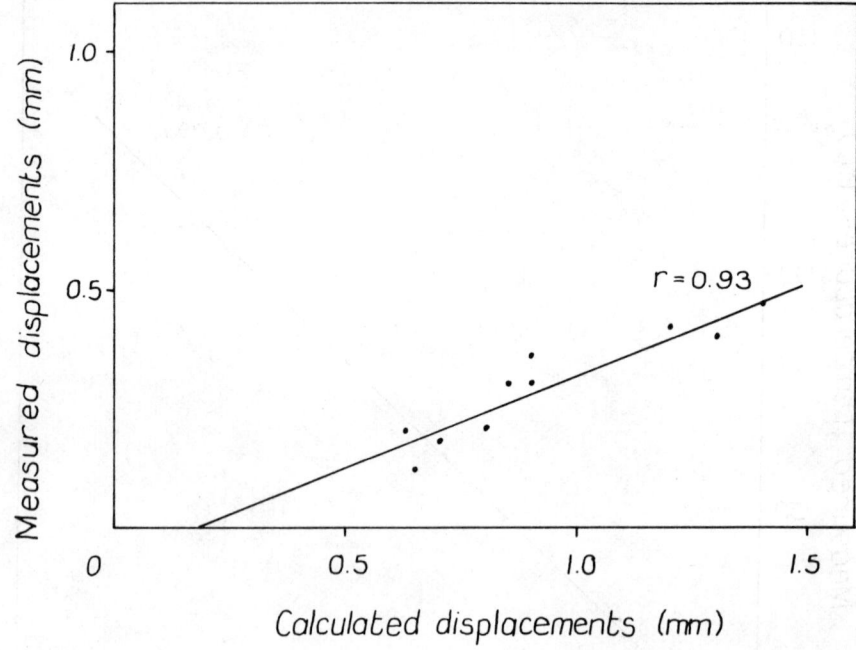

Figure 6. The relationship between calculated and measured thermal displacements in the rock adjacent to a 5kW heater at Stripa.

across the central drift at two cross sections separated by an axial distance of about 22 m. Mining of the central drift comprised the "mine-by" experiment, during which the effects of excavating this drift were measured by the extensometers and stress meters.

The displacements of the extensometer anchors relative to the extensometer sensing heads in the outside drifts, as a result of excavating the central drift, were predicted by calculation using an elastic, finite element model (ADINA). Measured and calculated displacements are as is illustrated in Figure 7. From this figure and Table 1 it can be seen that the rock appears to be significantly non linear, $b' = 0.28$, and that the predictive model is not applicable, $r = 0.13$. However, much of the disparity can be attributed to those extensometer measurements made horizontally across the two pillars between the outside and central drifts, for which the data lie in the lower, right hand quadrant of Figure 7. If these data are excluded, a least squares regression line can be fitted to the remaining values in the lower, left hand quadrant of this figure. The non linearity of the data in the left hand quadrant is much less than that of all the data, $b = 0.10$, and the applicability of the model, though not good, is much improved $r = 0.43$. It has been

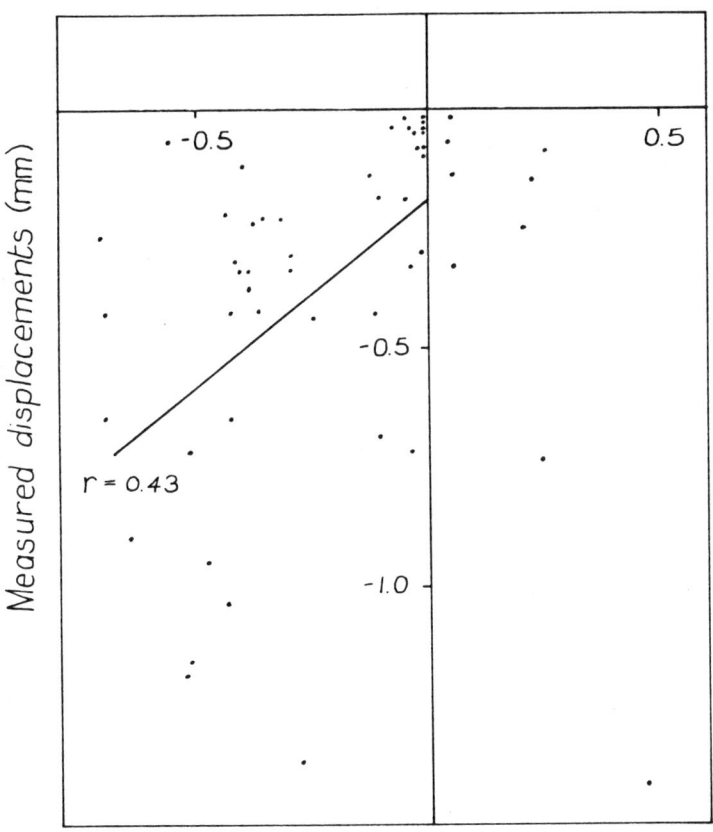

Figure 7. Measured and calculated displacements for the "mine by" experiment at Climax.

pointed out that the ADINA and other finite element models used to calculate the displacements are mutually consistent (Heuze, 1981). The question that arises is: How consistant are the measurements at each cross section with one another? If the relative displacements measured at the one cross section are compared with the corresponding displacements measured at the other cross section, as is illustrated in Figure 8 and as is shown in Table 1, the behavior of the rock at each cross section appears to be quite consistent, r = 0.82, but the magnitudes of the displacements differ significantly, m = 0.55, that is, the moduli of deformation at the two cross sections are significantly different from one another.

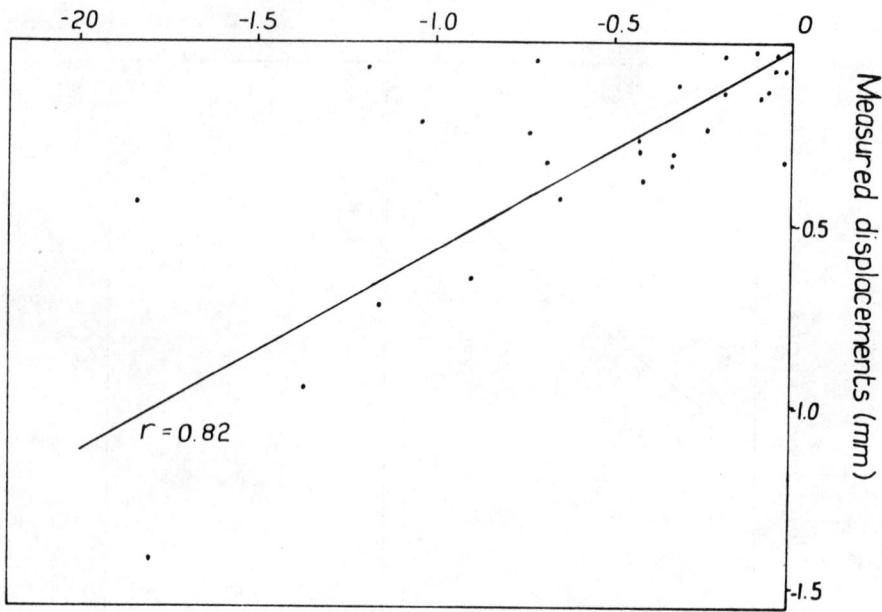

Figure 8. The correlation between like displacements measured at two cross sections for the "mine by" experiment at Climax.

Comparisons between predicted and measured values of the temperatures have been good (Patrick et al., 1981 and Ramspott et al., 1982). The relationship between calculated and measured temperatures at various locations in the rock, together with a linear, least squares regression fit to them, are as is illustrated in Figure 9. From this Figure and Table 1 it can be seen that b' = 0.08, r = 0.98, so that the model is quite good, and m = 0.85 which indicates that the coefficients of thermal conductivity and diffusivity are slightly too high.

DISCUSSION

The preceding analyses suggest that deformations of rock masses, as a result of excavation can be described well using the linear theory of elasticity, provided that the values of the mean compressive stresses are large and those of the deviatorial stresses are small.

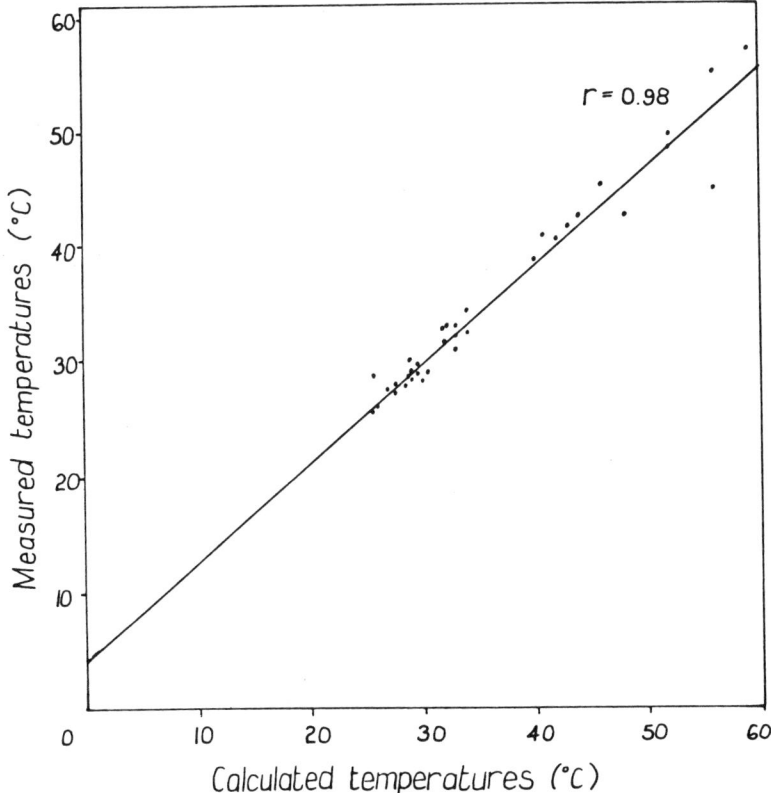

Figure 9. The relationship between calculated and measured temperatures in the rock at Climax.

High values for Young's modulus, determined from laboratory tests, are applicable to hard rock under high mean and low deviatorial stresses, whereas artificially low values are needed to model the laminated strata of soft rocks at low mean compressive and deviatorial stresses, such as those above a coal seam. High deviatorial stresses at low mean stresses result in a response of the rock mass to exavation that cannot be modeled using simple, linear elasticity. However, the response of the rock mass under low mean and high deviatorial stresses is systematic; measurements of like displacements at different cross sections show a significant correlation with one another.

Heat transfer in rock can be predicted well using the theory of linear heat conduction. However, the theory of linear thermoelasticity may not describe thermal deformations well. In particular, the coefficient of linear thermal expansion that is determined from laboratoary tests on specimens of rock <u>and is consistent with the mineral composition of such rocks</u> appears to be much too <u>large</u> in pervasively jointed rock.

In view of these conclusions, it would be interesting to establish the extent to which the deformations and stresses induced by making the outer two of the three co-planar holes on centers of three times their diameter, and then cooling one of them and heating the other, Figure 10 may be predicted for: (i) a test on a laboratory scale model (hole diameter, say, 50mm); (ii) an <u>in situ</u> block test (hole diameter, say, 250mm); (iii) a small field test (hole diameter, say, 1.5m), and (iv) a full scale field test (hole diameter, say, 7.5m).

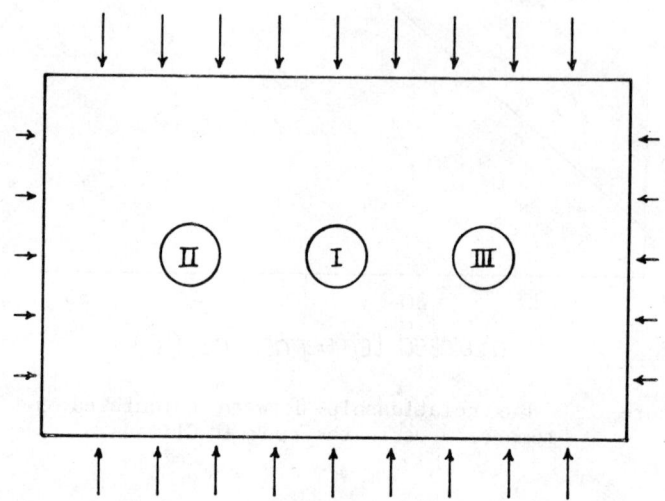

Figure 10. A sketch of the proposed laboratory, block and field tests. Hole I is drilled first; the effects of drilling holes II and III under stress are measured with intrumentation in the block and in hole I. Hole II is then cooled and hole III heated and the effects of cooling and heating are measured.

EXPERIMENTAL ROCK MECHANICS

ACKNOWLEDGEMENTS

The author is pleased to thank Ellen Klahn for compiling the typescript and Anna-Marie Cook for drafting the figures.

REFERENCES

Carter, N.L., Friedman, M., Logan, J.M., Sterns, D.W., (eds) 1981. Mechanical Behavior of Crustal Rocks. The Handin Volume, Geophysical Monograph 24, American Geophysical Union, Washington, D.C.

Chan, T., Cook, N.G.W., and Tsang, C.F. (1978). Theoretical Temperature Fields for the Stripa Heater Project, LBL 7082, SAC 09. Lawrence Berkeley Laboratory, Berkeley, California.

Cook, N.G.W., Hoek, E., Pretorius, J.P.G., Ortlepp, W.D. and Salamon M.D.G. 1966. Rock mechanics applied to the study of rockbursts, J. S. Afr. Inst. Min. Metall. 66, p. 435-528

Griggs, D.T. and Handin, J., (eds.), 1966. Rock Deformation, Geol. Soc. Am. Mem. 79. New York.

Heuze, F.H., 1981. Geomechanics of the Climax 'Mine-By', Nevada Test Site, Proceedings of the 22nd U.S. Symposium on Rock Mechanics. Rock Mechanics from Research to Application. Mass. Inst. Tech., June 28-July 2, pp. 428-434.

Ortlepp, W.D. and Cook, N.G.W. 1964. The measurement and analysis of the deformation around deep, hard-rock excavations, Proc. Fourth International Conference on Strata Control and Rock Mechanics, Henry Krumb School of Mines, Columbia University, New York, pp. 140-152.

Patrick, W., Montan, D. and Ballou, L. Near-Field Heat Transfer at the Spent Fuel Test-Climax: A Comparison of Measurements and Calculations, presented at the OECD Nuclear Energy Agency Workshop on Near-Field Phenomena in Geological Repositories For Radioactrive Waste, Seattle, Wa., August 31, 1981 and to be published in the Proceedings of the OECD Nuclear Energy Agency Workshop on Near-Field Phenomena in Geological Repositories.

Ramspott, L.D., Ballou L.B., and Patrick, W.C., 1982. Status Report on the Spent Fuel Test-Climax, Nevada Test Site: A Test of Dry Storage of Spent Fuel in a Deep Granite Location, UCRL-87448 Preprint, Lawrence Livermore National Laboratory.

Salamon, M.D.G., 1964. Elastic analysis of displacements and stresses induced by mining of reef deposits, Part II, J. S. Afr. Inst. Min. Metall., Vol 64, pp. 197-218.

Salamon, M.D.G., Ryder, J.A., and Ortlepp, W.D., 1964. An analogue solution for determining the elastic response of strata surrounding tabular mining excavations, J. S. Afr. Inst. Min. Metall Vol. 65, pp. 115-37.

Salamon, M.D.G., and Oravecz, K. I., 1970. The electrical resistance analogue as an aid to the design pillar workings, Proceedings Second Congress Int. Soc. for Rock Mechanics, Theme 4, Paper 18, Belgrade.

Thorpe, R., 1981. An example of fracture characteristics in granite rock, Proceedings of the 22nd U.S. Symposium on Rock Mechanics. Rock Mechanics from Research to Application. Cambridge, Mass., June 28-July 2.pp. 467-472.

Witherspoon, P.A., Cook, N.G.W., and Gale, J.E., 1981. Geologic storage of radioactive waste: Field studies in Sweden, Science, Vol. 211, pp. 894-900.

Chapter 76

EFFECTS OF BLOCK SIZE ON THE
SHEAR BEHAVIOR OF JOINTED ROCK

Nick Barton and Stavros Bandis

Geomechanics Division, Terra Tek, Inc.
Salt Lake City, Utah

Consultant
Thessaloniki, Greece

ABSTRACT

The descriptive term "rock mass" encompasses individual block dimensions ranging from centimeters to many tens of meters. Strength and deformability vary both qualitatively and quantitatively as a result of this size range. A key issue is therefore the appropriate size of the test sample. A large body of test data was reviewed to determine the influence of block size on the displacement required to mobilize peak strength. It is shown that the shear strength and shear stiffness reduce with increased block size due to reduced effective joint roughness, and due to reduced asperity strength. Both are a function of the delayed mobilization of roughness with increasing block size. A method of scaling shear strength and shear displacement from laboratory to in situ block sizes is suggested. It is based on the assumption that size effects disappear when the natural block size is exceeded. This simplification appears to be justified over a significant range of block sizes, but is invalidated when shearing along individual joints is replaced by rotational or kink-band deformation, as seen in more heavily jointed rock masses. Recent laboratory tests on model block assemblies illustrate some important effects of block size on deformability and Poisson's ratio.

INTRODUCTION

The wide range of natural block sizes found in nature has a strong and obvious influence on the morphology of a landscape. The contrast in natural slope angles and slope heights sustained by a ravelling "sugar cube" quartzite and a monolithic body of granite suggests that block size may be a controlling factor when compressive strength and slake durability are high in each case. In a tunnel, the contrast in behavior may produce more than an order of magnitude change in costs per meter. It is clear that the strength, the deformability and the mode of deformation (ravelling versus elastic) are strongly controlled by relative block size.

The mode of deformation cannot, however, be exclusively tied to block size. The loaded volume *relative* to block size, and the level of stress *relative* to the yield stress will each tend to control the mode of deformation. The above factors illustrate the difficulty that often arises in selecting the appropriate sizes of test sample. Usually, a jointed laboratory size sample will be small compared to the natural block size, and very small compared to the loaded volume in situ. Size effects will then be evident. On occasion, large size cores may be recovered which include a representative number of interlocked blocks, giving presumably a fair approximation to the strength and deformation behavior of a heavily jointed rock mass.

JOINT SAMPLE SIZE EFFECTS

Major through-going joint sets or individual discontinuities often dominate the stability and deformability of engineered structures in rock. Attempts to sample and test these surfaces are less successful than generally appears. This is because the size of the test sample often determines the magnitude of the strength data obtained.

Examples of joint sample size effects are illustrated in Figures 1 and 2. These shear tests were performed at such low normal stress (self-weight) that no shearing of asperities occurred. The marked difference in strength is strictly a function of different effective joint roughness. The small samples have the necessary degree of freedom to rotate slightly and "feel" the smaller, steeper asperities, while the monolithic blocks register only the flatter slopes of the major asperities, as clearly shown by Bandis et al (1981).

The combined effect of reduced effective roughness and increased individual contact areas causes a marked change in the shape of shear stress-shear displacement curves as sample size is increased. These effects are illustrated graphically in Figure 3. Extensive laboratory testing by Bandis (1980) suggests that the widely different shape of these illustrative strength-displacement curves is in no way exaggerated for the case of rough or moderately rough joints. However, smooth planar joints indicate only limited effects when sample size is increased.

A convenient way of interpreting the above size effects is to express shear strength in terms of its components. The peak drained friction angle (ϕ') developed by a rock joint can be expressed as:

$$\phi' = \phi_r + i \qquad (1)$$

where ϕ_r = the residual friction angle of a smoothed surface

i = total roughness component

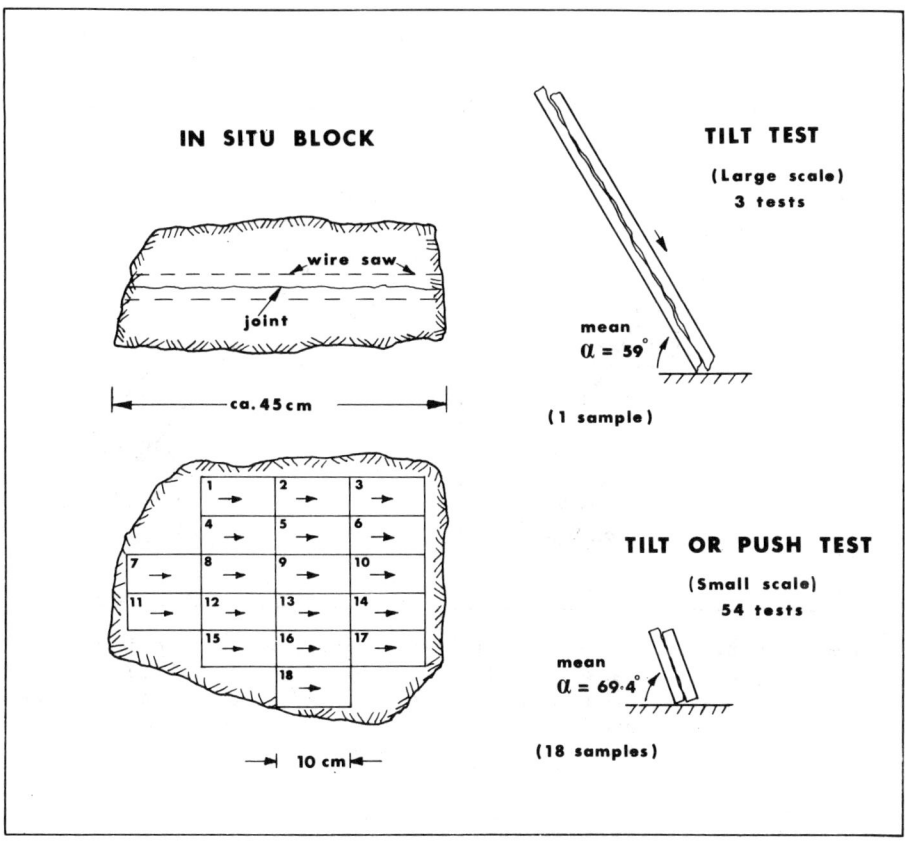

Figure 1. Tilt (self-weight sliding) tests of a natural joint in granite illustrate the exaggerated shear strength obtained when joint samples are too small. (Barton and Choubey, 1977).

The total roughness component (i) can be broken down as follows:

$$i = JRC \log (JCS/\sigma_n') \qquad (2)$$

where JRC = joint roughness coefficient

JCS = joint wall compression strength

σ_n' = effective normal stress

The joint wall compression strength can be measured with a Schmidt hammer (Barton and Choubey, 1977) while the joint roughness coefficient can be calculated by rearrangement of equations 1 and 2:

$$\text{JRC} = \frac{\phi' - \phi_r}{\log (\text{JCS}/\sigma_n')} \tag{3}$$

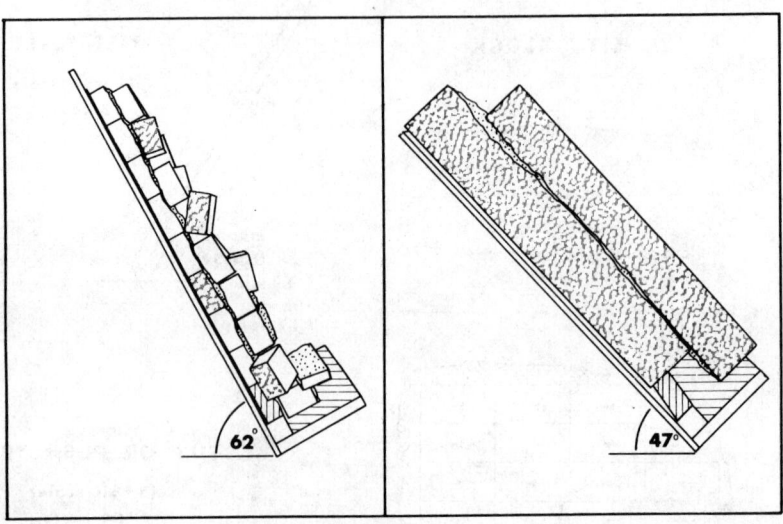

Figure 2. Contrasting shear strength of a large monolithic joint sample and an assembly of smaller blocks, each fabricated with the same batch of model material, and cast against the same joint surface. (Bandis et al. 1981).

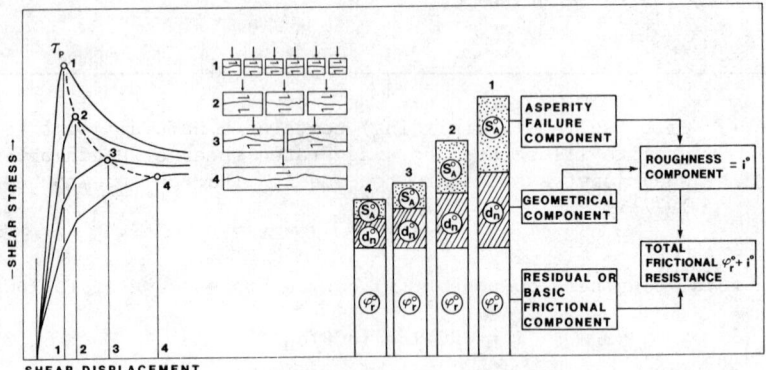

Figure 3. Increases in joint sample size cause the three fundamental changes in behavior; i.e. reduced asperity strength, reduced dilation, and increased displacement to mobilize peak strength. (Bandis et al. 1981).

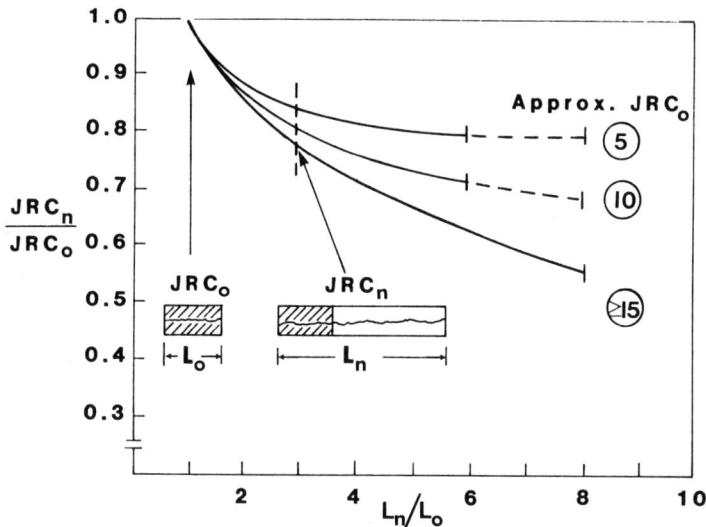

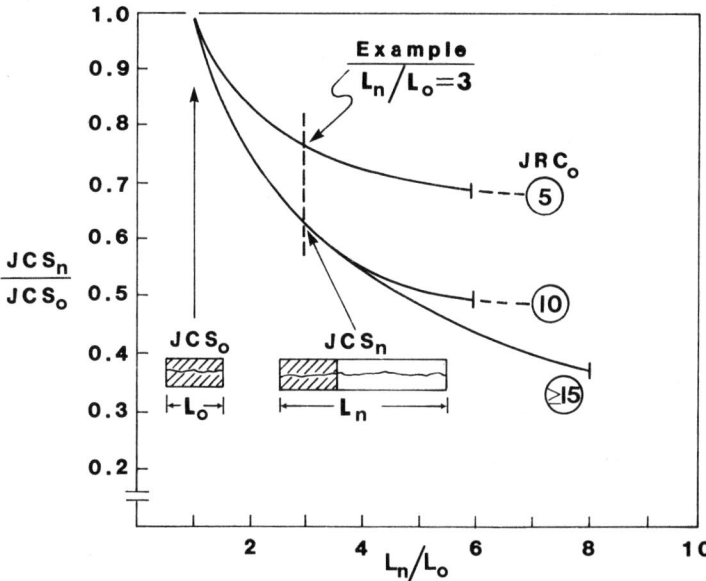

Figure 4. An approximate method for reducing joint roughness and wall strength parameters, to allow for size effects between laboratory and in situ sample sizes. (Bandis et al 1981).

Typical parametric values for a moderately rough joint in slightly weathered granite would be as follows: $\phi_r = 25°$, JRC = 10 and JCS = 100 MPa. An effective normal stress of 1 MPa would give a value of (ϕ') equal to 45°, while 10 MPa would give 35°.

Extensive size effects testing reported by Bandis et al (1981) suggest that both JRC and JCS reduce with increasing block size. Experimental data available at present are summarized in Figure 4. Greatest reductions in these parameters occur with the roughest joint surfaces due to the marked change in size of the individual contact points between opposed asperities.

SHEAR DISPLACEMENT SIZE EFFECTS

The block size effects discussed above are basically the result of the reduced degrees of freedom as block size is increased. The inability of a large block to slightly rotate and register all scales of roughness results in the situation depicted in Figure 5.

The parameter δ(peak) increases significantly as block size is increased. A wide ranging survey of laboratory and in situ test data, numbering approximately 650 tests, is summarized in Figure 6. For convenience, the data was grouped into three size categories: laboratory (30-300mm); in situ (300mm-3m); and novel (3m-12m). Table 1 summarizes mean values and number of tests for three categories of surface. The data has been expressed in a manner that minimizes the size effect, since all δ(peak) values are normalized by sample length. Nevertheless, significant size effects are evident.

TABLE 1

Summary of Mean Peak Shear "Strains" for
Joints and Clay-Filled Discontinuities

TYPE OF SAMPLE	LAB. SCALE (30-300mm)	IN SITU (0.3-3.0m)	NOVEL (3-12m)	ALL SIZES
(1) Filled discontinuities	1.31% (56)	0.55% (94)	0.13% (5)	0.81% (155)
(2) Rock joints	1.28% (224)	0.72% (71)	--	0.98% (295)
(3) Model joints	--	1.04% (96)	0.58% (66)	0.86% (162)

BLOCK SIZE EFFECT ON JOINTED ROCK

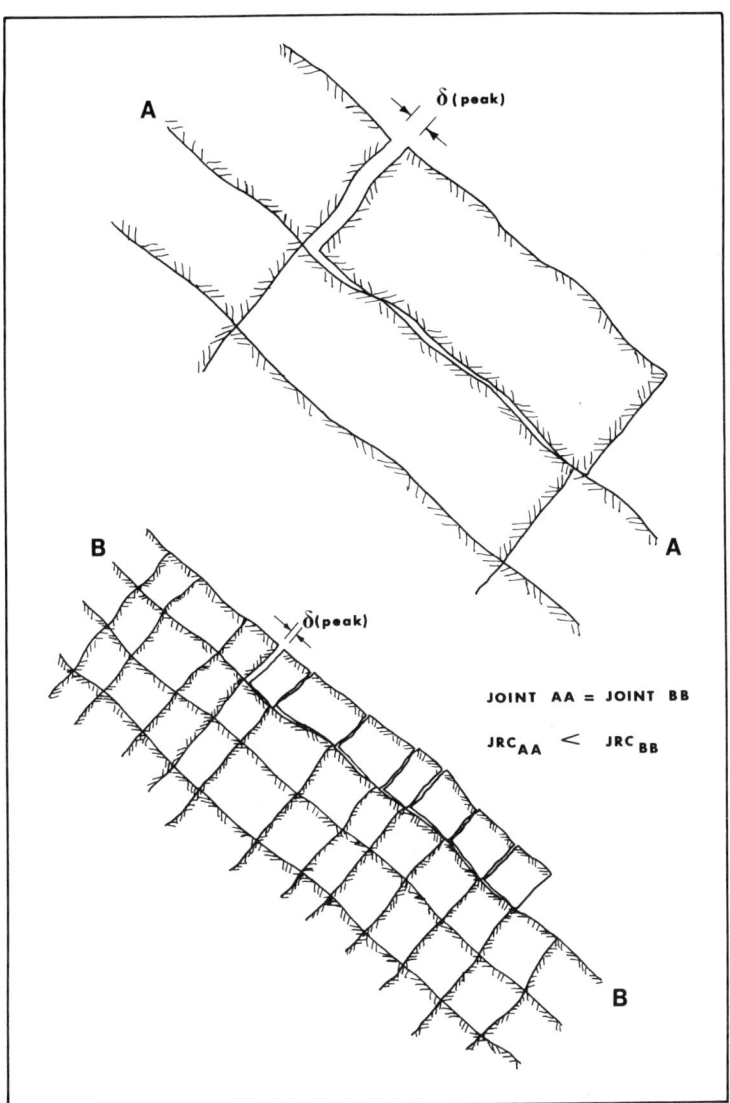

Figure 5. An illustration of the block size dependence of δ(peak), the shear displacement to mobilize peak strength.

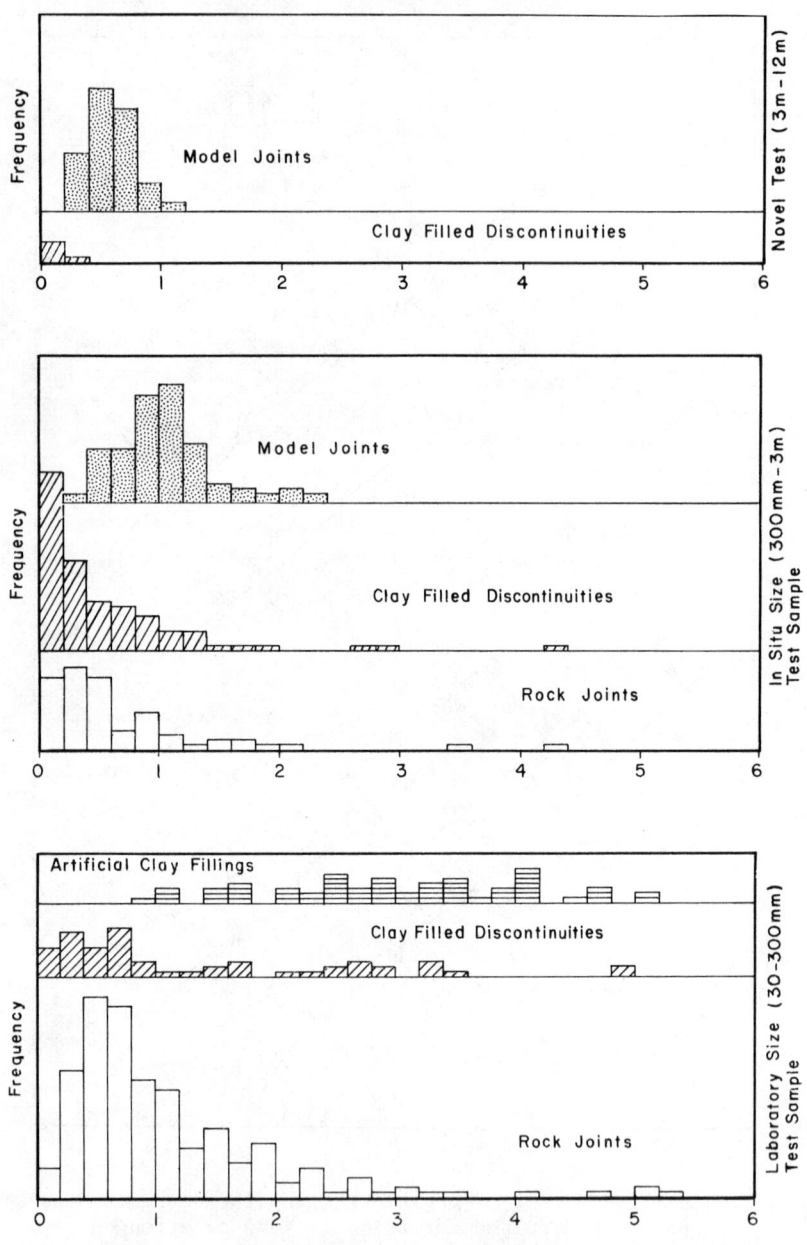

Figure 6. Distribution of δ(peak) as a function of sample length for rock joints, clay-filled discontinuities and model joints.

Figure 7 illustrates the apparent consistency in displacements observed in shear tests that involve *loading* in shear, and earthquake slip magnitudes which involve *unloading* in shear. An analysis of the data indicates that the following equation gives a reasonable approximation to the observed values:

$$\delta = \frac{L}{500} \cdot \left(\frac{JRC}{L}\right)^{0.33} \qquad (4)$$

where δ = slip magnitude required to mobilize peak strength, or that occurring during unloading in an earthquake

L = length of joint or faulted block (meters)

JRC = joint roughness coefficient (>0)

Example 1. <u>Laboratory Specimen</u>. Assume: JRC = 15 (rough), L = 0.1m
Equation 4 gives δ = 1.0mm

Example 2. <u>Natural Jointed Block</u>. Assume: JRC = 7.5, L = 1.0m
Equation 4 gives δ = 3.9mm

Example 3. <u>Earthquake Fault</u>. Assume: JRC = 0.5 (near-residual), L = 100km
Equation 4 gives δ = 3.6 meters

The above examples of size effects illustrate that equation 4 gives an acceptable degree of accuracy for most practical applications. The implication that (δ) is smaller when surfaces are smoother or closer to residual (JRC ≅ 0) also appears to be consistent with observations.

SHEAR STIFFNESS SIZE EFFECTS

Increased block size has been shown to:

increase δ(peak)

reduce   JRC

reduce   JCS

The combined effect is to noticably reduce the peak shear stiffness (K_S) which was defined by Goodman (1970) as:

$$K_S = \frac{\sigma_n' \tan \phi'}{\delta \text{ (peak)}} \qquad (5)$$

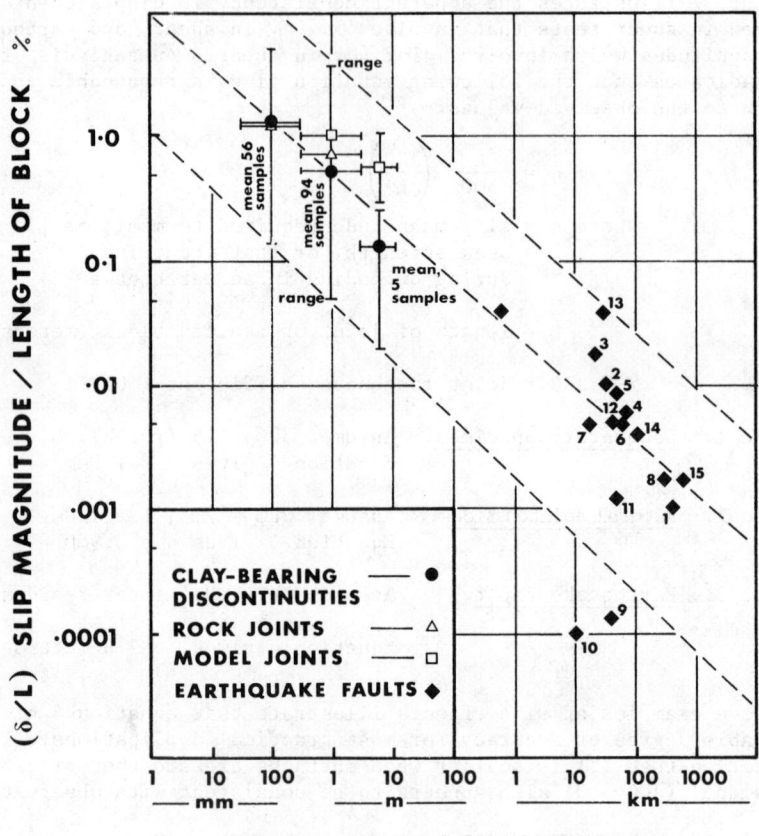

Figure 7. Slip magnitudes required to mobilize peak strength show a consistent effect of block length. Extrapolation to data for earthquake slip magnitudes (after Nur, 1974) suggests consistent trends between loading and unloading shear stiffness.

A useful approximation to K_s is given by rearrangement of equations 1, 2, 4 and 5:

$$K_s = \frac{\sigma_n' \tan [JRC \log (JCS/\sigma_n') + \phi_r]}{\frac{L}{500} \cdot \left(\frac{JRC}{L}\right)^{0.33}} \quad (6)$$

Measured values of K_s derived from a wide ranging review of test data are shown in Figure 8, each as a function of block size. The stippled lines representing normal stress levels were located using the mean

BLOCK SIZE EFFECT ON JOINTED ROCK

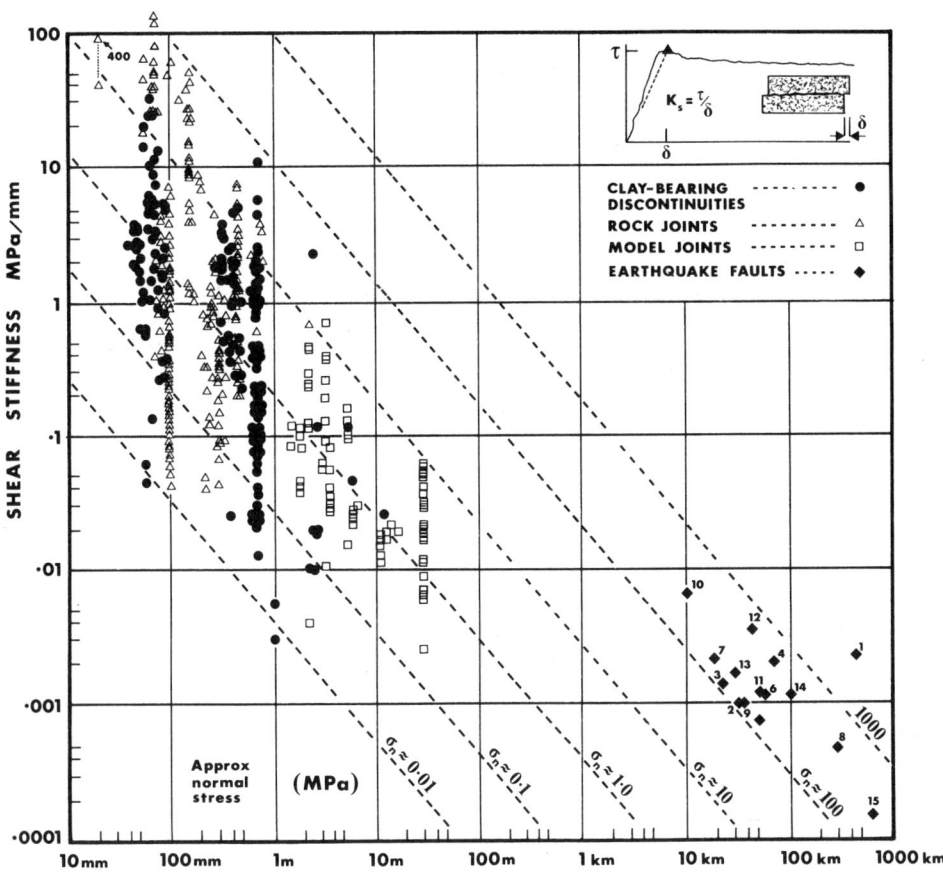

Figure 8. Laboratory and in situ shear stiffness data reported in the literature indicate the important influence of block size. Comparison is also made with the average values of stiffness derived for earthquake events, as reviewed by Nur (1974). The normal stress diagonals have been extrapolated linearly outside the range of typical test sizes of 100mm-1 meter.

values of JRC, JCS and ϕ_r obtained from the 137 shear tests on rock joints reported by Barton and Choubey (1977):

$$JRC = 8.9$$
$$JCS = 92 \text{ MPa}$$
$$\phi_r = 27.5°$$
$$L = 0.1m$$

The most frequently measured value of δ(peak) was 0.6mm, giving a peak shear stiffness value of 1.7 MPa/mm under a normal stress of 1 MPa. The gradient of the normal stress lines was derived from the best fit relationships to the data shown in Figure 4.

$$JRC_n \cong JRC_o \left(\frac{L_n}{L_o}\right)^{-0.02 \, JRC_o} \quad (7)$$

$$JCS_n \cong JCS_o \left(\frac{L_n}{L_o}\right)^{-0.03 \, JRC_o} \quad (8)$$

These equations were derived from shear tests over a ten-fold range of block sizes, and linear extrapolation outside the size range 100mm - 1 meter has been assumed when drawing the effective normal stress diagonals in Figure 8. It will be noted that the earthquake fault stiffnesses (mean values from Nur, 1974) are bracketed by the effective normal stress diagonals 100-1000 MPa (1-10 Kbars).

Tentative application of equations 6, 7 and 8 over a two order of magnitude range of block sizes shown in Figure 9 suggests a gradual flattening out of the normal stress diagonals with increasing block size. Tentative scaling of the same data to earthquake fault sizes indicates normal stress levels closer to the range 5-20 MPa (50-200 bars). This is conveniently close to the effective normal stress levels operating at depth in the vicinity of the San Andreas fault (Zoback et al. 1980).

It will be noticed that the stiffness of the rough, competent joint and that of the weaker, smooth joint (Figure 9) converges when either the stress level, or block size is increased. The above method of estimating peak shear stiffness for rock joints is specifically directed at clay-free discontinuities. When clay is present, preventing (to a greater or lesser extent) rock-to-rock contact, the peak shear stiffness tends not to be so size-dependent, and is also somewhat less stress dependent, due to the low shear strength (Infanti and Kanji, 1978).

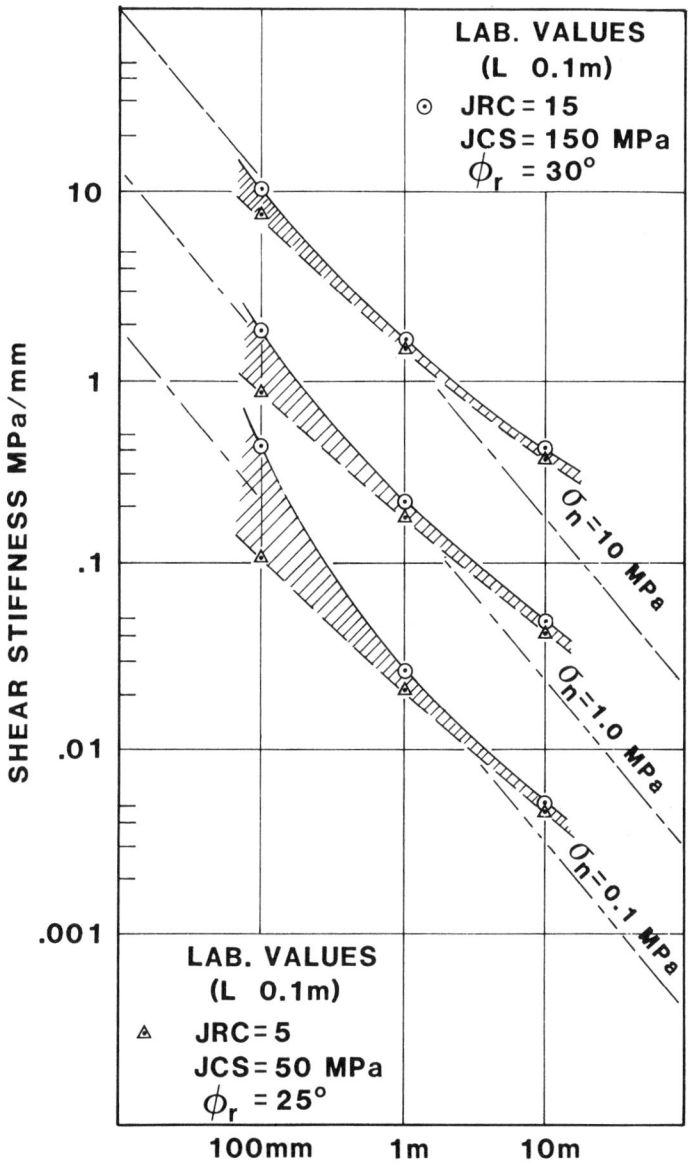

Figure 9. Application of equations 6, 7 and 8 to the scaling of typical laboratory test data. The diagonal normal stress lines match those in Figure 8.

SIZE EFFECTS IN BLOCK ASSEMBLIES

It has been shown earlier that joint properties are size-dependent when shear displacement is involved, due to the displacement-dependence of strength mobilization. However, the size dependency may apparently die out in an assembly of rock blocks when a "sample" exceeds the natural block size. Figure 10 shows schematically that, for unchanged roughness, the smaller the block size, the higher the shear strength of an assembly of blocks. The spacing of joints intersecting a potential shear plane also defines the distance between potential "hinges" in the assembly. The slightest block rotation allows the finer features of roughness to be felt as opposed to sheared over, hence the scale effect.

The biaxial samples depicted in Figure 10 were each fabricated with the same weak, brittle model material. Joint sets were formed using the same guillotine (Barton and Hansteen, 1979). Joint angles were the same in each case. Thus, the only difference was the joint spacing. Assembly No. 1 consisted of 4000 blocks; No. 2, 1000 blocks, and No. 3, 250 blocks. Joint spacing was doubled and quadrupled to produce these totals.

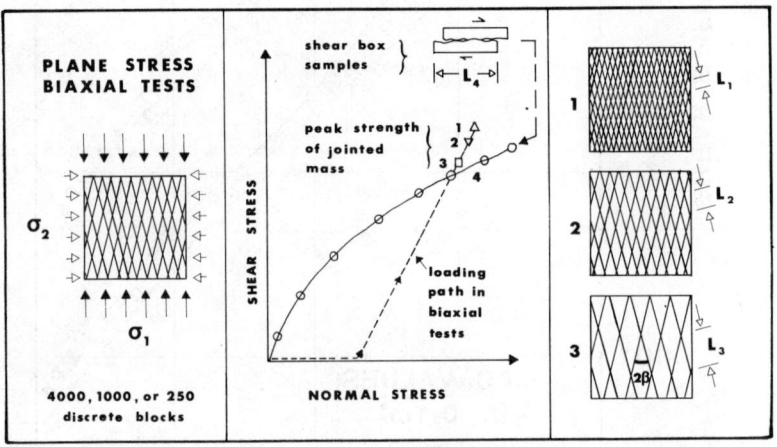

Figure 10. The size of individual blocks in a jointed assembly determines the shear strength of the assembly.

Sample Nos. 2 and 3 failed by shear along several of the primary continuous joints when the mobilized roughness JRC had reached values of 25.1 and 21.6 respectively. Higher differential stress was required to fail assembly No. 1, and failure did not occur by shear parallel to the weakest joint set. The contrasting failure modes are illustrated in Figure 11. Several tests were performed on assemblies of the smallest size blocks, using both diamond and square shaped blocks. In each case, failure occurred by rotation of blocks within a "kink" band at least eight blocks wide. Sequence photographs of such a development are given in Figure 12.

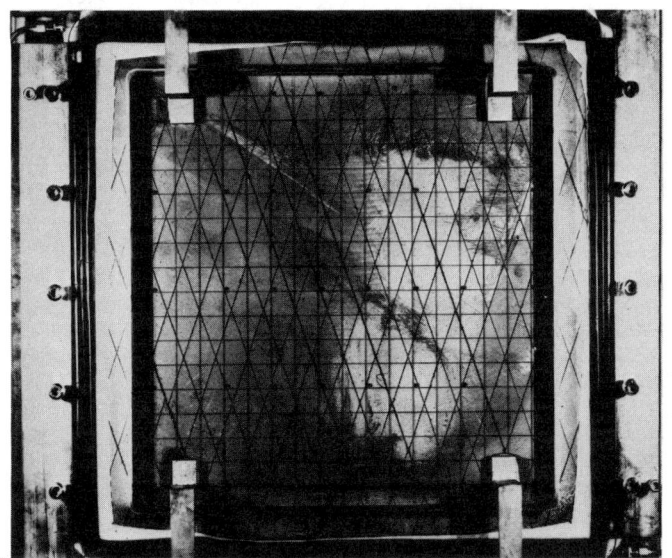

Test IV

↓ σ_1 = 1.75
← σ_2 = 0.2

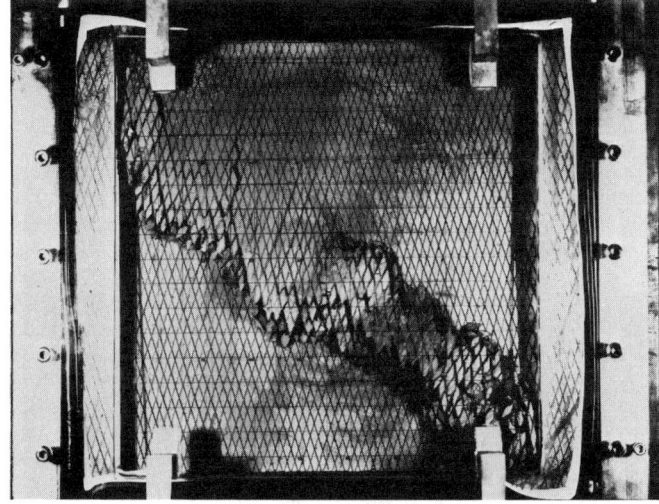

Test III

↓ σ_1 = 1.9
← σ_2 = 0.2

Figure 11. Block assemblies with large relative block size fail by shear along the weakest (continuous) joint set. All the assemblies with the smallest block sizes failed by rotation within a "kink" band.

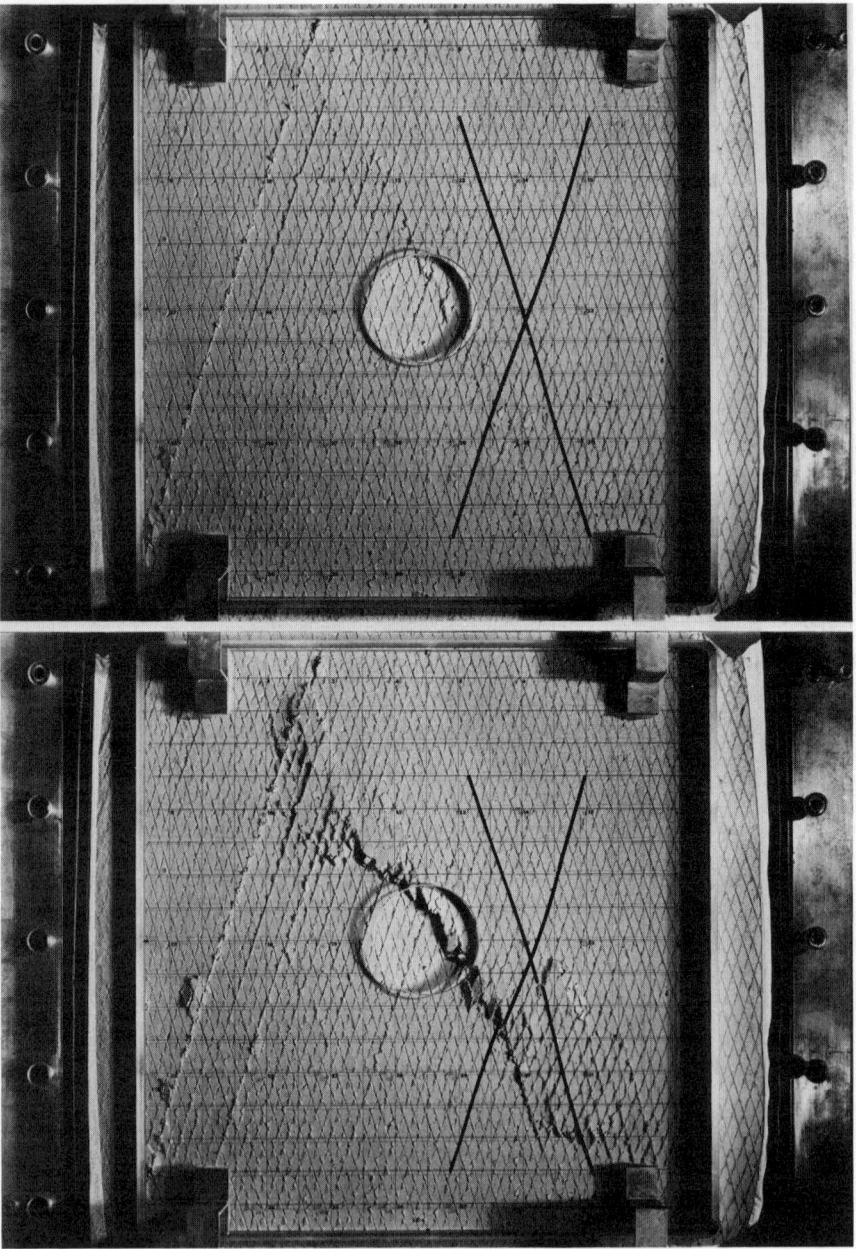

Figure 12. Sequence photographs illustrating the development of a rotational "kink" band concurrently with in-plane joint shear.

BLOCK SIZE EFFECT ON JOINTED ROCK

It appears that a fundamental change in deformation and strength behavior occurs when the number of blocks per loaded area exceeds some limit. Rotational modes of deformation have been observed in model studies by Ladanyi and Archambault (1972), and by Hayashi and Kitahara (1970). In 1974, Goodman made the following comments on the subject: "Rotational friction is important in view of the low shear strengths associated with instability by buckling and kinking of layers or rows of joint blocks. Unfortunately, our appreciation and understanding of these phenomena is only just beginning". These comments appear equally valid today.

Triaxial compression testing of jointed models reported by Baecher and Einstein (1981) indicate that both strength and deformation moduli decrease logarithmically with numbers of joints. The reported tests were performed with just one set of joints perpendicular to the major principal (axial) stress. As pointed out by these authors, changing scale may change the relative importance of the various possible failure mechanisms. In addition, changing the shape of blocks and the angle between joint sets may result in failure modes other than those commonly considered, especially when confining stresses are low or zero (Brown, 1970).

SIZE EFFECTS ON POISSON'S RATIO

An attempt to investigate the onset of different failure modes with different block sizes is shown in Figure 13. The axial and lateral strains were recorded using a photogrammetric technique. The most obvious difference in behavior is the ratio of axial to lateral strain, or Poisson's ratio. The stiffer, large-block assemblies deform by in-plane joint shear, and this results in rapid increases in Poisson's ratio due to the combined effect of joint shear and dilation caused by over-riding of roughness. The relatively moderate build-up of Poisson's ratio in the heavily jointed assembly (Figure 13) is probably a function of the large amount of axial consolidation that can be accomodated by numerous joints, before significant shear is apparent. At a later stage of loading when shear failure is imminent, Poisson's ratio is seen to increase up to, and beyond 1.0. An example of physical measurements of this phenomenon is shown in Figure 14.

Large increases in Poisson's ratio due to joint shear have been observed in model studies by Muller and Packer (1965), John (1970), and Barton and Hansteen (1979). They have also been measured in large-scale in situ block tests on jointed rock by John (1961), Lögters and Voort (1974), and Barton and Lingle (1982).

APPROPRIATE TEST SIZE

It is apparent from these tests on assemblies of blocks that the size of individual blocks controls both the shear strength of the assembly, and its deformation characteristics. It appears reasonable to assume that a test on one jointed block will give nearly the same result as a test on two adjacent blocks, if the pair of blocks are "hinged" (i.e. cross-jointed) so that the necessary freedom for rotation is present. If this is true, then a significant rock mechanics test size for jointed media will be the natural block size as depicted in Figure 15.

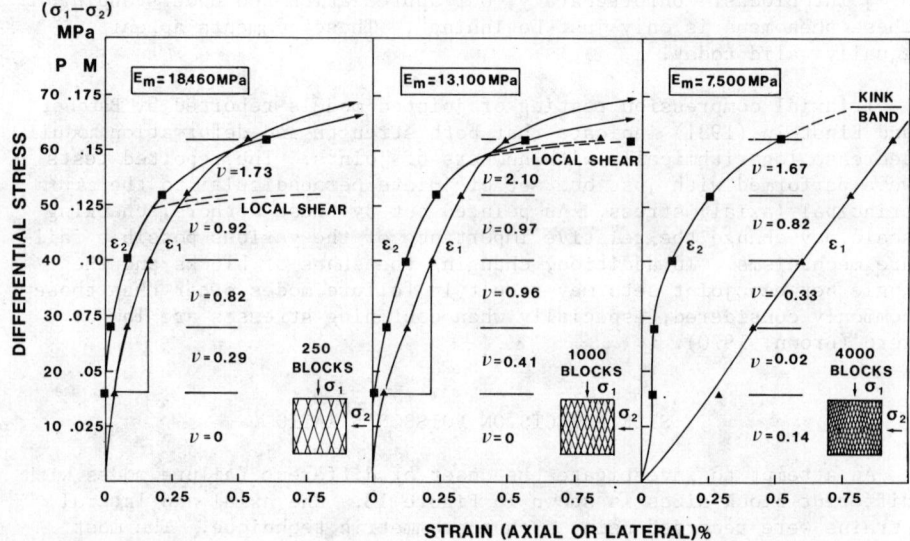

Figure 13. Contrasting deformation behavior exhibited by assemblies of different sized blocks.

Tests on smaller jointed blocks, for example on jointed drill core from the same rock mass, will automatically incur a size effect. The magnitude of this size effect may be significant if the particular joint set is non-planar. An example of measured size effects on shear stress-displacement behavior is given in Figure 16. A most significant point to note is that the assumed "residual" strength remaining at the end of the small-sample test is significantly higher than the peak strength of the full-size sample.

Tests on the natural-size blocks of a rockmass will obviously need to be performed in large numbers, before a statistically viable sample of test data is achieved. The combination of inexpensive tilt tests and Schmidt hammer tests is therefore attractive for obtaining the necessary shear strength data.

When deformability is also of concern, more expensive in situ block tests may need to be performed. Test volumes should then include a significant number of each of the joint sets thought to control the deformability of the rock mass. Suitable choice of instrumentation to span intact rock, single joints and multiple joints will then provide invaluable data on the magnitude of potential size effects.

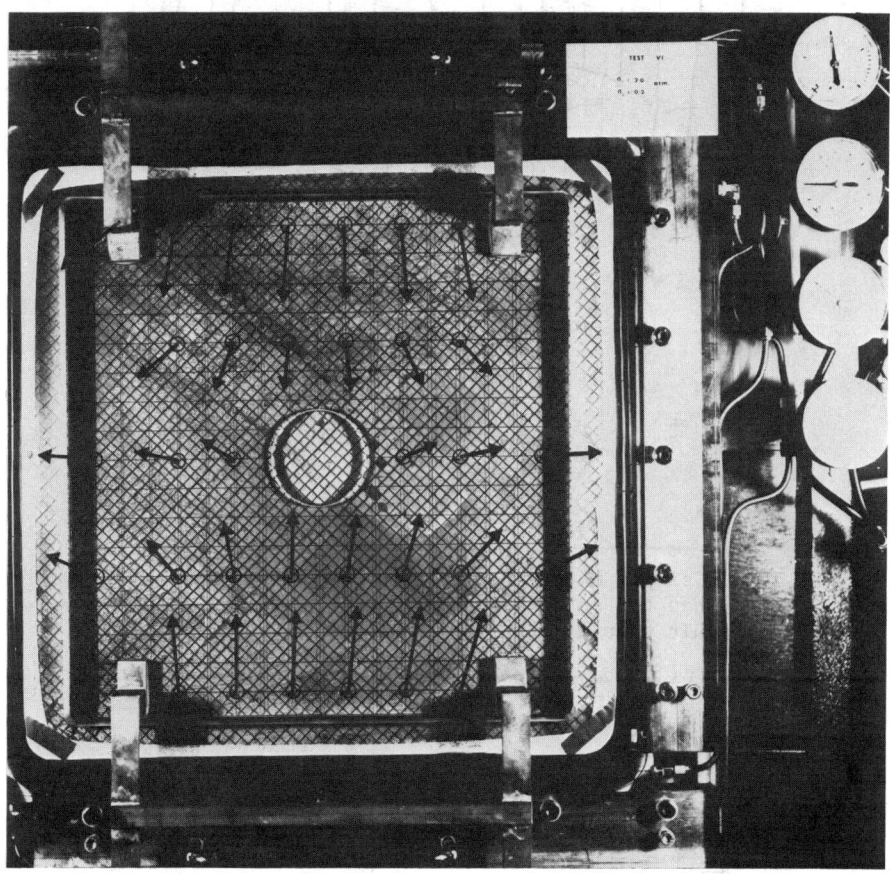

Figure 14. An example of the large values of Poisson's ratio (or transverse deformation) associated with shear and dilation of a jointed assembly of blocks. The displacement vectors were derived by photogrammetric analysis, and are relevant to the deformation occurring when the stress difference was increased from prototype (full-scale) values of 52 to 72 MPa.

Figure 15. A simple method for obtaining a scale-free value of JRC when the natural rock blocks are not too large or difficult to extract.

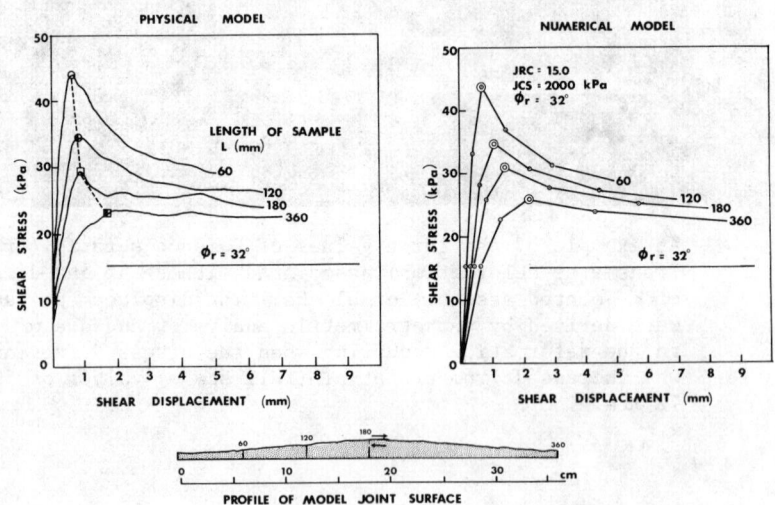

Figure 16. Measured size effects caused by testing jointed samples smaller than the full-size block. The physical model tests are reported by Bandis et al. (1981).

REFERENCES

Baecher, G.B., and Einstein, H.H., 1981, "Size Effect in Rock Testing", Geophysical Research Letters, Vol. 8, No. 7, pp. 671-674.

Bandis, S., 1980, "Experimental Studies of Scale Effects on Shear Strength and Deformation of Rock Joints", Ph.D. Thesis, University of Leeds, England.

Bandis, S., Lumsden, A.C. and Barton, N., 1981, "Experimental Studies of Scale Effects on the Shear Behavior of Rock Joints", Int. J. Rock Mech. Min. Sci. and Geomech. Abstr., Vol. 18, pp. 1-21.

Barton, N. and Choubey, V., 1977, "The Shear Strength of Rock Joints in Theory and Practice", Rock Mechanics, Vol. 10, pp. 1-54.

Barton, N. and Hansteen, H., 1979, "Very Large Span Openings At Shallow Depth: Deformation Magnitudes from Jointed Models and F.E. Analysis", Proc. Rapid Excavation and Tunneling Conf., Atlanta, Vol. 2, pp. 1331-1353.

Barton, N. and Lingle, R., 1982, "Rock Mass Characterization Methods for Nuclear Waste Repositories in Jointed Rock", Proc. ISRM Symposium, Rock Mechanics Related to Caverns and Pressure Shafts, Aachen, Germany.

Brown, E.T., 1970, "Modes of Failure in Jointed Rock Masses", Proc. 2nd Cong. of Int. Soc. Rock Mech., Belgrade, Vol. 2, 3-42.

Goodman, R.E., 1970, "The Deformability of Joints", Determination of the In Situ Modulus of Deformation of Rock, ASTM STP 477, pp. 174-196.

Goodman, R.E., 1974, "The Mechanical Properties of Joints", Panel Report, Proc. 3rd Cong. of Int. Soc. Rock Mech., Denver, Vol. 1A, pp. 127-140.

Hayashi, M. and Kitahara, Y., 1970, "Anisotropic Dilatancy and Strength of Jointed Rock Masses, and Stress Distribution in Fissured Rock Masses", Rock Mechanics in Japan, Vol. 1, pp. 85-87.

Infanti, N. and Kanji, M.A., 1978, "In Situ Shear Strength, Normal and Shear Stiffness Determinations at Agua Vermelha Project, Proc. 3rd Int. Cong. of Int. Assoc. of Eng. Geologists, Madrid, Vol. 2, pp. 175-183.

John, K.W., 1961, "Die Praxis der Felsgrossversuche Beschreiben am Beispielder Arbeiten au der Kurobe IV Staumauer in Japan", Geologie und Bauwesen, Vol. 27, No. 1.

John, K.W., 1970, "Civil Engineering Approach to Evaluate Strength and Deformability of Regularly Jointed Rock", Proc. 11th Symp. on Rock Mechanics, Berkeley, pp. 69-80.

Ladanyi, B. and Archambault, G., 1972, "Evaluation of Shear Strength of a Jointed Rock Mass", Proc. of 24th Int. Geological Congress, Montreal, Sect. 13D. (Translation from French.)

Lögters, G. and Voort, H., 1974, "In Situ Determination of the Deformational Behaviour of a Cubical Rock Mass Sample Under Triaxial Load", Rock Mechanics, Vol. 6, pp. 65-79

Muller, L. and Pacher, F., 1965, "Modellversuche zur Klärung der Bruchgefahr Geklüfteter Medien", Rock Mechanics and Engineering Geology, Suppl. II, pp. 7-24.

Nur, A., 1974, "Tectonophysics: The Study of Relations Between Deformation and Forces in the Earth", General Report Proc. 3rd Int. Cong. of Int. Soc. Rock Mech., Denver, Vol. 1A, pp. 243-317.

Zoback, M.D., Tsukahara, H. and Hickman, S., 1980, "Stress Measurements at Depth in the Vicinity of the San Andreas Fault: Implications for the Magnitude of Shear Stress with Depth, J. Geophys. Res., Vol. 85, pp. 6157-6173.

Chapter 77

THE COREJACKING TEST: AN ANALYSIS
OF THE COREJACK LOADING SYSTEM

by Douglas A. Blankenship and Randall G. Stickney

Staff Engineer, RE/SPEC Inc.
Rapid City, South Dakota

Manager of Field Engineering, RE/SPEC Inc.
Rapid City, South Dakota

ABSTRACT

The corejacking test is a field test designed to measure the response of salt to known boundary conditions. A 1.0-m-diameter salt core is externally pressurized using curved flatjacks placed in the annulus surrounding the core. Time-dependent deformation of the salt is measured by monitoring the diametrical closure of an inner borehole. A finite element elastic analysis of the jack loading system has been performed. The purpose of this analysis was to determine if the original assumption of axisymmetric loading is correct. Six loading cases were investigated, each representing either hypothetical or previously encountered jack configurations.

INTRODUCTION

The corejacking test has been developed to provide field data on the time-dependent response of rock salt when subjected to load and elevated temperature. The primary purpose of the test is to provide data that can be used for verification of the numerical modeling techniques used to predict the behavior of salt under various conditions. Therefore, the test was designed to be axisymmetric so that it could be easily modeled.

Small salt specimens have been extensively tested under controlled conditions in laboratory settings. Field tests have also been performed in salt to measure creep behavior, but have often necessitated broad assumptions with regard to boundary conditions. The boundary conditions of the corejacking test can be determined with only minor assumptions if axisymmetric loading is achieved. This paper is dedicated to the examination of the assumption of axisymmetric loading.

TEST DESCRIPTION

The corejacking test consists of a large hollow cylinder of rock salt to which constant external pressure and temperature are applied (Van Sambeek, 1981; Stickney, 1981). Response of the salt to this thermal and mechanical loading is monitored by measuring the closure of the inside of the cylinder as a function of time. The cylinder has an outside diameter of 1 m, an inside diameter of 0.2 m, and an overall length of 1 m.

The hollow cylinder is formed by drilling into the mine floor with a large diamond coring bit. The drilling results in a 1-m core in the mine floor encircled by an annulus 30-mm wide. A 0.2-m diameter borehole is then concentrically drilled in the 1-m core, also using a diamond coring bit.

Curved flatjacks, descriptively termed corejacks, are placed in the annulus surrounding the salt core and grouted in place. Constant external load is applied to the core by hydraulic pressurization of the jacks. A conceptual drawing of the core and corejack arrangement is shown in Figure 1. Tests currently in progress utilize jacks that are 1 m in length and 0.6 m along the arc. Five jacks of this size are necessary to surround the core. In practice two rows of these jacks have been used and staggered in such a way

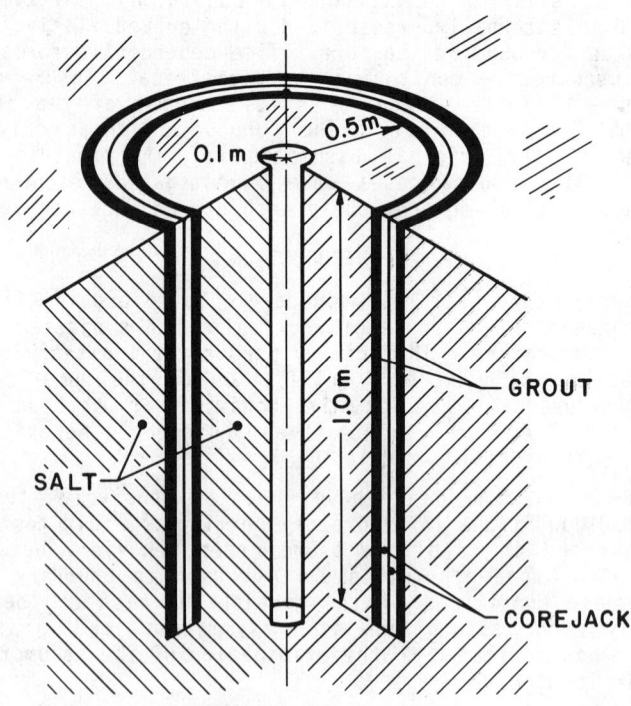

Figure 1. Conceptual drawing of corejack loading system.

that the center of the outside jack is along the same radial line as the gap between the edges of two inside jacks. A plan view of the corejack placement scheme is shown in Figure 2. The tests have been performed at both ambient and elevated temperatures.

Closure of the inner (0.2 m) borehole is measured for the test duration using a three-point inside micrometer. The borehole diameter changes are measured at nominal elevations of 0.25 and 0.5 meters below the top of the core. At both elevations three fixed measurement pads are anchored to the borehole wall at 120° azimuthal separation. The tool is self-centering in the borehole and reads the gross borehole diameter directly with a resolution of about 0.005 mm.

ANALYSIS OF THE LOADING SYSTEM

The purpose of this analysis was to determine the symmetry of the state of stress in the salt core due to corejack loading. The typical jack configuration shown in Figure 2 and variations of this system were investigated.

Using the finite element program, SPECTROM 11 (Yamada, 1981), elastic analyses of various corejack loading cases were performed. All analyses were performed on a segment of core "AOB" as seen in Figure 2. This 36° segment represents the minimum geometric symmetry for the typical loading case.

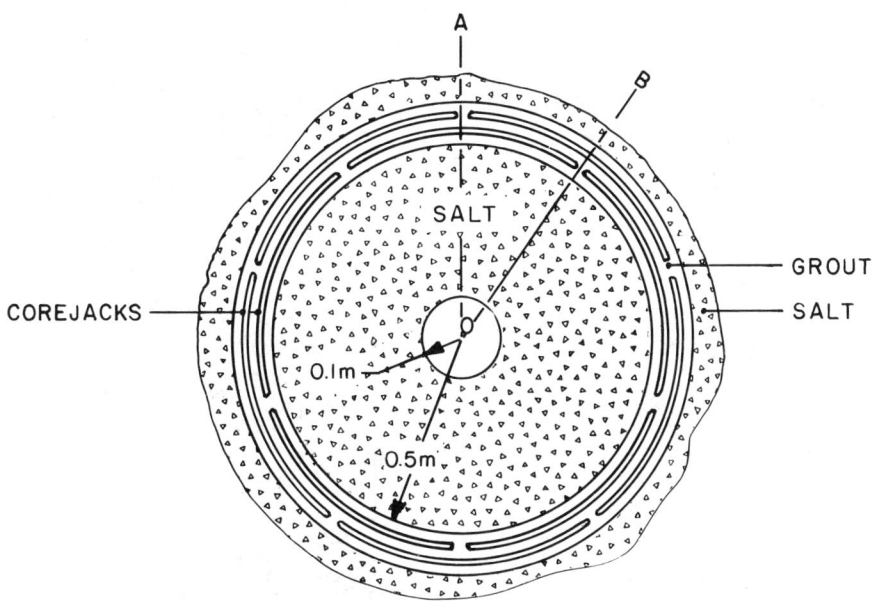

Figure 2. Plan view of typical corejack arrangement.

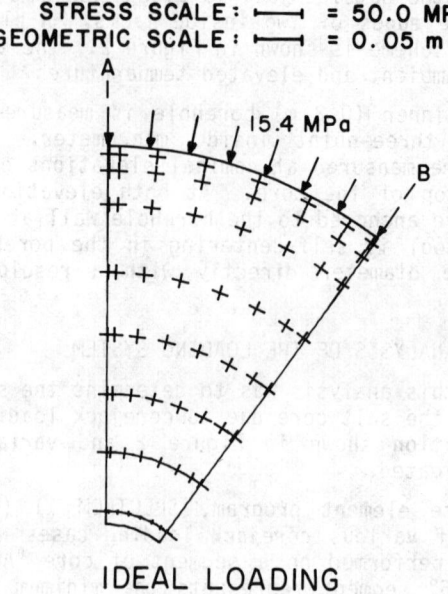
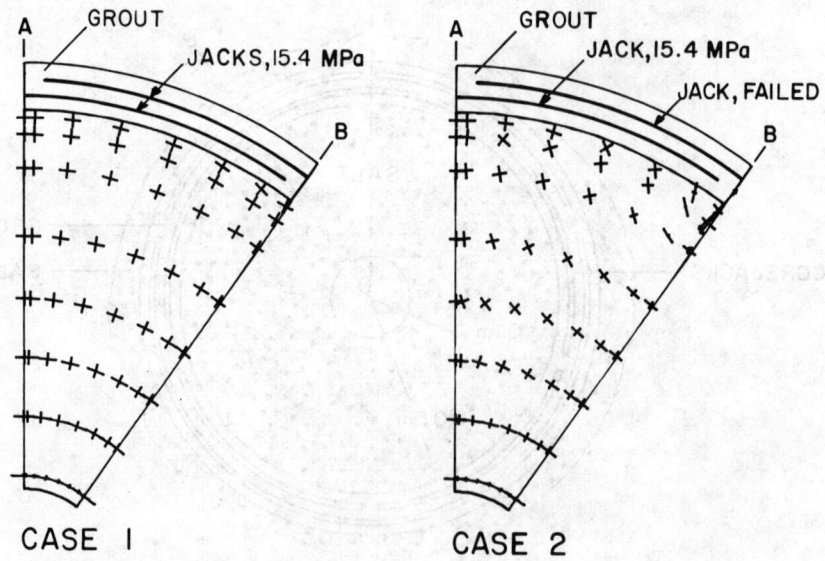

Figure 3. Symmetric segment "AOB" with jack configurations and associated principle stress trajectories.

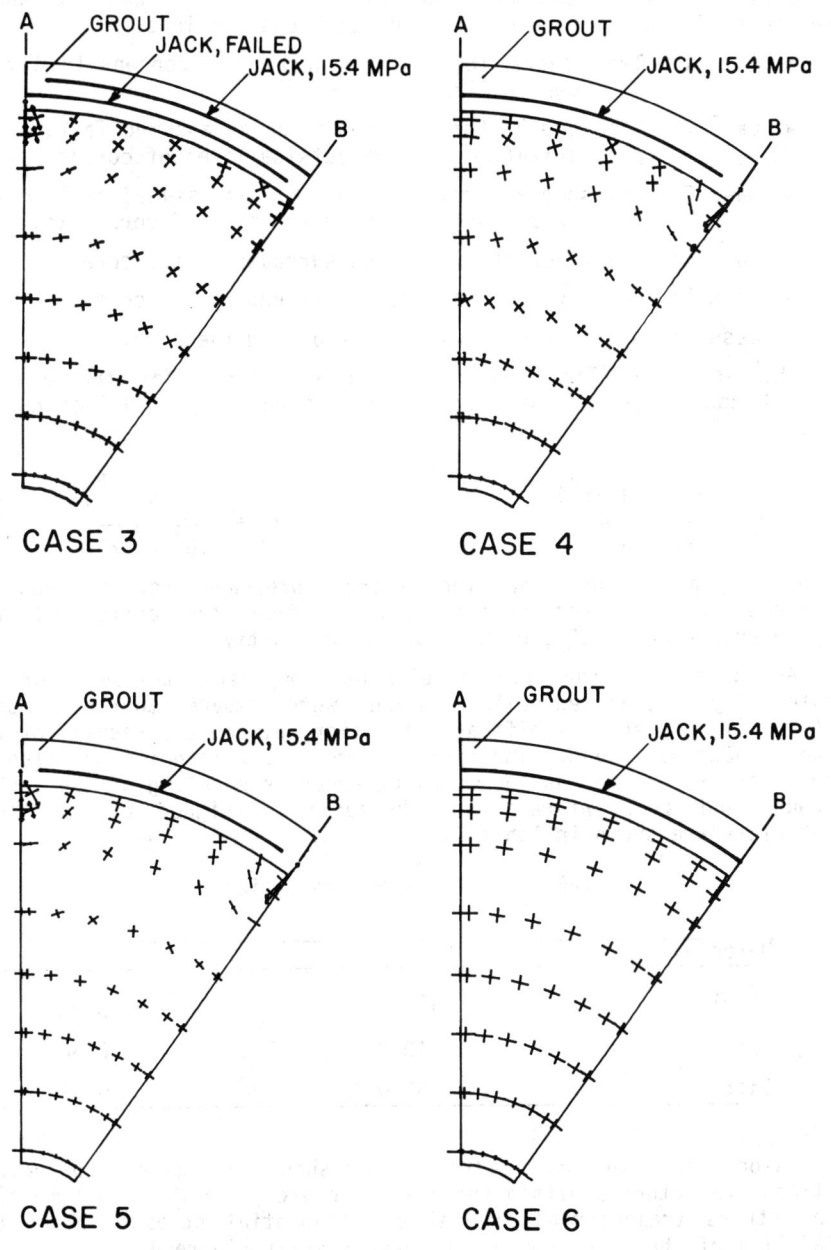

Figure 3. Continued.

In all, six jack configurations compatible with the symmetric segment "AOB" were examined. Each of the six loading cases is shown in Figure 3. A brief description of each case follows:

- Case 1: Two layers thick, five jacks per concentric layer, as shown in Figure 2.
- Case 2: The same as Case 1 except for an assumed failure (no pressurization) of the outside layer of corejacks.
- Case 3: The same as Case 1 except for an assumed failure (no pressurization) of the inner layer of corejacks.
- Case 4: One layer of five jacks surrounding the core.
- Case 5: One layer of ten jacks surrounding the core.
- Case 6: A single continuous jack around the core.

The solution (Timoshenko and Goodier, 1951) for the elastic radial and tangential stresses in an externally-pressurized hollow cylinder is:

$$\sigma_r = \frac{-P_o b^2 (a^2 - r^2)}{(b^2 - a^2) r^2} \qquad \sigma_\theta = \frac{P_o b^2 (a^2 + r^2)}{(b^2 - a^2) r^2} \qquad (1)$$

where P_o, a, b, and r are the external pressure, internal radius, external radius, and radial distance from the center of the cylinder, respectively, with compression positive.

As a check, the finite element computed stresses for an externally-pressurized hollow cylinder were compared to the stresses of Eq. (1) and an insignificant variation between the results of the two methods was found. The finite element mesh is quite detailed in the vicinity of the annulus enabling accurate modeling of the jacks, grout, and salt interaction. Properties assigned to the three materials are shown in Table 1.

TABLE 1. Material Properties

Material	E (GPa)	ν
Steel	180.0	0.28
Grout	10.0	0.30
Salt	31.0	0.31

Results

Associated with each loading system shown in Figure 3, principle stress trajectories within the salt core are plotted. Complementing the stress trajectories, radial and tangential stresses along the "A" line of the core segment are presented in Figure 4.

Figures 3 and 4 clearly show that Cases 1 and 6 approximate the ideal solution very well. In Case 1 a small perturbation in stress symmetry is noticed near the exterior of the core segment, but stresses along the "A" line show no graphic deviation from the ideal case. As expected, the stresses computed for Case 6 are all but identical to those resulting from ideal loading.

Cases 2, 3, 4, and 5 vary considerably from the ideal case. In Cases 2 and 3, where the outside and inside jacks fail respectively, the pressurized jack tends to straighten; thus, the edges of the jacks pull away from the core causing tensile stresses to develop in these areas. Figure 4 captures this drift toward tension for Case 3 because the edge of the outside jack is near the "A" line. The tensile drift is not shown for Case 2 because the edge of the inside jack is near the "B" line. The stresses in the salt core due to loading Case 4 are essentially equal to those for Case 2, indicating the additional stiffness of the failed jacks is of no consequence in the transfer of load. In Case 5 the jacks respond in the same manner as Cases 2, 3, and 4; that is, the edges of the jacks tend to pull away from the core. The stiffness of the half-sized jack is less than the full-sized jack, thus the stresses in the salt core are less than those included by one row of full-sized jacks.

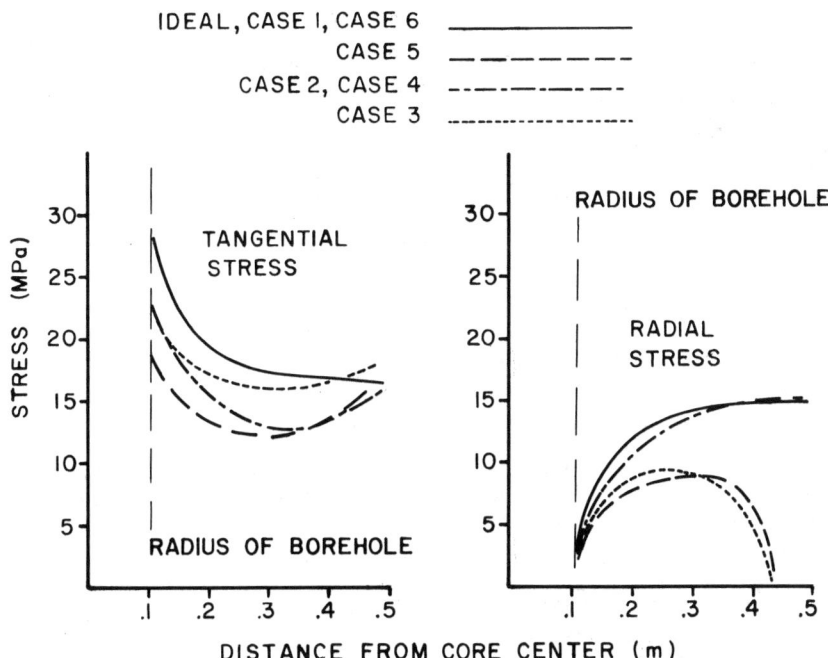

Figure 4. Radial and tangential stress along the "A" line.

CONCLUSIONS

Case 1, the typical loading system, has been shown to be more than adequate to provide load to the salt core in terms of symmetry and magnitude. Previously the typical two-layer, five jacks per layer loading system was thought to be somewhat of a fail-safe system. It was assumed that an entire layer of jacks could fail without significantly effecting the transfer of loads; the core would still be surrounded by active jacks. The problem with the fail-safe assumption is that the radius of curvature of a corejack will increase if not restrained and will no longer conform to the exterior of the salt core. When all jacks are pressurized, each acts to restrain the other. Cases 2 and 3 are extreme cases of entire layer failure, and it is open to question what effect several jack failures would have on the core stresses. However, each jack is seen to be an integral part of the system and axisymmetric loading requires all jacks to be in working order.

The corejacking test has been successful for pressures in the 10 MPa range at ambient temperatures. Existing corejacks have been unreliable when subjected to higher pressures and temperatures because of the large deformations associated with these conditions. More refined corejacks are currently under design. A jack described by Case 6 (continuous around the core) is also being investigated.

ACKNOWLEDGEMENTS

Financial support for the analysis of the corejacking system was provided by the Office of Nuclear Waste Isolation under Subcontract No. E512-02200, with Battelle Memorial Institute, Columbus, OH. The Technical Project Manager was Mr. William F. Ubbes. The authors wish to thank Mr. Marc C. Loken and Dr. Joe L. Ratigan for their technical guidance in the modeling performed.

REFERENCES

Stickney, R. G., 1981, "Engineering Test Plan for Phase II Accelerated Borehole Closure Experiments at Avery Island," Topical Report RSI-0146, ONWI- , Office of Nuclear Waste Isolation under Subcontract with Battelle Memorial Institute, Columbus, OH.

Timoshenko, S., and Goodier, J. N., 1951, Theory of Elasticity, 2nd ed., McGraw-Hill, New York, pp. 58-60.

Van Sambeek, L. L., 1981, "The Corejacking Test for Measuring Salt Response," Proceedings, First Conference on the Mechanical Behavior of Salt, to be published.

Yamada, S. E., 1981, "User's Manual for SPECTROM 11: A Finite Element Thermoelastic/Plastic Stress Analysis Program," Topical Report RSI-0158, ONWI- , Office of Nuclear Waste Isolation under Subcontract with Battelle Memorial Institute, Columbus, OH.

Chapter 78

EVALUATION OF OPENING AND HYDRAULIC CONDUCTIVITY OF ROCK DISCONTINUITIES

by Paulo T. Cruz, Eda F. Quadros, Diogo Correa F9 and Antonio Marrano

Consulting Engineer, Assistant Research Engineer and Geologists at the Hydrogeotechnical Division at Instituto de Pesquisas Tecnológicas de São Paulo - Brasil

ABSTRACT

For many years the water pressure test has been used as a normal procedure to investigate the hydraulic characteristics of rock masses and discontinuities. Although the test has given evidence of problematic areas, the nature of flow and the so called equivalent permeability are seldom seriously affected by the testing tecnique. In order to improve the information on these points the Instituto de Pesquisas Tecnológicas de São Paulo - IPT, has developed an extensive program at laboratory and field research on the hydraulics at rock fractures. This paper summarizes the main itens related to this research. Is shows that with only small improvements on field tests and some support of laboratory conductivity tests on natural rock joints it is possible to obtain a better understanding of the nature of flow in fractures, improve flow predictions and estimate the opening and hydraulic conductivity, as well as the transmissivity of natural joints or rock discontinuities.

INTRODUCTION

The research program consisted of simple conductivity tests performed on core boring rock samples (granite), with variable openings and rugosity. The field water pressure tests were performed in a known highly pervious discontinuity (fault joint) on the the foundation of the Nova Avanhandava Dam site (S.Paulo-Brasil), using the conventional equipment, but increasing the number of pressure stages both on the loading und unloading cycle. Infiltration tests were also performed in the low pressure range.

Test results both of field and laboratory were expressed as the relation of flow versus effective pressure.

A brief description of these tests and the analysis and conclusions follow.

LABORATORY TESTS

A simple device similar to a soil permeameter, was used to study flow characteristics and flow regimens within single rock fractures on three rock surfaces (a smooth surface, a semi-rough surface and a rough surface) em 2" core boring samples of granite. The variables were the opening, the rugosity and the gradient. Test results were compared with much more elaborated tests run by Lomize (1953), Louis (1969), Sharp (1970) and Rissler (1978) and the very good agreement was reached.

A typical test result, is shown in Fig. 1, relating flow with gradient. All test results were ploted on the same way.

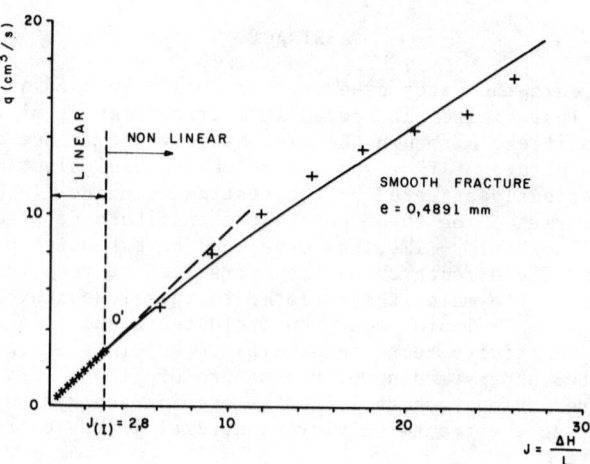

Fig. 1 - Relationship between flow and gradient. Results from laboratory test.

In order to simulate a field water pressure test, test results had to be transformed into relations of flow versus effective pressure.

In the laboratory test flow develops on a rectangular section of aproximately constant width, as shown in Fig. 2.

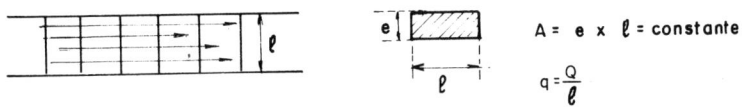

Fig. 2 - Single joint with a constant retangular section.

In a field water pressure test the flow is radial and in increasing areas. Fig. 3 shows the assumed radial distribution of flow.

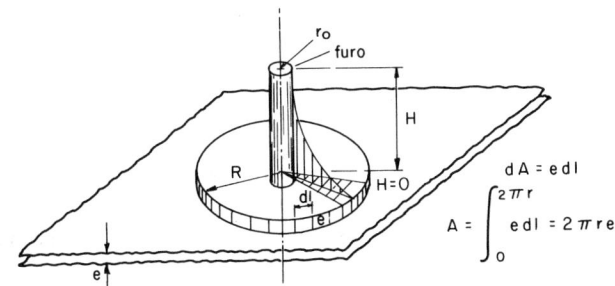

Fig. 3 - Radial distribution of flow.

Considering the above model and using the empirical equations derived from regression analysis relating unit flow (q), gradient (J), opening (e) and absolute rugosity (k), diagrams relating flow with effective pressure were prepared for the three types of rock surfaces tested on the laboratory.

Figure 4 shows test results for the smooth surface and Fig. 5 the test results for the three rock surfaces tested at the same opening (0,7 mm).

It is evident from the last figure the influence of the rock surface rugosity on the amount of flow for the same gradient (or the effective pressure).

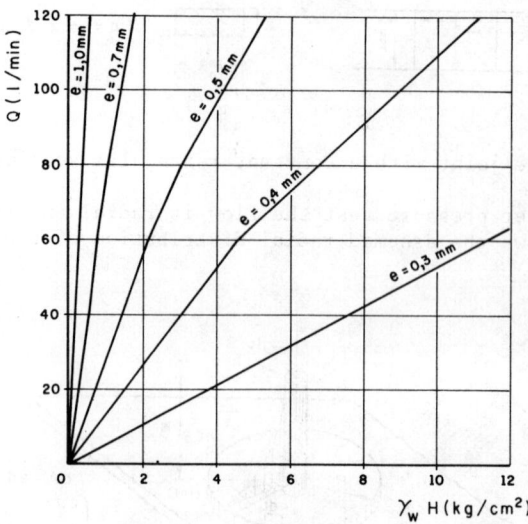

Fig. 4 - Relationship between flow and effective pressure.

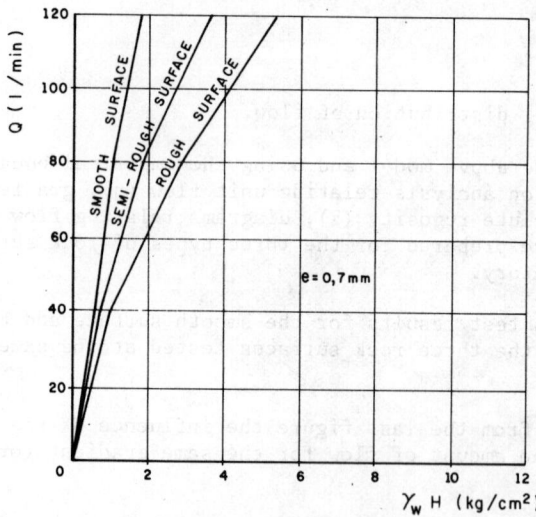

Fig. 5 - Relationship between flow and effective pressure for a constant opening.

ROCK DISCONTINUITIES

FIELD TESTS

Field water pressure tests were performed using a conventional equipment with the arrangment shown in Figure 6.

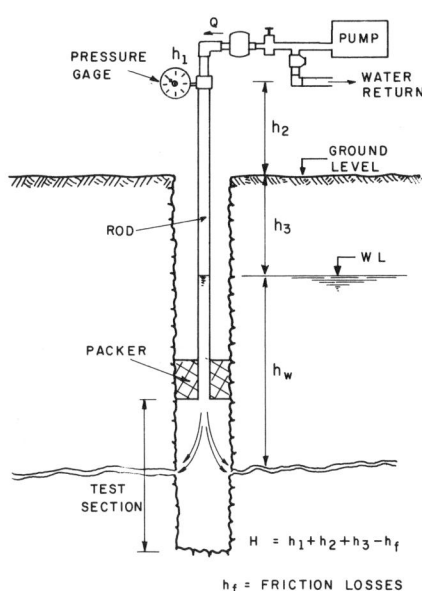

Fig. 6 - Arrangement of the water pressure test using a single packer. Case of a single facture in the test section.

Multiple pressure stages were used both on the loading and unloading cycles to improve the information on the nature of flow. Corrections were made for all the head losses, common in these tests, to compute the effective pressure $\gamma_w H$. Additional infiltration tests were also performed for the low pressure range.

Six water tests were performed on different areas of the basaltic joint.

A typical test result is shown in Figures 7 and 8 below relating flow with effective pressure.

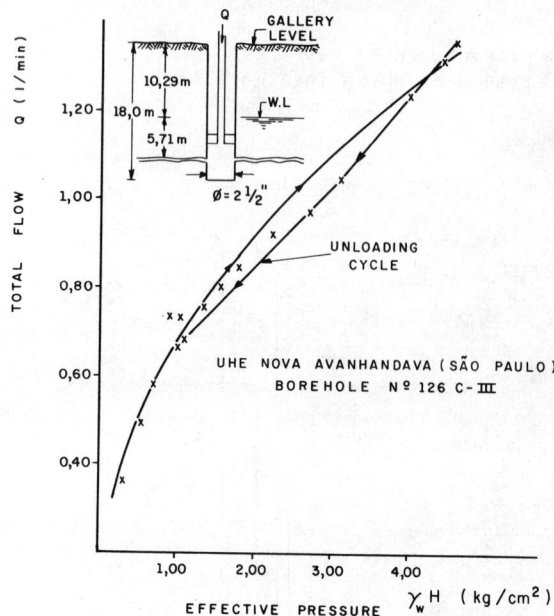

Fig. 7 - Relationship between Q and $\gamma_w H$ (field test).

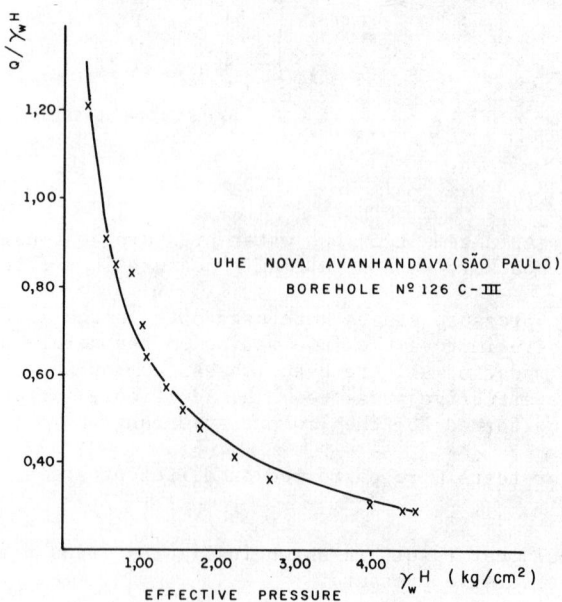

Fig. 8 - Relationship between $Q/\gamma_w H$ and $\gamma_w H$ (field test).

ROCK DISCONTINUITIES

When a linear relationship between flow and effective pressure exists the ratio $Q/\gamma_w H$ should be constant.

JOINT DESCRIPTION

The joint description was based on a carefull examination of bore hole samples, integral samples and extensive exposed areas where the escavations reached the joint level. This joint is composed of a portion of extremely fractured to very fractured rock (more than 10 fractures per meter). In the fractured zones the joints separate angular boulder-like fragments of compacted-textured basalt. The surface may be weathered or covered by rests and films of white and light yellow alochtonous clay. Some stretches of extremely fractured rock up to ten centimeters may be associated with the fractured zones. The walls are covered by thin films of white and light yellow clays or silty and clayey materials provenient of the rock weathering. In some areas the soft material is marked by striae similar to slickensides. Calcite may be present as a cement of the basaltic fragments.

The geometry of the joint is typical of the known fault-joints of basaltic rocks with a nominal width ranging from a few centimeters to 1 meter.

A model sketch of the joint is shown in Fig. 9.

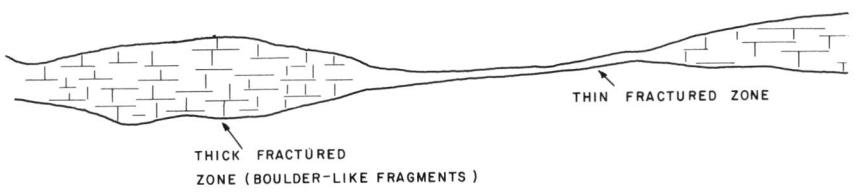

THICK FRACTURED ZONE (BOULDER-LIKE FRAGMENTS)

THIN FRACTURED ZONE

Fig. 9 - Model sketch of the joint.

ANALYSIS, INTERPRETATION AND CONCLUSIONS

Field water pressure test results, should be ploted in figures relating flow (Q) versus effective pressure ($\gamma_w H$) and the ratio of flow (Q) and effective pressure ($\gamma_w H$) versus pressure ($\gamma_w H$) as shown in Figures 7 and 8.

The nature of flow can be identified. A linear ratio of $Q/\gamma_w H$ versus $\gamma_w H$ indicates a "laminar flow". A transition type of flow follows up to a turbulent flow, when Q is related to the gradient J by an equation of the type $Q = f(J^\alpha)$ where α tends to 0.5. This was very clear on laboratory tests.

A confrontation of laboratory tests with field tests is possible, considering the joint description and assuming a pattern of rugosity for the joint. Compare Fig. 4 with Fig. 7 for example.

This comparison will lead to an "equivalent opening" if the field joint was continuous and open.

In the present case, an "equivalent opening" of 0,60 to 0,70 mm was obtained. Openings can be related to conductivity considering the flow regimen, but known formulae.

It is very important on the field tests to analyse the first stages of the curves were the flow regimen could aproach a linear laminar flow, because the normal technique leads very easily to either a transitional or turbulent flow with completely different characteristics.

Predictions of flows into large excavations, or into drains in the foundations of concrete structures, can be underestimated if the flow regimen is not corrected evaluated.

For flow prediction where well formulae, or modified well formulae are used, it could be usefull to introduce the conception of transmissivity (T = Ke) both for field and laboratory tests, due to the normal difficulties in determining both the opening and the conductivity of natural joints. Once flow is a function of T and the gradient (or the effective pressure), it could be enough to compute T for the different flow regimens that will develop in the water pressure test.

In conclusion, the proposed methodology seems to be useful in the interpretation of water pressure tests, but more research is under way to improve the present analysis.

Any discussions and contributions to this paper are very wellcome by the authors.

We want to express our acknowledgment for the IPT and CESP for the opportunity of presenting this small contribution on such a controversial area of water flow in rock discontinuities.

BIBLIOGRAPHY

ASSOCIAÇÃO BRASILEIRA DE GEOLOGIA DE ENGENHARIA (ABGE) - 1975 - Ensaio de perda d'água sob pressão; diretrizes. São Paulo. 16 p. (ABGE, Boletim 02).

CRUZ, P.T. - 1979 - Contribuição ao estudo do fluxo da água em meios contínuos. São Paulo, IPT (pré-print).

LOUIS, C. - 1969 - Étude des ecoulements d'eau dans les roche fissurées et de leurs influences sur la stabilité des massifs rocheurs. Bulletin de la Direction des Études et recherches. Série A (3): 5-132.

(Thèse presenteé a l'Université de Karlsruhe).

QUADROS,E.F. - 1982 - Determinação das características do fluxo de água em fraturas de rochas.São Paulo.(Dissert.de Mestrado,Escola Politécnica - Universidade de São Paulo).

RISSLER,P. - 1978 - Determination of the water permeability of jointed rock. Aachen, Institute for Foundation Engineering and Soil Mechanics.

SHARP,J.C. - 1970 - Fluid flow through fissured media.London (Ph D. Thesis, Imperial College of Sciences).

Chapter 79

EXTERNAL DISPLACEMENT METHOD FOR DETERMINING
THE IN-SITU DEFORMABILITY OF ROCK MASSES

Rodolfo V. de la Cruz

University of Wisconsin-Madison
Madison, Wisconsin

ABSTRACT

The in situ deformability of rock masses is determined by relating the applied load to the radial displacements of points on the borehole wall that are outside of the loaded surfaces.

The external displacement method includes the steps in which: a) radial displacement sensors are installed to measure the change in length of a diameter of a borehole; b) unidirectional self-equilibrating pair of forces perpendicular to the direction of radial displacement measurements are applied on opposite sectors of the borehole wall, and; c) using elastic theory, the in-situ deformability of the rock mass can be related with the applied loads and measured radial displacements.

The displacements outside the loaded surfaces are comparatively smaller in magnitude than displacements inside the loaded surfaces. These displacements, however, are sufficiently large and can be measured precisely by commercially available linear displacement sensors. Loads are applied by a conventional borehole jack.

Existing methods for measuring the in situ deformability of rock masses yields unreliable results due to complex boundary stresses, small and varying contact areas, and the existence of fractures. It was shown that in the external displacement method, the effect of boundary pressure distribution and contact area are negligible. Also, since the displacements measured in the external displacement method were due to materials that are deeper inside the rock mass, borehole fractures have less influence on the calculated deformation modulus.

EXTERNAL DISPLACEMENT METHOD

The external displacement method may be used by itself or combined in the same device with the NX borehole jack method and the modified Goodman jack method. As a multiple device, more accurate and reliable deformation modulus will be obtained, allows checking of other measurements, and provide a means of quantifying the degree of fissuring on the borehole walls.

INTRODUCTION

Information on the in situ deformability of rock masses is critical to the design and construction of structures on or in rock. In situ deformability values are necessary input for numerical modeling of rock structures. In situ deformability values are also required in calculating stresses from observed strains or deformations. In situ stresses, magnitudes and orientations of either absolute or change of stress, are vital in assessing the stability of underground openings.

There are a number of methods and techniques for measuring and/or estimating the in situ deformability of rock masses (1). Unfortunately, no one single method or even combinations of a number of methods are capable of obtaining reliable and accurate in situ deformability values (2). Each method yields a wide data scatter and substantial standard deviations (3). Variations between different methods are also significant. The primary reason for the discrepancies within and between methods is the presence of various types of discontinuities in the rock mass. These discontinuities affect the loading conditions, stress distributions, deformations, strains or any other parameters that are applied or measured in order to determine the in situ deformability of the rock mass. While discontinuities can be modeled and their effect on the in situ deformability can be determined, the mapping and defining of discontinuities especially those away from the borehole is difficult if not impossible (4).

This paper presents a new technique for determining the in situ deformability of rock masses by measuring the displacements of points on the borehole wall that are outside of the loaded areas. The instrumentation is described, the method of operation is discussed, and the theoretical basis of the method is derived using the principles of the classical theory of elasticity. Factors that may affect data interpretation such as boundary pressure distribution, contact area and existence of fractures were shown to have either negligible or minimum effects on the accuracy of the method. The advantages of the method were pointed out and its use in conjunction with other existing methods would allow duplication and checking of measurements.

EXTERNAL DISPLACEMENT METHOD

Principle of Operation

The in situ deformability of rock masses are usually determined by relating an applied load to the measured displacement of the loaded area. For example, in the NX-borehole jack method, unidirectional loads are applied on opposite sectors of a borehole through curved bearing plates (5). The displacements of these plates are measured and then related to the applied loads in order to determine the in situ deformability of the rock materials surrounding the borehole. Similarly, in the CSM cell method, the radial displacements of all points on the borehole wall are integrated to obtain an aggregate volume change of the borehole due to hydrostatic loading (6). Using elastic theory, the modulus of rigidity can be determined from the hydrostatic load and the measured volume change which is related to the displacements.

In the external displacement method, loads are applied in one area of the rock material while the displacements of the rock mass outside the loaded area are measured. The deformation modulus of the rock mass can be obtained by relating the displacements outside the loaded areas with the applied loads. The method maybe performed easily at the surface, e.g., in conjunction with conventional plate bearing test, Figure 1. Using a linear displacement sensor, the surface displacement away from the loaded surface can be measured and used to determine the deformation modulus of the material. A number of sensors located around and at various distances away from the load would yield multiple data for checking and/or profiling the surface.

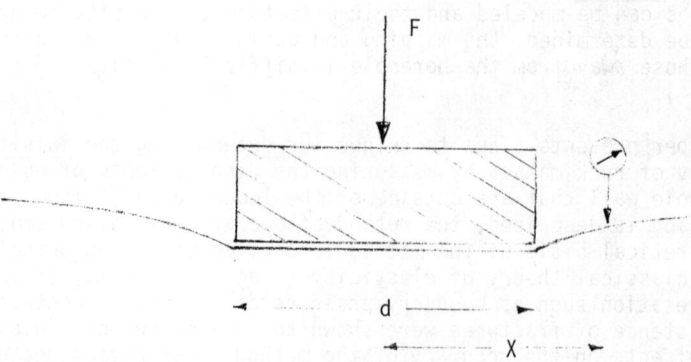

Fig. 1 External Displacements in the Plate Bearing Test.

EXTERNAL DISPLACEMENT METHOD

The external displacement method would be especially suitable for application in boreholes, thus allowing measurements of in-situ deformability at deeper locations and avoiding the effects of artifically induced fractures and other uncertainties at or near the surface. Figure 2 shows a transverse cross-section of a borehole and two curved bearing plates on opposite quadrants of the borehole. The bearing plates are forced apart hydraulically to load and deform the borehole approximately into an ellipse. Radial displacement sensors that are oriented perpendicular to the direction of the load and installed prior to jack loading, would measure the inward displacement of the walls during loading and the outward displacement during unloading. Using elastic theory, the deformation modulus of the rock mass can be determined from the measured displacements outside the loaded areas and the applied loads.

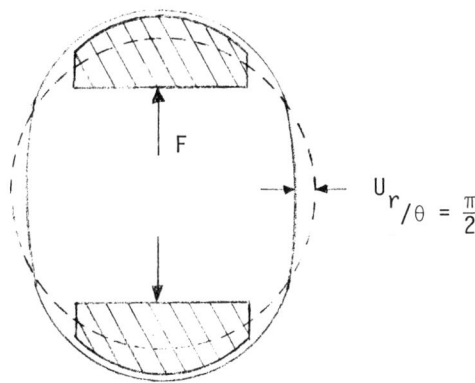

Fig. 2. Application of the External Displacement Method in a Borehole.

Instrumentation

<u>Borehole Jack</u>. The device for applying a unidirectional load on opposite quadrants of a borehole is similar to the well-known Goodman Jack with the bearing plates lengthened to six times the diameter of the borehole so that a state of plane strain may be assumed at the central region of the jack, Figure 3a. With the lengthened jack, the effects of the end of the jack on the stress distribution at the central region would be negligible.

Radial Displacement Sensor. The radial displacement of the borehole wall with jack loading may be measured by conventional linear displacement sensor. While any commercially available gage may be used, a miniature linear variable differential transformer (LVDT) is selected since it satisfies the size, range and sensitivity requirements expected in rock mass measurements, Figure 3b. With the LVDT core attached to one piston and the LVDT coil attached to the other, the pistons are retracted by air pressure during insertion in a borehole. At the point of measurement, the spring-loaded pistons are released until they contact the borehole walls at a uniform pressure. Any radial displacements of the borehole walls will be detected by the sensors. For more accurate measurements, the sensors are supported by flexible materials, such as by rubber O-rings, so that any motions of the jack during loading or unloading do not interfere with the measurements of the sensors.

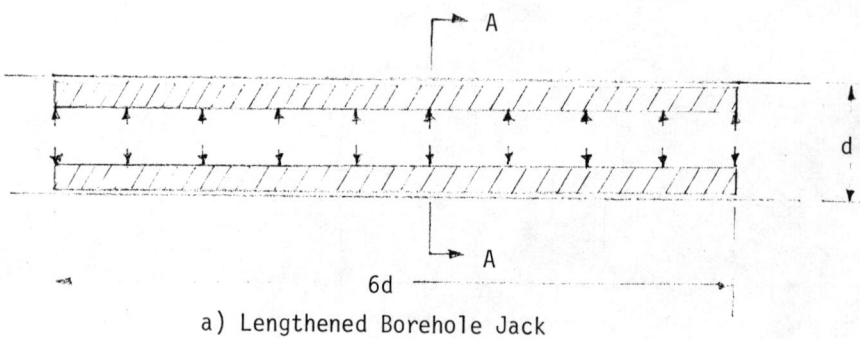

a) Lengthened Borehole Jack

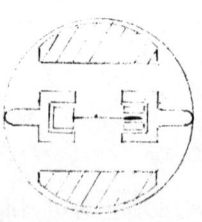

Section A-A
b) Radial Displacement Sensor

Fig. 3. Instrumentation for the External Displacement Method.

Theoretical Basis

The deformability of the rock mass surounding the borehole maybe determined from the applied pressure, Q, and the measured radial displacements, U_r, by using equations that have been derived previously (5,7). The boundary stress distribution with a uniform, unidirectional stress on opposite sectors of a borehole wall are shown in Figure 4.

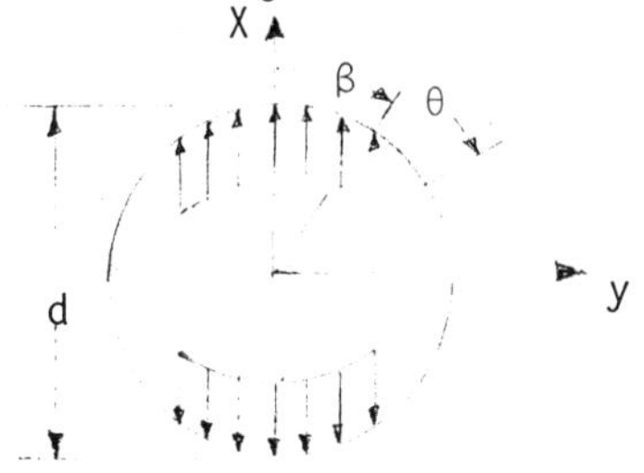

$$\sigma_x = \begin{cases} Q, & -\beta \leq \theta \leq \beta \\ & \pi - \beta \leq \theta \leq \pi + \beta \\ 0, & \beta < \theta < \pi - \beta \\ & \pi + \beta < \theta < 2\pi - \beta \end{cases}$$

Fig. 4 Uniform Boundary Stress Distribution at The Plate/Wall Interface.

Using the principles of the theory of elasticity, e.g., stress function method or complex variable method, the general expression for the radial displacements is given by (5):

$$U_d = -\frac{dp}{2G\pi} \{2\beta [1 + (3-4\nu) \cos 2\theta]$$

$$- \sum_{m=1}^{\infty} \frac{1}{m} \sin 2m\beta \; [\frac{(3-4\nu)}{2m+1} \cos 2(m+1) \theta$$

$$+ (\frac{3-4\nu}{2m-1} + \frac{1}{2m+1}) \cos 2m \theta$$

$$+ \frac{1}{2m-1} \cos 2(m-1) \theta]\} \tag{1}$$

where:

U_d = radial displacement of the borehole wall,

p = Q/2, half of the uniaxial pressure applied by the borehole jack,

2β = subtended angle of the bearing plate,

$G = \frac{E}{2(1+\nu)}$, modulus of rigidity of the rock mass where E and ν are the Young's modulus and Poisson's Ratio of the rock

θ = orientation of a point on the borehole measured from the direction of loading (x-axis),

d = diameter of the borehole.

When Equation (1) is evaluated at $\theta = \pm \pi/2$ corresponding to the points that are 90 degrees away from the direction of loading, the radial displacement becomes:

$$U_d \Big|_{\theta = \frac{\pi}{2}} = \frac{2d\ p(1-2\nu)}{G\ \pi} \left[\beta - \sum_{m=1}^{\infty} \frac{\cos m\pi \sin 2m\beta}{(m)(2m+1)(2m-1)} \right] \quad (2)$$

Equation (2) can now be used to determine the in-situ deformability of the rock mass by the external displacement method, or:

$$E = \frac{2(1+\nu)(1-2\nu)\ d\ Q}{\pi\ U_{d\big|_{\theta = \frac{\pi}{2}}}} \left[\beta - \sum_{m=1}^{\infty} \frac{\cos m\pi \sin 2m\beta}{(m)(2m+1)(2m-1)} \right] \quad (3)$$

For $\beta = \frac{\pi}{4}$, half of the included angle of the bearing plate adopted in the method, the equation becomes simplified:

$$E = \frac{2(1+\nu)(1-2\nu)d\ Q}{3\ U_{d/\theta = \frac{\pi}{2}}} \quad (4)$$

Magnitude of Displacements

The displacements of the rock mass away from the loaded areas are smaller than those in the loaded areas. If these displacements are extremely small then more sensitive instruments will be required to measure the displacements. This would mean less precision and accuracy in the measurements and the method will be more difficult to perform and less desirable. It is, therefore, instructive to compare the magnitudes of the displacements, that are measured in the external displacement method to those measured by other methods, e.g. the average unidirectional displacement under the bearing plate which is the theoretically predicted displacement for the NX borehole jack method. Previous calculation for the NX borehole jack method shows:

$$\bar{U} = K(\nu,\beta)\ d\ \frac{Q}{E} \quad (5a)$$

where: $\bar{U}$ = unidirectional displacement in the direction of the load.

$K(\nu,\beta)$ = stress factor

Assuming $\nu = 0.25$ and $\beta = \frac{\pi}{4}$;

$$\bar{U} = -1.254 \, d \, \frac{Q}{E} \qquad (5b)$$

The ratio of the displacements measured by the external displacement and the NX borehole jack method will then be:

$$\left| \frac{U_{d/\theta = \pm \frac{\pi}{2}}}{\bar{U}_{d/-\beta \leq \theta \leq \beta}} \right| = \frac{1}{3} \qquad (6)$$

Equation (6) indicates that the external displacement method measures one third of the displacement measured by the NX borehole jack method. Since the displacements measured in the NX borehole jack method are relatively large, existing commercially available sensors including those used in the NX borehole jack method would be fully adequate for use in the external displacement method.

FACTORS AFFECTING DATA INTERPRETATION

Pressure Distribution and Area of Contact

A primary concern with the NX borehole jack method is radius mismatch between the bearing plates and the borehole wall (7). Due to radius mismatch, the stress distribution at the plate/wall interface will not be uniform as required in the derivation of the theoretical basis of the method. Radius mismatch also mean that the entire area of the bearing plate will not be in full contact with the borehole wall. Laboratory tests in large blocks of aluminum showed that the included angle of the contact area vary from about 7° to about 17° for pressures varying from 1,000 to 10,000 psi applied hydraulic pressure. Thus, even at the maximum pressures, the contact area are considerably less than the assumed contact value of $2\beta = 90°$. Because of non-uniform pressure distribution and reduced area of contact, the displacements measured by the NX borehole jack method are, in general, much larger than the theoretically predicted values and the calculated deformation moduli are consistently much lower than the true values. Empirically and numerically derived correction factors have been applied in order to obtain more reasonable estimates of the *in situ* deformability values. The correction factors, however, are sensitive to moduli variations, the slope of the curve is very steep, and the reliability and accuracy of the NX borehole jack method is severely reduced.

The external displacement method utilizes a similar jack for loading the borehole walls. It is, therefore, informative to determine whether the radial displacements measured by the external displacement method are sensitive to non-uniform pressure distribution and variable contact area at the plate/wall interface. Previous analysis carried out for the modified Goodman jack method have shown that the stresses at points away from the loaded surfaces are essentially independent of the pressure distribution and area of contact at the plate/wall interface (8). Figure 5 shows a general (non-uniform) boundary pressure distribution at the plate/wall interface.

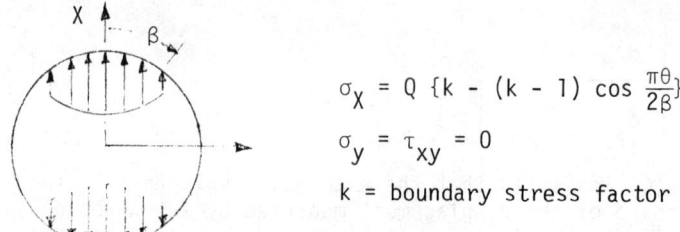

$$\sigma_x = Q \{k - (k-1) \cos \frac{\pi\theta}{2\beta}\}$$

$$\sigma_y = \tau_{xy} = 0$$

k = boundary stress factor

Fig. 5 Nonuniform Boundary Stress Distribution at the Plate/Wall Interface.

The pressure distribution is defined by a boundary stress distribution factor, k, which is the ratio of the stress at the edges ($\theta = \pm \beta$ and $\pi \pm \beta$) to the centers ($\theta = 0$ and π) of the bearing plates. The expression for the tangential stress, $\sigma_{\theta\theta}$, in terms of the total load, F, exerted by the bearing plates on the borehole wall is given by the following equation:

$$\sigma_{\theta\theta} = K(k, \beta, \theta) \frac{F}{a} \qquad (7)$$

Numerical analysis have shown that at $\theta = \pm \pi/2$ and when the contact angle $\beta \leq \pi/4$, the stress factor K (k, $\beta,\pi/2$) is practically constant for any value of the boundary stress distribution factor, k. Therefore, since K(k, $\beta,\pi/2$) is practically constant whatever the contact angle (less than or equal to that used in the borehole jack) and whatever the pressure distribution including concentrated line loading, then the tangential stress, $\sigma_{\theta\theta}$, at $\theta = \pm\pi/2$ is a function of the total load, F, only. With the tangential stress constant, the displacements at the same points will then be independent of the pressure distribution and contact area at the plate/wall interface.

Longitudinal Bending of the Bearing Plates

The location of measurement along the length of the bearing plate is critical since the stresses and displacements are affected by the ends and bending of the bearing plate. In the NX borehole jack method, the displacements are measured at the ends of the bearing plate and, therefore, correction factors are required. If measurements were conducted at the center of the bearing plates such corrections will not be necessary. In the external displacement method, the bearing plates were lengthened to six times the diameter of the borehole (in contrast, the length of the original NX borehole jack is only 2 2/3 times the diameter of the borehole) and measurements are also conducted at the midpoint of the bearing plates. For the external displacement method, therefore, end effects and longitudinal bending of the bearing plates will not be a factor.

Effect of Fractures on Measurements

Cracks or fractures at the borehole wall and in the rock mass invalidates idealized assumptions and results in large data scatter and substantial variations in the <u>in situ</u> deformability values obtained by existing methods. This is because rock mass response at the loaded surfaces are much more influenced by surface rocks compared with the deeper, less disturbed and less fractured rock materials (9). Previous studies in plate loading tests showed that about 80 percent of the plate displacement were due to materials within $Z/a = 4$, where Z is depth under the load and a is the radius of the plate, Figure 6. In contrast, the study showed that 80 percent of the displacements measured outside of the loaded surface were due to materials within $Z/a = 10$.

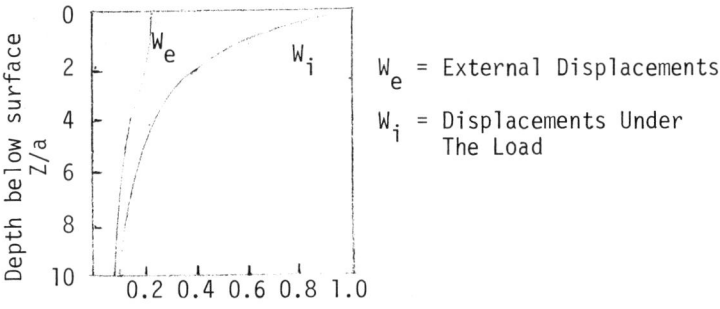

W_e = External Displacements

W_i = Displacements Under The Load

Relative displacement, W/W max

Fig. 6 Variation of displacement in the rock (after Waldorf, et al)

If this observation can be extrapolated to a borehole, then measurements involving the loaded surfaces are substantially influenced by surface fractures. In the external displacement method where measurements are outside of the loaded surfaces, the effect of surface fractures and irregularities on the calculated modulus would be much less. More important, since the displacements measured are contributed by much larger volume of rock, 10 d instead of 4 d only, the external displacement method provides a better assessment of the deformation modulus of the rock mass.

CONCLUSIONS

The instrumentation, measurement procedure, theoretical basis and the factors affecting the accuracy of measurements of the external displacement method for determining the in situ deformability of rock masses has been presented. Preliminary laboratory measurements in large aluminum blocks yielded accurate moduli values. Further tests in large rock specimens and various rock types in the field are required in order to fully validate the method.

While the external displacement method can be used by itself, more accurate and reliable measurements may be obtained when the method is combined in the same device with existing methods such as the NX borehole jack method and the modified Goodman jack method. The multiple simultaneous measurements involving the average displacements under the bearing plate, tangential strains and radial displacements outside the loaded areas would yield multiple, redundant data from which, using statistical techniques, a more precise estimate of the in situ rock mass deformability can be determined. The multiple measurements would also allow comparison and checking of the accuracies of each method as well as the validity of idealized elastic and continuity assumptions used in the analysis. The most important benefit from multiple measurements would be the quantification of the degree of fissuring in the rock mass. Current practice involves qualitative, empirical estimates of the degree of fissuring based on a number of geological observations. With multiple measurements, however, comparison of the various measurements may indicate the degree of fissuring in the rock mass.

REFERENCES

Schrauf, T.W., and H.R. Pratt, 1979, "Review of Current Capabilities for the Measurement of Stress, Displacement, and In Situ Deformation Modulus, ONWI Report 95.

Bieniawski, Z.T., 1978, "A Critical Assessment of Selected In Situ Tests for Rock Mass Deformability and Stress Measurement", Proc., 19th U.S. Symposium on Rock Mechanics.

de la Cruz, R.V., M. Karfakis and K. Kim, 1982, "Analysis of Displacement and Strain Data for the Determination of the In Situ Deformability of Rock Masses", AIME Annual Meeting, Dallas, Texas.

Goodman, R.E., 1967, "Analysis of Structures in Jointed Rock", U.S. Army Engineer District, Omaha Tech. Report No. 3.

Goodman, R.E., T.K. Van, and F.E. Heuze, 1968, "Measurement of Rock Deformability in Boreholes", Proc., 10th U.S. Symposium on Rock Mechanics.

Hustrulid, W., and A. Hustrulid, 1973, "The CSM Cell - A Borehole Device for Determining the Modulus of Rigidity of Rock", Proc., 15th U.S. Symposium on Rock Mechanics.

Hustrulid, W., 1976, "Analysis of the Goodman Jack", Proc., 17th U.S. Symposium on Rock Mechanics.

de la Cruz, R.V., 1978, "Modified Borehole Jack Method for Elastic Property Determination in Rocks", Rock Mechanics, vol. 10.

Waldorf, W.A., J. Veltrop and J. J. Curtis, 1963, "Foundation Modulus Tests for Karadj Arch Dam", J. Soil Mech. Found. Div., ASCE, vol. 89, July.

Chapter 80

SPATIAL DISTRIBUTION OF DEFORMATION MODULI
AROUND THE CSM/ONWI ROOM, EDGAR MINE, IDAHO SPRINGS, COLORADO

by A. Wadood M. A. El Rabaa[1], William A. Hustrulid[2], William F. Ubbes[3]

[1]Graduate Student, and [2]Professor, Department of Mining,
Colorado School of Mines, Golden, Colorado

[3]Program Manager, Office of Nuclear Waste Isolation (ONWI),
Battelle Memorial Laboratories, Columbus, Ohio

ABSTRACT

The spatial distribution of the deformation modulii around the CSM/ONWI test facility was determined using an NX version of the CSM cell. Approximately 840 modulus measurements were made in six rings (each ring containing seven holes each 6 m in length) of boreholes. These parallel rings were spaced at 2.4 m intervals along the room. Contour plots of the modulii revealed that the depth of the blast damage zone varied both with perimeter position and distance away from the opening. The maximum depth of disturbance appeared to be of the order of 60 cm. At locations removed from the opening, the variability in modulus was quite high reflecting the presence of joints, faults, shear zones and foliation. A comparison of rock mass modulii (E_R) with those obtained from the laboratory testing of core (E), revealed that $E_R \simeq 0.5\ E$.

INTRODUCTION

The Colorado School of Mines (CSM) under sponsorship of the Department of Energy (DOE) through the Office of Nuclear Waste Isolation (ONWI) has established a hard rock research facility at its Experimental Mine near Idaho Springs, Colorado. The facility is being used to develop mining, geologic and geotechnical procedures which can be applied to underground excavations made in hard rock as well as to provide a geotechnical data base for modelling studies. The scope of the program has been described by Hustrulid, et al. (1).

As part of the program, a room 30 m long, 5 m wide and 3 m high was driven in granitic gneiss using careful blasting techniques. The evaluation of the extent and degree of blast induced damage to the

DEFORMATION MODULI AT EDGAR MINE, CO

rock surrounding the room was of high priority. To accomplish this the pattern of NX (75 mm) diameter boreholes shown in Figure 1 were diamond drilled at six locations along the room. The spacing between rings was approximately 2.4 m (8 ft.) with each ring located about midway along each of six blast rounds. The cores obtained as well as the boreholes provided the opportunity to apply a number of blast damage techniques, including:

- core evaluation;
- borehole wall examination;
- permeability measurements;
- cross hole ultrasonic measurements;
- geophysical measurements;
- borehole modulus observations.

This paper presents some of the modulus results obtained using an NX version of the CSM cell. Another paper in this symposium will present some of the permeability results.

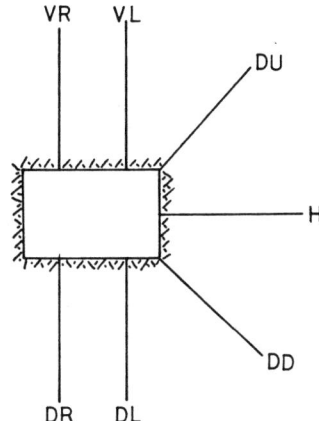

Figure 1. Diagrammatic representation of one of the rings of NX boreholes.

THE CSM CELL

The CSM cell (3, 4) is a borehole device for obtaining the modulus of rigidity (G) of rock.

Although originally developed for use in an EX (38 mm) diameter borehole, an NX (76 mm) version has been built and used for these tests. The active length of the two versions is the same (16.5 cm).

The basic elements of the system (Figure 2) are as follows.

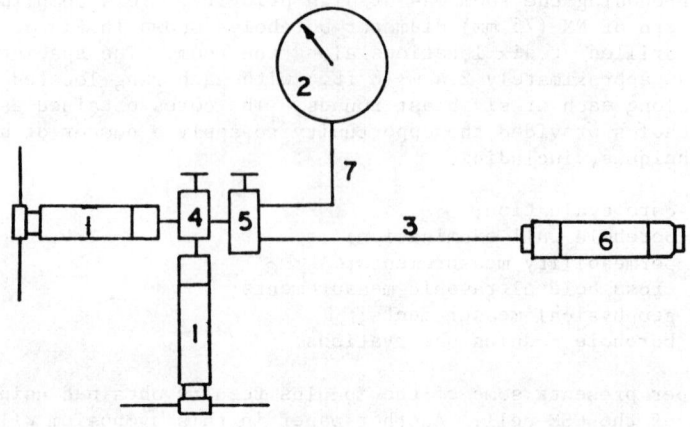

Figure 2. The components of the CSM-cell system.

1) Pressure generator (with vernier indicator) rated at 34 MPa (5,000 psi) with a fluid capacity of 60 cm^3 (3.66 in.^{3}).

2) Bourdon gauge rated at 0 to 41.3 MPa (6,000 psi) with a valve for bleeding the Bourdon tube. The pressure readings can be read with an accuracy of $\pm$ 35 kPa (5 psi).

3) Twelve (12) meters (40 ft.) of 3.18 mm (1/8") by 1 mm (0.14") ID high pressure stainless steel tubing, with short pieces of 6.35 mm OD stainless steel tubing 206 MPa (30,000 psi) silver soldered over the ends of the 3.18 mm (1/8") tubing in order to increase the strength and reliability of the connections.

4) Three-way high pressure valve rated at 206 MPa (30,000 psi).

5) Three-way high pressure valve rated at 206 MPa (30,000 psi).

6) CSM borehole cell.

7) Short stainless steel nipples 6.35 mm (1/4") OD and 2.11 mm (0.083") ID rated at 206 MPa (30,000 psi).

A diagrammatic representation of the NX version of the CSM cell is shown in Figure 3.

Two screw type pressure generators are used. Each pressure generator can displace 0.705 cm^3 (0.043 in.^{3}) of fluid per turn. A mixture consisting of one part of permanent antifreeze (ethyleen-glycol base) and five parts of water by volume has been used as pressurizing fluid.

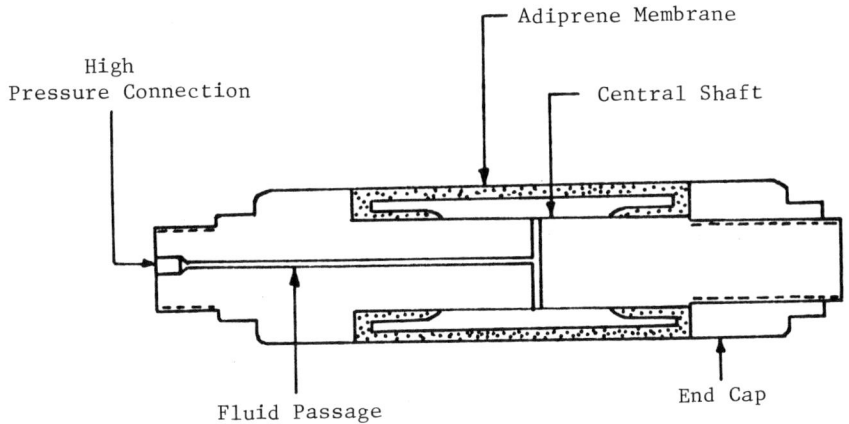

Figure 3. Diagrammatic representation of the CSM-cell.

Listed below are the procedures followed when evaluating rock mass modulii using the CSM cell.

1) The system should be at the test site long enough to reach ambient temperature.

2) The cell is calibrated in a calibration cylinder of known elastic properties before insertion in the borehole. This calibration gives the stiffness (pressure-volume relationship) of the pressurizing system plus the calibration cylinder. Since the stiffness of the calibration cylinder can be calculated, the stiffness of the pressurization system (M_s) alone can be found.

3) The cell is then inserted into the borehole with testing commencing at the hole bottom. As the cell is withdrawn from the hole tests are made at intervals of 30 cm (1 ft.). The relationship between the pressure and the volume (M_T) is determined for each position in the borehole.

4) After the borehole testing has been completed, the cell is inspected for fluid leakage or any air bubbles introduced into the cell membrane. The cell is recalibrated using the calibration cylinder.

5) The calibration slopes (before and after testing the rock) are compared. If the difference between the two slopes is within $\pm$ 70 kPa (10 psi)/turn, the averaged slope will be used in the calculations. A larger difference in the calibration slopes (yet within acceptable limits) will be

linearly distributed over the number of the tests to find the value of M_m which corresponds to each testing position.

6) The stiffness (M_R) of the rock surrounding the borehole is then calculated for each position using

$$M_R = \frac{M_T M_s}{M_s - M_T}$$

The modulus of rigidity (G_r) can be calculated by

$$G_r = M_R \frac{\pi L r_i^2}{\gamma}$$

where

M_R = rock mass stiffness (MPa/turn)
L = length of pressurization (16.5 cm)
r_i = borehole radius (cm)
γ = volume expelled/turn of the pressure generator (0.705 cm^3/turn)
G_R = modulus of rigidity (MPa)

If Poisson's ratio (ν_R) is known or can be estimated, the rock mass modulus (E_r) can be calculated by

$$E_r = 2(1 + \nu_R) G_R$$

where

E_r = rock mass modulus
ν_R = Poisson's ratio

CSM CELL RESULTS

Modulus measurements were made at 30 cm (1 ft.) intervals in each hole of every ring. Since the average hole depth was about 6 m, approximately 840 individual measurements were made. All the results are available in reference (5), and only the detailed results for Ring #1 (which are typical of all rings) will be presented here (Table 1 and Figure 4). Table 2 presents average results for each ring at each depth as well as overall averages. From the contour plots one can conclude that

- The room is surrounded by a zone of lower modulus of the order of 21 GPa (3 x 10^6 psi) which has an average thickness of 30 cm (1 ft.). This lower modulus zone was noticed to be thicker (max. 1.1 m (3.5)) near the corners of the room in

DEFORMATION MODULI AT EDGAR MINE, CO

Table 1. Rock Mass Modulii (10^6 psi) as a Function of Depth and Position in Ring No. 1.

Depth (ft)	H	VR	VL	DR	DL	DU	DD
19	-	-	-	-	-	-	-
18	-	4.01	-	-	-	-	-
17	-	3.52	-	-	-	-	-
16	-	4.87	3.65	-	-	-	-
15	-	4.90	5.09	5.41	-	-	-
14	5.2	4.89	6.80	5.42	-	6.32	5.3
13	6.19	4.90	5.35	5.01	5.22	6.13	4.29
12	5.35	5.31	4.83	4.9	4.78	4.23	4.68
11	5.32	5.21	2.34	5.78	4.12	6.5	5.29
10	5.31	5.22	5.48	7.2	4.18	6.4	4.63
9	5.94	5.01	5.35	5.90	6.65	6.41	3.64
8	3.90	4.70	2.84	5.8	3.5	6.33	4.28
7	6.3	4.20	7.40	4.9	2.82	5.38	4.87
6	5.46	3.4	8.00	5.82	3.5	5.64	5.39
5	5.03	4.86	8.20	5.90	5.9	6.37	3.45
4	5.91	4.63	5.90	4.55	7.5	4.23	7.2
3	4.97	2.81	4.30	4.61	3.04	6.13	7.09
2	6.80	5.80	2.06	4.75	5.02	2.87	4.38
1	6.10	3.95	3.07	4.56	5.02	3.17	1.66

*To obtain the modulii in GPa multiply the table values by 6.9. To obtain distances in meters multiply by 0.305. These conversions should also be used for Table 2 and Figure 4.

Table 2. Average Modulii as a Function of Ring Position and Depth.

Depth (ft)	Averaged E_r/ft/Ring						E_r average for each depth
	Ring #1	Ring #2	Ring #3	Ring #4	Ring #5	Ring #6	
19	-	3.34	3.01	-	-	4.19	3.51
18	4.01	2.88	2.63	5.95	3.00	4.25	3.79
17	3.52	3.14	2.25	5.39	3.00	3.94	3.54
16	4.26	3.42	2.27	5.24	2.50	3.80	3.58
15	5.13	3.66	2.83	4.89	3.06	3.85	3.90
14	5.65	4.05	4.44	4.56	4.04	4.27	4.50
13	5.29	4.93	4.54	5.07	4.16	4.18	4.70
12	4.86	5.98	4.08	4.91	3.65	4.68	4.69
11	4.94	5.34	3.99	5.10	4.66	4.07	4.58
10	5.49	4.80	3.76	5.22	4.18	4.08	4.59
9	5.47	5.05	3.34	5.64	4.42	3.83	4.63
8	4.55	5.00	3.09	5.08	4.56	3.91	4.37
7	5.12	5.38	3.82	4.55	4.12	3.98	4.50
6	5.32	4.95	4.14	4.68	4.36	4.01	4.58
5	5.67	5.04	4.34	4.58	3.30	4.50	4.57
4	5.70	5.66	4.17	6.99	3.54	4.16	5.04
3	4.71	4.63	4.87	5.42	4.15	4.14	4.65
2	4.52	4.58	4.34	4.28	3.95	3.96	4.27
1	3.93	4.46	3.07	5.34	3.05	3.83	3.95
E_r/Ring	4.89	4.54	3.63	5.16	3.73	4.09	4.34
St.Dev.	0.63	0.86	0.77	0.61	0.59	0.22	0.57

*If the values of E_r for deeper depths than 15 ft. are excluded, because only they represent the up vertical holes, then the averaged E_r for the whole test is $(4.54 \pm .23) \times 10^6$ psi.

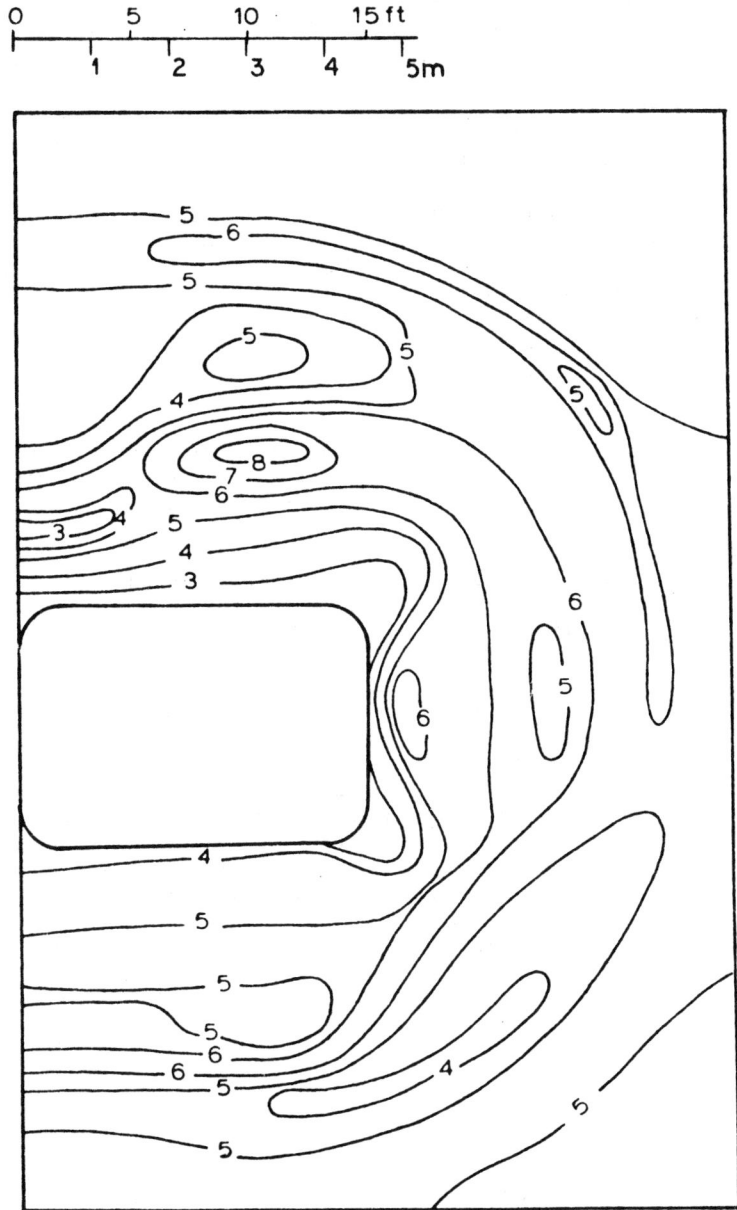

Figure 4. Rock Mass Mdoulus Distribution in Ring #1. The modulii are given in 10^6 psi. To convert to GPa multiply the contour values by 6.9.

some rings (ring 1, ring 2, ring 3) and the bottom corner in ring 4. One explanation for this phenomena can be attributed to drilling and blasting problems during the excavation. For rings 1 and 2 the hole deviations in the corner holes was large. For ring 3, a set of intersecting joints determined the blasting results. For ring 4, excessive hole deviation resulted in reshooting of the contour holes.

- A zone of higher modulus 41.4 GPa (6 x 10^6 psi) appears in the horizontal holes of rings 1, 3, 4, 5, and 6. This zone occurs at a distance from the opening ranging from 30 cm (1 ft.) (ring 1) to 1.5 m (5 ft.) (ring 5). As this higher modulus zone falls within the zone of increased stress due to the presence of the opening, it is suggested that the two are related. The higher stresses would close fractures and, hence the modulus would increase.

- A statistical evaluation reveals that rings no. 3, and no. 5, were located in more fractured sections than the others, and ring no. 4 is in a less fractured section.

- At locations removed from the opening, there is a considerable variability in modulus reflecting the jointed nature of the rock and the presence of structural features (shear zones, faults, foliation, etc.).

ROCK MASS MODULUS VS INTACT MODULUS

Samples having length-diameter ratios of 2:1 were prepared from the cores collected during the drilling of the ring holes. These samples were loaded in compression using a 100 tonne (220,000 lb.) capacity MTS servo-controlled universal testing machine. Values of the modulus of elasticity (E), Poisson's ratio, and the compressive strength were obtained. It was found that when using average values

$$E_R = 31 \text{ GPa } (4.5 \times 10^6 \text{ psi}) \text{ \{average of 640 values\}}$$

$$E = 61 \text{ GPa } (8.9 \times 10^6 \text{ psi}) \text{ \{average of 97 values\}}$$

the ratio E_R/E becomes

$$\frac{E_R}{E} = 0.51$$

Of the 97 samples tested in the laboratory, 60 were from locations with a corresponding CSM cell test result. The 60 pairs of E_r, E values have been plotted in Figure 5. As can be seen, all the points fall under the $E = E_r$ line which means that E_r was always less than its corresponding value of E. The average ratio of E_r/E for each pair, was found to equal to .51, with a standard deviation of 0.17.

A histogram representation of the E_r/E ratio was constructed using the above data. Figure 6 shows the frequency distribution (expressed as a percent) of the E_r/E ratio. It can be seen that, the ratio data was normally distributed around the mean. Twenty five (25 percent of the data fell between 0.5 and 0.6.

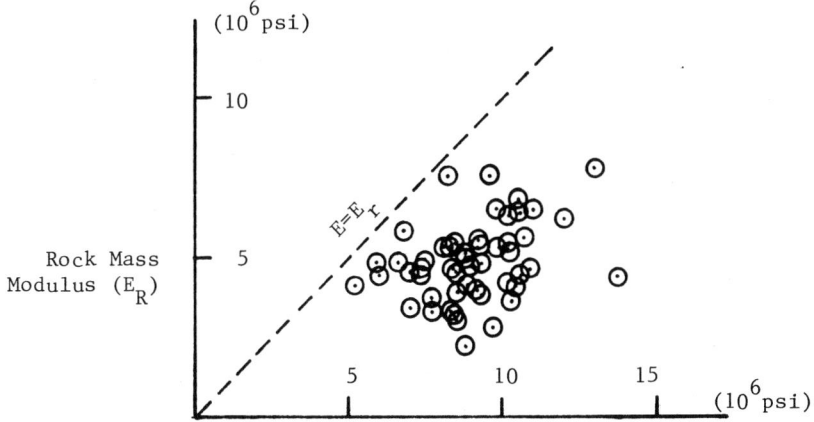

Figure 5. Rock mass modulus as a function of intact modulus.

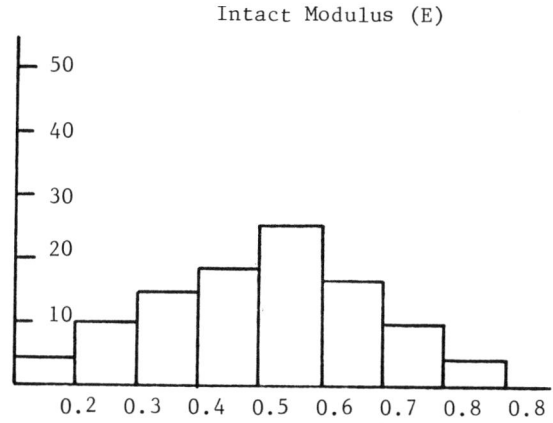

Figure 6. Histogram showing the percent of data pairs having a given E_r/E ratio.

CONCLUSIONS

The NX version of the CSM cell has proven to be a simple, reliable borehole device for determining the deformation modulus of the rock surrounding an underground opening. The results from the CSM/ONWI test facility suggest that

- the modulii in the near vicinity of the opening are affected both by blasting damage and an increase in the stress field. The first effect results in a modulus reduction, whereas the second through the closing of joints, cracks, and fissures would tend to increase the modulus. The superposition of these two effects would tend to off set one another.

- the near surface blast damage varies with position around the drift. The maximum thickness of the blast damaged zone appears to be of the order of 60 cm (2 ft.)

- away from the opening, there is considerable variability in modulus reflecting the influence of structure (joints, faults, shear zones, foliation).

- the rock mass modulus as determined using the CSM cell is of the order of one half that obtained from laboratory tests on cores.

NOTICE/DISCLAIMER

This report was prepared as an account of work sponsored by the United States Government. Neither the United States nor the Department of Energy, nor any of their employees, nor any of their contractors, subcontractors, or their employees, makes any warranty, express or implied, or assumes any legal liability or responsibility for the accuracy, completeness, or usefulness of any information, apparatus, product, or process disclosed, or represents that its use would not infringe privately-owned rights.

ACKNOWLEDGMENTS

The financial support for the work reported was through contract E512-04800 with the Project Management Division of the Battelle Memorial Institute, Columbus, Ohio.

REFERENCES

1. Hustrulid, W., Cudnik, R., Trent, R., Holmberg, R., Sperry, P. E., Hutchinson, R., and P. Rosasco, "Mining Technology Development for Hard Rock Excavation", Sub-Surface Space (Proceedings of Rockstore 80, Stockholm, Sweden, June 23-27, 1980), volume 2, pp. 916-926.

2. Montazer, P., Chitombo, G., King, R. M., and W. F. Ubbes, "Spatial Distribution of Permeability Around the CSM/ONWI Room, Edgar Mine, Idaho Springs, Colorado", Proceedings 23rd U.S. Symposium on Rock Mechanics, Berkeley, California, August, 1982.

3. Hustrulid, A. and W. Hustrulid, "CSM Borehole Device - Assembly, Calibration and Use", Final Report to U.S.B.M. on Extension to Contract HO 101705 (Parts I and II), July, 1972.

4. Hustrulid, William and Andrew Hustrulid, "The CSM Cell - A Borehole Device for Determining the Modulus of Rigidity of Rock", Applications of Rock Mechanics, ASCE, 1975.

5. El Rabaa, Abdel Wadood, M.A., "Measurements and Modelling of Rock Mass Response to Underground Excavation", M.S. Thesis, Colorado School of Mines (T-2470), December, 1981.

Chapter 81

MEASURING THE THERMOMECHANICAL AND TRANSPORT PROPERTIES
OF A ROCKMASS USING THE HEATED BLOCK TEST

E. Hardin*, N. Barton, M. Voegele*, M. Board*, R. Lingle, H. Pratt*,
W. Ubbes**

Terra Tek, Inc.
Salt Lake City, Utah

*Science Applications, Inc.
Salt Lake City, Utah

**Office of Nuclear Waste Isolation
Columbus, Ohio

A 2 m. cube of jointed, Pre Cambrian biotite gneiss was subjected to uniaxial and biaxial loading at ambient and elevated temperature. The effects of different boundary conditions on the following rockmass properties were investigated: deformation modulus, joint stiffness, dynamic modulus, coefficient of thermal expansion, thermal diffusivity and conductivity, and joint permeability. Test conditions ranged between 0 and 6.9 MPa uniaxial and biaxial load, and 12° to 74°C mean block temperature. The tests were performed as part of the Office of Nuclear Waste Isolation program for evaluating site charterization test methods for measuring the hydro-thermal-mechanical properties of rock masses. The fracture flow experiments discussed in this paper were described in detail in a similar paper given to this symposium in 1981 (Voegele, 1981).

DEFORMATION MODULUS

Horizontal deformation measurements oriented parallel to the loading direction were made at the block surface. Several instrument types were employed: surface short-rod extensometers, bonded strain gauges, vibrating wire strain gauges and the Whittemore caliper strain gauge. Two vertical and two slant borehole extensometers measured vertical displacement. Loading history, especially during uniaxial loading, was critical to interpretation of observed deformation and fracture flow data. This important tendency has been observed in other large-scale laboratory (Bandis, 1980) and field tests (Pratt, et al., 1977). Excavation of the block resulted in minimal disturbance to the intrinsic condition of the sample. Even so, permanent, unrecovered "set" was imparted to the block beginning with the first low-stress biaxial load cycle to 3.45 MPa (Table 1). Unloading moduli suggested a joint rebound effect, whereby joints remained interlocked until the applied stress fell below a threshold level. At elevated temperature, modulus was particularly high on unloading, as determined from measurements on the block surface. Deformation modulus measured across the full 2 m. block dimension by boundary crack monitoring (Figure 1) indicated similar values for ambient and elevated temperature. Improved closure of joints with temperature apparently counteracted the thermal expansion effect. The major difference in deformability associated with elevated temperature was increased loading hysteresis caused by coupled stress and thermal cycles (Table 1).

Table 1. Mean Values of Block Deformation Modulus (GPa) Derived from Boundary Crack Monitoring.

Test Conditions	Mean Block Temp. (C)	Loading 0-3.45	(MPa) 3.45-6.9	Unloading 6.9-3.45	(MPa) 3.45-0
Biax	12	28.1	(low stress test cycle)		37.6
Biax	12	50.2	18.6	34.7	17.8
NS-Unix	12	19.3	20.6	26.1	18.3
EW-Unix	12	14.7	35.6	28.8	17.1
Biax (load)	12	32.5	13.8	(load, then heatup)	----
Biax (cycle)	50	----	33.6(2)	47.3(1)	----
Biax (cycle)	74	----	52.2(2)	47.7(1)	----
Biax (unload)	16	(cooldown, then unload)		96.0	16.9
Biax (cycle)	50	32.6(1)	----	----	30.8(2)

A distribution of modulus values was obtained, as represented by the accompanying histogram (Figure 2). Observed modulus ranged widely with measurement gauge length and location on the block surface. This distribution agrees reasonably well with that proposed by Bieniawski (1978) which gives mean and minimum values of 31 and 12.5 GPa, respectively. The block was significantly stiffer parallel to foliation than perpendicular to it. Poisson ratio values obtained from measurements spanning several joints, during biaxial and uniaxial loading, ranged from -1 to +2 (Table 2). The negative values came from the second uniaxial load cycle which immediately followed and was transverse to the first such cycle.

Rockmass anisotropy in the well-foliated gneiss was particularly evident from the numerous Whittemore strain measurements and the USBM borehole deformation gauge. The ratio of EW (perpendicular to foliation) to NS (parallel) deformation moduli, determined from the average of 22 Whittemore strain measurements, was ~1.2. The maximum and minimum borehole deformations measured by the single USBM BDG, in response to biaxial loading at ambient temperature, were in the ratio of 6.5. Plane strain orthotropic analysis of this reponse (Hooker and Johnson, 1969) suggests that the ratio of the moduli is app. 8:1. Since this is highly unlikely, it must be assumed that the readings were affected by a structural feature associated with foliation, striking EW. Similar anisotropic behavior, of appreciable magnitude, was noted from the response of an IRAD stress meter rosette located 71 cm from the USBM BOG.

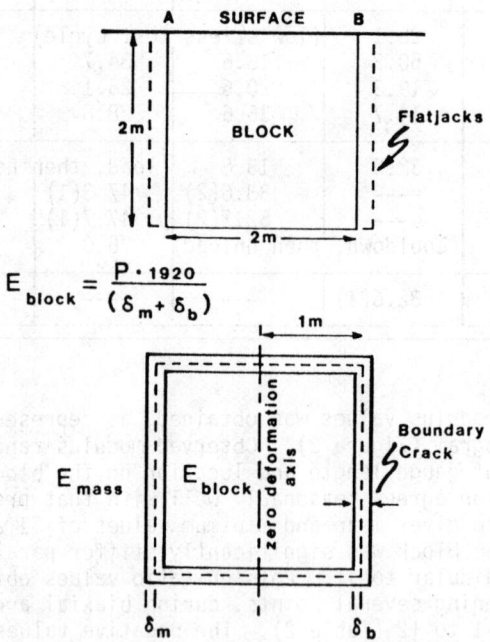

Figure 1. Computational scheme for reduction of block modulus from measurements of boundary-crack aperture (δ_m & δ_b) during load cycling. (Poulos and Davis, 1973)

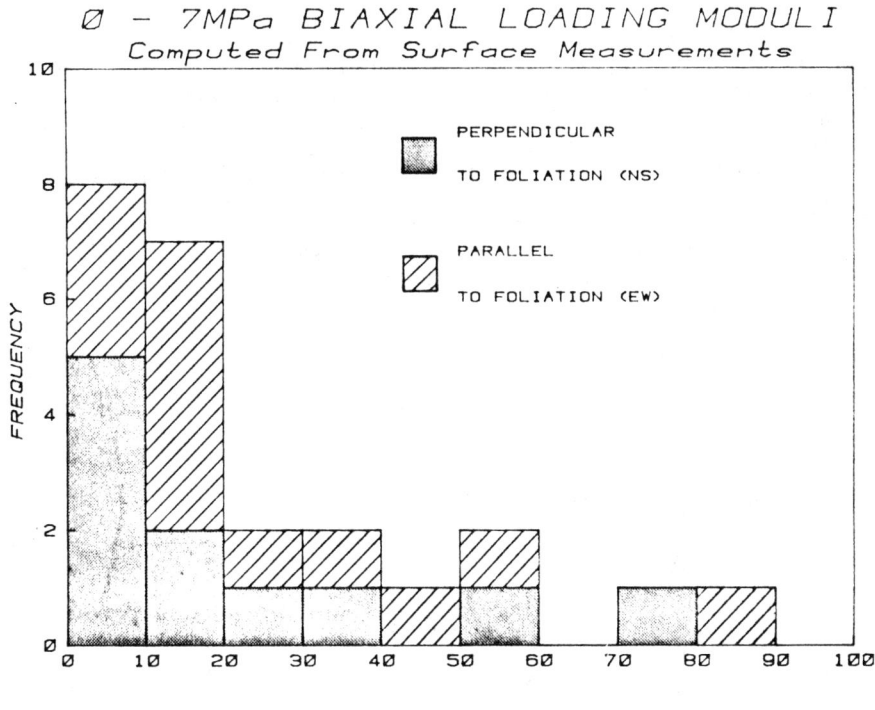

Figure 2. Distribution of modulus values, from 0-6.9 MPa biaxial loading at ambient temperature. (All oriented surface measurements combined)

Table 2. Mean Poisson Ratio Values - Strain Data from Whittemore Gauge Surface Measurements and Vertical Extensometers. Behavior Determined from Loading, 0 to 6.9 MPa.

Loading Direction	Poisson Ratio, From Transverse Strains Oriented:		
	NS	EW	Vertical
Biaxial	----	----	.246
NS Uniaxial	----	.245	.063
EW Uniaxial	-.441	----	.071

JOINT STIFFNESS

Joint deformation measurements were concentrated on the main diagonal "study" joint, where permeability experiments were conducted. Highly variable normal displacements were measured across the joint during biaxial loading. Shear displacement measured during uniaxial loading was more coherent, and successfully correlated with flow tests. Shear which occurred upon initial uniaxial loading was only partially recovered by unloading and subsequent shear reversal. The normal and shear displacements of the study joint induced by uniaxial loading are presented as Figures 3 and 4. Observations of normal joint closure showed that the foliation joints were more "gapped" and less stiff than the diagonal joints.

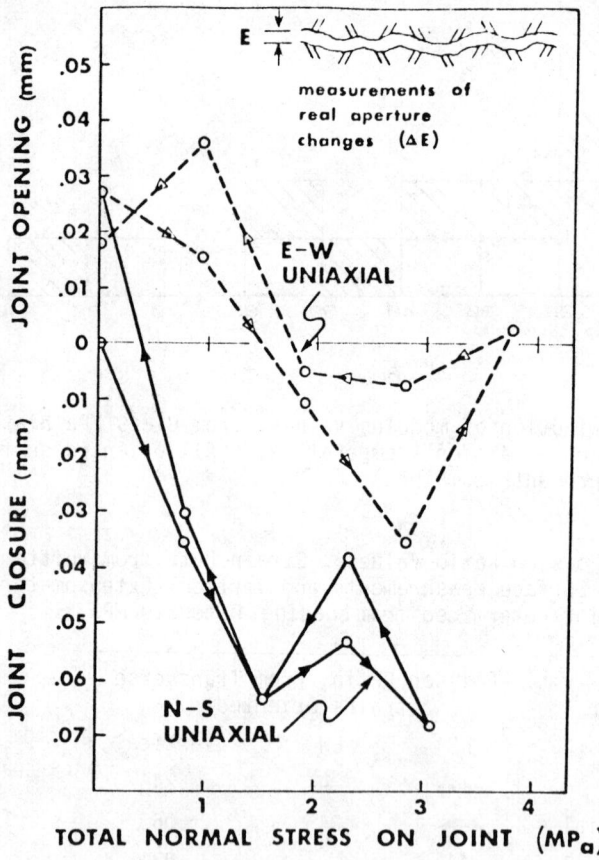

Figure 3. Measured changes in real aperture of the study joint (E), in response to uniaxial loading at ambient temperature. (Compare to Figures 4 & 9)

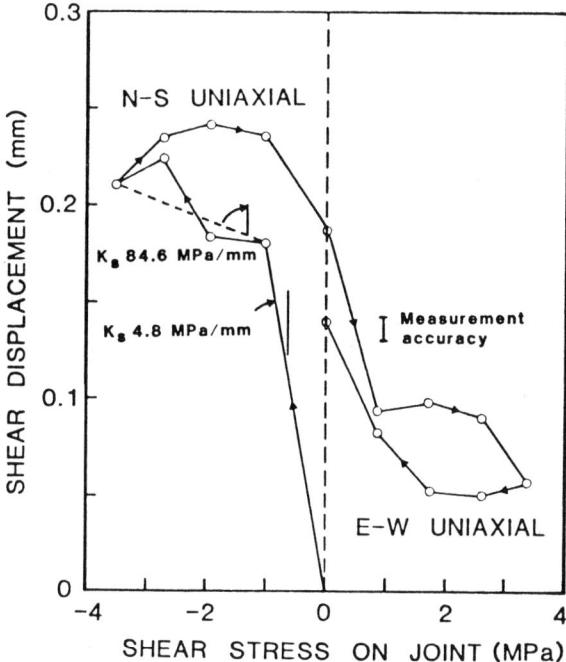

Figure 4. Measured shear displacement of the study joint, in response to uniaxial loading at ambient temperature.

Loading moduli were generally low (<25 GPa) at biaxial stress levels under 3.5 MPa. Unloading moduli above this stress level were characteristically high. The largest discrepancy between loading and unloading moduli was noted when a thermal cycle intervened. After cooldown, when the influence of expansion was absent, but the joints remained tightly closed, the modulus was exceptionally high (96 GPa) on initial unloading to 3.5 MPa, but was dramatically reduced (16.9 GPa) in the final stage of unloading. Numerous Whittemore strain measurements across mapped joints showed average overall ambient and elevated temperature moduli of 16.9 GPa perpendicular, and 19.3 GPa parallel to foliation. The NS uniaxial load cycle was the first shearing event, and it was resisted along the study joint by an initial shear stiffness of 4.8 MPa/mm. The stiffness increased by a factor of at least 20 as shear displacement increased to a maximum of about 0.2 mm.

A shear failure envelope for the study joint was developed from joint profile, mineralization and Schmidt hammer studies in the test room (Hardin, et al., 1981). Comparison with measured shear displacements showed the stiffness to exceed that predicted, suggesting that joint behavior was affected by the continuous base of the block.

DYNAMIC MODULUS

Biaxial loading to 6.9 MPa generally caused a 5 to 20 GPa increase in dynamic modulus. At elevated temperature, there was a marked increase of 10-30 GPa in the range of modulus values in the E-W direction, parallel to foliation. Modulus typically increased with depth, although not always. Upon the initial application of uniaxial load, a significant degradation of dynamic modulus in the transverse direction was observed. When the shear stress was reversed on the next load cycle, the average dynamic modulus in the transverse direction was only slightly lower than the loading direction. This is presumably due to the re-closing of joints during reversed shear.

THERMAL EXPANSION

The thermomechanical properties of the block were influenced by temperature in several ways. Thermal strains were measured by the surface and borehole instruments mentioned above, especially the short-rod extensometers, Whittemore strain gauge and vertical extensometers. The temperature field was measured by an array of 52 thermocouples in 16 boreholes, and an additional 25 thermocouples installed with the heaters and borehole extensometers (Figure 5).

Strain meters oriented parallel to the loading axes indicated some initial reductions in the thermal expansion coefficient (α) with increasing temperature, a tendency also noted in laboratory tests with confining pressure (Figures 6, 7). In the unconfined vertical direction "α" increased significantly with increasing temperature. The average value of "α" was approximately twice as high perpendicular to foliation as parallel to it, as determined from surface observations. Overall, measured thermal strains were more consistent with respect to measurement gauge length, instrument type and location and were easier to measure than deformation response to loading. The distribution of values obtained for "α", under 6.9 MPa biaxial confinement and over the entire temperature range is presented in Figure 8.

THERMAL DIFFUSIVITY AND CONDUCTIVITY

The temperature field produced by the line of borehole heaters which bisected the block was largely one-dimensional within the block itself. Best fit values for conductivity and diffusivity were obtained with three dimensional heat flow models based on the program JUDITH (St. John, 1977), using known values for density and specific heat of the gneiss. The borehole heaters were considered as an array of point heat sources for analysis, and heat flux at the block surface

HEATED BLOCK TEST

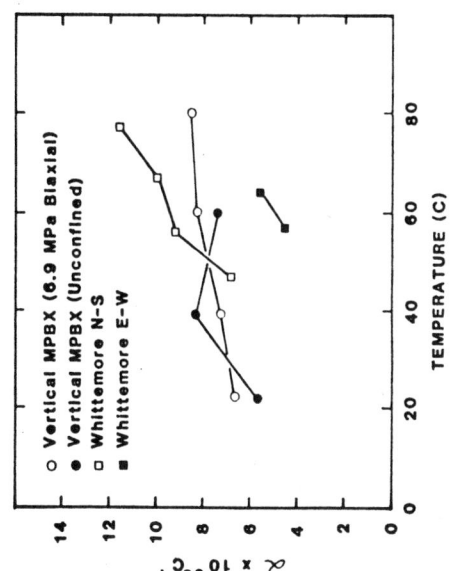

Figure 6. Thermal expansion coefficient as a function of temperature, for several orientations. Whittemore readings from all thermal cycles were averaged for this figure.

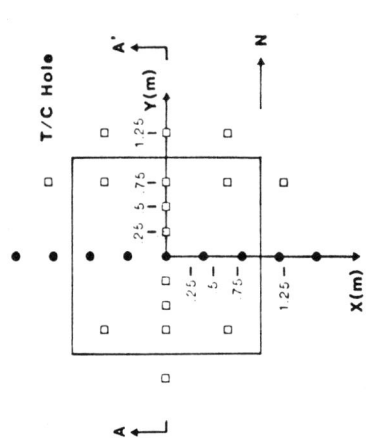

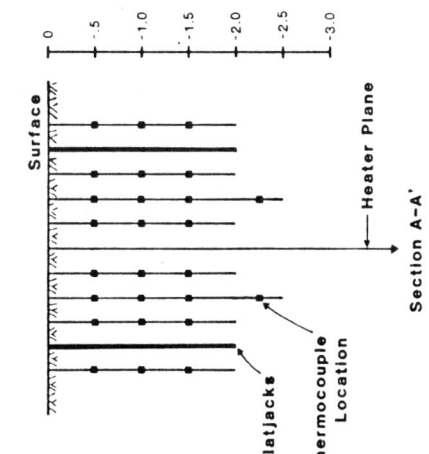

Figure 5. Schematic of heater boreholes and thermocouple array.

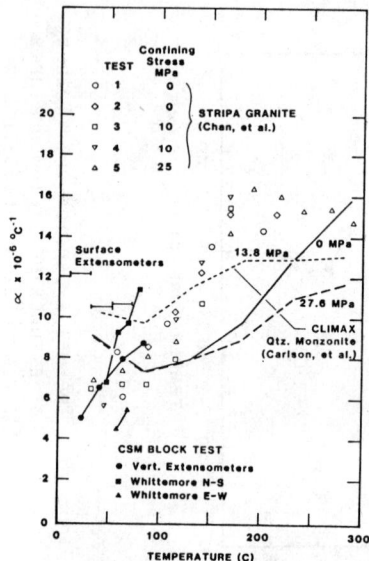

Figure 7. Thermal expansion coefficient as a function of temperature, comparing the heated block test to lab results from the Stripa and Climax programs.

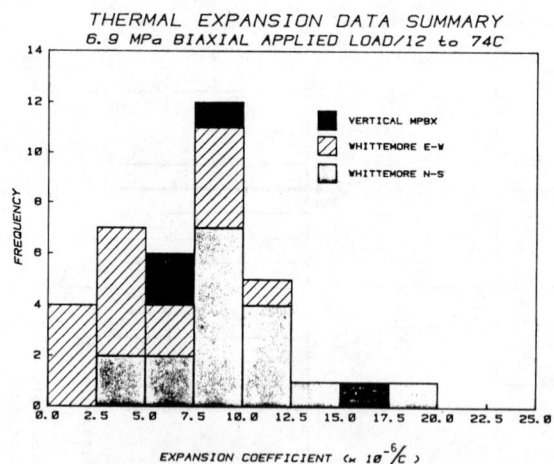

Figure 8. Distribution of thermal expansion measurements under 6.9 MPa biaxial confinement, and over the entire temperature range.

was assumed to be zero. The calculated temperature field within and around the block was forced into agreement with the measured temperatures by adjustment of diffusivity or conductivity, depending on the model used. In this manner the temperature of any part of the block could be calculated with an accuracy of $\pm 2C^o$; this error was attributed to water movement, rock inhomogeneity, heat loss to the test room and experimental uncertainty. The flatjacks caused no apparent perturbation of the temperature field.

Each of the three heating events (step function power increases) was continued until the temperature measured at any point in the block varied by no more than $0.5C^o$/day. The linear thermal gradient, as computed from a least squares fit to all the temperature data, stabilized within seven days after heater power adjustment in every case. The magnitude of this gradient was used to compute thermal conductivity for comparison to more detailed models; the results appear in Table 3. Some trends are present in this data, however, they may lie within the range of experimental uncertainty.

JOINT PERMEABILITY

The study joint was rough, mineralized, and intersected the block diagonally. During biaxial loading at ambient temperature, permeability was reduced by a factor of four at 6.9 MPa. When subjected to shear and reversed shear, the effect of dilation on the permeability of the joint was detected, even for shear displacements as small as 0.25mm. Rock and injection water temperatures were subsequently varied between 12^oC and 74^oC under biaxial load. The results show that coupling of stress and temperature reduced permeability as much as thirty-fold from the initial level. This is interpreted as improved asperity-interlock caused by thermal expansion along the joint profile. Where the joint was unconfined and therefore poorly interlocked, temperature had little effect on permeability. Comparison of measured changes in joint aperture (ΔE) with calculated changes in smooth wall aperture (Δe), suggested that $\frac{E}{e} > 1$ for rough, "closed" joints (Figures 3, 9). This is presumably due to the combined effects of toruosity and roughness, which may be particularly marked for the cased of mineralized joints.

REFERENCES

Bandis, S., 1980, "Experimental Studies of Scale Effects on Shear Strength and Deformation of Rock Joints," Ph.D. Thesis, University of Leeds, Department of Earth Sciences.

Bieniawski, Z.T., 1978, "Determining Rock Mass Deformability: Experience from Case Histories," Int. Journal of Rock Mech. and Min. Sci. and Geomech. Abstr., Vol 15, pp. 237-247.

Table 3. Thermal Conductivity As Calculated From the Transient and Steady State Analyses.

Test Conditions	Diffusivity ($\times 10^{-6}$ m^2/sec.)	K as Inferred from Diffusivity (W/m°K)	Thermal Gradient (°C/m)	K as Determined From 1-d Analysis (W/m°K)
6.9 MPa Biaxial Confinement, 500 W./heater	1.36	3.30	37.2	3.72
6.9 MPa Biaxial Confinement, 700 W./heater	1.61	3.89	51.1	3.90
Unconfined, 500 W./heater	1.55	3.75	36.8	3.77

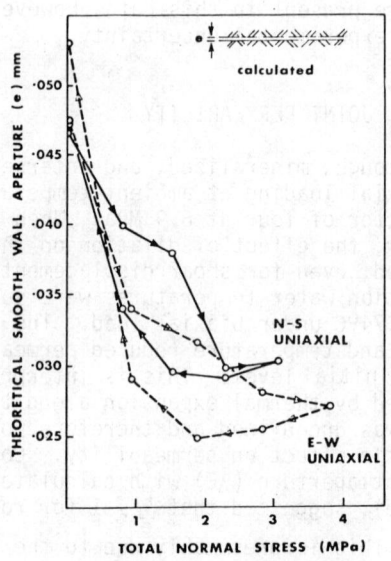

Figure 9. Calculated smooth-wall aperture (from fracture flow tests) vs. normal stress on the study joint, for uniaxial loading at ambient temperature.

Carlson, R.C., et al., 1980, "Spent Fuel Test-Climax: Technical Measurements Interim Report, FY 1980", Lawrence Livermore National Laboratory, UCRL-53064.

Chan, T., M. Hood and M. Board, 1980, "Rock Properties and Their Effect on Thermally Induced Displacements and Stresses," Proc. of Annual Meeting: American Society of Mechanical Engineers, (New Orleans, LA, February, 1980).

Hardin, E., et al., 1981, "A Heated, Flatjack Test Series to Measure the Thermomechanical and Transport Properties of In Situ Rock Masses," U.S. Department of Energy, Office of Nuclear Waste Isolation Publication ONWI-260.

Hooker, V.E. and C.F. Johnson, 1969, "Near Surface Horizontal Stresses Including the Effects of Rock Anisotropy," U.S. Bureau of Mines RI 7224, February.

Poulos, H.G. and E. H. Davis, 1973, Elastic Solutions for Soil and Rock Mechanics, John Wiley and Sons, New York.

Pratt, H.R., et al., 1977, "Elastic and Transport Properties of an In Situ Jointed Granite," Int. J. of Rock Mech. Min. Sci. and Geomech. Abstr., Vol. 14, pp. 35-45.

St. John, C.M., 1977, "Thermoelastic Analysis of Spent Fuel in High Level Radioactive Waste Repositories in Salt - A Semi-Analytical Solution," Office of Waste Isolation, Y-OWI-SUB7118-1, submitted by Univ. of Minn., April.

Voegele, M.D., et al., 1981, "Site Characterization of Joint Permeability Using the Heated Block Test," Proc. of the 22nd U.S. Symposium on Rock Mechanics, M.I.T., Cambridge, Mass.

Chapter 82

THE INFLUENCE OF TEST PLATE FLEXIBILITY ON THE
RESULTS OF CABLE JACKING TESTS

J.K. Jeyapalan
Asst. Prof. of Civil Engineering
Texas A&M University
College Station, Texas, U.S.A.

A.P.S. Selvadurai
Professor of Civil Engineering
Carleton University
Ottawa, Ontario, CANADA.

ABSTRACT

Plate loading tests which use surficial loading of a rock mass are employed quite extensively for the determination of in situ deformability characteristics of rock masses. The cable jacking method is essentially a surficial plate load test in which the test load is provided by reaction against a cable anchored at some depth below the rock surface. The results of the test are usually evaluated within the framework of the classical theory of elasticity to generate estimates for the bulk deformability properties of the rock mass. In the conventional evaluation of this test it is assumed that the test plate is rigid and that the embedded anchor load has no influence on the test plate settlement. The purpose of this paper is to examine the possible influences of the plate flexibility and the anchor load location on the effective settlement of the test plate. The variational estimates presented in the paper are valid for moderately flexible test plates which are characterized by a relative rigidity parameter. Numerical results are developed in the form of correction factors that should be applied to the cable jacking results to arrive at the deformability characteristics of the tested rock mass.

INTRODUCTION

It is generally recognized that in situ methods of testing of rock masses have a proven advantage over other techniques, chiefly those which involve laboratory measurement of deformability and strength properties. A variety of testing techniques such as plate loading, flat jack and borehole dilatometer tests have been used quite effectively for the determination of in situ properties of rock masses (Brown, 1981). In particular the plate load testing of a rock mass enables the determination of the bulk properties of the rock mass which could be influenced by small scale effects such as inhomogeneities, intrusions and microfissures (see,e.g., Jaeger, 1972; Coulson, 1979; Goodman,1980).

TEST PLATE FLEXIBILITY ON CABLE JACKING

As the sizes of these defects increase, the dimensions of the loading plate should be increased so that representative values of the bulk properties of the tested rock mass can be determined. With large plate sizes it becomes necessary to subject the plate to large loads so that the deformations induced in the rock mass can be measured with sufficient accuracy. The cable method of in situ testing provides an effective means for the application of such large loads. The cable jacking test is essentially a surficial plate load test in which the test load is provided by reaction against a steel cable anchored at some depth in a small diameter borehole. The effect of the borehole is usually assumed to be negligible. (When the diameter of the rigid plate is greater than three times the borehole diameter, the presence of the borehole has little effect on the plate settlement; see, e.g., Parlas and Michalopoulos, 1972 and Bandyopadhyay and Kassir, 1978.) The cable method of testing was formally proposed by Stagg and Zienkiewicz (1968). Recently, Selvadurai (1978, 1979a,b,c, 1980) and Selvadurai and Faruque (1981) have made an exhaustive study of the cable jacking test in which factors such as the depth of location of the anchor and its load distribution, transverse isotropy, interface friction and creep effects were investigated. A number of useful results including the critical depth of location of the anchor region (i.e. the depth of the anchor for which there is no interaction between the test plate and the anchor) have been developed in the above studies. These theoretical developments were initiated primarily as a prelude to the development of a standardized test (Selvadurai, 1982).

The present paper examines the problem related to the cable jacking test which is performed with a flexible plate. The problem represents a commonly occurring situation in which the relative rigidity of the test plate - rock mass system is such that the rigid plate approximation is clearly inappropriate. The test plate will experience flexural deflections which depend on the test plate - rock mass relative rigidity, $R = E_p h^3 / E a^3$ (where E_p is the modulus of elasticity of the test plate, E is the modulus of elasticity of the rock, h is the test plate thickness and a its radius). Investigations pertaining to soil-foundation interaction suggest that moderate values (such as $R \to 10^{-1}$) have an appreciable effect on the plate settlement.

The combined interaction problem related to the rock mass, a flexible circular test plate and a concentrated anchor load system is examined, in the context of the classical theory of elasticity, by using essentially a variational formulation. The contact region between the flexible test plate and the rock mass is idealized as a frictionless interface. The settlement of the test plate is evaluated in two parts; firstly a 'rigid' settlement of the test plate due to the internal anchor is evaluated in exact form. The settlement experienced by the test plate due to the flexibility effects are superposed on this rigid settlement. These latter effects are examined by employing a variational formulation in which the plate deflection is approximated by a polynomial in the radial coordinate. The solutions developed in the paper are expected to give accurate estimates when the test plate - rock mass relative rigidity (R) is of the order 10^{-1}. Alternatively, the results could be used to establish plate dimensions and anchor load locations which could minimize errors in interpretation of the test results.

ANALYSIS

We consider the problem of a flexible test plate which is in smooth contact with an isotropic elastic halfspace. The anchor load is represented by a concentrated force which acts along the axis of symmetry (Fig. 1). The analysis of the problem is reduced to the determination of the deflections of the test plate as a function of the location of the anchor load and relative flexibility of the test plate - elastic rock mass system. For the analysis of the problem it is convenient to visualize the settlement of the test plate ($w(r)$) as composed of two parts. First, a 'rigid settlement' w_a of the test plate which is assumed to be caused by the internal anchor load (P_0). The second is a non-uniform settlement, $w_p(r)$, which is caused by the action of the external load (of stress intensity p_0 and radius αa) on the flexible plate. Since the anchor load solution is derived for a rigid settlement of the test plate the perturbation solution for the flexibility effects can be superposed on a plane (horizontal) surface where the point of reference is altered only by a rigid body displacement. This superposition can be made for arbitrary P_0 and p_0. The solution for the cable jacking test will be obtained when $P_0 = \pi\alpha^2 a^2 p_0$.

Settlement of the test plate due to the internal anchor load.

For the analysis of this problem we represent the internal anchor load as a vertically directed Mindlin force (Mindlin, 1936) which acts at a point on the axis of symmetry (Fig. 1). When the force P_0 acts in the -ve z-direction, the surface displacement $u_z^M(r,0)$ of the traction free halfspace is given by

$$u_z^M(r,0) = \frac{P_0(1-\nu)}{2\pi G}\left\{\frac{1}{(r^2+c^2)^{1/2}} + \frac{c^2}{2(1-\nu)(r^2+c^2)^{3/2}}\right\} \quad (1)$$

where G and ν are respectively the shear modulus and Poisson's ratio of the elastic material. We now investigate the problem wherein the region $r \leq a$ and $z=0$ is enforced to undergo a rigid body displacement w_a in the -ve z-direction in the presence of the anchor load (Fig. 2a). This can be visualized as a frictionless bonding in the test plate region which enables the development of tensile contact stresses. The resulting axisymmetric mixed boundary value problem can be examined quite conveniently by employing the complex potential function approach of Green (1949). Owing to limitations of space the details of the method of analysis shall not be presented here. Details of the method are given in Selvadurai (1979a); it is sufficient to note the salient final results here. The rigid displacement in the foundation region is given by

$$w_a = -\frac{P_0(1-\nu)}{4aG}\left\{\frac{2}{\pi}\tan^{-1}\left(\frac{a}{c}\right) + \frac{ac}{\pi(1-\nu)(a^2+c^2)}\right\} \quad (2)$$

The above result represents the surface displacement of a region of a halfspace ($r \leq a$; $z=0$) which is enforced to undergo uniform axial displacement. This uniform settlement of the region $r \leq a$ takes place in the absence of any shear traction on $z=0$. This is a purely

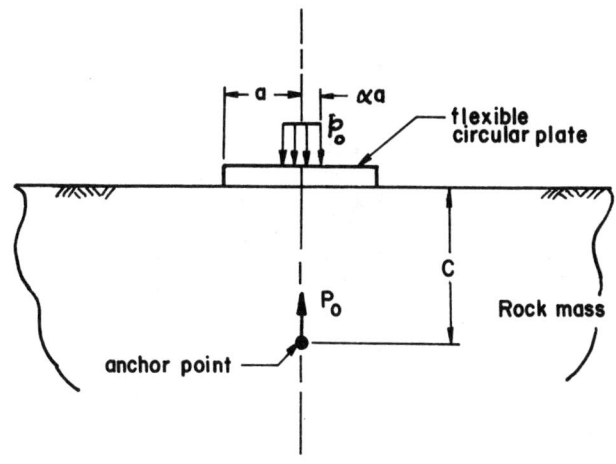

Fig. 1. Definition sketch of the plate-rock system.

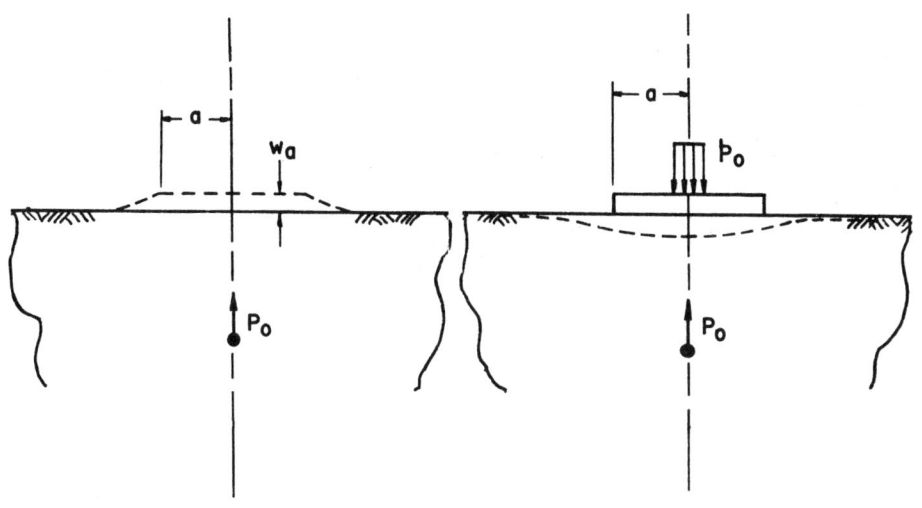

Fig. 2a. Rigid settlement due to P_o

Fig. 2b. Non uniform settlement due to p_o

mathematical constraint which can be physically realized only in the instance when the contact stresses within the test plate region (under the action of both the external and internal loads) are compressive.

Settlement of the flexible plate due to the applied loads.

We assume that the plate is subjected to the uniformly distributed load p(r) (Fig. 2b) when it is placed in the surface of the internally loaded halfspace which is subjected to the surface constraint $u_z(r,0) = w_a$ on $r \leq a$. The results for the flexural interaction analysis will be unaffected by the virtual rigid body displacement w_a. Furthermore, since the theoretical developments are restricted to the classical theory of elasticity, the boundary conditions corresponding to the flexural interaction problem can be assigned to an undeformed surface of the halfspace which does not contain any internal loads.

An extensive account of the flexural interaction problem related to a circular plate resting in smooth contact with an elastic halfspace is given by Selvadurai (1979d). More recently, Selvadurai (1979e) has shown that a solution to the flexural interaction between a uniformly loaded circular plate and an elastic halfspace can be obtained via a variational method. In these developments the flexural deflection of the plate is approximated by a power series of the form

$$w(r) = a \sum_{n=0}^{3} C_{2n} \left(\frac{r}{a}\right)^{2n} \tag{3}$$

The total potential energy functional (U) required for the variational formulation consists of the flexural energy of the plate (U_F), the strain energy of the halfspace region (U_E) and the potential energy of the external load (U_P). It may be noted that any work component associated with the internal load is neglected since an equilibrium solution for the internal force is already established. The total potential energy functional for the plate - elastic halfspace system U (=$U_E + U_F + U_P$) can be evaluated to within four arbitrary constants C_{2i} (i = 0,....3). Two of these constants (C_4 and C_6) can be eliminated by making use of the Kirchhoff boundary conditions applicable to a plate with a free boundary (Selvadurai, 1979d). The remaining constants C_0 and C_2 are determined from the equations generated from the minimization conditions

$$\frac{\partial U}{\partial C_0} = \frac{\partial U}{\partial C_2} = 0 \tag{4}$$

The details of the method of analysis will not be discussed here. The result of primary importance to this paper, namely the deflected shape of the plate, can be expressed in the form

$$w(r) = \frac{P_0(1-\nu)}{4aG} \left(\frac{2}{\eta} \{ \chi_1\chi_4 - 2\chi_2 - R^*\chi_3 - (\chi_1 - 2\chi_4)(\rho^2 + \lambda_1\rho^4 + \lambda_2\rho^6) \} \right.$$
$$\left. - \frac{2}{\pi}\tan^{-1}\left(\frac{a}{c}\right) - \frac{ac}{\pi(1-\nu)(a^2+c^2)} \right) \tag{5}$$

where

$$R^* = \frac{\pi(1-\nu^2)}{6(1-\nu_p^2)} \frac{E_p h^3}{E a^3} \quad ; \quad \rho = r/a \qquad (6)$$

and λ_n, χ_n, η etc., are defined in Appendix 1. Assuming that the settlement of the test plate is measured at the centre of the loaded area (w_0) we have

$$w_0 = \frac{P_0(1-\nu)}{4aG} \left\{ 2\{\frac{\chi_1\chi_4 - 2\chi_2 - R^*\chi_3}{\chi_1^2 - 4\chi_2 - 2R^*\chi_3}\} - \frac{2}{\pi}\tan^{-1}(\frac{a}{c}) - \frac{ac}{\pi(1-\nu)(a^2+c^2)} \right\} \qquad (7)$$

It may be observed that when the anchor location is remote (i.e. c → 0) (7) gives the solution to the problem of the indentation of the rock mass by a smooth flexible test plate. Putting $R^* \to \infty$ in the resulting expression yields Boussinesq's (1885) classical result for the rigid punch. Similarly by setting $R^* \to \infty$, (7) gives the result obtained previously for the interaction between a rigid plate and a Mindlin force. Since the solution to the interaction problem has been derived with w_a = const., for all (c/a), the result (7) is not applicable to extremely flexible plates (R → 0) (i) which undergo deflections that are not representative of (3), or (ii) which are subjected to concentrated loads that give logarithmic singularities in w(r). For this reason any numerical evaluation of (7) will be attempted only to examine the influence of moderate plate flexibility (R > 10^{-1}) on the inferred deformability parameters. The analysis can be extended to reduce these limitations but this can be achieved only at the expense of an inordinate amount of computational effort which may not be justified on practical grounds.

NUMERICAL RESULTS

The results of this study can be best presented as a correction factor Ω to the measured deformability parameter, $\{G\}_{measured}$, which takes into account the plate-foundation relative stiffness $R = E_p h^3/Ea^3$ and the depth of location of the anchor c/a. From (7) it can be shown that

$$\{G\}_{in\ situ} = \{G\}_{measured}\ \Omega(R,\nu,\nu_p,c/a,\alpha) \qquad (8)$$

where $\{G\}_{measured}$ is the value of G obtained by neglecting the plate flexibility and anchor load effects and

$$\Omega = \left\{ 2\{\frac{\chi_1\chi_4 - 2\chi_2 - R^*\chi_3}{\chi_1^2 - 4\chi_2 - 2R^*\chi_3}\} - \frac{2}{\pi}\tan^{-1}(\frac{a}{c}) - \frac{c^2}{\pi(1-\nu)(a^2+c^2)} \right\}^{-1} \qquad (9)$$

Clearly, to determine $\{G\}_{in\ situ}$, it is necessary to assign some values for the unknown parameters R and ν and perform an iterative procedure to arrive at the correction factor (i.e. assuming E_p, ν_p, α

and c/a are specified in a test). The iterative procedure can be summarized as follows: (i) Assume that the plate is rigid and compute G; (ii) Use this value of G to evaluate the relative rigidity R^A; (iii) Obtain Ω^A consistent with the estimated R^A and compute $\{G\}^A_{in\ situ}$ = G Ω^A (iv) Repeat procedure using $\{G\}^A_{in\ situ}$ to derive a new relative rigidity value R^B.

The variations of Ω with c/a, R, α, ν and ν_p are shown in Figs. 3-6. In order to illustrate the use of these charts in evaluating test results obtained by cable jacking tests, a sample application is given as follows. A cable jacking test was performed on a layer of sandstone with a test plate of effective diameter (2a) 0.4 m and thickness (h) 0.08 m (i.e. the plate may be composed of several elements). The load is localized over an area such that α = 0.2. The load-displacement curve at the end of four cycles records a central settlement (w_0) of 0.25 mm when the applied load (P_0) is 3150 kN. If the anchor region is located at a depth of 4 m (c), estimate $\{G\}_{in\ situ}$ for the sandstone assuming that ν = 0.3, ν_p = 0.3 and E_p = 210 x 10^6 kN/m^2. Neglecting the flexibility of the plate

$$\{G\}_{measured} = P_0(1-\nu)/4aw_0$$
$$\cong 11.0 \times 10^6 \text{ kN/m}^2.$$

This gives a relative rigidity R $\cong$ 0.47 ; a corresponding first estimate for Ω is 0.77 (i.e. in Fig. 5, c/a = 20, logR = - 0.33; ν = ν_p = 0.3). A corrected first estimate for G is

$$\{G\}_{in\ situ} = \Omega\{G\}_{measured} \cong 8.5 \times 10^6 \text{ kN/m}^2.$$

CONCLUSIONS

This paper examines the influence of test plate flexibility on the results of cable jacking tests. The approximate variational solution developed here is suitable for test plates which exhibit relative rigidities in the region R > 0.1. The influence of the anchor load is expressed as a far field corrective solution which is evaluated exactly. It is shown that the results can be used to assign a permissible depth for the anchoring point or to incorporate suitable modifications to the measured deformability properties by taking into account effects of plate flexibility. We have examined here only a relatively simple problem which involves a homogeneous rock mass. An extension of this work to include joint deformability characteristics, rock stratification, etc. (see, e.g., Goodman, 1970) merits further investigation.

ACKNOWLEDGEMENTS

The work described in this paper forms a part of the overall programme in In Situ Geomechanics currently in progress at Carleton University, Ottawa, Canada. It is supported by Natural Sciences and

Engineering Research Council of Canada Grant No. A3866. Specific examination of the Cable Method of In Situ Testing was undertaken with a view to formulating an ISRM Standard for this test. These standards are coordinated by Dr. J.A. Franklin (Canada) and Dr. Z.T. Bieniawski (U.S.A.).

REFERENCES

Bandyopadhyay, K.K. and Kassir, M.K., 1978,"Contact problem for solids containing cavities",J. Eng. Mech. Div., Proc. ASCE, Vol. 104, EM6 pp. 1389-1402.

Boussinesq, J., 1885, Application des potentiels, Gauthier-Villars, Paris.

Brown, E.T., 1981, Ed., Rock Characterization, Testing and Monitoring, ISRM Suggested Methods, Pergamon, N.Y.

Coulson, J.A., 1979,"Suggested methods for determining in situ deformability of rock" Int. J. Rock Mech. Min. Sci. & Geomech. Abstr., Vol. 16, pp. 195-214.

Goodman, R.E., 1970, Deformability of Joints, ASTM Spec. Tech. Publ. No. 477, pp. 174-196.

Goodman, R.E., 1980, Rock Mechanics, Wiley, N.Y.

Green, A.E., 1949,"On Boussinesq's problem and penny shaped cracks", Proc. Camb. Phil. Soc., Vol. 45, 251-257.

Jaeger, J.C., 1972, Rock Mechanics and Engineering, Cambridge University Press, London.

Mindlin, R.D., 1936, "Force at a point in the interior of a semi-infinite solid", Physics, Vol. 7, pp. 195-202.

Parlas, S.C. and Michalopoulos, C.D., 1972, "Axisymmetric contact problem for an elastic halfspace with a cylindrical hole", Int. J. Engng. Sci., Vol. 10, pp. 699-707.

Selvadurai, A.P.S., 1978, "The interaction between a rigid circular punch on an elastic halfspace and a Mindlin force", Mech. Res. Comm. Vol.5, pp.57-64.

Selvadurai, A.P.S., 1979a, "The cable method of in situ testing; Influence of the anchor region", Proc. 3rd Int. Conf. Num. Meth. Geomech., Aachen, Vol. 3, pp.1237-1243.

Selvadurai, A.P.S., 1979b, "The displacement of a rigid circular foundation anchored to an elastic halfspace", Geotechnique, Vol.29, pp.195-202.

Selvadurai, A.P.S., 1979c, "Some results concerning the viscoelastic relaxation of prestress in a near surface rock anchor", Int. J. Rock Mech. Min. Sci. & Geomech. Abstr., Vol. 16, pp.309-317.

Selvadurai, A.P.S., 1979d, Elastic Analysis of Soil-Foundation Interaction, Developments in Geotechnical Engineering Vol.17, Elsevier, Amsterdam.

Selvadurai, A.P.S., 1979e, "The interaction between a uniformly loaded circular plate and an isotropic elastic halfspace: A variational approach", J. Struct. Mech., Vol.7, pp.231-246.

Selvadurai, A.P.S., 1980, "The elastic settlement of a rigid circular foundation anchored to a transversely isotropic rock mass", Proc. Int. Conf. Struct. Fdns. on Rock, Vol. 1, pp.23-28.

Selvadurai, A.P.S., 1982, "Measuring rock mass deformability using the cable jacking test". ISRM Suggested Method (in preparation).

Selvadurai, A.P.S. and Faruque, O.Md., 1981, "The influence of interface friction on the performance of cable jacking tests of rock masses", Proc. Symp. on Implementation of Computing Procedures and Stress-Strain Laws in Geotechnical Engineering, Chicago.

Stagg, K. and Zienkiewicz, O.C., 1968, Rock Mechanics in Engineering Practice, Wiley, N.Y.

APPENDIX 1

$$\lambda_1 = -\frac{3(1-\nu_p)}{4(2+\nu_p)} \quad ; \quad \lambda_2 = \frac{1+\nu_p}{6(2+\nu_p)}$$

$$\chi_1 = \xi_0 + \frac{2}{3}(1+\xi_2) + \frac{8}{15}(\lambda_1+\xi_4) + \frac{16}{35}(\lambda_2+\xi_6)$$

$$\chi_2 = \frac{2}{3}\xi_0 + \frac{8}{15}(\lambda_1\xi_0+\xi_2) + \frac{16}{35}(\lambda_2\xi_0+\lambda_1\xi_2+\xi_4)$$
$$+ \frac{128}{315}(\xi_2\lambda_2+\xi_4\lambda_1+\xi_6) + \frac{256}{693}(\xi_4\lambda_2+\xi_6\lambda_1) + \frac{1024}{3003}\xi_6\lambda_2$$

$$\chi_3 = 8 + 32\lambda_1 + \frac{144}{3}\lambda_2 + \frac{128}{3}\lambda_1^2 + \frac{1188}{5}\lambda_2^2 + 144\lambda_1\lambda_2$$
$$- (1-\nu_p)(4+16\lambda_1+24\lambda_2+16\lambda_1^2+36\lambda_2^2+48\lambda_1\lambda_2)$$

$$\chi_4 = \frac{\alpha^2}{2} + \frac{\alpha^4\lambda_1}{3} + \frac{\alpha^6\lambda_2}{4}$$

$$\xi_0 = -2 - \frac{8}{9}\lambda_1 - \frac{16}{25}\lambda_2 \quad ; \quad \xi_4 = \frac{64\lambda_1}{9} - \frac{128\lambda_2}{25}$$

$$\xi_2 = 4 - \frac{32\lambda_1}{9} - \frac{32\lambda_2}{25} \quad ; \quad \xi_6 = \frac{256\lambda_2}{25}$$

$$\eta = \chi_1^2 - 4\chi_2 - 2R^*\chi_3$$

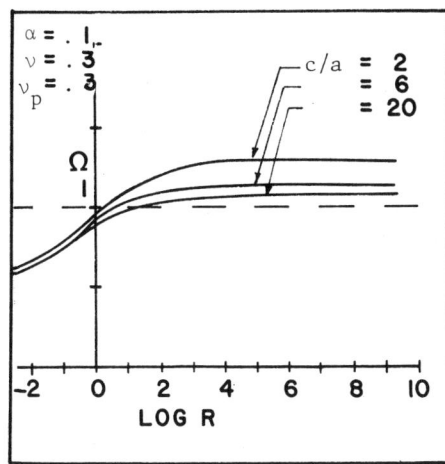

Fig. 3

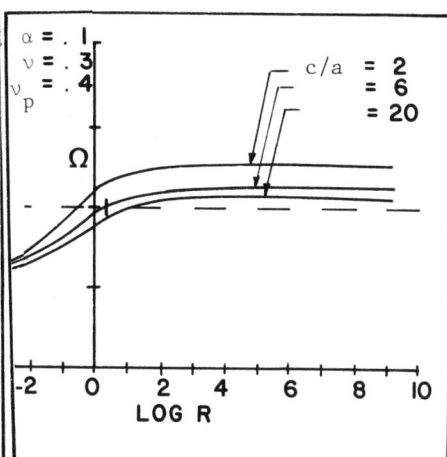

Fig. 4

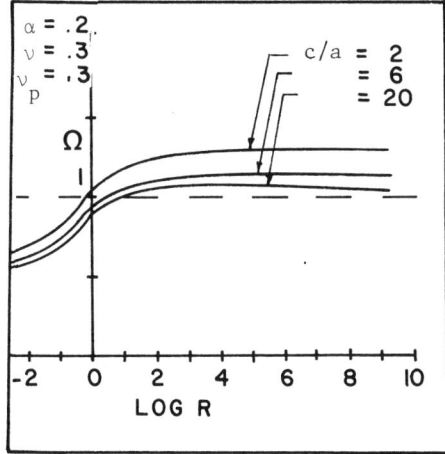

Fig. 5

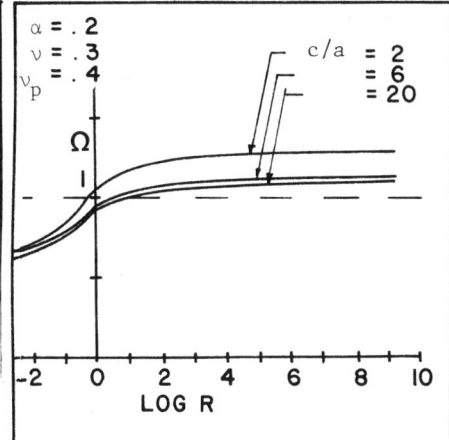

Fig. 6

Correction factor, Ω

Chapter 83

TUNNEL RESPONSE IN MODELED JOINTED ROCK

by Herbert E. Lindberg

Staff Scientist, SRI International
Menlo Park, California 94025

ABSTRACT

Laboratory-scale intact and jointed rock masses were tested in field and laboratory experiments to investigate tunnel response in high stress environments. In most of the experiments the tunnels were modeled by a rigid-plastic foam polyurethane with a crush strength of 3.4 MPa (500 psi) and emphasis was placed on jointed rock parameters: joint sets (single set of parallel plane joints versus a double set of orthogonal parallel plane joints), joint plane angle with respect to the direct loading direction, and joint spacing. Other parameters were loading stress, loading direction, and tunnel reinforcement strength. Results show quantitatively the increase in tunnel deformation in jointed compared with intact rocks, double versus single joint sets, and large versus small angle between the joint plane normal and the loading direction. A surprising result was the decreases in tunnel deformation with increases in the number of joints across a tunnel diameter.

The efficacy of scale modeling with laboratory-constructed rock was tested by modeling complex reinforced concrete and steel tunnel reinforcements in jointed rock masses and comparing the results with large scale tunnels previously fielded in highly jointed granite. Tunnel response was reproduced in surprising detail, including concrete fracture, steel liner bending and fracture, and critical load to produce these responses. The scale models also allowed sectioning of the jointed rock masses to observe gross sliding along the joints that accompanied tunnel failure.

INTRODUCTION

One of the most difficult aspects of underground structure design and analysis is the effect of rock jointing and planes of weakness on structure response. Most designs are based on limited field measurements and empirical rules for weakening caused by joints. Analysis of elastic response is reasonably accurate because stiffness can be measured in the field on rock masses comparable in size to the tunnel or gallery being constructed. Inelastic response and failure is much more difficult to predict because field measurement of strength properties is extremely difficult on large rock masses. Also, even where field experiments are performed on structures loaded to yield and failure, it is impractical to observe rock failure and sliding along joints surrounding the structure because of the difficulty in drilling to make observations.

To circumvent these difficulties, experiments on tunnels in jointed rock masses have been performed in the laboratory (Hendron and Engeling, 1973), and several investigations have been performed to determine the strength of laboratory constructed jointed rock masses (for example, Rosengren and Jaeger, 1968, Einstein and Hirshfeld, 1973, and Reik and Zacas, 1978). A fundamental limitation of such tunnel tests in the laboratory is the size and expense of testing machines required to test large rock masses at high stress levels, a combination that is required if realistic tunnel reinforcements are to be investigated. The research reported here demonstrated that this limitation can be overcome by testing laboratory constructed rock masses and tunnels as add-ons to underground nuclear tests.

MODEL CONSTRUCTION

The two basic types of jointed models tested are illustrated in Fig. 1. The first type was made of an assembly of 12.7-mm-thick plates of rock simulant through which a tunnel was drilled with a core drill. The second type was made of an assembly of 12.7 mm and 25.4 mm square bars. The bars were assembled around a tunnel reinforcement using the scheme shown in the figure. The smaller bars were used near the tunnel and the larger bars were used to complete the model in order to reduce expense in bar fabrication and assembly. At the rock-tunnel interface, some of the smaller bars were trimmed to a triangular cross section to match the contour of the tunnel. Small voids were filled with grout. Tunnel diameter was typically 76 mm and the rock mass diameter was 0.56 m.

Composition and properties of the rock simulants are given in Table 1. Two simulants were used, one to model granite as closely as possible with cement and sand materials, and the other to model weaker rocks. As indicated in the table, it was not possible to obtain the high friction angle and strength of granite, but it was felt that the values obtained were high enough to give tunnel response similar to that for granite.

We chose to model granite because previous underground tests had already been performed on reinforced tunnels in natural, highly jointed granite, and we wanted to test the efficacy of scale modeling

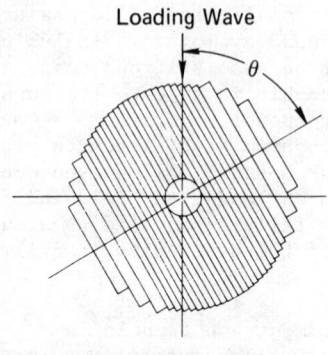
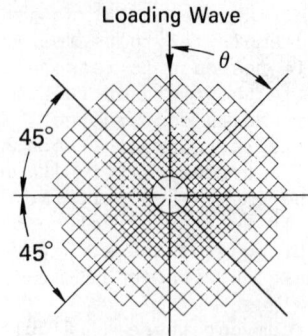

(a) Single Joint Set, Fielded at $\theta = 0°, 45°$ and $90°$

(b) Double Orthogonal Joint Sets, Fielded at $\theta = 0°$ and $45°$

MA-5088-7B

FIGURE 1   JOINT ARRANGEMENTS AND ORIENTATIONS

Table 1.
COMPOSITION AND PROPERTIES OF ROCK SIMULANTS

Composition (% by weight)			Constitutive Parameters			
Component	16A	GS	Parameter	16A	GS	Granite
Portland Cement, Type 1	18.61	34.97	Young's Modulus, E			
			(GPa)	25.1	34	69
Limestone Sand	61.37		(10^6 psi)	3.65	4.9	10
Granite Sand (Sherman granite, Wyoming)	6.62	55.33	Poisson's Ratio, ν	0.23	0.21	
CFR-2 (friction reducer)		0.26	Compressive Strength, σ_u			
			(GPa)	37.2	124	228
Melment L-10 (water reducer)		1.05	(10^3 psi)	5.4	18.0	33
Water	13.40	8.39	Friction Angle, ϕ (degrees)	29	38	55
Total:	100.00	100.00				

16A — Simulant used in parameter study and triaxial test machine.
GS — Granite simulant used in 1:28-scale models of tunnels in granite.

with laboratory constructed rock. These previous tests were with 2.82 m diameter rock openings, as shown in Figure 2. The jointing systems surrounding the tunnels were determined from wall observations and core drilling and are indicated in the figure. Also shown is the highly idealized modeled jointed rock used in our model tests, performed at 1:28 scale. The models were constructed on the assumption that the major joints in the natural granite surrounding the large tunnels consisted of two sets, one nearly parallel to the loading wave direction and the other at right angles. The number of major joints across the rock opening diameter was taken as 8 in the models.

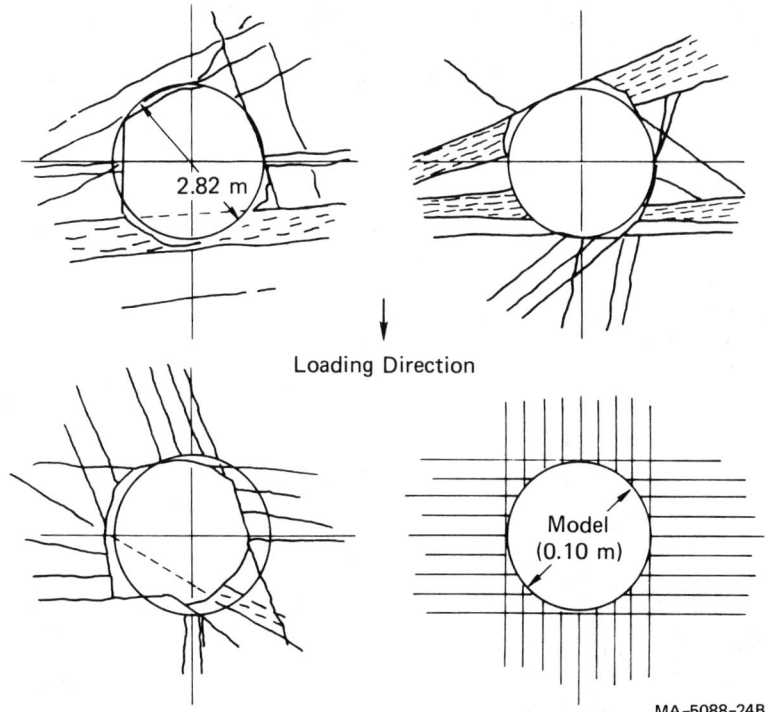

FIGURE 2   JOINTING IN FIELD STRUCTURES AND 1:28-SCALE MODEL

Figure 3 shows medium and close-up views of one of the three models of these reinforced tunnels in granite as sectioned after the test. This model was fielded at the lowest stress level, corresponding to the low stress level larger tunnel, and is essentially undamaged. It shows the details of rock mass and tunnel reinforcement construction. Figure 4 shows the rebar cage for the reinforced concrete liner of a similarly constructed model. An essential feature of the modeling and loading technique described here is that the model rock masses were large enough and the model rock strength high enough that the reinforced concrete tunnel liners could be modeled with full strength model concrete and steel rebars.

ISSUES IN ROCK MECHANICS

JOINTS NEAR TUNNEL

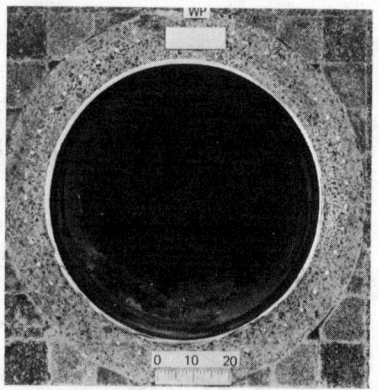
LINER CLOSE UP

MP-5088-70A

FIGURE 3 SECTIONED VIEWS OF 1:28-SCALE MODEL FIELDED AT LOW STRESS LEVEL

Figure 5 shows a completed assembly of a typical model. The jointed rock mass is held in place by casting a collar of grout around it, with the entire mass supported on a metal base plate and stand. The strength and stiffness of the grout was selected to approximately match the strength and stiffness estimated for the jointed rock mass. This assembly was then placed in an underground tunnel at an

MP-5088-80A

FIGURE 4 REBAR CAGE FOR SCALE MODEL

FIGURE 5 ASSEMBLY OF TYPICAL MODEL

MA-5088-8

appropriate location to be loaded at the desired direct stress (defined here as the stress in the direction of the loading wave).

The volume of the fielding tunnels (typically 3 m in diameter) in which the jointed rock models were placed was then filled with grout of the same type as in the grout collar. With this scheme the direct and lateral stresses on the model were approximately what they would be had the modeled jointed rock extend a great distance from the model tunnel. The actual stresses experienced by the models were measured by brinell-type gages (Peekna, 1976) bonded to the grout collars as shown in Fig. 5. Comparison of direct and lateral stresses as measured by these gages showed that the lateral stresses were about 60% of the direct stresses, as compared with 40% found to be required in the laboratory (Senseny and Lindberg, 1979) for uniaxial strain loading of similar jointed rock masses (uniaxial strain approximates the condition during the rising portion of a loading wave far from an underground explosion).

COMPARISON OF MODEL RESPONSE WITH FIELD STRUCTURES IN GRANITE

Figure 6 gives section views of a model tunnel and rock mass identical to that in Fig. 3 but loaded at a higher stress level. The tunnel is heavily damaged. The steel liner is severely bent and fractured over some distance along the bend. The surrounding reinforced concrete has fractured and substantial sliding has occurred along the joints in the rock mass. We speculate that the underlying cause of the failures is the fracture in the concrete, which consists mainly of shear fractures under the large radial deformation at this high load. After fracture, the liner system strength dropped and localized interior metal liner bending occurred as the deformation became even larger.

This speculation is based in part on observations in laboratory experiments on tunnels with steel liners with no surrounding concrete (Kennedy and Lindberg, 1978). It was found in these experiments that steel liners can sustain very large plastic deformations and buckling without precipitous failure, and that the liners continued to provide a reinforcing pressure against the rock opening equal to the yield pressure of the liner in simple hoop compression. Furthermore, the buckling was in a mode with many wrinkles around the circumference, rather than with a single bend as in Figure 6.

The joint sliding in Fig. 6 is mainly on the two vertical joints tangent to either side of the concrete reinforcement, although there is substantial sliding along all of the vertical joints between these two. There is little sliding along the horizontal joints because, after vertical sliding, horizontal sliding would have to be accompanied by fracturing of the rock blocks. This pattern is the result of the regular pattern of joints in this highly idealized model, but we feel that similar patterns also occur in natural jointed

Loading Wave

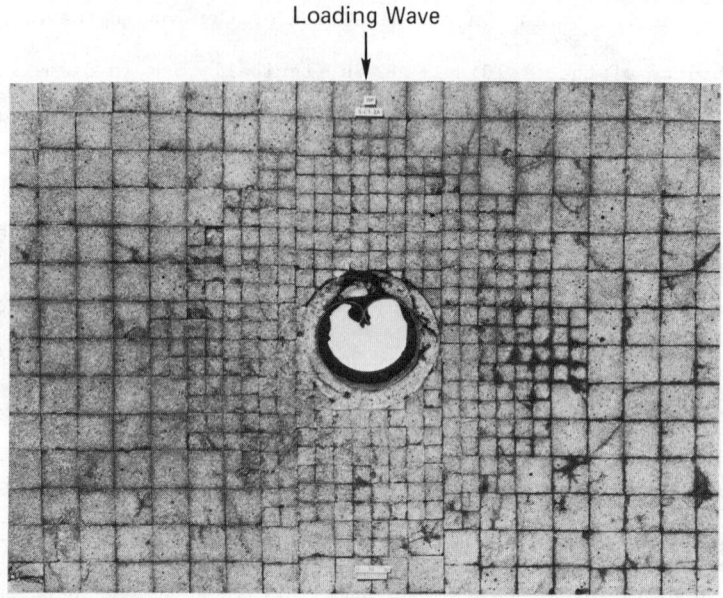

(a) JOINT OVERVIEW

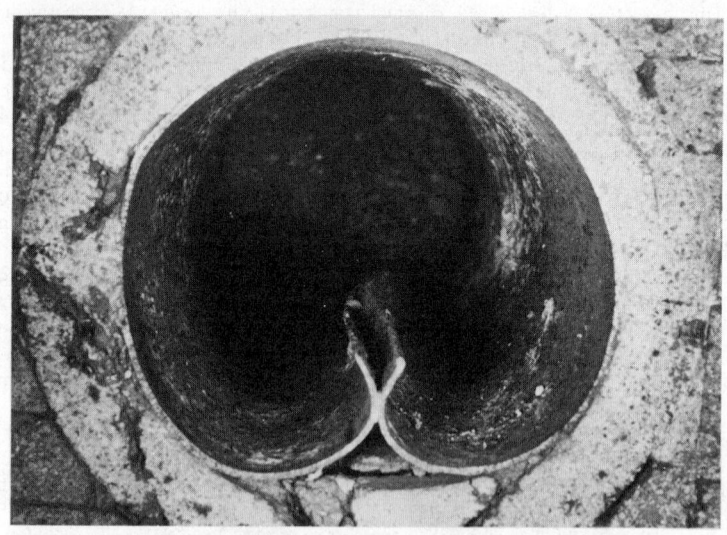

(b) INTERNAL TUNNEL DEFORMATION

MP-5088-90A

FIGURE 6 SECTIONED VIEWS OF 1:28-SCALE MODEL FIELDED AT HIGH STRESS LEVEL

TUNNEL RESPONSE IN MODELED JOINTED ROCK 831

rock because they also form interlocking systems after motion has occurred along one set of joints.

Figure 7 is an interior view of the larger tunnel fielded in natural jointed granite. Only the damage in the steel liner can be seen, but it has all the features of the severe bending and fracture seen in the small scale model in Fig. 6. The major difference is that in the natural granite the steel liner is bent at both "top" and bottom (i.e., along the angular orientation in the direction of the loading wave) rather than just at the top as in the small model. The fact that damage is concentrated in this direction suggests that joint sliding in the natural granite was similar to that in the model granite, that is, mainly along the set of joints nearly in line with the loading wave.

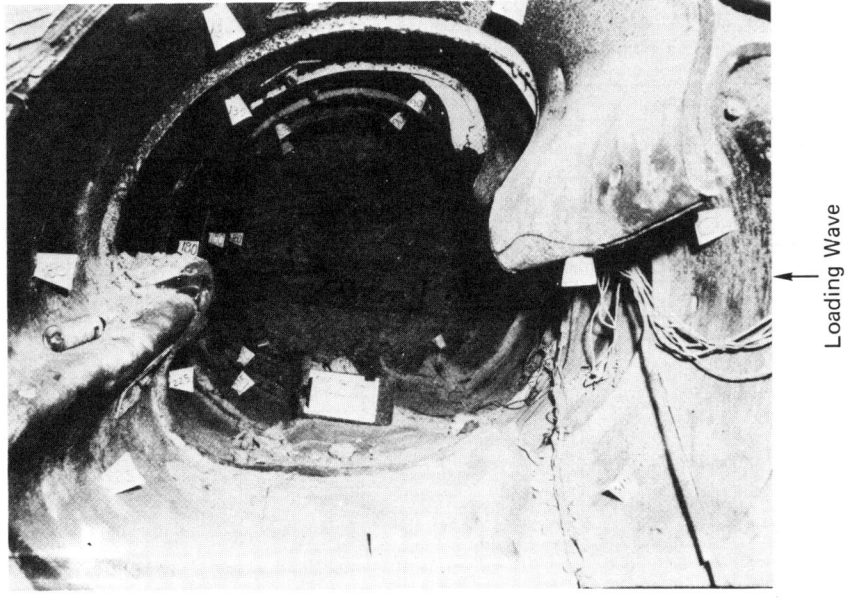

MP-5088-107A

FIGURE 7 LINER BUCKLING AND FRACTURE IN LARGE SCALE TUNNEL FIELDED IN GRANITE

This favorable comparison between model and natural rock-tunnel systems lends confidence to using model tunnel-rock systems to investigate jointed rock behavior and to screen various tunnel reinforcement concepts. The models give the further advantage of allowing the entire rock mass surrounding the tunnel to be sectioned so that the mechanisms of deformation can be directly observed. Response of the reinforced concrete and surrounding rock mass for the tunnel in Fig. 7 was largely conjecture because it cannot be seen.

JOINTED ROCK PARAMETER STUDY

Table 2 summarizes the results of 15 jointed rock models with parameters varied systematically to observe their effects on tunnel response. The most pertinent comparisons to be made are indicated by the bars and arrows in the lower half of the table. For example, to see the effect of loading stress with other parameters held fixed, one can compare model 1 with model 10 and model 6 with model 12. In intact models 1 and 10, peak inward deformation increased from 0.9% to 1.3% with an increase in direct stress from 59 to 65 MPa. In models 6 and 12, which have double sets of joints oriented at ±45 degrees from the loading direction, inward deformation increases from 1.3% at 46 MPa to 5.1% at 57 MPa.

Table 2.
SUMMARY OF PARAMETER STUDY FOR TUNNEL RESPONSE IN JOINTED ROCK

Model Number		1	2	3	4	5	6	7	8	9	10	11	12	13	14	15
Tunnel Reinforcement, P_i		\multicolumn{9}{c}{3.4 MPa (500 psi)}					10.3 MPa (1500 psi)		
Joints	Angle — degrees	Intact	(a)	90	45	0	±45	45	0	(b)	Intact	±45	±45	45	0	±45
	Spacing D/S		6	6	6	6	6	10	10	6		3	6	6	6	6
Load σ_x — MPa		59	65	63	63	67	57	60	58	59	65	47	46	51	59	53
Response W_{inward} — percent		0.9	1.0	4.1	4.0	2.3	5.1	2.4	1.8	1.5	1.3	2.3	1.3	0.5	1.0	1.6

Parameter Comparisons:
- Loading Stress
- Reinforcement Strength
- Joint Angle
- Joint Sets
- Joint Spacing
- Loading Direction

(a) Joint planes perpendicular to tunnel axis.
(b) Model axis along loading wave direction (axisymmetric loading).

The next parameter is reinforcement strength. Three pairs of models can be compared for this parameter. In pair 4 and 13, models with a single set of joints oriented at 45 degrees from the loading direction had a tunnel closure of 4.0% with a 3.4 MPa reinforcement and 0.5% with a 10 MPa reinforcement. For 0 degree joints (models 5

TUNNEL RESPONSE IN MODELED JOINTED ROCK 833

and 14) closure is 2.3% with a 3.4 MPa reinforcement compared with 1.0% with a 10 MPa reinforcement. Corresponding closures for models with a double set of orthogonal joints oriented ±45 degrees from the loading direction (models 6 and 15) were 5.1% and 1.6%, respectively. All these comparisons are what one would expect for an increase in reinforcement strength.

Four groups of comparisons can be made for the effect of joint angle with respect to the loading direction. Comparison of models 1 and 2 shows that there is little effect of joints if they are perpendicular to the tunnel axis; model 2, with these joints, has a deformation of only 1.0% at 65 MPa compared with 0.9% for intact model 1 at 59 MPa. Models 5, 4, and 3 show a systematic increase in deformation with increasing angle between the joint plane normal and the loading direction, from 2.3% at 0 degrees to 2.3% at 45 degrees to 4.1% at 90 degrees, respectively. These models all have 6 joints across the tunnel diameter. Models 7 and 8 have 10 joints across a tunnel diameter and also show the increase in tunnel deformation with increasing joint normal angle, from 1.8% at 0 degrees to 2.4% at 45 degrees. The only reversal in this trend is in models 13 and 14, which have D/S = 6 and a tunnel reinforcement pressure of 10.3 MPa. In these models, deformation is 1.0% for the 0 degree orientation and 0.5% for the 45 degree orientation. This reversal in trend is attributed to the higher load on the model with 0 degree joints, 59 MPa compared with 51 MPa for the model with 45 degree joints.

The effect of joint sets is seen by comparing model 4 with model 6 and model 13 with model 15. The former show an increase in tunnel deformation from 4.0% for a single set of 45 degree joints to 5.1% for a double set of joints at ±45 degrees. This difference is surprisingly small, even though the model with the single set has a higher pressure of 63 MPa compared with 57 MPa for the double set model. Models 13 and 15 show a similar comparison but with a tunnel reinforcement pressure of 10.3 MPa; the increase in closure is from 0.5% for the single set of joints to 1.6% for the double set.

The most surprising result is the effect of joint spacing. In all three comparisons the model with the larger number of joints across a tunnel diameter has the smaller deformation. Models 4 and 7, with single sets of joints at 45 degrees, show a deformation of 4.0% with 6 joints across a tunnel diameter and 2.4% with 10 joints. Models 5 and 8, with joints at 0 degrees, show 2.3% deformation with 6 joints and 1.8% with 10 joints. In this comparison, the 6 joint model has 67 MPa loading while the 10 joint model has only 58 MPa loading, which tends to give the smaller deformation in the 10 joint model. Finally, models 11 and 12, with double joint sets at ±45 degrees, show 2.3% deformation with 3 joints across the tunnel diameter compared with 1.3% with 6 joints.

These trends, in all three model pairs with different joint spacing, are opposite to the rule used to estimate closure of tunnels in jointed rock and to what one would predict based on measurements of modulus and strength in small samples of jointed rock. The difference may be the result of our model rock being more ductile than natural rocks, but the trend reversal certainly warrants further investigation to determine whether current empirical rules need to be reevaluated.

All of the results in Table 2 are very self consistent. They are also consistent with laboratory tests performed on jointed models made with the same 16A rock simulant (Senseny and Lindberg, 1979). The laboratory tests were performed with 0.3-m-diameter rock specimens and 50-mm-diameter tunnels cored along a diameter of the right circular cylindrical specimens. The jointed models were made with $D/S = 6$ by making the rock plates thinner than in the larger field models.

Figure 8 gives plots of rock cavity closure versus direct stress in these models for both mild steel liners and backpacked liners, with a backbacking crush stress of 3.4 MPa as in many of the field models. Four of the models in Table 2 have parameters consistent with the backpacked models in Fig. 8 and their closures are therefore plotted for comparison. In plotting the data for jointed models 4 and 5, 1% closure was added to account for the 1% initial closure seen in all of the laboratory jointed rock models because of initial gaps between the joints at zero stress.

The curves to which the four field data points are to be compared are indicated by light lines drawn from the points to the appropriate curves. The field data points are consistently below the laboratory curves. This shift is attributed to the difference in lateral stress condition in the two test types, as mentioned earlier. The lateral stresses were 60% of the direct stresses in the field models and 40% of the direct stresses in the laboratory tests. This greater confinement in the field models made the load more nearly symmetric so that oval closure was reduced (in all the tunnels, deformation is dominated by oval deformation).

CONCLUSIONS

It has been demonstrated that complex jointed rock masses and reinforced concrete tunnel liners can be successfully fabricated and tested as add-ons to underground tests. This allows testing of models small enough that the entire rock mass can be recovered and sectioned for inspection, but large enough to allow model complexity. Testing underground allows these relatively large models to be tested at high stress levels, which would require very large, special equipment in the laboratory. Using this modeling and testing technique, we showed the effect of several rock joint parameters on tunnel deformation. Favorable comparison of these results with laboratory tests gives further confidence in use of the field model technique for other tests on underground structures.

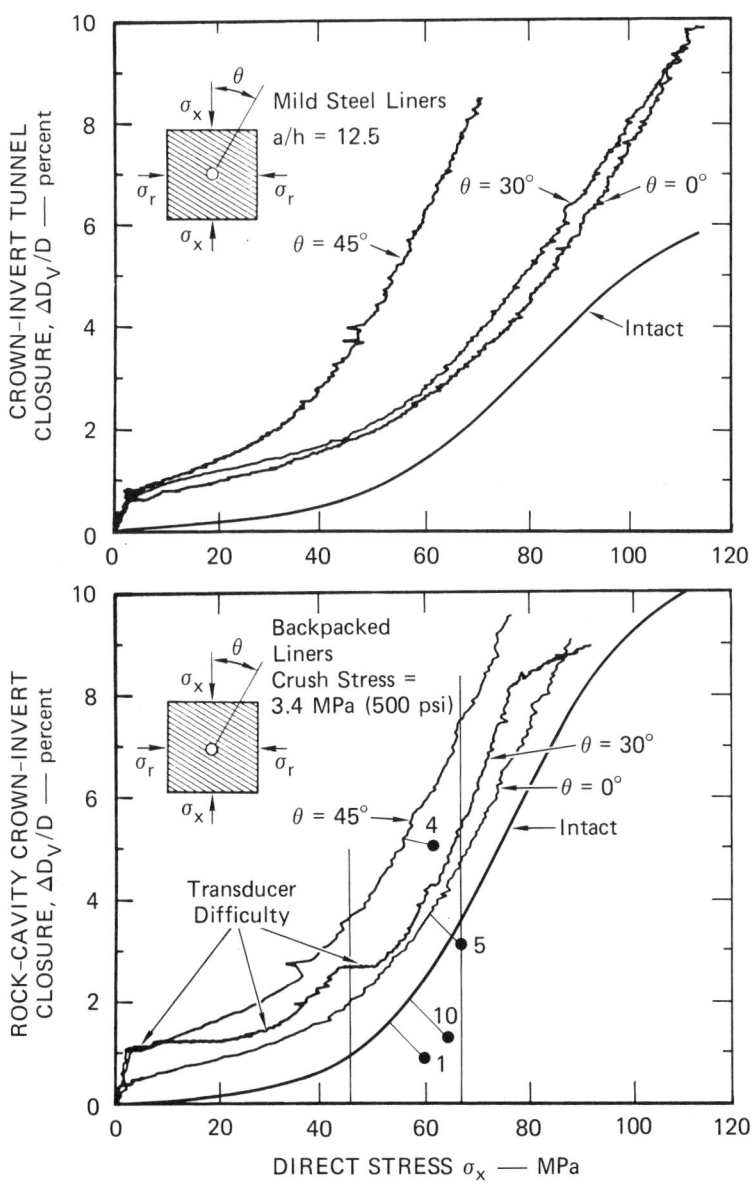

FIGURE 8 TUNNEL CLOSURE VERSUS DIRECT STRESS FOR JOINTED ROCK MODELS TESTED IN A TRIAXIAL LOADING MACHINE

(Rock specimen 0.3 m in diameter and 0.3 m tall; tunnel diameter 50 mm)

ACKNOWLEDGMENTS

This work was supported by Defense Nuclear Agency under Contracts DNA001-76-C-0292 and DNA001-76-C-0385. The technical monitors were Lt. Col. D. Burgess and Maj. R. Swedock. I am also indebted to P. E. Senseny, B. S. Holmes, and M. Sanai, who were leaders of the projects on which the paper is based, and to R. M. Stowe of Waterways Experiment Station, Vicksburg, Miss. for developing the rock simulants and directing fabrication of the jointed rock masses at WES.

REFERENCES

Einstein, H.H., and Hirschfeld, R.C., 1973, "Model Studies on Mechanics of Jointed Rock," Journal of the Soil Mechanics and Foundations Division of the ASCE, Vol. 99, pp. 229-248.

Hendron, A.J., and Engeling, P., 1973, "Model Tests of Lined Tunnels in a Jointed Rock Mass," TR M-41, Construction Engineering Research Laboratory, Champaign, IL.

Kennedy, T.C., and Lindberg, H.E., 1978, "Model Tests for Plastic Response of Lined Tunnels," Proc. ASCE, J. Eng. Mech. Div., Vol. 104, No. EM2, pp. 399-420.

Peekna, A., 1976, "Development of the Brinell Sandwich Passive Transducer," Proceedings of the Army Science Conference, Vol. III, pp. 43-56.

Reik, G., and Zacas, M., 1978, "Strength and Deformation Characteristics of Jointed Media in True Triaxial Compression," Int. J. Rock Mech. Min. Sci. & Geomech. Abstr., Vol. 15, pp. 295-303.

Rosengren, K.J., and Jaeger, J.C., 1968, "The Mechanical Properties of an Interlocked Low-Porosity Aggregate," Geotechnique, Vol. 18, pp. 317-326.

Senseny, P.E., and Lindberg, H.E., 1979, "Theoretical and Experimental Study of Deep-Based Structures in Intact and Jointed Rocks," DNA 5208F, Sept., for Defense Nuclear Agency, Washington, D.C.

Chapter 84

IN SITU MEASUREMENTS OF STRESS CHANGE INDUCED BY THERMAL LOAD:
A CASE HISTORY IN GRANITIC ROCK

by Richard Lingle and Philip H. Nelson

Terra Tek, Inc.
Salt Lake City, Utah

Lawrence Berkeley Laboratory
Berkeley, California

ABSTRACT

Vibrating wire stressmeters (VWS's) and borehole deformation gages (BDG) were deployed in two in-situ heater experiments at Stripa, Sweden to determine the thermally induced stress changes in the rock mass. The heater experiments ran for 18 months, including a six-month cool-down period after the heaters were turned off. Gages were installed in both horizontal and vertical boreholes and were subjected to temperatures ranging from 10°C to 120°C, depending upon the distance of a gage from the central heater. Most gages operated in a moist environment because groundwater flowed into many of the instrument boreholes during the experiments.

Of the 36 VWS's installed, six failed. Failures were attributed to corrosion of the vibrating wire due to moisture entering the gage body. Of the 30 BDG's installed at the beginning of the experiments, 22 failed, also as the result of water entering the gage body. Modification and reinstallation of the gages eliminated the problem of water infiltration, but much data from early stages of the experiments was lost due to gage failures.

Calibration of the VWS in blocks of Stripa granite showed a lack of repeatability. Variability in calibration results are attributed to seating effects, to variations in the elastic modulus of the rock, and to changes in temperature. These three effects combined to produce an error judged to be approximately ±33% of the stress changes measured at Stripa. In contrast, the major source of error in determining stress with the BDG appears to be the uncertainty in rock modulus and thermal expansion coefficient. Sample calculations based on data acquired near the end of the experiments (during cool-down) yield values of stress change which are in reasonable accord with theoretical values.

Off-the-shelf VWS's and BDG's installed at Stripa did not function well enough to provide reliable data on stress changes. Further effort is required to provide stress monitoring instrumentation which can function adequately in the environment of a nuclear waste repository.

INTRODUCTION

A joint USA-Swedish program was initiated in May 1977 to assess the response of a granitic rock mass subjected to thermal loading. The experiment was conducted at the 340 meter level in the Stripa mine, an inactive iron ore mine located in central Sweden. Electrically-powered heaters installed in boreholes were used to simulate the expected heat output of high-level nuclear waste canisters. The rock mass surrounding the heaters was instrumented to determine the temperature, stress and deformation changes induced by the heating. This report is confined to the efforts to determine stress changes.

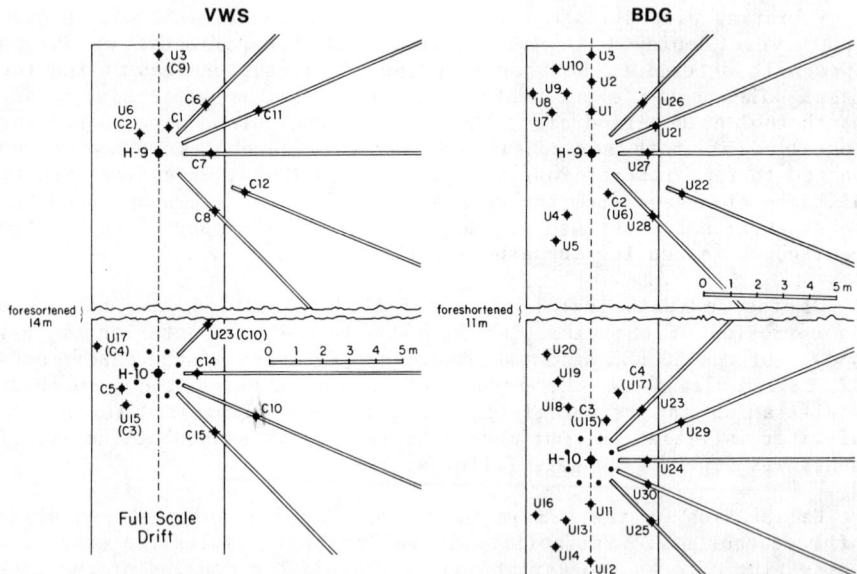

Fig. 1. Location (plan view) of vibrating wire stressmeters (VWS) and USBM borehole deformation gages (BDG) at the Stripa heater experiments. Horizontal boreholes entering from right are collared in an adjacent drift. Heater holes H9 and H10 are separated by a distance of 22 m.

STRESS CHANGE INDUCED BY THERMAL LOAD 839

Two instrument types were deployed in the two full-scale heating experiments to determine the stress change in the rock (Fig. 1). Vibrating wire stressmeters (VWS's) were installed in a total of 15 boreholes (two per borehole) and USBM type borehole deformation gages (BDG's) were installed in 30 boreholes (one per borehole); both vertical and horizontal boreholes were used. Gages were installed 3 to 5 meters below the floor of the drift and at radial distances from the central heaters ranging between 1 and 4 meters.

During the experiments, the gages were subjected to temperatures ranging from 10°C to as high as 120°C for gages closest to the H10 heater. The maximum thermomechanical stress at any gage location was about 60 MPa. Water seeped into the boreholes during the experiments; periodic dewatering of the vertical boreholes prevented submersion of the gages, but, nevertheless, the gages operated in a moist environment. Horizontal boreholes drained freely because they were open and were inclined slightly downwards towards the collars.

VIBRATING WIRE STRESSMETER

The vibrating wire stressmeter (VWS) consists of a hollow steel cylinder which is set into position in a 38 mm borehole by means of a sliding wedge and platen assembly. The sensing element is a highly stressed steel wire stretched across the diameter of the hollow cylinder. A small electromagnet is used both to excite and sense the natural vibration of the wire. A change of stress in the rock results in a change in the stress of the wire causing a shift in its fundamental vibratory period. More detail on the design and operation of the gage is given by Hawkes and Bailey (1973) and Dutta, et al. (1981). The gages used in the Stripa heating experiments were manufactured by Irad Gage, Inc., at Lebanon, New Hampshire and are the VBS-1HT type with the hard rock (HR) platen.

Calibration

The VWS's were calibrated by installing them in a borehole in a block of Stripa granite and subjecting the rock to a known stress field. The period of vibration (R) of the wire was recorded at increments of applied stress. The slope of the straight line fit of applied stress plotted as a function of the $1/R^2$ is taken as the gage calibration factor (C_1).

Gages recovered from boreholes when the experiments terminated were calibrated in two different facilities. The first was constructed at the site by line-drilling out a 38 cm block in the drift floor centrally located between the two experiments. Load was applied to the block by flatjacks grouted in the drilled slots. The second calibration facility was comprised of a 30 cm (flatjack-loaded) block of Stripa granite set up in the laboratory. To avoid the effects of crack closure at very low confining stress, either the block was preloaded to 6.9 MPa before gage installation, or else C_1

was computed using only data taken at loads in excess of 6.9 MPa. Comparison of room temperature calibrations from these two blocks showed a large amount of scatter, as illustrated in Fig. 2.

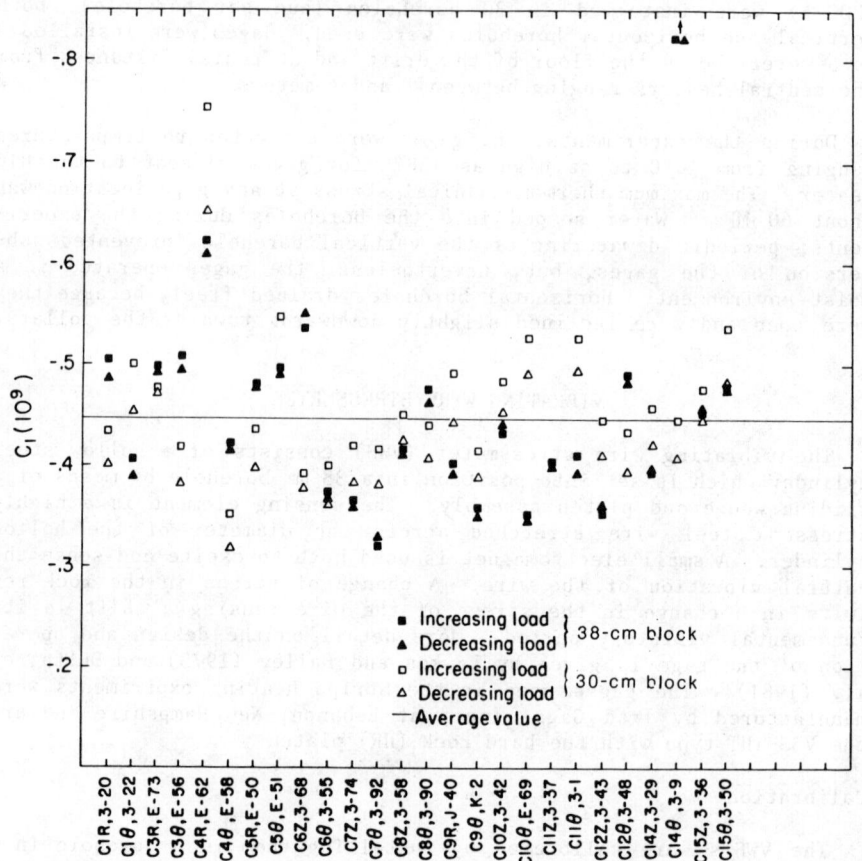

Fig. 2. Distribution of C_1 values from calibrations of 26 VWS in 30 cm and 38 cm blocks.

We found that the calibration factor C_1 was sensitive to gage seating. The C_1 value obtained from a series of repeater installations and loadings exhibited a scatter as great as ±20% from the mean. It was also found that C_1 varied systematically with setting pressure, as measured by the change in period from the unloaded state (R_o) to the installed state (R_s). As shown in Fig. 3, C_1 is higher (gage sensitivity is lower) at low setting pressures than at high setting pressures. Results by Dutta, et al. (1981) (large squares in Fig. 3) showed the same dependence, although their results were ob-

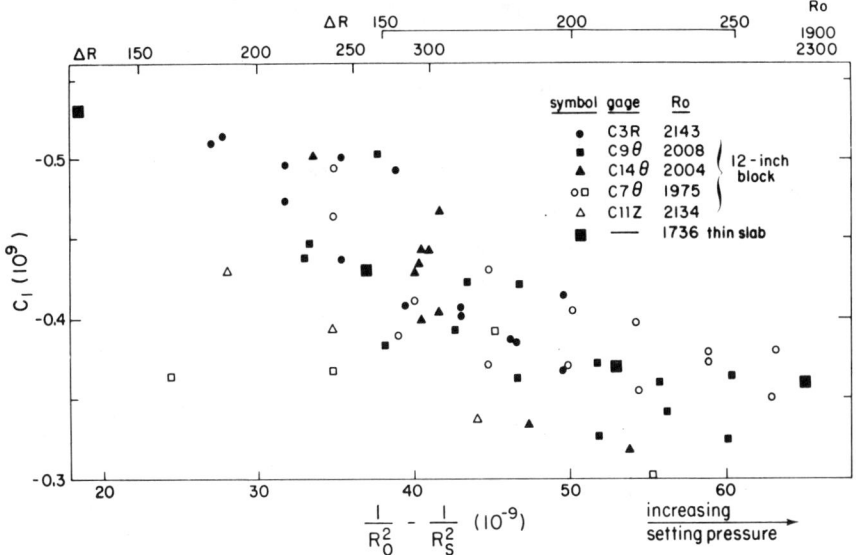

Fig. 3. VWS calibration results as a function of setting pressure. Thin slab calibrations taken from Dutta, et al. (1981).

tained in a thin slab rather than a block. Hence in practice, C_1 was dependent upon the force applied to the wedge when setting the gage.

Further testing showed that C_1 depends upon the elastic modulus of the rock and upon temperature. We were unable to separate satisfactorily the effects of temperature and rock modulus from gage seating effects. Therefore, for treatment of the Stripa data, the error bounds cannot be reduced below the sum of all effects, which is of the order of ±33% in terms of computed stress.

Operating History

Of the 36 VWS's used at Stripa, 6 failed and were replaced, 16 operated for the full 550 day experimental period, and another 14 operated satisfactorily although installed for less than the full term. Dissection of some of the failed gages revealed that the wire had corroded. This was attributed to water entering the gage body. Despite corrosion of some gages, long-term stability was quite good. The period of vibration of gages recovered after the experiment was compared with the period before installation. Only four gages had changed by more than 20 counts.

BOREHOLE DEFORMATION GAGE

The borehole deformation gage (BDG) measures diametral deformation of a 38 mm borehole by detecting the deflection of three pairs of strain-gaged cantilevers spaced 60° apart. A more detailed description of the BDG and its calibration for use at elevated temperature is given by Schrauf, et al. (1979). The gages were developed by the U.S. Bureau of Mines and were manufactured by Rogers Arms and Machine Company, Inc. at Grand Junction, Colorado. The standard gage was modified in order to improve its performance at elevated temperature. However, prior to the heater experiments, no special precautions were taken to protect the gage against infiltration of water.

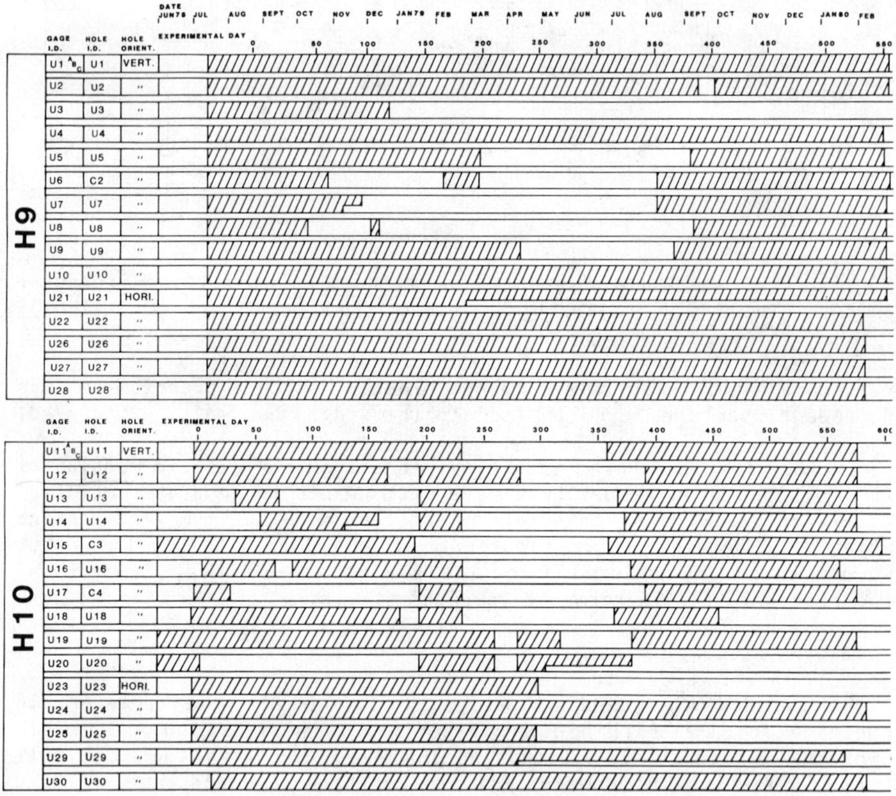

Fig. 4. Borehole deformation gage operating history at Stripa. Hachured areas denote time periods of successful operation.

Operating History

The failure rate of the BDG's used during the heating experiments at Stripa was extremely high, as shown in Fig. 4. At the end of 11 months, 57% (17 of 30) of the gages originally installed were inoperative and at the end of the experiment 73% of the original gages had failed. As with the VWS's, the failure rate was highest for gages in vertical boreholes, where 85% (17 of 20) failed. In all cases, the failures were attributed to internal corrosion problems as a result of water infiltration.

Because failures were so high, 15 gages were removed from boreholes while the heater experiments continued, and returned to the manufacturer for modifications. All modified gages were filled with silicone oil before reinstallation. The modifications eliminated the problem of water infiltration, although some gages did fail after reinstallation due to other causes.

Error In Computed Stress

A linear equation relates the measured strain-gage bridge voltages to diametral displacements. However, the offset voltage (voltage output with no cantilever deflection) and calibration coefficient are temperature-dependent and hence their contribution to total error varies with temperature. Thermal expansion of the gage body must be accounted for, the uncertainty in this value likewise contributes to the overall error in displacement. Assessment of these effects showed that the uncertainty in diametral displacement obtained during satisfactory gage operation of the BDG's used at Stripa was approximately ±10%.

Conversion of measured displacement to stress change requires a knowledge of the elastic properties and coefficient of thermal expansion of the rock. Uncertainty in these quantities contribute directly to the uncertainty in the computed stress change. It appears at this stage that the uncertainty in rock modulus of the Stripa granite constitute the largest source of error in the stress change values computed from BDG data.

Sample Calculations

Stress changes were made from the measurements taken by some of the gages for the cool-down period by taking values measured just before turning off the heaters and just before gage deactivation. These data are shown in Fig. 5, all gages shown were located in vertical boreholes. The results are compared with predictive computations by Chan and Cook (1979). Correspondence between stress values derived from BDG's and predicted values appears reasonable both in terms of magnitude and in terms of the decline of stress with increasing distance from the heater. Not enough VWS values are available to warrant comment.

SUMMARY

Two problem areas have been identified with the VWS's during the heating experiments at Stripa, gage failure and the problem of gage

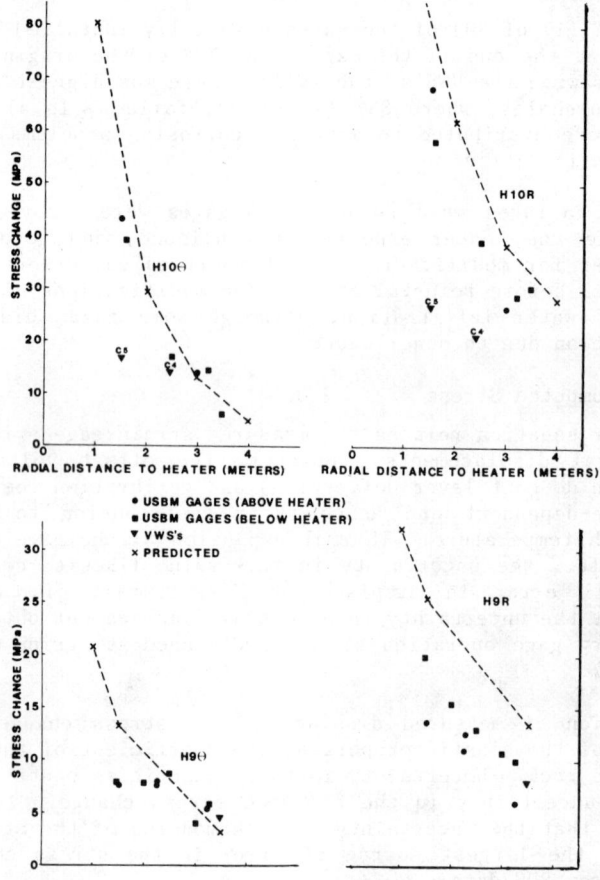

Fig. 5. Comparison of stress change from BDG and VWS measurements with predictive results from Chan and Cook (1979). Stress components are radial (R) and tangential (θ) to the central heater.

calibration. For the purpose of Stripa VWS data, the calibration problems have not been sufficiently defined or interpreted to enable the error band to be reduced below that obtained by combining all the effects, which is the order of ±33% of the indicated stress.

The operating performance record of the BDG's used at Stripa was worse than the VWS's, in all, 73% (22 of 30) of the gages that were installed at the beginning of the experiment failed. BDG measurements (when the gages were operating properly) were accurate to approximately ±10%. However, in computing stress from the deformation data, the uncertainty in the elastic and thermal properties of the rock must also be considered.

As a result of the experience gained from the heating experiments at Stripa, it is concluded that off-the-shelf instruments for monitoring in-situ stress <u>change</u> is not yet to the point of providing reliable long-term data in a nuclear waste repository environment. If this data will be required in the future in the repository licensing process or for model validation, a concerted effort involving the funding agencies, the users, and the manufacturers will be required to advance the state-of-the-art.

ACKNOWLEDGEMENTS

This work was performed under subcontract with Battelle Memorial Institute, Project Management Division, under Contract DE-AC06-76RL01830 with the Department of Energy. The contract was administered by the Battelle Project Management Division.

We wish to acknowledge the work of H. Sellden in gage installation and maintenance, A. Dubois in performance assessment, and P. A. Halen and H. Carlsson for support at the Stripa mine. M. Board, T. Schrauf and R. Clayton conducted many of the laboratory calibrations.

REFERENCES

Chan, T. and Cook, N.G.W., 1979, "Calculated Thermally Induced Displacements and Stresses for Heater Experiments at Stripa, Sweden," LBL-7061, SAC-22, Lawrence Berkeley Laboratory, Berkeley, Cal.

Dutta, P.K., Hatfield, R.W. and Runstadler, P.W., Jr., 1981, "Calibration Characteristics of IRAD Gage Vibrating Wire Stressmeter at Normal and High Temperature," Tech. Report 80-2, UCRL-15426, Lawrence Livermore National Laboratory, Livermore, Cal.

Hawkes, I. and Bailey, W.V., 1973, "Design, Develop, Fabricate, Test and Demonstrate Permissible Low Cost Cylindrical Stress Gages and Associated Components Capable of Measuring Change of Stress as a Function of Time in Underground Coal Mines," U.S. Bureau of Mines Contract Report (H0220050), Washington, D.C.

Schrauf, T., Pratt, H., Simonson, E. and Hustrulid, W., 1979, "Instrument Evaluation, Calibration and Installation for the Heater Experiments at Stripa," LBL-8313, SAC-25, UC-70, Lawrence Berkeley Laboratory, Berkeley, Cal.

Chapter 85

GEOTECHNICAL MONITORING OF HIGH-LEVEL NUCLEAR WASTE
REPOSITORY PERFORMANCE

by Christopher M. St.John and Michael P. Hardy

Consultant, Los Angeles, California

Principal, J.F.T. Agapito and Associates, Inc.
Grand Junction, Colorado

ABSTRACT

The paper discusses an approach to geotechnical monitoring of a geological repository constructed for the purpose of isolation of high-level nuclear waste, and describes the conceptual framework for development of an integrated monitoring scheme that is initiated during the earliest phases of site investigation and continues through facility decommissioning. The approach that is proposed recognizes the need to provide specific data to support the licensing process for the repository while avoiding adverse impact upon the system performance. Further, it recognizes present and inherent limitations in both the instrumentation and techniques that might be used within a monitoring scheme.

INTRODUCTION

It is the opinion of the authors that geotechnical monitoring of a high-level nuclear waste repository site should start during the site screening activities undertaken by the Department of Energy (DOE) and should be terminated in the post-decommissioning phase of the isolation facility. Precisely what measurements are undertaken at any time will depend upon the objectives and functions of the monitoring at any stage in the development of the facility. Some understanding of repository development and the licensing process is therefore a prerequisite for any discussion of repository monitoring.

Following selection of a number of sites for characterization, the DOE will present to the Nuclear Regulatory Commission (NRC) Site

GEOTECHNICAL MONITORING 847

Characterization Reports (SCR's) for each site. Initiation of site characterization, following comment on the SCR by the NRC, marks the start of a five phase program for the development of a waste isolation at that site. These five phases - site characterization, construction, waste emplacement and retrievable storage, decommissioning, and post-decommissioning - each have associated monitoring activities. The paper discusses the objectives of monitoring within each phase, and recommends geotechnical activities that should be undertaken to meet these objectives in a timely manner. The work upon which this paper is based is documented in a report prepared for the Nuclear Regulatory Commission (St.John et al, 1981)(1). However, the opinions expressed are those of the authors and should not be construed as representing the position of the NRC.

FUNCTIONS OF GEOTECHNICAL MONITORING

Geotechnical monitoring will have the ultimate objective of the determination of site and facility suitability for emplacement and isolation of nuclear waste. Intermediate objectives are characterization of the geologic setting, monitoring of operational safety, and design confirmation. The decisions that will be based on the results of monitoring activities are:

- o acceptance or rejection of a site - based on an assessment of beneficial and adverse site conditions;
- o requirements for design modification - based on encountered characteristics or unexpected performance of natural or engineered components of the repository system;
- o retrieval of emplaced waste - based on measured or predicted capability of the facility to meet specified performance criteria; and,
- o decommissioning of the subsurface facilities - based on measured performance of the as-built facility during the retrievable storage phase.

Specific objectives of monitoring during each phase of development of a facility are summarized in Table 1, together with the type of activities likely to be undertaken to meet these objectives. The types of activities may be broadly categorized as one or more of the following functions:

- o performance monitoring - to observe the response of the system to excavation, waste emplacement, backfilling, etc.;
- o design verification - to compare the response of the system to the assumed conditions and predicted response used as the basis for design decisions and in the license applications;
- o quality assurance - to confirm that the as-built repository is within specifications; and,
- o background monitoring - to monitor the baseline

conditions to assess annual variations and long-term trends that impact repository performance.

REGULATORY AND PRACTICAL CONSTRAINTS

There are both regulatory and practical constraints that are placed on the design of a repository geotechnical surveillance program. Regulatory constraints, defined in 10 CFR Part 60 (2, 3), consist of specifications of minimum monitoring procedures required for performance confirmation, and a time frame that is determined by the several steps of the licensing process. Practical constraints arise because of the requirement to not adversely affect the natural or engineered isolation barriers of the system and, because of technical limitations of monitoring capabilities.

The most important constraint of the geotechnical monitoring and surveillance system is the nature of the licensing process and the time frames within which decisions will be required to be made. The duration of various steps in the development and licensing of a repository are summarized in Table 2. The practical consequence is that extensive periods of time will be available during each phase, so that sufficient data may be gathered to move forward to the next phase with adequate assurance of an acceptable outcome. The long time frames require that instrumentation would have to be either replaceable or very durable.

Any potential for causing adverse effects to either the natural or engineered isolation barriers can be avoided by using methods of measurement that do not require penetration of the rock mass, backfill, or seals. Such measurement techniques may be referred to as a "non-incursive" means of measurement. Unfortunately, such measurements seldom permit direct measurement of the physical condition or response that is of concern. Most direct geotechnical measurements are incursive and will result in the creation of potential conduits for ground water flow, either through the rock mass or through backfill and seals. This potential can be minimized by careful sealing of instrumentation holes and by use of as few data transmission lines as possible. Alternatively, instrumentation could either be removed before decommissioning so that there would be no long-term impact, or incursive methods of measurement could be largely restricted to sections of the repository from which waste could be removed if any adverse effects of monitoring were detected.

An assessment of technology (St.John et al, 1981) revealed the important practical constraint that there is no immediate prospect of using completely remote monitoring stations. This situation arises because of a lack of power supplies capable of delivering the necessary electrical power over long time frames, limitations in

GEOTECHNICAL MONITORING   849

capabilities for through-the-earth data transmission, and the relatively short functional life span of most available geotechnical instrumentation. There are also practical limitations in measurement capability that may result from the resolution of the measuring technique used, unavoidable propagation errors during measurement, or disturbance of the condition being measured by the measurement or sampling process itself.

AN APPROACH TO MONITORING

Geotechnical monitoring is initiated during the site selection process. Before construction authorization is received, there must be reliable assurance that the site is suitable for waste isolation, based on site characterization and monitoring, and that the proposed design has an adequate measure of conservatism. Also, the layout of the repository, the distribution of waste, and the decommissioning plan should not be highly sensitive to expected variations in the geologic setting. During construction of the subsurface facility, the geotechnical monitoring scheme will provide evidence of variance in site conditions so that repository layout can be modified, or particular sections abandoned or sealed without emplacement of nuclear waste. Geotechnical monitoring of the performance of engineered components of the repository also will provide data required before the issuance of a license to receive and store high-level nuclear waste can be issued.

After waste emplacement, the geotechnical surveillance should continue to provide performance monitoring for design verification and quality assurance. The data gathered during this phase will be used to support the license amendment to allow decommissioning of the repository. During decommissioning further monitoring is required for quality assurance of the decommissioning procedures and for initial performance verification of the sealing components. Post-decommissioning monitoring and surveillance should be minimized because of the lack of institutional control of the facility after license termination.

The approach to monitoring throughout these phases must recognize the needs for, and constraints imposed upon, the geotechnical surveillance system. The approach must resolve fundamental issues such as the duration and density of observation, the parameters to be monitored, and the limitations of present technology. Development of the approach proposed by the authors is discussed in the remaining paragraphs of this section.

The main objective of repository design will be to provide a system that will contain the waste and prevent leakage to the biosphere. To satisfy this objective several redundant barriers are to be engineered into the repository system. The first barrier is the waste package; 10 CFR 60 requires that the package be designed to contain the waste for 1000 years. Backfilling and sealing of all excavations during decommissioning provides other engineered barriers which are designed to contain the waste within the repository after any releases from the canister subsequent to the first 1000 years. The natural geologic barriers provide the ultimate containment of the waste. Monitoring during the 50 years retrievable storage phase will provide a very limited performance sample of the first 1000 year period. Monitoring must concentrate, therefore, on identifying events or changes from the assumed design conditions that are precursors of early canister failure or breaches in the natural geologic barriers.

At some point in time, a decision will be made as to whether the waste package meets performance requirements. Once the decision has been made, all waste package related monitoring could cease. Attention is then directed towards the effectiveness of other engineered barriers and the natural barriers afforded by the geologic setting. Terminal isolation, marked by license termination, will only be possible after the performance of all components of the repository system have been demonstrated to be acceptable.

In the same way that the decision to decommission provides a natural duration for canister scale monitoring, the duration and scope of other monitoring activities can be defined by particular needs. Once sufficient data has been gathered to support that decision, further monitoring can be eliminated. Actions, such as backfilling and sealing, that are a consequence of that decision will require monitoring for the purposes of quality assurance.

A high density monitoring program is likely to be in conflict with the constraint of not jeopardizing the natural barriers. However, a wide range of coverage of the monitoring system is required because the geologic setting will exhibit significant spatial variability, and performance assessment and design verification must be demonstrated for the prevailing geologic conditions. One approach to such variability is to measure the same conditions or responses throughout the facility.

An alternative approach is to quantify the variability of geotechnical parameters throughout the repository by careful characterization, and to monitor the performance by non-incursive, domain-type measuring techniques that provide condition or response measurement for a relatively large volume of the repository system. Detailed performance assessment could then be carried out in a special design verification facility. This approach has the advantages of allowing testing to occur much sooner than might be practical during normal facility operation, and also removes the potential for breach of the natural and engineered barriers as all waste could be recovered from the Design Verification Facility (DVF) and placed in a routine disposal area, if necessary. This approach is the one that has been used as a basis for development of the surveillance program proposed by the authors.

The conditions and responses to be monitored depend on the objectives and development phase of the monitoring system. Conditions and responses to be monitored must be measurable quantities which provide early warning of departure of the repository performance from that predicted during design analysis. Such early warning is best provided by domain-type response measurements since point observations have to be fortuitously located to detect important changes in a heterogeneous system.

SUMMARY OF RECOMMENDED GEOTECHNICAL MONITORING

Following the logic discussed above, repository monitoring has been divided into two distinct activities. The first concerns routine surveillance during all phases of construction, operation and decommissioning of the facility. This monitoring is designed to provide widespread coverage of the repository without compromising the natural or engineered barriers. The second concerns monitoring within a Design Verification Facility. This special facility, located at the same horizon and with the same canister placement configuration as in the repository, is highly instrumented. Following the decision to decommission the repository the canisters in the DVF may be retrieved for permanent disposal elsewhere.

Routine Repository Monitoring

Conditions and responses to be measured as part of the routine monitoring activities are summarized in Table 3. Within the table, a distinction is drawn between monitoring on canister, room and repository or regional scales. On the smallest scale, measurements are made in the immediate vicinity of the waste package, including physical inspection for signs of accelerated corrosion or damage. Canister scale monitoring starts when the emplacement holes are prepared and terminates when the emplacement rooms, or corridors, are backfilled. Room scale monitoring embraces all measurements within and around the emplacement rooms and emphasizes non-incursive, domain-type observations. Some discrete or point observations of physical response are required to aid in interpretation of domain measurements. For example, infrared thermal scanning techniques provide evidence of superficial variations, while discrete measurements enable this to be related to the rock mass temperature. Room scale monitoring starts during excavation of the emplacement room and is terminated as those rooms are sealed. Repository scale monitoring again emphasizes domain response measurements. Monitoring activities start during site characterization and development and are phased out during facility decommissioning.

Monitoring Within the Design Verification Facility

The DVF should consist of a complete panel in which phases of the repository development, operation and decommissioning are represented. The panel should be identical to all others in the repository, with the exception that it is fully instrumented, and should be filled with waste of the same form and at the same density as in regular panels. Approximately one-third of the rooms should be open, and one-third should be backfilled and sealed and in a semi-decommissioned state. The remaining one-third of the panel should be in a completely decommissioned state. The facility should operate on an accelerated time frame so that data is provided sufficiently early to influence decisions made in regard to the rest of the repository.

Measurements to be undertaken within the DVF are summarized in Table 4, where the subdivisions are in terms of components of the repository system and major stages during its development. Responses monitored are essentially the same as during routine repository surveillance but the density of instrumentation, the frequency of observation, and the precision of observation would all be greater because the objective of monitoring is design verification. The difficulties of achieving such a high level of surveillance will be somewhat offset by the possibility of instrument replacement, the accelerated time frame, and the possibility of removing all waste after the monitoring period should the instrumentation have had a locally adverse effect on engineered or natural components of the repository system.

REFERENCES

1. St. John, C.M., Aggson, J.R., Hardy, M.P., and Hocking, G., 1981. Evaluation of Geotechnical Surveillance Techniques for Monitoring High-Level Waste Repository Performance. Prepared for U.S. Nuclear Regulatory Commission, NUREG/CR-2547.

2. NRC, 1981. Disposal of High-Level Radioactive Wastes in Geologic Repositories: Licensing Procedures. Federal Register, 46 FR 13971, February 25.

3. NRC, 1981. Disposal of High-Level Radioactive Wastes in Geologic Repositories. Federal Register, 46 FR 35280, July 8.

GEOTECHNICAL MONITORING

TABLE 1. Timetable for Repository Development

PHASE	ACTIVITY	START	END	APPROXIMATE DURATION
Site Characterization	Define site Establish initial conditions Monitor stability	Submission of Site Characterization Report	Receipt of construction authorization	10 years
Construction	Develop underground facility for waste emplacement	Receipt of construction authorization	Completion of all development	40-50 years
Operation and Retrieval	Emplace waste Monitor for 50+ years	Receipt of operating license	Receipt of license to decommission	80 years
Retrieval	Retrieve waste	Decision to retrieve	Completion of retrieval	30 years*
Decommissioning	Backfill and seal all excavations	Receipt of license modification to decommission	Completion of decommissioning	30 years
Post-decommissioning	Remote monitoring	Completion of decommissioning License termination	-	-

*Occurs only if the performance requirements for the natural or engineered components of the system are not satisfied.

TABLE 2. Summary of Geotechnical Surveillance of a Repository

PHASE	MAJOR OBJECTIVES	RELATION TO LICENSING DECISIONS	ACTIVITIES
Site Characterization	Establish baseline conditions Set-up long term surface monitoring	Support license application -before construction authorization	Surface & limited subsurface investigation Investigation surface monitoring networks for seismic, microseismic, piezometric head.
Construction	Demonstrate site homogeneity Performance verification	Support license application -required before issue of license	Detailed subsurface investigation Establish room and canister scale monitoring. Extend surface microseismic. Establish surface displacement scheme.
Emplacement and Retrievable Storage	Monitor waste package performance Performance verification	Support application for amendment to decommission	Monitor response of waste package and rock mass. Thermal scanning in emplacement rooms.
Decommissioning	Quality assurance for backfill and sealing Performance Assessment	Support application for license termination	Monitor behavior of backfill and and seals. Continue repository/regional monitoring.
Post-Decommissioning	Demonstrate compliance with environmental requirements	None	Continue long term monitoring from the surface.

TABLE 3. Routine Repository Surveillance - Summary of Conditions and Responses Measured

PHASE / SCALE	SITE CHARACTERIZATION	CONSTRUCTION	OPERATION AND RETRIEVABLE STORAGE	DECOMMISSIONING	POST-DECOMMISSIONING
Canister	None	Rock mass deformability	Waste package temperature Corrosion rates Ground water Chemistry	None	None
Room	None	Global Observations Microseismicity Sonic profile Ground water inflow Point Observations Deformation Temperature Piezometric head	Global Observations Thermal imagery Microseismicity Sonic profile Ground water inflow Ventilation air Point Observations Deformation Temperature Piezometric head	Phase out all previous monitoring progressively. Geophysical logging of backfill and seals for temperature; density moisture content closure/voids, etc.	None
Repository/ Regional	Characterization of site: Initial stress, Temperature, Ground water age/chemistry, Piezometric head, Permeability, Ground water velocity, Seismicity, Microseismicity, etc.	Continue characterization measurements. Gobal Observations Surface and underground microseismic, Surface absolute displacements, Point Observations Deformation Temperature Piezometric head	As during construction	Phase out underground microseismic network and spot observations Continue surface characterization Surface seismic and microseismic Absolute displacement	Continue: ground water sampling; Seismic and microseismic; Absolute displacement

TABLE 4. Monitoring Within a Design Verification Facility

COMPONENT OF REPOSITORY DESIGN / TEST SECTION OF DVF	RETRIEVABLE STORAGE	PARTIALLY DECOMMISSIONED	FULLY DECOMMISSIONED
Waste Package and Waste Package Environment	Waste package temperature Deformation Induced stress Acoustic emission Geophysical logging of open holes Ground water chemistry Radiation Waste package inspection	None	None
Emplacement Room	Microseismic monitoring Seismic profiling Infrared scanning Ventilation air Temperature Deformation Piezometric head Rock support monitoring	Microseismic monitoring Seismic profiling Temperature Induced stress Deformation Piezometric head Geophysical logging of backfill and seals	Microseismic monitoring Geophysical logging of backfill and seals
Access Rooms and Shafts	Temperature Deformation Induced stress Piezometric head Rock support monitoring	N/A	Temperature Deformation Induced stress Piezometric head Geophysical logging of backfill and seals

Chapter 86

CALCULATED AND MEASURED DRIFT CLOSURE DURING THE
SPENT FUEL TEST IN CLIMAX GRANITE

Jesse L. Yow Jr. and Theodore R. Butkovich

Lawrence Livermore National Laboratory
Livermore, California

Geological storage of spent-fuel assemblies from an operating nuclear reactor has been underway since the Spring of 1980 at the U.S. Department of Energy's Nevada Test Site. The primary objective of this generic test is to evaluate granite as a medium for deep geological storage of high level reactor waste and to provide data on thermal and thermomechanical behavior of granite from imposed heat loads.[1]

The underground installation in the Climax stock granite was sited at about 420 m below the surface and 145 m above the existing water table. As shown in Figure 1, three parallel drifts were excavated spaced on approximately 10 m centers. Seventeen storage holes were drilled at the central drift on 3 m centers in which eleven spent fuel canisters and six thermally identical electrical simulators were emplaced. Electrical resistance heaters were emplaced in vertical holes in the floor of the side drifts on 6 m centers. The thermal output of these heaters is being periodically adjusted to simulate the thermal response of a large storage array. Rock and ventilation air temperatures are being measured continuously in the drifts and throughout the rock surrounding the excavations.

Measurements of drift deformation have been made routinely since the emplacement of the spent fuel. Both vertical and horizontal measurements are being taken at five locations along the heater drifts and at six locations along the central canister drift (Fig. 1). Two types of instrumentaton are being used: convergence wire extensometers which are monitored automatically, and a manually operated tape extensometer with which measurements are taken periodically. The tape extensometer measurements were initiated about 6 weeks following fuel emplacement. The lack of a temperature correcting capability for the convergence wire extensometers make that data unavailable for analysis at this time.

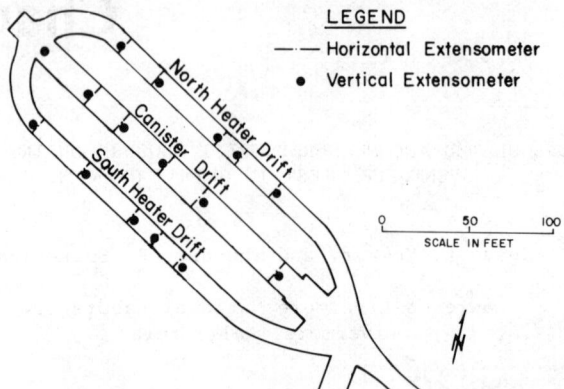

Figure 1. Plan of Spent Fuel Test with Convergence Measurement Locations.

A series of finite element calculations were run with as-built drift geometry and measured physical, mechanical, and thermal properties of Climax Stock granite.[2] The ADINA[3] structural analysis and compatible ADINAT[4] heat flow codes were chosen for this because of their ability to handle diverse factors such as heat flow by conduction, radiation and convection, thermoelasticity, and excavation. ADINAT was adapted to model both internal radiative heat transfer within the drifts and ventilation.[5] The ADINA calculations were run with an isotropic thermal-elastic model. Separate ADINA calculations were made using different elastic moduli for the rock. Averages of laboratory measurements on small samples of rock taken from the site give a value of 48 GPa, while the average value from field determinations of effective elastic modulus is about 27 GPa. In addition, an 0.5 m region around each opening that was damaged by explosives during excavation was shown to have an effective elastic modulus of about 13 GPa.[6]

As expected, the thermally induced closures of the drifts are different in each calculation. Figure 2 shows as an example the horizontal closure of the canister drift from the thermal load for each calculation. The smallest closures are for the stiffest laboratory measured modulus. Values are larger by a factor of about 2 for the field determined value, and are about 10% still larger when the low modulus damaged region is included. When plotted as a function of time, the closure of the drifts is in each case parallel for the three calculations. The calculations show that the vertical closure of the canister drift and the horizontal closure of the heater drift occurs within about 6 months after start-up, followed by a small divergence after maximum closure. The horizontal closure of the canister drift and the vertical motion of the side drifts are similar in that most of the closure occurs early, but in these cases the closures continue at a much slower rate without reversal.

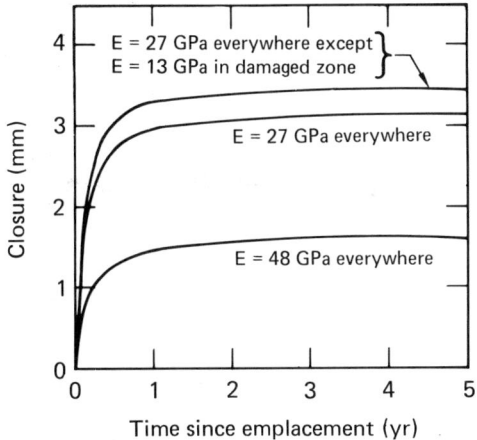

Figure 2. Calculated Horizontal Closure of Canister Drift.

Drift closure measurements reported here were made manually with a model 51855 Tape Extensometer, manufactured by Slope Indicator Co., of Seattle, Washington. A set of measurements were made at each location in the canister and heater drifts at about one month intervals. Potential sources of error in the final values obtained from given readings include thermal expansion effects and operator influences. Measurements are made to the nearest 0.025 mm (0.001 in.). Thermal correction errors are no more than ± 2% of these values if the instrument temperature is known. Of greater significance are operator errors, since measurements are made by several operators. The variation between operators repeating the same measurement falls between ± 0.1 mm.[7]

Measurements from five locations in the canister drift, five locations in the north heater drift, and four locations in the south heater drift were used in the analysis. Those in the curved part of the south heater drift which may show tunnel end effects and the northwest end of the canister drift were not used. Temperature corrected results from redundant measurements and different operators were arithmatically averaged at given locations and times to produce single values for closure with time in a given drift.

Figure 3 shows an example of sets of measurements of horizontal closure of the north heater drift, including one plot for each location where measurements were made. The variations between locations do not appear to be systematic, that is, values do not increase or decrease from the center of the drift to either end. The variation between locations can be as high as ±0.2 mm. Again, these curves were averaged arithmatically to produce an average curve for each drifts' horizontal and vertical closure for comparison with calculational results.

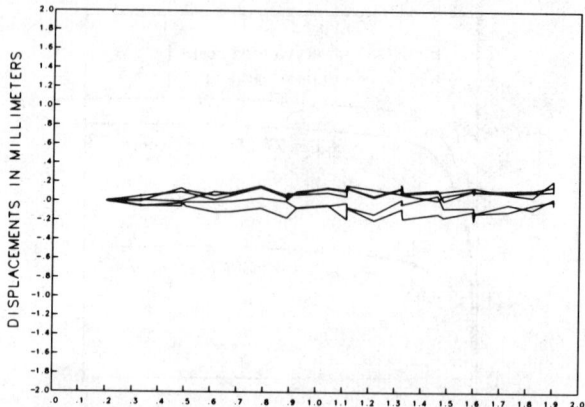

Fig. 3. Measured Horizontal Closure of North Heater Drift.

Tape extensometer results averaged in this manner are compared with the calculated displacements of the drifts at floor and ceiling centerlines and at the mid-height of the walls in each drift. The calculation used is that which has no explosive-damaged region, that is, the modulus is 27 GPa throughout the finite element mesh. This calculation was chosen for the comparison because prior to the installation of the anchor points for drift closure measurements, most of the blast-damaged rock was scaled from the walls. Figures 4a and 4b show the results compared for the horizontal and vertical closure of the canister drift respectively; Figures 5a and 5b for the north heater drift; and 6a and 6b for the south heater drift. The solid lines are the averages of the measured results, and the lines with cross marks are the calculational results. In each case, the total calculated closure since emplacement of the spent fuel (May 6, 1980) is plotted. The tape extensometer values are assumed to agree with the calculated result when the first set of measurements were made. The calculational results show that most of the closure occurred in the interval between the time of emplacement of the spent fuel and the first set of tape extensometer measurements.

The calculations and data track each other very well. For the measurements showing the closure to be continuing throughout the total time, the calculational results also show this effect. Where the measurements show closure followed by dilation, the calculation shows this effect as well. In general, considering the extremely small closures that were measured in the time interval starting at about six weeks after emplacement of the spent fuel canister, the agreement with calculations is very good. When considering the total closure the measured numbers are by comparison extremely small and in all cases are less than one millimeter since the start of tape extensometer measurements. When considering the difference between calculation and measurement, the fractional difference should be based on the total closure. In all cases the difference is less than 30%. When the total closure is almost zero as in the case of the horizontal closure of the heater drifts, this analysis doesn't hold because the calculated displacements are within the range of error for the tape extensometer measurements. Table I shows the results.

Table 1: Difference Analysis between Measurements and Calculations
1.54 Years Since Start

		Calculated Closure since Start (mm)	Closure since 1st meas. (mm)	Measured Closure (mm)	Diff.	% Diff.
Canister drift	H	2.60	1.13	0.38	0.75	28
	V	2.76	0.60	0.10	0.5	18
S. heater drift	H	----	0.11	0.05	0.06	--
	V	1.88	0.79	0.60	0.19	10
N. heater drift	H	----	0.11	0	0.11	--
	V	1.88	0.79	0.60	0.19	10

SUMMARY

Horizontal and vertical measurements of drift closures have been made with a manually operated tape extensometer since about 6 weeks after the emplacement of the spent fuel at various locations along the length of the drifts. The averaged closures are less than 0.6 mm from the onset of measurements through about two years after the spent fuel emplacement.

These results have been compared with thermo-elastic finite element calculations using measured medium properties. The comparisons show that most of the closure of the drifts occurred between the time the spent fuel was emplaced and the time of first measurement. The comparisons show that the results track each other, in that where closure followed by dilation is measured, the calculations also show this effect. The agreement is excellent, although where closures of less than 0.2 mm are measured the comparison with calculations is limited by measurement reproducability. Once measurements commenced the averaged measured closures remain to within 30% of the calculated total closure in each drift.

ACKNOWLEDGMENT

Work performed under the auspices of the U.S. Department of Energy by the Lawrence Livermore National Laboratory under contract number W-7405-ENG-48.

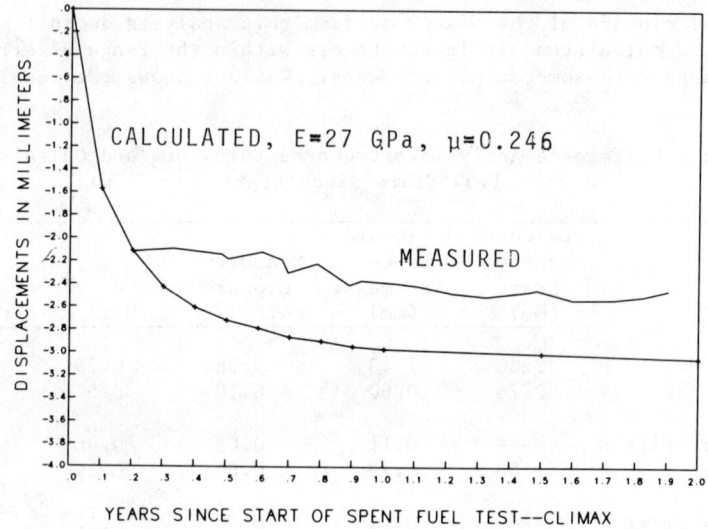

Figure 4a. Calculated and Measured Horizontal Canister Drift Closure.

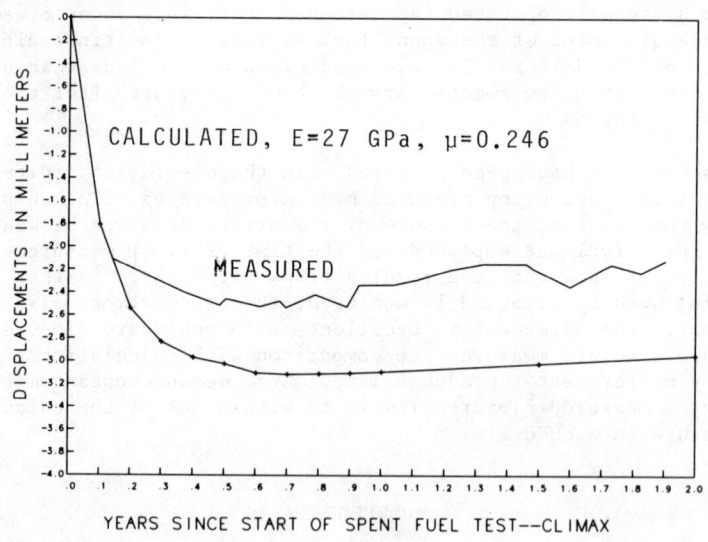

Figure 4b. Calculated and Measured Vertical Canister Drift Closure.

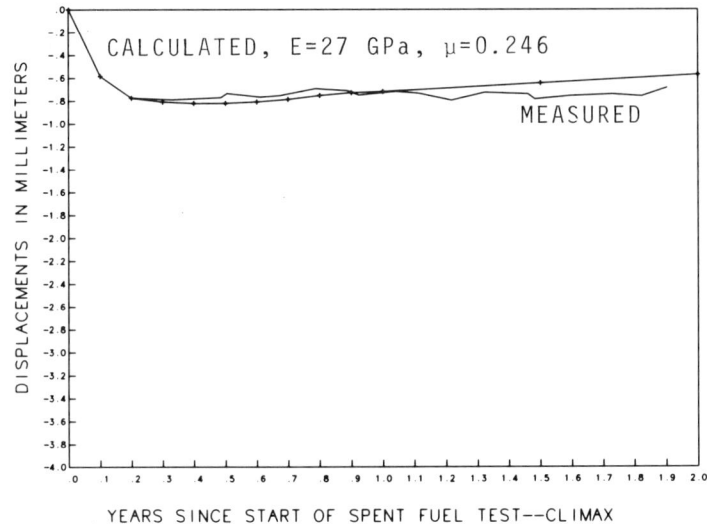

Figure 5a. Calculated and Measured Horizontal North Heater Drift Closure.

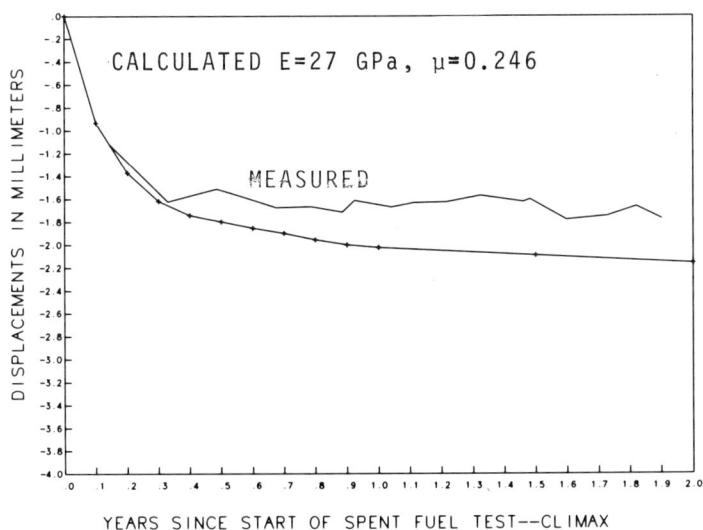

Figure 5b. Calculated and Measured Vertical North Heater Drift Closure.

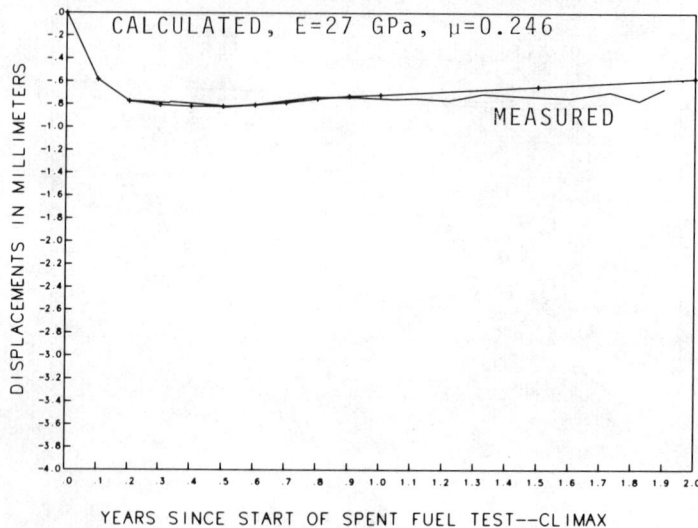

Figure 6a. Calculated and Measured Horizontal South Heater Drift Closure.

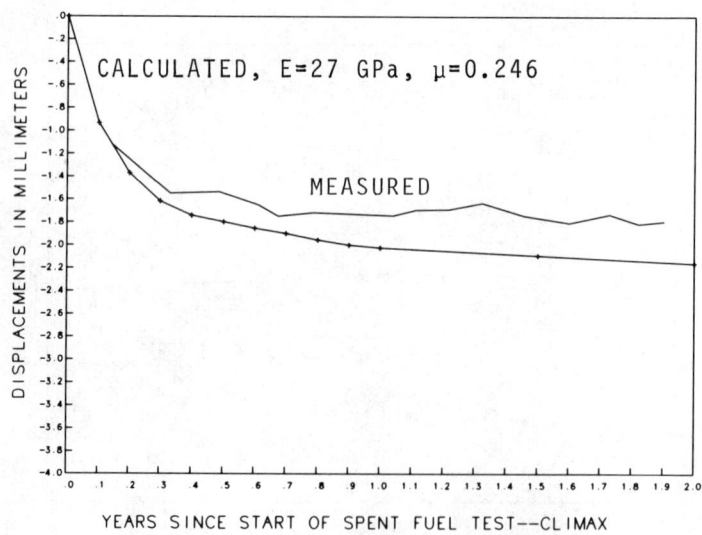

Figure 6b. Calculated and Measured Vertical South Heater Drift Closure.

BIBLIOGRAPHY

1. Ramspott, L.D., Ballou, L.B., Carlson, R.C., Montan, D.N., Butkovich, T.R., Duncan, J. B., Patrick, W.C., Wilder, D.G., Brough, W.G., and Mayr, M.C., Technical Concept for a Test of Geologic Storage of Spent Reactor Fuel in Climax Granite, Nevada Test Site, Lawrence Livermore National Laboratory Report UCRL-52796 (1979).

2. Butkovich, T.R., As-Built Mechanical and Thermomechanical Calculations of a Spent-Fuel Test in Climax Stock Granite, Lawrence Livermore National Laboratory Report UCRL-53198 (1981).

3. Bathe, K.J., ADINA, a Finite Element Program for Automatic Dynamic Incremental Nonlinear Analysis, Massachusetts Institute of Technology, Cambridge, Mass., 82448-1 (1978).

4. Bathe, K.J., ADINAT, A Finite Element Program for Automatic Dynamic Incremental Analysis of Temperature, Massachusetts Institute of Technology, Cambridge, Mass., 82448-5 (1977).

5. Butkovich, T.R., and Montan, D.N., A Method for Calculating Internal Radiation and Ventilation with the ADINAT Heat Flow Code, Lawrence Livermore National Laboratory Report UCRL-52918 (1980).

6. Heuze, F., Patrick, W., DeLaCruz, R., and Voss, C., In-Situ Deformability, In-situ Stresses and In-situ Poisson's Ratio, Climax Granite, Nevada Test Site, Lawrence Livermore National Laboratory Report UCRL-53076 (1980).

7. Yow, J.L., and Wilder, D.G., Tape Extensometer Sensitivity and Reliability, Lawrence Livermore National Laboratory Report UCRL-86100 (1981).

Chapter 87

SOME CONSIDERATION OF IN-SITU TESTING
ON MECHANICAL PROPERTIES OF ROCK MASS

by Zhu Weishen and Xu Dongjun

Associate Professor and Doctor

Assistant Professor

Institute of Rock and Soil Mechanics
Wuhan, China

ABSTRACT

In the 1st part of this paper a comparassion of the different methods for determining the modules of elasticity and deformation of hard rock is made. It is found that the plate bearing method has a serious shortcoming. The slit flat-jack method seems to be a more reasonable technique, which is recommended as a desirable method for in situ testing. In the 2nd part, a new test method to determine the properties of deformation and strength of weak plane in rock mass is recommended. Then the relation of weak plane character to the character of stress-strain curve, the relation of normal stress to the character of shear displacement curve and the yielding criterion of weak plane are discussed.

INTRODUCTION

In situ mechanical testings of rock mass are generally the most important content for the research of the stability of rock engineering. However it is a very expensive and time-consuming job as well. Therefore choosing the reasonable testing method is very important. The following paragraphs will discuss some current methods of in situ testing for rock mass.

SOME PROBLEMS OF IN SITU TESTING METHODS RELATING TO DETERMINING DEFORMATION PROPERTIES IN HARD ROCK

A series of rock mechanics in situ testing have been conducted by our institute in an engineering site in 1974 and 1975. Several conventional testing methods to determine modules of elasticity and deformation were comparaed. It has been found that the testing results were quite different for the different methods even for in situ testings. Now we shall mainly discuss these methods, their results and analyze the cause of difference in testing results.

These testings were made in a testing tunnel. Rock stratum was a relatively homogeneous dolomitic limestone with large thickness, obviousless anisotropy and undevelopment of joints.

1. <u>True Triaxial Compression Testing.</u> The specimens were prepared by pneumatic drill. The vertical loading was supplied by 6 hydrolic jacks with the max. loading up to 900t in total and the two lateral loadings σ_2 and σ_3 were supplied by flat jacks with the max. pressure of 100kg/cm. The dimesion of specimen was 60cm high with the cross section 50 x 50cm². The test layout is shown as fig.1 and the testing curves were drawn according to the relation of the deviatoric stress ($\sigma_i - \sigma_0$) to the deviatoric strain ($\varepsilon_i - \varepsilon_0$). It can be found that the cycle of loading and unloading had obvious hysteresis. To estimate the modules elasticity and deformation the following formula was adopted:

$$E_i = (1 + \mu) \frac{\sigma_i - \sigma_0}{\varepsilon_i - \varepsilon_0}$$

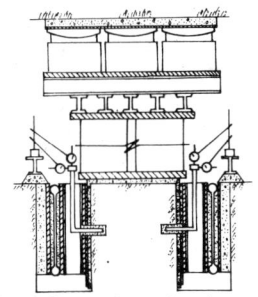

Fig.1 The equipement scheme for in-situ triaxial compression testing

2. <u>The Uniaxial Compression Test.</u> The specimens were cut on the floor with dimension 45 x 45 x 50 cm³ or so (Fig. 2). The vertical loading was supplied by two flat jacks with capacity up to 800T in total. The unit pressure on the specimen achieves 460 kg/cm² in max.. The complete stress-strain curves were drawn as fig. 3. It can be seen that the character of these curves are very similar to the character of curves for small rock sample in lab..

3. <u>Stiff Plate Bearing Test.</u> The specimens were prepared on the floor by clearing away the disturbed surface layer of the rock. The dimension of the plate was 30 x 30 cm² and the max. loading capacity was over 260 kg/cm² .

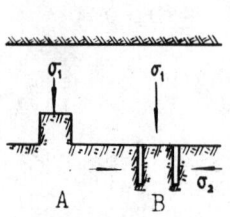

Fig.2 Methods A and B

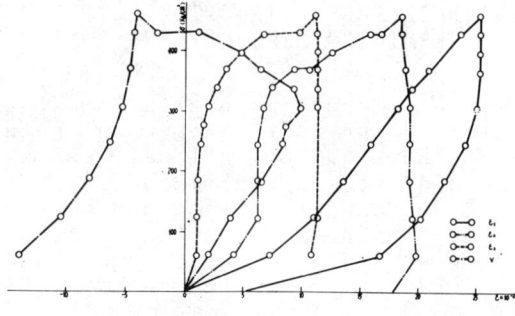

Fig.3 Complete stress-strain curve

4. **Slit Flat-jack Test Method.** It is the case that the specimen were formed by cutting a narrow vertical or horizontal slit in the wall with length about 4 times longer than that of flat jack (190 x 200 x 20 cm^3). Then the loading was supplied by means of the flat jack and it was treated as a semiplane problem to determine the E_e and E_d. The calculating formula is as follows (Fig. 4):

$$E = \frac{q}{\pi \varepsilon} \sin2\theta + 2\theta + \mu (\sin2\theta - 2\theta)$$

in which, q—supplied loading, μ —Poisson ratio, θ —the angle between the central line of loading and the line connected measuring point and edge point of that loading.

5. **Acoustic Wave Method and Laboratory Testing.** The results of above mentioned testing methods are shown in table 1:

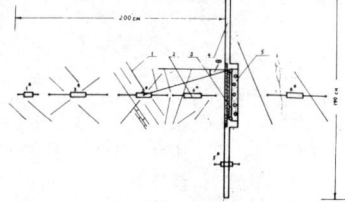

Tab.1 Variation of E_e and E_d for different methods (10^4 kg/cm^2)

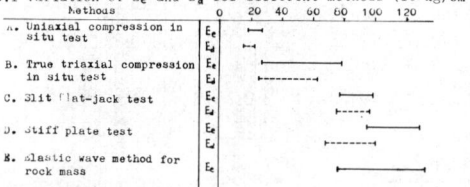

1. Joint 2. Strainmeter 3. Flat jack 4. Slit 5. Cement

Fig.4 Test layout of slit flat jack Method

6. **Discussions.** Now, we are going to discuss some problems from the above mentioned tests: 1) The Comparassion of Different methods. It can be seen from table 1 that the difference between the max. and min. values of E_e for different methods can be achieved up to 6.7 times and for E_d up to 7.3 times (methods A and D). The values E obtained from the testing methods with confining rock like C and D are much greater (1 to 5 times) than those without confining rock like A and B. There were obvious hysteresis phenomena in the cycles of loading and unloading for methods A, B and D (see fig. 8). However, in the method of C (fig. 4) the testing curves of left and right

side of the loading plane are quite different. On the right side, it has a hysteresis, because the measuring points cut across a big joint. On the contrary the curves of loading and unloading process are very closed each other on the left side (Fig. 5). It has been proved that the phenomenon of hysteresis is created just because of the presence of big joint, by which the internal resistance of recovering deformation is increased. It can be considered that during excavating the specimens of methods A and B undergo more vibration of pneumatic drilling or blasting than another's, by which much more new joints created in the specimens, because they have more free surfaces. Then in the case of method D having the loading direction perpendicular to the floor or the wall of tunnel (Fig. 6), the test loading just make a compaction to the unloading joints, which have been formed around the tunnel in rock mass due to the excavation. Moreover the deeper part of rock mass including the unloading joints up to a certain depth will be compacted continuously with the increasing of supplied loading and values of E_e and E_d will be decreased as well (Fig. 7). From above discussion it shows that the value of E_e and E_d obtained by means of methods A and B will be too low certainly, however the method D as a current one in the world has a very serious shortcoming as well. By this technique people can always obtain the results of E_e to be too high and E_d to be too low (Fig. 8).

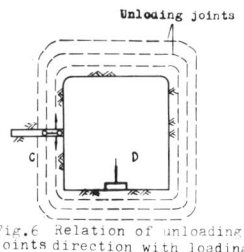

Fig.5 stress-strain curve (C)

Fig.6 Relation of unloading joints direction with loading direction of methods C and D

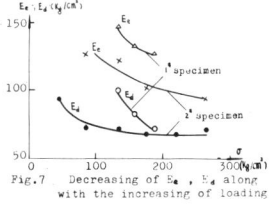

Fig.7 Decreasing of E_e, E_d along with the increasing of loading

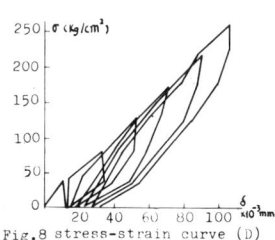

Fig.8 stress-strain curve (D)

2) Recommended Method. It is considered that the current method D to determine the E_e and E_d is not desirable. According to the opinions of the authers, the more reasonable one among above mentioned would be the method C. As the loading direction of that is approximately parallel with the main direction of unloading joints in confining rock, latter will have a small influence on the test results only. So it can be considered that the test curve of this method can indicate the property of deeper part of rock mass better than others. The another merit of this method

is that it can be used to detect the deformation behavior of deeper part of rock in detail and directly.

TEST STUDY OF MECHANICAL PROPERTIES OF WEAK PLANE IN ROCK MASS

<u>1. Shear Test Method Of Weak Plane In Rock Mass.</u> Generally the stability of engineering rock mass mainly depends on the deformation and strength properties of weak plane in rock mass. A new method to carry out this kind of testing will be recommended in this paragraph.

The specimens shown in fig. 9 a, b and test layout shown in fig. 10 a, b, may be used to determine the deformation and strength properties of weak plane, which is of

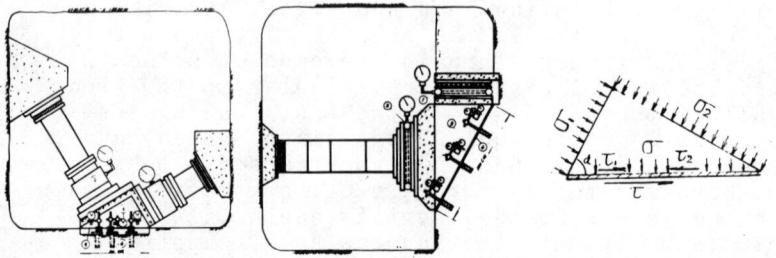

Fig.9 Specimens with weak plane for shear test

Fig.10 Shear test layout of weak plane in rock mass

following advantages:
1). The stresses on shear plane are distributed uniformly.
2). It is applicable for weak plane of any occurence.
3). The velocity of shear displacement can be controlled to be constant by means of two jacks, which is convenient for obtaining the complete process of stress-strain relation.
4). It is convenient for using the single specimen method to determine the strength properties by one specimen and one test only.

The plane, on which σ_1 act, must be perpendicular to that of σ_2, and it is necessary to make $\sigma_1 \sin\alpha$ greater than $\sigma_2 \cos\alpha$. The shear area is about 3000--$15000 cm^2$.

According to Mohr-Comlomb strength criterion, on shear plane,

$$\tau = \sigma tg\phi + c$$

$$\tau = 1/2 (\sigma_1 - \sigma_2) \sin(2\alpha)$$

$$\sigma = \tfrac{1}{2}(\sigma_1 + \sigma_2) + \tfrac{1}{2}(\sigma_1 - \sigma_2)\cos(2\alpha)$$

Loading is supplied according to the stress path, which is shown in fig. 11, the test way and steps are following:

1). σ_1 and σ_2 are supplied synchronously to make shear stress on weak plane to be $\tau = 0$, and normal stress increases up to given value σ_0 gradually.
2). By increasing the σ_1 with simultaneous decreasing the σ_2 to keep the σ_0 to be a

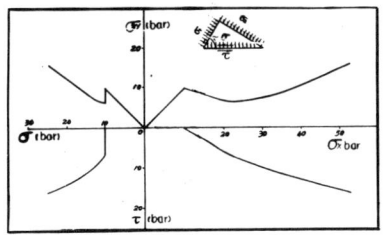

Fig.11 Stress path of shear test

constant the shear stresson weak plane increase gradually. In this way the test is continued until the weak plane yields or approaches to shear strength and the curvature of τ-U_s curve is to be closed to zero.

In order to obtain the τ-σ curve it is necessary generally to cut more than three specimens at least on the same weak plane and to carry out above mentioned shear test under variety of constant σ_0. But in case that it is difficult to cut more than three specimens along the same weak plane, this test may be carried out by a new technique, which is called as single specimen method and detailed as following.

At first, the above mentioned conventional shear test should be conducted until the curvature of σ_1 -- U_s curve approaches to zero, as shown in fig. 12. Then it is necessary to abandon the restraint of keeping the σ_0 to be a constant, and to increase the σ_1.
Meantime let the σ_2 increases relatively with increasing of shear displacement U_s. This time both τ and σ on the weak plane increase systematically with U_s, which is shown as the case over points F and F' in fig. 12. Corresponding to the same U_s values, the values of σ'_1 -- σ'_2, σ''_1 -- σ''_2, σ'''_1 -- σ'''_2, etc. are obtained and the curve of τ -- σ relationship is determined by Mohr-circle method.

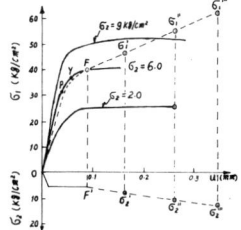

Fig.12. The stress-strain curves of granodiorite weak plane with fault mud. The solid lines show the curves of conventional method; The dotted lines show the curves of new single sample method.

2. <u>The Characters Of Stress-Strain Curve Of Weak Plane In Rock Mass.</u> a) The properties of weak plane and its stress-strain relation. On basis of data collected by the authors, the relationship curves of shear stress τ and shear

displacement U_S of various weak plane may be classified into three classes as shown in fig. 13. Under the same normal stress σ_0, the stress-strain relationship of different kinds depends on the thickness, mineral composition, filling condition irregularity of weak intercalation and the property of rock on both sides of weak plane, etc..
b) The relationship of normal stress and characters of τ - U_S curve. As shown in fig. 14, the curve of shear stress τ and shear displacement U_S of the same weak plane in rock mass may change from one kind to another with increasing normal stress . It can be often met in the case of the waek plane filled with thicker pelitic intercalation. Fig. 14 shows the test curve of pelitic intercalation of clay rock. It is resulted obviously by further consolidating of weak plane with increasing normal stress σ. As shown in fig. 15, when one or both side wall of the weak plane are very hard and with increasing of σ the material in weak intercalation is easy to be crushed, the τ - U_S curve transforms from class I to class II or III. c) The yielding criterion of weak plane. With the test layout shown in fig. 10 the shear displacement U_S and normal displacement U_V of weak plane are measured at the same time. It may be appropriate to determine proportion limit of τ - U_S curve according to the pushing end A of specimen changes from compression ($-U_V$) to tension ($+U_V$), because point P represents the value of shear stress, at which one end of weak plane begins to crack. It may be feasible to consider the cracking of whole weak plane

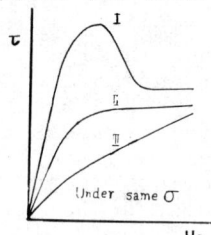

Fig.13 $\tau - U_s$ curve of various weak planes under the same normal stress σ

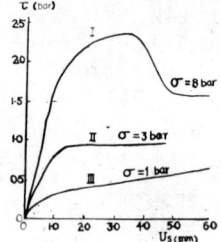

Fig.14 The curves of $\tau - U_s$ relationship of the same weak intercalation under different normal stresses

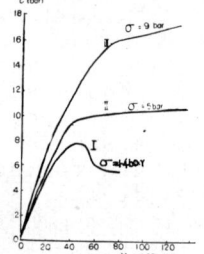

Fig.15 Shear test curve of cementation plane between the concrete and clay siltstone

as yielding criterion and the point Y as yielding limit value, which corresponds to point Y', at which the back end B of specimen begins to crack.(Fig.16)

As has been said above, the deformation and strength properties of weak plane in rock mass must be determined with correct test method. The characters of weak plane and normal stress supplied on it would affect its stress -strain relation strongly. It is necessary to choose the range of normal stress correctly and make it compatible with the consolidation condition of weak intercalation during the testing process.

Fig.16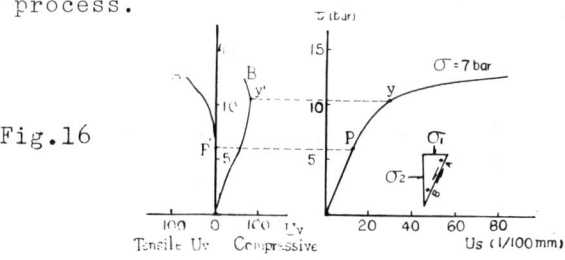

ACKNOWLEDGMENT

The authors wish to express their gratitude to Mr.Wu Yushan, Chen Shengqiang, Dai Guanyi and others for taking part in the perfoming in situ tests.

REFERENCES

Zhu Weishen, 1979, "Stability of Large-Span Cavern in Jointed Rock Mass", Proceedings of 4th Congress of ISRM, T. 3, Montreux.

Xu Dongjun, 1980, "The Rheological Behavior of the Weak Rock Mass and the Methed of Determining Long-Term Strength", Rock and Soil Mechanics, No. 1 (in Chinese).

Chapter 88

ISSUES RELATED TO FIELD TESTING IN TUFF

R. M. Zimmerman

Sandia National Laboratories, Albuquerque, NM

INTRODUCTION

Tuff is being considered as a possible geologic medium for the underground storage of commercial high level radioactive wastes by the Department of Energy (DOE). DOE has the responsibility for developing or improving the technology for safely and permanently isolating these wastes. DOE programs must satisfy requirements established by the Nuclear Regulatory Commission (NRC) for the licensing and regulating of these storage facilities (NRC 1981). Pertinent draft NRC requirements are: 1) releases of radioactivity to the environment must be within prescribed limits and 2) emplaced canisters containing wastes must be retrievable up to 50 years following emplacement. High level radioactive wastes produce heat in storage and it is essential that the thermal/mechanical/hydrological response of the host rock will not significantly degrade either the structural performance of the repository or the ability of the natural or engineered barriers to retard radionuclide migration.

The Nevada Nuclear Waste Storage Investigations (NNWSI) was established by DOE to evaluate the suitability of disposing of high level nuclear wastes on or near the Nevada Test Site (NTS) (NNWSI 1980). NNWSI has focused exploration and site evaluation on the tuffs of Yucca Mountain. Field testing is essential to support site characterization and repository design efforts. Preliminary studies have been conducted indicating that tuffs in G-tunnel on the NTS contain welded and nonwelded tuffs that have similar properties to those in Yucca Mountain. Thus, field testing can be accelerated. This paper discusses the issues supporting the planned field experiments in G-tunnel. Issues not related to rock mechanics are not included.

BACKGROUND

Tuff is a form of volcanic rock. Ash-flow tuffs are deposited as sheets from a gas-charged density cloud that rolls down the slope of a volcano; ash-fall tuffs are emplaced by the aerial settling of particulates and may vary greatly in thickness (Smith 1960). Welding of tuffs results when the emplacement temperature (>500°C) and lithostatic pressure are sufficiently high to result in plastic compaction and cohesion of the glass fragments. The degree of welding ranges from incipient, where there is simply bonding of contact points of individual glassy fragments or shards, to complete, which is marked by the cohesion of the surfaces of these fragments, coupled with plastic deformation, so that the pore space between the fragments is practically eliminated. Slightly welded tuffs may have a porosity near 0.3 by volume; moderately welded from 0.2-0.3; and densely welded from 0 to 0.2. For convenience, nonwelded tuffs are considered to have a porosity more than 0.25 (Johnstone and Wolfsberg 1980). The cooling processes of the tuff deposits affect the resulting porosity grain size, chemical and/or mineralogical composition, and structural continuity (Smith 1960). The latter is of particular importance. Welded tuff is more brittle and, as a result of cooling shrinkage, is more jointed than nonwelded tuff.

Zeolites usually form in the pore space of nonwelded tuffs when they are deposited at depth and are in the presence of groundwaters. This is considered secondary mineralogical alteration. The zeolites have the effect of changing the porous nonwelded tuff into a relatively impermeable continuum.

TUFF ISSUES

This section outlines the major rock mechanics related issues and subissues associated with the use of tuffs as the host rock for a nuclear waste repository. These are defined and discussed in the following paragraphs and are summarized in Table 1.

The first major issue is the effect of TEMPERATURE rise on the rock mass due to the emplacement of heat producing wastes. Temperature rises are distributed through the rock mass by thermal conductivity and result in thermal stresses due to thermal expansion/contraction phenomena (Johnstone and Wolfsberg 1980). A subissue identified with temperature is that of tuff dehydration. Nonwelded tuffs can dehydrate at moderate temperatures and thermal contraction can occur even below 100°C due to complex behavior in and around the pore space (Lappin 1978). Such effects might lead to the destabilization or spalling of excavated surfaces, or possibly unwanted internal rock stress redistributions. A second subissue associated with temperature is that increased temperatures may increase the tendency for time-dependent viscous deformations, or creep of nonwelded tuffs. Creep could lead to the geometrical distortion of excavated shapes. Possible impacts are that required canister

Table 1
Summary of Issues, Subissues, Responses, and Impacts

Issue*	Tuff** Type	Subissue	Potential Response/ Possible Impacts
T	N	Dehydration	Thermal Contraction/ o Surface destabilization or spalling o Stress redistributions
T	N	Creep	Geometrical Distortions/ o Retrieval problem o Sealing problem
T,P	A	Pore Pressure	Pore Pressure Changes/ o Mechanical strength changes
T,P	A	Pressure Gradient	Water Migration/ o Flooding of emplacement hole o Emplacement hole sealing problems o Water borne heat transmission
T,P	A	Vaporization	Pore dewatering/ o Increase in canister temperatures
T,J	W	Joint Characteristics	Water Transmission/ o Preferential pathways
T,J	W	Jointed Rock Behavior	Nonuniform or Anisotropic Properties o Reduced nonlinear, or nonuniform modulus of deformation o Nonuniform or time-dependent thermal properties
T,D	W	Inhomogeneity	Stress Concentrations/ o Localized failures o Inaccurate model representations
T,D	A	Stratigraphic Variation	Nonuniform Rock Response/ o Inaccurate model representations o Shear failures at interfaces

*T - temperature
P - pore water
J - joints
D - depositional patterns

**N - nonwelded tuff
W - welded tuff
A - all tuffs

retrievability could be affected and/or borehole sealing problems could be intensified.

A second major issue is that of PORE WATER. Pore water can exist in either the liquid or vapor state. Pore pressure is identified as a subissue. Pore pressures can exist due to hydraulic head or be generated due to thermal stresses when the pore water is in the liquid form. Increases in effective pore pressure can lead to decreases in mechanical strengths. The second subissue is pore pressure gradient. Pore water under pressure gradients results in water migration. Pore pressure gradients can be established due to excavations below the static water level under ambient temperatures. Tunnel water accumulation problems can result. Emplacement hole sealing problems could result. Pore water can also migrate if a gradient is formed by differential pore pressures caused by thermal stresses. Pore water can also migrate in vapor form if the gradient is caused by differential vapor pressures (Johnstone 1979). The possible impact of the latter two cases is that there might be water migration into canister emplacement holes or accompanying drifts. Such water migration could lead to problems in designing engineered barriers required to contain radioactive wastes (NRC 1981). A secondary effect associated with water migration is that heat can be transported with the water. Such action could alter expected thermal conditions around emplacement holes and drifts. The third subissue also relates to problems in defining thermal properties around emplacement holes. This is the subissue of pore water vaporization. Pore water can vaporize due to increased temperatures. The result is that the thermal properties can change significantly. Both thermal conductivity and heat capacity can be reduced due to vaporization (Lappin, 1980). The potential impact is that canister temperatures can be increased.

A third major issue relates to the potential effects of JOINTS on welded tuffs. Joints are significant because they limit the usefulness of continuum theories (Johnstone and Wolfsberg 1980). The first subissue relates to joint characteristics. Joint dimensions and frequencies within a stratum of joints and the possible extension of joints into adjoining strata govern water migration phenomena. They become preferential pathways for water migration. The second subissue covers jointed rock behavior. Jointed rock tends to have nonuniform or anisotropic mechanical and possibly thermal properties. In the mechanical case, joints tend to reduce the bulk rock modulus of deformation. Parallel joint sets can cause the modulus to be somewhat anisotropic. Joints can also cause the modulus to be nonlinear and this complicates modeling representations. The effects of joints on thermal properties needs attention. It is speculated that the presence or absence of moisture in joints may cause the thermal properties to be nonuniform and possibly anisotropic. There is also a possibility that joints might partially heal under increased temperatures and the thermal and mechanical properties could be time dependent.

The fourth major issue deals with the DEPOSITIONAL PATTERNS associated with tuffs. Ash-flow tuffs travel over the terrain before settling and cooling (Smith 1960). They may collect rubble during their travels and it can become entrapped in pockets or layers. This leads to a subissue of inhomogeneity. The impact is that stress concentrations may occur and localized failures might result. These pockets can complicate constitutive developments for modeling. Another factor causing inhomogeneities in tuff is the entrapment of air or gas during the deposition process (lithophysae). Lithophysae zones in welded tuff units may have properties different from more uniform deposits of welded tuff. A second subissue is that of stratigraphic variation. Depositional processes lead to horizontal and vertical variations within single cooling units. Variable boundary conditions due to flow frequencies and contents of preceding and following flows cause the welding processes to vary. Therefore, there may be thermal and mechanical variations through the thickness of a unit. There may also be significant variations between adjoining units as welded tuffs may be located between nonwelded tuffs. Special efforts could be required to describe these interfaces when conducting computer modeling studies. In particular, potential shear failures may be difficult to define.

EXPERIMENTAL PROGRAMS

The issues and subissues outlined in this paper have evolved from NNWSI studies, which have included laboratory and limited field investigations. Tuff site characterization efforts require baseline information and analyses. To obtain a meaningful evaluation of the Yucca Mountain site, the pertinent geotechnical properties must be established. Repository conceptual designs require an understanding of the basic phenomena relevant to tuff. Behavior of tuff in response to issues such as temperature, pore water, joints and depositional patterns must be understood. There is a need to develop confidence in predictive modeling capabilities as well as develop instrumentation and control system experiences. Based on these needs, an assessment of pragmatic program directions, site limitations, and cost considerations, a plan for field testing in G-tunnel has been formulated (Zimmerman 1982). Experiments that address to these issues and subissues are listed and summarized in Table 2. The table shows the more descriptive subissues identified with each experiment. The following paragraphs briefly discuss these experiments. Planned dates to initiate testing are included.

One limitation of the G-tunnel site should be mentioned as it has a direct impact on extrapolations to Yucca Mountain. G-tunnel is located in nonwelded and welded tuffs above the static water table; however, a perched water table causes the exposed, nonwelded tuffs to be very nearly saturated and the saturation of the welded tuffs ranges to greater than 0.9 (Warpinski, et al, 1978). Thus, water related experiments with a hydraulic head cannot be conducted there, but temperature-dependent water migration can be studied. The impact of this limitation is not clear at this time. Four potential reposi-

Table 2

Summary of Planned Field Experiments

Experiment: Small Diameter Heater (SDH) - 1982
  Brief Description: Small-scale canister simulation
  Measurements: Temperatures, amount of water and vapor
  Media: Welded and nonwelded tuff
  Subissues: Dehydration Pore Pressure, Pressure Gradient, Vaporization

Experiment: Unit Cell Canister Scale (UC) - 1983
  Brief Description: Canister scale heater simulation with 8 supplemental guard heaters in axisymmetric arrangement
  Measurements: Temperatures, deformations, stress changes, pore pressure
  Medium: Welded tuff
  Subissues: Pore Pressure, Pressure Gradient, Vaporization, Jointed Rock Behavior, Inhomogeneity, Stratigraphic Variation

Experiment: Heated Block (HB) - 1982
  Brief Description: Two-meter cubed prismatic block defined by 4 vertical slots and heated by 2 parallel lines of 7 guard heaters
  Measurements: Temperatures, deformations, stress changes, pore pressure, hydraulic conductivity, ultrasonic properties
  Medium: Welded tuff
  Subissues: Pore Pressure, Pressure Gradient, Vaporization, Joint Characteristics, Jointed Rock Behavior, Inhomogeneity

Experiment: Rocha Slot (RS) - 1983
  Brief Description: Flatjack placed in close fitting slot and pressurized
  Measurements: Flatjack pressure, slot surface deflection
  Media: Welded and nonwelded (includes creep) tuff
  Subissues: Creep, Jointed Rock Behavior

Experiment: Heated Slot (HS) - 1983
  Brief Description: Rocha Slot with addition of 2 parallel lines of guard heaters
  Measurements: Flatjack pressure, slot surface deflection, temperatures
  Medium: Nonwelded (includes creep) tuff
  Subissues: Dehydration, Creep, Pore Pressure, Pressure Gradient, Vaporization

Experiment: Thermal Properties (TP) - 1983
  Brief Description: Apply thermal probe techniques
  Measurements: Temperature in and around transient line heat source
  Medium: Welded tuff
  Subissue: Jointed Rock Behavior

tory horizons are being evaluated at Yucca Mountain. Two of these are welded tuff units below the static water table. Another welded tuff unit located above the water table, and a nonwelded tuff unit that straddles it are being evaluated. The final repository horizon is scheduled to be selected in December 1982 and input from field tests will be used as appropriate.

The Small Diameter Heater (SDH) Experiment is planned for welded and nonwelded units. The purpose of the experiment is to evaluate temperature distributions and moisture states in a relatively small emplacement hole, 13 cm in diameter and 3.1 m deep. The experiment will provide data for thermal model evaluations of 2-phase hydrological phenomena. The experiment will provide an understanding of water migration towards the heater source.

The Unit Cell (UC) Canister Scale Experiment is designed to evaluate thermal, mechanical, and hydrological responses of welded tuff due to heating on a canister scale. The experiment will provide for thermal and mechanical model evaluations of a jointed rock mass. Hydrological phenomena are to be monitored. High moisture contents of the relatively porous tuff provide a potential for pore water vaporization and resulting changes in the thermal and mechanical properties. The experiment is designed as a relatively long-term (a maximum of 2 years) experiment to evaluate time-dependent factors. A sufficient volume of rock is heated such that stratum inhomogeneities and variations can be evaluated.

The Heated Block (HB) Experiment is designed so that geotechnical measurements may be taken on a significant volume of jointed welded tuff (8 m^3). The block is free on five sides and boundary conditions can be controlled. Two lines of guard heaters provide a one-dimensional temperature field through the block. Coupled or uncoupled stress and temperature fields can be imposed on four of the exposed sides. The constitutive bulk behavior of the block can be measured and joint behavior can be evaluated. Effects of temperature and stress on joint permeability can be evaluated.

The Rocha Slot (RS) experiment is planned for welded and nonwelded tuffs. The experiment involves cutting a thin slot (~7 mm wide) in the rock and inserting a flatjack that contains pressure and deformation measuring sensors. Pressure-deformation responses of excavated rock surfaces can be measured. The responses can be related to joint behavior in welded tuff and to creep in nonwelded tuff. The latter requires time-dependent measurements.

The Heated Slot (HS) is a combination of the flatjack-slot measurements of the RS experiment and the 2 guard heater lines of the HB experiment. Time-dependent creep at controlled temperatures is to be evaluated. Dewatering and dehydration problems associated with nonwelded tuffs are to be evaluated.

The final experiment is the Thermal Property (TP) experiment in welded tuff. A small heater is to be inserted in tuff near a joint and the thermal conductivity and heat capacity measured. A transient line source thermal pulse is to be used. The joint is to be evaluated both saturated and dry.

SUMMARY AND CONCLUSIONS

This paper has brought out the unique properties of tuffs and related them to needs associated with their use as a host rock for a high level nuclear waste repository. Major issues of temperature, pore water, joints, and depositional patterns have been identified and related responses and impacts outlined in Table 1. Planned experiments have been outlined and their relationships to the rock mechanics issues summarized in Table 2.

The conclusions from this paper are:

(1) tuff is a complex rock and basic phenomenological understanding is incomplete, and

(2) available field test facilities will be used for a series of experiments designed to improve phenomenological understanding and support repository design efforts.

REFERENCES

1) U. S. Nuclear Regulatory Commission, Disposal of High-Level Radioactive Wastes in Geologic Repositories, Proposed Rule 10CFR60, Federal Register, Vol. 46, No. 130, July 8, 1981.

2) Nevada Nuclear Waste Storage Investigations; NVO-196-13 FY 1980 Project Plan and FY 1981 Forecast, Las Vegas, Nevada, February 1980.

3) Smith, M. L., Zones and Zonal Variations in Welded Ash Flows; Geological Survey Professional Paper 354-F, U.S. Geological Survey, Washington, D. C. 1960.

4) Johnstone, J. K. and Wolfsberg, K., editors, Evaluation of Tuff as a Medium for Nuclear Waste Repository Interim Status Report on the Properties of Tuff; SAND80-1464, Sandia National Laboratories, Albuquerque, NM, July 1980.

5) Lappin, A. R., Preliminary Thermal Expansion Screening Data for Tuffs, SAND78-1147, Sandia National Laboratories, Albuquerque, NM (1978).

6) Lappin, A. R., *Thermal Conductivity of Silicic Tuffs: Predictive Formalism and Comparison with Preliminary Experimental Results*: SAND80-0769, Sandia National Laboratories, Albuquerque, NM, July 1980.

7) Johnstone, J. K., *In-Situ Tuff Water Migration/Heater Experiment: Experiment Plan*; SAND79-1276, Sandia National Laboratories, Albuquerque, NM, August 1980.

8) Zimmerman, R. M., *Test Plan for Rock Mechanics Field Experiments in Tuff in G-Tunnel*, NTS; SAND82-0108, Sandia National Laboratories, Albuquerque, NM, in preparation.

9) Warpinski, N. R., Schmidt, R. A. et.al., *Hydraulic Fracture Behavior at a Geologic Formation Interface: Pre-Mineback Report*, SAND78-1578, Sandia National Laboratories, Albuquerque, NM, October 1978.

REINFORCEMENT

Chairman
Jose F.T. Agapito
Agapito and Associates
Grand Junction, Colorado

Co-Chairman
Francis S. Kendorski
Engineers International Inc.
Downer's Grove, Illinois

Keynote Speaker
Richard E. Goodman
University of California
Berkeley, California

Chapter 89

CALCULATION OF SUPPORT FOR HARD, JOINTED
ROCK USING THE KEYBLOCK PRINCIPLE

By Richard E. Goodman, Gen-hua Shi, and William Boyle

University of California, Berkeley, Department of Civil Engineering

ABSTRACT

This keynote paper calls attention to two critically important issues relating to selecting supports for excavations in hard rock. The first is the proportion of the ultimate sliding volume that needs to be stabilized by the support system. Some excavations can be imagined to liberate very large rock volumes, whose weight can not be assigned to supports without severe economic penalty. Using the keyblock principle, it is argued that only the most critically oriented and finite blocks right on the periphery of the opening have to be supported; if these are retained by the support system, or by self resistance due to friction on the sliding faces, then the blocks behind will remain in place. As all concepts, this one has bounds and limits. Since the keyblock analysis for rock engineering is very new, these limits are not yet understood... The second issue is the degree to which possibly stabilizing forces associated with the in-situ stress field ought to be accepted as support for potential keyblocks. It is shown that the initial stress can be effective in stabilizing rock wedges whose faces are nearly perpendicular to the excavation surface. It is too early to decide this issue but important to consider it.

INTRODUCTION

Even though hard rocks may possess considerable specimen strength, they can fail around excavations by block movement. The blocks are defined by systematic joints which can usually be described in advance of excavation. Therefore an analysis of the directions and properties of the joint sets should be able to determine the support requirements. That is the object of this paper.

When an excavation is made, the new surfaces, together with the system of pre-existing discontinuities, determine some critically located blocks around the surface of the excavation. In soft or very highly stressed rocks, the stresses flowing around the opening can cause additional cracks to form and to propagate so that the problem of block morphology is dependent upon the stress field. We are not ready to handle that problem and will confine ourselves to a discussion of critical blocks in rock masses unlikely to fracture.

The keyblock theory provides an optimum solution to the support problem by focusing on the "weak links" in the rock mass. When the excavation is made, many shapes and sizes of blocks are created but only some are shaped and oriented in a potentially unsafe manner. If the construction has preserved the basic integrity of the rock mass, these blocks will be knit together in an interlocking mosaic and only the loss of keyblocks will allow larger movements. Whether or not greater movements will follow the loss of a keyblock is also subject to analysis. But that analysis is unnecessary if the keyblocks are prevented from movement by timely support, or if they are shown to be safe without support. If all potential keyblocks are restrained, the rocks that abut against them may never know that an excavation has been made. It is very unlikely that movements of enormous multi-block volumes of rock, as experienced in a number of documented failures, initiate simultaneously over the entire rock volume. If so, attempts to use supports to prevent large failures need not be based upon the requirements for support of the entire weight of the ultimate failing mass. Discretion is important in deciding this issue.

In this paper we review the keyblock principle, previously explained by Shi and Goodman (1981) and Goodman and Shi (1981) and extend its application to excavations with compound shape, including complete tunnel sections, and edges and corners of underground chambers. We also discuss the stability analysis of keyblocks including allowance for possibly stabilizing influence of initial stress acting on the faces of the keyblock.

As discussed in the references cited, the keyblock theory makes use of a rigorous topological theorem that optimizes the search of all possible blocks formed by the intersection of a system of joints and excavation planes. Consider the joint system of Table 1, with four different sets. In Figure 1, each set has been plotted in an upper hemisphere (lower focal point) stereographic projection yielding four circles. The dashed circle -- the reference circle -- is the projection of the equator and the region within it is the upper hemisphere. The intersections of the four joint circles define 14 regions, each identified by a four digit number, e.g., 1100. The digit "0" means "the region inside a circle" and the digit "1" identifies a region outside a circle. The digits are arranged in order so that 1100 identifies the region outside circles one and two and inside circles three and four. In this upper hemisphere stereographic projection, with the focus at the bottom of the reference sphere, the region <u>inside</u> a circle corresponds to the half-space <u>above</u> the corresponding joint.

TABLE 1

Joint Set	Dip	Dip Direction
1	60	270°
2	60	130
3	60	50
4	30	90

Thus region 1100 corresponds to the block created by the intersection of half spaces below joints 1 and 2 and above joints 3 and 4. Such an intersection of half spaces will not in general be finite but it can become finite when cut by one or more excavation surfaces.

Shi's theorem allows one to determine which of these regions will be finite when intersected by a set of excavation surfaces. The excavation planes are plotted on the stereographic projection to define an "excavation pyramid". This construction creates a "target" figure. Block 1100 is finite only if it plots completely within the target region; if block 1100 hits any part of the target boundary, it is infinite. If new rock fracture is barred, then a block can be a "keyblock" only if it is finite. The construction of target diagrams for several examples will explain the method further.

APPLICATION OF KEYBLOCK PRINCIPLE FOR SURFACES, EDGES AND CORNERS OF A CHAMBER

Consider the box-shaped excavation of Figure 2a, with vertical walls striking NW and NE, a flat floor, and a roof inclined 10 degrees to the East. For the given joint system, keyblocks can intersect the floor or the roof, any of the walls, any of the edges and any of the corners. There are 6 faces, 12 edges, and 8 corners, so 26 different target diagrams can be constructed and compared with Figure 1. We will illustrate the process with three of these -- for the roof, for the Northwest edge, and for the upper Northwest corner.

Figure 2b shows the excavation pyramid corresponding to the roof. The roof plane dips 10 degrees to the East and projects as a circle as shown. The region inside the circle corresponds to the halfspace above the roof. The space liberated with the roof is the unshaded area outside the circle. A region of Figure 1 corresponds to a finite block if it plots entirely in the unshaded area. Similarly, Figure 2c shows the target area for the Northwest edge of the excavation, defined by vertical walls to the Northwest and to the Northeast. (In the stereographic projection, a vertical plane maps as a straight line, in the direction of the strike, drawn through the center of the reference circle). The space defined by this intersection of edges -- the

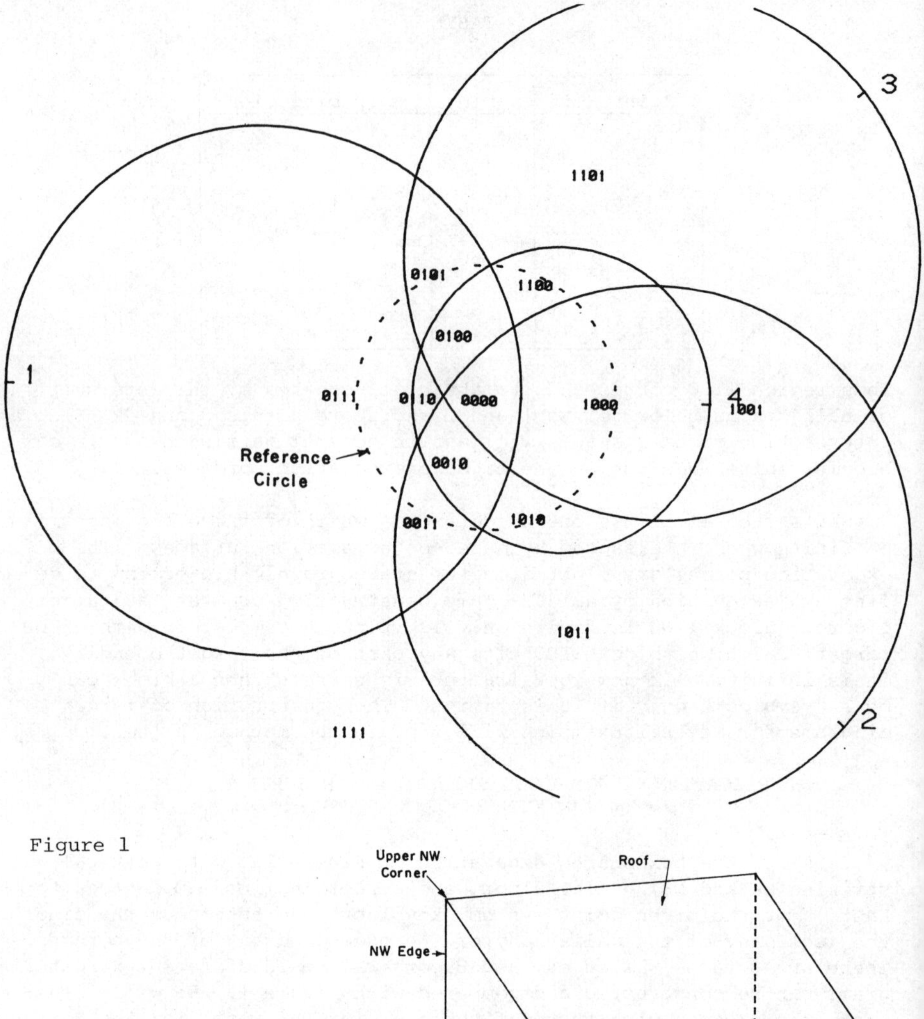

ISSUES IN ROCK MECHANICS

Figure 1

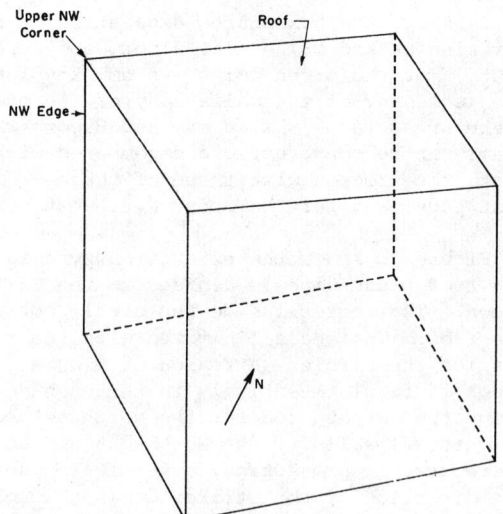

Figure 2a

ROCK SUPPORT USING KEYBLOCK PRINCIPLE 887

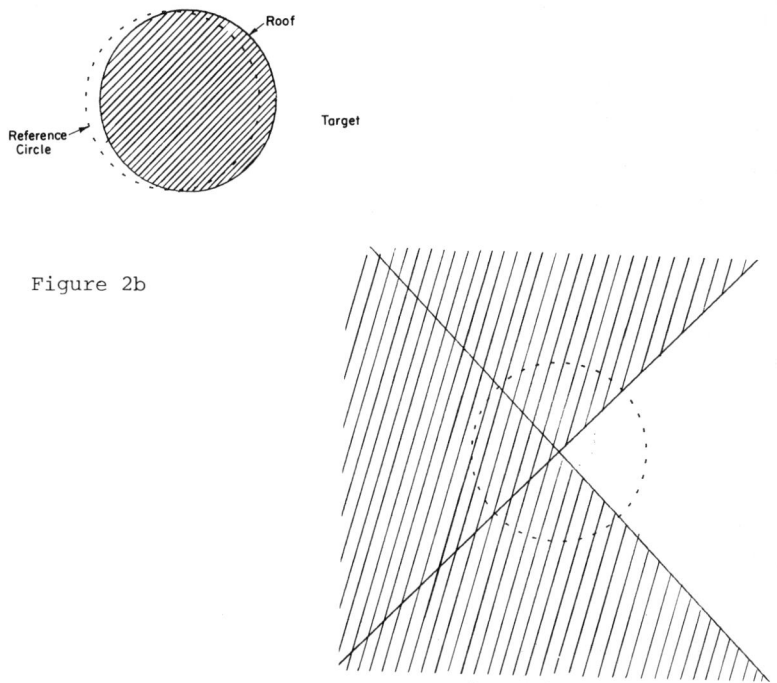

Figure 2b

Figure 2c

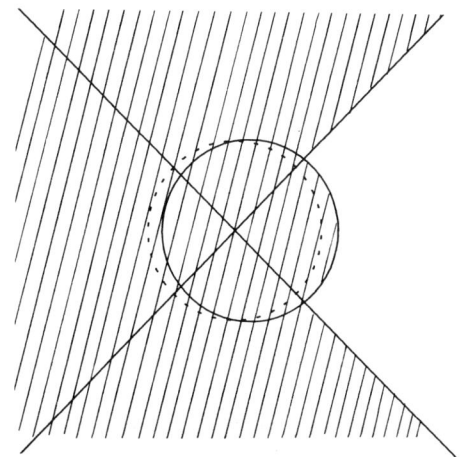

Figure 2d

target -- is the unshaded area of Figure 2c. Similarly Figure 2d shows the target area for the upper Northwest corner, which lies simultaneously below the roof and in the space of the Northwest edge.

Comparing the joint map of Figure 1 with the target area for the upper Northwest corner shows that block 1001 is a potential keyblock of the corner because it plots entirely within the target area (Figure 3a). It is also a potential keyblock of the roof alone, and of the Southwest and Northwest walls of the gallery. Figure 3b shows this block as it appears in the Northwest wall, and Figure 3c, shows how the same keyblock would appear in the roof.

KEYBLOCKS OF TUNNELS AND SHAFTS

A cylinder can be represented on the stereographic projection by the family of planes tangent to its cross section. These will intersect at a single point representing the axis (generator) of the cylinder. Thus in Figure 4, A is the axis of a horizontal cylinder -- a tunnel; B is the axis of an inclined shaft; and C is the axis of a vertical shaft. Any region on the sphere that does not contain or touch the axis of the tunnel or shaft is a potential keyblock somewhere around the section. The limits of the region within which the largest corresponding keyblock can lie are easily determined by the construction shown in Figure 5a. In this Figure block 1001 is being investigated for a horseshoe tunnel having its axis at A of Figure 4a. Two great circles are drawn through A to enclose extreme points of region 1001. The smaller of these circles rises 26.9 degrees above horizontal from west in the cross section of the tunnel, and the keyblock is below it. The larger of these circles rises 53.9 degrees from the west side of the tunnel section; the keyblock is above it. Figure 5b transfers this information to a section perpendicular to A and thereby establishes the limits of the largest keyblock of type 1001 as seen in the tunnel cross-section.

A similar construction can be made for an inclined shaft. In figure 6a, two great circles are drawn through B, the axis of the shaft so as to envelope block 1001. These great circles intersect the plane perpendicular to the tunnel as shown, permitting construction of Figure 6b, which shows the size of the region in which the keyblock of type 1001 can lie. If the keyblock lacks sufficient friction, or is subject to additional forces it may require support, which can be directed to these regions enclusively.

STABILITY ANALYSIS

Having identified an important potential keyblock, and its position in an excavation, the issue to be addressed is its support. Given the dimensions of the excavation, the largest keyblock of the type identified can be drawn and its volume and weight determined. Then, for each keyblock a stability analysis can be made using the technique introduced by John (1968), Wittke (1965), and Londe et al (1969) and (1970). (This procedure was discussed fully by Goodman (1976)). We

ROCK SUPPORT USING KEYBLOCK PRINCIPLE 889

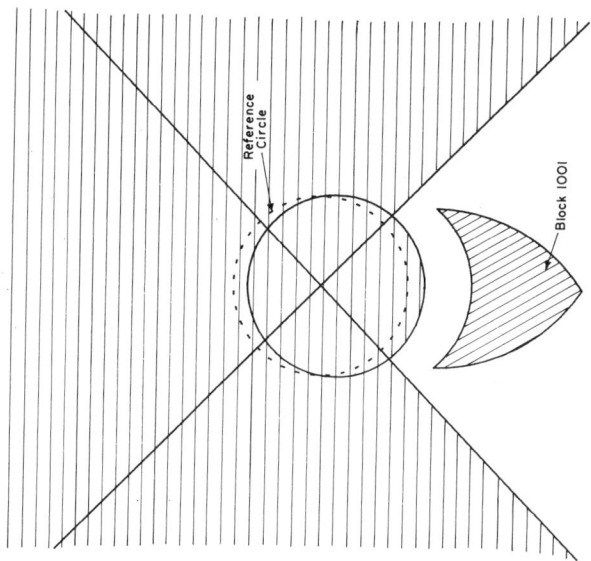

Figure 3a

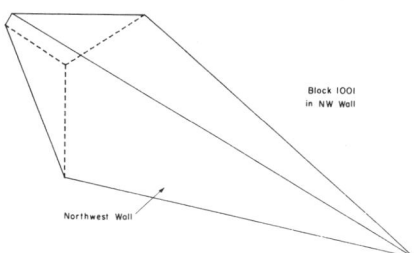

Figure 3b

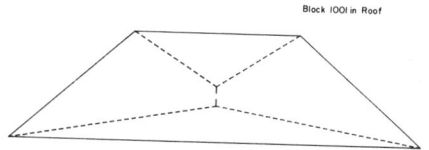

Figure 3c

Figure 4

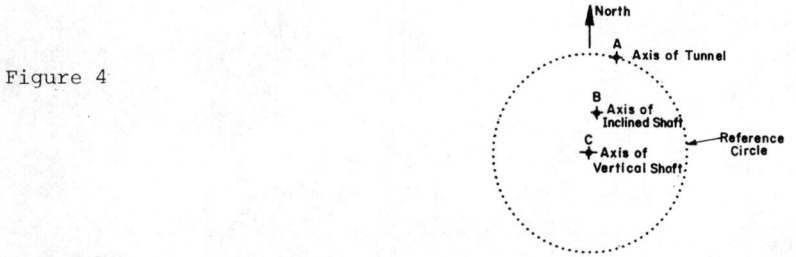

Figure 5a

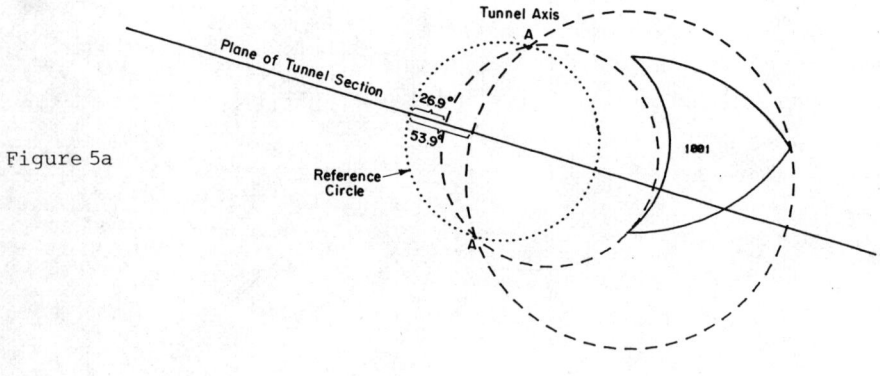

Figure 5b

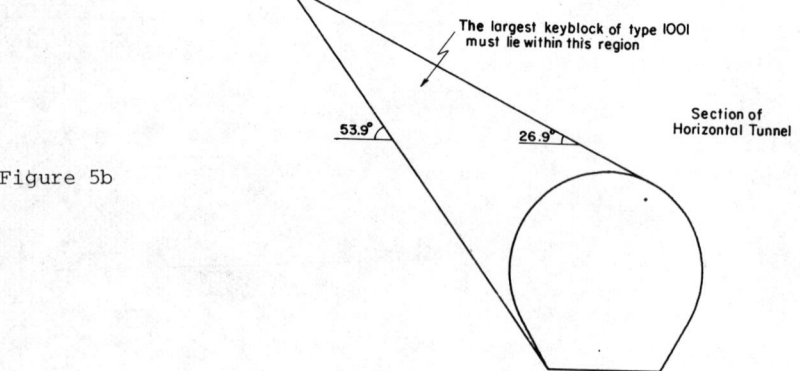

ROCK SUPPORT USING KEYBLOCK PRINCIPLE

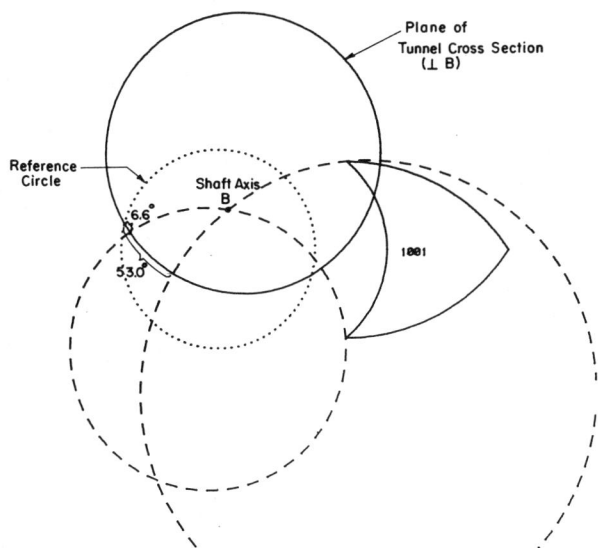

Figure 6a

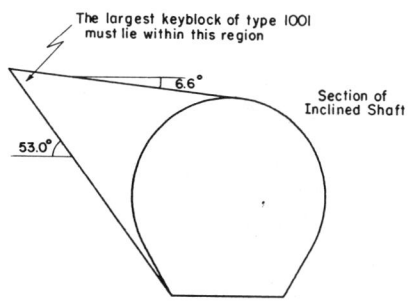

Figure 6b

have recently extended this analysis for blocks with any number of faces.

Figure (7a) shows the lower hemisphere stereographic projection of the results of a stability analysis for block 1001 relevant to sliding from the Northwest or Southwest walls of the gallery of Figure 2a. The dashed lines are contours of the friction angle required for stability corresponding to all possible orientations for the resultant force on the key block. This was drawn assuming all the friction angles to be the same on each face; however, variable friction can be input without added difficulty. For weight alone, the resultant force is in the center of the reference circle at position R; hence the block is safe without support only if the friction angle on the joint surfaces is more than 48 degrees. To find the required support corresponding to any smaller value of friction, $\phi_{required}$, the support force must be added vectorially to R to shift the resultant direction onto the contour for $\phi = \phi_{required}$ as discussed by Goodman (1976).

EFFECTS OF INITIAL STRESS ON EQUILIBRIUM

A rock wedge that intersects an excavation such that its faces are inclined at a small angle with the excavation surface will tend to pop out under the tangential stresses flowing around the opening. Conversely, a wedge with faces joining the excavation surface at angles approaching ninety degrees may be considerably reinforced by friction along the potentially sliding faces owing to the action of initial normal stress. In examining the possible stabilizing influence of initial stress on the faces of a two dimensional wedge in the roof of an opening, John Bray suggested the following approach: Calculate the normal and shear forces on the faces of the wedge that are in equilibrium with the tangential-stress-flow around the excavation, ignoring the weight of the wedge. Then switch on the weight and support force simultaneously and determine the resulting change in normal and shear forces on the face of the wedge corresponding to its displacement in a known direction and assigned joint stiffness values. According to the initial stress conditions, the geometry, of the wedge, and the friction, stiffness, and roughness properties of the bounding discontinuities, an equilibrium may be established with support considerably smaller than that required to support the full weight of the wedge when initial stress is absent. The final traction on the faces of the wedge after switching on gravity, will be smaller and differently oriented than the initial value. Continued support of the wedge will require that this final traction be maintained despite blasting, weathering, and relaxation. Whether or not it will be deemed acceptable to use initial stress contribution to underground support remains to be seen. For the present, it will be instructive to consider a particular 3 dimensional case.

The particular problem to be studied is a convex polyhedral block with n planar sides, one of which is the roof of an excavation. The remaining (n-1) faces are oriented so that all will open if the block displaces vertically downward into the excavation. It is assumed that

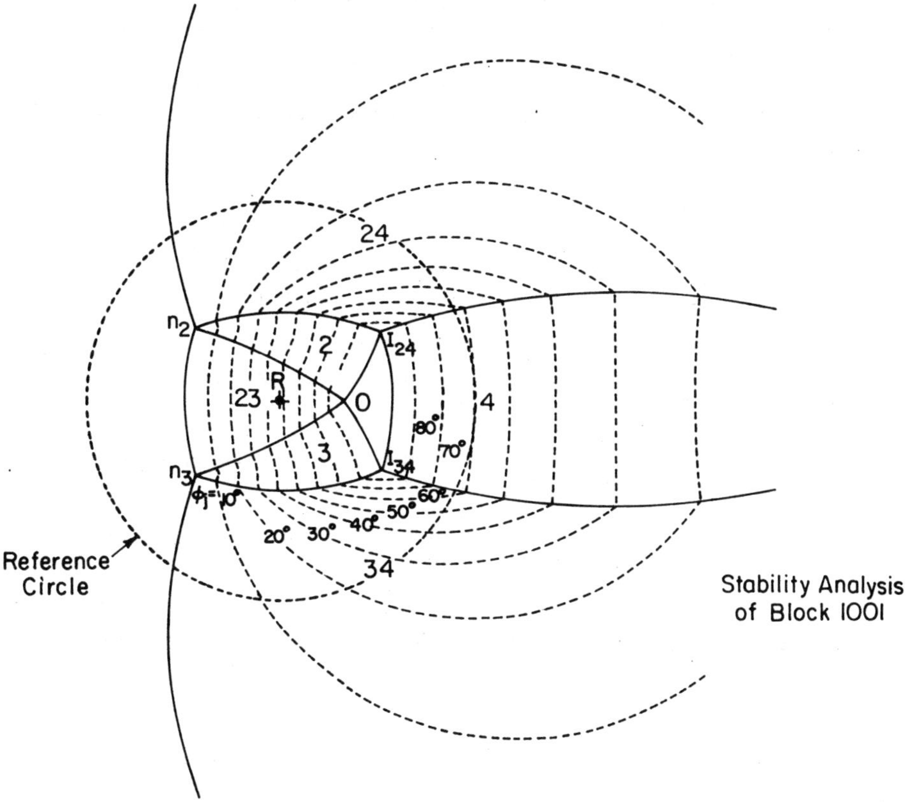

ROCK SUPPORT USING KEYBLOCK PRINCIPLE

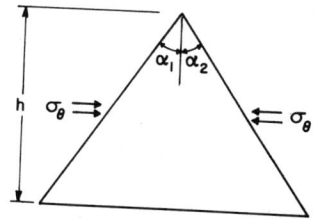

Figure 7

Figure 8

a vertical support force and gravity are applied instantaneously resulting in a small vertical displacement. Also it is assumed that the stress state before any movement due to gravity is characterized by average principal stresses $\sigma_{1,0}$ and $\sigma_{2,0}$ parallel to the block's base and $\sigma_{3,0}$ which is taken to be zero. The amount of support necessary to prevent collapse of the block can be determined as follows. Let $(\sigma_{n,i})$ be the initial normal stress on face i of the wedge. It can be computed by

$$(\sigma_{n,i})_0 = \sigma_{1,0}(\cos \alpha_i \cos(\theta_{P,1} - \theta_i))^2 + \sigma_{2,0}(\cos\alpha_i \cos(\theta_{P,2} - \theta_i))^2 \quad (1)$$

where: $\sigma_{1,0}$ and $\sigma_{2,0}$ are the greatest and least principal stress, respectively, in the plane of the roof (horizontal) before turning on gravity;

$\theta_{P,1}$ and $\theta_{P,2}$ are the clockwise bearings from North to the directions of $\sigma_{1,0}$ and $\sigma_{2,0}$ respectively;

θ_i is the dip direction of plane i (clockwise from North) and α_i is the complement of the dip angle of plane i.

The proportion of the weight of the wedge that must be supported for equilibrium is then determined by

$$\frac{A_0}{W} = 1 - \sum_{i=1}^{n-1} \frac{(\sigma_{n,i})_0 A_i \sin(\phi_i - \alpha_i) k_{s,i}/k_{n,i}}{W\cos\phi_i} \frac{[\cos\alpha_i \cos i_i + \tan\alpha_i \sin(\alpha_i - i_i)]}{k_{s,i}/k_{n,i} [\cos\alpha_i \cos i_i + \tan\phi_i \sin(\alpha_i - i_i)]} \quad (2)$$

where: A_0 = vertical support force; W = the wedge's weight;

A_i is the area of face i; ϕ_i is the friction angle on face i; and i_i is the roughness angle for plane i; and $k_{s,i}$ and $k_{n,i}$ are the shear and normal stiffnesses of joint plane i. In equation (2), $\phi_i \geq \alpha_i$.

For a 2-dimensional case, with a horizontal roof (Figure 8) the stress state is completely described by σ_θ acting in the plane of the roof before movement. Equation 2 then simplifies to:

$$\frac{A_o}{W} = 1 - \frac{(2\sigma_\theta)}{\gamma h (\tan \alpha_1 + \tan \alpha_2)} (M_2 + M_1) \quad (3)$$

where:

$$M_1 = \frac{[(k_{s,1}/k_{n,1})\cos^2\alpha_1 \cos i_1 + \sin\alpha_1 \sin(\alpha_1 - i_1)]}{[(k_{s,1}/k_{n,1})\cos\alpha_1 \cos i_1 + \tan\phi_1 \sin(\alpha_1 - i_1)]} \frac{\sin(\phi_1 - \alpha_1)}{\cos\phi_1}$$

and

$$M_2 = \frac{\left[(k_{s,2}/k_{n,2}) \cos^2\alpha_2 \cosi_2 + \sin\alpha_2 \sin(\alpha_2-i_2)\right] \sin(\phi_2-\alpha_2)}{\left[(k_{s,2}/k_{n,2}) \cos\alpha_2 \cosi_2 + \tan\phi_2 \sin(\alpha_2-i_2)\right] \cos\phi_2}$$

where γ = unit weight of the rock material
       h = height of the intersection of the bounding planes above the block base (see Figure 8)
       σ_θ = stress parallel to the excavation roof before movement

note $\phi_1 \geq \alpha_1$ and $\phi_2 \geq \alpha_2$

Consider again the rock mass having discontinuities of Table 1. This system of joints was previously plotted in Figure 1. For a horizontal roof, the target diagram (the excavation pyramid) is the region outside the reference circle. Region 1111 is in the target area and is therefore a potential keyblock of the roof. Block 1111 is defined by the intersection of half-spaces below each of the joint planes. To simplify the discussion, we will delete reference to joint plane 4. Figure 9 shows the block defined by 111, together with the horizontal roof of an underground opening. Equations 1 and 2 can be used to analyze the support requirements for this block under initial stress. Let the major and minor horizontal initial stresses above the roof of the gallery be East-West and North-South respectively. First, we examine a case in which the available friction angle on all planes is 35° which includes a roughness angle, i, equal to 5° on each plane. The ratio of shear to normal stiffness is 0.1 on each of the three joint faces of the block. Figure 10 shows the ratio of support to weight A_0/W, for equilibrium as a function of the greatest horizontal initial stress, with the minor initial horizontal stress equal to 0, one-half, and one times the major horizontal initial stress. Even without any minor horizontal principal stress, it can be seen that an average horizontal stress of 27 psi (186 kN/m^2) or less will stabilize this wedge without any external support, i.e. A_0/W equals zero. This initial stress is assumed to be constant over the full height of the wedge; the scale of the block is such that the trace length of plane 3 on the roof is 17 feet (5.2 m).

Figure 11 shows the influence of the stiffness ratio k_s/k_n on the previous calculation, with both horizontal principal stresses equal to 10 psi (69.kN/m^2). For the stiffness ratio on all joint planes equal to 0, the support for equilibrium must be 0.538 times the weight of the wedge. This number is only slightly affected by the stiffness ratio. Figure 12 shows the results of a third set of calculations with the stresses as in the previous example but with the friction angle on all planes equal to 30 degrees plus i. The roughness angle i is seen to be an important parameter in regard to the stabilizing effect of the initial stress. As i varies from zero to eleven degrees, the support required for equilibrium of the wedge descends almost linearly from full wedge-weight to zero.

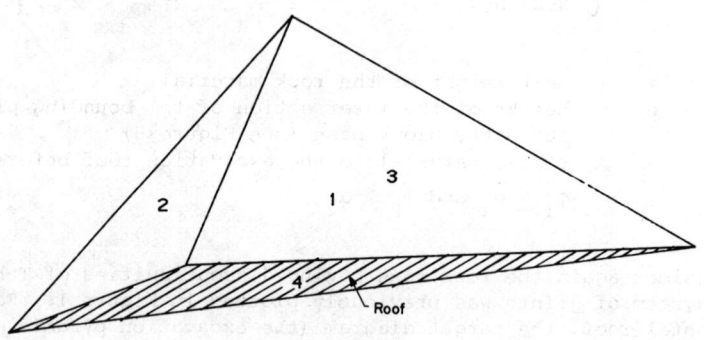

Figure 9

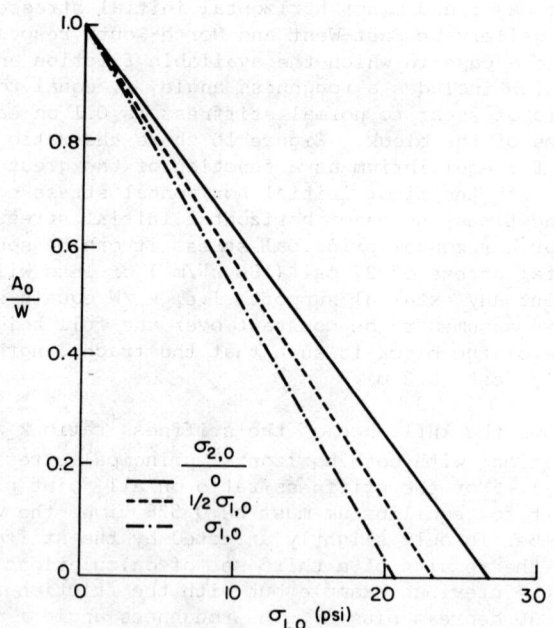

Figure 10

Figure 11

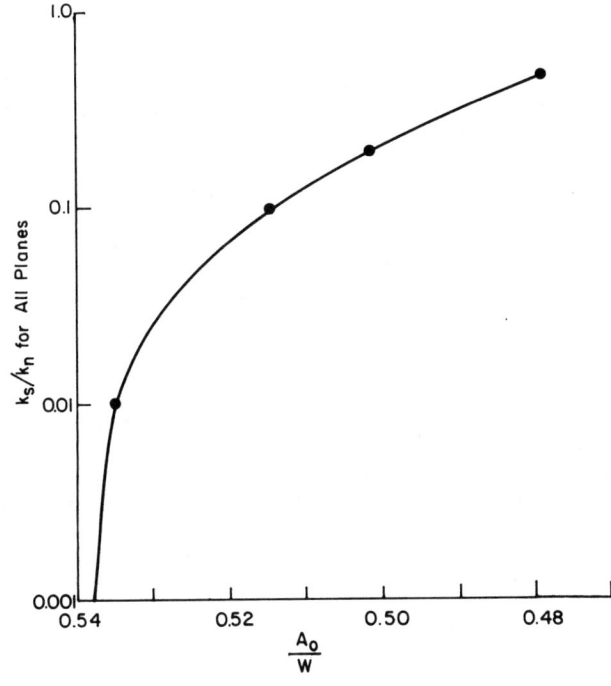

Figure 12

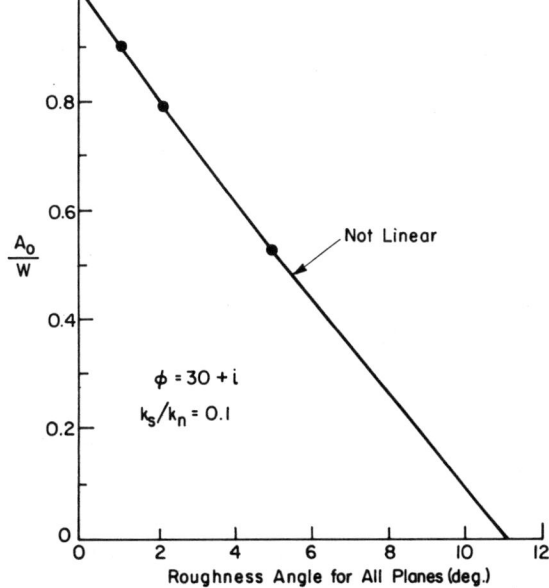

CONCLUSION

The design of support for hard, jointed rock is seen to depend heavily on the intersection of the joint system and the excavation surfaces. The keyblock theory displayed here permits all the critical blocks to be identified. Stability analysis then permits the required support to be calculated. If initial tangential stress can be shown to flow around the excavation, it is possible to consider reducing the support requirement from that calculated by limit equilibrium methods. A formula for determining the strengthening of three dimensional wedges by the action of initial tangential stress has been presented. The advisability of relying on initial stress support for engineering practise is currently under investigation.

ACKNOWLEDGEMENT

The analysis of a two dimensional, symmetrical wedge in the roof of an opening with in-situ tangential stress was described in unpublished notes by John Bray, Imperial College, London. The authors thank Dr. Bray for providing this valuable start on the initial stress problem. The support of the California Institute for Mining and Mineral Resources Research, Douglas Fuerstenau, Director, is greatly appreciated.

REFERENCES

Goodman, R.E., 1976, "Methods of Geological Engineering in Discontinuous Rocks", (West Publishing Co., St. Paul).

Goodman, R.E., and Shi, Gen-hua, 1982, "Geology and Rock Slope Stability - Application of the Keyblock Concept for Rock Slopes", Proc. 3rd Int. Conf. on Stability in Surface Mining, SME (AIME) pp 347-373.

John, K.W., 1968, "Graphical Stability Analysis of Slopes in Jointed Rock", J. Soil Mech. & Found. Div., ASCE, V. 94, No. SM2.

Londe, P., Vigier, G., and Vormeringer, 1969 and 1970, "Stability of Rock Slopes, a Three Dimensional Study", J. Soil Mech. & Found. Div., ASCE, Vol. 95, No. SM1, and continued "Graphical Methods", Vol. 96, No. SM 4.

Shi, Gen-hua and Goodman, R.E., 1981, "A New Concept for Support of Underground and Surface Excavations in Discontinuous Rocks Based On a Keystone Principle", Proc. 22nd Symposium on Rock Mechanics, MIT, Boston, pp. 290-296.

Wittke, W., 1965, "Methods to Analyze the Stability of Rock Slopes with and without Additional Loading", (in German) Rock Mechanics and Engineering Geol. Supplement II, p. 52.

Chapter 90

CASE STUDIES OF ROCK SLOPE REINFORCEMENT

Peter N. Calder

Professor and Head
Department of Mining Engineering
Queen's University

ABSTRACT

This paper deals with the reinforcement of igneous rock slopes at three open pit iron ore mines in northern Canada. The pits involved utilize overall slope angles in excess of 60° with ultimate depths in the order of 250 m. Three case studies are described.

The first concerns the reinforcement of a 300,000 Ton undercut rock wedge, approximately 70 m in height. This was accomplished using 20 cm diameter steel bales, installed in 25 cm diameter vertical holes, and a cement-sand grout.

The second involves the reinforcement of a 40 m wide diabase dyke which had been continuously failing as the pit deepened. A unique form of rock reinforcement was used in that the entire mass of the undercut dyke was not pinned to the wall. Rather an interlocking effect between individual blocks within the dyke was relied upon.

A third concerns the reinforcement of a berm which was undercut by the contact between the ore and waste. This reinforcement was also accomplished using a combination of steel bales, rods and cement grout.

INTRODUCTION

In designing open pit mines in hard igneous and other competent rock materials, the strength parameters of the rock and the stress levels present throughout the area are generally not significant factors. It is well recognized now that the structural conditions are far more important. Much attention has been paid to detailed methods of structurally mapping rock masses as an aid to mine design. However there are often structural problems encountered which are not generally apparent at the conclusion of a preliminary conventional structural analysis. For example the presence of a single fault, which may not even be evident during the exploration and planning stage, may be the single most important structural feature present. The same can be true of major dykes and the zones of alteration which often occur in the vicinity of the ore waste contact. The problems described in this paper deal with structural problems of this nature. The pits involved are very steep, however the problems encountered would not have been greatly alleviated by a flatter slope design.

INTERSECTING DYKES

Numerous dykes of various types are present throughout the mines concerned. In this particular example the dykes involved are less than 1 m in thickness. They initially appeared to be relatively competent when viewed during the mining of the first several benches and did not appear to present a stability problem. As the pit deepened, however, it became apparent that the surface of one of the dykes was highly altered and it also became evident that the two dykes together would form a wedge which would be undercut by the pit design. Figure 1 is a view of the area taken towards the end of the life of the mine. Figure 2 is a three dimensional sketch indicating the two intersecting dykes and their relative position to the pit face.

The orientation of the two dykes was mapped using terrestrial photogrammetry. A good estimation of the angle of friction of the rocks involved was known from previous experience and testing. It was decided to assume a zero value of cohesive strength along the base of the wedge, and to calculate the factor of safety for that condition. A conventional sliding wedge analysis was used, (Calder, 1970). This indicated a factor of

Figure 1: Photograph of two intersecting dykes forming an undercut wedge.

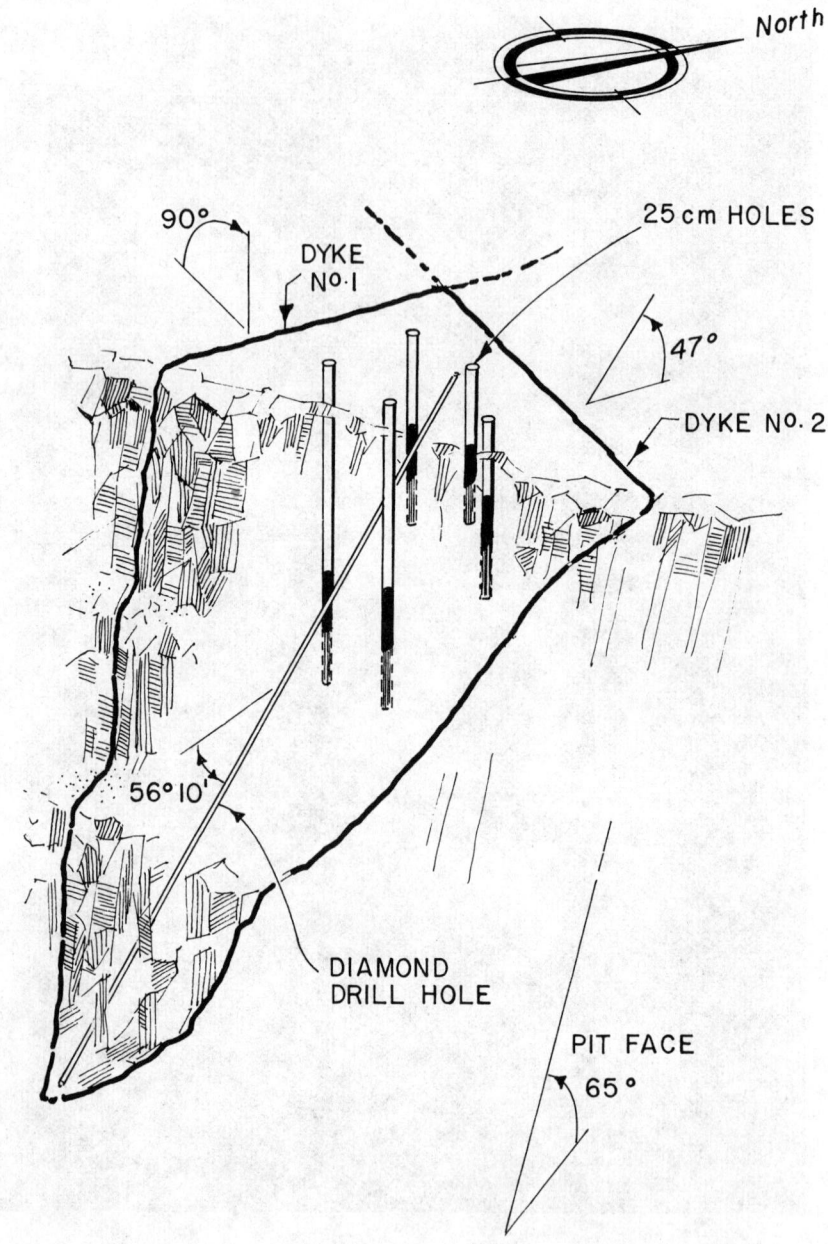

Figure 2: Isometric view of the intersecting dykes.

safety less than 0.8. It was obvious that the dyke did have some cohesion but it was felt that this could not be relied upon over the long term. It was therefore decided to increase the factor of safety to 1.1 assuming a zero value of cohesion, by adding artificial support.

The use of cable bolts was considered but this was impractical due to difficulties of accessing the vertical face. This method also would have been much more costly than the method selected. It was decided to drill vertical blast holes from the upper surface of the wedge to penetrate the surface of the dykes. These would then be reinforced using bales of reinforcing rod supplemented with grout injection. It was concluded that five 25 cm diameter boreholes would be required, located as indicated on Figure 3. It was thought that the use of pumps to inject grout under pressure was impractical due to the fact that this might have precipitated a failure. However, as it was necessary to use grout to encase the rods ridgidly in the boreholes it was also decided to pump grout using the rods in the boreholes as pistons. Figure 6 is a photograph illustrating the installation procedure. The procedure consisted of filling the borehole, which was designed to penetrate 5 m below the dyke surface, with grout. The 10 m long, 20 cm diameter, steel bales were then lowered into the holes and grout was forced into the open fractures within the wedge. This grout could visibly be observed on the face as much as 200 m away from the location of the actual boreholes. Although it was estimated that the bonding strength between the grout and the rock would add significant cohesion (approximately 1000 kPa) it was decided not to rely on this form of reinforcement because of the difficulty in accurately assessing the the distribution of the grout along the potential failure surface. It was therefore decided to add sufficient steel so that the shear strength of the steel across the failure surface would raise the factor of safety to a level of 1.1. The action of the grout would then add an additional measure of support. An extensometer was also installed in the wedge.

The method of support used would be classified as a passive system similar to a grouted rockbolt. Figure 4 illustrates the principle involved, the steel rods grouted into the boreholes form a rigid column which is much stiffer than the rock mass itself. Any small movement of the rock wedge which would be a necessary prelude to failure would immediately mobilize the shearing resistance of the passive reinforcement system and at that point it would become active. Figure 5 is a cross-

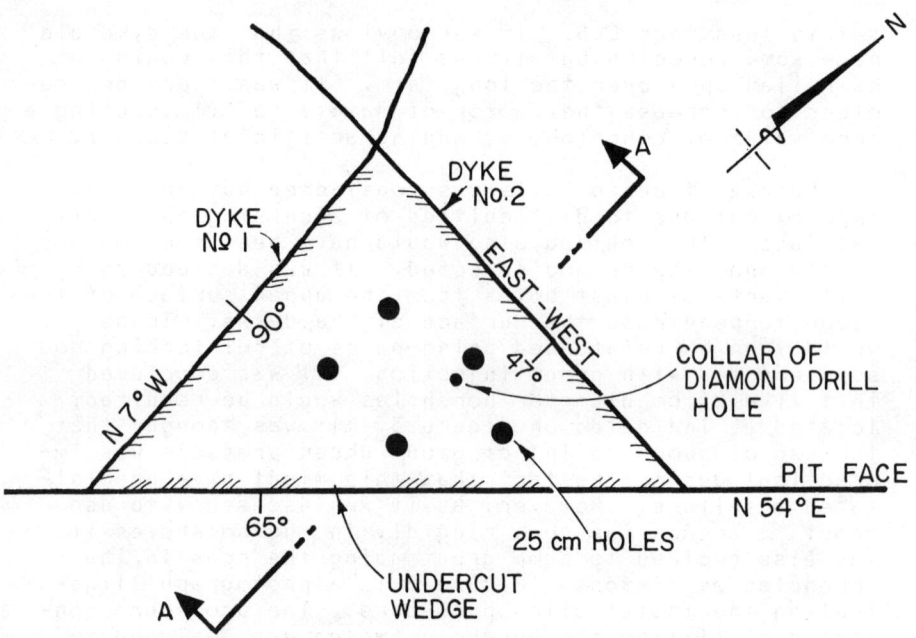

Figure 3: Plan view of the intersecting dykes showing borehole locations.

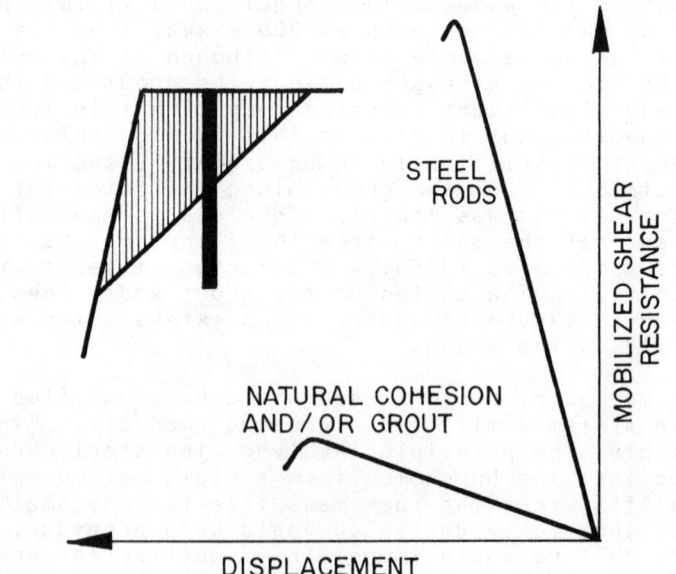

Figure 4: Relationship between mobilized shear resistance and displacement.

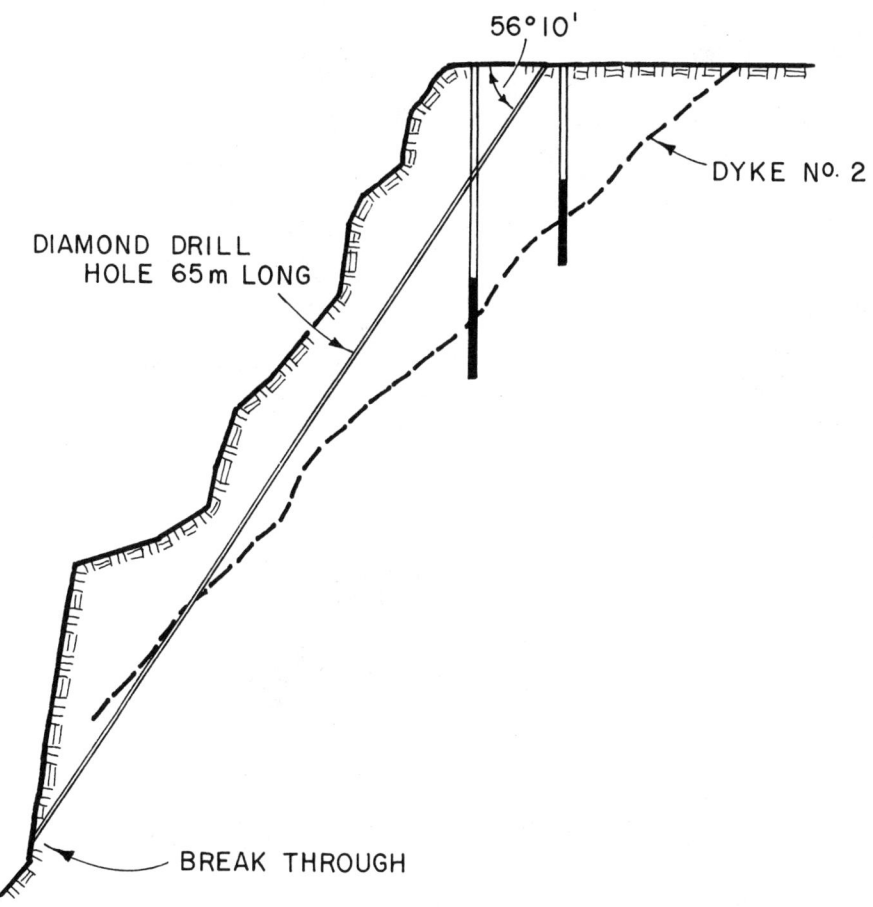

Figure 5: Cross-section A-A (Figure 3) through undercut wedge.

Figure 6: Installation of 20 cm steel bales.

Figure 7: Diabase dyke - failed upper section.

section through the wedge at the position indicated on
Figure 3. A 65 m inclined diamond drill hole was used
to confirm the presence of dyke #2 at depth and was also
used to inject grout into the nose of the wedge area.
The diamond drill hole was allowed to break through onto
the pit face so that it could be surveyed in to confirm
its' location. The bottom of this hole was then sealed
with bentonite prior to grouting the nose area. Following installation of the steel bales this hole was used
as an extensometer installation. This extensometer was
monitored on a routine basis and would have provided
warning if the wedge had started to fail.

This pit has since been deepened an additional 200 m
and mining is now terminated. No difficulties with this
wedge were encountered during mining.

REINFORCEMENT OF A MAJOR DIABASE DYKE

The second case history deals with the reinforcement
of a 40 m wide diabase dyke. This dyke contained
structure which was significantly flatter (56°) than the
face slope angle of 80° which was being utilized in that
area. Each time this dyke had been exposed on the upper
levels, it would eventually fail back to the natural
structure causing considerable difficulties in the operating pit. The unreinforced failed section of the dyke
is visible in the upper portion of Figure 7. It was
apparent that the dyke was not failing on a smooth continuous failure surface but rather was ravelling. The
blocky structure of the dyke was such that there was
considerable interlocking of individual blocks. To assure that no weight was transferred vertically to successive berms, the face slope angle was reduced to 71°
as illustrated in Figure 8. It was then felt that if
the face of the dyke could be supported so that unravel-
ling could not begin, a massive block-type slide would
not be possible because of the interlocking. Therefore
it was decided to reinforce only the face of the dyke
and not to attempt to pin the entire undercut mass to
the wall. This was accomplished using 3 m long rockbolts which were sealed in the hole with epoxy resin.
Steel straps and heavy steel mesh were pinned to the
face by the rockbolt anchor plates. The face of the
dyke was formed using pre-splitting (Calder, 1976) and
the blasts in the vicinity were lightly loaded. The
quality of the work is evident in Figure 9. This pit
has since been deepened an additional 50 m and no further problems have been encountered.

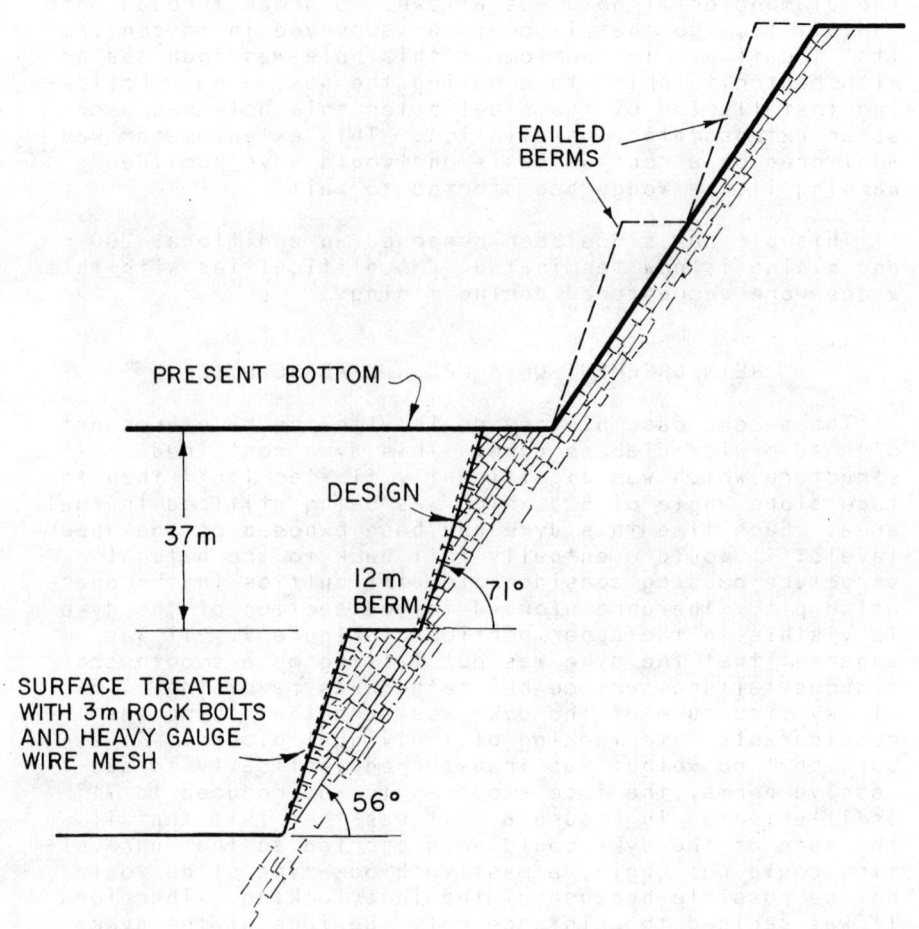

Figure 8: Reinforcement of a major diabase dyke.

Figure 9: View of reinforced diabase dyke.

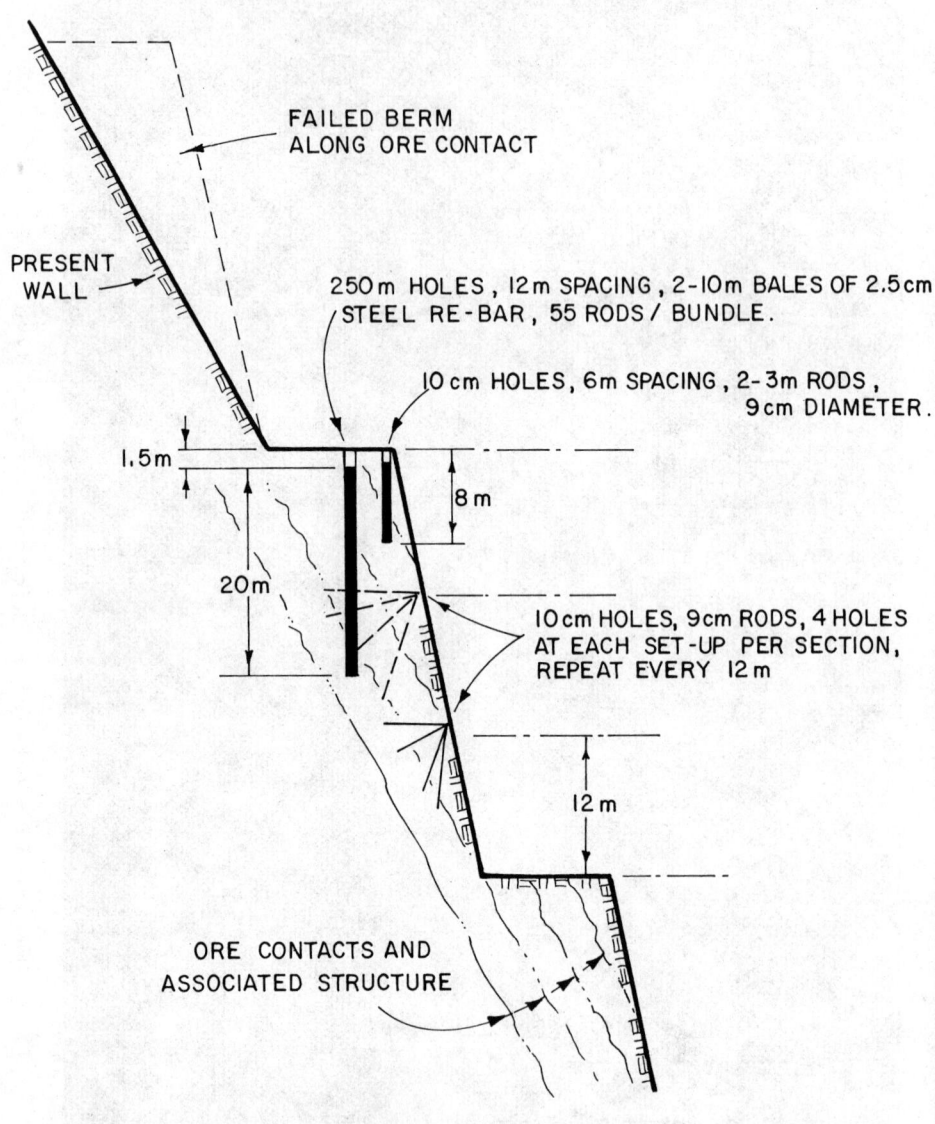

Figure 10: Reinforcement of berm undercut by ore contact.

REINFORCEMENT OF A BERM UNDERCUT BY AN ORE CONTACT

The stability of berms along the footwall side of an open pit is often a problem as the wall closely parallels the ore waste contact. This particular contact is severely altered and represents an extremely weak surface in comparison to the host rock. A method currently being used to deal with this problem is illustrated in Figure 10. This procedure actually evolved from the wedge reinforcement method described in the first section, the underlying principles remain the same. Several key points to note on Figure 10 are the location of the short crest holes containing the 6 m steel rods, and the location of the 3 inclined holes drilled from each working level to provide support to the toe area.

This reinforcement system has presently been installed to the bottom of the first working bench in the series. The inclined fan of holes used for toe support will be added shortly. Some time will be necessary to fully evaluate the effectiveness of this method but based on past experience a successful result is expected.

ACKNOWLEDGEMENT

The author wishes to thank the management of Cliffs of Canada Limited for permission to publish this paper, which involves projects undertaken at their open pit mines.

REFERENCES

Calder, P.N., 1970, "Slope Stability in Jointed Rock," The Canadian Institute of Mining and Metallurgical Bulletin, Vol. 63, No. 683, May.

Calder, P.N., 1977, "Pit Slope Manual, Perimeter Blasting," Canada Centre for Mineral & Energy Technology (CANMET), Report 77-14, May.

Chapter 91

ROCK SUPPORT AT PINE FLAT
A CASE HISTORY

by John Cogan and P. M. Gomez

Principal Geotechnical Engineer and Project Geologist,
International Engineering Company
San Francisco, California

ABSTRACT

A typical problem in applied rock mechanics is an accurate prediction of support requirements for deep rock cuts. This paper presents experience with this problem as obtained from excavations for a powerhouse at the Pine Flat Dam near Fresno, California. The project, sponsored by the Kings River Conservation District is still under construction. Highly jointed and weathered rock was discovered during initial excavation. Wedge failures along clay filled joints and bedding planes presented potential support problems. Detailed mapping of joints and beds was undertaken to establish basic patterns of all rock discontinuities including mean orientations and an estimate of orientation variations. Rock bolting patterns were devised to meet these patterns as excavation proceeded. Shear friction angles across joints and beds were estimated from rock wedge fallouts. Friction angles as low as 27° were identified on highly weathered joints. Weathering strongly influenced stability conditions. Examples of bolting patterns to stabilize critical slopes are presented.

INTRODUCTION

One of the most persistent problems encountered in geotechnical engineering is the problem of accurately predicting support requirements for rock excavations. This problem is usually expressed as an underestimate of rock bolt lengths, the total number of rock bolts required, the total amount of shotcrete required to stabilize slope faces, and the amount of wire mesh needed to prevent rock ravelling and/or to secure shotcrete facing. These underestimates often result in significant cost additions to the project and may even result in slowing expected excavation work progress. This paper presents experience with this problem as encountered during the excavations for the Pine Flat Powerhouse on the Kings River near Fresno, California.

PROJECT DESCRIPTION

The Pine Flat Hydroelectric Project consists of the construction of a powerhouse at the toe of the existing Pine Flat Dam, and then interconnecting it to three penstocks included in the dam during original construction in the early 1950's. The powerhouse is being built into a hole excavated into the right bank of the Kings River. As shown in Figure 1, the dam is a concrete gravity dam with an overflow spillway aligned along river center, but shifted slightly toward the right bank. In addition, the existing penstocks connections built into the dam are located on the left bank side. For this reason the powerhouse had to be located in the left bank in such a position that a cofferdam could be constructed between the excavation and the spillway stream. This dictates a site with very little room for changes in location or orientation in order to minimize possible adverse rock conditions. It also dictates that a cut may be made into the natural hillside on the left bank.

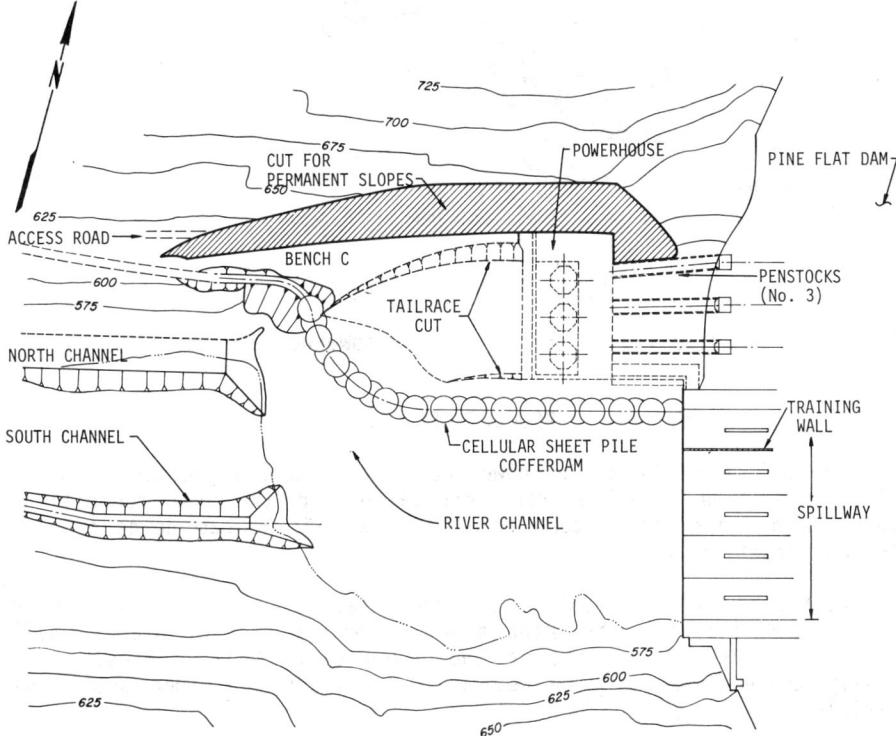

Fig. 1. Plan view of Pine Flat powerhouse excavation.

Figure 2 shows the north slope of the excavation. On the slope there are three benches, benches A, B, and C. All excavations above bench C are permanent slopes against which no facilities will be constructed. On the original design only benches B and C were planned. All major slopes were to be cut to 1/4 H:1 V. Below bench C vertical cuts were made in the powerhouse area and a 1/4 H:1 V in the tailrace area. The maximum height of this cut is 94 feet with a 9 ft wide bench approximately at mid height. The tailrace wall was to be lined with a 6 inch thick facing of shotcrete applied over wire mesh.

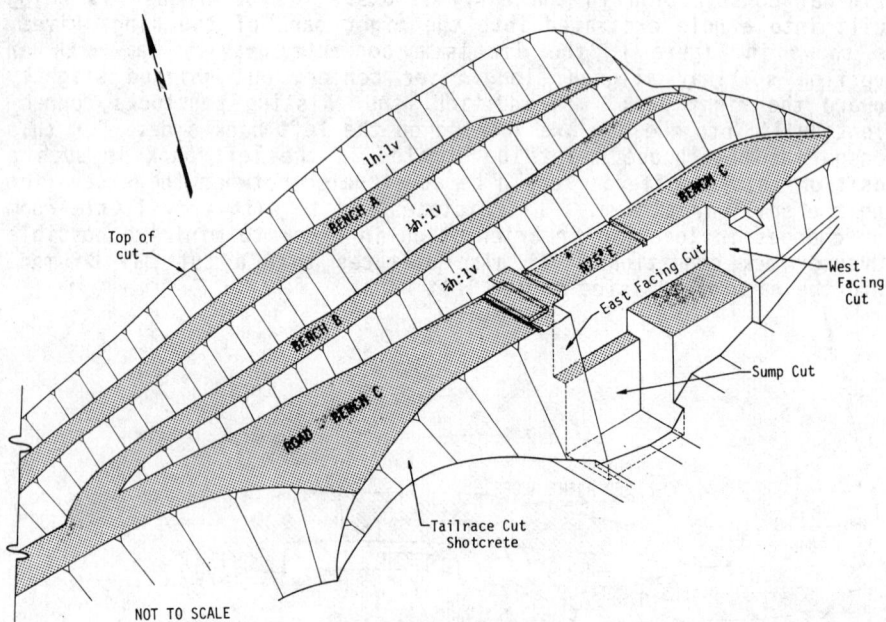

Fig. 2. Isometric view of north side of powerhouse excavation.

PROJECT GEOLOGY

The rock throughout the powerhouse site is plagioclase amphibolite, the same type of rock excavated for the dam foundation. Outcrops of this rock extend from the base of the dam downstream to the construction camp.

Fresh rock samples are hard and moderately strong but they tend to be brittle, sometimes breaking along microscopic joint planes which frequently are not visible to the naked eye. Many samples, especially those from the more weathered locations contained small but visible micro cracks.

Structural Features

The main structural features are bedding, jointing, dikes, and shear zones. All of these features coupled with surface weathering played a role in slope stability. No major faults cross the site, but locally dikes and shear zones cut across bedding. Mean orientations for these features are listed in Table I.

TABLE I
PINE FLAT PROJECT
SCHMIDT PLOT OF
STRUCTURAL DISCONTINUITIES

Feature	Orientation	Notes
Primary Bedding (B_1)	N35°W 35°NE	Consistent throughout excavation.
Secondary Bedding (B_2)	N41°W 54°NE	Present at all elevations but becomes slightly more distinct with depth.
Cross Bed Joints (J_1) (J_1)	N27°W 44°SW N41°W 71°SW	Both sets highly variable.
High Angle Joints (J_2)	N56°E 86°SE	Strikes to east and steepens slightly with depth.
Flat Joints (J_3) (J_3)	N56°E 05°NW N18°W 15°SW	Variable, generally dips less than 20°.

The bedding is generally uniform throughout the site, following a consistent orientation across the site, striking almost perpendicular to the river bank and dipping upstream between 32° and 38° (Figure 3). This layering may reflect the original bedding produced by the basaltic flows, but because of metamorphism all of the original rock fabric was destroyed, making it impossible to recognize any basaltic flow structure at all.

Jointing is extremely prominent in all areas. There are three main sets as shown in Table I, cross-bed joints, J_1, near vertical high angle joints, J_2, and flat joints, J_3.

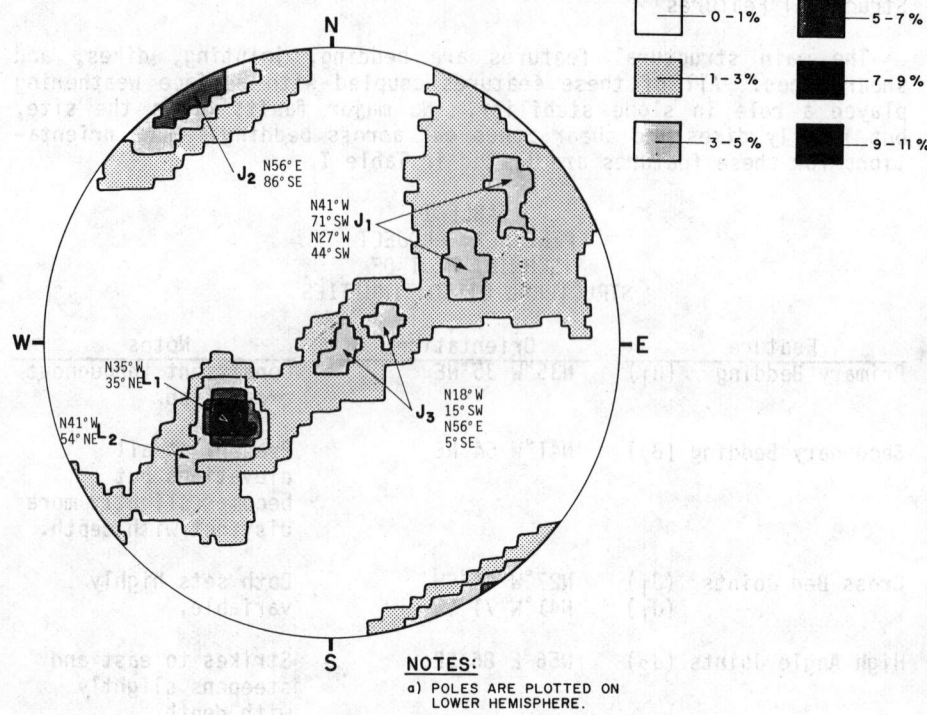

Fig. 3. Stereo plot of bedding and jointing.

Joint set J_1 is complementary to the bedding as it tends to strike parallel to it but dip normally. From Figure 3 it can be seen that a wide scatter exists in these joints and there are actually two orientations around which these joints cluster. The most prominent cluster strikes N27°W and dips 44°SW. The variability of joint set J_1 is probably due to the jointing following the long axes of the amphibole crystals which aligned themselves in a variable manner during metamorphism.

Joint Set J_2, the high angle set, acts as a slip plane for block fallouts in all east or west facing slopes. It also acts as a slip plane in slopes parallel to the river when it dips toward the opening. In stereo plots of orientations taken at differing elevations, a shift to the east is noted with depth. Dips also become slightly steeper. This could be seen on the north cut as well as the stereo plots. The variation in strike may be sampling error (Figure 4).

ROCK SUPPORT AT PINE FLAT

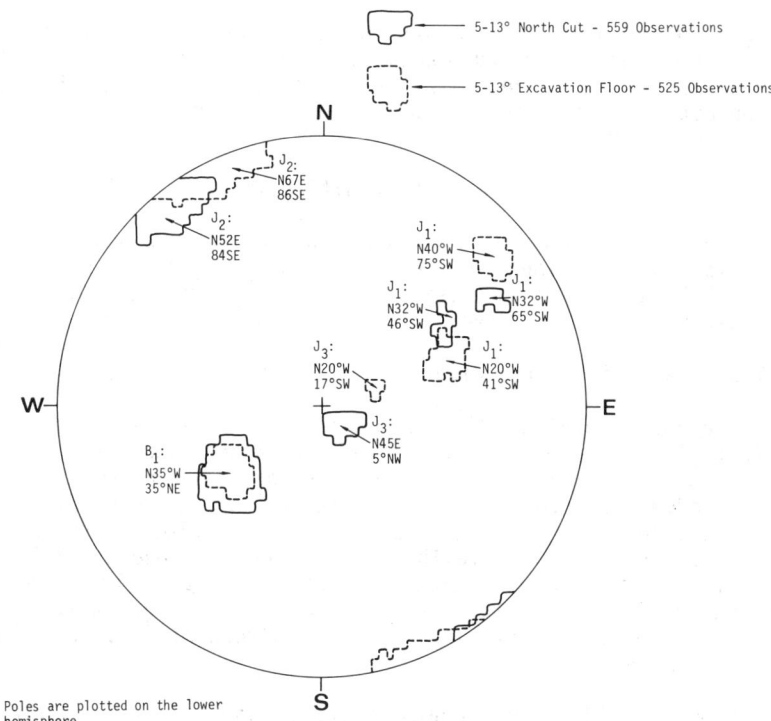

Fig. 4. Stereo plot comparing orientations on the north cut and the excavation floor.

Joint set J₃ is a flat lying set. As seen on figure 3 the pole directions vary, but generally dips are less than 20°. These variations may be present simply because flat lying joints are likely to be more numerous than can be seen on a flat or nearly flat surface. Individual joints can be traced across the entire north cut.

Weathering

The site has a very pronounced weathering pattern. The weathering has been controlled by water flowing along bedding planes and joints. In the hillside, run off water not only drains along the surface but also filters downward along joints and bedding planes working its way through these openings toward river level. This seepage pattern was especially evident after excavation for the powerhouse cut was well underway. During storms water seepage from all exposed joints and bedding planes was visible on steep slopes long after rainfall ended. Percolation down primary bedding planes and certain near vertical joints was especially pronounced.

In upper slope areas clay filled joints and bedding planes were expecially prominent, and the potential for rock wedge fallouts high. Down lower, near and below river level identical planes exhibited little or no weathering and appeared to be fresh with interlocking joint surfaces. In many cases a degree of cohesion was also noticed across the fresher joint surfaces.

SLOPE STABILITY

The general pattern of slope behavior can be represented by two excavation directions; the cuts on the north side of the powerhouse parallel to the river striking N75°E and cuts for the sump walls perpendicular to the river striking N15°W.

Slopes Parallel to the River

The cuts parallel to the river should be discussed first. At first glance the average orientations of beds and joints as shown in Figure 5 are such that the only kinematically favorable slip planes for wedge failures are those formed by joints J_1 and J_2 in combination with the slope face. Since J_2 almost parallels the cut and dips very steeply, failures should be confined to slabbing on the face. On the 1/4 h:1v slopes the average dip is less than the slip of the face and slabbing should be suppressed, but on vertical slopes slabbing should occur. This general pattern of failure did hold true during excavation.

However, all joints showed distinct variations in strike and dip as shown in the stereo plot of Figure 3. In addition, intermediate joints not following the general pattern also existed, and it is these joints which led to the larger wedge failures during construction. For the most part joints striking between due north and N45°E and dipping 50° to 65° outward from the hillside combined with cross bed joints, J_1, to yield failures.

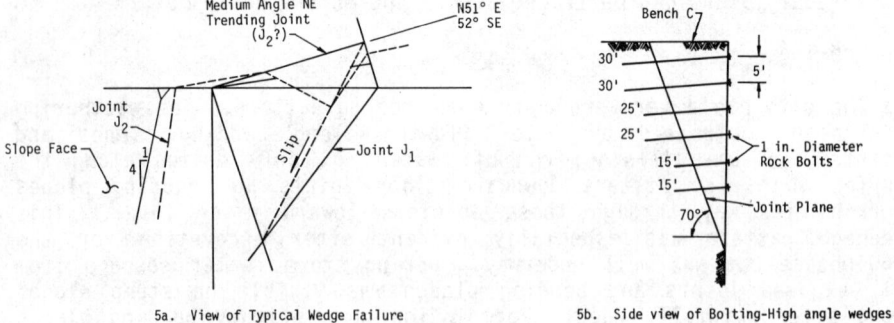

5a. View of Typical Wedge Failure Case 6, Table I

5b. Side view of Bolting-High angle wedges.

Fig. 5. Wedge conditions on north side cut slope.

Table II lists the results of graphical analyses (Goodman, 1976) of the stability of some face wedges including some which failed and others which did not. All of these analyses are taken from selected rock wedges in benches B and C as shown in Figure 5.

TABLE II

ROCK WEDGE STABILITY
IN NORTH SIDE CUT

CASE	LOCATION	ORIENTATIONS OF WEDGE PLANES/OPEN FACE	$\emptyset$	NOTES
1.	15+20 Bench B	N51°E 52°SE/N63°W 59°SW N75°E 76°SE	36.4°	No failure
2.	15+25 Bench B	N39°E 68°SE/N74°W 54°SW N75°E 76°SE	34.3°	No failure
3.	15+60 Bench B	N51°E 62°SE/N62°W 59°SW N75°E 76°SE	52.4°	Failed
4.	16+70 Bench B	N04°E 45°SE/N58°W 63°SW N75°E 76°SE	27.0°	Failed
5.	15+20 Bench C	N42°W 35°NE/N50°E 86°SE N15°W Vert.	27.7°	Failed
6.	16+25 Bench C	N64°E 62°SE/N06°W 72°SW N75°E 76°SE	55.9°	Failed

Listed in the table are the wedge slip planes orientations, the location of the wedge, notes on behavior, and an angle $\emptyset$. The tangent of this angle equals the total shear load acting down a slip direction divided by the total normal load on each slip face. This angle is a rough measure of frictional resistance if a wedge is just about to fail. Thus if a slope fails the friction angle, $\emptyset_f$, should be less than $\emptyset$.

It can be seen from Table II that failures occurred at angles as low as 27° implying frictional resistance is less, say 25°. In these areas weathering on joints was severe. At two other locations wedges did not fail with $\emptyset$ values of 34° and 36° implying the $\emptyset_f$ is larger, say 40°, or some cohesion exists across joint planes. Only moderate weathering existed in these locations. Finally one failure occured with a $\emptyset$ of 55.9°, but this failure actually occurred in two stages; a lower wedge falling from the face first, followed by a delay of a few days, followed by the upper wedge failure. It is hard to imagine this delay being caused by a $\emptyset_f$ of 55° in the weathered material. Puzzling situations like this were common.

Slopes Perpendicular to the River

The vertical east and west facing slopes were the most hazardous. For east facing slopes bedding planes dip downward and outward from the face. The cross bed joints J_1 acted to form wedges while the high angle joints released these wedges as shown in Figure 6. Progressive loosening and wedge slip-out became a major problem. The analysis in Table II showed a $\emptyset$ value of 27.7° along the weathered bedding plane for Case 6 in Table II. As the excavation went deeper, however, bedding and joint spacing again increased. The slope remained unstable but the instability appeared to confine itself more and more to just a 10 ft surface thickness.

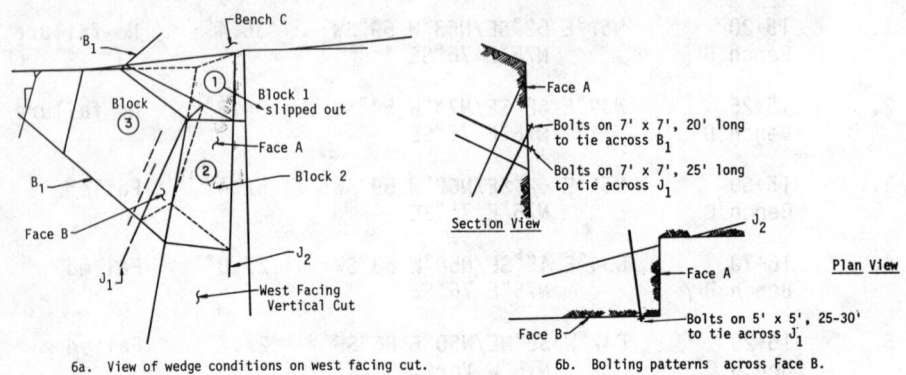

6a. View of wedge conditions on west facing cut.   6b. Bolting patterns across Face B.

Fig. 6. Wedge conditions for west facing slope.

A similar situation existed for the west facing joints, except J_1 was the slip plane and the bedding plane was the wedge former. Again the slope improved with depth.

ROCK SUPPORT

All slopes were stabilized with shotcrete and/or rock bolts. Rock bolts were 1 inch and 1-1/4 inch high strength steel bars grouted into 1-5/8 inch diameter drill holes. Fast set chemical grout was used to cover bolt ends for tensioning. Slow set grout was used to fill the remainder of the hole, thus insuring the entire bolt was covered. The bolts were tensioned before the slow set resin hardened. All shotcreting was done with wire mesh attached to bolt heads.

The philosophy behind support application was as follows:

ROCK SUPPORT AT PINE FLAT

- o Bolting patterns are designed to stabilize rock wedges in the outer 5-10 feet of rock slope faces. Alter the pattern and/or add bolts if potential for large wedge failures exist.

- o All bolts should be tensioned to 20,000 lbs and be fully covered by grout.

- o Bolts lengths should exceed 1.7 times bolt pattern spacing to insure the development of a compression zone behind the rock face.

- o Shotcrete with wire mesh should be applied to all permanently exposed slopes where severe ravelling and erosion is probable.

- o If the potential for very large wedge failures appears during excavation, perform analysis and design a seperate support system to stabilize that wedge.

This philosophy assumes that large slope failures are initiated by smaller ones, and that stabilizing these smaller ones will prevent the larger ones. For the slopes paralleling the river, this approach is correct because large failures could only occur with infrequent northeast striking joints. During construction extensive joints of this orientation with severe deep weathering were not observed. For slopes striking normal to the hillside, the potential for large wedge failures did exist, and bolting patterns had to be redesigned during excavation. It should be noted, however, that even where the potential for large failures existed, this step by step loosening action of smaller near-surface wedge failures seemed to predominate.

SUPPORT PROBLEMS

The following notes are comments on support problems during construction with a brief descriptions of their solution.

In the slopes above benches B and C, 12 ft bolts on a 7 ft by 7 ft pattern were used. In many instances the rock was so weathered that it was feared the bond between the rock and the grout would not hold. Thus bolt lengths were increased until rock hardness increased as indicated by drilling rates. Then the bolt length became the hole length. Bolts up to 20 ft long were installed. Most of the longer bolts were placed in the first several rows below bench floors.

When the excavation dropped below bench C, two major problems arose. First, the integrity of bench C had to be insured because of requirements for construction access by a 200 ton crane for handling steel penstock sections. Secondly the potential for large wedge failures appeared in the east and west facing slopes as described earlier. Allowing these slopes to fail to any great degree would have required extensive backfilling with concrete.

Joint mapping for the first case showed that the only potential failure was along wedges controlled by joints sets J_1 and J_2 and the high vertical face as shown in Figure 5. Joints were mapped and graphical analyses were performed. Dips on the J_2 set varied between 70° and 90° SE. Because of the height of the cut some very large wedges were possible on the order of 15 feet deep by 20 feet wide by 50 ft high. The weight of these wedges was on the order of 600 kips. It was found in the analysis that placing bolts through all potential failure planes on 5 ft centers would give a safety factor of 1.8 if a shear friction angle of 30° was used. Thus the bolting pattern was changed to that shown in Figure 5b.

For the second case, sliding down the bedding plane was the major problem. The configuration of block slippage is shown in Figure 6a. Blocks sliding down the bedding plane are released by cross bed joints J_1 and high angle joints J_2. To stabilize block 2 in the figure bolting can be done in three ways:

o   Bolt upward through face A into block 3.

o   Bolt laterally through face B across joint J_2 into the hillside.

o   Bolt downward through face A across bedding plane B, into lower rock.

For the first case block 2 is tied to block 3. If block 3 also fails so will block 2. However, there are many cross bed joints cutting through block 2 so that these bolts will serve to hole block 2 together, and prevent ravelling. This type of bolting was done on the loose upper portions of the cuts where joint spacing was on the order of 1 to 2 feet.

For the second case, the bolts will pull block 2 flush against the joint J_2 and more competent rock further in the hillside. This yeilds sliding resistance in the form of increase normal loads across J_2 and the bolt shear strength. Long 25 and 30 ft bolts were placed on face B as shown in Figure 6b.

For the third case, the bolts will increase frictional resistance to sliding along the bedding plane through increase normal loads and the shear strength of the bolts steel. This type of bolting was also done on the upper slope area. However, as the excavation proceeded below rock conditions improved, and case 1 bolting was merged with case 2 bolting.

Graphical analyses of block stability showed that the most effective pattern to prevent sliding was pattern 2. Because of a slight component of bedding dip into the hillside, block sliding tended to push up against joint J_2, thus increasing frictional resistance.

Thus bolts up to 30 ft long were placed in face A on 5 ft by 5 ft centers. To further secure loose rock on bolted faces wire mesh and shotcrete were also applied.

SUMMARY AND CONCLUSIONS

Some interesting conclusions can be obtined from the excavation experience at Pine Flat. These are:

o   A very limited amount of surface mapping did identify all major bedding and jointing orientations. However, it did not give an indication of the variability of these orientations.

o   Limited surface observations did identify the existence of weathering but in no way did it identify how strongly weathering would influence slope stability.

o   All slope stability problems were directly related to bed and joint orientations. Simple wedge failures predominated.

o   Weathering reduced joint friction angles from values of larger than 34° to 36° down to values of less than 27°.

o   Some cohesion was noticed across some fresh joints at depth below river level.

o   Weathering was controlled by the seepage of run off water downward through joints and bedding planes.

o   All slopes were supported by shotcrete and, or rock bolts. The limited size of most potential wedge failures allowed for rock bolt stabilization. Excessive bolting was required only where blocks reached dimensions of 30 to 40 feet in depth, width or height.

REFERENCES

Goodman, R.E., 1976, <u>Methods of Geological Engineering</u>, West Publishing Company, San Francisco

Chapter 92

GENERALIZATION OF THE GROUND REACTION CURVE CONCEPT

by Emmanuel Detournay[1] and Charles Fairhurst[2]

[1] Research Fellow, Department of Civil and Mineral Engineering
University of Minnesota

[2] Professor and Head, Department of Civil and Mineral Engineering
University of Minnesota

ABSTRACT

A method is proposed for computing (a possible solution of) the ground reaction curve, for use in tunnel support analysis, in cases where the initial stress field is not hydrostatic. Rock around a circular tunnel is assumed to behave as an elasto-plastic material with a Mohr-Coulomb yield envelope.

INTRODUCTION

Tunneling in squeezing rock conditions is usually characterized by large ground displacements or "convergences". These result from the development, around the tunnel, of a failed zone produced by stress changes due to excavation.

Computation of the so-called "Ground Reaction Curve" (GRC), i.e., the relationship between the ground displacement and the tunnel support pressure p is usually based on the assumption of an infinite plane strain model with the excavation sequence simulated by "unloading" of the tunnel boundary. The simplest such model assumes a circular tunnel and an initial homogeneous hydrostatic stress field in the plane. Determination of the ground reaction curve and its application to tunnel design have been the subject of numerous investigations (Fenner, 1938; Labasse, 1949; Salencon, 1969; Egger, 1973; Ladanyi, 1974; Rechsteiner and Lombardi, 1974; Daemen, 1975; Panet, 1976; Nguyen Minh and Berest, 1979; Berest, Habib, and Nguyen Minh, 1980; Kaiser, 1981; Vardoulakis and Detournay, 1982).

The object of this paper is two-fold: (1) generalize the concept of the GRC for a non-hydrostatic initial stress field, and (2) provide a semi-analytic means to compute the GRC, assuming the rock to have an elastoplastic material behavior, with a Mohr-Coulomb yield envelope.

PROBLEM STATEMENT

Consider an infinite plane, loaded in plane strain, containing a circle of radius a. The reference cartesian coordinates system (x,y) is defined in the plane, with origin at the center of the hole, see Fig. 1. The plane is initially subjected to a uniform compressive stress field ($\underline{\tau}^0$) with the major initial compressive stress τ_1^0 parallel to the y-axis.

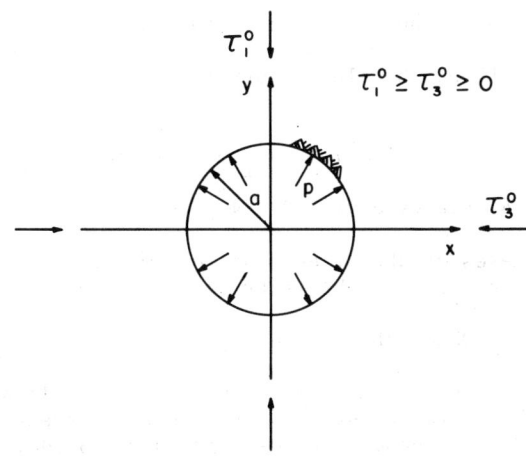

Figure 1 : Problem Definition

We seek to determine the displacement at the circular boundary (r = a) caused by quasi-static "unloading" of that boundary. It is required that:

1. the initial state (Σo) be elastic;
2. the final state (Σ_1) corresponding to zero stress at r = a, may be characterized by a plastic zone fully surrounding the hole.
3. the elastic limit of the system (Σpo) be reached when:

$$\tau_{rr} = p_p \; ; \; \tau_{r\phi} = 0 \text{ at } r = a. \qquad (1)$$

4. after attaining the limit elastic state Σpo, further unloading be proportional, i.e.,

$$\tau_{rr} = (1-\lambda)p_p \; ; \; \tau_{r\phi} = 0, \; \lambda \in [0,1] \qquad (2)$$

The material is assumed to be homogeneous, isotropic, linearly elastic – perfectly plastic, with a Mohr-Coulomb yield criterion that may be expressed as:

$$\tau_1 = K_p \tau_3 + q \tag{3}$$

where,

τ_1, τ_3 are the principal stresses in the plane of deformation ($\tau_1 \geq \tau_3 \geq 0$)

K_p is the passive coefficient $(1+\sin\Phi)/(1-\sin\Phi)$; Φ, friction angle

q is the unconfined strength of the material.

During plastic flow, the elastic strain increments are neglected, thus:

$$d\varepsilon_3 = \mu(1-\sin\Phi^*) \; ; \; d\varepsilon_1 = -\mu(1+\sin\Phi^*) \tag{4}$$

where, Φ^* is the dilatancy angle ($0 \leq \Phi^* \leq \Phi$).

The principal axes of the stress and incremental strain tensor are coincident on account of the isotropy assumption.

ADMISSIBLE INITIAL STRESS FIELD

The requirements 1–3 on the loading path at $r = a$ impose some restrictions on the characteristics of the initial stress field $\underline{\tau}^o$. Thus, if we let P and S denote the mean pressure and deviatoric stress invariant respectively in the plane of deformation:

$$P = \frac{\tau_1 + \tau_3}{2} \; ; \; S = \frac{\tau_1 - \tau_3}{2} \tag{5}$$

the parameter m defined as follows:

$$m = S^o/S_\ell^o \; ; \; \text{with } S_\ell^o = \frac{K_p - 1}{K_p + 1}(P^o + \frac{q}{K_p - 1}) \tag{6}$$

is a measure of the "obliquity" of the initial stress in the space (P^o, S^o). $m < 1$ corresponds to an initial elastic state.

The set of initial stress states compatible with the hypotheses 1–4 is indicated by the hatched region IIb (case $\Phi = 30°$) in Fig. 2. This set is obtained as follows:

1. The restriction $m \leq 0.5$ is a necessary condition for requirement 3 on the loading path. If $m \leq 0.5$, the boundary conditions Eq. (1), with p_p defined by

$$p_p = (2P^o + 4S^o - q)/(K_p + 1) \tag{7}$$

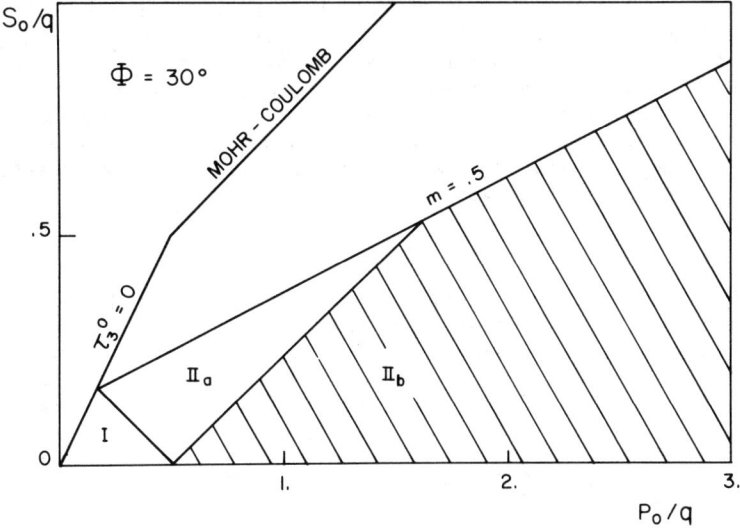

Figure 2 : Admissible Initial Stress Conditions ($\Phi=30°$)

correspond to an elastic limit of the system, with the points (a,o) and (a,II) on the boundary in a state of impending failure.

2. Kirsch's solution (e.g., Sokolnikoff, 1956) for the elastic stress distribution around a circular hole in a loaded plate indicates that the final state (Σ_1) is elastic if the initial stress point lies in region I (Fig. 2) of the stress space (P^o, S^o), the line of separation between I and II is given by:

$$4S^o = q - 2P^o \quad ; \quad 0 \leq P^o \leq q/2 \tag{8}$$

3. If the initial stress point falls within the limits of region IIb (Fig. 2), the hole is completely surrounded by a plastic zone at the end of the loading history ($\lambda=1$). The demarcation line between IIa and IIb is obtained on the basis of a statical solution of the elasto-plastic interface. The parametric equations of this line are:

$$P^o = \frac{q}{2\alpha} h(m)(1-m)^{-2\alpha} - \frac{q}{K_p-1} \quad ; \quad S^o = \frac{mq}{2} h(m)(1-m)^{-2\alpha} \quad ; \quad 0 \leq m \leq .5 \tag{9}$$

where, $\alpha = (K_p-1)/(K_p+1)$; $h(m) = F(-\alpha, -\alpha, 1, m^2)$ (Gauss hypergeometric series).

GENERALIZATION OF THE GROUND REACTION CURVE

The loading path[*] followed to reach the limit elastic state Σpo may be arbitrary, provided only that Σpo is reached by a sequence of elastic states only. Beyond the elastic limit the loading is proportional (see Eq. (2)), the loading parameter is assumed to increase monotonically with time. The intermediate plastic states are denoted by $\Sigma p\lambda$, $0 \leq \lambda \leq 1$. In the transition from Σpo to $\Sigma p1 \equiv \Sigma 1$, the configuration of the plastic region, the stress $\underline{\tau}$ and a displacement field $\underline{u}$ are all functions of λ, which plays the role of a kinematic parameter. We may therefore define the "velocity" field associated with the state $\Sigma p\lambda$ as follows:

$$\underline{v}(x,y;\lambda) = \frac{D}{D\lambda} \underline{u}(x,y;\lambda) \tag{10}$$

Let $\underline{u}_a^e(\phi)$, $\underline{u}_a(\phi,\lambda)$, and $\underline{v}_a(\phi,\lambda)$ denote, respectively, the elastic displacement at $\lambda=0$, the displacement at the state $\Sigma p\lambda$ for a point (a,ϕ) of the circular boundary and the velocity at the same point. The displacement $\underline{u}(\phi,\lambda)$ must be determined incrementally by integrating the velocity $\underline{v}_a(\phi,\lambda)$ associated with each plastic state $\Sigma p\lambda$, i.e.,

$$\underline{u}_a(\phi,\lambda) = \underline{u}_a^e(\phi) + \int_0^\lambda \underline{v}_a(\phi,\mu)\,d\mu \; ; \; 0 \leq \lambda \leq 1 \tag{11}$$

The vectorial function $\underline{u}_a(\phi,\lambda)$ of λ is defined as the ground reaction "curve" at a point (a,ϕ) of the boundary. It must be emphasized that this definition applies only to the plastic part of the loading history.

The elastic displacement $\underline{u}_a^e(\phi)$ and the limiting value of the velocity $\underline{v}_a(\phi,\lambda)$ at $\lambda=0$ are obtained from the Kirsch's solution. In the cylindrical coordinate system, the components of $\underline{u}_a^e(\phi)$ and $\underline{v}_a(\phi,0)$ are as follows:

$$\begin{aligned} u_{ar}^e(\phi)/\ell &= (3-4\nu)\,S^{o'}\cos 2\phi + p_p' - p^{o'} \\ u_{ar}^e(\phi)/\ell &= -(5-4\nu)\,S^{o'}\sin 2\phi \end{aligned} \tag{12}$$

where,

$\ell = aq/2G$ is a characteristic length
G, ν the shear modulus and Poisson's Ratio respectively
' denotes division of the stress quantity by q

and
$$v_{ar}(\phi,0)/\ell = -p_p' \; ; \; v_{a\phi}(\phi,0)/\ell = 0 \tag{13}$$

Unfortunately, the elasto-plastic analysis needed to determine the velocity $\underline{v}_a(\phi,\lambda)$ cannot, at the present time, be performed analytically except for the case of an initial hydrostatic stress field. The remainder of the paper discusses the construction of a (statically) possible solution of $\underline{v}_a(\phi,\lambda)$.

(*) "unloading" of the circular boundary causes loading of the plane.

OUTLINE OF THE METHOD OF SOLUTION

As the loading parameter λ increases monotonically from 0 to 1, the plastic zones initiated at (a,o) and (a,π) spread around and away from the hole towards the interior of the plane. If the initial stress conditions allow, the failed zones will eventually coalesce for $\lambda = \lambda_\ell$ so as to form a continuous plastic region around the hole. Three stages can thus be distinguished in the loading history $0 \leqslant \lambda \leqslant 1$: (i) a stage of impending failure ($\lambda=0$), (ii) a first plastic stage where the hole is partially surrounded by a plastic zone ($0<\lambda \leqslant \lambda_\ell$), and (iii) a second plastic stage where the hole is fully surrounded by a failed zone ($\lambda_\ell \leqslant \lambda \leqslant 1$).

For $\lambda > 0$, the velocity $\underline{v}_a(\phi,\lambda)$ is determined as follows:

1. $\lambda_\ell \leqslant \lambda \leqslant 1$: Computation of the velocity is based on the construction of a statically admissible solution of the elasto-plastic interface Γ. For λ increasing from λ_ℓ to 1, the static solution provides a continuous sequence of configurations for the plastic and elastic regions, thus allowing for the calculation of the velocity field.

2. $0 \leqslant \lambda \leqslant \lambda_\ell$: a solution for the interface during the first plastic stage is not yet available. It is assumed that, at a given point (a,ϕ) of the boundary, the function $\underline{v}_a(\phi,\lambda)$ evolves in a continuous fashion from $\lambda=0$ (elastic solution) to $\lambda=\lambda_\ell$ (statically admissible solution).

COMPUTATION OF THE VELOCITY FIELD, $\lambda_\ell \leqslant \lambda \leqslant 1$

A possible solution for the velocity field during the second plastic stage is calculated from a sequence of statically admissible configurations for the plastic and elastic regions.

Statical Configuration of the Elasto-plastic Interface Γ

A statical solution for Γ is constructed on the assumption that the stress field in the plastic region is (i) continuous and (ii) statically determined by the boundary conditions at $r=a$. The unknown boundary Γ between the plastic and elastic regions must satisfy the requirements that (i) all plane components of the stress tensor are continuous across Γ, and (ii) the stress field in the elastic region is everywhere admissible (i.e. the yield criterion is not violated).

The solution for the interface Γ was obtained using the Kolossov-Muskhelishvili complex variable method by determining the unknown mapping function which transforms the infinite elastic region bounded by Γ (physical plane z) onto the domain exterior to the unit circle in the reference plane ζ (Detournay and Fairhurst, 1982). It was found (Detournay, 1982) that an explicit approximate solution for the mapping function $z/a = \omega(\zeta;\lambda)$ is given by:

$$z/a = R(\lambda)\zeta(1 + \frac{m}{\zeta^2})^{2/(K_p+1)} \qquad (14)$$

where,

$z = x+iy$ is the complex variable in the physical plane
$\zeta = \xi+i\eta$ is the complex variable in the reference plane
$R(\lambda) = R_o(\lambda)/[h(m)]^{1/(K_p-1)}$ [cf Eq. (9) for $h(m)$].

$aR_o(\lambda)$ is the radius of the elasto-plastic interface for the limiting case of an initial hydrostatic stress field of magnitude P^o (m=0).

It should be noted that there is an error associated with the approximate solution Eq. (14). This error, measured by the maximum stress discontinuity across Γ, increases with m and K_p but, even so, remains less than $10^{-3}S_\ell^o$ for m=0.3. The stress field in the elastic region defined by Eq. (14) is admissible as indicated by the contour levels of the yield function $F \equiv (S-S_\ell)/S_\ell^o$ in Fig. 3.

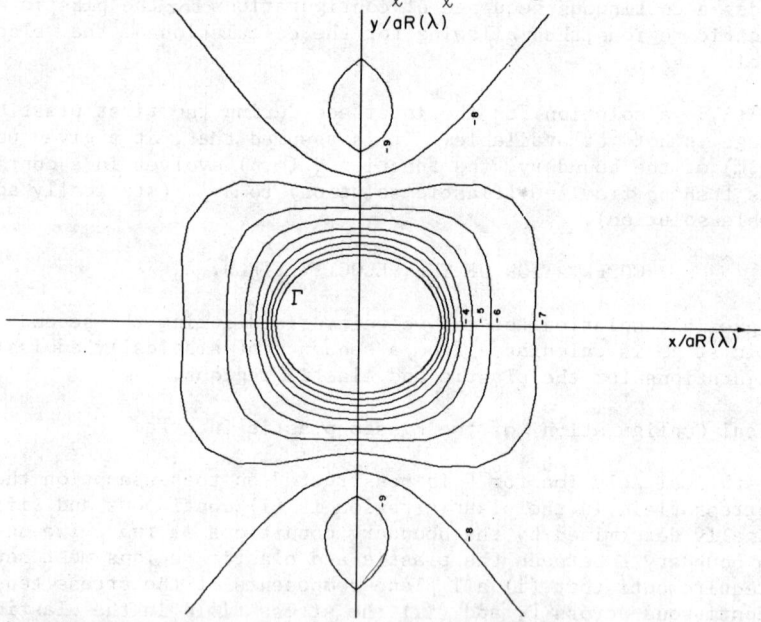

Figure 3. Contours of the Yield Function in the Elastic Zone ($\phi=30°$, m=0.2)

The statical solution of the interface Γ has some remarkable properties: (i) Γ has an oval shape symmetric about the x and y coordinate axes; (ii) the direction of largest expansion of the plastic zone is perpendicular to the direction of the initial maximum compression stress; (iii) the plastic region grows with λ, but the configuration of the elastic region remains self-similar.

Velocity Field

The induced displacement field in the elastic region can be computed for each configuration defined by Eq. (14). That solution, in conjunction with Eq. (14) is used to determine the velocity field in the elastic zone for each state $\Sigma p\lambda$ $\lambda_\ell \leq \lambda \leq 1$. It can be shown that the elastic velocity solution is of the form:

$$v_x + iv_y = \ell \, S_\ell^{o'} \frac{dR(\lambda)}{d(\lambda)} \, \hat{v}(\zeta) \; ; \; |\zeta| \geq 1 \qquad (15)$$

where,

v_x, v_y are the cartesian components in the x and y directions respectively of the induced velocity in the elastic zone.

$\hat{v}(\zeta)$ is a complex function of the reference variable ζ depending upon m, K_p, and the value of Poisson's ratio ν.

The velocity at the elasto-plastic interface Γ, obtained from (15) with $|\zeta|=1$, is then used as "boundary conditions" for the calculation of the velocity in the plastic zone. Since elastic strain increments during plastic flow are neglected, the plastic velocity may be considered to be governed by hyperbolic partial differential equations (Salencon, 1972), which can be solved numerically by the method of characteristics (Masseau, 1889).

In order to take advantage of the self-similar configuration of the elastic region during "unloading" ($\lambda_\ell \leq \lambda \leq 1$), computation of the plastic velocity is carried out in the unit plane $\hat{z} = z/aR(\lambda)$, with the normalized velocity $\hat{\underline{v}} = \underline{v}/\ell S_\ell^{o'} \frac{dR(\lambda)}{d\lambda}$. In the unit plane $\hat{z}$, the elasto-plastic interface is represented by a fixed curve $\hat{\Gamma}$, the boundary r=a by circles of varying radius. The transformation $\hat{z} = z/aR(\lambda)$ considerably reduces the computational effort required to calculate the velocity at the circular boundary by making use of the correspondence between the velocity at (a,ϕ) in the loading range $(\lambda_\ell,1)$ and the velocity along the radial segment of line $([aR(1)]^{-1},\phi)$, $([aR(\lambda_\ell)]^{-1},\phi)$ in the unit plane.

COMPUTATIONAL RESULTS

To illustrate the computational process for determining the ground reaction curves we will consider the following case: $\tau_{xx}^o/q=1.85$, $\tau_{xy}^o/q=0$, $\tau_{yy}^o/q=2.15$, $\Phi=30°$, $\nu=0.5$, and $\Phi^*=0°$ or $30°$. This problem is characterized by a small obliquity of the initial stress field (m=0.12). Fig. 4 illustrates the variation of the computed velocity-normalized with respect to the characteristic length ℓ-with the loading parameter λ at the points 1 and 2 of the boundary (curves a: $\Phi^*=30°$; curves b: $\Phi^*=0$). The velocity at 1 and 2 has a radial component only because these points are on the axes of symmetry of the problem (see insert of Fig. 4). The portion of the curve between

$\lambda=0$ and $\lambda=\lambda_\ell$ is a quadratic interpolation between the elastic solution ($\lambda=0$) and the statical solution ($\lambda_\ell \leqslant \lambda \leqslant 1$). The velocity curves are

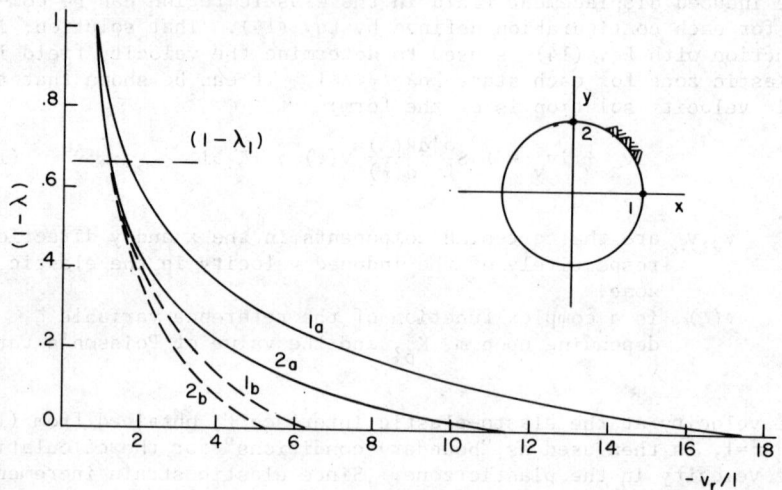

Figure 4. Variation of the Velocity at two Points of the Boundary

integrated to calculate the displacements for the ground reaction curves (Fig. 5). The GRC of Fig. 5 show two types of behavior depending upon the value of dilatancy angle. If $\phi^*=30°$ the direction of greatest convergence changes, during the loading history ($\lambda=0.4$), to become perpendicular to the major initial compressive stress τ_1^o. With no dilatancy assumed during plastic flow, the direction of largest displacement remains vertical, i.e. parallel to τ_1^o; however,

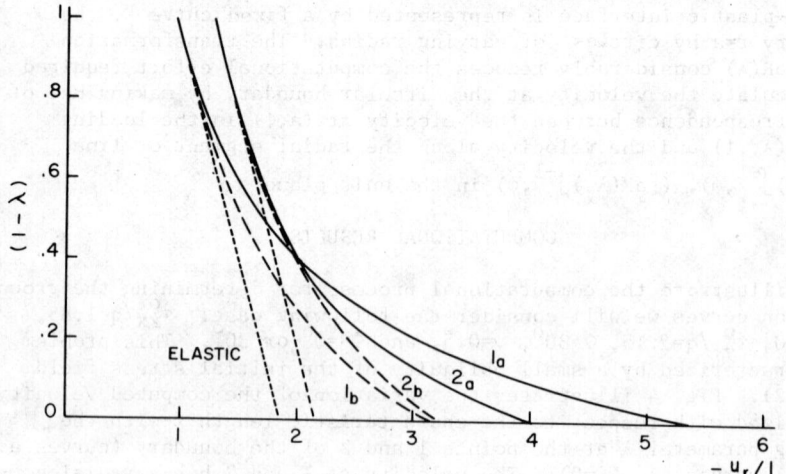

Figure 5. Ground Reaction Curves

the difference in the magnitude of the displacement at 1 and 2 decreases with λ, suggesting that a change in the initial conditions (e.g., increase of the ratio P^o/q but same obliquity m) could eventually produce a rotation of the direction of maximum convergence.

CONCLUSIONS

The concept of ground reaction curve, used in the design of deep tunnels, has been defined for cases of non-hydrostatic initial stress fields. A method for developing possible solutions for these curves has also been described. The main advantage of the approach proposed is that it is able to generate these curves at negligible computational cost, thus permitting inexpensive parametric analysis.

On the basis of a single simulation it was shown that: (i) a small anisotropy in the initial stress field may induce considerable anisotropy in the displacement at the tunnel wall, and (ii) the direction of greatest convergence in the tunnel may rotate during unloading. One of the sensitive factors controlling this "rotation" is the dilatancy of the material during plastic flow.

REFERENCES

Berest, P., Habib, P., Nguyen Minh, D., 1980,"Tentative d'interpretation des deformations observees aux tunnels du Frejus et du Grand Sasso," Revue Francaise de Geotechnique, 12:44-55.

Daemen, J.J.K., 1975, Tunnel Support Loading Caused by Rock Failure, Ph.D. Thesis, University of Minnesota.

Detournay, E., 1982, Contribution to the design of deep tunnels, Ph.D. Thesis, To be submitted to the University of Minnesota.

Detournay, E., Fairhurst, C., 1982, "Approximate Statical Solution of the Elastoplastic Interface in the Problem of an Infinite Medium with a Circular Hole," In 4th Int. Conf. Numerical Methods in Geomechanics, Edmonton, Canada.

Egger, P., 1975, Einsfluss des post-failure verhaltens von fels, Technical Report 57, IBF, Karlsruhe.

Fenner, R., 1938, "Untersuchungen zur erkenntis des gebiegsdruckes," Glueckauf, 681-695, 705-715.

Kaiser, P.K., 1981, "A new concept to Evaluate Tunnel Performance - Influence of Excavation Procedure," 22nd U.S. Symp. on Rock Mechanics, M.I.T., pgs. 264-271.

Labasse, H., 1949, "Les pressions de terrains autour des puits," Revue Universelle des Mines, 5:78-88.

Ladanyi, B., 1974, "Use of Long-Term Strength Concept in the Determination of Ground Pressure on Tunnel Linings," Proc. 3th Congress of ISRM, Vol. II-B, pgs. 1150-1156, Denver.

Masseau, J., "Memoire sur l'integration graphique des equations aux derivees partielles", Ann. Ass. Ing. Gand, 1899.

Nguyen Minh, D., Berest, P., 1979, "Etude de la stabilite des cavites souterraines avec un modele de comportement elastoplastique radoucissant," In Proc. 4th Congress of ISRM, Vol. 1, pgs. 249-256, Montreux.

Panet, M.,1976, "Analyse de la stabilite d'un tunnel creuse dans un massif rocheux en tenant compte du comportement apres la rupture," Rock Mechanics 8:209-223.

Rechsteiner, G.F., Lombardi, G., 1974, "Une methode de calcul elastoplastique de l'etat de tension et de deformation autour d'une cavite souterraine," In Proc. 3th Congress of ISRM, Vol. II-B, pgs. 1049-1054, Denver.

Salencon, J., 1969, "Contraction quasi-statique d'une cavite a symetrie spherique ou cylindrique dans un milieu elastoplastique," Annales des Ponts et Chaussees 4:231-236.

Sokolnikoff, I.S., 1956, Mathematical Theory of Elasticity, McGraw-Hill.

Vardoulakis, I., Detournay, E., 1982, "Determination of the Ground Reaction Curve in Deep Tunnels Using Biot's Hodograph Method," In 4th Int. Conf. Numerical Methods in Geomechanics, Edmonton, Canada.

Chapter 93

STABILIZATION OF ROCK EXCAVATIONS
USING ROCK REINFORCEMENT

by Thomas A. Lang and John A. Bischoff

Senior Consultant and Associate

Woodward-Clyde Consultants
San Francisco, California

ABSTRACT

The use of rock bolts for the support and stabilization of coal mining excavations was introduced over 30 years ago and, since that time, has progressively increased until today it is the primary means of support and stabilization of large underground openings. Early analytical studies and work with photoelastic and physical models established that even highly fragmented rock could be stabilized by rock bolting and form a load-carrying structural member that would span an opening. For stability, interaction of the bolts is essential and it depends on the bolt tension, the length and spacing of the bolts, the characteristics of the rock, and the applied load. However, the development of an explicit functional relationship embracing all these factors proved to be elusive.

A review of the early rock bolting investigations led to the concept that the basic element of a rock bolted roof is the reinforced rock unit (RRU), consisting of an individual bolt and the rock immediately surrounding and adjacent to it, and to the development of mathematical equations which give the minimum bolt tension required to ensure that the RRU's are stable relative to one another and act together as a structural member. Analytical procedures have been developed for integrating RRU's into a reinforced rock structural (beam-arch) member which spans an underground opening and providing a rational basis for the design and installation of rock reinforcement systems.

INTRODUCTION

The use of rock bolts for the support and stabilization of coal mining excavations was introduced over 30 years ago (Weigel 1943, Thomas, et al. 1949) and, since that time, has progressively increased until today it has become the primary support system in the mining industry and for the stabilization of large excavations for underground power stations and other purposes. In the early rock bolt research by the U. S. Bureau of Mines (USBM), the analytical studies and laboratory investigations with physical models were directed primarily towards beam action in which the strata were clamped together by tensioned bolts to form a laminated beam with enhanced bending strength.

Using the results of the USBM research, a more general approach was taken by the Snowy Mountains Hydroelectric Authority (SMA) in Australia because its underground works were in igneous and metamorphic rocks as well as stratified rocks. The SMA investigators used photoelastic and physical models and elastic analysis to show that even highly fragmented rock could be stabilized by rock bolting and form a load-carrying structural member that would span an opening. It was found that, for stability, interaction of the bolts was essential and that it depended on bolt tension, length/spacing ratio and time of installation of the bolts, physical characteristics of rock and depth of distressed rock or applied load (Lang 1957, 1962). However, although semi-empirical rules for safe rock reinforcement practice were developed, the development of explicit functional relationships embracing all of these factors proved to be elusive.

Recently, a review of these early investigations led to the concept that a reinforced rock structure was a system of reinforced rock units (RRU), each of which consisted of a bolt and the rock immediately surrounding and adjacent to it, CJEF, Fig. 1. Using this concept, equations have been developed that express the relative stability of the

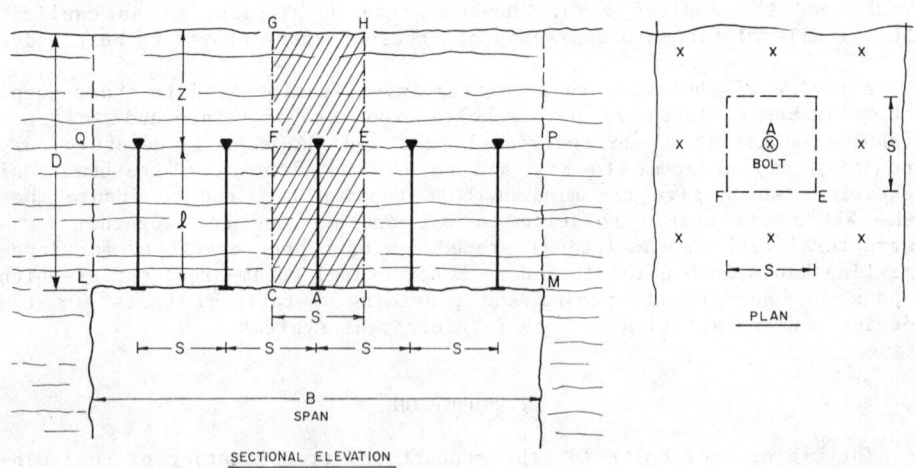

Figure 1. Reinforced Rock Unit.

STABILIZATION OF EXCAVATIONS

RRU's, unify the results of past research and experience and provide a rational basis for quantifying the contribution of rock reinforcement to the stability of an underground excavation (Lang, et al. 1979, 1981).

REINFORCED ROCK UNIT

Consider first the limiting equilibrium of an unreinforced rock unit, CJEF, Fig. 2(a), relative to its neighbors (List of Symbols is at the end of this paper).

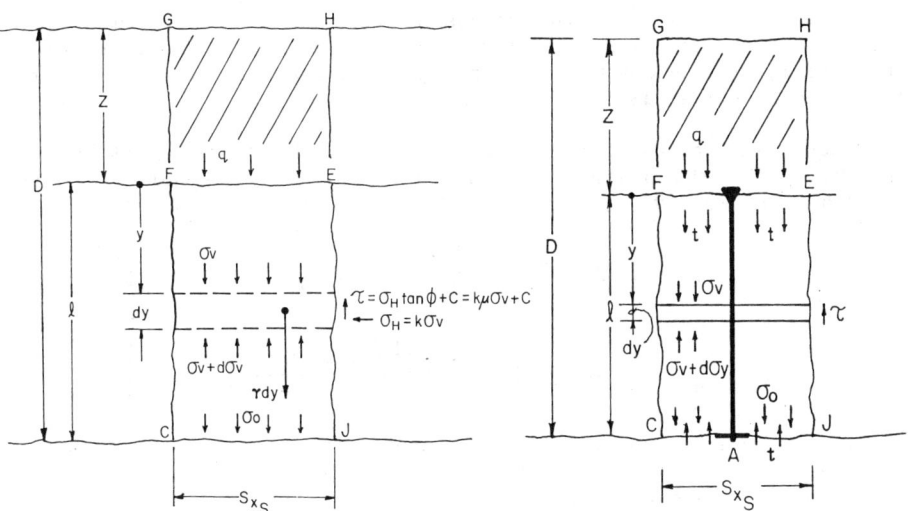

(a) Unreinforced Rock Unit   (b) Reinforced Rock Unit

Figure 2. Stability of Rock Unit.

For a given value of y, say $y = \ell$, the vertical stress σ_o at CJ, if it was not a free surface, can be shown to be:

$$\sigma_o = (\gamma R - c)\frac{1}{k\mu}(1 - e^{-k\mu \ell/R}) + qe^{-k\mu \ell/R} \quad (1)$$

If the superimposed stress q at FE results from a depth of distressed rock z and $D = z + \ell$, then for the limiting condition (i.e., shear is fully developed along the periphery of the unit), Equation (1) becomes:

$$\sigma_o = (\gamma R - c)\frac{1}{k\mu}(1 - e^{-k\mu D/R}) \quad (2)$$

In the effect, $(\gamma R - c)$ is the passive direct support that a prop under the rock unit would have to provide to prevent fallout of the unit relative to the surrounding rock.

If a stabilizing pressure, t, equal to σ_o, is provided by a rock bolt installed at the center of the rock unit, Fig. 2(b), then the total tensile load, T, in the rock bolt will be:

$$T = tA = ts^2 \text{ (for a square pattern)} \qquad (3)$$

Assuming that the bolting stress is uniformly distributed over the section of the rock unit and the effective length of the rock bolt is equal to ℓ, the rock bolt supplies a passive support pressure, t, at CJ and an equal stress is added at FE.

Under these conditions, it can be shown that, for equilibrium of the RRU, CJEF, the passive pressure, t, applied by the bolt at CJ must equal σ_o, and we have:

or

$$\left. \begin{array}{c} t(1 - e^{-k\mu\ell/R}) = \dfrac{\alpha}{k\mu} (\gamma R - c)(1 - e^{-k\mu D/R}) \\[6pt] \dfrac{t}{\gamma R} = \dfrac{T}{\gamma A R} = \dfrac{\alpha}{k\mu}\left(1 - \dfrac{c}{\gamma R}\right)\left[\dfrac{1 - e^{-k\mu D/R}}{1 - e^{-k\mu\ell/R}}\right] \end{array} \right\} \qquad (4)$$

In the above equations, it is assumed that a limiting condition has been reached where the downward deformation of the rock unit has proceeded to the stage where the full shear strength of the rock, in accordance with Coulomb's criteria, has been developed. In other words, the bolt is applying a "passive" pressure at A, Fig. 2(b), which is just adequate to provide support for that portion of the total weight of distressed rock, D, which is not carried by shear on the sides of the rock unit. For these conditions, $\alpha = 1.0$. If the rock reinforcement is installed either before or immediately after excavation exposes the surface CJ, Fig. 2(b), that is, before significant deformation has taken place, then the pressure at CJ will "actively" retard deformation of the unit and "actively" contribute to its stability. Under these conditions the bolts could be described as providing "active" reinforcement and it can be shown that $\alpha = 0.5$. In practice, the time of installation of the bolts is generally intermediate between these two extremes and the use of $\dot{\alpha} = 1.0$ is conservative. It may be noted that the use of fully grouted untensioned rebar as rock reinforcement, where the tension in the rebar is developed by the rock deformation, would also be intermediate between the "active" and "passive" conditions defined above and would provide a confining stress, t, similar to that provided by tensioned bolts.

Field stresses parallel to the free surface (CJ, Fig. 2) can also be provided for in Equation (4) but, for simplicity, have not been included

STABILIZATION OF EXCAVATIONS

in these notes. Parametric and sensitivity studies (Lang, et al. 1981) as well as experience have shown that values for the intrinsic shear strength (cohesion) which result from presently available methods of testing can be misleading and should be used with caution in making stability appraisals. Accordingly c should be taken as zero in initial designs and investigations. Appropriate modifications can be made as information is obtained regarding the behavior charactersitics of the in situ rock.

The variation of $T/\gamma AR$ with D/R and ℓ/R is shown in Fig. 3 for $\phi = 30$ degrees. As the ℓ/R and D/R ratios increase, asymptotic values for $T/\gamma AR$ or $t/\gamma R$ are approached and the $\Delta = 5$ and 1 percent curves shown when the variation in $T/\gamma AR$ will be less than these values.

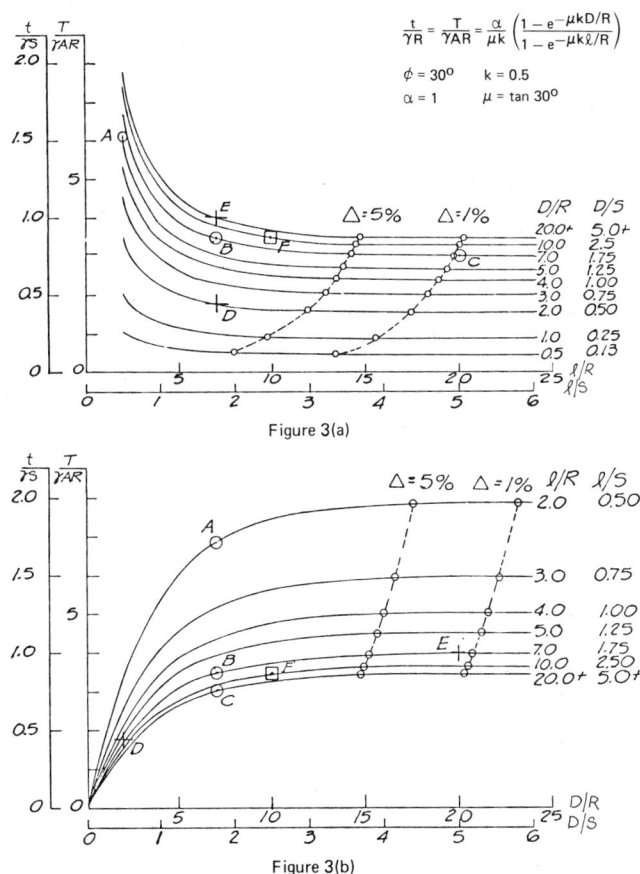

Figure 3. Variation of $T/\gamma AR$ with D/R and ℓ/R.

A clearer appreciation of these inter-relationships can be obtained from the 3-D diagram, Fig. 4, by comparing the curves DBE and ABC in the two figures. To illustrate the effects of the variation of ϕ, Fig. 5 shows a range of values with ϕ varying from 20 to 60 degrees and assuming that D/R approaches infinity. This shows that for given rock conditions there is a practical limit to the ℓ/R or ℓ/s ratio beyond which there is little decrease in the required bolt tension. It also shows clearly that as the ratio ℓ/s decreases below two, the required bolt tension increases dramatically, particularly for the smaller values of ϕ.

REINFORCED ROCK ARCH

The RRU's which form the reinforced zone in the roof of an excavation are analogous to the voussoirs in masonry arch structures. This is the case whether the reinforced roof is flat, Fig. 6(b), as is common in sedimentary strata, or excavated as an arch, Fig. 6(a).

In Fig. 6, abce may be taken as representing a segment of a voussoir arch with an effective depth, d, and an area of (s x s) at the intrados. Although the reinforced rock arch is not free-standing, it is convenient to treat it as a free-standing arch loaded with its own weight and the superincumbent distressed rock and then consider the constraints that are imposed on it by the surrounding rock mass with which it is integrated.

An initial approach to defining the loading on such a reinforced rock arch can be made by applying Equation (2) to the full span, B, and taking the attenuated stress, σ_o, corresponding to a depth, D, of distressed rock as being the distributed load on the arch. The load is attenuated only by shear in the abutment areas. Using an appropriate shear radius to reflect this case will enable the determination of a suitable pattern of the minimum rock reinforcement that is required to stabilize the roof (Lang, et al. 1981). This pattern would need modification to take into account the effects of joints and other discontinuities in the rock mass.

CONCLUSION

The concepts and equations that have been developed quantitatively describe the conditions for stability of the rock bolt reinforced zone of the roof of an underground opening as a structural (beam-arch) member consisting of reinforced rock units. The equations provide, for the first time, a rational basis for the optimum design and installation of rock reinforcement systems. The equations take into account the rock bolt tension, length, spacing, and orientation of the bolts, size of the opening, height of distressed rock above the roof, and the strength and other physical properties of the rock. The use of a rational design procedure for a rock bolted roof will reduce the uncertainties that are inherent in the present semi-empirical methods and thus provide enhanced safety for underground excavation operations.

STABILIZATION OF EXCAVATIONS

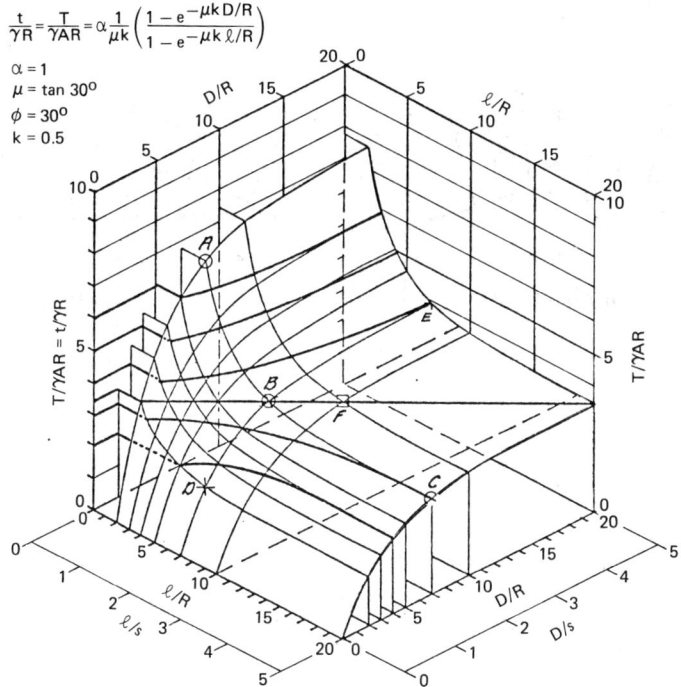

Figure 4. 3-Diagram of $T/\gamma AR$ vs. ℓ/R and D/R for $\phi = 30°$.

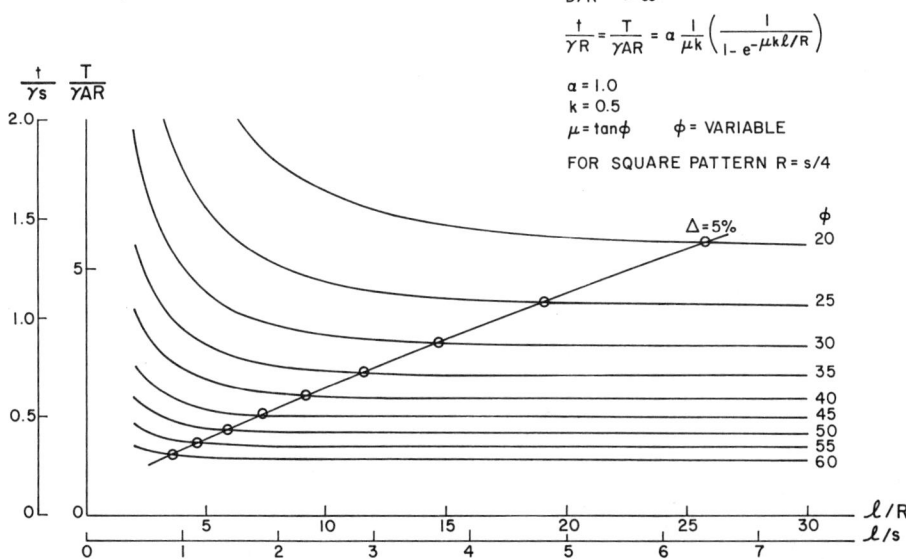

Figure 5. Variation of $T/\gamma AR$ with ℓ/R for Values of ϕ from 20° to 60°.

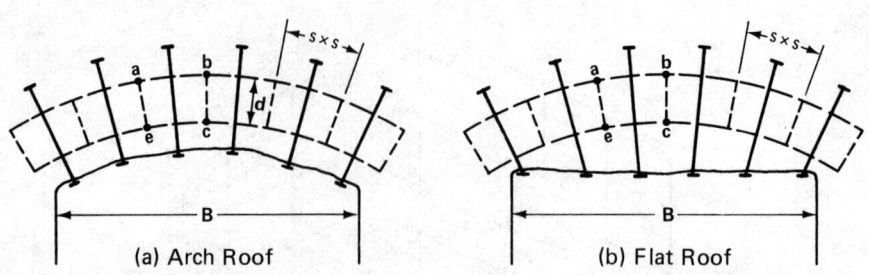

Figure 6. RRU Arches.

REFERENCES

Lang, T. A., 1957, "Rock Behavior and Rock Bolt Support in Large Excavations. Snowy Mountains Scheme--Australia, T1 Power Station." Symp. on Underground Power Stations, ASCE, New York, October. (Abstract, Lang, T. A., 1958, "Rock Bolting Speeds Snowy Mountains Project." Civil Engineering, Vol. 28, No. 2, February, pp. 40-42.)

Lang, T. A., 1962, "Theory and Practice of Rock Bolting." Trans., AIME (Mining), Vol. 223.

Lang, T. A., Bischoff, J. A., and Wagner, P. L., 1979a, "A Program Plan for Determining Optimum Roof Bolt Tension--Theory and Application of Rock Reinforcement Systems in Coal Mines," Leeds, Hill and Jewett, Inc., San Francisco, Vol. 1, 256 p.

Lang, T. A., Bischoff, J. A., and Wagner, P. L., 1979b, "A Program Plan for Determining Optimum Roof Bolt Tension--Theory and Application of Rock Reinforcement Systems in Coal Mines," Leeds, Hill and Jewett, Inc., San Francisco, Vol. 2, Bibliography, 43 p.

Lang, T. A. and Bischoff, J. A., 1981, "A Research Study of Coal Mine Rock Reinforcement," Leeds, Hill and Jewett, Inc., San Francisco, 224 p.

Thomas, E., Barry, A. J., Metcalfe, A., 1949, "Suspension Roof Support," BuMines IC 7533, September.

Weigel, W. W., 1943, "Channel Irons for Roof Control," Engineering and Mining Journal, Vol. 144, No. 5, May, pp. 70-72.

LIST OF SYMBOLS

T = Bolt tension

t = Average bolting stress = $\dfrac{T}{A}$

α = Factor depending on time of installation of bolts after excavation (probable variation of 0.5 to 1.0)

γ = Unit weight of the rock

μ = Tan ϕ where ϕ is the angle of internal friction of the rock mass

k = Ratio of average horizontal to average vertical stresses

A = Area of reinforced rock unit (e.g., s^2 for a bolt pattern with s x s spacing)

P = Shear perimeter of reinforced rock unit (e.g., 4s for a bolt pattern with s x s spacing)

R = Shear radius = $\dfrac{A}{P}$

c = Apparent cohesion of rock mass (i.e., intrinsic strength at zero normal stress)

ℓ = Length of rock bolts

D = Height of distressed rock above the surface of an opening

σ_v = Average vertical stress

σ_h = Horizontal stress conjugate to σ_v

   = $k\sigma_v$ where k is assumed to be constant

τ = Shear strength

  = $\sigma_h \tan\phi + c = k\sigma_v \mu + c$

   where ϕ is the angle of internal friction

Chapter 94

ROCK SLOPE REINFORCEMENT
WITH PASSIVE ANCHORS

By Dennis P. Moore and Michael R. Lewis

Senior Rock Engineer, B.C. Hydro
Vancouver, British Columbia

Rock Engineer, B.C. Hydro
Revelstoke, British Columbia

ABSTRACT

Untensioned, fully grouted, steel bars (dowels) up to 30 m long and 45 mm in diameter were used to reinforce rock slopes several hundred metres high, excavated for a hydroelectric project near Revelstoke, British Columbia. These dowels were successfully used to prevent surface rock from loosening, to prevent sliding on interconnected discontinuities and to prevent sliding of relatively intact rock on a thin shear zone. Because the interaction between rock slopes and dowels is not well understood, critical slopes were extensively monitored. Based on the experience from Revelstoke and elsewhere, it appears to be worthwhile, in most circumstances, to reconsider the need for tensioning rock anchors.

INTRODUCTION

Untensioned, grouted, steel bars (dowels) up to 30 m long were used to reinforce rock slopes several hundred metres high, excavated for a hydroelectric project near Revelstoke, British Columbia. Dowels were used to prevent surface rock from loosening, to prevent sliding along interconnected discontinuities and to prevent sliding of relatively intact rock along a thin shear zone.

During the early years of construction, tensioned rock bolts and high capacity tensioned anchors were the primary means of rock reinforcement (Moore and Imrie, 1982). Dowels were first used to prevent further loosening of fractured and weathered surface rock in the excavations for a highway to bypass the dam on the east bank. The ease of installation and obvious effectiveness of the dowels in the bypass road excavation encouraged the use of high capacity dowels in the east abutment and powerhouse excavations and later in the west wall of the powerhouse.

Because the interaction between rock slopes and passive anchors is not well understood (Azuar et al, 1979), these critical slopes were extensively monitored. The results of this monitoring, brief descriptions of the rock reinforcement system and some general considerations about the use of passive anchors are contained in this paper.

EAST ABUTMENT AND POWERHOUSE EXCAVATION

The east abutment and powerhouse excavation is an example of where dowels were used to prevent deep seated sliding along interconnected discontinuities in a highly fractured rock mass. Although no outward dipping shears existed in the east bank rock, foliation shears dipping gently into the slope were common, as were steep fractures parallel to the slope. A major gouge-filled shear zone, dipping into the slope as well as many other smaller, cross-cutting shears, existed in the slope (Fig. 1). This major shear zone is the most obvious expression of a regional fault which occurs at the damsite. Early movements on this fault resulted in a zone of mylonitized rock

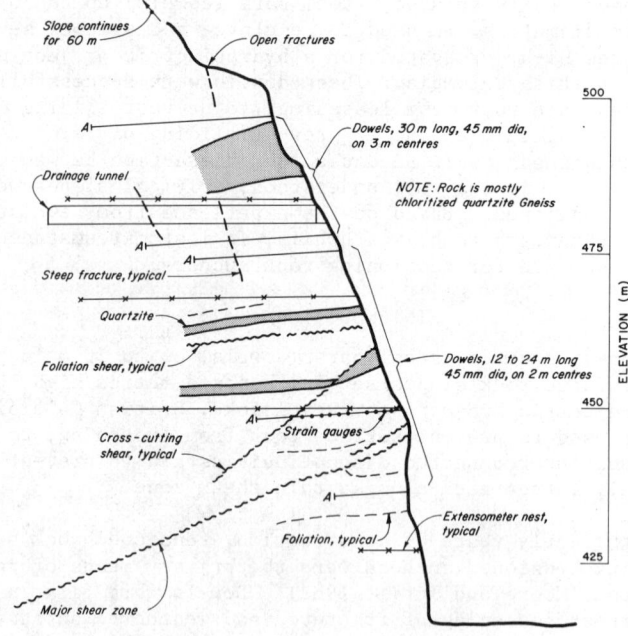

FIG.1 GEOLOGY & ROCK MOVEMENT - POWERHOUSE EAST WALL

hundreds of metres thick, and the most recent movement about 40 million years ago resulted in the fractures and shears, as well as, pervasive chloritization of the gneissic rocks at the site.

By the time the east abutment excavation was started it was clear from these geological conditions that the 100 m high, 0.5H:1V slopes would require substantial rock support. The slopes could not be flattened without undercutting the newly completed highway. Construction was behind schedule and the excavations were on the critical path for completion of the project.

Despite our incomplete understanding of passive reinforcement, a passive system was adopted because of the potential savings in time over a tensioned system. The passive system consisted of about 23,000 m of 45 mm diameter, Grade 60 rebar dowels between 12 and 30 m long, installed in grout-filled holes spaced 2.1 to 5 m apart. In addition, 8500 m of 22 mm diameter dowels were installed. This reinforcement system was coupled with a movement monitoring system, a program to investigate the interaction between rock and dowel and a program to investigate the possibility of corrosion. Grout temperatures in the rock were measured and grout samples were routinely cured in the laboratory at the same temperature. Laboratory strengths were correlated with dowel-grout bond strengths to ensure that excavation was allowed to proceed as soon as sufficient bond strength was available.

The behaviour of the slopes was closely monitored with particular emphasis on frequent visual observations by the rock engineering staff and on good communication with the inspectors and construction workers. Rock movement of about 5 cm was measured at the top of the powerhouse wall excavation. Steep cracks parallel to the top of the slope opened up a few millimeters and plaster tell-tales in a drainage tunnel near midslope became cracked where they crossed inward dipping shear zones. Only minor creep movements have occurred since the excavation was completed.

Strain gauges welded along the axes of three dowels near midslope, showed that the outer 10 to 15 m of dowel picked up load from the rock and the inner 10 to 15 m transferred load to deeper rock (Fig. 2). The displacement of the rock measured in a nearby extensometer compares well to that of the dowels except in the outer 5 m where displacement of the dowel was far less; probably because the dowel was not fitted with a plate at the surface. A recent design for a dowel system to be used at another B.C. Hydro damsite includes a $90°$ end hook to be embedded in shotcrete at the surface. With this modification or with a surface plate the load carrying capacity of the system should develop closer to the surface.

At several locations the dowels were strained close to or just above the yield strain but no strains were measured near the failure strain and no load losses were detected. Pull tests carried out in a separate test program at the site showed that the grout-bar bond fails as the steel begins to yield in the grout thereby lowering the load in the bar for a given displacement. This mechanism probably operated in the powerhouse wall as well.

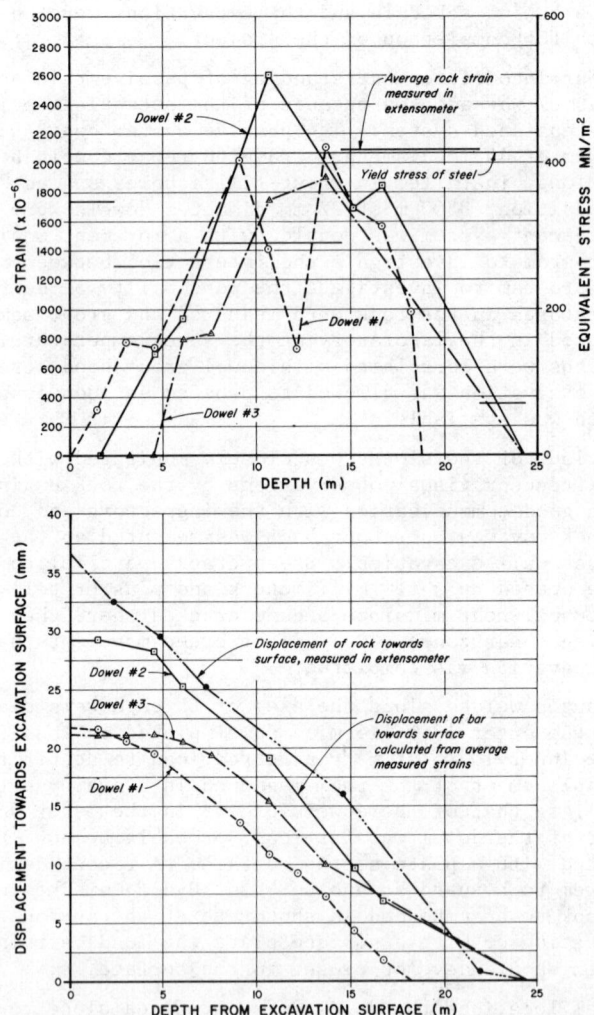

FIG. 2 DOWEL STRAIN & DISPLACEMENT - POWERHOUSE EAST WALL

POWERHOUSE WEST WALL

Dowels were also used at Revelstoke to prevent sliding along a thin shear zone. Movement of a very large block of rock occurred when a foliation shear containing about 2 cm of gouge was daylighted along the powerhouse excavation for a length of about 200 m (Fig. 3). Drainholes were drilled and a tunnel was excavated along the shear and backfilled with concrete to stop this movement. When a small blast triggered some rapid movements totalling several centimetres just downstream from the shear key tunnel, it became evident that **additional stabilization was required.** A toe berm was constructed to control movements and provide access for remedial measures. Almost 10,000 m of Grade 60 rebar dowels 45 mm in diameter were installed along the toe of the block to resist further movement (Fig. 3). Rock movement was monitored with extensometers at numerous locations and strain gauges were installed on three dowels. As soon as the berm was removed from the toe of the slope, rock movement accelerated and load was transferred to the dowels (Fig. 4). After about 5 mm of movement the rates of movement and load increase slowed down. Although these movements have resulted in a somewhat uneven distribution of strain between the dowels there is no evidence of progressive failure of the system or of failure of any of the individual dowels.

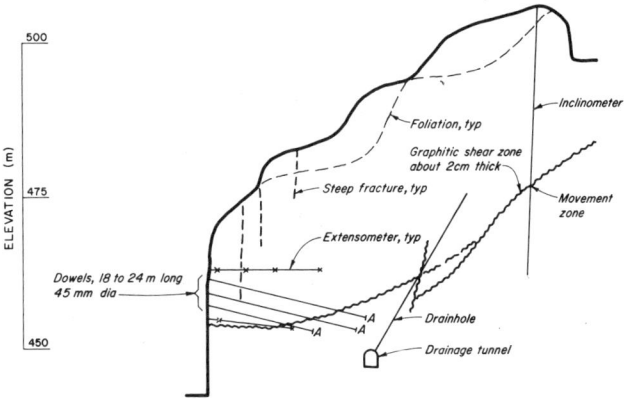

FIG. 3 GEOLOGY - POWERHOUSE WEST WALL

DISCUSSION

Progressive failure of the reinforcement system resulting from the lack of control over the time, location and magnitude of the load application is considered to be the most serious possible problem with dowel systems. If dowels are installed as excavation progresses, those installed first could become overloaded before those installed later become fully loaded. The same situation could occur if rock movement is not uniform or bond shear stresses develop unevenly. In addition, if the initial installation of dowels is found to be insufficient, supplemental dowels would not start to share the load until additional movements occur. These problems would not occur where the system has sufficient capacity for strain prior to loss in load carrying ability. This appears to have been the case of Revelstoke where relatively ductile steel was used. Some dowels were clearly yielding before others become loaded but no failures were apparent. Yielding probably occurred uniformly over a reasonable length along the bar, therefore no loss in load carrying capacity occurred.

Although the movement necessary to develop load in a dowel system could lead to a reduction in rock strength, this does not appear to have been the case at Revelstoke. Rock movements of only a few millimetres were necessary to develop loads in the dowels. These loads developed first where the movements occurred first and they generally developed most where the movements were the largest. This is an important characteristic of passive systems, since they tend to allow slope readjustments to occur while at the same time they tend to resist concentrated movements on any particular fracture.

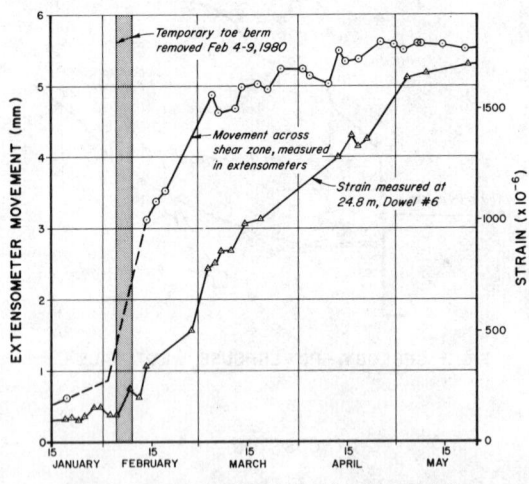

FIG. 4 ROCK MOVEMENT & DOWEL STRAIN

Even when shear movements were concentrated along a thin weak zone in relatively intact rock, such as on the powerhouse west wall, shear failure of the dowels did not occur. Fractured rock masses, such as at Revelstoke, usually dilate prior to failure, creating tension in the rock anchors. Numerous failures of rock masses were observed at Revelstoke and in almost every case rock anchors either stretched and bent until the rock fell apart around the anchor or the anchors pulled out completely. Shear failure of a dowel prior to mobilization of the full shear strength on a joint appears to be unlikely in rocks such as those at Revelstoke.

Cracks formed in the grout during loading of a dowel could reduce the resistance of the system to corrosion. At Revelstoke measurements of electric potential and groundwater chemistry showed that corrosion would not be a problem. In addition, the use of Grade 60 steel which is less susceptible to corrosion than the stronger steels and the use of solid bars which have a high ratio of cross-sectional area to surface area, reduced the susceptibility of the anchor system to corrosion.

With all rock reinforcement systems there is difficulty in determining from rock movements when additional rock support is required. With a dowel system it is even more difficult since the load on the dowel caused by rock movement is not usually known and cannot be calculated. Usually breaks or load losses can only be detected indirectly and could go unnoticed until a slope failure is well underway. Thus, it is even more important to appropriately monitor a passive system than it is to monitor a tensioned system.

CONCLUSIONS

Although serious problems with passive reinforcement systems can be theorized, at Revelstoke they never materialized. The uses of dowels to prevent surface rock from loosening, to prevent sliding on interconnected discontinuities and to prevent sliding of a relatively intact block on a thin shear zone are considered to have been successful and to have cost about 50% less than tensioned systems. Based on the experience at Revelstoke and elsewhere, it appears to be worthwhile in most circumstances, to reconsider the need for tensioning rock anchors. It is important, however, to evaluate the geological conditions at any particular site before choosing between a tensioned or untensioned system and to provide an appropriate monitoring system.

ACKNOWLEDGEMENTS

The authors appreciate the permission given by B.C. Hydro to write this paper and acknowledge the valuable assistance given by staff members during the course of the excavation work, especially B.J. Hutchison, T.E. Little, L. Lane and C.D. Martin.

The authors would also like to express appreciation to Dr. R.P. Benson the rock mechanics consultant, to Dr. L.T. Jory the geological consultant and to Mr. A.S. Imrie the geological supervisor all of whom were closely involved with the work and provided competent guidance.

The opinions expressed in this paper are those of the authors and do not necessarily represent the views of any other individual or organization.

REFERENCES

Azuar, Debreuille, Habib, Londe, Panet, Rochet and Salembier, 1979, "Le Renforcement Des Massifs Rocheux Par Armatures Passives", Proceedings, ISRM, 4th International Congress, Balkema, Rotterdam, Vol. I, pp. 23-25.

Moore, D.P. and Imrie, A.S., 1982, "Rock Slope Stabilization at Revelstoke Damsite", Transactions of the 14th International Congress on Large Dams, ICOLD, Paris, Vol. II, pp. 365-385.

Chapter 95

REVISING TERZAGHI'S TUNNEL
ROCK LOAD COEFFICIENTS
By
Don Rose
Tudor Engineering Company
San Francisco, California

In the USA the cost of steel ribs for tunnels approximates $100 million each year. European practice has long since abandoned heavy steel ribs for tunnel support, and tunnel costs in Europe (normalized for wages) are about half the costs incurred in the USA. A feeling has existed for some time that the USA practices of supporting tunnels with heavy steel ribs have been overly conservative. The USA practice is based on rock load coefficients by Karl Terzaghi.

In 1946, Karl Terzaghi published a tunnel rock classification system (Terzaghi, 1946), with the rock loads on tunnels predicted for each class of rock. No data on real rock loads were presented by Terzaghi and his rock load coefficients may have been largely intuitive. His classification system and associated predicted rock loads have been widely used in the USA. Each rock class, taken from Terzaghi's classification system, is assumed to generate loads of loosened rock resting as dead load. The dimensions of the loosened rock are described by rock load coefficients multiplied by the tunnel diameter. The resulting rock loads are used by design engineers, contractors and agencies to size supporting steel ribs.

Terzaghi clearly stated that he had conservatively assumed high rock loads for his classes 4, 5 and 6 to take a presumed effect of groundwater into account. If the tunnel was <u>not</u> affected by groundwater, Terzaghi stated those rock loads should be reduced by 50 percent. Twenty two years later Tor Brekke (1968) wrote a paper clearly showing that based on observation of more than 250 tunnels groundwater does not affect tunnel rock loads. In 1969, Deere, et al., published measurements made on a number of tunnels, and a well-known curve was drawn by Deere to illustrate Terzaghi's coefficients graphically. Deere's text stated that Terzaghi's predicted rock loads were too high and should be reduced. Cording and Deere (1972) more recently published data indicating that very large real rock caverns and chambers were being successfully built using rock bolts, which can take only a very limited rock load, thus indicating that real rock load is much smaller than Terzaghi or even Deere predicted.

953

Real rock load is actually dependent on how discreet blocks of rock actually interlock and create an "arch action" which supports the roof of tunnels. Voegele (1978) published a two-dimensional computer program which showed this effect. More recently, Goodman and Hua Shi (1981) published what may be a definitive explanation of this arch action, showing in 3-D how certain "key blocks" serve to hold the rock mass intact in a tunnel roof preventing failure and acting to self-stabilize the tunnel. Provided that normal care is taken not to allow the rock to unravel, real rock loads are not large, and heavy support systems are not required.

Rose, et al.,(1981) parametrically showed how Terzaghi's 1946 rock classification system leads to given predicted rock loads, which then require that certain sizes of steel rib be provided. The paper showed the approximate costs of these steel rib supports for each class of rock. Costs are extraordinarily and unreasonably high, especially for larger tunnels in moderately poor rock, using Terzaghi's 1946 system.

To summarize, field experience and measured rock loads have shown that the massive steel ribs required by Terzaghi's coefficients are overly heavy, conservative and costly. The steel ribs used in Europe in association with the NATM shotcrete and rock bolts are very slender and lightweight. It appears necessary to revise Terzaghi's 1946 tunnel rock load coefficients.

A review of Figure 1 shows that Deere's curve "b" suggesting Terzaghi's loads could be reduced by 20 percent, enveloped 90 percent of the data points while remaining roughly parallel to Terzaghi's curve "a". Inspection of the data indicates that the Revised curve "R" (Figure 2) proposed here also envelopes 90 percent of the data and virtually coincides with Terzaghi's own recommendations for rock _not_ affected by groundwater. This paper recommends the use of the Revised curve "R" and the Revised coefficients in Table I for engineers, contractors and agencies designing steel rib supports for tunnels.

Table II indicates the Revised calculated rock loads and the resulting more slender, lighter steel ribs required by the Revised coefficients. As can be seen, the lighter ribs are still within the range of conventional practice and conventional spacing, but the resulting savings in weight and cost of steel are substantial. Figure 3 summarizes steel ribs weight and cost for various rock classes, using the Revised coefficients.

The risk associated with revising Terzaghi's 1946 coefficients and providing more slender, lightweight steel ribs in tunnels is clearly not significant. European practice has long proven that heavy steel ribs are not needed and that they only increase tunnel costs. Theoretical studies by Voegele and Goodman, et al., show how real rock supports itself and rarely loosens and lies as dead weight on the tunnel support. Lightweight steel ribs will adequately support most tunnel loads. Should an unfavorable combination of geological features require a local, unusually sturdy tunnel support, the lightweight ribs on hand could simply be very closely spaced by the miners to span the local problem area. The miners would still be protected from sudden catastrophic trouble by the steel ribs overhead.

In those few tunnels where local overloading of the lightweight steel ribs occurred, remedial measures at the local problem area could be taken, without agencies, engineers and/or contractors overdesigning steel ribs for the entire length of the tunnel.

Using the Revised curve (which changes only rock classes 4, 5 and 6 and does not apply to the special cases of squeezing or swelling ground) costs might be reduced on USA tunnels to approach costs of tunnels now being built in Europe and Japan. This might encourage the USA's ailing tunnel industry to renewed vigor, while preserving safety for the miners, a development of which Terzaghi himself might well heartily approve.

REFERENCES

Brekke, T.L. (1968). "'Blocky and Seamy Rock' in Tunneling," Bull., Association of Engineering Geologists, Vol. V, Number 1.

Cording, E.J., and D.U. Deere. (1972), "Rock Tunnel Supports and Field Measurements," Proceedings, Rapid Excavation and Tunneling Conference, American Institution of Mining Engineers, New York, pp. 601-622.

Deere, D.U., R.B. Peck, B. Schmidt and J.E. Monsees (1969), Design of Tunnel Liners and Support Systems, U.S. Department of Transportation, Washington, D.C.

Goodman, R.E. and Gen Hua Shi (1981), "A new concept for support of underground and surface excavations in discontinuous rocks based on a Keystone Principle" 22nd U.S. Symposium on Rock Mechanics, Mass. Inst. Technology, June 1981, pp. 290-296.

Rose, D., P. Kaboli and R. Mayes (1981) "Influence of Geologic Logs and Descriptions on Tunnel Design and Costs", 22nd U.S. Symposium on Rock Mechanics, Mass. Inst. Technology, June 1981, pp. 443-448.

Terzaghi, K. (1946). "Introduction to Tunnel Geology" in Rock Tunneling with Steel Supports (1968 revised edition: R.V. Proctor and T.L. White, editors), Commercial Shearing and Stamping Company, Youngstown, Ohio.

Voegele, M.D. (1978). An interactive graphics based analysis of the support requirements of excavations in jointed rock masses, Ph.D. Thesis, University of Minnesota.

TABLE I

Rock load H_p in feet of rock on roof of support in tunnel with width B (ft) and height H_t (ft) at depth of more than $1.5(B + H_t)$

ROCK CONDITION	RQD	ROCK LOAD H_p (ft)	REMARKS
1. Hard and intact	95-100	Zero	Same as Terzaghi (1946)
2. Hard stratified or schistose	90-99	0 to 0.5 B	Same as Terzaghi (1946)
3. Massive, moderately jointed	85-95	0 to 0.25 B	Same as Terzaghi (1946)
4. Moderately blocky and seamy	75-85	0.25 B to $0.20(B+H_t)$	Types 4, 5, and 6 reduced by about 50% from Terzaghi values because water table has little effect on rock load (Brekke, 1968 & Terzaghi, 1946)
5. Very blocky and seamy	30-75	$(0.20$ to $0.60)(B+H_t)$	
6. Completely crushed but chemically intact	3-30	$(0.60$ to $1.10)(B+H_t)$	
6a. Sand and gravel	0-3	$(1.10$ to $1.40)(B+H_t)$	
7. Squeezing rock, moderate depth	Not applicable	$(1.10$ to $2.10)(B+H_t)$	Same as Terzaghi (1946)
8. Squeezing rock, great depth	Not applicable	$(2.10$ to $4.50)(B+H_t)$	Same as Terzaghi (1946)
9. Swelling rock	Not applicable	Up to 250 ft irrespective of value of $(B+H_t)$	Same as Terzaghi (1946)

ROCK LOADS - REVISED DESIGN COEFFICIENTS

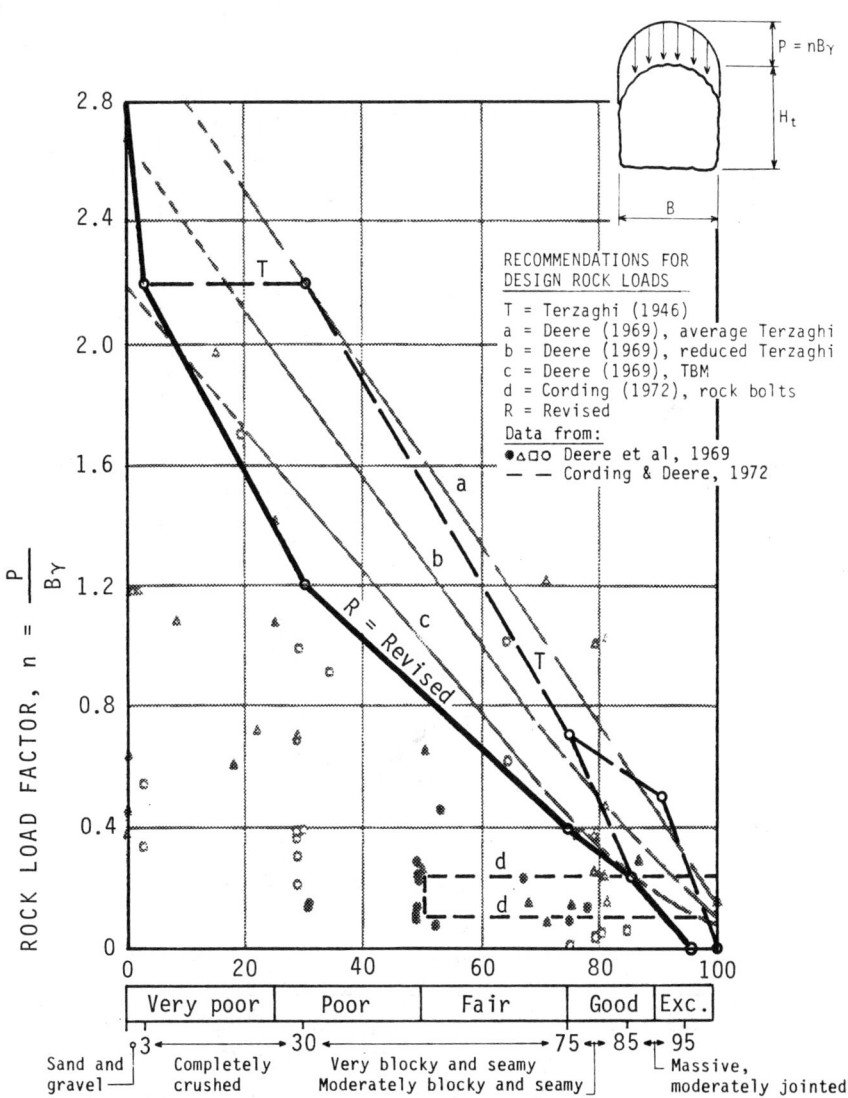

ROCK LOADS - MEASURED DATA

Figure 1

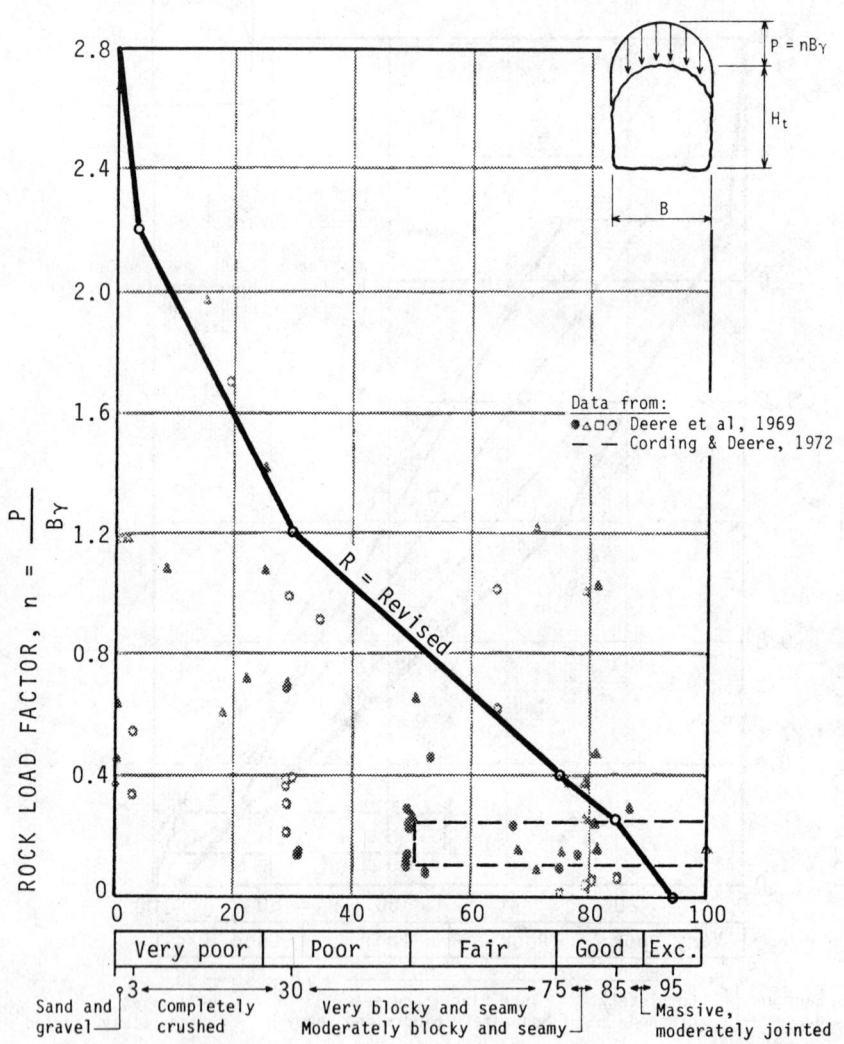

ROCK LOADS - REVISED DESIGN CURVE

Figure 2

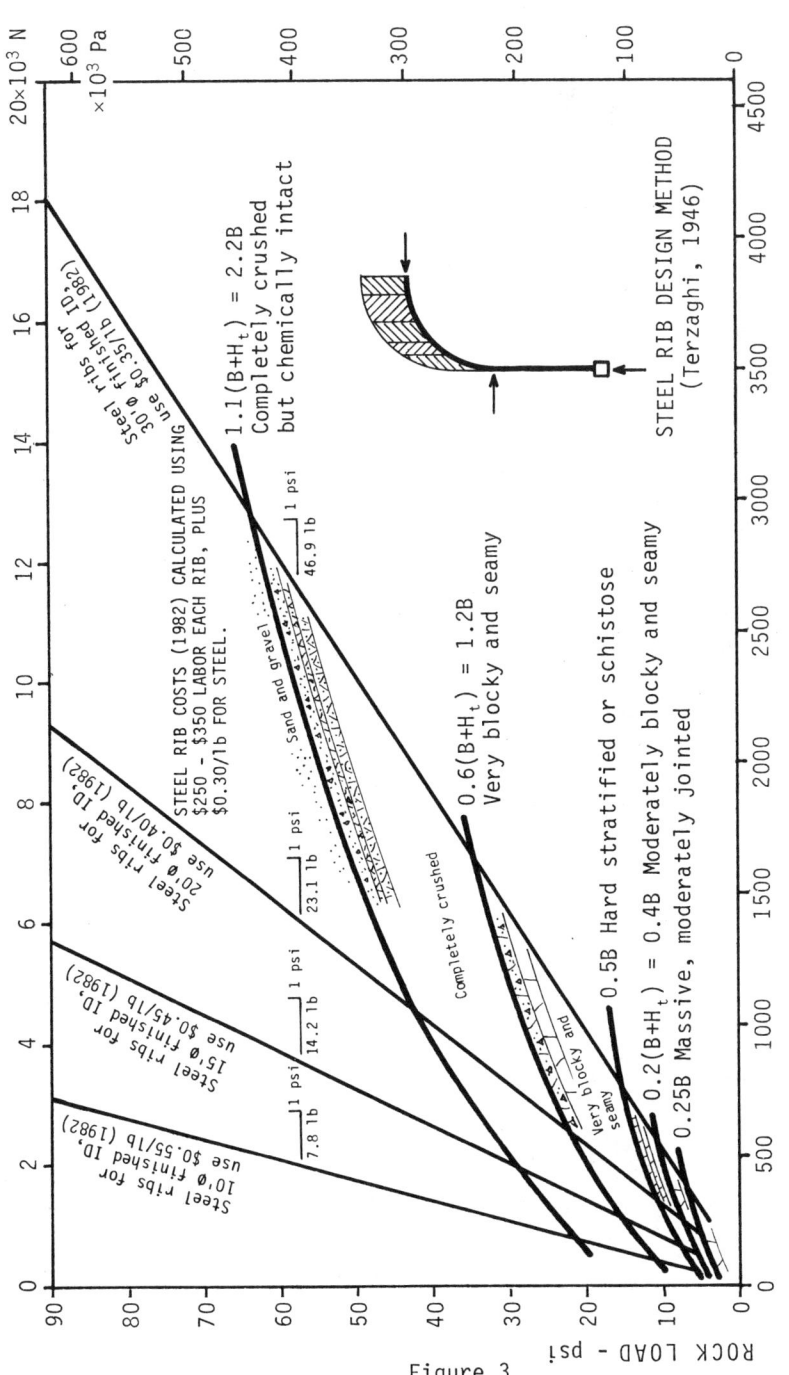

Figure 3

WEIGHT AND COST OF STEEL RIBS VS ROCK LOADS

TABLE II

SAVINGS POSSIBLE WITH REVISED ROCK LOADS

	5. Very Blocky and Seamy — RQD = 30 - 75			4. Moderately Blocky and Seamy — RQD = 75 - 85		
	10'∅	20'∅	30'∅	10'∅	20'∅	30'∅
Terzaghi Load Factor	2.2-0.7	2.2-0.7	2.2-0.7	0.7-0.25	0.7-0.25	0.7-0.25
Revised Load Factor	1.2-0.4	1.2-0.4	1.2-0.4	0.4-0.25	0.4-0.25	0.4-0.25
Terzaghi Loads, psi	22-7	44-14	66-21	7-3	14-5	21-8
Revised Loads, psi	12-4	24-8	36-12	4-3	8-5	12-8
Load Reduction, psi	10-3	20-6	30-9	3-0	6-0	9-0
Cost of Steel Ribs $/psi	$ 4.29	$ 9.24	$ 16.42	$ 4.29	$ 9.24	$ 16.42
Range of Cost Savings $/LF	$42.90–$12.87	$184.80–$ 55.44	$492.60–$147.78	$12.89–$ 0.00	$55.44–$ 0.00	$147.78–$ 0.00
Average Cost Saving $/LF	$28	$120	$320	$6	$28	$74
Terzaghi Rib Size (Typ)	W 6×20	W 8×35	W 14×68	W 4×13	W 6×20	W 8×35
Revised Rib Size (Typ)	W 4×13	W 6×20	W 8×35	W 4×13	W 4×13	W 6×20

Chapter 96

ROCK BOLT REINFORCEMENT SYSTEM TO STABILISE SHAFT INTERSECTIONS
AND PIT BOTTOM ROADWAYS DURING UNDERGROUND RECONSTRUCTION

by Raghu N. Singh and Ali M. Heidarieh Zadeh

Department of Mining Engineering, University of Nottingham,
Nottingham, NG7 2RD, U.K.

ABSTRACT

The paper presents a stability investigation of shaft intersections and pit bottom roadways in the vicinity of upcast and downcast shafts at a colliery in the U.K. A scheme of instrumentation was aimed at monitoring the strata displacement surrounding the pit bottom roadways and shaft intersections with a view to ascertaining the origin of the strata movement and effectiveness of the control measures. The borehole instrumentation results together with a theoretical analysis suggested that the cause of instability could be attributed to the drivage of a roadway within the shaft pillar.

A semi-empirical technique was used to calculate the amount of support resistance required to reinforce the pillar to take additional abutment load. A rock bolting system was designed to provide a required increase in thrust capacity of the roadway support. The results indicated that roadway closure was effectively controlled after installation of bolts indicating efficacy of the control measures adopted.

INTRODUCTION

Recent events in the international energy market have culminated in placing more emphasis on the exploitation of indigenous energy resources and revitalising coal industry, in the U.K. In order to increase the potential recovery of coal the existing shafts must be utilised to their maximum capacity. Most shafts are constructed to have a life span of at least 50 years, and in some cases in excess of 100 years. During this period the coal is won from several coal seams which often necessitates the modification of the pit bottom layouts whenever an access is desired to a new coal seam from the existing shaft bottom. These modifications may be enlargement of shaft intersections to accommodate loading facilities, drivage of vertical bunker for storing additional output from the new seam or drivage of new roadways. The most critical part of such modification of shaft structure is design and location of these intersections specially in view of their size, geometry, geological environment and interaction.

Interaction of new construction is of vital importance and may cause major instability of shaft intersections and pit bottom roadways. The paper describes the pit bottom instability caused by the interaction of a roadway drivage in a vicinity of access shafts at a colliery in South Yorkshire in the U.K. The application of rock bolt reinforcement system, as a successful remedial measure for restoration of stability is discussed.

SITE DESCRIPTION

The coal mine involved is situated in South Yorkshire coalfield, in the United Kingdom, some 36 miles north of Nottingham. The present mining operations are concentrated to Haigmoor Seam, producing about 0.6 million tonnes per annum from two mechanised working districts. The access to Haigmoor coal seam was obtained by sinking two shafts in 1914. Figure 1 shows the pit bottom layout. The downcast shaft, was originally sunk up to a depth of 660 m which was filled up to 652 m level in 1976 and the winding inset was maintained at 623 m level below the surface. The shaft was equipped with skip, transporting 10 tonnes of coal per wind. Shaft intersection was enlarged to house a skip-bunker, automatic skip loading pockets and control instrumentation.

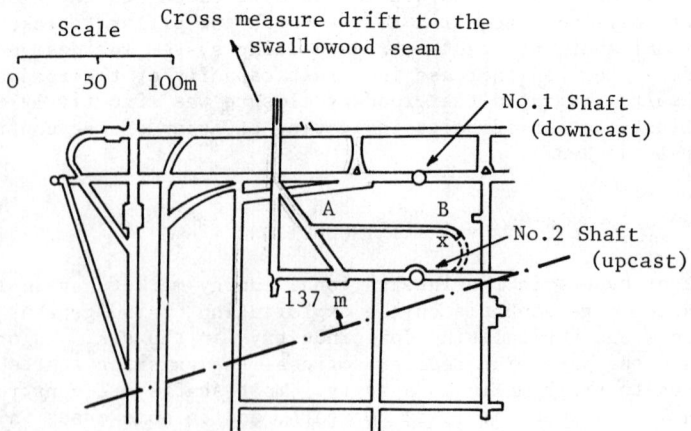

Fig.1. Shaft bottom layout of the colliery ('x' shows the location of the drivage which triggered the shaft bottom movements)

ROCK BOLT REINFORCEMENT SYSTEM 963

The upcast shaft had a winding inset at about 595 m below surface and was equipped with a pair of double deck cages and was mainly used for men and material winding. The strata in the immediate vicinity of Downcast shaft inset consisted of coal, blackshale, silty mudstone, seatearth, fine grain sandstone and sandstone + seatearth. The detailed geology of the upcast shaft section shows that the strata consists of interbedded seatearth and sandstone. A survey of the area shows location of a down throw fault with slight roadway deformation across the fault. The whole length of the fault plane shears through three sequences of strata comprising seatearth, sandstone and siltstone.

STATEMENT OF PROBLEM

In order to obtain access to Swallowood seam a pair of cross-measure drifts were driven from the present pit bottom. On account of dirt disposal problems associated with driving the cross-measure drifts, a roadway (AB shown in Figure 1) was driven in close proximity to the downcast shaft to improve the efficiency of the pit bottom mine car circuit. This effectively reduced the width of the shaft pillar from 60 to 27 m and resulted in instability of pit-bottom roadways, shafts and shaft intersections. Redistribution of strata pressures as a consequence of drivage of roadway AB, resulted in a crushing of the remnant shaft pillar. Types of instability observed were, side closure, cracking of support walls, buckling of the support girders and distortion of steel girder supports in the vicinity of the downcast shaft.

RESEARCH TECHNIQUE

Detailed investigation was designed to assess the extent of instability and obtain the following information.

(i) To determine the cause and origin of the strata movement.

(ii) Laboratory assessment of strength and deformation parameters of rocks in the vicinity of No. 1 and No. 2 shafts.

(iii) To obtain 'in situ' roadway closure data to assess severity of the problem.

(iv) To suggest the remedial measures to control the strata movement and calculate the necessary design parameters for any additional support measures.

(v) Monitoring the effectiveness of the control measures installed, to evaluate efficiency of the system.

The research programme included the following main investigations: (a) In situ strata closure measurements incorporating a scheme of surface and borehole instrumentation for underground stability evaluation, (b) Determination of the strength and deformation parameters of rock by laboratory tests, and (c) A theoretical or semi-analytical technique for estimating the degree of lining reinforcement required to restore stability.

Instrumentation for investigation of excavation stability :-

The favoured method for investigating excavation stability of major drivages within shaft pillars is based on the observation of displacement changes particularly within the shaft, tunnel or room. In addition observations were made of wall/side movement relative to a datum located in a borehole remote from the excavation.

The instrumentation methods used by the authors to investigate stability of supported excavations in the shaft pillar are as follows;

(a) telescopometer for observation of height and width changes to an accuracy of 1 mm over measured lengths in the range 4 - 7 m;

(b) dial gauge extension rod method which consists of an engineering dial gauge calibrated to 0.01 mm and mounted on screwed rods of standard length to permit height and width measurements to be made in the range 4 - 7 m;

(c) rod-type borehole extensometer which permitted insertion to a depth of up to 6 m into the solid rock from the roadway side.

The extensometer design was based on placing 4 rods within a 43 mm diameter borehole, each rod being positioned freely in a plastic sleeve at the end of which was a grouted anchor to provide a fixed reference point. Figure 2 shows the instrumentation scheme used in the project. The types of instrumentation described here proved sufficiently robust and accurate for the relatively difficult measuring conditions encountered close to the bottom of the main shaft. It was important that the measuring programme did not hinder production operations but would provide adequate indication of the state of stability of the lining of the excavations being investigated.

LABORATORY EVALUATION OF STRENGTH AND DEFORMATION PARAMETERS OF ROCKS OBTAINED FROM THE VICINITY OF THE ACCESS SHAFTS

Samples of rocks collected from the roadway in the vicinity of the access shafts were brought to the laboratory for assessing (i) Triaxial compressive strength up to confining pressures of 35 MN/m^2, (ii) Uniaxial stress/strain characteristics and Poisson's ratio of rock, (iii) Uniaxial unconfined compressive strength, (iv) Indirect tensile strength, and (v) Rock density evaluation.

The results of the physical properties of rocks are summarised in Table 1.

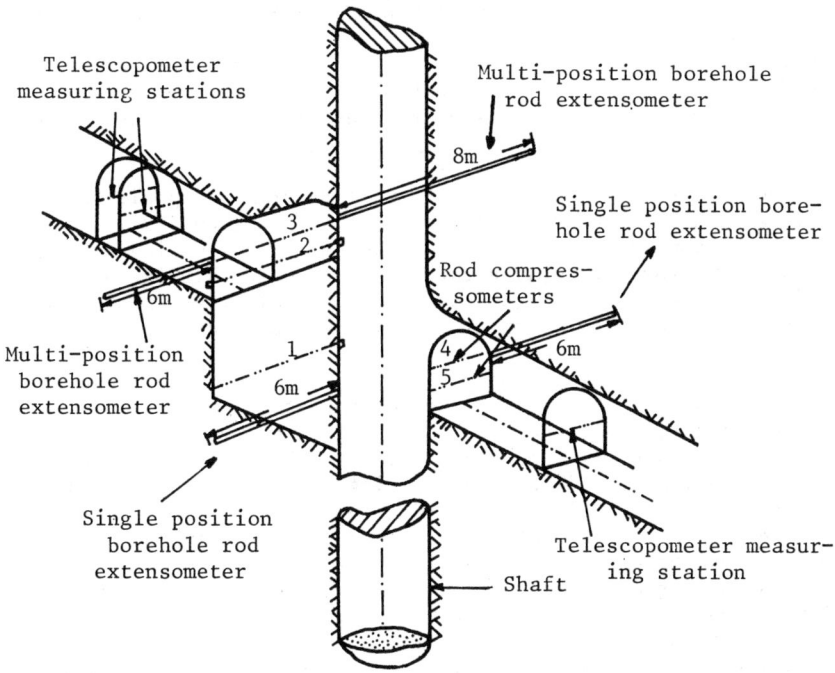

Fig.2. Instrumentation scheme, used in the project

Table 1. Summary of Strength and Deformation Parameters of Rock in the vicinity of access shaft

Rock Type	Bed Thickness m	Tri-axial factor k	Compressive Strength MN/m^2	Tensile Strength MN/m^2	Cohesive Strength MN/m^2	Internal Angle of Friction ϕ^o
F.G. Sandstone	1.36	3.13	107.27	47.30	41.88	13.05
Coal	1.16	3.45	21.95	4.29	3.98	33.00
Silty Mudstone	0.54	3.17	95.98	30.28	26.95	31.36
Mean	1.02	3.25	75.27	27.29	24.27	32.13
Weighted Mean		3.26	72.93	27.99	24.87	31.84

THEORETICAL ANALYSIS

The theoretical analysis was necessary to calculate stress acting on the pillar surrounding pit bottom roadways and to design rock bolting systems for stabilising roadways in the vicinity of the access shafts.

The stress acting on the shaft pillar surrounding pit bottom roadways was calculated using elastic analysis (Wilson, 1977). This analysis assumes that rocks associated with coal mining are relatively soft and that vertical strata pressure under virgin conditions arises from the overburden and the horizontal strata pressure of coal measures is equal to vertical stress on account of rock creep.

When an excavation is made in virgin soft rock the tangential stress around the excavation exceeds the stress field beyond the failure limit, and the strata surrounding the roadway fails gently to form a broken or yield zone. If the roadway is supported by a support system providing a restraining force, the friction present in the failed rock allows a build up of load bearing capacity as the distance from the excavation increases. Eventually the resistance offered by the broken material is sufficient to prevent further extension of yield zone. At this point an abutment peak will be formed and the stress will further diminish according to an exponential function of the distance. Wilson (1977) studied the stress/deformation around an excavation in soft rock using elastic analysis. This analysis permitted the evaluation of the abutment pressure width of yield zone and stress distribution on the pillar (see Table 2).

Figure 3(a) shows the stress distribution on the shaft pillar which was 60 m wide. Driving a roadway in shaft pillar effectively reduced the width of shaft pillar to 27 m. Figure 3(b) shows stress distribution on shaft pillar after reducing the shaft pillar area by drivage of a new roadway. It can be seen that the heavy pillar stress in the second case was of the order of 120 MN/m^2.

Design of Rock Bolting System to Stabilize the Roadway

This problem can be resolved in two stages

(a) Calculations of the support resistance requirement to reinforce the pillar so as to prevent further extension of yield zone, and

(b) Design of rock bolting system to provide the design degree of reinforcement.

The analysis assumes that the optimum width of a pillar between 2 roadways in solid ground should be twice the width of the yield zone surrounding the roadway plus an optimum width of a solid core capable of supporting strata stresses. The further assumptions made are :

(i) The extent of yield zone depends upon the depth, size of excavations, triaxial stress concentration factor (K) and restraining pres-

Table 2. Calculation of stress acting on the pillar surrounding pit bottom

(a) Abutment pressure, Wilson (1977) $\hat{\sigma} = Kq + \sigma_c/F$ $\hat{\sigma} = 66.01$ MPa	Notations $\hat{\sigma}$ = strata abutment pressure H = depth below surface = 625 m q = vertical strata pressure due to depth = γH = 15.625 MPa (for 625 m depth) σ_c = compressive strength of rock = 75.27 MPa
(b) Stress distribution in yield zone $\sigma_y = P.K \left(\dfrac{X+\frac{1}{2}M}{\frac{1}{2}M}\right)^{K-1}$ $= 0.047 \times 3.25 \left(\dfrac{X+1.5}{1.5}\right)^{2.25}$ =3m	f = scale factor = 5 M = height of the roadway X = distance within yield zone where stress is to be calculated K = triaxial stress concentration factor = 3.25 P = restraining pressure of roadway support, for the existing support
(c) Width of yield zone $XB = \dfrac{M}{2} \left\| \left(\dfrac{q}{P}\right)^{1/K-1} -1 \right\|$ $XB = 11.88$ m	= 0.047 MPa W = width of the roadway = 4.27 m $Aw = \frac{1}{2} W(H - \dfrac{W}{1.2}) \ \gamma = 33.17$
(d) Stress decay in elastic zone $\Delta\sigma_y = (\hat{\sigma}-q) \ e^{XB-X/C}$ Also $F = \left(\dfrac{K-1}{K}\right)+\left(\dfrac{K-1}{K}\right)^2 \ Tan^{-1} K$ $C = 3.148$	$Ab = \dfrac{M}{F} \cdot K(q-p) = 60.2$ $C = \dfrac{Aw+qXB-Ab}{(\hat{\sigma}-q)}$

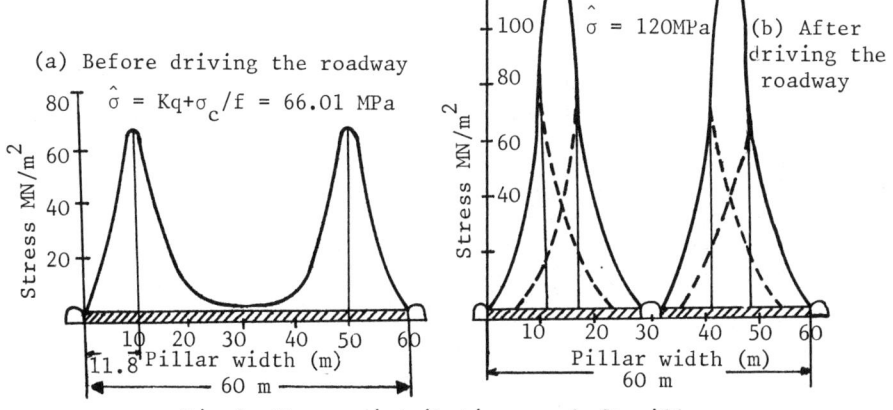

Fig.3. Stress distribution on shaft pillar

(a) Before driving the roadway $\hat{\sigma} = Kq+\sigma_c/f = 66.01$ MPa

(b) After driving the roadway $\hat{\sigma} = 120$ MPa

sure applied by the roadway support.

(ii) The reinforcement does not permit the width of the yield zone to extend further as a consequence of reducing the size of the pillar.

(iii) When the support resistance is increased by installing a rock bolt system the resulting pillar is capable of supporting the superincumbent strata loading without further crushing of the pillar.

Table 3 shows the calculation of the required support resistance to prevent further deterioration of the pillar.

Table 3. Calculation of the required support resistance

	Notations
Width of the yield zone $= XB = \frac{M}{2}[(\frac{q}{p})^{1/K-1} - 1]$ Optimum width of the pillar surrounding the roadway $= Q = \frac{1}{K-1}(W - \frac{W^2}{1.2H}) + 2XB$ substituting the values in the equations, XB = 11.88 m, P = 0.21 MPa	XB=Width of the yield zone M=Height of the roadway q=Vertical strata pressure $=\gamma H=0.025 \times 625 = 15.625$ MPa H=Depth below surface=625 m γ=Average rock density=0.025 MPa W=Width of the roadway=4.27 m K=Triaxial stress factor=3.25

The results in Table 3 show that XB = 11.88 m, and p = 0.21 MN/m^2. The resistance offered by the existing support system was estimated as 0.047, so the required resistance offered by the reinforcement will be 0.16 MN/m^2 and the increase in unit thrust capacity of reinforced rock mass per metre length was 1.03 MN.

The design of a rock bolting system for stabilising the roadway surrounding a shaft pillar was affected by using the technique proposed by Bischoff and Smart, 1975. From this approach, it can be shown that

$$\Delta ta \, S^2 + \frac{1}{F} \sigma_b A \, Tan^2 (\frac{\phi}{2} + 45) S - \sigma_b A (Tan^2 (\frac{\phi}{2} + 45) \frac{L}{F} = 0$$

where
- Δta = increase in unit thrust capacity of the rock mass per unit length of tunnel
- $\Delta \sigma_1$ = effective increase in the allowable stress of the reinforced rockmass
- D = bolt diameter (29 mm)
- t = effective thickness of reinforced rockmass = (L-S)
- ϕ = angle of internal friction of rockmass = 31.5°
- $\Delta \sigma_3$ = increase in rockmass confinement provided by bolting
- F = factor of safety = 1.3
- σ_b = yield stress of the bolt = 414 MN/m^2
- A = effective x-section area of the bolt = 0.00066 m^2
- S = spacing of bolt (square grid)
- L = length of bolt (3 m)

ROCK BOLT REINFORCEMENT SYSTEM

Substituting the above values in equation, we obtain

$$1.339 \, S^2 + 0.082 \, S - 2.646 = 0$$

$$S = 1.11 \text{ m approximately}$$

Based on the above calculations rock bolts on the shaft inset were required to be installed in a square grid.

DISCUSSION

Based on the design calculation a full remedial reinforcement programme was implemented using 3 m long 29 mm diameter steel bars fully resin grouted throughout their length. Figure 4 shows the influence rock bolts had on reducing the rate of closure at a number of measuring stations close to the shaft side. The results indicate that the rock bolts required a period over which supporting resistance needed to be built up before their full beneficial effect could be realised. Figure 4 also shows the duration of the drivage which initiated the shaft bottom movements and the 84 m of drivage took place between 20 and 30 days before the commencement of the measurements shown plotted in this illustration.

Drivage of the curved connecting roadway was suspended in view of the large scale movements encountered in the shaft bottom and within the shaft pillar area. As can be seen from Figure 4 it has taken about 1 year for the movements to become of little significance. Drilling and blasting in a drivage in such a situation should be avoided; conversely formation of the heading by machine cutting would have most probably reduced this problem in the first place.

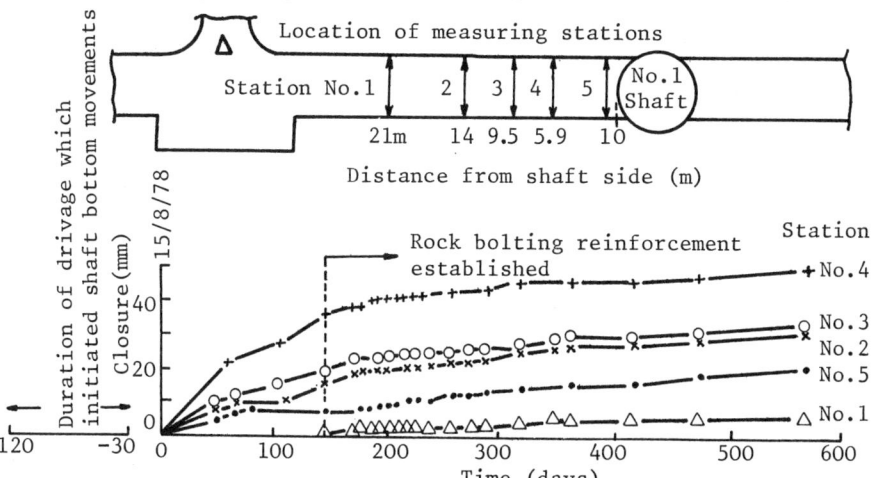

Fig.4. Comparison of closure data showing effect of introducing rock bolting remedial reinforcement at Downcast shaft

1. Shafts associated with coal mining usually provide services for multiple seam operations. When a new seam is planned to be exploited from an existing shaft it is often necessary to modify the pit bottom and shaft intersection layouts. Great care should be exercised in locating the position of new mining excavations so as to obviate the interaction effects of new excavations on the shaft intersections.

2. It is now possible to design a rock bolting system which is capable of offering a desired resistance to the strata. However, effectiveness of any rock bolting system in soft coal measures rock should be monitored over a long period to evaluate long term efficiency of the system.

3. Simple rock displacement measuring instrumentation is a powerful took for any strata mechanics investigation.

ACKNOWLEDGEMENTS

The authors record their gratitude to the National Coal Board for financial and practical support given to the research described here. Any views expressed are those of the authors and not necessarily those of the Board.

REFERENCES

Wilson, A., 1977, "The effect of yield zone on the control of ground", 6th International Strata Control Conference, Banff Spring, Canada, September.

Bischoff, J. A., Smart, J. D., 1975, "A method of computing a rock reinforcement system which is structurally equivalent to an internal support system", Proceedings, Sixteenth Symposium of rock mechanics, September 22-24, University of Minnesota, Minneapolis, U.S.A., pp.179-184.

Heidarieh Zadeh, A. M., 1982, "Support characteristics of underground mining excavations with special reference to rock reinforcement techniques", Ph.D. thesis, University of Nottingham, 249 pp.

Chapter 97

EQUIPMENT, AUTOMATION, ROCK MECHANICS PRINCIPLES AND SAFETY
INTERFACES IN THE CONTROL OF ROOF AND RIBS OF MINES

by Dr. James J. Scott

President, Scott Mine Technical Services, Inc.
Rolla, Missouri

Adjunct Professor of Mining
University of Missouri-Rolla

INTRODUCTION

This paper presents the basic principles which must be followed to create a truly inherently safe mining system. The need to relate the support mechanism to be employed to the ground reaction curve and basic rock mechanics principles will be pointed out. The support system must function in concert with the deformations that are inherent in the selected mining system. Following the support selection, the mechanics for installing support must be developed. This is best done through a cooperative effort between mine engineers, mine operators, safety personnel, equipment suppliers, and government inspectors, both state and federal.

Accident statistics, near miss statistics, review of equipment used by others, geologic features, etc. will necessarily have to be studied. The goal of equipment development and selection should be to develop inherent safety by removing the miner from the hazard area or by providing absolute protection for the miner if he must work in close proximity to the potential fall of ground. The role of good productivity in reducing the numbers of men exposed to hazards is emphasized as a good safety measure.

ROCK MECHANICS PRINCIPLES

A thorough understanding of the ground reaction curve surrounding openings to be created is essential in the process of designing ground control methods. See figure 1. It is first necessary to investigate the geologic conditions at the site as this is the one thing that cannot be changed.[1]

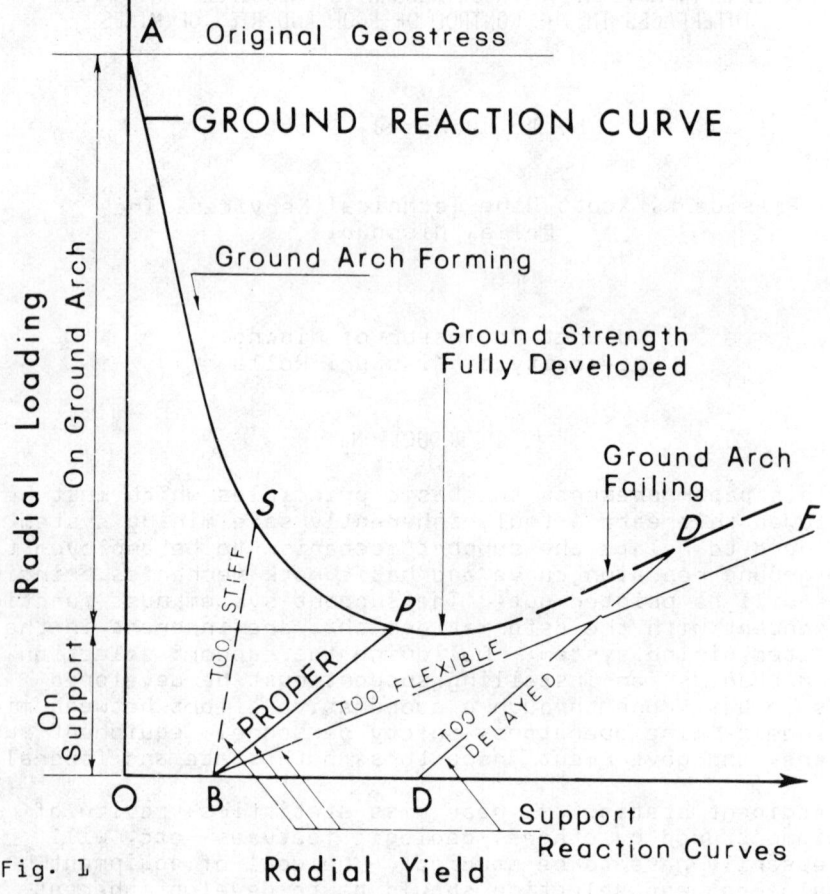

Fig. 1

Secondly, the deformation associated with the presently mined openings should be measured and plotted on time deformation plots. It may also be necessary to measure absolute stress conditions to determine whether or not regional tectonics play a role in ground stability.

Third, based on the above investigations, the best shape of opening should be decided upon. The shape being predicated upon location and orientation of geologic flaws, the effects of loading, time dependent creep of the opening and the shape necessary to allow mining equipment to operate.

CONTROL OF MINE ROOF, RIBS

Lastly, with the above information, the mining system can best be designed. The orientation and span of openings, pillar sizes and shapes can be determined.

The major goal in the design process will be to produce a stable opening which uses no artificial support. A field example of this is the use of the boring machines in potash or trona which cut an ovaloidal opening and often can be mined without use of roof fixtures. In the same formation, if conventional ripper miners are used to cut rectangular openings with sharp corners, which produce stress concentrations, roof fixtures are often necessary to keep the openings safe.

The first support used in a mine opening should be installed quickly as the face advances and may be temporary in nature. If interior rock supports are to be used, the host rock must have sufficient strength so that the fixture placed can reinforce the media. The simplest form of quick support may be mechanical roof bolts, Friction Rock Stabilizers, resin or cemented rebars. Figure 2 is a table showing the commonly available fixtures.[1]

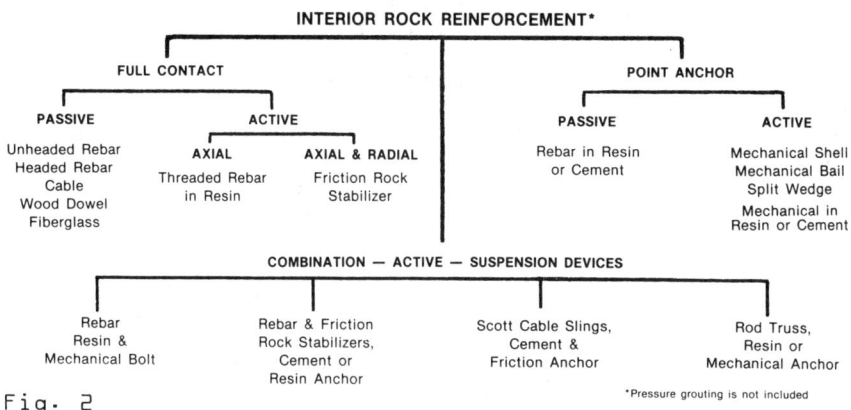

Fig. 2

Rock reinforcement in whatever form it takes should be placed in such a manner that it matches the yield of the load reaction curve. A first consideration should be whether or not the fixture should be active or passive. An active fixture is one which loads the geologic mass in which it is placed immediately upon installation. A passive fixture is one that when placed in the geologic mass does not play a role in ground support until the mass moves and, subsequently, loads the fixture. For example, a standard mechanical bolt is an active fixture when tightened in place. The leaves of the bale or shell act radially outward at their point of contact with the rock, while tension in the bolt creates a clamping action in the rock between the plate and the anchor. At the other extreme is the completely passive condition found when the rebar is hand inserted into a borehole filled with wet flowable cement. This device as placed fills the borehole but in no way loads the geologic mass.

A second consideration in matching the load reaction curve is to eliminate stress risers in the deformation system. The anchors of mechanical bolts often over stress rock. It has been estimated that the rock stress under the leaves of an anchor may be as high as 125 000 psi {862 x 10^{12} Pa}.[2] Such pressure causes rock fracture and creep or slippage of the anchor. Another form of stress concentrator not readily recognized is the mismatch which exists between high modulus resin rebar and full contact fixtures when they are used in soft viscoelastic medium. The modulus of the fixture is very high, while the modulus of the rock may be very low. This mismatch causes a breakdown at the metal, resin, rock interface due to a radical difference in deformation characteristics.

For the above reasons, the yield of interior rock support fixtures should match the load reaction curve as nearly as possible. The fixtures obviously need to be strong to reinforce the formation and thereby retard deformation, but they should not produce such massive point loads that they break the medium surrounding the fixture and the mine opening. Consider some practical examples. It has not been recognized by investigators that some slippage of mechanical anchors may be desirable and serve to enhance stability. Anchors can load up quickly to high loads when they are placed at the face, stretch some due to that load, and then over a period of time the anchor may creep under constant load, causing a bleed-off of tension in the bolt, but at the same time, strengthening the rock in a gradually decreasing manner which allows the rock fabric and fractured pieces to shift and adjust to the best position for long term stability. This condition is often found where mechanical bolts are used in limestone mines. Fixtures that are a long distance back from the mining face quite often are found to not be carrying any load or, if they are loaded, the load may be as little as a ton {907 kg}. The author believes that this is due to time dependent creep of the anchor under load and to the fact that the rock finally adjusts until it reaches a state of unstable equilibrium so that the roof bolts are not loaded and are essentially nonfunctional. This does not mean that they are not needed or that they have not served their purpose. If mining is carried on again in the area away from the face they may be reloaded and repeat a somewhat similar cycle due to the change in loading conditions produced.

Another example of the need for the fixtures to match the deformation curve is evidenced by the manner in which grouted cables are used in large stope operations throughout the world.[3] See figure 3. Long cables, placed in

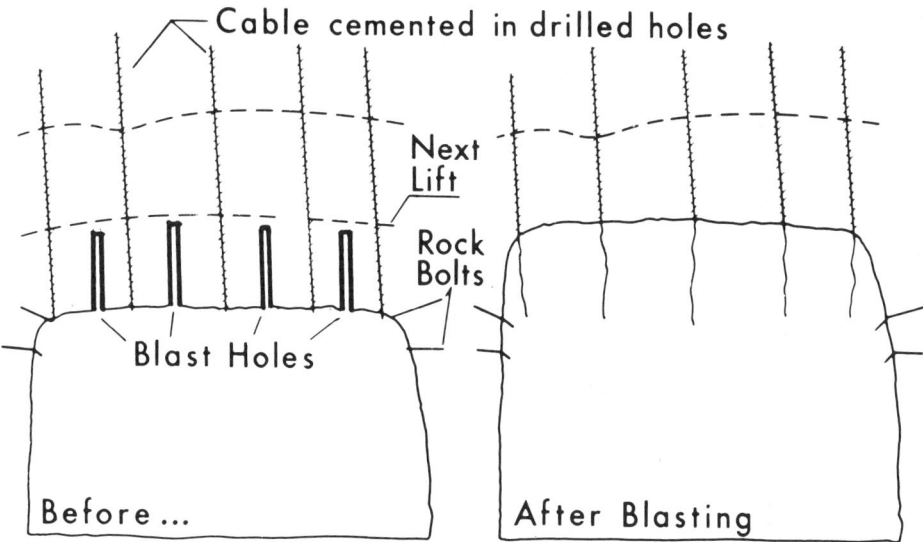

Fig. 3

holes up to 100 ft long {30.5 m} in the backs of stopes, were prestressed and grouted in place. Blast holes were placed between the cable locations and eight to ten ft {2 - 3 m} of the stope back was blasted out. The new stope back was thus presupported by the grouted cables. It was found that as the rock deformed in the grout arch the cables became overloaded, broke and the arch failed. The problem was solved by placing the cables loosely in the hole, eliminating prestressing when grouting them in place. Thus the cables were placed in passive form and, as the ground moved they were loaded, but not beyond their ultimate load carrying capabilities. Thus the need for a passive fixture under some conditions of rock deformation is evidenced. The modulus of the rock in these stopes was quite high so that the modulus of the grout and steel matched it quite well.

For long term stability and life of a mine opening, more permanent forms of lining should be considered. Timber, shotcrete, steel sets, liner plates, poured concrete, etc. can all be used. These supports may be placed late in the mining cycle when the remaining deformation associated with the ground reaction curve may be quite small. If a rigid support is to be used, this would be the time to place it. It is also important to decide whether or not these supports should be active or passive. To be active, the support will have to be loaded in some manner. For example, timber can be wedged, steel sets can be loaded by placing wedges in blocking or pressure grouting, and concrete can be prestressed by pressure grouting or use of pressure cells. The end use of the opening most often dictates what lining will be used. But for it to be fully successful it will have to match or resist ground deformations which may be present over its entire life.

MINER SAFETY

Repetitious mining systems such as those used in room and pillar operations in coal have an inherent safety factor built into the mining process. As shown in figure 4, the bituminous coal industry of the United States had a near identical fatal safety history for four consecutive years, 1968 through 1971, with the only deviations being due to work stoppages and major disasters from dust and gas explosions.[4] During this period, 4500 to 5000 separate underground mines were in operation. The period of 1972 through 1980 shows an improvement in fatalities but also shows a marked similarity in rate from year to year. How is this possible? It is the author's opinion it is due to the fact that all manufacturers constructing coal mining equipment produced equipment very similar to their competitors which meant that all men were exposed to near the same hazards. For example, roof bolters consist of two fundamental types for coal. One is the boom type bolter and the other the mast type. Both units have controls within easy reach of the operator so that he can feed steel and roof fixtures at the drill head and at the same time reach controls of the machine. Many machines have a dual set of controls so that the operator can move back away from the bolting site and finish placement of the fixture. Prior to the late 1960's and early 1970's the use of temporary roof support was not always mandatory and canopies were not required by law. Falls of ground contributed to a high percentage of the fatal accidents. Much of the reduction in recent years of fatal accidents in coal can be attributed to the use of canopies.

To accomplish further reductions in loss of life from

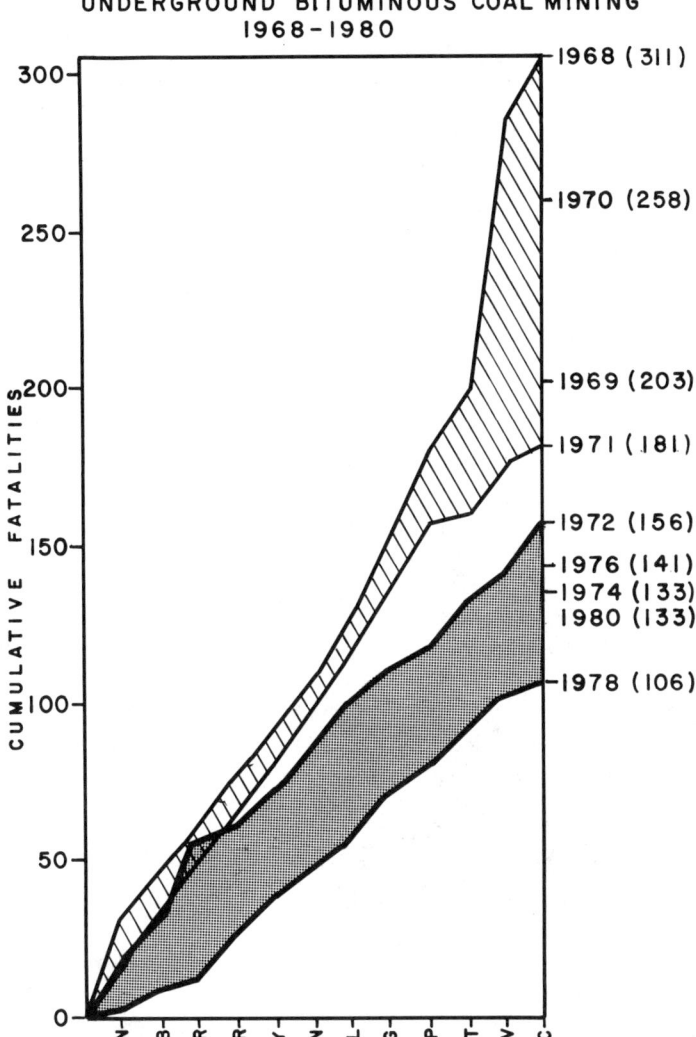

Fig. 4

falls of ground, additional inherent safety can be achieved by removing the man from the hazard through automation. The U.S. Bureau of Mines and equipment manufacturers have ongoing programs to develop just such equipment. Onboard bolters for continuous miners have been used with limited success. Fully automated systems that place roof fixtures by remote control are becoming popular in metal mines where more room is often available which makes the design of the machines easier. It is expected that the new thrust in oil shale mining in the western United States will result in the development and deployment of fully automated systems to place mechanical, resin or Friction Rock Stabilizer roof fixtures. Figures 5 and 6 show two operational systems that are now being tested at these mines. It seems reasonable to expect that the development of Robotics by the automotive industry may in someway spill over into mining to allow machines to do more of the tasks in hazardous areas.

PRODUCTIVITY AND AUTOMATION

Safety and productivity go hand in hand. A truly productive mine can be and should be a safe mine. Dr. Michael Zabetakis has documented the history of accidents in the coal industry including fatalities, disabling injuries and accident frequency.[5] Since the passage of the Coal Mine Act of 1970, fatal accidents have continued to decrease but accident frequency and disabling injuries have increased. See figures 7, 8, and 9. The increases coincide with the enforcement of the safety act. During the period 1970 to 1979 the productivity of coal mines

Fig. 5

Fig. 6

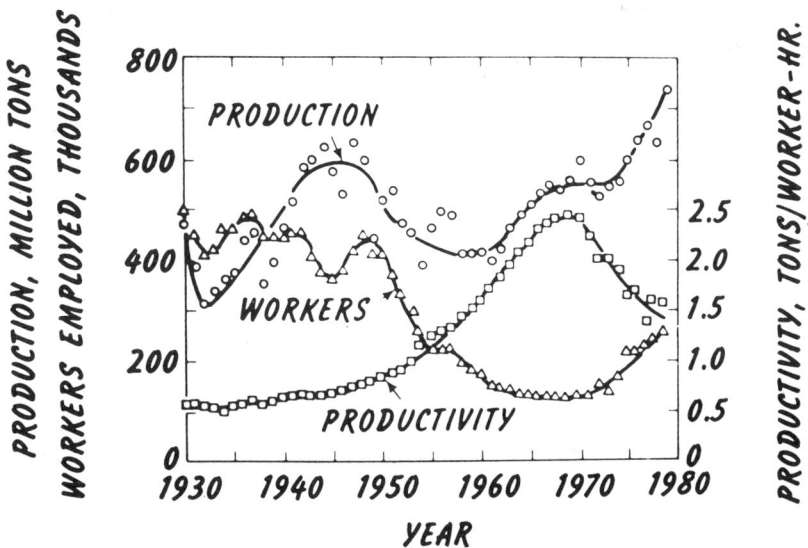

Fig. 7

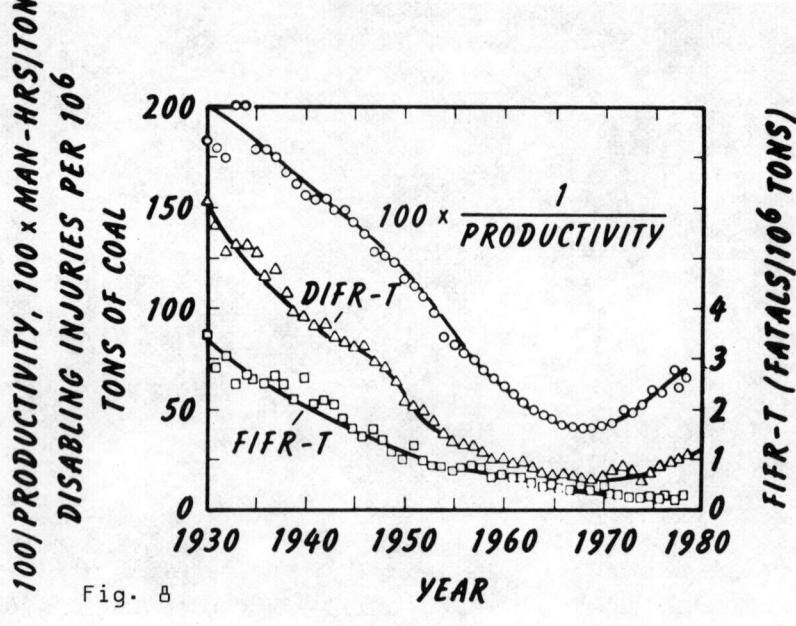

Fig. 8

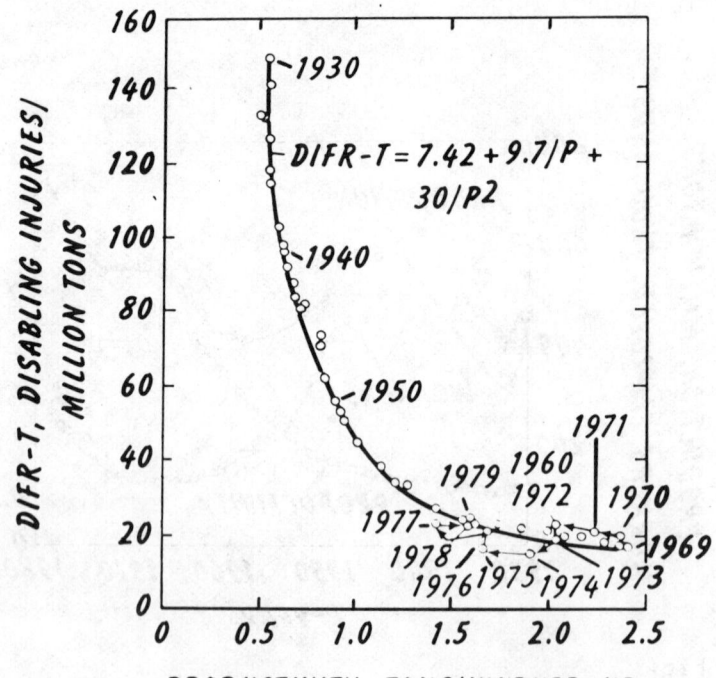

Fig. 9

decreased by almost 50%. This loss of productivity necessitated employment of many more workers to obtain the same tonnage and increased the workers exposure to hazards. Thus it is not surprising that good productivity that leads to a lesser number of workers being exposed is a good safety measure.

The simple mechanics of placing roof fixtures would seem to dictate that the one with the fewest parts and the greatest simplicity will be easiest to place in a remote manner. This has proven to be true. The Friction Rock Stabilizer which has no moving parts and consists of only a tube and a plate, can be placed faster than any other fixture and best lends itself to automated placement. Standard mechanical bolts are also quite simple and once assembled only require insertion and rotation for placement.

Resin and cementeous anchor devices are more difficult to automate. The resin and cement components must be placed in the borehole in a separate step and their capsules are quite fragile. Breakage often occurs and, in the case of resin especially, the broken capsules are quite messy and may prove toxic to the skin. There is also the belief in some circles that the resin component is a fire hazard in mines. One of the cementeous capsule systems that seems to have the greatest chance of success is the U.S. Bureau of Mines water-microcapsule system now under development.

Pumpable systems, where wet cementeous materials or resin are pumped into a borehole, are being investigated by the U.S. Bureau of Mines and others. These systems have some advantages for automation as pumping of materials is a well developed technology but at the same time it has several disadvantages. The systems are not good for intermittent use, clean up of hoses and pumps is a problem, keeping the fluid substances in the borehole may prove difficult, and all of the systems require special equipment not normally found in mines. Capital costs for such equipment may be high.

Shotcrete linings do lend themselves to semi-remote operator placement. The nozzle for the shotcrete can be extended out under the unsupported roof to spray the shotcrete. By using accelerators, quick set times and high strength can be achieved. It is possible to use a thin layer of shotcrete in conjunction with wire mesh attached to the surface of the opening as a flexible lining early in the ground reaction curve. The roof fixtures, wire mesh and thin shotcrete strengthen the ground arch and protect the workmen and equipment.

After the face has advanced and the area has stabilized through the final deformation, a thick, rigid layer of shotcrete can be placed and provide long term stability to the opening.

IMPLEMENTATION

To apply basic rock mechanic principals, develop automated equipment, improve safety, improve productivity and finally lower costs, the cooperative effort between theoreticians, academics, mining engineers, mine operators, safety personnel, equipment manufacturers and government inspectors, both state and federal, will be required. Top management of the mining operation must be committed to improvement in all of these areas. The objectives of any phase of the projects must be clearly stated and the scope of work must be defined in such a way that it interfaces well with other aspects of the program. Time studies and operational analysis can be used for optomizing the mining plan to reduce delays in roof support, materials handling and storage. Quick placement of roof fixtures should minimize roof falls and thus lower costs. Operator training and improvement of the mining cycle may be keys to success. Many new people are entering the mineral industry and this presents a great opportunity to mine management to train them adequately and thus eliminate bad habits sometimes inherent in existing working forces.

A concerted effort on the part of management in the field of rock mechanics can and will pay rich dividends. Early warning systems which alert workmen to hazards can be employed. Thorough examination of areas that do fail will assist in future planning and assure cooperation between the rock mechanics engineer and the mine operators. The end result of the thoroughly integrated study of the mining operation should be to improve mine design, improve roof control procedures, improve safety, increase productivity and lower costs.

A TOTAL COST APPROACH TO ROOF CONTROL

Studies conducted by Wesley C. Patrick in his Ph.D. dissertation at the University of Missouri-Rolla, showed conclusively that roof fall related costs have a substantial effect on the overall profitability of the mine. Support systems having a higher initial installation cost may be less expensive when viewed on a total cost basis.[6] Falls of ground generate many costs which are often not readily recognized. There is, of course, the tragedy associated with personal injury, loss of wages, hospital-

ization costs, etc. There are also other costs to the
company such as high level supervisors and office personnel which are involved in the investigation and documentation of the event, and another workman must be transferred to take the injured persons place on the job.
Overtime work may result from the accident. Equipment
may be damaged and taken out of service so that repair
costs are produced but loss of production of the equipment
may not be so evident. A more subtle and therefore more
difficult cost to evaluate are the losses incurred in
decreased productivity of an injured worker after he returns to work and also of the loss of productivity of his
fellow workmen who are on the job and are disrupted due
to the accident. Figure 10 shows the effect of a roof

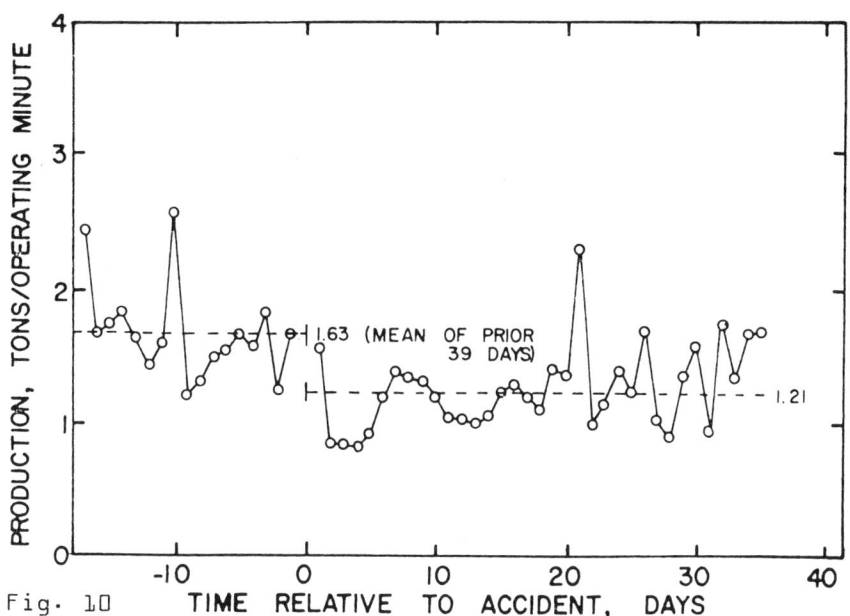

Fig. 10 TIME RELATIVE TO ACCIDENT, DAYS

fall fatality on production per net operating shift.
Before the accident the section averaged 572 tons per
shift and after the accident, averaged 392 tons per shift
for a period in excess of 30 days. Also the effect of
falls of ground even when injuries do not occur may have
profound effects on total mining costs if they occur in
travel roads or main ventilation roads. Rock cleaned up
with coal mining equipment may prove extremely costly
per ton as it damages the equipment. The normally
productive areas of the mine may be isolated and contract
commitments may not be met.

Finally, it should be realized that losses generated from one major fall of ground on a main travel way which disrupts production for an extended period of time could well equal the total yearly expenditure for a well staffed rock mechanics department.

In 1975 industry studies indicated that it costs approximately $4.50 per cubic ft to clean up roof falls in bituminous coal mines.[7] A fall four ft thick by 16 ft by 20 ft would cost $5,800 to clean up. The elimination of one fall of this type per month could well pay for an engineer and equipment for a full time study effort on ground control problems.

CONCLUSIONS

Investigators in the field of rock mechanics have in the author's opinion a humanitary responsibility to do everything possible to make working places in underground operations safe for workers. It is not enough to just come up with a new theory on a better way to support or stabilize ground around an opening without putting our very best thinking into the problems of application of the methods. Great strides have been made in the past 25 years in acquiring new knowledge about rock mechanics and some application of this knowledge has been transfered to operations. I look forward to the period of the 1980's and the 1990's to being the period of maturing so that theories of the present and the future will be more quickly transferred into engineering practice.

REFERENCES

1. Scott, James J., 1980, Interior Rock Reinforcement Fixtures - State of the Art, 21st Symposium on Rock Mechanics, University of Missouri-Rolla.
2. Lane, K.S., 1975, Field Test Sections Save Cost in Tunnel Support. This work was funded by the National Science Foundation with support of the American Society of Civil Engineers.
3. Schmuck, C.H., Dec. 1979, Cable Bolting at the Homestake Gold Mine. Mining Engineering.
4. MSHA Reports, 1968 - 1980, Tables for Coal Fatalities, Health and Safety Analysis Center, Denver, Colorado
5. Zabetakis, M., 1981, "Productivity and Safety in U.S. Bituminous Coal Mines", Mining Congress Journal, Vol. 67, pp. 19-21.
6. Patrick, W.C., 1978, A Total Cost Approach to Mine Roof Support Selection, Ph.D. Thesis, University of Missouri-Rolla.
7. Burggraf, C., 1981, "Roof Control at Republic's Kitt Mine", Mining Congress Journal, December, 1981, p. 37.

Chapter 98

NEW LABORATORY INSTRUMENTATION FOR THE EVALUATION
OF ROCK BOLT BEHAVIOR

E. Unal, H. Reginald Hardy, Jr. and Z. T. Bieniawski

Department of Mineral Engineering
The Pennsylvania State University
University Park, PA 16802

INTRODUCTION

Rock bolts are a major means of support in underground coal mines in the United States. However, two of the most crucial problems facing the mining engineer today are that of finding a well defined and standard method for testing the anchoring ability of rock bolts, and of selecting the most suitable type of rock bolt and bolting pattern for a particular mine or specific mine section (Anon., 1982). Such selection requires not only an understanding of the behavior of the rock mass and the rock-bolt unit itself, but also a knowledge of the basic mechanisms by which rock bolts provide support while they interact with the rock mass during various field loading conditions. A recent study in this area has shown the inadequacy of present methods used in selecting and testing of rock bolts and the lack of clearly defined guidelines for safe and economical design of bolting patterns (Bieniawski, et al., 1980).

The present paper describes a research study aimed at developing laboratory instrumentation and associated techniques for the evaluation of bolt-rock interaction. The instrumentation system developed, procedures followed during various experiments, and typical results from a series of studies carried out on mechanical bolts are presented.

DEVELOPMENT OF THE ROCK BOLT TESTING FACILITY

General

A block diagram of the experimental facilities are shown in Fig. 1. The system consists of three main sections. The first

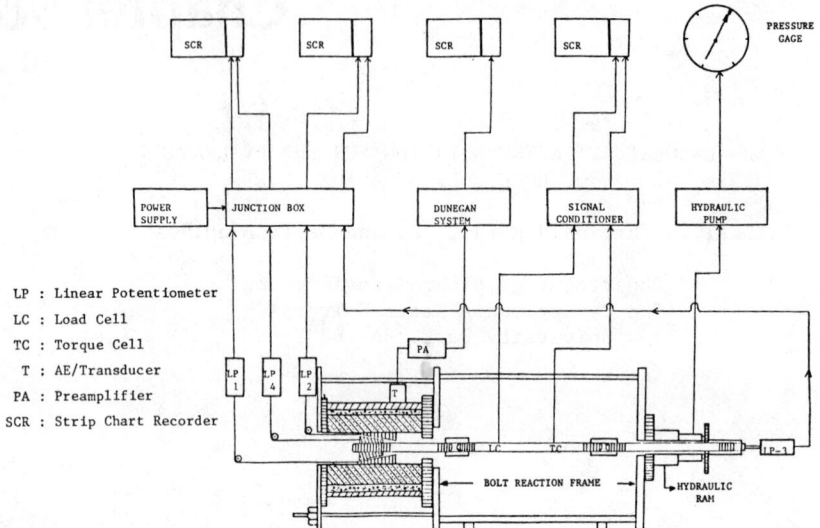

Fig. 1. Block diagram of experimental facilities for evaluation of rock-bolt stability.

includes the rock specimen, the rock bolt and the hydraulic ram, all of which are attached to the rock-bolt reaction frame. Also included is a hydraulic pump and a torque wrench which are used to apply tension and torque respectively to the system. The second section includes various transducers, including four linear potentiometers, an acoustic-emission (AE) transducer, and a torque-tension load cell. The third section includes two groups of signal conditioners, the first a Dunegan Acoustic-Emission System which is used for AE signal conditioning, and the second a system associated with the torque-tension load cell. Also included are various readout devices including pressure gages and four strip-chart recorders. Fig. 2 provides a general view of the rock bolt testing facility utilized in the study. Further details of specific components of the testing facility are described in the following section.

Details of Specific Components

Test Specimens. The test specimens for the experiments were thick-walled hollow cylinders prepared from 254x254x178 mm rectangular blocks of commercially available Berea Sandstone. The blocks were first cored in the center using a 32 mm diameter diamond bit. The desired hollow cylinders were then obtained by overcoring the central hole using a 152 mm diameter diamond bit. In order to simulate the effect of lateral confinement, and to prevent the movement of the rock specimens during the experiments, each specimen was encased in a 203 mm diameter cast iron pipe (jacket), 250 mm long and 9.5 mm thick. The annular void between the rock and the pipe was filled

Fig. 2. Experimental facilities for evaluation of rock-bolt stability.

with a commercially available quick-setting expanding-type cement known as "water plug." Following preparation, the test specimens were stored in the laboratory at room temperature for a minimum of four weeks to insure complete curing of the cement. Fig. 3 shows two of the prepared test specimens ready for subsequent tests.

Fig. 3. Prepared specimens ready for subsequent tests.

Rock-Bolt Assembly. The rock-bolt assembly consisted of a bail-type expansion shell, a 15.9 mm diameter steel bolt-rod, a standard steel bearing plate and a hardened washer. A torque-tension load cell was also incorporated in this assembly.

Test Frame. The test (reaction) frame used during the experiments consisted of a specially fabricated rectangular steel box with centrally located holes on opposite faces. This frame, bounded at one end by the rock specimen, represents an equivalent thickness of rock and due to its rigidity eliminates bending and other effects which would otherwise complicate the analysis. The hollow test specimen was attached to one end of the test frame by means of two threaded "U-bolts." The hydraulic ram was fixed to the opposite end of the frame. The mechanical details of these arrangements are presented elsewhere (Unal, 1982).

Linear Potentiometers. Four linear potentiometers (LP) were used to continuously monitor the movement of the rock-bolt shell (cone and leaves) and the elongation of the rock-bolt rod. The arrangement of the linear potentiometers has been shown earlier in Fig. 1. Two of these (LP-1 and LP-2) were connected to the leaves of the rock-bolt shell and a third one (LP-4) was connected to the rock-bolt cone. The mechanical connections between the linear potentiometers (mounted on a rigid base plate) and the various sections of the rock bolt (located inside the hollow test specimen) were provided by means of thin steel wires passing around brass pulleys. In order to obtain a low hysteresis movement in both directions a suitable spring was mounted between the shaft and the body of each potentiometer. Bolt-rod movement was monitored by a fourth linear potentiometer (LP-3) attached directly to the head of the rock-bolt assembly where the torque was applied to the system. The connection between this linear potentiometer and the bolt head was provided by means of a magnet mounted on the linear-potentiometer shaft.

The electronic circuits associated with the four linear potentiometers were located in an associated junction box. The input voltage to the system was provided by means of a dc power supply, and the output signals from each linear potentiometer were connected to an associated strip-chart recorder. Calibration studies indicated that the resolution of the linear potentiometers were 0.025 mm.

Torque-Tension Load Cell. As illustrated earlier in Fig. 1, continuous monitoring of tension and torque in the bolt-rod during various phases of experiments was provided by means of a torque-tension load cell specifically designed for this research work. Fig. 4 provides a general view of this unit. The cell contained various sets of electrical resistance strain gages mounted on a 22.9 mm thick and 330 mm long section of the Inconel-625 high strength steel rod. This rod was attached to the bolt-rod by means of couplers.

The tension-monitoring section of the load cell consisted of two axial and two transverse strain gages spaced at 90° apart around the circumference of the steel rod. These gages were connected in a full wheatstone bridge configuration and powered by a suitable dc supply. This section of the load cell was designed for a maximum tensile load capacity of 110 kN. Subsequent calibration indicated that the sensitivity of this section of the cell was 1.72 mV/V/1100 kN, and during rock-bolt experiments the tensile load could be resolved to 500 N.

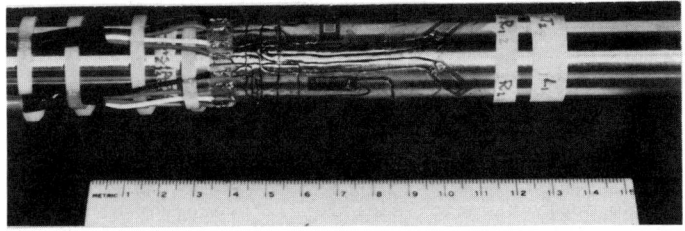

Fig. 4. A general view of the torque-tension load cell utilized during the experiments.

The torque-monitoring section of the load cell consisted of four strain gages bonded to the steel rod at a 45° angle with respect to the rod axis. These gages were spaced at 90° apart around the circumference of the steel rod. The four gages were connected in a full wheatstone bridge configuration eliminating the effects of all strains other than torsional ones. The torque section of the load cell was calibrated in the load frame, over the range 0-160 Nm, using a torque wrench, and the sensitivity of the cell was found to be 0.9 mV/V/ 160 Nm. During subsequent rock bolt tests torque could be resolved to 1.5 Nm.

AE Monitoring. A suitable AE transducer (Dunegan/Endevco, type D140B) was used to detect AE signals occurring during rock bolt tests. The transducer was attached to the surface of the specimen jacket using vacuum grease and masking tape. A Dunegan-3000 Acoustic-Emission System was utilized to process the signals detected by the AE transducer, and to provide a suitable analog output, in terms of total AE events, for input to one of the associated strip-chart recorders.

EXPERIMENTAL PROCEDURES

The eight parameters monitored during the experiments were:
(1) Applied torque
(2) Torque in the bolt-rod
(3) Tension in the bolt-rod
(4) Displacement of the shell (LP-1)
(5) Displacement of the shell (LP-2)
(6) Elongation of the steel rod (LP-3)
(7) Displacement of the bolt cone (LP-4)
(8) Acoustic Emission (AE)

The first seven of these parameters, which are referred to in this paper as anchorage-testing parameters and the associated acoustic emission (AE) were monitored during nine detailed final experiments carried out in this study. Furthermore, a large number of initial tests were also carried out prior to the final experiments in order to gain experience in the use of the testing facility and to investigate the form of data from such experiments.

In general, the experiments carried out in this research each involved five different steps, as shown in Fig. 5, namely:

<u>Torquing</u>: Application of torque to the rock-bolt assembly, thus increasing the tension in the bolt rod. This step represents the actual rock-bolt installation.

<u>Bleed-off I</u>: No application of additional torque, or other external forces to the rock-bolt assembly. Meanwhile monitoring the changes in the various parameters under study. This step constitutes the anchorage-efficiency test and demonstrates the time-dependent behavior of the bolt after installation.

<u>Pull-out</u>: Application of an increasing bolt-tension by means of the associated hydraulic ram. This step represents the pull-out or anchorage-capacity test. It illustrates the effect of rock strata exerting extra weight on the rock bolt.

<u>Bleed-off II</u>: No application of additional bolt tension, or other external forces to the rock-bolt assembly. Meanwhile monitoring the changes in the various parameters under study. This step constitutes an anchorage-efficiency test during the post-failure stage.

<u>Tension Release</u>: Reduction of the bolt tension to zero by releasing the pressure in the hydraulic ram. This step represents the equivalent of complete loss of anchorage capacity.

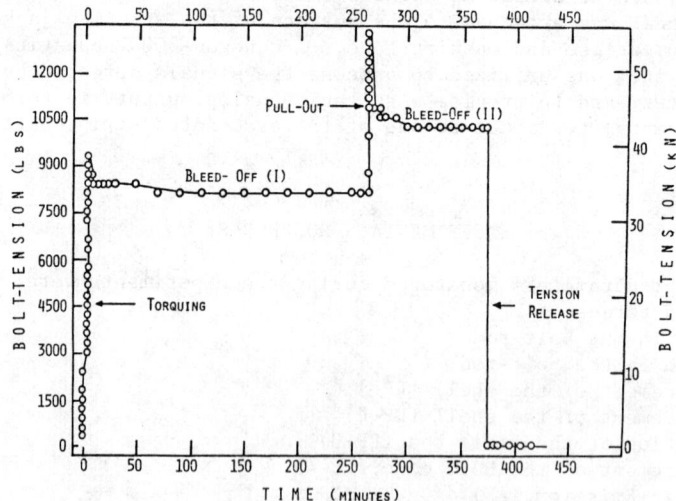

Fig. 5. A typical bolt-tension versus time curve showing the five steps involved in a typical experiment.

NEW LABORATORY INSTRUMENTATION

EXPERIMENTAL RESULTS

General

During the nine final laboratory experiments carried out to date, the anchorage-testing parameters and the associated acoustic emission have been recorded. A detailed program of data analysis has been carried out in order to establish general relationships between a number of the experimental parameters, namely:
(1) Applied and measured torque
(2) Torque and tension
(3) Tension and AE
(4) Tension and shell displacement
(5) Tension and cone displacement
(6) AE and shell displacement
(7) AE and cone displacement

Due to limited space available in this paper, only the highlights of some of the more important results obtained will be included here. The detailed results of the analysis related to acoustic emission has been published elsewhere (Unal and Hardy, 1981), and those related to the establishment of complete support characteristic curves and analysis of support and ground interaction will be the subject of a future paper.

Applied versus Measured Torque

The clamping force or supporting action of a rock bolt depends on many factors but primarily on the torque applied to the bolt head. The current study has clearly shown that the applied torque was much larger than the torque actually transmitted to the bolt. The analyses of the test results indicate a linear relationship between these two parameters. The following equation was derived as a result of the least-squares regression analysis performed on the data:

$$TC = 0.586(\pm 0.012)TW + 0.678(\pm 0.973) \qquad \text{(Eq. 1)}$$

where TC is the torsion developed in the bolt rod in Nm, and TW is the applied torque indicated on the torque wrench in Nm. The numbers enclosed in brackets ( ) are the standard deviations associated with the regression determined coefficients. The results indicate that the torque measured on the bolt is approximately 59 percent of the applied torque. This value is slightly higher than the earlier predictions made by other workers.

Torque versus Tension

One of the basic problems involved in any rock bolt installation is the determination of the quantity of torque transmitted to the bolt in the form of tension. The most common method of tensioning rock bolts is the application of torque to the bolt head, and using this method it is not possible to predict the actual values of tension attained in bolts with any degree of reliability.

In the present study a series of statistical analyses were undertaken on the bolt-tension versus applied torque and the bolt-tension versus bolt-rod torsion data. These analyses indicated a linear relationship between the bolt-tension and applied torque with the associated regression equation being as follows:

$$LC = 276.2(\pm 5.9)TW - 267.8(\pm 476.0) \qquad (Eq. 2)$$

Furthermore, the relationship between bolt-tension and bolt-rod torque was found to be as follows:

$$LC = 428.1(\pm 17.4)TC + 2620.4(\pm 810.9) \qquad (Eq. 3)$$

In Eqs. 2 and 3, LC is the tension in the bolt rod in N, TW is the applied torque in Nm, TC is the torsion in the bolt rod (Nm).

Acoustic Emission versus Anchorage Testing Parameters

It is a known fact that acoustic emission (AE) is often a very useful parameter for evaluating the stability of geologic structures (Hardy, 1981). For example, when a structure, such as bolted-roof in an underground mine, changes its stage of mechanical equilibrium during a mining cycle, some of the associated energy should be released in the form of AE. If AE could be correlated to one or more of the rock bolt anchorage-testing parameters discussed earlier in this paper, then it would be possible to develop an early warning system capable of continuously monitoring of the stability of sections of bolted mine roof.

In the present study attempts were made to correlate AE with various anchorage-testing parameters. This section of the paper presents some of the results of the associated analyses; however, further details on this aspect are presented elsewhere (Unal and Hardy, 1982). Fig. 6 illustrates the variation of accumulated AE activity and various anchorage-testing parameters as a function of time. In general it appears that there is a significant correlation between the various anchorage testing parameters and AE.

One of the most significant findings of these analyses was the observed sudden increase in the rate of AE associated with changes in the behavior of the rock mass and/or the rock-bolt assembly. The first warning point was observed as the yield strength of the bolt-rod steel was reached (see region A-A', Fig. 6). The second warning point was associated with the movement of the bolt anchor in the bore hole and observed as the load bearing capacity of the bolt was reached (see region B-B', Fig. 6). Finally, the third warning point was observed when the complete loss of anchorage capacity occurred (see Fig. 6, point C).

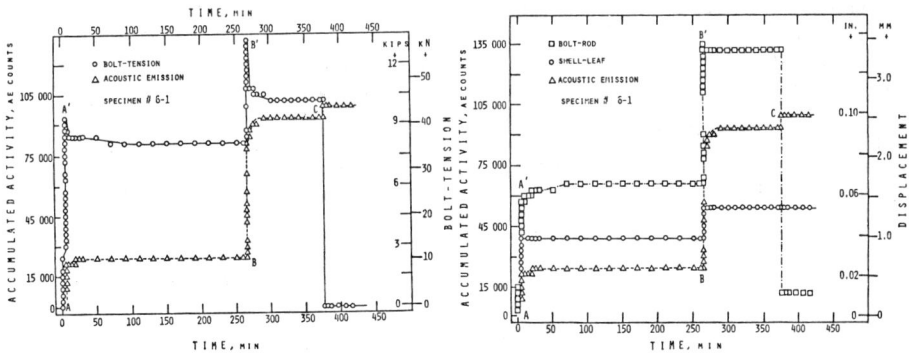

Fig. 6. Variation of accumulated acoustic emission and anchorage-testing parameters as a function of time.

The variation of accumulated AE was analyzed as a function of a number of anchorage testing parameters and the regression analysis based curves were determined from the data points associated with the first step of the experiments (i.e., torquing). In all cases the analysis yielded positive correlation coefficients indicating an increasing accumulated AE activity with increasing tension or movement of the rock-bolt assembly. The best correlation obtained was that associated with shell-leaf and shell-cone movement and AE activity. The regression equations and the associated statistical data from these analyses are presented elsewhere (Unal and Hardy, 1982).

Development of a Complete Support Characteristic Curve for Rock Bolts

In general the behavior of any support element is characterized by its support characteristic curves. Such curves can be obtained by plotting the displacement of the support-assembly components as a function of applied load. From its installation stage through various mining cycles the components of a rock-bolt assembly react to changes in loading conditions in the form of displacements. The two most significant displacements are the rigid body motion of the shell leaves with respect to rock, and the deformation of the bolt. In order to determine the complete support-characteristic curves, the behavior of these two components should be monitored during various experimental steps, including: torquing, bleed-off, and tension release. The resulting curves would provide information about load bearing capacity (maximum load which can be carried by the support assembly) and load bearing efficiency (change of load with time) of the rock-bolt assembly.

Fig. 7 shows an example of the complete support-characteristic curves obtained for one of the bail-type mechanical bolts used in the current study. The curves indicate the higher load bearing capacity of the bolt leaves (point B) relative to the bolt-rod yield-point (point A). This result suggests the necessity of using a higher strength steel and/or more suitable thread design for the bolt rod since the present bolt-rod design did not make full use of the anchorage capacity of the shell-leaves. If in contrast point A were much higher than point B in Fig. 7, this would indicate an unnecessary strong bolt-rod. Ideally the anchorage capacity of the bolt should match the yield strength of the associated bolt rod.

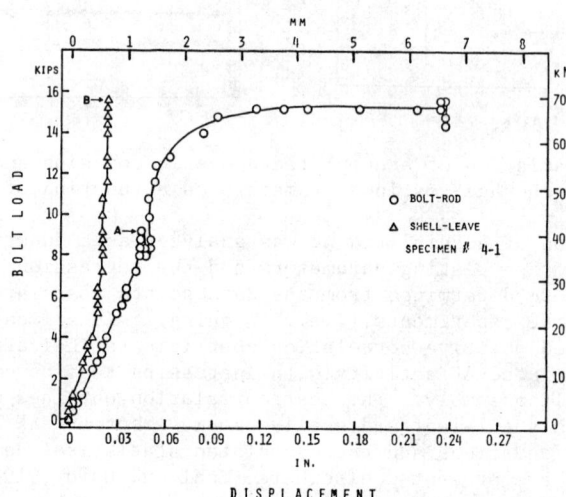

Fig. 7. Typical support characteristic curves for rock bolts. (A - Rod yield-point, B - Load bearing capacity of shell-leaf.

DISCUSSION

The following accomplishments and conclusions are based on the work carried out during the recent study.

(1) Laboratory instrumentation was developed in order to evaluate in detail the mechanical behavior of rock bolts.

(2) An experimental method was developed to evaluate the stability of rock-bolt assemblies based on a variety of in-situ loading conditions.

(3) A detailed analysis of rock bolt behavior was carried out considering various anchorage-testing parameters and acoustic emission (AE) activity.

(4) AE techniques were used for the first time in the evaluation of rock bolt anchor stability.

(5) A significant relationship was found between various anchorage-testing parameters and AE activity, thus the possibility of developing an early warning system for monitoring the instability of bolted underground structures appears feasible.

(6) The complete support-characteristic curves for one type of rock bolt were developed considering both the motion of the shell and deformation of the bolt-rod. Such curves are considered necessary in order to indicate the true behavior of a rock bolt unit.

REFERENCES

Anon., 1982, "Federal Register, Proposed Rules," Vol. 47, No. 46, March 9, Part III, Department of Labor, Mine Safety and Health Administration, Section 3, pp. 10191-10192.

Bieniawski, Z. T., Cincilla, W. A., and Unal, E., 1981, "A Design Approach for Coal Mine Tunnels," Proceedings, 1981 Rapid Excavation and Tunneling Conference, San Francisco, California, AIME, ASCE, V. 1, pp. 147-164.

Hardy, H. R., Jr., 1981, "Application of Acoustic Emission Techniques to Rock and Rock Structures: A State of the Art Review," Acoustic Emissions in Geotechnical Engineering Practice, STP 750, V. P. Drnevich and R. E. Gray, Eds., ASTM, pp. 4-92.

Unal, E., 1982, "Assessment of Design Needs and Support Guidelines for Coal Mine Tunnels," Ph.D. Thesis, The Pennsylvania State University, 175 pp.

Unal, E. and Hardy, H. R., Jr., 1982, "Application of AE Techniques in the Evaluation of Rock Bolt Anchor Stability," Proccedings, 3rd Conference on Acoustic Emission/Microseismic Activity in Geologic Structures and Materials, University Park, Pennsylvania, The Pennsylvania State University, in press.

CASE HISTORIES

Chairman
John E. O'Rourke
Woodward-Clyde Consultants
San Francisco, California

Co-Chairman
Donald C. Rose
Tudor Engineering Co.
San Francisco, California

Keynote Speakers
Chikaosa Tanimoto
Kyoto University
Kyoto, Japan

Issa S. Oweis
Converse Consultants
Caldwell, New Jersey

Chapter 99

ENGINEERING EXPERIENCE WITH WEAK ROCKS IN JAPAN

by   Chikaosa Tanimoto

Lecturer in
Construction Engineering
Department of Civil Engineering
Kyoto University
Sakyo, Kyoto, Japan

INTRODUCTION

The committee on Rock Mechanics, Japanese Society for Civil Engineers, has been discussing 'soft rock engineering in Japan' and presented several papers concerning dam, tunnel, bridge and slope constructions to International Symposium on Weak Rock, Tokyo 1981. This is the summarized paper based on References of Okamoto et al (1981), Iida et al (1981), Working Group on Tunneling-JSCE (1981), Working Group on Bridge Foundations-JSCE (1981) and Kikuchi et al (1981), which can convey general view on engineering experiences with weak rocks in Japan.

From the point of view to the origin and/or distinctive physical properties, weak rocks in Japan may be classified into (1) sedimentary, (2) weathered, (3) low welded-pyroclastic and (4) fractured ones. Their typical characteristics are summarized in the following section.

Concerning dam foundations on soft rock, several hundreds have already been constructed and, at present, nearly one hundred are under construction or at planning stage. These dams, almost without exception, have some problems such as site selection, material selection for fill-type and concrete dams, design of dam and its foundation, and construction procedures, requiring special ingeneous and cautious measures to cope with the local geotechnical difficult situations.

In tunneling, for the decade of 1970s, the number of tunnels in soft rock either completed or under construction was at least 160, including Seikan Tunnel of 53.9 km long, Nakayama Tunnel of 14.8 km long and so on. Most of these tunnels are distributed in the northern half of Mainland, consisting of a basement of Paleozoic strata and Plutonic rocks, overlayed by Neogene Tertiary formations which include large quantities of ejecta from volcanoes active from th Tertiary to the Holocene periods.

Accompanying the development of such major traffic networks as super-express railways, free ways, and trans-strait bridges, many long span bridges have been constructed on weak rock. It has become necessary to carry out surveys and tests on the strength and deformation characteristics of weak rock. In Japan a standard survey and testing method for weak rock as bridge foundation has not yet been established. In this paper, case studies, concerning five bridges built on weak rock such as Tertiary sedimentary rock, weathered granite, and pyroclastic rock are mentioned. As a special case, the elevated bridge of ShinKobe station, which has been built on an active fault, is included.

Landslides and slope failures are frequently attributable to environmental conditions and additional human interferences to ground. Based on the past experiences through disasters caused by mass movement, considerable improvement has been made in analysis, investigation and controlling technique.

Respective experiences in different fields may have some disagreements more or less due to various geological and mechanical conditions, situations, interaction between rock in-situ and man-made structure, highly developed techniques, and so on. The author tried to stand on the neutral position in this paper.

GEOLOGICAL AND GEOTECHNICAL FEATURES ON WEAK ROCKS IN JAPAN

The geological structure of Japan composed of island arcs is complicated. Japan Islands covered an area of about 370,000 square kilometers had several times of orogenic movement in geological history. The amount of upheaval and subsidence of ground occurred since Quaternary period is estimated about $\pm 1,500$ m. At present, the geological phenomena (earthquake, volcanism etc.) are active and many active faults are found in this land. Fig. 1 shows the outline of geological structure of Japan Island Arcs.

Weak rocks, which are of low strength due to loose texture, low induration or poor cementation, are broadly grouped as follows:
(1) Sedimentary rocks of the Neogene Tertiary
    a. Sedimentary rocks in non-Green Tuff region
    b. Sedimentary rocks in Green Tuff region
(2) Weathered granite
(3) Low welded volcanic rocks of the Quaternary
The distributions of respective rocks are shown in Fig. 2 - 4.

The regions occupied by Neogene rocks can be subdivided into non-Green Tuff and Green Tuff regions. The latter region, which has geotechnical features somewhat different from the former, occupies Northeast Japan (including the Fossa Magna), Southwest Japan on the side of Japan Sea, and eastern Hokkaido. In this region, submarine volcanic activity was violent and extensive from early to middle Miocene epoch, hence the formations up to the middle Miocene are mainly composed of volcanic rocks and pyroclastic rocks such as lava, tuff and tuff breccia. As this region underwent subsidence from late Oligocene or early Miocene and remained under deep sea in the latter half of Miocene, upheaval to land taking place from the end of Pliocene to the beginning of Pleistocene, most of the formations are the products of submarine eruptions.

As the volcanic activity was somewhat calm and local in the latter half of the Miocene, the deposits in this age are mostly mudstone and sandstone with some intercalations of volcanic and pyroclastic rocks. The sedimentary rocks of this age are poorly indurated in general and low in strength. While sedimentary rocks of the Neogene including those of non-Green Tuff region, have a common tendency to deteriorate and disintegrate by drying and wetting cycles, most of sedimentary rocks in Green Tuff region show a remarkably high tendency of slaking and swelling due to the high content of swelling clay minerals such as montmorillonite group. In the lower formations of Neogene deposits of Green Tuff region consisting mainly of pyroclastic rocks, which are mostly altered due to intense submarine volcanic activity, the sedimentary rock of this kind occurs either as a contained mass or as interbeddings of several centimeters to several decimeters thick. Upper formations of Neogene deposits in Green Tuff region are mainly composed of sandstone and mudstone which shows typical characteristics of slaking and swelling. In contrast, as the volcanic activity in non Green Tuff region was calm in the Neogene, the Neogene deposits in this region mostly consist of sandstone and mudstone with scarce

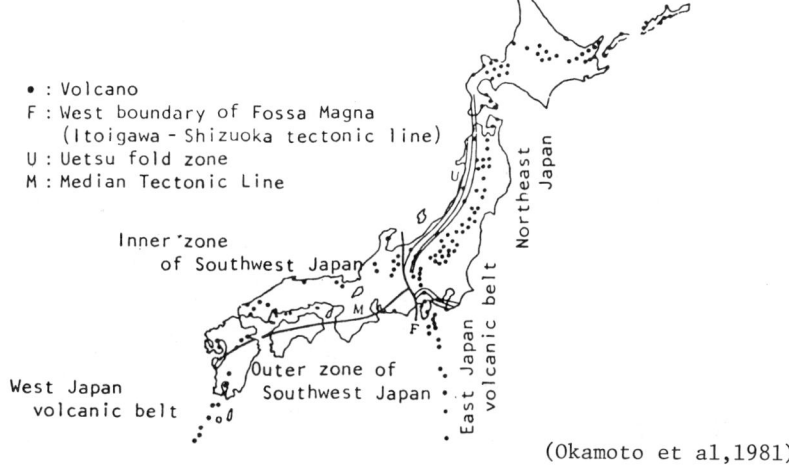

(Okamoto et al,1981)

Fig. 1 Outline of geological structure of Japan Island Arcs

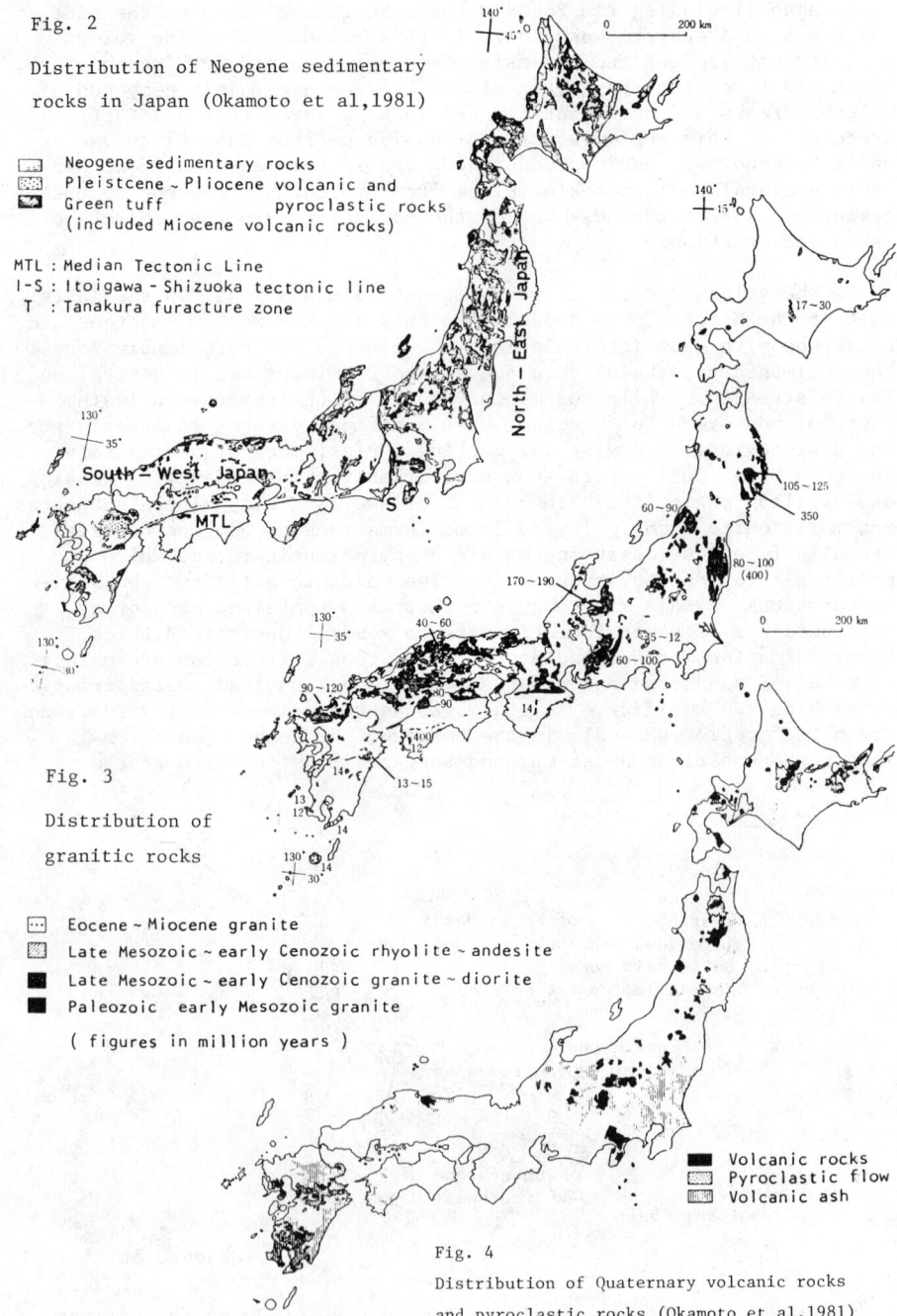

Fig. 2 Distribution of Neogene sedimentary rocks in Japan (Okamoto et al, 1981)

- Neogene sedimentary rocks
- Pleistcene ~ Pliocene volcanic and pyroclastic rocks
- Green tuff (included Miocene volcanic rocks)

MTL : Median Tectonic Line
I-S : Itoigawa - Shizuoka tectonic line
T   : Tanakura furacture zone

Fig. 3 Distribution of granitic rocks

- Eocene ~ Miocene granite
- Late Mesozoic ~ early Cenozoic rhyolite ~ andesite
- Late Mesozoic ~ early Cenozoic granite ~ diorite
- Paleozoic early Mesozoic granite

(figures in million years)

- Volcanic rocks
- Pyroclastic flow
- Volcanic ash

Fig. 4 Distribution of Quaternary volcanic rocks and pyroclastic rocks (Okamoto et al, 1981)

pyroclastic rocks. The degree of induration of these rocks is generally lower than that of the pyroclastic rocks in the lower formations of Green Tuff region although it increases as depth does.

In the Pleistocene epoch, the sea, once transgressed covering all over Japan, largely regressed and Japanese islands took the shape as they have at present but volcanism has continued mainly in Green Tuff region constructing numerous volcanic cones, calderas and lava plateaus on land. Among the volcanic rocks in Quaternary system, the typical soft rocks are low welded tuff and "Shirasu", defined in a Japanese term to mean white pyroclastic sediment such as non-welded pumice flow and pumice fall deposits including the secondary deposit transported from the original pumice deposits, which widely occur in Southern part of Kyushu Island.

These soft and weak rocks also raise various intricate problems not only due to their peculiar engineering properties (weak and very susceptible to erosion) but also due to their peculiar pattern of occurrence (almost always adjoined by highly welded jointed volcanic rocks).

ENGINEERING PROPERTIES OF WEAK ROCKS IN JAPAN

Sedimentary Rocks of Neogene Tertiary

Geological Structures. Sedimentary rocks of the Neogene Tertiary period are largely divided into green tuffs and non-green tuffs. Green tuff was effused from underwater volcanoes during the Miocene period and is one of the most outstanding rocks typically found in northwest Japan where recent orogeny, though to be still progressing, is the most active. The green tuff regions cover over half of Japan's surface area and the thickness can reach up to several thousand meters. Non-green tuffs are common in the Miocene-Pliocene sedimentary rocks lying on top of the green tuffs and in the sedimentary rocks on the Pacific Ocean side of Japan. The sedimentary rocks of Neogene Tertiary have relatively simple geological structures, but in the Uetsu fold zone in north-east Japan the anticlines are still upheaving and synclines subsiding, in other words, it is an active fold zone. The rate of rise and fall is approximately 1 mm per year.

Characteristics. Green tuffs have various physical properties ranging from hard gravels to clays with high swelling properties when immersed in water. Sometimes altered clay, that is solfatalic clay, has been produced by hydrothermal alteration. Solfatalic clay contains large amounts of natrium montmorillonite, and distinctly expands with water. This clay frequently found in Quaternary volcanic areas too. On the other hand, non-green tuffs are mainly sandstones and mudstones, and in general they are relatively softer but those which are sandy have low cohesion and collapse easily when exposed to water. Tuffaceous mudstones often undergo rapid decomposition due to repeated wetting and drying. In Table 1, the engineering geological characteristics of soft rocks, based mainly on sedimentary rocks of Neogene Tertiary, are compared with soils and hard rocks.

Lithification and Mechanical Properties. Lithification is the process by which "sedimentary deposits are hardened by the actions of compaction and cementation over a geological period". It sets the particles into a stable arrangement by means of accumulating loads due to continued sedimentation. The dehydration and the volume reducing which dominantly takes place during the initial stage of sedimentation. This stage is called consolidation. The latter is the 'bonding process' due to precipitation into the pores of cementing materials such as carbonates which dissolve into the water contained in the sedimentary deposits, or for deep locations the replacement and crystallization of minerals. With reference to the muddy sediments formed since the Miocene period to recent and found in the Boso Peninsula, the variation of properties due to increasing load is shown in the relation among porosity, unconfined compressive strength q_u, and maximum depth of burial. Lithification of strata is dependant on the past maximum depth of burial. The maximum depth of burial for sedimentary deposits older than diluvium differs to the existing depth due to the effects of crustal movements and it is estimated from the thickness of strata determined stratigraphically. Representing the lithification process from soil to rock as a relationship between porosity and strength, since the maximum depth of burial is difficult to measure, Fig. 5 is obtained. In Stage 1, there is a large change in volume due to dehydration and the rate of dehydration is almost the same as the rate of

Table 1 Change in geological elements of rocks with lithification
(Okamoto et al,1981)

Lithification		SOIL ①	SOFT ROCK ②	HARD ROCK ③
Stage Geologic age		Alluvium - Dilvium (upper)	Dilvium (lower) - Neogene	Paleogene - Palaeozoic
Characteristics of strata		Horizontal beds. The buried topographic surface formed during the ice age makes the discontinuous surface. Existing overburden and mechanical properties correspond.	Generally inclined beds. Mechanical properties are governed by depth of burial rather than present overburden.	Stratified but mechanical properties governed by geological discontinuities such as fault and joint rather than lithification.
Geologic discontinuities	Bedding plane	Usually no problems.	Anisoreopic but weakens as it becomes softer. Homogeneiety becomes a problem with alternating thin layers and steep slopes.	Anisotropy and heterogeneity develops in strength and deformation characteristics. Bedding planes are usually dealt with in relation to joints.
	Fault & shear zone	No deterioration of ground due to fault.	The scale of shear is generally small. Lithofacies variation due to slipping has greater effects on strength and deformation.	The scale of shear is sometimes large. Fractuation and agrilization are remarkable in sheared zone.
	Joints & crack	No joints and mechanical properties governed by lithology and/or soil quality.	Generally few joints and has little effect in mechanical properties and permeability.	Strength and deformation characteristic of rockmass are greatly affected by nature and amount of joints.
Weathering & alteration		Affects strength but deterioration minimal.	Accompanies deterioration but not apparent except for those due to land slides and solfataric alteration. Rapid decomposition after excavation.	Weathering and alteration are remarkable.

reduction of volume, but the strength does not increase appropriately. It can be considered as the stage when consolidation progresses. This stage is evident for muds but is not so obvious for sands. During Stage 2, the intergranular bond due to compaction increases for mud and the strength distinctly increases. As time passes, cementation is added. In case of sand, slow compaction continues. In this stage, the amount of mud and the degree of cementation sometimes causes extreme dispersion to the extent of lithification. In Stage 3, strength ceases to increase by compaction for both sand and mud, and cementation becomes the main lithification process. According to Hoshino (1979) the porosity become about 30% (void ratio, $e \cong 0.43$), and the volumetric compressibility becomes almost uniform and compaction terminates. Lithification of sediments reaches the limit during this stage and the properties of rock mass are controlled by geological discontinuities such as faults and joints.

MODULUS OF DEFORMATION: The properties of rock fragments which from rock masses have so far been discussed, but in case of properties of soft rock masses, it can be dealt with in a similar way to soils since it can be estimated to a certain extent from the nature of the rock fragments which form it. For example, in Fig. 6, the moduli of deformation for the rock masses and the rock fragments sampled from the test site are compared, and though joints and cracks in hard rocks cause a wide dispersion, the values are relatively close to each other

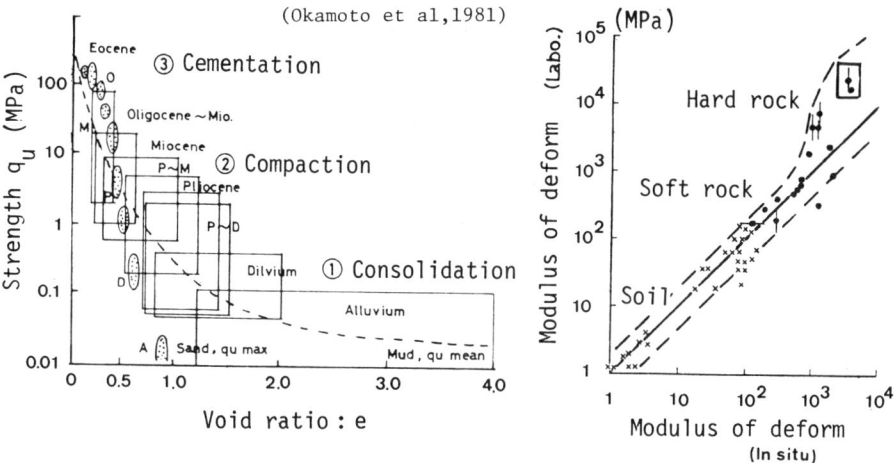

Fig. 5

Change in physical properties of Cenozoic sediments in Japan with lithification (q_u : unconfined compressive strength)

Fig. 6

Variation of elastic moduli in rock and rock mass with lithification of sediments

(x : by borehole test, • : by jack test)

in cases of soils and soft rocks. Also, concerning permeability of rock masses in Stage 1 - 2, clear tendency has been observed. The influence of pore water pressure, which extends to the mechanical properties of soft rocks, can be treated in the same way as for soils, but in case of having various water content, it is needless to say that, since the texture of the particles have been solidified by compaction, once this is broken down by swelling or slaking, the strength and the modulus of deformation distinctly decrease.

Weathered Granite

Distribution and Structure of Granite Rocks. Granitic rocks covers a wide area in Japan and makes up approximately 13% of the total land surface. The largest distribution of granitic rocks is found on the north side of the Median Tectonic Line in south-west Japan. The granitic rocks mentioned here include granite, granodiorite, granite porphyry, quartz porphyry and gneiss. Those formed during the late Mesozoic to the Cenozoic era are most common. The structure varies according to the formation process and the subsequent influence of diastrophism (Koide,1973).

Weathering Characteristics of Granite. Weathering of granite differs depending on the meteorological conditions such as rainfall and temperature as well as the degree of discontinuity developing in rocks. Also, thickness of weathered zone formed in this way varies with the intensity of erosion. The Japanese archipelago has experienced temperate to subtropical climates since the Tertiary period to date, and also much rain. For these reasons, weathered zone with thickness reaching 100 m are not uncommon. A typical example of mineral changes in granite is shown in Table 2 (Kinomiya,1975). For this, the thickness of weathered zone is 20 - 40 m, and this is classified into 7 zones by degree of weathering (rigidity,variation in mineral qualities etc.). Zone I is a typical fresh granite consisting of quartz, potash feldspar, plagioclase and biotite. Plagioclase and biotite become clayish as weathering progresses, transforming into chrorite, illite and hydrobiotite in Zone IV where hammering produces sandy materials, and into gibbsite in the final zone VII. The proportion of clay minerals increases as weathering progresses. Consequently, the crystallized water and the interlayer water contained in clay minerals also increase and variation such as percentage ignition loss are measured. The relation between porosity and ignition loss of weathered granite is given.

Mechanical Properties of Weathered Granite. As well as the relationship between RQD and deformation modulus obtained by pressure-meter (E_b), the one between porosity, which is an important index of the degree of weathering, and the deformation modulus E_{d1} (from the virgin curve) and E_b, obtained from plate loading test is shown in Fig. 7.

Weak Volucanic Rocks

The Origin and Characteristics of Weak Volcanic Rocks. The total number of volcanoes which erupted during the Quaternary period was about 200, and 64 are still regarded as active volcanoes. The lava, pyroclastic rocks and sediments produced by the volcanic activities include those rocks which are mechanically treated as weak rocks as follows. (1) Rapid chill facies of lava, (2) Pyroclastic rocks, especially loosely welded tuffs and pumice flows, and (3) Alteration facies. Pyroclastic rocks are widely distributed on a large scale, representing the characteristics of weak volcanic rocks. In particular they are often related to the volcanic actions which took place in the Quaternary period, discharging a massive amount of pyroclastic materials later forming calderas. There is not much variation of grain size distribution in respect to distance, but its character lies in the vertical variations of structure and strength due to the difference in welding intensities caused by a varying chill rate. The "Shirasu" in southern Kyushu is a typical pumice flow. These lithologic characteristics correspond to the mechanical properties, and the mechanical parameters such as strength and coefficient of permeability steeply change in the vertical direction although not in horizontal.

Mechanical Properties of Weak Volcanic Rocks. One of the mechanical properties of weak volcanic rocks is the horizontal uniformity and vertical variation of strength and permeability in respect to regional pyroclastic rocks, and the other is the severe lowering of strength and slaking due to alterations. The relationships among structure, strength and permeability cause a great deal of difficulty in cut-off operations. It is well known that strength of rock diminishes in becoming clay-like when affected by alterations. The results of tests to determine a limit of dehydration on instantaneous collapse under drying show that critical dehydration time upto immersion is 6 - 9 hours, and it means that slope protection works should be carried out as early after excavation as possible.

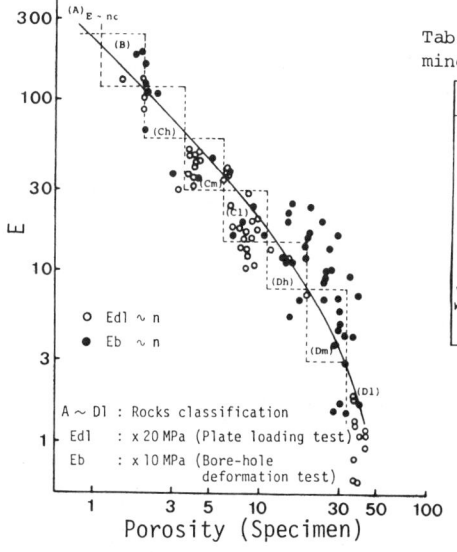

Table 2   Weathering and the change of mineral composition of granite

	zone I	zone II	zone III	zone IV	zone V	zone VI	zone VII
quartz							
K-feldspar							
plagioclase							
biotite							
chlorite							
illite							
hydro-biotite vermiculite							
kaolinе-minerals							
gibbsite							

(Kinomiya, 1975)

Fig. 7   Deformation modulus and porosity of granite (Okamoto et al, 1981)

PROBLEMS AND EXPERIENCES
IN DAM CONSTRUCTION

On Sedimentary Rocks of the Neogene

In Non-Green Tuff Region. The sedimentary rocks in this region are mostly sandstone and mudstone which are not well indurated even in the lower formations. As the shear strengths of those rocks are 0.3 - 0.4 MPa at maximum, the heights of concrete gravity dams are in the range of 20 - 30 meters (e.g. Toyofusa,Kameyama dams,etc. on early or middle Miocene deposits). The strength parameters required for design were obtained by in-situ direct shear tests so as to keep foundation rocks wet and fresh during tests. In construction stage the rocks disintegrated by alternate wetting and drying were removed with care. Dams 40 - 50 m high in this region are fill-typed generally. In the formations younger than the late Miocene, the degree of induration of sandstone is very low, scarcely being cemented and mudstone is also less indurated than older ones, allowing only earth and earth-rock dams to be constructed in most cases (e.g. Kori,Kuriyama dams). The permeability of these soft/weak rocks is generally low and the improvement (to reduce permeability) by grouting is not easily attained. Sedimentary rocks of the Neogene are in general not used as embankment materials or as aggregates for concrete dams as they are mostly unstable. Hence, except for homogeneous embankment dams, it is often to transport rock materials from relatively remote areas. However, in case of Azuma Dam (central-core-type,completed in 1970) in non-Green Tuff region of Hokkaido, mudstone was successfully employed in shell zones, although the upstream slope had to be covered with cobbles and boulders fixed with concrete frames to prevent the mudstone near the surface from disintegration.

In Green Tuff Region. Rocks are highly indurated and the topography where they prevail are precipitous, thus allowing relatively high dams to be built (e.g.Muromaki,Hoheikyo,Miyagase,Kawaji dams,etc.). Even relatively high arch dams have been constructed on this rock (e.g. Kawaji,Okumiomote,Hoheikyo,Shimouke,Susobana,Muromaki dams). The rocks in the formations of the early Miocene are fairly well indurated and show the characteristics more of hard rock rather than of soft rock, whilst those of middle Miocene are less indurated and, as is often the case with sedimentary rocks generally in Green Tuff region, mudstone with high slaking-swelling characteristics is intercalated. In case of Susobana Dam, the mudstone was fairly well indurated and showed a high Young's modulus ($E = 20,000$ MPa) in an intact condition, but it was quite easily deteriorated into mud once exposed to drying/wetting cycles. It is also common in Green Tuff region that mudstone susceptible to slaking/swelling occurs as interbedding of several centimeters to decimeters (e.g.Konade,Tatsumi,Aokata,Enaga,Arakawa dams). As mudstone of this type above ground water level is often deteriorated and aggravated by large deformation, area of high ground water level is preferred in dam site selection. It is not so difficult to find such area as sedimentary rocks in Neogene Tertiary are of low permeability in general. In Green Tuff region, a fill-typed dam is usually chosen.

WEAK ROCKS IN JAPAN

In contrast with those in non-Green Tuff region, the rocks mostly have a high tendency of slaking/swelling, hence it is a usual practice to take such special precautions as sprinkling water or covering the rock with wet mats during rock works to keep the rock as intact as possible.

As is the same with non-Green Tuff region, good concrete aggregates and competent rock fill materials are generally not easily available near dam sites in Green Tuff region. Even the highly indurated sedimentary rocks in older formations of Neogene Tertiary are not reliable in their durability, hence lava or intrusive rocks are mostly utilized as aggregates or rock fill materials. However lava is often altered into incompetent material. Although an intrusive rock is most competent and stable, it is not rare that its volume is not enough. Installation of a concrete grouting gallery underneath a high fill dam is strongly recommended in Japan but this is giving a new different problem on stress and deformation of a gallery, specially in mudstone.

On Weathered Granite

Granitic rocks are wide occurrence in Japan and one of suitable bed rocks for high concrete dams (e.g.Kurobe IV,Nagawado,Okutadami dams). Portions of abutment in high elevations are often weathered and sometimes behave like a weak rock. In this case, however, stresses and water pressures due to dam and reservoir are small in high elevations so that difficulties seldom occur. But if the granite is of coarse grained type, weathering often develops to deeper portion, and measures are sometimes required for such problems as altering a dam type and seepage control as are the cases with Hirose, Yamagami, and Naramata Dams. When a fill dam is selected for the weathered granite, seepage problem is far more important than the consideration of deformation and strength. Generally, permeability is higher in fairly weathered portions rather than heavily weathered ones, but usually fairly weathered portions are more easily improved in its permeability by grouting than heavy ones. Grout materials of fine particle such as colloid cement are often used for heavily weathered granite.

On Quaternary Volcanic Regions

Most of Quaternary volcanic rocks are terrestrial products. Therefore, of these Quaternary volcanic rocks, poorly welded volcanic rocks belong to soft/weak rocks, represented by poorly welded tuff and pumice fall deposits "Shirasu". Welded volcanic rocks ranges from highly welded one which is of high strength and has abundant cooling joints to poorly welded one which is of low strength and has few cooling joints. These rocks are usually low in specific gravity and porous. These rocks have not been consolidated enough by overburden so that the permeability of these rocks are generally higher than the Neogene sedimentary rocks. However, seepage flow through cooloing joints and nonconformable portions become problems more important than the seepage thro' rock itself. The maximum depth of a flow unit of Quaternary volcanic rocks seldom exceeds several tens of meters, being overlain or underlain by another flow unit or nonconformable deposit, which often gives difficulty in construction.

Cooling joints are open in highly welded volcanic rocks of the Quaternary. If the damsite is composed of highly welded tuff, it is general that cooling joints of high permeability exist even in deeper part of the foundation and abutment. The velocities of several to more than ten cm/sec (Midorikawa, Shimoyu, and Izarigawa Dams) are often measured at the field survey of seepage flow. Accordingly, in the highly welded volcanic rock regions of the Quaternary, grouting curtain is necessary for all the portions where their elevation is lower than the maximum normal water level.

On the contrary, poorly welded volcanic rocks show low permeability compared to highly welded ones. But it is difficult to improve their permeability by grouting, so that it is important to investigate permeability of rock itself in intact condition. The foundation of Izarigawa Dam consists of poorly welded tuff, and its abutment is composed of highly indurated welded tuff. The maximum normal water level was restricted below the lowest elevation of highly welded tuff. The permeability of the poorly welded portion showed the coefficient of around 10^{-4} cm/sec. In Iwase and Hikumi Dam, there are pumice fall deposits in the higher elevation of the dam. In these cases, little leakage was observed after impounding if the ground water table in the abutment had existed approximately as high as the maximum normal water surface before impounding, and a fair amount of leakage was often observed if the ground water table was below the maximum normal water level before impounding.

THE CURRENT STATE OF TUNNELING IN WEAK ROCKS

Formerly, the hilly regions, composed of soft/weak rocks, which is overlayed by distributed formations after the Tertiary, have been detoured around, or passed through by cuttings. However, with the upgrading of routes and in order to avoid the densely populated areas of the plain rims, recent years have seen a large increase in the number of tunnels constructed in these areas. Design and construction of tunnels in weak rock meets with various problems and requires a high degree of investigative work. For instance, there is the result of heavy rock load and large deformation encountered in tunnels with high overburden. There are also such matters to be dealt with as the liquefaction of soft/weak rock under high water pressure, or subsidence occurring with low heights of overburden.

Survey and Testing

Tunnels differ from other structures because of their underground siting, and geological survey with high precision cannot be expected, for survey and test concerning tunnels is done intermittently from design stage to construction stage, with a degree of accuracy which is intended to increase gradually. At planning stage a most favorable route is chosen and geological feasibility surveys are carried out in order to avoid major changes in cost or schedule to be needed later on. At construction stage, investigation including field measurement and monitoring are conducted so as to meet actual conditions. Seismic prospecting by refraction method is employed very conveniently through design and construction stages, and it should be pointed out that rock classification based on seismic rate is quite common in Japan. Since liquefaction of rock generated by ground water gives difficulties in case of confronting sand beds and sedimentary rocks with low solidity factor, tests and surveys on permeability, grain size constituents and inflow water are given a great attention. Where expansive rock pressure is anticipated, analysis and test on mechanical properties, consistency, slaking, clay minerals etc. are executed. If ground water inflow develops, tests on mechanical properties, unit weight and permeablity are undertaken. According to situations with geological difficulties, advance boring, drainage and/or grouting are employed, but works for examining galleries or large-scale in-situ testing is rare except tunnels accompanied by particularly large cross section or considerable expansive behavior.

Difficulties and Rock Classification

In particular, major difficulties are encountered during construction due to 'plastic ground pressure' found in regions of Tertiary mudstone and tuffaceous layers, as well as the outflow and extrusions from layers of Quaternary volcanic pyroclastics which have plentiful ground water. Difficulty is found in clearly explaining such problems on rock mechanics encountered in design and construction. This is because that actual behaviors of those base rocks produce complex interactions with supports. Also, troublesome characteristics of soft/ weak rock include as well, chemical changes caused by weathering and hydrothermal alterations, swelling in clay minerals, joints and bedding planes. Furthermore, the fault/highly fractured zones, overall shearing and inflow with abundant water are still major difficulties. A combination of all these factors is the cause behind failures, and it is necessary to establish geological survey with much higher accuracy and to advance such practical a research as to provide not only qualitative but also quantitative informations regarding behaviors of weak rock in-situ in an adequate way.

The major factors used in rock classification in Japanase tunneling are as follows. (1) Kind of rock, (2) Hardness/strength, (3) Weathering /hydrothermal alteration, (4) Degree and condition of fissuring, (5) Seismic velocity. The current classification has been applicable mainly to medium to hard rocks.

Drainage and Grouting Operations

Drainage Operation. In the past, drainage detour galleries and boring have been used as the most effective methods of dealing with the high pressure inflow which occurs fractured zones. Drainage galleries have a cross section of 4 - 10 m^2 and are situated on the outside of a face. Boring is a very common method. It frequently involves advance boring into a tunnel face to 50 m deep with a diameter of 100 mm. Further, 'deep well' and 'well point' are employed conveniently in case of having aggressive inflow. However, they cause environmental disruption such as drying up and subsidence.

<u>Grouting Operation.</u> Grouting is used effectively to obstruct water and to strengthen underwater tunnels where water pressure is extremely high. Cement products and cement waterglass products (LW) are availble grouting materials. The rocks surrounding the Seikan Tunnel were reinforced with LW-grouting to ensure safety of mining and to reduce inflow. Since the grouting area reached to 1 - 2 times of the diameter the distance of 5 - 10 m between a face and a grouting front was utilized as a protector. In this situation, the pressure of the final grouting took into account the ground and water pressures, that is 6 - 7 MPa.

Expansive Ground at Tunneling

In rock composed of Tertiary mudstone which contains large quantities of clay minerals, and in altered rocks, it causes extremely heavy load and expansive behavior. In tunneling through this rock, the side drifting and the bottom heading have often been carried out. Although the suitability of the side-drifting is adjudged through experiences, it is considered most effective to mobilize natural ground arch (bearing ring) successfully around an opening. Recently there has been a trend to employ rockbolts, shotcrete and flexible ribs as principal components for treating expansion. The Shin-Usami Tunnel is a recent and successful example concerning this method. The tunnel was constructed in parallel, at a distance of 50 m, with the old Usami Tunnel, subjected to extremely heavy expansion in 1930s.

BRIDGE FOUNDATION ON WEAK ROCKS

The foundation rock of the majority of the long-span bridge is soft, weak, sedimentary one, followed by weathered or altered rock. In planning stage, states of consolidation and weathering are discussed. Core drilling, in addition to standard penetration tests, is also employed in order to predict strength and deformation characteristics. Specially in case of underwater foundation, borehole-loading tests such as pressuremeter or Goodman jack tests are conveniently carried out. Following must be taken into account in geological investigations.
(1) As it is considered that weak rock is anisotropic and of a heterogeneous character, to constitute accurate geomechanical models is required based on appropriate geological surveys. (2) Especially where rock consists of Tertiary sedimentary rock, a distinct relation exists between laboratory tests and in-situ test results. (3) Elasto-plastic behavior of weak rock shows a remarkable time dependency.

In designing, physical properties such as static and dynamic deformation, shear strength and creep deformation are required. Estimating deformation characteristics depends mainly on plate loading test. For shear strength (C and ϕ), triaxial tests at laboratory and/or rock shear tests in exploratory adits. The strength parameters of weathered or altered rock, mainly granite, is generally C = 0 - 0.5 MPa, and ϕ = 30°- 35°. With regard to soft sedimentary rocks, they are C = 0 MPa, ϕ = 5°- 20°, and parameters for sandstone are C = 0 - 1 MPa, $\phi \geq$ 30°.

Bridge foundation types can be classified into three:'spread','pile' and 'caisson' foundations. As large foundations are situated on slopes in many cases, assessment of the lateral contact pressure of substructure under eccentric load (subjected to earthquakes) is important. Therefore, depth of foundation is extended by using either the Shinso piles (large cast-in-place piles) or the caisson by employing the reverse lining method. The special features of strait bridges, on alluvial deposits consisting of remarkably weathered and poorly consolidated rocks, are to have a large center span and high contact pressure. Generally, piles (e.g.a ganged-pier-typed) or laid down caisson foundations are selected for submarine.

Spread-typed Hirado Strait Bridge of 665 m long was built on Neogene sedimentary rocks covered by Quaternary volcanic rocks. In case of Kawaguchi highway bridge on Neogene sedimentary rock, tests by pressuremeter was conveniently employed to obtain the relation among reductions of C, ϕ and crack factor C_r (defined by $C_r = 1 - (Vp/Vpo)^2$, where Vp: P-wave velocity in rock mass, Vpo: one in rock specimen). Daikoku B. on sandy silt with N-value of 0 - 8 required to put steel pile foundations of large diameter because the bearing stratum lay at 60 m deep. Innoshima Bridge of 1,270 m long took a spread foundation on weathered granite. In case of Roppozawa B. on complicated formations consisting of Quaternary volcanic rocks, andesitic lava, pyroclastic rock, tuff breccia and tuff, because of a lack of sufficient bearing capacity at the abutments, movable and hinged supports were employed. For ShinKobe station, as an elevated bridge situated above 'Suwayama' active fault (with a 5 m wide band of fault clay),the alluvial gravel deposits have been dragged up about 70 cm over the fault plane, and assuming a uniform fault movement a rate of 0.1 - 1.0 mm/year was estimated.

SLOPE ENGINEERING

Landslides in Japan are often characterized according to three types: (1) 'Tertiary' type landslide refers to those found in regions of Tertiary strata including Green tuff, weakly consolidated rocks and coal strata. This is densely distributed in the northern part of Mainland along Japan Sea. (2) 'Fracture zone'(or,'Mesozoic,Paleozoic and metamorphic') occurs mainly in areas where rocks are shattered along a tectonic or fault line (i.e. Fossa Magna, Median Tectonic Line). Most cases of this type occur in Sambagawa cryatalline schist and Mikabu green rocks with intrusions of serpentinite, distributed along Median Tectonic Line. (3) 'Hydrothermal'(or,'Volcanic rock') type occurs in areas where volcanic rocks are altered and argillized to form a special clay layer called "Onsen-yodo" by hydrothermal process.

In 1977, the Ministry of Public Works investigated dangerous areas for landslide. Of 5,600 sites investigated, 'Tertiary type' areas accounted for about 68%, 'Fracture zone type', 28% and the remaining was 'Hydrothermal' one. Tertiary type trends to occur frequently around anticlinal axes and dome structures, coming folds caused by tectonic movement accompanying tension fractures.

It is possible to estimate the position of the potential sliding surface, while strength and pore pressure at the sliding surface can be evaluated. Also, slope stability can be judged accurately by measurement. However, it is difficult now to forecast landslides and slope failures caused artificially in Tertiary areas. In such areas, a sliding surface may be formed on a bending plane, shattered fault zone, in rocks deteriorated by weathering.

The most important objectives of investigations of slope stability are strength, permeability of ground and underground water conditions. It is also essential to confirm location and configuration of weak strata or sliding surface.

Works for landslide control and slope failure prevention are generally classified into two categories: 'suppression' and 'inhibition' works. Suppression works are intended to remove causes of a possible landslide or slope failure. These are : surface drain, drain work, interceptor drain/wall, groundwater drain(horizontal/oblique drain drilling,drainage wells of liner plates, drainage wells of ferroconcrete, drainage tunnels), river structure(consolidation dam,revetment,groin), slope concrete frame, grouting, rockbolting, and so on. Inhibition works are done for providing stress against a thrust caused by rock and soil so as to prevent a landslide or a slope failure. Those are : pile works (steel/PS piles), cast-in-place piles, anchoring, excavation, counterweight filling, retaining wall (concrete retaining wall, crib, gabion, tied-back wall by combination of steel piles and steel sheets), rock shade, and so on.

REFERENCES

Hoshino,K.,1979,'Change of mechanical properties during solidification of sedimentary rocks," Rock Mechanics in Japan,Vol.III, Japanese Committee for ISRM, pp.23-25.

Iida,R.,Nakano,R.,and Matsumoto,N.,1981,"Experiences and problems of construction of dams on soft rock foundation in Japan," Proc. of Int.Symp. on Weak Rock,Tokyo, Vol.5, pp.104-111.

JSCE,Working Group on Tunneling,1981,"The current state of soft rock tunneling in Japan," Proc. of Int.Symp. on Weak Rock,Tokyo,Vol.5, pp.112-122.

JSCE,Working Group on Bridge Foundations,1981,"Recent experience in bridge construction on weak foundation rock in Japan," Proc. of Int.Symp. on Weak Rock,Tokyo,Vol.5,pp.123-134.

Kikuchi,K. et al,1981,"Recent development of slope engineering of soft rock in Japan," Proc. of Int.Symp. on Weak Rock,Tokyo,Vol.5, pp.135-146.

Koide,H.,1973,"Land of Japan," Tokyo University Press, pp.15-23.

Kinomiya,K.,1973,"Tensile strength as a physical scale of weathering of granitic rocks," J. of Geol.Soc. of Japan, 81, pp.349-364. (in Japanese)

Okamoto,R.,Kojima,K.,and Yoshinaka,R.,1981,"Distribution and engineering properties of weak rocks in Japan," Proc. of Int.Symp. on Weak Rock,Tokyo,Vol.5, pp.89-101.

Chapter 100

SOME APPLICATIONS OF ROCK - ENGINEERING TO GEOTECHNICAL PRACTICE

Issa S. Oweis, Converse Consultants, Inc., Caldwell, NJ
Walter W. Lilly, The Austin Company, Roselle, NJ

Application of rock engineering in geotechnial consulting practice considered herein is in areas of: (a) bedrock verification for foundation support, (b) bearing capacity and settlements, and (c) excavations.

A routine geotechnical investigation usually provides data on depth to rock, rock classification, and rock quality. In cases of rock outcrop, such outcrops are mapped for geologic structures and rock mass description. In addition, the regional geology and the general aspects of the groundwater regime are usually known from either published literature or previous experience. The predominant technique for ground and site investigation are usually drilling and geologic reconnaissance. Other techniques involving seismic and geophysical soundings, borehole camera, and oriented cores are used by by no means routine. Performing laboratory strength and compressibility tests is often made to be considered as routine. The use of the point load test is gaining increasing use in the United States and would, perhaps, become a routine field or laboratory test due to its simplicity and, more importantly, its relatively low cost. The purpose of this paper is to describe cases where mostly routine investigation and testing were used to arrive at engineering decisions.

Bearing Capacity and Settlements

The allowable bearing values for foundation support on rock is usually dictated by local building codes which presumably should reflect local experience. The interpretation of the code, however, can present a problem, especially if the interpretation is made by a designer under the delusion of knowledge in rock engineering. For example, the provisions of the New York City Code allow 5750 kN/m^2 (60 tons per square foot TSF) bearing on Hard Sound Rock, 3830 kN/m^2 (40 TSF) on Medium Hard Rock, 1916 kN/m^2 (20 TSF) on Intermediate Rock, and 766 kN/m^2 (8 TSF) on Soft Rock.

The above bearing values are in many cases based on examination on exposed rock at footing bottom without consideration for possible presence of softer rock at some depth below footing, and the criterion for acceptance for 40 TSF bearing is contingent on whether the rock "rings." In some cases, this approach may be safe, especially when rock is excavated over an area much larger than the footing and rock walls can be examined. However, in many cases, this approach may be too simplistic as discussed below.

Figure 1 shows a subsurface profile at the location of one of the footings for a proposed 40-story structure in New York City hereinafter called Site A. Rock is Manhattan Schist, and the foundations for the building had to be installed in the basement of an existing building. There are no rock outcrops nearby, and the contractor elected to excavate locally for footing with no blasting. Thus, borings were drilled and NX rock cores were recovered at essentially all locations of footings. In addition to RQD, it was possible to assign a rock mass rating as proposed by Bieniawski (1974, 1976). For example, the hard Schist from 3.1 m - 4.33 m (14 - 18 feet) in Figure 1 was assigned a rating of 11 for hardess, 17 for RQD, 10 for jointing, 12 for conditions of the jointing, and 0.0 for water conditions. Thus, the overall rating is 50, and according to Figure 2, the in situ modulus is essentially zero based on extrapolation of the linear relationship shown on the Figure. Thus, more data are needed to revise Figure 2 for application to rocks with ratings of 0 to 50. Raphael and Goodman (1981) reported a modulus of 6 GPa and a rating of 33. For Site A, the Young's modulus for hard rock based on uniaxial compression tests or intact specimens varied from 6.9 GPa (10^6 psi) to 20.7 GPa (3×10^6 psi). From Figure 3, a reduction factor of 0.15 was selected to estimate the in situ modulus, and based on this, a modulus range 1.0 GPa to 3.0 GPa with an average of 2 GPa were used for estimating settlement. In order to attempt to improve the estimate of in situ modulus, the rock modulus was backcalculated from four pile load tests on the Manhattan Schist at other sites within the City. The backcalculated modulus varied from 0.5 GPa to 2.34 GPa. Based on this, laboratory compression tests on intact specimens, RQD, rock core examination and classification, modulus values were assigned depending on rock conditions. For example, layers similar to that shown in Figure 1 (4.33 m - 7.33 m) were judged to behave in a soil-like manner with modulus varying from 0.02 GPa - 0.08 GPa (200 TSF - 800 TSF) depending on location.

Based on the above analysis, settlements were evaluated for many footings using Figure 4. At footing locations where the profile is depicted in Figure 1, it was concluded that bearing pressures of 3,830 kN/m2 (40 TSF) would result in settlements larger than 12.9 mm (0.5 inch). The criterion for settlement was dictated by the

APPLICATIONS OF ROCK-ENGINEERING

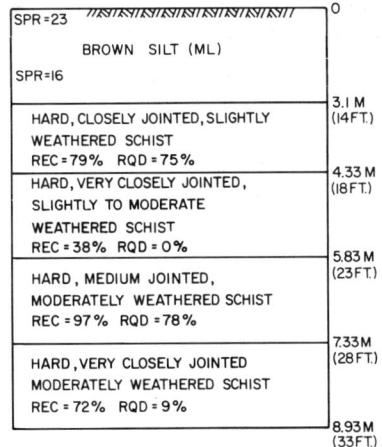

Fig. 1. Subsurface Profile, Site A

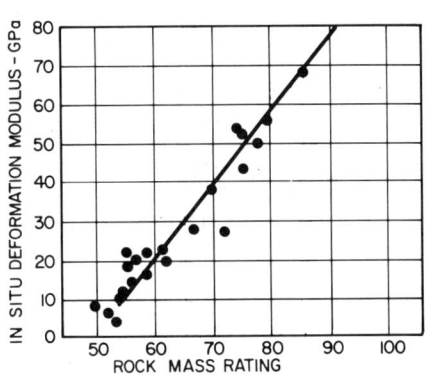

Fig. 2. Rock Mass Quality and in situ Deformation Modulus (Bieniawski, 1976)

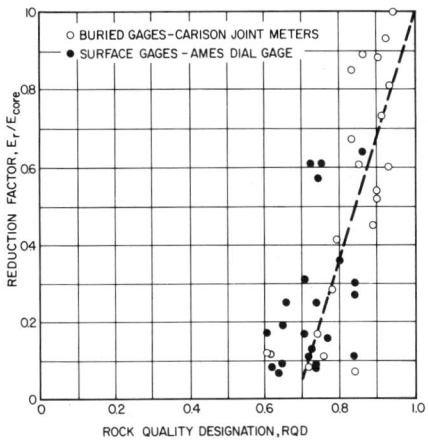

Fig. 3. Reduction Factor and Rock Quality from Plate Jacking Tests (Hall, et al, 1974)

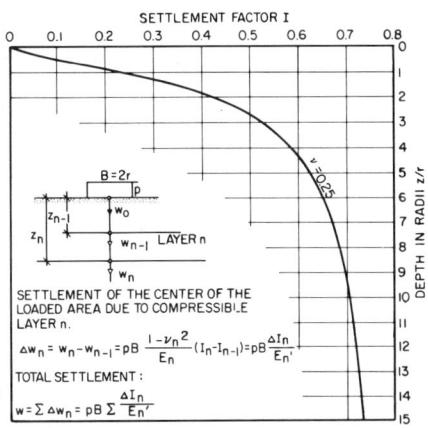

Fig. 4. Analysis of Deflection (Vesic, 1963)

designer. Had the footing been exposed without the benefit of borings and analysis, it is most likely that the rock at footing location depicted by Figure 1 would have been judged suitable for 3,830 kN/m² (40 TSF) bearing.

It is usual in geotechnical practice to assign a performance criterion for foundations in terms of allowable bearing dictated by anticipated settlements. The authors see no reason why this same practice should not be applicable to foundations on rock. This may be complicated by the fact that case histories for settlement on rock are meager and the reliability of estimating settlement on rock is perhaps less than that for soils.

Another important aspect of rock engineering which is in a somewhat state of confusion is the allowable loads for piers (or piles) socketed into rock. The usual practice is to compute the capacity as the sum of end bearing and shaft resistance, and appropriate factors of safety are applied to both components of resistance. The New York City Code allows 1.38 MPa (200 psi) as allowable adhesion between rock and concrete for Sound Hard Rock.

Horvath and Kenney (1979) proposed the following relationship for estimating the average ultimate shaft resistance (S_r) and unconfined compressive strength f'_w.

$$S_r \text{ (MPa)} = b \sqrt{f'_w} \text{ (M Pa)}$$

b = 0.2 - 0.25 for diameters larger than 400 mm (16 inches)

  = 0.25 - 0.33 for diameters smaller than 400 mm

f'_w = unconfined compression strength for rock or concrete, whichever is smaller

The unconfined compressive strength for the Sound Manhattan Schist can be in the range 60 - 150 MPa (8,700 psi - 22,000 psi). Based on the above expression, the average ultimate shaft resistance may range from 1.55 MPa to 4.0 MPa (225 psi - 580 psi). Thus, the factor of safety can be as high as 3 and as low as 1.1, depending on how sound the rock is.

Another problem in socket design is to decide on the portion of the applied load carried by friction and end bearing. Several investigators pointed out that the portion of the load carried by end bearing decreases as length of socket increases (see for example Figure 5 and Donald, Chiu and Sloan, 1979). The authors used Figure 5 for evaluating caisson socket requirements for Site D (see following section). It is not clear how fractures and weathered seams in rock affect the local transfer depicted in Figure 5. The authors interpret E_r to mean the mass (in situ) modulus of rock (see Figure 3), and socket areas determined based on actual rock contact with stringent requirements on cleaning the socket before concreting. Williams and Pells (1981) contented that sub-horizontal seams of extremely weathered rock or clay within the length of a socket would not significantly affect the interface shear resistance along the rest of the socket provided that total thickness of such seams is less than 15% of socket length, and seam material is not smeared over the remainder of socket wall.

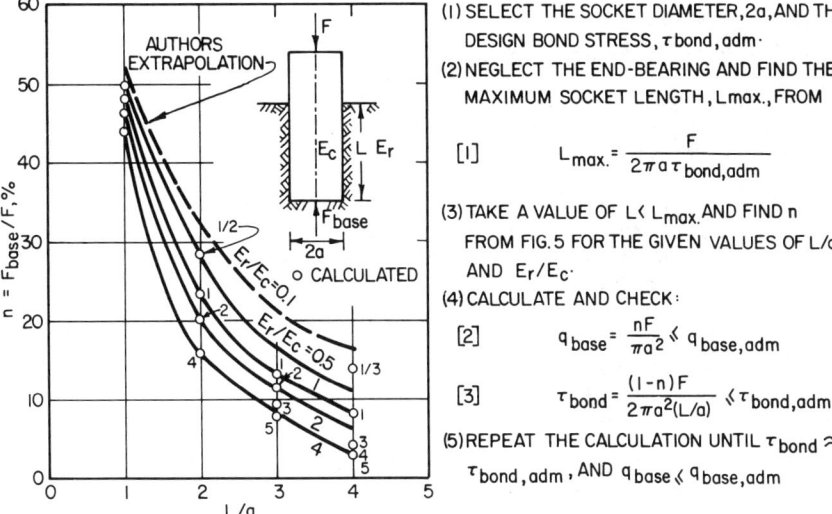

Fig. 5. End Bearing Ratio, for an Elastic Behavior of Rock and Concrete. (After Ladanyi, 1977)

It is clear from the above that the selection of design parameters for footing and socket design should be based on properties of site specific geologic materials, and selection of a parameter specified by the Code without verification may lead to a design with unknown margins of safety. Furthermore, such estimates can be made based on routine investigations and testing involving (a) rock description and classification, (b) RQD, and (c) compressive strength and Young's modulus of intact rock on NX size cores. If the point load test (Broch and Franklin, 1972) (Wijk, 1980) is used in lieu of the uniaxial compressive test, few uniaxial compression tests should be made to establish a correlation of the ratio of the compressive strength to the point load splitting strength. Average ratio of 24 have been reported (Bieniawski, 1974). However, the ratio is expected to depend on rock type. Ratios near 12 were determined for the Manhattan Mica Schist.

Excavations

The primary considerations for excavation in rock are:

a. Evaluating ease of excavation (rippability) and need for blasting.

b. Evaluating the stability of cuts and need for excavation support, and

c. Effect on nearby or adjacent facilities.

Ease of Excavation

Ease of excavation is usually evaluted in terms of one or combination of three broad techniques.

1. Trenching: Test trenching in rock is attempted at several locations using heavy equipment anticipated for use during construction.

2. Seismic Velocity Measurements: Many equipment manufacturers correlated the ripping capacity of their equipment with seismic velocity usually obtained from seismic refraction (e.g., Caterpillar Tractor Co., 1975).

3. Retrieval of rock cores and evaluating ease of excavation in terms of rock classification and testing: This is based on the observation that the tensile strength of rock and fracture frequency influences the rippability of rock (e.g., Dubbe, 1974).

Figure 6, allows assessment of rock rippability based on data which can be obtained from routine investigations using rock coring and point load testing either in the field on the laboratory. It is always preferable to combine this technique with other techniques such as trenching and/or seismic refraction.

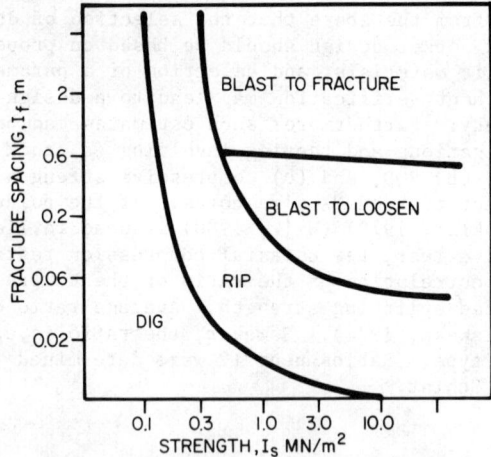

Fig. 6. Ease at Excavation Based on Fracture Spacing I_f (Meters) and Point Load Strength Index I_s (MN/M^2) (Franklin, et al 1971)

Stability of Cuts

In cases where rock outcrops can be found near the site, or where experiences for nearby sites are available, a preliminary pre-excavation evaluation can be made of the staility of the cut. Based on this, a preliminary design of the support anchoring system is made.

APPLICATIONS OF ROCK-ENGINEERING

As each lift of rock is blasted and excavated, the characteristics of rock exposed, the nature of discontinuities exposed from previous blast are observed and mapped. Figure 7 presents the attitude of foliational and joint/fracture as mapped after the first lift for an excavation in the Manhattan Schist rock in New York City, hereinafter referred to as Site B.

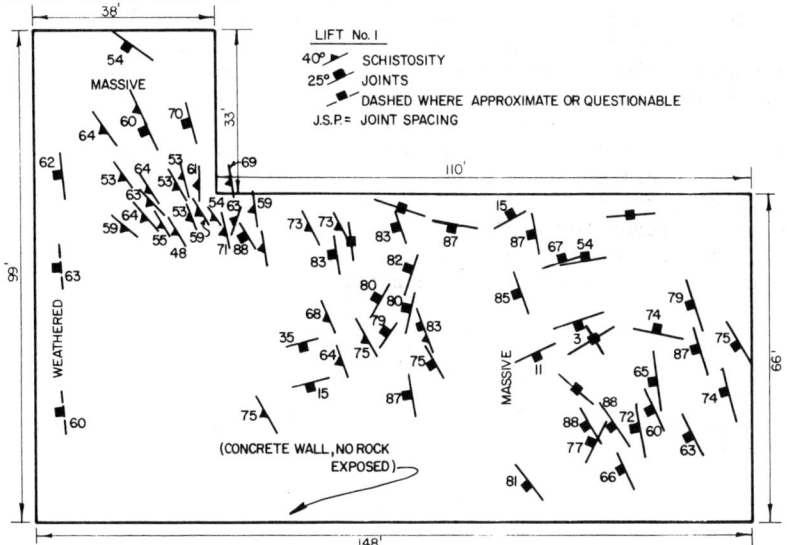

Fig. 7. Geologic Mapping-First Lift, Site-B

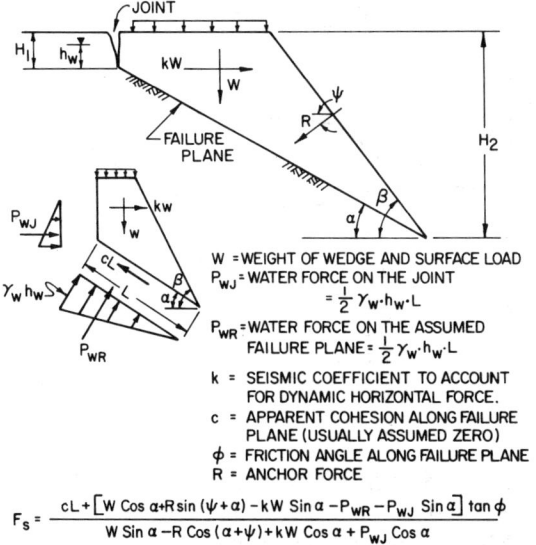

W = WEIGHT OF WEDGE AND SURFACE LOAD
P_{WJ} = WATER FORCE ON THE JOINT
$= \frac{1}{2} \gamma_w \cdot h_w \cdot L$

P_{WR} = WATER FORCE ON THE ASSUMED FAILURE PLANE = $\frac{1}{2} \gamma_w \cdot h_w \cdot L$

k = SEISMIC COEFFICIENT TO ACCOUNT FOR DYNAMIC HORIZONTAL FORCE.
c = APPARENT COHESION ALONG FAILURE PLANE (USUALLY ASSUMED ZERO)
ϕ = FRICTION ANGLE ALONG FAILURE PLANE
R = ANCHOR FORCE

$$F_s = \frac{cL + \left[W\cos\alpha + R\sin(\psi+\alpha) - kW\sin\alpha - P_{WR} - P_{WJ}\sin\alpha \right] \tan\phi}{W\sin\alpha - R\cos(\alpha+\psi) + kW\cos\alpha + P_{WJ}\cos\alpha}$$

Fig. 8. Stability of Anchored Rock Slope

Analysis for rock anchoring requirements is usually made utilizing a two-dimensional model (Figure 8) if the strike of particular joint is reasonably parallel to the side of excavation. A three-dimensional model (Hendron, Cording and Aiyer, 1980) (Hoek, Bray and Body, 1973) is usually necessary when additional anchors are needed to stabilize a rock wedge before excavating the next lift after rock mapping similar to Figure 7 is made and mapped joints are anticipated to "daylight" upon removing the next lift. The analysis should be made after mapping following each lift to evaluate if more anchors are needed to stabilize a particular area of the rock face.

Because of the uncertainties in parameter selection and the behavior of rock wall during excavation and blasting, it is advisable to monitor rock movement during construction. This can become especially important for a site such as Site B where multi-story structure abut the excavation from two sides, and major streets exist on the two sides. Figure 9 shows a partial record of extensometers

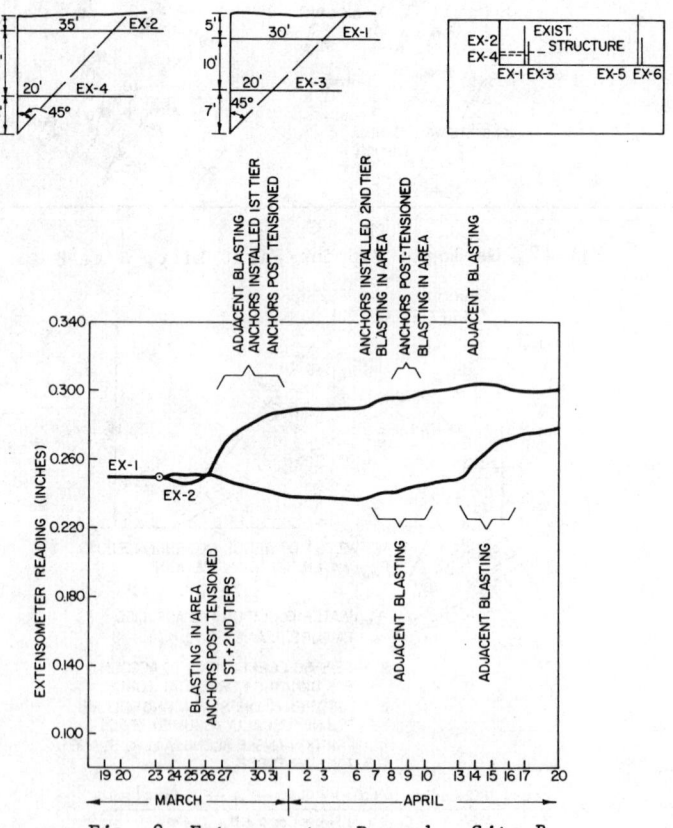

Fig. 9. Extensometer Record - Site B

installed to provide some warning of impending rock movement. As indicated, blasting contributed to movement of rock at two locations of the extensometers, and rock was reinforced in time to mitigate further movements. Figure 10 shows the record of the peak particle velocities measured during blasting of Site B and two other sites for comparison. All values for Site B were recorded in basements of structures abutting the excavation. The upper bound curve in Figure 10 is more conservative than compared to the relationship presented by Hendron (1980) for the World Trade Center. This is perhaps due to the more confined configuration at Site B. The three data points presented for Brunswick shale in Central New Jersey are for blasting in partially excavated small basement areas. The data for blasting in dolomites are for trenches in an open site in Western New Jersey (identified as Site D in this paper).

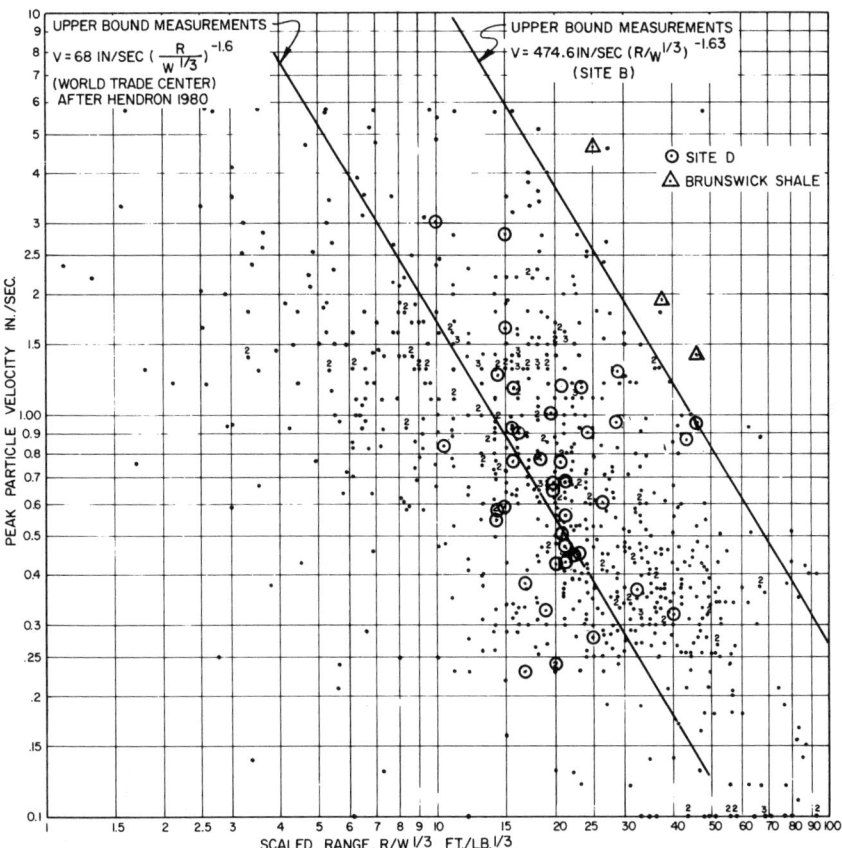

Fig.10. Peak Particle Velocity Record–Site B and other Sites

As evident from Figure 10, many particle velocities exceeded the specified 2-inch/second for Site B. No damage was observed. Most of the high velocities were measured when blasting was at a close distance from the structures, and the spectral velocity was perhaps below the level necessary to cause damage. Figure 11 shows average velocity response spectra for blasting rock motion. It is clear that for a short distance from the source of blasting and for low predominant structural frequency, the amplification for a nearby blast is less than blasts far away. Thus, a distance dependent blasting criterion compatible with structures to be protected is sensible and can be economical. Establishing such criterion requires non-routine investigations and analysis to determine the response characteristics of structures from field measurement.

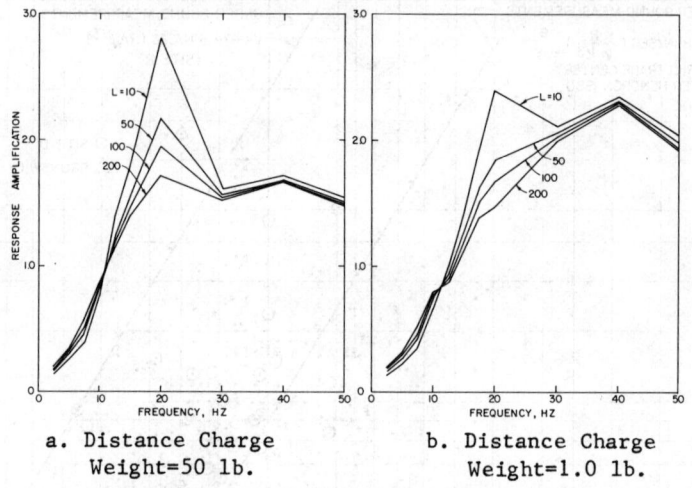

a. Distance Charge Weight=50 lb.

b. Distance Charge Weight=1.0 lb.

Fig.11. Variation of 5 Percent Pseudo Relative Velocity Response Spectrum with distance (Deduced from Madearis 1979)

Probabilistic Assessment - Cavernous Limestone

While the probabilistic approach is not used by the authors as a design tool, it has been used in the planning of exploration programs and indirectly, to interpretate field data. Probabilistic models, such as Poisson's distribution, can be used to make rough and preliminary judgements based upon geological maps, aerial photographs and field mapping programs. For example, considering investigation of a limestone terrain where a number of discrete events (sinkholes) occur randomly over a large area, with the random variable being the frequency of event occurrence; the distribution of local active site populations on a land surface with respect to Poisson's equation can be examined:

x = the number of events occurring in the unit area

u = the average number of events

$$P_u(x) = \frac{e^{-u} u^x}{x!}$$

where $P_u(x)$ is the probability that event "x" will occur when the mean or expected value of x is u.

At several construction site locations this model was used based upon surface features indicating subsurface solution activity. The model results (Expected Events) are given in Table 1, together with actual occurrences.

TABLE 1 - Preliminary Probabilistic Assessment for Sinkhole Development

Site	Location	Active Construction Area	Expected Sinkholes	Actual Occurrences of Sinkholes
C	Allentown, PA	2 acres (0.81 hectares)	0	1
D	Western NJ	7 acres (2.8 hectares)	1	8
E	Allentown, PA	33 acres (13.4 hectares)	4	3

Another distribution free model arises in the estimation of the abundance and size of cavities from a series of borehole logs obtained during an investigation. An estimate can be made of the probability of a boring missing sinkholes completely with the following known conditions (Lumb, 1967): boreholes of length L are located at rectangular grid points at spacings B times D. The expected numbers of cavities in the volume V = BDL is λV where λ is the mean number of cavities per unit volume. The probability of one drilled hole missing all sinkholes is:

$$P = (1-a/BD)^{\lambda V}$$

The possibility of missing one sinkhole is $(1-a/BD)$ where a is the projected area of one cavity and for less than 50% probability of missing a cavity, the grid spacing for borings should be less than 1.5 times the diameter of the cavity which is, in most cases, impractical. Because a/BD will always be a very small quantity. The above equation can be approximated by:

$$P = \exp(-3pL/2d)$$

where: d = diameter of assumed spherical cavity
L = length of borehole
$p = \lambda v$ = mean number of sinkholes per unit volume
v = mean volume of one cavity

Thus, the probability of missing all the sinkholes is independent of the borehole spacing but dependent on the ratio of borehole length to void size.

Site D offered the opportunity to utilize this model within a selected construction area of 4.4 acres (1.8 hectares) and where 33 boreholes were drilled into residual soil and cored into carbonate bedrock with NX diameter cores. From the borehole logs, mean volume of cavity per unit volume was estimated as p = 23.2/1,584 = 1.4%

where: total length of boreholes = 1,584 ft (483.1 m)
total length of sinkholes encountered = 23.2 ft (7.1 m)

The esimated cavity size $d = \dfrac{(3 \times 23.2)}{(2 \times 5)} = 6.96$ ft (2.1 m).

where: 23.2 = total length of cavities
5 = total number of cavities encountered

Therefore, the probability of not encountering a sinkhole is:

$$P = \exp \dfrac{(-3 \times .014 \times 48)}{2 \times 6.96} = 86.5\%$$

Because of extensive use of protective casing during drilled caisson construction, it was not possible to measure, in detail, the sinkholes encountered during construction. Since one caisson in 1.23 did not encounter a sinkhole, the actual probability was 81%, which gives a reasonable agreement with the predicted.

The probablistic approach based on surface features used at construction Sites C, D, and E provided an economical method when, limited time was essential, for planning foundations and site preparation, however, these approaches were of little value in determining conditions during construcion as indicated by the variations of predicted to that encountered.

Bedrock Verification

In areas where rock is not cavernous, assignment of bearing capacity and prediction of settlement can be usually made by techniques described under "Bearing Capacity and Settlements." However, in cavernous rock, the presence of cavities should be accounted for.

APPLICATIONS OF ROCK-ENGINEERING

Some of the simple analytical models usually considered are shown in Figure 12. The apparent crude models provides, in most cases, reasonable tools for assessing safe bearing on cavernous rock. The models help explain the fact that large sinkholes can be masked if the soil cover possesses significant cohesion (e.g., glacial till).

As shown in Figure 12a, the beam bending theory is used for evaluating the minimum rock cover over a cavity for foundation support. In fractured rock the bending resistance may be relatively small and roof support is provided by arch action, as illustrated in Figure 12e. It is usual to consider safety factors of 5 or more with the type of analysis depicted in Figures 12a and 12e.

Rock probe programs often are used in solution susceptible bedrock in conjunction with core borings to verify the presence of cavities.

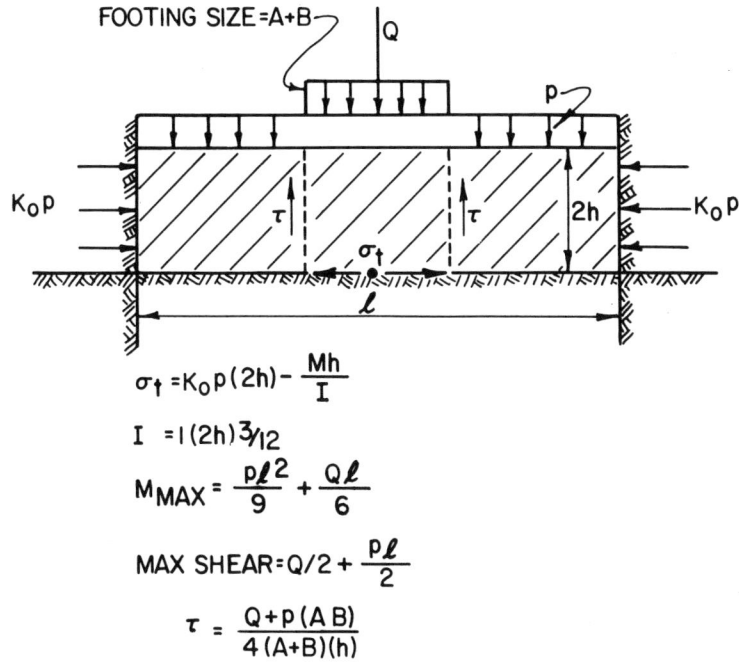

$\sigma_t = K_o p(2h) - \dfrac{Mh}{I}$

$I = 1(2h)^3/12$

$M_{MAX} = \dfrac{p\ell^2}{9} + \dfrac{Q\ell}{6}$

MAX SHEAR $= Q/2 + \dfrac{p\ell}{2}$

$\tau = \dfrac{Q + p(AB)}{4(A+B)(h)}$

a. Evaluating Cover Requirements Based on Beam Theory

Fig.12. Simple Analytical Models for Evaluating Foundation Performance on Caverneous Rock (Cont.)

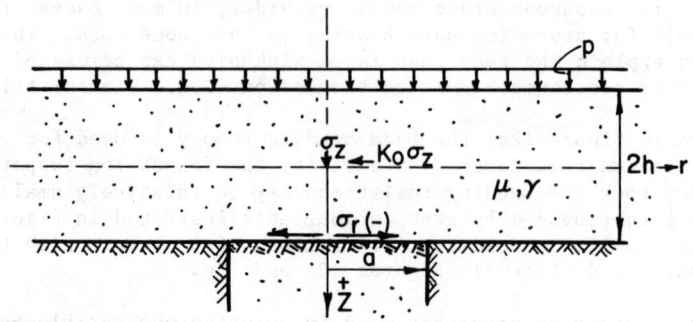

FOR $\mu = .3$

$(\sigma_r)_{r=0} = p(K_0 + 0.17\, Z/h - 0.29\,(Z/h)^3 - 0.31\, \dfrac{a^2 Z}{h^3})$

b. Solution for Simply Supported Circular Plate (Timoshenko, 1934) with Addition of in situ Lateral Confining Stress

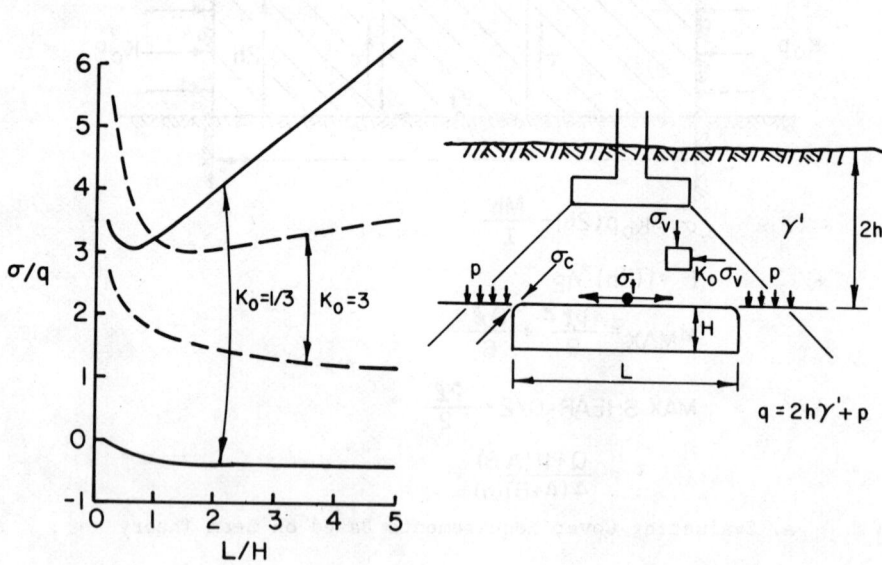

c. Roof Stress Based on Elastic Theory (Coates, 1978)

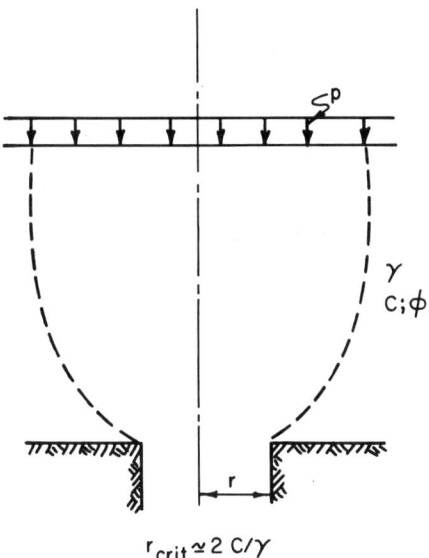

d. Arching of Soil (Terzaghi, 1943)

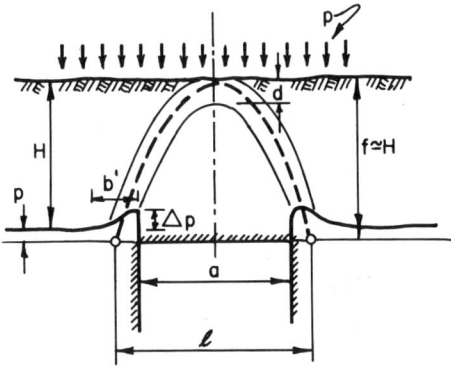

(f) REQUIRED $\approx (1.25 \ell/\sqrt{n})\, F_s$

n = RATIO OF ROCK COMPRESSIVE STRENGTH TO ARCH LOAD p
F_s = FACTOR OF SAFETY
ℓ = 1.25 a FOR HARD ROCK TO 2.5 a FOR SOFT ROCK

e. Arching Action of Rock Over a Cavity (Vassilopoulos, 1973)

The authors, as have many others, found this method to be economical, especially beneath founding levels of footings or caissons to identify soft rock, clay pockets and voids. In addition, probes generally have the advantage of speed.

At Site D, bedrock was identified as Cambro-ordovician carbonates, consisting primarly of dolomites. The correlations between air track probe penetration rates and rock conditions in situ were conducted using a Gardner-Denver air-track drilling rig at selected locations where the in situ conditions of the rock could be visually inspected. On this basis, the in situ rock conditions were classified for the purpose of caission construction. Figure 13a shows a correlation of drilling rate (95 psi at 1,800 rpm) with RQD. The correlation is based on probe rates determine during caisson construction and borings located not more than 5 feet away from probe location. Criterion for proof testing of rock below the foundation element ranged between a minimum of 10 feet to 20 feet to determine the presence of voids or seams of soft loose materials.

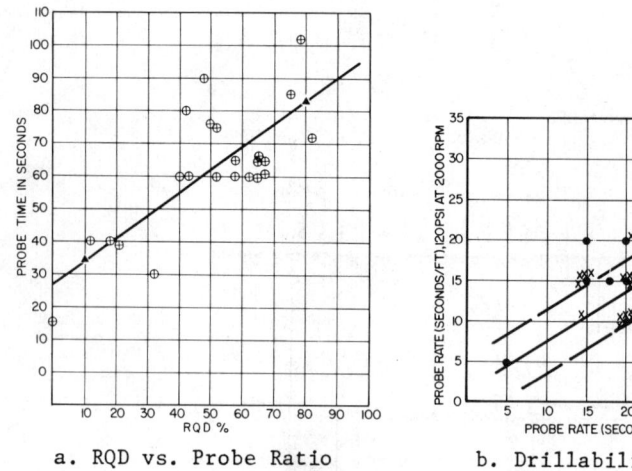

a. RQD vs. Probe Ratio

b. Drillability vs. Probe Pressure and RPM

Fig.13. Evaluation of Rock Drillability vs. Quality-Site D

Penetration rates can vary depending on drilling conditions. The authors encountered this problem when difficulties arose at Site D due to the presence of water and mud. Because of the difficult condition encountered the pressure was increased to 120 psi and the speed to 2,000 RPM to improve operating performance. This resulted in lower probe rates masking the previously defined correlation for rock quality. The two rates were then evaluated at selected locations under identical simulated conditions to reduce the introduction of variables. The new time rates were logged. A regression analysis was performed correlating 40 identical sets of probes (Figure 13b). The result of the analysis indicated that lower rates at higher operating pressure and RPM produces a rock quality reliability equal to or exceeding that corresponding to lower pressure and RPM.

APPLICATIONS OF ROCK-ENGINEERING

During the caisson construction for Site D, continuous observations of subsurface conditions were conducted, logged and selected manifestions were photographed. There were numerous voids encountered during the drilling, most all of which were tubular in shape, ranging from 3 inches in diameter (7.62 cm) to 12 inches (30.5 cm) in diameter. Experience at this site indicates that many of these voids and subsequent sinkhole collapse features were enhanced by on-going drilling activity for the caissons. The complex rock structure also caused drifting during socket and rock drilling which was corrected by revising the drilling procedure to include a smaller diameter pilot hole.

Techniques other than probes have been used with mixed success for cavity detection in cavernous rock. These include gravity and magnetic methods, electric resistivity and cross hole seismic velocity measurements (Rubin and Fowler, 1978) (Higginbottom, 1970) (Millet and Moorhouse, 1973). The purpose of these are usually to detect anamolies which are in turn interpreted for cavity presence. At Site D, referred to above, a subsurface interface radar survey was conducted. The survey was limited to a portion of the active construction site containing 0.31 acres (0.13 hectares) where a high density of solution activity was suspected. The results of the survey when compared with actual construction experiences identified only 39% of the cavities encountered. It is believed that the presence of wet clayey soil above bedrock made interpretation uncertain.

SUMMARY

Several project case histories are presented where some aspects of rock mechanics in practice are presented with regard to bearing capacity and settlements, excavations, bedrock verification, and application of probabilistic models to predict sinkhole development. From these applications, several issues are identified which are of significant practical interest. Among them are: (a) the need to link allowable bearing on rock to permissible settlement and reduce guess work and narrow interpretation of building codes, (b) improve techniques for estimating in situ rock modulus using routine testing and, (c) improve techniques to characterize the frequency content of rock motion generated by construction blasting.

REFERENCES

Bieniawski, Z.T., 1974, "Geomechanics Classification of Rock Masses and Its Application to Tunnel Support," Rock Mechanics, Volume 6, No. 4, pp 189-226.

Bieniawski, Z.T., 1976, "Rock Mass Classification in Rock Engineering," Proceedings, Symposium on Exploration for Rock Engineering, Johannesburg, Volume 1, pp 97-106.

Bieniawski, Z.T., 1978, "Determining Rock Mass Deformability: Experience for Case Histories," International Journal on Rock Mechanics and Mining Science, Volume 15, pp 237-248.

Broch, E. and Franklin, J.A., 1972, "The Point Load Strength Test," International Journal on Rock Mechanics and Mining Science, Volume 9, pp 669-697.

Catepillar Tractor Company, 1975, "Handbook of Ripping," Fifth Edition, Peoria, Illinois.

Coates, D.F., 1978, "Rock Mechanics Principles," Canadian Department of Energy, Mines Branch Nomograph 87A, pp. S-2 and S-3.

Donald, I.B., Chiu, H.K., and Sloan, S.W., May 1979, "Theoretical Analyses of Rock Socketed Piles in Structural Foundation on Rock," Proceedings, International Conference on Structural Foundations on Rock, Sydney, pp 303-316.

Dubbe, R.E., 1974, "A Fundamental Study of Prediction & Rock Rippability," Misc. Thesis, Duke University.

Franklin, J.A., Borch, E., and Walton, G., 1971, "Logging the Mechanical Character of Rock," Transaction, Institute of Mining and Metallurgy (Section A), A-1 - A-9.

Hall, W.J., Newmark, N.M., and Hendron, A.J., May 1974, "Classification Engineering Properties and Field Exploration of Soils, Intact Rock and In-situ Rock Masses," prepared for U.S. Atomic Energy Commission, p. 116.

Hendron, A.J., December 1980, "Rock Engineering of High Vertical Rock Excavations in an Urban Environment," ASCE Seminar, New Development in Earth Support System, New York.

Hendron, A.J. and Dowding, C.H., 1974, "Ground and Structural Response Due to Blasting," Advances in Rock Mechanics, Volume II, Part B, National Academy of Sciences, Washington, D.C.

Hendron, A.J., Cording, E. and Aiyer, A., 1980, Analytical and Graphical Methods for the Analysis of Slopes in Rock Masses," Technical Report No. GL80-2, U.S. Army Waterways Experiment Station, Vicksburg, MS.

Higginbottom, I.E., May 1976, "The Use of Geophysical Methods in Engineering Geology, Part II, Electrical Resistivty, Magnetic and Cavity Methods," Ground Engineering, Volume 9, No. 2.

Hoek, E.H., Bray, J.W. and Boyd, J.M., 1973, "The Stability of Rock Slope Containing a Wedge Resting on Two Intersecting Discontinuities," Engineering Geology, Volume 6, pp 1-55.

Horvath, R.G. and Kenney, T.C., 1979, "Shaft Resistance of Rock-Socketed Drilled Piers," preprint 3698, ASCE Convention and Exposition, Atlanta, October 23-25.

Ladanyi, B., 1977, "Discussion, Friction and End Bearing Tests on Bedrock for High Capacity Socket Design," Canadian Geotechnical Journal, 14, 153, pp 153-155.

Lumb, P., 1967, "Statistical Methods in Soil Investigation," Proceedings, Fifth Australian-New Zealand Conference on Soil Mechanics and Foundation Engineering, Auckland, New Zealand.

Medearis, K., April 1979, "Dynamic Characteristics of Ground Motions Due to Blasting," Bulletin of the Seismological Society of America, Volume 69, No. 2, pp. 627-639.

Millet, R.A. and Moorhouse, D.C., 1973, "Bedrock Verification Program for Davis-Besue Nuclear Power Station," Proceedings, Specialty Conference on Structural Design of Nuclear Plant Facilities, ASCE.

Raphael, J.M. and Goodman, R.E., March 1981, "Strength and Deformation of Highly Fractured Rock," Journal of the Geotechnical Engineering Division, ASCE, Volume 107, GT3, pp 366-367.

Rubin, L.A. and Fowler, J.C., 1978, "Ground Probing Radar for Delineation of rock Features," Engineering Geology, Volume 12, pp 163-170.

Terzaghi, K., 1943, Theoretical Soil Mechanics, Wiley.

Timoshenko, S., 1934, Theory of Elasticity, McGraw-Hill.

Vassilopoulos, E., 1973, "Vorschlag Zum Entwurf von Grundungen auf verkarsteten Felsformationen ("Proposals for the Design of Foundations on Karstic Rock"), Proceedings, Symposium, Sinkholes, and Subsidence, International Association of Engineering Geology, Hanover.

Vesic, A.B., 1963, "The Validity of Layered Solid Theories for Flexible Pavements," Proceedings, International Conference on the Structural Design of Asphalt Pavements, University of Michigan.

Wijk, G., June 1980, "Point Load Test for the Tensile Strength of Rock," Geotechnical Testing Journal, GTJODJ, Volume 3, No. 2, pp 49-54.

Williams, A.F. and Pells, P.J.N., 1981, "Side Resistance Rock Sockets in Sandstone, Mudstone and Shale," Canadian Geotechnical Journal, 18, pp 502-513.

Chapter 101

TOPPLING INDUCED MOVEMENTS
IN LARGE, RELATIVELY FLAT
ROCK SLOPES

By
Adrian Brown, Principal
Golder Associates
Denver, Colorado

ABSTRACT

Over the last decade, a number of large slopes have been excavated in rock which contains families of joints or faults which dip steeply into the cut slope. Despite the absence of "adverse" structures, these slopes have exhibited alarming movement rates at very modest heights (as little as 100 meters) and slope angles (as little as 20°). Classical stability analysis suggests that such slopes should be unconditionally stable.

This paper describes the toppling mechanism which has been found to be the cause of the observed movements and develops a general method for analysis of it. Simple cases are analyzed in order to provide charts which allow quick evaluation of the likelihood of this phenomenon occurring. The safety implications of such movements in slopes are also examined.

INTRODUCTION

Many large rock slopes are in a continual state of movement, particularly when excavation takes place at the toe, as is usually the situation in active open pit mines. There is a class of such moving slopes which are characterized by the following conditions:

(1) Relatively flat slopes (20° to 30°)
(2) Relatively high slopes (300 meters or more)
(3) Relatively good quality rock
(4) High fracture intensity
(5) No weak planes which dip into the pit
(6) A family of weak planes dipping steeply into the wall
(7) Partial or total saturation by ground water

Classical slope analysis techniques show such slopes to be highly stable, yet surface movements of up to 10 meters per year have been observed to occur on them.

OBSERVED SLOPE MOVEMENTS

Detailed long term instrumentation of several mine slopes which exhibit this movement phenomenon produce the following results:

(1) Displacements accelerate during periods when there is mining going on at the toe, when the slope is steepened, or when the water pressure in the slope rises.

(2) Displacements slow and eventually cease when the slope conditions mentioned above are held constant.

(3) With depth, displacement of points below the slope surface decreases linearly to zero with depth.

(4) Disturbance is deep seated; typically almost all the material beneath the slope which is above the elevation of the toe is involved.

(5) There is very little toe heave.

(6) There is usually substantial cracking at the crest.

(7) The slope generally exhibits echelon "obsequent scarps"-- that is, reverse fault scarps--all the way from crest to toe.

(8) Originally vertical posts on the slope move outwards and tilt towards the pit; little vertical movement generally occurs.

After such movements have occurred in a slope for a period of years, the slope is gradually reduced to rubble. Benches become impossible to maintain and haul roads require constant rebuilding. If movement is allowed to continue, deep seated catastrophic failure can occur in the weakened slope.

MOVEMENT MECHANISM

The mechanism of this kind of movement is clearly a classical toppling failure. Shear occurs on the weak, in-dipping planes, allowing them to rotate. This process is illustrated in Figure 1.

The "basal plane" shown on Figure 1 defines the limit of material disturbance in the slope. It is not a geologically determined plane, but is usually normal to the weak plane inclination and passes approximately through the toe. Rotation of each slab appears to occur about its point of intersection with this plane.

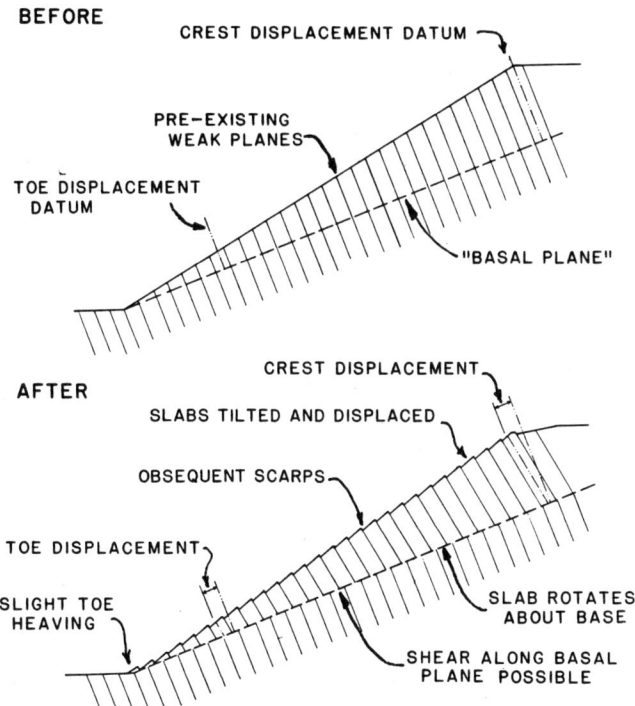

Figure 1. Mechanism of flat slope toppling

The kinematics of this process are important. Initially, rotation of each slab of rock occurs, causing shear failure of the material between each slab. As this process continues, the material in the slope either expands or contracts (depending on the angle between the weak planes and the basal plane), causing shear on the basal plane material. As this material is not as weak as the pre-existing weak planes, the system "locks up" until additional driving forces are mobilized. Examination of Figure 1 shows why the various phenomena observed in actual slope movements of this type (listed above) occur.

ANALYSIS

Toppling failure analysis has been attempted by a number of workers, notably Goodman and Bray (1976). An excellent discussion of the mechanics of toppling and a summary of (then) current knowledge on the subject is presented in Hoek and Bray (2nd edition, 1977). In the past, analyses have depended upon limit force equilibrium methods. However, these methods have serious limitations for analyzing large, shallow slopes because they require the selection or evaluation of lines of action of forces between "slices" in the toppling system.

For large assemblages of toppling slabs, it was necessary to move away from these methods and look more directly at the system using the principles of energy minimization. An elegant method of analysis of mechanical and civil engineering systems and structures, that of the principle of virtual work, was apparently particularly applicable to this system.

This principle was first enunciated by John Bernoulli in 1717 (Bernoulli, 1742) and is described in many texts (see for example, Wang, 1953, p. 144). In this approach, the entire system is subjected to a small displacement and the work done by all the forces acting within the system is computed. The system is stable if the work done by the resistive forces exceeds that done by the driving forces. For toppling systems, it is attractive to subject the entire system to a small rotational displacement in order to evaluate the net work.

Using this method, generalized analysis of this kind of slope movement has been performed, taking account of a wide range of geometric configurations. The basic system analyzed includes material toppling in two zones of a slope. Figure 2 defines these zones and other terms used in this paper.

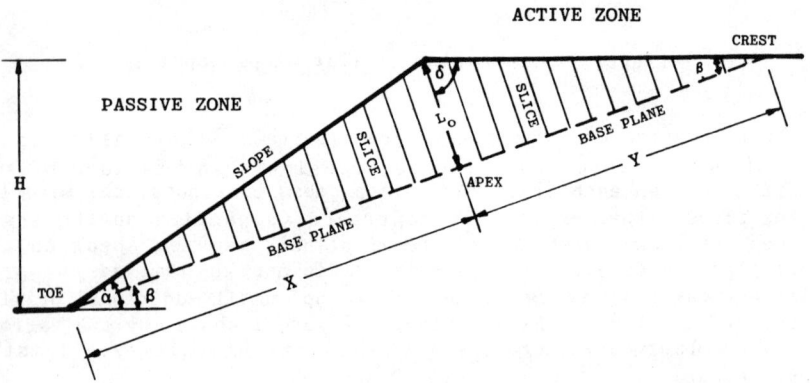

Figure 2. Definitions of terms used in analysis

The selection of the zones is primarily for analytical convenience.

The primary difference between virtual work and force balance analysis is that in the former it is necessary to know the stress distribution in the slope. A series of finite element and boundary integral analyses were performed to evaluate stresses in typical mine slopes. The results of these studies indicated that the computed post-mining stress field was very close to lithostatic for slopes up to about 60°. Accordingly, a linearly varying stress field was used in the analysis, as shown on Figure 3.

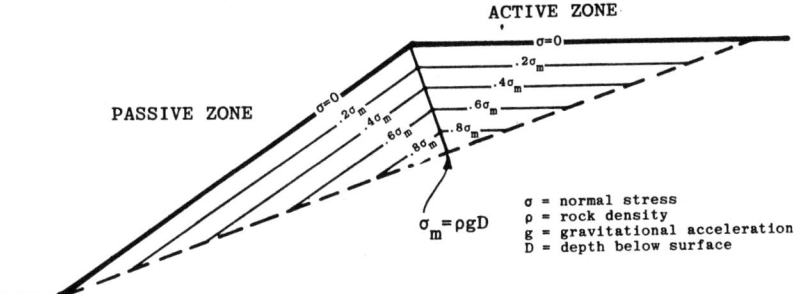

Figure 3. Normal stress field used in analysis

Any analysis of this system must include the effects of seepage. Because may slopes of this type are of low permeability, they tend to be fully saturated. A convenient method to describe water pressure is in an analogous way to that used for stress as shown on Figure 4.

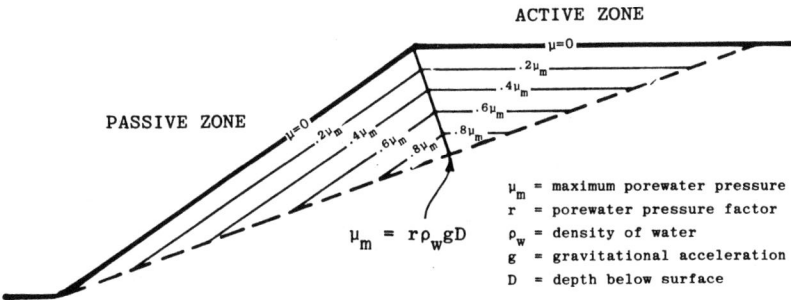

Figure 4. Water pressure regime used in analysis

Heads can be computed using the familiar formula:

$$h = \frac{\mu}{\rho_w g} + z$$

where Z defines the distance above a fixed datum. Pressure reductions in the slope (due to depressurization for example) are achieved analytically by reducing the value of the porewater pressure factor (r), while leaving the slope saturated. Desaturation is achieved by effectively reducing the head gradients (and hence the seepage forces) which remain when the porewater pressure factor has been reduced to zero.

Analysis is performed by considering a small virtual rotation of a typical slice about its base. The force system which does work when this occurs is shown on Figure 5.

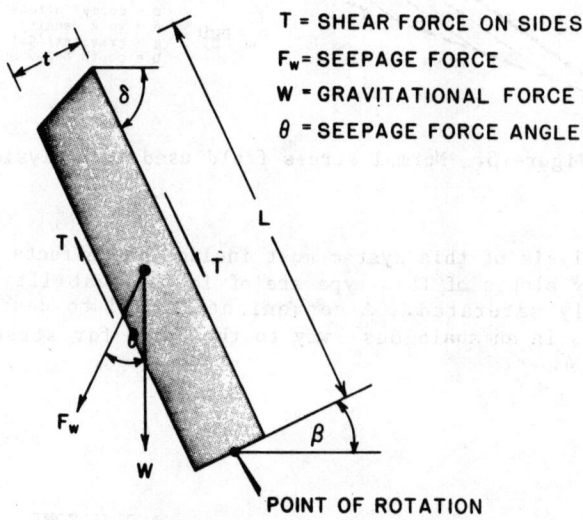

Figure 5. Forces on a typical slice

Results are integrated over all slices in the active and passive zones. The resulting work elements are:

W_R = Work done by gravitational forces due to rotation of the slice about the base.

W_{WR} = Work done by seepage forces due to rotational movement of the slice.

W_{SR} = Work done by shear forces on the sides of slices as rotation takes place.

Equations for these work elements for both zones are given in Table 1.

Table 1. Work Element List

	PASSIVE		ACTIVE
W_R	$\frac{1}{6} \rho g X L_o^2 \sin(\beta+\delta) \cos\delta \, d\psi$	W_R'	$\frac{1}{6} \rho g Y L_o^2 \sin(\beta+\delta) \cos\delta \, d\psi$
W_{WR}	$\frac{1}{6} \rho_w g X L_o^2 \dfrac{\sin\alpha \sin(\beta+\delta) \cos(\delta-\theta)}{\sin(\alpha+\theta)} d\psi$	W_{WR}'	$\frac{1}{6} \rho_w g (1-r) Y L_o^2 \sin(\beta+\delta) \cos\delta \, d\psi$
W_{SR}	$\frac{1}{6} \dfrac{X L_o}{\sin(\beta+\delta)} \left[(\sigma_m - \mu_m)\tan\phi + 3c \right] d\psi$	W_{SR}'	$\frac{1}{6} \dfrac{Y L_o}{\sin(\beta+\delta)} \left[(\sigma_m - \mu_m)\tan\phi + 3c \right] d\psi$

$L_o = H \sin(\alpha-\beta)/\sin\alpha$

$X = L_o \sin(\alpha+\beta)/\sin(\alpha-\beta)$

$Y = L_o \sin\delta/\sin\beta$

$\theta = \tan^{-1}\left[\dfrac{r}{(1-r)/\tan\alpha + 1/\tan\delta}\right]$

$\sigma_m - \mu_m = \rho g D (1 - r \rho_w/\rho)$

$D = L_o \sin\delta$

Note: This Table valid only for $\delta + \beta \geq 90°$.

For those who utilize a factor of safety approach to stability analysis, the factor of safety against toppling is given by:

$$FS = \frac{W_{SR} + W_{SR}'}{W_R + W_R' + S(W_{WR} + W_{WR}')}$$

where S = saturation factor

This is a "genuine" factor of safety in the sense that it is equal to the total resisting work divided by the total driving work. For those who favor a probabilistic approach, the probability of failure is given by:

$$PF = p\left\{\left[W_R + W_R' + S(W_{WR} + W_{WR}')\right] > (W_{SR} + W_{SR}')\right\}$$

It should be noted that this version of the analysis is only valid for cases where the weak planes are normal to the base plane, or where they dip more steeply than this. (This is because the equations do not consider shear along the base plane, which is covered in a future paper.)

RESULTS

The equations have been used to back analyze a number of actual toppling slopes where geometry, water pressure, and weak plane strength parameters were known. In each case, an excellent match between predicted and actual behavior has been found to occur, providing some validation for the approach.

The equations have also been used to investigate the conditions under which toppling can occur, and the effectiveness of water pressure reduction as a control strategy. Figure 6 shows the results of analyses of cohesionless slopes under various geometric and ground water conditions.

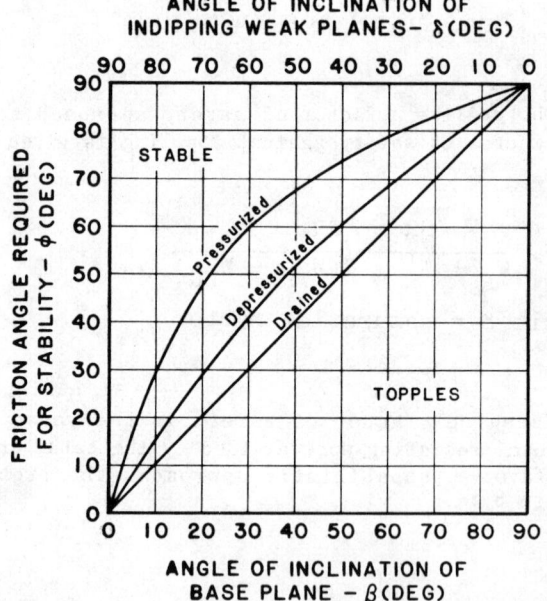

Figure 6. Onset of toppling of cohesionless slopes

This figure assumes that the base plane is normal to the weak planes (the least stable condition). The results are independent of slope height and slope angle, although the slope angle must, by definition, exceed the base plane angle. As can be seen from the figure:

(1) For reasonable values of friction angle on the weak plane (10° to 30°) slope angles in excess of 10° will likely topple if the ground water system is fully pressurized and an appropriate weak plane set is available.

(2) Even for a fully drained slope, toppling will likely take place if the slope angle exceeds 30°, providing an appropriate in-dipping joint or fault set is available.

This figure makes it clear why this phenomenon is probably common in heavily jointed rock slopes. The author suspects that many slope failures are in fact non-catastrophic toppling of this type. Almost any excavated rock slope over 20° should be checked for this possible movement mechanism.

Cohesion on the weak planes has the effect of making the results very sensitive to the height, the slope, and the base plane inclination. An example of the effects of cohesion on the friction angle required for stability is shown on Figure 7.

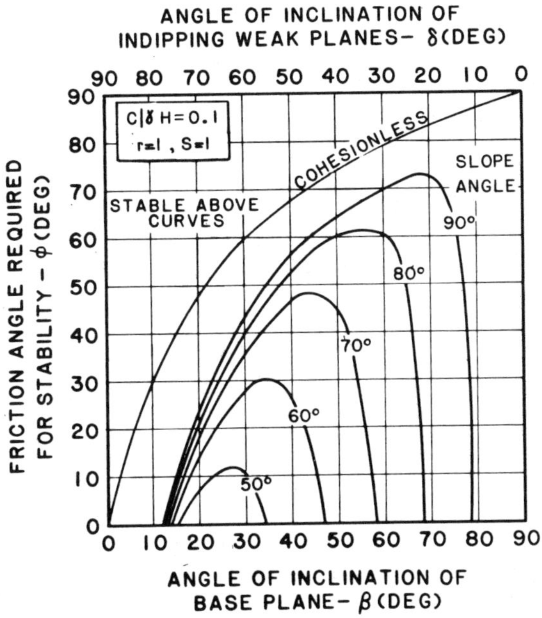

Figure 7. Effect of cohesion on onset of toppling

Clearly, the exact inclination of the weak planes now becomes important to the evaluation of the likelihood of toppling in any given slope. For slope evaluations it is often difficult to establish the joint inclination and continuity before excavation. A useful design tool can, therefore, be developed by assuming that a weak plane set exists at the most unfavorable inclination.

It is then possible to develop stability charts for toppling which are analogous to those for circular failure (Hoek and Bray, 1977). The toppling charts for fully drained conditions, (Hoek and Bray, Water Case #1), fully depressurized conditions, and fully pressurized conditions (Hoek and Bray, Water Case #5) are given on Figures 8, 9, and 10, respectively. These charts can be used for quick feasibility evaluations in place of the more cumbersome equations.

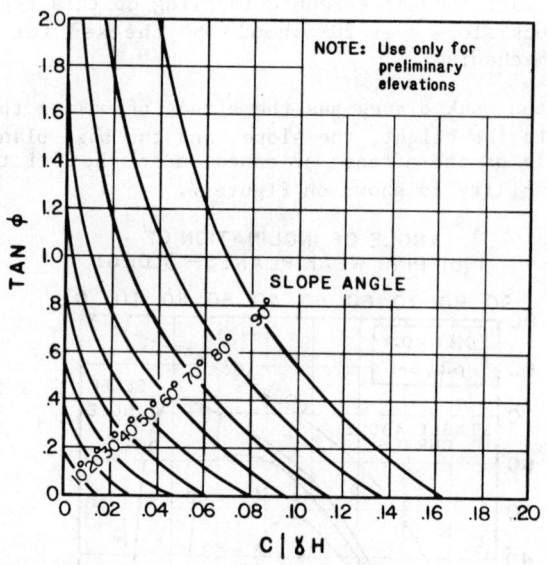

Figure 8. Toppling chart - fully drained case

DISCUSSION

In flat rock slopes, the occurrence of major displacements accompanied by relatively rapid surface degradation is usually regarded as a procursor of catastrophic failure. Toppling has rarely been considered as a possible cause of such movements because it tends to be associated with very steep slopes in very hard rock.

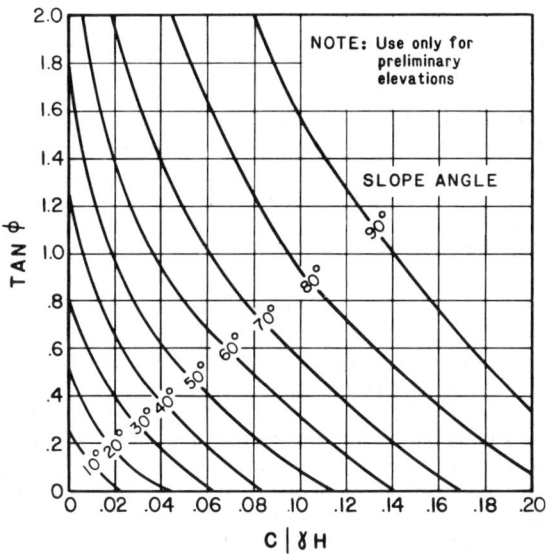

Figure 9. Toppling chart - saturated/depressurized case

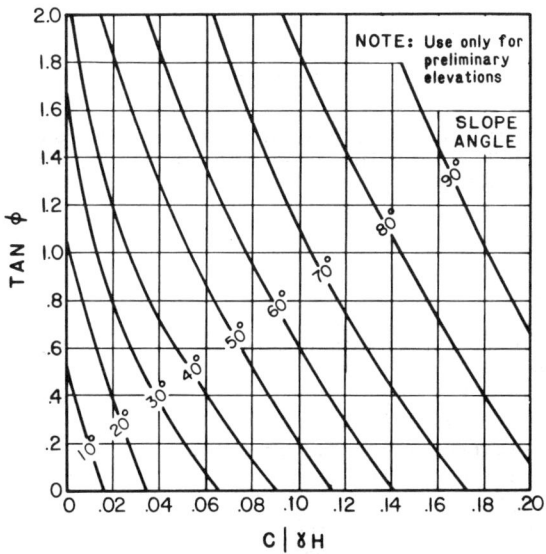

Figure 10. Toppling chart - saturated/pressurized case

This study shows that:

(1) Toppling can take place in very flat slopes and may disturb material to considerable depths.

(2) Movements caused by toppling in flat slopes cannot of themselves lead to catastrophic slope failure.

(3) Toppling can lead to translational failure of a slope by weakening the slope materials to the point where circular failure is possible.

These findings suggest that excavation can proceed safely in a mine where toppling-induced movements are occurring, provided that monitoring shows that the toppling is not causing catastrophic failure by weakening the slope materials.

The rate of toppling-induced movement in a flat slope is strongly dependent upon the slope geology, slope geometry, the weak plane strength, and the water pressure conditions in the slope. Given a desired geometry, the major movement control strategy which is available is ground water pressure reduction. This strategy is at least as effective in controlling toppling movements as it is in the control of traditional slope failure mechanisms.

NOMENCLATURE

c	=	effective stress cohesion of weak plane material
D	=	depth of apex below ground surface
FS	=	factor of safety
F_w	=	seepage force on slice
g	=	acceleration due to gravity
H	=	slope height
h	=	head
L	=	length of slice
L_o	=	length of common plane between active and passive zone
p	=	probability
PF	=	probability of failure
r	=	porewater pressure factor ($0 \geqslant r \geqslant 1$)
S	=	saturation factor ($0 \geqslant s \geqslant 1$)
T	−	shear force along slice side
t	=	thickness of each slice
W	=	gravitational force on slice
X	=	length of base of active zone
Y	=	length of base of passive zone
z	=	distance above an arbitrary datum

α = slope angle
β = base plane inclination
δ = angle of inclination of in-dipping weak planes
ρ = density of rock
ρ_w = density of water
ϕ = effective stress friction angle of weak plane material
σ = normal stress
σ_m = maximum normal stress
θ = seepage force inclination
μ = porewater pressure
μ_m = maximum porewater pressure

REFERENCES

Goodman, R.E. and Bray, J.W., 1967, "Toppling of Rock Slopes," Proceedings of the Specialty Conference Rock Engineering and Slopes, Boulder, Colorado, ASCE, Vol. 2.

Hoek, E. and Bray, J.W., 1977, Rock Slope Engineering, 2nd edition, I.M.M., London.

Bernoulli, John, 1742, Opera Omnia, 4 volumes, Lausanne and Geneva.

Wang, C.T., 1953, Applied Elasticity, McGraw Hill, New York.

Chapter 102

SUBSIDENCE MONITORING - CASE HISTORY

Dr. Peter J. Conroy - Partner
Julianne H. Gyarmaty - Staff Engineer

DAMES & MOORE
1550 Northwest Highway
Park Ridge, Illinois 60068

INTRODUCTION

The current study is part of the U.S. Department of Energy's (DOE) on-going subsidence research program. The long-term objective of the DOE program is to develop analytical methods of subsidence prediction applicable over a broad range of geologic and mining conditions encountered in the United States. The DOE currently has several contracts in progress to develop a sound data base that can be used in the development of predictive models. Dames & Moore is currently under contract to DOE to collect and analyze detailed subsidence and hydrologic data from a coal mine employing longwall mining methods in the Eastern Coal Province.

Work under Contract No. DE-AC22-80PC 30335 began in October of 1980 and will continue through September of 1984. The contract includes monitoring of two adjacent longwall panels. This paper summarizes the surface monitoring work that was performed through February of 1982. Mining of the first of the two panels (3 off 3 north) was completed on February 3, 1982. Results of the surface monitoring through completion of this panel are presented and discussed.

SITE DESCRIPTION

The mine being studied is located in eastern Ohio. The site is west of the Ohio River, approximately 56 km (35 miles) south of Wheeling, West Virginia. The topography of the unglaciated region is characterized by closely spaced valleys situated between steep, narrow ridges. Depth of cover in the study area ranges from approximately 102 m (310 feet) to 249 m (760 feet) above the coal, which is at an elevation of 161 m (490 feet) above mean sea level (MSL). Two longwall panels are being studied, each approximately 1640 m (5000 feet) long by 157 m (480 feet) wide.

The geologic strata of the study area are composed of interbedded shale, siltstone, sandstone, and limestone units primarily from the Greene, Washington, and Monongahela formations of Permian and Pennsylvanian age. The mine is producing from the Pittsburgh No. 8 seam. The Pittsburgh No. 8 seam at the project site is 1.6 m (4.95 feet) thick and is overlain by 0.6 m (1.85 feet) of roof coal with interbedded shale. Mined thickness ranges from 2.3 m (85 inches) to 2.4 m (88 inches).

The two panels selected for study, PANEL 3 OFF 3 NORTH and PANEL 4 OFF 3 NORTH, are shown in relation to surface topography on Figure 1. The mine employs a retreat longwall mining method. The direction of mining is from east to west. The panel sequence is from south to north. The panel directly to the south of PANEL 3 OFF 3 NORTH had been mined prior to the start of the current project.

INSTRUMENTATION

Surface instrumentation includes the installation of 338 surface monuments and the installation of full-profile borehole inclinometers (FPBI) and extensometers (PFBX) in two boreholes at the locations shown on Figure 1. Installation of all surface monuments has been completed. Initial elevations and horizontal distances have been taken for all monuments. Installation of full-profile borehole inclinometers and full-profile borehole extensometers has been completed at the locations shown on Figure 1.

An automatic data acquisition system (ADAS) developed by Dames & Moore in conjunction with the Department of Energy has been used to monitor strain and tilt development as the face passed beneath the house located as shown on Figure 1. In addition, the ADAS will be installed at the four locations over panel 4 OFF 3 NORTH, also shown on Figure 1.

The ADAS system is composed of a tilt measurement unit and a linear displacement measurement unit. Both units use Transtek angular transducers. The transducers are connected to a 16 channel Datel data logger and are read hourly. Readings are recorded as voltages on a cassette tape in the data-logger. The transducers may be mounted on monuments or they may be mounted directly on a

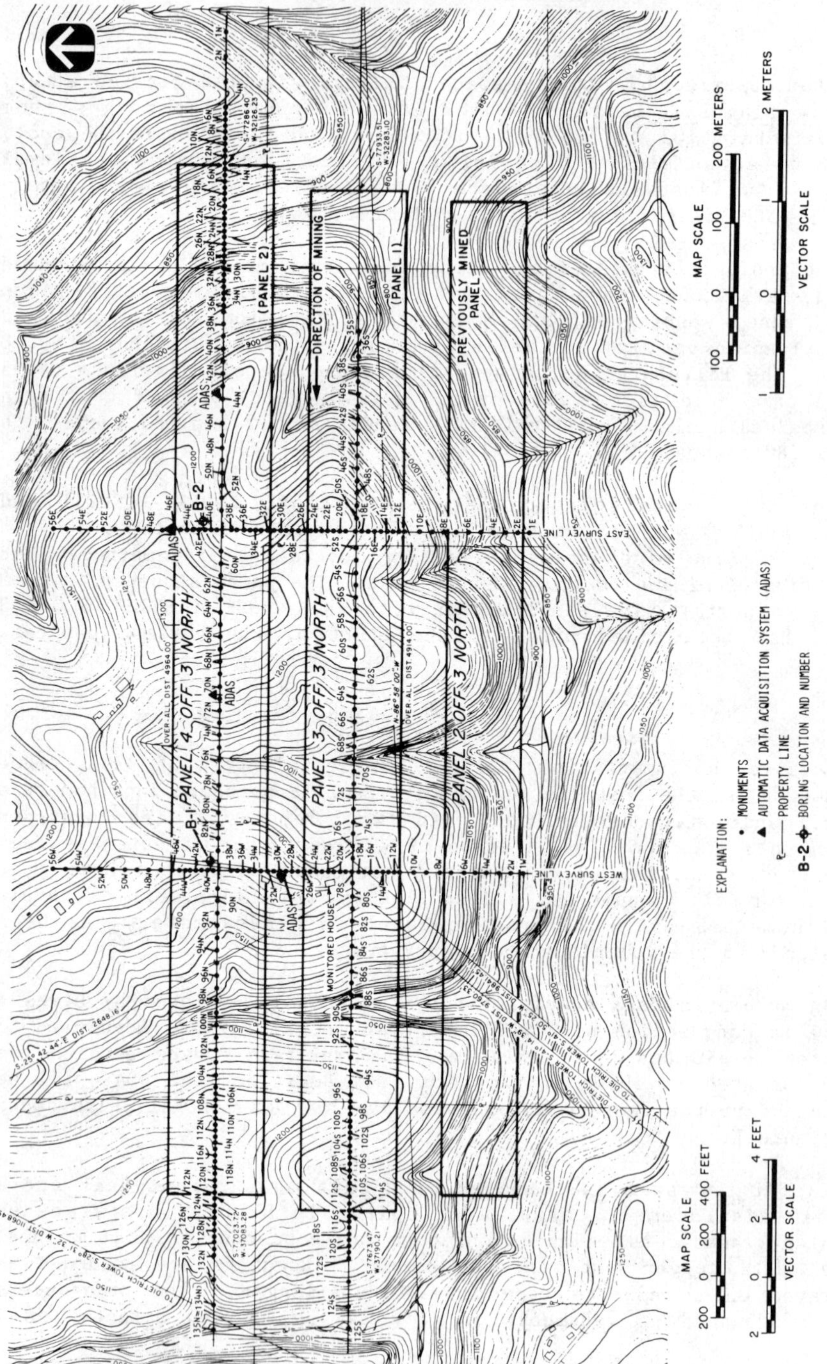

FIGURE 1 SURFACE INSTRUMENTATION

structure wall as was done in the house over Panel 3 OFF 3 NORTH. The cassette data tape is read directly into a computer where readings are converted to strain and tilt.

RESULTS

The results of the subsidence monitoring on Panel 3 OFF 3 NORTH are shown on the subsidence development curves on Figures 2 and 3. It should be kept in mind that mining was in progress when these curves were developed.

Figure 2 presents the subsidence development curve plotted along the centerline of the panel. Maximum subsidence at the time the figure was prepared was 4.7 feet or 65 percent of seam height. For comparison purposes the subsidence development curve from the Subsidence Engineers Handbook (NCB, 1975) is plotted on the same figure. The comparison indicates that the subsidence at this mine in the eastern U.S. lags behind that which would be predicted from the NCB curve.

Figure 3 is a subsidence development curve prepared from data obtained along the east survey line, perpendicular to the centerline of the panel. The curve plotted from data obtained north of the centerline reflects the presence of the unmined panel to the north. The curve of the data obtained south of the centerline demonstrates the effect of the previously mined panel on the subsidence of the panel being mined.

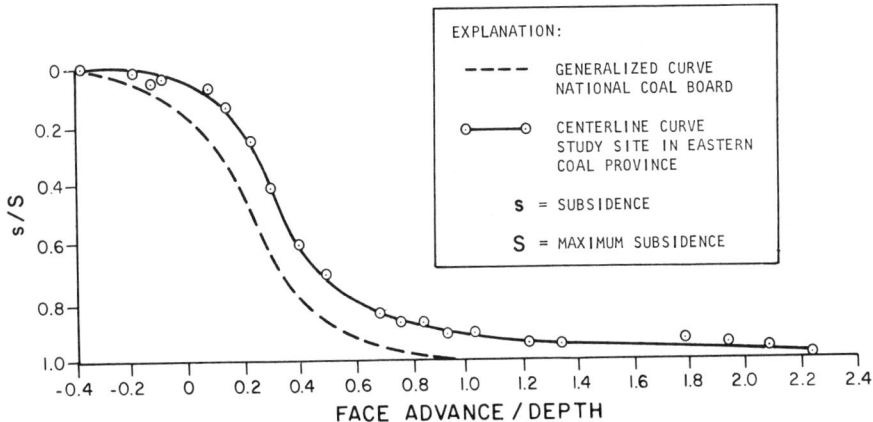

FIGURE 2 SUBSIDENCE DEVELOPMENT CURVE ALONG CENTERLINE

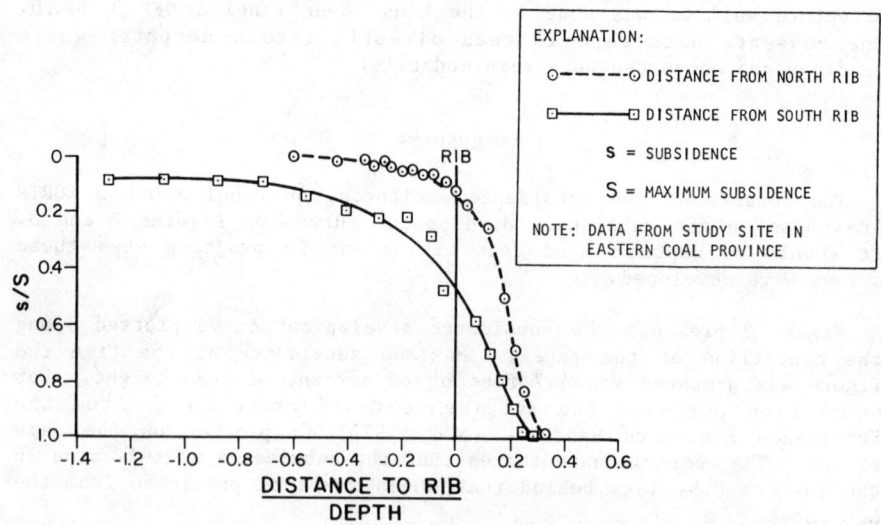

FIGURE 3 SUBSIDENCE DEVELOPMENT CURVES PERPENDICULAR TO CENTERLINE

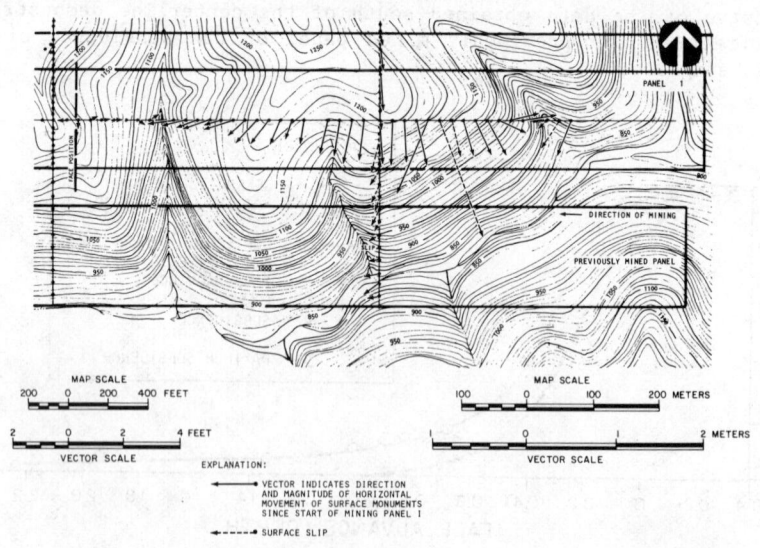

FIGURE 4 HORIZONTAL MOVEMENT OF SURFACE MONUMENTS

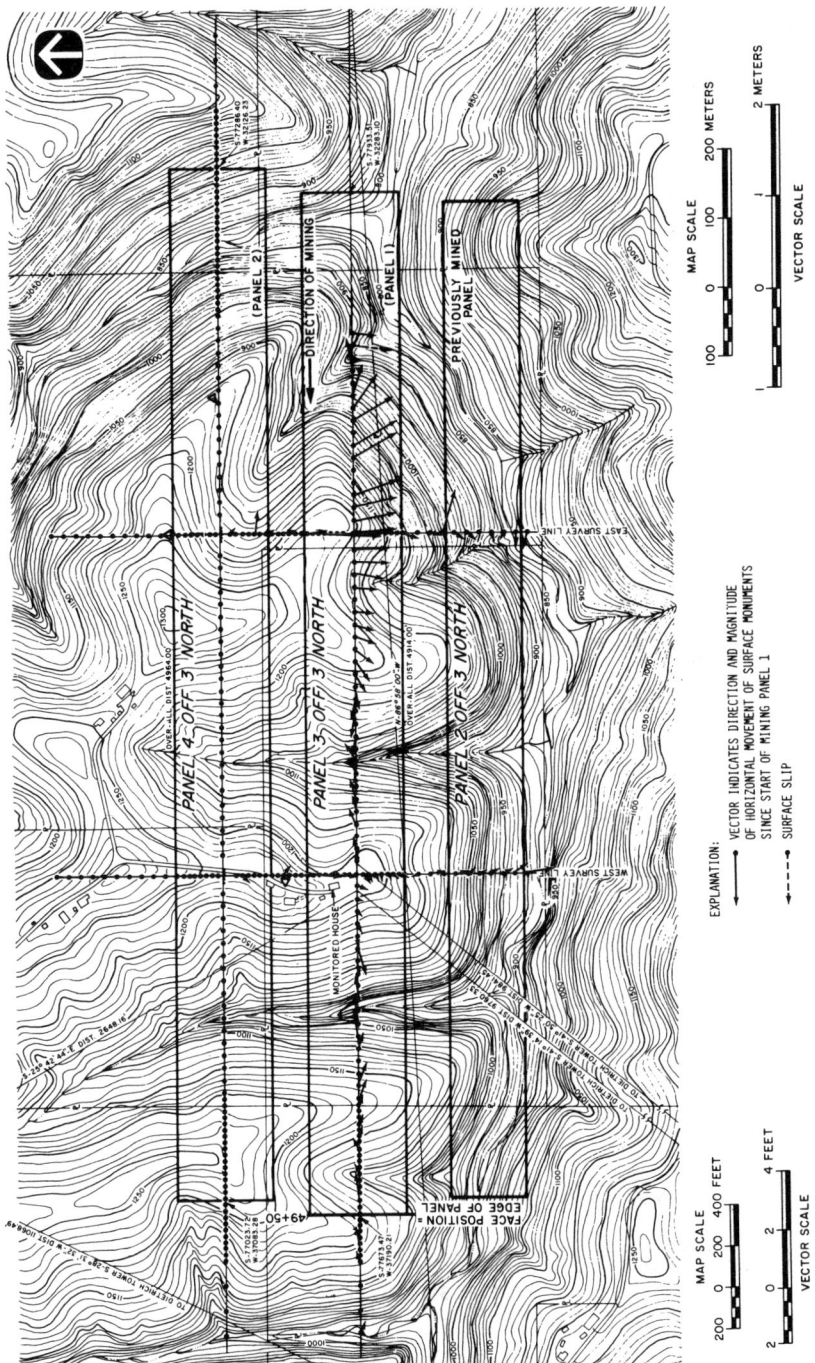

FIGURE 5 HORIZONTAL MOVEMENT OF SURFACE MONUMENTS

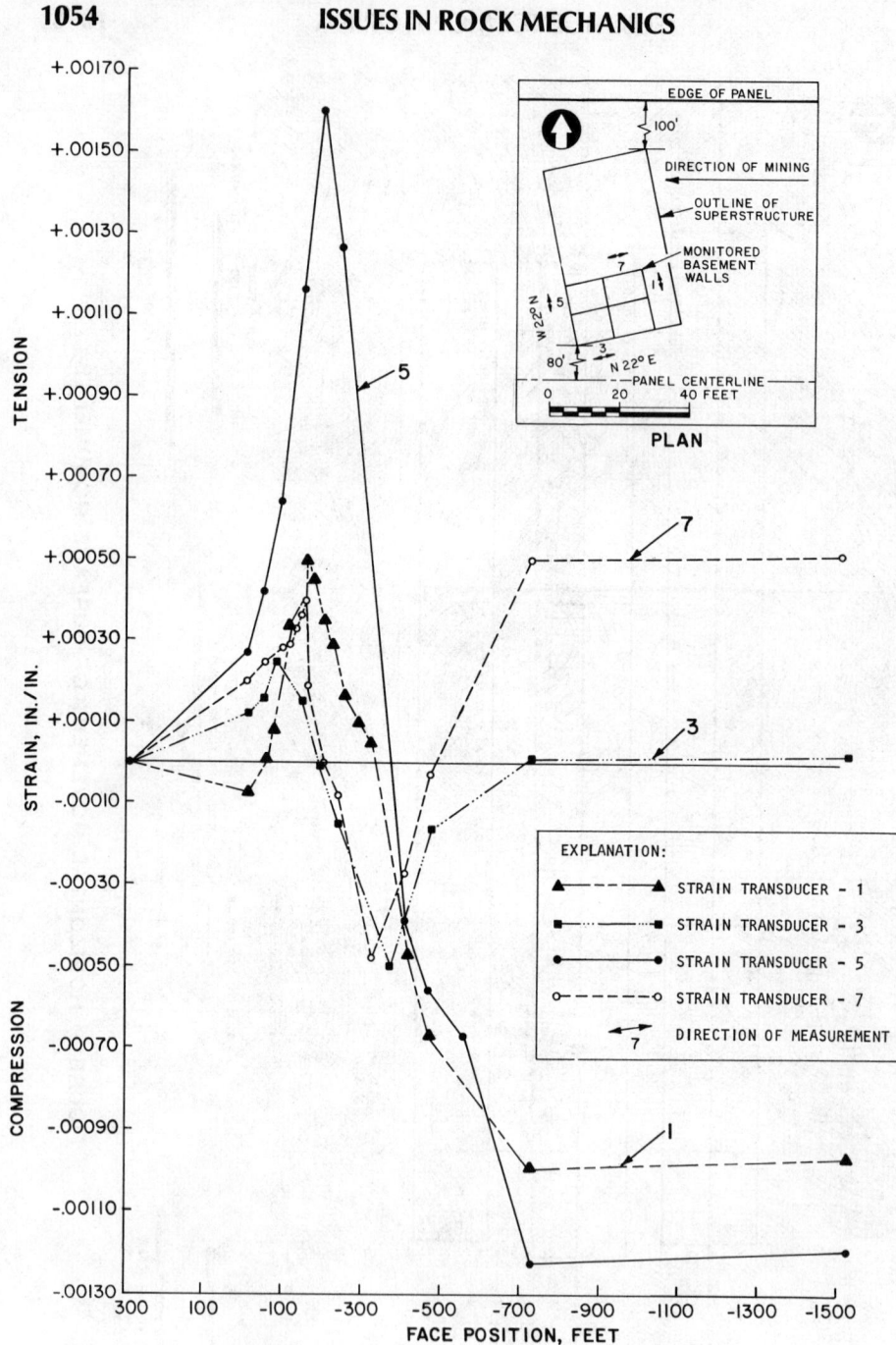

FIGURE 6   STRUCTURE STRAIN DATA

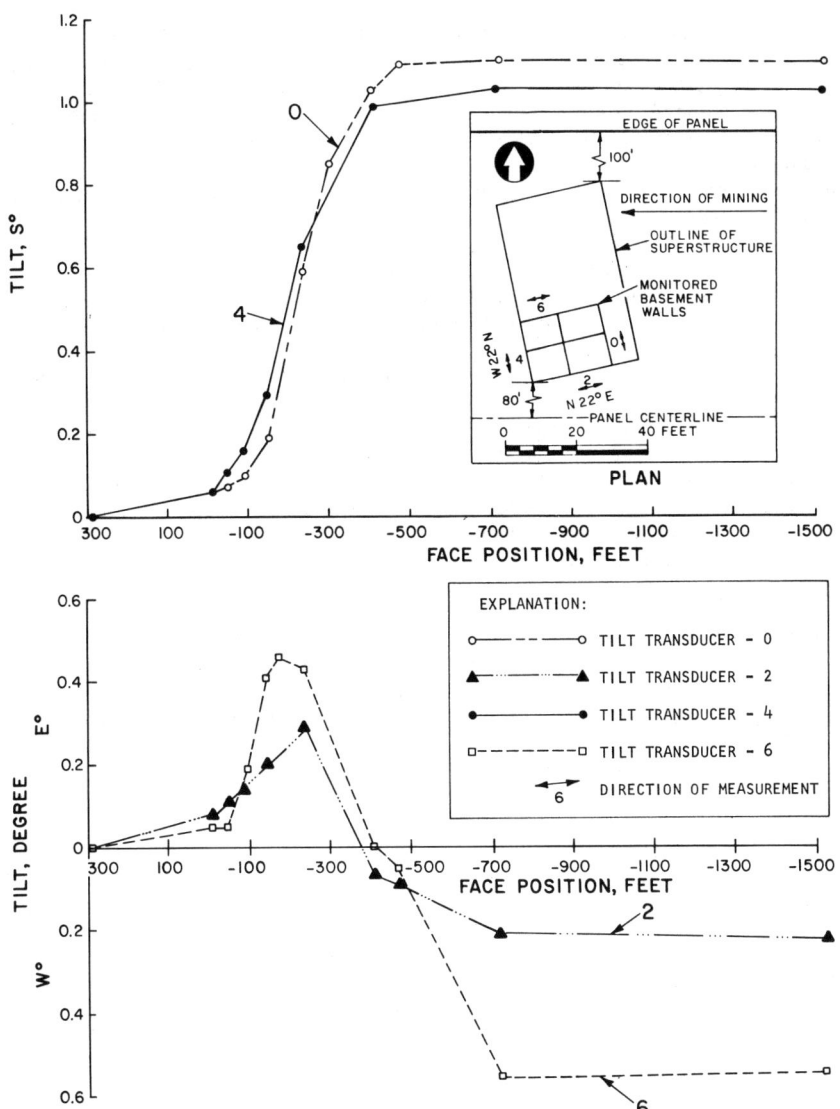

FIGURE 7 STRUCTURE TILT DATA

Figures 4 and 5 are plots of horizontal movements of surface monuments. The movement is that which has occurred during the time from the start of mining of Panel 3 OFF 3 NORTH to the time the face was at the position indicated. Movements where surface slumps or slips are apparent have been indicated. These movements have been affected by surficial slumping which was observed in the form of tension cracks or in tilting of the monuments. In general, the data indicates that topography may be a pronounced factor in subsidence related surface movement.

The results of strain and tilt measurements taken on the masonry walls of the basement of the house are shown on Figures 6 and 7. Hourly readings were taken automatically using the ADAS.

Strain data are plotted on Figure 6. Transducers 3 and 7 monitored strain in a direction 22° from the direction of mining. The data from these transducers exhibit the classical development of tension and compression due to the subsidence wave with the exception that transducer 7 indicated residual tension after the wave passed. Transducers 1 and 5 monitored strain on the walls oriented 68° from the direction of mining. The data from these transducers indicated tension as the subsidence wave passed and indicated a residual compression after mining.

Corresponding tilt data are shown on Figure 7. Residual tilt to the south was indicated from the data from transducers 0 and 4 which measured tilt in a direction 68° from the direction of mining. This is consistent with the horizontal movement indicated on Figures 4 and 5. Transducers 2 and 6 indicated an initial easterly tilt and finally a westerly residual tilt.

CONCLUSIONS

The data presented herein is from an on going study. This data and data to be obtained during the mining of the second longwall panel will be utilized to develop relationships for subsidence prediction in the region. Measured strain and tilt data obtained from the ADAS will be used to verify strain and tilt computed from subsidence curves developed in the conventional manner. The topographic effects will also be considered in developing predictive relationships.

REFERENCES

National Coal Board, Subsidence Engineers Handbook, National Coal Board, London, 1975, 111 p.

Chapter 103

ROCK WEDGE STABILITY

Adrian M. Crawford

Mineral Resources Engineering
Imperial College
London, U.K.

INTRODUCTION

A study has been made of the stability of blocks and wedges of rock subjected to gravity forces in the roofs of underground excavations. Rock wedges may become wholly or partly self supporting with the mobilization of shear resistance of discontinuities bounding such wedges. The influence of in-situ stresses and the relative stiffnesses of the intact rock, and shear and normal stiffnesses of the discontinuities has been investigated.

A two dimensional, plane strain model of symmetric rock wegdes in a horizontal stress field has been utilized. A comparison between the results of an extension of a closed form solution proposed by Bray(1981), and numerical analyses with Displacement Discontinuity elements (Crouch, 1976), are presented. These analyses indicate that the stability of the wedge and the stress redistribution around the wedge upon excavation of an opening is markedly influenced by the ratio of intact rock stiffness to seam normal and shear stiffness.

It is shown that the analytical solution provides a reasonable measure of the lower and upper bounds of the loads necessary to cause failure of the wedge.

MODELS

The problem considered (Fig. 1) is a two dimensional plane strain model containing two discontinuities. The discontinuities are symmetrically disposed with respect to a horizontal excavation which forms a wedge of rock bounded by these discontinuities. The angle of the intersecting discontinuities at the apex of the wedge is 2α. Mechanical properties of the two discontinuities are identical, being represented by linear normal and shear joint stiffnesses K_n and K_s and angle of friction ϕ.

The rock wedge is loaded by an in-situ horizontal stress field σ_h

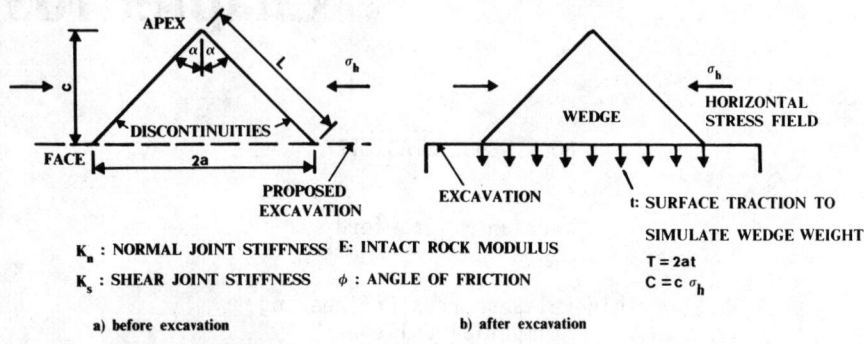

Fig. 1. Model of rock wedge

and a surface traction t which represents the self weight of the rock wedge or an external force applied to the wedge.

Only cases in which the half wedge apex angle (α) is less than the discontinuity friction angle ϕ have been considered. Thus, prior to excavation of the opening, the discontinuity is able to transmit the horizontal stress field without the discontinuity being in a state of limiting equilibrium. With excavation of the opening, the wedge is stabilized by the horizontal stress field. A downward force T (=2at) must be applied to cause the wedge to fail.

NUMERICAL TECHNIQUE

The Displacement Discontinuity (DD) method is an indirect boundary element formulation based on a solution which expresses displacements and stresses resulting from constant normal and shear displacement discontinuities over a finite line segment in an elastic body. The method comprises placing displacement discontinuities along the boundary region being analyzed. Seam or discontinuity elements may be represented as displacement discontinuity elements, the opposite faces of which are connected by elements representing the shear and normal stiffnesses (Ks and Kn) of the seam.

In the formulation used here the normal and shear stiffnesses of the discontinuities have been modelled by linear elastic elements, but in general, non-linear discontinuity elements may be modelled with the programme. Seam failure is incorporated by imposing a constraint on the maximum shear stress that can be developed in a seam. A Mohr-Coulomb failure criterion is used (Crouch, 1976).

The resulting system of algebraic equations is solved to find the unknown discontinuity displacements that are consistent with the prescribed boundary tractions and displacements.

With this analysis the stress distribution on the seam elements may be determined both before and after an excavation has been made. Prior to excavation, provided the shear resistance of the discontinuities has not been exceeded, the normal and shear stresses on the discontinuities are equal to the in-situ stress field values. Upon excavation and loading of the wedge the joints may load or unload elastically or fail depending on the changes in normal and shear stresses on the discontinuities. Hence with this analysis it is possible to determine the complete load-deformation relationship for a loaded wedge.

ANALYTICAL SOLUTION

In this model, modified from Bray(1981), the influences of the horizontal stress field and joint stiffnesses on the stability of a symmetrical rock wedge under a vertical load are considered. Unlike the DD model, the effect of the deformability of the intact rock is not accounted for.

Taking a unit length of excavation, static equilibrium of a symmetrical, rigid wedge of rock (Fig. 1) is considered. Prior to excavation of the opening the horizontal stress field σ_h results in a horizontal force C being transmitted across the wedge. Thus

$$C = \sigma_h c \tag{1}$$

After excavation, under the influence of the vertical force T, the wedge will be displaced downward by an amount δ. The effect of such a deformation is to cause a change in the normal and shear forces N and S on the discontinuity faces bounding the wedge. Thus, the new values of normal and shear forces N' and S' are given by

$$N' = N - K_n\delta\sin\alpha = C\cos\alpha - K_n\delta\sin\alpha$$
$$S' = S + K_s\delta\cos\alpha = C\sin\alpha + K_s\delta\cos\alpha \tag{2}$$

At limiting equilibrium the net vertical displacement of the wedge δ and T, the vertical force applied to the wedge may be found. The solution is given for two limiting cases, when $K_s=K_n$ and $K_s<<K_n$.

$K_s=K_n$

$$\delta = \frac{C(\tan\phi - \tan\alpha)}{K_n(1 + \tan\alpha\tan\phi)}$$

$$T_{yield} = \frac{2C(\tan\phi - \tan\alpha)}{(1 + \tan\alpha\tan\phi)} \tag{3}$$

For this case, the full in-situ stress field is transmitted across the wedge resulting in a value for T which may be taken as an upper bound to the failure load. For positive α, T/C is a maximum when $\alpha=0$.

Ks<<Kn

$$\delta = \frac{C(\tan\phi - \tan\alpha)}{K_n \tan\alpha \tan\phi}$$

$$T_{yield} = \frac{2C\sin\alpha\cos\alpha(\tan\phi - \tan\alpha)}{\tan\phi} \qquad (4)$$

With Ks<<Kn, the shear forces and stresses along the discontinuity are the same both before and after excavation, loading and deformation of the wedge. This expression for T approximates a lower bound for the failure load. T/C has a maximum value when $\alpha=\phi/2$, that is when the angle at the apex of the symmetric wedge equals the angle of friction.

The relationship between surface traction t, horizontal stress σh and forces T and C is given by

$$\frac{t}{\sigma h} = \frac{1}{2\tan\alpha} \frac{T}{C} \qquad (5)$$

Note that the values of joint normal and shear stiffness Kn and Ks in the analytical solution are related to those of the numerical solution by the length L of the discontinuities bounding the wedge (Fig. 1). Thus

$$K_{numerical} = K_{analytical}/L \qquad (6)$$

The ratios Ks/Kn are identical.

RESULTS OF ANALYSES

Discontinuity Parameters

In general, the shear and normal stiffnesses of a discontinuity are not related to each other, hence a range of values of Ks/Kn have been considered to determine its effect on the failure load.

For medium and high stress levels, Barton(1981) and Ludvig(1981) suggest that the ratio of Ks/Kn, assuming Kn to be independent of scale, is in the range of 1/41 to 1/91. Alternatively, if the shear and normal stiffnesses are related as shear and elastic modulii, then $Ks=\frac{1}{2}Kn$ for a Poisson's ratio of zero.

In the analyses, values of Ks/Kn from 1.0 to 1/1000 were adopted, encompassing the most likely and realistic range of estimates of Ks/Kn.

The ratio E/Kn was varied from between 1 and 100. E/Kn equal to unity represents a seam with stiffness per unit thickness identical to the intact rock. The upper limit of 100 was judged to afford a sufficient range to the values of E/Kn to illustrate the influence of this ratio on the stability of rock wedges. It is possible that for gouge filled joints, the ratio E/Kn may be considerably greater than the upper limit chosen for this analysis.

Two friction angles ϕ equal to 45 and 30 degrees have been considered.

Model Discretization

One symmetric wedge has been studied with half apex angle α equal to 25 degrees. Both discontinuities and the excavated face have been modelled with ten DD elements each.

For the numerical model, the failure load T (or equivalently surface failure traction t) was taken to be the load at which the downward deformations of the wedge became large as both bounding discontinuities achieved a state of limiting equilibrium. Practically T was determined by gradually incrementing the stress t applied to the wedge face, and plotting the average deformation δ of the wedge face. Near the failure load the load-deformation curve was observed to become non-linear. Failure of the wedge was indicated by an instability in the iterative solution of the DD equations as the deformations became large.

Results

For the symmetrical wedge problem the half wedge apex angle has a value of 25 degrees. Two friction angles of ϕ equal to 45 and 30 degrees have been adopted.

The results of the analyses are presented in normalized form (Figs. 2, 3 and 4). Thus the surface traction t at failure is given by $t/\sigma h$, where σh is the horizontal stress field. The normal and shear joint stiffnesses (Kn and Ks) and modulus of elasticity of the intact rock (E) are given as Ks/Kn and E/Kn.

The influence of normal and shear seam stiffnesses, intact rock modulus of elasticity and angle of friction on the stability of the rock wedge is shown in Fig. 2.

For the value of ϕ approaching the wedge angle ($\phi=30°$, $\alpha=25°$), the effects of seam stiffness and intact rock modulus on the failure stress ratio ($t/\sigma h$) is quite small. The numerical solution plots within the closed form solution.

An increase in ϕ ($\phi=45°$) shows the failure stress ratio ($t/\sigma h$) to be sensitive to the values of Ks, Kn and E. For both Ks=Kn and

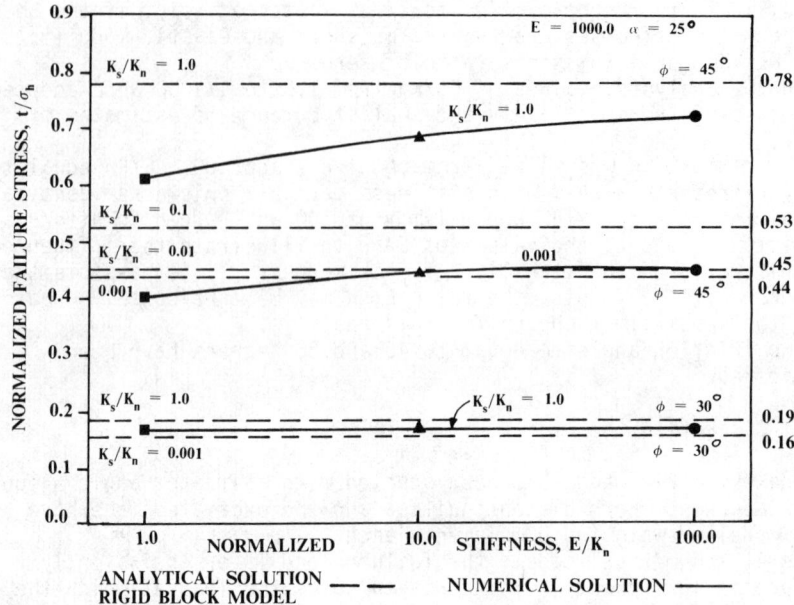

Fig. 2. Rock wedge failure stress as a function of intact rock modulus, seam normal and shear stiffnesses, apex angle and angle of friction.

$K_s<<K_n$, the value of ($t/\sigma h$) increases as the seam becomes softer relative to the intact rock.

For the stiff seam (E/Kn=1.0, Ks/Kn=1.0, ϕ=45°) the pattern of the shear and normal stresses along the discontinuities bounding the wedge are non-uniform due to a redistribution of stresses around the wedge (Fig. 3). With E/Kn=100 the seam is so soft that deformations of the intact rock surrounding the wedge have little influence on the stress distribution within the discontinuity. This results in a quite uniform stress pattern within the discontinuity (Fig. 3) and a greater failure load than for E/Kn=1.0. A similar though less pronounced effect is evident for Ks<<Kn (Fig. 3) and for ϕ=30 degrees.

Load-deformation curves for Ks=Kn, ϕ=45 degrees, are plotted in Fig. 4. A comparison is made between the numerical and analytical solutions. For Ks and Kn<<E the wedge deforms essentially as a rigid block and the analytical solution is a good representation of the curve predicted by numerical analysis. With increased stiffness of the seam relative to the intact rock, deformation of the intact rock forming the wedge and redistribution of stresses around the wedge cause the analytical solution to underestimate deformations of the face of the wedge. Similar observations were made for

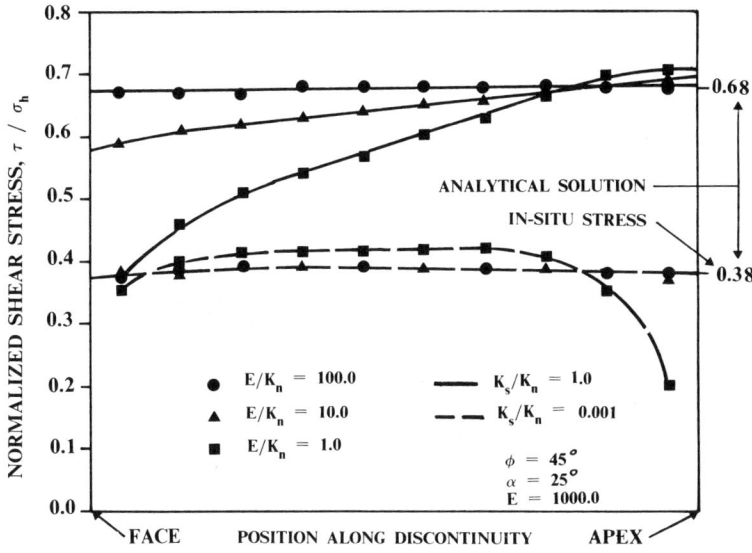

Fig. 3. Shear stress distribution along discontinuity.

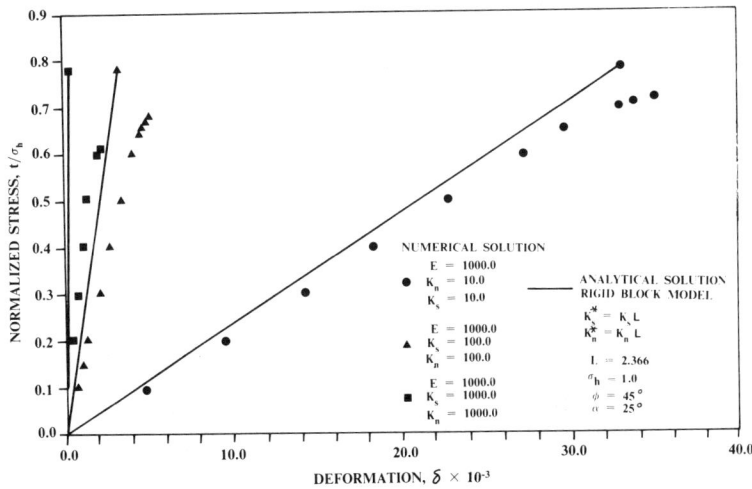

Fig. 4. Deformation behaviour of rock wedge.

Ks<<Kn, φ=45 degrees and for the φ=30 degrees case.

CONCLUSIONS

This work has demonstrated that the presence of a stress field tangential to the face of an excavation can have an important stabilizing influence on rock wedges.

The most significant factor determining the load necessary to cause failure of symmetric wedges was found to be the relative normal to shear stiffnesses of the discontinuities. A comparison of numerical and analytical solutions showed that a reasonable measure of an upper bound to the failure load is given by the analytical solution with equal joint normal and shear stiffnesses (that is Ks=Kn). A lower bound may be obtained using the same solution with Ks<<Kn.

There are some limitations to the models presented here.

Underground rock wedge stability problems are most likely to be concerned with three dimensional wedges in three dimensional stress fields. The load-deformation relationships of discontinuities are known to be non-linear, and the effects of the excavation processes, such as drilling and blasting on the stability of wedges needs to be considered.

Nevertheless, it is believed that these models give a useful insight into the phenomenon.

ACKNOWLEDGEMENTS

The work of Dr. J.W. Bray is acknowledged.

REFERENCES

Barton, N., 1981, "Estimation of in-situ joint properties, Näsliden mine", <u>Application of Rock Mechanics to Cut and Fill Mining</u>, Stephansson, O., and Jones, M.J., ed., IMM, London, pp 186-192.

Bray, J.W., 1981, Personal Communication.

Crouch, S.L., 1976, "Analysis of Stresses and Displacements around Underground Excavations: An application of the Displacement Discontinuity Method", Dept. Civil and Mineral Engineering, University of Minnesota, Minnesota.

Ludvig, B., 1981, "Direct Shear Tests of Filled and Unfilled Joints", <u>Application of Rock Mechanics to Cut and Fill Mining</u>, Stephansson, O., and Jones, M.J., ed., IMM, London, pp 179-185.

Chapter 104

FIREFLOOD MICROSEISMIC MONITORING: ROCK MECHANICS IMPLICATIONS

Maurice B. Dusseault, PhD, P.Eng., and Edo Nyland, PhD, P.Geoph.

Faculties of Engineering and Science
Departments of Mineral Engineering and Physics
University of Alberta
Edmonton, Alberta, Canada, T6G 2G6

ABSTRACT

Numerous consistent seismic signals are being generated in a pilot fireflood in a 750 m deep high permeability unconsolidated channel sand in Eastern Alberta. The pilot has a central air injection well, eight production wells on artificial lift, and two internally placed observation wells. Surface seismometers and a 12-hydrophone downhole string were deployed to detect seismics signals. Date were collected for 22 hours employing signal-triggered event detectors recording on cassette tapes.

Monitoring during twelve hours of steady air injection at 625 m^3/s yielded a clear event frequency of about 50 events/h. The events were mostly clear compressional wave arrivals originating in the vicinity of the firefront. Stopping air injection caused a gradual drop in event frequency, and the rate was reestablished when air injection was reinitiated. Spectral frequency analysis and knowledge of material properties indicate that the events are shears of a stick-slip nature associated with the strain-weakening behavior of the over-compacted reservoir sands. The event amplitude indicates small-scale shearing, and different amplitudes suggest that planes re-shear with lower energy release levels. The triggering mechanism for shear rupture is high shear stress created by thermal expansion in the area of highest thermal gradient.

Other enhanced recovery techniques may also produce useful signals for process control. In particular, steam flooding, hydraulic fracturing, and high-pressure injection operations likely generate mappable microseismic events which can be applied directly to production/injection strategy.

RESERVOIR CHARACTERISTICS

The reservoir is a 750 m deep channel sand of Cretaceous age averaging 35 m in thickness with a width of 1.5 km and a length of over 15 km. The Bodo Pilot reservoir properties are given in Table 1.

Table 1. Reservoir Characteristics

Geometry: 750 m deep, channel sand averaging 35 m in thickness, 1.5 km in width, with an oil-containing extent of at least 15 km along the channel axis.

Porosity: 30% average from geophysical interpretation.

Lithology: Sublitharenitic, fine- to medium-grained, relatively homogeneous but fining upwards, a few stringers of silt, clay, or sharpstone conglomerates.

Mineralogy: Mostly quartz, some lithics and feldspars, minor clays.

Oil Type: Heavy oil averaging 13^0 API gravity.

Permeability: 1-3 Darcy absolute permeability.

Saturation: 0.27 water, 0.73 oil, no free gas in the pilot area. A water leg exists in the bottom few metres of the pilot.

Natural Radioactivity: Very low, averaging 30 API units in the sands.

Sonic Velocity: 2.65-2.75 km/s for P-waves, interpreted from borehole sonic velocity logs.

Diagenetic State: Slightly overcompacted due to deeper burial in geological history, authigenic kaolinite clays are observed, no cementitious material in the clean reservoir sands, mild diagenetic fabric.

In summary, the reservoir is a high-porosity cohesionless sublitharenitic sand.

Alberta's heavy oil deposits, and most other heavy oils, are found in cohesionless sands. The oil contains dissolved gases and there are serious problems of sample disturbance because the fabric of the sands is destroyed as the gas evolves during core recovery (Dusseault, 1980; Dusseault and Van Domselaar, 1982). Core samples of the test reservoir show expansions of about 6% of total volume, or 20% of the pore volume. Porosities of 35% to 37% are calculated in the laboratory, whereas values of about 30% are determined from geophysical bulk density logs and specific gravity data on the rock components (Collins, 1977). The problems of sample disturbance and restoration of in situ properties are common to friable reservoir sands (Donaldson and Staub, 1981). Comparison to tests on similar materials (Dusseault and Morgenstern, 1979; Barnes and Dusseault, 1982) suggest that the Bodo sand is overconsolidated with respect to present stress state; therefore the sand will behave as a strain-weakening material in its undisturbed state at depth under confining stresses.

The in situ stress state is not explicitly known for this region, but data from other areas (Dusseault, 1980; Gough and Bell, 1981) and analysis of geophysical density logs suggest that the vertical total stress is 16.5 MPa at a depth of 750 m, and that this stress is the principal stress. The intermediate and minor principal stresses are horizontal, and the minor stress is probably 70% to 80% of overburden and is oriented at Az 160°. The initial pore pressures in the reservoir are less than hydrostatic, but the present pore pressures are a function of the geometry of the fireflood as the air is being injected at a pressure below the hydraulic fracture pressure, and the production wells are produced with bottom-hole pumps which create a condition of low pressure to maximize the flow gradients. Because the burned reservoir rock is extremely permeable, the pressure drop takes place in the fire zone and the unburned region. The failure criterion is considered to be appropriate for materials of this type.

Figure 1 is a stress-strain curve for a sample of Athabasca oil sands under two different effective confining stresses. The plot shows that the materials are dense, dilatent, have relatively small strains to failure, and are strain-weakening. The sands have no stress-independent cohesion, therefore they have no tensile strength. As the Bodo sands are similar in porosity, mineralogy, and geological history to the Athabasca and Cold Lake oil sands deposits, those materials are considered suitable model materials for behavior in the absence of tests on undisturbed core from this pilot. Therefore, the range of initial compressibility is estimated at $0.2-0.8 \times 10^{-6}$ kPa^{-1}. Elastic moduli can be estimated from the sonic transit time of 2.65-2.75 km/s by assuming a reasonable value of Poisson's Ratio of 0.3.

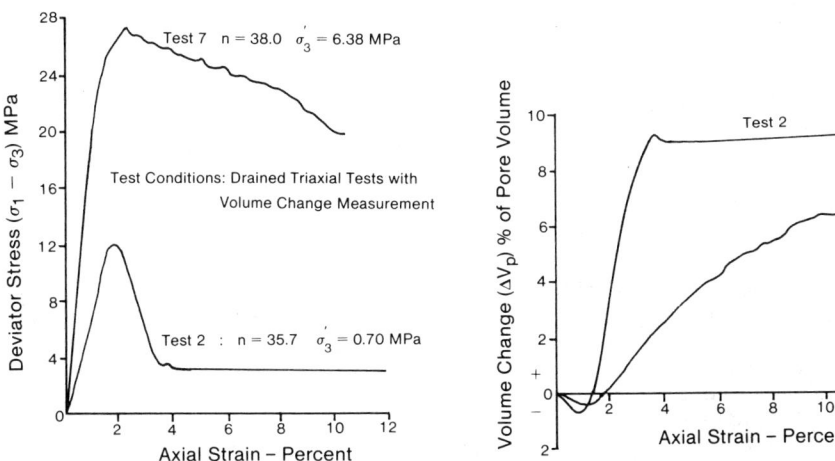

Figure 1: Drained Triaxial Test Behavior of Athabasca Oil Sand.

Since sonic moduli are higher than "static" moduli by factors of 1.3 to 3 (Lama and Vutukuri, 1978), moduli derived in this way must be reduced for "static" analysis (Dusseault and Simmons, 1980).

EXPERIMENTAL STRATEGY

Three surface seismometers and 12 hydrophones in a single water-filled borehole were deployed. The hydrophones were spaced at 15 m intervals along a seismic velocity string lowered into an observation well, and the lower two hydrophones were located within the reservoir body. Three event detector cassette-recording units collected data at a sample rate of 200 samples/s on three channels each. The recorders were triggered by a short-term average to long-term average amplitude ratio arrived at by trial and error in the initial stages of the field experiment. Upon triggering, the individual recording units acquired data in fixed blocks of 2.16 s length. The first block for a trigger contains about 1.5 s of pre-event buffer, and, in practice, a post-event buffer of about 0.4 s was acquired, although a specific post-event buffer was not set on the detector. Few events lasted long enough to result in more than one data block. The three units (three channels each) could not be slaved to one another for uniform triggering of all nine channels. Data on cassettes were transferred to computer tape for plotting, statistical analysis, and spectral density analysis. The recording strategy was determined in the field, and the gain, bandpass, and trigger level were set on the basis of the observed responses. Continuous recording or low trigger levels would have resulted in huge amounts of data, on the order of 150×10^6 sample points for 24 hours continuous recording. This translates into 600 cassettes, whereas 35 cassettes were actually filled.

Among the equipment problems encountered in the field were leaking connectors on the velocity string and a relatively high noise level within the reservoir as the result of the processes taking place.

RESULTS

Over 400 excellent quality compressional wave arrivals were recorded during a 22 h recording interval on the hydrophones. No useful data were recorded at any time on the surface seismometers because of the relatively small amplitude of the events and the presence of a glacial low-velocity layer about 75 m thick at this site. No teleseisms or activity extraneous to the pilot were recorded.

Figure 2 shows the logical flow of events for data processing. The human input is necessary to assess whether real events are being picked up, or whether the triggering was initiated by noise or extraneous events. Figure 3 shows a plot of a poor quality event registered on one recording unit coupled to two hydrophones and a surface seismometer. The lowest trace in these diagrams is the trigger channel. Figure 4 contains traces of several events of varying quality, from excellent to poor. Figure 5 shows a data trace with arrival times picked on a CRT terminal display, and peak arrivals determined by a

FIREFLOOD MICROSEISMIC MONITORING

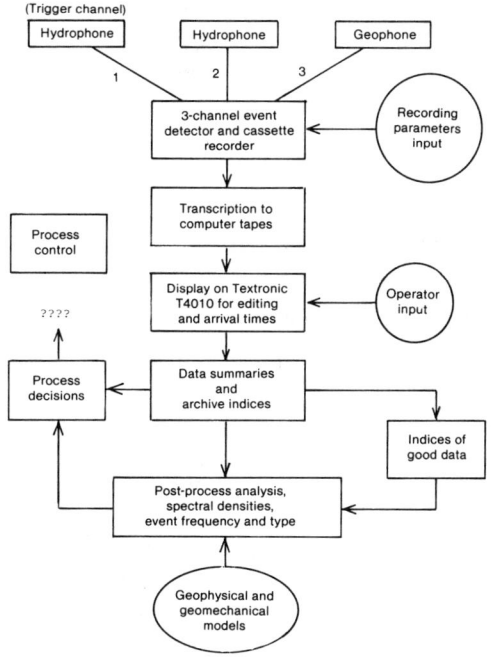

Figure 2. Logical Flow Chart for Data Analysis.

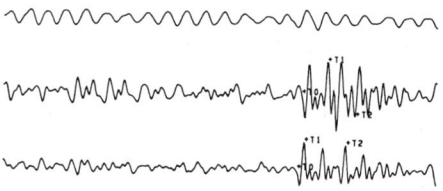

NOTE: The signal is two superposed doublets, indicating near-simultaneous events of a similar nature.

Figure 3. Trace of a Poor-Quality Event.

programmed algorithm. Arrival times for the P-wave can be consistently picked to an accuracy of ±0.005 s. All the good quality events show a wave doublet, each wave consisting of one to two cycles, and the doublets separated by about 0.2 seconds. Some traces showed a second doublet which registered about 0.5 seconds after the first. This may be a reflection or a second event initiated within the reservoir in some way by the first event. The noise preceding and following the events is of seismic origin associated with the process. This has been confirmed by a second monitoring period in a quiet reservoir situation. Figure 6 contains an example of two doublets as well as a

ISSUES IN ROCK MECHANICS

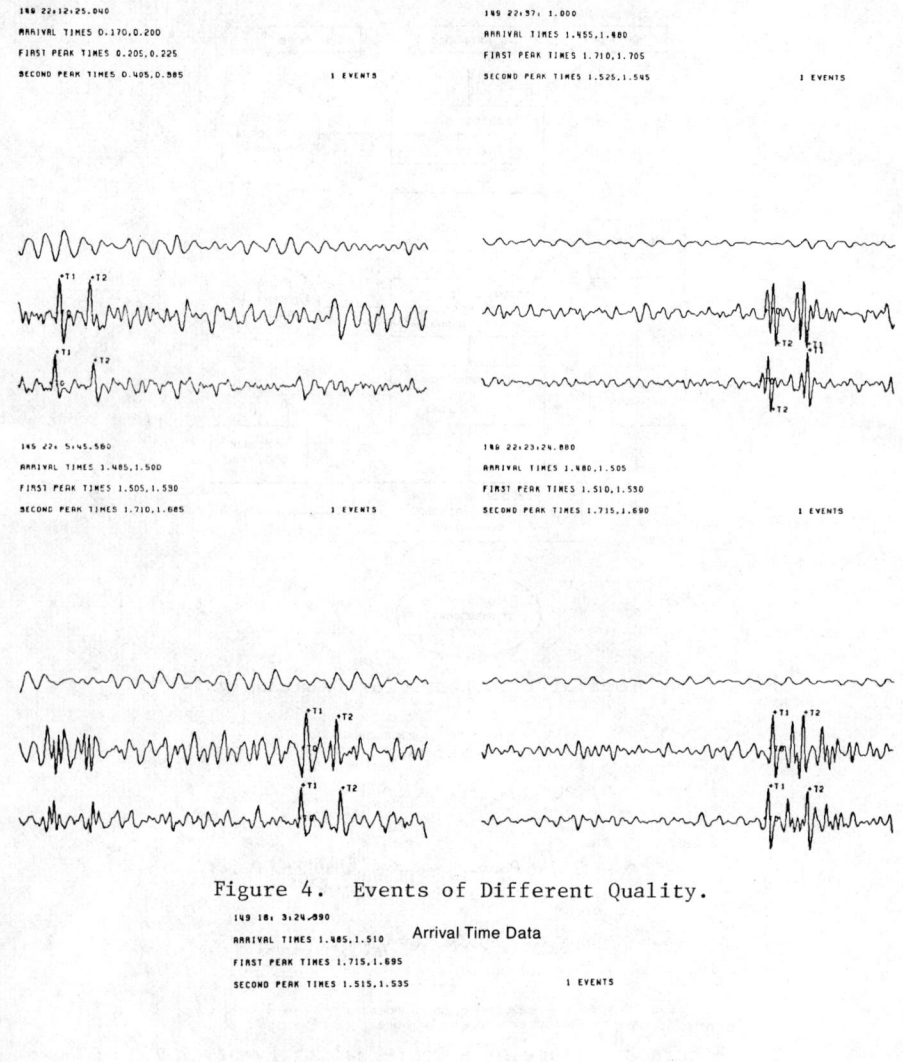

Figure 4. Events of Different Quality.

Figure 5. CRT Display of Three Channels with Arrival Times Determined.

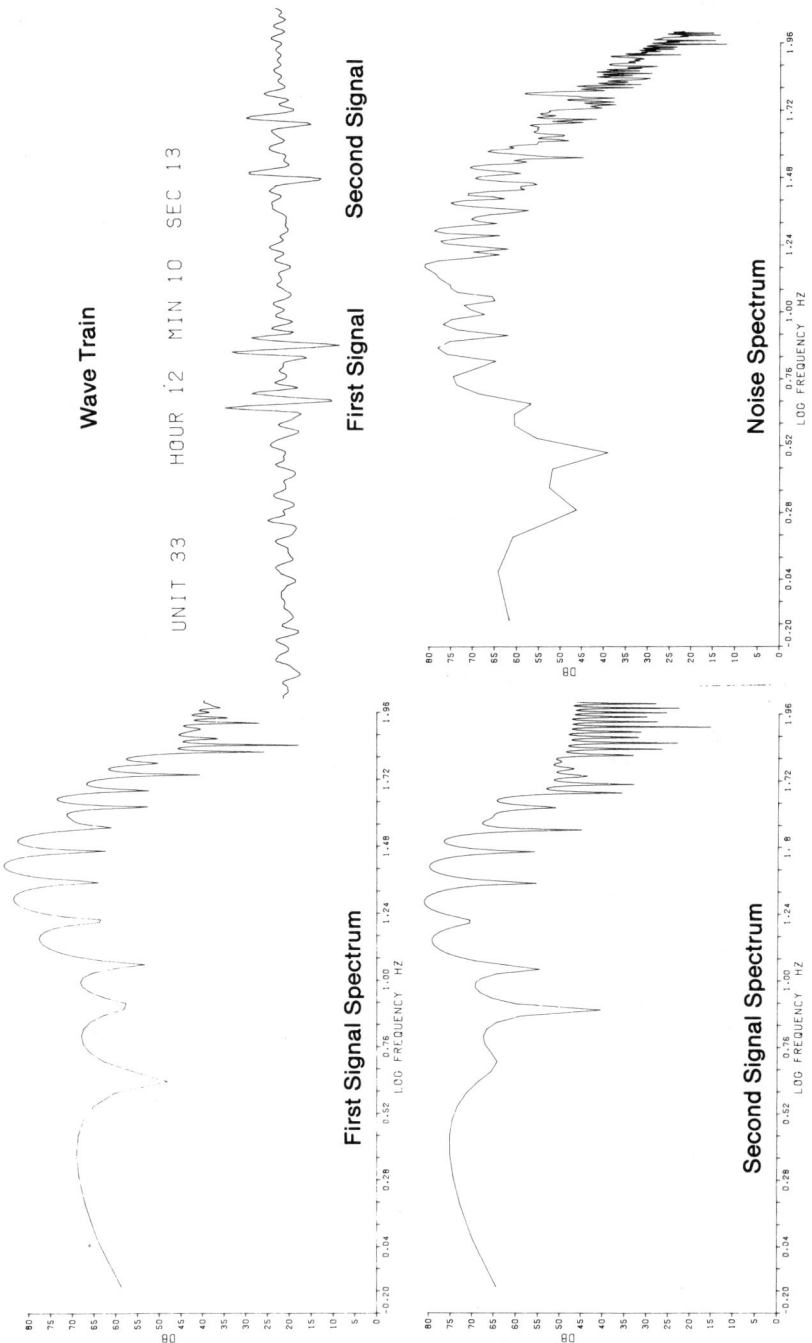

Figure 6. Frequency Spectral Density of Two Doublets.

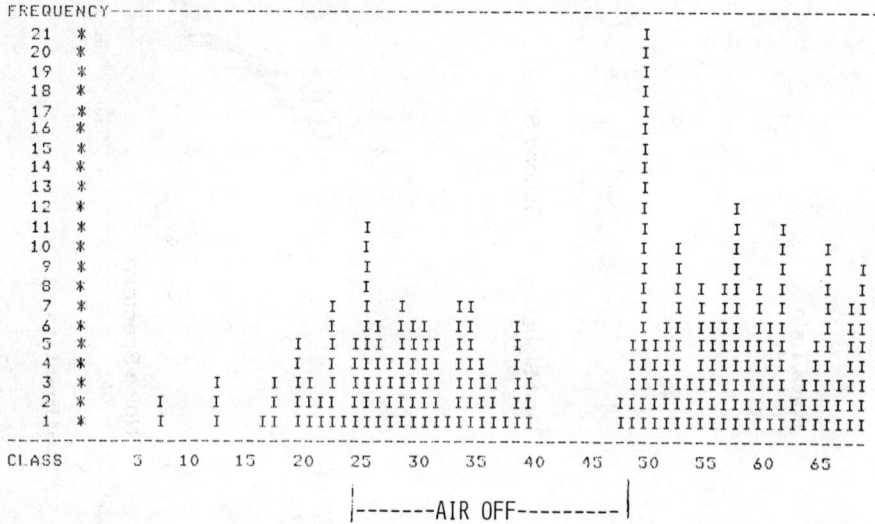

Figure 7. Event Frequency versus Time for Period of Air Shut-Off.

frequency spectral analysis of the noise, the first (typical) doublet, and the second doublet. These data have been chosen to be representative of the information collected.

During the experiment, the air injection was stopped for a period to assess the effect on event frequency. Figure 7 shows how the event frequency declined gradually from the level of about 50 recorded events/h. After recommencing air injection, the pilot displayed a sudden burst of activity, then settled down to the previous rate.

GEOPHYSICAL INTERPRETATION

The seismic signal is characterized by a very simple pulse form. This and the lack of significant local activity (Milne et al., 1978) indicates that the events are clear compressional wave arrivals originating from the vicinity of the burn. The spectral density analyses are all similar to those in Figure 6 and are characterized by a well-defined plateau at low frequencies, a relatively uniform falloff at high frequencies, and a corner frequency in the range of 25 to 40 Hz. This is the classic signature of an earthquake (Brune, 1970; 1971) and is interpreted in terms of shear failure. The only interpretation of the doublets is that they are the result of a wave propagating up and down the borehole, and move-out analysis shows that the apparent velocity is indeed that of sound in water, 1.5 km/s.

The spectra must be treated with caution. The spiky holes in the spectra result from interference between the two similar parts of the signature. The corner frequency is too low to accurately reflect the postulated source characteristics, as a source size on the order of 50 to 70 m is implied, which is too large for the dimensions of the fireflood. The contamination of the spectrum may be explained by the presence of the observation well, which, in its role as wave guide, also acts as a signal filter.

It is uncertain whether the dropoff in high frequency content is the result of attenuation or is a source characteristic. The spectral difference for the two signals in Figure 7 indicates a different source location with different slip velocity, stress drop, or area. The spectral similarity to the noise shows that the noise is also seismically generated in the reservoir. Calculations of transmission and reflection coefficients (Aki and Richards, Volume 1, 1980) suggest that the reservoir traps the seismic energy except where a low reflectance path can be found. The observation well is such a path.

GEOMECHANICAL MODEL

The geomechanical model chosen to explain the behavior is one of shear rupture (Mode II fracture) in the zone of highest thermal gradient, which is in the fireflood front. The events cannot be hydraulic fracture features as the injection pressure is below the fracture pressure and no mechanism can be postulated in a permeable reservoir to give rise to the high pressures required. The sands are also cohesionless and no significant energy release mechanism is available from tensile phenomena (Mode I fracture). The slow dropoff in event frequency when air injection ceases implies a thermally-activated process for event generation. The actual stress path to failure is not explicitly known. It is a combination of these simple stress paths with the added complexity that material properties, and therefore stress path, are changing with temperature. A thermally expanding zone will tend to increase the stress normal to the zone and decrease the stresses parallel to the front. This thermoelastic stress field, as yet unmodeled, must be added to the initial stress fields to determine the true principal stress orientations in and around the thermal front. Because the greatest stress changes occur in the zone of highest thermal gradient, this zone is the likely source of the shear events.

Figure 8 is a conceptual sketch of the stresses across an event-generating shear plane. Shear stress accumulates until the material strength is locally exceeded, failure occurs, and the sand loses some of its capacity to withstand shear stress because it is strain-weakening. The shear stress is then redistributed to the "shoulders" of the initially ruptured area, where rupture also occurs if the failure criterion is exceeded. The process propagates outwards extremely rapidly until sufficient lower stressed area is available for redistribution without failure, and the rupture then terminates. The unfailed region is now at a higher shear stress level than before rupture of the adjacent plane, and will be a preferred zone for the next failure. Repeated activation of failed planes may be the cause of much of the noise and of some of the lower amplitude and "smeared" events.

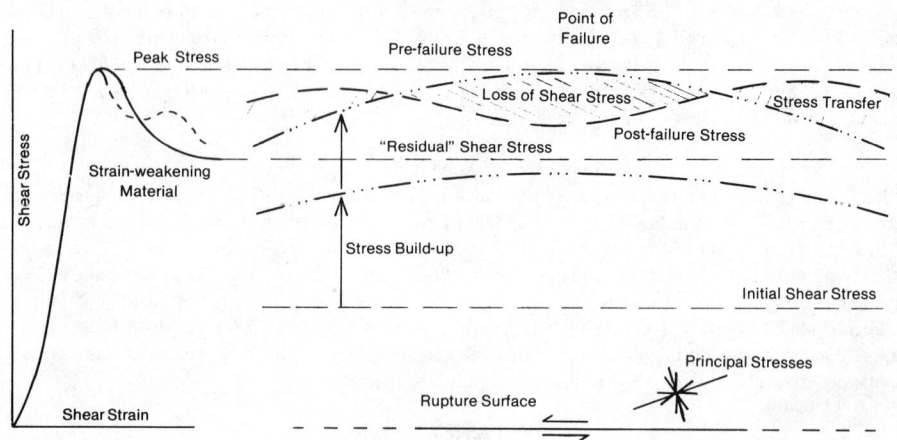

Figure 8. Geomechanical Model of Thermally Activated Rupture.

Although it is not directly relevant to rock mechanics, a fireflood such as this may provide a useful scale model of the physics of earthquake generation in the much larger tectonic systems involved in large magnitude seismic events. Earthquakes are generally not thermally activated, but their driving mechanism must involve an external energy source which may be similar to the driving mechanism postulated for these data.

IMPLICATIONS FOR PROCESS MONITORING

As a result of the wave consistency and the reliability of arrival time determination, a spatial discrimination of 10 to 15 m can be achieved with a triple borehole strategy. The signal pick-up is remote, which is not the case in temperature and pressure monitoring. The accuracy is high as the sensors are within the reservoir, as opposed to magnetotelluric or tilt measurement methods. New devoted wells are not necessary; production or injection wells which are temporarily inactive may be used, providing they have a blind tubing string or can retain water for hydrophone coupling. The use of wall-coupled triaxial seismometers to collect the full P and S wave trains will allow superior spatial resolution and analysis.

The technique seems to have application in fireflood mapping to determine the areas of highest burn rate, front location, and front velocity. These data, in conjunction with other process data, may allow the operator to adjust injection/production strategy to avoid premature well burn-through, and to maximize the areal sweep in a fireflood. Finally, other in situ processes may generate mappable microseismic events of use to operations.

CONCLUSIONS

1. Consistent interpretable mappable microseismic events are generated during fireflooding.

2. Surface seismometers are of no value if event amplitudes are small, particularly in areas with a thick low-velocity layer.

3. The events recorded are shear ruptures energized by thermal expansion acting on cohesionless, over-compacted sand.

4. The events are associated with the fire front and may be used to map its presence and velocity.

ACKNOWLEDGMENTS

We sincerely thank Norcen Energy Resources Ltd. and their partners in the Bodo Pilot for their support in this project. The Natural Sciences and Engineering Research Council of Canada provides support to both of us. The Alberta Oil Sands Technology and Research Authority supports MBD.

REFERENCES

Aki, K., and Richards, P.G., 1980, Quantitative Seismology Theory and Methods, Vols. I, II, W.H. Freeman & Co., San Francisco, CA.

Barnes, D.J., and Dusseault, M.B., 1982, "The Influence of Diagenetic Micro-Fabric on Oil Sands Behavior," Canadian Journal of Earth Sciences, 19, pp. 804-818.

Brune, J.N., 1970, "Tectonic Stress and the Spectra of Seismic Shear Waves from Earthquakes," Journal of Geophysical Research, 75, pp. 4997-5009.

Brune, J.N., 1971, Correction, op. cit., Journal of Geophysical Research, 76, p. 5002.

Collins, H.N., 1977, "Log-Core Study of Bodo Upper Mannville 'B' Pool," Interim Report IR-4, Petroleum Recovery Institute, Calgary.

Donaldson, E.C., and Staub, H.L., 1981, "Comparison of Methods for Measurement of Oil Saturation," Preprint 10298, Society of Petroleum Engineers of AIME.

Dusseault, M.B., 1980, "Sample Disturbance in Athabasca Oil Sands," The Journal of Canadian Petroleum Technology, 20, pp. 85-92.

Dusseault, M.B., 1980, "The Behaviour of Hydraulically Induced Fractures in Oil Sands," Underground Rock Engineering, 13th Canadian Rock Mechanics Symposium, CIM Special Volume 22, pp. 36-41.

Dusseault, M.B., Morgenstern, N.R., 1979, "Locked Sands," Quarterly Journal of Engineering Geology, 12, pp. 117-131.

Dusseault, M.B., and Simmons, J.V., 1980, "Fracture Orientation Changes during Injection," 33rd Canadian Geotechnical Conference, Calgary, proceedings published.

Dusseault, M.B., and Van Domselaar, H.R., 1982, "Canadian and Venezuelan Oil Sand: Problems and Analysis of Uncemented Gaseous Sand Sampling," Updating Subsurface Sampling of Soils and Rocks and Their In-Situ Testing, Engineering Foundation Conference, Santa Barbara, CA, in press.

Gough, D.I., and Bell, J.S., 1981, "Stress Orientations from Oil Well Fractures in Alberta and Texas," Canadian Journal of Earth Sciences, 18, pp. 638-645.

Lama, R.D., and Vutukuri, V.S., 1978, Handbook on Mechanical Properties of Rocks, Vol. II, Trans Tech Publications, Clausthal, Germany.

Milne, W.G., and Rogers, R.P., Riddihough, McMechan, G.A., and Hyndman, R.D., 1978, "Seismicity of Western Canada," Canadian Journal of Earth Sciences, 15, pp. 1170-1193.

Chapter 105

SHEAR STABILITY OF MINE PILLARS IN DIPPING SEAMS

By William G. Pariseau

Department of Mining and Fuels Engineering
University of Utah
Salt Lake City, Utah

ABSTRACT

The extraction ratio approach to pillar design in flat seams is based on a mathematically exact analysis that is a reasonable physical approximation to room-and-pillar layouts in many instances. The essentials of the approach are reviewed and then mathematically approximate formulas for pillar design in inclined seams of arbitrary dip are proposed. Detailed numerical analysis of stress about a typical 5 entry, 4 pillar panel at dips of 0, 15, 30, 45, 60, 75 and 90 degrees at low and high extraction ratios show that the proposed inclined seam extraction ratio formulas for pillar normal and shear stresses, although conservative, have an accuracy quite adequate for engineering purposes. Appropriate pillar safety factors for compressive and shear failure and a safety criterion for overturning are specified.

INTRODUCTION

In flat, tabular ore bodies mined by room-and-pillar methods, the extraction ratio formula[1] provides a reasonable basis for pillar design. This approach which is also known as the tributary area method is based on an exact equilibrium analysis of the forces acting on a typical pillar in a largy array of similar pillars. The extraction ratio formula provides a satisfactory approximation to actual room-and-pillar mines in many cases including gently dipping seams, and thus allows one to estimate a pillar safety factor based on average pillar stress and strength.

---

1 Obert, L. and Duvall, W. I., 1967, Rock Mechanics and The Design of Structures in Rock, Wiley, N. Y., p. 541.

However, when the dip is arbitrary, the presence of shear forces complicates the situation. No formula comparable to the extraction ratio formula appears to have been developed for pillar design in dipping seams. In fact, there appears to be very little information available concerning the shear stability of pillars in underground mines of either the soft or hard rock type. At present, a detailed analysis of the stress changes induced by mining is required for the calculation of the forces, average stresses and safety factors for pillars in dipping seams. Although such calculations may be considered routine for the rock mechanics specialist, they are generally beyond the resources of the engineer assigned to ground control at an operating mine. Elaborate tables of results of detailed stress analyses or intricate plots of the results are also unlikely to be used without special training or additional simplification. This paper examines the possibility of simplification by extending the extraction ratio approach from flat seams to seams of arbitrary dip.

A brief review of the extraction ratio formula is presented in the second part of this paper. A hypothesis for extending the extraction ratio formula to dipping seams is presented in part three and tested in a series of numerical experiments in part four of this paper. Some additional results and conclusions are presented in part five.

EXTRACTION RATIO FORMULA REVIEW

The extraction ratio formula for pillar design is based on two essential assumptions:

(i) the pre-mining stress field is characterized by a vertical stress S_v equal to the average unit weight of overburden times the depth h, a horizontal stress S_h equal to a constant K_o times the vertical stress, and no shear stresses.

(ii) a large array of pillars is formed, so that vertical planes through the adjacent entry and crosscut centerlines are planes of symmetry.

A consequence of these two assumptions that is of critical importance to the development of the extraction ratio formula is that the forces acting on the overburden block after mining are equal to the corresponding forces before mining. Fig. 1 shows a typical overburden block before and after mining. Both sets of forces must individually satisfy equilibrium, of course.

The forces before mining are calculated from the assumed pre-mining stress field. The forces after mining are then expressed exactly as an average stress times an area. Equating the two sets of forces results in the extraction ratio formula for the average pillar stress. A pillar safety factor can then be computed.

Accordingly, the pre-mining stress field is given by

$$S_v = \gamma h \quad \text{and} \quad S_h = K_o S_v \qquad (1)$$

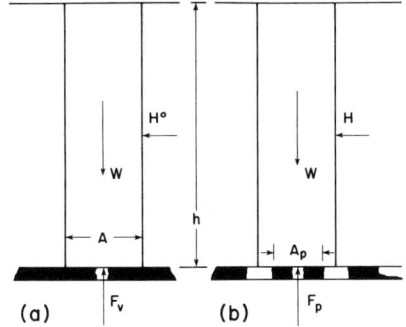

FIG. 1 FLAT SEAM OVERBURDEN BLOCK FORCES
(a) PRE- AND (b) POST-MINING.

where S_v = vertical stress at depth H, γ = average unit weight of overburden, S_h = horizontal stress (any direction), and K_o = constant.

With reference to Fig. 1, the pre-mining forces in the vertical direction are

$$F_v = \int_A S_v dA = \gamma h A \quad \text{and} \quad W = \int_V \gamma dV = \gamma A h \qquad (2)$$

where dA = element of area and dV = Adh = volume element. The average pre-mining vertical stress at seam level is by definition

$$S_v = F_v/A = \gamma h.$$

Equilibrium of the vertical pre-mining forces requires that

$$W - F_v = 0. \qquad (3)$$

Substitution from Eq. 2 into the first of Eq. 3 shows that the assumed pre-mining stress field is in fact an equilibrium stress field.

Equilibrium after mining requires that

$$W - F_p = 0. \qquad (4)$$

Comparison with Eq. 3 shows that indeed the post-mining forces are equal to the pre-mining forces, that is, $F_p = F_v$.

The average vertical pillar stress is by definition

$$S_p = F_p/A_p$$

where A_p = the pillar cross-sectional area. Hence $S_p A_p = S_v A$ and

$$S_p = S_v A/A_p. \tag{5}$$

The extraction ratio is by definition

$$R = A_m/A = 1 - A_p/A \tag{6}$$

where A_m = area mined = $A - A_p$. The extraction ratio formula for average pillar stress is thus

$$S_p = S_v/(1-R). \qquad \text{(flat seam formula, 7)}$$

A pillar safety factor F_c with respect to compression normal to the seam is

$$F_c = C_p/S_p \tag{8}$$

where C_p = pillar "strength."

Equilibrium of horizontal forces in the flat seam case is of no consequence for the pillar; details are therefore unnecessary.

HYPOTHESIS FOR INCLINED SEAMS

Extension of the applicability of the extraction ratio formulation to inclined seams of arbitrary dip leads to a hypothesis worthy of testing. Justification lies in the expectation that the approximation involved is well within the accuracy required for engineering design.

Statement of Formulas and Hypothesis

The essential steps followed in arriving at the flat seam extraction ratio formula for average pillar stress (Eq. 7) when closely examined provide a basis for extending the formula to seams of arbitrary dip. In the flat seam case, the vertical pre-mining stress S_v is also the direct stress S_n acting normal to the seam. Thus, in the general case

$$S_p = S_n/(1-R) \qquad \text{(inclined seams, 9)}$$

where S_p is the average pillar stress in the normal direction and the extraction ratio R is determined by areas measured in the plane of the seam.

The average shear stress in the pillar after mining is given by an analogous formula

$$T_p = T_n/(1-R) \qquad \text{(inclined seams, 10)}$$

where T_n = average pre-mining shear stress acting tangential to the seam and T_p' = average pillar shear stress. Eq. 9 and 10 reduce to the flat seam case when the dip angle is zero, since in that case $S_n = S_v$ and $T_n = 0$.

A factor of safety F_s with respect to shear along the dip is

$$F_s = (S_p \tan\phi + K)/T_p \tag{11}$$

where the contributions to shear strength in the numerator are from the frictional resistance mobilized by the force normal to the dip acting through the coefficient of friction, $\tan\phi$, and the force of cohesive resistance K. A factor of safety with respect to compression across the seam normal to the dip is given by Eq. 8.

The inclined seam extraction ratio formulas (Eq. 9 and 10) constitute a hypothesis to be tested. The hypothesis is that their accuracy is quite adequate for engineering purposes.

Justification of the Hypothesis

Although the pre-mining stress field assumed in the flat seam case is also an equilibrium stress field in the inclined seam case, the symmetry of the flat seam case is lacking in the inclined seam case. Thus only the first assumption underlying the traditional extraction ratio formula is applicable.

The forces before and after mining must still satisfy equilibrium. However, they are not generally equal. If they were, then the extended extraction formulas (Eq. 9 and 10) would be exact. To see that this is the case, one notes that if N^o and N are the forces normal to the seam before and after mining as shown in Fig. 2 and 3, and if $N = N^o$, then $S_p A_p = S_n A$. Formula 9 follows immediately. Similarly, if the tangential forces are $T^o = T_n A$ and $T = T_p A_p$ before and after mining, and if $T = T^o$, then $T_p A_p = T_n A$. Formula 10 follows.

An analysis of force equilibrium shows that before mining

$$\begin{aligned} N^o &= [W - (V_2^o - V_1^o)]c + (H_2^o - H_1^o)s \\ T^o &= [W - (V_2^o - V_1^o)]s - (H_2^o - H_1^o)c. \end{aligned} \tag{12}$$

After mining,

$$\begin{aligned} N &= [W - (V_2 - V_1)]c + (H_2 - H_1)s \\ T &= [W - (V_2 - V_1)]s - (H_2 - H_1)c \end{aligned} \tag{13}$$

where $c = \cos(\alpha)$, $s = \sin(\alpha)$, and α = dip angle.

As in the flat seam case, explicit expressions for the pre-mining forces follow from the pre-mining stress field. All quantities in Eq. 12 are thus known. The steps and results are summarized in Tables

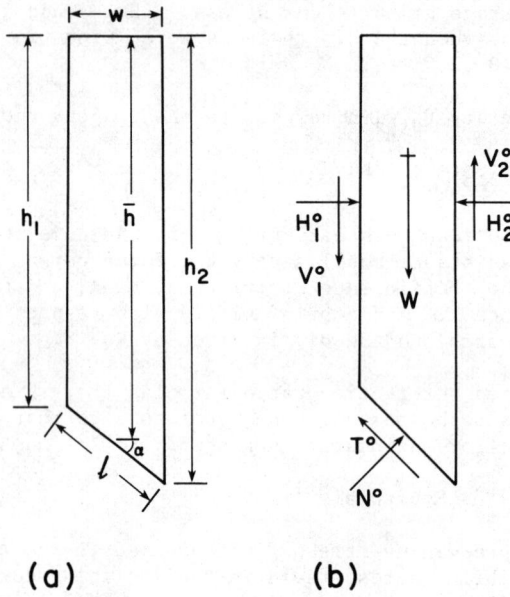

FIG. 2. INCLINED SEAM OVERBURDEN BLOCK PRE-MINING GEOMETRY (a) AND FORCES (b).

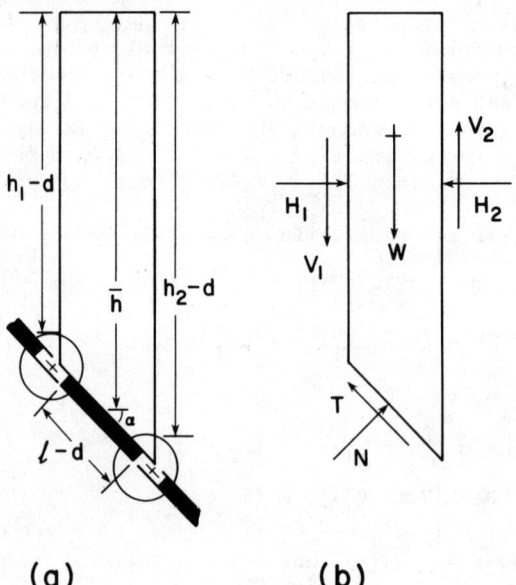

FIG. 3. POST-MINING OVERBURDEN BLOCK GEOMETRY (a) AND FORCES (b).

SHEAR STABILITY OF MINE PILLARS

TABLE 1. OVERBURDEN BLOCK STRESSES.

Stress	Time	Pre-mining	Post-mining
Vertical		$S_v = h\gamma$	$\sigma_v = S_v + \Delta\sigma_v$
Horizontal		$S_h = K_o S_v$	$\sigma_h = S_h + \Delta\sigma_h$
Shear		$S_{hv} = 0$	$\sigma_{hv} = S_{hv} + \Delta\sigma_{hv}$
Normal		$S_n = S_v c^2 + S_h s^2$	$S_p = \sigma_v c^2 + \sigma_h s^2 + 2\sigma_{hv} sc$
Shear		$S_{nt} = -(S_v - S_h)sc$	$T_p = -(\sigma_v - \sigma_h)sc + \sigma_{hv}(c^2 - s^2)$

v = vertical direction, h = horizontal direction, Δ = change,

γ = average unit weight of overburden, h = depth to point of interest

K_o = ratio of horizontal to vertical pre-mining stresses (a constant),

c = cos(α), s = sin(α), α = dip angle, -α = angle of rotation v,h to n,t

n = direction normal to the seam, t = direction tangential to the seam

TABLE 2. OVERBURDEN BLOCK FORCE FORMULAS.

Force	Time	Pre-mining	Post-mining
Weight		$W = \int_V \gamma dV$	$W = \int_V \gamma dV$
Horizontal		$H_1^o = \int_0^{h_1} bS_h dh$	$H_1 = \int_0^{h_1} b\sigma_h dh$
		$H_2^o = \int_0^{h_2} bS_h dh$	$H_2 = \int_0^{h_2} b\sigma_h dh$
Vertical		$V_1^o = \int_0^{h_1} bS_{hv} dh$	$V_1 = \int_0^{h_1} b\sigma_{hv} dh$
		$V_2^o = \int_0^{h_2} bS_{hv} dh$	$V_2 = \int_0^{h_2} b\sigma_{hv} dh$
Normal		$N^o = \int_{\ell_1}^{\ell_2} bS_n d\ell$	$N = \int_{\ell_1}^{\ell_2} b\sigma_n d\ell$
Shear		$T^o = \int_{\ell_1}^{\ell_2} bS_{nt} d\ell$	$T = \int_{\ell_1}^{\ell_2} b\sigma_{nt} d\ell$

b = breadth of overburden block (into the page), dV = volume element

h = depth from surface, ℓ = entry spacing on the dip, down dip distance

TABLE 3. OVERBURDEN BLOCK FORCES AFTER INTEGRATION.

Force	Pre-mining	Post-mining
Weight	$W = \gamma b \bar{h} \ell c$	$W = \gamma b \bar{h} \ell c$
Horizontal	$H_1^o = b K_o \gamma h_1^2 / 2$	$H_1 = H_1^o + \int_0^{h_1} b \Delta \sigma_h \, dh$
	$H_2^o = b K_o \gamma h_2^2 / s$	$H_2 = H_2^o + \int_0^{h_2} b \Delta \sigma_h \, dh$
Vertical	$V_1^o = 0$	$V_1 = 0 + \int_0^{h_1} b \Delta \sigma_{hv} \, dh$
	$V_2^o = 0$	$V_2 = 0 + \int_0^{h_2} b \Delta \sigma_{hv} \, dh$
Normal	$N^o = b \gamma \bar{h} \ell (c^2 + K s^2)$	$N = N^o + \int_D b \Delta \sigma_n \, d\ell$
Shear	$T^o = b \gamma \bar{h} \ell (1 - K_o) s c$	$T = T^o + \int_D b \Delta \sigma_{nt} \, d\ell$

$\bar{h}$ = average depth = $(h_1 + h_2)/2$ = depth to pillar center at top,

D = bd = area of influence of entries = (h_1-d to h_1 and h_2-d to h_2 on sides, ℓ_1 to ℓ_1-d and ℓ_2-d to ℓ_2 on the block bottom)

1, 2 and 3. However, none of the quantities in Eq. 13 can be considered known. A detailed analysis of the stress changes induced by mining is required for their evaluation. The forces before and after mining are thus not expected to be equal. Evaluation of their exact differences again would require a detailed analysis of stress.

There are, however, reasons to believe that the differences between pre- and post-mining forces are small provided that the depth is large compared with the opening width. Before mining, the vertical shears $V_1^o = V_2^o = 0$. After mining, $V_2 - V_1$ is expected to be negligible. The reason for this is illustrated in Fig. 3 which shows "zones" of influence about the entries adjacent to the pillar. A zone of influence defines a region of significant stress concentration or change. Beyond the zone of influence, the stresses are close to their pre-mining values. Stress concentrations decay rapidly with distance from the periphery of an opening. As a rule of thumb, the influence of the opening extends as high as the opening is wide and as wide as the opening is high. One may double these distances, but the height of influence would be only 12.2 m (40 ft) for a 6.1 m (20 ft) wide entry. In a 1.5 m (5 ft) thick seam, the pre-mining stress would be reached 3 m (10 ft) into the pillar. These distances are but small fractions of the depth of many underground mines.

The vertical shear forces are of particular interest. They are given as the product of the average shear stress times the appropriate area. Beyond the zone of influence of the openings, the vertical shear stress is zero and so therefore is its contribution to the shear force.

Within the zone of influence the shear stress is not likely to reach a substantial value because it is zero at the roof line as well as at the limit of the zone of influence. V_1 and V_2 are thus expected to be small (relative to the weight of the overburden block). Their difference which enters the post-mining equations of equilibrium (Eq. 13) should be even less. Hence, it appears that a reasonable approximation to post-mining equilibrium can be obtained by setting $V_2 - V_1 = 0$. ($V_2^o - V_1^o = 0$.) A similar line of reasoning suggests setting $H_2 - H_1 = H_2^o - H_1^o$, which generally is not zero. It is then approximately true that $N = N^o$ and $T = T^o$. The extended extraction ratio formulas (Eq. 9 and 10) follow.

NUMERICAL TESTS

A series of finite element analyses were carried out in order to quantitatively test the accuracy of the proposed extraction ratio formulas for inclined seams. The analyses were two-dimensional (plane strain) and were performed within the elastic range. A homogenous body was assumed in order to ascertain the effects of seam dip alone. The mining geometry modeled is shown in Fig. 4 and consists of a panel of five 6.4 m (20 ft) wide entries driven on strike on 30.4 m (100 ft) centers. Entry height is 3.1 m (10 ft). The extraction ratio is low at 20 percent. Dips were 0, 15, 30, 45, 60, 75 and 90 degrees. A second series at a relatively high extraction ratio of 60 percent was also run. Fig. 5 shows the seam level geometry of the two panels. In all runs the roof centerline of the central entry was held constant at a depth of 305 m (1000 ft).

In order to compare the results of the numerical analyses with the proposed extraction ratio formula predictions, Eq. 9 and 10 are rewritten as

$$S_p = K_p S_n \quad \text{and} \quad T_p = K_t T_n \qquad (14)$$

The factors K_p and K_t in Eq. 14 are pillar stress concentration factors for the normal and shear stress in the pillar. According to Eq. 9 and 10, K_p and K_t depend only on the extraction ratio and are independent of dip and depth. In the low extraction ratio case

$$K_p = 1.25 \quad \text{and} \quad K_t = 1.25. \qquad (15)$$

The stresses S_n and T_n are the pre-mining normal and shear stresses and are thus known:

$$S_n = \gamma \bar{h}[(1 + K_o) + (1 - K_o)\cos(2\alpha)]/2$$
$$T_n = \gamma \bar{h}[(1 - K_o)\sin(2\alpha)]/2 \qquad (16)$$

where $\bar{h}$ is the depth to the considered pillar center. The stresses S_p and T_p are obtained from the finite element analysis output by an

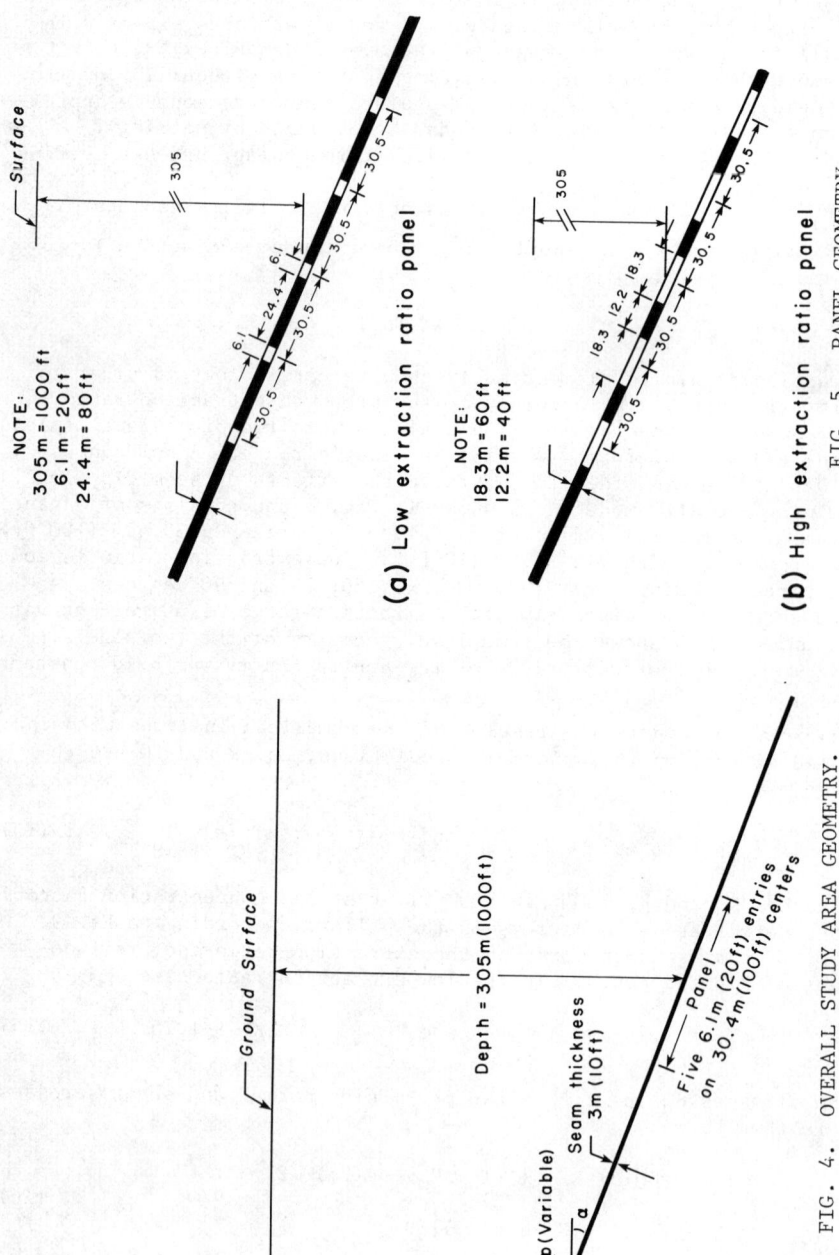

FIG. 5. PANEL GEOMETRY.

FIG. 4. OVERALL STUDY AREA GEOMETRY.

element stress averaging process over all elements in a pillar. The results are summarized in Table 4. Inspection of Table 4 shows that essentially constant values of K_p and K_t of 1.24 is obtained in all pillars at all dips. The values calculated from the suggested extensions of the extraction ratio formula (Eq. 9 and 10) are thus within 1 percent of the actual values. In this regard, the extraction ratio values are expected to be somewhat higher than the five entry panel values obtained here and are therefore conservative.

The values for the stress concentration factors obtained from the extended extraction ratio formulas in the high extraction ratio case are

$$K_p = 2.50 \quad \text{and} \quad K_t = 2.50 \tag{17}$$

The actual values obtained from the numerical analyses are presented in Table 5 which also contains a summary of the average stresses in each of the four pillars for all seven values of dip. The results in Table 5 are generally within 5 to 10% of the extraction ratio values. The latter are again higher and thus conservative with respect to the actual values contained in Table 5. The accuracy of the extended extraction ratio formulas (Eq. 9 and 10) for inclined seams thus appears to be quite adequate for engineering purposes.

ADDITIONAL RESULTS AND CONCLUSION

Additional results of interest concern pillar overturning and a simple approximate formula for pillar stress. Equilibrium of forc or equivalently a safety factor greater than unity based on forces does not guarantee moment equilibrium. There is a possibility of a rotational or kinematic instability if the pillar is high relative to its width. Consider the pillar shown in Fig. 6. Summation of moments about the pillar center requires

$$(T_1 + T_2)(H_p/2) = (N_1 + N_2)(X/2) \tag{18}$$

The normal forces must shift up and down dip in order to balance the couple associated with the shear forces. This tends to load the pillar diagonally from updip roof to downdip floor. Force equilibrium requires $T_1 = T_2 = T$ and $N_1 = N_2 = N$, so that $TH_p = NX$. Thus,

$$(T/N) = (X/H_p) \leqq (W_p/H_p). \tag{19}$$

The inequality (19) is a well known elementary criterion for preventint overturning of blocks and so forth. Within the present context and in view of Eq. 14

$$(W_p/H_p) \geqq (T_n/S_n), \tag{20}$$

that is, the pillar width-to-height ratio measured parallel and per-

TABLE 5. SUMMARY OF NUMERICAL RESULTS (R=60%).

Pillar	S_n	S_p	K_p	T_n	T_p	K_t
			$\alpha = 0°$			
1	995	2370	2.38	0	—	—
2	995	2413	2.43	0	—	—
3	995	2413	2.43	0	—	—
4	995	2370	2.38	0	—	—
			$\alpha = 15°$			
1	918	2165	2.36	181	412	2.28
2	942	2261	2.40	186	435	2.34
3	967	2318	2.40	191	444	2.33
4	991	2334	2.36	196	407	2.23
			$\alpha = 30°$			
1	755	1781	2.36	302	686	2.27
2	796	1909	2.40	318	742	2.33
3	836	2004	2.40	334	776	2.32
4	877	2059	2.35	350	783	2.24
			$\alpha = 45°$			
1	561	1324	2.36	337	769	2.29
2	605	1448	2.39	363	847	2.33
3	649	1550	2.39	390	903	2.32
4	694	1623	2.34	416	944	2.24
			$\alpha = 60°$			
1	382	898	2.35	283	649	2.29
2	420	1002	2.39	311	727	7.34
3	458	1089	2.38	340	797	2.32
4	495	1153	2.33	367	822	2.24
			$\alpha = 75°$			
1	257	599	2.33	161	368	2.29
2	286	680	2.38	179	418	2.34
3	315	747	2.37	197	456	2.32
4	344	796	2.31	215	480	2.23
			$\alpha = 90°$			
1	213	492	2.31	0	—	—
2	238	561	2.36	0	—	—
3	263	619	2.35	0	—	—
4	288	661	2.30	0	—	—

TABLE 4. SUMMARY OF NUMERICAL RESULTS (R=20%).

Pillar	S_n	S_p	K_p	T_n	T_p	K_t
			$\alpha = 0°$			
1,2,3,4	995	1238	1.24	0	—	—
			$\alpha = 15°$			
1	918	1142	1.24	181	225	1.24
2	942	1171	1.24	186	231	1.24
3	967	1202	1.24	191	236	1.24
4	991	1234	1.24	196	241	1.23
			$\alpha = 30°$			
1	755	939	1.24	302	374	1.24
2	796	988	1.24	318	395	1.24
3	836	1040	1.24	334	413	1.24
4	877	1091	1.24	350	432	1.23
			$\alpha = 45°$			
1	561	697	1.24	337	417	1.24
2	605	751	1.24	363	450	1.24
3	649	805	1.24	390	482	1.24
4	694	862	1.24	416	513	1.23
			$\alpha = 60°$			
1	381	473	1.24	283	351	1.24
2	420	520	1.24	311	387	1.24
3	458	568	1.24	340	419	1.24
4	495	615	1.24	367	453	1.24
			$\alpha = 75°$			
1	257	318	1.24	161	199	1.24
2	285	354	1.24	179	221	1.23
3	315	391	1.24	197	244	1.24
4	344	426	1.24	215	266	1.24
			$\alpha = 90°$			
1	213	262	1.24	0	—	—
2	238	294	1.24	0	—	—
3	263	325	1.24	0	—	—
4	288	355	1.24	0	—	—

pendicular to the dip must exceed the pre-mining shear-to-normal stress ratio referred to the same direction. Inequality (20) in terms of the pre-mining horizontal to vertical stress ratio, K_o, and seam dip is

$$(W_p/H_p) \gtreqless [\frac{(1-K_o)\sin(2\alpha)}{(1+K_o) + (1-K_o)\cos(2\alpha)}]. \tag{21}$$

A simple "correction" to the flat seam pillar stress formula is suggested by writing Eq. 7 as

$$S_p = \gamma h/(1-R)$$

and then using the component of γ across the dip for the inclined seam case. Instead of Eq. 9, one has the "corrected" formula

$$S_p = \gamma h \cos(\alpha)/(1-R) \tag{22}$$

Eq. 22 is surprisingly accurate up to about 60 degrees as can be seen in Fig. 7 which shows the cosine "correction" and the actual pillar stresses. Eq. 22 should not be used in place of Eq. 9, of course.

The main conclusion reached in this paper is that the proposed inclined seam extraction ratio formulas, Eq. 9 and 10, are as adequate for pillar design in dipping seams as is the well-known flat seam extraction ratio formula. Figure 8 shows the accuracy of the proposed extraction ratio formulas compared with the finite element results in the high extraction ratio case (R=60%). The average normal and shear stress computed by the extraction ratio formulas are seen to be somewhat higher than the actual stresses (averages) obtained from the finite element analyses. In the low extraction ratio case, the differences are less. The slight overestimation of pillar stresses by the proposed formulas for inclined seams (Eq. 9 and 10) should be of no real consequence for the extraction ratio approach to the design of mine pillars. The extraction ratio formulas for pillar design and a knowledge of stress concentration about the entries and crosscuts complete what is essentially a "strength of materials" approach to room-and-pillar mine design.

Acknowledgment

The assistance of students S. Smith, M. Messick and B. McGavin in performing much of the computer work in this study is gratefully acknowledged. Financial assistance was provided in part by the Utah Mineral Leasing Fund.

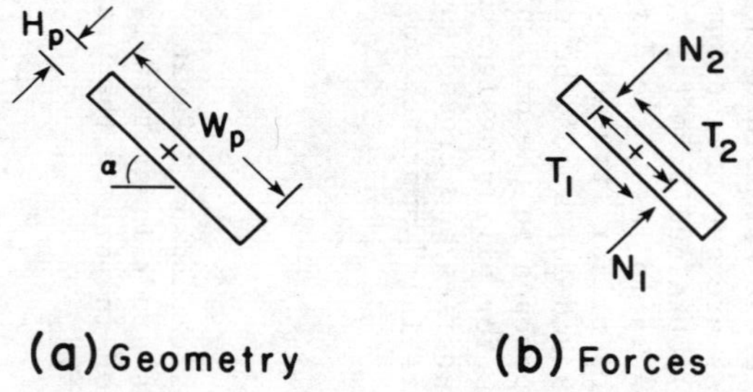

1090 ISSUES IN ROCK MECHANICS

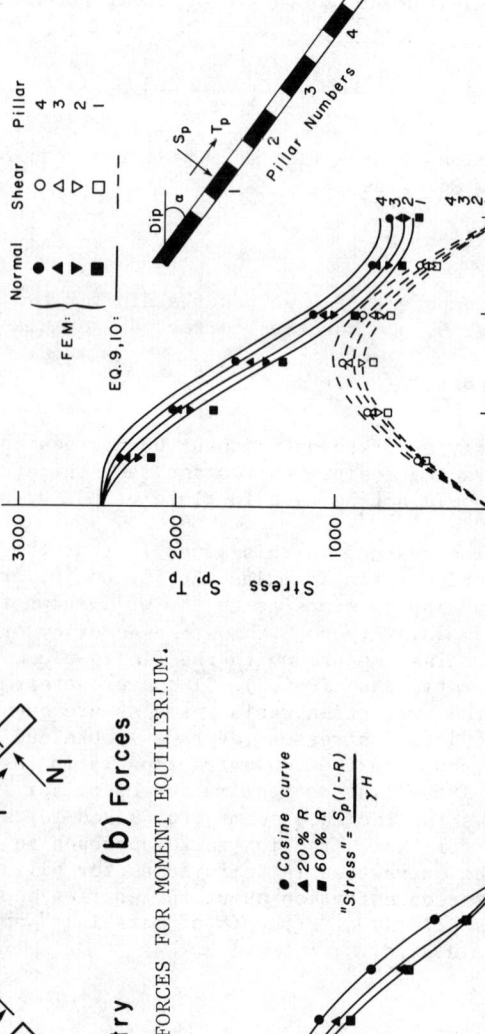

FIG. 6. PILLAR FORCES FOR MOMENT EQUILIBRIUM.

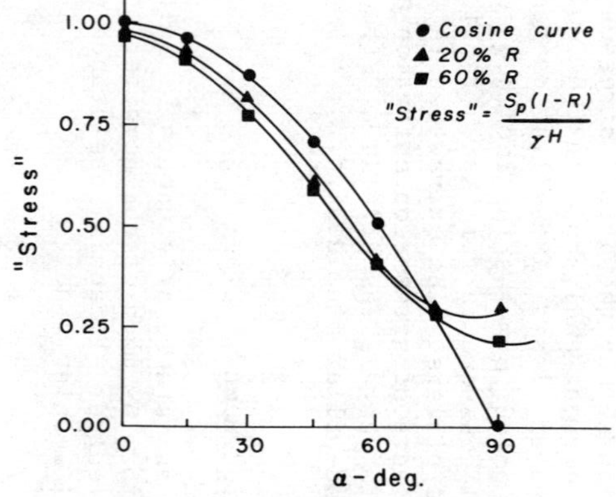

FIG. 7. COSINE "CORRECTION" FORMULA (EQ. 22) FOR PILLAR NORMAL STRESS.

FIG. 8. PILLAR STRESS COMPARISONS. FEM IS FINITE ELEMENT METHOD. EQ. 9 AND 10 ARE PROPOSED EXTRACTION RATIO FORMULAS.

Chapter 106

FACTORS GOVERNING THE STABILITY OF ROCK SLOPES IN
BRITISH SURFACE COAL MINES

by M. J. Scoble and W. J. P. Leigh

Mining Engineering Department, Nottingham University,
England.

National Coal Board, Opencast Executive,
Newcastle, England

ABSTRACT

This paper aims to characterise the forms of slope instability incident in surface coal mines in Britain. These are related to observed controlling factors, with particular reference being given to slope geometry and structure. The work is based on specific data collection programmes considering actual stable and unstable mine slopes. Reference is made to field examples.

INTRODUCTION

The Opencast Executive of the National Coal Board has an annual production target of 15 million tonnes of coal which entails the operation of around 75 multi-seam mines. The average national seam width is 1.1 m and the maximum number of seams per mine is 20. The average vertical stripping ratio is 15:1 (maximum 35:1), whilst the average working depth is 50 m with a planned ultimate of 214 m at one open pit (Lindley 1980).

This paper is based upon the analysis of data generated by two research projects: a National Slopes Survey (NSS) and an Instability Data Base (IDB). The NSS was conducted over a three year period - the geotechnical and mining features exposed on each mine slope were logged for all producing mines, regardless of stability history. The IDB holds available geotechnical and mining data relating to past national mine slope instability cases. It is an on-going project and currently holds 120 case histories. The influence exerted by slope geometry, structure and rock mass quality on mine slope stability is examined with reference to both projects.

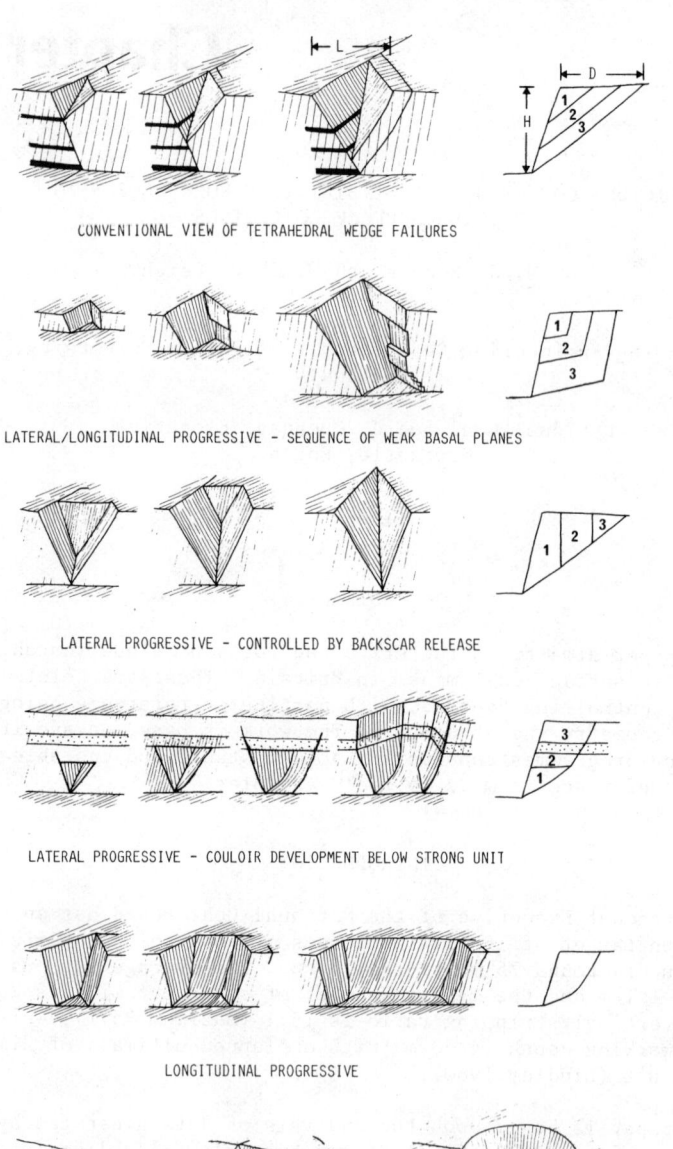

Fig. 1 Characteristic forms of instability development observed in mine rock slopes.

ROCK SLOPE STABILITY IN BRITAIN 1093

Groundwater is recognised as an important factor (Norton, 1982) but lack of piezometric data for most IDB cases has hindered any detailed analysis. Mines are tending to work deeper deposits in often more complex geotechnical environments and guidelines for a sounder engineering approach to slope design are desirable.

INSTABILITY CHARACTERISTICS AND SLOPE GEOMETRY

The relationship between height (H), depth (D), length (L), slope angle (A) and volume (V) of each IDB case study was analysed in order to establish the geometric pattern of instability types and any influence exerted by slope geometry on stability. H and A appear to exert an important influence on the stability of loose-wall and pavement slopes. In high, side and low wall slopes, however, their influence appears to relate only to their control over kinematic feasibility and exposure of weak and adversely-orientated structures.

The IDB indicated that 52% of cases were under 10000 cu m. in volume. Any ability to predict the limits to lateral and longitudinal extension of an instability, i.e. D and L, based upon a developed or assumed H would be of value in stability assessment and control. Fig. 1 presents the characteristic forms of progressive development observed in mine rock slopes. 9% of IDB cases were observed to develop significant lateral extension (H:D ratios under 0.6). All of these were based on planar, low shear strength horizons with dips under $12°$ into the excavation. Three cases of such acute lateral extension are illustrated: one with basal shear along a rockhead interface with boulder clay on a valley side (Fig. 2A), and two others based on weak lithological units (Figs. 2B, 2C). When block release of translational slides occurs by parting and shear along joint and intact rock bridges then H:D values lie in the range 1 to 4. Rock slope instability not controlled by bedding shear, i.e. the multi-planar or quasi-circular mode, involving joint shear and intact rock breakage in slopes of low rock mass quality (dominated by U1-U4 rock units, Table 1) also exhibit similar H:D ratios, Fig. 2D. The longitudinal extension of IDB cases was controlled by rock mass quality and release afforded by faults and underground mine subsidence fractures.

Slope curvature in plan and section was also evident as a release factor. Fig. 2E shows the plan promontory excavated on a 100 m highwall which resulted in a 2000000 cu m slide, released by reverse faulting aligned with the NW-SE tension cracks shown. Fig. 2F shows three cross-sections through a pavement failure of 65000 cu m. The slab slide was witnessed to initiate overnight by buckling and breakage at the mid-face point of greatest curvature. An upper slab was initially released in the northern area to override and motivate the lower slab.

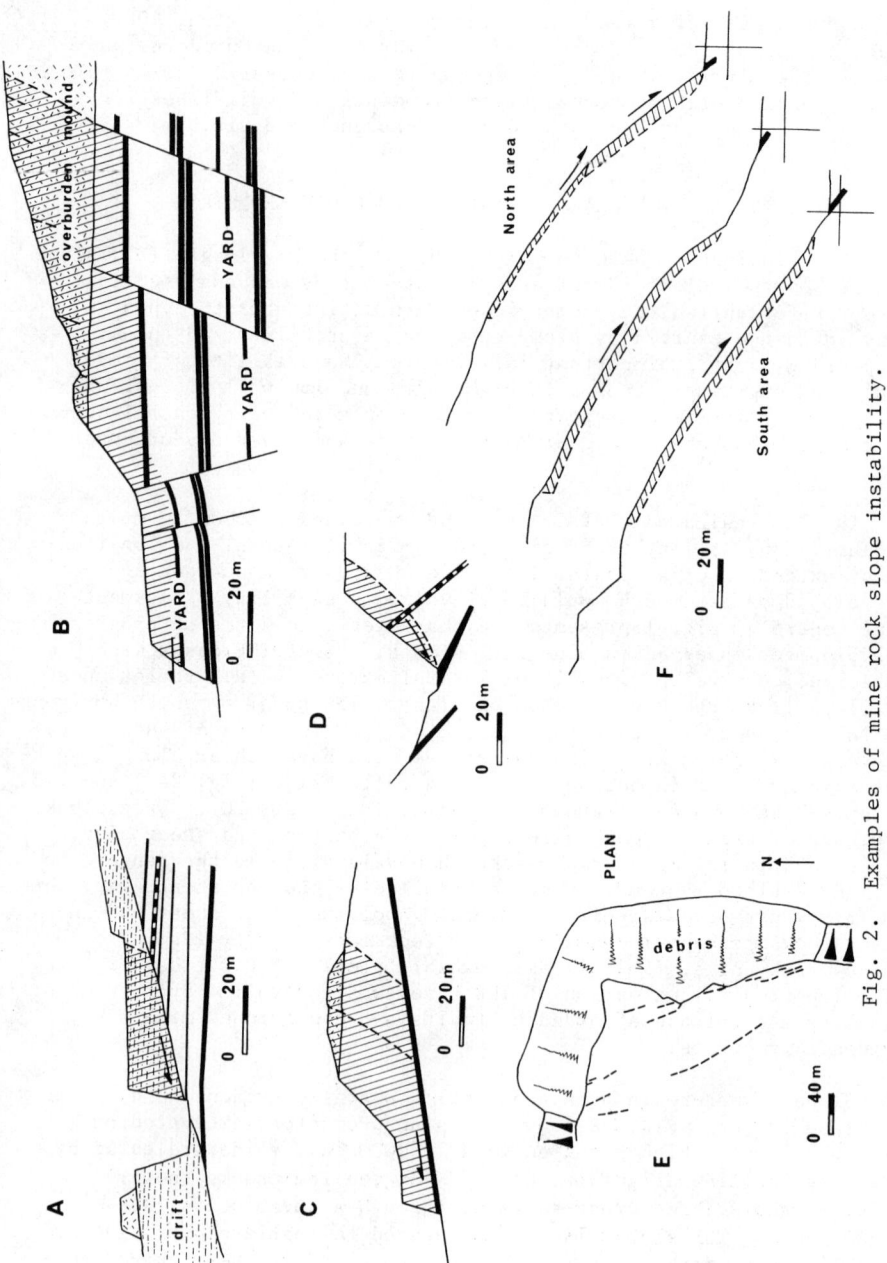

Fig. 2. Examples of mine rock slope instability.

ROCK SLOPE STABILITY IN BRITAIN 1095

STRUCTURAL CONTROLS

Unstable blocks frequently exhibited asymmetry in plan with rotation accompanying deformation, reflecting the structural control over basal and release faces of unstable blocks. 71% of IDB cases were in slopes with <u>strata dips</u> into the excavation, in contrast to only 33% of all mine slopes, as mapped in the NSS. In 26% of IDB cases the dip component corresponded to the regional strata dips but the remainder arose from localised folding, mainly drag folding associated with faulting.

In 52% of IDB cases <u>faulting</u> was a contributory factor by provision of one or more of the following:

- peripheral release planes (22% of fault-related IDB cases).
- low shear strength basal planes comprising the fault itself or a weak clay band (65% of fault-related IDB cases): intra-formational shear zones (ISZ) have been genetically linked with Carboniferous faulting (Salehy et al, 1977).
- adverse steepening of strata dips due to drag folding (73% of fault-related IDB cases).

Fig. 2B demonstrates the risk inherent in placing a mine boundary and overburden mound along a fault zone. The antithetic fault zone was approached up dip by a strike cut, along which a sequence of planar slides occurred on ISZ horizons in mudstone and seatearth units; total affected volume 540000 cu m. Fig. 3 shows the release of a 200000 cu m block by a single fault traversing a 60 m boundary sidewall (Leigh et al, 1980). This incident is also an example of the cyclic pattern of ground deformation commonly observed, where the displacement-time function consists of a series of exponential decay curves, similar to deformation characteristics observed in hard rock open pit incidents (Broadbent and Ko, 1971). Both examples demonstrate the need to define faulting features (drag folding, throws, shear zones) in mine peripheral zones with care.

<u>Jointing</u> contributed to 54% of IDB cases by lowering rock mass quality (controlling quasi-circular and erosional instability) or delineating toppling or sliding blocks. Subsidence zones associated with shallow and deep <u>underground mine workings</u> contributed to 25% of IDB cases by reducing rock mass quality, providing block release features and modifying groundwater regimes.

ROCK MASS QUALITY

Table 1 presents an engineering classification of the component rock units encountered in British mine slopes. The Rock Mass Quality (RMQ) is determined by the inherent strength and durability of these discontinuous units together with modifications arising from contained faulting, folding and underground workings. RMQ

Table 1: An Engineering Classification of Coal Measures Slope Units

	ROCK UNIT No	U1	U2	U3	U4	U5	U6	U7	U8	U9	U10	U11
	ROCK UNIT DESCRIPTION	MASSIVE MUDSTONE	FISSILE MUDSTONE	LAMINATED MUDSTONE	INTERBEDDED SILTSTONE AND MUDSTONE	MASSIVE SILTSTONE	LAYERED SILTSTONE	COMPLEX INTERBEDDED SILTSTONE AND SANDSTONE	LAYERED SILTSTONE AND SANDSTONE	MASSIVE SANDSTONE	LAYERED SANDSTONE	SEATEARTH
PHYSICAL PROPERTIES	DRY DENSITY t/m^3	2.1 ±0.2	2.1 ±0.2	2.1 ±0.2	2.3 ±0.2	2.6 ±0.2	2.6 ±0.2	2.5 ±0.2	2.5 ±0.2	2.4 ±0.2	2.4 ±0.2	2.4 ±0.1
	SLAKE DURABILITY Id_2 %	<80	<80	<80	70±20	75±10	85±10	90±10	90±10	95±5	95±5	40±20
	POROSITY n %	4±2	4±2	4±2	4±2	8±3	8±3	12±2	12±2	13±3	13±3	–
	P-WAVE VELOCITY (IN SITU) m/s	1150 ±250	950 ±250	950 ±250	1250 ±350	1850 ±650	1600 ±400	1600 ±400	1600 ±400	1850 ±650	1600 ±400	–
STRUCTURAL PROPERTIES	BEDDING SPACING m	>0.6	<0.1	<0.01	0.1 TO 0.3	>0.6	0.1 TO 0.5	0.1 TO 0.5	0.05 TO 0.3	>0.6	0.1 TO 0.5	–
	JOINT SPACING m	>0.6	<0.6	<0.6	<0.6	>0.6	<1.0	<0.6	<1.0	>0.6	<1.0	–
	R.Q.D. %	>50	<30	<50	<60	70 TO 100	30 TO 100	30 TO 100	30 TO 100	70 TO 100	50 TO 100	<30
	FRACTURE FREQUENCY per m	<10	>20	<20	>10	<2	<15	<15	<15	<2	<12	>20
	BLOCK SHAPE	BLOCKY OR IRREG	FLAKY	FLAGGY	BLOCKY OR FLAGGY	BLOCKY OR IRREG	FLAGGY	BLOCKY OR FLAGGY	FLAGGY	BLOCKY OR COLUMN	FLAGGY	IRREG
	BLOCK VOLUME m^3	>0.2	<0.01	<0.01	<0.01	>0.2	<0.3	<0.2	<0.2	>0.2	<0.3	<0.01
INTACT STRENGTH	COMPRESS STRENGTH MN/m^2 SAT		5 ± 2		25±3	45±4	45±4	48±12	48±12	112 ±65	112 ±65	1±0.2
	COMPRESS STRENGTH MN/m^2 DRY		40 ± 10		50±7	63±7	63±7	64±11	64±11	163 ±89	163 ±89	27±7
	TENSILE STRENGTH MN/m^2 SAT		0.6 ± 0.2		2±1	4±1	4±1	5±1	5±1	9±5	9±5	0.1 ±0.01
	TENSILE STRENGTH MN/m^2 DRY		4 ± 1		6±1	8±1	8±1	7±2	7±2	12±7	12±7	3±0.5
	POINT LOAD INDEX DRY I_{S50}		1.4 ± 0.3		1.7 ±0.2	2.2 ±0.2	2.2 ±0.2	2.2 ±0.3	2.2 ±0.3	5.6 ±3.0	5.6 ±3.0	0.9 ±0.2
	SCHMIDT HAMMER (TYPE N)		20 TO 40		20 TO 40	30 TO 50	30 TO 50	30 TO 40	30 TO 50	30 TO 60	30 TO 50	20 TO 40
DISCONTINUITY STRENGTH	APPARENT COHESION kN/m^2		0 TO 200		0 TO 200	100 TO 300	100 TO 300	100 TO 300	100 TO 300	100 TO 400	100 TO 400	0 TO 30
	FRICTION ANGLE degrees BASIC		27 TO 29		27 TO 33	28 TO 33	28 TO 33	25 TO 33	25 TO 33	25 TO 37	25 TO 37	6 TO 24
	FRICTION ANGLE degrees PEAK		21 TO 33		21 TO 33	28 TO 33	28 TO 33	28 TO 33	28 TO 33	32 TO 37	32 TO 37	15 TO 24
	$\tau = A\sigma_n^{0.8}$ INDEX A SAT		2.1		2.1	2.4	2.4	2.4	2.4	3.0	3.0	1.8
	$\tau = A\sigma_n^{0.8}$ INDEX A DRY		2.4		2.4	2.8	2.8	2.8	2.8	3.5	3.5	2.0

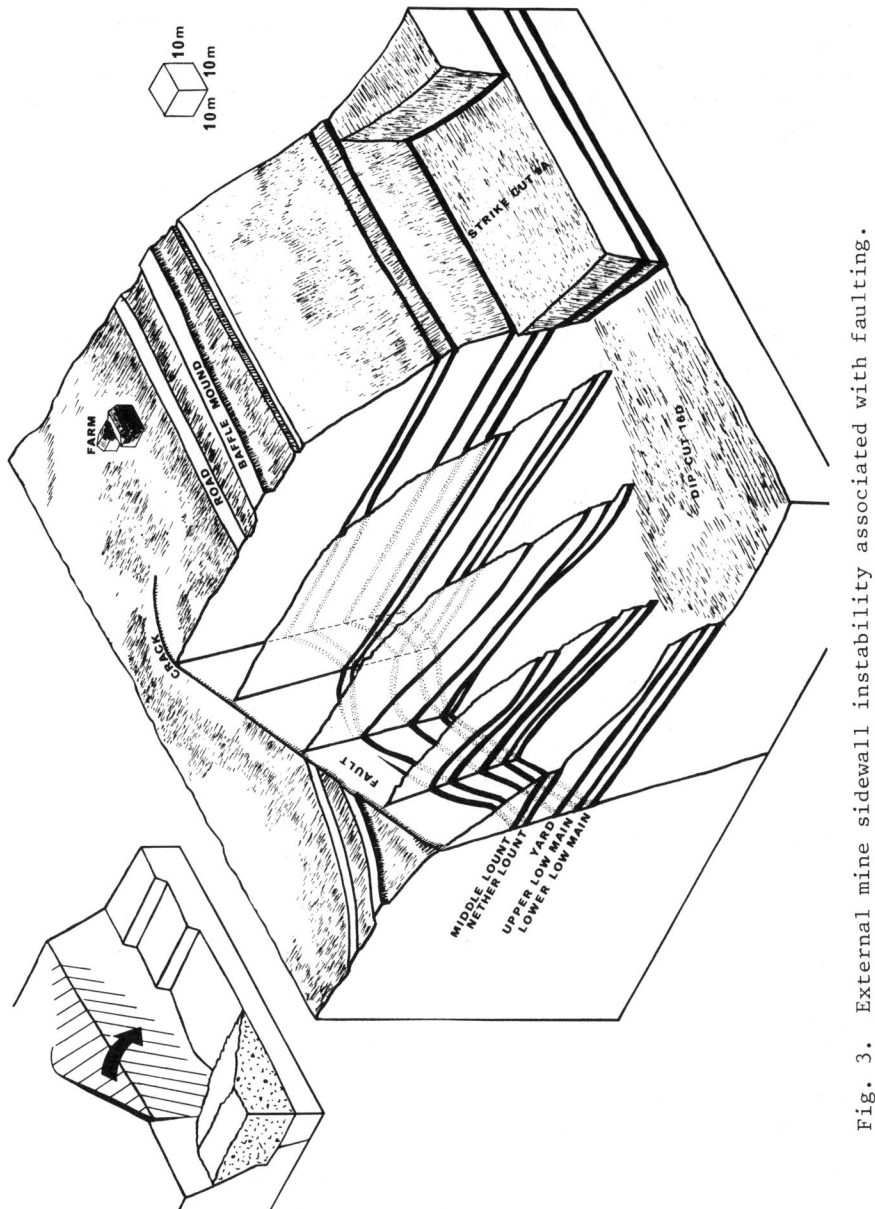

Fig. 3. External mine sidewall instability associated with faulting.

influences the development of quasi-circular, toppling, wedge and rock fall instabilities. These generally involve volumes under 10000 cu m but pose serious hazards if bench design and dressing is inadequate. The dominance of planar and bi-planar instability (34% and 40% of IDB cases respectively) reflects the key role played by weak zones, particularly ISZ bands, within rock units serving as the loci of translational slides (Hassani and Scoble, 1981).

CONCLUSION

This work has highlighted the particular influence exerted by adverse structure and weak lithological zones on mine slope stability in Coal Measures. Effective mine design to minimise instability must be based upon reliable and adequate detail of structure and weak zones in addition to rock mass quality and groundwater.

ACKNOWLEDGEMENTS

The authors wish to acknowledge the assistance and support of the N.C.B. Opencast Executive in the research programmes underlying this work. The views expressed are entirely those of the authors and not the N.C.B.

REFERENCES

Broadbent, C. D. and Ko, K. C., 1971, "Rheological aspects of rock slope failures", Proc. 13th Symp. Rock Mech., Illinois, 573-594.

Hassani, F. P. and Scoble, M. J., 1981, "Properties of weak rocks, with special reference to the shear strength of their discontinuities, as encountered in British surface coal mining", Proc. Int. Symp. on Weak Rock, Tokyo.

Leigh, W. J. P., Scoble, M. J. and Young, G. J., 1980, "Ground movement adjacent to a deep excavation", 2nd Conf. Ground Movement and Structures, Brit. Geotech. Soc., Cardiff.

Lindley, G. F., 1980, "Opencast Mining in the United Kingdom", Proc. Symp. Opencast Coal and Quarrying , Ass. Min. El. Mech. Eng.

Norton, P. J., 1982, "Groundwater problems in surface coal mining in Scotland", Int. Jnl. Mine Water, No. 1.

Salehy, M. R., Money, M. S. and Dearman, W. R., 1977, "The occurrence and engineering properties of intraformational shears in Carboniferous rocks", CORE-UK Conf., Newcastle upon Tyne.

Chapter 107

PRELIMINARY FOUNDATION STUDIES FOR RAISING A GRAVITY-ARCH DAM

by Gregg A. Scott, Karl J. Dreher, and Charles C. Hennig

Geotechnical Engineer, Bureau of Reclamation, Denver, Colorado

Formerly with the Bureau of Reclamation, Denver, Colorado

Civil Engineer, Bureau of Reclamation, Denver, Colorado

INTRODUCTION

Theodore Roosevelt Dam is a cyclopean-masonry, gravity-arch dam located on the Salt River northeast of Phoenix, Arizona. Construction of the dam began in 1903 and was completed in 1911. The dam is about 85 m high with a total reservoir capacity of about 1.7×10^9 m^3. Raising the dam is now being considered to provide additional water storage and flood control. This paper describes preliminary analyses performed to assess foundation stability for increased loading resulting from raising the dam and reservoir water surface about 12 m.

The foundation of Theodore Roosevelt Dam consists of variably metamorphosed limestones, sandstones, mudstones, and shales. Bedding uniformly strikes nearly normal to the river (N 40° W) and dips upstream at 25 to 30°. The rock is cut by four prominent joint sets as indicated by the joint contour diagram shown in figure 1A. The diagram was developed from surface observations and vertical drill core oriented relative to the bedding. Sampling biases account for the differences in concentrations of various joint sets. However, field observations indicate that all sets are relatively prominent, planar, and continuous.

SEEPAGE ANALYSES

Consistent with practice at the time of construction, no foundation drainage was installed at Theodore Roosevelt Dam. Finite element seepage analyses were therefore performed to study the potential for large uplift pressures in the foundation. The dam has steep abutments and is curved in plan as shown in figure 1B. The effects of modeling the seepage two-dimensionally, radially symmetric about the dam's axis center, and three-dimensionally were therefore com-

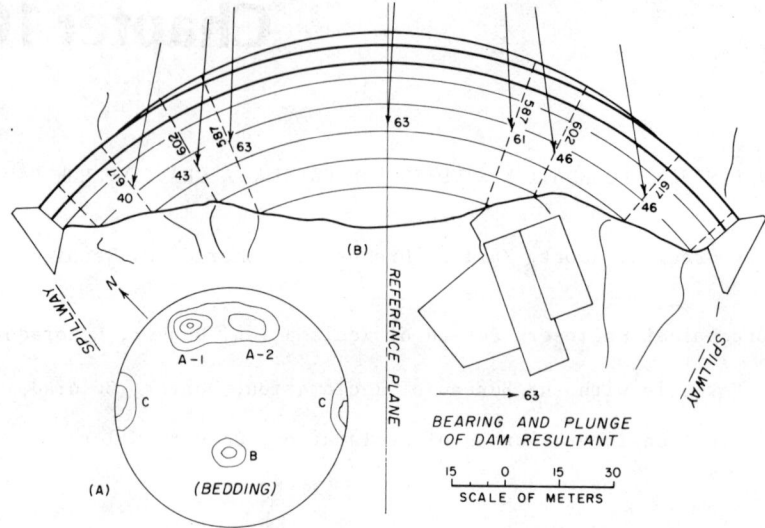

Fig. 1(A). Joint contour diagram, poles on lower hemisphere of equal area stereonet (B) Plan of raised dam.

pared. Two-dimensionally, the shortest seepage path is radial to the dam. Therefore, two-dimensional finite element meshes were developed radial to the dam at the reference plane and two locations on the left abutment (looking downstream). The meshes were configured such that they could be connected to form a three-dimensional mesh, allowing for direct comparison of results. Seepage measurements indicate that the flow changes almost simultaneously with a change in reservoir water surface elevation. Modeling the flow as confined steady-state seepage was therefore considered appropriate.

Results are presented in figure 2A at the reference plane assuming the foundation to be homogeneous and isotropic. As expected, the two-dimensional analysis shows the equipotential lines to be evenly distributed throughout the flow regime. However, the axisymmetric and three-dimensional analyses show the equipotential lines concentrated near the downstream portion of the foundation as a result of the flow converging toward the stream channel. The calculated uplift pressure at any point is largest from the axisymmetric model and smallest from the two-dimensional model. An intermediate value is indicated by the three-dimensional model which is considered to be the most correct. Similar results were observed at the other two sections studied.

Studying an existing dam affords the opportunity to observe prototype behavior through instrumentation. Piezometers were installed in the foundation at the reference plane and two locations on each abutment. By assuming a permeability 10 times greater parallel to the bedding than perpendicular to the bedding,

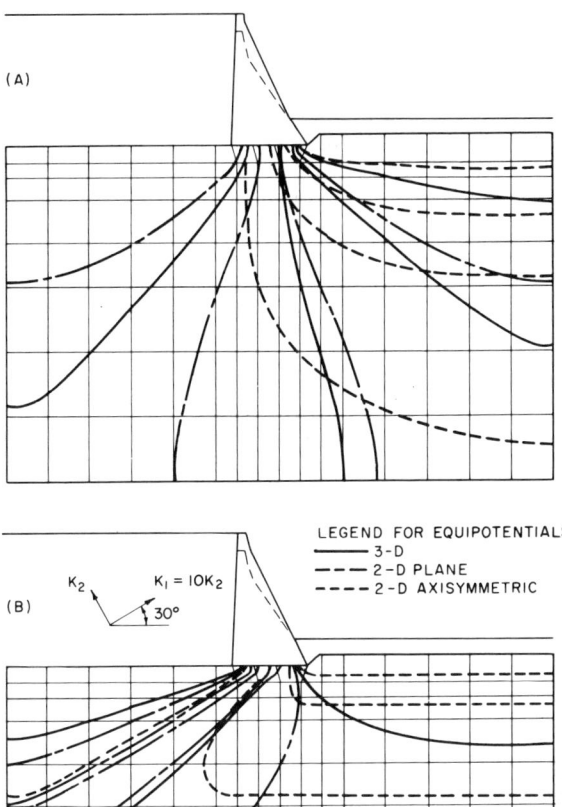

Fig. 2. Comparison of seepage models (A) isotropic (B) anisotropic

close agreement was obtained between the three-dimensional seepage model and the piezometer data at the reference plane. Assuming symmetry of the model about the reference plane, all other piezometer data indicated lower pressures than the model, except data from a piezometer located at a depth of about 110 m on the right abutment. However, foundation stability at this depth is not considered to be of concern. A comparison of the seepage models for a permeability ratio of 10:1 is shown in figure 2B. The results are similar to the homogeneous case, except that uplift pressures are generally reduced due to the bedding orientation.

STABILITY ANALYSES

Potentially unstable blocks are formed in the foundation of Theodore Roosevelt Dam by upstream dipping bedding planes, rear release planes consisting of A-1 or A-2 joints, and side planes consisting of C joints. A stepped surface could occur at any location, and definition of three-dimensional wedges is difficult. Since joint set C is generally parallel to the loads from the dam, as shown graphically in figure 1, the potential sliding resistance offered by the side planes is considered small. Consequently, assessment of numerous two-dimensional wedges was considered to approach a stepped surface without being overly conservative. Two-dimensional geologic sections (see fig. 3) were developed in the directions of resultant dam loading at selected elevations on both abutments. Rigid block limit equilibrium analyses were conducted for sliding on the bedding at various depths beneath the dam. The rear release planes of the wedges were assumed to coincide with the upstream face of the dam.

Laboratory direct shear tests were conducted on 28 samples of open bedding joints obtained from Nx and 102 mm diameter drill core. The resulting strength envelopes were nonlinear as expected. To simplify the analyses, zero strength was assumed at zero normal stress and applicable friction angles were estimated based on normal stresses of interest to this problem. This was considered to be reasonable since uncertainties due to scale effects (Bandis, Lumsden, and Barton, 1981) may be large compared to differences introduced by this approach.

Water forces acting on the wedges were determined from the anisotropic three-dimensional seepage model, assuming symmetry about the reference plane, with boundary conditions consistent with the raised dam. Dam loads, water forces, and dead loads were resolved into forces normal and tangential to the bedding, and factors of

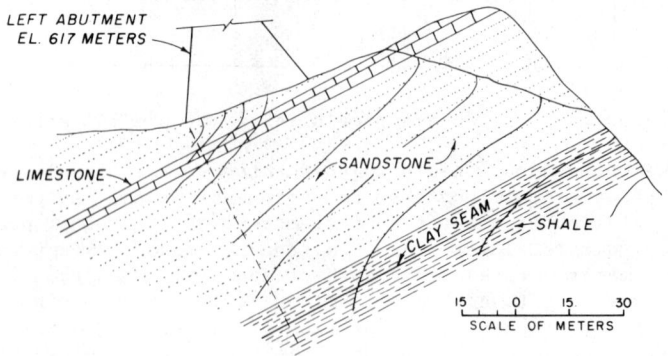

Fig. 3. Typical 2-D analysis section, geology and equipotentials.

safety were calculated by considering the appropriate friction angle. The sliding plane for each section with the smallest factor of safety was determined by examining all clay seams and geologic contacts. It was also assumed that a continuous bedding joint could occur at any depth below the dam. The results are shown in table 1.

Table 1. - Summary of minimum safety factors

Elevation (m)/ abutment	Minimum F.S.	90° wedge corner elevation (m)	Bedding discontinuity
617 left	1.4	604	joint-limestone
602 left	1.7	587	joint-sandstone
587 left	2.0	545	clay seam
Ref. plane	7.6	541	clay seam
587 right	4.4	575	joint-sandstone
602 right	1.8	587	joint-sandstone
617 right	1.0	606	joint-sandstone

FOUNDATION TREATMENT

The results of the analyses indicate that the foundation of Theodore Roosevelt Dam is basically stable for the assumptions and loading conditions analyzed. However, current Bureau of Reclamation criteria require a factor of safety of at least 1.3 for loading conditions associated with the Maximum Credible Earthquake. Although the seismic input had not been formulated at the time of this study, experience suggests that a factor of safety of 2.7 for the conditions analyzed would satisfy the requirement for earthquake stabilty. Four foundation treatment alternatives were therefore considered.

Steel bars grouted into vertical holes drilled through the toe of the dam could be used to reinforce the foundation. Downstream and upward movement along the bedding discontinuities in the foundation would cause elongation, shearing, and bending of the bars over a short interval across the zone of sliding, thus producing a stabilizing effect. The behavior of passive reinforcement is still not completely understood as evidenced by the varied results reported in the literature. Laboratory tests performed by various investigators are summarized in figure 4. Only results from direct shear tests performed on smooth discontinuities with securely anchored reinforcement are shown. The thickness of the discontinuity and normal dilation rate are therefore approximately constant. The maximum reinforcement contribution to shear resistance is normalized with respect to the tensile yield force to facilitate comparisons of various sizes and grades of reinforcing steel. When the yield strength of the steel was not clearly stated, reasonable assumptions were made. Bjurstrom (1974) indicates that the steel fails in a combined shear and tension mode for the reinforcement angles plotted. Generally, increasing resistance is shown with decreasing reinforcement angle, which is indicative of increased frictional resistance due to tension in the steel. However, the reinforcement contribution does not appear to decrease significantly with decreasing friction angle indicating a resisting component due to the shear stress in the steel.

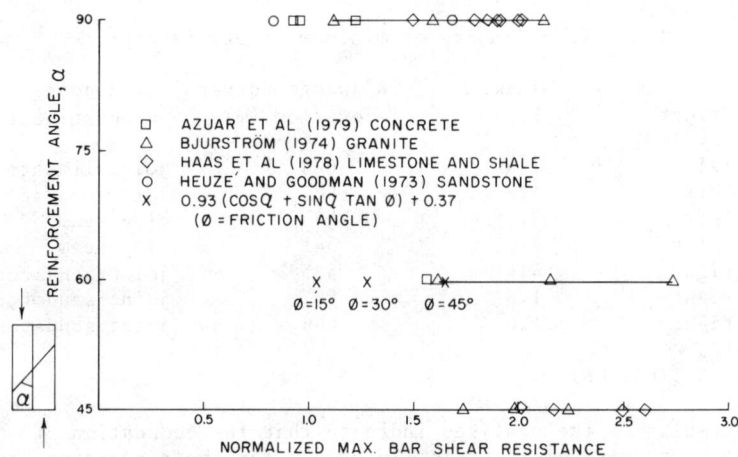

Fig. 4. Laboratory tests on passive reinforcement.

Linear elastic finite element studies were performed to better understand the behavior of the steel reinforcement. The rock mass was modeled with two-dimensional continuum elements in rows parallel to the direction of the bedding. Orthotropic material properties were used to model a 64-mm-thick discontinuity and beam elements were used to represent a 114-mm-diameter bar at an angle of 60° to the bedding. Displacement along the potential sliding zone was simulated using boundary elements. For the purposes of this study, the steel was considered to have a yield strength of 248 MPa and an ultimate strength of 450 MPa. The displacement was adjusted until the combined axial and bending tensile stresses in the steel were slightly less than ultimate strength. At this displacement, the average tensile force in the bar is approximately 93 percent of yield, and the mobilized shear force in the steel is about 37 percent of the tensile yield force. The resulting resistance, shown in figure 4, was used for this study.

Installation of foundation drainage would produce a stabilizing effect by reducing potential uplift pressures. Relief wells installed from the toe of the dam would likely be of limited effectiveness because holes drilled to penetrate the upstream third of the foundation would intercept very few bedding planes. In addition, potential reduction in uplift pressures would be limited by drill hole collar elevations. Therefore, a drainage gallery system located above maximum tailwater under the upstream third of the dam is considered to be a more effective alternative. To determine the potential merits of this type of system, factors of safety were computed assuming a 50 percent reduction in water forces.

Many of the critical potential sliding planes are relatively shallow, which would facilitate installation of mass concrete shear keys constructed in excavated trenches at the toe of the dam. The exception is at elevation 587 m on the left abutment where the potential sliding planes are too deep. The potential mobilized shear strength of the concrete was considered to be about 2.8 MPa under zero normal stress.

Post-tensioned cables could be installed in a manner similar to the passive reinforcement. However, only the contribution due to tensile forces in the cables can be considered since they are flexible in shear. Additional resistance would be provided without requiring movement in the foundation. However, the possibility of load relaxation due to creep effects must be considered. Cable strands of 1860 MPa ultimate and 1120 MPa working stress were considered for the purposes of this study.

The results of the foundation treatment studies for a required factor of safety of 2.7 are shown in table 2. Embedment lengths were calculated by estimating bond strengths and requiring anchorage below the deepest potential sliding plane with a factor of safety less than 2.7. Mobilized strength of the passive steel and concrete was assumed to occur at small displacements compatible with peak joint strengths. Peak joint strengths of foundation laboratory samples occurred at shear displacements of about 1 mm. The test results of Azuar et al., (1979), indicate compatible displacements (less than 2 mm) for maximum reinforcement contribution at angles less than 90°. The finite element studies indicate a slightly smaller displacement required to mobilize maximum reinforcement resistance. However, non-linear effects which would occur near the sliding plane were neglected. Further studies of displacement and scale effects related to treatment design are required.

Table 2. - Summary of foundation treatment studies

Elevation (m)/ abutment	Area passive steel m^2/m	F.S. - 50% water forces	Shear area conc. m^2/m	Area cable mm^2/m
617 left	0.015	3.01	3.02	4090
602 left	0.013	2.84	2.87	3910
587 left	0.011	*	**	5880
Ref. plane		(no treatment required)		
587 right		(no treatment required)		
602 right	0.011	4.00	2.26	2670
617 right	0.018	***2.15	3.81	4450

* Negative driving force
** Not considered to be feasible at this location
*** Additional treatment required at this location

Chemical analyses of ground water indicate that encasing the passive steel or cables in cement grout should be sufficient corrosion protection. Steel connections must also be considered for the passive reinforcement. Strict excavation control would be required for installing concrete shear keys or drainage galleries. Grouting and pumping would likely be required to handle inflowing water. The concrete shear keys would need to be installed in short segments to avoid undercutting too much of the toe of the dam at once. In all cases, care must be taken to avoid sealing open bedding joints at the toe, which would create the potential for full uplift pressures beneath the dam.

CONCLUSIONS

Preliminary studies indicate that the foundation of Theodore Roosevelt Dam can withstand increased loading from raising the dam and reservoir water surface by 12 m. Discontinuities are present which form potentially unstable foundation blocks. The results of finite element seepage studies and static rigid block stability studies indicate that remedial foundation treatment will probably be required. Four types of treatment are considered viable at this stage of study. The final selection and design of treatment should be based on seismic loading conditions and cost considerations. Additional studies and tests are required prior to final design to investigate displacement and scale considerations, refine estimates of treatment effectiveness, and investigate three-dimensional stability effects.

REFERENCES

Azuar, et al., 1979, "Le Renforcement des Massifs Rocheux par Armatures Passives," Proceedings, 4th ISRM Congress, Montreaux, Switzerland, vol. 1, pp. 23-30

Bandis, S., Lumsden, A. C., and Barton, N. R., 1981, "Experimental Studies of Scale Effects on the Shear Behaviour of Rock Joints," International Journal of Rock Mechanics and Mining Sciences and Geomechanics Abstracts, vol. 18, No. 1, Feb., pp. 1-21

Bjurstrom, S., 1974, "Shear Strength of Hard Rock Joints Reinforced by Grouted Untensioned Bolts," Proceedings 3rd ISRM Congress, Denver, Colorado, vol. 2-B, pp. 1194-1199

Hass, C. J., et al., 1978, "An Investigation of the Interaction of Rock and Types of Rock Bolts for Selected Loading Conditions," PB293988, May, for U.S. Bureau of Mines, available from NTIS, Springfield, Virginia, pp. 101-168

Heuze, F. E., and Goodman, R. E., 1973, "Numerical and Physical Modeling of Reinforcement Systems in Jointed Rock," AD-766833, August, for U.S. Corps of Engineers, available from NTIS, Springfield, Virginia, pp. 43-47 some concern was expressed about pillar instability. Large scale pillar failure could cause breaks up to the base of a Permian aquifer, 50 m above the Main seam, and cause excessive strains on the sea bed. This aquifer is exposed on the sea bed, so any fractures reaching it, could cause water inflows large enough to close the mine. Due to this danger, Newcastle University was requested to monitor pillar stability as the underlying Brass Thill was extracted.

In January, 1977, extraction of the Brass Thill by longwall mining was begun. The first longwall face, KS4, was a 180 m wide retreat unit, taking the full seam height of 1.6 m. This face only undermined room and pillar workings in the Main seam and some areas of High Main extraction. The observations taken in the room and pillar workings indicated no signs of pillar instability.

On completion of KS4 face, KS3 was worked by advancing longwall, also undermining Main and High Main seam workings only. This was followed by KS6, which some 300 m from the start line undermined an area in which both Main and Yard seams had been worked, with High Main coal also having been taken in some roadways. Since potential pillar and junction failures in the Yard seam could affect the overlying Main seam pillars, the University was again requested to monitor pillar stability in the Main seam. This was made worse by there being no consistency in the relative positioning of pillars in the Main and Yard seams.

INSTRUMENTATION

In the two investigations at Lynemouth Colliery above KS4 and KS6 faces, pillars in the Main seam were instrumented using the well known anchor/strainwire/extensometer technique. Convergence stations were established at the junctions in the instrumented areas to monitor vertical closure. These stations consisted of roof and floor bolts with steel tapes attached to locate a reversed extensometer.

Above KS4 face, two rows of pillars extending from above the solid to the face center-line were instrumented to monitor pillar deformation perpendicular to direction of advance, Figure 1. Holes were drilled through the full pillar width of 23 m, and anchors were installed from each side of the pillar at depths of 10.7m, 6.1m, 3m, and 1.5m. In addition a wire was passed through the full pillar width, enabling lateral pillar expansion to be measured.

Above KS6 face four rows of pillars extending from above the solid to the face center-line were instrumented, Figure 2. Horizontal boreholes were drilled parallel to the direction of face advance, to a depth of 17 m, 6 m from the rear faces of the pillars. Anchors were installed at depths of 17m, 6m, 3m and 1.5m. In both the KS4 and KS6 investigations, vertical boreholes were drilled down from the Main seam to monitor inter-strata settlements caused by longwall extraction. The results of these investigations have been described previously by Styler and Dunham (1), Tubby and Farmer (2).

Chapter 108

INTERACTION EFFECTS ASSOCIATED WITH LONGWALL COAL MINING

by Dr. A. N. Styler* and Dr. R. K. Dunham*
Assistant Professor of Mining Engineering
University of Pittsburgh
Associate Director, British Mining Consultants, Inc.
*Formerly with the University of Newcastle-Upon-Tyne, England

ABSTRACT

The results of two studies into the effects of undermining offshore room and pillar workings by longwall faces are presented. The first study indicated no potential instability from undermining one level of room and pillar workings. On the basis of this it was decided to undermine two levels of room and pillar workings, again monitoring pillar stability. The results showed no signs of increased instability.

INTRODUCTION

At Lynemouth colliery, Northumberland, England, all the coal workings are under the North Sea, and extend approximately 8 km from the coast. The mine is divided in two by an East-West fault with a throw of approximately 200 m to the South. Prior to 1977 all extraction in the Northern part of the mine had been in the Main, High Main, and Yard seams by room and pillar. In the Main seam, approximately 85 m below the sea bed, the initial extraction height was 2.5 m, in some areas this was followed by removing the supports, and blasting down the overlying 1.5 m of High Main coal. In the Yard seam, between 3 m and 15 m below the Main seam, the full height of 2.4 m was extracted. All the workings consisted of 23 m square pillars, and 5.5 m wide roadways giving an extraction of 35%. The main seam in the instrumented area was extracted between 1964 and 1966, and the overlying Yard seam was extracted between 1972 and 1974.

The Brass Thill seam which underlies these workings, was 1.6 m thick at a depth of 160 m below the sea bed in areas suitable for extraction. As this seam was more than 105 m below the sea bed, British mining regulations permitted longwall extraction providing the tensile strain on the sea bed was kept below 10 mm/m. However, as extraction of the Brass Thill would mean undermining the Main and Yard seam workings,

INTERACTION EFFECTS OF MINING 1109

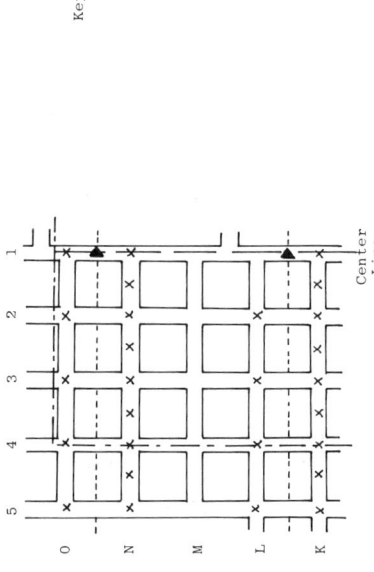

Key:
× Horizontal boreholes
--- Vertical boreholes
● Convergence stations
—·— Limits of KS4 in the Brass Thill

Figure 1. KS4 Instrumentation Layout

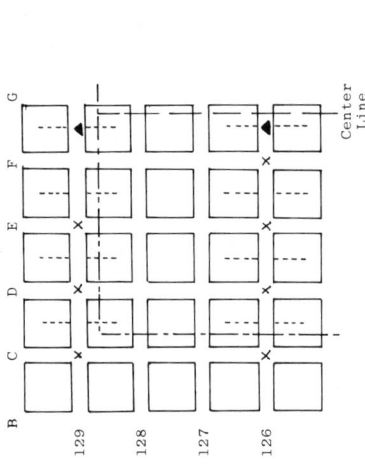

Figure 2. KS6 Instrumentation Layout

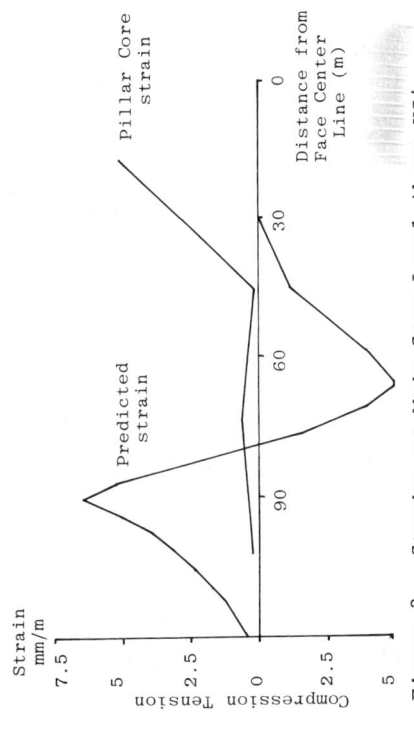

Figure 3. Strains at Main Seam Level Above KS4

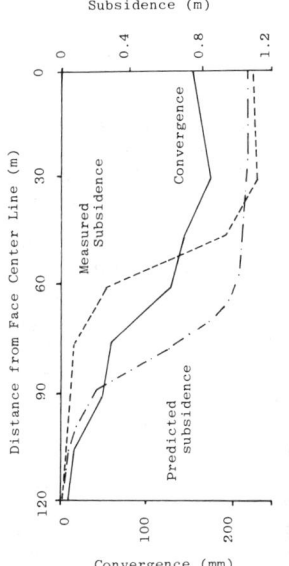

Figure 4. Convergence, and Subsidence at Main Seam Level Above KS4

RESULTS OF KS4 INVESTIGATION

Unfortunately due to roof falls, access was lost to some of the instrumentation soon after the face had passed. Therefore, only the results of pillar deformation between entries N and O, and convergence and settlement along entry N are discussed here. A detailed description of this investigation has been given by Hodkin (3).

Measurements taken close to the face center line showed that there was a delay in pillar deformation until the face had advanced approximately 40 m past the pillars, or a total face advance of 60 m. Visual observations of the gob showed that the first major break occurred after a total advance of 36 m, coinciding with the first significant movement measured at the anchor located 6.5 m above the face in the vertical borehole between entries O and N. Movement of the higher anchors in this borehole occurred in sequence, with the top anchor, 13.7 m below the Main seam, not moving until the face had advanced a total of 60 m. This indicates that the delay in pillar deformation was due to the delay in subsidence close to the face start line. As the face advanced, pillar deformation increased until the face was 120 m past the pillar, after which no further deformation occurred. This pattern of movement was fairly typical for all the instrumented pillars, with a decrease in the deformation towards the rib side. In all cases the pillar deformations reached a stable value, with no indication of failure.

From the measurements of total pillar deformation, and movement of the anchors at 6.6 m from the pillar edge, the strains over the central 10.7 m core of the instrumented pillars were calculated. The maximum pillar core strains are plotted in Figure 3, against distance from the face center line. Also shown on this figure is the horizontal strain profile at the Main seam level, corrected for proximity to the face start line, as predicted using the Subsidence Engineers Handbook (4). Comparison of the two curves shows that there is little correlation between pillar core dilation, and predicted horizontal strain. The maximum pillar core strain occurs towards the face center line, which for a supercritical panel width is the region where subsidence is purely vertical, theoretically inducing no horizontal strain.

The final roof to floor convergence profile along entry N, and the measured and predicted subsidence corrected for proximity to the face start line, are shown in Figure 4. This diagram shows that there is a fairly uniform increase in convergence from above the ribside to 30 m from the face center line, which follows the general trend of the predicted subsidence profile. What portions of the convergence is due to floor heave as opposed to roof lowering is unknown. However, it is interesting to note that towards the ribside the props became loose and fell out as the face undermined the entry, indicating that the floor was lowering faster than the roof. The new props installed in this area were broken by the time the face had advanced a further 30 m.

Comparison of the predicted and measured subsidence profiles in Figure 4, shows fairly good agreement over the center of the face, and above the ribside. However, between 45 m and 90 from face center line there is considerable disagreement between the two curves. The measured results indicate that the subsidence is more confined to the central portions of the face, than predicted. It can also be seen that the volume of the measured subsidence though, is much less than that of the predicted subsidence trough.

RESULTS OF KS6 INVESTIGATION

Unfortunately, it was not possible to obtain a complete set of results from the first line of instrumentation, due to the collapse of junction 126D, Figure 2, 30 m behind the face line. Due to operational difficulties, the instrumented pillars adjacent to roadway 129 were not undermined, hence no significant deformations were measured. Therefore, only the results from the first line of instrumentation, along roadway 126 will be presented. A detailed description of this investigation has been given by Styler (5).

The plot of anchor movements against face position, for the pillar closest to the center line between roadways 125 and 126 are shown in Figure 5. This diagram shows that pillar expansion started to occur

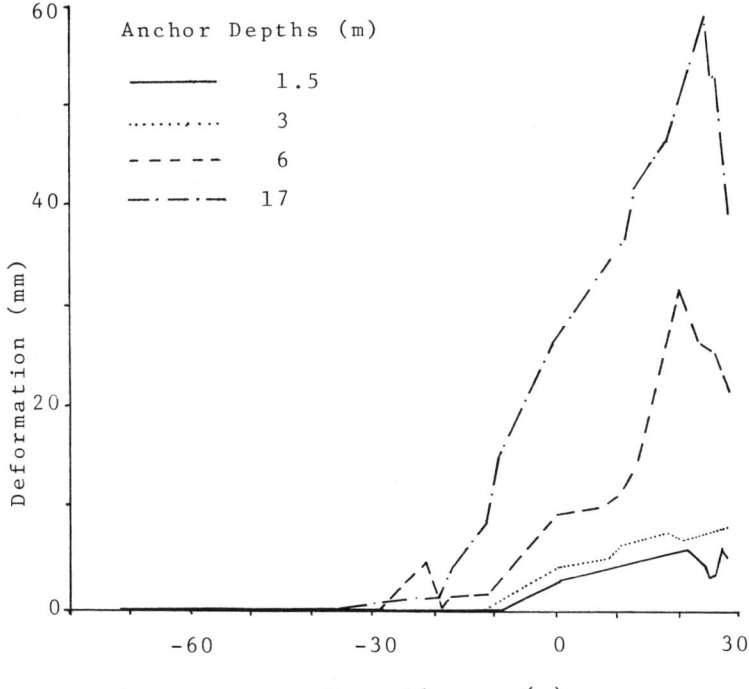

Figure 5. Pillar Anchor Movements

about 5 m in advance of the face, reaching to a peak 35 m behind the face, after which the pillar started to contract. Similar behavior was exhibited by the pillars between cross cuts D and F, but with a decrease in maximum deformation towards the ribside. The maximum deformation of the pillar above the solid, between cross cuts C and D, was an order of magnitude smaller, with no discernable pattern of deformation.

The anchor movements were converted into bay strains by calculating the differential movements between anchors, and dividing by the distance between them. The anchors were positioned in the boreholes to enable the strains across the central core of the pillars to be determined.

The plot of bay strain against face position are shown in Figure 6 for the same pillar as in Figure 5. With respect to pillar stability, the most important plot is the strain between the anchors at 6 m and 17 m, as this represents strain across the core of the pillar. For the pillars between roadways 125 and 126, the peak tensile core strain occurs between 25 m and 39 m behind the face, with maximum values of: 3mm/m between crosscuts F and G, 2.4 mm/m between E and F, and 1.5mm/m

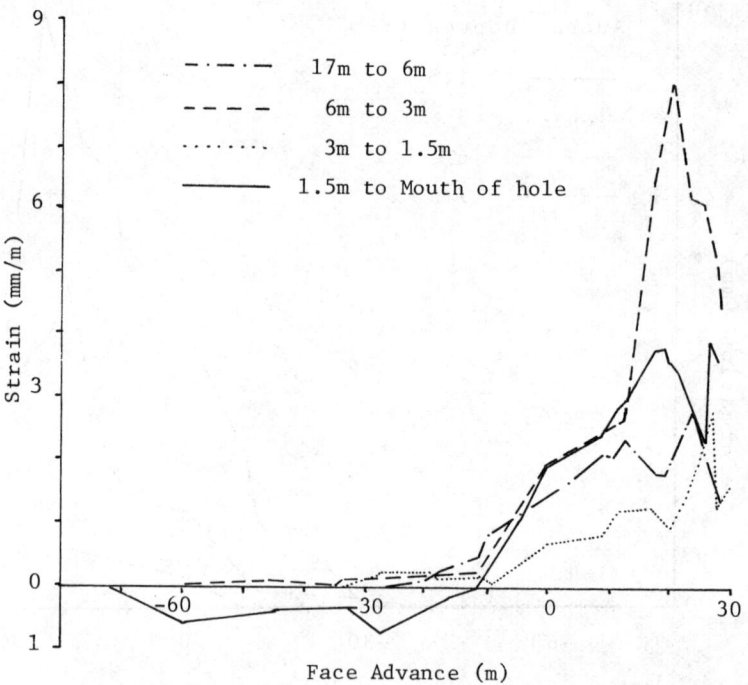

Figure 6. Pillar Bay Strains

between D and E, with no significant strain being recorded between C and D. As can be seen from Figure 6, the pillar core moves into tension above the face-line, with strain reaching a peak approximately 35 m behind the face. The Subsidence Engineers Handbook (4) predicts similar behavior with an increasing tensile strain in advance of the face which reaches a peak above the face-line, and then decreases, eventually changing to compression. Therefore, it could be expected that the pillar cores would eventually go into compression, thus causing an increase in pillar strength. Unfortunately, due to the collapse of junction 126D, access was lost, and it was not possible to confirm this hypothesis by measurement.

The only indication of possible instability caused by undermining both the Yard and Main seams, was a junction collapse in the Yard seam extending through to the Main seam. The fall occurred underneath crosscut D, between roadways 127 and 128, approximately 10 m behind the face. The separation between the Main and Yard seams in this area was approximately 13 m. This was the only fall creating an observed connection between the two sets of room and pillar workings, and was therefore probably structurally controlled. As there was little consistency in the relative positioning of pillars between the Main and Yard seams, the effects of a junction collapse beneath a Main seam pillar was analyzed. This resulted in a safety factor of over 2 for the worst possible case.

CONCLUSIONS

The principal conclusions that can be drawn from these two studies are:

Comparison of the results from above KS4 and KS6, indicate that the maximum pillar core strains, and the roof to floor convergences, were similar for both faces. This indicates that extraction of the Yard seam by room and pillar had little effect on the stability of the Main seam workings, when they were undermined by a longwall face.

The pillar core strains above KS6 showed some correlation with the dynamic strain profile predicted from the Subsidence Engineers Handbook (4), suggesting that the pillars may eventually go into compression.

Comparison of the measured pillar core strains, and the predicted horizontal strains, indicated that the position of maximum tensile strain was displaced from above the edge of the face, to above the excavated area for both faces.

Pillar dilation was approximately an order of magnitude higher above the excavation, than the solid. This indicates that deformation due to undermining was confined to a zone directly above the face.

ACKNOWLEDGEMENTS

The authors wish to thank the National Coal Board for financial and practical support during the Lynemouth investigation.

REFERENCES

1. Styler, A.N., Dunham, R.K., 1980, "Strata Deformation Above Longwall Faces," 21st U.S. Symposium on Rock Mechanics, Rolla, Mo.

2. Tubby, J.E., Farmer, I.W., 1981, "Stability of Undersea Workings at Lynemouth and Ellington Collieries." Transactions Institute of Mining Engineers, vol. 141, pp. 87-97.

3. Hodkin, D.L., 1978, "Interaction between Longwall and Pillared Workings at Lynemouth Colliery," Ph.D. Thesis, University of Newcastle-Upon-Tyne.

4. National Coal Board, 1975, "Subsidence Engineers Handbook."

5. Styler, A.N., 1980, "Inter Burden Strata Deformation and Interaction Effects Associated with Longwall Coal Mining," Ph.D. Thesis, University of Newcastle-Upon-Tyne.

Chapter 109

COMPLEMENTARY INFLUENCE FUNCTIONS FOR
PREDICTING SUBSIDENCE CAUSED BY MINING*

by H. J. Sutherland and D. E. Munson

Geotechnical Engineering Division
and
Experimental Programs Division
Sandia National Laboratories
Albuquerque, New Mexico 87185

ABSTRACT

Surface subsidence caused by underground mining is described through complementary influence functions. The complementary functions developed here differ from the simple functions previously used in that the surface displacement is the result of the combined contributions of the mined and unmined zones. This eliminates computational difficulties experienced with the simple functions in determining the deflections above the rib side and in the eventual application of influence functions to complex room-and-pillar configurations. Although the analysis framework presented is intended for predicting subsidence over complex mine configurations, use of the complementary functions is illustrated adequately by application to a longwall panel of the Old Ben No. 24 coal mine.

INTRODUCTION

In the first part of this century, mining engineers realized that they needed the capability to predict ground displacements and strains caused by subsurface mine workings. This need to predict surface movements, commonly termed subsidence, is especially critical in Europe because of the extensive surface utilization over economic coal and ore bodies. A similar need is now developing in the United States.

Early considerations of subsidence, before the advent of computer analysis, led to empirical functions that describe mathematically the

---

*This work supported by the U. S. Department of Energy.

observed surface displacements. Two classes of empirical functions are commonly used: profile functions and influence functions (Brauner, 1973 and Hood et al, 1981). Profile functions are direct fits to empirical data and are typically used for subsidence predictions over long wall panels (Munson and Eichfeld, 1980). These functions cannot describe geometrically complex mining areas such as found in room-and-pillar mining. Influence functions are more applicable to these complex mining areas and have been previously developed from the viewpoint of the excavated material. Each unit element of the ground surface above the mined volume is assigned the same response, and integration of this elemental influence over all the elements in the mined area yields the subsidence prediction. These influence functions are widely used, with considerable success. There is, however, a major problem with this formulation because it significantly overpredicts the subsidence directly over the rib side. Typically the problem is handled by integrating the elements to an imaginary rib location within the mined area rather than to the actual rib location. Solution of the rib side subsidence problem is crucial in developing the capability to predict subsidence over room-and-pillar, as well as over longwall, mines.

In recent analyses, numerical computer methods have been used to predict subsidence. Instead of analyzing the mined material, these techniques analyze the behavior of the coal (ore) and overburden layers remaining after mining. Calculations of considerable detail are possible and have shown how elastic bending, breaking and bulking and void volume are transmitted through the overlying strata to cause surface subsidence ((Munson and Benzley, 1980 and Sutherland and Schuler, 1982). Normally, such large scale analyses are beyond the means of mine operators, and a simpler analysis method is required, especially for room-and-pillar configurations.

In this work, we examine the motivating concepts for influence functions in order to remove the inadequacy of the prediction at the rib side. In the examination, it is apparent that the remaining material in the seam is as influential as that of the open volume left by the excavation. This viewpoint leads to a formulation based on the concept of complementary influence functions.

COMPLEMENTARY INFLUENCE FUNCTIONS

The fundamental concept of complementary influence functions is that the separate influence functions describing the response of mined and unmined zones act together to produce the observed subsidence. Each influence function is defined by the response of a unit element: an "unmined" element for the coal (ore) left by the mining process, and a "mined" element for the open volume created by the removal of the coal (ore). These elements are shown in Figure 1. When these elements are integrated (appropriately summed) over the entire seam (both the mined and the unmined zones) a subsidence prediction results. Thus, the subsidence is the sum of the influence of the mined and unmined response.

FUNCTIONS FOR PREDICTING SUBSIDENCE

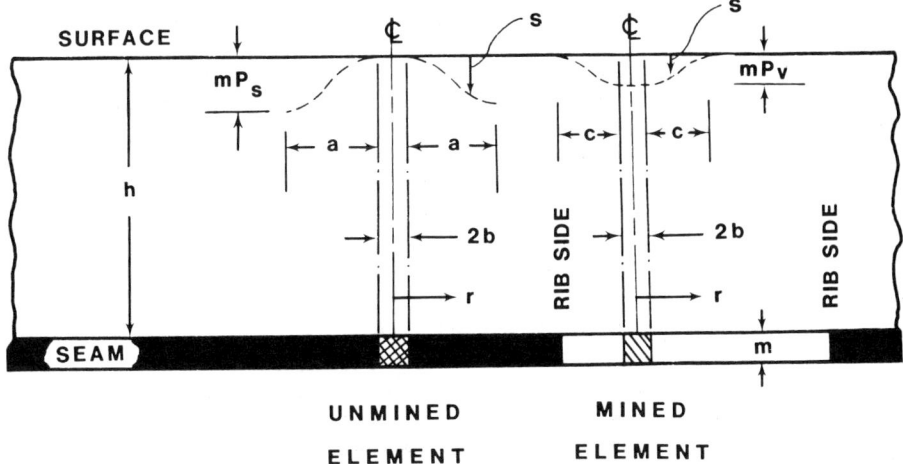

Fig. 1. Complementary Elements

Response of the Unmined Element

The elemental response of the unmined element is based on the elastic response of a plate supported by the element (Timoshenko and Woinowsky-Krieger, 1959) and is given by the function

$$s_s/m = P_s \begin{cases} 0 & , 0 \leq r \leq b \\ \{I\}\{\Delta\}\{r_s^2[1 - 2\ln(r_s)]\} & , b \leq r \leq a + b \\ 1 & , a + b \leq r \end{cases}, \qquad (1)$$

where

$$r_s = \frac{r}{a} - \frac{b}{a} ,$$

s_s is the vertical subsidence (positive down) around the unmined element, m is the mined height, P_s is the proportion of the maximum subsidence attributed to the unmined element, r is the radial coordinate from the centerline of the element, b is the half-width of the element, and a is the radial extent of influence outside the element (see Figure 1). $\{I\}$ and $\{\Delta\}$ are functions that express the effects of the moment of inertia of the overlying strata and the crushing of unmined material on the element response, respectively. As this solution suggests, the unmined elements hold the surface above them at its original position, but allow the surface above neighboring mined zones to move down according to the elastic solution.

The elastic solution, however, must be modified to account for possible variation in the thickness of the elastic beam representing the unfractured overlying strata. Physical and numerical models of the response of the overlying strata (Sutherland and Schuler, 1982) have shown that the failure zone above a longwall panel produces a thinning of the remaining elastic beam as one moves away from the rib. This thinning can be accommodated in Equation 1 by using the moment of inertial function $\{I\}$. If $t(r_s)$ is the thickness of the plate as a function of the radius r_s and t_o is the center thickness of the plate, then for $b \leq r \leq a + b$

$$\{I\} = \left[\frac{t_o}{t(r_s)}\right]^3 . \tag{2}$$

The elastic solution must also be modified to incorporate the effect of possible crushing of the unmined element $\{\Delta\}$. Analysis of this effect is the subject of a forthcoming paper and will not be treated further. This crushing effect, while very important for room-and-pillar mines, is less essential to the illustrations involving longwall panels discussed here.

Response of the Mined Element

As a portion of the overlying strata progressively breaks and falls into the mined cavity, vertically nonuniform voids are left throughout the caved overburden. Description of this distribution of void in the panel center has led to the prediction of maximum subsidence (Munson and Benzley, 1980). Near the rib side, both the horizontal and vertical distribution of residual void is non-uniform. The horizontal distribution is simply related to the probability of lateral migration of a void as it moves to the surface. By assuming a Gaussian probability distribution for this migration process, the integrated effect at the surface has the form of an error function; namely,

$$s_v/m = P_v \begin{cases} 1 & , 0 \leq r \leq b \\ \text{erfc}(2r_v) = 1 - \frac{2}{\sqrt{\pi}} \int_0^{2r_v} \exp(-\xi^2) d\xi & , b \leq r \leq c + b \\ 0 & , c + b \leq r \end{cases} , \tag{3}$$

where

$$r_v = r/c - r/b ,$$

and s_v is the vertical subsidence due to the mined element, P_v is the proportion of the maximum subsidence due to the mined element, c is the radial extent of influence outside the element, and ξ is an integration parameter. This probability distribution appears similar to

FUNCTIONS FOR PREDICTING SUBSIDENCE 1119

other subsidence analyses (Brauner, 1973); but contrary to the other analyses, we are concerned only with the residual void in the overburden and not the total mined volume. Thus, the mined element response is based on the horizontal distribution of part of the excavated volume (the bulking) in the overlying strata.

APPLICATION TO A LONGWALL PANEL

According to the functions just presented, the parameter set that governs subsidence is a, c, P_s and P_v. In terms of a longwall panel, a and c can be interpreted as the half range of the profile function and the draw angle, and P_s and P_v are related to the centerline displacement and the displacement above the rib side, respectively. Specific values of these parameters, together with simple computational techniques, are applied to the analysis of subsidence over the longwall panel 2N at the Old Ben No. 24 coal mine. This panel is located at a depth h of 189 m and the excavated height m is nominally 2.1 m (Edl and Eichfeld, 1978). In nondimensional form, the parameters a/h, b/h, P_s, c/h and P_v were taken to be 0.48, 0.004, 0.5, 0.48, and 0.08. The resulting displacements and their sum (i.e., the total subsidence) are plotted in Figure 2 for the case where $\{I\}$ is a constant. The measured subsidence data, also plotted in Figure 2, show an influence of a variable beam thickness. Consequently, a calculation was made that assumed a moment-of-inertia term of the form

$$\{I\} = \left[\frac{t_o}{t(r_s)}\right]^3 = \left[\frac{1}{1 + p(1 - r_s)^3}\right]^3 . \qquad (4)$$

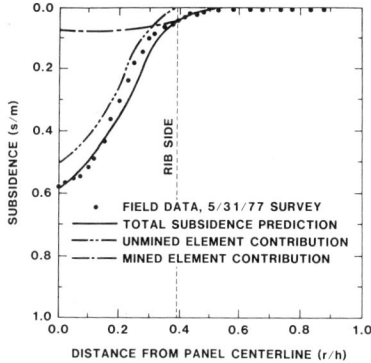

Fig. 2. Subsidence Prediction (Without Variation in in Beam Thickness)

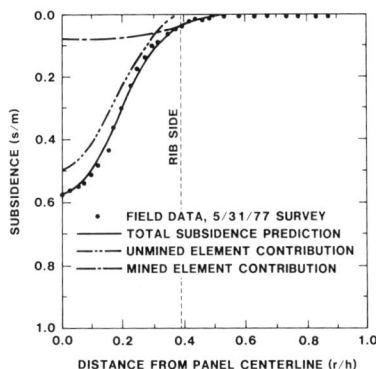

Fig. 3. Subsidence Prediction (With Variation in in Beam Thickness)

This equation implies that the beam forms a "stress arch" above the mined elements, where the ratio of the beam thicknesses at the ends of the span to that at the center of the span is p + 1. The result for p = 0.3 is shown in Figure 3 and is in much better agreement with the observed profile.

The application of complementary functions to three-dimensional panel geometry of Old Ben is illustrated in Figure 4. Superimposed on a quadrant of a longwall panel are the predicted subsidence contours for the cases of the influence of the unmined element alone, the mined element alone, and the total subsidence of the complementary functions. As with the subsidence profiles, the contours are in quantitative agreement with field measurements (Edl and Eichfeld, 1978).

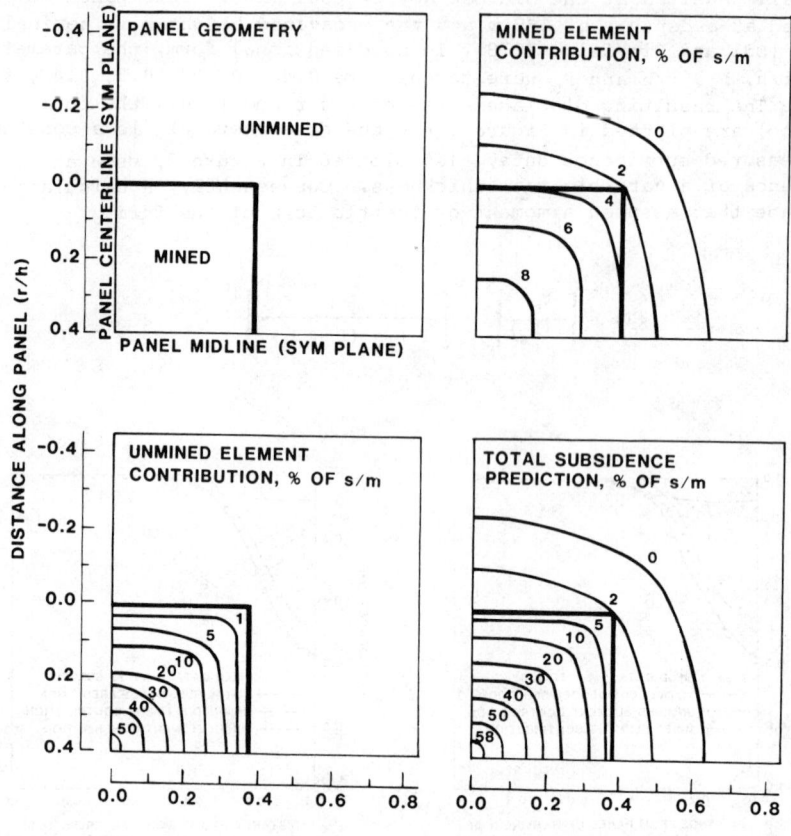

Fig. 4. Subsidence Contours Over a Longwall Panel

CONCLUSIONS

A new approach is developed for the use of influence functions in the prediction of mine subsidence. In this approach, complementary influence functions representing the response of both mined and unmined elements are integrated over the area and summed. Both elements strongly affect the subsidence. Development of complementary influence functions is essential in advancing the state-of-the-art in subsidence analysis of complicated room-and-pillar mines. Limited comparisons between field data and predictions are very encouraging.

REFERENCES

Brauner, G., 1973, Subsidence Due to Underground Mining (Part 1) USBM Information Circular 8571, 56 pp.

Edl, J.N., Jr., and Eichfeld, W.F., 1978, Subsidence Related Data from Four Representative United States Coal Mines, Laboratory Report S&SE 78-2, Carbondale Mining Technology Center, U.S. Dept. of Energy, 154 pp.

Hood, M., Ewy, R.T., Riddle, L.R., & Daeman, J.J.K., 1981, "Empirical Methods for Subsidence Prediction and Their Applicability to U.S. Mining Conditions," Final Report, Contract No. 62-0200, Dept. of Materials Science and Mineral Engineering, University of Calif., Berkeley, 241 pp.

Munson, D.E., and Benzley, S.E., 1980, "Analytic Subsidence Model Using Void-Volume Distribution Functions," Proc. 21st U.S. Symposium on Rock Mechanics, Rolla, MO, p 102.

Munson, D.E., and Eichfeld, W.F., 1980, Evaluation of European Empirical Methods for Subsidence in U.S. Coal Fields, SAND80-0537, Sandia National Laboratories, Albuquerque, NM.

Sutherland, H.J., and Schuler, K.W., (in press) 1982, "A Review of Subsidence Prediction Research Conducted at Sandia National Laboratories," Proceeding of the Workshop on Surface Subsidence Due to Underground Mining, S.S. Peng, ed.

Timoshenko, S., and Woinowsky-Krieger, S., 1959, Theory of Plates and Shells, McGraw-Hill, New York, p 68.

Chapter 110

DESIGN AND ANALYSIS OF A CIRCULAR UNDERGROUND POWERHOUSE

by D. Zayakov, G. Yoshikado* and P.R. Kneitz*

Project Engineer, Pacific Gas and Electric Company, S.F.
Principal Engineer, International Engineering Company, S.F.
Civil Engineer, Tudor Engineering, S.F.

ABSTRACT

The design and analysis of a circular, domed powerhouse for Pacific Gas and Electric (PG&E) Company's Kerckhoff 2 Project is described. Exploration, geologic stress determinations, mechanical properties, reinforcement and instrumentation are addressed briefly. The analyses of the circular, domed cavern and the comparison of predicted and field deformations are presented in somewhat more detail.

The behavior of the completed powerhouse cavern has been found to be substantially as predicted by the analysis.

INTRODUCTION

The Kerckhoff 2 Hydroelectric Project is located on the San Joaquin River about 48 Km (30 miles) northeast of Fresno, California. Water will be conveyed to an underground powerhouse through a 6.4 Km (4 mile) long, machine bored, 7.3 m (24 ft.) diameter tunnel from PG&E's existing Kerckhoff Lake. A single 140 MW unit will be housed in a circular, 26 m (85 ft) diameter cavern topped by a 28.3 m (93 ft) diameter dome; one of the worlds largest. The main features are shown on Figure 1.

*Formerly of PG&E

CIRCULAR UNDERGROUND POWERHOUSE

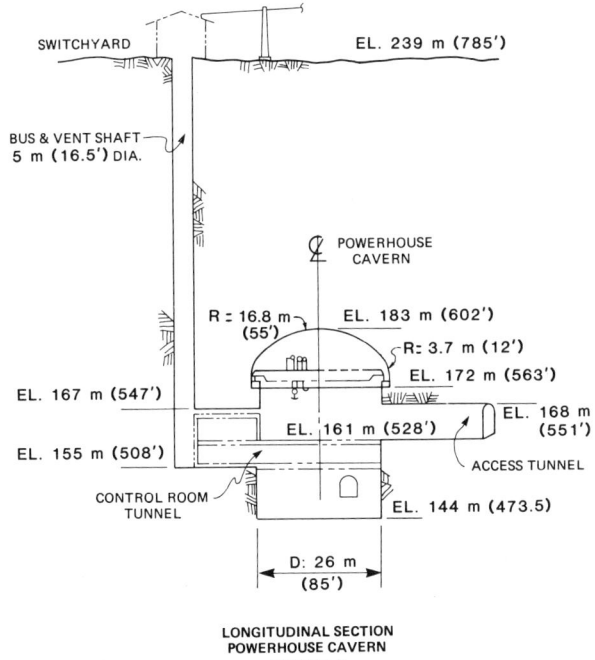

LONGITUDINAL SECTION
POWERHOUSE CAVERN
FIGURE 1

EXPLORATION AND ROCK PROPERTIES

The Project is located in the granitic rock of the western foothills of the Central Sierra Nevada Mountains. The Powerhouse Cavern is located in fresh granodiorite with core recovery and Rock Quality Designation (RQD) generally at about 90%.

Exploration for the cavern consisted of a series of cored 76 mm (3 in.) borings. A set of borings was made at a prospective site and severe artesian conditions were encountered. A new location was selected and a second set of borings was made. The second site was more favorable and therefore selected. A 213 m (700 ft.) horizontal boring was also made to monitor the ground water. All rock cores were logged, sketched, photographed in color, bound in a formal report and made available to the contractors.

Surface geology and joint patterns were thoroughly mapped. A model of the Powerhouse was made and projected joint patterns and borings were shown on it to aid in design. A video tape recording of a borehole was also made by using a down-hole television camera in order to check joint orientations.

Young's Modulus (E) of 48,263 MPa (7,000,000 psi) and Poisson's Ratio of 0.19 were established from seismic refraction survey data. Laboratory tests on solid rock cores gave an average "E" of 55,158 MPa (8,000,000 psi) and an average Poisson's Ratio of 0.17.

An "E" of 48,263 MPa (7,000,000 psi) and a Poisson's Ratio of 0.17 were chosen for use in the Finite Element Method (FEM) analysis. However, the elements that would be within 6.1 m (20 ft.) of a blasted surface were assigned an "E" of 34,474 MPa (5,000,000 psi) to compensate for the disturbance caused by excavation.

In-situ stresses of the site were determined by borehole "hydrofracturing" (Haimson, 1977). These stresses were found to increase with depth, as expected, but the major principal horizontal stresses were 3 to 6 times larger than the vertical stresses. The minor principal horizontal stresses were only 2 to 3 times larger than the vertical stresses.

DESIGN OF A CIRCULAR UNDERGROUND POWERHOUSE

An underground powerhouse was chosen rather than a conventional surface design for environmental as well as economic reasons. An exposed powerhouse and penstock would have been highly visible from nearby recreation areas. Also, space for an outdoor plant was extremely limited. Another consideration was the 18 m (60 ft.) fluctuation in tailwater elevation. These factors resulted in the economics favoring an underground powerhouse.

A circular plan was adopted when it became apparent that a single, large unit could be used rather than multiple units. A 2,090,664 N (235 ton) polar crane was provided to fit this arrangement. A rock bench topped with a concrete runway was found to be the most economic way to support this crane.

A circular powerhouse was found to have the following advantages over a rectangular one: its volume of excavation will be about 25% less; its wall area will be about 20% less (permitting fewer rock bolts and less shotcrete); it takes advantage of the inherent strength of a circular section in ring compression and the multidirectional action of a dome.

FINITE ELEMENT ANALYSES

Introduction

In-situ stresses, seismic accelerations and rock weights were used as the loads in analyses of the powerhouse cavern by the

Finite Element Method (FEM). The results were used to check the design for stability and stress concentrations. A comparison was also made between a circular and a rectangular cavern to evaluate the advantages of a circular scheme. An instrumentation program was implemented to check actual deformations. The measurements in the completed cavern have confirmed the results of the analyses.

The analyses were made with the SAP IV FEM Program (Bathe, et al, 1973). The following separate analyses were made: 3-dimensional (3-D) static; 2-dimensional (2-D) static; 2-D pseudostatic seismic load; and 2-D stage of construction.

The 2-D model was considered to be representative of a long, rectangular cavern. It was therefore compared with the 3-D case to demonstrate the advantages of the circular plan.

The "stage of construction analysis" examined the cavern when excavation reached the crane girder support bench.

The 2-D static analysis was independently "checked" by using STRUDL (Logcher, et al, 1968). The close comparison of results indicated that a reasonable study had been made.

SAP IV Program

SAP IV is a structural analysis program for obtaining the static and dynamic responses of linear systems. Its 3-D, 8-node, isoparametric element with 3 translational degrees of freedom per node was used. Isotropic material properties were assumed and stresses were computed at the element centroids.

Finite Element Models and Loadings

The models included only the main powerhouse geometry and omitted the connecting tunnels. This simplification was made because the estimated small increase in accuracy that could have been gained by including such openings did not justify the huge increase in complexity.

The 3-D model consisted of 1488 "brick" elements that had a total of 1952 nodes. Smaller elements were used to model the curved surfaces of the cavern and larger ones to represent undisturbed rock. The horizontal boundaries of the model were set at 1-1/2 diameters and the vertical boundaries at a distance equal to the height of the cavern. These boundaries proved to be remote enough when computed stresses were found to approach in-situ values at the boundary elements. The boundary nodes were considered to be fixed points.

The 2-D model was simply a cross-section of the 3-D model. It consisted of 130 rectangular, plain-strain elements with 170 total nodes.

Both models were "loaded" with in-situ stresses and the weights of the elements. The 2-D model was also subjected to seismic accelerations.

The rock was assumed to be homogeneous in these models and elastic analyses were made. Therefore, the true, jointed character of the rock and any disturbances during excavation were not modeled. This will be partly offset by the rockbolts and the concrete crane-support girder.

Three-Dimensional Static Analysis

Only one analysis was required in this case because it was possible to consider the in-situ stresses and the rock weights simultaneously. The resulting stresses and deformations were evaluated by plotting them on cross-sections parallel with both principal horizontal stress axes. The results were also plotted on several plan views taken at various elevations. Stress contours were drawn on these plots. An example is given on Figure 2.

The maximum vertical compressive stress computed was 5412 KPa (785 psi) and the maximum horizontal stress 15,169 KPa (2200 psi). These stresses are an order of magnitude less than the laboratory unconfined compressive strengths that average about 117,212 KPa (17,000 psi). The only tensile stress computed was at the center of the floor and can be attributed to the removal of rock loads.

One unusual result found was the upward movement of 0.3 mm (0.01 inch) at the crown of the dome. This is due to the squeezing effect of the large horizontal in-situ stresses. The maximum wall deflection was 2.8 mm (0.11 inch) in the direction of the maximum horizontal in-situ stress.

Two Dimensional Static Analysis

Cross-sections parallel with both the major and minor horizontal principal stress axes were analyzed. Vertical stress contours on the major axis are shown on Figure 3. The behavior of the rock wall was found to be similar to that of a beam in bending with tension on the extreme fibres. Since tension indicates potential spalling, the walls were "pattern bolted" although the circular plan was expected to offset such problems.

The maximum inward deflection on the major axis was found to be 5.3 mm (0.21 inch), almost twice that computed in the 3-D analysis.

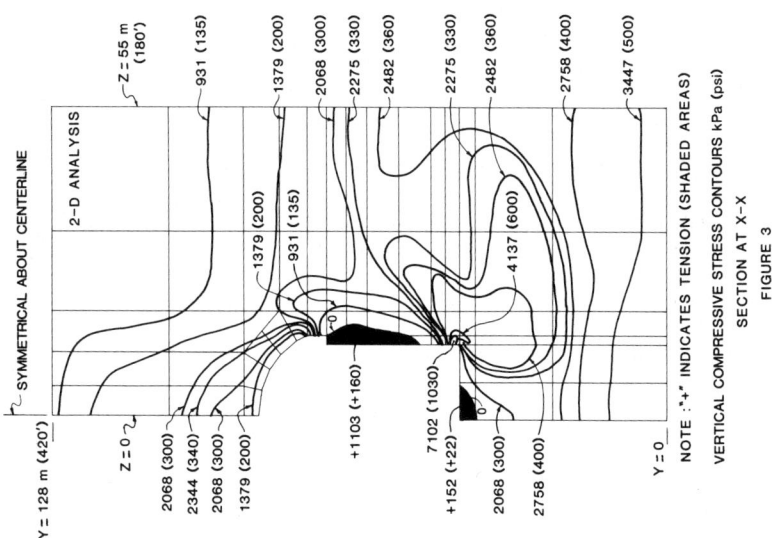

VERTICAL COMPRESSIVE STRESS CONTOURS kPa (psi)
SECTION AT X-X
FIGURE 3

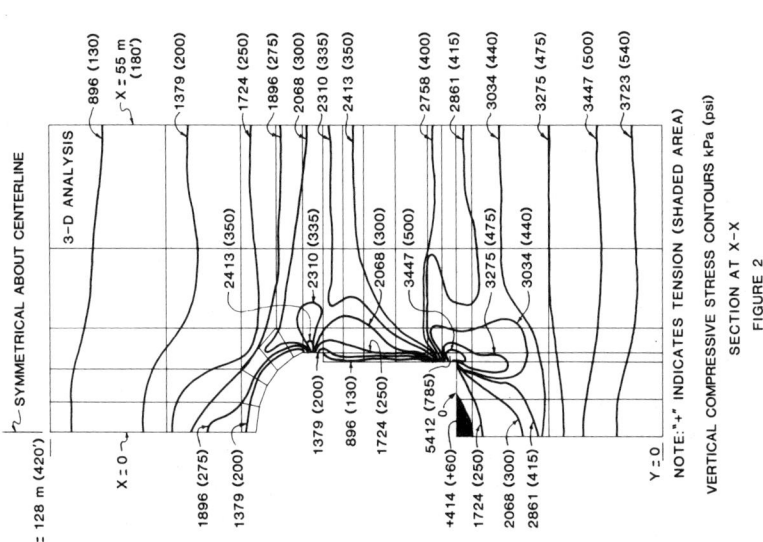

VERTICAL COMPRESSIVE STRESS CONTOURS kPa (psi)
SECTION AT X-X
FIGURE 2

Two-Dimensional "Pseudostatic" Seismic Load Analysis

Horizontal accelerations of 0.15g and vertical accelerations of 0.10g were superimposed on the in-situ conditions. The results found were that an earthquake producing such accelerations would have little effect on the underground cavern.

Two-Dimensional Stage of Construction Analysis

The behavior of the dome when excavation had reached the elevation of the crane bench was investigated. The crown was found to move downward 0.3 mm (0.01 inch). As excavation proceeded, the inward movement of the walls forced the crown back up.

Comparison of Circular (3-D) and Rectangular (2-D) Caverns

The maximum compressive stress in the 3-D dome was 11,239 KPa (1630 psi) whereas the maximum for the 2-D arch was 14,065 KPa (2040 psi), a 25% increase. Therefore, the circular cavern was designed with a smaller rise to span ratio and resulted in less excavation and rock bolting than a long, rectangular cavern. The circular shape was found particularly superior in cases where high horizontal in-situ stresses were present. The ring compression in the circular walls overcame the tendency for tensile stresses to form. Therefore, wall reinforcement was less costly for the circular cavern.

FIELD MEASUREMENTS

Rod Type extensometers with remote readouts and rock bolts with strain gages were installed. Tape extensometers were also used. The instruments were read about once a week but this varied with the construction activities.

The deflections at the crown can be summarized as follows: the movement was downward during the first two months of excavation; there was little movement during the next 7 months (the reading stayed at about 0.5 mm (0.02 inch); the movement was upward during the next 1-1/2 months and the crown stabilized at about 0.3 mm (0.01 inch) of deflection. The 7 months of little movement corresponds to the period when excavation progressed downward to the level of the access tunnel, El. 161 m (529 ft.) and was held there for about 1-1/2 months while work progressed elsewhere. The upward movement began when the excavation was progressing downward from El. 161 m (529 ft.) to the bottom of the sumps. Apparently, the high horizontal stresses caused this as predicted by the analysis.

A 1.3 mm (0.05 inch) inward deflection of the wall was recorded in the minor horizontal principal stress direction. This compares very favorably with the 1.0 mm (0.04 inch) predicted.

Inexplicably, no wall movement was detected in the major horizontal principal stress direction. An inward deflection of 2.8 mm (0.11 inch) had been predicted. The two large unmodeled openings and the heavy concrete crane girder may be contributing to this anomaly.

The instrumented rock bolts showed little load as would be expected with such small deflections. The maximum stress was about 34,474 KPa (5000 psi).

CONCLUSION

The FEM analysis was successfully used to predict the order of magnitude of very small rock movements and also their direction. It appears that considerable refinement in the analyses would be required before the absolute values of such small deformations can be predicted.

The field measurements have confirmed that the circular, domed, powerhouse cavern is very stable. The dome and cavern walls deflected very little, as was expected.

The small loads on the instrumented rock bolts indicated that the short, 4.6 m (15 ft.), bolts installed on a 1.2 m (4 ft.) by 1.2 m (4 ft) pattern were more than adequate.

REFERENCES

Bathe, K.J, Wilson, E.L., and Peterson, F.E. 1973
"SAP IV, a Structural Analysis Program For Static and Dynamic Response of Linear Systems," June, University of California, Berkeley, CA.

Haimson, B.C., 1977, "Hydrofracturing Stress Measurements, Kerckhoff 2 Powerhouse Site", November, Pacific Gas and Electric Company, San Francisco, CA.

Logcher, et al, 1968, "ICES STRUDL-II", November, Massachusetts Institute of Technology, Cambridge, MA.

AUTHORS' INDEX

Aboustit, B. L., 587
Adams, T. F., 637
Advani, S. H., 315, 471, 587
Alm, O., 261, 542
Amadei, B., 157
Anderson, G. D., 551

Baecher, G. B., 67
Bandis, S., 739
Barker, D. B., 441
Barton, C. C., 449
Barton, N., 739, 802
Bauer, S. J., 279
Beech, J. F., 463
Beloff, W. R., 211
Bieniawski, Z. T., 985
Bischoff, J. A., 935
Blackwood, R. L., 168
Blankenship, D. A., 761
Board, M., 802
Borschel, T. F., 341
Boyle, W., 883
Brady, B. H. G., 571, 628, 692
Brown, A., 1035
Butkovich, T. R., 855

Calder, P. N., 899
Cardenas-Garcia, J. F., 441
Carlsson, H. S., 23
Chitombo, G., 47
Chu, C.-1., 270
Cogan, J., 912
Conroy, P. J., 1048
Cook, N. W. G., 723
Correa F^o, D., 769
Crawford, A. M., 1057
Crouch, S. L., 704
Cruz, P. T., 769

Dawson, P. R., 299
Deadrick, F. J., 57
de la Cruz, R., 778
Demou, S., 341
Detourney, E., 924
Dienes, J. K., 86
Doe, T. W., 30, 645
Dongjun, X., 864
Dorwart, B. C., 211
Dreher, K. J., 1099
Duddeck, H. W., 596
Dunbar, W. S., 604
Dunham, R. K., 1107
Dusseault, M. B., 1065

El Rabaa, A. W. M. A., 790
Endo, H. K., 30

Fairhurst, C., 673, 924
Fourney, W. L., 523
Friedman, M., 279

Gale, J. E., 290
Gay, N. C., 176
Ghosh, S., 95
Glynn, E. F., 95
Goldsmith, W., 488
Gomez, P. M., 912
Goodman, R. E., 157, 883
Gronseth, J. M., 183
Guangyu, L., 230
Gunsallus, K. L., 463

Hadizadeh, J., 372
Haimson, B. C., 143, 190

Handin, J., 279
Hanson, M. E., 551
Hardin, E., 802
Hardy, H. R., Jr., 985
Hardy, M. P., 846
Heard, H. C., 249
Hennig, C. C., 1099
Hennig-Michaeli, C., 380
Herget, G., 203
Heuze, F. E., 612
Holloway, D. C., 441, 523
Hume, H. R., 104
Hustrulid, W. A., 790

Ingraffea, A. R., 463

Jaworski, G. W., 211
Jeyapalan, J. K., 814
Judd, W. R., 104

Kenner, V. H., 471
Keough, D., 504
Kerrich, R., 389
King, M. S., 39
King, R., 47
Kneitz, P. R., 1122
Kobayashi, T., 480
Kolluru, S., 333
Kumano, A., 488

Ladanyi, B., 3
Lajtai, E. Z., 496
Lang, T. A., 935
LaTour, T. E., 389
Lee, J. K., 587
Leigh, W. J. P., 1091
Lewis, M. R., 945
Lilly, W. W., 1015
Lindberg, H. E., 824

Lingle, R., 802, 837
Lo, K. Y., 620
Long, J. C. S., 30
Lorig, L. J., 628
Lytle, R. J., 17, 57

Mahtab, M. A., 116
Margolin, L. G., 637
Marrano, A., 769
McClennan, J. D., 219
McGroarty, S., 531
McHugh, S., 504
Mellegard, K. D., 684
Miller, S. M., 124
Montazer, P., 47
Moore, D. P., 945
Munson, D. E., 299, 1115

Nelson, P., 463
Nelson, P. H., 837
Nipp, H.-K., 596
Noorishad, J., 645
Nyland, E., 1065

Ouchterlony, F., 515
Oweis, I. S., 1015

Pariseau, W. G., 1077
Paulsson, B. N. P., 39
Pleifle, T. W., 307
Pratt, H., 802
Preece, D. S., 655

Quadros, E. F., 769

AUTHORS' INDEX

Ramirez, A. L., 57
Ratigan, J. L., 423
Richard, T. G., 315, 471
Robertson, E. C., 397
Roegiers, J.-C., 219
Rose, D., 953
Rowe, R. G., 324
Rowe, R. K., 620
Rutter, E. H., 372

Ubbes, W., 47, 790, 802
Unal, E., 985
Upadhyay, P. C., 664

van der Heever, P. J., 176
Voegele, M. D., 673, 802

St. John, C. M., 571, 846
Sandhu, R. S., 587
Schatz, J. F., 341
Schultz, L. D., 219
Schwartz, C. W., 333
Scoble, M. J., 1091
Scott, G. A., 1099
Scott, J. J., 971
Selvadurai, A. P. S., 814
Senseny, P. E., 307, 684
Shaffer, R. J., 551
Shi, G.-h., 883
Shiwie, B., 230
Sibson, R. H., 361
Siemes, H., 380
Simha, K. R. Y., 523
Singh, J. G., 664
Singh, R. N., 961
Sinha, K. P., 341
Stavropoulou, V. G., 351
Stickney, R. G., 761
Stone, C. M., 655
Styler, A. N., 1107
Summers, D. A., 531
Sutherland, H. J., 1115
Swan, G., 542

Wagner, R. A., 684
Wai, R. S. C., 620
Wang, C.-Y., 270
Warren, N., 132
Weishen, Z., 864
Wenk, H. R., 405
West, T. R., 104
White, W. F., 211
Wilson, C. R., 30
Witherspoon, P. A., 560

Yegulalp, T. M., 116
Yeung, D., 692
Yongjia, W., 704
Yoshikado, G., 1122
Yow, J. L., Jr., 855

Zadeh, A. M. H., 961
Zayakov, D., 1122
Zimmerman, R. M., 872
Zimmerman, R. W., 712
Zoback, M. D., 143

Tanimoto, C., 999
Teufel, L. W., 238
Tham, L., 620
Thorpe, R. K., 551
Tsang, Y. W., 560